HADRON SPECTROSCOPY

Previous Proceedings in the Series of Conferences on Hadron Spectroscopy

Conference		Held in	Publisher	ISSN/ISBN
HADRON99	8th	Beijing, China	North-Holland Nuclear Physics A, Vol. 675	0375-9474
HADRON97	7th	Upton, New York, USA	AIP Conf. Proceedings Vol. 432	1-56396-765-0

Other Related Titles from AIP Conference Proceedings

603 Mesons and Light Nuclei: 8th Conference
Edited by Jiri Adam, Petr Bydzovsky, and Jiri Mares, December 2001, 0-7354-0047-4

602 QCD@Work: International Workshop on Quantum Chromodynamics: Theory and Experiment
Edited by Pietro Colangelo and Giuseppe Nardulli, December 2001, 0-7354-0046-6

594 Hadrons and Nuclei: First International Symposium
Edited by Il-Tong Cheon, Taekeun Choi, Seung-Woo Hong, and Su Houng Lee, November 2001, 0-7354-0037-7

549 Intersections of Particle and Nuclear Physics: 7th Conference, CIPANP2000
Edited by Zohreh Parsa and William J. Marciano, December 2000, 1-56396-978-5

540 Particle Physics and Cosmology: Second Tropical Workshop
Edited by José F. Nieves, October 2000, 1-56396-965-3

508 Hadron Physics: Effective Theories of Low Energy QCD
Edited by A. H. Blin, B. Hiller, M. C. Ruivo, C. A. Sousa, and E. van Beveren, March 2000, 1-56396-927-0

496 Workshop on Instabilities of High Intensity Hadron Beams in Rings
Edited by T. Roser and S. Y. Zhang, December 1999, 1-56396-910-6

494 New Directions in Quantum Chromodynamics
Edited by Chueng-Ryong Ji and Dong-Pil Min, November 1999, 1-56396-908-4

488 High Energy Physics at the Millennium: MRST '99
Edited by Pat Kalyniak, Stephen Godfrey, and B. Kamal, October 1999, 1-56396-902-5

482 RHIC Physics and Beyond: Kay Kay Gee Day
Edited by Berndt Müller and Robert Pisarski, July 1999, 1-56396-878-9

To learn more about these titles, or the AIP Conference Proceedings Series, please visit the webpage **http://proceedings.aip.org**

HADRON SPECTROSCOPY

Ninth International Conference on
Hadron Spectroscopy

Protvino, Russia   25 August–1 September 2001

EDITORS
Dmitry Amelin
Alexander M. Zaitsev
Institute for High Energy Physics
Protvino, Russia

SPONSORING ORGANIZATIONS
Ministry of Atomic Energy
Ministry of Industry, Science and Technologies
Russian Academy of Sciences
Russian Fund of Fundamental Research

Melville, New York, 2002
AIP CONFERENCE PROCEEDINGS ■ VOLUME 619

Editors:

Dmitry Amelin
Alexander M. Zaitsev

Institute for High Energy Physics (OAF)
RU-142284 Protvino (Moscow Region)
RUSSIA

E-mail: amelin@mx.ihep.su
         zaitsev@mx.ihep.su

Authorization to photocopy items for internal or personal use, beyond the free copying permitted under the 1978 U.S. Copyright Law (see statement below), is granted by the American Institute of Physics for users registered with the Copyright Clearance Center (CCC) Transactional Reporting Service, provided that the base fee of $19.00 per copy is paid directly to CCC, 222 Rosewood Drive, Danvers, MA 01923. For those organizations that have been granted a photocopy license by CCC, a separate system of payment has been arranged. The fee code for users of the Transactional Reporting Service is: 0-7354-0067-9/02/$19.00.

© 2002 American Institute of Physics

Individual readers of this volume and nonprofit libraries, acting for them, are permitted to make fair use of the material in it, such as copying an article for use in teaching or research. Permission is granted to quote from this volume in scientific work with the customary acknowledgment of the source. To reprint a figure, table, or other excerpt requires the consent of one of the original authors and notification to AIP. Republication or systematic or multiple reproduction of any material in this volume is permitted only under license from AIP. Address inquiries to Office of Rights and Permissions, Suite 1NO1, 2 Huntington Quadrangle, Melville, N.Y. 11747-4502; phone: 516-576-2268; fax: 516-576-2450; e-mail: rights@aip.org.

L.C. Catalog Card No. 2002105262
ISBN 0-7354-0067-9
ISSN 0094-243X
Printed in the United States of America

CONTENTS

Preface xix
Committees xxi
Group Photo xxiii

PLENARY SESSIONS

VECTOR MESON STATES

Light-Quark Vector-Meson Spectroscopy 5
    A. Donnachie and Y. S. Kalashnikova
Precise Measurement Hadronic Cross Sections with CMD-2 Detector at VEPP-2M 15
    R. R. Akhmetshin, E. V. Anashkin, V. M. Aulchenko, V. S. Banzarov,
    L. M. Barkov, S. E. Baru, N. S. Bashtovoy, A. E. Bondar, D. V. Bondarev,
    A. V. Bragin, D. V. Chernyak, S. I. Eidelman, *G. V. Fedotovitch*,
    N. I. Gabyshev, A. A. Grebeniuk, D. N. Grigoriev, V. W. Hughes,
    P. M. Ivanov, S. V. Karpov, V. F. Kazanin, B. I. Khazin, I. A. Koop,
    M. S. Korostelev, P. P. Krokovny, L. M. Kurdadze, A. S. Kuzmin,
    I. B. Logashenko, P. A. Lukin, A. P. Lysenko, K. Y. Mikhailov,
    I. N. Nesterenko, V. S. Okhapkin, A. V. Otboev, E. A. Perevedentsev,
    A. A. Polunin, A. S. Popov, T. A. Purlatz, N. I. Root, A. A. Ruban,
    N. M. Ryskulov, A. G. Shamov, Y. M. Shatunov, A. I. Shekhtman,
    B. A. Shwartz, A. L. Sibidanov, V. A. Sidorov, A. N. Skrinsky,
    V. P. Smakhtin, I. G. Snopkov, E. P. Solodov, P. Y. Stepanov,
    A. I. Sukhanov, J. A. Thompson, V. M. Titov, A. A. Valishev, Y. V. Yudin,
    and S. G. Zverev
Review of Results from SND Detector 30
    M. N. Achasov, V. M. Aulchenko, K. I. Beloborodov, A. V. Berdyugin,
    A. G. Bogdanchikov, A. V. Bozhenok, A. D. Bukin, D. A. Bukin,
    S. V. Burdin, T. V. Dimova, A. A. Drozdetski, V. P. Druzhinin,
    D. I. Ganushin, V. B. Golubev, V. N. Ivanchenko, P. M. Ivanov,
    A. A. Korol, S. V. Koshuba, I. N. Nesterenko, E. V. Pakhtusova,
    A. A. Polunin, A. A. Salnikov, S. I. Serednyakov, V. V. Shary,
    Y. M. Shatunov, V. A. Sidorov, Z. K. Silagadze, A. N. Skrinsky,
    A. G. Skripkin, Y. V. Usov, and A. V. Vasiljev

LIGHT MESONS FROM CHARM DECAYS

Recent Results from Belle 43
    S. R. Hou (for the Belle Collaboration)
Three Body Decays of D^0 and D_s Mesons 53
    A. Palano

Italicized name indicates the author who presented the paper.

Light Meson Physics from Charm Decays at Fermilab E791 63
    C. Göbel (for the E791 Collaboration)
Heavy Quark Production and Spectroscopy at HERA 73
    U. Karshon (for the H1 and ZEUS Collaborations)

HEAVY Q SPECTROSCOPY

Understanding B Meson Branching Fractions 85
    B. Gittelman
Charmonium Spectroscopy from Fermilab E835 95
    G. Garzoglio, K. Gollwitzer, M. Hu, S. Pordes, S. Werkema, W. Baldini,
    D. Bettoni, R. Calabrese, G. Cibinetto, P. Dalpiaz, E. Luppi, M. Negrini,
    G. Stancari, M. Stancari, S. Bagnasco, A. Buzzo, M. LoVetere, M. Macrì,
    M. Marinelli, M. Pallavicini, C. Patrignani, E. Robutti, A. Santroni,
    G. Lasio, M. Mandelkern, J. Schultz, W. Roethel, M. Graham, R. Rusack,
    S. H. Seo, T. Vidnovic III, D. Joffe, J. Kasper, Z. Metreveli, J. Rosen,
    P. Rumerio, K. K. Seth, *A. Tomaradze*, P. Zweber, G. Borreani, R. Cester,
    F. Marchetto, E. Menichetti, R. Mussa, M. Obertino, and N. Pastrone
Aspects of Charmonium Physics 105
    S. F. Tuan
Analysis of Nature of $\phi \to \gamma \pi \eta$ and $\phi \to \gamma \pi^0 \pi^0$ Decays 112
    N. N. Achasov

EXOTICS

Search for Exotic Baryons with Hidden Strangeness in Proton Diffractive Production at the Energy of 70 GeV 125
    Y. M. Antipov, A. V. Artamonov, V. A. Batarin, O. V. Eroshin,
    S. V. Golovkin, Y. P. Gorin, A. P. Kozhevnikov, V. P. Kubarovsky,
    V. Kurshetsov, L. G. Landsberg, V. A. Medovikov, V. V. Molchanov,
    V. A. Mukhin, D. I. Patalakha, S. V. Petrenko, A. I. Petrukhin, V. S. Vaniev,
    D. V. Vavilov, V. A. Victorov, S. A. Zimin, V. Z. Kolganov, G. S. Lomkatsi,
    A. F. Nilov, and V. T. Smolyankin
Recent Results from Brookhaven E852 Experiment 135
    A. V. Popov (for the E852 Collaboration)
The $J^{PC}=1^{-+}$ Hunting Season at VES 143
    D. V. Amelin, *V. Dorofeev*, R. I. Dzhelyadin, Y. P. Gouz, I. A. Kachaev,
    A. N. Karyukhin, Y. A. Khokhlov, A. K. Konoplyannikov,
    V. F. Konstantinov, S. V. Kopikov, V. V. Kostyukhin, V. D. Matveev,
    V. I. Nikolaenko, A. P. Ostankov, B. F. Polyakov, D. I. Ryabchikov,
    A. A. Solodkov, O. V. Soloviianov, and A. M. Zaitsev
The Resonance Structures of $K_s K_s$ and $\Lambda \bar{\Lambda}$ Spectrum at MIS ITEP 155
    I. A. Erofeev, O. N. Erofeeva, V. K. Grigoriev, Y. V. Katinov, V. I. Lisin,
    V. N. Luzin, *V. N. Nozdrachev*, Y. P. Shkurenko, V. V. Sokolovsky,
    G. D. Tikhomirov, and V. V. Vladimirsky

Italicized name indicates the author who presented the paper.

SCALARS

Light Scalar Meson Spectrum 167
    W. Ochs

Investigating the Light Scalar Mesons 178
    D. Black, S. Moussa, S. Nasri, A. H. Fariborz, and *J. Schechter*

Covariant Classification Scheme of Hadrons 187
    S. Ishida and M. Ishida

Systematics, $q\bar{q}$ States Scalar Mesons, and Glueball 197
    V. V. Anisovich

On the Dominance of Nonexotic Meson-Meson Scattering by S-Channel $q\bar{q}$ Confinement States and the Classification of the Scalar Mesons 209
    E. van Beveren and G. Rupp

NONPERTURBATIVE QCD

Light Hadron Spectroscopy from Lattice QCD 221
    T. Kaneko

Gluonic Excitations in Lattice QCD: A Brief Survey 231
    C. Morningstar

QCD String Model for Hybrid Adiabatic Potentials 241
    Y. S. Kalashnikova and D. S. Kuzmenko

Remembering Nathan 251
    T. Barnes

HISTORY

Hadronic Octaves: Symphony in Treble Clef 259
    Y. Ne'eman

BARYONS, QG PLASMA

Open Questions in Baryon Spectroscopy 277
    V. Credé

A Review of Baryon Resonance Analysis and Comparisons with Meson Resonance Analyses 287
    S. A. Dytman

The New Crystal Ball Experimental Program 297
    W. J. Briscoe (for the Crystal Ball Collaboration)

J/ψ Attenuation and the Quark-Gluon Plasma 306
    K. K. Seth

Italicized name indicates the author who presented the paper.

The Sigma Meson and Chiral Transition in Hot and Dense Matter 315
   T. Kunihiro

PHENOMENOLOGY

The Constituent Quark Model: A Status Report 327
   E. S. Swanson
Heavy Quark Potential and Mass Spectra of Heavy Mesons 336
   D. Ebert, R. N. Faustov, and V. O. Galkin
**Recent Results of $\psi(2S)$ Decay Branching Ratios and Decay Widths
from BES** 346
   Y. Zhu (for the BES Collaboration)
$q\bar{q}$ **and "Extra" Mesons up to 2400 MeV** 356
   D. V. Bugg

STANDARD MODEL

**A New Measurement of $\mathrm{Re}(\varepsilon'/\varepsilon)$ in $K^0 \to 2\pi$ Decays by the
Experiment NA48 at CERN** 367
   M. Holder and *M. Ziolkowsky* (for the NA48 Collaboration)
W Mass and Width Determination at LEP II 375
   F. Ligabue
Recent QCD Results from CDF 385
   I. V. Gorelov

NEW FACILITIES

Recent J/ψ Physics Results from BES 399
   Z. Gou (for the BES Collaboration)
The Hall D Detector at Jefferson Lab 408
   C. A. Meyer
The Antiproton Project at GSI 414
   H. Koch
The KLOE Physics Program 424
   A. Aloisio, F. Ambrosino, A. Antonelli, M. Antonelli, C. Bacci,
   G. Barbiellini, F. Bellini, G. Bencivenni, S. Bertolucci, C. Bini, C. Bloise,
   V. Bocci, F. Bossi, P. Branchini, S. A. Bulychjov, G. Cabibbo, R. Caloi,
   P. Campana, G. Capon, G. Carboni, M. Casarsa, V. Casavola, G. Cataldi,
   F. Ceradini, F. Cervelli, F. Cevenini, G. Chiefari, P. Ciambrone, S. Conetti,
   E. De Lucia, G. De Robertis, P. De Simone, G. De Zorzi, S. Dell'Agnello,
   A. Denig, A. Di Domenico, C. Di Donato, S. Di Falco, A. Doria,
   M. Dreucci, O. Erriquez, A. Farilla, G. Felici, A. Ferrari, M. L. Ferrer,
   G. Finocchiaro, C. Forti, A. Franceschi, P. Franzini, C. Gatti, P. Gauzzi,
   A. Giannasi, S. Giovannella, E. Gorini, F. Grancagnolo, E. Graziani,

Italicized name indicates the author who presented the paper.

S. W. Han, M. Incagli, L. Ingrosso, W. Kluge, C. Kuo, V. Kulikov,
F. Lacava, G. Lanfranchi, J. Lee-Franzini, D. Leone, F. Lu,
M. Martemianov, M. Matsyuk, W. Mei, A. Menicucci, L. Merola, R. Messi,
S. Miscetti, M. Moulson, S. Mueller, F. Murtas, M. Napolitano,
A. Nedosekin, F. Nguyen, M. Palutan, L. Paoluzi, E. Pasqualucci,
L. Passalacqua, A. Passeri, V. Patera, E. Petrolo, D. Picca, G. Pirozzi,
L. Pontecorvo, M. Primavera, F. Ruggieri, P. Santangelo, E. Santovetti,
G. Saracino, R. D. Schamberger, B. Sciascia, A. Sciubba, F. Scuri,
I. Sfiligoi, J. Shan, P. Silano, T. Spadaro, E. Spiriti, G. L. Tong, L. Tortora,
E. Valente, *P. Valente*, B. Valeriani, G. Venanzoni, S. Veneziano,
A. Ventura, Y. Wu, G. Xu, G. W. Yu, P. F. Zema, and Y. Zhou

Exploring the Charm Sector with CLEO-c 434
D. Urner

SUMMARY

Hadron 2001 Conference Summary: Theory 447
T. Barnes

Hadron 2001 Summary: Experiment 463
E. Klempt

PARALLEL SESSIONS

SCALARS, THEORY

Mixing among Scalar Mesons and Scalar Glueball 487
T. Teshima, I. Kitamura, and N. Morisita

The f_0 Mesons in Processes $\pi\pi \to \pi\pi$, $K\bar{K}$ 491
Y. S. Surovtsev, D. Krupa, and M. Nagy

Chiral Symmetry and Scalars 495
S. F. Tuan

The σ-Meson Production in Excited Υ Decay Processes 499
T. Komada, M. Ishida, and S. Ishida

BARYONS, EXPERIMENT

**High-Lying N* Studies in Electromagnetic Double Charged
Pion Production** 505
V. I. Mokeev, M. Ripani, M. Anghinolfi, M. Battaglieri, R. De Vita,
G. V. Fedotov, E. N. Golovach, B. S. Ishkhanov, M. V. Osipenko, G. Ricco,
V. Sapunenko, M. Taiuti, and the CLAS Collaboration

Strangeness Production at CLAS 510
K. H. Hicks (for the CLAS Collaboration)

Italicized name indicates the author who presented the paper.

Experimental Check of the GDH Sum Rule at MAMI and ELSA 514
    J. Krimmer (for the GDH-Collaboration)

CENTRAL PRODUCTION, HEAVY IONS

Preliminary Results of a PWA of the Centrally Produced $\phi\phi$ System. 521
    M. A. Reyes, M. C. Berisso, D. C. Christian, J. Felix, A. Gara,
    E. E. Gottschalk, G. Gutiérrez, E. P. Hartouni, B. C. Knapp, M. N. Kreisler,
    S. Lee, K. Markianos, G. Moreno, M. H. L. S. Wang, A. Wehman, and
    D. Wesson

Glueball Candidates Production in Peripheral Heavy Ion Collisions
at ALICE 525
    M. Bashkanov

J/Ψ Production in the Peripheral Collisions of Heavy Ions 529
    S. Timoshenko

A Comment on the New Low-Lying States in the Recent
Experimental Results 533
    H. Noya and H. Nakamura

CHARM AND BEAUTY DECAYS

Measurement of HQET Parameters and CKM Matrix Elements 539
    D. Urner

Study of Ω_c^0 Production and New Results on Λ_c^+ Decays at Belle 543
    R. Chistov (for the Belle Collaboration)

Production and Decay of the Λ_c Charmed Baryon
from Fermilab E791 547
    B. Meadows (for the E791 Collaboration)

Dalitz Analysis of $D_S^\pm \to K_S^0 K_S^0 \pi^\pm$ and $D_S^\pm \to \pi^+ \pi^- \pi^\pm$ 551
    T. Deppermann, K. Peters, and H. Schmüker

Measurement of Inclusive f_1 (1285) and f_1 (1420) Production in
Z Decays with the DELPHI Detector 555
    D. Ryabchikov (for the DELPHI Collaboration)

EXOTICS

Exotic $J^{PC}=1^{-+}$ Mesons at the Present Time (Discussion
of Some Problems) 561
    L. I. Sarycheva and V. L. Korotkikh

A Study of the Reaction $\pi^- p \to \omega \pi^- \pi^0 p$ at 18 GeV/c 565
    A. V. Popov (for the E852 Collaboration)

Partial Wave Analysis of $\pi^- \pi^- \pi^+ \eta$ in the Reaction
$\pi^- p \to \pi^- \pi^- \pi^+ \eta p$ at 18 GeV/c 569
    J. Kuhn (for the E852 Collaboration)

Italicized name indicates the author who presented the paper.

A Study of the $\eta\eta\pi^-$ System Produced in the Reaction
$\pi^- p \to p\pi^+\pi^-\pi^-4\gamma$ at 18 GeV/c 573
    P. Eugenio (for the Brookhaven E852 Collaboration)

Study of Reaction $\pi^- A \to \pi^+\pi^-\pi^- A$ at VES Setup 577
    D. V. Amelin, R. I. Dzhelyadin, V. A. Dorofeev, Y. P. Gouz, *I. A. Kachaev*,
    A. N. Karyukhin, Y. A. Khokhlov, A. K. Konoplyannikov,
    V. F. Konstantinov, S. V. Kopikov, V. V. Kostyukhin, V. D. Matveev,
    V. I. Nikolaenko, A. P. Ostankov, B. F. Polyakov, D. I. Ryabchikov,
    A. A. Solodkov, O. V. Soloviyanov, and A. M. Zaitsev

Study of Isospin 2 States in $\bar{n}p$ Annihilations 582
    A. Filippi (for the Obelix (PS201) Experiment)

MODELS

Instanton Effects in the Meson and Baryon Spectrum 589
    B. Metsch

Light-Light and Heavy-Light Mesons in the Model of QCD String
with Quarks at the Ends 595
    A. V. Nefediev

Effective Lagrangians Induced by the Anomalous Wess-Zumino
Action and $I^G(J^{PC}) = 1^-(1^{-+})$ Exotic States 599
    N. N. Achasov and *G. N. Shestakov*

Rare Radiative B Decays to Excited K Mesons 603
    D. Ebert, R. N. Faustov, and *V. O. Galkin*

Wavelet Analysis in Physics of Resonances: Application to Spectrum
of ρ' and ω' Excitations and Ratio $R_{e^+e^-}$ 607
    V. K. Henner, P. G. Frick, and T. S. Belozerova

Scalar Strong Interaction Hadron Theory—An Alternative to QCD 611
    F. C. Hoh

Wavelet Analysis of E852 Experimental Data 615
    V. L. Korotkikh and L. I. Sarycheva

Quasirotational States in Various String Hadron Models and
Regge Trajectories 619
    G. S. Sharov

A Light Meson Translatable Template 623
    C. E. Allgower and *D. C. Peaslee*

STATES WITH OPEN/HIDDEN STRANGENESS

Some Features of $\Lambda\bar{\Lambda}$ System in $\pi^-(p,C) \to \bar{\Lambda}\Lambda + A^*$ at 40 GeV 629
    I. A. Erofeev, O. N. Erofeeva, V. K. Grigoriev, Y. V. Katinov, V. I. Lisin,
    V. N. Luzin, V. N. Nozdrachev, Y. P. Shkurenko, *V. V. Sokolovsky*,
    G. D. Tikhomirov, and V. V. Vladimirsky

Italicized name indicates the author who presented the paper.

Observation of Narrow States of $K_s^0 K_s^0$ System on 6-m Spectrometer
and Comparison with L3 Results.. 633
        I. A. Erofeev, O. N. Erofeeva, V. K. Grigoriev, Y. V. Katinov, *V. I. Lisin*,
        V. N. Luzin, V. N. Nozdrachev, Y. P. Shkurenko, V. V. Sokolovsky,
        G. D. Tikhomirov, and V. V. Vladimirsky

Study of the $\eta\eta$ System in the $\pi^- p$ Charge Exchange Reaction
at 32 GeV/c with the GAMS-4π Spectrometer.................................... 637
        A. M. Blick, F. G. Binon, A. V. Dolgopolov, S. V. Donskov, S. Inaba,
        Y. Fujii, G. V. Khaustov, *V. N. Kolosov*, A. A. Kondashov, A. A. Lednev,
        V. A. Lishin, J. P. Peigeneux, V. A. Polyakov, S. A. Sadovsky,
        V. D. Samoylenko, P. M. Shagin, H. Shimizu, A. V. Singovsky,
        A. E. Sobol, J. P. Stroot, V. P. Sugonyaev, K. Takamatsu, T. Tsuru, Y. Yasu,
        and A. Y. Zvyagin

A Study of Reaction $K^- N \to (K^- \pi^+ \pi^-) N$ at 28 GeV/c.......................... 641
        V. Nikolaenko (for the VES Experiment)

PHENOMENOLOGY

Relative Phase between the Three-Gluon and One-Photon Amplitudes
of the J/ψ Decays 649
        N. N. Achasov

Sum Rules for Total Hadronic Widths of Mesons 653
        M. Majewski

Possible Evidence for a Chiral Axial-Vector State in
the D Meson System... 657
        K. Yamada, M. Ishida, S. Ishida, D. Ito, T. Komada,
        and H. Tonooka

Weak Decays of Heavy Mesons in a Covariant Quark Model................. 661
        D. Merten

Potential of the Rare Heavy Quark Decay Studies at
the ATLAS Experiment 665
        N. Nikitine

The Heavy Baryons in the Nonperturbative String Approach................ 669
        I. M. Narodetski and M. A. Trusov

Hadron-Hadron Scattering in the Nonrelativistic Quark Model............. 673
        T. Barnes

Radial Excitations of Pseudoscalar Mesons 678
        J. S. Suh

Higher Vector Meson States 683
        B. Pick (for the Crystal Barrel Collaboration)

On Spectroscopy of ρ', ρ'' and ω', ω'' Resonances 687
        N. N. Achasov and *A. A. Kozhevnikov*

Italicized name indicates the author who presented the paper.

LOW ENERGY

**New Results on Baryon Spectroscopy with the
Crystal Ball Spectrometer** 693
    D. Isenhower, M. Sadler, C. E. Allgower, H. Spinka, J. R. Comfort,
    K. Craig, A. F. Ramirez, M. Clajus, A. Marušić, S. McDonald,
    B. M. K. Nefkens, N. Phaisangittisakul, S. Prakhov, J. W. Price,
    A. Starostin, W. B. Tippens, J. Peterson, W. J. Briscoe, A. Shafi,
    I. I. Strakovsky, H. M. Staudenmaier, *D. M. Manley*, J. Olmsted,
    D. C. Peaslee, V. V. Abaev, V. Bekrenev, A. A. Kulbardis, N. G. Kozlenko,
    S. Kruglov, I. V. Lopatin, N. Knecht, G. Lolos, Z. Papandreou, I. Supek,
    D. Grosnick, D. D. Koetke, R. Manweiler, and T. D. S. Stanislaus

**Measurement of Differential Cross Sections of the Reaction $\pi^- p \rightarrow \eta n$
Using the Crystal Ball Detector** 697
    B. Draper, S. Hayden, J. Huddleston, D. Isenhower, C. Robinson,
    M. Sadler, C. Allgower, R. Cadman, H. Spinka, J. Comfort, K. Craig,
    A. Ramirez, T. Kycia, M. Clajus, A. Marusic, S. McDonald,
    B. M. K. Nefkens, N. Phaisangittisakul, S. Prakhov, J. W. Price,
    W. B. Tippens, J. Peterson, W. Briscoe, A. Shafi, I. Strakovsky,
    H. Staudenmaier, D. M. Manley, J. Olmsted, D. Peaslee, V. Abaev,
    V. Bekrenev, A. Koulbardis, *N. G. Kozlenko*, S. Kruglov, I. Lopatin,
    A. Starostin, N. Knecht, G. Lolos, Z. Papandreou, I. Supek, A. Gibson,
    D. Grosnic, D. D. Koetke, R. Manweiler, S. Stanislaus, H. Calen,
    A. Kupse, T. Johanson, and U. Wiedner

**Differential Cross Sections of the Charge Exchange Reaction
$\pi^- p \rightarrow \pi^0 n$ in the Momentum Range from 148 to 323 MeV/c** 701
    B. Draper, S. Hayden, J. Huddleston, D. Isenhower, C. Robinson,
    M. Sadler, C. Allgower, R. Cadman, H. Spinka, J. Comfort, K. Craig,
    A. Ramirez, T. Kycia (deceased), M. Clajus, A. Marusic, S. McDonald,
    B. M. K. Nefkens, N. Phaisangittisakul, S. Prakhov, J. W. Price,
    W. B. Tippens, J. Peterson, W. Briscoe, A. Shafi, I. Strakovsky,
    H. Staudenmaier, D. M. Manley, J. Olmsted, D. Peaslee, V. Abaev,
    V. Bekrenev, *A. Koulbardis*, N. Kozlenko, S. Kruglov, I. Lopatin,
    A. Starostin, N. Knecht, G. Lolos, Z. Papandreou, I. Supek, A. Gibson,
    D. Grosnic, D. D. Koetke, R. Manweiler, S. Stanislaus, H. Calen,
    A. Kupse, T. Johanson, and U. Wiedner

**LEPS Experiment at SPring-8: Detector Status and
Preliminary Results** 705
    P. Shagin (for the LEPS Collaboration)

RADIATIVE DECAYS

**Detection of $\phi \rightarrow \pi^0 \gamma$, $\phi \rightarrow \eta \gamma$ and $\phi \rightarrow \eta' \gamma$ with KLOE Detector
at DAΦNE** 711
    A. Aloisio, F. Ambrosino, A. Antonelli, M. Antonelli, C. Bacci,
    G. Barbiellini, F. Bellini, G. Bencivenni, S. Bertolucci, C. Bini, C. Bloise,

Italicized name indicates the author who presented the paper.

V. Bocci, F. Bossi, P. Branchini, S. A. Bulychjov, G. Cabibbo, R. Caloi,
P. Campana, G. Capon, G. Carboni, M. Casarsa, V. Casavola, G. Cataldi,
F. Ceradini, F. Cervelli, F. Cevenini, G. Chiefari, P. Ciambrone, S. Conetti,
E. De Lucia, G. De Robertis, P. De Simone, G. De Zorzi, S. Dell'Agnello,
A. Denig, A. Di Domenico, C. Di Donato, S. Di Falco, A. Doria,
M. Dreucci, O. Erriquez, A. Farilla, G. Felici, A. Ferrari, M. L. Ferrer,
G. Finocchiaro, C. Forti, A. Franceschi, P. Franzini, C. Gatti, P. Gauzzi,
A. Giannasi, S. Giovannella, E. Gorini, F. Grancagnolo, E. Graziani,
S. W. Han, M. Incagli, L. Ingrosso, W. Kluge, C. Kuo, V. Kulikov,
F. Lacava, G. Lanfranchi, J. Lee-Franzini, D. Leone, F. Lu,
M. Martemianov, M. Matsyuk, W. Mei, A. Menicucci, L. Merola, R. Messi,
S. Miscetti, M. Moulson, S. Müller, F. Murtas, M. Napolitano,
A. Nedosekin, F. Nguyen, M. Palutan, L. Paoluzi, E. Pasqualucci,
L. Passalacqua, A. Passeri, V. Patera, E. Petrolo, D. Picca, G. Pirozzi,
L. Pontecorvo, M. Primavera, F. Ruggieri, P. Santangelo, E. Santovetti,
G. Saracino, R. D. Schamberger, B. Sciascia, A. Sciubba, F. Scuri,
I. Sfiligoi, J. Shan, P. Silano, T. Spadaro, E. Spiriti, G. L. Tong, L. Tortora,
E. Valente, P. Valente, B. Valeriani, G. Venanzoni, S. Veneziano, A. Ventura,
Y. Wu, G. Xu, G. W. Yu, P. F. Zema, and Y. Zhou

Study of ϕ Decays to $f_0(980)\gamma$ and $a_0(980)\gamma$ with KLOE at DAΦNE 716

A. Aloisio, F. Ambrosino, A. Antonelli, M. Antonelli, C. Bacci,
G. Barbiellini, F. Bellini, G. Bencivenni, S. Bertolucci, C. Bini, C. Bloise,
V. Bocci, F. Bossi, P. Branchini, S. A. Bulychjov, G. Cabibbo, R. Caloi,
P. Campana, G. Capon, G. Carboni, M. Casarsa, V. Casavola, G. Cataldi,
F. Ceradini, F. Cervelli, F. Cevenini, G. Chiefari, P. Ciambrone, S. Conetti,
E. De Lucia, G. De Robertis, P. De Simone, G. De Zorzi, S. Dell'Agnello,
A. Denig, A. Di Domenico, C. Di Donato, S. Di Falco, A. Doria,
M. Dreucci, O. Erriquez, A. Farilla, G. Felici, A. Ferrari, M. L. Ferrer,
G. Finocchiaro, C. Forti, A. Franceschi, P. Franzini, C. Gatti, P. Gauzzi,
A. Giannasi, S. Giovannella, E. Gorini, F. Grancagnolo, E. Graziani,
S. W. Han, M. Incagli, L. Ingrosso, W. Kluge, C. Kuo, V. Kulikov,
F. Lacava, G. Lanfranchi, J. Lee-Franzini, D. Leone, F. Lu,
M. Martemianov, M. Matsyuk, W. Mei, A. Menicucci, L. Merola, R. Messi,
S. Miscetti, M. Moulson, S. Müller, F. Murtas, M. Napolitano,
A. Nedosekin, F. Nguyen, M. Palutan, L. Paoluzi, E. Pasqualucci,
L. Passalacqua, A. Passeri, V. Patera, E. Petrolo, D. Picca, G. Pirozzi,
L. Pontecorvo, M. Primavera, F. Ruggieri, P. Santangelo, E. Santovetti,
G. Saracino, R. D. Schamberger, B. Sciascia, A. Sciubba, F. Scuri,
I. Sfiligoi, J. Shan, P. Silano, T. Spadaro, E. Spiriti, G. L. Tong, L. Tortora,
E. Valente, P. Valente, *B. Valeriani*, G. Venanzoni, S. Veneziano,
A. Ventura, Y. Wu, G. Xu, G. W. Yu, P. F. Zema, and Y. Zhou

Study of the Process $e^+e^- \to \eta\gamma \to 7\gamma$ in the Energy Region $\sqrt{s} < 1.4$ GeV 721

M. N. Achasov, S. E. Baru, K. I. Belkoborodov, *A. V. Berdyugin*,
A. G. Bogdanchikov, A. V. Bozhenok, A. D. Bukin, D. A. Bukin,
S. V. Burdin, T. V. Dimova, A. A. Drozdetski, V. P. Druzhinin,
V. B. Golubev, V. N. Ivanchenko, P. M. Ivanov, I. A. Koop, A. A. Korol,

Italicized name indicates the author who presented the paper.

M. S. Korostelev, S. V. Koshuba, A. V. Otboev, E. V. Pakhtusova,
E. A. Perevedentsev, A. A. Salnikov, S. I. Serednyakov, V. V. Shary,
Y. M. Shatunov, V. A. Sidorov, Z. K. Silagadze, A. G. Skripkin, and
A. V. Vasiljev

CHIRAL THEORY

Chirality and the Quark Model 727
    E. S. Swanson and A. P. Szczepaniak
Property of Chiral Scalar and Axial-Vector Mesons in Heavy-Light Quark Systems 731
    M. Ishida and S. Ishida
Confirmation of $\sigma(450-600)$-Meson in $Y' \to Y\pi\pi$ and Other $\pi\pi$-Production Processes 735
    M. Ishida, S. Ishida, T. Komada, and S. I. Matsumoto
Complete Meson-Meson Scattering within One Loop in Chiral Perturbation Theory: Unitarization and Resonances 739
    J. R. Peláez and A. Gómez Nicola

VERY LOW ENERGY, DEUTERON

Detection of Atoms Consisting of π^+ and π^- Mesons at PS CERN 745
    B. Adeva, *L. Afanasyev*, M. Benayoun, Z. Berka, V. Brekhovskikh,
    G. Caragheorgheopol, T. Cechak, M. Chiba, S. Constantini,
    S. Constantinescu, A. Doudarev, D. Dreossi, D. Drijard, M. Ferro-Luzzi,
    T. Gallas Torreira, J. Gerndt, R. Giacomich, P. Gianotti, F. Gomez,
    A. Gorin, O. Gortchakov, C. Guaraldo, M. Hansroul, R. Hosek, M. Iliescu,
    M. Jabitski, N. Kalinina, V. Karpukhin, J. Kluson, M. Kobayashi,
    P. Kokkas, V. Komarov, A. Koulikov, A. Kouptsov, V. Krouglov,
    L. Krouglova, K.-I. Kuroda, A. Lamberto, A. Lanaro, V. Lapshin,
    R. Lednicky, P. Leruste, P. Levi Sandri, A. Lopez Aguera, V. Lucherini,
    T. Maki, I. Manuilov, L. Montanet, J.-L. Narjoux, L. Nemenov, M. Nikitin,
    T. Nunez Pardo, K. Okada, V. Olchevskii, A. Pazos, M. Pentia, A. Penzo,
    J.-M. Perreau, C. Petrascu, M. Plo, T. Ponta, D. Pop, G. F. Rappazzo,
    A. Riazantsev, J. M. Rodriguez, A. Rodriguez Fernandez, V. Rykalin,
    C. Santamarina, J. Saborido, J. Schacher, C. Schuetz, A. Sidorov, J. Smolik,
    F. Takeutchi, A. Tarasov, L. Tauscher, M. J. Tobar, S. Trousov, P. Vazquez,
    S. Vlachos, V. Yazkov, Y. Yoshimura, and P. Zrelov
Deeply Bound $1s$ and $2p$ Pionic States in ^{205}Pb and Effective Pion Mass in Nuclear Matter 749
    H. Geissel, H. Gilg, A. Gillitzer, R. S. Hayano, S. Hirenzaki, *K. Itahashi*,
    M. Iwasaki, P. Kienle, M. Münch, G. Münzenberg, W. Schott, K. Suzuki,
    D. Tomono, H. Weick, T. Yamazaki, and T. Yoneyama

Italicized name indicates the author who presented the paper.

Search for Supernarrow Dibaryons Production in $pd \to p+pX_1$
and $pd \to p+dX_2$ Reactions 753
    L. V. Fil'kov, V. L. Kashevarov, E. S. Konobeevski, M. V. Mordovskoy,
    S. I. Potashev, V. A. Simonov, and V. M. Skorkin

Study of Two Photon Production Process in Proton-Proton Collisions
at 216 MeV 761
    A. S. Khrykin

Are the Exotic Mesons and Baryons, Recently Observed, a Signature
of Quark-Hadron Duality? 765
    B. Tatischeff

On a Manifestation of Dibaryon Resonances in the Structure
of Proton-Proton Total Cross Section at Low Energies 771
    A. A. Arkhipov

$N\bar{N}$

A High-Resolution Search for the Tensor Glueball Candidate
$\xi(2230)$ 779
    K. K. Seth (for the Crystal Barrel Collaboration)

Antiproton Annihilation on Nuclei 783
    K. Protasov and R. Duperray

$\rho\pi$-States in the Antiproton-Neutron-Annihilation into $\pi^-3\pi^\circ$ 787
    F. Meyer-Wildhagen (for the Crystal Barrel Collaboration)

Search for an Exotic Partial Wave in $\pi\eta'$ 792
    J. Reinnarth (for the Crystal Barrel Collaboration)

$\gamma\gamma$ AND INCLUSIVE REACTIONS

Inclusive D*-Meson Production in Two-Photon Collisions at LEP 799
    A. A. Sokolov

Inclusive J/ψ Production in Two-Photon Collisions at LEP II with the
DELPHI Detector 803
    M. Chapkine

Relating Production and Masses of the Vector and P-Wave Mesons
for Light and Heavy Flavours 807
    P. V. Chliapnikov

Do the Strange Quarks in Z^0 Decays and Kaon Valence $\bar{s}$ Quark in
K^+p Reactions Fragment Differently? 811
    P. V. Chliapnikov

Determination of the Strangeness Content of Light-Flavour Isoscalars
from Their Production Rates in Hadronic Z Decays at LEP 815
    V. A. Uvarov

Italicized name indicates the author who presented the paper.

Measurements of the Vector and Tensor Analyzing Powers of the Inelastic Scattering of Deuterons on Nuclei in the Vicinity of Baryonic Resonances Excitation 819
    L. S. *Azhgirey*, V. P. Ladygin, S. V. Afanasiev, V. V. Arkhipov,
    V. K. Bondarev, G. Filipov, A. Y. Isupov, V. I. Ivanov, A. A. Kartamyshev,
    V. A. Kashirin, A. N. Khrenov, V. I. Kolesnikov, V. A. Kuznetsov,
    N. B. Ladygina, A. G. Litvinenko, S. G. Reznikov, P. A. Rukoyatkin,
    A. Y. Semenov, I. A. Semenova, G. D. Stoletov, V. N. Zhmyrov, and
    L. S. Zolin

RARE DECAYS, SEARCHES FOR NEW PHENOMENA

New Results on Rare Decays and on Future NA48 825
    A. *Zinchenko* (for the NA48 Collaboration)

Search for the Standard Model Higgs Boson at the ALEPH Detector 829
    R. R. *White*

RADIATIVE DECAYS

Radiative Decay Width of the $a_2(1320)$ Meson 835
    V. V. *Molchanov* (for the SELEX Collaboration)

Radiative Transitions in Mesons within a Nonrelativistic Quark Model 839
    R. Bonnaz, B. *Silvestre-Brac*, and C. Gignoux

Hadron 2001 Participants' List 845

Author Index 849

Italicized name indicates the author who presented the paper.

Preface

The Ninth International Conference on Hadron Spectroscopy (HADRON01) took place at the Institute for High Energy Physics, Protvino, Russia, from 25 August 2001 to 1 September 2001. Some 145 scientists from 15 countries attended the Conference.

Conferences in this series take place every two years, starting with 1985. The previous Conference was held in 1999 in Beijing, China. The scope of these conferences is hadron spectroscopy and related aspects of hadron dynamics. At present this extensive field of high energy physics demonstrates significant progress both in the experimental and in theoretical aspects.

HADRON01 started with registration and reception on Sunday, 25 August 2001, and ended with two summary talks, experimental and theoretical, Saturday morning, 1 September. In the organization of the Conference, we chose the format where some of the reports were presented at plenary sessions and some at three parallel sessions. The special after dinner session was devoted to the history of hadron spectroscopy.

The efforts of the International Advisory Committee and Program Committee in shaping the program and the high quality of the speakers guaranteed a high scientific level of the conference. The social program, such as concerts and tours, as well as good weather also helped to reach a positive result. It is a pleasure for the Chairman to thank all the participants and all those who have contributed to the success of this Conference.

Thanks are due to the sponsors of the conference:

- Ministry of Atomic Energy;
- Ministry of Industry, Science and Technologies;
- Russian Academy of Sciences;
- Russian Fund of Fundamental Research.

I would like to thank the members of the Local Organizing Committee - Yu. Ryabov, I. Kostyuhina, V. Komarova, M. Lissina, G. Solodkova, and M. Pirogova for their hard and effective work

Finally, it is a great pleasure to wish full success to HADRON03, which will be held in Germany.

Alexander M. Zaitsev

INTERNATION ADVISORY COMMITTEE

C. Amsler (Zurich)
A. Astbury (TRIUMF)
J. Appel (FNAL)
P. Barnes (Los Alamos)
E. Berger (Argonne)
T. Bressani Torino)
S. Chung (BNL)
F. Close (CERN & Rutherford)
S. Denisov (Protvino)
A. Donnachie (Manchester)
W. Dunwoodie (SLAC)
A. Dzierba (Indiana)
S. Gershtein (Protvino)
C. Guaraldo (Frascati)
N. Isgur (Jefferson Lab)
E. Klempt (Bonn)
H. Lipkin (Weizmann)
V. Matveev (Moscow)
L. Montanet (CERN)
S. Nagamiya (KEK)
S. Paul (CERN & Munich)
D. Peaslee (Maryland)
A. Skrinsky (Novosibirsk)
K. Takamatsu (KEK)
W.G. Li (IHEP,Beijing)

INTERNATION PROGRAM COMMITTEE

N. Achasov (Novosibirsk)
S. Dytman (Pittsburg)
T. Barnes (Oak Ridge)
K. Kilian (Julich)
A. Kirk (Birmingham)
S. Narison (Montpellier)
M. Pennington (Durham)
K.K. Seth (Northwest Univ)
T. Tsuru (KEK)
L. Landsberg (Protvino)
Z.G. Zhao (IHEP, Beijing)
P. Franzini (University of Rome)

CONFERENCE CHAIRMAN

A.M. Zaitsev, IHEP

LOCAL ORGANIZING COMMITTEE

Y. Ryabov
D. Amelin
A. Karpenko
V. Komarova
I. Kostyukhina
M. Lissina
G. Malitskij
V. Nikolaenko
M. Pirogova
G. Solodkova
V. Ugarov
O. Zdorovenkov

PLENARY SESSIONS

VECTOR MESON STATES

Light-Quark Vector-Meson Spectroscopy

A. Donnachie* and Yu S Kalashnikova[†]

*Department of Physics and Astronomy, University of Manchester,
Manchester M13 9PL, England*
[†] *ITEP, 117259 Moscow, Russia*

Abstract. The current situation for vector meson spectroscopy is outlined, and it is shown that the data are inconsistent with the generally accepted model for meson decay. A possible resolution in terms of exotic mesons is given. Although this resolves some of the issues, fresh theoretical questions are raised.

INTRODUCTION

It is now 15 years since it was first suggested [1,2] that the $\rho'(1600)$, as it was then known, is in fact a composite structure, consisting of at least two states: the $\rho(1450)$ and $\rho(1700)$. Their existence, and that of their isoscalar counterparts, the $\omega(1420)$ and $\omega(1650)$, and of an associated hidden-strangeness state, the $\phi(1680)$, is now well established [3]. So it is pertinent to ask why the light-quark vector mesons remain an important field of study. The answer is straightforward. Although there is general consensus on the existence of these states, there is considerable disparity on their masses and widths. Further what is known about the composition of their hadronic decays raises fundamental questions about the nature of these states and our understanding of the mechanism of hadronic decays.

A further complication has been highlighted in two of the talks [4,5] at this meeting, with indications of an isovector vector meson at a mass of around 1200 to 1250 MeV. This revives an old controversy. Many years ago evidence was presented [6] for two vector states with masses 1097 ± 19 MeV and 1266 ± 35 MeV in the reaction $\gamma p \to e^+ e^- p$. The evidence for these two states was obtained from the interference between the Bethe-Heitler amplitude and the real part of the hadronic photoproduction amplitude. Additionally in $\omega\pi$ photoproduction [7,8,9], $\gamma p \to (\omega\pi)p$, the $\omega\pi$ system is dominated by a low-mass enhancement with a peak at about 1250 MeV and it seemed natural to associate this with a vector state. However it appears that this enhancement is dominated by the $J^P = 1^+$ $b_1(1235)$ meson. The evidence for this comes from the analysis [8,9] of the decay angular distributions of the $\omega\pi$ system. The conclusion of [8] is that the data are best described by production of the $b_1(1235)$ together with a small $J^P = 1^-$ contribution.

Additionally the production mechanism does not appear to conserve s-channel helicity. This latter conclusion is confirmed in the experiment of [9] which had the benefit of a linearly-polarised beam. The angular distribution of the production plane relative to the photon polarisation vector has structure which is inconsistent with s-channel helicity conservation. The decay angular distributions of the two experiments agree and it was also concluded by [9] that the data favour a $b_1(1235)$ interpretation over a vector-meson interpretation. Preliminary data [10] on $\omega\pi$ photoproduction at high energy indicate a cross section for the 1250 MeV enhancement which is comparable to or larger than the cross section at much lower energy [8,9], with the natural inference that it is being produced diffractively. However diffractive photoproduction of the $b_1(1235)$ is inconsistent with all we know about diffraction from other reactions. It violates the Gribov-Morrison rule [11,12] and, more seriously, at the parton level it requires both spin flip and angular-momentum flip. The photon is some combination of $^3S_1$ and $^3D_1$ $q\bar{q}$ states and the $b_1(1235)$ is $^1P_1$. To avoid conflict with diffraction phenomenology, a possible interpretation of the $\omega\pi$ photoproduction data is that it is a mix of 1^- and 1^+ at the lower energies, and entirely 1^- at high energy, requiring a vector meson at about 1250 MeV.

Obviously the light-quark vector mesons present an exciting theoretical challenge.

Information on the vector states comes principally from e^+e^- annihilation, and also τ decay for the isovector states, but there are problems with much of the data:
- inconsistencies, even in recent high-statistics data
- restricted energy ranges, e.g. Novosibirsk and CLEO
- poor statistics in some channels and missing channels
- inadequate knowledge of multiparticle final states

Fortunately this is set to change with a range of possible new facilities for e^+e^- annihilation and τ decay:
- upgrade of Novosibirsk to higher energy
- the PEP-N proposal at SLAC
- the use of initial state radiation (ISR) at BABAR
- emphasis on τ and charm at CLEO

There is also the possibilty of complementary data on vector meson photoproduction:
- from the upgrade of CEBAF
- from real photon radiation in proton-ion and ion-ion collisions at RHIC
- photo- and electroproduction at HERA

So the future study of vector mesons looks healthy.

THE DATA

The key data in determining the existence of the two isovector states were $e^+e^- \to \pi^+\pi^-$ [13] and $e^+e^- \to \omega\pi$ [14]. These original data sets have subsequently been augmented by data on the corresponding charged channels in τ decay

[15,16], to which they are related by CVC. These new data confirm the earlier conclusions. The data on $e^+e^- \to \pi^+\pi^-\pi^+\pi^-$ [17] and $e^+e^- \to \pi^+\pi^-\pi^0\pi^0$ [17,18] (excluding $\omega\pi$) and the corresponding charged channels in τ decay [16] are consistent with the two-resonance interpretation [19,20], although they do not provide such good discrimination. It was also found that the $e^+e^- \to \eta\pi^+\pi^-$ cross section is better fitted with two interfering resonances than with a single state [21]. Independent evidence for two $J^P = 1^-$ states is provided in a high statistics study of the $\eta\pi\pi$ system in π^-p charge exchange [22]. Decisive evidence for both the $\rho(1450)$ and $\rho(1700)$ in their 2π and 4π decays has come from the study of $\bar{p}p$ and $\bar{p}n$ annihilation [23]. The data initially available for the study of the $\omega(1420)$ and $\omega(1600)$ were $e^+e^- \to \pi^+\pi^-\pi^0$ (which is dominated by $\rho\pi$) and $e^+e^- \to \omega\pi^+\pi^-$ [24]. The latter cross section shows a clear peak which is apparently dominated by the $\omega(1600)$. The former cross section is more sensitive to the $\omega(1420)$. However the only channel in e^+e^- annihilation and τ decay with really consistent data sets over a wide energy range is the $\pi\pi$ channel, and that runs out of statistics at the upper end of the relevant energy range.

In addition to the direct experimental problems there are theoretical uncertainties which affect the analysis of e^+e^- annihilation and τ decay, and which present data are insufficiently precise to resolve. Firstly there is the "tail-of-the-ρ" problem. In some channels, most notably $\pi\pi$ and $\pi\omega$, there is strong interference between the high-energy tail of the ρ and the higher-mass resonances. The magnitude and shape of this tail are not known with any precision. They can only be specified in models and strictly should be part of the parametrisation. Different models yield different results for the masses and widths of the resonances. A related problem is the question of the relative phases. These can be specified in simple models, but we know that these models are not precise and leaving the phases as free parameters has a major effect on the results of any analysis.

The experimental challenge is easily stated; high-statistics excitation curves for a wide range of hadronic final states:

$$\pi\pi \quad \omega\pi \quad a_1\pi \quad h_1\pi \quad \rho\rho \quad \rho(\pi\pi)_S \quad K\bar{K} \quad K^*\bar{K} \cdots \qquad (1)$$

Note that the $n\bar{n}$ states can decay to $K\bar{K}$, $K^*\bar{K}$ etc. with significant partial widths, so isospin separation is necessary in these channels, and there can be mixing between the isoscalar $n\bar{n}$ states and the $s\bar{s}$ states.

THE THEORETICAL PROBLEM

Despite these various difficulties, an apparently natural explanation for the higher-mass vector states is that they are the first radial, 2^3S_1, and first orbital, 1^3D_1, excitations of the ρ and ω and the first radial excitation of the ϕ, as the generally-accepted masses [3] are close to those predicted by the quark model [25]. However this argument is suspect as the masses of the corresponding $J^P = 1^-$

strange mesons are less than the predictions, particularly for the 2^3S_1 at 1414 ± 15 MeV [3] compared to the predicted 1580 MeV [25]. Quite apart from comparing predicted and observed masses, one would expect the $n\bar{n}$ mesons to be 100 to 150 MeV lighter than their strange counterparts, putting the 2^3S_1 at less than 1300 MeV and the 1^3D_1 below 1600 MeV. Also this interpretation faces a more fundamental problem. The data on the 4π channels in e^+e^- annihilation are not compatible with the $^3P_0$ model [26,27,28,29] which is accepted as the most successful model of meson decay. The model works well for decays of established ground-state mesons:
- widths predicted to be large, are found to be so
- widths predicted to be small, are found to be so
- calculated widths agree with data to $25-40\%$
- signs of amplitudes are correctly predicted

As far as one can ascertain the $^3P_0$ model is reliable, but it has not been seriously tested for the decays of excited states.

The $^3P_0$ model predicts that the decay of the isovector 2^3S_1 to 4π is extremely small:

$$\Gamma_{2S\to a_1\pi} \sim 3\text{MeV} \qquad \Gamma_{2S\to h_1\pi} \sim 1\text{MeV} \tag{2}$$

and for the isovector 1^3D_1 the $a_1\pi$ and $h_1\pi$ decays are large and equal:

$$\Gamma_{1D\to a_1\pi} \sim \Gamma_{1D\to h_1\pi} \sim 105\text{MeV} \tag{3}$$

As $h_1\pi$ contributes only to the $\pi^+\pi^-\pi^0\pi^0$ channel in e^+e^- annihilation, and $a_1\pi$ contributes to both $\pi^+\pi^-\pi^+\pi^-$ and $\pi^+\pi^-\pi^0\pi^0$, then after subtraction of the $\omega\pi$ cross section from the total $\pi^+\pi^-\pi^0\pi^0$ the $^3P_0$ model predicts:

$$\sigma(e^+e^- \to \pi^+\pi^-\pi^0\pi^0) > \sigma(e^+e^- \to \pi^+\pi^-\pi^0\pi^0) \tag{4}$$

This contradicts observation over most of the available energy range. Further, and more seriously, it has been shown recently by the CMD collaboration at Novosibirsk [30] and by CLEO [31] that the dominant channel by far in 4π (excluding $\omega\pi$) up to ~ 1.6 GeV is $a_1\pi$. This is quite inexplicable in terms of the $^3P_0$ model. So the standard picture is wrong for the isovectors, and there are serious inconsistencies in the isoscalar channels as well. One possibility is that the $^3P_0$ model is simply failing when applied to excited states, which is an intriguing question in itself. An alternative is that there is new physics involved.

POSSIBLE SOLUTIONS

A favoured hypothesis is to include vector hybrids [32,33], that is $q\bar{q}g$ states. The reason for this is that, firstly, hybrid states occur naturally in QCD, and secondly, that in the relevant mass range the dominant hadronic decay of the isovector vector hybrid ρ_H is believed to be $a_1\pi$ [33]. The masses of light-quark hybrids have been obtained in lattice-QCD calculations [34,35,36,37], although with quite large errors. Results from lattice QCD and other approaches, such as the bag model [38,39], flux-tube models [40], constituent gluon models [41] and QCD sum rules [42,43], show considerable variation from each other. So the absolute mass scale is somewhat imprecise, predictions for the lightest hybrid lying between 1.3 and 1.9 GeV. However it does seem generally agreed that the mass ordering is $0^{-+} < 1^{-+} < 1^{--} < 2^{-+}$.

Evidence for the excitation of gluonic degrees of freedom has emerged in several processes. Two experiments [44,45] have evidence for an exotic $J^{PC} = 1^{-+}$ resonance, $\hat{\rho}(1600)$ in the $\rho^0\pi^-$ channel in the reaction $\pi^-N \to (\pi^+\pi^-\pi^-)N$. A peak in the $\eta\pi$ mass spectrum at ~ 1400 MeV with $J^{PC} = 1^{-+}$ in $\pi^-N \to (\eta\pi^-)N$ has also been interpreted as a resonance [46]. Supporting evidence for the 1400 state in the same mode comes from $\bar{p}p \to \eta\pi^-\pi^+$ [47]. There is evidence [48] for two isovector 0^{-+} states in the mass region 1.4 to 1.9 GeV; $\pi(1600)$ and $\pi(1800)$. The quark model predicts only one. Taking the mass of the $1^{-+} \sim 1.4$ GeV, then the 0^{-+} is at ~ 1.3 GeV and the lightest 1^{--} at ~ 1.65 GeV, which is in the range required for the mixing hypothesis to work. Of course if hybrids are comparatively heavy, that is the $\hat{\rho}(1600)$ is the lightest 1^{-+} state, and the $\pi(1600)$ presumably the corresponding 0^{-+} hybrid (or at least with a significant hybrid component) then the vector hybrid mass ~ 2.0 GeV making strong mixing with the radial and orbital excitations unlikely.

Two specific models for the hadronic hybrids are the flux-tube model [33,40] and the constituent gluon model [49,50]. There are some substantial differences in their predictions for hybrid decays. For the isovector 1^{--} the flux-tube model predicts $a_1\pi$ as essentially the only hadronic mode, and a width of ~ 100 MeV. The constituent gluon model predicts dominant $a_1\pi$, but with significant $\rho(\pi\pi)_S$ and $\omega\pi$ components, and a larger width. For the isoscalar 1^{--} the flux-tube model predicts $\rho\pi$ as essentially the only hadronic mode, with a width of ~ 20 MeV. The constituent gluon model predicts dominant $\rho\pi$, a significant $\omega(\pi\pi)_S$ component and a larger width.

An alternative explanation could be to invoke the old concept of multiquark states. These are defined as solutions of the multiquark Hamiltonian with totally confined boundary conditions. In the pioneering paper on bag-model four-quarks [51], the states were considered with all interquark orbital momenta equal to zero. Such states easily decay into mesons, so that these multiquarks usually do not exist as relatively narrow resonances. The $q^2\bar{q}^2$ states with vector quantum numbers necessarily contain an extra unit of orbital momentum between constituents, which could reduce the amplitude of their "superallowed" decays. Namely, for the four-quark configurations corresponding to the $(3\bar{3})$ diquark-antidiquark colour representation, $J^{PC} = 1^{--}$ quantum numbers are achieved if orbital excitation $L = 1$ is taken between the diquark and the antidiquark. Extra suppresion of superallowed decay happens if the string model for the multiquark state is adopted, in which a string with junction and antijunction points is formed between the diquark and the antiquark.

In the bag model the masses of such vector states [52] lie well above 1.7 GeV. The string model with junctions [53] lowers the mass to 1.5 GeV giving the possibility for $q^2\bar{q}^2$ states to participate in the higher vector meson phenomena. The detailed structure of $q^2\bar{q}^2$ vector states was considerd in [54]. The peculiar feature of the multiquark scenario is that it is necessary to take into account three lowest states with different total quark spins. Another interesting feature is that the lowest isovector state is about 200 MeV higher than the lowest isoscalar one. Similarly to the hybrid case, selection rules for the multiquark superallowed decay exist which forbid the decay into a pair of S-wave mesons [54]. The main decay modes of $q^2\bar{q}^2$ states are to S-wave plus P-wave mesons, and, in principle, mixing between $q\bar{q}$ and $q^2\bar{q}^2$ states does the same job as mixing between $q\bar{q}$ states and hybrids. The resulting mixing scheme should include five states, and is much more complicated than in the hybrid case. On the other hand it offers new opportunities, as the low-lying four-quark $\rho(1250)$ and $\omega(1100)$ might be responsible for the photoproduction data and the former be the "new" vector meson at about 1250 MeV.

The general conclusion is that the e^+e^- annihilation and τ-decay data require the existence of a "hidden" vector exotics in the isovector and isoscalar channels (assuming that the $^3P_0$ results are qualitatively reliable). The mixing required is non-trivial, although schemes can be devised which are qualitatively compatible with the data [54,55]. The unseen physical states are "off-stage", in the 1.9 to 2.1 GeV mass region. Nonetheless, it appears difficult to achieve quantitative agreement with data (within the constraint of specific models) unless the exotics and the 1^3D_1 states have direct electromagnetic coupling. At the simplest level hybrids do not, but these couplings can be generated by relativistic corrections at the parton level [25] or via intermediate hadronic states, for example hybrid $\to a_1\pi$ $\to$ "ρ" $\to e^+e^-$.

RADIATIVE DECAYS: AN ALTERNATIVE

Radiative decays offer several theoretical advantages. They are a much better probe of wave functions, and hence of models, than are hadronic decay modes because of the direct coupling to the charges and spins of the constituents. This can be particularly relevant, for example, in distinguishing gluonic excitations from conventional radial and orbital excitations as in a 1^{--} hybrid the $q\bar{q}$ are in a spin-singlet state which is the reverse of the usual $q\bar{q}$ configuration. The results of detailed calculation are encouraging [56]. Crucial channels can be specified and, importantly from the practical point of view, it is found that interesting channels should be easily identified. The widths for radiative decays to pseudoscalar states are generally small, but some of those to the $1P$ states are large. Some preliminary results are given in Table 1.

TABLE 1. Preliminary radiative widths in keV [56].

	$\Gamma(\rho_S)$	$\Gamma(\omega_S)$	$\Gamma(\rho_D)$	$\Gamma(\omega_D)$
$a_0(1300)\gamma$	~ 15	~ 140	~ 110	~ 990
$a_1(1260)\gamma$	~ 45	~ 420	~ 80	~ 740
$a_2(1320)\gamma$	~ 75	~ 695	~ 12	~ 110
$f_0(1300)\gamma$	~ 140	~ 15	~ 990	~ 110
$f_1(1285)\gamma$	~ 420	~ 45	~ 740	~ 80
$f_2(1270)\gamma$	~ 695	~ 75	~ 110	~ 12

The larger partial widths should be measurable at the new high-intensity facilities. In some cases they may be measurable in the data from present experiments. We give two specific examples of quarkonia decay and a comment on hybrid radiative decay.

The $\omega\eta$ decay of the $\omega(1650)$ has been observed in the E852 experiment [57]. If the $\omega(1650)$ is the $1D$ $q\bar{q}$ excitation of the ω, then the $^3P_0$ model gives the partial width for this decay as 13 MeV [29]. The partial width for the radiative decay $\omega(1650) \to a_1(1250)\gamma$ is of the order of 1 MeV, that is about 8% of the $\omega\eta$ width. The E852 experiment has several thousand events in the $\omega\eta$, so we may expect several hundred events in the $a_1\gamma$ channel. Similarly both the $\rho(1450)$ and $\rho(1700)$ are seen by the VES collaboration [58] in the $\rho\eta$ channel with several thousand events. Both these states have strong radiative decays, the $\rho(1450)$ to $f_2(1270)\gamma$ and the $\rho(1700)$ to $f_1(1285)\gamma$ both of the order of 1 MeV. Assuming that the $\rho(1450)$ and the $\rho(1700)$ are respectively the $2S$ and the $1D$ excitations of the ρ, then the $^3P_0$ model gives the partial widths for the $\rho\eta$ decays of the $\rho(1450)$ and $\rho(1700)$ as 23 MeV and 25 MeV respectively [29], so the radiative decays should again be present at the level of a few hundred events.

The $\pi(1800)$ is interesting as it could be a conventional $\pi(2S)$ or $\pi(3S)$, the latter being the more natural if the $\pi(1300)$ is interpreted as the $\pi(2S)$ as in the conventional quark model, or it could be a π_g hybrid. The $\omega(1420)$ and $\omega(1650)$ could be conventional $2S$ and $1D$ or a hybrid ω_g. The widths of the radiative decays $\pi(1800) \to \omega(1420)$ or $\omega(1650)$ depend sensitively on which of these configurations the mesons are in, and potentially can discriminate among them. For example, if the $\pi(1800)$ is $2S$ then the width of the radiative decay to $\omega(1420)$ will be large, 0.5 to 1.0 MeV. If it is $3S$ then the radiative decay will be strongly suppressed because of the orthogonality of the wave functions, unless the $\omega(1420)$ is $3S$, which is highly unlikely. Equally, if the $\pi(1800)$ is a hybrid, then the width of the radiative decay to the hybrid ω_g will again be large. Thus if a radiative width of ~ 1 MeV is found then the two states must be siblings, which would be most natural for hybrids.

SUMMARY

Despite 15 years of work we do not yet understand the light-quark vectors. The data raise tantalising questions which go to the heart of nonperturbative QCD:
- How many light-quark vector mesons are there?
- Are there exotic states hiding in there?
- What are the masses, widths, decay channels?
- Does the $^3P_0$ model fail?
- If there are exotics, where are the corresponding states in the strange and charm sectors?

The e^+e^- partial widths and the radiative decay widths of the light-quark vector mesons provide a particularly sensitive probe of wave functions, and the nature of the states involved. The hadronic channels test our understanding of decay mechanisms for radial and orbital excitations, and for exotic states if they are present.

We are in the intriguing position of having sufficient information to realise that the light-quark vector mesons present an exciting challenge, but have insufficient information to solve it. Whatever the answers, new physics is guaranteed!

ACKNOWLEDGEMENTS

This work was supported in part by grant INTAS-RFBR 97-232

REFERENCES

1. Erkal, C. and Olsson, M. G., *Z.Phys.* **C31**, 615 (1986)
2. Donnachie, A. and Mirzaie, H., *Z.Phys.* **C33**, 407 (1987)
3. Particle Data Group, *European Physical Journal* **C15**, 1 (2000)
4. Achasov, M. (SND Collaboration), "*Review of experimental results from SND*"
5. Pick, B. (Crystal Barrel Collaboration), "*Higher vector meson states*"
6. Bartalucci, S. et al., *Nuovo Cimento* **49A**, 207, (1979)
7. Ballam, J. et al., *Nucl.Phys.* **B76**, 375 (1974)
8. Atkinson, M. et al., *Nucl.Phys.* **B243**, 1 (1984)
9. Brau, J. E. et al., *Phys.Rev.* **D37**, 2379 (1988)
10. H1 Collaboration, Paper submitted to *XX International Symposium on Lepton and Photon Interactions*, Rome, 2001
11. Gribov, V. N., *Sov.J.Nucl.Phys.* **5**, 138 (1967)
12. Morrison, D. R. O., *Phys.Rev.* **165**, 1699 (1968)
13. Barkov, L. M. et al., *Nucl.Phys.* **B256**, 365 (1985)
Bisello, D. et al., *Phys.Lett.* **B220**, 321 (1989)
14. Dolinsky, S. I. et al., *Phys.Lett.* **B174**, 453 (1986)
Bisello, D. et al., *Proc. XXV ICHEP*, edited by K.K. Phua and Y. Yamaguchi, World Scientific,

Singapore, 1992
15. R. Barate et al (ALEPH Collaboration), Z.Phys. **C76**, 15 (1997)
Perera, L. P. (CLEO Collaboration), Proc. Hadron'97, edited by S-U Chung and H. J. Willutski, American Institute of Physics, New York, 1998, 595
16. Albrecht, H. et al. (ARGUS Collaboration), Phys.Lett. **B185**, 223 (1987)
17. Dolinsky, S. I. et al., Phys.Rep. **202**, 99 (1991)
Stanco, L. (DM2 Collaboration), Proc. Hadron'91, edited by Y. Oneda and D. Peaslee, World Scientific, Singapore, 1992, 84
18. Bacci, C. et al., Nucl.Phys. **B184**, 31 (1981)
Cosme, G. et al., Nucl.Phys. **B152**, 215 (1979)
19. Clegg, A. B. and Donnachie, A., Z.Phys. **C62**, 455 (1994)
Donnachie, A. and Clegg, A. B., Phys.Rev. **D51**, 4979 (1995)
20. Achasov, N. N. and Kozhevnikov, A. A., Phys.Rev. **D55**, 2663 (1997)
21. Antonelli, M. et al., Phys.Lett. **B212**, 133 (1988)
22. Fukui, S. et al., Phys.Lett. **B202**, 133(1988)
23. Abele, A. et al. (Crystal Barrel Collaboration), Phys.Lett. **B391**, 191 (1997)
24. Antonelli, M. et al., Z.Phys. **C56**, 15 (1992)
25. Godfrey, S. and Isgur, N., Phys.Rev. **D32**, 189 (1985)
26. Busetto, G. and Oliver, L., Z.Phys. **C20**, 247 (1983)
Geiger, P. and Swanson, E. S., Phys.Rev. **D50**, 6855 (1994)
Blundell, H. G. and Godfrey, S., Phys.Rev. **D53**, 3700 (1996)
27. Kokoski, R. and Isgur, N., Phys.Rev. **D35**, 907 (1987)
28. Ackleh, E. S., Barnes, T. and Swanson, E. S., Phys.Rev. **D54**, 6811 (1996)
29. Barnes, T., Close, F. E., Page, P. R. and Swanson, E. S., Phys.Rev. **D55**, 415 (1997)
30. Akhmetshin, R. R. et al., Phys.Lett. **B466**, 392 (1999)
31. Anderson, S. et al., Phys.Rev. **D61**, 112002 (2000)
32. Donnachie, A. and Kalashnikova, Yu. S., Z.Phys. **C59**, 621 (1993)
33. Close, F. E. and Page, P. R., Phys.Rev. **D56**, 1584 (1997)
34. Lacock, P., Michael, C., Boyle, P. and Rowland, P., Phys.Lett. **B401**, 308 (1997)
35. Bernard, C. et al., Phys.Rev. **D56**, (1997) 7039 and hep-lat/9809087
36. Lacock, P. and Schilling, K., hep-lat/9809022
37. McNeile, C., hep-lat/9904013
38. Barnes, T. and Close, F. E., Phys.Lett. **116B**, 365 (1982); ibid **123B**, 89 (1983)
Barnes, T., Close, F. E. and de Viron, F., Nucl.Phys. **B224**, 241 (1983)
39. Chanowitz, M. and Sharpe,S., Nucl.Phys. **B222**, 211 (1983)
40. Isgur, N. and Paton, J. E., Phys.Lett. **124B**, 247 (1983)
Isgur, N., Kokoski, R. and Paton, J. E., Phys.Rev.Lett. **54**, 869 (1985)
Isgur, N. and Paton, J. E., Phys.Rev. **D31**, 2910 (1985)
Barnes, T., Close, F.E. and Swanson, E. S., Phys.Rev. **D52**, 5242 (1995)
41. Kalashnikova, Yu. S. and Yufryakov, Yu. B., Phys. Lett. **B359**, 175 (1995)
Kalashnikova, Yu. S. and Yufryakov, Yu. B., Phys. At. Nucl. **60**, 307 (1997)
42. Balitsky, I. I., Dyakonov, D. I. and Yung, A.V., Z.Phys. **C33**, 265 (1986)
43. Latorre, J. I., Pascual, P. and Narison, S., Z.Phys. **C34**, 347 (1987)

44. Weygand, D. P. (E852 Collaboration), *Proc. HADRON'97*, edited by S-U Chung and H. J. Willutski, American Institute of Physics, New York, 1998, 313
45. Gouz, Yu. P. (VES Collaboration), *Proc. XXVI ICHEP* edited by J. R. Sanford, 572
46. Thompson, D. R. et al. (E852 Collaboration), *Phys.Rev.Lett.* **79**, 1630 (1997)
47. Abele, A. et al. (Crystal Barrel Collaboration), *Phys.Lett.* **B423**, 175 (1998)
48. Zaitsev, A. (VES Collaboration), *Proc. Hadron'97*, edited by S-U Chung and H. J. Willutski, American Institute of Physics, New York, (1998), 461
Amelin, D. V. (VES Collaboration), *ibid* 770
49. Le Yaounac, A., Oliver, L., Pène, O., Raynal, J. C. and Ono, S., *Z.Phys.* **C28**, 309 (1985)
Iddir, F., Le Yaouanc, A., Oliver, L., Pène, O, and Raynal,J. C., *Phys.Lett.* **B205**, 564 (1988)
50. Kalashnikova, Yu. S., *Z.Phys.* **C62**, 323 (1994)
51. Jaffe, R. L., *Phys.Rev.* **D15**, 267 (1977)
52. Mulders, P.J., Aerts, A.T. and de Swart, J. J., *Phys.Rev.* **D21**, 1370 (1980)
53. Badalyan, A. M., LNF-91/017(R) (1991)
54. Donnachie, A. and Kalashnikova, Yu. S., *Z.Phys.* **C59**, 621 (1993)
55. Donnachie, A. and Kalashnikova, Yu. S., *Phys.Rev.* **D60**, 114011 (1999)
56. Close, F.E., Donnachie, A. and Kalashnikova, Yu. S., in preparation
57. Eugenio, P. et al. (E852 Collaboration), *Phys.Lett.* **B497** 190 (2001)
58. D.V. Amelin et al. (VES Collaboration), *Nucl.Phys.* **A668**, 83 (2000)

Precise measurement hadronic cross sections with CMD-2 detector at VEPP-2M

R.R. Akhmetshin*, E.V. Anashkin*, V.M. Aulchenko*, V.Sh. Banzarov*,
L.M. Barkov*, S.E. Baru*, N.S. Bashtovoy*, A.E. Bondar*,
D.V. Bondarev*, A.V. Bragin*, D.V. Chernyak*, S.I. Eidelman*,
G.V. Fedotovitch*, N.I. Gabyshev*, A.A. Grebeniuk*, D.N. Grigoriev*,
V.W. Hughes[†], P.M. Ivanov*, S.V. Karpov*, V.F. Kazanin*, B.I. Khazin*,
I.A. Koop*, M.S. Korostelev*, P.P. Krokovny*, L.M. Kurdadze*,
A.S. Kuzmin*, I.B. Logashenko*, P.A. Lukin*, A.P. Lysenko*,
K.Yu. Mikhailov*, I.N. Nesterenko*, V.S. Okhapkin*, A.V. Otboev*,
E.A. Perevedentsev*, A.A. Polunin*, A.S. Popov*, T.A. Purlatz*,
N.I. Root*, A.A. Ruban*, N.M. Ryskulov*, A.G. Shamov*,
Yu.M. Shatunov*, A.I. Shekhtman*, B.A. Shwartz*, A.L. Sibidanov*,
V.A. Sidorov*, A.N. Skrinsky*, V.P. Smakhtin*, I.G. Snopkov*,
E.P. Solodov*, P.Yu. Stepanov*, A.I. Sukhanov*, J.A. Thompson**,
V.M. Titov*, A.A. Valishev*, Yu.V. Yudin* and S.G. Zverev*

Budker Institute of Nuclear Physics, Novosibirsk 630090, Russia
[†]*Yale University, New Haven, CT 06511, USA*
**University of Pittsburgh, Pittsburgh, PA 15260, USA*

Abstract.
The Cryogenic Magnetic Detector (CMD-2) and main it's parameters are shortly described. The results for the cross sections of e^+e^- annihilation into hadrons are presented in the c.m. energy range from 0.37 to 1.39 GeV. The total integrated luminosity of about 31 pb^{-1} has been collected. The new results for the ρ and ω meson parameters were obtained. The pion form factor was determinated with a 0.6% systematic uncertainty. The major decay modes of the φ meson as well as multihadron final states have been studied in a broad energy range.

INTRODUCTION

Despite thirty years of experimental studies e^+e^- annihilation into hadrons at low energies is still rather far from the complete understanding. More precise measurements are needed to determine ρ, ω and φ meson parameters as well as continuum properties providing the unique information about interactions of light quarks.

Exact data on the pion form factor is necessary for the precise determination of the ratio $R = \sigma(e^+e^- \to hadrons)/\sigma(e^+e^- \to \mu^+\mu^-)$, which in the VEPP-2M energy range is dominated by the $e^+e^- \to \pi^+\pi^-$ channel. Knowledge of R with high accuracy is required to evaluate the hadronic contribution to the (g-2) of muon and to the running electromagnetic constant $\alpha(M_Z^2)$. In case of the muon (g-2) the energy range of VEPP-2M

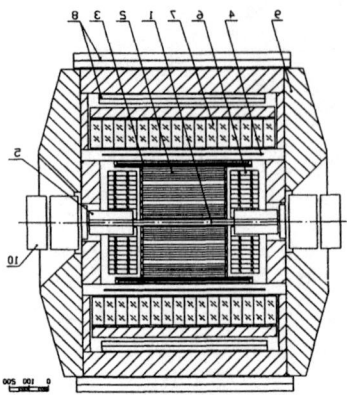

FIGURE 1. Cross sections of the CMD-2 detector. 1 - vacuum chamber, 2 - drift chamber, 3 - Z-chamber, 4 - main solinoid, 5 -compensating solinoid, 6 - BGO endcap calorimeter, 7 - CsI barrel calorimeter, 8 - muon range system, 9 - magnet yoke, 10 - collider lenses

gives the major contribution both to the hadronic contribution itself and to its uncertainty.

Assuming concervation of the vector current (CVC) and isospin symmetry, the spectral function of $\tau^- \to \pi^-\pi^0\nu_\tau$ decay can be related to the isovector part of the pion form foctor. The comparison of the pion form factor measured at e^+e^- colliders with the spectral function of the $\tau^- \to \pi^-\pi^0\nu_\tau$ decay provides a test of CVC. If CVC holds with high accuracy, τ-lepton decay data can be also used to improve the accuracy of the calculations mentioned above.

E821 has recently published the result of its measurement of $(g-2)_\mu$ with an relative accuracy of 1.3 ppm [1]. The measured value of $(g-2)_\mu$ is 2.6 standard diviations higher than the SM prediction. This observation makes new high precision measurement of the $e^+e^- \to hadrons$ and particularly of the pion form factor extremely important.

The precise measurement of the hadronic cross section was one of the main goal CMD-2 detector running at the VEPP-2M e^+e^- collider in Novosibirsk since 1992. A short description of the most interesting results is presented in this paper.

CMD-2 DETECTOR

The CMD-2 is a general purpose detector. The cross section of the detector is shown in Fig. 1 in outline. The drift chamber of the detector (DC) [3] has 80 jet type drift cells. The resolution in the plane transverse to the beam axis is about of 250 μ. The double layer multiwire proportional chamber (ZC) [4] is installed just after DC. ZC has an accurate resolution (~ 0.5 mm) of the z-coordinate of particle track along the beam direction. The time resolution of ZC is near 5 ns. The beam cicle time at VEPP-2M is 60 ns. Both chambers are inside a thin (0.38 X_0) superconducting solenoid with a field of 1 T.

The barrel calorimeter is placed outside of the solenoid and consists of 892 CsI crystals [5] of $6 \times 6 \times 15$ cm^3 size. The crystals are arranged in eight octants. The light readout is performed by PMTs. The energy resolution is about of 8% for photons with the energy more than 100 MeV. The both azimuthal and polar angle resolution is about of 0.02 radian.

The endcap calorimeter [6] consists of 680 BGO crystals of $2.5 \times 2.5 \times 15$ cm^3 size. The light readout is performed by vacuum phototriodes placed on the crystals. The energy and angular resolution were found to be $\sigma_E/E = 4.6\%/\sqrt{E(GeV)}$ and $\sigma_{\phi,\theta} = 2 \cdot 10^{-2}/\sqrt{E(GeV)}$ radians respectively. The solid angle covered by both parts of the calorimeter is about of 96% of 4π.

The muon range system consists of two double layers of the streamer tubes operating in a self-quenching mode and is aimed to separate pions and muons. The inner and outer parts of this system are arranged in 8 modules each and cover 55% and 48% of the solid angle respectively. More details on the detector can be found elsewhere [2].

MEASUREMENT OF $E^+E^- \to \pi^+\pi^-$ CROSS SECTION AROUND ρ MESON

A large data sample of about 2 million $e^+e^- \to \pi^+\pi^-$ events was collected by CMD-2 detector in the whole energy range available at VEPP-2M. Analysis of the data is completed for 10% of the statistic only. The beam energy was measured by the resonant depolarization technique at almost all energy points. The special kinematic cuts were applied to events in order to separate the collinear ones. The signature of two particles of opposite charge and nearly back-to-back momenta originating from the interaction point.

The selected sample of collinear events contains $e^+e^- \to e^+e^-$, $e^+e^- \to \pi^+\pi^-$, $e^+e^- \to \mu^+\mu^-$ events (below refered to as beam originating) as well as the background of cosmic muons which pass near the interaction region and are misidentified as collinear events. The number of cosmic background events N_{cosmic} was determined by the analysis of the spatial distribution of the vertex. Both distributions - along coordinate z and transverse to the beam axis (distance ρ from the beam) are peaked around zero for the beam originating events and are very broad for the cosmic events. Typical ρ and z-distributions are shown in Fig. 2, 3. The open histogram corresponds to beam originating collinear events, the filled one shows the subset of background events.

The energy deposition of the charged particles in the calorimeter is quite different for e, μ and π. The typical energy deposition of two particles (E^+ vs E^-) for experimental events selected for the beam energy of 400 MeV is shown in Fig. 4.

The high deposition spot corresponds to e^+e^- pairs, where both particles leave almost all their energy in the calorimeter. The low deposition spot represents $\mu^+\mu^-$ pairs, cosmic muons and part $\pi^+\pi^-$ pairs when both particles interact as minimum-ionizing. The long tails correspond to $\pi^+\pi^-$ pairs where one or both particles undergo nuclear interactions inside the calorimeter.

The energy deposition of the particles was used for the separation of the beam

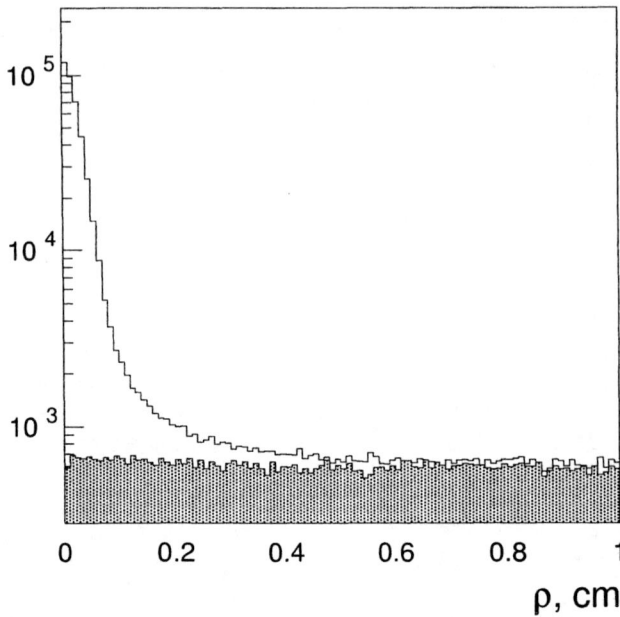

FIGURE 2. The distribution of the distance from the vertex to the beam axis.

originating events. Since the overlap of the distributions for e^+e^- and $\pi^+\pi^-$ pairs is small, this approach gives stable results with a small systematic error.

The number of $\mu^+\mu^-$ pairs was derived from the number of e^+e^- pairs according to QED, taking into account radiative corrections and detection efficiencies. Since in this energy range the number of $\mu^+\mu^-$ pairs is small with respect to $\pi^+\pi^-$ the systematic error caused by the corresponding calculation is negligible (less than 0.03%).

The likelihood function was written with the following global fit parameters: $(N_{ee} + N_{\mu\mu})$, $N_{\pi\pi}/(N_{ee}+N_{\mu\mu})$, N_{cosmic}. The ratio $N_{\mu\mu}/N_{ee}$ was fixed according to QED calculation. The number of cosmic events was determined before the fit and was fixed during the fit. The radiative corrections (RC) for these channels were calculated according to works [7, 8]. The estimated systematic error for RC is better than 0.2%. The emission of the second photon inside a narrow cone along the electron and positron (initial and final) was taken into account. The lepton and hadron contributions to vacuum polarization were included in the RC for $e^+e^- \to e^+e^-$, $\mu^+\mu^-$, but they were excluded for the channel $e^+e^- \to \pi^+\pi^-$.

Special studies were performed to estimate the systematic error of the separation procedure. The dominant effect was produced by the small nonuniformity of the calorimeter response. Due to forward-backward asymmetry of the e^+e^- cross section, that leads to the slight difference between e^+ and e^- energy depositions. The corresponding error was found to be less than 0.2%.

Other factors possibly influencing the results are varying energy calibrations as well

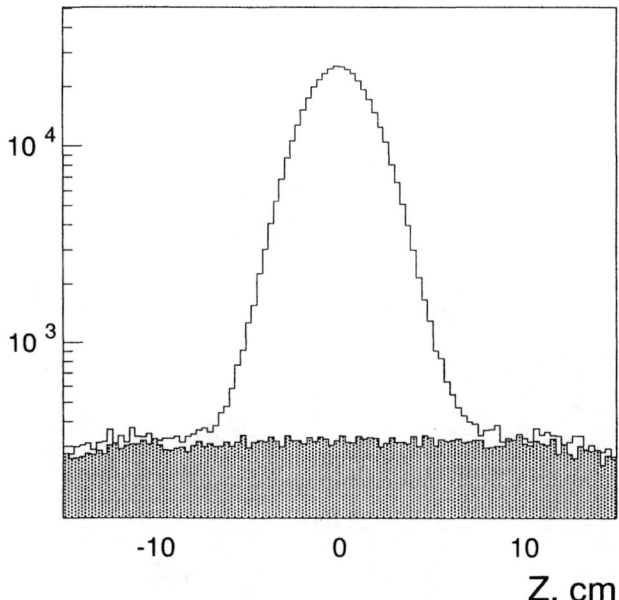

FIGURE 3. The distribution of the distance between the vertex and the center of the detector along the beam axis (z)

as the energy resolution of the calorimeter. Their effect was also below 0.1%. Finally, the large amount of $e^+e^- \to e^+e^-\gamma$, $\mu^+\mu^-\gamma$, $\pi^+\pi^-\gamma$ was generated in a proper proportion with the help of full detector simulation in several energy points covering the whole energy range under study. The simulated events were processed with the same procedure as for the data. The reconstructed number of the events is always consistent with the input one and the average difference is below 0.1% or consistent with zero. Thus, the systematic error because of the event separation taking into account the above effects is estimated to be 0.2%.

The pion form factor presented in Fig. 5 is based on the data sample at 53 energy points in the energy range from 0.37 to 0.96 GeV. The obtained ρ meson parameters based on Gounaris-Sakurai parameterization were found to be:

$$M_\rho = 776.2 \pm 0.8 \text{ MeV}, \qquad \Gamma_\rho = 144.6 \pm 1.6 \text{ MeV},$$
$$\Gamma_{\rho \to e^+e^-} = 6.86 \pm 0.12 \text{ keV}, \qquad Br(\omega \to \pi^+\pi^-) = (1.34 \pm 0.25)\%.$$

The radiative corrections for the process $e^+e^- \to \pi^+\pi^-$ include the effects of both initial (ISR) and final state radiation (FSR) and do not include the vacuum polarization terms (both leptonic and hadronic) since the latter are considered to be an intrinsic part of the hadronic cross section. However, for various applications based on dispersion relations and involving the total cross section of $e^+e^- \to $ hadrons, the radiation by the final pions is no longer a radiative correction, so that $\pi^+\pi^-\gamma$ with photon radiate by

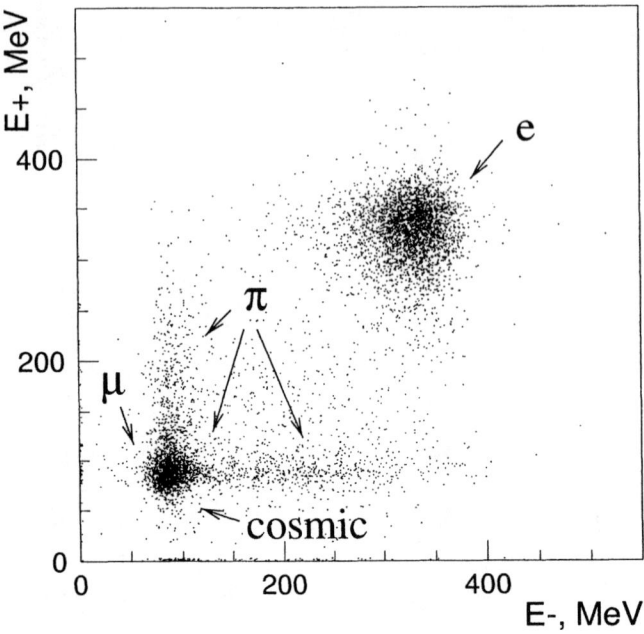

FIGURE 4. The energy deposition of collinear events for the beam energy of 400 MeV

one of the final pions should be considered as one of the possible hadronic final states contributing to the total cross section. Therefor, the bare cross section $e^+e^- \to \pi^+\pi^-(\gamma)$, below refered to as $\sigma^0_{\pi\pi(\gamma)}$, was also calculated as

$$\sigma^0_{\pi\pi(\gamma)} = \frac{\pi\alpha^2}{3s} \cdot \beta_\pi^3 \cdot |F_\pi(s)|^2 \cdot |1 - P(s)|^2 \cdot \left(1 + \frac{\alpha}{\pi}\Lambda(s)\right). \quad (1)$$

The factor $|1 - P(s)|^2$ with a polarization operator P(s) excludes the effect of leptonic and hadronic vacuum polarization and we obtained the bare hadronic cross section required for various applications. A quantity $\Lambda(s)$ represents contribution comes from final state radiation and was calculated according to [10].

The measured values of the pion form factor as well as the bare $e^+e^- \to \pi^+\pi^-(\gamma)$ cross section are shown in Table 1. The main sources of the systematic error are listed in Table 2.

The overall systematic error obtained by summing individual contributions in quadrature is about 0.6% - the best result with respect to all previous experiments.

The energy range around the ω, ϕ and upper ϕ meson was scanned and detail analysis of the data is done only for ω and ϕ meson. The high precision measurements of the ω meson parameters were performed using the dominant decay mode: $\omega \to \pi^+\pi^-\pi^0$. Events with two tracks originating from the same vertex, each with a polar angle $0.85 < \theta < \pi - 0.85$ within fiducial volume of the detector were selected only.

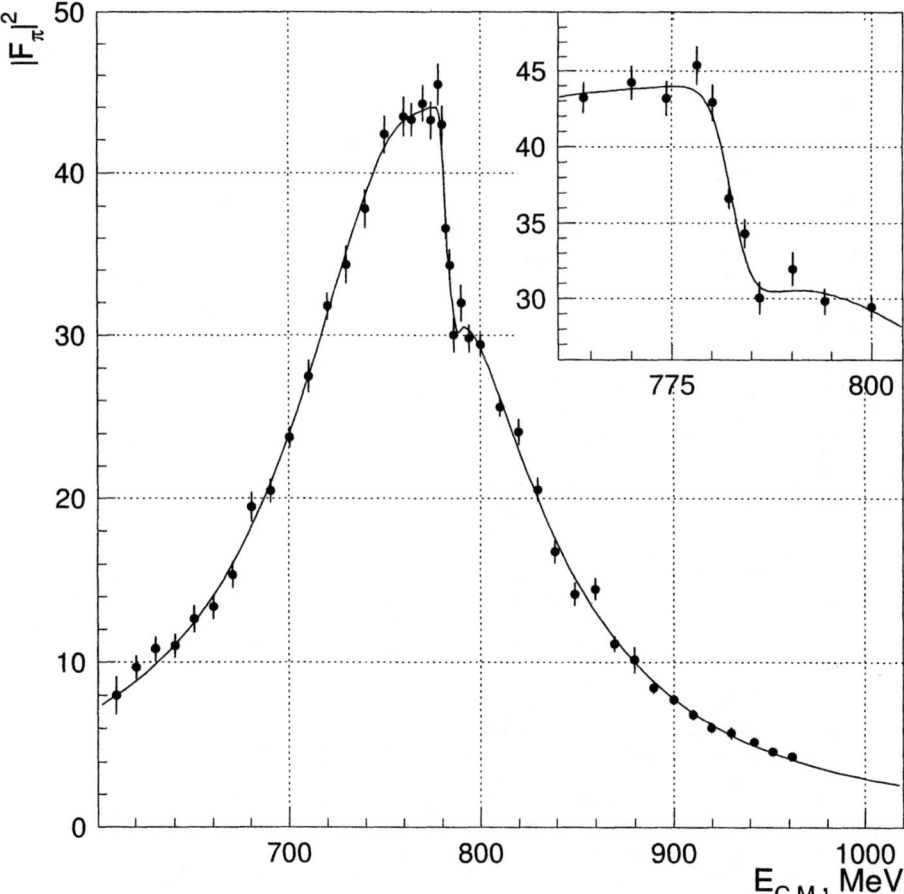

FIGURE 5. The pion form factor vs c.m. energy and the fit with Gounaris-Sakurai parameterization.

Most of the background comes from the processes with the hard photon emission: $e^+e^- \to e^+e^-\gamma, \pi^+\pi^-\gamma, \mu^+\mu^-\gamma$. These processes have the same signature as the reaction $e^+e^- \to \pi^+\pi^-\pi^0$, except for the very different distribution on acollinearity angle $\triangle\phi$ between two charged particles which peaked near $\triangle\phi = 0$. Thus, the rejection of events with a small $\triangle\phi$ drastically reduces the background.

The resonant depolarization method was used for the precise calibration of the beam energy at each point. The production cross section was calculated according to the expression:

$$\sigma(e^+e^- \to \pi^+\pi^-\pi^0) = (N_{3\pi}/L) \cdot \varepsilon_{trig} \cdot \varepsilon_{MC} \cdot \varepsilon_{M^2_{miss}} \cdot (1+\delta_{rad}),$$

where $N_{3\pi}$ is the number of events; L is the integrated luminosity determined from large angle Bhabha events; δ_{rad} is the radiative correction calculated according to [9] with an accuracy better than 0.2% by the Monte Carlo integration of the differential cross

TABLE 1. The measured values of the pion form factor and $e^+e^- \to \pi^+\pi^-(\gamma)$ cross section. Only statistical errors are shown. The systematic error is estimated to be 0.6%.

2E (MeV)	$|F_\pi|^2$	$\sigma^0_{\pi\pi(\gamma)}$, nb	2E (MeV)	$|F_\pi|^2$	$\sigma^0_{\pi\pi(\gamma)}$, nb
610.50	7.98 ± 1.12	327.1 ± 46.1	784.24	34.30 ± 0.97	936.0 ± 26.5
620.50	9.69 ± 0.72	389.7 ± 29.0	786.04	30.02 ± 1.07	809.6 ± 28.9
630.50	10.80 ± 0.72	426.1 ± 28.6	790.10	31.96 ± 1.11	859.4 ± 30.0
640.51	10.99 ± 0.72	425.4 ± 27.9	794.14	29.82 ± 0.85	799.8 ± 22.7
650.49	12.64 ± 0.83	479.5 ± 31.6	800.02	29.41 ± 0.68	783.0 ± 18.1
660.50	13.38 ± 0.75	498.0 ± 27.8	810.14	25.61 ± 0.55	670.0 ± 14.4
670.50	15.34 ± 0.82	559.6 ± 29.9	820.02	24.11 ± 0.78	619.4 ± 20.0
680.59	19.47 ± 0.90	696.3 ± 32.2	829.97	20.53 ± 0.73	517.9 ± 18.5
690.43	20.47 ± 0.73	717.7 ± 25.5	839.10	16.76 ± 0.70	415.7 ± 17.4
700.52	23.78 ± 0.61	817.2 ± 21.1	849.24	14.18 ± 0.70	345.1 ± 17.1
710.47	27.49 ± 0.99	925.2 ± 33.4	859.60	14.48 ± 0.67	346.1 ± 16.0
720.25	31.77 ± 0.82	1047.2 ± 26.9	869.50	11.12 ± 0.46	261.1 ± 10.7
730.24	34.32 ± 1.18	1106.5 ± 38.0	879.84	10.15 ± 0.79	234.0 ± 18.1
740.20	37.80 ± 1.18	1191.3 ± 37.0	889.72	8.45 ± 0.34	191.4 ± 7.7
750.28	42.31 ± 1.18	1302.4 ± 36.2	900.04	7.75 ± 0.29	172.4 ± 6.4
760.18	43.43 ± 1.21	1306.5 ± 36.5	910.02	6.83 ± 0.31	149.4 ± 6.8
764.17	43.23 ± 1.03	1289.4 ± 30.6	919.56	6.04 ± 0.30	129.9 ± 6.5
770.11	44.23 ± 1.14	1304.5 ± 33.5	930.11	5.71 ± 0.36	120.6 ± 7.7
774.38	43.18 ± 1.15	1265.6 ± 33.7	942.19	5.17 ± 0.25	107.1 ± 5.1
778.17	45.41 ± 1.29	1321.6 ± 37.7	951.84	4.57 ± 0.23	93.2 ± 4.8
780.17	42.91 ± 1.21	1235.0 ± 34.8	961.52	4.30 ± 0.23	86.4 ± 4.6
782.23	36.59 ± 0.66	1026.6 ± 18.5			

TABLE 2. Main sources of the systematic errors

Source	Contribution
Beam energy determination	0.1%
Fiducial volume	0.2%
Reconstruction efficiency	0.2%
Event separation	0.2%
Correction for pion losses	0.2%
Radiative corrections	0.4%
Total	0.6%

section under all selection criteria.

The efficiencies were calculated by simulation and their systematic errors were estimated with the help of special "test" events obtained as a result of the constrained fit based on information from the ZC and CsI calorimeter only. "Test" events with the neutral trigger were also used to determine the charged trigger efficiency.

The effects of vacuum polarization by leptons and hadrons were excluded from RC. The following parameters have been obtained from the fit:

$$M_\omega = 782.71 \pm 0.07 \pm 0.04 \text{ MeV}, \quad \sigma_0 = 1457 \pm 23 \pm 19 \text{ nb},$$
$$\Gamma_\omega = 8.68 \pm 0.23 \pm 0.10 \text{ MeV}, \quad \Gamma_{e^+e^-} = 0.595 \pm 0.014 \pm 0.009 \text{ keV}.$$

These results are more precise than those from previous experiments. The mass value

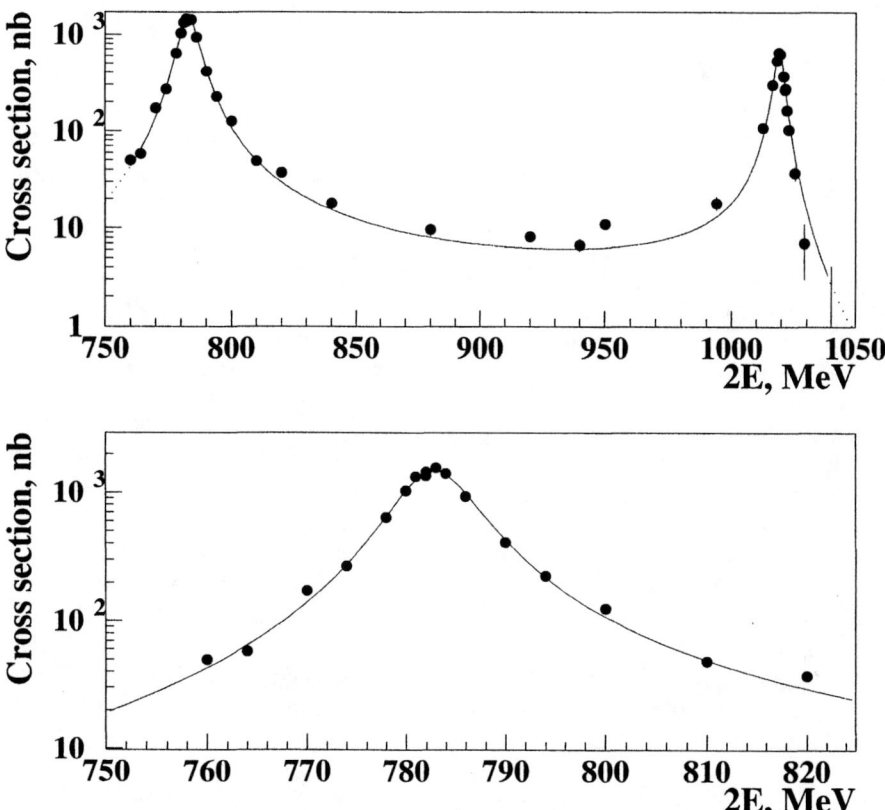

FIGURE 6. The upper figure - cross section vs c.m.energy and common fit for the ω, φ mesons. The lower figure - cross section in the energy range around ω-meson.

differs (930 keV) more than seven standard deviations in CMD87 [11]. Due to the present more thorough study of systematic errors, our mass measurement supersedes that of [11]. The common excitation curve for the ω and φ meson is presented in Fig. 6.

MEASUREMENTS OF φ MESON PARAMETERS

The φ resonance is a copious source of charged and neutral kaon pairs and therefor the large data sample of φ meson allowed a number of results on kaon physics to be obtained, first of all results on the study of $e^+e^- \to K^+K^-$, $K_L^0 K_S^0$.

The new more precise results were obtained for the channel $\phi \to K_L^0 K_S^0$ when K_S^0 decays into a $\pi^+\pi^-$ pairs. The data sample was collected in four scans in the energy range from 984 to 1040 MeV with the integrated luminosity of 2.37 pb^{-1} and contains

2.97×10^5 of selected $K_L^0 K_S^0$. The following parameters have been obtained from the fit:

$$\sigma_0(\phi \to K_L^0 K_S^0) = 1376 \pm 6 \pm 23 \text{ nb},$$
$$M_\phi = 1019.483 \pm 0.011 \pm 0.025 \text{ MeV},$$
$$\Gamma_\phi = 4.280 \pm 0.033 \pm 0.025 \text{ MeV},$$
$$Br(\phi \to ee) \cdot Br(\phi \to K_L^0 K_S^0) = (0.975 \pm 0.004 \pm 0.017) \times 10^{-4}.$$

The data sample collected in c.m.energy range from 1050 to 1380 MeV were used to study the cross section of the production charged and neutral kaon pairs. The preliminary results for the cross section with neutral kaons in the final state are presented in Fig. 7 obtained by CMD-2, SND, OLYA and DM2 detectors. The curve is a fit in VDM frame where ρ, ω and ϕ mesons are taken into account only. It is clear seen disagreement between data and VDM prediction above 1100 MeV. But if $\phi'(1680)$ meson is added to fit in this case VDM describes cross section very well in a broad energy range.

The systematic error change from 4.9% (at 1050–1125 MeV) to 9.6% (at 1270–1380 MeV). The main source of systematic error comes from background substraction and selection criteria and more detail can be found elsewhere [13].

The preliminary results for the cross section with charged kaons in the final state obtained with CMD-2 and SND as well as other detectors are shown in Fig. 8. A good agreement between the data from the different experiments is observed. The curve is a fit with the contribution of ρ, ω and ϕ in VDM frame. The deviation between experimental data and VDM prediction above 1100 MeV little by little becomes significant. Again if $\phi'(1680)$ meson is added to fit VDM describes data well enough.

STUDY OF THE REACTION $E^+E^- \to 4\pi$

Recently the investigation of e^+e^- annihilation into hadrons was restricted by measurements of the cross section only. Operating at high luminosity colliders VEPP-2M has been provided a large data sample and opened qualitatively new possibilities for the investigation of the multihadronic production in e^+e^- annihilation.

Production of four pions is one of the dominant process of e^+e^- annihilation into hadrons in the energy range from 1.05 to 2.5 GeV. One of the main difficulties in the experimental studies of four pions production is caused by the existence of different intermediate states providing production of the final pions.

The analysis is based on 5.8 pb^{-1} data collected at center-of-mass energies from 1.05 to 1.38 GeV. The energy range was scanned twice up and down with a step of 10 MeV. The analysis is based on completely reconstructed events: one primary vertex with two or four oppositely charged tracks come from interaction region, distance from the track to the beam axis should be less than 0.3 cm, vertex position along the beam axis should be inside ± 10 cm. The collinear events were rejected by the acollinearity angle between any tracks in R - Φ plane should be greater than 0.1 radians. The clusters in EM calorimeter which one do not match with a charged track projection are paired to form π^0 candidates. The energy threshold for cluster was set 20 MeV.

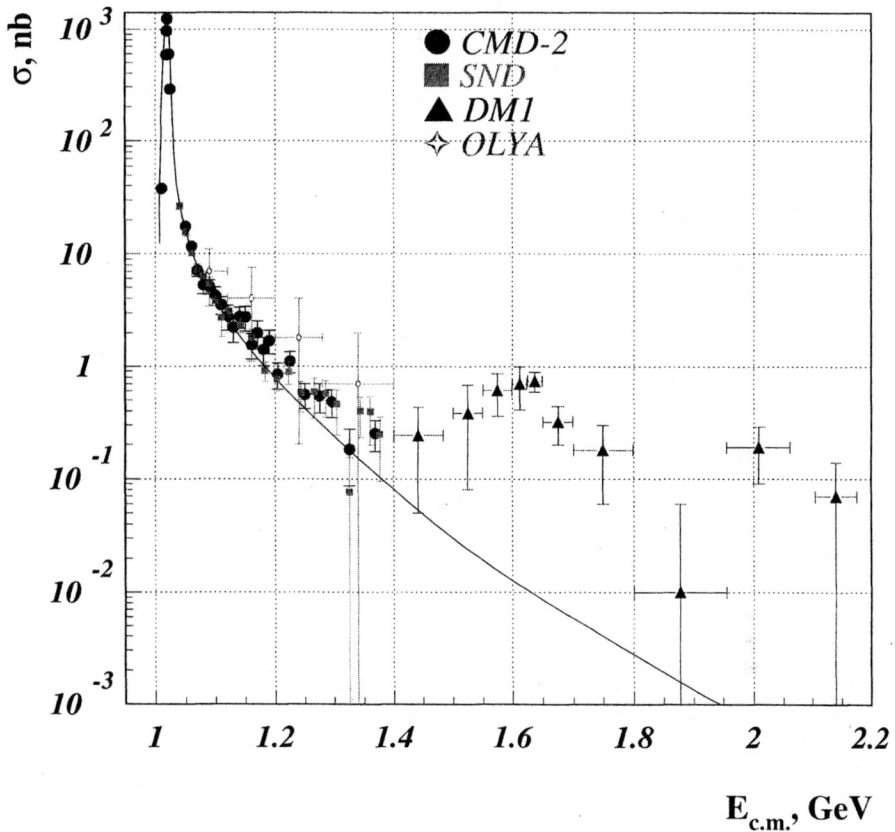

FIGURE 7. Cross section of $e^+e^- \to K_L^0 K_S^0$ vs energy. The dot line - fit with ρ, ω and ϕ, solid line - fit with ρ, ω, ϕ and $\phi'(1680)$

The distribution of the recoil mass for one of the neutral pions was studied and we defined - the signal from the $\omega\pi^0$ intermediate state gives only $\sim 60\%$ of the observed events that points to the existence of additional intermediate states. The further analysis of the angular distributions for the recoil pion definitely confirmed existence $a_1(1260)\pi$ intermediate state, although a small admixture of other states is not excluded.

The analysis of the process $e^+e^- \to \pi^+\pi^-\pi^+\pi^-$ was done too in the energy range from 0.6 to 0.97 GeV. About 153 $\pi^+\pi^-\pi^+\pi^-$ events were observed and the first measurement of the corresponding cross section at a level as low as several tenths of pb was done. The kinematic fit was performed for the selected events assuming that all tracks are pions. The detection efficiency was determined from MC assuming four different quasi-twobody production mechanisms: $a_1(1260)\pi$, $\rho\sigma$, $\pi(1300)\pi$ and $a_2(1320)\pi$. The

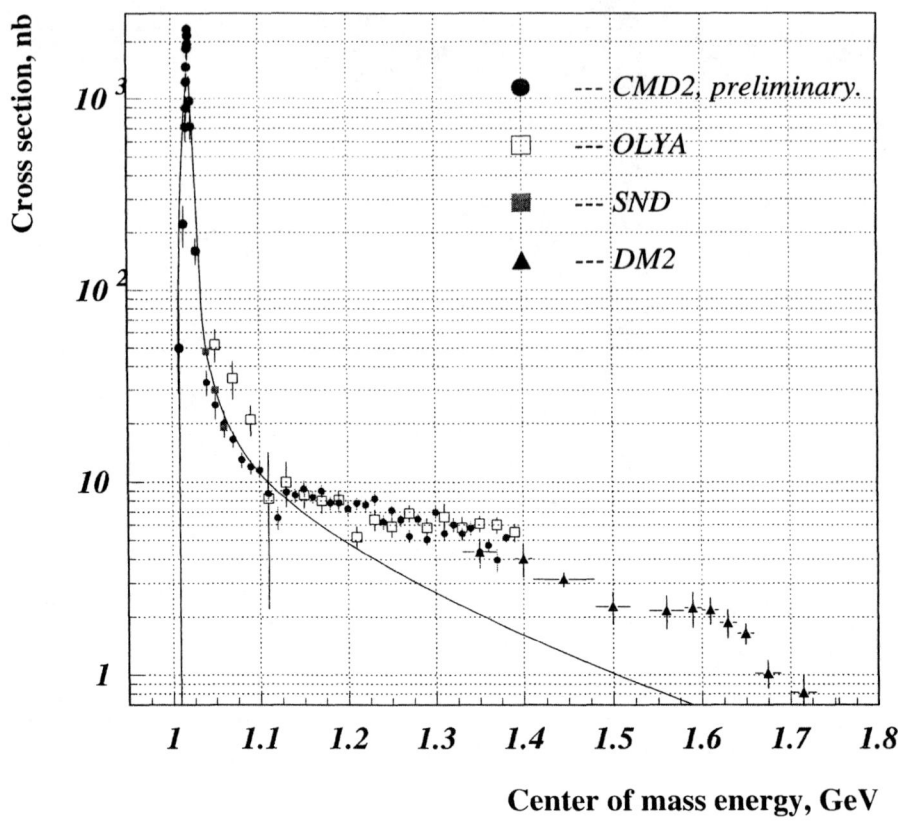

FIGURE 8. Cross section of $e^+e^- \to K^+K^-$ vs energy. The dot line - fit with ρ, ω and ϕ, solid line - fit with ρ, ω, ϕ and $\phi'(1680)$

fit of invariant mass and angular distributions showed that $\pi(1300)\pi$ and $a_2(1320)\pi$ intermediate states can not well describe the data whereas $a_1(1260)\pi$ and $\rho\sigma$ are almost indistinguishable in this energy range. For the final fit the $a_1(1260)\pi$ production mechanism which dominates at higher energy has been chosen. The following values for width and branching ratio were found to be:

$$\Gamma(\rho^0 \to \pi^+\pi^-\pi^+\pi^-) = (2.8 \pm 1.4 \pm 0.5) \text{ keV}$$
$$Br(\rho^0 \to \pi^+\pi^-\pi^+\pi^-) = (1.8 \pm 0.9 \pm 0.3) \cdot 10^{-5}.$$

STUDY OF THE PROCESS $E^+E^- \to \eta\gamma$

Radiative magnetic dipole transitions of ρ, ω and ϕ mesons to $\eta\gamma$ have traditionally been a good laboratory for various tests of theoretical concepts from the quark model and SU(3) symmetry to Vector Dominance Model (VDM).

Large integrated luminosity collected by CMD-2 in recent experiments allows qualitatively new analysis of the $\eta\gamma$ final state produced in e^+e^- annihilation. The analysis is based on the data sample corresponding to 26.3 pb^{-1}.

To study the process $e^+e^- \to \eta\gamma$ the decay mode $\eta \to 3\pi^0$ was chosen when seven photons in final state are. At the first stage of analysis events were selected which have no tracks in the DC, the number of photons from 6 to 8, the total energy deposition in calorimeters more than $1.5E_{beam}$, the total momentum $P_{tot} < 0.4 \cdot E_{beam}$. Among of these photons there is one monochromatic with the energy 362 MeV. Other photons have been pairing to form π^0 by combinatorial way and the combinations with the best χ^2 were chosen. After kinematic reconstruction requiring energy-momentum conservation events were remained which had good reconstruction quality. The detection efficiency for this process was determined by simulation and was found to be $(10.8 \pm 0.1)\%$.

The maximum likelihood method was applied to fit the experimental data. Two different models were considered for the fit: the first one is VDM with the ρ, ω, ϕ mesons and in the second one an additional ρ' meson was included. Results of the fits are shown in Fig. 9.

Although the fits quality (χ^2) are good for both models, but it is clear seen that at high energies the measured cross section is not described by one VDM. This fact is a hint for possible additional contributions coming from the higher resonance. Another words we consider our observation as evidence for the existence for $\rho' \to \eta\gamma$ decay. The following results were obtained from the fit:

$$Br_{\rho \to e^+e^-} \cdot Br_{\rho \to \eta\gamma} = (1.46 \pm 0.17 \pm 0.10) \cdot 10^{-7},$$
$$Br_{\omega \to e^+e^-} \cdot Br_{\omega \to \eta\gamma} = (3.60 \pm 0.51 \pm 0.22) \cdot 10^{-7},$$
$$Br_{\phi \to e^+e^-} \cdot Br_{\phi \to \eta\gamma} = (3.849 \pm 0.040 \pm 0.158) \cdot 10^{-6},$$
$$Br_{\rho' \to e^+e^-} \cdot Br_{\rho' \to \eta\gamma} = (9.4 \pm 2.0) \cdot 10^{-9},$$
$$m_\phi = 1019.40 \pm 0.04 \pm 0.05 \text{ MeV},$$
$$m_{\rho'} = 1497 \pm 14 MeV, \Gamma_{\rho'} = 226 \pm 44 \text{ MeV}.$$

CONCLUSION

For the first time the pion form factor is measured with the systematic error 0.6%. This accuracy is adequate to the requirements of $(g-2)_\mu$ experiments at BNL.

The accuracy of measurement others hadronic cross sections is enough to get the systematic error in (g-2) of muon less than 0.15 ppm. These results are outlined in the Table 3. It is clear seen the main contribution to the error of a_μ comes again from two pions channel in the energy range under study. But nevertheless after our result the energy range from 1.4 GeV to 2 GeV will be a main source of the total uncertainty in a_μ

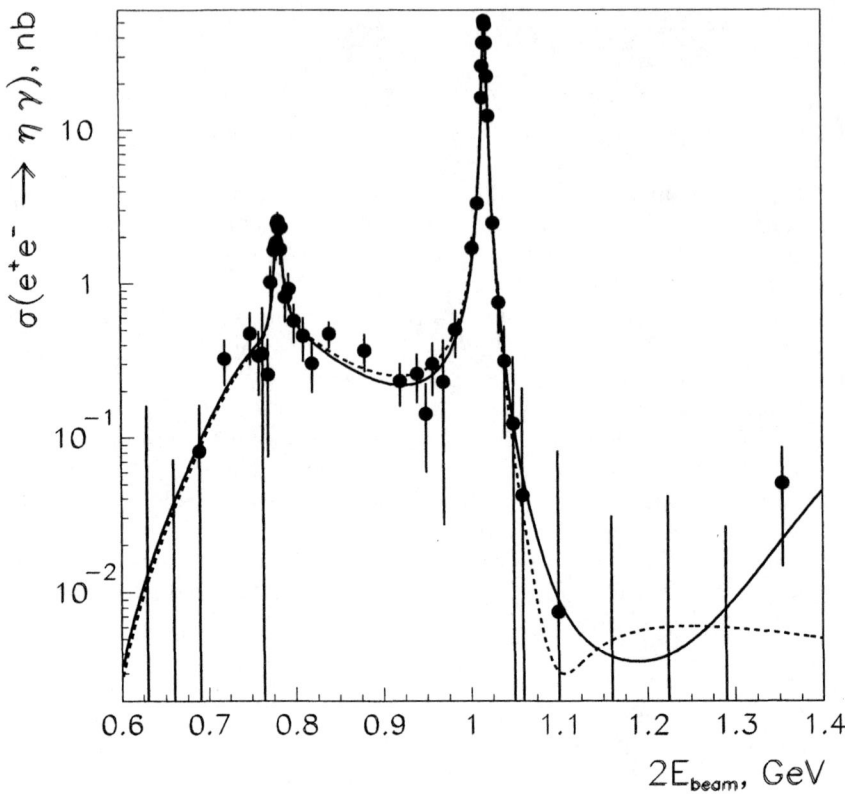

FIGURE 9. Cross section of the $e^+e^- \to \eta\gamma$ process in the optimal fits: the solid curve is VDM, the dashed one is VDM + ρ'.

and further improvement will be possible when VEPP-2000 e^+e^- collider begin to run.

This work is supported in part by the grants: RFBR-98-02-17851, RFBR-99-02-17053, RFBR-99-02-17119, INTAS 96-0624.

REFERENCES

1. PRL, v.86, 11(2001)2227-2231
2. G.A.Aksenov et al., Preprint BINP 85-118, Novosibirsk, 1985.
3. F.V.Ignatov et al., Preprint BINP 99-64, Novosibirsk, 1999.
4. E.V.Anashkin et al., Nucl. Instr. and Meth., A323 (1992) 178.
5. V.M.Aulchenko et al., Nucl. Instr. and Meth., A336 (1993) 53.

TABLE 3. The relative precision hadronic cross sections measurements

Channel	Precision of measurements, %	Contribution to the error of a_μ, ppm
$\pi^+\pi^-$	0.6	0.26
$\pi^+\pi^-\pi^0$	1.5	0.06
K^+K^-	5.2	0.09
$K_L K_S$	1.9	0.02
$\pi^+\pi^-\pi^+\pi^-$	7	0.04
$\pi^+\pi^-\pi^0\pi^0$	7	0.05
$\omega \to \pi^0\gamma, \phi \to \eta\gamma$	6	0.02
Total		0.29

6. D.N.Grigoriev et al., IEEE Trans. Nuc. Sci., 42, N4 (1995) 505.
7. A.B.Arbuzov et al., JHEP 10 (1997) 006,
8. A.B.Arbuzov et al., JHEP 10 (1997) 001
9. E.A.Kuraev and V.S.Fadin, Sov. J. Nucl. Phys., 41 (1985) 466
10. K.Melnikov, SLAC-PUB-8844, hep-ph/0105267, May 2001.
11. L.M.Barkov et al., JETP Lett., 46 (1987) 164
12. R.R.Akhmetshin et al., Phys.Lett., B364 (1995) 199.
13. E.V.Anashkin et al., Preprint BINP 01-58, Novosibirsk, 2001.
14. Review of Particle Physics, The European Physical Journal, C3 (1998).
15. *S.I.Dolinsky et al.* Sov. J. Nucl. Phys., 48 (1988) 277.

Review of results from SND detector[1]

M. N. Achasov*, V.M.Aulchenko*, K.I.Beloborodov*, A.V.Berdyugin*,
A.G.Bogdanchikov*, A.V.Bozhenok*, A.D.Bukin*, D.A.Bukin*,
S.V.Burdin*, T.V.Dimova*, A.A.Drozdetski*, V.P.Druzhinin*,
D.I.Ganushin[†], V.B.Golubev*, V.N.Ivanchenko*, P.M.Ivanov*, A.A.Korol*,
S.V.Koshuba*, I.N.Nesterenko*, E.V.Pakhtusova*, A.A.Polunin*,
A.A.Salnikov*, S.I.Serednyakov*, V.V.Shary*, Yu.M.Shatunov*,
V.A.Sidorov*, Z.K.Silagadze*, A.N.Skrinsky*, A.G.Skripkin[†], Yu.V.Usov*
and A.V.Vasiljev*

*Budker Institute of Nuclear Physics, Siberian Branch of the Russian Academy of Sciences,
Laurentyev 11, Novosibirsk, 630090, Russia
[†]Novosibirsk State University, Novosibirsk, 630090, Russia

Abstract. The review of experimental results obtained with SND detector at VEPP-2M e^+e^- collider in the energy region $\sqrt{s} = 0.36 - 1.38$ GeV is given. The presented results include the following items: studies of the light vector mesons radiative decays, OZI-rule and G-parity suppressed ϕ-meson rare decays, ϕ-meson parameters measurements, studies of $e^+e^- \to \pi^+\pi^-\pi^0$ process dynamics, η and K_S mesons rare decays, η and ϕ mesons conversion decays, and study of the e^+e^- annihilation into hadrons.

The Spherical Neutral Detector (SND) operated since 1995 up to 2000 at VEPP-2M [1] e^+e^- collider in the energy range from 0.36 to 1.38 GeV. SND was described in detail in Ref.[2]. During six experimental years SND had collected data with integrated luminosity about 30 pb^{-1}.

Radiative decays of the ϕ, ω, ρ mesons

Electric dipole transitions of the ϕ meson. Till recently ϕ meson electric dipole transitions were not observed. A search for such decays was first performed with ND detector at VEPP-2M and the upper limits of about 10^{-3} were obtained [3]. About the same time the theoretical proposal of the $\phi \to f_0\gamma$, $a_0\gamma$ decays search appeared [4]. In 1997 the $\phi \to \pi^0\pi^0\gamma$, $\eta\pi^0\gamma$ decays were observed with SND [5]. The SND results based on the full data sample look as follows [6]: $B(\phi \to \pi^0\pi^0\gamma) = (1.22 \pm 0.10 \pm 0.06) \cdot 10^{-4}$, $B(\phi \to \eta\pi^0\gamma) = (0.88 \pm 0.14 \pm 0.09) \cdot 10^{-4}$. Studies of the $\pi^0\pi^0$ and $\eta\pi$ invariant mass spectra (Fig.1) demonstrate that $f_0\gamma$ and $a_0\gamma$ mechanisms dominate in these decays [6]. So the following branching ratios were obtained: $B(\phi \to f_0\gamma) = (3.5 \pm 0.3^{+1.3}_{-0.5}) \cdot 10^{-4}$,

---

[1] Presented by M.N.Achasov, e-mail:achasov@inp.nsk.su

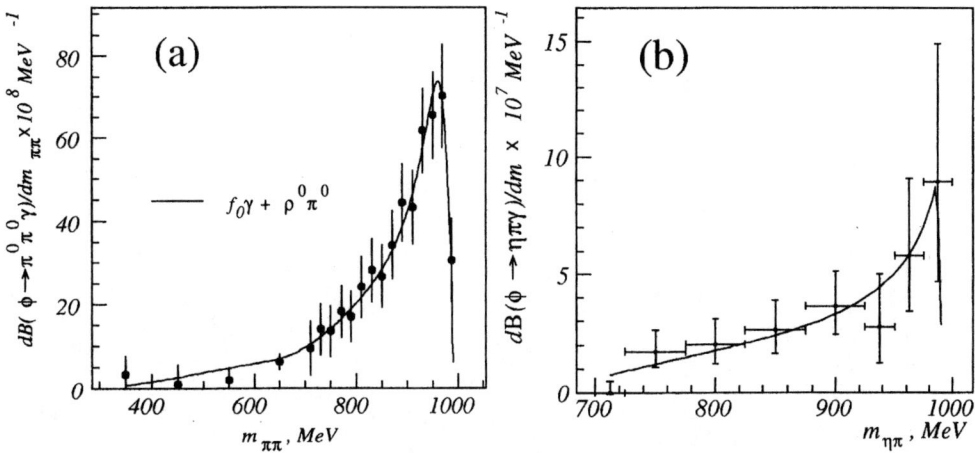

FIGURE 1. The $\pi^0\pi^0$ mass in the decay $\phi \to \pi^0\pi^0\gamma$ (a) and the $\eta\pi^0$ mass in the decay $\phi \to \eta\pi^0\gamma$ (b)

$B(\phi \to a_0\gamma) = (0.88 \pm 0.14 \pm 0.09) \cdot 10^{-4}$. These relatively large values point out the exotic four-quark structure of a_0 and f_0 mesons [7]. CMD2 measurements reported in Ref. [8] agree with SND results. Also results of such measurements were recently reported by KLOE [9].

The decays $\rho, \omega \to \pi^0\pi^0\gamma$. In VDM model these decays proceed through the $\rho \to \omega\pi^0 \to \pi^0\pi^0\gamma$ and $\omega \to \rho\pi^0 \to \pi^0\pi^0\gamma$ transitions with the relative probability about 10^{-5} [10]. The same final state is also possible through the vector mesons radiative transitions to the $\pi^0\pi^0$ scalar state with expected branching ratio about $1.4 \cdot 10^{-5}$ [11, 12]. The only measurement of $\omega \to \pi^0\pi^0\gamma$ decay by GAMS [13] gives value of $(7.2 \pm 2.5) \cdot 10^{-5}$. The SND studies of these decays based on the one third of the accumulated statistics were already reported in Ref. [14]. The results of a new analysis based on the full data sample of about 9 pb^{-1} are presented here.

We cannot extract any information about ω decay mechanisms from the energy or angular distributions due to insufficient statistics. The photon energy spectrum shape agrees with VDM as well as with the sum of VDM and $V \to S\gamma$ (V denotes vector meson and S – the scalar one, for example σ meson) radiative decays mechanisms in case of constructive interference between them (Fig.2 (a)). The destructive interference is ruled out experimental distribution.

The fit of the cross section (Fig.2 (b)) included $\rho, \rho' \to \omega\pi^0$ transitions and $\omega \to \pi^0\pi^0\gamma$ decay in VDM model and through $V \to S\gamma$ transitions. The strong difference in the energy dependences of the phase space for $\rho \to \omega\pi^0$ and $\rho \to S\gamma$ mechanisms allows to distinguish the different models. The model without $\rho \to S\gamma$ contribution gives $P(\chi^2) \simeq 1\%$. Inclusion of the scalar mechanism to the fit improves $P(\chi^2)$ to 30%. The results of the fit follows: $B(\omega \to \pi^0\pi^0\gamma) = (6.3 \pm 1.4 \pm 0.8) \cdot 10^{-5}$, $B(\rho \to \pi^0\pi^0\gamma) = (4.0 \pm 0.9 \pm 0.4) \cdot 10^{-5}$, $B(\rho \to S\gamma \to \pi^0\pi^0\gamma) = (2.0 \pm 0.7 \pm 0.3) \cdot 10^{-5}$. So we confirm the value of ω decay obtained by GAMS. The ρ meson decay to $\pi^0\pi^0\gamma$ was observed for the first time.

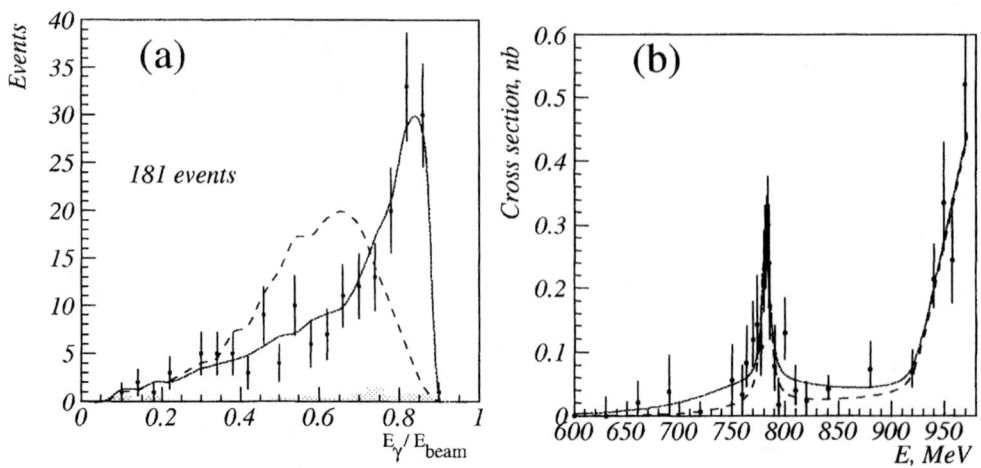

FIGURE 2. (a) – the photon energy spectrum in the reaction $e^+e^- \to \pi^0\pi^0\gamma$ in the energy range near ω meson mass. Solid curve – VDM model, dashed curve – sum of VDM and $\sigma\gamma$ mechanisms in case of destructive interference between them. (b) – cross-section for the $e^+e^- \to \pi^0\pi^0\gamma$ reaction. Solid line – fit in VDM model, dashed line – fit with sum of VDM and $\rho \to S\gamma$ decay.

The magnetic dipole transitions of the light vector mesons. The magnetic dipole radiative decays are traditional objects in the light meson spectroscopy. Only one decay of this type $\phi \to \eta'\gamma$ was not observed till recently. This decay was observed with CMD2 detector [15] and then confirmed by SND [16]. The results of SND studies of the $\phi \to \eta'\gamma$ in comparison with CMD2 and KLOE measurements are listed in Table 1.

TABLE 1. The comparison of the $B(\phi \to \eta'\gamma)$ obtained with SND and results of the other experiments [17, 18]

	SND [16] $\eta' \to \pi^+\pi^-\eta$	SND $\eta' \to \pi^0\pi^0\eta$	SND (average)	CMD2 [17]	KLOE [18]
$B(\phi \to \eta'\gamma) \cdot 10^5$	$6.7^{+3.4}_{-2.9}$	$4.3 \pm 1.6 \pm 0.9$	$4.9^{+1.6}_{-1.5}$	6.4 ± 1.6	6.8 ± 0.8

The process $e^+e^- \to \eta\gamma$ in the seven photon final state was studied in full available energy region. The results based on the part of accumulated statistics were already published [19]. Here we present the result obtained using full data set. It was found that cross section (Fig.3) can be described by sum of ρ, ω and ϕ resonances contributions only. The branching ratios obtained from the fit are presented in Table 2. The experimental ratio of the partial width $\Gamma_{\omega\eta\gamma} : \Gamma_{\rho\eta\gamma} : \Gamma_{\phi\eta\gamma} = 1 : (11.7 \pm 1.9) : (15.9 \pm 1.9)$ is consistent with a prediction of the simple quark model 1:8:12.

The probability of the $\phi \to \eta\gamma$ decay was measured by SND in two other η meson decay modes: $\eta \to \pi^+\pi^-\pi^0$ [21] and $\eta \to \gamma\gamma$ [22]. Combining the results of the three different modes the SND average was obtained: $B(\phi \to \eta\gamma) = (1.310 \pm 0.045)\%$. It is the most precise measurement of this value.

The process $e^+e^- \to \pi^0\gamma$ was studied in the vicinity of ϕ-meson [22] and in the ρ, ω energy region [23]. As in previous case the cross section can be described by sum of ρ, ω and ϕ mesons only. The obtained branching ratios are listed in Table 2. The ρ and ω

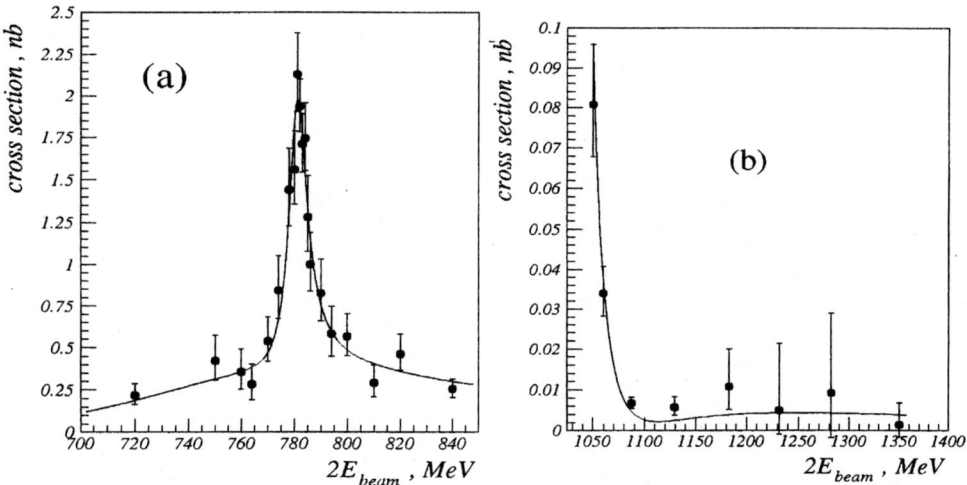

FIGURE 3. The cross section of the reaction $e^+e^- \to \eta\gamma$ in the ρ and ω energy region (a) and above ϕ meson (b)

branching ratios are in good agreement with both PDG values and prediction of a simple quark model. These results are based on a one third of available statistics. For full data sample we expect improvement of accuracy of the ρ meson branching ratio. We also hope that combined analysis of data from ϕ and ρ, ω energy regions could reduce the systematic error of $\phi \to \pi^0\gamma$ branching ratio caused by the model dependence of $\phi - \omega$ interference description.

ϕ meson energy region study

OZI rule and G-parity suppressed $\phi \to \omega\pi^0$ and $\pi^+\pi^-$ decays. Till recently only one decay of this type $\phi \to \pi^+\pi^+$ was observed by detectors OLYA [24] and ND [27]. Such decays are possible through the $\omega - \phi$ mixing or direct transition [25, 26]. In the SND experiment the $\phi \to \pi^+\pi^+$ decay was studied [28] and the decay $\phi \to \omega\pi$ was observed for the first time [29].

These decays are seen as interference patterns around ϕ-resonance in the energy dependence of the cross section. The Born cross section can be written as follows [25]:

$$\sigma(s) = \sigma_0(s) \times \left|1 - Z\frac{m_\phi \Gamma_\phi}{D_\phi(s)}\right|^2,$$

where σ_0 is nonresonant cross section and Z is complex interference amplitude. The measured branching ratios are listed in Table 2. The imaginary parts of Z amplitudes: $\mathrm{Im}(Z_{\pi\pi}) = -0.041 \pm 0.007$, $\mathrm{Im}(Z_{\omega\pi}) = -0.125 \pm 0.020$ agree with theoretical predictions based on standard $\omega - \phi$ mixing, while the expected values of the real parts exceed our results: $\mathrm{Re}(Z_{\pi\pi}) = 0.061 \pm 0.006$, $\mathrm{Re}(Z_{\omega\pi}) = 0.108 \pm 0.16$. The possible cause of this disagreement could be a nonstandard $\omega - \phi$ mixing or direct decays.

TABLE 2. The results of the $\rho,\omega,\phi \to \eta\gamma$ decays studies using the seven photon final state, results of $\rho,\omega,\phi \to \pi^0\gamma$, $\phi \to \pi^+\pi^-$, $\omega\pi$ decays measurements, the obtained ϕ-meson parameters, results on η and K_s mesons rare decays and conversion decays of η and ϕ mesons

	SND	Other data	
$B(\rho \to \eta\gamma) \cdot 10^4$	$2.77 \pm 0.26 \pm 0.16$	$3.28 \pm 0.37 \pm 0.23$	(CMD2 [20])
$B(\omega \to \eta\gamma) \cdot 10^4$	$4.22 \pm 0.47 \pm 0.17$	$5.10 \pm 0.72 \pm 0.34$	(CMD2 [20])
$B(\phi \to \eta\gamma) \cdot 10^2$	$1.34 \pm 0.01 \pm 0.05$	$1.287 \pm 0.013 \pm 0.063$	(CMD2 [20])
$B(\rho \to \pi^0\gamma) \cdot 10^4$	$5.03 \pm 1.17 \pm 0.83$	6.8 ± 1.7	(PDG-2000)
$B(\rho \to \pi^\pm\gamma) \cdot 10^4$		4.5 ± 0.5	(PDG-2000)
$B(\omega \to \pi^0\gamma) \cdot 10^2$	$9.17 \pm 0.16 \pm 0.46$	8.5 ± 0.5	(PDG-2000)
$B(\phi \to \pi^0\gamma) \cdot 10^3$	$1.23 \pm 0.04 \pm 0.09$	1.26 ± 0.10	(PDG-2000)
$B(\phi \to \pi^+\pi^-) \cdot 10^5$	7.1 ± 1.4	8^{+5}_{-4}	[24, 27]
$B(\phi \to \omega\pi^0) \cdot 10^5$	$5.2^{+1.3}_{-1.1}$		
m_ϕ, MeV	$1019.42 \pm 0.02 \pm 0.04$	1019.417 ± 0.014	(PDG-2000)
Γ_ϕ, MeV	$4.21 \pm 0.03 \pm 0.02$	4.458 ± 0.032	(PDG-2000)
$B(\phi \to e^+e^-) \cdot 10^4$	$2.93 \pm 0.02 \pm 0.14$	2.91 ± 0.07	(PDG-2000)
$B(\phi \to K^+K^-)$, %	$47.6 \pm 0.3 \pm 1.6$	49.2 ± 0.7	(PDG-2000)
$B(\phi \to K_S K_L)$, %	$35.1 \pm 0.2 \pm 1.2$	33.8 ± 0.6	(PDG-2000)
$B(\phi \to \pi^+\pi^-\pi^0)$, %	$15.9 \pm 0.2 \pm 0.8$	15.5 ± 0.6	(PDG-2000)
$B(\phi \to \eta\gamma)$, %	$1.33 \pm 0.03 \pm 0.05$	1.297 ± 0.033	(PDG-2000)
$B(\eta \to \pi^0\pi^0) \cdot 10^4$	< 6 [38]	< 4.3	(CMD2 [8])
$B(\eta \to \pi^0\gamma\gamma) \cdot 10^4$	< 8.4 [39]	7.1 ± 1.4	(PDG2000)
$B(K_S \to 3\pi^0) \cdot 10^5$	< 1.4 [40]	< 1.9	(CPLEAR [41])
$B(\phi \to \eta e^+e^-) \cdot 10^4$	1.19 ± 0.22 [42]	1.17 ± 0.12	(CMD2 [43])
$B(\eta \to e^+e^-\gamma) \cdot 10^3$	5.15 ± 0.96 [42]	7.10 ± 0.79	(CMD2 [43])
$B(\phi \to \pi^0 e^+e^-) \cdot 10^5$	1.05 ± 0.37	1.22 ± 0.40	(CMD2 [44])

ϕ meson parameters study. The main parameters of the ϕ meson were measured through studies of the processes $e^+e^- \to K^+K^-, K_S K_L$ and $\pi^+\pi^-\pi^0$ [30]. The measured cross sections were approximated within the VDM, taking into account ρ, ω and ϕ mesons. Contributions from higher resonances ρ', ω', ϕ' were included in each cross section as constant terms. The K^+K^- and $K_S K_L$ cross sections can be fitted by a sum of ρ, ω and ϕ contributions only, while for a good approximation of the $e^+e^- \to \pi^+\pi^-\pi^0$ cross section the additional contribution, which can be attributed to the higher resonances, is strongly required. The obtained ϕ-meson parameters (Table 2) mainly agree with PDG data and have accuracies comparable with the world averages. The only measured value which is in conflict with the now days world average is ϕ-meson width. The world average Γ_ϕ value is strongly based on CMD2 measurement $\Gamma_\phi = 4.477 \pm 0.036 \pm 0.022$ MeV [31] which contradict to the SND one. But the recent CMD2 result $\Gamma_\phi = 4.280 \pm 0.033 \pm 0.025$ MeV [32] agreed with SND measurement.

The ϕ-meson leptonic branching ratio was also measured using $e^+e^- \to \mu^+\mu^-$ reaction [33]: $\sqrt{B(\phi \to e^+e^-)B(\phi \to \mu^+\mu^-)} = (2.93 \pm 0.11) \cdot 10^{-4}$, which is in a good agreement with branching value of $\phi \to e^+e^-$ decay. Using SND value of $\phi \to e^+e^-$ decay width we obtained the following leptonic branching ratio: $B(\phi \to l^+l^-) = (2.93 \pm 0.09) \cdot 10^{-4}$.

The $e^+e^- \to \pi^+\pi^-\pi^0$ dynamics study and other results. In SND experiment the dipion mass spectra were studied in the $e^+e^- \to \pi^+\pi^-\pi^0$ process in the energy region around ϕ-meson [34]. Such studies provide the information about reaction dynamics as well as about ρ-meson parameters – mass and width, $\rho^\pm$ and ρ^0 mass difference [35]. Spectra were analyzed within the VDM framework taking into account $\rho\pi$ transition, $\rho - \omega$ mixing and possible transition through intermediate states different from $\rho\pi$ (for example, via $\rho'\pi$). It was found that the experimental data can be described as a pure $\rho\pi$ transition. Upper limit on the branching ratio of the non $\rho\pi$ $\phi(1020) \to 3\pi$ decay was obtained: $B(\phi \to \pi^+\pi^-\pi^0) < 6 \cdot 10^{-4}$. This result agrees with CMD2 similar studies [36]. Also the result of such studies was reported by KLOE [37], but unfortunately the information given there is insufficient to do the comparison of the results. Neutral and charged ρ-mesons mass difference was found to be consistent with zero: $m_{\rho^\pm} - m_{\rho^0} = -1.3 \pm 2.3$ MeV. The ρ-meson mass and width were measured equal to $m_\rho = 775.0 \pm 1.3$ MeV, $\Gamma_\rho = 150.4 \pm 3.0$ MeV. The ρ mass values obtained by using different reactions contradict each other. SND ρ-mass value support the results of the e^+e^- annihilation and τ decay experiments: $m_\rho = 776 \pm 0.9$ MeV. But the PDG value 769.3 ± 0.8 MeV, which takes into account all experiments in which the ρ-meson mass was measured, contradicts our result.

Some other results obtained using statistics collected in the ϕ-meson energy region are presented in Table 2.

e^+e^- annihilation into hadrons above 1 GeV

The light vector mesons are studied rather well. They are 2 quark states, their masses, widths and the main decays are measured with high accuracy. The experimental data also point out the existence of the states with vector meson quantum numbers $I^G(J^{PC}) = 0^+(1^{--}), 0^-(1^{--})$ and masses above 1 GeV. Parameters of these states are not well established due to the poor accuracy and conflicting of experimental data. The nature of these states is not clear. They are considered as a mixture of two quark, four quark and hybrids states [45] or as a two quark states – radial and orbital excitations of the ρ, ω and ϕ mesons [46]. In this context the main experimental task is the improvement of the cross sections measurement accuracy. In SND experiment the following processes were studied.

The $e^+e^- \to K_S K_L$ cross section was measured using $K_S \to \pi^0\pi^0$ decay mode. Our measurements in comparison with OLYA and DM1 results are shown in Fig.4 (a). The curve is theoretical cross section with ρ, ω and ϕ contributions only. Experimental data above 1.2 GeV exceed the conventional VDM prediction.

The process $e^+e^- \to \omega\pi$ was studied in the $\pi^0\pi^0\gamma$ final state [47]. Measured cross section in comparison with the other results is shown in Fig. 4 (b). The systematic error of SND measurement is about 5%. The CLEO2 results[2] are in good agreement with ours, while CMD2 measurements are about 10% lower, but this difference is smaller

---

[2] e^+e^- annihilation cross section was calculated from $\tau \to 3\pi\pi^0$ decay data using CVS hypothesis

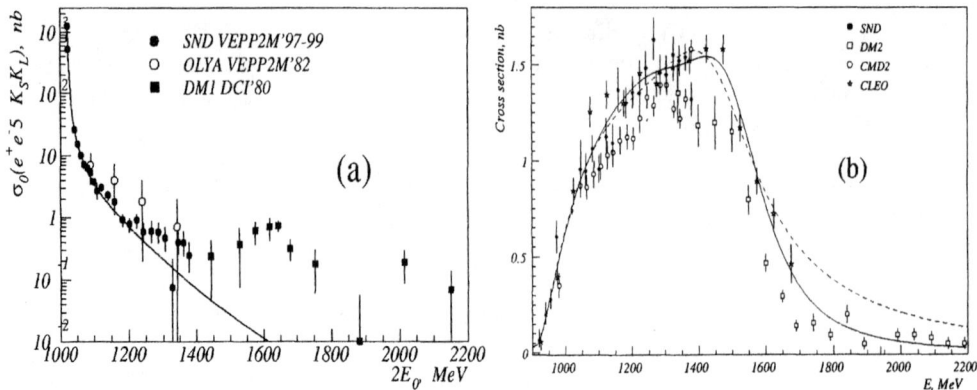

FIGURE 4. (a) – the cross section of the reaction $e^+e^- \to K_S K_L$. The results of the SND, OLYA [53] and DM1 [54] are shown. Curve is theoretical cross section in conventional VDM. (b) – the cross section of the reaction $e^+e^- \to \omega\pi \to \pi^0\pi^0\gamma$. The results of the SND [47], DM2 [48], CMD2 [49] and CLEO2 [50] are shown. Curves are results of fitting in models described in Ref.[47]

than the 15% systematic error quoted in Ref.[49]. The same process was also studied in the $\pi^+\pi^-\pi^0\pi^0$ final state [51]. Obtained cross section agrees with our result in $\pi^0\pi^0\gamma$ mode. Its systematic error was estimated to be 20% for $\sqrt{s} < 1150$ MeV and 15% at $\sqrt{s} > 1150$.

The $e^+e^- \to \pi^+\pi^-\pi^0\pi^0$ cross section with subtracted contribution from $\omega\pi^0$ is shown in Fig.5 (a). The systematic error of SND measurements is about 20%. The SND result is compared with CLEO2, CMD2 and DM2 data[3]. The two groups are seen: CMD2 dots better agree with DM2 while SND ones – with CLEO2. But the difference is smaller than systematic errors. The measured cross section of the e^+e^- annihilation into four charged pions is shown in Fig.5 (b). Here SND dots better agree with DM2 result. The systematic error of the $e^+e^- \to \pi^+\pi^-\pi^+\pi^-$ cross section was estimated to be 12% for $\sqrt{s} < 1150$ MeV and 8% at $\sqrt{s} > 1150$.

Both $e^+e^- \to \rho\pi$ and $e^+e^- \to \omega\pi^0$ mechanisms contribute to the $\pi^+\pi^-\pi^0$ final state. The $\omega\pi$ contribution was predicted in Ref.[55] and observed by SND [56]. SND already reported the total cross section measurements based on the part of the accumulated statistics [57]. The new result of the cross section measurement is presented here. For data analysis we use the following theoretical model, taking into account the $\rho - \omega$ mixing [55]:

$$\frac{d\sigma}{dm_0 dm_+} = \frac{4\pi\alpha}{s^{3/2}} \frac{|\vec{p}_+ \times \vec{p}_-|^2}{12\pi^2 \sqrt{s}} m_0 m_+ \cdot \left| A_{\rho\pi}(s) \sum_{i=+,0,-} \frac{g_{\rho^i\pi\pi}}{D_\rho(m_i)} + A_{\omega\pi}(s) \frac{\Pi_{\rho\omega} g_{\rho^0\pi\pi}}{D_\rho(m_0) D_\omega(m_0)} \right|^2$$

---

[3] While extracting cross section data from CLEO2 results the ratio $\sigma_{e^+e^-\to\pi^+\pi^-\pi^+\pi^-}/\sigma_{e^+e^-\to\pi^+\pi^-2\pi^0} = 2$ was assumed, which is confirmed by SND measurements [51]

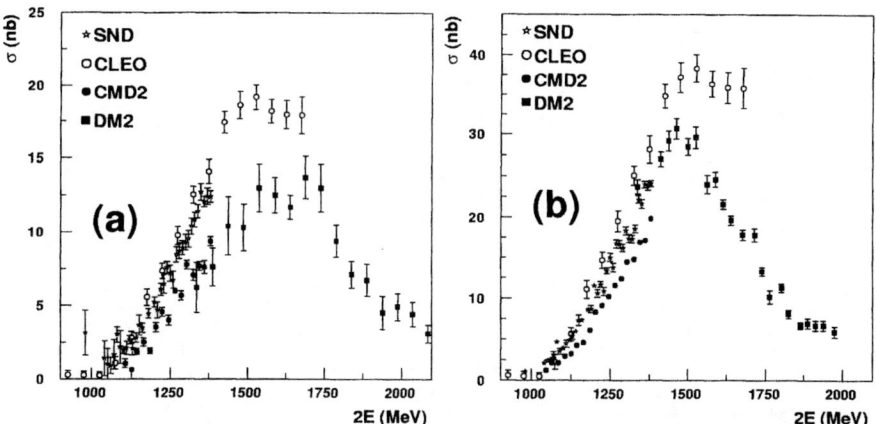

FIGURE 5. (a) – the cross section of the reaction $e^+e^- \to \pi^+\pi^-\pi^0\pi^0$ with subtracted contribution from $\omega\pi^0$. (b) – the cross section of the reaction $e^+e^- \to \pi^+\pi^-\pi^+\pi^-$. The results of the SND [51], DM2 [52], CMD2 [49] and CLEO2 [50] are shown.

TABLE 3. ω^i parameters obtained from the fit. Here ϕ denotes a relative phases between ω and ω primes.

	ω^1	ω^2	ω^3
m, MeV	1250±29	1400±19	1771±28
Γ, MeV	426±135	626±89	473±76
$\sigma(V \to \rho\pi)$, nb	0.56±0.25	3.90±0.39	2.28±0.46
$\sigma(V \to \omega\pi\pi)$, nb	0	0.046±0.039	2.49±0.33
ϕ	π	π	0
$\Gamma(V \to e^+e^-)$, eV	~25	~300	~470

$$A_{\rho\pi} \sim \sum_{V=\omega,\phi,\omega',...} \frac{\Gamma_V m_V^2 \sqrt{m_V \sigma(V \to \rho\pi)}}{D_V(s)} \qquad A_{\omega\pi}(s) = \sum_{V=\rho,\rho',...} \frac{g_{\gamma V} g_{V\omega\pi^0}}{D_V(s)}$$

$$\text{Im}(\Pi_{\rho\omega}) \ll \text{Re}(\Pi_{\rho\omega}), \quad \text{Re}(\Pi_{\rho\omega}) = 2m_\omega\delta, \quad \delta = 2.3 \text{ MeV}, \quad \delta \sim \sqrt{B(\omega \to \pi^+\pi^-)}$$

The combined studies of the total cross section and dipion mass spectra provide the information about relative phase between $A_{\rho\pi}$ and $A_{\omega\pi}$ amplitudes and $\omega \to \pi^+\pi^-$ branching ratio. We performed the combined fit of $\pi^+\pi^-\pi^0$ (Fig.6) and $\omega\pi^+\pi^-$ cross sections. The cross section measured by SND in the ϕ-meson energy region was also included in the fit. The best description of the data was obtained when the cross sections were fitted by a sum of ω, ϕ and three ω^i amplitudes. The obtained ω^i parameters are listed in Table 3.

To obtain relative phase between $A_{\rho\pi}$ and $A_{\omega\pi}$ amplitudes and $\omega \to \pi^+\pi^-$ branching ratio the invariant mass distribution and the ratio of $\rho\pi$ to 3π cross sections were fitted together in each energy point. The energy dependence of the relative phase is shown in Fig.6. The obtained branching ratio $B(\omega \to \pi^+\pi^-) = 2.46 \pm 0.42 \pm 0.15$ agrees with world average value.

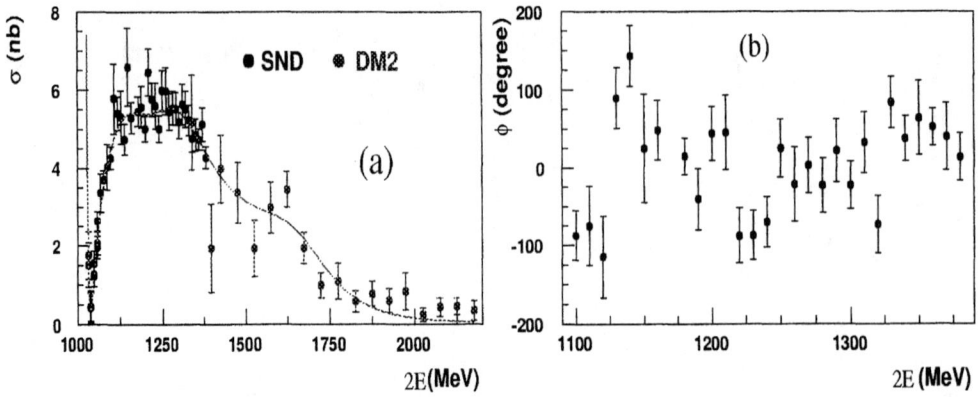

FIGURE 6. (a) – the cross sections of the reactions $e^+e^- \to \pi^+\pi^-\pi^0$. The results of the SND and DM2 [58] are shown. Curve is the fit by a sum of ω, ϕ and three ω' amplitudes. (b) – the energy dependence of the relative phase between $A_{\rho\pi}$ and $A_{\omega\pi}$ amplitudes.

A search for direct production of a_2 and f_2 mesons in e^+e^- annihilation was performed with SND [59]. The following upper limits were obtained $\Gamma(a_2 \to e^+e^-) < 0.56$ eV and $\Gamma(f_2 \to e^+e^-) < 0.11$ eV. These upper limits are only four times higher than unitarity limit [60].

Conclusion

The SND detector operated since 1995 up to 2000 at VEPP-2M collider in the energy range $360 < \sqrt{s} < 1380$ MeV and had collected data with integrated luminosity of about 30 pb^{-1}. The ρ, ω, ϕ mesons decays and e^+e^- annihilation into hadrons were studied. New rare decays $\phi \to \pi^0\pi^0\gamma$, $\eta\pi^0\gamma$, $\omega\pi^0$ and $\rho \to \pi^0\pi^0\gamma$ were observed. Many other results were obtained.

The present work was supported in part by grants RFBR 00-15-96802, 01-02-22003, 01-02-16934-a, 00-02-17478, 00-02-17481, grant no. 78 1999 of Russian Academy of Science for young scientists.

REFERENCES

1. A.N. Skrinsky, in Proc. of Workshop on physics and detectors for DAΦNE, Frascati, 1995, p.3
2. M.N. Achasov et al., Nucl. Instr. and Meth. A449, 125 (2000)
3. S.I. Dolinsky et al., Phys. Rep. 202 (1991) 99
4. N.N. Achasov, V.N. Ivanchenko, Nucl. Phys. B315 (1989) 465
5. M.N.Achasov et al., in Proc. of the 7th Int. Conf. on Hadron Spectroscopy, Brookhaven (BNL), 1997, p.783; Phys. Lett. B436 (1998) 441; Phys. Lett. B440 (1998) 442; Yad. Fiz., 62 (1999) 484
6. M.N.Achasov, et al., Phys Lett. B485 (2000) 349; Phys Lett. B479 (2000) 53
7. N.N.Achasov and V.V. Gubin, Phys. Rev. D 63 (2001) 094007
8. R.R.Akhmetshin et al., Phys. Lett. B 462 (1999) 380; Phys. Lett. B 462 (1999) 371
9. A.Aloisio, hep-ex/0107024

10. A.Bramon et al., Phys. Lett. B283 (1992) 476
11. E.Marco et al., Phys. Lett. B470 (1999) 20
12. A.Bramon et al., Phys. Lett. B289 (1992) 97
13. D.Albe et al., Phys. Lett. B340 (1994) 122
14. M.N.Achasov et al., JETP Letters, 71(9) (2000) 355
15. R.R. Akhmetshin et al., Phys. Lett. B415 (1997) 445
16. V.M.Aulchenko, et al., JETP Letters 69 (1999), 97
17. R.R. Akhmetshin et al., Phys. Lett. B494 (2000), 26
18. A.Aloisio, et al., hep-ex/0107022
19. M.N.Achasov et al., JETP Letters 72 (2000) 282
20. R.R. Akhmetshin et al., Phys. Lett. B509 (2001), 217
21. M.N.Achasov et al., JETP 90 (2000) 17
22. M.N. Achasov et al., Eur.Phys.J. C12 (2000)
23. M.N.Achasov et al., Preprint, Budker INP 2001-54, Novosibirsk, 2001 (in Russian)
24. I.B.Vasserman et al., Phys. Lett. B99 (1981) 62
25. N.N.Achasov, A.A.Kozhevnikov, Int. J. Mod. Phys. A7 (1992) 4825
26. J.A. Oller et al., Phys. Rev. D62 (2000) 114017
27. V.B.Golubev et al., YAF 44 (1986) 633; Sov. JNP 44 (1986) 409
28. M.N.Achasov et al., Phys. Lett. B474 (2000) 188
29. M.N.Achasov et al., Phys. Lett. B449 (1999) 122; Nuc. Phys. B569 (2000) 158; V.M.Aulchenko et al., JETP 90 (2000) 927
30. M.N.Achasov et al., Phys. Rev. D 63, (2001) 072002
31. R.R.Akhmetshin et al., Phys. Lett. B466 (1999) 385
32. R.R.Akhmetshin et al., Phys. Lett. B508 (2001) 217
33. M.N.Achasov et al., Phys. Lett. B456 (1999) 304; Phys. Rev. Lett. 86 (2001) 1698
34. M.N.Achasov et al., hep-ex/0106048
35. M.N.Achasov, N.N.Achasov, Pis'ma Zh. Eksp. Teor. Fiz. 69 (1999) 8; JETP Lett. 69 (1999) 7
36. R.R.Akhmetshin et al., Phys. Lett. B434 (1998) 426
37. M.Adinolfi et al., hep-ex/0006036
38. M.N.Achasov et al, Phys. Lett. B425 (1998) 388
39. M.N.Achasov et al, Nucl. Phys. B600 (2001) 3
40. M.N.Achasov et al, Phys. Lett. B459 (1999) 674
41. A.Angelopoulos et al., Phys. Lett. B425 (1998) 391
42. M.N.Achasov et al, Phys. Lett. B504 (2001) 275
43. R.R.Akhmetshin et al., Phys. Lett. B501 (2001) 191
44. R.R.Akhmetshin et al., Phys. Lett. B503 (2001) 237
45. A.Donnachie, Yu.S.Kalashnikova Phys. Rev. D60 (1999) 114011; Z. Phys. C60 (1993) 187; Z. Phys. C59 (1993) 621; A.B.Clegg, A.Donnachie Z. Phys. C62 (1994) 455
46. N.N.Achasov, A.A. Kozhevnikov Phys. Rev. D55 (1997) 2663; Phys. Rev. D57 (1998) 4334; Phys. Rev. D62 (2000) 117503
47. M.N.Achasov et al., Phys. Lett. B486 (2000) 29
48. D.Bisello et al., Nucl. Phys. Proc. Suppl. 21 111 (1991)
49. R.R.Akhmetshin et al., Phys. Lett. B466 (1999) 392
50. K.W.Edwards et al., Phys. Rev. D61 (2000) 072003
51. M.N.Achasov et al., Preprint, Budker INP 2000-34, Novosibirsk, 2000 (in Russian)
52. L.Stanco, in Proc. of the Int. Conf. on Hadron Spectroscopy, College Park, Maryland, 1991, p.84
53. P.M.Ivanov et al., Pis'ma Zh. Eksp. Teor. Fiz. 39 (1982) 91
54. F.Mane et al., Phys. Lett. B99 (1981) 261
55. N.N. Achasov, A.A.Kozhevnikov, G.N.Shestakov, Phys. Lett. B50 (1974) 448
56. M.N. Achasov et al., Preprint Budker INP 98-65 (1998); hep-ex/9809013
57. M.N. Achasov et al., Phys. Lett. B462 (1999) 365
58. A.Antonelli et al., Z. Phys C56 (1992) 15
59. M.N.Achasov et al., Phys. Lett. B492 (2000) 8
60. A.I.Vainshtein, I.B.Khriplovich, Yad. Fiz. 13 (1971) 620

LIGHT MESONS FROM CHARM DECAYS

Recent results from Belle

S. R. Hou   for the Belle collaboration

Department of Physics, National Taiwan University, Taipei 106, Taiwan

Abstract. Using data collected by the Belle detector at KEKB, we report preliminary results on resonance formation in two-photon collisions. Tensor meson production has been studied in the K^+K^-, $K_S^0 K_S^0$ and $\pi^+\pi^-\pi^0$ final states. χ_{c2} production is measured in the $\gamma J/\psi$ decay channel. We also report results on the charmed meson decay lifetimes, D^0-$\bar{D}^0$ mixing, and prompt charmonia production.

Introduction

The Belle detector [1] is a general purpose spectrometer designed and constructed mainly for observing CP violation in B decay. With the high luminosity provided by the KEKB [2] storage ring, the Belle experiment is currently accumulating data at a rate of more than 1.5 fb^{-1} per month, corresponding to 1.6 million $B\bar{B}$ and 2 million $c\bar{c}$ events per month produced at the $\Upsilon(4S)$ resonance. With the advantage of large statistics, precision measurements are performed for charmed hadron properties and resonance formation in two-photon collisions.

The Belle detector is equipped with a 1.5 T superconducting solenoid magnet. Charged tracks are reconstructed in a 50 layer Central Drift Chamber (CDC) and in three concentric layers of double sided silicon strip detectors. The charged particle acceptance of the spectrometer covers the laboratory polar angles of $17° < \theta < 150°$ which corresponds to 92% of the full center-of-mass (CM) solid angle. Photons and electrons are identified using the CsI(Tl) Electromagnetic Calorimeter (ECL) located inside the magnet coil. Muons and K_L^0's are detected using resistive plate chambers embedded in the iron magnetic flux return (KLM). Charged particles are identified using specific ionization measurements in the Central Drift Chamber, pulse heights from the Aerogel Cherenkov Counters (ACC) and timing information from the Time of Flight (TOF) Counters. By combining the information from these three detector systems, K/π separation is achieved over the momentum range from about 0.2 to 3.5 GeV/c.

Production of $\gamma\gamma \to K\bar{K}$

Resonance formation in the kaon-pair final states in two photon collisions is dominated by the $f_2'(1525)$ meson. A higher mass resonance state has been reported by the L3 experiment for the $K_S^0 K_S^0$ final state at 1750 MeV/c^2 [3]. We have conducted measurements for both the K^+K^- and the $K_S^0 K_S^0$ channels.

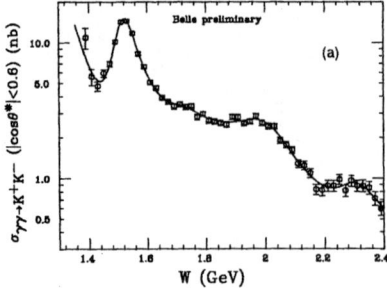
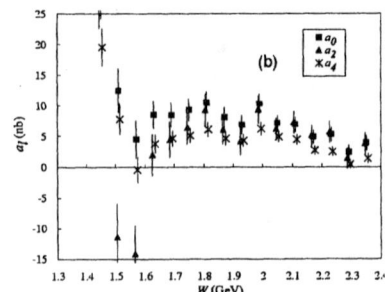

FIGURE 1. (a) The cross section of $\gamma\gamma \to K^+K^-$ in the polar angle range $|\cos\theta| < 0.6$. The curve is the fit by a function including resonances and continuum production. (b) The moments a_0(squares), a_2(triangles) and a_4(crosses) of the fit to Legendre polynomials.

Candidate K^+K^- events in two-photon collisions are selected from pairs of charged particles with the scalar sum of track momenta smaller than 6 GeV/c. The transverse momenta of the charged particles are required to be larger than 400 MeV/c. To enhance events induced by collisions of quasi-real photons, a strict p_t-balance cut is imposed on the final state, $|\sum p_t|^2 < 0.01$ GeV$^2/c^2$. Charged kaons are identified using information from the CsI calorimeter, TOF counters, ACC counters and dE/dx in the CDC. The data sample analyzed corresponds to an integrated luminosity of 32.6 fb^{-1}.

The cross section for the $\gamma\gamma \to K^+K^-$ process is calculated as a function of the two-photon invariant mass W (Fig. 1(a)). In addition to the clean peak observed for the $f_2'(1525)$, enhancements are also seen in the vicinities of 1.75 GeV, 2.0 GeV and 2.3 GeV. The W dependence is fitted by a function including resonances and continuum production. The resonance parameters obtained by the fit are summarized in Table 1. The systematic error of around 10% is mainly from uncertainties in kaon identification. The angular distributions at several W intervals are fitted by a sum of Legendre polynomials, $d\sigma/d|\cos\theta| = \sum_{l=0,2,4} \sqrt{2l+1}\, a_l P_l(|\cos\theta|)$. The moments a_l obtained are shown in Fig. 1(b). The fact that all the three moments have enhancements at 1.75 GeV and 2.0 GeV suggests that these resonances have $J = 2$.

Events in the $K_S^0 K_S^0$ final states are identified from $K_S^0 \to \pi^+\pi^-$ decays. Events produced by quasi-real photons are selected by requiring the scalar sum of the track momenta be smaller than 6 GeV/c and the total momentum imbalance $|\sum p_t|^2 < 0.1$ GeV$^2/c^2$. The analysis is conducted on a 28.1 fb^{-1} data sample. The mass distribution of the two pion pairs is shown in Fig. 2(a). $K_S^0 K_S^0$ candidates are

TABLE 1. The resonance parameters of the fit in Fig. 1. σ_{peak} is the cross section for $|\cos\theta| < 0.6$ in the CM frame. The errors are statistical only.

Resonance	M(MeV/c^2)	Γ (MeV)	σ_{peak} (nb)	Relative phase (rad)
$f_2'(1525)$	1525(fixed)	98 ± 4	9.9 ± 1.1	-0.2 ± 0.2 to "$f_2 + a_2$"
1.75 GeV	1742 ± 11	257 ± 46	1.21 ± 0.25	2.8 ± 0.2 to "2.0 GeV"
2.0 GeV	2048 ± 8	272 ± 22	1.71 ± 0.28	
2.3 GeV	2317 ± 11	157 ± 41	0.39 ± 0.10	incoherent

TABLE 2. Resonance parameters of the fit to the $K_S^0 K_S^0$ mass spectrum. The errors are statistical only.

Resonance	M (MeV/c^2)	Γ (MeV)	$\Gamma_{\gamma\gamma}BR(R \to K_S^0 K_S^0)$ (eV)
1750 MeV	1768 ± 9.6	323 ± 29	
2300 MeV	2261 ± 16	91 ± 34	
χ_{c0}	3412.5 ± 6.8	28.8 ± 18.8	8.2 ± 3.5

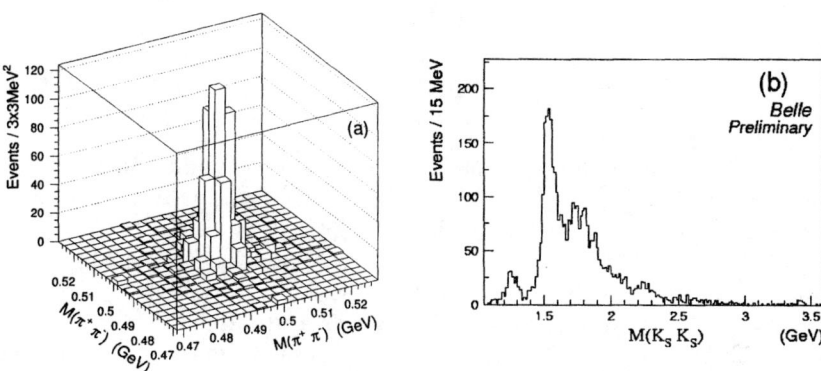

FIGURE 2. (a) Distribution of the two K_S^0 masses, and (b) invariant mass distribution for the $K_S^0 K_S^0$ candidates.

selected by requiring the masses of both pion pairs be within 10 MeV of the nominal K_S^0 mass. The invariant mass distribution for the $K_S K_S$ candidates is shown in Fig. 2(b). The $f_2'(1525)$ meson is dominant. Events in the $f_2(1270)$-$a_2(1320)$ interference region are suppressed by the track momentum threshold. The resonance structure at 1700 MeV/c^2 is consistent with the L3 measurement. The enhancement at 2.3 GeV/c^2 is similar to the K^+K^- observation. The signal for χ_{c0} at 3.41 GeV/c^2 is also seen. The resonance parameters obtained are summarized in Table 2. The χ_{c0} radiative width obtained is consistent with the CLEO report [4]. A preliminary study for the angular distribution for events in the 1700 MeV/c^2 region had shown a significant $J = 0$ contribution. The determination of the spin-parity and helicity state is not yet conclusive.

Resonance production in $\gamma\gamma \to \pi^+\pi^-\pi^0$

The exclusive $\pi^+\pi^-\pi^0$ final state produced in two-photon collisions is dominated by the formation of the $a_2(1320)$ meson. In addition, a resonance state in the mass region around 1750 MeV/c^2 has been reported. The L3 result [5] indicates a new radial excitation state, $a_2(1750)$, which is predominantly of $J^P = 2^+$, helicity 2 wave.

We have performed a spin-parity analysis [6] using a 6.54 fb^{-1} data sample. Two-photon events are selected for the total energy sum of observed tracks and photons below 5 GeV. To eliminate contamination from beam related background, the photons from π^0 decay are required to have energies larger than 150 MeV. Candidate π^0's

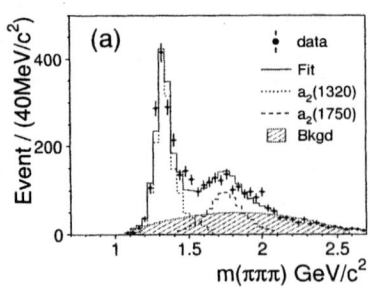
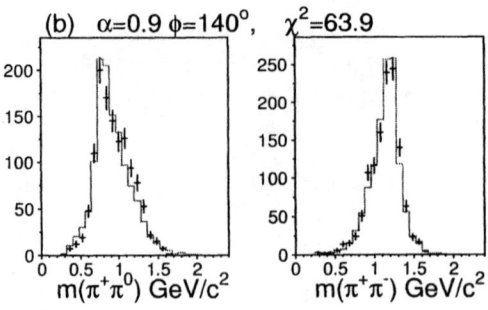

FIGURE 3. (a) Invariant mass distribution for the $\pi^+\pi^-\pi^0$ candidates of $p_t^2 < 0.0002$ GeV$^2/c^2$. (b) The di-pion mass distribution is compared to the Monte Carlo of interference amplitude $\alpha = 0.9$ and phase angle $\phi = 140°$.

are selected with the invariant mass of the pair of photons within 3σ of the nominal π^0 mass. The mass resolution for π^0's is 6 MeV/c^2. The low p_t^2 events correspond to interactions of quasi-real photons with $Q^2 \sim 0$, and are the least contaminated by beam background or other two-photon interactions. Therefore a stringent requirement of $p_t^2 < 0.0002$ GeV$^2/c^2$ is applied to select the final $\pi^+\pi^-\pi^0$ sample.

The invariant mass spectrum obtained (Fig. 3(a)) is fitted to the Monte Carlo distributions of $a_2(1320)$ and $a_2(1750)$ with the background parameterized by a second order polynomial. The Monte Carlo sample for the $a_2(1320)$ is normalized to the nominal two-photon radiative width and serves as a sideband constraint. The background is also constrained by the sideband above 2.0 GeV/c^2.

The resonance mass and width are determined by minimizing the χ^2 between the data and the Monte Carlo distributions generated at several mass positions and widths. The dominant systematic uncertainties come from the estimation of background and selection efficiency. The selection efficiency differs for the decay intermediate state and spin-parity. For a pure $J^P = 2^+$ helicity 2 state with equal decay amplitude into the interference of $\rho\pi$ and $f_2\pi$, the parameters obtained are

mass $m = 1740 \pm 10(\text{stat.}) \pm 10(\text{syst.})$ MeV/c^2;
width $\Gamma = 290 \pm 30(\text{stat.}) \pm 20(\text{syst.})$ MeV;
radiative width $\Gamma_{\gamma\gamma} \cdot BR(\pi^+\pi^-\pi^0) = 0.27 \pm 0.02(\text{stat.}) \pm 0.04(\text{syst.})$ keV.

Events in the mass region corresponding to the $a_2(1750)$ have been studied for spin-parity and decay modes. The decay amplitude is a sum of orthogonal spherical harmonics multiplied by the di-pion Breit-Wigner terms. The interference between $\rho\pi$ and $f_2\pi$ is present for $J^P = 2^+$. The decay matrix element is given by

$$\mathcal{T}_\lambda = \left| \left(\frac{T_\lambda(\rho^+\pi^-)}{m_\rho^2 - s_{\pi^+\pi^0} - im_\rho\Gamma_\rho} + \frac{T_\lambda(\rho^-\pi^+)}{m_\rho^2 - s_{\pi^-\pi^0} - im_\rho\Gamma_\rho} \right) + \alpha e^{i\phi} \frac{T_\lambda(f_2\pi^0)}{m_{f_2}^2 - s_{\pi^+\pi^-} - im_{f_2}\Gamma_{f_2}} \right|^2,$$

where λ denotes the helicity state, α^2 is the branching ratio of $f_2\pi$ to $\rho\pi$, and ϕ the relative phase angle of the two decay amplitudes. The angular distributions for the di-pions and the pions from di-pion decay are described by the transition amplitude T_λ.

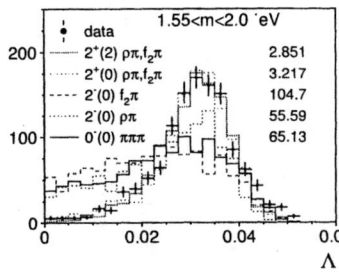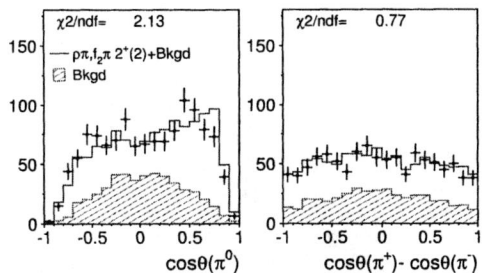

FIGURE 4. Λ and $\cos\theta$ distributions and the χ^2 tests to Monte Carlo (histograms) for events in the 1750 MeV/c² mass region.

For $J^P = 2^+$, the interference phase angle has an effect of shifting the di-pion mass peaks of ρ and f_2 toward each other as ϕ gets smaller. The two interference parameters are determined by χ^2 tests for the di-pion mass spectra and the Monte Carlo predictions of pure $J^P = 2^+$ helicity 2 state. Shown in Fig. 3(b) is a test to the Monte Carlo of $\alpha = 0.9$ and $\phi = 140°$. The phase angle determined by the χ^2 tests is 125 ± 5 degrees. The relative amplitude obtained is $\alpha = 0.9 \pm 0.1$.

The resonance spin-parity effects the magnitude of the norm of the decay plane. The distribution for an optimized Λ parameter, $\Lambda = |\vec{p}(\pi^+) \times \vec{p}(\pi^-)/Q|^2$, where Q is the kinetic energy of the pions, is shown in Fig. 4. The χ^2 tests for Monte Carlo samples of various spin-parity and decay modes are also shown. The angular dependence of the spin-parity and helicity state can be directly seen in the polar angle distributions for the final state pions. The $\cos\theta$ distributions are also shown in Fig. 4. In both the Λ and the $\cos\theta$ distributions, the $J^P = 2^+$ helicity 2 wave gives the best agreement with data.

The Monte Carlo samples are generated assuming independent resonance formations for the $a_2(1320)$ and a resonance at 1750 MeV/c². The dominant $a_2(1320)$ sample serve as a reference for studying events in the 1750 MeV/c² region. The di-pion mass spectra and the angular distributions of final state pions are consistent with the hypothesis for the $a_2(1750)$ of $J^P = 2^+$ helicity 2 wave in decays to $\rho\pi$ and $f_2\pi$.

Charmonium production of $\gamma\gamma \to \chi_{c2}$

The two-photon decay widths of charmonium states give valuable information for testing models describing the nature of heavy quarkonium system. Using a large data sample of 32.6 fb^{-1}, we have measured the production of χ_{c2} in two-photon collisions with the decay channel $\chi_{c2} \to \gamma J/\psi$, $J/\psi \to l^+l^-$.

The event selection criteria are: two oppositely-charged tracks in the Central Drift Chamber of transverse momentum $p_t \geq 0.4$ GeV/c and one photon with energy $E_\gamma \geq 0.2$ GeV. To ensure that the events originate from two-photon collisions, the total momentum of the $l^+l^-\gamma$ system is required to be less than 6 GeV/c. Events contaminated with background photons and inclusive two-photon interactions are suppressed by requiring the transverse momenta of the $l^+l^-\gamma$ system be smaller than 0.2 GeV/c, and the

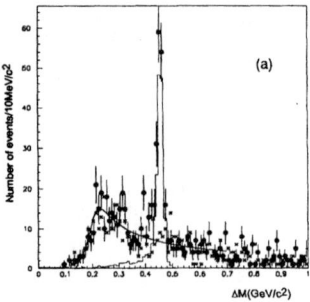
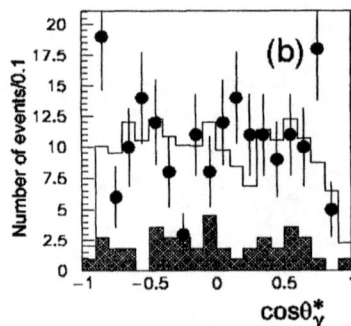

FIGURE 5. (a) Distribution of the mass difference $\Delta m = m(\gamma l^+ l^-) - m(l^+ l^-)$. Dot points are events of $m(l^+l^-)$ within the $m(J/\psi)$ mass window, asterisk points are those of the sidebands. Histogram is the Monte Carlo for signal and the curve is the fit to background. (b) Polar angle distribution of photon in the rest frame of χ_{c2}. Signal and background of Monte Carlo predictions are shown in open and hatched histogram.

sum of the l^+l^- pair be larger than 0.1 GeV/c.

Candidate $J/\psi \to l^+l^-$ decays are selected with $3.06 < m(l^+l^-) < 3.13$ GeV. The mass difference distribution for the selected $l^+l^-\gamma$ events is shown in Fig. 5(a), where $\Delta m = m(l^+l^-\gamma) - m(l^+l^-)$. The number of the signal events obtained is 151 ± 14. The angular distributions are investigated for events within the Δm peak. Plotted in Fig. 5(b) is the polar angle distribution of the photon in the $l^+l^-\gamma$ rest frame. The photon distribution is sensitive to the helicity state and is consistent with the Monte Carlo prediction of a helicity 2 state.

The two-photon decay width is derived for the number of events obtained, $\Gamma_{\gamma\gamma}(\chi_{c2}) = 0.84 \pm 0.08(\text{stat.}) \pm 0.07(\text{syst.}) \pm 0.07(\text{B.R.})$ keV for a pure helicity 2 state. The systematic error comes from the trigger efficiency(3%), event selection cuts (7%), and background subtraction(3%). This result is consistent with those reported by [4, 7], however, it is larger than the results from $p\bar{p}$ experiments [8].

Charmed meson Lifetimes

Measurement of charmed meson lifetimes provide useful information for the theoretical understanding of the heavy flavor decay mechanisms. Moreover, the D^0-$\bar{D}^0$ mixing parameters y_{CP} can be determined from the lifetime difference of the D^0 meson decaying into the state $K^-\pi^+$, and the CP-eigenstate K^-K^+. The y_{CP} parameter is given by

$$y_{CP} \equiv \frac{\Gamma(CP_{even}) - \Gamma(CP_{odd})}{\Gamma(CP_{even}) + \Gamma(CP_{odd})} \approx \frac{\tau(D^0 \to K^-\pi^+)}{\tau(D^0 \to K^-K^+)} - 1.$$

Lifetime measurements were conducted on a 23.4 fb^{-1} data sample [9] for the D^0, D^+ and D_s^+ in the fully reconstructed decay chains, $D^0 \to K^-\pi^+$, $D^0 \to K^-K^+$, $D^+ \to K^-\pi^+\pi^+$ (from $D^{*+} \to D^+\pi^0$), $D^+ \to \phi\pi^+$ (with $\phi \to K^+K^-$), $D_s^+ \to \phi\pi^+$, and $D_s^+ \to \bar{K}^{*0}K^+$ (with $\bar{K}^{*0} \to K^-\pi^+$). The invariant mass distribution of selected $K^-\pi^+$ events is plotted in Fig. 6. D^0 candidates are selected within 6σ of the measured peak value.

The decay vertex of the charmed meson is determined and then the production vertex is obtained by extrapolating the charmed meson flight path to the interaction region. The

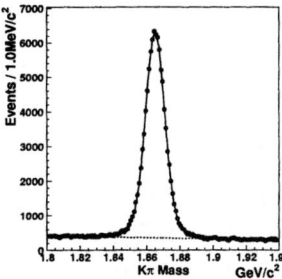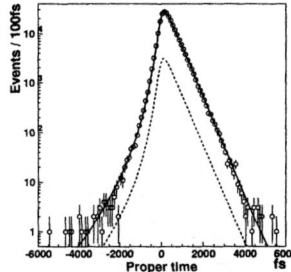

FIGURE 6. The mass distribution for $D^0 \to K^-\pi^+$ and the proper time distribution for events selected within 3σ of the mass peak. The dashed curve in the proper-time distribution shows the contribution of the background function in the fit.

projected decay length and the proper-time are then obtained. An unbinned maximum likelihood fit is performed to extract the charmed meson decay lifetime. The probability density function for each event consists of a signal term, and two background terms to represent background components with non-zero and zero lifetimes, respectively. The probability density function takes into consideration the signal and background fractions and the resolution for the measured decay time. The result of the fit to the proper time distribution for the $D^0 \to K^-\pi^+$ decay is shown in Fig. 6. The D^0 lifetime obtained is $414.6 \pm 1.7^{+1.9}_{-1.8}$ fs. Similarly, the lifetimes obtained for D^+ is $1037 \pm 12^{+5}_{-6}$ fs, and for D_s^+, $485.4^{+7.9}_{-7.7}{}^{+2.9}_{-4.2}$ fs. The main systematic uncertainties are in the resolution function, the proper time dependence of the reconstruction efficiency, and the reconstruction of the decay vertex. The ratios of the lifetimes of D^+ and D_s^+ with respect to D^0 are measured to be $\tau(D^+)/\tau(D^0) = 2.50 \pm 0.03^{+0.01}_{-0.02}$ and $\tau(D_s^+)/\tau(D^0) = 1.17 \pm 0.02 \pm 0.01$ The D^0-$\bar{D}^0$ mixing parameter y_{CP}, obtained from the lifetimes of $D^0 \to K^-\pi^+$ and $D^0 \to K^-K^+$ decays is $y_{CP} = -0.5 \pm 1.0^{+0.7}_{-0.8}$ %, corresponding to a 95% confidence interval of $-3.0 < y_{CP} < 2.0$ %.

D^0 decays to $K_L^0\pi^0$ and $K_S^0\pi^0$

An important quantity in the interpretation of D^0-$\bar{D}^0$ mixing is the strong phase difference $\delta_{K\pi}$ between the decays $D^0 \to K^-\pi^+$ (Cabibbo favored) and $D^0 \to K^+\pi^-$ (doubly Cabibbo suppressed). New information on this quantity may be obtained by measuring the asymmetry between the decay rates of D^0 into $K_L^0\pi^0$ and $K_S^0\pi^0$, where the effect may be as large as $O(\tan^2\theta_c)$, or 5% [10].

The analysis [11] is performed on a 23.6 fb^{-1} data sample. The K_L^0 flight direction is reconstructed using showers in the KLM system. The precision is further improved if the energy deposition is also present in the CsI calorimeter. The magnitude of the K_L^0 momentum is calculated using a D^0 mass constraint for the $K_L^0\pi^0$ combination. The D^0's are selected from the decay of $D^{*\pm} \to D^0\pi^\pm$. D^* candidates are required to have a scaled momentum $x_P > 0.6$, to reject D^* from B-meson decays and combinatoric background. The K_S^0's are selected in the $K_S^0 \to \pi^+\pi^-$ mode. The relative detection efficiency for K_L^0/K_S^0 can also be determined from the D^0 decay $D^0 \to K^{*-}\pi^+$, where K^{*-} decays into

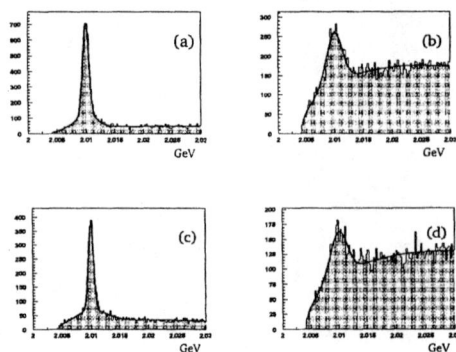

FIGURE 7. Mass distributions for D^* with D^0 decays into (a) $K_S^0\pi^0$, (b) $K_L^0\pi^0$, (c) $(K_S^0\pi)\pi$, and (d) $(K_L^0\pi)\pi$.

$K_S^0\pi^-$ and $K_L^0\pi^-$. The K_0 momentum in both the $K^0\pi$ and $(K^0\pi)\pi$ modes are selected in a range from 0.6 to 2.5 GeV/c. The resulting D^* mass distributions in the four D^0 decay modes are shown in Fig. 7. The yield of events of the four decay modes are used to calculate the ratio of branching fractions of $D^0 \to K_L^0\pi^0$ and $D^0 \to K_S^0\pi^0$,

$$\frac{\mathcal{B}(D^0 \to K_L^0\pi^0)}{\mathcal{B}(D^0 \to K_S^0\pi^0)} = \frac{N(K_L^0\pi^0)/N(K_S^0\pi^0)}{N((K_L^0\pi)\pi)/N((K_S^0\pi)\pi)} = 0.88 \pm 0.09 \text{(stat)}$$

The kinematic properties for the $K^0\pi$ and the $(K^0\pi)\pi$ modes are similar, therefore, by dividing the ratios of them the systematic uncertainties are largely canceled out. The remaining systematic errors are from imperfections of the fitting model, estimated to be 10%. Expressing the result in terms of the rate asymmetry defined in [10], we find

$$\mathcal{A} = \frac{\Gamma(D^0 \to K_S^0\pi^0) - \Gamma(D^0 \to K_L^0\pi^0)}{\Gamma(D^0 \to K_S^0\pi^0) + \Gamma(D^0 \to K_L^0\pi^0)} = 0.06 \pm 0.05 \text{(stat)} \pm 0.05 \text{(syst)},$$

which is consistent with unity. At the current level of precision we are not able to place any strong constraint on the parameters of $D \to K\pi$ decays, and the strong phase $\delta_{K\pi}$. However, the statistical error will soon be improved as more candidates are accumulated. Changes of technique to reduce systematic error are also being actively evaluated.

Prompt Charmonia

The production of prompt charmonia is important for studying the perturbative QCD and non-perturbative effects. Studies of prompt charmonia in e^+e^- collisions at the $\Upsilon(4S)$ resonance can provide tests of the recently developed non-relativistic QCD (NRQCD) theory [12] and the estimates of some of its non-perturbative matrix elements.

We report results of a measurement [13] using data recorded at the $\Upsilon(4S)$ (21.3 fb^{-1}) and in the continuum 60 MeV below the resonance (2.3 fb^{-1}). Candidate J/ψ mesons are reconstructed using the leptonic decays $J/\psi \to \mu^+\mu^-$ and e^+e^-. For $J/\psi \to \mu^+\mu^-$, both charged tracks are identified as muons in the KLM. For $J/\psi \to e^+e^-$, oppositely charged track pairs are identified as electrons based on a combination of CDC dE/dx, ACC response, and associated ECL shower parameters. The signal region is defined by the mass window $-93 < M_{l^+l^-} - M_{J/\psi} < 33$ MeV/c^2.

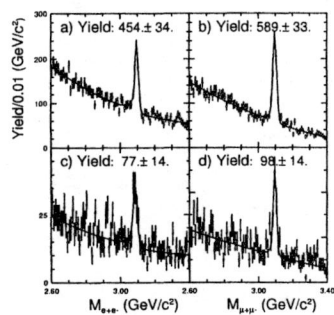
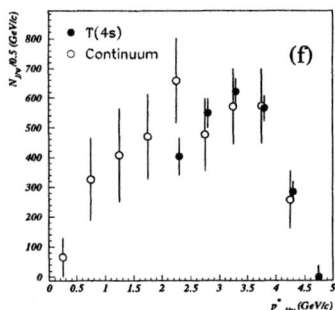

FIGURE 8. Mass distributions for the $J/\psi \to e^+e^-$ (a,c) and $\mu^+\mu^-$ (b,d) candidates; (a) and (b) are for $\Upsilon(4S)$ data with $p^* > 2$ GeV/c, (c) and (d) are for off-resonance data. In (f) the momentum distributions of the J/ψ are shown for $\Upsilon(4S)$ data with $p^* > 2$ GeV/c (dots) and for off-resonance data (circles).

For background, the secondary charmonia from the B meson decays is eliminated by requiring the charmonium CM momentum p^* to be above the B decay kinematic limit, $p^* > 2.0$ GeV/c. This value is robust against the effects of momentum measurement errors and motion of the B-meson in the CM. This requirement is not applied to off-resonance data. The background also comes from initial state radiation with a hard photon and higher order QED processes $e^+e^- \to J/\psi\gamma^*, J/\psi l^+ l^-$. These are suppressed by requiring the number of charged tracks $N_{ch} > 4$.

The mass distributions for prompt J/ψ candidates in the $\Upsilon(4S)$ and off-resonance data samples are shown in Fig. 8. The momentum distribution of the J/ψ in the CM system is determined by a sideband subtraction method. The acceptance-corrected distributions are shown in Fig. 8(f). In the region of $p^* > 2$ GeV/c where we can directly compare the $\Upsilon(4S)$ data with the off-resonance data, agreement is good indicating that all the J/ψ production at $\Upsilon(4S)$ can be explained by continuum production. The branching fraction for the direct decay $\Upsilon(4S) \to J/\psi X$ can be estimated from the difference between these two yields, which is found to be -27 ± 132. The upper limit for this process is thus determined to be $\mathcal{B}(\Upsilon(4S) \to J/\psi X) < 1.9 \times 10^{-4}$ (90%CL).

The off-resonance data is combined with the $\Upsilon(4S)$ data for $p^* > 2$ GeV/c to extract charmonium production cross-sections and angular distributions. To minimize model dependence for detection efficiency, the data is corrected using a two-dimensional weight matrix as a function of p^* and $\cos\theta^*$, where θ^* is the J/ψ production angle in the CM system. The resulting cross-section is $\sigma(e^+e^- \to J/\psi X) = 1.57 \pm 0.13(\text{stat.}) \pm 0.35(\text{syst})$ pb. The statistical error comes mostly from the limited statistics in the off-momentum data. The partial cross section for $p^* > 2$ GeV/c, determined from the large $\Upsilon(4S)$ sample only, is $1.08 \pm 0.05(\text{stat.})$ pb. The systematic errors estimated include the uncertainty in the efficiency determination, selection cuts, lepton identification, luminosity measurement, contamination from QED processes, subtraction for $\psi(2S)$ feed-down, and errors of the respective branching fractions.

In the momentum distributions (Fig. 8(f)) the J/ψ distribution is softer than the NRQCD prediction for color-singlet $J/\psi gg$, and agrees qualitatively with the predicted

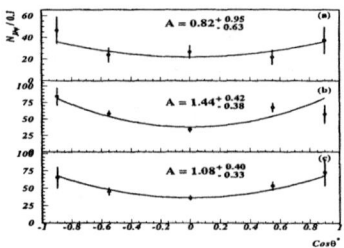 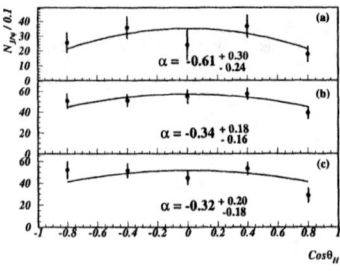

FIGURE 9. The $\cos\theta^*$ and $\cos\theta_H$ distributions in three momentum intervals (a) $2.0 < p^* < 2.6$ GeV/c, (b) $2.6 < p^* < 3.4$ GeV/c, and (c) $3.4 < p^* < 5.0$ GeV/c. The curves represent fits to $1+A\cos^2\theta^*$ for cos^* distributions, and $1+\alpha\cos^2\theta_H$ for cos_H distributions, respectively.

shape of the color-singlet $J/\psi c\bar{c}$ component [14, 15]. Some NRQCD calculations also predict a dramatic rise in the cross-section at the end-point due to color-octet $e^+e^- \to J/\psi g$ [15, 16]. Assuming a 1 pb cross-section for this process, we expect > 300 events in the last two bins of Fig. 8(b). Such events are not observed.

The distributions of the prompt J/ψ CM production angle θ^*, and the helicity angle θ_H (the angle between the positive lepton daughter momentum vector in the J/ψ rest frame, and the J/ψ momentum vector in the CM system), have also been studied in different momentum intervals. The distributions (Fig. 9) are fitted with the parameterizations $1+A\cos^2\theta^*$ and $1+\alpha\cos^2\theta_H$. No statistically significant p^* dependence is seen for either A nor α parameters.

REFERENCES

1. K. Abe *et al.* (Belle Collaboration), KEK Report 2000-4, to be published in Nucl. Inst. and Methods.
2. KEKB B Factory Design Report, KEK Report 95-7, 1995, unpublished.
3. L3 Collab., M. Acciarri et al., Phys. Lett. B 501 (2001) 173.
4. CLEO Collab., B.I. Eisenstein et al., Phys. Rev. Lett. 87 (2001) 061801.
5. L3 Collab., M. Acciarri et al., Phys. Lett. B 413 (1997) 147.
6. Belle Collab., K. Abe et al., BELLE-CONF-0117 (2001).
7. CLEO Collab., J. Dominick et al., Phys. Rev. D 50 (1994) 4265; OPAL Collab., K. Ackerstaff et al., Phys. Lett. B 439 (1998) 197; L3 Collab., M. Acciarri et al., Phys. Lett. B 453 (1999) 73.
8. E760 Collab., T.A. Armstrong et al., Phys. Rev. D 54 (1993) 7067; E835 Collab., M. Stancari et al., Nucl. Phys. B (Proc. suppl.) 82 (2000) 300.
9. Belle Collab., K. Abe et al., BELLE-CONF-0131 (2001).
10. E.Golowich and S. Pakvasa, Phys. Lett. B 505 (2001) 94.
11. Belle Collab., K. Abe et al., BELLE-CONF-0129 (2001).
12. G.T. Bodwin, E. Braaten and G.P. Lepage, Phys. Rev. D 51 (1995) 1125.
13. Belle Collab., K. Abe et al., BELLE-CONF-0128 (2001).
14. S. Baek, P. Ko, J. Lee, and H.S. Song, J. Kor. Phys. Soc. 33 (1998) 97; hep-ph/9804455.
15. P. Cho and A.K. Leibovich, Phys. Rev. D 54 (1996) 6690.
16. E. Braaten and Yu-Qi Chen, Phys. Rev. Lett. 76 (1996) 730.

Three body decays of D^0 and D_S mesons

Antimo Palano

INFN and University of Bari, Italy
representing the BaBar Collaboration

Abstract. Results are presented on the study of three body decays of $D^0 \to K_S^0 h^+ h^-$, where $h = \pi/K$ and $D_S^\pm \to K_S^0 K_S^0 \pi^\pm$. The data have been collected by the BaBar experiment at SLAC and are extracted from continuum $e^+ e^-$ annihilations at the $\Upsilon(4S)$ energy.

Introduction

New generation experiments are providing large data sets for charm physics with statistics which supersede most previous measurements. The Dalitz plot analyses of 3-body charm decays have been performed in the past but these new large and clean samples will allow high precision measurements that were never before possible.

The Dalitz plot analysis of three-body decays is a relatively new technique in development for charm physics studies. This method of analysis is the most complete way of analyzing the data since it allows measurement of both decay amplitudes and phases. The final state is the result of the interference of all intermediate states. The significant results provided by these studies are:

- Accurate measurements of branching fractions.
- A study of Final State Interactions.
- A study of CP violation in rates and decay amplitudes.
- New input to several old unsolved problems in light meson spectroscopy, in particular to the scalar mesons puzzle.

Factorization models assume the weak decay amplitudes to be real. The fact that the observed amplitudes have a relative complex phase is a consequence of final state interaction.

CP violation is expected to be small in charm decays ($\approx 10^{-3}$) [1]. Two amplitudes with different phases are needed:

$$Ae^{i\delta_A} + Be^{i\delta_B}$$

In singly Cabibbo-suppressed decays penguin terms may provide a weak phase, while Final State Interactions provide a strong phase shift. Under CP the weak phases change sign but the strong ones do not. Any difference between D and $\bar{D}$ in the Dalitz plot would be evidence for CP violation.

Throughout this paper charge conjugate modes, where not explicit, are implied.

The BaBar Experiment

The PEPII Collider is an asymmetric storage ring which collides 9 GeV electrons with 3.1 GeV positrons with a peak luminosity of 4.2×10^{33} $cm^{-2}s^{-1}$. The $\Upsilon(4S)$ resonance is produced with $\beta\gamma = 0.56$ in the laboratory frame at zero crossing angle. Details on the layout of the apparatus, trigger conditions and data processing can be found in previous publications [2].

The $\Upsilon(4S)$ resonance sits on a large continuum background with a contribution from $e^+e^- \to c\bar{c}$ of 1.30 nb. The power of BaBar for studying charm physics is based on:

- Relatively small combinatorics because of e^+e^- interactions.
- Good vertexing.
- Good Particle Identification.
- Detection of all possible final states, with charged tracks and γ's.
- Very high statistics.

The BaBar experiment is continuously collecting data. This work is based on the data taken during 1999/2000 and corresponds to an integrated luminosity of $\approx 23 fb^{-1}$ unless otherwise specified.

Study of $D_S^+ \to K_S^0 K_S^0 \pi^+$

The resonance $f_J(1710)$ has been measured with spin 0 or 2 in different experiments. It has been a candidate for being the lowest lying scalar or tensor glueball. It has been observed in J/ψ decay, central production and $\gamma\gamma$ collisions [3]. Details on the selection criteria used to isolate the $K_S^0 K_S^0 \pi^+$ final state can be found in ref. [4]. The $K_S^0 K_S^0 \pi^+$ mass spectrum (relative to an integrated luminosity of 18.4 fb^{-1}) is plotted in fig. 1 and shows signals from D^+ and D_S^+.

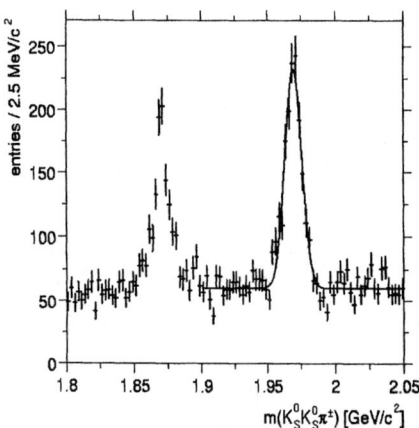

FIGURE 1. $K_S^0 K_S^0 \pi^+$ mass spectrum.

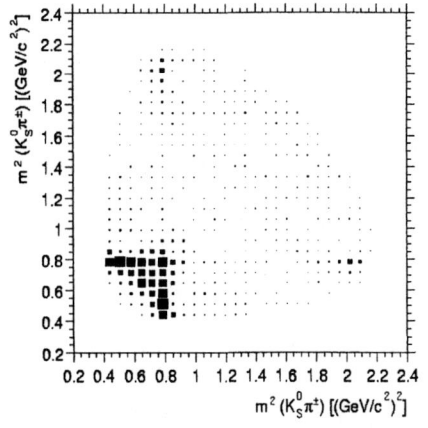
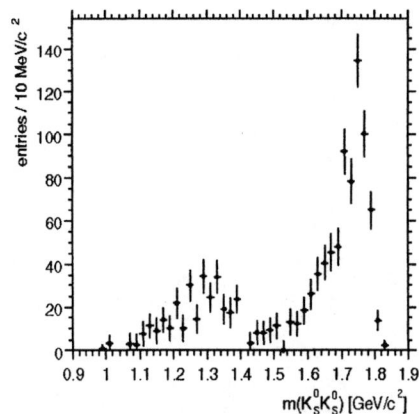

FIGURE 2. Background subtracted Dalitz plot and $K_S^0 K_S^0$ projection for $D_S^+ \to K_S^0 K_S^0 \pi^+$.

The background subtracted Dalitz plot and its $K_S^0 K_S^0$ projection are shown in fig. 2. Strong accumulations in the region of the $K^*(892)$ can be seen. A preliminary Dalitz plot analysis shows that the data cannot be described with $K^*(892)$ alone but require the presence of the $D_S^+ \to f_J(1710)\pi^+$ final state.

Selection of $D^0 \to K_S^0 h^+ h^-$

D^0's are required to come from a D^{*+} decay:

$$D^{*+} \to D^0 \pi^+$$
$$\to \bar{K}^0 \pi^+ \pi^-$$

$$D^{*-} \to \bar{D}^0 \pi^-$$
$$\to K^0 \pi^+ \pi^-$$

Therefore the charge of the slow π gives the flavour of the D^0 and that of the K^0 (apart from the contribution from DCSD, which is expected to be $\approx 10^{-4}$).

The selection of the channel starts with the reconstruction of the K_S^0 and D^0 vertices. The slow π are refitted using the beam spot constraint to improve the resolution, (beam spot size: $\sigma_x = 0.15mm, \sigma_y = 0.05mm, \sigma_z = 8mm$). The center of mass momentum of the D^0 (p^*) has been required to be greater then 2.2 GeV/c.

The mass difference:

$$\Delta m = m(K^0 \pi^+ \pi^- \pi_s) - m(K^0 \pi^+ \pi^-)$$

where the slow pion π_s has a momentum below 0.6 GeV/c is plotted in fig. 3. Fig. 3a) and fig. 3b) show respectively the Δm distribution before and after having required a 2.5 σ cut around the D^0 mass.

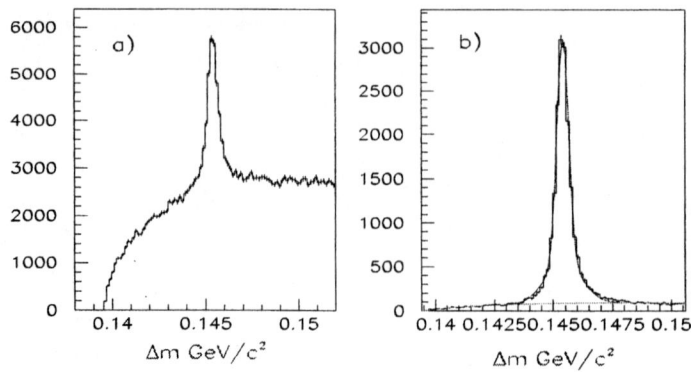

FIGURE 3. Δm distributions for $K_S^0 \pi^+ \pi^-$ a) before and b) after a D^0 mass cut at 2.5σ.

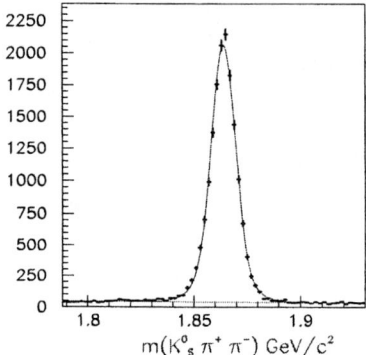

FIGURE 4. $K_S^0 \pi^+ \pi^-$ mass distribution after a D^{*+} cut at 2σ.

Fitting the Δm width using a single Gaussian and a threshold function $(m-m_{th})^{\alpha} e^{-\beta m - \gamma m^2}$, we obtain $\sigma = 326 \pm 10$ keV/c^2. Performing now a 2.0 σ cut on Δm we obtain the $m(K_S^0 \pi^+ \pi^-)$ shown in fig. 4. The D^0 width obtained fitting only one Gaussian and a linear background is: $\sigma = 6.3 \pm 0.1$ MeV/c^2.

The Dalitz plot of $D^0 \rightarrow K_S^0 \pi^+ \pi^-$, obtained by selecting events within 2.5 σ of the D^0 mass, is shown in fig. 5 (15753 events) and its projections are shown in fig. 6. The background fraction, estimated to be 4.1 %, has not been subtracted.

This Dalitz plot reveals a complex structure. Several intermediate resonant states involving $K^{*+}(892)$, $K_0^{*+}(1430)$, $\rho^0(770)$, $f_0(980)$, $f_0(1370)$ resonances can be seen.

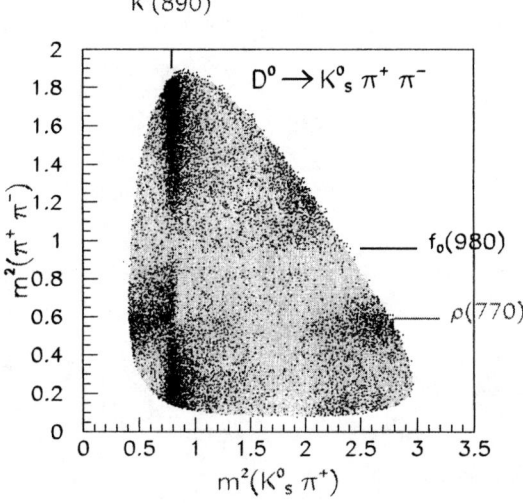

FIGURE 5. $K_S^0 \pi^+ \pi^-$ Dalitz plot with no background subtraction.

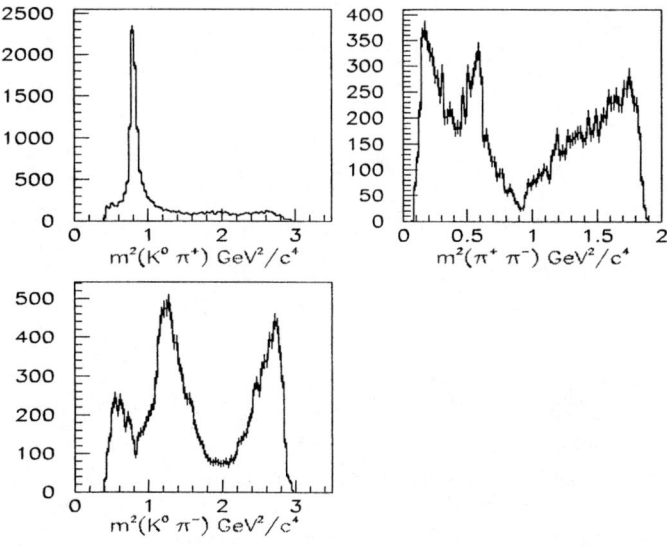

FIGURE 6. $K_S^0 \pi^+ \pi^-$ Dalitz plot projections.

Strong interferences can also be seen. In particular, $f_0(980)$ resonance shows up as a uniform horizontal depletion suggesting interference with a broad scalar resonance.

Selection of $D^0 \to K_S^0 K^\pm \pi^\mp$

The decay $D^0 \to K_S^0 K\pi$ contains two possible D^0 decay modes:

$$D^0 \to K^0 K^- \pi^+ \quad (a)$$

$$D^0 \to \bar{K}^0 K^+ \pi^- \quad (b)$$

The charge of the pion separates the two decay modes. Diagrams contributing to the two decay channels are shown in fig. 7.

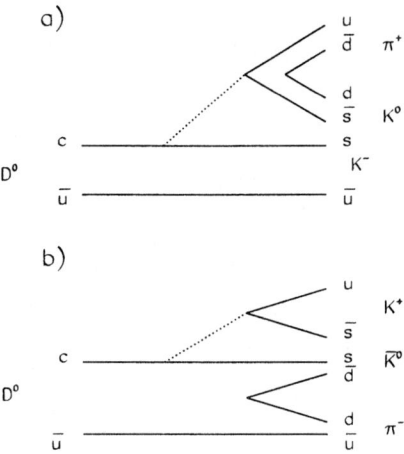

FIGURE 7. Diagrams which contribute to the two different D^0 decay modes to $K_S^0 K\pi$.

The channels have been isolated using similar cuts on the corresponding Δm distribution. Requiring one of the two charged tracks to be positively identified as a kaon we obtain the $K_S^0 K\pi$ mass distributions for the two decay modes shown in fig. 8.

The fitted widths and yields obtained using only one Gaussian and a linear background are the following:

(a) $D^0 \to K^0 K^- \pi^+$: (6.0 ± 0.1) MeV/c^2 2335 events

(b) $D^0 \to \bar{K}^0 K^+ \pi^-$: (5.1 ± 0.1) MeV/c^2 731 events

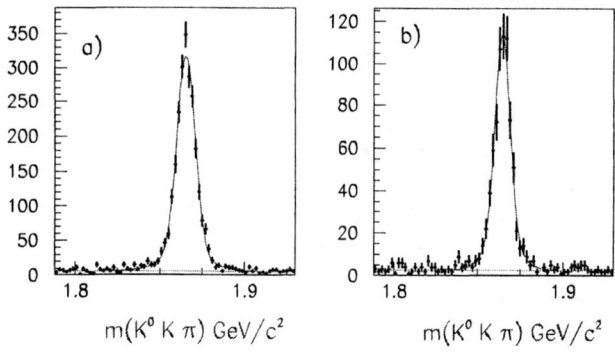

FIGURE 8. a) $D^0 \to K^0 K^- \pi^+$ and b) $D^0 \to \bar{K}^0 K^+ \pi^-$ signals.

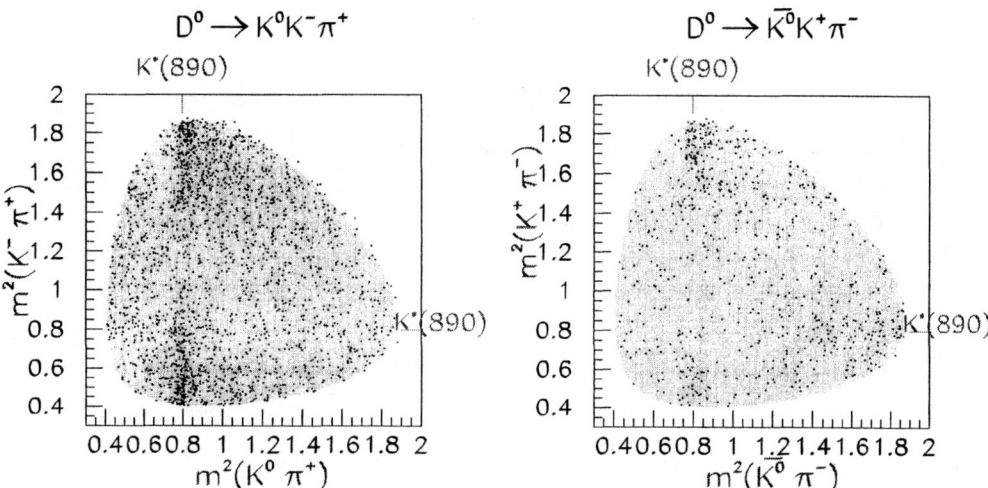

FIGURE 9. a) $D^0 \to K^0 K^- \pi^+$ and b) $D^0 \to \bar{K}^0 K^+ \pi^-$ Dalitz plots with no background subtraction.

The above yields indicate that branching fractions for the above D^0 decay channels are different.

The two Dalitz plots are shown in fig. 9 and their projections are shown in fig. 10. The background fractions for decays a) and b) have been estimated to be 4 % and 5% respectively and have not been subtracted.

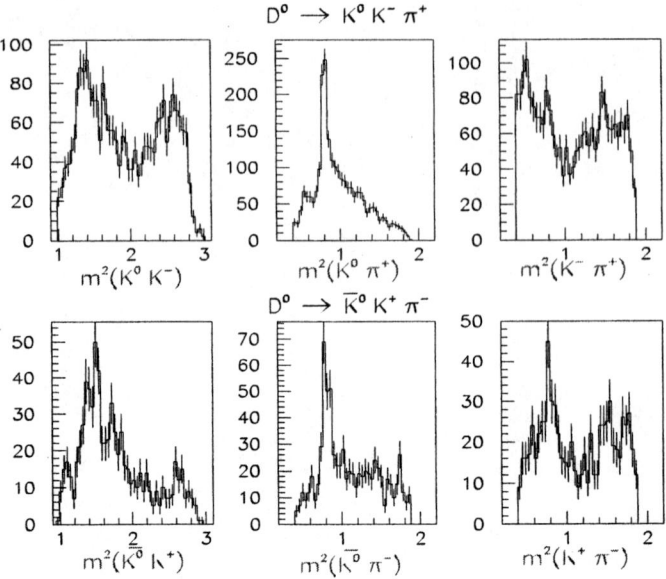

FIGURE 10. First row: $D^0 \to K^0 K^- \pi^+$ Dalitz plot projections. Second row: $D^0 \to \bar{K}^0 K^+ \pi^-$ Dalitz plot projections

The decay $D^0 \to K^0 K^- \pi^+$ is dominated by $K^{*0} K^-$ with a small contribution from $\bar{K}^{*0} K^0$. We also observe the presence of broad structure. The decay $D^0 \to \bar{K}^0 K^+ \pi^-$ shows more symmetric $K^{*-} K^+$ and $K^{*0} \bar{K}^0$ contributions.

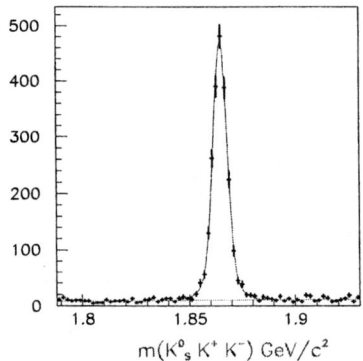

FIGURE 11. $K_S^0 K^+ K^-$ mass spectrum

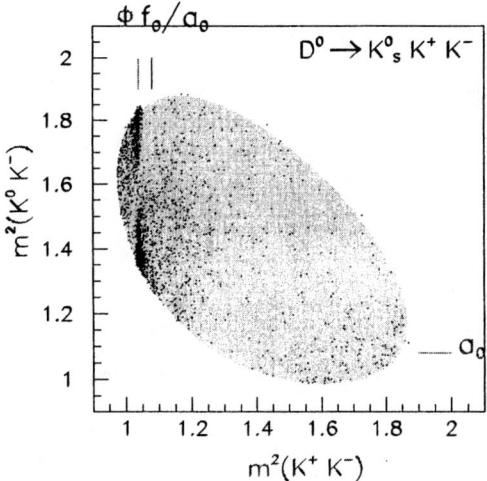

FIGURE 12. $D^0 \to \bar{K}^0 K^+ K^-$ Dalitz plot.

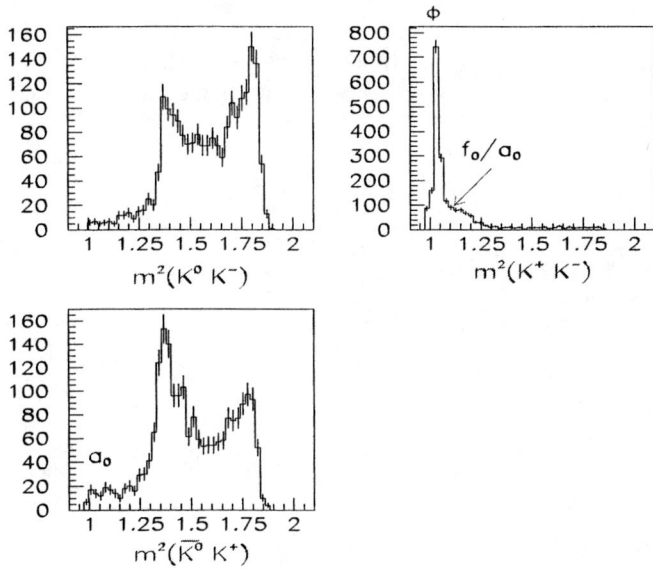

FIGURE 13. $D^0 \to \bar{K}^0 K^+ K^-$ Dalitz plot projections.

Selection of $D^0 \to K_S^0 K^+ K^-$

The channel has been isolated performing a similar Δm cut and requiring at least one of the two charged tracks to be positively identified as a kaon. The $K_S^0 K^+ K^-$ mass spectrum is shown in fig. 11.

The D^0 width obtained fitting the mass spectrum with one Gaussian and a linear background is (3.7 ± 0.1) MeV/c^2 and the yield is 2089 events.

The Dalitz plot of $D^0 \to K_S^0 K^+ K^-$ is shown in fig. 12 and contains a background fraction of 3 % (not subtracted). The corresponding projections are shown in fig. 13.

The presence of intermediate states involving ϕ, $f_0(980)$ and $a_0(980)$ resonances is clearly visible.

Conclusions

Charm Physics can be performed at B-factories with high statistics and small backgrounds. The Dalitz plot analysis as well as measurements of CP asymmetries in the decay amplitudes and rates for different charmed mesons are in progress.

In the near future, Charm Physics will be dominated by B-factories and τ/charm factories. Present available statistics for Dalitz charm decays from fixed target and B-factories are: 1–5×10^4 events for Cabibbo allowed, 1–10×10^3 events for Cabibbo suppressed and 50–300 events for double Cabibbo suppressed decays.

Given the large data samples being accumulated, we expect in the next few years an increase of these yields by a factor 20. This will allow charm physics to be rewritten with errors on branching fractions reduced by more than a factor 10.

REFERENCES

1. M. Golden and B. Grinstein, Phys. Lett. **B222** 501 (1989).
2. B. Aubert et al., The BaBar Detector, SLAC-PUB-8569, hep-ex/0105044, to appear in NIM.
3. D.E. Groom et al., The European Physical Journal C15 (2000) 1.
4. T. Deppermann (BaBar Collaboration), Dalitz Analyses of $D_S^\pm \to K_S^0 K_S^0 \pi^\pm$ and $D_S^\pm \to \pi^+ \pi^- \pi^\pm$, these proceedings.

Light Meson Physics from Charm Decays at Fermilab E791

Carla Göbel [1]

Instituto de Física, Facultad de Ingeniería, Universidad de la República, C.C. 30, CP 11300, Montevideo, Uruguay

Abstract. We present recent results on light mesons based on Dalitz plot analyses of charm decays from Fermilab experiment E791. Scalar mesons are found to have large contributions to the decays studied, $D^+ \to K^-\pi^+\pi^+$ and $D^+, D_s^+ \to \pi^-\pi^+\pi^+$. From the $K\pi\pi$ final state, we find good evidence for the existence of the light and broad κ meson and we measure its mass and width. We also discuss recently published results on the 3π final states, especially the measurement of the f_0 parameters and the evidence for the σ meson from $D^+ \to \sigma\pi^+$. These results demonstrate the importance of charm decays as a new environment for the study of light meson physics.

INTRODUCTION

The decays of charm mesons are currently a new source of information for the study of light meson spectroscopy, with the advantages of having well defined initial state (the D meson, a 0^- state with defined mass). This new information is complementary to that from scattering experiments and can be particularly relevant to the understanding of the scalar sector.

Here we present preliminary results for the Dalitz-plot analysis of the Cabibbo-favored decay $D^+ \to K^-\pi^+\pi^+$ using data from Fermilab E791 experiment. We also present an overview of our results for $D_s^+ \to \pi^-\pi^+\pi^+$ [1] and $D^+ \to \pi^-\pi^+\pi^+$ [2] Dalitz-plot analyses. The E791 data was collected in 1991/92 from 500 GeV/c π^--nucleon interactions. For details see [3].

For the $D^+ \to K^-\pi^+\pi^+$ analysis, when we include all known $K\pi$ resonant channels plus a non-resonant (NR) contribution, we find that the NR decay is dominant. This is unusual in D decays. Moreover, the fit model has important discrepancies with respect to the data. By including an extra scalar resonant state, with unconstrained mass and width, we obtain a fit which is substantially superior to that without this state. The values for its mass and width are found to be $797 \pm 19 \pm 42$ MeV/c^2 and $410 \pm 43 \pm 85$ MeV/c^2 respectively. We refer to this state as the κ. The existence of such a state has been greatly discussed in the literature in recent years [4]–[14]. We also obtain new measurements for the mass and the width of the $K_0^*(1430)$ resonance: $1459 \pm 7 \pm 6$ MeV/c^2 and $175 \pm 12 \pm 12$ MeV/c^2 respectively.

From our analysis of $D_s^+ \to \pi^-\pi^+\pi^+$ decays, we find that the dominant decay fraction

---

[1] For the E791 Collaboration.

comes from $f_0(980)\pi^+$. We obtain new measurements for the $f_0(980)$ and $f_0(1370)$ masses and widths. From $D^+ \to \pi^-\pi^+\pi^+$ decays, we find that a model with only known $\pi\pi$ resonances plus a NR channel is not able to describe the data adequately. We find strong evidence for the presence of a light and broad scalar resonance, the $\sigma(500)$, the $\sigma\pi^+$ channel being responsible for half of the decay rate. We measure the mass and the width of this scalar meson to be $478^{+24}_{-23} \pm 17$ MeV/c^2 and $324^{+42}_{-40} \pm 21$ MeV/c^2, respectively.

THE $D^+ \to K^-\pi^+\pi^+$ DALITZ-PLOT ANALYSIS

From the original 2×10^9 events collected by E791, and after reconstruction and selection criteria, we obtained the $D^+ \to K^-\pi^+\pi^+$ sample shown in Figure 1(a). The filled area represents the level of background; besides the combinatorial, the other main source of background comes from the reflection of the decay $D_s^+ \to K^-K^+\pi^+$ (through $\bar{K}^*K^+$ and $\phi\pi^+$). The crosshatched region contains the events selected for the Dalitz-plot analysis. There are 15090 events in this sample, of which 6% are background.

Figure 1(b) shows the Dalitz-plot for these events. The two axes are the squared invariant-mass combinations for $K\pi$, and the plot is symmetrized with respect to the two identical pions. The plot presents a rich structure, where we can observe the clear bands from $\bar{K}^*(890)\pi^+$, and an accumulation of events at the upper edge of the diagonal, due to heavier resonances. To study the resonant substructure, we perform an unbinned maximun-likelihood fit to the data, with probability distribution funtions (PDF's) for both signal and background sources. In particular, for each candidate event, the signal PDF is written as the square of the total physical amplitude $\mathcal{A}$ (defined below) and it is weighted for the acceptance across the Dalitz plot (obtained by Monte Carlo (MC)) and by the level of signal to background for each event, as given by the line shape of Figure 1(a). The background PDF's (levels and shapes) are fixed for the Dalitz-plot fit, according to MC and data studies.

We begin describing our first approach to fit the data, which represents the conventional Dalitz-plot analysis including the known $K\pi$ resonant amplitudes ($\mathcal{A}_n$, $n \geq 1$), plus a constant non-resonant contribution. The signal amplitude is constructed as a coherent sum of the various sub-channels:

$$\mathcal{A} = a_0 e^{i\delta_0}\mathcal{A}_0 + \sum_{n=1}^{N} a_n e^{i\delta_n}\mathcal{A}_n(m_{12}^2, m_{13}^2) \tag{1}$$

Each resonant amplitude is written as

$$\mathcal{A}_n = BW_n \, F_D^{(J)} \, F_n^{(J)} \, \mathcal{M}_n^{(J)} \,. \tag{2}$$

where BW_n is the relativistic Breit-Wigner propagator

$$BW_n = \frac{1}{m_0^2 - m^2 - im_0\Gamma(m)} \tag{3}$$

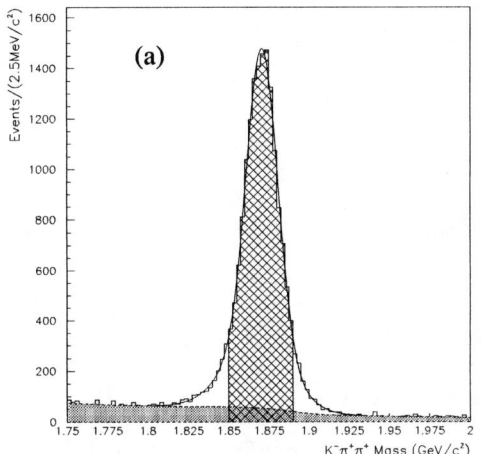 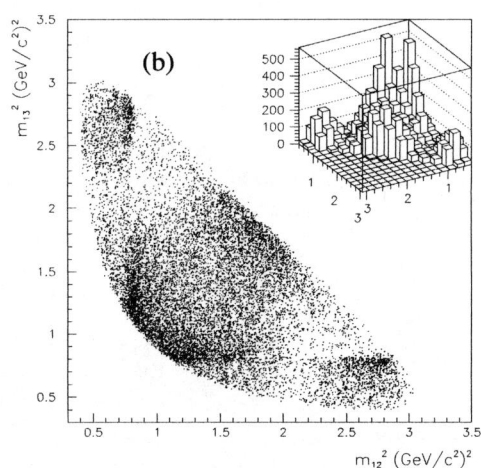

FIGURE 1. (a) The $K^-\pi^+\pi^+$ invariant mass spectrum. The filled area is background; (b) Dalitz plot corresponding to the events in the dashed area of (a).

with mass dependent width,

$$\Gamma(m) = \Gamma_0 \frac{m_0}{m} \left(\frac{p^*}{p_0^*}\right)^{2J+1} \left(\frac{F_n^{(J)}(p^*)}{F_n^{(J)}(p_0^*)}\right)^2. \tag{4}$$

The quantities F_D and F_R are the Blatt-Weisskopf damping factors respectively for the D and the $K\pi$ resonances, they depend on the radii of the decaying meson and are set to $r_D = 3.0$ GeV^{-1} and $r_R = 1.5$ GeV^{-1} [18]; p^* is the pion momentum in the resonance rest frame at mass m_{12} ($p_0^* = p^*(m_0)$). $\mathcal{M}_n^{(J)}$ describes the angular distribution due to the spin J of the resonance. See details in [2]. Finally each amplitude is Bose symmetrized $\mathcal{A}_n = \mathcal{A}_n[(\mathbf{12})\mathbf{3}] + \mathcal{A}_n[(\mathbf{13})\mathbf{2}]$.

Using this model (Model A), we find contributions from the following channels: the non-resonant, responsible for more than 90% of the decay rate, followed by $\bar{K}_0^*(1430)\pi^+$, $\bar{K}^*(892)\pi^+$, $\bar{K}^*(1680)\pi^+$ and $\bar{K}_2^*(1430)\pi^+$. The decay fractions and relative phases are shown in Table 1. These values are in accordance with previous results from E691 [15] and E687 [16]. We thus confirm a high non-resonant contribution according to this model, which is totally unusual in D decays. Besides, there is an important destructive interference pattern, since all fractions add up to 140 %.

To evaluate the fit quality of Model A, we compute a two-dimensional χ^2 in the Dalitz plot, from the difference in densities for the model (from a fast-MC algorithm) and the data. We obtain $\chi^2/\nu = 2.7$ (ν being the number of degrees of freedom), with a corresponding confidence level (CL) of 10^{-11}. In Figure 2(a) we show the $K\pi$ low and high squared-mass projections for data (error bars) and model (solid line). The discrepancies are evident at the very low-mass region for $m^2(K\pi)_{low}$ and near

TABLE 1. Results without κ (Model A) and with κ (Model B). *Preliminary.*

Decay Mode	Model A: No κ		Model B: With κ	
	Fraction (%)	Phase	Fraction (%)	Phase
NR	90.9 ± 2.6	$0°$ (fixed)	$13.0 \pm 5.8 \pm 2.6$	$(349 \pm 14 \pm 8)°$
$\kappa\pi^+$	–	–	$47.8 \pm 12.1 \pm 3.7$	$(187 \pm 8 \pm 17)°$
$\bar{K}^*(892)\pi^+$	13.8 ± 0.5	$(54 \pm 2)°$	$12.3 \pm 1.0 \pm 0.9$	$0°$ (fixed)
$\bar{K}_0^*(1430)\pi^+$	30.6 ± 1.6	$(54 \pm 2)°$	$12.5 \pm 1.4 \pm 0.4$	$(48 \pm 7 \pm 10)°$
$\bar{K}_2^*(1430)\pi^+$	0.4 ± 0.1	$(33 \pm 8)°$	$0.5 \pm 0.1 \pm 0.2$	$(306 \pm 8 \pm 6)°$
$\bar{K}^*(1680)\pi^+$	3.2 ± 0.3	$(66 \pm 3)°$	$2.5 \pm 0.7 \pm 0.2$	$(28 \pm 13 \pm 15)°$

2.5 $(\text{GeV}/c^2)^2$ for $m^2(K\pi)_{high}$. These regions of disagreement are the same observed previously by E687 [16]. We thus conclude that a model with the known $K\pi$ resonances, plus a non-resonant amplitude, is not able to describe the $D^+ \to K^-\pi^+\pi^+$ Dalitz plot satisfactorily.

A similar pattern – bad fit quality with large NR fraction – is found in the analysis of the decay $D^+ \to \pi^-\pi^+\pi^+$ when allowing only the established $\pi\pi$ resonances [2]. There we find that the inclusion of an extra scalar resonance improves the fit substantially, giving strong evidence for the $\sigma(500)$. See the section on $D^+ \to \pi^-\pi^+\pi^+$ below. Thus, we are lead to try an extra scalar resonance in our fit model here. The possible existence of a light and broad $K\pi$ scalar state has been suggested by many authors [4, 5, 6, 7, 8], some of them believing it would be a member of a light scalar nonet [9, 10, 11]; however, its existence has been the subject of some controversy also [12, 13, 14].

A second fit model, Model B, is constructed by the inclusion of an extra scalar state, with unconstrained mass and width. For consistency, the mass and width of the other scalar state, the $K_0^*(1430)$, are also free parameters of the fit. We adopt a better description for these scalar states by introducing gaussian-type form-factors [12] to take into account the finite size of the decaying mesons. Two extra floating parameters are the meson radii r_D and r_R introduced above.

Using this model, we obtain the values of $797 \pm 19 \pm 42$ MeV/c^2 for the mass and $410 \pm 43 \pm 85$ MeV/c^2 for the width of the new scalar state (first error statistical, second error systematic), referred to here as the κ. The values of mass and width obtained for the $K_0^*(1430)$ are respectively $1459 \pm 7 \pm 6$ MeV/c^2 and $175 \pm 12 \pm 12$ MeV/c^2, appearing heavier and narrower than presented by the PDG [17]. The decay fractions and relative phases for Model B, with systematic errors, are given in Table 1. Compared to the results of Model A (without κ), the non-resonant mode drops from over 90% to 13%. The $\kappa\pi^+$ state is now the dominant channel with about 50%. The meson radii r_D and r_R are found to be respectively 5.0 ± 0.5 GeV^{-1} and 1.6 ± 1.3 GeV^{-1}, in complete agreement with previous estimates [18, 19].

Moreover, the fit quality of Model B is substantially superior to that of Model A. The χ^2/ν is now 0.73 with a CL of 95%. The very good agreement between the model and the data can be seen in the projections of Figure 2(b).

A number of studies were done to check these results. For example, we replaced the complex κ Breit-Wigner by a real Breit-Wigner, with no phase variation. In this case, we got similar mass and width for this extra state, but with unphysical fractions for this state and the NR, and a worse fit quality. We also replaced the κ by a hypothetical

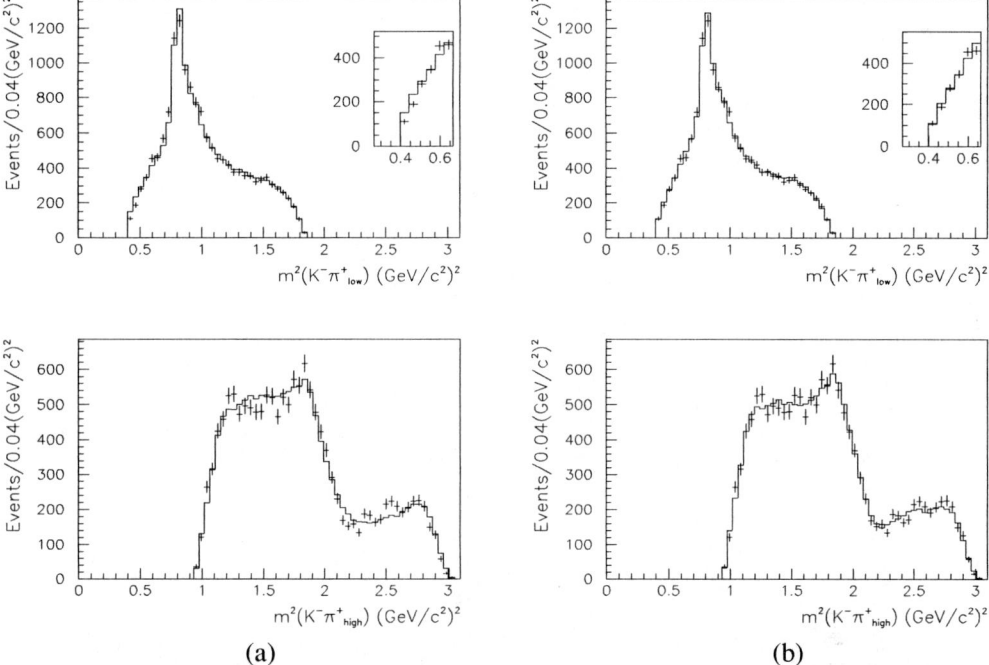

FIGURE 2. $m^2(K\pi_{\text{low}})$ and $m^2(K\pi_{\text{high}})$ projections for data (error bars) and fast MC (solid line): (a) fit to Model A, without κ, and (b) fit to Model B, with κ.

vector state, with unconstrained mass and width, but it appears with a small fraction, the fit quality being comparable to the model without it, and all fractions and phases remaining unchanged. A tensor model was also tried without convergence, the width being driven to large, negative values. Other models with the κ were also tried. For example, modifications to the scalar Breit-Wigner amplitude and to the form-factors were introduced. A number of studies for the parameterization of the NR amplitude were tried [22], with and without the κ. No model without the κ was able to describe our data satisfactorily. All variations of models with κ gave similar results for the κ mass and width (within errors) although the fractions for $\kappa\pi$ and NR showed correlations.

Thus, from the results above, we find strong evidence that a light and broad scalar $K\pi$ resonance gives an important contribution to the $D^+ \to K^-\pi^+\pi^+$ decay.

THE $D_S^+ \to \pi^-\pi^+\pi^+$ RESULTS

In Figure 3 we show the $\pi^-\pi^+\pi^+$ invariant mass distribution for the sample collected by E791 after reconstruction and selection criteria [1, 2]. Besides combinatorial background, reflections from the decays $D^+ \to K^-\pi^+\pi^+$, $D^+ \to K^-\pi^+$ (plus one extra

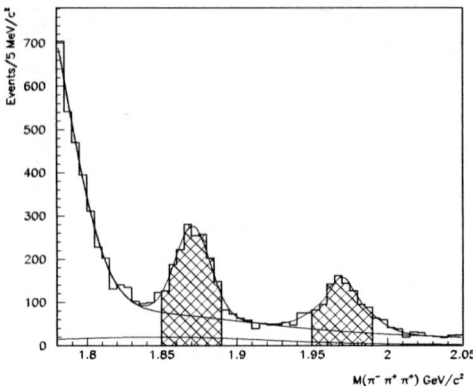

FIGURE 3. The $\pi^-\pi^+\pi^+$ invariant mass spectrum. The dotted line represents the $D^0 \to K^-\pi^+$ plus $D_s^+ \to \eta'\pi^+$ reflections and the dashed line is the total background. Events used for the Dalitz analyses are in the hatched areas.

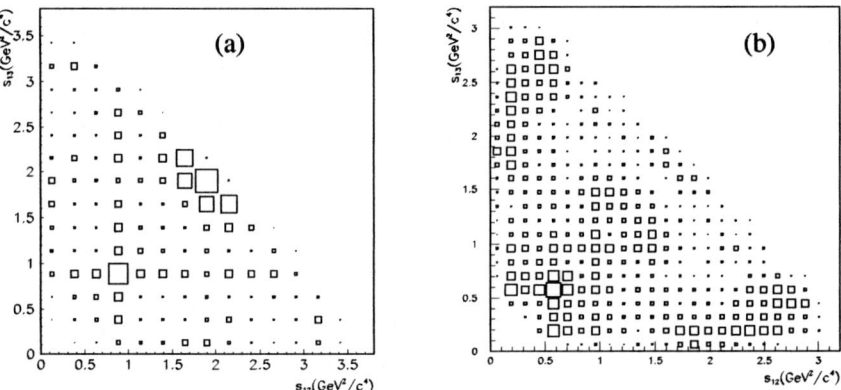

FIGURE 4. (a) The $D_s^+ \to \pi^-\pi^+\pi^+$ Dalitz plot and (b) the $D^+ \to \pi^-\pi^+\pi^+$ Dalitz plot. Since there are two identical pions, the plots are symmetrized.

track) and $D_s^+ \to \eta'\pi^+$, $\eta' \to \rho^0(770)\gamma$ are all taken into account. The hatched regions in Figure 3 show the samples used for the Dalitz-plot analyses. There are 1686 and 937 candidate events for D^+ and D_s^+ respectively, with a signal to background ratio of about 2:1. The Dalitz plots for these events are shown in Figure 4, the axes corresponding to the two $\pi^-\pi^+$ invariant-masses squared. For the Dalitz-plot fits of both $D^+ \to \pi^-\pi^+\pi^+$ and $D_s^+ \to \pi^-\pi^+\pi^+$ decays, we use essentially the same formalism as for the $D^+ \to K^-\pi^+\pi^+$ decays. See details in [1, 2].

TABLE 2. Dalitz fit results for $D_s^+ \to \pi^-\pi^+\pi^+$.

Decay Mode	Fraction (%)	Phase
$f_0(980)\pi^+$	$56.5 \pm 4.3 \pm 4.7$	$0°$ (fixed)
NR	$0.5 \pm 1.4 \pm 1.7$	$(181 \pm 94 \pm 51)°$
$\rho^0(770)\pi^+$	$5.8 \pm 2.3 \pm 3.7$	$(109 \pm 24 \pm 5)°$
$f_2(1270)\pi^+$	$19.7 \pm 3.3 \pm 0.6$	$(133 \pm 13 \pm 28)°$
$f_0(1370)\pi^+$	$32.4 \pm 7.7 \pm 1.9$	$(198 \pm 19 \pm 27)°$
$\rho^0(1450)\pi^+$	$4.4 \pm 2.1 \pm 0.2$	$(162 \pm 26 \pm 17)°$

For the $D_s^+ \to \pi^-\pi^+\pi^+$ events in Figure 4(a), the signal amplitude includes the following channels: $\rho^0(770)\pi^+$, $f_0(980)\pi^+$, $f_2(1270)\pi^+$, $f_0(1370)\pi^+$, $\rho^0(1450)\pi^+$ and the non-resonant, assumed constant across the Dalitz plot.

For the $f_0(980)\pi^+$ amplitude, instead of a simple Breit-Wigner of Eq. 3 [2], we use a coupled-channel Breit-Wigner function [23],

$$BW_{f_0(980)} = \frac{1}{m_{\pi\pi}^2 - m_0^2 + im_0(\Gamma_\pi + \Gamma_K)}, \qquad (5)$$

$$\Gamma_\pi = g_\pi \sqrt{m_{\pi\pi}^2/4 - m_\pi^2}, \quad \Gamma_K = \frac{g_K}{2}\left(\sqrt{m_{\pi\pi}^2/4 - m_{K^+}^2} + \sqrt{m_{\pi\pi}^2/4 - m_{K^0}^2}\right). \qquad (6)$$

The $D_s^+ \to \pi^-\pi^+\pi^+$ Dalitz plot is fit to obtain not only the decay fractions and phases of the possible sub-channels, but also the parameters of the $f_0(980)$ state, g_π, g_K, and m_0, as well as the mass and width of the $f_0(1370)$. The other resonance masses and widths are taken from the PDG[17]. The resulting fractions and phases are given in Table 2.

The measured $f_0(980)$ parameters are $m_0 = 977 \pm 3 \pm 2$ MeV/c^2, $g_\pi = 0.09 \pm 0.01 \pm 0.01$ and $g_K = 0.02 \pm 0.04 \pm 0.03$. Our value for g_π is in very good agreement with OPAL and MARKII results [24], but WA76 [23] found a much larger value, $g_\pi = 0.28 \pm 0.04$. Our value of g_K indicates a small coupling of $f_0(980)$ to $K\bar{K}$. The values of the $f_0(980)$ mass and of g_π, as well as the magnitudes and phases of the resonant amplitudes, are relatively insensitive to the value of g_K. Both OPAL and MARKII results are also insensitive to the value of g_K. WA76, on the contrary, measured $g_K = 0.56 \pm 0.18$.

By fitting the Dalitz plot using for the $f_0(980)$ a simple Breit-Wigner function, we find $m_0 = 975 \pm 3$ MeV/c^2 and $\Gamma_0 = 44 \pm 2 \pm 2$ MeV/c^2, and the results for fractions and phases are indistinguishable.

The confidence level of the fit for $D_s^+ \to \pi^-\pi^+\pi^+$ is 35% [1]. In Figure 5 we show the $\pi^-\pi^+$ mass-squared projections for data (points) and model (solid lines, from fast-MC).

As we can see by the results of Table 2, approximately half of the $D_s^+ \to \pi^-\pi^+\pi^+$ rate is via $f_0(980)\pi^+$. If the spectator amplitude is dominant in this decay, this would support the interpretation of the $f_0(980)$ as an $s\bar{s}$ state. On the other hand, the large contribution from $f_0(1370)\pi^+$ indicates the presence of either W-annihilation amplitudes or strong rescattering in the final state. In fact, the $f_0(1370)\pi^+$ is not observed in the $D_s^+ \to$

---

[2] For both $D, D_s \to 3\pi$ analyses, the relativistic Breit-Wigner for each resonant amplitude is defined with a factor (-1) with respect to Eq. 3.

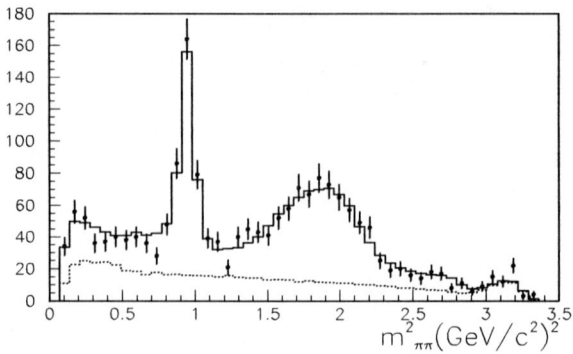

FIGURE 5. s_{12} and s_{13} ($m_{\pi\pi}^2$) projections for $D_s^+ \to \pi^-\pi^+\pi^+$ data (dots) and our best fit (solid). The hashed area corresponds to background.

$K^+K^-\pi^+$ final state[25], pointing to the $f_0(1370)$ being a non-$s\bar{s}$ state, as suggested by the naive quark model[17]. There is no evidence in the D_s^+ decay for a low-mass broad scalar particle as seen in the $D^+ \to \pi^-\pi^+\pi^+$ decay, discussed below.

THE $D^+ \to \pi^-\pi^+\pi^+$ DECAY

In a first approach, we try to fit the $D^+ \to \pi^-\pi^+\pi^+$ Dalitz plot of Figure 4(b) with the same amplitudes used for the $D_s^+ \to \pi^-\pi^+\pi^+$ analysis. Using this model, the non-resonant, the $\rho^0(1450)\pi^+$, and the $\rho^0(770)\pi^+$ amplitudes are found to dominate, as shown in Table 3, and in agreement with previous reported analyses [26, 27]. However, this model does not describe the data satisfactorily, especially at low $\pi^-\pi^+$ mass squared, as can be seen from Fig. 6(a). The χ^2/ν obtained from the binned Dalitz plot for this model is 1.6, with a CL less than 10^{-5}.

To investigate the possibility that another $\pi^-\pi^+$ resonance contributes to the $D^+ \to \pi^-\pi^+\pi^+$ decay, we add an extra scalar resonance amplitude to the signal PDF, with mass and width as floating parameters in the fit.

We find that this model improves our fit substantially. The mass and the width of the extra scalar state are found to be $478^{+24}_{-23} \pm 17$ MeV/c^2 and $324^{+42}_{-40} \pm 21$ MeV/c^2, respectively. Refering to this state as the σ, we obtain that the $\sigma\pi^+$ chanell produces the largest decay fraction, as shown in Table 3; the non-resonant amplitude, which is dominant in the model without $\sigma\pi^+$, drops substantially. This model describes the data much better, as can be seen by the $\pi\pi$ mass squared projection in Fig. 6(b). The χ^2/ν is now 0.9, with a corresponding confidence level of 91%.

The existence of a light $\pi\pi$ state, or the σ, has been the subject of a long-standing controversy [28, 29]. Various experiments have presented inconsistent evidence for this state [30], yielding conflicting results [17, 29].

TABLE 3. Dalitz fit results for $D^+ \to \pi^-\pi^+\pi^+$. First errors are statistical, second systematics (only for fit with $\sigma\pi^+$ mode).

Decay Mode	Fit without $\sigma\pi^+$ Fraction (%)	Phase	Fit with $\sigma\pi^+$ Fraction (%)	Phase
$\sigma\pi^+$	–	–	$46.3 \pm 9.0 \pm 2.1$	$(206 \pm 8 \pm 5)°$
$\rho^0(770)\pi^+$	20.8 ± 2.4	$0°$ (fixed)	$33.6 \pm 3.2 \pm 2.2$	$0°$ (fixed)
NR	38.6 ± 9.7	$(150 \pm 12)°$	$7.8 \pm 6.0 \pm 2.7$	$(57 \pm 20 \pm 6)°$
$f_0(980)\pi^+$	7.4 ± 1.4	$(152 \pm 16)°$	$6.2 \pm 1.3 \pm 0.4$	$(165 \pm 11 \pm 3)°$
$f_2(1270)\pi^+$	6.3 ± 1.9	$(103 \pm 16)°$	$19.4 \pm 2.5 \pm 0.4$	$(57 \pm 8 \pm 3)°$
$f_0(1370)\pi^+$	10.7 ± 3.1	$(143 \pm 10)°$	$2.3 \pm 1.5 \pm 0.8$	$(105 \pm 18 \pm 1)°$
$\rho^0(1450)\pi^+$	22.6 ± 3.7	$(46 \pm 15)°$	$0.7 \pm 0.7 \pm 0.3$	$(319 \pm 39 \pm 11)°$

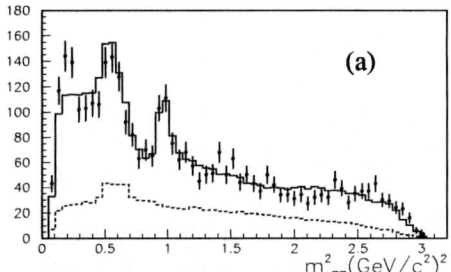
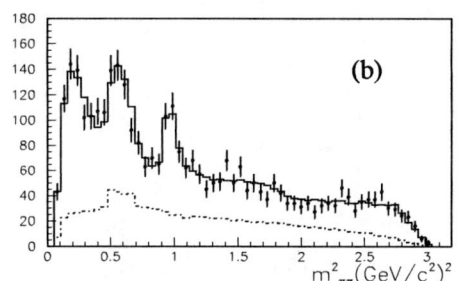

FIGURE 6. s_{12} and s_{13} ($m_{\pi\pi}^2$) projections for $D^+ \to \pi^-\pi^+\pi^+$ data (dots) and our best fit (solid) for models (a) without and (b) with $\sigma\pi^+$ amplitude. The dashed distribution corresponds to the expected background level.

To test the model above, we replace the scalar amplitude by vector and tensor states, and also by a real Breit-Wigner, with no phase variation (as also done in the $D^+ \to K^-\pi^+\pi^+$ analysis). All these alternative models fail to describe the data as well as the scalar (regular) Breit-Wigner amplitude. See detailed discussion in [2].

CONCLUSION

From the data of the Fermilab E791 experiment, we studied the Dalitz plots of the decays $D^+ \to K^-\pi^+\pi^+$, $D_s^+ \to \pi^-\pi^+\pi^+$ and $D^+ \to \pi^-\pi^+\pi^+$. In these three final states, the scalar intermediate resonances were found to give the main contribution to the decay rates. We obtained strong evidence for the existence of the σ and κ scalar mesons, measuring their masses and widths. We also obtained new measurements for masses and widths of the other scalars studied, $f_0(980)$, $f_0(1430)$ and $K_0^*(1430)$.

The results presented here show the potential of D meson decays for the study of light meson espectroscopy, in particular in the scalar sector.

REFERENCES

1. E791 Collaboration, E.M. Aitala *et al.*, Phys. Rev. Lett. **86** 765 (2001).
2. E791 Collaboration, E.M. Aitala *et al.*, Phys. Rev. Lett. **86** 770 (2001).
3. J.A. Appel, Ann. Rev. Nucl. Part. Sci. **42**, 367 (1992); D. Summers *et al.*, hep-ex/0009015; S. Amato *et al.*, Nucl. Instr. Meth. A **324**, 535 (1993); E791 Collaboration, E.M. Aitala *et al.*, Eur. Phys. J. direct C **4**, 1 (1999).
4. E. van Beveren *et al.*, Z. Phys. C **30**, 615 (1986).
5. S. Ishida *et al.*, Prog. Theor. Phys. **98**, 621 (1997).
6. D. Black *et al.*, Phys. Rev. D **58**, 054012 (1998).
7. J.A. Oller, E. Oset, J.R. Peláez, Phys. Rev. D **59** 074001 (1999); M. Jamin, J.A. Oller, and A. Pich, Nucl. Phys. **B587**, 331 (2000).
8. C.M. Shakin, H. Wang, Phys. Rev. D **63**, 014019 (2001).
9. M. Ishida, Prog. Theor. Phys. **101**, 661 (1999).
10. D. Black *et al.*, Phys. Rev. D **59**, 074026 (1999).
11. J.A. Oller and E. Oset, Phys. Rev. D **60**, 074023 (1999).
12. N.A. Törnqvist, Z. Phys. C **68**, 647 (1995).
13. A.V. Anisovitch and A.V. Sarantsev, Phys. Lett. B **413**, 137 (1997).
14. S.N. Cherry and M.R. Pennington, Nucl. Phys. **A688** 823 (2001).
15. E691 Collaboration, J.C. Anjos *et al.*, Phys. Rev. D **48**, 56 (1993).
16. E687 Collaboration, P.L. Frabetti *et al.*, Phys. Lett. B **331**, 217 (1994).
17. Particle Data Group, D.E. Groom *et al.*, Eur. Phys. Jour. C **15**, 1 (2000).
18. ARGUS Collaboration, H. Albrecht *et al.*, Phys. Lett. B **308**, 435(1993).
19. CLEO Collaboration, S. Kopp *et al.*, Phys. Rev. D **63**, 092001 (2001).
20. J.M. Blatt and V.F. Weisskopf, Theoretical Nuclear Physics, John Wiley & Sons, New York, 1952.
21. LASS Collaboration, D. Aston *et al.*, Nucl. Phys. **B296**, 493 (1988).
22. I. Bediaga, C. Göbel, and R. Méndez-Galain, Phys. Rev. Lett. **78**, 22 (1997) and Phys. Rev. D **56**, 4268 (1997); C. Göbel, Ph.D. Thesis, Centro Brasileiro de Pesquisas Físicas, Rio de Janeiro, Brazil (1999).
23. WA76 Collaboration, T.A. Armstrong *et al.*, Z. Phys. C **51**, 351 (1991).
24. OPAL Collaboration, K. Ackerstaff *et al.*, Eur. Phys. J. C **4**, 19 (1998); MARKII Collaboration, G. Gidal *et al.*, Phys. Lett. B **107**, 153 (1981).
25. E687 Collaboration, P.L. Frabetti *et al.*, Phys. Lett. B **351**, 591 (1995).
26. E691 Collaboration, J.C. Anjos *et al.*, Phys. Rev. Lett. **62**, 125 (1989).
27. E687 Collaboration, P.L. Frabetti *et al.*, Phys. Lett. B **407**, 79 (1997).
28. M.R. Pennington, in *Proceedings of the Workshop on Hadron Spectroscopy*, Frascati Physics Series Vol. XV (Laboratory Nazionali de Frascati, Frascati (Roma), Italy, 1999), p. 95.
29. N. Törnqvist, in *Proceedings of the Workshop on Hadron Spectroscopy*, Frascati Physics Series Vol. XV (Laboratory Nazionali de Frascati, Frascati (Roma), Italy, 1999), p. 237, and N. Törnqvist, hep-ph/0008135.
30. WA102 Collaboration, D. Barberis *et al.*, Phys. Lett. B **453**, 316 (1999); CLEO Collaboration, D.M. Asner *et al.*, Phys. Rev. D **61**, 012002 (2000); GAMS Collaboration, D. Alde *et al.*, Phys. Lett. B **397**, 350 (1997).

Heavy Quark Production and Spectroscopy at HERA

Uri Karshon [1]

Weizmann Institute of Science, Israel

Abstract. Production of final states containing open charm (c) and beauty (b) quarks at HERA is reviewed. Photoproduction (PHP) of the charm meson resonances D^*, D^0 and D_s, as well as D^* production in the deep inelastic scattering (DIS) regime, are measured and compared to QCD predictions. The excited charm mesons $D_1^0(2420)$, $D_2^{*0}(2460)$ and $D_{s1}^\pm(2536)$ have been observed and the rates of charm quarks hadronising to these mesons were extracted. A search for radially excited charm mesons has been performed. PHP and DIS beauty cross sections are higher than expected in next-to-leading order (NLO) QCD.

INTRODUCTION

The HERA e-p collider accelerates electrons (or positrons) and protons to energies of $E_e = 27.5\,\text{GeV}$ and $E_p = 920\,\text{GeV}$ ($820\,\text{GeV}$ until 1997), respectively. The H1 and ZEUS experiments are located at two collision points along the circulating beams. The incoming $e^\pm$ interacts with the proton by first radiating a virtual photon. The photon is either quasi-real with $Q^2 < 1\,\text{GeV}^2$ and $Q^2_{median} \approx 3 \cdot 10^{-4}\,\text{GeV}^2$ (PHP regime) or highly virtual ($Q^2 > 1\,\text{GeV}^2$ - DIS regime).

The large masses of the heavy quarks (HQ) c and b provide a "hard" scale needed for the comparison of data to QCD predictions. In leading-order (LO) QCD, two types of processes are responsible for the PHP of HQ's: Direct photon processes, where the photon interacts as a point-like particle with a parton from the incoming proton, and resolved photon processes, where a parton from the photon partonic structure scatters off a parton from the proton. Heavy quarks (Q) present in the parton distributions of the photon lead to LO resolved processes, such as $Qg \to Qg$ (where g is a gluon), which are called heavy flavour excitation. In NLO calculations, only the sum of direct and resolved processes is unambiguously defined.

Two different NLO calculations are available for comparison with measurements of HQ PHP at HERA: 1) a fixed-order ("massive") approach [1, 2], where HQ's are produced only dynamically in the hard subprocess. This calculation is expected to become less accurate when $p_\perp^2 >> m_Q^2$, where $p_\perp$ and m_Q are the transverse momentum and mass of the HQ; 2) a resummed ("massless") approach [3, 4], where the massless HQ's from the photon and proton parton distributions are used explicitly. This calculation is expected to yield better results as $p_\perp^2 >> m_Q^2$.

---

[1] On behalf of the H1 and ZEUS Collaborations

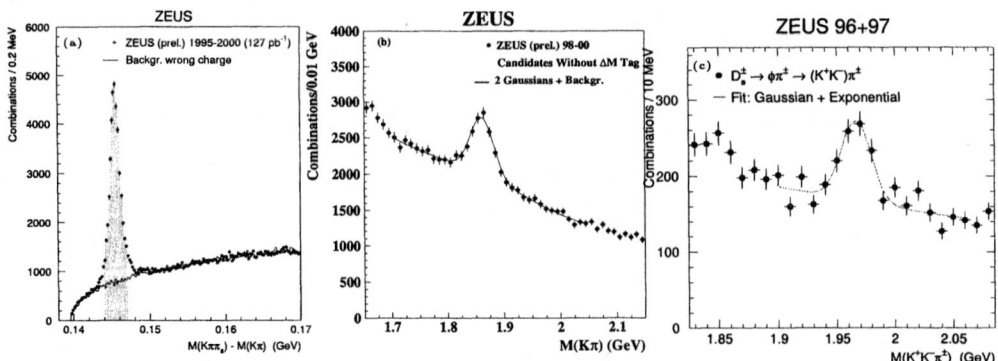

FIGURE 1. (a) $M(K\pi\pi_S) - M(K\pi)$ distribution in the D^0 mass region (full dots). The histogram is the distribution for wrong charge combinations. (b) $M(K^\pm\pi^\mp)$ distribution for events excluding the $D^{*\pm}$ region. The solid curve is a fit to two Gaussian shapes for the right and wrong K mass assignments plus a sum of exponential and linear backgrounds. (c) $M(K^+K^-\pi^\pm)$ distribution for events inside the ϕ mass range $(1.0115 < M(K^+K^-) < 1.0275\,\text{GeV})$. The solid curve is a fit to a Gaussian resonance plus an exponential background.

PRODUCTION OF $D^{*\pm}$, D^0 AND D_S MESONS

The charmed meson $D^{*\pm}$ has been reconstructed via its decay chain $D^{*+} \to D^0\pi_S^+ \to (K^-\pi^+)\pi_S^+$ (+ c.c.). Fig. 1(a) shows the mass difference distribution, $\Delta M = M(K\pi\pi_S) - M(K\pi)$, in the D^0 mass region $1.83 < M(K\pi) < 1.90\,\text{GeV}$ in the kinematic range $p_\perp^{D^*} > 2\,\text{GeV}$ and $|\eta^{D^*}| < 1.5$, where $p_\perp$ is the transverse momentum and $\eta = -\ln\tan(\theta/2)$ is the pseudorapidity. The polar angle, θ, is defined with respect to the proton beam direction. The plot includes PHP and DIS ZEUS data collected during 1995-2000 [5]. A clear $D^{*\pm}$ signal is seen (dots) on top of a small combinatorial background, estimated by wrong charge combinations (histogram), where both D^0 tracks have the same charge and π_S has the opposite charge. Defining $D^{*\pm}$ candidates as events with $0.144 < \Delta M < 0.147\,\text{GeV}$, a signal of 31350 ± 240 $D^{*\pm}$ mesons was found after background subtraction.

Inclusive production of charm hadrons other than the $D^{*\pm}$ have also been observed. The $M(K^\pm\pi^\mp)$ distribution for PHP events excluding the $D^{*\pm}$ region $(0.143 < \Delta M < 0.148\,\text{GeV})$ is shown in Fig. 1(b) for a restricted kinematic region, using a ZEUS event sample with integrated luminosity $\mathcal{L} = 66\,\text{pb}^{-1}$ [6]. A clear D^0 signal is seen. The production ratio, P_v, of vector to pseudoscalar+vector ground state (orbital angular momentum $L = 0$ of the $c\bar{q}$ system) charm mesons can be approximated by $P_v = (\sigma(D^0)/\sigma(D^{*+}) - B_{D^{*+}\to D^0\pi})^{-1}$, where $\sigma(D^0)$ and $\sigma(D^{*+})$ are, respectively, the inclusive D^0 and $D^{*\pm}$ cross sections and $B_{D^{*+}\to D^0\pi}$ is the D^{*+} branching ratio to $D^0\pi^+$. Using also the D^0 signal from events in the D^* region with the same cuts yields the preliminary result $P_v = 0.546 \pm 0.045(\text{stat.}) \pm 0.028(\text{syst.})$. This measurement is in good agreement with the e^+e^- annihilation results, 0.57 ± 0.05 and 0.595 ± 0.045 [7], supporting the universality of charm fragmentation.

Using a ZEUS PHP sample with $38\,\text{pb}^{-1}$ in a restricted kinematic region, a

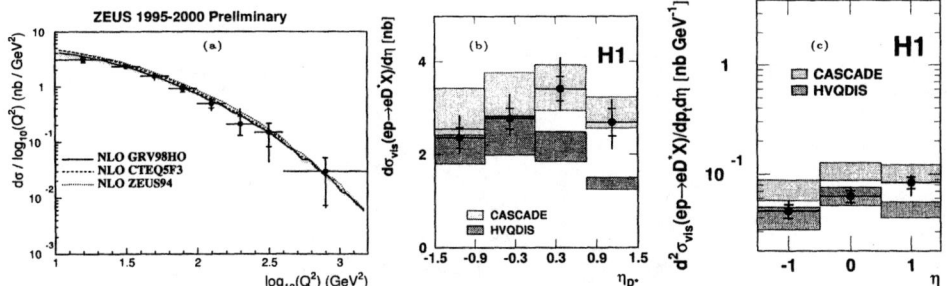

FIGURE 2. (a) Differential D^* cross section in $\log_{10}(Q^2)$ compared to the NLO HVQDIS calculations with different structure function parametrisations. (b) Differential D^* DIS ($1 < Q^2 < 100\,\text{GeV}^2$) cross section in η^{D^*} compared with HVQDIS (lower shaded band) and CASCADE (upper shaded band) predictions. (c) Same as b) for $4 < p_\perp^{D^*} < 10\,\text{GeV}$.

D_s signal is seen (Fig. 1(c)) in the $K^+K^-\pi^\pm$ mass distribution for events where $M(K^+K^-)$ is in the ϕ mass range [8]. The ratio of the D_s to D^* cross sections in identical kinematic regions was measured to be $\sigma_{ep \to D_s X}/\sigma_{ep \to D^* X} = 0.41 \pm 0.07\,(\text{stat.})^{+0.03}_{-0.05}\,(\text{syst.}) \pm 0.10\,(\text{br.})$, where the last error is due to the uncertainty in the $D_s \to \phi\pi$ branching ratio. This result is in good agreement with the ratio $f(c \to D_s^+)/f(c \to D^{*+}) = 0.43 \pm 0.04 \pm 0.11\,(\text{br.})$ obtained [8, 9] from e^+e^- experiments, again confirming the universality of charm fragmentation.

$D^{*\pm}$ Production in DIS

Open charm production in the DIS regime is dominated by boson-gluon fusion (BGF) processes, where the boson (gluon) is emitted from the incoming electron (proton). Fixed-order NLO perturbative QCD (pQCD) calculations are available in the form of a Monte Carlo (MC) integrator (HVQDIS) [10]. The ZEUS preliminary D^* differential cross section in Q^2, using a sample of $\mathcal{L} = 82.6\,\text{pb}^{-1}$, is shown in Fig. 2(a) for the kinematic region $Q^2 > 10\,\text{GeV}^2$, $p_\perp^{D^*} > 1.5\,\text{GeV}$ and $|\eta^{D^*}| < 1.5$ [11]. The distribution compares well with the HVQDIS calculations, using 3 different parton distribution functions in the proton and a charm quark mass range $1.3 < m_c < 1.6\,\text{GeV}$. For $Q^2 \gg m_c^2$, resummed NLO calculations should be superior. However, up to $Q^2 \approx 1000\,\text{GeV}^2$, the data is nicely described by the fixed-order scheme. The $D^{*\pm}$ cross sections were also measured separately for e^+ and e^- beams [12]. Integrated over $Q^2 > 20\,\text{GeV}^2$, the e^-p cross section is higher than that for e^+p by ≈ 3 standard deviations. Both results are compatible with the predictions within the theoretical uncertainties.

The H1 Collaboration has measured $D^{*\pm}$ production in DIS, using $\mathcal{L} = 18.6\,\text{pb}^{-1}$, in a kinematic region $1 < Q^2 < 100\,\text{GeV}^2$, $p_\perp^{D^*} > 1.5\,\text{GeV}$ and $|\eta^{D^*}| < 1.5$ [13]. The differential cross section in η^{D^*} is compared in Figs. 2(b-c) to HVQDIS [10] and to CASCADE [14] calculations, which implement a version of the CCFM evolution scheme [15]. The shaded bands reflect the uncertainties in the predictions due to m_c (1.3 − 1.5 GeV) and the allowed fragmentation parameters. CASCADE shows

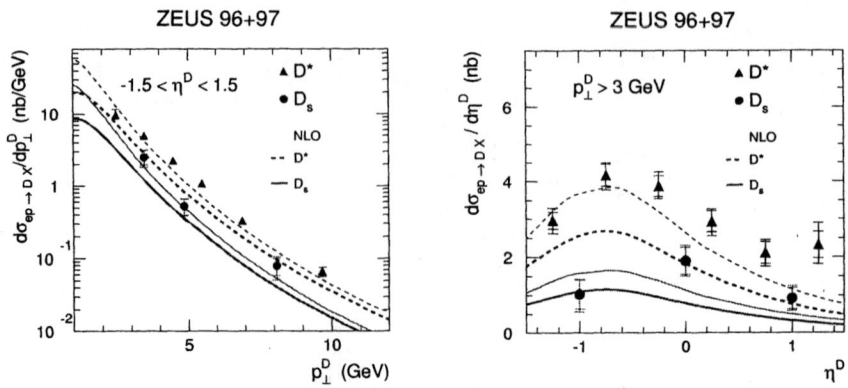

FIGURE 3. Differential cross sections in $p_\perp^D$ and η^D, where D stands for D^* or D_s. The D_s (dots) and D^* (triangles) data are compared with NLO predictions for D_s (full curves) and D^* (dashed curves) with two parameter settings: $m_c = 1.5\,\text{GeV}$, $\mu_R = m_\perp$ (thick curves) and $m_c = 1.2\,\text{GeV}$, $\mu_R = 0.5 m_\perp$ (thin curves). Here μ_R is the renormalisation scale and $m_\perp = \sqrt{m_c^2 + p_\perp^2}$.

better agreement with the overall η^{D^*} distribution (Fig. 2(b)), while the HVQDIS prediction is too low for the forward region. CASCADE agrees poorly with the data at the low-η^{D^*} high-$p_\perp^{D^*}$ region (Fig. 2(c)).

Photoproduction of charm mesons

Differential $D_s^\pm$ cross sections in $p_\perp$ and η [8] were compared to those for $D^{*\pm}$ [16] in the same kinematic region $Q^2 < 1\,\text{GeV}^2$, $130 < W < 280\,\text{GeV}$, $p_\perp^D > 3\,\text{GeV}$ and $|\eta^D| < 1.5$, where W is the γp centre-of-mass energy, and with fixed-order NLO calculations [17] (Fig. 3). The cross sections for both cases are above the predictions, in particular for η in the forward (proton) direction. NLO resummed predictions [3] are closer to the D^* data [16], but still too low for high η. Using different photon parton density functions shows some sensitivity to the parton density parametrisation of the photon. A tree-level pQCD calculation [18], where the $c\bar{q}$ state is hadronised rather than the c quark, gives a better agreement with the data compared to the NLO calculations.

$D^{*\pm}$ and Associated Dijets

Events with a reconstructed $D^{*\pm}$ and at least 2 hadron jets ("dijet event") enable one to study the photon structure, in particular its charm content. $D^{*\pm}$ candidates with $p_\perp^{D^*} > 3\,\text{GeV}$ and $|\eta^{D^*}| < 1.5$ have been selected. The two jets with the highest transverse energy, E_T, in the pseudorapidity range $|\eta^{jet}| < 2.4$ are required to have E_T^{jet1} and E_T^{jet2} above certain values. The fraction of the photon momentum participating in the dijet production, x_γ^{OBS}, is defined as $x_\gamma^{\text{OBS}} = \frac{\Sigma_{\text{jets}} E_T e^{-\eta}}{2 y E_e}$, where y is approximately the fraction of incoming electron energy carried by the photon. In LO QCD, direct processes have $x_\gamma^{\text{OBS}} = 1$ while

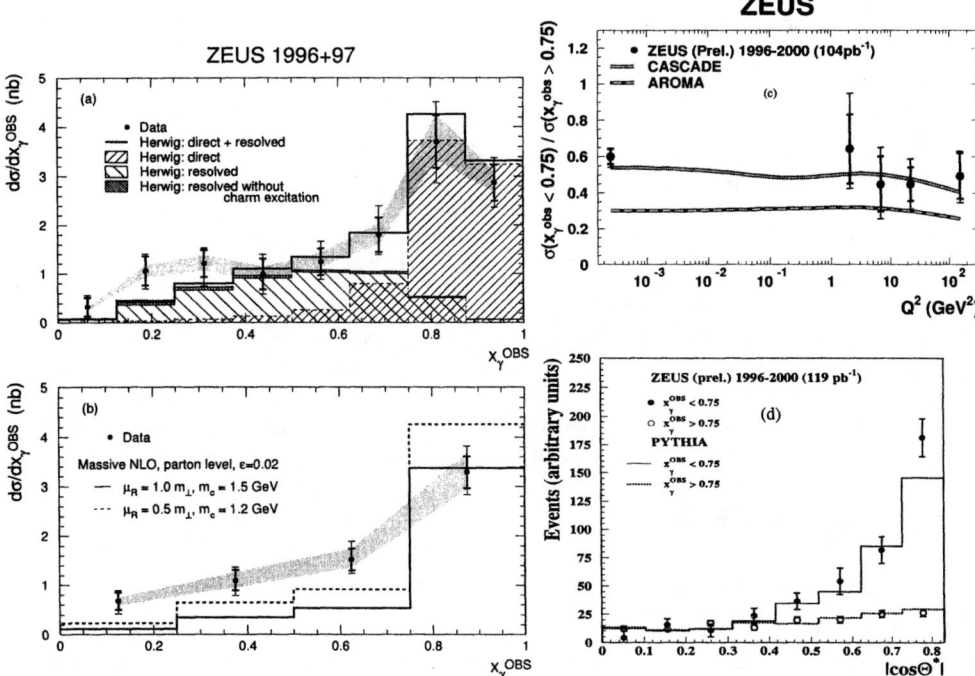

FIGURE 4. (a-b) Differential cross sections $d\sigma/dx_\gamma^{\mathrm{OBS}}$ for dijets with an associated D^* in the PHP range with $E_T^{jet1} > 7\,\mathrm{GeV}$ and $E_T^{jet2} > 6\,\mathrm{GeV}$. In (a), the experimental data (dots) are compared to the expectations of the HERWIG MC simulation, normalised to the data, for LO-direct (right hatched), LO-resolved (left hatched), LO-resolved without charm excitation (dense hatched) and the sum of LO-direct and LO-resolved photon contributions (full histogram). In (b), the data are compared with a parton level NLO fixed-order calculation. (c) Ratio of low to high x_γ^{OBS} for D^* dijet events with $E_T^{jet1,2} > 7.5$, $6.5\,\mathrm{GeV}$ vs. Q^2 compared to the AROMA and CASCADE MC's. The left-hand point is due to PHP events. (d) Differential distributions in $|\cos\theta^*|$ for D^* dijet events with $E_T^{jet1,2} > 5\,\mathrm{GeV}$. Data (dots) and PYTHIA MC (lines) are shown separately for direct- (open dots/dashed lines) and resolved- (black dots/full lines) photon events. All distributions are normalised to the resolved data in the lowest 4 bins.

resolved processes have $x_\gamma^{\mathrm{OBS}} < 1$. Samples enriched with direct (resolved) events are separated by a cut $x_\gamma^{\mathrm{OBS}} > 0.75 (< 0.75)$.

Differential cross sections in x_γ^{OBS} for ZEUS measurements with $\mathcal{L} = 37\,\mathrm{pb}^{-1}$ are shown in Fig. 4(a-b) [16] and compared with LO MC simulation and NLO fixed-order calculation. The peak at high x_γ^{OBS} is due to the LO-direct BGF process. The low x_γ^{OBS} tail comes from LO-resolved processes, dominated by photon charm excitation. The shape of the x_γ^{OBS} distribution is in good agreement with the MC simulation with $\approx 40\%$ resolved contribution (Fig. 4(a)). The fixed-order NLO calculation lies below the data at low x_γ^{OBS} values (Fig. 4(b)). This could be due to the fact that no explicit charm excitation component exists in this calculation.

The dependence of the virtual-photon structure on its virtuality, Q^2, has been studied with ZEUS dijet events containing a D^* [19]. The cross section ratio $R = \sigma(x_\gamma^{\text{OBS}} < 0.75)/\sigma(x_\gamma^{\text{OBS}} > 0.75)$ as a function of Q^2 is shown in Fig. 4(c) to be approximately constant up to $Q^2 \approx 200\,\text{GeV}^2$, contrary to the no-charm-tag case, where R falls with increasing Q^2. The data are compared to two MC models which implement no specific partonic structure for the photon, generating all low x_γ^{OBS} events from parton showers in two different schemes. The AROMA [20] model, which implements the DGLAP evolution scheme [21], lies below the data. The CASCADE results [14] are much closer to the data.

The angular distributions, $dN/d|\cos\theta^*|$, of dijet events containing a D^* in the PHP regime have been measured by ZEUS [22]. Here θ^* is the angle between the jet-jet axis and the beam direction in the dijet rest frame. The results are shown in Fig. 4(d) separately for direct and resolved photon events. The resolved processes peak at high $|\cos\theta^*|$, in agreement with LO MC predictions, as expected for dominant gluon exchange. The direct processes are much flatter in $|\cos\theta^*|$, consistent with quark exchange. The steep rise towards high $|\cos\theta^*|$ of the resolved charm events provides evidence that the bulk of the resolved contribution is due to charm excitation in the photon.

EXCITED CHARM MESONS

P-wave charm mesons (L=1 of the $c\bar{q}$ system) can decay into L=0 states plus a π or a K [23]. They are predicted [24] to appear in two doublets with total angular momentum j=3/2 (narrow states) or j=1/2 (broad states). Narrow states, $D_1(2420)$ and $D_2^*(2460)$, were observed in the $D^*\pi$ decay mode and identified as members of the j=3/2 doublet with spin-parity $J^P = 1^+$ and 2^+, respectively [23]. A charm-strange excited meson, $D_{s1}^\pm(2536)$, was found in the $D^{*\pm}K_S^0$ final state [23].

$D_1^0(2420)$ and $D_2^{*0}(2460)$ production

D_1^0 and D_2^{*0} mesons were reconstructed by ZEUS [25] via their decays to $D^{*\pm}\pi_4^\mp$, followed by the $D^{*\pm}$ decays, $D^{*+} \to D^0\pi_S^+ \to (K^-\pi^+)\pi_S^+(+c.c.)$. Fig. 5(a) shows the "extended" mass difference distribution, $M(K\pi\pi_S\pi_4) - M(K\pi\pi_S) + M(D^*)$, where $M(D^*)$ is the nominal $D^{*\pm}$ mass [23] (full dots). A clear excess is seen around the mass region of the $D_1^0(2420)$ and $D_2^{*0}(2460)$ mesons. No enhancement is seen for wrong charge combinations (dashed histogram), where D^* and π_4 have the same charges. The solid curves in Figs. 5(a-b) are an unbinned likelihood fit to two Breit-Wigner shapes with masses and widths fixed to the nominal D_1^0 and D_2^{*0} values [23], convoluted with a Gaussian function and multiplied by helicity spectrum functions for $J^P = 1^+$ and 2^+ states, respectively. The background shape was parametrised by the form $x^\alpha \cdot exp(-\beta \cdot x + \gamma \cdot x^2)$, where $x = M(K\pi\pi_S\pi_4) - M(K\pi\pi_S) - M(\pi)$. The fitted curves describe the distribution reasonably well, except for a narrow enhancement near 2.4 GeV (Fig. 5(b)). In Fig. 5(c), a similar fit is shown with an additional Gaussian-shaped resonance with free mass and width. The fit yielded 211 ± 49 entries for the narrow enhancement with mass value

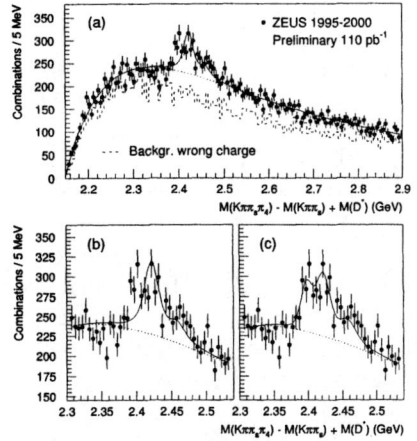
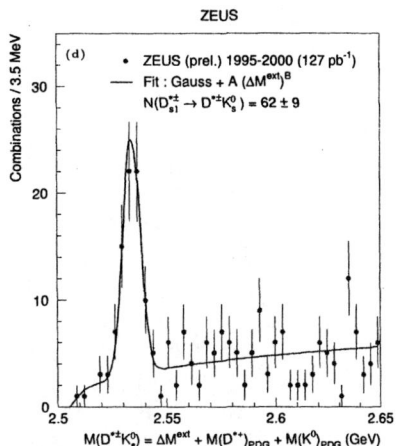

FIGURE 5. (a-c) Extended mass difference distribution for $D^{*\pm}$ candidates (full dots). The dashed histogram is for wrong charge cominations. The curves are the results of unbinned likelihood fits. In (a) and (b) the solid curves are a fit to background parametrisation and two Breit-Wigner distributions convoluted with a Gaussian function. In (c) an additional Gaussian-shaped resonance near 2.4 GeV is assumed in the fit. The dotted curves are fitted shapes of the combinatorial background. (d) Effective $M(D^{*\pm}K_S^0)$ distribution (full dots). The solid line is a fit to a Gaussian resonance plus background of the form $A(\Delta M^{ext})^B$.

$2398.1 \pm 2.1\,(\text{stat.})^{+1.6}_{-0.8}\,(\text{syst.})$ MeV. The width was consistent with the resolution expected from the tracking detector. The enhancement may indicate a new excited charm meson, a result of an interference effect or a statistical fluctuation. The number of reconstructed D_1^0 and D_2^{*0} mesons in the 3-resonance fit are 526 ± 65 and 203 ± 60, respectively.

The acceptance-corrected fractions of $D^{*\pm}$ mesons originating from D_1^0 and D_2^{*0} in the measured kinematic range were found to be $R_{D_1^0 \to D^{*\pm}\pi^{\mp}/D^{*\pm}} = 3.40 \pm 0.42\,(\text{stat.})^{+0.78}_{-0.63}\,(\text{syst.})\%$ and $R_{D_2^{*0} \to D^{*\pm}\pi^{\mp}/D^{*\pm}} = 1.37 \pm 0.40\,(\text{stat.})^{+0.96}_{-0.33}\,(\text{syst.})\%$. Extrapolating to the full kinematic phase space by a MC simulation and using the partial width ratio, $\Gamma(D_2^{*0} \to D^+\pi^-)/\Gamma(D_2^{*0} \to D^{*+}\pi^-) = 2.3 \pm 0.6$ [23], the rate of c quarks hadronising as D^{*+} mesons, $f(c \to D^{*+}) = 0.235 \pm 0.007 \pm 0.007$ [9], and isospin conservation, the rates of c quarks hadronising as D_1^0 and D_2^{*0} mesons are found to be: $f(c \to D_1^0) = 1.46 \pm 0.18\,(\text{stat.})^{+0.33}_{-0.27}\,(\text{syst.}) \pm 0.06\,(\text{ext.})\%$ and $f(c \to D_2^{*0}) = 2.00 \pm 0.58\,(\text{stat.})^{+1.40}_{-0.48}\,(\text{syst.}) \pm 0.41\,(\text{ext.})\%$. The third errors arise from uncertainties in $f(c \to D^{*+})$ and the $D_2^{*0} \to D^{*+}\pi^-$ branching ratio. The results are consistent with the e^+e^- rates measured by CLEO [26]: $f(c \to D_1^0) = 1.8 \pm 0.3\%$ and $f(c \to D_2^{*0}) = 1.9 \pm 0.3\%$.

Search for radially excited $D^{*'\pm}$

Radially excited charm mesons with mass around 2.6 GeV are predicted [27] to decay into $D\pi\pi$ or $D^*\pi\pi$. A narrow resonance in the $D^{*\pm}\pi^+\pi^-$ final state at 2637 MeV, interpreted as the radially excited $D^{*'\pm}$, was reported by DELPHI [28]. No evidence for this state has been found by OPAL and CLEO [29].

$D^{*'\pm}$ candidates were reconstructed by ZEUS [25] from their decays to $D^{*\pm}\pi_4^+\pi_5^-$. No narrow resonance is seen in the extended mass difference $M(K\pi\pi_S\pi_4\pi_5) - M(K\pi\pi_S) + M(D^*)$. An upper limit of $R_{D^{*'+} \to D^{*+}\pi^+\pi^-/D^{*+}} < 2.3\%$ (95% C.L.) is obtained in the measured kinematic region for the fraction of $D^{*\pm}$ originating from $D^{*'\pm}$ decays within a signal window $2.59 < M(D^{*'\pm}) < 2.67$ GeV, which covers theoretical predictions [27] and the DELPHI measurement [28]. Extrapolating by a MC simulation to the full kinematic phase space and using the known $f(c \to D^*)$ value, a $D^{*'\pm}$ production limit of $f(c \to D^{*'+}) \cdot B_{D^{*'+} \to D^{*+}\pi^+\pi^-} < 0.7\%$ (95% C.L.) is obtained. A similar limit of 0.9% has been reported by OPAL [30].

Production of the charm-strange meson $D_{s1}^\pm(2536)$

$D_{s1}^\pm$ mesons were reconstructed by ZEUS [5] via the $D^{*\pm}K_S^0$ decay mode with $K_S^0 \to \pi^+\pi^-$. K_S^0 candidates were identified by using pairs of oppositely charged tracks with $p_\perp > 0.2$ GeV. A clean $K_S^0 \to \pi_3\pi_4$ signal was extracted after applying standard V^0-finding cuts. K_S^0 candidates with $0.480 < M(\pi_3\pi_4) < 0.515$ GeV were kept for the $D_{s1}^\pm$ reconstruction. Fig. 5(d) shows the effective $M(D^{*\pm}K_S^0)$ distribution in terms of $\Delta M^{ext} + M(D^{*+})_{PDG} + M(K^0)_{PDG}$ (solid dots), where $\Delta M^{ext} = M(K\pi\pi_S\pi_3\pi_4) - M(K\pi\pi_S) - M(\pi_3\pi_4)$ and $M(D^{*+})_{PDG}$ ($M(K^0)_{PDG}$) is the nominal $D^{*\pm}$ (K^0) mass [23]. A clear signal is seen at the $M(D_{s1}^\pm)$ value. The solid curve is an unbinned likelihood fit to a Gaussian resonance plus background of the form $A(\Delta M^{ext})^B$. The fit yielded 62.3 ± 9.3 $D_{s1}^\pm$ mesons. The mass value was found to be $M(D_{s1}^\pm) = 2534.2 \pm 0.6 \pm 0.5$ MeV, in rough agreement with the PDG value [23]. The last error is due to the uncertainty in $M(D^{*+})_{PDG}$.

The angular distribution of the D_{s1} signal was studied via the helicity angle, α, between the K_S^0 and π_S momenta in the $D^{*\pm}$ rest frame. The $dN/d\cos\alpha$ distribution was fitted to $(1 + R\cos^2\alpha)$. An unbinned likelihood fit yielded $R = -0.53 \pm 0.32$ (stat.)$^{+0.05}_{-0.14}$ (syst.), consistent with the CLEO value [31] $R = -0.23^{+0.40}_{-0.32}$. Both measurements are consistent with $R = 0$, i.e. $J^P = 1^+$ for the D_{s1} meson. However, our result is not inconsistent with $R = -1$, i.e. $J^P = 1^-$ or 2^+ [32].

The fraction of $D^{*\pm}$ mesons originating from $D_{s1}^\pm$ in the measured kinematic region is $R_{D_{s1}^\pm \to D^{*\pm}K^0/D^{*\pm}} = 1.77 \pm 0.26$ (stat.)$^{+0.11}_{-0.09}$ (syst.)%. The rate of c quarks hadronising as D_{s1}^+ mesons, after a MC extrapolation to the full kinematic phase space, is $f(c \to D_{s1}^+) = 1.24 \pm 0.18$ (stat.)$^{+0.08}_{-0.06}$ (syst.) ± 0.14 (br.)%. The third error is due to uncertainties in $f(c \to D^{*+})$ and the $D_{s1}^+ \to D^{*+}K^0$ branching ratio. The rate agrees with the OPAL value [33] $1.6 \pm 0.4 \pm 0.3\%$. This rate is about twice that expected, assuming $f(c \to D_1^0) \approx 2\%$ [25] and $\gamma_s \approx 0.3$ [8], where γ_s is the strangeness suppression factor in charm production.

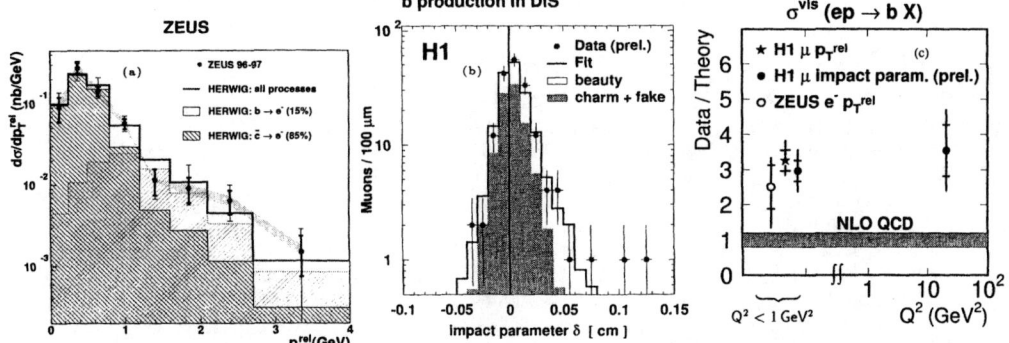

FIGURE 6. (a) Differential cross section in p_T^{rel} for the production of electrons in dijet PHP events compared to HERWIG MC shape prediction (solid line). The MC beauty (charm) component is given by the forward- (backward-) diagonally hatched histogram. (b) Muon impact parameter, δ, distribution for DIS events, with a likelihood fit decomposition into beauty (white region) and charm + fake (shaded region). (c) Ratio of measured b production cross sections over theoretical expectation, as a function of Q^2. The shaded band represents the theoretical uncertainty.

OPEN BEAUTY PRODUCTION

The production rates for beauty at HERA are about two orders of magnitude smaller than for charm. Theoretical uncertainties are expected to be smaller due to the high b quark mass. Enriched b samples have been obtained [34, 35, 36, 37] by studying electrons or muons from semileptonic (SL) b decays. Due to the heavy b quark, the high tail of the lepton p_T with respect to the axis of the closest jet, p_T^{rel}, provides a b signature. H1 has also exploited, using their micro-vertex detector, the long lifetimes of b hadrons to extract b production cross sections by measuring the impact parameter, δ, which is the distance of closest approach of SL muons to the primary vertex in the plane transverse to the beam axis.

The ZEUS differential cross section, $d\sigma/dp_T^{rel}$, for a PHP dijet event sample of $\mathcal{L} = 38.5\,\mathrm{pb}^{-1}$ with identified electrons is compared in Fig. 6(a) [35] with a MC c and b production simulation. The shape is fitted by varying the relative contributions of c and b. A b fraction of $\approx 15\%$ is obtained, consistent with the MC expectation.

For a PHP dijet sample with an identified muon, H1 has shown [36] that the p_T^{rel} and δ methods provide independent and consistent b production results. The H1 analysis of a smaller DIS sample from $\mathcal{L} = 10.5\,\mathrm{pb}^{-1}$ [37] uses the combination of the two observables. A likelihood fit of $b\bar{b}$, $c\bar{c}$ and fake muon spectra to a two-dimensional distribution in δ and p_T^{rel} adjusts the relative weights of all three components in the data. It yields a $b\bar{b}$ fraction of $(43 \pm 8)\%$. The δ projection of this distribution is shown in Fig. 6(b) together with the fit decomposition.

The ratio of the measured visible b cross sections over theoretical expectations [10, 17] is shown in Fig. 6(c) as a function of Q^2 [37]. The ratio is roughly constant with Q^2 and the discrepancy between data and NLO calculations is quite significant. The DIS case is theoretically cleaner, since at high Q^2 the resolved contribution is expected to be suppressed.

REFERENCES

1. S. Frixione et al., Nucl. Phys. B412 (1994) 225;
   M. Mangano et al., Nucl. Phys. B373 (1992) 295.
2. R.K. Ellis and P. Nason, Nucl. Phys. B 312 (1989) 551;
   P. Nason, S. Dawson and R.K. Ellis, Nucl. Phys. B 303 (1988) 607;
   J. Smith and W.L. van Neerven, Nucl. Phys. B 374 (1992) 36.
3. B.A. Kniehl et al., Z. Phys. C76 (1997) 689;
   J.Binnewies et al., Z. Phys. C76 (1997) 677; Phys. Rev. D58 (1998) 014014.
4. M. Cacciari et al., Z. Phys. C69 (1996) 459; Phys. Rev. D55 (1997) 2736;
   M. Cacciari and M. Greco, Phys. Rev. D55 (1997) 7134.
5. ZEUS Collaboration, paper 497 submitted to EPS Conf. on HEP2001, Budapest, Hungary.
6. ZEUS Collaboration, paper 501 submitted to EPS Conf. on HEP2001, Budapest, Hungary.
7. OPAL Collaboration, K. Ackerstaff et al., Eur. Phys. J. C 5 (1998) 1;
   ALEPH Collaboration, R. Barate et al., Eur. Phys. J. C 16 (2000) 597.
8. ZEUS Collaboration, J. Breitweg et al., Phys. Lett. B481 (2000) 213.
9. L. Gladilin, hep-ex/9912064.
10. B.W. Harris and J. Smith, Phys. Rev. D57 (1998) 2806;
    E. Laenen et al., Nucl. Phys. B392 (1995) 162.
11. ZEUS Collaboration, paper 449 submitted to XXX ICHEP2000, Osaka, Japan.
12. ZEUS Collaboration, paper 493 submitted to EPS Conf. on HEP2001, Budapest, Hungary.
13. H1 Collaboration, C. Adloff et al., hep-ex/0108039, submitted to Phys. Lett. B (2001).
14. H. Jung and G.P. Salam, Eur. Phys. J. C19 (2001) 351; H. Jung, hep-ph-0109146.
15. M. Ciafaloni, Nucl. Phys. B296 (1988) 49;
    S. Catani et al., Phys. Lett. B234 (1990) 339; Nucl. Phys. B 336 (1990) 18;
    G. Marchesini, Nucl. Phys. B445 (1995) 49.
16. ZEUS Collaboration, Breitweg J. et al., Eur. Phys. J. C6 (1999) 67.
17. S. Frixione et al., Nucl. Phys. B454 (1995) 3 ; Phys. Lett. B348 (1995) 633.
18. A.V. Berezhnoy, V.V. Kiselev, and A.K. Likhoded, hep-ph/9901333, hep-ph/9905555, Yad. Fiz. [Phys. At. Nucl.] in print (2000).
19. ZEUS Collaboration, paper 495 submitted to EPS Conf. on HEP2001, Budapest, Hungary.
20. G. Ingelman et al., Comp. Phys. Comm. 101 (1997) 135.
21. V. Gribov and L. Lipatov, Sov. J. Nucl. Phys. 15 (1972) 438; ibid. 15 (1972) 675;
    L. Lipatov, Sov. J. Nucl. Phys. 20 (1975) 94;
    G. Altarelli and G. Parisi, Nucl. Phys. B126 (1977) 298;
    Y. Dokshitser, Sov. Phys. JETP 46 (1977) 641.
22. ZEUS Collaboration, paper 499 submitted to EPS Conf. on HEP2001, Budapest, Hungary.
23. Particle Data Group, D. E. Groom et al., Eur. Phys. J C15 (2000) 1.
24. N. Isgur and M.B. Wise, Phys. Lett. B232 (1989) 113;
    M. Neubert, Phys. Rep. A245 (1994) 259.
25. ZEUS Collaboration, paper 448 submitted to XXX ICHEP2000, Osaka, Japan.
26. CLEO Collaboration, Avery P. et al., Phys. Lett. B331 (1994) 236.
27. S. Godfrey and N. Isgur, Phys. Rev. D32 (1985) 189;
    D. Ebert et al., Phys. Rev. D57 (1998) 5663.
28. DELPHI Collaboration, P. Abreu et al., Phys. Lett. B426 (1998) 231.
29. OPAL Collaboration, XXIX ICHEP1998, Vancouver, Canada;
    CLEO Collaboration, hep-ex/9901008.
30. OPAL Collaboration, hep-ex/0101045, submitted to Eur. Phys. J. C (April 2001).
31. CLEO Collaboration, J.P. Alexander et al., Phys. Lett. B303 (1993) 303.
32. S. Godfrey and R. Kokoski, Phys. Rev. D43 (1991) 1130.
33. OPAL Collaboration, K. Ackerstaff et al., Z. Phys. C76 (1997) 425.
34. H1 Collaboration, C. Adloff et al., Phys. Lett. B467 (1999) 156.
35. ZEUS Collaboration, J. Breitweg et al., Euro. Phys. J. C18 (2001) 625.
36. H1 Collaboration, abstract 982 submitted to XXX ICHEP2000, Osaka, Japan.
37. H1 Collaboration, paper 807 submitted to EPS Conf. on HEP2001, Budapest, Hungary.

HEAVY Q SPECTROSCOPY

IHEP, Protvino

Understanding B Meson Branching Fractions

Bernard Gittelman

Cornell University, Ithaca, NY 14853-5001

Abstract

This is a discussion of the main branching fractions of B Meson decays ($b \to cW^-$). What has been measured and what remains unmeasured is summarized, and a plan to measure charged track multiplicity accompanying the $D^{(*)}$ meson in B decay is presented.

Introduction

The existence of the B Meson was infered from the discovery of the $\Upsilon(1S)$ and $\Upsilon(2S)$ mesons at Fermilab in 1977[1], and the measurement of the electromagnetic production cross section, $e^+e^- \to \Upsilon(1S)$ and $e^+e^- \to \Upsilon(2S)$, in 1978 at the DORIS storage ring of DESY[2]. From the magnitude of these cross-sections, it became clear that the Υ resonances are to be described as $b\bar{b}$ quark pairs, not $t\bar{t}$. The Cornell-Electron-Storage-Ring, CESR, started operating in late 1979, and the two Collaborations there, CLEO and CUSB published observation of the $\Upsilon(1S)$, $\Upsilon(2S)$, $\Upsilon(3S)$[3], and $\Upsilon(4S)$[4] in 1980. The $\Upsilon(4S)$ is a much wider resonance and is believed to decay exclusively to $B\bar{B}$ mesons. Over the past 20 years, the general properties of the B meson decays have been studied by many different groups[5]. For reasons having to do with minimizing backgrounds, CLEO has studied B mesons by taking data at the $\Upsilon(4S)$. For every data sample recorded at the $\Upsilon(4S)$, CLEO has collected a data sample of approximately half as much integrated luminosity at an energy of 40 to 80 MeV below the $\Upsilon(4S)$ to measure the background from the continuum. This energy is below $B\bar{B}$ threshold. The original CLEO detector, CLEO-I, was used from 1979 until 1988. It was then replaced by the CLEO-II detector, which consisted of a system of tracking chambers with much better resolution, and a Cesium-Iodide crystal electromagnetic calorimeter (see figure 1). In 1995, the CLEO-II detector was upgraded, by replacing the inner-most tracking chamber, (the PTL), with a silicon vertex tracker. The newer version of CLEO-II was named CLEO-II.V. In 1998, CESR shut down for upgrading, and the CLEO-II.V detector was upgraded further and renamed CLEO-III. CLEO-III contains a new drift chamber, a new silicon detector, and a Ring-Imaging-CHerenkov detector, RICH, for better pion-kaon identification, but the time-of-flight scintillation counters were removed. Data taking with CLEO-III began in 2000. Figures 1 and 2 illustrate the CLEO-II and CLEO-III detectors. Table 1 provides a summary of the integrated luminosity recorded on the $\Upsilon(4S)$ for each of the CLEO detectors.

Inclusive Branching Fractions

B meson decay is described by 4 basic diagrams, as shown in Fig. 3. The diagrams describing $b \to cW^-$ and $b \to uW^-$ are shown in Figs 3a and 3b. The magnitude of $b \to cW^-$ branching fractions versus $b \to uW^-$ branching fractions have been estimated by the study of inclusive production of D mesons in non-leptonic and semi-leptonic final states. Evidence for the existence of diagrams 3c and 3d (the penquins), comes from the observation of $b \to s\gamma$ final states such as $B \to K^*\gamma$. An estimate of the importance of the different contributions to B meson decays as measured from the inclusive branching fractions is given in Table 2.

Semi-leptonic Decay

The existence of the B meson was first verified by the observation of leptons (electrons and muons) coming from $\Upsilon(4S)$[6] decays, with a momentum spectrum expected from B meson semi-leptonic decays[6]. The semi-leptonic decay of B mesons has been studied in detail. A summary of what is known today about the branching fractions is given in Table 3[7]. The total branching fraction for semi-leptonic decay is approximately 24%. This comes from the almost equal branching fraction for inclusive electrons and muons, and an estimate of the expected branching fraction for τ leptons using the phase space factor. As shown in Table 3, some of the exclusive final state branching fractions have been measured. The sum of the measured exclusive semi-leptonic branching fractions is approximately 72% of the inclusive semi-leptonic branching fraction. From the fact that the $b \to uW^-$ semi-leptonic final state branching fractions are approximately 100 times smaller than the $b \to cW^-$ states, we estimated the total inclusive $b \to uW^-$ branching fraction shown in Table 2.

Hadronic Final States

From inclusive measurements of $B \to DX$, 87%; $B \to \Lambda_c X$, ≈6%; and $B \to (c\bar{c})X$, ≈2%; one concludes that $b \to cW^-$ is responsible for most, (95%), of B decays. The challenging problem is to measure the exclusive branching fractions. The measured two body exclusive branching fractions for hadronic decay of B mesons are listed in Table 4. One observes for $B \to D^{(*)}X$, total measured for $B^- = 10\%$, and for $B^o = 6.9\%$; for $B \to (c\bar{c})X$, total measured is 0.41% and 0.33%; for $b \to uW^-$ plus $b \to sg$ plus $b \to s\gamma$, the total branching fractions are 0.017% and 0.012%. This again leads to the conclusion, understanding the $b \to cW^-$ decays will enable us to describe most (more than 90%) of B meson decays. The measured, three, four, and five body final state branching fractions for the hadronic decay of B mesons via $b \to cW^-$ are listed in Table 5. The measured exclusive branching fractions for 2 body final states is much larger than for 3, 4, or 5 body states, but that probably has to do with complexity and larger combinatoric backgrounds for multi-body (>2) final states. We know of no reason to expect the total branching fraction for 3 body, 4 body or 5 body hadronic states ($D^{(*)}$ meson plus 4 particles) should be so much smaller than that of 2 bodies states.

A Proposed Method to measure "Exclusive" B Decays.

From the decay modes listed in Table 3, the total exclusive semi-leptonic branching fractions that have been measured is 7.5%; i.e., we have measured $7.5/10.5 = 72\%$ of the semi-leptonic decays. Adding up the various hadronic decay modes listed in Tables 4 and 5 and averaging over B^- and B^o, we find only 14.8% for the hadronic final states branching fractions. A summary of the major hadronic decay branching fractions listed by multiplicity is given in Table 6. Assuming the total hadronic branching fraction is $(100 - 24)\% = 76\%$, we have measured only $14.8/76 = 19\%$ of the exclusive hadronic decay modes. It is interesting to speculate on why we have measured 72% of the semi-leptonic final states where we are always missing the energy and momentum of the neutrino, but only measured 19% of the hadronic final states where for most modes we should be able to detect every final state particle (photons, charged pions, kaons, and protons, but not neutrons). The only explanation is based on higher multiplicity of the hadronic final state.

As mentioned earlier, CLEO has collected approximately 15 fb^{-1} of data on the $\Upsilon(4S)$ resonance. This provides $\sim 15 \times 10^6$ $B\bar{B}$ pairs. If one were to try reconstructing the B meson in one of the many clean easy to reconstruct decay modes, and the D (or Ψ meson) from the $\bar{B}$ decay in the same event, then one would have a very useful sample of $\bar{B}$ decays. In this reconstruction process, we will be removing events in which the sum of the charge of the tracks does not add up to zero, and those events in which the total momentum vector of the D meson, plus the charged tracks, and the photons does not point in "exactly" the direction opposite to the momentum vector of the reconstructed B meson. We have estimated having 15,000 events in which we have reconstructed The B meson and the D meson from the $\bar{B}$ decay. This sample of 15,000 events might be useful to let us know how often the B meson decays into

$$B \to D + 1track; D + 2tracks; ....; D + 9tracks; ...$$
$$B \to D + \pi^o; D + \eta^o; D + \pi^o + 1track; D + \eta^o + 1track; ....; D + \eta^o + 9tracks; ...$$

Perhaps we will also be able to learn how often the B meson decays into

$$B \to D^* + ntracks; \quad D^* + \pi^o + ntracks; \quad D^* + \eta^o + ntracks$$
$$B \to D^{(*)} + 2\pi^o + ntracks; \quad D^{(*)} + 3\pi^o + ntracks$$

Given a sample of $B \to D^{(*)} + n$ tracks, one can investigate the invariant mass distribution of the n tracks. We would also study the invariant mass distribution of all two track combinations for each set of n charged tracks accompanying the $D^{(*)}$ mesons.

Summary

Evidence for B mesons was discovered in 1980 and after 20 years, the decay modes are understood at the 40% level (See the last line of Table 6.). The D mesons were discovered in 1975, but 90% of their decay modes are already measured. This enormous difference is due to the three times higher mass of the B meson leading to more complicated, higher multiplicity states. However, we know from inclusive measurements that more than 90% of B mesons decay to a DX (87.6%), ΨX (1.15%), or ΛX (6.4%). It will be interesting if we are able to measure the charged track multiplicity of the X part of $B \to DX$ final states, to learn whether the Standard Model with QCD is able to describe (predict) these multiplicities. Measuring the same for $B \to D^*X$ (and also for $B \to D\pi^o X$ and $B \to D^*\pi^o X$) would be very useful. If our 15 fb^{-1} data sample is large enough (?), we should be able to complete this over the next two years.

References

1. Observation of a new group of resonances at 10 GeV decaying to muons pairs
   S.W. Herb et al., Phys. Rev. Lett. $\underline{39}, 252(1977)$.

2. Confirmation of the existence of the $\Upsilon(1S)$ and $\Upsilon(2S)$
   C. Berger et al., Phys. Lett. $\underline{76B}, 243(1978)$.
   C.W. Darden et al., Phys. Lett. $\underline{78B}, 246(1978)$.
   J.K. Bienlein et al., Phys. Lett. $\underline{78B}, 360(1978)$.
   C.W. Darden et al., Phys. Lett. $\underline{80B}, 419(1979)$.

3. Observation of Three Upsilon States
   D. Andrews et al., Phys. Rev. Lett. $\underline{44}, 1108(1980)$.
   T. Bohringer et al., Phys. Rev. Lett. $\underline{44}, 1111(1980)$.

4. Observation of a Fourth Upsilon State in e^+e^- Annihilations
   D. Andrews et al., Phys. Rev. Lett. $\underline{45}, 219(1980)$.
   G. Finnocchiaro et al., Phys. Rev. Lett. $\underline{45}, 222(1980)$.

5. Collaborations studying B meson decays:
   CLEO, ARGUS, ALEPH, DELPHI, OPAL, SLD, CDF

6. M.E.Nelson et al., Phys. Rev. Lett. $\underline{50}, 1542(1983)$.
   E.Fernandaz et al., Phys. Rev. Lett. $\underline{50}, 2054(1983)$.

7. The branching fractions quoted in this paper are approximate values. Most of these numbers have been checked for consistency with the Review of Particle Physics, The European Physical Journal, $\underline{15}, 1-4(2000)$.

Table 1, The CLEO Detectors Used to Study B Meson Decay

CLEO has been studying B Decay since 1980

Detector		$\Upsilon(4S)$ Luminosity
CLEO_I	1980 - 1989	295pb^{-1}
CLEO_II	1990 - 1995	3136pb^{-1}
CLEO_II.V	1996 - 1998	6064pb^{-1}
CLEO_III	2000 - 2001	$\sim$ 6000pb^{-1}

Table 2, Status of Our understanding of B Meson Decays

$b \to cW^-$	$\sim (92 - 95)\%$
$b \to uW^-$	$\sim (1.2 - 2)\%$
$b \to sg$	??
$b \to s\gamma$	$\sim (3 \times 10^{-2})\%$

Table 3, B Meson Semi-Leptonic Decays $\quad \bar{B} \to l^- \bar{\nu}_l X$

(From inclusive measurements of Semi-Leptonic Decays)

The leptons, l,

$l^- = e^-$	10.5%
$l^- = \mu^-$	10.4%
$l^- = \tau^-$	$\sim$ 3% (from theory)
Total 'measured'	10.5 + 10.4 + 3.0 = 24%

X, the hadronics, for	B^-	$\bar{B}^o$
$X = D$	2.15%	2.10%
$X = D^*$	5.3%	4.6%
$X = D^{**}$	0.6%	??
$X = \pi$	??	$(1.8 \times 10^{-2})\%$
$X = \rho$	??	$(2.6 \times 10^{-2})\%$
Total measured	$\sim$ 8%	$\sim$ 7%

Table 4. Two Body B Meson Hadronic Decays

From $b \to c W^-$, for	$B^- \to D^{(*)o}X^-$	$\bar{B}^o \to D^{(*)+}X^-$
$D\pi$	$(5.3 \pm 0.5) \times 10^{-3}$	$(3.0 \pm 0.4) \times 10^{-3}$
DK	$(2.9 \pm 0.8) \times 10^{-4}$	??
$D\rho$	$(1.34 \pm 0.18) \times 10^{-2}$	$(7.9 \pm 1.4) \times 10^{-3}$
$D^*\pi$	$(4.6 \pm 0.4) \times 10^{-3}$	$(2.76 \pm 0.21) \times 10^{-3}$
$D^*\rho$	$(1.55 \pm 0.31) \times 10^{-2}$	$(6.8 \pm 3.4) \times 10^{-3}$
DD_s	$(1.3 \pm 0.4) \times 10^{-2}$	$(0.8 \pm 0.3) \times 10^{-2}$
DD_s^*	$(0.9 \pm 0.4) \times 10^{-2}$	$(1.0 \pm 0.5) \times 10^{-2}$
D^*D_s	$(1.2 \pm 0.5) \times 10^{-2}$	$(0.96 \pm 0.34) \times 10^{-2}$
$D^*D_s^*$	$(2.7 \pm 1.0) \times 10^{-2}$	$(2.0 \pm 0.7) \times 10^{-2}$
D^*D^*	??	$(6.2^{+4.1}_{-3.1}) \times 10^{-4}$
Total measured	$(10.0 \pm 1.3)\%$	$(6.9 \pm 1.0)\%$

From $b \to c W^-$, for	$B^- \to \Psi^o X^-$	$\bar{B}^o \to \Psi^o X^o$
$\Psi(1S)\bar{K}$	$(1.0 \pm 0.1) \times 10^{-3}$	$(0.89 \pm 0.12) \times 10^{-3}$
$\Psi(1S)\bar{K}^*$	$(1.48 \pm 0.27) \times 10^{-3}$	$(1.50 \pm 0.17) \times 10^{-3}$
$\Psi(1S)\pi$	$(0.051 \pm 0.015) \times 10^{-3}$	??
$\Psi(2S)\bar{K}$	$(0.58 \pm 0.10) \times 10^{-3}$	??
$\Psi(2S)\bar{K}^*$	??	$(0.93 \pm 0.23) \times 10^{-3}$
$\chi_{c1}(1P)\bar{K}$	$(1.0 \pm 0.4) \times 10^{-3}$	??
Total measured	$(0.41 \pm 0.05)\%$	$(0.33 \pm 0.03)\%$

From $b \to uW^-$ and/or $b \to sg$ for	B^-	$\bar{B}^o$
$\pi^o\pi^-$, $\pi^+\pi^-$	??	$(4.3^{+1.6}_{-1.4}) \times 10^{-6}$
$\pi^o K^-$, $\pi^+ K^-$	$(1.2^{+0.4}_{-0.3}) \times 10^{-5}$	$(1.7^{+0.4}_{-0.3}) \times 10^{-5}$
$\pi^- K^o$, $\pi^o K^o$	$(1.8^{+0.5}_{-0.5}) \times 10^{-5}$	$(1.4^{+0.7}_{-0.6}) \times 10^{-5}$
$\eta' K^-$, $\eta' K^o$	$(6.5 \pm 1.7) \times 10^{-5}$	$(4.7^{+2.8}_{-2.2}) \times 10^{-5}$
ωK^-, ωK^o	$(1.5^{+0.7}_{-0.6}) \times 10^{-5}$	??
γK^{*-}, γK^{*o}	$(5.7 \pm 3.3) \times 10^{-5}$	$(4.0 \pm 1.9) \times 10^{-5}$
Total measured	$(0.017 \pm 0.004)\%$	$(0.012 \pm 0.004)\%$

Table 5. Multi-Body B Meson Hadronic Decays

Three Body Hadronic Final States
From $b \to c W^-$, for

$B^- \to D^{(*)}X$		$\bar{B}^o \to D^{(*)}X$	
$D^o \pi^- \rho^o$	$(4.2 \pm 3.0) \times 10^{-3}$	$D^+ \pi^- \rho^o$	$(1.1 \pm 1.0) \times 10^{-3}$
$D^{*o} \pi^- \pi^o$	??	$D^{*+} \pi^- \pi^o$	$(1.5 \pm 0.5) \times 10^{-2}$
$D^{*+} \pi^- \pi^-$	$(2.1 \pm 0.6) \times 10^{-3}$	$D^{*o} \pi^+ \pi^-$	??

$B^- \to \Psi X^-$		$\bar{B}^o \to \Psi X^o$	
$\Psi(1S) K^- \pi^o$	??	$\Psi(1S) K^- \pi^+$	$(1.2 \pm 0.6) \times 10^{-3}$

$B^- \to N\bar{N}'X$		$\bar{B}^o \to N\bar{N}'X$	
$\Lambda_c^+ \bar{P} \pi^-$	$(6.2 \pm 2.7) \times 10^{-4}$	$\Lambda_c^+ \bar{N} \pi^-$	??

Total measured $(0.69 \pm 0.31)\%$ $\qquad$ $(1.7 \pm 0.5)\%$

==

Four Body Hadronic Final States
From $b \to c W^-$, for

$B^- \to D^{(*)}X$		$\bar{B}^o \to D^{(*)}X$	
$D^o \pi^- \pi^- \pi^+$	$(5 \pm 4) \times 10^{-3}$	$D^+ \pi^- \pi^- \pi^+$	$(8.0 \pm 2.5) \times 10^{-3}$
$D^{*o} \pi^- \pi^- \pi^+$	$(9.4 \pm 2.6) \times 10^{-3}$	$D^{*+} \pi^- \pi^- \pi^+$	$(6.8 \pm 3.4) \times 10^{-3}$
$D^{*+} \pi^- \pi^- \pi^o$	$(1.5 \pm 0.7) \times 10^{-2}$	$D^{*o} \pi^- \pi^+ \pi^o$	??

$B^- \to \Psi X^-$		$\bar{B}^o \to \Psi X^o$	
$\Psi(1S) K^- \pi^+ \pi^-$	$(1.4 \pm 0.6) \times 10^{-3}$	$\Psi(1S) K^- \pi^+ \pi^o$	??
$\Psi(2S) K^- \pi^+ \pi^-$	$(1.9 \pm 1.2) \times 10^{-3}$	$\Psi(1S) K^- \pi^+ \pi^o$	??

$B^- \to N\bar{N}'X$		$\bar{B}^o \to N\bar{N}'X$	
$\Lambda_c^+ \bar{P} \pi^- \pi^o$	??	$\Lambda_c^+ \bar{P} \pi^+ \pi^-$	$(1.3 \pm 0.6) \times 10^{-3}$

Total measured $(3.3 \pm 0.9)\%$ $\qquad$ $(1.6 \pm 0.4)\%$

==

Five Body Hadronic Final States
From $b \to c W^-$, for

$B^- \to D^{(*)}X$		$\bar{B}^o \to D^{(*)}X$	
$D^o \pi^- \pi^- \pi^+ \pi^o$	$(0.41 \pm 0.09)\%$	$D^+ \pi^- \pi^- \pi^+ \pi^o$	$(0.28 \pm 0.06)\%$
$D^{*o} \pi^- \pi^- \pi^+ \pi^o$	$(1.8 \pm 0.4)\%$	$D^{*+} \pi^- \pi^- \pi^+ \pi^o$	$(1.72 \pm 0.28)\%$
$D^{*o} \pi^- \pi^- \pi^+ \pi^o$	??	$D^{*o} \pi^- \pi^+ \pi^- \pi^+$	$(0.3 \pm 0.1)\%$

Total measured $(2.2 \pm 0.4)\%$ $\qquad$ $(2.3 \pm 0.3)\%$

Table 6, B Meson Branching Fractions

	B^-	$\bar{B}^0$
Inclusive Semi-Leptonic		
electrons, e^-	$(10.9 \pm 0.6)\%$	$(10.2 \pm 0.6)\%$
muons, μ^-	$(10.8 \pm 0.6)\%$	$(10.1 \pm 0.6)\%$
tau, τ^-	$(3.1 \pm 0.6)\%$	$(2.9 \pm 0.6)\%$
	$(24.7 \pm 1.1)\%$	$(23.1 \pm 1.1)\%$
Two Body Hadronic Final States with D or D^* or D^{**}		
	$(10.0 \pm 1.3)\%$	$(6.9 \pm 1.0)\%$
Two Body Hadronic Final States with $c\bar{c}$. (i.e., Ψ's)		
	$(0.41 \pm 0.05)\%$	$(0.33 \pm 0.03)\%$
Three Body Hadronic Final States		
	$(0.69 \pm 0.31)\%$	$(1.7 \pm 0.5)\%$
Four Body Hadronic Final States		
	$(3.3 \pm 0.8)\%$	$(1.6 \pm 0.4)\%$
Five Body Hadronic Final States		
	$(2.2 \pm 0.4)\%$	$(2.3 \pm 0.3)\%$
Total, $b \to c$, Inclusive Semi-Leptonic Decays plus Exclusive Hadronics Decays		
	$(41.3 \pm 1.8)\%$	$(35.9 \pm 1.7)\%$

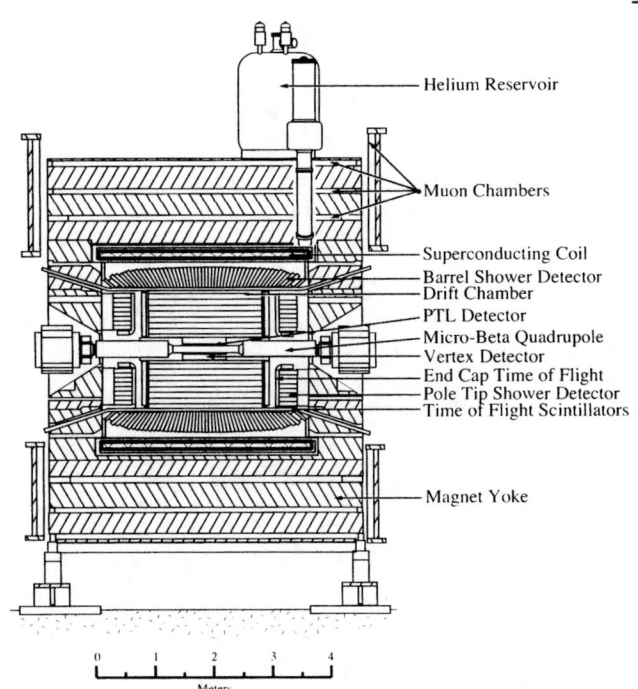

FIGURE 1. The CLEO II Detector

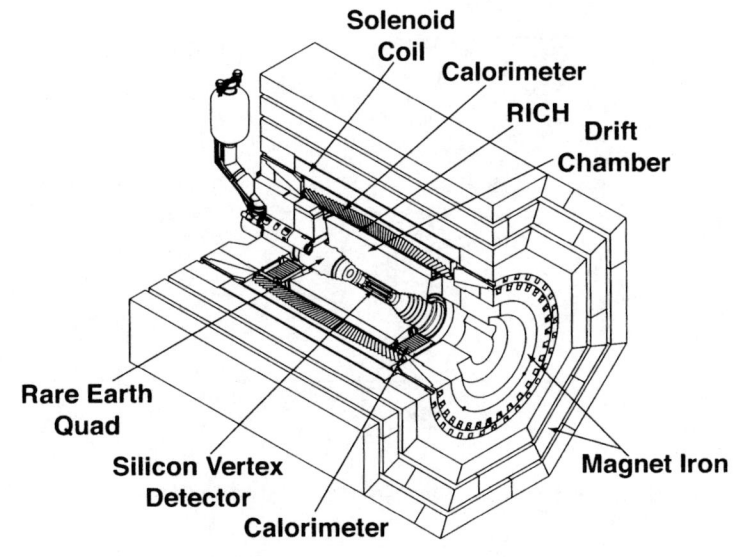

FIGURE 2. The CLEO III Detector

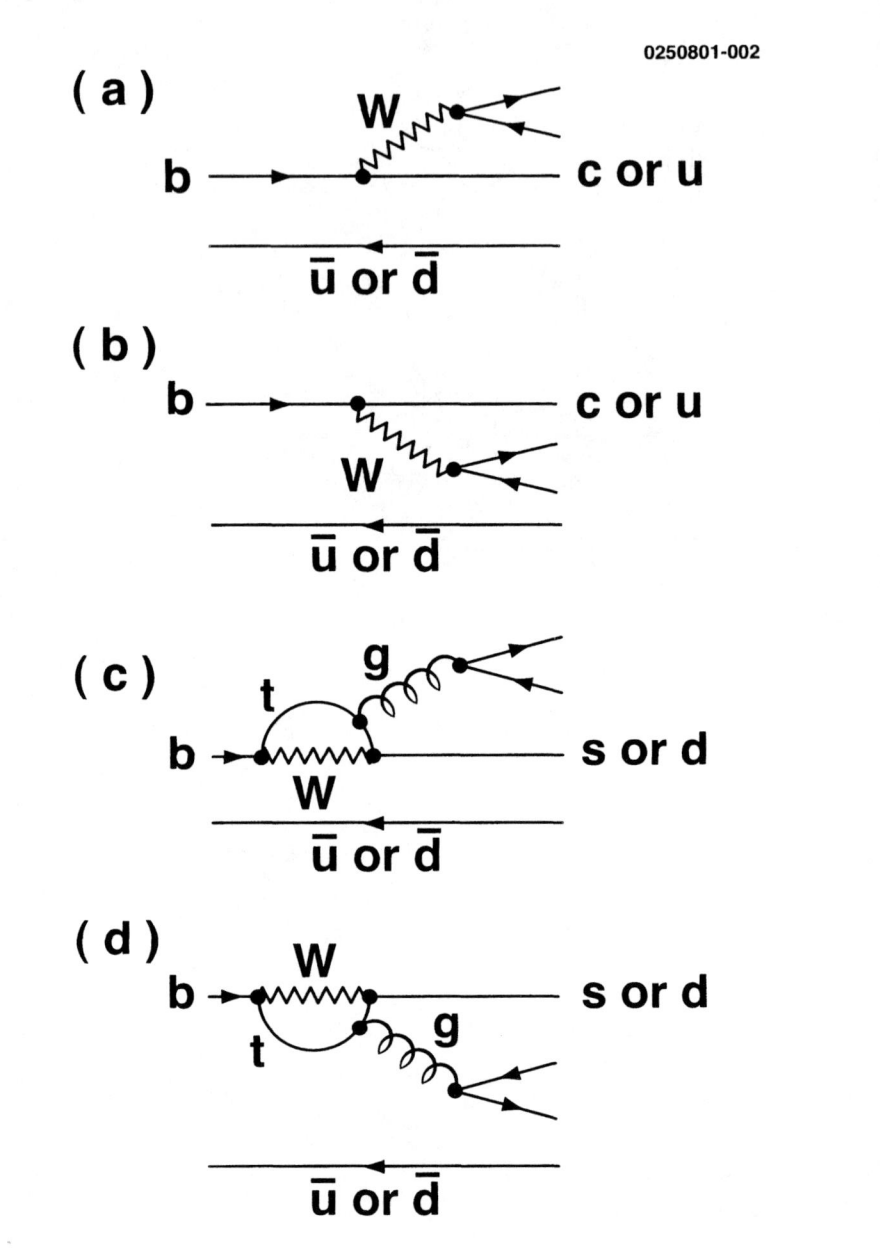

FIGURE 3. Feynman Diagrams describing B Meson decays

Charmonium Spectroscopy From Fermilab E835

Amiran Tomaradze[1]

Department of Physics, Northwestern University, Evanston, IL 60208

Abstract. The Fermilab experiment E835 is dedicated to the study of charmonium spectroscopy via direct formation of charmonium resonances in proton-antiproton annihilations. The latest results from this experiment and their comparison to the world data are presented. The physics implications are discussed.

INTRODUCTION

Precision spectroscopy of the charmonium system can yield quantitative results concerning the nature of QCD in the non–perturbative regime, where it is not well understood. The mass of the charm quark is small enough that $c\bar{c}$ cross sections are reasonably large, yet it is large enough that its motion within charmonium is nearly non–relativistic. The spectrum of charmonium states is shown in Fig. 1.

From the discovery of J/ψ in 1974 until the mid–1980's, nearly all charmonium spectroscopy was studied via e^+e^- annihilations at facilities like SLAC and DESY. Because e^+e^- annihilations proceeds via a virtual photon, only $J^{PC} = 1^{--}$ states could be formed directly. The study of other states is possible only through the observation of radiative decays of $J/\psi, \psi' \rightarrow (c\bar{c})_R \gamma$. Thus mass and width measurements for these states were limited by the photon energy resolutions of the detector system. While most of the charmonium states were discovered in such experiments, their masses, and especially widths, were not well measured. In order to do precision spectroscopy of charmonium, a new technique was required. Such a technique, charmonium via $p\bar{p}$ annihilations, was pioneered by R704 at CERN [1]. Fermilab E760 and E835 have fully exploited this technique, and have obtained results which represent improvements in precision, in some cases by an order of magnitude.

The advantage of the technique of charmonium formation in $p\bar{p}$ annihilations is that all charmonium states can be formed directly; the $p\bar{p}$ annihilation process occurs

---

[1] On behalf of the Fermilab E835 Collaboration: G. Garzoglio, K. Gollwitzer, M. Hu, S. Pordes, S. Werkema (Fermi National Accelerator Laboratory), W. Baldini, D. Bettoni, R. Calabrese, G. Cibinetto, P. Dalpiaz, E. Luppi, M. Negrini, G. Stancari, M. Stancari (INFN and University of Ferrara), S. Bagnasco, A. Buzzo, M. LoVetere, M. Macrì, M. Marinelli, M. Pallavicini, C. Patrignani, E. Robutti, A. Santroni (INFN and University of Genova), G. Lasio, M. Mandelkern, J. Schultz, W. Roethel (University of California at Irvine), M. Graham, R. Rusack, S. H. Seo, T. Vidnovic III (University of Minnesota), D. Joffe, J. Kasper, Z. Metreveli, J. Rosen, P. Rumerio, K. K. Seth, A. Tomaradze, P. Zweber (Northwestern University), G. Borreani, R. Cester, F. Marchetto, E. Menichetti, R. Mussa, M. Obertino, N. Pastrone (INFN and University of Torino).

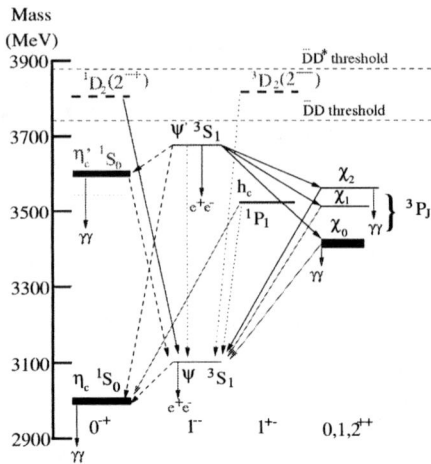

FIGURE 1. The spectrum of charmonium states.

through two or three gluons in the intermediate state, which may have any value of J^{PC}. Since every charmonium state can be formed directly, its resonance parameters may be determined with a precision dependent only on the knowledge of the beam energy and energy distribution, and event statistics.

FERMILAB EXPERIMENT E835

The E835 experiment is located in the Fermilab Antiproton Accumuliator (see Fig. 2). The circulating beam of antiprotons intersects a hydrogen jet target to produce $p\bar{p}$ interactions. During Fermilab fixed target operations, the Accumulator stores antiprotons at 8.9 GeV, and decelerates them to the momentum required for the resonant formation of the $c\bar{c}$ state under investigation.

The stochastically cooled $\bar{p}$ beam consists of $\sim 10^{12}$ $\bar{p}$ which orbit the 474.0454 m path length of the accumulator with frequency of ~ 0.62 MHz. This frequency is measured using the Schotty noise spectrum of the beam to about 2 parts in 10^7. The beam intersects the gas jet target [2], whose density in the interaction region can be varied between 10^{12} and 10^{14} atoms/cm^3, in order to keep the instantaneous luminosity nearly constant (typically, 2×10^{31} cm^{-2}sec^{-1}).

The center of mass energy $\sqrt{s}$ is determined by the revolution frequency and the orbit length, which is measured accurate to ± 7 mm by a calibration scan at the ψ', whose mass is 3685.96$\pm$0.09 MeV [3]. Deviations of the beam from this reference orbit length are measured by beam position monitors to ± 1 mm, so that $\sqrt{s}$ may be known to $\sim \pm 50$ keV. The beam momentum spread, which is directly related to the width of the Schotty noice spectrum, is $\Delta p/p \sim 1-2 \times 10^{-4}$. This Δp corresponds to the center of mass energy spread $\sigma(\sqrt{s}) \approx 200$ keV at the ψ'.

The E760 detector system, which is optimized for the detection and identification

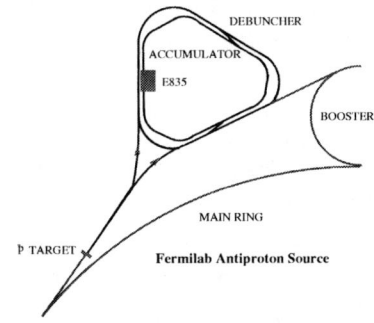

FIGURE 2.

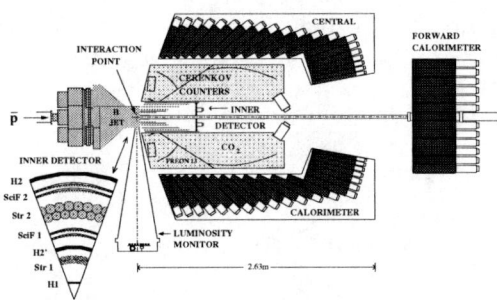

FIGURE 3. Schematic of the E835 detector system.

of photons and electrons, has been described before [4–6]. The E835 detector system (See Fig. 3) is an upgrate of the E760 detector [7]. The apparatus has full acceptance in azimuth (ϕ), with a cylindrical central system covering the polar angle $11° < \theta < 70°$, and a planar forward system covering $3° < \theta < 12°$. The central system contains three azimuthally segmented scintillator hodoscopes for triggering and measuring dE/dx, straw tubes for tracking in ϕ, scintillating-fiber trackers for tracking in θ, a 16 cell threshold gas Čerenkov counter for electron identification, and a 1280 element lead-glass calorimeter (CCAL) for measuring the direction and energy of photons and electrons. The forward system consists of a scintillator hodoscope and a 144 element lead-glass array (FCAL).

The luminosity monitor [8] counts proton recoils near $90°$ from elastic $p\bar{p}$ scatters and provids absolute luminosity with a statistical precision of 0.1% and systematic uncertainty of $\pm 2.5\%$.

In the E835 experiment, charmonium states are studied by decelerating the $\bar{p}$ beam so that $\sqrt{s}$ is varied in small steps across the resonance under study. At each step, the yield of the final state particles from the electromagnetic decays is measured. The number of events divided by luminosity taken at each point, plotted versus the center of mass energy, gives the resonance excitation curve. This is just the convolution of the measured

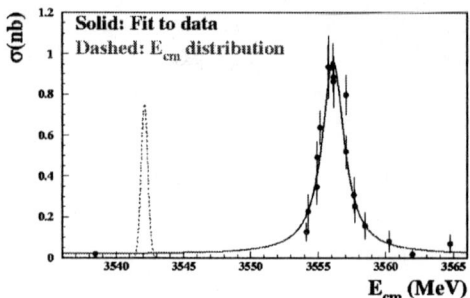

FIGURE 4. Example of the scan of χ_{c2} state. The solid curve is a fit to the acceptance and efficiency–corrected data, while the dotted curve shows the beam energy spread.

beam energy distribution function and the Breit–Wigner resonance cross section (which is proportional to the product of the branching ratios for the formation and the decay of the resonance, $B_{in} \times B_{out}$). Fig. 4 shows an example of such an excitation curve from the scan of χ_{c2} in E760 [4].

DETECTION of $(c\bar{c})_R$ via $p\bar{p} \to (c\bar{c})_R \to e^+e^-X$

The almost background free channels for the investigation of $(c\bar{c})$ states in E835 are those which include a single high–mass e^+e^- pair. These events are triggered by a high mass pair of clusters in CCAL, assocoated with two tracks in the inner charged tracking detectors and at least one associated signal in the Čerenkov detector.

Measurement of χ_{c0} resonance parameters via $p\bar{p} \to \chi_{c0} \to J/\psi\gamma \to (e^+e^-)\gamma$ detection

Our new measurements of the χ_{c0} from the year 2000 run [9] are based on a total luminosity of 32.8 pb^{-1}. Three cluster events consisting of an e^+e^- pair and a photon, and having a fit probability meeting the $J/\psi\gamma$ hypothesis $\geq 0.1\%$, lead to a very clean signal of χ_{c0}, as shown in Fig. 5 (left). The resonance parameters are determined to be
$M(\chi_{c0}) = 3415.4 \pm 0.4 \pm 0.2$ MeV,
$\Gamma_{tot}(\chi_{c0}) = 9.8 \pm 1.0 \pm 0.1$ MeV, and
$B_{p\bar{p}}(\chi_{c0}) = (4.1 \pm 0.3^{+1.5}_{-0.9}) \times 10^{-4}$.
Althought the hadron helicity conservation rule for massless quarks forbids the formation of zero spin states in $p\bar{p}$ annihilations, our value of $B_{p\bar{p}}(\chi_{c0})$ is actually about four times larger than $B_{p\bar{p}}(\chi_{c1})$ and $B_{p\bar{p}}(\chi_{c2})$. The values for the mass and width of the χ_{c0} from this and previous experiments are shown in Fig. 5 (right). The present results represent a significant improvement in precision for both mass and width.

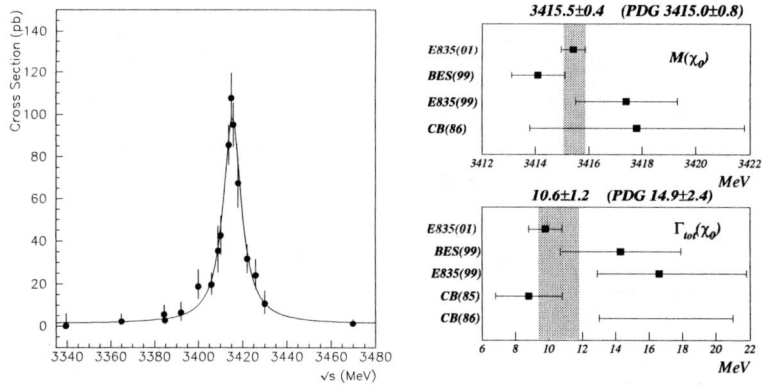

FIGURE 5. (left) $J/\psi\,\gamma$ cross section in the χ_{c0} mass region; (right) the mass and width of the χ_{c0} from the present and other experiments; the new world average values are shown as shaded bands and are compared with the value in PDG–2000 [3].

Angular distribution analysis for $\chi_{c1,c2} \to J/\psi\gamma \to (e^+e^-)\gamma$

The electron gamma angular correlation for the reaction $\chi_{c1,c2} \to J/\psi\gamma \to (e^+e^-)\gamma$ was measured using the data collected in 1996-97 run. The radiative transitions are dominantly $E1$ nature, with small $M2$ components allowed. The fractional magnetic quadrupole amplitudes $a_2 = M2/E1$ are found to be [10]:

$a_2(\chi_{c1})=0.002\pm0.032$, $\quad a_2(\chi_{c2})=-0.093\pm0.040$.

While the $a_2(\chi_{c2})$ is in agreement with what is expected, the near zero value of $a_2(\chi_{c1})$ is surprising.

Branching ratios from the Reactions $p\bar{p} \to \psi' \to J/\psi\gamma\gamma$

During both 1996–97 and 2000 runs of E835 the ψ' resonance was scanned. The nearly 24 pb^{-1} of ψ' data have been analyzed for the reaction $\psi' \to J/\psi\gamma\gamma$ to determine $B_{J/\psi\eta}(\psi')$, $B_{J/\psi\pi^0}(\psi')$ and $B_{J/\psi\gamma}(\chi_{c0,c1,c2})$. In Fig. 6 (left) the Dalitz plot for $M(\gamma\gamma)$ versus $M(J/\psi\gamma)$ is shown. Appropriate cuts in this plot determine the desired branching ratios. As an example, the projection for $M(J/\psi\gamma_{high})$ is shown in Fig. 6 (right). The χ_{c1} and χ_{c2} resonances are well resolved, while χ_{c0} is barely visible.

The data are still being analyzed.

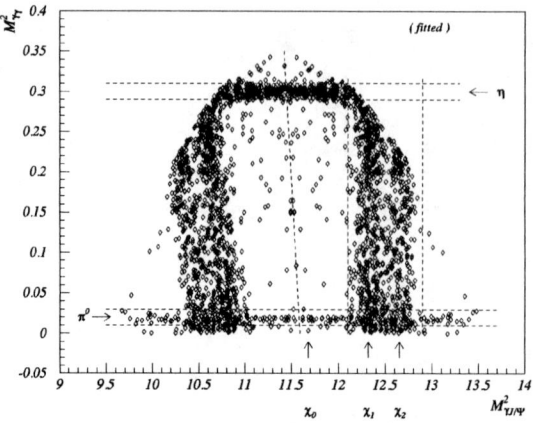 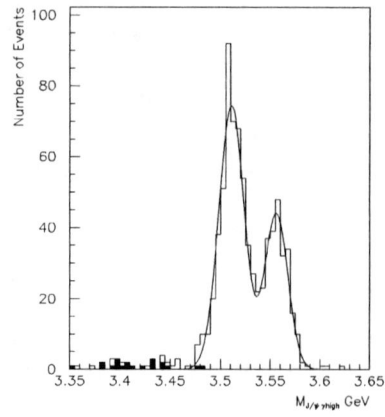

FIGURE 6. (left) the Dalitz plot, $M^2(\gamma\gamma)$ versus $M^2(J/\psi\gamma)$, in reaction $\psi' \to J/\psi\gamma\gamma$; (right) the invariant mass of J/ψ and fast γ, the $\chi_{c1}(3510)$ and $\chi_{c2}(3556)$ resonances are well resolved.

DETECTION of $(c\bar{c})_R$ via $p\bar{p} \to (c\bar{c})_R \to \gamma\gamma$

The two photon event selection used in our measurements has been described in detail in two recent publications [11,12].

Measurement of two–photon width of χ_{c2} and χ_{c0}

In 1993 E760 reported a measurement of $\Gamma_{\gamma\gamma}(\chi_{c2}) = 0.321 \pm 0.078 \pm 0.054$ [13] which was smaller than all previous measurements and upper limits by factors 5 to 30. Most of the previous determinations of the $\gamma\gamma$ partial width of χ_{c2} came from $\gamma\gamma$-fusion experiments and all such measurements appeared to be consistent with each other. It was therefore considered important to measure $\Gamma_{\gamma\gamma}(\chi_{c2})$ again.

A measurement of $\Gamma_{\gamma\gamma}(\chi_{c2})$ was made in E835 with approximately $8.4 pb^{-1}$ of integrated luminosity near the χ_{c2} peak [12], compared to $\sim 2.6 pb^{-1}$ in E760. The cross sections for $p\bar{p} \to \gamma\gamma$ in this region are shown in Fig. 7 (left). Since the total width $\Gamma_{tot}(\chi_{c2}) = 1.98 \pm 0.18$ MeV, the $p\bar{p}$ branching fraction $B_{p\bar{p}}(\chi_{c2}) = 1.0 \pm 0.1 \times 10^{-4}$ and the mass $M(\chi_{c2}) = 3556.16 \pm 0.14$ MeV were each well measured in E760, the data were analyzed using these values as constants. The new result obtained is

$\Gamma_{\gamma\gamma}(\chi_{c2}) = 0.270 \pm 0.049 \pm 0.033$ keV.

Thus the present measurement confirms the E760 result, and has nearly factor two smaller errors. Our measurement of $\Gamma_{\gamma\gamma}(\chi_{c2})$ is shown for comparison with other experimental results in Fig. 7 (right). It is encouraging to note that the CLEO results appear to be converging to our results, although their errors remain much larger.

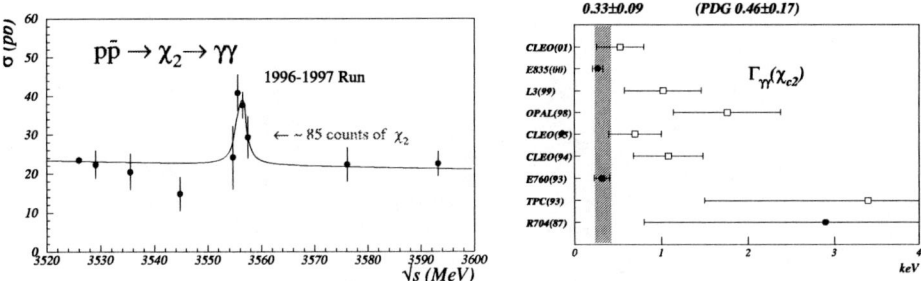

FIGURE 7. (left) γγ cross section in the χ_{c2} mass region; (right) the two photon partial width of the χ_{c2} from the present and other experiments. The new world average value is shown as shaded band and is compared with the value in PDG–2000 [3].

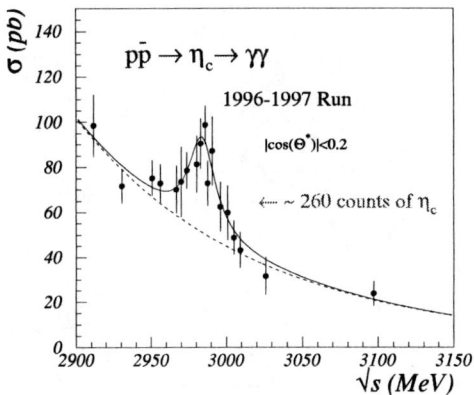

FIGURE 8. γγ cross section in the η_c mass region.

No statistically significant signal for χ_{c0} was found in the reaction $p\bar{p} \to \chi_{c0} \to \gamma\gamma$ in the E835 data collected in the 1996–97 run. We set a 95% upper limit of $\Gamma_{\gamma\gamma}(\chi_{c0}) \leq 3.47$ keV. We have collected ~10 times more data at χ_{c0} peak in the year 2000 run, which should lead to more definite results on $\Gamma_{\gamma\gamma}(\chi_{c0})$. We note, in the meantime, that CLEO has reported $\Gamma_{\gamma\gamma}(\chi_{c0}) = 3.8 \pm 0.7 \pm 0.4 \pm 1.7$ keV [14].

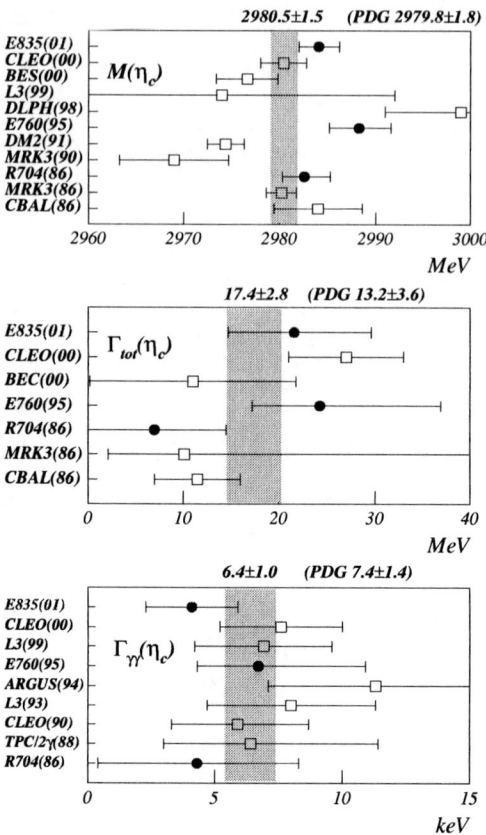

FIGURE 9. The mass, width, and two photon partial width of the η_c from the present and other experiments. The new world average values are shown as shaded bands and are compared with the values in PDG–2000 [3].

Measurement of η_c resonance parameters

The E760 experiment reported on the reaction $p\bar{p} \to \eta_c \to \gamma\gamma$ in 1995 [15]. The most significant results of these measurements, in which a total luminocity of 3.5 pb^{-1} was invested, were that the width of η_c was nearly twice as large ($\Gamma_{tot}(\eta_c) = 23.9^{+12.6}_{-7.1}$ MeV) as the average of previous measurements ($\Gamma_{tot}(\eta_c) = 10.3^{+3.8}_{-3.4}$ MeV [16]), and that $\Gamma_{\gamma\gamma}(\eta_c) = 6.7^{+2.4}_{-1.7} \pm 2.4$ keV.

The E835 experiment has made a new study of the $p\bar{p} \to \eta_c \to \gamma\gamma$ reaction with a much larger integrated luminosity of 24.4 pb^{-1}. The measured cross section are shown in Fig. 8.

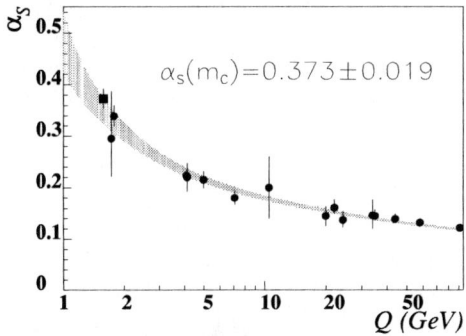

FIGURE 10. Measurements of $\alpha_s(Q)$ from various experiments. The shaded band indicates the evaluation of the PDG average value of $\alpha_s(m_Z)$ down to $Q=1$ GeV.

The resonance was parametrized by the Breit–Wigner formula with M, Γ_{tot} and $B_{in} \times B_{out}$ (or $B_{in} \times \Gamma_{out}$) as fit parameters. The background evaluation was done in a two ways:
• parametrized to a power law curve $A(\sqrt{s})^B$.
• using 'feeddown calculation': the shape of background was evaluated starting from $p\bar{p} \to \pi^0\gamma$ and $p\bar{p} \to \pi^0\pi^0$ cross sections, and taking account of one and two missing photons. The two methods give consistent results:
$M(\eta_c) = 2984.1 \pm 2.1$ MeV,
$\Gamma_{tot}(\eta_c) = 21.6 \pm 7.5$ MeV, and
$\Gamma_{\gamma\gamma}(\eta_c) = 4.1 \pm 1.3 \pm 1.4$ keV.
These results are preliminary. The value of mass, width and two photon partial width of the η_c from this and the other experiments are shown in Fig. 9. It should be noted that except for the BES result, all recent measurements give $\Gamma_{tot}(\eta_c) \approx 20$–$25$ MeV, which is significantly different from the current PDG average of 13.2 ± 3.6 MeV [3].

Determination of α_s

We can determine the strong coupling constant $\alpha_s(m_c)$ by our measurements of the total and partial widths of η_c and χ_{c2}. The pQCD formulas with first order radiative corrections give:

$$\frac{\Gamma_{gg}(\eta_c)}{\Gamma_{\gamma\gamma}(\eta_c)} = \frac{9\alpha_s^2}{8\alpha^2} \times \frac{(1+4.8\alpha_s/\pi)}{(1-3.4\alpha_s/\pi)}$$

since $\Gamma_{gg}(\eta_c) \approx \Gamma_{tot}(\eta_c)$, we obtain $\alpha_s = 0.33 \pm 0.05$.

$$\frac{\Gamma_{gg}(\chi_{c2})}{\Gamma_{\gamma\gamma}(\chi_{c2})} = \frac{9\alpha_s^2}{8\alpha^2} \times \frac{(1-2.2\alpha_s/\pi)}{(1-5.3\alpha_s\pi)}$$

since $\Gamma_{gg}(\chi_{c2}) \approx \Gamma_{tot}(\chi_{c2}) - \Gamma_{J/\psi\gamma}(\chi_{c2})$, we get $\alpha_s = 0.38 \pm 0.02$.

We note, however, that both determination are strongly affected by the first order radiative corrections, the determination at η_c much more so than determination at χ_{c2}.

In Fig. 10 we show the average value of $\alpha_s(m_c)$ determined from our data with the other experimental results.

SUMMARY

E760 made precision measurements of the resonance parameters of several charmonium resonances. E835 has continued the program, with
- First observation of χ_{c0} in $p\bar{p}$ annihilations and precision measurements of mass, total width and $p\bar{p}$ partial width of χ_{c0}.
- Branching ratios from the analysis of ψ' decay into $J/\psi\gamma\gamma$ final states.
- Angular correlation analysis for the $\chi_{c1} \to J/\psi\gamma$ and $\chi_{c2} \to J/\psi\gamma$ decays.
- Branching ratios and two photon partial widths for χ_{c0} and χ_{c2}.
- Precision measurements of mass, total width and two photon partial width of η_c.

ACKNOWLEDGMENTS

The E835 Collaboration wishes to acknowledge the important contribution made by the Antiproton Source Department of the Fermilab Beams Division in making the E835 experiment successful. We also acknowledge the support of the U.S. Department of Energy and the Italian Istituto Nazionale di Fisica Nucleare. The speaker wishes to thank Prof. A. Zaitsev and all organizers of the conference for their hospitality and assistance. Special thanks to Prof. K. Seth for his help in preparation of the talk and manuscript.

REFERENCES

[1] C. Baglin et al., Nucl. Phys **B286**, 592 (1987).
[2] D. Allspach et al., Nucl. Inst. Meth. A**410**, 195 (1998).
[3] D.E. Groom *et al.*, PDG–2000, The Eur. Phys. Jour. C**15**, 653, (2000).
[4] T. A. Armstrong et al., Nucl. Phys **B373**, 35 (1992).
[5] L. Bartozek et al., Nucl. Inst. Meth, **A301** 47 (1991).
[6] C. Biino et al., Nucl. Inst. Meth, **A317** 135 (1992).
[7] M. Ambrogiani et al., Phys. Rev. D **60**, 032002 (1999).
[8] S. Trokenheim et al., Nucl. Inst. Meth. A**355**, 308 (1995).
[9] S. Bagnasco et al., submitted to Phys. Rev. Lett.
[10] M. Ambrogiani et al., Phys. Rev. D. Accepted for publication.
[11] M. Ambrogiani et al., Phys. Rev. D **64**, 052003 (2001).
[12] M. Ambrogiani et al., Phys. Rev. D **62**, 052002 (2000).
[13] T.A. Armstrong et al., Phys. Rev. Lett. **70**, 2988 (1993);
[14] B.I. Eisenstein et al., Phys. Rev. Lett. **87**, 061801 (2001);
[15] M. Ambrogiani et al., Phys. Rev. D **52**, 4839 (1995).
[16] PDG–1994, Phys. Rev. D **49**, 1 (1994).

Aspects of Charmonium Physics

S.F. Tuan

*Department of Physics, University of Hawaii at Manoa
Honolulu, HI 96822-2219, U.S.A.*

Abstract. I review possible resolution of the $J/\psi(\psi') \to \rho - \pi$ puzzle based on two inputs: the relative phase between the one- photon and the gluonic decay amplitudes, and a possible hadronic excess in the inclusive nonelectromagnetic decay rate of ψ'. The status of a universal large phase here is examined for its meaning and implications (including those for B-physics). Since the future of tau/charm facility(s) are again under consideration together with a future anti-proton facility at GSI, I propose to extend my review to include a broader discussion of charmonium physics. Outstanding questions like the status of the $^1P_1$ state of charmonium, measuring $D^0 - \bar{D}^0$ mixing and relative strong phases, status of molecular P(S) - wave charmonia will also be discussed amongst others.

The status of the famous $J/\psi(\psi') \to \rho - \pi$ puzzle has been summarized well [1] elsewhere, including the pertinent experimental features. By the time of the Hadron'99 Conference in Beijing it was known [2] that almost all theoretical models aiming to explain the puzzle, were found to be inadequate. Gu and Li [3] stressed that a key premise for physical considerations is to establish whether the J/ψ decays or the $\psi(2S)/\psi'$ decays are anomalous. This is particularly relevant relative to a more recent model [4] where it was proposed that the $J/\psi \to \rho\pi$ decay is *enhanced* (rather than $\psi' \to \rho\pi$ being suppressed) by mixing of the J/ψ with light-quark states, notably ω and ϕ. Arguments against the model are given by Rosner [1] include the appearance of certain unsuppressed light-quark decay modes of the ψ' and the lack of evidence for helicity suppression in J/ψ decays involving a single virtual photon. Indeed an amplitude analysis made for the two-body decays of J/ψ to VP [5] has shown that nothing anomalous is found in the magnitudes of the three-gluon and one-photon amplitudes. Hence those arguments presupposing the J/ψ as the origin of the puzzle anomaly should be disregarded, including another model presented at Hadron'99 [6].

In order to search for a clue to solve the $\rho\pi$ puzzle, Suzuki [7] argued with conviction that we must locate the source of the problem before offering a final solution of the problem. In other words we must first identify the correct inputs towards resolution. Two threads have been exposed clearly since Hadron'99 which may eventually lead to a solution of the $\rho\pi$ puzzle. Motivated by the results in the amplitude analyses [5,7] of the two-body decays of J/ψ, the following two input postulates seem eminently reasonable. (1) The relative phases between the gluon and the photon decay amplitudes are *universally* large (close to $\pi/2$) for all two body decays of J/ψ. The photon decay amplitudes are predominantly real and consequently the gluon decay amplitudes are imaginary. The same pattern holds for $\psi(2S)$ decay as well. (2) A possible hadronic excess in the inclusive nonelectromagnetic decay rate of $\psi(2S)$.

The large phase assumption (1) is based on the amplitude analysis of the J/ψ decay where the relative phase of the gluonic and the one-photon decay amplitude is close to $90°$ for all two-body decay channels so far studied [7]: 1^-0^-, 0^-0^-, 1^-1^-, 1^+0^-, and $N\bar{N}$. The appropriate references are given in ref. [7], to which we add the almost model independent work of Achasov and Gubin [8] on large (nearly $90°$) relative phase between the one-photon and the three-gluon decay amplitudes in $J/\psi \to \rho\eta$ and $\omega\eta$ decays. Of particular interest is that the nucleon-antinucleon FSI large phase is based purely on *experiment* of the FENICE Collaboration. As an historical note it was pointed out to me by Achasov [9] that more than 10 years ago MARK III [10] and DM2 [11] teams as well as numerous theoretical users missed a top level result by failing to decode their data on the strong evidence (if not discovery) of a large (near $90°$) phase between the $J/\psi \to \rho\eta$ (one-photon) and the $\omega\eta$ (three-gluon) amplitudes! Assumption (2) is based on experimental information relevant to the issue. That is the hadronic decay rate of $\psi(2S)$ which is normally attributed to $\psi(2S) \to ggg$. When we compute the current data for the inclusive gluonic decay rate of $\psi(2S)$ by subtracting the cascade and the electromagnetic decay rate from the total rate, it is 60-70% larger, within experimental uncertainties, than what we expect from short-distance (i.e. computable from perturbative QCD) gluonic decay alone - the celebrated 12-14% rule. Smaller errors have been attached to this discrepancy [3] with a different error estimate, but the conclusion remains basically the same.

On the matter of 'universal large phases', this was argued recently by Gérard and Weyers [12] in their model. Some of the experimental and theoretical difficulties for $\rho-\pi$ puzzle resolution via this model have already been pointed out by Gu and Tuan [2]. While agreeing that there could be a more universal nature to the conclusion that the photon and three-gluon amplitudes for J/ψ are out of phase with one another, Rosner [13] found their argument that the three-gluon and photon final states are orthogonal somewhat curious. They can populate the same hadronic final states (with the same spin and parity). Nothing then in principle would prevent a real relative phase between a three-gluon and a one-photon amplitude. Thus the argument (based on incoherence) for this large relative phase leaves something to be desired. Nevertheless, without over emphasizing the large universal phase, Rosner [14] pointed out that in the charm decays themselves, even the $\rho - K$ final state had large relative phases among different amplitudes, despite the fact that these did not show up in the isospin triangle discussed earlier [15] as an aftermath of Suzuki's analysis [5].

In point of fact, the decay branching fractions of $\psi(2S) \to 1^-0^-$ clearly show a suppression of the gluon amplitude and favor a *small* relative phase between the gluon and photon amplitudes [7]. A small phase seems likely for $\psi(2S) \to 1^+0^-$ also. At first sight this seems a violation of assumption (1) above. However taking into account assumption (2) about the possible excess in the inclusive hadronic decay rate of $\psi(2S)$, it is proposed that this excess is related to both the suppression and the *small* relative phase of the 1^-0^- amplitude which would otherwise have the *large* relative phase of assumption (1). The proposition is then that an additional decay process generating the excess should largely cancel the short-distance gluon amplitude in the exclusive decay into 1^-0^- and that the resulting small residual amplitude is not only *real* but also destructively interferes with the photon amplitude. Remember [16] the picture is still that the three- gluon decay amplitudes of J/ψ and ψ' have large phases because the three gluons are on mass shell. In contrast, one-photon annihilation amplitudes are real because the photon is off shell. Suzuki [7] then examined two scenarios which may possibly generate the excess inclusive hadronic decay. They seem to be among a very few possibilities that have not yet been ruled out by experiment. We summarize these model scenarios in the sections below.

The $\psi(2S) \to$ *resonance* $\to$ *hadrons* is a twist of an old one: A noncharm resonance may exist near the $\psi(2S)$ mass and give an extra contribution to the hadronic decay rate. A glueball was proposed earlier at the J/ψ mass to boost the $\rho\pi$ decay of J/ψ [17]. However we now want it near $\psi(2S)$, not near J/ψ. Suzuki [7] argued against this resonance, call it R, being a $q\bar{q}$ or four-quark resonance. For a 1^{--} glueball, the lattice calculations [18] indeed support such a state at 3.7 GeV or even higher. However [16] it is always nice to get a physical picture from some analytic calculation. Lattice calculation may be fine but only after one gets a reasonable quantitative understanding without it. It is reassuring that a very recent work by Hou et al. [19] using the constituent gluon model estimate the mass of 1^{--} glueball to be 3.1 - 3.7 GeV, which is close to the mass of the J/ψ and ψ'. We look into the possibility that R around the $\psi(2S)$ mass destructively interferes with the perturbative $\psi(2S) \to ggg \to 1^-0^-$ decay. Note unlike the glueball proposed near J/ψ mass to enhance $J/\psi \to \rho\pi$ [17] and hence the unnatural property that it decays predominantly into 1^-0^-, in our case, since glueball R is introduced to account for the hadronic excess, it should couple not primarily to the 1^-0^- channels, but to many other channels and hence is quite natural. If R has total width Γ_R is as narrow as 100 MeV, for instance, the mixing $|\epsilon| = O(10^{-2})$ would be able to account for the excess in the inclusive hadron decay of $\psi(2S)$. It was also argued [7] that the amplitude for $\psi(2S) \to R \to$ hadrons has automatically a large phase when mass difference between R and $\psi(2S)$ is smaller than Γ_R. Hence the resonant amplitude can interfere strongly with the three-gluon amplitude in two-meson decays. However Hou [20] has argued that the glueball could be significantly narrower than 100 MeV using the "[OZI]$^{1/2}$" rule [21] that a typical glueball R total width is $\Gamma_R = [\Gamma_{ordinary} \times \Gamma_{OZI-violating}]^{1/2}$. In terms of $\psi(2S)$ where $\Gamma_{ordinary}$ could be generously up to 500 MeV in width at this mass, while $\Gamma(\psi(2S))$ is about 277 KeV, the square root formula gives for a gluonium R degenerate with $\psi(2S)$ a width of order 11.8 MeV. Using Meshkov's [21] benchmark, something like 35.4 MeV would be a conservative estimate. It would be interesting for BES to proceed from their 3.96 million $\psi(2S)$ to search for indirect evidence of R via scanning across $\psi(2S)$ for shape distortion, but even the setting of a limit on the gluonium width will be of interest. We need of course to work with unsuppressed $\psi(2S) \to PA$ modes ($\pi b_1(1235)$, $K_1(1270)\bar{K}$) or even the unsuppressed SV mode $f_0\phi$ of $\psi(2S)$. If the glueball is significantly narrower than 100 MeV, the mixing parameter ϵ must be adjusted accordingly. The adjustment depends on how closely $\psi(2S)$ and the glueball R are degenerate. How much this affects the $\psi(2S)$ resonance shape depends sensitively on how closely $\psi(2S)$ and the glueball are degenerate. To observe a distortion of the resonance shape of $\psi(2S)$, they would have to be degenerate nearly to the $\psi(2S)$ width, 277 KeV!

The $\psi(2S) \to D\bar{D} \to$ hadrons model [7] has been much expanded by Rosner [22] very recently. He noted that if a ψ' decay amplitude due to coupling to virtual (but nearly on shell) charmed particle pairs interferes destructively with the standard three-gluon amplitude, the suppression of these (and possibly other) modes in ψ' final states can be understood. However Rosner goes further and noted effects of the proximity of the $D\bar{D}$ threshold can mix the ψ' and the $\psi(3770)$. Perhaps the missing partial width of $\psi' \to \rho\pi$ (less than half a KeV) is showing up in the $\psi(3770)$. If so, since the latter state has a total width nearly 100 times that of the ψ', it would correspond to a very tiny branching ratio yet to be detected. As pointed out [7] the $D\bar{D}$ amplitude needs to have a large final-state phase, in order to interfere destructively with the perturbative 3g contribution in the $\rho\pi$ and $K\bar{K}^*(892)$ + c.c. channels. If this new contribution is due to rescattering into non-charmed final states through charmed particle pairs, it is exactly the type of contribution proposed by many and quoted in Ref. [22], in which the decay $\bar{b} \to \bar{c}c\bar{s}$ or $\bar{b} \to \bar{c}cd$ contributes to the penguin amplitude *with a large phase*. The rescattering of $D\bar{D}$ states into non-charmed final states (like $\rho - \pi$) could also be responsible for the larger-than-expected penguin amplitude in B decays about which the Rome people [23] have been writing about, and for the large $B \to K\eta'$ branching ratio. If the phase is large, as needed for a suppression of $\psi' \to \rho\pi$, this again would be good news for the observation of a large CP asymmetry in B decays with both tree and penguin contributions [15]. Hence there is cautious optimism that the amplitude for $\psi(2S) \to D\bar{D} \to mesons$ can have the requisite large size and phase to help resolve the $\rho - \pi$ puzzle.

Remarks: (a) It must be recognized that both models above, resonance R or $D\bar{D}$ scattering, involves long distance effects not easily computable, unlike the short distance perturbative QCD. (b) The situation for $\psi(2S) \to VT$ (suppressed), $\psi(2S) \to PA$ (unsuppressed for $\pi b_1(1235)$, $K_1(1270)\bar{K}$), and $\psi(2S) \to SV$ (unsuppressed for $f_0\phi$) exclusive channels deserve further study as the first step (we recognize other anomalies, e.g. anomalous enhancement of $\psi(2S) \to K_1(1270)\bar{K}$, different isospin violations in $K^*\bar{K}$ decays for J/ψ and $\psi(2S)$, flavor-SU(3)-violating $K_1(1270) - K_1(1400)$ asymmetries with opposite character for J/ψ and $\psi(2S)$, which need to be taken up later). At present we have no way to relate among two-body charmonium decay amplitudes of different spin-parities. Probably we shall not have one for a long time. This is a weak point of the argument of "long-distance physics" in contrast to the short-distance argument such as "helicity suppression" of perturbative QCD. It would be good to have some systematic "long-distance" argument which covers two-body decays of different spin-parities in one shot! (c) The decay angular distributions for 1^-0^- and 0^-0^- should be tested. A large interference can occur only when the dynamical mechanisms of the two processes are similar. When a large disparity is observed between the corresponding two-meson decay rates of J/ψ and $\psi(2S)$, the decay angular distribution of this channel will also be very different between J/ψ and $\psi(2S)$. This will give a good test of the idea of interference with an additional amplitude. However in the PP and VP decays, the decay angular distribution is unique, independent of dynamics, because there is only one relative orbital angular momentum involved. That is not the case for the decays such as VT: the final orbital angular momentum can take different values (0, 2, or 4 for VT). This actually opens up the possibility to test Suzuki's [7] dynamical assumptions by concentrating on PP decays especially. As pointed out [3] since $\psi(2S) \to PV$ strong decay is much suppressed, measurement of its decay angular distribution would be extremely difficult. However $\psi(2S) \to PP$ is already giving tantalizing hint of constructive interference in (large) decay rates [7] and should be checked in decay angular distribution. (d) Unlike for the charmonium case where hadron helicity conservation (HHC) and perturbative QCD (PQCD) is of doubtful validity [24], HHC/PQCD at the $\Upsilon(nS)$ mass are more immune from breakdown. Mechanisms for possible breakdown at $\psi(nS)$ energies are minimized at the heavier $b\bar{b}$ mass. For $\Upsilon(nS)$ exclusive decays to light hadrons, violations are of order m_h/m_Q (light hadron mass/heavy quark mass) in matrix element; or of order $(m_h/m_b)^2$ if the light hadron has intrinsic $b\bar{b}$ content in decays [16]. The $c\bar{c}$ contribution from light hadrons in upsilon exclusive decays should also be not very important, such as for $\Upsilon(nS) \to \rho\pi$, since the probability of the intrinsic bottom in the ρ and π is suppressed by $(m_c/m_b)^2$ relative to the intrinsic charm probability. This is shown rigorously using the operator product expansion OPE [25]. Both these corrections are negligible for $\Upsilon(nS)$. Nevertheless it would be of interest to test $\Upsilon(nS) \to \omega - \pi^0$ (I=1 electromagnetic) decay strength, to reassure us of the absence of long distance effects at $\Upsilon(nS)$. (e) At a deeper level Kochelev [26] has advanced the instanton approach (shared by M.A. Shifman) for understanding the $\rho - \pi$ puzzle. He noted that if one consider the decay $J/\psi(\psi(2S)) \to ggg$ then the average virtuality of each gluon is approximately $(2m_c/3)^2 \approx 1 GeV^2$. It is difficult to believe that at such virtuality one can apply perturbative QCD. He has estimated the direct instanton contribution to nucleon sum rules [27] where he showed that at scale $M_P^2 \approx 1 GeV^2$ the instanton contribution is very important. But the virtuality of each gluon for the $\Upsilon(nS)$ will be even larger. An interesting question is whether the relative phases are close to 90°, with photon decay amplitudes real and consequently the gluon decay amplitudes are imaginary for $\Upsilon(nS)$ as once conjectured by Suzuki [16].

Since the future of tau/charm facility(s) are again under consideration together with a future anti-proton facility at GSI, I shall very briefly list some broader aspects of charm/charmonium physics which can be accomplished when such facilities are available. (i) The study of $\psi(^1P_1)$ state should be vigorously pursued. Rosner's mixing solutions [22] are in tantalizing agreement with those of Kuang-Yan [28] and make the case for charmonium $^1P_1$ even more exciting. Other details are given elsewhere [29]. (ii) The measuring of $D^0 - \bar{D}^0$ mixing and relative strong phases at a Charm Factory has been discussed recently by Gronau et al. [30]. (iii) The Martenelli lattice group in Rome (with their more powerful computing facility) should be encouraged to reach a conclusion whether the Alford/Jaffe [31] prediction of a $J^{PC} = 0^{++}$ stable S-wave four-quark bound state, with non-exotic flavor quantum numbers, just below the threshold for $D\bar{D}$ in the charmonium spectrum, is sustainable or not. Though P-wave molecular charmonium states [32], because of centrifugal barrier and hence less quark wavefunction overlap for formation [33], should also be tested by the lattice group for sustainability.

I wish to thank my scientific colleagues Kolia Achasov, Jon Rosner, and Mahiko Suzuki for very helpful communications and discussions. This work was supported in part by the U.S. Department of Energy under Grant DE-FG-03- 94ER40833 at the University of Hawaii at Manoa.

REFERENCES

1. See for instance F.A. Harris, hep-ex/9903036, March 1999; J.Z. Bai et al., Phys. Rev. Lett. **83**, 1918 (1999); Y.F. Gu, Proc. of DPF '96, **Vol. 2**, p. 986 [World Scientific Publishing (1998)]; J.L. Rosner, see Ref. 22 below.
2. Y.F. Gu and S.F. Tuan, Nucl. Phys. **A675**, 404 (2000); S.F. Tuan, Commun. Theor. Phys. **33**, 285 (2000).
3. Y.F. Gu and X.H. Li, Phys. Rev. **D63**, 114019 (2001).
4. T. Feldmann and P. Kroll, Phys. Rev. **D62**, 074006 (2000).
5. M. Suzuki, Phys. Rev. **D57**, 5717 (1998).
6. C.T. Chan and W.S. Hou, Nucl. Phys. **A675**, 367(2000).
7. M. Suzuki, Phys. Rev. **D63**, 054021 (2001).
8. N.N. Achasov and V.V. Gubin, Phys. Rev. **D61**, 117504 (2000).
9. N.N. Achasov, private communication.
10. MARK III Collaboration, Phys. Rev. **D38**, 2695 (1988).
11. DM2 Collaboration, Phys. Rev. **D41**, 1389 (1990).
12. J.-M. Gérard and J. Weyers, Phys. Lett. **B462**, 324 (1999).
13. J.L. Rosner, private communication.
14. J.L. Rosner, Phys. Rev. **D60**, 114026 (1999).
15. J.L. Rosner, Phys. Rev. **D60**, 074029 (1999).
16. M. Suzuki, private communication.
17. S.J. Brodsky et al., Phys. Rev. Lett. **59**, 621 (1987).
18. M. Peardon, hep-lat/9710029; C. Morningstar and M. Peardon,hep-lat/9901004 v.2; C. Morningstar, these Proceedings.
19. W.S. Hou et al., Phys. Rev. **D64**, 014028 (2001).

20. W.S. Hou, private communication.
21. D. Robson, *Nucl. Phys.* **B130**, 328 (1977); S. Meshkov et al., *Phys. Lett.* **99B**, 353 (1981); S. Meshkov, CALT-68-923, UCI Technical Report 82/35 (1982).
22. J.L. Rosner, hep-ph/0105327.
23. M. Ciuchini et al., *Nucl. Phys.* **B501**, 271 (1997); M. Ciuchini, R. Contino et al., *Nucl. Phys.* **B512**, 3 (1998).
24. V. Chernyak, hep-ph/9906387.
25. S.J. Brodsky, private communication concerning the work of M. Franz, M.V. Polyakov, and K. Goeke, *Phys. Rev.* **D62**, 074024 (2000).
26. N. Kochelev, private communication.
27. N. Kochelev, *Zeit. Phys.* **C46**, 281 (1990).
28. Y.P. Kuang and T.M. Yan, *Phys. Rev.* **D41**, 155 (1990).
29. T. Barnes et al., UH-511-868-97, Ms. # 8921 at PLB.
30. M. Gronau et al., hep-ph/0103110.
31. M. Alford and R.L. Jaffe, hep-lat/0001023, MIT-CTP-2940.
32. A. De Rújula et al., *Phys. Rev. Lett.* **38**, 317 (1977); S.F. Tuan, *Phys. Lett.* **B473**, 136 (2000).
33. A. De Rújula and R.L. Jaffe, *Experimental Meson Spectroscopy* (1977), p. 83 [Published by Northeastern University Press, Boston, Mass 02115].

Analysis of nature of $\phi \to \gamma\pi\eta$ and $\phi \to \gamma\pi^0\pi^0$ decays

N.N. Achasov

Laboratory of Theoretical Physics, Sobolev Institute for Mathematics, Academician Koptiug prospekt, 4, 630090 Novosibirsk, Russia

Abstract. We study interference patterns in the $\phi \to (\gamma a_0 + \pi^0\rho) \to \gamma\pi\eta$ and $\phi \to (\gamma f_0 + \pi^0\rho) \to \gamma\pi^0\pi^0$ reactions. Taking into account the interference, we fit the experimental data and show that the background reaction does not distort the $\pi^0\eta$ spectrum in the decay $\phi \to \gamma\pi\eta$ everywhere over the energy region and does not distort the $\pi^0\pi^0$ spectrum in the decay $\phi \to \gamma\pi^0\pi^0$ when the invariant mass $m_{\pi^0\pi^0} > 670$ MeV. We discuss the details of the scalar meson production in the radiative decays and note that there are conclusive arguments in favor of the one-loop mechanism $\phi \to K^+K^- \to \gamma a_0$ (or γf_0). We discuss also distinctions between the four-quark, molecular, and two-quark models and argue that the establishment of the scalar meson production mechanism in the ϕ radiative decays gives new strong evidence in favor of the four-quark nature of the scalar $a_0(980)$ and $f_0(980)$ mesons.

INTRODUCTION

As was shown in a number of papers, see Refs. [1, 2, 3, 4, 5, 6] and references therein, the study of the radiative decays $\phi \to \gamma a_0 \to \gamma\pi\eta$ and $\phi \to \gamma f_0 \to \gamma\pi\pi$ can shed light on the problem of the scalar $a_0(980)$ and $f_0(980)$ mesons. These decays have been studied not only theoretically but also experimentally [7]. Present time data have already been obtained from Novosibirsk with the detectors SND [8, 9, 10, 11] and CMD-2 [12], which give the following branching ratios : $BR(\phi \to \gamma\pi\eta) = (0.88 \pm 0.14 \pm 0.09) \cdot 10^{-4}$ [10], $BR(\phi \to \gamma\pi^0\pi^0) = (1.221 \pm 0.098 \pm 0.061) \cdot 10^{-4}$ [11] and $BR(\phi \to \gamma\pi\eta) = (0.9 \pm 0.24 \pm 0.1) \cdot 10^{-4}$, $BR(\phi \to \gamma\pi^0\pi^0) = (0.92 \pm 0.08 \pm 0.06) \cdot 10^{-4}$ [12]. DAΦNE also confirms the Novosibirsk results [13].

These data give evidence in favor of the four-quark ($q^2\bar{q}^2$) [1, 14, 15, 16, 17, 18, 19, 20, 21, 22] nature of the scalar $a_0(980)$ and $f_0(980)$ mesons. Note that the isovector $a_0(980)$ meson is produced in the radiative ϕ meson decay as intensively as the well-studied η' meson involving essentially strange quarks $s\bar{s}$ ($\approx 66\%$), responsible for the decay.

As shown in Refs. [1, 3, 23], the background situation for studying the radiative decays $\phi \to \gamma a_0 \to \gamma\pi^0\eta$ and $\phi \to \gamma f_0 \to \gamma\pi^0\pi^0$ is very good. For example, in the case of the decay $\phi \to \gamma a_0 \to \gamma\pi^0\eta$, the process $\phi \to \pi^0\rho \to \gamma\pi^0\eta$ is the dominant background. The estimation for the soft, by strong interaction standard, photon energy, $\omega < 100$ MeV, gives $BR(\phi \to \pi^0\rho^0 \to \gamma\pi^0\eta, \omega < 100 \text{ MeV}) \approx 1.5 \cdot 10^{-6}$. The influence of the background process is negligible, provided $BR(\phi \to \gamma a_0 \to \gamma\pi^0\eta, \omega < 100 \text{ MeV}) \geq 10^{-5}$. In Sec. II we show that for the obtained experimental data the influence of the background processes is negligible everywhere over the photon energy region [19].

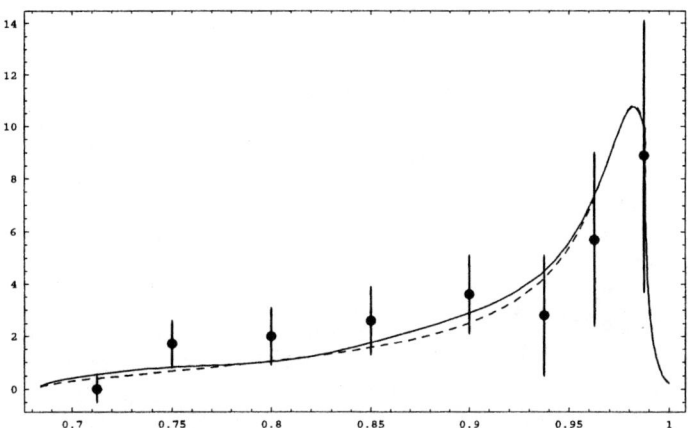

FIGURE 1. Fitting of $dBR(\phi \to \gamma\pi\eta)/dm \times 10^4 \text{GeV}^{-1}$ with the background is shown with the solid line, the signal contribution is shown with the dashed line.

The situation with $\phi \to \gamma f_0 \to \gamma\pi^0\pi^0$ decay is not much different. As was shown in [1, 3, 23] the dominant background is the $\phi \to \pi^0\rho^0 \to \gamma\pi^0\pi^0$ process with $BR(\phi \to \pi^0\rho^0 \to \gamma\pi^0\pi^0, \omega < 100 \text{ MeV}) \approx 6.4 \cdot 10^{-7}$. The influence of this background process is negligible, provided $BR(\phi \to \gamma f_0 \to \gamma\pi^0\pi^0, \omega < 100 \text{ MeV}) \geq 5 \cdot 10^{-6}$.

The exact calculation of the interference patterns between the decays $\phi \to \gamma f_0 \to \gamma\pi^0\pi^0$ and $\phi \to \rho^0\pi \to \gamma\pi^0\pi^0$ [19], which we present in Sec. III, shows that the influence of the background in the decay $\phi \to \gamma\pi^0\pi^0$ for the obtained experimental data is negligible in the wide region of the $\pi^0\pi^0$ invariant mass, $m_{\pi\pi} > 670$ MeV, or in the photon energy region $\omega < 300$ MeV.

In Sec. IV we discuss the mechanism of the scalar meson production in the radiative decays and show that experimental data obtained in Novosibirsk give the conclusive arguments in favor of the one-loop mechanism $\phi \to K^+K^- \to \gamma a_0$ and $\phi \to K^+K^- \to \gamma f_0$ of these decays [19]. We explain also why this circumstance gives new strong evidence in favor of the four-quark nature of the scalar $a_0(980)$ and $f_0(980)$ mesons.

INTERFERENCE BETWEEN THE REACTIONS $\phi \to \gamma a_0 \to \gamma\pi^0\eta$ AND $\phi \to \pi^0\rho^0 \to \gamma\pi^0\eta$

The fit [19] of the experimental data from the SND detector [10] is shown in Fig. 1. The total branching ratio, taking into account the interference, is $BR(\phi \to (\gamma a_0 + \pi^0\rho) \to \gamma\pi\eta) = (0.79 \pm 0.2) \cdot 10^{-4}$, the branching ratio of the signal is $BR(\phi \to \gamma a_0 \to \gamma\pi\eta) = (0.75 \pm 0.2) \cdot 10^{-4}$ and the branching ratio of the background is $BR(\phi \to \rho^0\pi^0 \to \gamma\pi^0\eta) = 3.43 \cdot 10^{-6}$. So, the integral part of the interference is negligible. The influence of the interference on the mass spectrum of the $\pi\eta$ system is also negligible, see Fig. 1.

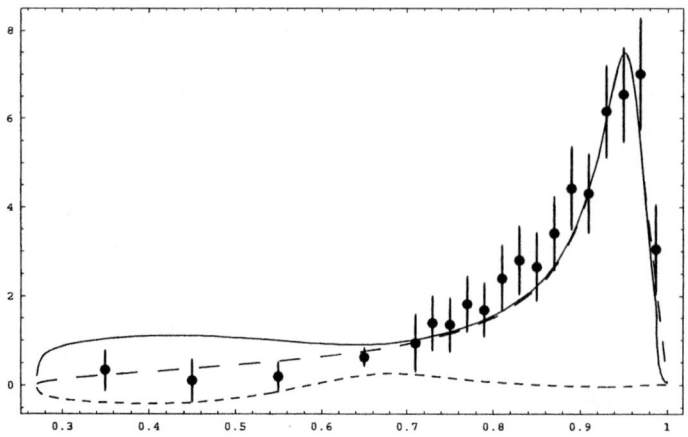

FIGURE 2. Fitting of $dBR(\phi \to \gamma\pi^0\pi^0)/dm \times 10^4 \text{GeV}^{-1}$ with the background is shown with the solid line, the signal contribution is shown with the dashed line. The dotted line is the interference term. The data are from the SND detector.

INTERFERENCE BETWEEN THE $e^+e^- \to \gamma f_0 \to \gamma\pi^0\pi^0$ AND $e^+e^- \to \phi \to \pi^0\rho \to \gamma\pi^0\pi^0$ REACTIONS

The fit [19] of the experimental data [11], obtained using the total statistics of SND detector, is shown in Fig. 2.

The total branching ratio, with interference being taken into account, is $BR(\phi \to (\gamma f_0 + \pi^0\rho) \to \gamma\pi^0\pi^0) = (1.26 \pm 0.29) \cdot 10^{-4}$, the branching ratio of the signal is $BR(\phi \to \gamma f_0 \to \gamma\pi^0\pi^0) = (1.01 \pm 0.23) \cdot 10^{-4}$, the branching ratio of the background is $BR(\phi \to \rho^0\pi^0 \to \gamma\pi^0\pi^0) = 0.18 \cdot 10^{-4}$.

One can see from Fig. 2 that the influence of the background process on the spectrum of the $\phi \to \gamma\pi^0\pi^0$ decay is negligible in the wide region of the $\pi^0\pi^0$ invariant mass, $m_{\pi\pi} > 670$ MeV, or when photon energy less than 300 MeV.

The difference from the experimental data, observed in the region $m_{\pi\pi} < 670$ MeV, is due to the fact that in the experimental processing of the $e^+e^- \to \gamma\pi^0\pi^0$ events the background events $e^+e^- \to \omega\pi^0 \to \gamma\pi^0\pi^0$ are excluded with the help of the invariant mass cutting and simulation, in so doing the part of the $e^+e^- \to \phi \to \rho\pi^0 \to \gamma\pi^0\pi^0$ events is excluded as well.

CONCLUSION

The experimental data give evidence [19] not only in favor of the four-quark model but in favor of the dynamical model suggested in Ref. [1], in which the discussed decays proceed through the kaon loop, $\phi \to K^+K^- \to \gamma f_0(a_0)$, see Fig. 3.

Indeed, according to the gauge invariance condition, the transition amplitude $\phi \to \gamma f_0(a_0)$ is proportional to the electromagnetic field strength tensor $F_{\mu\nu}$ (in our case to the

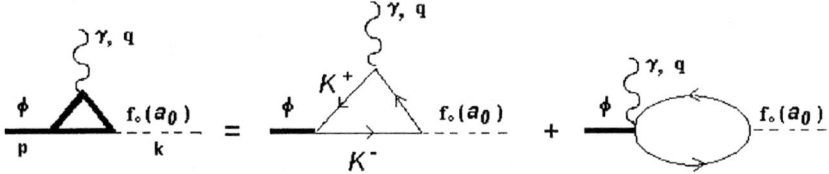

FIGURE 3. Diagrams of the K^+K^- loop model.

electric field). Since there are no pole terms in our case, the function $g(m)$ in

$$\frac{d\Gamma(\phi \to \gamma a_0 \to \gamma\pi\eta\, m)}{dm} = \frac{2}{\pi}\frac{m^2 \Gamma(\phi \to \gamma a_0(m)) \Gamma(a_0(m) \to \pi\eta)}{|D_{a_0}(m)|^2}$$

$$= \frac{2|g(m)|^2 p_{\eta\pi}(m_\phi^2 - m^2)}{3(4\pi)^3 m_\phi^3} \left|\frac{g_{a_0 K^+ K^-} g_{a_0 \pi\eta}}{D_{a_0}(m)}\right|^2 \quad (1)$$

and [1]

$$\frac{d\Gamma(\phi \to \gamma f_0 \to \gamma \pi^0 \pi^0\, m)}{dm}$$

$$= \frac{|g(m)|^2 \sqrt{m^2 - 4m_\pi^2}(m_\phi^2 - m^2)}{3(4\pi)^3 m_\phi^3} \left|\sum_{R,R'} g_{RK^+K^-} G^{-1}_{RR'} g_{R'\pi^0\pi^0}\right|^2 \quad (2)$$

is proportional to the energy of photon $\omega = (m_\phi^2 - m^2)/2m_\phi$ in the soft photon region. To describe the experimental spectra in Figs. 1 and 2, the function $|g(m)|^2$ should be smooth (almost constant) in the range $m \leq 0.99$ GeV, see Eqs. (1) and (2). Stopping the function ω^2 at $\omega_0 = 30$ MeV, using the form-factor of the form $1/(1 + R^2\omega^2)$, requires $R \approx 100$ GeV^{-1}. It seems to be incredible to explain the formation of such a huge radius in hadron physics. Based on the large, by hadron physics standard, $R \approx 10$ GeV^{-1}, one can obtain an effective maximum of the mass spectra under discussion only near 900 MeV. In the meantime, the K^+K^- loop, see Fig. 3, gives the natural description to this threshold effect, see Fig. 4.

To demonstrate the threshold character of this effect we present Fig. 5 and Fig.6 in which the function $|g(m)|^2$ is shown in the case of K^+ meson mass is 25 MeV and 50 MeV less than in reality.

One can see from Figs. 5 and 6 that the function $|g(m)|^2$ is suppressed by the ω^2 low in the region 950-1020 MeV and 900-1020 Mev respectively.

In the mass spectrum this suppression is increased by one more power of ω, see Eqs. (1) and (2), so that we cannot see the resonance in the region 980-995 MeV. The

---

[1] Eq. (2) takes into account the mixig of $f_0(980)$ meson with other scalar resonances, see Refs. [3, 19].

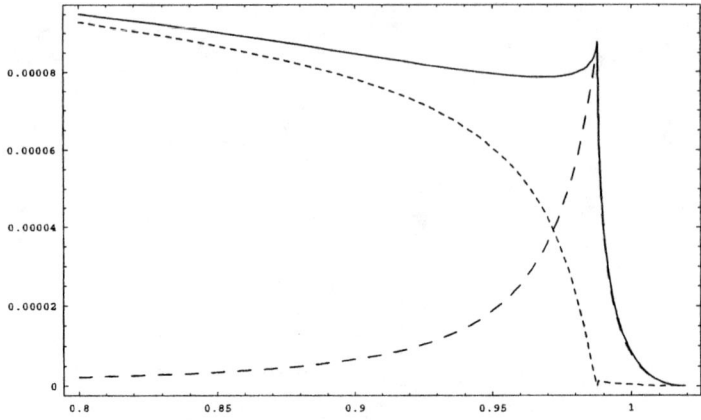

FIGURE 4. The function $|g(m)|^2$ is drawn with the solid line. The contribution of the imaginary part is drawn with the dashed line. The contribution of the real part is drawn with the dotted line.

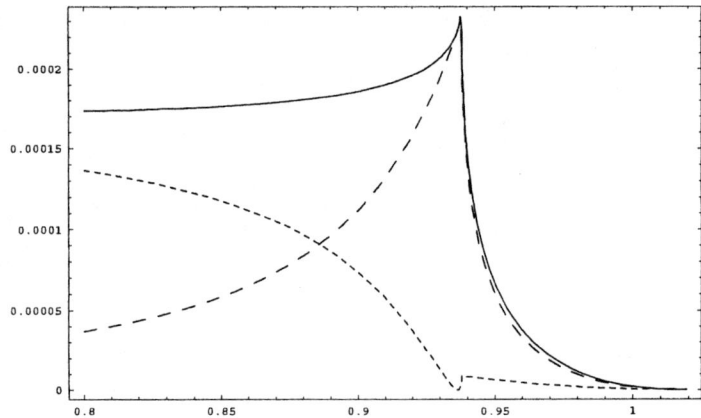

FIGURE 5. The function $|g(m)|^2$ for $m_{K^+} = 469$ MeV is drawn with the solid line. The contribution of the imaginary is drawn with the dashed line. The contribution of the real part is drawn with the dotted line.

maximum in the spectrum is effectively shifted to the region 935-950 MeV and 880-900 MeV respectively.

In truth this means that $a_0(980)$ and $f_0(980)$ resonances are seen in the radiative decays of ϕ meson owing to the K^+K^- intermediate state, otherwise the maxima in the spectra would be shifted to 900 MeV.

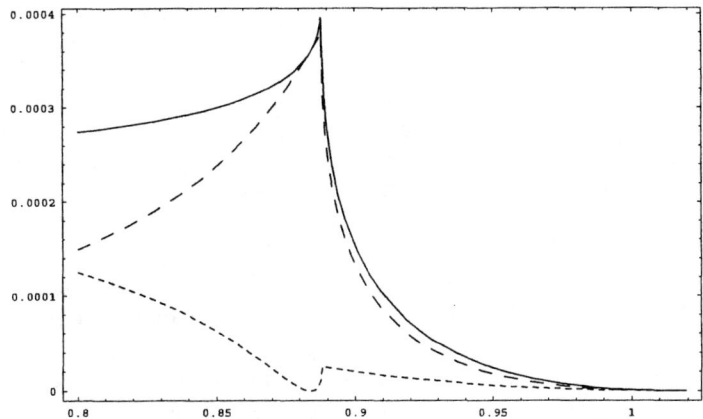

FIGURE 6. The function $|g(m)|^2$ for $m_{K^+} = 444$ MeV is drawn with the solid line. The contribution of the imaginary part is drawn with the dashed line. The contribution of the real part is drawn with the dotted line.

Thus the mechanism of production of the scalar mesons in the ϕ radiative decays is established at a physical level of proof. It is the rarest case in hadron physics. [2]

Both real and imaginary parts of the $\phi \to \gamma a_0(f_0)$ amplitudes are caused by the K^+K^- intermediate state. The imaginary parts are caused by the real K^+K^- intermediate state while the real parts are caused by the virtual compact K^+K^- intermediate state, that is, we are dealing here with the four-quark transition [3]. Needless to say radiative four-quark transitions can happen between two two-quark states as well as between two-quark and four-quark states but their intensities depend strongly on a type of the transitions. A radiative four-quark transition between two two-quark states requires creation of an additional $q\bar{q}$ pair, that is, such a transition is forbidden according to the Okuba-Zweig-Izuka (OZI) rule, while a radiative four-quark transition between two-quark and four-quark states happens without creation an additional $q\bar{q}$ pair, that is, such a transition is allowed according to the OZI rule.

Let us discuss this problem from two point of views: i) from point of view of intermediate states and ii) from point of view of the $1/N_c$ expansion.

i) It was noted already in paper [1] that the imaginary part of the K^+K^- loop is calculated practically in a model independent way making use of the coupling constants $g_{\phi K^+K^-}$ and $g_{a_0(f_0)K^+K^-}$ due to the Low's theorem [27] for the photons with energy $\omega < 100$ MeV which is soft by the standard of strong interaction.

In the same paper it was noted that the real part of the loop (with accuracy up to 20% in the width of the $\phi \to \gamma f_0(a_0)$ decay) is practically not different for the point-like particle and the compact hadron with form-factor which has the cutting radius in the

---

[2] It is worth noting that the K^+K^- loop model is practically accepted by theorists, compare, for example, Ref. [24] with Ref. [25]; true there is exception [26].

[3] It will be recalled that the imaginary part of every hadronic amplitude describes a multi-quark transition.

momentum space about the mass of ρ meson ($m_\rho = 0.77$ GeV).

In contrast to the four-quark state which is the compact hadron [14], the bound $K\bar{K}$ state is the extended state with the spatial radius $R \sim 1/\sqrt{m_K \varepsilon}$, where ε is the binding energy. Corresponding form-factor in the momentum space has the radius of the order of $\sqrt{m_K \varepsilon} \approx 100$ MeV for $\varepsilon = 20$ MeV, [28]. The more detailed calculation [2] gives for the radius in the momentum space the value $p_0 = 140$ MeV. As a result, the contribution of the virtual intermediate K^+K^- states in the K^+K^- loop is suppressed by the momentum distribution in the molecule, and the real part of the loop amplitude is negligible [4]. It leads to the branching ratio much less than the experimental one. In addition, the spectrum is much narrower in the $K\bar{K}$ molecule case that contradicts to the experiment, see the behavior of the imaginary part contribution in Fig. 4 and in corresponding figures in [4].

Of course, the two-quark state is as compact as four-quark one. The question arises, why is the branching ratio in the two-quark model suppressed in comparison with the branching ratio in the four-quark model? There are two reasons. First, the coupling constant of two-quark states with the $K\bar{K}$ channel is noticeably less [3, 15] and, second, there is the OZI rule that is more important really.

If the isovector $a_0(980)$ meson is the two-quark state, it has no strange quarks. Hence [1, 3, 17], the decay $\phi \to \gamma a_0$ should be suppressed according to the OZI rule. On the intermediate state level, the OZI rule is formulated as compensation of the different intermediate states [29, 30, 31]. In our case these states are $K\bar{K}$, $K\bar{K}^* + \bar{K}K^*$, $K^*\bar{K}^*$ and so on. Since, due to the kinematical reason, the real intermediate state is the only K^+K^- state, the compensation in the imaginary part is impossible and it destroys the OZI rule. The compensation should be in the real part of the amplitude only. As a result, the $\phi \to \gamma a_0$ decay in the two-quark model is mainly due to the imaginary part of the amplitude and is much less intensive than in the four-quark model [1, 3]. In addition, in the two-quark model, $a_0(980)$ meson should appear in the $\phi \to \gamma a_0$ decay as a noticeably more narrow resonance than in other processes, see the behavior of the imaginary part contribution in Fig. 4.

As regards to the isoscalar $f_0(980)$ state, there are two possibilities in the two-quark model. If $f_0(980)$ meson does not contain the strange quarks the all above mentioned arguments about suppression of the $\phi \to \gamma a_0$ decay and the spectrum shape are also valid for the $\phi \to \gamma f_0$ decay. Generally speaking, there could be the strong OZI violation for the isoscalar $q\bar{q}$ states (mixing of the $u\bar{u}$, $d\bar{d}$ and $s\bar{s}$ states) with regard to the strong mixing of the quark and gluon degree of freedom which is due to the nonperturbative effects of QCD [32]. But, an almost exact degeneration of the masses of the isoscalar $f_0(980)$ and isovector $a_0(980)$ mesons excludes such a possibility. Note also, the experiment points directly to the weak coupling of $f_0(980)$ meson with gluons, $B(J/\psi \to \gamma f_0 \to \gamma \pi\pi) < 1.4 \cdot 10^{-5}$ [33].

If $f_0(980)$ meson is close to the $s\bar{s}$ state [17, 34], there is no suppression due to the the OZI rule. Nevertheless, if $a_0(980)$ and $f_0(980)$ mesons are the members of the same multiplet, the $\phi \to \gamma f_0$ branching ratio, $BR(\phi \to \gamma \pi^0 \pi^0) = (1/3) BR(\phi \to \gamma \pi\pi) \approx 1.8 \cdot 10^{-5}$, is significantly less than that in the four-quark model, due to the relation between the coupling constants with the $K\bar{K}$, $\pi\eta$ and $K\bar{K}$, $\pi\pi$ channels inherited in the two-quark model, see Refs. [1, 3]. In addition, in this case there is no natural explanation of the a_0 and f_0 mass degeneration.

Only in the case when the nature of $f_0(980)$ meson in no way related to the nature of $a_0(980)$ meson (which, for example, is the four-quark state) the experimentally observed branching ratio of the $\phi \to \gamma f_0$ decay could be explained by $s\bar{s}$ nature of $f_0(980)$ meson. But, from the theoretical point of view, such a possibility seems awful [17].

ii) What is more, the OZI allowed transition is bound to have a small weight in the $1/N_c$ expansion in case of $s\bar{s}$ nature of $f_0(980)$ meson. Indeed, the main term of the $1/N_c$ expansion of the $\phi \to \gamma f_0$ amplitude, i.e, the OZI allowed transition, has the order of N_c^0 but does not contains the K^+K^- intermediate state. This state emerges only in the next to leading term of the $1/N_c$ expansion, i.e., in the OZI forbidden transition, which has the order of $1/N_c$.

If $f_0(980)$ meson is the two-quark state without the strange quarks, the $1/N_c$ expansion of the $\phi \to \gamma f_0(980)$ amplitude starts with the OZI forbidden transition of the order of $1/N_c$. But a weight of this term is bound to be small, because it does not contain the K^+K^- intermediate state, which emerges only in the next to leading term of the order of $1/N_c^2$.

In the two-quark model of $a_0(980)$ meson the $1/N_c$ expansion of the $\phi \to \gamma a_0(980)$ amplitude starts also with the OZI forbidden transition of the order of $1/N_c$, whose weight is bound to be small, because this term does not contain the K^+K^- intermediate state, which emerges only in the next to leading term of the order of $1/N_c^2$ too [4].

In the meantime, if $a_0(980)$ and $f_0(980)$ mesons are compact $K\bar{K}$ states, i.e., four-quark states, the $1/N_c$ expansions of the $\phi \to \gamma a_0(980)(f_0(980))$ amplitudes start with the OZI allowed transitions of the order of $N_c^{-1/2}$, which contain the K^+K^- intermediate state.

As we see, the knowledge of the mechanism of the scalar meson production in the ϕ radiative decays gives the new very strong (if not crucial) evidence in favor of the four-quark nature of the scalar $a_0(980)$ and $f_0(980)$ mesons.

ACKNOWLEDGEMENT

I thank very much Organizers for the invitation and the complete financial support.
I gratefully acknowledge the V.V. Gubin help too.
This work was supported in part by INTAS-RFBR, grant IR-97-232.

REFERENCES

1. N.N. Achasov and V.N. Ivanchenko, Nucl. Phys. B315, 465 (1989).
2. F.E. Close, N. Isgur and S. Kumano, Nucl. Phys. B389, 513 (1993).
3. N.N. Achasov and V.V. Gubin, Phys. Rev. D 56, 4084 (1997).
4. N.N. Achasov, V.V. Gubin and V.I. Shevchenko, Phys. Rev. D 56, 203 (1997).
5. J.L. Lucio, M. Napsuciale, Contribution to 3rd Workshop on Physics and Detectors for DAPHNE (DAPHNE 99), Frascati, Italy, 16-19 Nov 1999. hep-ph/0001136.

---

[4] It will be recalled that the OZI allowed $\phi \to \gamma \eta'$ amplitude has the order of N_c^0.

6. N.N. Achasov and V.V. Gubin, Phys. Rev. D 57, 1987 (1998).
7. M.N. Achasov, Plenary session, these Proceedings, hep-ex/0109035.
8. M.N. Achasov et al., Phys. Lett. B438, 441 (1998).
9. M.N. Achasov et al., Phys. Lett. B440, 442 (1998), hep-ex/9807016.
10. M.N. Achasov et al., Phys. Lett. B479, 53 (2000), hep-ex/0003031.
11. M.N.Achasov et al., Phys. Lett. B485, 349 (2000), hep-ex/0005017.
12. R.R. Akhmetshin et al., Phys. Lett. B462, 380 (1999).
13. Barbara Valeriani, Parallel session, these Proceedings.
14. R.L. Jaffe, Phys. Rev. D15, 267, 281 (1977).
15. N.N. Achasov, S.A. Devyanin and G.N. Shestakov, Usp. Fiz. Nauk. 142, 361 (1984) [Sov. Phys. Usp. 27, 161 (1984)].
16. N.N. Achasov and G.N. Shestakov, Usp. Fiz. Nauk. 161, 53 (1991)[Sov. Phys. Usp. 347, 471 (1991)].
17. N.N. Achasov, Usp. Fiz. Nauk. 168, 1257 (1998) [Phys. Usp. 41, 1149 (1998)]; hep-ph/9904223; Nucl. Phys. A 675, 279c (2000).
18. O. Black, A. Fariborz, F. Sannino, and J. Schechter, Phys.Rev. D 59, 074026 (1999).
19. N.N. Achasov and V.V. Gubin, Phys. Rev. D 63, 094007 (2001).
20. J. Schechter, Plenary session, these Proceedings.
21. S.F. Tuan, Parallel session, these Proceedings, hep-ph/0109191.
22. T. Teshima, Parralel session, these Proceedings.
23. A. Bramon, A. Grau, G. Pancheri, Phys.Lett. B 289, 97 (1992).
    A. Bramon, A. Grau, and G. Pancheri, Phys.Lett. B 283,416 (1992).
24. E. Marco, S. Hirenzaki, E. Oset, and H. Toki, Phys. Lett. B470, 20 (1999).
25. A. Bramon et al., hep-ph/0008188.
26. A.V. Anisovich, V.V. Anisovich, and V.A. Nikonov, hep-ph/0011191,
    Alexei Anisovich, Parralel session, these Proceedings.
    The authors of this paper use the amplitude of the $\phi \to \gamma(f_0 + \text{background}) \to \gamma \pi^0 \pi^0$ decay which does not vanish when $\omega \to 0$, i.e. which does not satisfy the gauge invariance condition. This amplitude is not adequate to the physical problem since the mass spectrum under discussion should have the behavior ω^3 at $\omega \to 0$ and not ω as in hep-ph/0011191. With the same result one can study the electromagnetic form-factor of π meson in the $e^+e^- \to \pi^+\pi^-$ reaction near the threshold considering that the cross-section of the process is proportional to the momentum of π meson while it is proportional to the momentum in the third power.
    Hereinafter the comment could not be included in Physical Review D 63, 094007 (2001)(published 30 March 2001) for the temporal reasons.
    To provide the spectrum behavior ω^3 at $\omega \to 0$ the authors, correcting some typos and undoing some references in hep-ph/0011191 v4 27 Mar 2001, inserted a crazy common factor $F_{thresh}(\omega) = \sqrt{1 - \exp\{-(\omega/36 \text{ MeV})^2\}}$ **in the** $\phi \to \gamma(f_0 + \text{background}) \to \gamma\pi^0\pi^0$ **amplitude without any explanations, see Eq. (39) in hep-ph/0011191. But the real trouble is that the calculation in hep-ph/0011191 is not gauge invariant. The calculation of the** $\phi \to q\bar{q} \to \gamma f_0$ **amplitude requires a gauge invariant regularization (for example, the substraction at $\omega = 0$) in spite of the integral convergence. A text-book example of such a kind is the** $\gamma\gamma \to e^+e^-(\text{or} q\bar{q}) \to \gamma\gamma$ **scattering. The authors of the paper under discussion obtained that** $A_{\phi \to \gamma f_0} = $ **(in our symbols)** $g(m)(g_{f_0 K^+ K^-}/e) \neq 0$ **at** $\omega = (m_\phi^2 - m^2)/2m_\phi = 0$ **($A_{\phi \to \gamma f_0}$ does not depend on m at all), see Eq. (30) in hep-ph/0011191. This means that the authors created the false pole in the invariant amplitude free from kinematical singularities:** $\left(eA_{\phi \to \gamma f_0}/\left(m_\phi^2 - m^2\right)\right) \times$ $(\phi_\mu p_\nu - \phi_\nu p_\mu)(\varepsilon_\mu q_\nu - \varepsilon_\nu q_\mu)$, compare with Eq. (9) in hep-ph/0011191. So, once again, the calculation of hep-ph/0011191 is not adequate to the physical problem!
27. F.E. Low, Phys. Rev. 110, 574 (1958).
28. Unfortunately, in the interesting paper V.E. Markushin, Eur. Phys.J., A8, 389 (2000), the potential in the momentum space was taken as the momentum distribution in the molecule instead of the wave function in the momentum space. But the momentum distribution radius of the potential is 5-8 times as large as one of the wave function, that was the reason for the misleading conclusion on the possibility to explain the Novosibirsk reasults in the molecule case.
29. H. Lipkin, Nucl. Phys. B 291, 720 (1987).
30. P. Geiger and N. Isgur, Phys. Rev. D 44, 799 (1991).

31. N.N. Achasov and A.A. Kozhevnikov, Phys. Rev. D 49, 27 (1994).
32. A.I. Vainshtein, V.I. Zakharov, V.A. Novikov and M.A. Shifman, Fiz. Elem. Chastits At. Yadra 13, 542 (1982) [Sov. J. Part. Nucl., 13, 224 (1982)].
33. G. Eigen, Proc. of the XXIV Int. Conf. on High Energy Phys., Munich, August 4-10, 1988 (Eds. R. Kotthaus and J.H. Kuhn) Session 4(Berlin: Springer-Verlag, 1988) p. 590.
34. N.A. Törnqvist, Z. Phys. C 68, 647 (1995).

EXOTICS

Search for exotic baryons with hidden strangeness in proton diffractive production at the energy of 70 GeV

Victor Kurshetsov

IHEP, Protvino, 142284, Russia

On behalf of the SPHINX Collaboration[1]

Abstract. First preliminary results from upgraded SPHINX spectrometer, working in the proton beam with the energy of 70 GeV of IHEP accelerator, are presented. The data for the reaction $p + N \rightarrow [\Sigma^0 K^+] + N$ based on a new statistics are in a good agreement with our previous data and strongly supports the existence of X(2000) state (with the increase of statistics for this state by a factor of ~ 5). We also observed radiative decay of $\Lambda(1520) \rightarrow \Lambda + \gamma$. The significant increase of statistics for many diffractive production reactions will allow us to study them in great detail.

EXOTIC BARYONS AND THEIR PRODUCTION PROCESSES

Extensive studies of the diffractive baryon production and search for cryptoexotic pentaquark baryons with hidden strangeness ($B_\phi = |qqqs\bar{s}>$, here $q = u, d$ quarks) are being carried out by the SPHINX Collaboration at IHEP accelerator. This program was described in detail in reviews [1, 2].

The cryptoexotic B_ϕ baryons do not have external exotic quantum numbers and their complicated internal valence quark structure can be established only indirectly, by examination of their dynamic properties which can be quite different from those for ordinary $|qqq>$ baryons. Examples of such anomalous features are listed below (see [1, 2] for more details):

1. The dominant OZI allowed decay modes of B_ϕ baryons are the ones with strange particles in the final state (for ordinary baryons such decays have branching ratios at the per cent level).

2. Cryptoexotic B_ϕ baryons can possess both large masses ($M > 1.8 - 2.0$ GeV) and narrow decay widths ($\Gamma \leq 50 - 100$ MeV). This is due to a complicated internal color structure of these baryons which leads to a significant quark rearrangement of color clusters in the decay process and due to a limited phase space for the

---

[1] Yu.M.Antipov, A.V.Artamonov, V.A.Batarin, O.V.Eroshin, S.V.Golovkin, Yu.P.Gorin, A.P.Kozhevnikov, V.P.Kubarovsky, V.F.Kurshetsov, L.G.Landsberg, V.A.Medovikov, V.V.Molchanov, V.A.Mukhin, D.I.Patalakha, S.V.Petrenko, A.I.Petrukhin, V.S.Vaniev, D.V.Vavilov, V.A.Victorov, S.A.Zimin, IHEP, Protvino, Russia.
V.Z.Kolganov, G.S.Lomkatsi, A.F.Nilov, V.T.Smolyankin, ITEP, Moscow, Russia.

OZI allowed $B \to YK$ decays. At the same time, typical decay widths for the well established $|qqq>$ isobars with similar masses are ≥ 300 MeV.

As was emphasized in a number of papers (see reviews [1, 2] and the references therein), diffractive production processes with Pomeron exchange offer new tools in searches for the exotic hadrons. In modern notion Pomeron is a multigluon system which allows for the production of the exotic hadrons in gluon-rich diffractive processes.

The Pomeron exchange mechanism in diffractive production reactions can induce the coherent processes on the target nucleus. In such processes the nucleus acts as a whole. Owing to the difference in the absorptions of singe-particle and multiparticle objects in nuclei, coherent processes could serve as an effective tool for separation of resonance against non-resonant multiparticle background.

The SPHINX spectrometer was working in the proton beam of IHEP accelerator with the energy $E_p = 70$ GeV and intensity $I \simeq (2-3) \cdot 10^6$ protons/spill. The experiments on the SPHINX facility can be divided into two stages:

a) First generation measurements — with "old" SPHINX setup (the runs of 1990-1994). The main results of these measurements were published in 1994-2000 [3-16]. The most sensitive data were obtained in 1999-2000 [15, 16] ("previous data").

b) Second generation measurements — with completely upgraded SPHINX setup. With the modified setup more than 10^9 events were recorded in 1996-1999. Preliminary results of these measurements will be presented at the conference "Hadron2001" for the first time ("new data").

The "old" SPHINX and upgraded one had the same structure, but after the upgrade the facility was equipped with the new tracking system, new hodoscopes, hadron calorimeter and modernized RICH spectrometer, new electronics, DAQ and online computers (which increased the maximum flux of data per spill by an order of magnitude). As the result of this upgrade we have obtained practically new setup.

Let us briefly summarize some results of the searches for cryptoexotic baryon states which were obtained earlier in the experiments of the SPHINX Collaboration.

MAIN RESULTS OF THE PREVIOUS MEASUREMENTS ON THE SPHINX FACILITY

In the previous measurements on the SPHINX spectrometer several unusual baryonic states were observed in the study of diffractive production reactions (see [1-7,10,11,15,16]. The most interesting information was obtained in the study of the reaction

$$p + N(C) \to [\Sigma^0 K^+] + N(C) \qquad (1)$$

(here C corresponds to the coherent reaction on carbon nuclei). The key element of the analysis of this reaction is the selection of $\Sigma^0 \to \Lambda\gamma$ decay which is rather complicated problem due to the soft character of photon spectrum in the laboratory frame ($E_\gamma < 6$ GeV) and significant background. A detailed GEANT-based Monte-Carlo simulation of the setup was done for efficiency calculations and cross section evaluations.

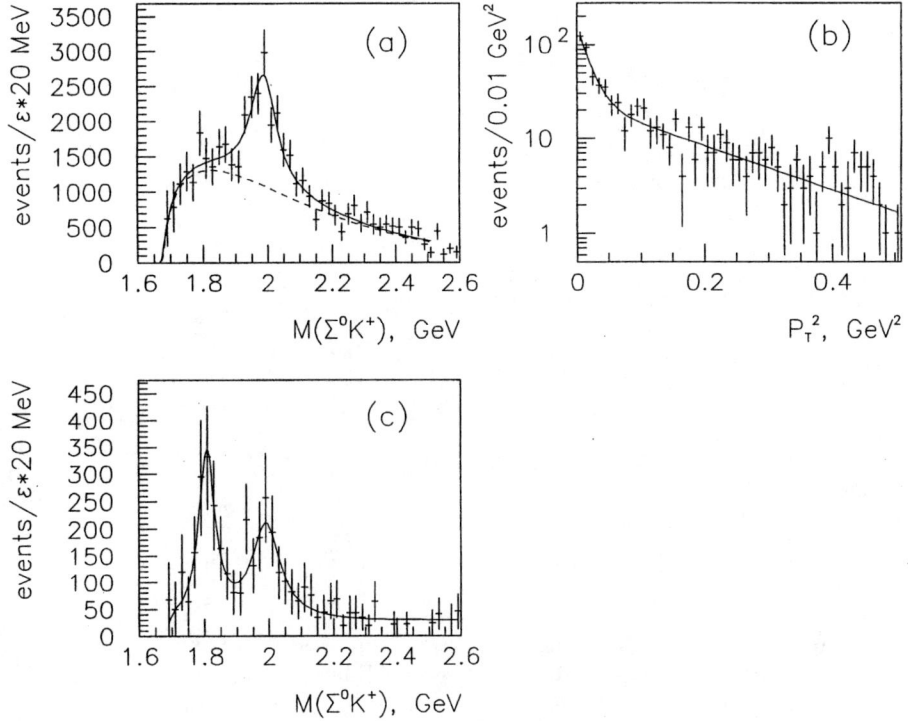

FIGURE 1. Distributions for diffractive reaction $p + N \to [\Sigma^0 K^+] + N$. a) Corrected mass spectrum $M(\Sigma^0 K^+)$ for all P_T^2 (soft photon cut). b) transverse momentum distribution dN/dP_T^2. c) Corrected mass spectrum $M(\Sigma^0 K^+)$ for the region of very small $P_T^2 < 0.01$ GeV^2 (strong photon cut).

The reaction (1) was studied in our previous works [6, 10] and [15] in different experimental and kinematic conditions, with successively improved separation of Σ^0 signal due to improvement in the measurements and data analysis. The results of all these studies are in a good agreement which support our conclusion of the observation of two new baryonic states:

a) the state $X(2000)^+ \to \Sigma^0 K^+$ with the mass $M = 1989 \pm 6$ MeV and the width $\Gamma = 91 \pm 20$ MeV;
b) the state $X(1810)^+ \to \Sigma^0 K^+$ with $M = 1807 \pm 7$ MeV and $\Gamma = 62 \pm 19$ MeV.

The effective mass spectrum $M(\Sigma^0 K^+)$ in the reaction (1) for all values of the square of transverse momentum(P_T^2) is presented in Fig. 1a. The peak of X(2000) is seen very clearly in this spectrum with a good statistical significance. Thus, the reaction

$$p + N \to X(2000) + N \qquad (2)$$

is well separated in the SPHINX data. We estimated the cross section for X(2000) production in (2):

$$\sigma[p+N \to X(2000)+N] \cdot BR[X(2000) \to \Sigma^0 K^+] = 95 \pm 20 \text{ nb/nucleon} \quad (3)$$

(assuming $\sigma \propto A^{2/3}$, e.g. for the effective number of nucleons in carbon nucleus equal to 5.24). The parameters of X(2000) peak are not sensitive to different photon cuts (see Table 1).

TABLE 1. Data on $M(\Sigma^0 K^+)$ in reaction $p+N \to [\Sigma^0 K^+]+N, \Sigma^0 \to \Lambda\gamma$ with different photon cuts[15] (for all P_T^2)

Photon cut	Soft	Intermediate	Strong
N events in X(2000) peak	430 ± 89	301 ± 71	190 ± 47
Correction factor for photon efficiency	1.0	1.4	2.25
Parameters of X(2000)			
M (MeV) weighted spectrum	1986 ± 6	1991 ± 8	1988 ± 6
Γ (MeV) weighted spectrum	98 ± 20	96 ± 26	68 ± 21
$\sigma[p+N \to X(2000)+N] \cdot BR[X(2000) \to \Sigma^0 K^+]$ (nb/nucleon)	100 ± 19	93 ± 25	91 ± 21

The transverse momentum distribution dN/dP_T^2 for reaction (2) is shown in Fig. 1b. From this distribution the coherent diffractive production reaction on carbon nuclei is identified as a diffractive peak with the slope $b \simeq 63 \pm 10$ GeV^{-2}. The cross section for coherent reaction is determined as

$$\sigma[p+C \to X(2000)^+ + C]_{\text{Coherent}} \cdot BR[X(2000)^+ \to \Sigma^0 K^+] = 260 \pm 60 \text{ nb/C nuclei}. \quad (4)$$

The errors in (3) and (4) are statistical only. Additional systematic errors are about ±20% due to uncertainties in the cuts, in the Monte Carlo efficiency calculations and in the absolute normalization. In the study of coherent reaction (1) (with $P_T^2 < 0.075 \text{GeV}^2$) in the mass spectrum $M(\Sigma^0 K^+)$ we observed not only the peak of X(2000), but another state X(1810). Study of the yield of X(1810) as function of P_T^2 demonstrates that this state is produced only in the region of very small $P_T^2 (\lesssim 0.01 \text{GeV}^2)$ where it is well defined (see Fig. 1c). From this data parameters of X(1810) are determined as well as the coherent cross section

$$\sigma[p+C \to X(1810)+C]_{P_T^2 < 0.01 \text{GeV}^2} \cdot BR[X(1810)^+ \to \Sigma^0 K^+] = 215 \pm 44 \text{ nb} \quad (5)$$

In the mass spectrum $M(\Sigma^0 K^+)$ in Fig. 1a there is only a slight indication for X(1810) structure which is seen very clearly in the coherent reaction (1). This difference is caused by a large background in this region for the events in Fig. 1a (for all P_T^2 values).

To explain the production of X(1810) state only at a very small P_T^2, the hypothesis of the electromagnetic production of this state in the Coulomb field of carbon nucleus was proposed [17] and it seems to be in no contradiction with the experimental data.

X(2000) as a candidate for pentaquark baryon

In a comparative study of coherent reactions $p+C \to p\pi^+\pi^- +C$ and $p+C \to \Delta^{++}\pi^- +C$ under the same kinematics as $p+C \to \Sigma^0 K^+ +C$ a search for other decay modes of the X(2000) was performed. No peaks in 2 GeV mass range were observed in $M(p\pi^+\pi^-)$ and $M(\Delta^{++}\pi^-)$ mass spectra and lower limits for the ratios

$$R[X(2000)] = BR[X(2000) \to (\Sigma K)]/BR[X(2000) \to (\Delta \pi); p\pi^+\pi^-] \gtrsim 1 \qquad (6)$$

were obtained. Thus two unusual properties of X(2000) state were found:

- Anomalously large branching ratios for decay channels with strange particle emission ($R[X(2000)] \gtrsim 1$). At the same time for ordinary isobars R ≤ few percents.
- Small enough decay width of heavy X(2000) state. For well established isobars in this mass region $\Gamma \gtrsim 300 - 400$ MeV.

These anomalous dynamical properties of X(2000) baryon are the reasons to consider it as a serious candidate for pentabaryon with hidden strangeness $|X(2000) = |uuds\bar{s}>$ (see more details in [2] and [6]).

The reality of X(2000)

We have obtained some additional data to support the reality of X(2000) baryon state

1. In the experiments with the SPHINX setup we studied the reaction

$$p+N(C) \to [\Sigma^+ + K^0]+N(C) \qquad (7)$$

In spite of a limited statistics, we observed the X(2000) peak and the indication for X(1810) structure in this reaction which are quite compatible with the data for reaction (1) [16].

2. In the experiment at the SELEX (E781) spectrometer [18] with the Σ^- hyperon beam of the Fermilab Tevatron, the diffractive production reaction

$$\Sigma^- + N \to [\Sigma^- K^+ K^-] + N \qquad (8)$$

was studied at the beam momentum $P_{\Sigma^-} \simeq 600$ GeV. In the invariant mass spectrum $M(\Sigma^- K^+)$ for this reaction a peak with parameter $M = 1962 \pm 12$ MeV and $\Gamma = 96 \pm 32$ MeV was observed (see [2, 19]). The parameters of this structure are very close to the parameters of $X(2000)$. Thus, the real existence of X(2000) baryon seems to be supported by the data from another experiment and in another process.

PRELIMINARY RESULTS FROM UPGRADED SPHINX FACILITY

Let us now present some preliminary results from the upgraded spectrometer, final version of which includes:

- A wide aperture magnetic spectrometer with proportional chambers, drift tubes and scintillator hodoscopes.
- Multichannel lead glass γ-spectrometer with 1052 $5\times5\times42$cm^3 counters.
- System of Cherenkov counters for identification of secondary particles (including RICH spectrometer with photomatrix of 736 small phototubes — the first RICH device of this type (see [3, 20]).
- Hadron calorimeter with 96 total absorption detectors.
- Guard system of scintillator counters and lead-scintillator sandwiches for separation of exclusive reactions.
- Trigger and front-end electronics, DAQ and fast on-line computers.

New front-end electronics and DAQ-system allowed us to record up to ~ 3000 triggers per 10s accelerator cycle. This in turn gave us the possibility to loose old trigger requirements and to introduce new types of triggers. During the runs in 1996-1999, more than 10^9 events were recorded, corresponding to approximately 10^{12} protons passing through (C/Cu) target.

Status of new data processing

Up to now almost 70% of data were processed by a tracking reconstruction program which was completely rewritten. We also finished the preliminary calibration of RICH and γ-detector. These detectors were used in the analysis, but there is a place for improvements. New Geant based Monte-Carlo program is under development. Based on preliminary analysis we can estimate the possible increase in statistics for some reactions, studied with "old" SPHINX(see Table 2).

TABLE 2. Estimated increase in statistics for diffractive production reactions with the upgraded SPHINX facility(relative to the previous data)

reaction		relative factor
$p+N$	$\to [\Sigma^0 K^+] + N$	5-7
	$\to [\Sigma^+ K^0] + N$	10-15
	$\to [\Sigma^*(1385)K^+] + N$	5-7
	$\to [p\eta] + N$	~ 15
	$\to [p\eta'] + N$	~ 15

Two new types of trigger were introduced in the runs with upgraded SPHINX. One of them (meson trigger) was designed to continue our investigations of the quasiexclusive meson production in the deep fragmentation region ([14]). The other (multiparticle trigger) was developed to search for possible narrow exotic baryons, in particular for

SPHINX PRELIMINARY

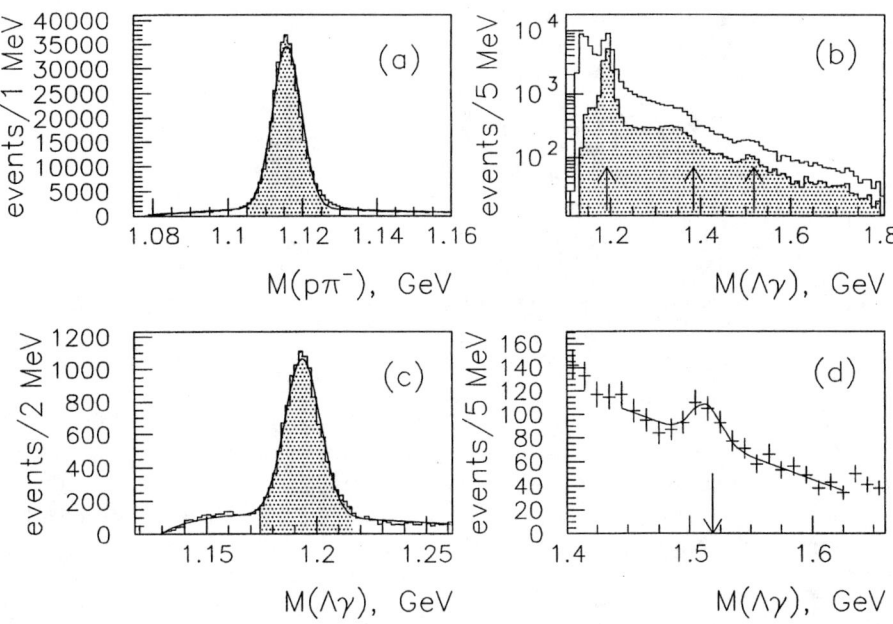

FIGURE 2. a) Effective mass distribution M($p\pi^-$) for the reaction (9); b) Effective mass distribution M($\Lambda\gamma^-$) for the reaction (10): unshaded histogramm – all events, shaded – soft photon cut, the arrows show the positions of $\Sigma^0(1192),\Sigma^0(1385)$ and $\Lambda(1520)$; c) and d): the same as b) but in the regions of $\Sigma^0(1192)$ and $\Lambda(1520)$(soft photon cut).

$Z^+(1530), Z^+ \to nK^+(pK^0)$, predicted in [21]. The data from these types of the trigger have also passed track reconstruction stage and preliminary results will be available in the near future.

The reactions with $\Lambda(1115)$ in the final state

Currently we are concentrating on studying the reactions with $\Lambda(1115)$ in the final state

$$p+N \to (\Lambda+K^+ +n\gamma)+N, \Lambda \to p+\pi^- \qquad (9)$$

The identification of Λ in the reaction (9), and more generally, the identification of the ΛK^+-system in the final state, was done using the combined information from RICH and tracking system. This results in very clean signal for Λ, which presented in Fig. 2a.

The decomposition of the sample (9) into the the reactions with the different number of photons($E_\gamma \geq 1$ GeV) is presented in Table 3.

TABLE 3. Number of events with different topology of photons for the reaction $p + N \to (\Lambda + K^+ + n\gamma) + N (\sim 70\%$ of data)

reaction	number of events
$p + N \to [\Lambda K^+] + 0\gamma + N$	171k
$\to [\Lambda K^+] + 1\gamma + N$	57k
$\to [\Lambda K^+] + 2\gamma + N$	31k
$\to [\Lambda K^+] + 3\gamma + N$	12k
$\to [\Lambda K^+] +$ any $\gamma + N$	322k

Further discussion will be devoted to the reaction with single photon in the final state

$$p + N \to (\Lambda + K^+ + \gamma) + N \qquad (10)$$

General spectrum of the $\Lambda\gamma$-effective mass for this reaction is shown in Fig. 2b for all events and the events with special cuts for the selection of "real" photons. There are three distinct structures in this distribution: the decay $\Sigma^0(1192) \to \Lambda\gamma$, the decay $\Sigma^0(1385) \to \Lambda\pi^0$ with one missing photon, and the decay $\Lambda(1520) \to \Lambda\gamma$. The signal for Σ^0 is shown in more detail in Fig. 2c, for $\Lambda(1520)$ -in Fig. 2d.

X(2000) and X(1810) in the "new" data

Using the cuts shown in Fig. 2c, the reaction (1) in the data from upgraded SPHINX-spectrometer was finally selected. The results are presented in Fig. 3 together with the distribution from "previous" data.

Note that the effective mass distribution $M(\Sigma^0 K^+)$ from first generation experiment (Fig. 3a) is not corrected for the efficiency, thus allowing to compare it directly with the same distribution from new data(Fig. 3b). The distributions are very similar with the evident increase of statistics in new data and can be easily fitted by the same function. In fact, the fit of the distribution Fig. 3b was done using the parameters of X(2000) from the previous one with two free parameters for normalization. The effective mass distributions $M(\Sigma^0 K^+)$ with our standard P_T^2-cuts are shown in Fig. 3c(coherent region) and Fig. 3d(the region of very small P_T^2). Note that in the last two figures an additional cut on the kaon momentum $p_{K^+} \leq 25$ GeV was introduced to ensure the identification capabilities of RICH. As can be seen from the comparison of Figs.1 and 3, the results for the reaction $p + N \to [\Sigma^0 K^+] + N$ based on a new statistics are in a good agreement with our previous data and strongly supports the existence of X(2000) state (with the increase of statistics for this state by a factor of ~ 5).

With new statistics we hope to receive quantitative information for X(2000) state (cross sections, angular decay distributions, quantum numbers, branching ratios for different channels) and study the features of X(1810) production in the small P_T^2 region with C and Cu targets.

SPHINX PRELIMINARY

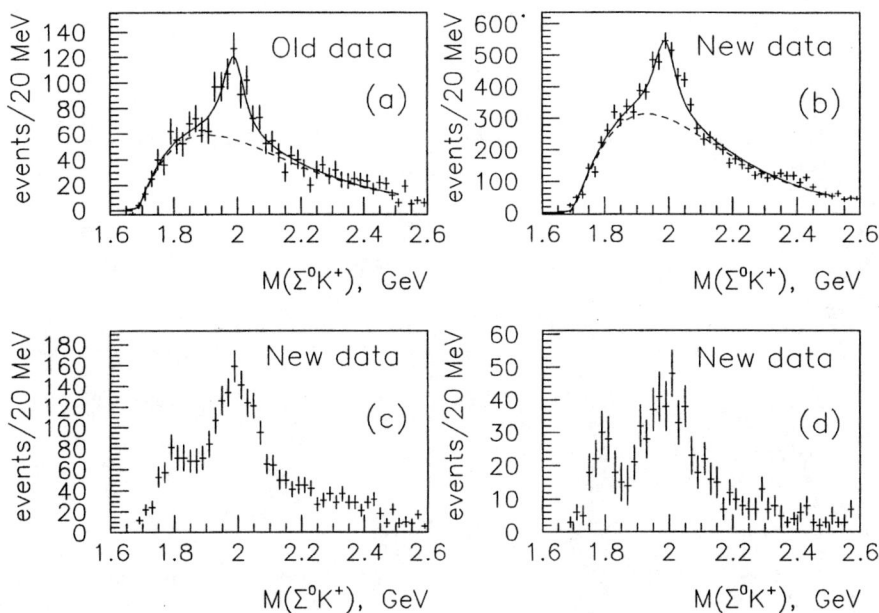

FIGURE 3. Invariant mass spectra $M(\Sigma^0 K^+)$ in the diffractive reaction $p + N \to [\Sigma^0 K^+] + N$ (with soft photon cut): a) "old" data, all P_T^2; b) "new" data, all P_T^2; c) "new" data, $P_T^2 \leq 0.75$ GeV^2, $P_{K^+} \leq 25$ GeV; d) "new" data, $P_T^2 \leq 0.01$ GeV^2, $P_{K^+} \leq 25$ GeV;

CONCLUSIONS

In the experiments with "old"SPHINX we investigated the reaction $p + N \to [\Sigma^0 K^+] + N$ and observed the new baryon state $X(2000) \to [\Sigma^0 K^+]$ with $M = 1989 \pm 6$ MeV, $\Gamma = 91 \pm 20$ MeV and with anomalous dynamical properties. This X(2000) state is a serious candidate for pentaquark baryon with hidden strangeness $|uuds\bar{s}\rangle$.

In the new runs with completely upgraded SPHINX facility (practically new setup) a large statistics for many proton induced reactions was obtained. First preliminary results for the reaction $p + N \to [\Sigma^0 K^+] + N$ based on a new statistics are in a good agreement with our previous data and strongly supports the existence of X(2000) state (with the increase of statistics for this state by a factor of ~ 5).

Radiative decay of $\Lambda(1520) \to \Lambda + \gamma$ is observed in a new statistics. There is also a hope that we can see the decay $\Lambda(1520) \to \Sigma^0 + \gamma$. The data about these decays can be very important for the investigation of the mechanisms of the SU(3) symmetry breaking (see [22] for a review).

We have a large program for further analysis of a new statistics and first of all for quantitative data on several interesting objects which were indicated in our old measurements.

ACKNOWLEDGMENTS

This work was supported in part by the Russian Foundation for Basic Research, grant 99-02-18251.

REFERENCES

1. L. G. Landsberg, UFN **164** (1994) 1129 [Physics Uspekhi (Engl. Transl.) **37** (1994) 1043];
   V. F. Kurshetsov, L. G. Landsberg, Yad. Fiz. **57** (1994) 2030 [Phys. At. Nucl. (Engl. Transl.) **57** (1994) 1954];
   L. G. Landsberg, Yad. Fiz. **60** 1997 1541 [Phys. At. Nucl. (Engl. Transl.) **60** (1997) 1397].
2. L. G. Landsberg, Phys. Rep. **320** 223 (1999);
   L. G. Landsberg, Yad. Fiz. **62** 2167 (1999).
3. D. V. Vavilov et al. (SPHINX Collab.). Yad. Fiz. **57** (1994) 241 [Phys. At. Nucl. (Engl. Transl.) **57** (1994) 227];
   M. Ya. Balatz et al. (SPHINX Collab.). Z. Phys. **C61** (1994) 220.
4. M. Ya. Balatz et al. (SPHINX Collab.). Z. Phys. **C61** (1994) 399.
5. D. V. Vavilov et al. (SPHINX Collab.). Yad. Fiz. **57** (1994) 253 [Phys. At. Nucl. (Engl. Transl.) **57** (1994) 238].
6. L. G. Landsberg et al. (SPHINX Collab.). Nuov. Cim. **A107** (1994) 2441.
7. D. V. Vavilov et al. (SPHINX Collab.). Yad. Fiz. **57** (1994) 241 [Phys. At. Nucl. (Engl. Transl.) **57** (1994) 1970].
8. D. V. Vavilov et al. (SPHINX Collab.). Yad. Fiz. **57** (1994) 1449; **58** (1995) 1426 [Phys. At. Nucl. (Engl. Transl.) **57** (1994) 1376; **58** (1995) 1342].
9. S. V. Golovkin et al. (SPHINX Collab.). Z. Phys. **C68** (1995) 585.
10. S. V. Golovkin et al. (SPHINX Collab.). Yad. Phys. **59** (1996) 1395 [Phys. At. Nucl. (Engl. Transl.). **59** (1996) 1336].
11. V. A. Bezzubov et al. (SPHINX Collab.). Yad. Phys. **59** (1996) 2199 [Phys. At. Nucl. (Engl. Transl.). **59** (1996) 2117].
12. L. G. Landsberg, Hadron Spectroscopy ("Hadron 97"). Seventh Intern. Conf. Upton, NY, August 1997 (ed. S.-U. Chung, H. J. Willutzki), p. 725.
13. V. A. Victorov et al. Yad. Fiz. **59** (1996) 1229 [Phys. At. Nucl. (Engl. Transl.). **59** (1996) 1175];
    M. Ya. Balatz et al. (SPHINX Collab.). Yad. Fiz. **59** (1996) 1242 [Phys. At. Nucl. (Engl. Transl.). **59** (1996) 1186];
    S. V. Golovkin et al. (SPHINX Collab.). Z. Phys. **A359** (1997) 435.
14. S. V. Golovkin et al. (SPHINX Collab.). Z. Phys. **A359** (1997) 327.
15. S. V. Golovkin et al. (SPHINX Collab.). Eur. Phys. J. **A5** (1999) 409.
16. D. V. Vavilov et al. (SPHINX Collab.). Yad. Fiz. **63** (2000) 1469.
17. D. V. Vavilov et al. Yad. Fiz. **62** (1999) 501.
18. R. Edelstein et al. Fermilab Proposal P781, 1987 (revised in 1993).
19. L. G. Landsberg, Proc. of 4th Workshop on Small-X and Diffractive Physics, Fermilab, Batavia, 17-20 September 1998, p.189;
    L. G. Landsberg, Proc. of Hyperon-99, Fermilab, Batavia, September 1999, p.29.
20. A. Kozhevnikov, V. Kubarovsky, V. Molchanov, V. Rykalin and V. Solyanik, Nucl. Instrum. Meth. A **433**, 164 (1999).
21. D. Diakonov, V. Petrov and M. V. Polyakov, Z. Phys. A **359**, 305 (1997) [hep-ph/9703373].
22. L. G. Landsberg, Phys. Atom. Nucl. **59**, 2080 (1996) [Yad. Fiz. **59**, 2161 (1996)].

Recent Results from Brookhaven E852 Experiment

Presented by Alexei V. Popov on behalf of the E852 Collaboration

Institute for High Energy Physics, Protvino, Russia

Abstract. Experiment E852 at Brookhaven National Laboratory is an experiment in meson spectroscopy configured to detect both neutral and charged final states of $\pi^- p$ collisions at 18 GeV/c. The main goal of this experiment is to search for the signs of mesons which cannot be described in the scope of the constituent quark model, with an emphasis on the meson with the manifestly exotic quantum numbers. In this talk the description of the experimental apparatus along with a general overview of the latest E852 results are presented.

INTRODUCTION

The objective of this talk is to give an overview of the recent results obtained by E852 experiment. QCD demands a much richer spectrum of meson states than a constituent quark model which includes extra states such as hybrids ($q\bar{q}g$), multiquarks ($q\bar{q}q\bar{q}$), and glueballs (gg or ggg). The main goal of E852 experiment is to search for meson states incompatible with the constituent quark model, with an emphasis on the mesons with the manifestly exotic $J^{PC} = 0^{--}, 0^{+-}, 1^{-+}, 2^{+-}, etc.$ quantum numbers. Observation of a state with such quantum numbers would directly indicate that this state is an exotic meson and not a conventional $q\bar{q}$ pair. The final states being searched for the possible exotic meson candidates include $\eta\pi^-$, $\eta'\pi^-$, $\pi^+\pi^-\pi^-$, and $b_1(1235)\pi$. In addition to the final states mentioned above the $\eta\pi^+\pi^-$, $\omega\eta$, $\omega\rho$, $K^+K^-\pi^0$, $\omega\pi^-$ final states were studied and provide a valuable information about a higher spin states as well as the excitations of the known ground states.

EXPERIMENTAL APPARATUS

The apparatus was located at the Multi-Particle Spectrometer (MPS) [1] of Brookhaven's Alternating Gradient Synchrotron (AGS). A diagram of the experimental apparatus is shown in Fig. 1.

A tagged π^- beam of momentum 18.3 GeV/c and a 30.5 cm liquid hydrogen target were used. The target was placed at the center of the MPS magnet with the field of 1 Tesla. The target was surrounded by a four-layer cylindrical drift chamber (TCYL) [2] used to trigger on the recoil particle, and a 198-element cylindrical thallium-doped cesium-iodide array (CsI) [3] to reject events with wide-angle photons. The downstream part of the MPS magnet housed the main components of the charged-tracking region. It consisted of 3 proportional wire chambers (TPX1-3) and 6 drift chamber modules

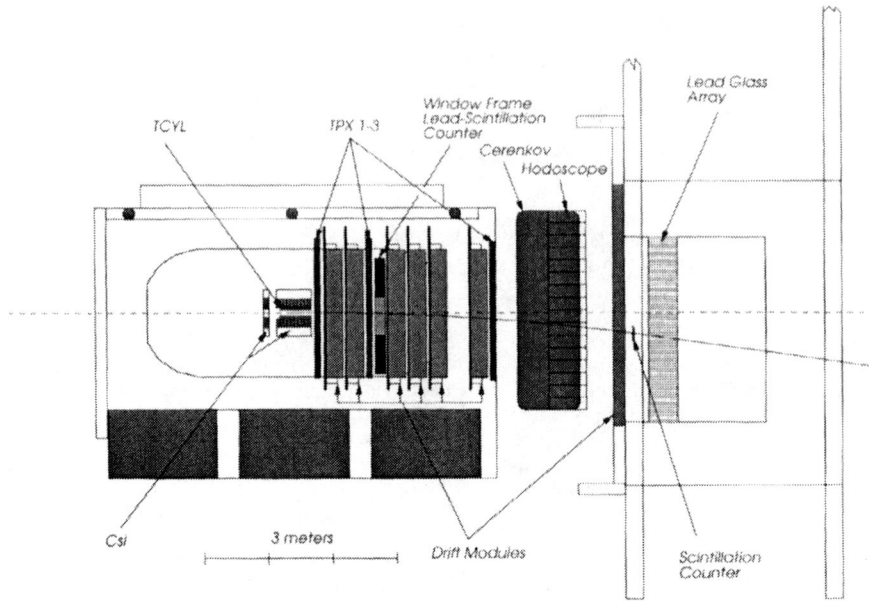

FIGURE 1. The plan view of E852 apparatus.

(DM1-6) [4] each with seven-layers. Photon hermeticity was ensured by a window-frame lead scintillator photon veto counter (DEA) in combination with an upstream segmented scintillator counter (CPVC) to identify charged tracks entering DEA. A forward scintillator counter (CPVB) was used to veto charged tracks for neutral triggers. A 96-segment threshold Cherenkov counter was used to distinguish charged kaons from pions. Non interacting beam and elastic scattering events were rejected with the help of two forward scintillator counters (EV and BV). Forward photons were detected by a 3045-element lead-glass electromagnetic calorimeter (LGD) [5]. Large drift chamber (TDX4) located directly in front of the LGD for tagging charged particles entering the LGD and to improve the momentum resolution. Data acquisition system was running at a rate of about 700 events per AGS spill. A more detailed description of the E852 apparatus is given in [6]. The E852 experiment had four data-taking runs and collected about 2 billion triggers. At this time detector is disassembled and data-taking part of the experiment is over.

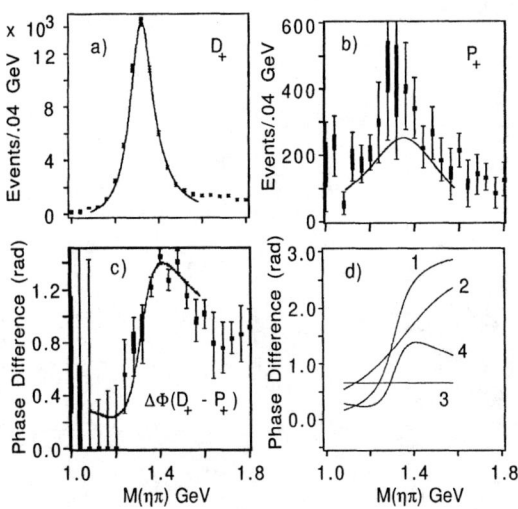

FIGURE 2. Results of the Partial Wave amplitude analysis for $\eta\pi^-$ system. Shown a) the fitted intensity distribution for the D_+ and b) the P_+ partial waves, and c) their phase difference. The range of values for the eight ambiguous solutions is shown by the central bar and the extent of the maximum error is shown by the error bars. Also shown as curves in a), b), and c) are the results of the mass-dependent analysis. The lines in d) correspond to (1) the fitted D_+ phase, (2) the fitted P_+ phase, (3) the relative production phase, and (4) the overall phase difference.

EXOTIC MESONS

$\pi_1(1400)$

In previous publications [7, 8] E852 presented evidence for an exotic meson produced in the reaction $\pi^- p \to \eta\pi^- p$ at 18 GeV/c from an analysis of the 1994 E852 data set. Interference between D - wave ($J^{PC} = 2^{++}$) and P - wave ($J^{PC} = 1^{-+}$) amplitudes produced with natural parity exchange is required in order to explain the data. The $a_2(1320)$ was observed in the D - wave and there is a broad enhancement between 1200 and 1600 GeV/c^2 in the P - wave. Using this observed interference it was shown that the P - wave phase has a rapid variation with mass and that this variation coupled with the fitted P - wave intensity distribution is well-fitted by a Breit-Wigner resonance with mass and width of $1370 \pm 16^{+50}_{-30}$ MeV/c^2 and $385 \pm 40^{+65}_{-105}$ MeV/c^2 respectively. Since a P-wave resonance in $\eta\pi$ system has $J^{PC} = 1^{-+}$ it is manifestly exotic. The results of the PWA and mass-dependent analysis for $\eta\pi^-$ system are shown in Fig. 2.

Preliminary results from an analysis of the 1995 data set confirmed the existence of this exotic object and studies of the mass-dependence show that the data is well described by the $a_2(1320)$ interfering with an exotic $J^{PC} = 1^{-+}$ resonance with mass and width of 1369 ± 11 MeV/c^2 and 257 ± 19 MeV/c^2 respectively.

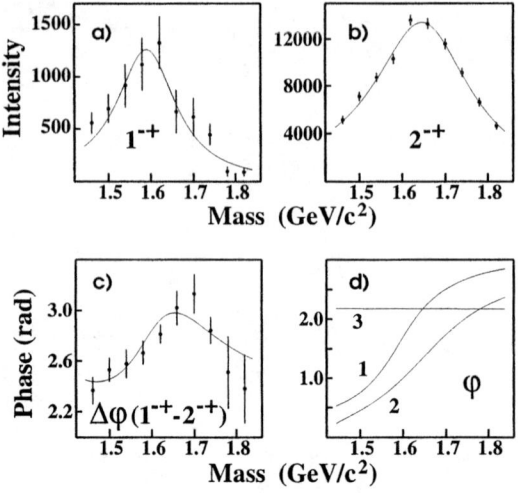

FIGURE 3. Results of the mass-dependent analysis for exotic wave in $\rho\pi$ system: a) 1^{-+} $\rho\pi$ $P1^+$ wave intensity; b) 2^{-+} $f_2\pi$ $S0^+$ wave intensity; c) phase difference between these waves; d) phase motion of the 1^{-+} wave (1), 2^{-+} wave (2), and the production phase between them (3).

$\pi_1(1600)$

The first system in which this $J^{PC} = 1^{-+}$ exotic resonance was observed is $\rho\pi^-$. A partial wave analysis of the reaction $\pi^- p \to \pi^+\pi^-\pi^- p$ [9] has been performed on a data sample of 250000 events obtained in the E852. The well-known $a_1(1260), \pi_2(1670), a_2(1320)$ resonant states are observed. In addition, the natural parity exchange wave with manifestly exotic quantum numbers $J^{PC} = 1^{-+}$ shows broad enhancements in the 1.1-1.3 GeV/c^2 and 1.6-1.7 GeV/c^2 regions. It was found that the first enhancement produced mostly by the leakage from nonexotic waves (especially $J^{PC} = 1^{++}$ wave) but the second structure (at 1.5-1.7 GeV/c^2 region) has no impact from this leakage. The phase difference between 1^{-+} $\rho\pi$ wave and all other significant natural parity exchange waves indicates a rapid increase in the phase of the 1^{-+} wave across 1.5-1.7 GeV/c^2 region. This is consistent with resonance behaviour. To determine the resonance parameters a mass-dependent analysis was made and the fitted mass and width of the 1^{-+} state, obtained from this analysis, are $M = 1593 \pm 8^{+29}_{-47}$ MeV/c^2, $\Gamma = 168 \pm 20^{+150}_{-12}$ MeV/c^2. The systematic errors were estimated by fitting the PWA results obtained for different sets of partial waves and a different rank of the PWA fit. The results of the PWA fit for exotic wave as well as the mass-dependent fit results are shown in Fig. 3.

Another system in which this $\pi_1(1600)$ exotic meson was observed is $\eta'\pi^-$ [10]. A partial wave amplitude analysis of an exclusive sample of 5765 events from the reaction $\pi^- p \to \eta'\pi^- p$ was performed. The $\eta'\pi^-$ production is dominated by natural parity exchange and by three partial waves with $J^{PC} = 1^{-+}, 2^{++}, 4^{++}$. A mass-dependent

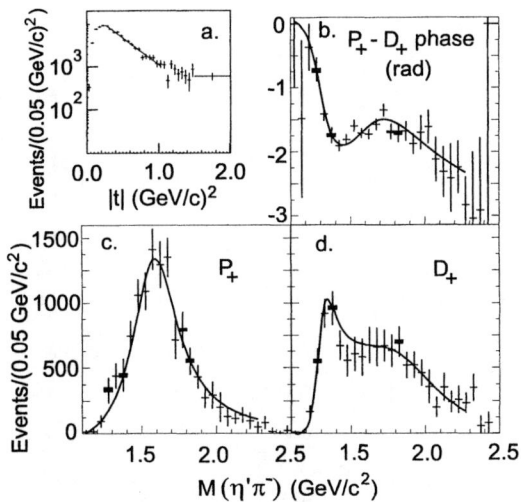

FIGURE 4. a) The acceptance corrected $|t|$ distribution fitted with the function $f(t) = ae^{b|t|}$. b),c),d) The results of the mass-independent PWA (horizontal lines with error bars) and mass-dependent fit (solid curve). The range of the ambiguous solutions is plotted with black rectangles. b) The $(P_+ - D_+)$ phase difference. c) The intensity distribution of the P_+ partial wave. c) The intensity distribution of the D_+ partial wave.

analysis of the partial-wave amplitudes indicates the production of the $a_2(1320)$ meson as well as the $a_4(2040)$ meson, observed for the first time decaying to $\eta'\pi^-$. Between 1.5 and 1.8 GeV/c^2 the exotic wave is the dominant wave. Its intensity distribution consists of a broad structure peaked near 1.6 GeV/c^2. The phase difference between P_+ ($J^{PC} = 1^{-+}$) and D_+ ($J^{PC} = 2^{++}$) waves exhibits a clear phase motion from $a_2(1320)$ in 1.1-1.4 GeV/c^2 region followed by the phase motion from P_+ wave which indicates a resonant behaviour in this wave. Mass-dependent analysis gave the parameters of this exotic object: $M = 1597 \pm 10^{+45}_{-10}$ MeV/c^2, $\Gamma = 340 \pm 40 \pm 50$ GeV/c^2. The mass and width of the P_+ state are consistent with those of the $\pi_1(1600)$ exotic state observed in the $\pi^+\pi^-\pi^-$ system and can be interpreted as the observation of a second decay mode of the $\pi_1(1600)$. The results of the PWA fit for exotic wave as well as the mass-dependent fit results are shown in Fig. 4.

Another analysis currently underway studies the $b_1(1235)\pi$ system produced in the reaction $\pi^-p \to \omega\pi^-\pi^0 p$. There is a significant 1^{-+} wave in $b_1(1235)\pi$ system which peaks near 1.6 GeV/c^2 and has a shoulder at 1.9 - 2.0 GeV/c^2. The relative phase between this wave and 2^{++} $\omega\rho$ wave is rising in the interval 1.4 - 1.9 GeV/c^2, which gives a strong evidence of an exotic resonance in this wave. The parameters of this object from the mass-dependent analysis are: $M = 1582 \pm 10(stat.) \pm 20(syst.)$ MeV/c^2, $\Gamma = 289 \pm 16(stat.) \pm 27(syst.)$ MeV/c^2 which are consistent with those obtained from $\pi^+\pi^-\pi^-$ and $\eta'\pi^-$ system analysis for $\pi_1(1600)$ exotic meson. To describe a shoulder at 1.9 - 2.0 GeV/c^2 the mass-dependent fit with two resonances was performed. This fit gives a better description of the data than a fit with one resonance, which gives some

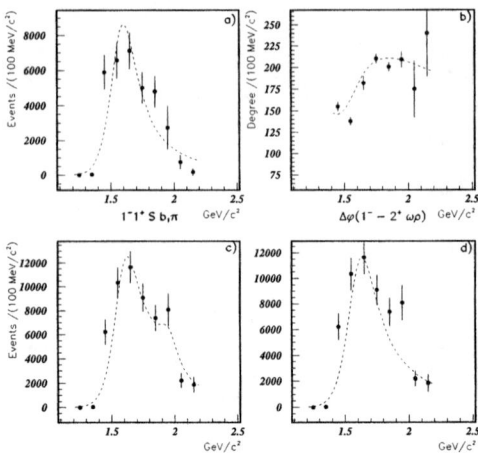

FIGURE 5. a) Acceptance corrected partial wave intensity for the 1^{-+} wave in the $b_1(1235)\pi$ system; b) relative phase difference between 1^{-+} and 2^{++} waves; c) and d) overall intensity for 1^{-+} waves and their mass-dependent fit description in a model with two resonances and in a model with one resonance.

evidence of another object existence at 2.0 GeV/c^2, but at this time it is hard to make any conclusions about existence and properties of this object. The results of the mass-independent and mass-dependent PWA are shown in Fig. 5.

NON EXOTIC MESONS

A partial wave analysis of the $\eta\pi^+\pi^-$ [11] has been performed. The data are dominated by $J^{PC} = 0^{-+}$ partial waves consistent with observation of the $\eta(1295)$ and $\eta(1440)$. Evidence of $J^{PC} = 1^{++}$, consisted with the $f_1(1285)$ is also seen. Results of the partial wave analysis were combined with the results of other experiments to estimate $f_1(1285)$ branching fractions. This values are considerably different from current values determined without the aid of amplitude analysis.

A partial wave analysis of the $\omega\eta$ final state ($\omega \to \pi^+\pi^-\pi^0$, $\eta \to \gamma\gamma$) [12] was performed. We observe the previously unreported decay mode $\omega(1650) \to \omega\eta$ and a new $J^{PC} = 1^{+-}$ state, $h_1(1595)$, with a mass $M = 1594 \pm 15^{+10}_{-60}$ MeV/c^2 and width $\Gamma = 384 \pm 60^{+70}_{-100}$ MeV/c^2. The $h_1(1595)$ exhibits resonance-like phase motion relative to the $\omega(1650)$. The results of the partial wave analysis and mass-dependent analysis of the $\omega\eta$ system are presented in Fig. 6.

Both mass-independent and mass-dependent PWA's of the reaction $\pi^- p \to \omega\pi^-\pi^0 p$ were performed. In the $\omega(782)\rho(770)$ intermediate state the resonance behavior was observed in the $J^{PC} = 0^{-+}, 4^{++}$ waves which is consistent with the existence of the $\pi(1800)$ and $a_4(2040)$ mesons decaying into $\omega(782)\rho(770)$. For the $J^{PC} = 2^{-+}$ waves

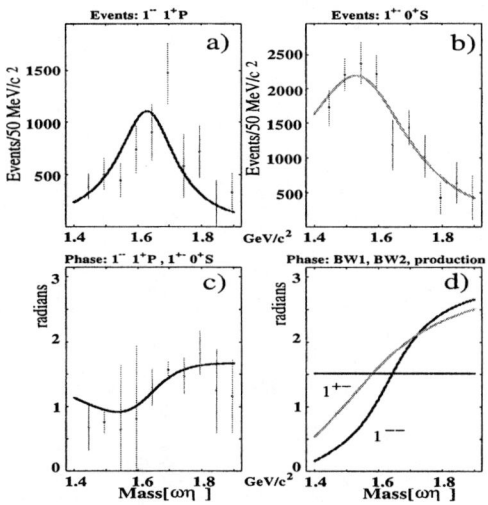

FIGURE 6. The results of the mass-dependent analysis for $\omega\eta$ system. a) $J^{PC} = 1^{--}$ state ($\omega(1650)$); b) new $J^{PC} = 1^{+-}$ $h_1(1595)$ state; c) the relative phase of 1^{+-} and 1^{--} waves; d) the individual Breit-Wigner phases and the overall production phase.

the results of the mass-independent and mass-dependent analysis can be described as a superposition of two resonances - the well known $\pi_2(1670)$-meson and another object at 1.9 GeV/c^2. In the $b_1(1235)\pi$ intermediate state resonant behavior was observed in $J^{PC} = 3^{++}$ which is consistent with $a_3(1900)$ decaying into $b_1(1235)\pi$. Details of this analysis can be found in these proceedings [13].

A partial wave analysis of the mesons produced in the reaction $\pi^- p \to K^+ K^- \pi^0 n$ was performed [14]. The intensity and phase of the resulting waves show clear excitation of several previously identified states: $f_1(1285), \eta(1295), \eta(1416), f_1(1420)$ and $\eta(1485)$. The existence of three low-mass pseudoscalars is confirmed. The $\eta(1416)$ and $\eta(1485)$ are distinguished by their decay properties: $\eta(1416)$ decays primarily to $a_0\pi^0$ but has a small $K^*\bar{K}$ branch as well. The $\eta(1485)$ was observed only in $K^*\bar{K}$ decay. The results of the mass-independent PWA and mass-dependent analysis are shown in Fig. 7 and 8.

TABLE 1. Resonance parameters and decay modes of the observed states in $K^+K^-\pi^0$ system

Resonance	M (GeV/c^2)	Γ (GeV/c^2)	Decay Modes
$f_1(1285)$	$1288 \pm 4 \pm 5$	$45 \pm 9 \pm 7$	$a_0\pi^0$
$\eta(1295)$	$1302 \pm 9 \pm 8$	$57 \pm 23 \pm 21$	$a_0\pi^0$
$\eta(1416)$	$1416 \pm 4 \pm 2$	$42 \pm 10 \pm 9$	$a_0\pi^0, K^*\bar{K}$
$f_1(1420)$	$1428 \pm 4 \pm 2$	$38 \pm 9 \pm 6$	$K^*\bar{K}$
$\eta(1485)$	$1485 \pm 8 \pm 5$	$98 \pm 18 \pm 3$	$K^*\bar{K}$

A partial wave analysis of the $\omega\pi^-$ system produced in the reaction $\pi^- p \to \omega\pi^- p$ was used to determine the D-wave to S-wave decay amplitude ratio for $b_1^- \to \omega\pi^-$

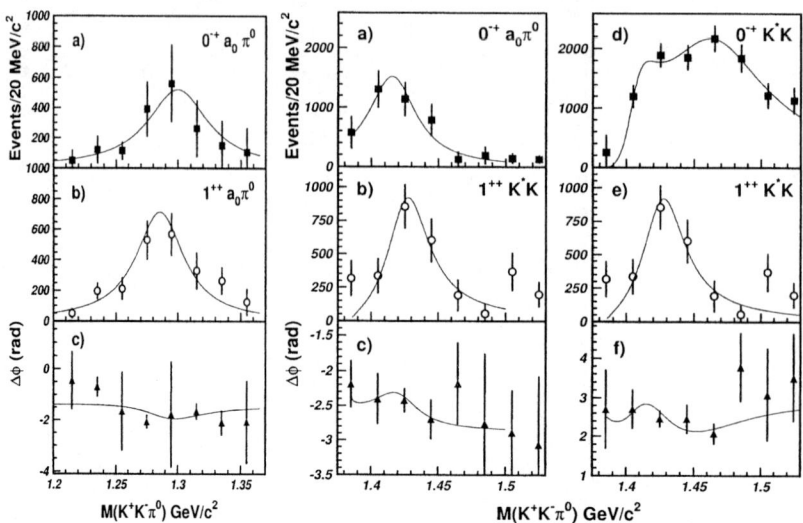

FIGURE 7. First column: results of mass-independent fit (points) and mass-dependent fit (lines) for $K^+K^-\pi^0$ system below $K^*\bar{K}$ threshold. a) 0^{-+} intensity; b) 1^{++} intensity; c) Phase difference between them. **Second and third column**: results of mass-independent fit (points) and mass-dependent fit (lines) for $K^+K^-\pi^0$ system above $K^*\bar{K}$ threshold. a) 0^{-+} $a_0\pi^0$ intensity; b) and e) 1^{++} $K^*\bar{K}$ intensity; c) relative phase between (a) and (b); f) relative phase between (d) and (e).

($\omega \to \pi^+\pi^-\pi^0$). A strong $b_1(1235)$ production was observed through natural parity exchange. A grid search in $|D/S|$ was performed in the PWA fits of the events in the b_1 region to find the best values corresponding to the maximum likelihood from the PWA fits. We find $|D/S| = 0.281 \pm 0.015(stat.) \pm 0.003(syst.)$ which is consistent with the current PDG average.

REFERENCES

1. S. Ozaki, *Abbreviated Description of the MPS*, Brookhaven MPS note 40 (1978)
2. Z. Bar-Yam et al., *Nucl. Instr. & Meth.* **A386**, 253 (1997)
3. T. Adams et al., *Nucl. Instr. & Meth.* **A368**, 617 (1996)
4. S.E. Eiseman et al., *Nucl. Instr. & Meth.* **A217**, 140 (1983)
5. R.R. Crittenden et al.,*Nucl. Instr. & Meth.* **A387**, 377 (1997)
6. S. Teige et al.,*Phys. Rev.* **D59**, 12001 (1999)
7. D.R. Thompson et al.,*Phys. Rev. Lett.* **79**, 1630 (1997)
8. S.U. Chung et al.,*Phys. Rev.* **D60**, 92001 (1999)
9. G.S. Adams et al.,*Phys. Rew. Lett.* **81**, 5760 (1998)
10. E. Ivanov et al.,*Phys. Rev. Lett.* **86**, 3977 (2001)
11. J.J. Manak et al.,*Phys. Rev.* **D62**, 012003 (2000)
12. P. Eugenio et al.,*Phys. Lett.* **B497**, 190 (2001)
13. A. Popov, *A Study of the Reaction $\pi^-p \to \omega\pi^-\pi^0 p$ at 18 GeV/c*, these proceedings
14. G.S. Adams et al.,*Phys.Lett.* **B516**, 264 (2001)

The $J^{PC} = 1^{-+}$ hunting season at VES

Valery Dorofeev (for VES Collaboration [1])

Department of Hadron Physics, IHEP, Protvino, Russia, 142284

Abstract. We present preliminary results of study of the $\eta\pi^-$, $\eta'\pi^-$ and $b_1\pi$-systems produced in the π^-Be-interaction at 28 GeV/c. $J^{PC} = 2^{++}$ and 1^{-+}-waves resulted from the PWA have been fitted in each system separately to establish the nature of the 1^{-+}-wave. A hypothesis of the 1^{-+}-wave resonant nature in the $\eta\pi^-$ and $\eta'\pi^-$ has no statistically significant preference over the non-resonant one. The $b_1\pi$-system analysis confirms the results of the 37 GeV/c-beam data analysis in favor of the resonant treatment of the bump at 1.6GeV.

INTRODUCTION

At present several groups have evidence for 1^{-+} meson production, which is forbidden for ordinary quarkonia and hence might be a good candidate for a hybrid [1], [2]. The $\pi_1(1400)$-meson shows up in the $\eta\pi$ final state in two experiments [3],[4],[5]. VES group found that the results of the $\eta'\pi^-$ and $b_1\pi$-system Partial-Wave Analysis(PWA) at 37GeV/c agree with the production of the higher mass $\pi_1(1600)$ [6], evidence for which in the $\eta'\pi^-$ has been also confirmed by E852 at BNL [7]. A broad bump observed by VES in the 1^{-+}-wave intensities of the $\eta'\pi$, $b_1\pi$, $\rho\pi$ at 1.6GeV was assumed to be a single object. The resonant nature results from the analysis of the $J^{PC} = 2^{++}$ and 1^{-+}-waves in the $\omega\pi^-\pi^0$ [8].

A data sample was collected in the 1996 data run at VES spectrometer exposed by the 28GeV/c momentum π^--meson beam. A detailed description of setup can be found elsewhere [9]. Over 5×10^8 triggers were recorded during the data taking period, which is ~ 2.5 more than at 37GeV/c. A trigger is an event with at least two charged tracks from the beam interaction moving in the beam direction and falling into the spectrometer aperture. Events of the studied final states production form subsets of reaction $\pi^-Be \to \pi^+2\pi^- + k\pi^0(\eta) + Be$, where $k = 1, 2$. To select events of this reaction the following criteria were applied to the reconstructed events:

- 3 reconstructed tracks should form the $\pi^+\pi^-\pi^-$-system;
- an interaction vertex must be inside the target;
- we require for 2-4 γ-clusters in the calorimeter;

---

[1] Amelin D.V., Dorofeev V.A., Dzhelyadin R.I., Gouz Yu.P., Kachaev I.A., Karyukhin A.N., Khokhlov Yu.A., Konoplyannikov A.K., Konstantinov V.F., Kopikov S.V., Kostyukhin V.V., Matveev V.D., Nikolaenko V.I., Ostankov A.P., Polyakov B.F., Ryabchikov D.I., Solodkov A.A., Solovianov O.V., Zaitsev A.M.

- a 2γ-system with mass closest to the mean $\pi^0(\eta)$-meson mass within 30(78)MeV is identified as the $\pi^0(\eta)$-meson. A 1C-fit to the $\pi^0(\eta)$ mass is applied to the identified γγ pairs;
- total visible energy E_{tot} must satisfy the following in-equation: $25 GeV < E_{tot} < 30 GeV$.

STUDY OF THE $\eta\pi^-$-SYSTEM

A decay chain $\eta \to \pi^+\pi^-\pi^0$, $\pi^0 \to 2\gamma$ was chosen for the $\eta\pi^-$ production study. A clear peak of the η-meson corresponding to the $\eta\pi^-$ production is observed in an invariant mass spectrum of the $\pi^+\pi^-\pi^0$-system in fig. 1a. The shape of the peak is parametrized with a sum of two Gaussian functions and resolution of the narrow one is equal to $6 MeV$. Shown in fig. 1b is the invariant mass distribution of the $\pi^+2\pi^-\pi^0$ for events from the η region ($530 < M_{\pi^+\pi^-\pi^0} < 566 MeV$) with superimposed plot for sidebands ($502 < M_{\pi^+\pi^-\pi^0} < 520 MeV$ and $576 < M_{\pi^+\pi^-\pi^0} < 594 MeV$) which are used for the background estimation. The background subtracted $\eta\pi^-$ invariant mass spectrum (fig. 1c) shows a dominant $a_2(1320)$-meson production with a small bump near $M_{\eta\pi^-} = 1 GeV$ of a possible $a_0(980)$-meson production.

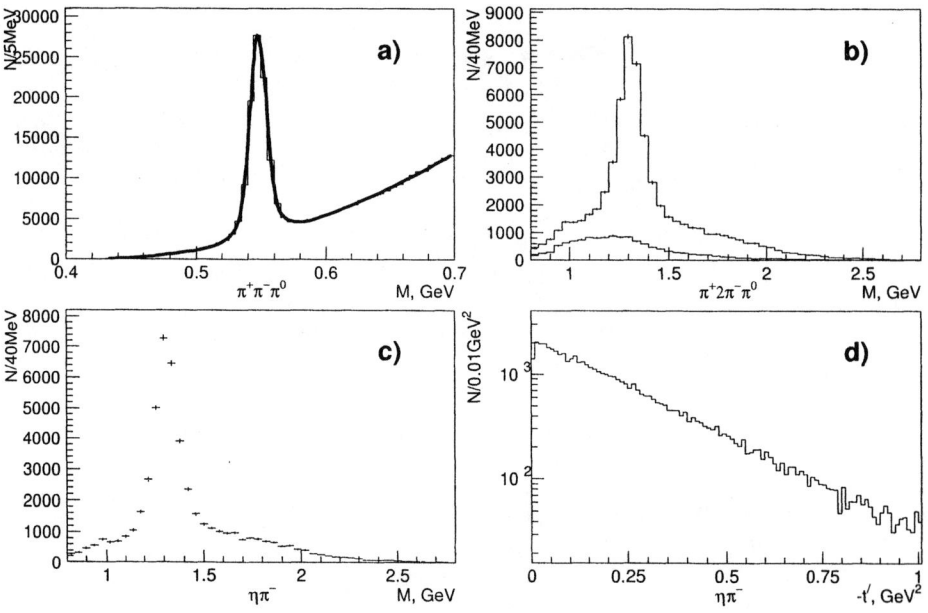

FIGURE 1. Effective mass of the $\pi^+\pi^-\pi^0$ a), $\pi^+2\pi^-\pi^0$ b) and $\eta\pi^-$ c). $-t'$ distribution d).

A negative value of the invariant four-momentum-transfer between the beam and the $\eta\pi^-$ squared $-t'$-distribution in fig. 1d shows the characteristic features of the projective wave production. Here $t' = t - t_{min}$, where t_{min}- is a minimum transfer squared for a given $\eta\pi^-$ mass.

The $\eta\pi^-$-system PWA

We applied the PWA procedure to expand the data in terms of the partial waves. An event is considered in the spirit of the isobar model [10] as a production process followed by a chain of the subsequent decays into the $\eta\pi^-$ and $\eta \to \pi^+\pi^-\pi^0$. The production process of the $\eta\pi^-$ is assumed to be described by a rank one density matrix for each naturality [11]. An event probability for each naturality is equal to an average over possible non-interfering $\pi^+\pi_i^-\pi^0$ combinations N_{comb} of a product: $P_{ev} = \frac{1}{N_{comb}} \sum_i P_i D_i(m_{3\pi}) G_i(m_{3\pi})$ due to presence of two π^-'s, where: D_i - is a probability of the $\eta \to \pi^+\pi_i^-\pi^0$ decay which corresponds to the decay Dalitz-plot distribution, G_i - is the η-meson shape parameterization being used to take into account the $\pi^+\pi_i^-\pi^0$ mass resolution. Here P_i is an expansion squared of the product of the two pseudo-scalar production and decay amplitudes in terms of the partial wave decay amplitudes. The decay amplitudes are defined by the following set of quantum numbers $J^P M^\eta$, where J stands for a total angular momentum and is equal to an orbital momentum L in the system of two pseudo-scalars. With P we denote parity, M - an absolute value of the J_z projection. Here η is exchange naturality. The decay amplitude is parametrized with spherical harmonics [12] multiplied by a breakup momentum involved in the L-th degree. Under the assumption about the rank of the density matrix the production amplitudes are denoted as: L_0 for the $M=0$ and L_η for the $M=1$ waves.

The complex production amplitudes are determined from the extended maximum likelihood fit [13].

A wave set includes 3 subsets of partial waves non-interfering with each other. The first is composed of waves with negative naturality, the second - with positive one and a subset made of a single wave FLAT which describes the background in the $\pi^+2\pi^-\pi^0$-system.

The ambiguous solutions [14] were found by repeating 100 times the fit from the random starting values of the fit parameters.

50300 events in the range $0.8 < M(\pi^+2\pi^-\pi^0) < 2.4 GeV$, $|t'| < 1 GeV^2$, with the mass of at least a single $\pi^+\pi^-\pi^0$ combination in the η region or the sidebands were subjected to the PWA. The PWA has been carried out independently in each of the 40MeV bins. To make the data set cleaner the following additional cuts were applied: $\cos\Theta^{hel}_{\pi^0} < 0.8$ and the track angular separation in projections $\Delta AX(AY) > 10(6) mrad$, where $\Theta^{hel}_{\pi^0}$ is the π^0 polar angle in the overall CM helicity reference frame.

In fig. 2 are shown predicted by the PWA fit acceptance-corrected wave intensities and a relative phase between the P_+ and D_+ waves as a function of the $\eta\pi^-$ mass. With a thick line is shown a range of possible ambiguous solutions and with a thin one the maximum extent of errors.

The intensities of the S and P waves with the negative naturality are small and consistent with zero in all the $M_{\eta\pi^-}$ range. The D_0 wave has a non-zero contribution in the $M_{\eta\pi^-} > 2 GeV$. We found a significant D_- intensity in the whole mass range of

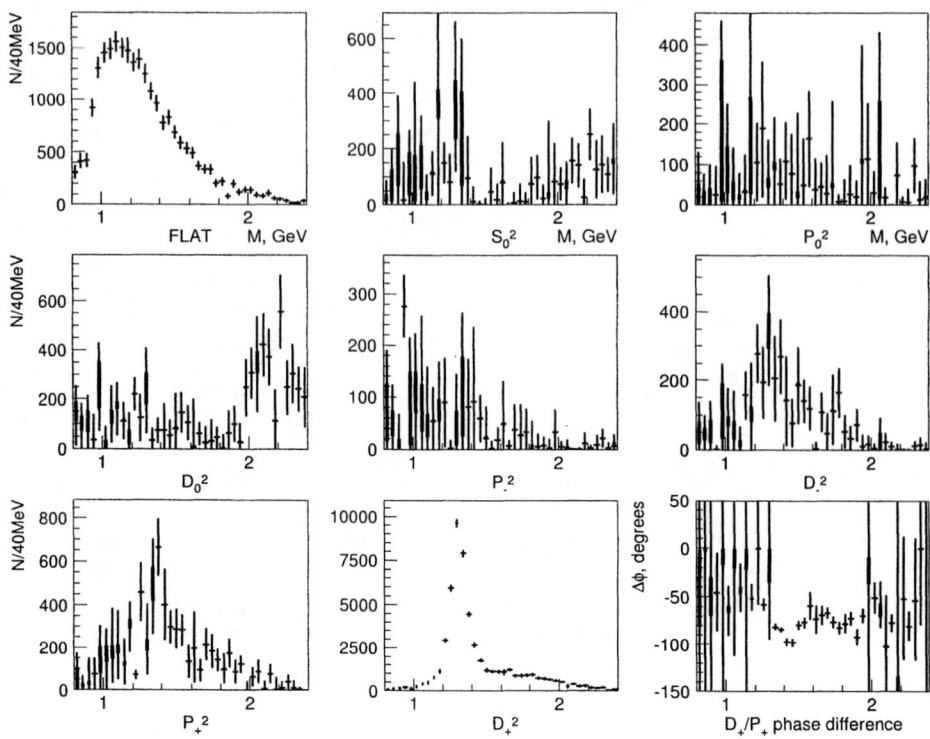

FIGURE 2. The $\eta\pi^-$ wave intensities and the D_+/P_+ relative phase.

unclear nature.

The D_+-wave is the dominantly produced wave in the $\eta\pi^-$, where a huge peak corresponding to the $a_2(1320)$ production is observed. A broad bump with a spike at $M_{\eta\pi^-} = 1.4 GeV$ is observed in the P_+. The spike might results from the fit quality worsening in the $a_2(1320)$ region and where one should notice the increase of errors. A sign ambiguous relative phase shows a rapid motion in the $a_2(1320)$ region.

We have tried also to include the waves with the orbital momentum higher than 2 and the waves with $M > 1$, but they were found to be not significant. The quality of the fit is controlled by comparison of distributions of variables which describe the $\eta\pi^-$-system for the experimental and Monte-Carlo data. The Monte-Carlo events were generated according to the described above model with the values of the production amplitudes resulted from the best fit solution in a mass bin.

The $\eta\pi^-$-system mass-dependent fit

In order to establish the nature of the P_+-wave we have carried out a combined mass-dependent(MD) fit of the expected production amplitudes to the P_+ and D_+ resulted from the PWA in the whole mass range. Fitted with the χ^2 method are the D_+ and P_+ intensities and the real and imaginary parts of the production of both amplitudes with the corresponding errors in each mass bin.

We assume two general production manners of a resonant and a non-resonant type. The resonant amplitude is parametrized with a Breit-Wigner form:

$$A_R = \frac{m_0\sqrt{\Gamma_0\Gamma_f}}{m_0^2 - m^2 - im_0\Gamma_{tot}}, \qquad \Gamma_f = BR_f\Gamma_0\frac{m_0}{m}\frac{q}{q_0}\left[\frac{B_L(q)}{B_L(q_0)}\right]^2$$

where: m - the $\eta\pi$ mass, m_0 - a resonance mass, Γ_0 - a nominal width, Γ_f - a partial width in the final channel, Γ_{tot} - a total width, $q(q_0)$ - a breakup momentum of the $\eta\pi$ with the $m(m_0)$ mass, $B_L(q)$ - a barrier factor [15], BR_f - a branching ratio into the f-th final state.

By the notion background we just mean the non-resonant production manner without going into the nature of the process. The main difference of the background type from the resonant one is absence of phase motion. The background amplitude is parametrized with the following free form $A_B = LIPS_l \times (m - m_t)^\alpha \exp(-\beta(m - m_t))$, where: $LIPS_l(m)$ - a phase space, m_t - a threshold mass, α and β are shape parameters.

A procedure of the wave construction from the objects is in the following. At first a wave is formed of evident structures such as the $a_2(1320)$ in the D_+-wave. If there are no such objects we take an object of any production type arbitrarily. After doing the fit to thus constructed wave we then necessarily try to describe the data with the object of the alternative type. At the next steps we add consistently another objects of both types to both waves until we will succeed in the description of the wave intensities and their interference in the whole mass region. All objects are added coherently with their own mass-independent phase. The parameters of the MD fit are: the mass, width and the number of the resonance events, the shape parameters and the number of the background events, the phase relative to a reference object.

TABLE 1. Comparison of the MD fit variants

D_+ bkg	a_2'	P_+ bkg	π_1	χ^2/NDF
+		+		244/149
	+	+		422/149
+			+	224/149
	+		+	219/149
+		+	+	176/145
+		+	E852	187/147

To begin with we have fitted the $a_2(1320)$ Breit-Wigner to the D_+ intensity and failed to describe the D_+ wave at masses $> 1.5 GeV$. Then we have applied the described above procedure. At first the D_+ was fitted to a sum of the $a_2(1320)$ and the background and the P_+ to the background. The second fit was carried out with the D_+ wave background replaced by the resonance, which will be called further as an a_2'-meson for convenience. There have just been reported the evidence for the $a_2(1750)$ [16]. The next steps were to fit the P_+ wave to the resonance, called a $\pi_1(1400)$. The evidence for it have been reported by E852 and CBAR groups [3], [4], [5] and which production did not contradict to VES results [17]. The results of the MD fits in the case, when the D_+ has been fitted to a sum of the $a_2(1320)$ and the background and the P_+ to the background or to the π_1 are shown in fig. 3 with thick solid and dashed curves respectively.

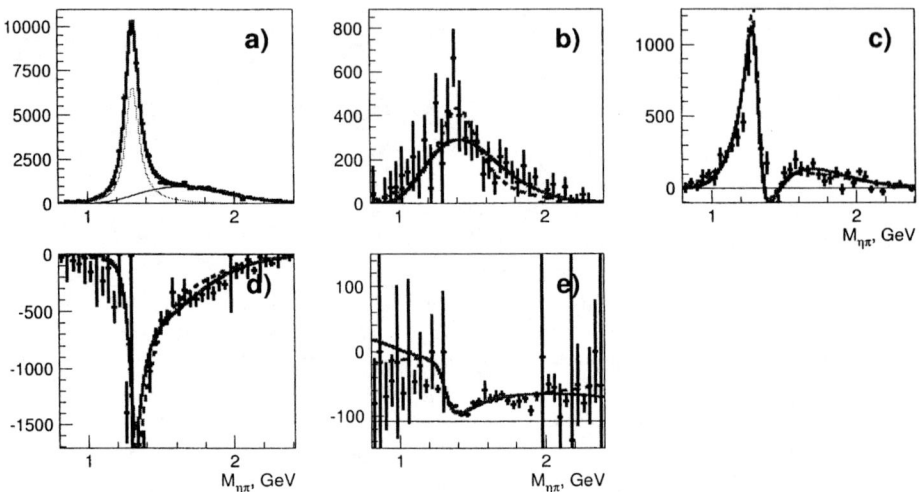

FIGURE 3. The $\eta\pi^-$ D_+ a) and P_+ b) intensities. The real c) and imaginary d) parts of D_+P_+ product. The D_+/P_+ relative phase e). The smooth curves are the MD fit results.

The summary of the MD fit results are presented in tab. 1. The objects which make up the fitted partial waves are marked by the plus sign in a row. A χ^2 over the number of degrees of freedom of the corresponding MD fit is shown in the fifth column.

From the comparison of the first four rows one can conclude that there are no statistically significant difference between a single resonant and non-resonant object description of the P_+.

The fits of the P_+ to a sum of the π_1 and the background are shown in the last two rows. A fit with free π_1 parameters resulted in the following values: $M_{a_2(1320)} = 1316 \pm 1 MeV$, $\Gamma_{a_2(1320)} = 113 \pm 3 MeV$, and $M_{\pi_1} = 1316 \pm 12 MeV$, $\Gamma_{\pi_1} = 287 \pm 25 MeV$. The quoted errors are statistical only. Statistical significance is nearly the same as for the fit with the π_1 parameters fixed to the values reported by E852 [3]. A closeness of the π_1 mass to the $a_2(1320)$ mass may indicate that there is an underestimated influence of the $a_2(1320)$ on the P_+.

STUDY OF THE $\eta'\pi^-$-SYSTEM

A decay chain $\eta' \to \eta\pi^+\pi^-$, $\eta \to 2\gamma$ was chosen for the $\eta'\pi^-$ production study. A clear peak of the η' corresponding to the $\eta'\pi^-$ production is observed in the invariant mass spectrum of the $\eta\pi^+\pi^-$ in fig. 4b. Shown in fig. 4c is the invariant mass distribution of the $\eta\pi^+2\pi^-$ for events from η' region ($940 < M_{\eta\pi^+\pi^-} < 976 MeV$) with superimposed plot for the η' sidebands ($912 < M_{\eta\pi^+\pi^-} < 930 MeV$ and $986 < M_{\eta\pi^+\pi^-} < 1004 MeV$) which are used for the background estimation. The background subtracted $\eta'\pi^-$ invariant mass spectrum (fig. 4d) shows a clear $a_2(1320)$ peak and a broad large bump at $M_{\eta\pi^-} = 1.6 GeV$, which strongly differs from the corresponding distribution of the $\eta\pi^-$.

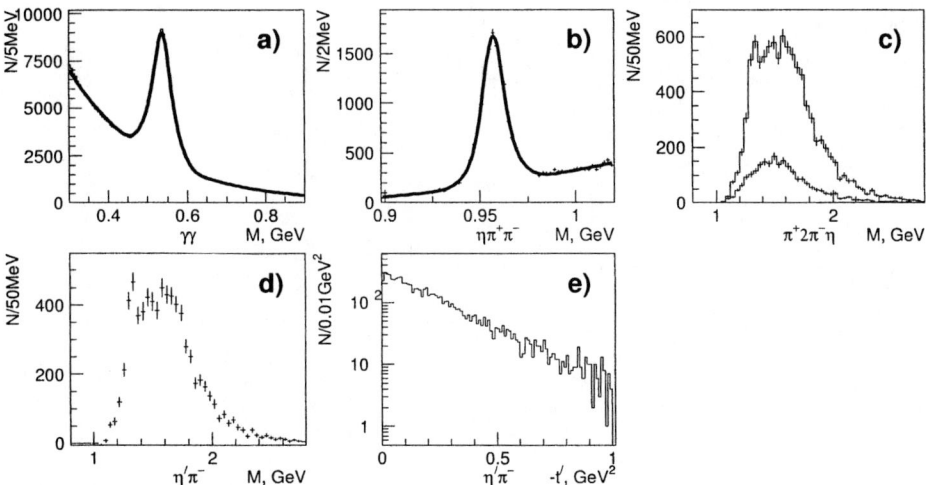

FIGURE 4. Effective mass of the $\gamma\gamma$ a), $\eta\pi^+\pi^-$ b), $\eta\pi^+2\pi^-$ c) and $\eta'\pi^-$ d). $-t'$ distribution e).

A $-t'_{\eta'\pi^-}$ distribution in fig. 4e shows characteristic features of the projective wave production like for the $\eta\pi^-$.

10700 events in the range $1.15 < M(\eta\pi^+2\pi^-) < 2.15 GeV$, $|t'| < 1 GeV^2$, with the mass of at least a single $\eta\pi^+\pi^-$ combination in the η'-meson region or the sidebands were subjected to the PWA. The PWA has been carried out independently in each of the $50 MeV$ wide bins.

We have applied the same procedure as for the $\eta\pi^-$-system PWA, described in the appropriate section. In fig. 5 are shown predicted by the PWA fit acceptance-corrected wave intensities and a phase difference of the P_+ and D_+ as a function of the $\eta'\pi^-$ mass. With a thick line is shown a range of possible ambiguous solutions and with a thin one the maximum extent of errors.

The intensities of the negative naturality waves are small and nearly consistent with zero in all the $M_{\eta'\pi^-}$ range. The D_0 wave only shows some non-negligible contribution in the regions $1.4 \div 1.6 GeV$ and $> 1.9 GeV$.

The $a_2(1320)$ peak is observed in the D_+ intensity. But contrary to the $\eta\pi^-$-system

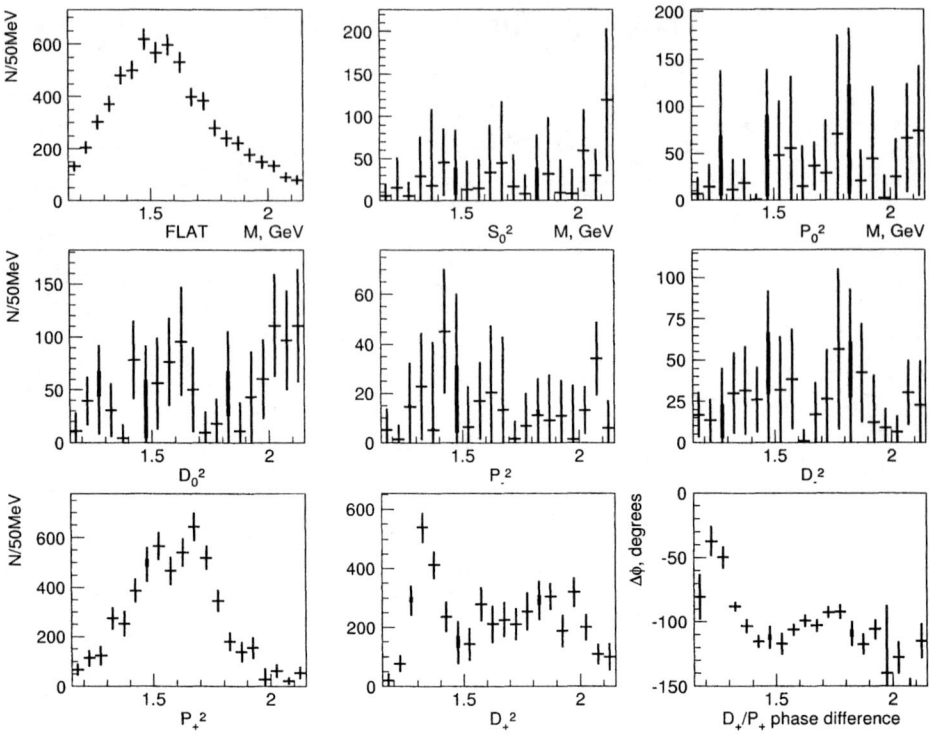

FIGURE 5. The $\eta'\pi^-$ wave intensities and the D_+/P_+ relative phase.

behavior this wave shows a broad structure in the region $M_{\eta'\pi^-} > 1.5 GeV$. The rate of this high mass production relative to the $a_2(1320)$ is larger for the $\eta'\pi^-$ than for $\eta\pi^-$.

A broad bump of a $\sim 350 MeV$ width is observed in the P_+ at $M_{\eta'\pi^-} = 1.6 GeV$. A sign ambiguous relative phase shows a rapid motion in the $a_2(1320)$ mass region.

We have tried also to include the waves with the orbital momentum higher than 2 but they were found to be not significant. The quality of the fit is controlled by comparison of distributions of variables which describe the $\eta'\pi^-$-system for the experimental and Monte-Carlo data.

To establish the nature of the P_+-wave we have carried out the MD fit following the procedure which had been applied for the $\eta\pi^-$ study. The D_+ partial wave amplitude was parametrized with a sum of the $a_2(1320)$ Breit-Wigner function and the background or an a_2'. This additional object was included to describe the broad bump in the region $M_{\eta'\pi^-} > 1.5 GeV$. The P_+ partial wave was parametrized with the background or with the $\pi_1(1600)$. The indication to the possible resonance nature of the bump at $1.6 GeV$ was reported by our group as a result of the $\eta'\pi^-$ PWA at $37 GeV/c$ [6]. Later the observation of the resonance have been reported by E852 [7]. The result of the MD fit in the case, when the D_+ is fitted to a sum of the $a_2(1320)$ and the a_2' and the P_+ to the resonance

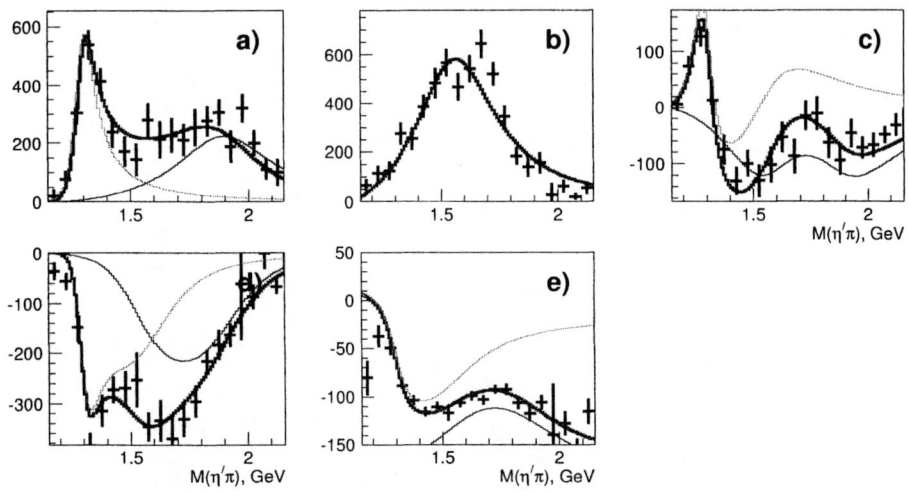

FIGURE 6. The $\eta'\pi^-$ D_+ a) and P_+ b) intensities. The real c) and imaginary d) parts of D_+P_+ product. The D_+/P_+ relative phase e). The smooth curves are the MD fit results.

only is shown in fig. 6.

We have found no statistically significant preference of the fit with the P_+ saturated with the resonant over the non-resonant description. In the case of the P_+ partial wave amplitude parameterization with a sum of the resonance and the background the statistical significance is larger than in the previous cases. However the $a_2(1320)$ width was found to be somewhat larger than it could be expected from the resolution study. In the cases, when the D_+ is composed of the $a_2(1320)$ and a a'_2, while the P_+ includes the π_1, the a'_2 mass was found to be somewhat $1.8 \div 1.9 GeV$ and the width $0.4 \div 0.5 GeV$, while one might expect to find the same parameters as for the $a'_2(1750)$.

The above consideration of results prevents us from drawing the unambiguous conclusion about the nature of the P_+-wave in the $\eta'\pi^-$.

STUDY OF THE $\omega\pi^-\pi^0$-SYSTEM

A decay mode $\omega \to \pi^+\pi^-\pi^0$ was chosen for the $\omega\pi^-\pi^0$ production study. A clear peak of the ω meson corresponding to the $\omega\pi^-\pi^0$ production is observed in the invariant mass spectrum of the $\pi^+\pi^-\pi^0$ in fig. 7a. Shown in fig. 7b is the invariant mass distribution of the $\pi^+2\pi^-2\pi^0$ for events from the ω region ($758 < M_{\pi^+\pi^-\pi^0} < 808 MeV$) with superimposed plot for the ω sidebands ($684 < M_{\pi^+\pi^-\pi^0} < 719 MeV$ and $846 < M_{\pi^+\pi^-\pi^0} < 871 MeV$) which are used for the background estimation. The background subtracted $\omega\pi^-\pi^0$ invariant mass spectrum (fig. 7c) shows an $a_2(1320)$ shoulder and a broad large bump at $M_{\omega\pi^-\pi^0} = 1.8 GeV$.

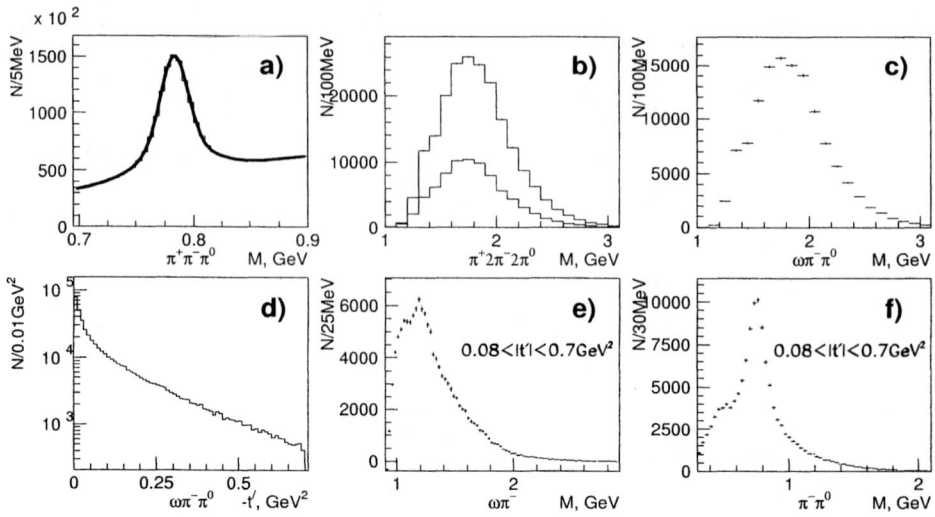

FIGURE 7. Effective mass of the $\pi^+\pi^-\pi^0$ a), $\pi^+2\pi^-2\pi^0$ b) and $\omega\pi^-\pi^0$ c). $-t'$ distribution d). Effective mass of the $\omega\rho^-$ e) and $\pi^-\pi^0$ f) from the $\omega\pi^-\pi^0$ for $0.08 < |t'| < 0.7 GeV^2$.

A $-t'_{\omega\pi^-\pi^0}$-distribution in fig. 7d shows a sharp peak in the low $|t'| < 0.08 GeV^2$ (LT) region, corresponding to the coherent production on a nucleus, which flattens in the high $|t'|$ region ($0.08 < |t'| < 0.7 GeV^2$) and which will be called further the HT. This behavior in the HT-region corresponds to the incoherent production on a nucleon. The projective wave production dominates in the HT-region. Therefore we have divided a data sample into the LT- and HT-sets, which were analyzed separately. A large ρ^--meson peak is observed in the invariant mass spectrum of the $\pi^-\pi^0$, produced with the ω(see fig. 7f). This means the dominant production of the $\omega\rho^-$. There is a peak of the $b_1(1235)^-$-meson in the $\omega\pi^-$ invariant mass spectrum shown in fig. 7e, corresponding to the $b_1(1235)\pi$ production. The same is for the $b_1(1235)^0$.

A PWA of 284000 events in the range $1.2 < M(\pi^+2\pi^-2\pi^0) < 3GeV$, $0.08 < |t'| < 0.7 GeV^2$, $\max_{\pi^\pm}(\cos\Theta^{hel}) < 0.92$ with the mass of at least one $\pi^+\pi^-\pi^0$-combination in the ω region or in the sidebands has been carried out. Here Θ^{hel} is a track polar angle in the overall helicity CM reference frame. Events in each of the $50MeV$ or $100MeV$ wide bins were fitted separately. The Illinois method [11] was used in the PWA, however with an important improvement. We added the possibility to restrict a rank of the density matrix [18]. The partial waves are denoted by the $J^P M^\eta LS(isobar - bachelor)$, where S stands for a total spin in the isobar-bachelor system. The partial waves of the $\omega\rho^-$, $b_1(1235)\pi$, $\rho_3(1690)\pi$, $\rho_1(1450)\pi$ are included in the wave set. The results of the PWA are consistent with our previous analysis of the $\omega\pi^-\pi^0$ at $37 GeV/c$ [19]. A $2^+1^+S2(\omega\rho)$ shown in fig. 8a was found to be a dominant wave with a clear $a_2(1320)$ peak and a broad bump at $M_{\omega\pi^-\pi^0} = 1.7GeV$. A significant $1^-1^+S1(b_1\pi)$ wave shown in fig. 8b is observed with a broad bump at $M_{\omega\pi^-\pi^0} = 1.6GeV$, which is in the maximum $\sim 15\%$ of the $a_2(1320)$ peak height. The important feature of the combined 2^+ and 1^- wave behavior demonstrated in fig. 8c is a region bounded to $1.5 < M_{\omega\pi^-\pi^0} < 2 GeV$, where the coherence is significantly non-zero. It should be noticed in fig. 8f an $80°$ rise of the 1^--wave phase relative to the 2^+-wave phase right in this region and which may be attributed to a 1^- resonance.

The MD fit has been carried out to establish the origin of the 1^-1^+-wave in the $b_1\pi$. The high intensity $2^+1^+S2(\omega\rho)$ was selected to be a reference wave. The fit has been done in the same way like the $\eta\pi^-$ fit. However absence of unambiguous knowledge about the nature of the bump at $M_{\omega\pi^-\pi^0} = 1.7 GeV$ and an extension of the density matrix rank to three in the PWA requires inclusion of additional fitting parameters in the MD fit. The naïve model is to describe the 2^+ wave by a sum of the $a_2(1320)$ Breit-Wigner and something, which has a form of the bump at $1.7 GeV$, while the 1^- may be constructed from a broad incoherent background and from a resonance, which is coherent to the 2^+ bump.

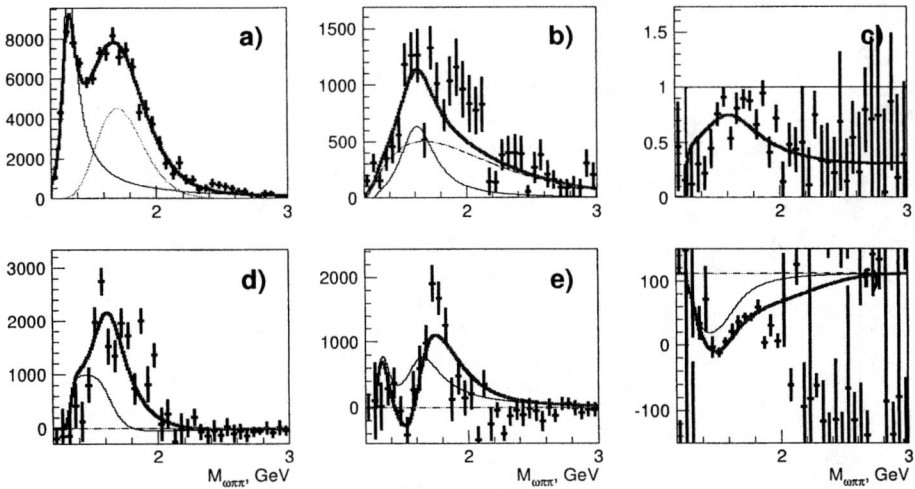

FIGURE 8. The $2^+1^+S2(\omega\rho)$ a) and $1^-1^+S1(b_1\pi)$ b) intensities. A coherence parameter c) [11]. The real d) and imaginary e) parts of their non-diagonal ρ-matrix element. The 1^- phase relative to 2^+ f). The smooth curves are the MD fit results.

The fit describes the interference pattern satisfactorily as seen in fig. 8. We have tried another hypotheses: to fit the 1^- wave to a partially coherent background or to make up the 2^+ wave from a a'_2 in addition to the $a_2(1320)$. However we failed to find qualitatively better solution. The data are consistent with the resonant description of the $1^-1^+S1(b_1\pi)$ with the mass $1.6 GeV$ and the width $0.33 GeV$.

CONCLUSIONS

We have carried out the PWA of the $\eta\pi^-$, $\eta'\pi^-$, $b_1\pi$ systems produced in π^-Be interaction at $28 GeV/c$.

We have tried to understand the nature of the observed partial waves in each system by the fit to the resonant and non-resonant description.

- We observed the P_+ in the $\eta\pi^-$ which exhibits the phase motion with respect to the D_+ in the coherent model. We could not unambiguously establish the nature of the P_+ in the $\eta\pi^-$.
- The production of the $\eta'\pi^-$ is dominated by the P_+ and D_+. The common analysis of the observed $\eta'\pi^-$ partial waves gives no preference to the hypothesis of the $\pi_1(1600)$ production.
- We confirmed the results of the $b_1\pi$ analysis of the $37GeV/c$-beam data. The broad bump has been observed in the $1^-1^+S1(b_1\pi)$ of the $\omega\pi^-\pi^0$. The simultaneous fit with the $2^+1^+S2(\omega\rho)$ favors the existence of a resonance in the $b_1\pi$ with the mass $\sim 1.6GeV$ and the width $\sim 330MeV$.
- The PWA of the $37GeV/c$-beam data sample in the several t' intervals shows a wide bump in the $J^PM^\eta(isobar - bachelor) = 1^-1^+(\rho\pi)$-wave at $M_{\rho\pi} < 2GeV$ [20].

ACKNOWLEDGMENTS

This is supported, in part, by INTAS-RFBR 97-02-71017, INTAS-RFBR 00-02-16555, RFBR 00-15-96689 grants.

REFERENCES

1. Isgur, N., and Paton, J., *Phys. Rev.*, **D31**, 2910 (1985).
2. Close, F. E., and Page, P. R., *Nucl. Phys.*, **B443**, 233 (1995).
3. Chung, S. U., et al., *Phys. Rev.*, **D60**, 092001 (1999).
4. Abele, A., et al., *Phys. Lett.*, **B423**, 175 (1998).
5. Abele, A., et al., *Phys. Lett.*, **B446**, 349 (1999).
6. Khokhlov, Y., et al., *Nucl. Phys.*, **A663**, 596 (2000).
7. Ivanov, E. I., et al., *Phys. Rev. Lett.*, **85**, 3977 (2001).
8. Dorofeev, V. A., et al., "New results from VES", in *Workshop on Hadron Spectroscopy*, edited by B. T., A. Feliciello, and A. Filippi, Frascati Phys. 15, 1999, p. 999.
9. Beladidze, G., et al., *Zeit. fur Phys.*, **C54**, 235 (1992).
10. Herndon, D. J., Söding, P., and Cashmore, R., *Phys. Rev.*, **D11**, 3165 (1975).
11. Hansen, J. D., Jones, G., Otter, G., and Rudolph, G., *Nucl. Phys.*, **B81**, 403 (1974).
12. Groom, D. E., et al., *Eur. Phys. J.*, **C15**, 1–878 (2000).
13. Orear, J., Notes on statistics for physicists (1958), in UCRL-8417.
14. Sadovsky, S. A., On the ambiguities in the partial-wave analysis of $\pi^-p \to \eta\pi^0 n$ reaction (1991), preprint IHEP-91-75.
15. von Hippel, F., and Quigg, C., *Phys. Rev.*, **D5**, 624 (1972).
16. Hou, S., in this proceedings (2001).
17. Zaitsev, A. M., et al., "Search for exotics in $I^G J^P = 1^-1^-$, 1^-0^- and 0^+2^+ waves", in *CP432, Hadron Spectroscopy: Seventh Internaltional Conference*, edited by S. U. Chung and W. H. J., 1997, pp. 461–470.
18. Chung, S. U., and Trueman, T. L., *Phys. Rev.*, **D11**, 633 (1975).
19. Amelin, D. V., et al., *Phys. Atom. Nucl.*, **62**, 445–453 (1999).
20. Kachaev, I. A., in this proceedings (2001).

The resonance structures of $K_S K_S$ and $\Lambda \bar{\Lambda}$ spectrum at MIS ITEP

V. N. Nozdrachev, SERP-E173 experiment[1]

ITEP, Moscow, Russia

Abstract. Experimental data (~ 40000 $K_S K_S$ events and ~ 2300 $\Lambda \bar{\Lambda}$ events) were obtained on the ITEP 6-meter spectrometer with a 40 GeV π^- beam of the IHEP U-70 accelerator. The experimental setup is being upgraded now.

The analysis of the 1200–1300 MeV/c^2 effective mass region of the $K_S K_S$-meson system was carried out for events obtained with the neutral trigger with 40 cm H_2 target. Irregularity near 1240 MeV/c^2 and width about 30 MeV/c^2 is clearly observed. Similar irregularity is observed on another sample of 15000 event on carbon target. A resonance states with mass near 1800 MeV/c^2 and 2230 MeV/c^2 are observed at selection of $\cos\theta_{GJ}$.

The results of studying the $\Lambda \bar{\Lambda}$ system produced in the reaction $\pi^- p \to \Lambda \bar{\Lambda} n$ at a π^--meson energy of 40 GeV are reported. The invariant-mass spectra for the events dominated by the singlet or triplet $\Lambda \bar{\Lambda}$ states were found to differ considerably from each other. The data give evidence for the existence of resonance states of the $\Lambda \bar{\Lambda}$ system in the mass regions near 2.3, 2.5 and 2.8 GeV.

INTRODUCTION

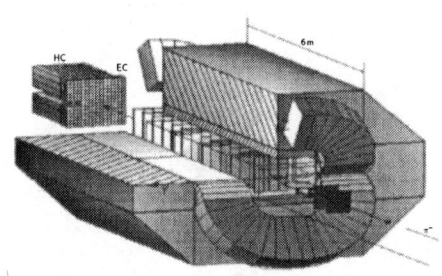

FIGURE 1. MIS ITEP experimental setup.

We present here recent results on reactions

$$\pi^- p \to K_S K_S n, \quad K_S \to \pi^+ \pi^- \quad (1)$$

$$\pi^- C \to K_S K_S A^* \quad (2)$$

$$\pi^- p \to \Lambda \bar{\Lambda} n, \quad \Lambda \to p\pi^-, \quad \bar{\Lambda} \to \bar{p}\pi^+ \quad (3)$$

$$\pi^- C \to \Lambda \bar{\Lambda} A^*, \quad (4)$$

at incident π^- momentum 40 GeV/c on MIS ITEP spectrometer (Fig. 1) at the 70 GeV IHEP accelerator. The target was surrounded by the system of scintillator veto counters. The counters form double shielding layer around the target. To suppress not only charged particles but also γ rays emanating from the target Pb convertors were placed between the layers. The coordinate detector consists of 62 planes of electrodynamic spark chambers producing 124 measurement planes in the magnetic field volume $6 \times 0.8 \times 1.54$ m^3. The events with strange particle production were identified on the basis of χ^2 values fitting the neutral V^0 events

---

[1] I.A. Erofeev, O.N. Erofeeva, V.K. Grigoriev, Y.V. Katinov, V.I. Lisin, V.N. Luzin, V.N. Nozdrachev, Y.P. Shkurenko, V.V. Sokolovsky, G.D. Tikhomirov, V.V. Vladimirsky. ITEP, Moscow.

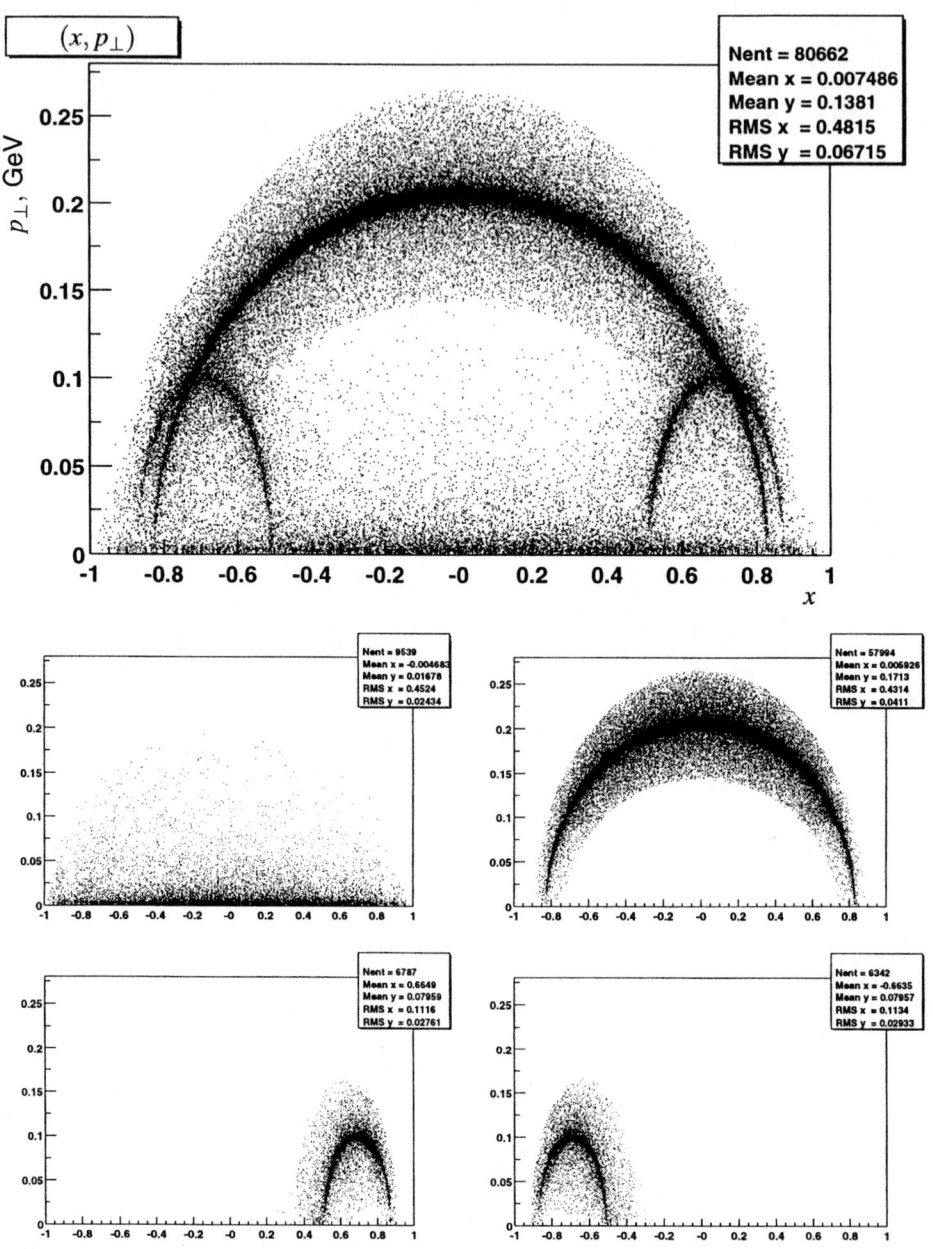

FIGURE 2. 2-D plot of $(x, p_\perp)$, $x = (p_{1\|} - p_{2\|})/(p_{1\|} + p_{2\|})$, where p_1, p_2 are the moments of positive and negative V_0-components. Points near large arc with $p_{\perp max} = 206$ MeV/c are $K_S \to \pi^+\pi^-$ decays. Points from the right small arc are $\Lambda \to p\pi^-$ decays, left — $\bar{\Lambda} \to \pi^+ \bar{p}$. Events near x-axis correspond to $\gamma \to e^+e^-$. Small regions of higher density near $p_\perp = 100$ Mev/c corresponding to $K_S - \Lambda$ and $K_S - \bar{\Lambda}$ ambiguity are clearly seen.

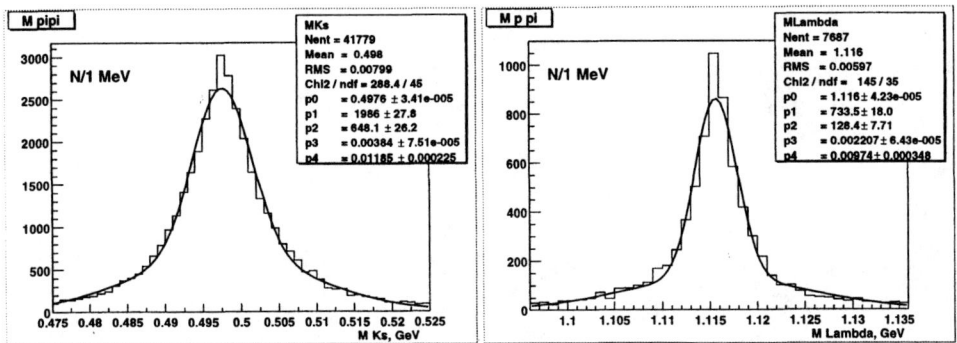

FIGURE 3. Measured $\pi^+\pi^-$ (left) and $p\pi^-$ (right) invariant mass. Both distributions are fitted by two gauss functions.

to the hypotheses of γ, K_S, Λ and $\bar{\Lambda}$ particles. For the cases where χ^2 values for different hypotheses were close, we took into account, in addition to χ^2, the probability of a given decay length to occur and the *a priori* probabilities of the K_S and $\Lambda(\bar{\Lambda})$ fluxes. For some events, K_S and Λ were identified ambiguously (Fig. 2), producing a background generated by pairs of K_S-mesons in hyperon sample, and vice versa. Resulting resolution is shown on Fig. 3.

RESONANCE STATES IN $K_S K_S$ SYSTEM

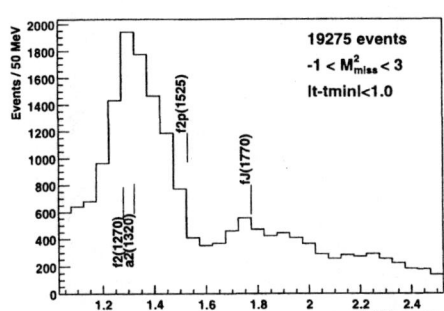

FIGURE 4. $K_S K_S$ effective mass distribution for events of reaction (1).

Large sample of 40000 events (19275 after cuts) obtained with neutral trigger on liquid hydrogen target in the reaction (1) starting from the run described in [1] up to our recent runs was used to analyze the 1200–1300 MeV/c^2 region of $K_S K_S$ effective mass, Fig. 4. K_S mesons were identified by the secondary decay vertex reconstruction as described earlier. The coordinates of the reconstructed primary vertex were checked to be compatible with the target position. Effective mass resolution for $K_S K_S$-system in this region is better than 5 MeV/c^2. Cuts on missing mass of the reaction select events of reaction (1) with small background. Fitted slope parameter of transverse momentum distribution $p_0 = 7.2$ GeV^{-2} demonstrates OPE-exchange domination in this reaction.

Threshold behavior of $K_S K_S$ distribution may be explained by roughly equal contributions of scalars $f_0(a_0)$.

The distribution of $K_S K_S$ mass in the reaction (1) shows (Fig. 4) clear peak from $f_2(1270)$ with small contribution of $a_2(1320)$, which has already been investigated in our previous works [2, 3] and also presented to this conference [4].

The system of two neutral kaons may be observed in even $J^{PC} = 0^{++}, 2^{++} \ldots$ While D_0 wave (Fig. 6) is rather small near the threshold, it is rapidly increasing at the left shoulders of $f_2(1270)$ and $a_2^0(1320)$ mesons. Clear interference type signal in D_0 wave near 1240 MeV/c^2 is shown in Fig. 6.

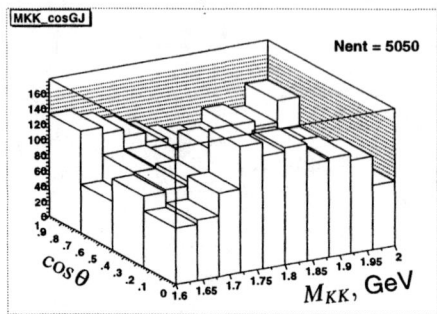

FIGURE 5. 2-D plot of $(M_{K_S K_S}, \cos\theta_{GJ})$ in the region 1600–2000 MeV/c^2.

An interesting interference phenomenon in the region 1200–1300 MeV/c^2 was also observed in two photon collisions studied with L3 detector at LEP [5] which is treated by authors in the text of paper [5] as $f_2(1270)$ and $a_2^0(1320)$ destructive interference, but the distribution near 1240 is fitted by the authors [5] by Breit-Wigner function with mass 1239 ± 6 MeV/c^2 and width 78 ± 19 MeV/c^2. On the other hand, our data show rather the interference effect with comparable width about 30 MeV/c^2 in the D_0 wave where $f_2(1270)$ and $a_2^0(1320)$ dominate. The difference in width may be due to the difference in experimental resolution: 29 MeV/c^2 in the case of L3 and bettter than 5 MeV/c^2 in our case.

Our recent results on reaction (2) show the same irregularity near 1240 MeV/c^2 on the left shoulder of $f_2(1270)$, $a_2^0(1320)$ which is shown on Fig. 7. This effect is seen even in the mass distribution of $K_S K_S$ system. Unfortunately available statistics is not still sufficient for more detailed PWA analysis.

2-dimensional distribution of $(M_{K_S K_S}, \cos\theta_{GJ}$ in Fig. 5 clearly demonstrates the resonant behavior of $K_S K_S$ spectrum in the region 1600–2000 MeV/c^2. The same distribution integrated over $\cos\theta_{GJ}$ is rather smooth, see Fig. 4.

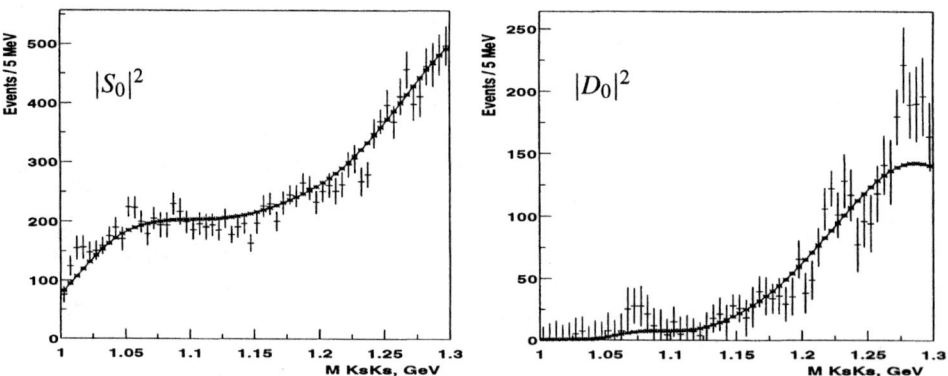

FIGURE 6. $|S_0|^2$ and $|D_0|^2$ as a function of $K_S^0 K_S^0$ effective mass near 1240 MeV/c^2 with step 5 MeV/c^2. Solid lines show results of 30 MeV/c^2 smoothing.

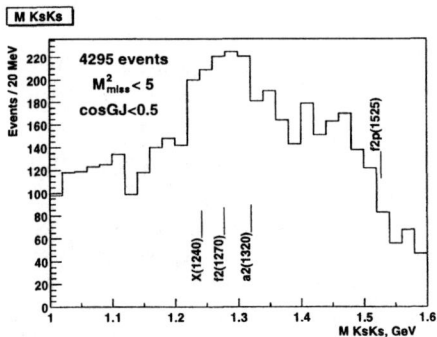
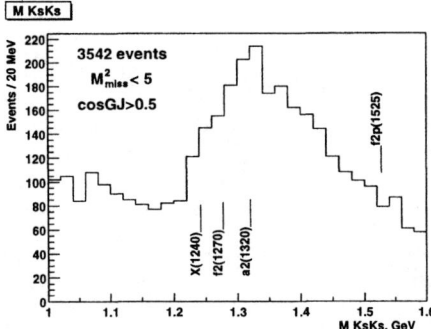

FIGURE 7. $K_S K_S$ effective mass distribution for events of reaction (2) with different selections on Gottfried-Jackson angle.

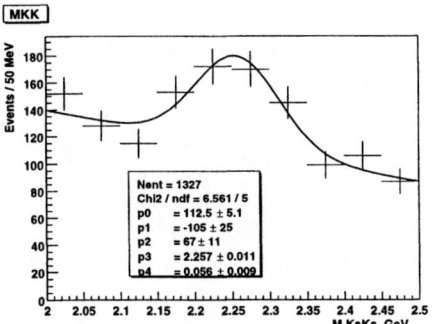

FIGURE 8. $K_S K_S$ effective mass distribution for events of reaction (1) near 2250 MeV/c^2 with selection $\cos\theta_{GJ} < 0.5$.

In $K_S K_S$ system produced in two photon collisions [5] there is no significant signal near 2230 MeV/c^2. In hadron collision in our data there is clear enhancement with mass 2257 MeV/c^2 and relatively small width 56 MeV/c^2 (Fig. 8). Our mass resolution in this region is better than 15 MeV/c^2. There is no simple way to explain this small width in the terms of quark-antiquark nature of this possible resonance. Even at mass of 1.3 GeV/c^2 width of $f_2(1270)$ is 185 MeV/c^2 and width should increase with increasing of resonance mass. On the other hand, the signal of $\xi(2230)$ is observed [6] in J/ψ radiative decays with mass 2232 MeV/c^2 and width 20 MeV/c^2.

The narrow interference phenomenon in the D_0-wave near $K_S^0 K_S^0$ mass 1240 MeV/c^2 exists in the reaction $\pi^- p \to K_S^0 K_S^0 n$. The same irregularity is observed in effective mass distribution of $K_S^0 K_S^0$ system from the reaction $\pi^- C \to K_S^0 K_S^0 A^*$ on the left shoulder of $f_2(1270)$, $a_2^0(1320)$.

Statistically significant bump in $K_S^0 K_S^0$ mass distribution near 2250 MeV/c^2 and relatively small width 56 MeV/c^2 is observed under cut $\cos\theta_{GJ} < 0.5$.

RESONANCE STATES IN $\Lambda\bar{\Lambda}$ SYSTEM

Interest in the investigation of baryon-antibaryon states is mainly caused by the fact that these states can be produced in the decay of usual mesons, thus providing a natural opportunity to study resonance states with very large masses. A number of broad resonances were found in the $\bar{p}p$ system [7] and in the $\Lambda\bar{p}$ ($\bar{\Lambda}p$) systems with strangeness [8, 9]. The $\Lambda\bar{\Lambda}$ system is also of interest because it may contain states with hidden strangeness, e.g., $s\bar{s}$ states of the ϕ- and f'_2-meson type. Relatively high mass of the $\Lambda\bar{\Lambda}$ system and the presence of strange quarks are the main reasons for experimental study of meson decay modes into this system [7, 8, 9]. Like the ordinary ($q\bar{q}$) states, quantum numbers of such mesons are: $P=(-1)^{l+1}$, $C=(-1)^{l+s}$. Isospin of the system I=0. For singlet states $s = 0$, $J^{PC} = 0^{-+}, 1^{+-}, ...$, for triplet P=C.

Quantum numbers of the $\Lambda\bar{\Lambda}$ system. The $\Lambda\bar{\Lambda}$ system consisting of a fermion and an antifermion has spin S = 0 or 1; total angular momentum **J = S + L**, where **L** is the orbital angular momentum; isospin $I = 0$; and negative intrinsic parity.

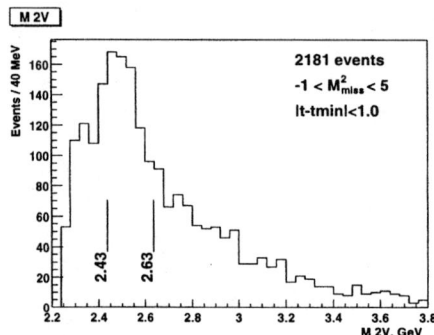

FIGURE 9. $\Lambda\bar{\Lambda}$ effective mass distribution for events of reaction (3).

The spatial parity is determined as $P = -(-1)^L = (-1)^{L+1}$, the charge parity as $C = (-1)^{L+S}$, and the combined PC parity depends on the spin S of the $\Lambda\bar{\Lambda}$ system. The singlet ($S = 0$) and triplet ($S = 1$) states have, respectively, opposite and identical signs of the P and C parities.

The system $\Lambda\bar{\Lambda}$ have been studied in [10, 11, 12] with statistics about 100 events and consequently without detailed analysis. Some experiments with larger statistics near the threshold [13, 14] showed polarization properties, but could not observe resonance behavior. Some evidence of $\Lambda\bar{\Lambda}$ resonance states was reported in our previous papers [15, 16, 17], more details are given in the report on this conference [18].

In this work, the production of pairs of Λ and $\bar{\Lambda}$ hyperons (Fig. 9,10) was analyzed on statistics of 2308 events detected in five exposures at 40 GeV measured at 6-meter spectrometer ITEP [19] in experiment SERP-E-173. These events are generated mainly in the "elastic" reaction (5).

$$\pi^- p \to \Lambda \bar{\Lambda} n, \quad \Lambda \to p\pi^-, \quad \bar{\Lambda} \to \bar{p}\pi^+ \tag{5}$$

The contribution of the "inelastic" channel

$$\pi^- p \to \Lambda \bar{\Lambda} X^0, \tag{6}$$

where X^0 is a neutral system with zero strangeness and mass larger than the neutron mass, is ~30%. The experimental sensitivity was ~70 events/nb.

Dozens of resonances are expected in the region 2.23–3.0 GeV/c². Our 2300 events of reactions (3) and (4) which are considerable part ($\sim$ 70 ev/nb) of fixed target world statistics, is still not enough for PWA.

For some events, K_S and Λ were identified ambiguously, resulting in a background generated by pairs of K_S-mesons. To suppress this background, the invariant masses of both forks were calculated on the assumption that they decay into π-mesons. The events for which both masses fell within the K_S-meson mass range (497±12 MeV/c²) were rejected. As a result, 10±2 % of the events were lost. Nevertheless, the final

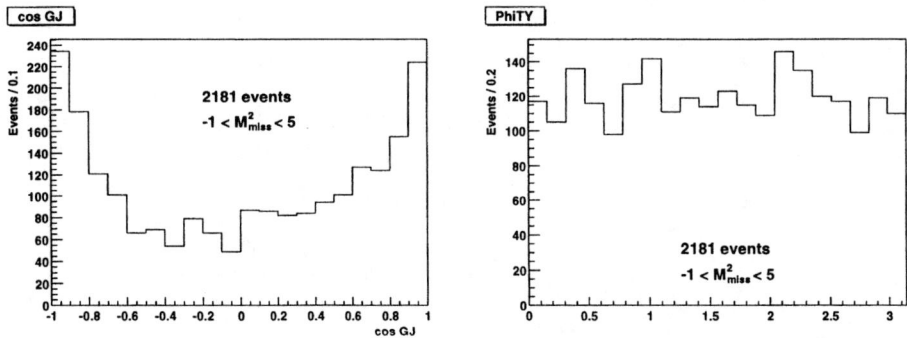

FIGURE 10. Gottfried-Jackson $\cos\theta_{GJ}$ (left) and Treiman-Yang angle ϕ_{TY} (right) distributions for full sample of $\Lambda\bar{\Lambda}$ events.

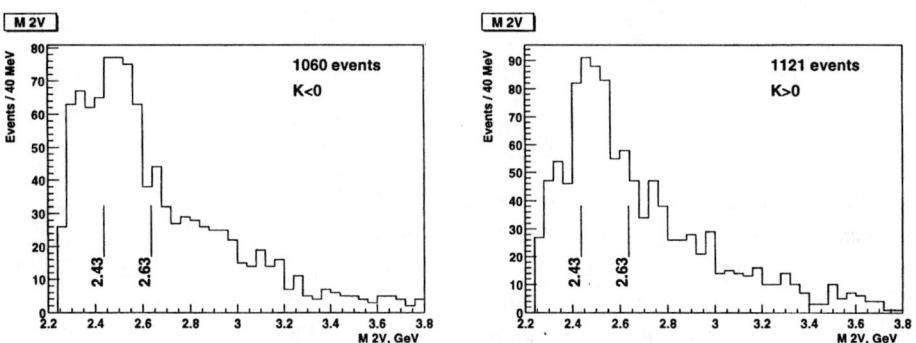

FIGURE 11. $\Lambda\bar{\Lambda}$ effective mass distributions for mainly triplet (left, $C > 0$) and singlet (right, $C < 0$) states.

samples of reactions (3) and (6) contained a 12±3% background. This background was estimated from the number of $2V^0$ events fitted to the $\Lambda\Lambda$ and $\bar{\Lambda}\bar{\Lambda}$ pairs forbidden by the conservation laws in the experimental topology.

High precision of coordinate detector of the spectrometer makes it possible to identify $K_S, \Lambda, \bar{\Lambda}$ and γ by kinematical properties of their decays into charged particles (Fig. 2) without identification of charged particle type.

The main part of background from $K_S \to \pi^+\pi^-$ decays is rejected by these cuts. By combining these cuts, $\Lambda\bar{\Lambda}$ sample is selected with purity 88%. The selection efficiency is estimated as 92%. Geometrical and track selection cuts were described in our previous papers [1, 3].

There is a sharp near-threshold rise which can be assigned to the resonances in this region and to the kinematic effect. A maximum at 2.5 GeV and a shoulder near 2.8 GeV may also be manifestations of the resonance structures. In our previous work [17], evidence was obtained for the existence of either resonances or sets of close resonances in these mass regions.

TABLE 1. Resonance parameters.

Mass (MeV)	Width (MeV)	Cut	A_p(norm.)	σ^* (nb)
2320 ± 20	65 ± 20	1	42 ± 15	1.8 ± 0.4
		2	21 ± 8	
2500 ± 10	205 ± 50	1	71 ± 12	2.9 ± 0.5
		2	41 ± 10	
2805 ± 15	90 ± 25	1	31 ± 10	1.0 ± 0.4
		2	6 ± 4	

Weak decays $\Lambda \to p\pi^-$ and $\bar\Lambda \to \bar p\pi^+$ give information on polarization and spin of the $\Lambda\bar\Lambda$ system. Parameter

$$C = -\frac{9}{\alpha^2} \overline{(\cos\theta_{p\bar p})} \qquad (7)$$

equals $C = 1$ for triplet state of $\Lambda\bar\Lambda$, $C = -3$ for singlet. In the definition (7) α is the polarization parameter $\alpha = \pm 0.647$, the angle $\theta_{p\bar p}$ is the angle between the directions of p and $\bar p$ moments.

The $\Lambda\bar\Lambda$ effective mass M distribution has three bumps near 2.3, 2.5 and 2.8 GeV/c^2. Figure 11 shows the same mass distribution, but for $C < 0$ and $C > 0$ samples separately.

Like $(q\bar q)$ states the $\Lambda\bar\Lambda$ system has the following quantum numbers: $P=(-1)^{l+1}$, $C=(-1)^{l+s}$, but isospin of the $\Lambda\bar\Lambda$ I=0. Bumps reported here can be considered as ordinary $(q\bar q)$ meson states or exotic one with $(q\bar q)$ admixture. We report here the preferable spin of $\Lambda\bar\Lambda$ only, but other quantum numbers remain unknown.

- Bumps near 2.3, 2.5 and 2.8 GeV/c^2 $\Lambda\bar\Lambda$ mass are clearly seen in the sample of 2300 events. Isospin I=0.
- State at 2.3 GeV/c^2 is mainly singlet. $J^{PC} = 0^{-+}, 1^{+-}, ...$
- State at 2.8 GeV/c^2 is mainly triplet.

MIS ITEP UNDER RECONSTRUCTION

The goal of the experiment SERP-E173 is the investigation of meson resonances decaying into neutral strange particles at Serpukhov energies. Our recent experimental results on narrow states in $K_S K_S$-system at Serpukhov energies showed great demand for more precise coordinate detector for experimental setup capable of measuring effective mass of this system with accuracy of few MeV. Upgraded coordinate detector for charged particles (Fig. 12) consists of 2680 30 mm drift tubes (Fig. 13) uniformly distributed in large workspace $6.0 \times 1.54 \times 0.8$ m^3 of 1.0 Tesla magnetic field. Tubes are grouped into perpendicular to the beam layers (Fig. 14). The number of charged particles is such that several tracks often pass through the same tubes of first layers. Reconstruction of track parameters is not a simple task for coordinate detector of any kind, because considerable

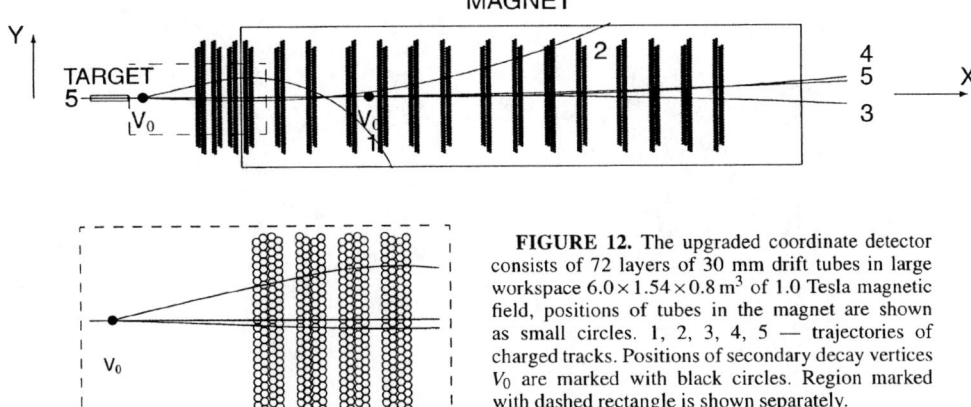

FIGURE 12. The upgraded coordinate detector consists of 72 layers of 30 mm drift tubes in large workspace $6.0 \times 1.54 \times 0.8\,\text{m}^3$ of 1.0 Tesla magnetic field, positions of tubes in the magnet are shown as small circles. 1, 2, 3, 4, 5 — trajectories of charged tracks. Positions of secondary decay vertices V_0 are marked with black circles. Region marked with dashed rectangle is shown separately.

part of neutral strange particles decays into secondary particles in nonuniform magnetic field on large distances from the primary interaction vertex. The distance between high energy secondary tracks is no more then few millimeters along first 30–80 centimeters of the track. Decay vertices are positioned near the beam axis and often are overlapped with other high energy charged tracks and noise beam particles. Additional complexity with drift tubes comes from left-right ambiguity of coordinate restored by drift time, and use of one-signal front electronics which makes tube insensitive to tracks passing the signal shadow.

Tracks in our experimental setup are overlapped for considerable part of it's length (Fig. 12), so reconstruction of event can not be reduced to selection of subset of fired wires, attribution it to a track and subsequent definition of track parameters according to this subset of measured coordinates [20]. Rather track must be localized in very small volume of it's parameter space. Precision of the setup permits to distinguish between $\simeq 10^{15}$ possible tracks.

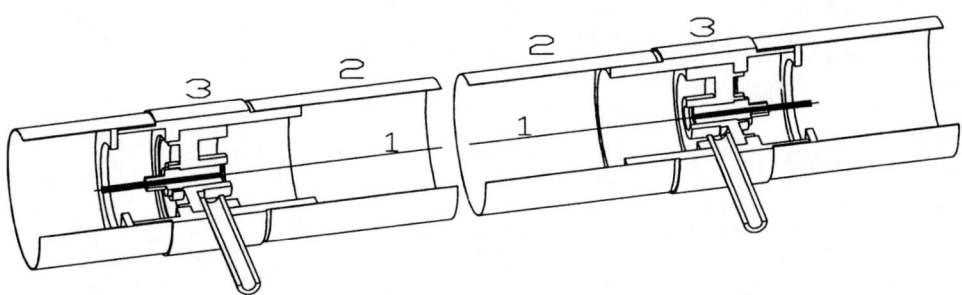

FIGURE 13. Drift tube (diameter 30-mm).

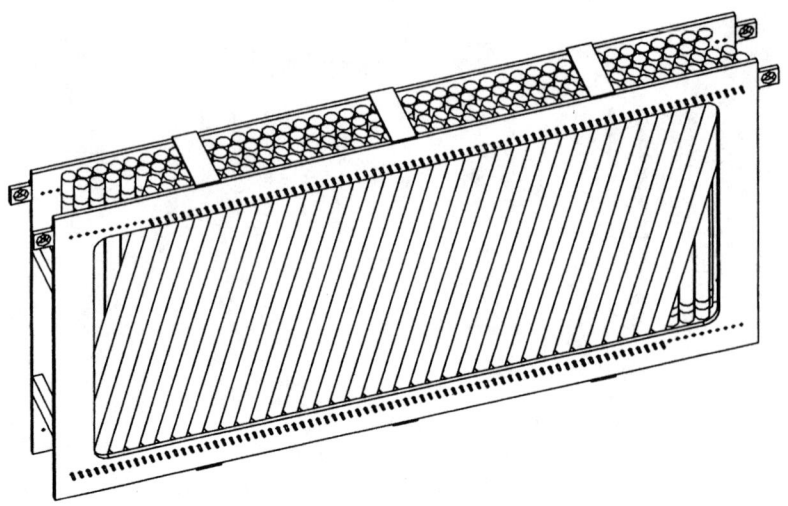

FIGURE 14.   4-plane module (162 drift tubes).

R&D are completed. Module assembling is in progress. We have proposed new algorithm [20] based on scanning of the whole space of track parameters, which is performed by decision tree. We hope to have first run before 2003.

REFERENCES

1. Bolonkin B.V. et al., *Nucl. Phys.* **B309** 426 (1988).
2. Barkov B. P. *et al.*, *in Proc. of HADRON91, College Park*, 47 (1991).
3. Bolonkin B.V. et al., *Yad. Phys.* **58** 1628 (1995); *Phys. Atom. Nucl.* **58**, 1535 (1995)
4. Lisin V.I, "Observation of narrow states of $K_S^0 K_S^0$ system on 6–m spectrometer and comparison with L3 results", this conference.
5. L3 Collab., Acciarri M. et al., *Phys. Lett.* **B501** 173 (2001).
6. BES Collab., Bai J.Z. et al., *Phys. Rev. Lett.* **76** 3502 (1996).
7. Rosanska M. et al., *Nucl. Phys.* **B162**, 505 (1980).
8. Baubillier M. et al., *Nucl. Phys.* **B183**, 1 (1981).
9. Armstrong T. et al., *Nucl. Phys.* **B227**, 365 (1983).
10. Beusch W. et al., *Phys. Lett.* **28B** 211 (1968).
11. Lichtman S. et al., *Nucl. Phys.* **B105** 229 (1976).
12. Anderson S. et al., *Preprint CLEO 96-19*, 1997.
13. Alexander G. et al., *CERN PPE/96-068*, 1996.
14. Barnes P.D. et al., *CERN PPE/96-58*, 1996.
15. Barkov B.P. et al., *Yad. Phys.* **22**, 223 (1975); *Sov. J. Nucl. Phys.* **22**, 113 (1975).
16. Bolonkin B.V. et al., *ITEP preprint* ITEP-86, Moscow, 1973.
17. Baloshin O.N. et al., *ITEP preprint* ITEP-2, Moscow, 1982.
18. Sokolovsky V.V., "Some Features of $\Lambda\bar{\Lambda}$ system in $\pi^-(p,C) \to \Lambda\bar{\Lambda} + A^*$ at 40 GeV", this conference.
19. Baloshin O.N. et al., *ITEP preprint* ITEP-154, Moscow, 1981.
20. Lisin V.I., Nozdrachev V.N., "Dichotomic Adaptive Algorithm for Track Reconstruction", in *New Computing Techniques in Physics Research VI*, edited by G.Athanasiu and D.Perret-Gallix, Parisianou, Athena, 2000, pp. 105-109.

SCALARS

Light scalar meson spectrum

Wolfgang Ochs

Föhringer Ring 6, D-80805 München, Germany

Abstract. We discuss the classification of the light scalar mesons with mass below 2 GeV into $q\bar{q}$ nonets and glueballs. The information on production and decay of these states, in particular recent information on the $f_0(980)$, $f_0(400-1200)$ (or $\sigma(600)$) and $f_0(1500)$ is considered. Although the data are not yet very precise the recent information is in favour of the previously developed scheme which includes $f_0(980)$, $a_0(980)$, $K_0^*(1430)$, $f_0(1500)$ into the lightest scalar nonet. The glueball in this approach appears as broad object around 1 GeV. Alternative schemes find the glueball at somewhat higher mass or suggest his mixing with $q\bar{q}$ states spread over a similar mass range. We do not see sufficient evidence yet for a light scalar nonet below 1 GeV around a $\sigma(600)$ resonance.

INTRODUCTION

Among the light mesons the scalars ($J^{PC} = 0^{++}$) are quite resistant against any classification which would be generally acceptable. In part this comes from the difficulty to identify very broad states, like $f_0(400-1200)$ and to determine their parameters or even to establish their presence, in part also from the possible existence of different types of mesons, namely besides the usual $q\bar{q}$ quarkonia, four quark or molecular states and glueballs. At the same time these possibilities are at the origin of the large interest in the scalar states as the lightest glueball is generally expected with these quantum numbers. It is therefore important to bring order into the scalar spectrum.

The existence of glueballs is confirmed by the nonperturbative QCD calculations and their discovery is a challenge for theory and experiment. In the quenched approximation of lattice QCD, i.e. neglecting sea quarks, one finds the mass of the scalar glueball in the range 1400-1800 MeV (recent reviews [2, 3]). There are still considerable uncertainties concerning the effects of sea quarks and the masses of the light quarks. The glueballs with tensor and pseudoscalar quantum numbers are next heavier in mass and already close to or above 2 GeV. At this higher mass it is certainly more difficult to establish the nature of the observed resonances, therefore the scalar glueball is the primary target for glueball searches. An alternative approach to glueball masses is based on QCD sum rules. In these calculations a light gluonic state near 1 GeV is demanded from a particular sum rule [4, 5]. According to these QCD results it seems plausible to search for the lightest scalar glueball in the range 1 - 2 GeV and not in a much narrower region.

In a first step it should be clarified which states with $J^{PC} = 0^{++}$ are really established and what their internal flavour structure is. Then one can attempt to group them into $q\bar{q}$ nonets. The existence of glueballs is indicated if there are supernumerary states. In the following we discuss in particular results from Ref. [6, 7] including new analyses [8] and compare with other approaches.

LIGHT SCALAR MESONS: EVIDENCE AND FLAVOUR STRUCTURE

First we attempt to identify the $q\bar{q}$ meson nonet(s). The scalar states listed by the Particle Data Group (PDG) are shown in Table 1. In this talk we concentrate our attention to the isoscalar (f_0) states of lowest mass and the construction of the lightest nonet.

In order to establish a resonance in a general environment with background, which case is relevant to our discussion, we request as necessary condition for a state to be acceptable a definite evidence for the movement of the partial wave amplitude in both magnitude and phase according to a local Breit Wigner representation, i.e. a pole in the complex energy plane in general above some backgound. We begin our discussion with two well established isoscalars where the debate concerns their intrinsic structure and continue with two others whose very existence we consider in doubt.

The $f_0(980)$ meson

The existence of this state is well established by early phase shift analyses [9, 10]. There is a continuing debate on whether its internal structure corresponds to a quarkonium or rather to a 4-quark or molecular state. We follow here the standard quark model for simplicity as far as possible in the hope that some problems may disappear with improved calculations. The $q\bar{q}$ assumption is supported by various observations which are not natural for a complex 4q state: the close similarity in various production properties with other quarkonia of similar mass (like ρ, $\phi(1020)$, $a_0(980)$, η', in both e^+e^- annihilation [11] and νp interactions [12], see recent review [13]); the dominance of $f_0(980)$ production at large momentum transfer $|t| \sim 0.5$ GeV^2 in πp collisions [14] which suggests a $f_0 \pi A_1$ coupling [6]; the strong production in D, D_s decays (see below) through intermediate $d\bar{d}$ and $s\bar{s}$ states. This discussion will have to continue until a consistent description of all phenomena is achieved.

In the following we discuss the predictions from the quarkonium model for various ratios of observables, where the dependences on the less known intrinsic structures are expected to cancel. These ratios depend only on the mixing angle φ_s in the scalar nonet which we define through the amplitudes into strange and non-strange components

$$f_0(980) = \sin\varphi_s n\bar{n} + \cos\varphi_s s\bar{s} \quad \text{with} \quad n\bar{n} = (u\bar{u} + d\bar{d})/\sqrt{2}. \qquad (1)$$

The results can ultimately be compared with corresponding predictions from molecular models if available. A consistent description of data in terms of only one parameter φ_s is then a crucial test of the quarkonium model.

TABLE 1. Scalar mesons below 2 GeV according to Particle Data Group [1]

I = 0	$f_0(400-1200)$ (or σ)	$f_0(980)$	$f_0(1370)$	$f_0(1500)$	$f_0(1710)$	$f_0(2020)$?
I = $\frac{1}{2}$				$K_0^*(1430)$		$K^*(1950)$?
I = 1		$a_0(980)$		$a_0(1450)$		

TABLE 2. Summary of scalar mixing angle φ_s from three ratios of branching fractions involving the state $f_0(980)$.

	$R_1 = \dfrac{J/\psi \to \phi f_0}{J/\psi \to \omega f_0}$	$R_2 = \dfrac{f_0 \to \gamma\gamma}{a_0 \to \gamma\gamma}$	$R_3 = \dfrac{f_0 \to K\bar{K}}{f_0 \to \pi\pi}$
R_{exp}	2.3 ± 1.1	1.3 ± 0.6	$g_K^2/g_\pi^2 = 2.0 \pm 0.6$
R_{theor}	$\dfrac{p_\phi}{p_\omega} \cot^2 \varphi_s$	$\dfrac{2}{9}\left(\dfrac{5}{\sqrt{2}}\sin\varphi_s + \cos\varphi_s\right)^2$	$\dfrac{2}{3}\left(\cot\varphi_s + \dfrac{S}{\sqrt{2}}\right)^2$
φ_{s1}	$33 \pm 7°$	$25 \pm 12°$	$41° \pm 8°$
φ_{s2}	$147 \pm 7°$	$123 \pm 12°$	$153.8° \pm 3.4°$

The definition (1) is in analogy to the common definition for pseudoscalars

$$\eta' = \sin\varphi_p n\bar{n} + \cos\varphi_p s\bar{s} \tag{2}$$
$$\eta = \cos\varphi_p n\bar{n} - \sin\varphi_p s\bar{s}. \tag{3}$$

A recent determination of the pseudoscalar mixing angle yielded [15]

$$\varphi_p = 39.3° \pm 1.0°. \tag{4}$$

This result corresponds approximately to components $(u\bar{u}, d\bar{d}, s\bar{s})$

$$\eta' = (1,1,2)/\sqrt{6} \quad \text{near singlet} \quad (1,1,1)/\sqrt{3} \tag{5}$$
$$\eta = (1,1,-1)/\sqrt{3} \quad \text{near octet} \quad (1,1,-2)/\sqrt{6} \tag{6}$$

with mixing angle $\varphi_p = 35.3°$ near singlet-octet angle $\varphi_p = 54.7°$.

We consider here three ratios of branching ratios as in [6] but now express them in terms of the mixing angle φ_s. Then we can get a quantitative measure of the consistency of the approach. We summarize here only the final results in Table 2.

The experimental values for the ratios R_i are determined from the PDG results where available. The ratio R_1 estimates the ratio of strange and nonstrange components of the f_0. This ratio has been determined for the pseudoscalars and yielded a mixing angle consistent with all other determinations [15]. Remarkably the J/ψ-branching ratios entering R_1 are very similar for η' and $f_0(980)$ which gives the first hint towards the close similarity of the scalar and pseudoscalar multiplets. The ratio $R_1 = 2$ would correspond to the quark composition $\eta' = (1,1,2)/\sqrt{6}$ in (5). Accordingly, the mixing angle $\varphi_s \sim \varphi_p$, in addition a second solution φ_{s2} is possible.

The ratio R_2 is calculated from the $q\bar{q}$ annihilation amplitudes which are proportional to the squares of quark charges Q_q^2. It is assumed here that the $a_0(980)$ is a quarkonium as well with wave function $a_0 = (u\bar{u} - d\bar{d})/\sqrt{2}$.

For the ratio R_3 we assume a strange quark suppression amplitude $S = 0.8 \pm 0.2$ close to Ref. [16]. The reduced branching ratio g_K^2/g_π^2 is taken from determinations based on measurements of both $K\bar{K}$ and $\pi\pi$ final states in central production [17] and with low background in large t πp collisions [18].

The results for the mixing angles from the measured ratios are nicely compatible for the small angle solution ($\chi^2 = 1.4$) with

$$\varphi_s = 35° \pm 4° \tag{7}$$

whereas the large angle solution $\varphi_s = 154° \pm 3°$ closer to octet is disfavoured ($\chi^2 = 6.4$, Prob $\sim 1\%$). These results are based on three ratios and have little model dependence. Our favoured solution φ_s is similar to the pseudoscalar mixing angle φ_p in (4) as already suggested in [6].

Our two solutions are similar to those of Ref. [16] (using $g_K^2/g_\pi^2 \sim 1.5$) whereas in Ref. [19] calculations based on two absolute rates yielded the small angle solution $\varphi_s = 4° \pm 3°$ which is rejected in favour of the large angle solution $\varphi_s = 138° \pm 6°$.

Independent information on the relative phase of the $s\bar{s}$ vs. $n\bar{n}$ components in the $f_0(980)$ wave function is accessible from D and D_s charmed meson decays. Consider first the decay $D_s^+ \to \pi^+ K^+ K^-$ which shows a strong $\phi(1020)$ signal overlapping with $f_0(980)$. The dominant decay of D_s^+ proceeds through the emission of a π^+ and formation of an intermediate $s\bar{s}$ state subsequently decaying into ϕ and f_0 states with amplitudes 1 and $\cos\varphi_s$ according to (1). The absolute phases of ϕ and f_0 have been determined by the E687 Collaboration [20] as $(178 \pm 20 \pm 24)°$ and $(159 \pm 22 \pm 16)°$, i.e. the relative phase is consistent with zero degrees. Therefore

$$\cos\varphi_s > 0 \quad \to \quad 0 < \varphi_s < 90° \tag{8}$$

in agreement with the small phase solution (7).

The determination of the relative phases depends on the definition of the scattering angle. Under an exchange of the two particles of the decay the S-P wave interference term considered here would change sign. As a check we therefore studied a similar situation in the decay $D^+ \to \pi^+\pi^-\pi^+$ with the relative phase between the amplitudes $D^+ \to \pi^+\rho^0$ and $D^+ \to \pi^+ f_0(980)$. According to the dominant mechanism these resonances are produced through intermediate $d\bar{d}$ states with amplitudes $-1/\sqrt{2} < 0$ and $\sin\varphi_s/\sqrt{2} > 0$ respectively, so one expects a 180° phase difference. This expectation is in fact verified by the measurements by both E687 [20] and E791 Collaborations [21, 22] which confirms that the standard definition of the angles yields consistent results.

Another interesting ratio is $R = (\phi \to a_0(980)\gamma)/(\phi \to f_0(980)\gamma)$. As the decays involve two decay mechanisms with or without $s\bar{s}$ annihilation we have no straightforward prediction. An explanation is possible in terms of $a_0 - f_0$ mixing [23].

The $f_0(1500)$ meson

This state can be considered as well established by now also. In $p\bar{p} \to 3\pi$ the Dalitz plot has been fitted with some phase sensitivity and the S wave nature has been demonstrated (CBAR Collaboration [24]). Meanwhile various branching ratios became known.

The phase movement has also been seen in the Argand diagrams of $\pi\pi \to K\bar{K}$ and $\pi\pi \to \eta\eta$ as obtained from the πN production experiments which have been reconstructed using the data on $|S|$, $|D|$ and ϕ_{SD} together with Breit Wigner fits to the tensor mesons which provide the absolute phase [6]. The comparison of both reactions has also demonstrated through its interference with the tensor mesons that

$$T(\pi\pi \to f_0(1500) \to K\bar{K}) = -T(\pi\pi \to f_0(1500) \to \eta\eta) \tag{9}$$

which implies that the $f_0(1500)$ has an opposite sign of the $n\bar{n}$ and $s\bar{s}$ components [6]. A similar Argand diagram has been obtained for $\pi\pi \to K\bar{K}$ [25]; the opposite orientation of both amplitudes in (9) is also visible in the energy dependent fits in [26] although with different overall phase. Unfortunately, the elastic $\pi\pi$ scattering is not yet uniquely determined in this region.

Further interesting information on this meson can again be obtained from decays of D and D_s charmed mesons. In the decay $D_s \to \pi\pi\pi$ the scalars f_0 can be produced through the intermediate process $s\bar{s} \to \pi\pi$. This favours intermediate states with large $s\bar{s}$ - $n\bar{n}$ mixing. One observes a strong signal from $f_0(980)$ which proves again its strong $s\bar{s}$ component and a higher mass state related to $f_0(1500)$ by E687 [20] and to $f_0(1370)$ by E791 [21]. The signal is strongest near the edge of phase space in the Dalitz plot where the two resonance bands cross but the mass and width appear to be closer to $f_0(1500)$. Ultimately the study of branching ratios of this state has to decide. For the time being we take this state as $f_0(1500)$.

An interesting feature common to both experiments is the large relative phase consistent with 180° between the production amplitudes of $f_0(980)$ and "$f_0(1500)$" which we interpret as

$$T(s\bar{s} \to f_0(980)) = -T(s\bar{s} \to f_0(1500)). \tag{10}$$

This negative phase in the fit to the Dalitz plot obviously corresponds to a lack of enhancement at the off diagonal crossing point of the two resonance bands in this plot which contrasts the strong enhancement in the diagonal crossing points.

The mass of $f_0(1500)$ is close to the glueball mass obtained in quenched approximation of lattice QCD. This at first has lead to models with close connection between these two states [27]; further studies now prefer mixing models where the superposition of the glueball and nearby $q\bar{q}$ mesons correspond to the physical states $f_0(1370)$, $f_0(1500)$ and $f_0(1710)$ (for overview, see [13]). As an example we quote a recent result motivated by lattice calculations on mixing [28, 3]

$$f_0(1500) = -0.36\,n\bar{n} + 0.91\,s\bar{s} - 0.22\,\text{glueball}. \tag{11}$$

In this example the glueball component of the $f_0(1500)$ has a weight of only ~5%.

Contrary to a single glueball which would mix with the flavour singlet we see in (11) $n\bar{n}$ and $s\bar{s}$ with opposite sign. This octet type flavour mixing is in line with our findings (9),(10) and appears as "robust result" in fits of the above kind [29]. On the other hand, our finding (9) not only requires an octet type flavour mixing but also provides an upper limit to the glueball contribution; this contribution would add with the same sign to all pairs of pseudoscalars. This is an important additional limitation to such fits not yet taken into account so far.

If we take these observations together, especially the large components of both $n\bar{n}$ and $s\bar{s}$ (suggested from D_s decays) and their negative relative sign, then $f_0(980)$ and $f_0(1500)$ look like the orthogonal isoscalar members of the $q\bar{q}$ nonet. We will come back to this idea below.

The $f_0(400-1200)$ and the $\sigma(600)$ meson

This entry in the PDG refers to results from $\pi\pi$ phase shift analyses and from the observation of peaks in mass spectra around 400-600 MeV. We give a short account of these observations.

$\pi\pi$ phase shifts and $f_0(400-1200)$

It is a common feature of fits to $\pi\pi$ scattering that there is one broad object where the width is comparable to the mass (see [6], for example). The $\pi\pi$ scattering amplitudes are rather well known by now up to ~ 1.4 GeV, from single pion production with and without polarized target. Recent studies have removed remaining ambiguities [30] below 1 GeV in favour of a slowly rising S wave phase shift in the ρ region, excluding in particular a rapidly varying phase and resonance under the ρ, in essential agreement with the old results which were obtained using particular assumptions on the production mechanism [10]. A theoretical analysis [31] based on the constraints from S matrix theory provides a good description of the observed low energy (below 1 GeV) pion-pion interactions with slowly varying S wave.

The interpretation of the $I = 0$ S wave in terms of resonant states is less straightforward. The phase shifts pass through $90°$ at ~ 1000 MeV once the $f_0(980)$ effects are subtracted. This suggests a state at 1000 MeV [32, 6] with a large width of 500-1000 MeV. With a negative background phase added the resonance position can be shifted towards lower values and this has been considered as state $\sigma(600)$ in [33]. Fits over a large mass region including a background term yield resonance poles in the scattering amplitude around 1300 MeV or higher, again with a large width [34]. With such broad states the determination of the resonance mass depends on the assumed background in an essential way. There is a strong $\pi\pi$ interaction around 1 GeV and beyond but not necessarily and exclusively a broad $\sigma(600)$.

Peaks in mass spectra and $\sigma(600)$

There are a number of effects which have been related to $\sigma(600)$.

1. Decay $J/\psi \to \omega\pi\pi$

There is a peak around 500 MeV in the $\pi\pi$ mass spectra besides a strong signal from $f_2(1270)$ [35]. For a Breit Wigner σ resonance at 500 MeV the interference term $Re(SD^*)$ between the (almost real) D wave and the resonant S wave would change sign at the pole position and so the angular distribution $d\sigma/d\Omega \sim |S|^2 + (3\cos^2\vartheta - 1)Re(SD^*)/2 + O(|D|^2)$ would vary accordingly with a sign change of the $\cos^2\vartheta$ term (from + to −). The data [35] do not show any sign change below 750 MeV and therefore there is no indication for a Breit Wigner resonance below this mass.

2. $Y', Y'' \to Y\pi\pi$ and similar decays of J/ψ

Mass peaks are observed here as well. Unfortunately the angular distributions are not measured as function of the mass. Hopefully such measurement will be provided in the future to study possible phase variations and resonant behaviour.

3. Central production $pp \to p(\pi\pi)p$

At small momentum transfers between the protons this process is assumed to be dominated by double Pomeron exchange. The centrally produced $\pi\pi$ system peaks shortly above threshold below 400 MeV [36, 37, 38, 39] and has been related to the $\sigma(600)$ as well [37]. There are some other remarkable features in this process. Quite unusually, there is a strong D wave near threshold as well which peaks near 500 MeV; the total D wave contributions $\sum_\lambda |D_\lambda|^2$ at their peak are about five times larger than the $f_2(1270)$ contribution and about one third of the S wave contribution at its peak.

These observations are very similar to findings in $\gamma\gamma \to \pi\pi$ which suggest a close relation between the processes [8]

$$\text{Pomeron Pomeron} \to \pi\pi \quad \leftrightarrow \quad \gamma\gamma \to \pi\pi \qquad (12)$$

In fact, the $I = 0$ S wave component obtained from a fit to $\gamma\gamma \to \pi\pi$ for charged and neutral pions [40] peaks below 400 MeV and the D wave near 500 MeV with similar ratio 1/3; the origin of this unusual behaviour is the contribution of one-pion-exchange to $\gamma\gamma \to \pi^+\pi^-$. Therefore we propose that one-pion-exchange dominates the double Pomeron reaction at small $\pi\pi$ masses as well [8].[1] It reproduces the main characteristics. As the pion pole is near the physical region the $\pi\pi$ angular distribution is very steep, steeper than in more typical interactions mediated by vector (ρ) exchange. Therefore one estimates that the D wave becomes important not at m_{f_2} but already at $m_{f_2} \times (m_\pi/m_\rho) \sim 0.3$ GeV. This mechanism also explains the low mass peak of the S wave without associated phase variation. In fact, in the region of the peak no strong S-D phase variation is observed [38, 39] as would be expected for a $\sigma(600)$ resonance. The presence of the one-pion-exchange process does not exclude the presence of broad states as in $\pi\pi$ elastic scattering either from $\pi\pi$ rescattering or by direct formation, very much as it is discussed in the $\gamma\gamma$ process [40]; but there is no evidence for an additional low mass $\sigma(600)$ resonance near the peak.

4. Decay $D^+ \to \pi^-\pi^+\pi^+$

The $\pi^+\pi^-$ mass spectrum presented by the E791 Collaboration [21, 22] shows three prominent peaks, one just above $\pi\pi$ threshold, one related to ρ and one to $f_0(980)$. Only fits including a light σ particle have been found successful according to their analysis. In principle, the low mass peak could be due to the decays of resonances with higher spin $J \geq 1$ in the crossed channel.

Again, one would like to see the related Breit Wigner phase motion in a more direct way. There should be a large term $Re(SP^*)$ from the interference $\sigma - \rho$ which changes sign near the σ resonance in $s_{12}(\pi^-\pi_1^+)$ and therefore changes sign of the forward backward asymmetry in the σ decay angle $\cos\vartheta$ which is linearly related to the mass variable $s_{13}(\pi^-\pi_2^+)$ along the σ resonance (s_{12}) band in the Dalitz plot. Such an effect is actually visible in the Dalitz plot at the ρ resonance (presumably from the ($\rho - f_0(980)$) interference): the sign change of the asymmetry causes the appearance of the tails towards lower and higher mass (s_{12}) at the upper and lower part (s_{13}) of the ρ band respectively. A study with sufficiently fine binning should reveal this effect for the σ

---

[1] Some global properties of this process have been considered already long ago [41].

in the resonance Monte Carlo for the $\sigma + \rho$ superposition and prove or disprove the presence of the interference effect in the data (a sensitive observable is $\langle\cos\vartheta\rangle \frac{d\sigma}{ds_{12}}$ together with $\frac{d\sigma}{ds_{12}}$).

The $f_0(1370)$ meson

There is strong interaction in $\pi\pi$ and other channels in this mass region but again the clear evidence for a localized Breit Wigner phase motion is missing to support the resonance hypothesis. The $f_0(1370)$ and $f_0(400-1200)$ look like parts of a broader state in the channels with two pseudoscalars [6]. The missing information could be provided by phase shift analysis of recent high statistics $\pi^0\pi^0$ data [14, 42]. Furthermore, the resonance interpretation is found not consistent [13] with the different branching ratios observed in the 4π channel; there could be a broad background state interfering with the narrow $f_0(1500)$ to cause a peak near 1370 MeV. Again a phase shift analysis is necessary to clarify the situation.

CONSTRUCTION OF THE LIGHTEST QUARK ANTI-QUARK NONET

For the time being neither the $\sigma(600)$ nor the $f_0(1370)$ are acceptable for us as genuine resonant states. There are peaks at these masses but the associated Breit Wigner phase variation has not yet been established. Then the most natural candidates for the lightest $q\bar{q}$ nonet are

$$f_0(980), \quad f_0(1500), \quad K_0^*(1430), \quad a_0(980) \tag{13}$$

with mixing pattern very similar to the one observed in the pseudoscalar nonet

$$f_0(980) \sim \eta' \sim \text{singlet} \qquad f_0(1500) \sim \eta \sim \text{octet}. \tag{14}$$

A solution like this has been proposed on the basis of a renormalizable sigma model with instanton interactions [43]. It also has $f_0(1500)$ as octet member but would prefer $a_0(1450)$ over $a_0(980)$ as isovector. It explains why the octet state is above the singlet for the scalars and vice versa for the pseudoscalars. Similar models have been studied in [44, 45].

Independently, the correspondence (14) and the nonet (13) have been proposed on the basis of the phenomenological analysis [6]. These results were found consistent with a general QCD potential model for sigma variables; in this analysis the choice $m(a_0(980)) \approx m(f_0(980))$ is possible although not required or explained. Furthermore, it has been shown that the octet in (13),(14) fulfills approximately the Gell Mann Okubo formula, and from a_0 and K_0^* one predicts

$$3(m_{f_8}^2 - m_a^2) = 4(m_{K_0^*}^2 - m_a^2) \qquad \rightarrow \qquad m_{f_8} = 1550 \text{ MeV} \tag{15}$$

a good result for the octet isoscalar.

In the low mass region we are now left with $f_0(400-1200)$ and $f_0(1370)$ to which we come back below. It is interesting to note that the remaining states in the PDG below 2 GeV (Table 1) can be grouped together into a second nonet which includes

$$f_0(1720), \quad f_0(2020), \quad K_0^*(1950), \quad a_0(1450). \tag{16}$$

In this case the Gell Mann Okubo formula predicts for the octet scalar $m_{f_8} = 2.080$ GeV which fits to the highest state in (16) and therefore this nonet would repeat the mixing pattern of the lowest nonet. However as there is little further information on these states this assignment is rather speculative.

CANDIDATE FOR LIGHTEST GLUEBALL

After having selected the lightest nonet from well established resonances we are left with $f_0(400-1200)$ and $f_0(1370)$. The $\pi\pi$ data are consistent with the view that both states correspond to the low and high mass tails of a single resonance ("red dragon" [6]) and this state we take as the glueball

$$f_0(400-1200) \quad \text{and} \quad f_0(1370) \quad \rightarrow \quad gb(1000). \tag{17}$$

This mass corresponds to a resonance fit of $\pi\pi$ elastic scattering without background, other fits have lead to higher masses with the option of a broad glueball near 1400 MeV [34]. These results are a bit lower than expected from the lattice results in quenched approximation but looking at the large width and the still approximate nature of QCD results there is not necessarily a contradiction.

Our detailed arguments in favour of this glueball assignment have been summarized elsewhere [7, 8], together with plausible arguments for a large width of an S wave binary glueball. Here we only recall the most relevant observations.
1. The state $gb(1000)$ is produced in most reactions which are considered as gluon rich:
a. central production $pp \rightarrow pXp$;
b. Decays of radially excited heavy quarkonia like $Y', Y'' \rightarrow Y(\pi\pi)$;
c. $p\bar{p} \rightarrow 3\pi$
d. There is no prominent signal however in $J/\psi \rightarrow \gamma\pi\pi$ for $m_{\pi\pi} < 1$ GeV, this could possibly be due to instrumental problems at small masses.
2. The production in $\gamma\gamma$ collisions is untypically small [7] (based on fits [40]).
3. The decays of $f_0(1370)$ (part of glueball) favour the gb over the $n\bar{n}$ assignment [6, 46].

Alternatively one may attempt to explain the strong $\pi\pi$ interaction in the 1 GeV region without direct channel resonance in terms of $\rho - f$ exchange processes (ρ alone would not explain the strong $\pi^0\pi^0$ interaction) and to obtain the moving phase from a unitarization procedure (see review [13]). In considering this proposal we note that a t-channel analysis of $\pi\pi$ scattering [47] indicates a large component in the $I_t = 0$ channel which is not related to $q\bar{q}$ Reggeon exchange but indicates Pomeron exchange or gluonic interactions already for $m_{\pi\pi} < 1$ GeV [7, 8]. So in this explanation one has to take into account the presence of non-resonant effects and avoid double counting of direct and crossed channel exchanges where the fit has to be done to all isospin amplitudes.

SUMMARY AND CONCLUSIONS

We have presented the arguments to include the isoscalars $f_0(980)$ and $f_0(1500)$ in the lightest scalar nonet. Various ratios of branching fractions as well as new results on relative phases between different $q\bar{q}$ components and between the production amplitudes could be explained consistently in terms of one parameter, the scalar mixing angle $\varphi_s = 35° \pm 4°$. This is found close to the pseudoscalar equivalent but very different from "ideal mixing" with $\varphi_s = 0°, 180°$. The nonet (13) fulfills the Gell Mann Okubo formula which relates the masses of octet particles assuming symmetry breaking by quark mass terms. It will be important to improve the accuracy of these measurements as test of this classification scheme. Also it will be interesting to see whether the $K\bar{K}/$ 4q model for $f_0/a_0(980)$ can explain the data discussed here.

We do not see evidence yet for the Breit-Wigner resonance nature of peaks related to $\sigma(600)$. It will be important to investigate the phase motion in $D \to $ "σ"π. In some cases there is counter evidence for the expected phase motion (like $J/\psi \to \omega\pi\pi$ and $pp \to p(\pi\pi)p$). Therefore the existence of a meson multiplet below 1 GeV is not apparent. Also there is a lack of evidence for $f_0(1370)$ so far. There are data available which could be analysed in this respect (for example $\pi^- p \to \pi^0\pi^0 n$).

The PDG listing allows for a second scalar nonet below 2 GeV with similar mixing.

This leaves the broad state around 1 GeV (built from $f_0(400-1200)$ and $f_0(1370)$ + more (?)) with the large width of 500 - 1000 MeV as a candidate for the lightest scalar glueball. This assignment is in agreement with most phenomenological expectations. Also we do not see an obvious disagreement with QCD results taking into account the approximate nature of the calculations.

In alternative approaches a superposition of glueball and two neighbour scalar quarkonia (from nonet) builds up the physical states $f_0(1370)$, $f_0(1500)$ and $f_0(1710)$. In effect this would imply that the glueball is not localized in a narrow mass interval but contributes to scattering processes in quite a large mass range of $\sim$500 MeV as it happens in our "broad glueball" scheme. Major differences are in the $q\bar{q}$ sector, especially concerning the existence of $f_0(1370)$ and treating the left alone states $f_0, a_0(980)$ as molecules or forming an additional nonet around "$\sigma(600)$". Therefore improved experimental data and analyses in the low energy region are of great importance.

Of course it would be nice to have a more direct evidence for the gluonic nature of a particular candidate. A promising tool is the comparative study of glueball candidates in the fragmentation region of both quark and gluon jets [48, 49]. First results [50] look promising in indicating an extra neutral component in the gluon jet. There is something to look forward to.

ACKNOWLEDGMENTS

I would like to thank Peter Minkowski for discussions and collaboration on the problems presented in this talk.

REFERENCES

1. Particle Data Group, D.E. Groom et al., *Eur. Phys. J.* C**15**, 1 (2000).
2. G.S. Bali, "Glueballs: Results and Perspectives from the Lattice", hep-ph/0110254.
3. C.J. Morningstar, this conference.
4. S. Narison, *Nucl. Phys. B* **509** (1998) 312.
5. T.G. Steele and D. Harnett, "Two Topics in QCD Sum-Rules", hep-ph/0108232.
6. P. Minkowski and W. Ochs, *Eur. Phys. J. C* **9**, 283 (1999).
7. P. Minkowski and W. Ochs, *Proc. Workshop on Hadron Spectroscopy*, Frascati, March 1999, Italy, Eds. T. Bressani et al., Frascati Physics Series XV, p.245 (1999).
8. P. Minkowski and W. Ochs, in preparation.
9. S.D. Protopopescu et al., *Phys. Rev. D* **7**, 1279 (1973).
10. B. Hyams et al., *Nucl. Phys. B* **64**, 4 (1973); W. Ochs, thesis 1973 (unpublished).
11. OPAL Collaboration, K. Ackerstaff et al., *Eur. Phys. J. C* **4**, 19 (1998).
12. NOMAD Collaboration, P. Astier et al., *Nucl. Phys. B* **601**, 3 (2001).
13. E. Klempt, "*Meson Spectroscopy: Glueballs, Hybrids and $Q\bar{Q}$ Mesons*", hep-ex/0101031.
14. GAMS Collaboration: D. Alde et al. *Z. Phys. C* **66**, 375 (1995); *Phys. Atom. Nucl.* **62** 1993 (1999).
15. T. Feldmann, P. Kroll and B. Stech, *Phys. Rev. D* **58** 114006 (1998).
16. V.V. Anisovich, L. Montanet and V.N. Nikonov, *Phys. Lett. B* **480** 19 (2000).
17. WA102 Collaboration: D. Barberis et al., *Phys. Lett. B* **462**, 462 (1999).
18. D.M. Binnie et al., *Phys. Rev. Lett* **31**, 1534 (1973).
19. A.V. Anisovich, V.V. Anisovich and V.A. Nikonov, hep-ph/0011191.
20. E687 Collaboration, P.L. Frabetti et al., *Phys. Lett. B* **351**, 591 (1995); **407**, 79 (1997).
21. E791 Collaboration, E.M. Aitala et al., *Phys. Rev. Lett.* **86**, 770 (2001); **86**, 765 (2001).
22. Carla Göbel, this conference; hep-ex/0110052.
23. F.E. Close and A. Kirk, *Phys. Lett. B* **489**, 24 (2000).
24. Crystal Barrel Collaboration: V.V. Anisovich et al., *Phys. Lett. B* **323**, 233 (1994).
25. D. Cohen et al., *Phys. Rev.* **D22**, 2595 (1980).
26. V.V. Anisovich, Yu.D. Prokoshkin and A.V. Sarantsev, *Phys. Lett.* **B389**, 388 (1996).
27. C. Amsler and F.E. Close, *Phys. Rev.* **D53**, 295 (1996); *Phys. Lett.* **B353**, 385 (1995).
28. W. Lee and D. Weingarten, *Phys. Rev. D* **61**, 014015 (2000).
29. F.E. Close, *Acta Phys. Polon. B* **31**, 2557 (2000).
30. R. Kamiński, L.Leśniak and K. Rybicki, hep-ph/0109268.
31. B. Ananthanarayan, G. Colangelo, J. Gasser and H. Leutwyler, *Phys. Rep.* **353**, 207 (2001).
32. D. Morgan and M. R. Pennington, *Phys. Rev.* **D48** (1993) 1185.
33. S. Ishida et al., *Prog. Theor. Phys.* **95**, 745 (1996).
34. V.V. Anisovich et al., *Phys. Atom. Nucl.* **60**, 1410 (2000).
35. DM2 Collaboration, J.E. Augustin et al., *Nucl. Phys. B* **320**, 1 (1989).
36. Axial Field Spectrometer Collaboration, T. Akesson et al., *Nucl. Phys. B* **264**, 154 (1986).
37. GAMS Collaboration, D. Alde et al., *Phys. Lett. B* **397**, 350 (1997).
38. WA102 Collaboration, D. Barberis et al. *Phys. Lett. B* **453**, 325 (1999); **453**, 316 (1999).
39. GAMS Collaboration, R. Bellazzini et al., *Phys. Lett. B* **467**, 296 (1999).
40. M. Boglione and M.R. Pennington, *Eur. Phys. J. C* **9**, 11 (1999).
41. Ya.I. Azimov, E.M. Levin, M.G. Ryskin and V.A. Khoze, *Yad. Fiz.* **21**, 413 (1975).
42. E852 Collaboration: J. Gunter et al., *Phys. Rev. D* **64**, 072003 (2001).
43. E. Klempt, B.C. Metsch, C.R. Münz and H.R. Petry, *Phys. Lett. B* **361**, 160 (1995).
44. V. Dmitrašinović, *Phys. Rev. C* **53**, 1383 (1996).
45. L. Burakovsky and T. Goldmann, *Nucl. Phys. A* **628**, 87 (1998).
46. K.K. Seth, *Nucl. Phys. B (Proc. Suppl.)* **96**, 205 (2001).
47. C. Quigg, in: *Proc. 4th Int. Conf. on Experimental Meson Spectroscopy*, Boston, 1974 (AIP Conf. Proc. no. 21, particles and fields subseries no. 8) p. 297.
48. P. Minkowski and W. Ochs, *Phys. Lett. B* **485**, 139 (2000).
49. P. Roy and K. Sridhar, JHEP 9907, 013 (1999).
50. B. Buschbeck and F. Mandl (DELPHI Collaboration), "Study of Gluon Fragmentation and Color Neutralization", Intern. Symposium of Multiparticle Dynamics Sept. 2001, Datong, China.

Investigating the Light Scalar Mesons

D.Black[a], S. Moussa[a], S. Nasri[a], A. H. Fariborz[b] and J. Schechter[a]

[a] *Department of Physics, Syracuse University, Syracuse, New York 13244-1130, USA.*

[b] *Department of Mathematics/Science, State University of New York Institute of Technology, Utica, New York 13504-3050, USA.*

Abstract.
  We first briefly review a treatment of the scalars in meson meson scattering based on a non-linear chiral Lagrangian, with unitarity implemented by a "local" modification of the scalar propagators. It is shown that the main results are confirmed by a treatment in the SU(3) linear sigma model in which unitarity is implemented "globally". Some remarks are made on the speculative subject of the scalars' quark structure.

I INTRODUCTION

The "hydrogen atom" problem of meson spectroscopy is the study of the pion in terms of its fundamental constituents. Typically, this difficult problem is finessed by using an effective Lagrangian treatment of the composite field which includes the important feature of (almost) spontaneously broken chiral symmetry. Then one explains the presumed next highest mass meson– the rho– as a $q\bar{q}$ bound state and continues up the spectrum. But nowadays there is increasing support for the existence of the old "sigma" resonance which may be lighter than the rho. If this is true it certainly seems worthwhile to pause and examine the issue in detail. It is also a difficult problem because the sigma is in an energy range just above where one expects chiral perturbation theory to be practical but well below where asymptotic freedom permits a systematic perturbative QCD expansion.

In this talk a recent paper [1] on the subject will be discussed. Other work is referenced in that paper and in other contributions [2] to this conference. First, a brief review of our previous results based on the non linear chiral Lagrangian will be given. Then we try to check the form of these results by using the linear sigma model. This model, while less general, provides the usual physical intuition about the problem as it contains a scalar nonet linked to the pseudoscalars.

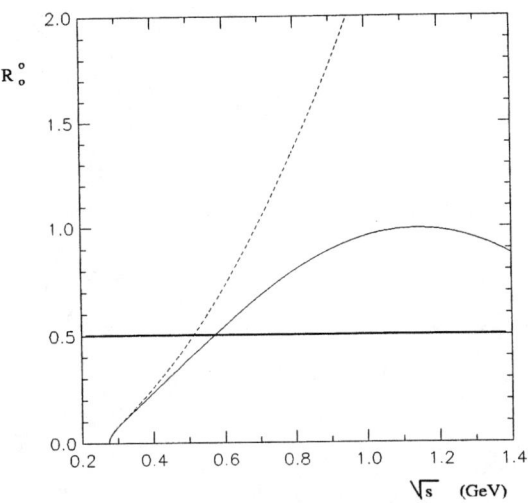

FIGURE 1. The solid line which shows the current algebra $+\rho$ result is much closer to the unitary bound of 0.5 than the dashed line which shows the current algebra result alone.

II BRIEF REVIEW OF OUR PREVIOUS WORK USING THE NONLINEAR CHIRAL LAGRANGIAN

Pi pi scattering [3]. It was noted that the $I = J = 0$ partial wave amplitude up to about 1 GeV could be simply explained as a sum of four pieces: i. the current algebra "contact" term, ii. the ρ exchange diagram iii. a non Breit Wigner σ pole diagram and exchange, iv. an $f_0(980)$ pole in the background produced by the other three. This is illustrated in a step by step manner for the real part R_0^0 in Figs. 1, 2 and 3.

We see in Fig. 1 that the "current algebra" piece starts violating the unitarity bound, $|R_0^0| \leq 1/2$ at about 0.5 GeV and then runs away. However the inclusion of the ρ meson exchange diagrams turns the curve in the right direction and improves, but does not completely cure, the unitarity violation. These pieces, which do not involve any unknown parameters, give encouragment to our hope that the cooperative interplay of various pieces can explain the low energy scattering. In order to fix up Fig. 1 we note that the real part of a resonance contribution vanishes at the pole, is positive before the pole and *negative* above the pole. Thus a scalar resonance with a pole roughly about 0.5 GeV (above which R_0^0 in Fig. 1 needs a negative contribution to stay below 1/2) should do the job. The result of including such a σ pole , with three parameters, is shown in Fig. 2. Now note that the predicted $R_0^0(s)$ in Fig. 2 vanishes around 1 GeV. Thus the phase δ at 1 GeV (assumed to keep rising) is about 90°. Considering this as a background phase for the known $f_0(980)$, the real part of the $f_0(980)$ contribution will get reversed in sign (Ramsauer–Townsend effect). As Fig. 3 shows this is the missing piece needed to give a simple explanation of the $J = I = 0$ $\pi\pi$ scattering up to about 1 GeV.

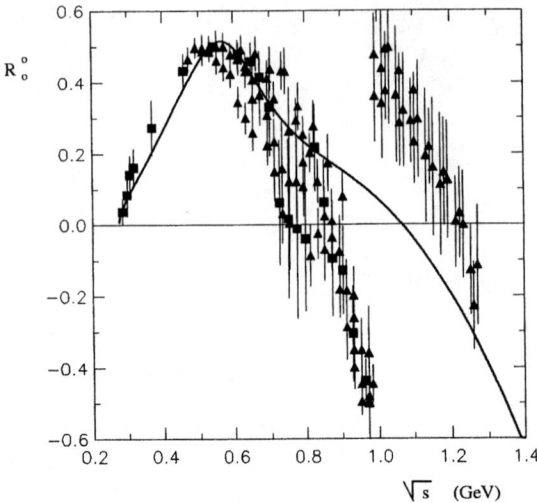

FIGURE 2. The sum of current algebra $+\rho + \sigma$ contributions compared to data.

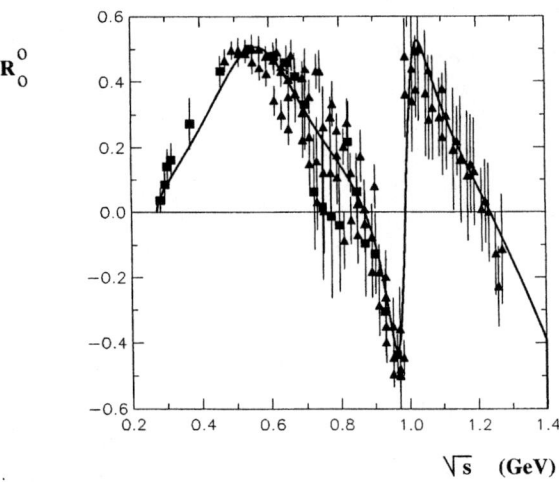

FIGURE 3. The sum of current algebra $+\rho + \sigma + f_0(980)$ contributions compared to data.

Pi K scattering [4]. In this case the low energy amplitude is taken to correspond to the sum of a current algebra contact diagram, vector ρ and K^* exchange diagrams and scalar $\sigma(560)$, $f_0(980)$ and $\kappa(900)$ exchange diagrams. The situation in the interesting $I = 1/2$ s-wave channel turns out to be very analogous to the $I = 0$ channel of s-wave $\pi\pi$ scattering. Now a non Breit Wigner κ is required to restore unitarity; it plays the role of the $\sigma(560)$ in the $\pi\pi$ case. We found that a satisfactory description of the 1-1.5 GeV s-wave region is also obtained by including the well known $K_0^*(1430)$ scalar resonance, which plays the role of the $f_0(980)$ in the $\pi\pi$ calculation.

Putative light scalar nonet [5]. The nine states associated with the $\sigma(560)$, $\kappa(900)$, $f_0(980)$ and $a_0(980)$ are required in order to fit experiment in our model. What do their masses and coupling constants suggest about their quark substructure? Clearly the mass ordering of the various states is inverted compared to the "ideal mixing" [6] scenario which approximately holds for most meson nonets. This means that a quark structure for the putative scalar nonet $N_a^b \sim q_a \bar{q}^b$ is unlikely since the mass ordering just corresponds to counting the number of heavier strange quarks. Then the degenerate $f_0(980)$ and $a_0(980)$ which must have the structure $N_1^1 \pm N_2^2$ would be lightest rather than heaviest. However the inverted ordering will agree with this counting if we assume that the scalar mesons are schematically constructed as $N_a^b \sim T_a \bar{T}^b$ where $T_a \sim \epsilon_{acd} \bar{q}^c \bar{q}^d$ is a "dual" quark (or anti diquark). This interpretation is strengthened by consideration [5] of the scalars' coupling constants to two pseudoscalars. That shows $\sigma \sim N_3^3 +$ "small", so it is a predominantly non-strange particle in this picture. Furthermore the states $N_1^1 \pm N_2^2$ now would each have two strange quarks and would be expected to be heaviest. The four quark picture was first suggested a long time ago [7] on dynamical grounds.

Mechanism for next heavier scalar nonet [8]. Of course, the success of the phenomenological quark model suggests that there exists, in addition, a nonet of "conventional" p-wave $q\bar{q}$ scalars in the 1+ GeV range. The experimental candidates for these states are $a_0(1450)(I = 1)$, $K_0^*(1430)(I = 1/2)$ and for $I = 0$, $f_0(1370)$, $f_0(1500)$ and $f_0(1710)$. These are enough for a full nonet plus a glueball. However it is puzzling that the strange $K_0^*(1430)$ isn't noticeably heavier than the non strange $a_0(1450)$ and that they are not lighter than the corresponding spin 2 states. These and another puzzle may be solved in a natural way [8] if the heavier p-wave scalar nonet mixes with a lighter $qq\bar{q}\bar{q}$ nonet of the type mentioned above. The mixing mechanism makes essential use of the "bare" lighter nonet having an inverted mass ordering while the heavier "bare" nonet has the normal ordering. A rather rich structure involving the light scalars seems to be emerging. At lower energies one may consider as a first approximation, "integrating out" the heavier nonet and retaining just the lighter one.

III THE PICTURE IN A 3 FLAVOR LINEAR SIGMA MODEL

In [1] we employed the conventional chiral field $M = S + i\phi$, where S is a hermitian matrix of nine scalars and ϕ is a hermitian matrix of nine pseudoscalars. The Lagrangian density is,

$$\mathcal{L} = -\frac{1}{2}Tr(\partial_\mu M \partial_\mu M^\dagger) - V_0 + \sum A_a(M_{aa} + M_{aa}^\dagger). \tag{1}$$

The three A_a's are numbers proportional to the (current) quark masses. V_0 may be considered to be an arbitrary function of the chiral $SU(3) \times SU(3)$ invariants constructed from M and $M^\dagger$. Note that most consequences at tree level follow just from chiral symmetry, irrespective of the form of V_0.

Pi pi scattering amplitude. The computed $I = J = 0$ partial wave amplitude at tree level has the form,

$$T^0_{0tree}(s) = cos^2\psi[\alpha(s) + \frac{\beta(s)}{m^2_{BARE}(\sigma) - s}] + sin^2\psi[\tilde{\alpha}(s) + \frac{\tilde{\beta}(s)}{m^2_{BARE}(\sigma') - s}], \tag{2}$$

where $\alpha(s), \beta(s)$ etc. are given in connection with Eq (3.2) of [1]. ψ is a mixing angle between the two $I = 0$ scalars, denoted as σ and σ'. The subscript "BARE" on their masses means the value at tree level. If V_0 is general, the three quantities ψ, $m_{BARE}(\sigma)$ and $m_{BARE}(\sigma')$ may be chosen at will. However if V_0 is taken to be renormalizable there is only one arbitrary parameter (say $m_{BARE}(\sigma)$) in the theory when the input set (say $m_\pi, m_K, m_\eta, m_{\eta'}, F_\pi$) is fixed.

It is instructive to first go back to the widely treated two flavor case. This corresponds to choosing $\psi = 0$ in (2). Then $m_{BARE}(\sigma)$ is the only unknown parameter. Near threshold, if $m_{BARE}(\sigma)$ is not too low, the amplitude is the "current algebra" result which agrees fairly well with experiment. It is a small quantity which emerges from an almost complete cancellation of the pole and non pole terms in (2). One would like to keep this result and utilize the effect of the sigma at higher energies. Since there is a true pole in (2) it seems reasonable to regulate this in the usual way by adding a term $-im\Gamma$ in the pole denominator. However, as Achasov and Shestakov [9] pointed out, this regulation completely destroys the good current algebra result. They instead adopt the K matrix approach (whereas the usual solution is to adopt the non linear model instead since the derivative coupling of the σ there suppresses the pole contribution near threshold). In this way the tree amplitude is not only regularized but made exactly unitary. One calculates the amplitude in terms of its tree value as:

$$T = \frac{T_{tree}}{1 - iT_{tree}}. \tag{3}$$

When T_{tree} is small, $T \approx T_{tree}$ so the behavior near threshold will now not be spoiled. At the other extreme, when T_{tree} gets very large $T \to i$.

Note that the pole position, z of the unitarized T will typically correspond to mass and width (via $z = m^2 - im\Gamma$) which differ from m_{BARE} and the starting perturbative width. Which one should be chosen? Since T in (3) evidently has the structure of a "bubble sum" in field theory it seems reasonable to regard the K matrix unitarization as an approximation to including the "radiative corrections". Then, as in usual field theory, the pole found is interpreted as giving the physical mass and width while the values of m_{BARE} and Γ_{BARE} would have no special significance. For the two flavor model we verified the result of [9] that a choice of $m_{BARE}(\sigma)$ around 0.8 to 1.0 GeV would result in a physical $m(\sigma)$ around 0.45 GeV and fit the first bump in Fig. 3. The physical mass is not very sensitive to the exact choice of bare mass and also the physical width is very greatly reduced. The predicted mass in this model is a bit less than the one we found in the non linear model reviewed in the previous section, but this is readily understandable as being due [10] to the neglect of vector mesons in the present model.

Three flavor linear model amplitude. The procedure was simply to use the full two pole tree amplitude (2) in the unitarization formula (3). We were not able to fit the entire T_0^0 amplitude shown in Fig. 3 up to about 1.2 GeV in the renormalizable model (which contains only the single unknown parameter $m_{BARE}(\sigma)$). However it is easy to find a fit in the chiral model with general V_0, in which we were able to choose the three parameters $m_{BARE}(\sigma) = 0.847$ GeV, $m_{BARE}(\sigma') = 1.300$ GeV and $\psi = 48.6^o$. The physical isoscalar masses (after unitarization) turned out to be 0.457 GeV and 0.993 Gev associated with respective widths 0.632 GeV and 0.05 GeV. Again these represent large shifts from the bare values. For illustrative purposes the unitarized amplitude is reasonably approximated as

$$T_0^0 \approx const. + \frac{0.167 + 0.210i}{s - (0.209 - 0.289i)} + \frac{0.053 + 0.005i}{s - (0.986 - 0.051i)}. \tag{4}$$

Neither the first (σ) pole nor the second ($f_0(980)$) pole is precisely of Breit Wigner type. However the $f_0(980)$ pole approximates a Breit Wigner except for an overall minus sign, which corresponds to the well known "flipping" of this resonance.

We similarly studied the $I = 0, J = 1/2$ scattering amplitude to find the properties of the κ resonance in the linear model. The bare mass of the κ is fixed once the input parameters are given. To allow us to vary this quantity (in a range where the $\pi\pi$ scattering is not much affected) we chose the alternative input set $(m_\pi, m_K, m_{\eta'}, F_\pi, F_K)$ and varied F_K. In this case, because the $K_0^*(1430)$ is not included in the model we can only fit the data up to about 1 GeV. It was found that the best fit corresponded to the bare κ mass about 1.3 GeV. After unitarization the physical kappa mass turned out to be about 0.800 GeV and this didn't change much as the bare value was varied from 0.9 to 1.3 GeV. Unitarization also substantially narrowed the physical kappa width. Furthermore, as for the case of the σ the κ pole is not of Breit Wigner type. An analogous calculation was carried out to study the properties of the $a_0(980)$ as observed in $\pi\eta$ scattering. A summary shown in Table 1 compares the physical widths obtained in this linear model with

	σ	f_0	κ	a_0
Present Model				
mass (MeV), width (MeV)	457, 632	993, 51	800, 260-610	890-1010, 110-240
Comparison				
mass (MeV), width (MeV)	560, 370	980±10, 40-100	900, 275	985, 50-100

TABLE 1. Predicted "physical" masses and widths in MeV of the nonet of scalar mesons contrasted with suitable (as discussed in the text) comparison values.

those obtained in the non linear model. In the cases of the $f_0(980)$ and the $a_0(980)$ the entries were taken from the Particle Data Group [11], with which the non linear model calculations agree.

Clearly, the complex pole positions and nature of the poles (non Breit Wigner for σ and κ and "Ramsauer Townsend" for $f_0(980)$) of the scalar nonet in the linear sigma model are similar to those obtained previously (putative scalar nonet) using a non linear chiral Lagrangian with a different "local" regulation. This statement makes heavy use of the unitarization of the three flavor linear model; otherwise the $f_0(980)$ and κ might be considered too high and wide to belong to a light scalar nonet. In particular, the κ clearly cannot be identified with the $K_0^*(1430)$ in this unitarized linear sigma model.

Speculation on scalar quark structure (Section V of [1]). At an intuitive level one might expect the scalar nonet, being the "chiral partner" of the light pseudoscalar nonet, to have a quark- anti quark structure. It was stressed [5] however that in the more general non linear Lagrangian approach (e.g. [12]) the scalar and pseudoscalar transformation properties are decoupled. Only the flavor SU(3) transformation property, not the chiral one, of the scalars is fixed in the effective non-linear Lagrangian treatment. Features, mentioned above, like isoscalar mixing angle and mass ordering suggest in fact the $qq\bar{q}\bar{q}$ structure for the light scalars as an initial approximation.

How might this kind of scenario play out in the linear model where the chiral properties of the scalars and the pseudoscalars are clearly linked? Even there, the quark substructure implied by the $SU(3) \times SU(3)$ transformation properties of the chiral matrix M in (1) is not unique [1] (However the $U(1)_A$ transformations do distinguish between $q\bar{q}$ and $qq\bar{q}\bar{q}$). There are three different "four quark" structures with the same transformation properties. Physically, they correspond to making the chiral mesons as a) meson meson "molecule" b) spin 0 diquark -spin 0 anti diquark and c) spin 1 diquark - spin 1 anti diquark. Actually these three are not linearly independent. Thus the molecule [13] and diquark- anti diquark [7] pictures are not clearly distinguished at the effective Lagrangian level. Presumably, large changes in the properties of the scalars due to unitarization in the effective theory must be counted as "four quark" admixtures at the underlying level.

In detail, the schematic structure for the matrix $M(x) = S + i\phi$ realizing a $q\bar{q}$ composite in terms of quark fields $q_{aA}(x)$ can be written

$$M_a^{(1)b} = (q_{bA})^\dagger \gamma_4 \frac{1+\gamma_5}{2} q_{aA}, \tag{5}$$

where a and A are respectively flavor and color indices. For the "molecule" model a) the schematic quark structure with the same $SU(3)_L \times SU(3)_R$ transformation property is,

$$M_a^{(2)b} = \epsilon_{acd}\epsilon^{bef}\left(M^{(1)\dagger}\right)^c_e\left(M^{(1)\dagger}\right)^d_f. \tag{6}$$

In the spin 0 diquark - spin 0 anti diquark case the same transformation property is realized with,

$$M_g^{(3)f} = \left(L^{gA}\right)^\dagger R^{fA}, \tag{7}$$

where

$$L^{gE} = \epsilon^{gab}\epsilon^{EAB} q_{aA}^T C^{-1} \frac{1+\gamma_5}{2} q_{bB},$$
$$R^{gE} = \epsilon^{gab}\epsilon^{EAB} q_{aA}^T C^{-1} \frac{1-\gamma_5}{2} q_{bB}, \tag{8}$$

in which C is the charge conjugation matrix of the Dirac theory. Finally the spin 1 diquark - spin 1 anti diquark case c) has the schematic structure,

$$M_g^{(4)f} = \left(L_{\mu\nu,AB}^g\right)^\dagger R_{\mu\nu,AB}^f, \tag{9}$$

where

$$L_{\mu\nu,AB}^g = L_{\mu\nu,BA}^g = \epsilon^{gab} q_{aA}^T C^{-1} \sigma_{\mu\nu} \frac{1+\gamma_5}{2} q_{bB},$$
$$R_{\mu\nu,AB}^g = R_{\mu\nu,BA}^g = \epsilon^{gab} q_{aA}^T C^{-1} \sigma_{\mu\nu} \frac{1-\gamma_5}{2} q_{bB}, \tag{10}$$

and $\sigma_{\mu\nu} = \frac{1}{2i}[\gamma_\mu, \gamma_\nu]$.

Now, as discussed before, the realistic situation is likely to contain substantial mixing between scalar $q\bar{q}$ and $qq\bar{q}\bar{q}$ nonets. To explore this we formulated a linear sigma model containing both a $q\bar{q}$ chiral matrix $M = S + i\phi$ and a $qq\bar{q}\bar{q}$ chiral matrix $M' = S' + i\phi'$. This is a very complicated system so we started with a "toy model" in which all current quark masses are neglected and only a minimum number of non derivative terms are included. In addition to two minimal kinetic terms as in (1), we took the simplified potential,

$$V_0 = -c_2 \text{Tr}\left(MM^\dagger\right) + c_4 \text{Tr}\left(MM^\dagger MM^\dagger\right) + d_2 \text{Tr}\left(M'M'^\dagger\right) + e\text{Tr}\left(MM'^\dagger + M'M^\dagger\right). \tag{11}$$

Here c_2, c_4 and d_2 are positive real constants. The M matrix field is chosen to have a wrong sign mass term so that there will be spontaneous breakdown of chiral symmetry. A pseudoscalar octet will thus be massless. The mixing is controlled by the parameter e. It is amusing to note that there will then be an induced $qq\bar{q}\bar{q}$ condensate $\langle S' \rangle$ in addition to the usual $q\bar{q}$ condensate $\langle S \rangle$.

We found that it is easy to obtain a situation where the the next highest state above the predominantly $q\bar{q}$ Nambu Goldstone pseudoscalar octet is a predominantly $qq\bar{q}\bar{q}$ scalar octet. Still heavier is the predominantly $qq\bar{q}\bar{q}$ pseudoscalar while heaviest of all is the $q\bar{q}$ scalar octet. Of course, SU(3) symmetry breaking and unitarization would be expected to modify this picture. It seems very interesting to further pursue a model of this type. There is evidently a possibility of learning a lot about non perturbative QCD from the light scalar system.

We would like to thank Francesco Sannino and Masayasu Harada for fruitful collaboration. One of us (J.S.) would like to thank the organizers for arranging a stimulating and enjoyable conference. The work has been supported in part by the US DOE under contract DE-FG-02-85ER40231.

REFERENCES

1. D. Black, A. H. Fariborz, S. Moussa, S. Nasri and J. Schechter, Phys. Rev.**D64**, 014031(2001).
2. See the write-ups of N. Achasov, C. Gobel, S. Ishida, M. Ishida, T. Kunihiro, J. L. Lucio-Martinez, G. Moreno, W. Ochs, J. Pelaez, Yu. Surovtsev, K. Teshihiko, T. Teshima, S. F. Tuan and E. van Beveren.
3. M. Harada, F. Sannino and J. Schechter, Phys. Rev. **D54**,1991(1996).
4. D. Black, A. H. Fariborz, F. Sannino and J. Schechter, Phys. Rev. **D58**, 054012 (1998).
5. D. Black, A. H. Fariborz, F. Sannino and J. Schechter, Phys. Rev. **D59**, 074026 (1999).
6. S. Okubo, Phys. Lett. **5**, 165 (1963).
7. R. Jaffe, Phys. Rev. **D15**, 267 (1977).
8. D. Black, A. H. Fariborz and J. Schechter, Phys. Rev. **D61**, 074001 (2000).
9. N. N. Achasov and G. N. Shestakov, Phys. Rev. **D 49**, 5779 (1994).
10. M. Harada, F. Sannino and J. Schechter, Phys. Rev. Lett., **78**, 1603 (1997).
11. Review of Particle Physics, Euro. Phys. J. **C3** (1999).
12. C. Callan, S. Coleman, J. Wess and B. Zumino, Phys. Rev. **177**, 2247 (1969).
13. N. Isgur and J. Weinstein, Phys. Rev. **D27**, 588 (1983).

Covariant Classification Scheme of Hadrons

S. Ishida* and M. Ishida[†]

*Atomic Energy Research Institute, College of Science and Technology, Nihon University, Tokyo 101-0062, JAPAN
[†]Department of Physics, Tokyo Institute of Technology, Tokyo 152-8551, JAPAN

Abstract. Starting from the multi-local Klein-Gordon equations with Lorentz-scalar squared-mass operator we give a covariant quark representation of the general composite mesons and baryons with definite Lorentz transformation property. The mass spectra satisfy the approximate symmetry under the $\tilde{U}(4)$ transformation group, including the chiral transformation as a subgroup, concerning the spinor freedom of light constituent quarks, and this symmetry predicts the existence of new type of chiral mesons and baryons out of the conventional framework in non-relativistic quark model: For example, for light $q\bar{q}$ systems, the scalar σ- and axial-vector a_1-nonets, and for heavy-light $Q\bar{q}$ and $q\bar{Q}$ systems the scalar and axial-vector mesons are predicted to exist as relativistic S-wave states besides the ordinary P-wave state mesons. The existence of two "exotic" 1^{-+} meson nonets is predicted as the relativistic P-wave states in $q\bar{q}$ systems. For light quark baryons the extra **56** with positive parity and the extra **70** with negative parity of the static $SU(6)$ are predicted to exist as the ground state chiral particles.

INTRODUCTION

There exist the two contrasting, non-relativistic and relativistic, viewpoints of level-classification. The former is based on the non-relativistic quark model (NRQM) with the approximate LS-symmetry and gives a theoretical base to the PDG level-classification. The latter is embodied typically in the NJL model with the approximate chiral symmetry. It is widely accepted that π meson nonet has the property as a Nambu-Goldstone boson in the case of spontaneous breaking of chiral symmetry.

Owing to the recent progress, both theoretical and experimental, the existence of light σ-meson as chiral partner of $\pi(140)$ seems to be established[1] especially through the analysis of various $\pi\pi$-production processes. This gives further a strong support to the relativistic viewpoint.

Thus, the hadron spectroscopy is now confronting with a serious problem, existence of the seemingly contradictory two viewpoints, Non-relativistic and Extremely Relativistic ones. The purpose of this talk is to present an attempt for a new level-classification

TABLE 1. Two Contrasting Viewpoints of Level Classification

	Non-Relativistic	Relativistic
Model	Non Relat. Q. M.	NJL model
Approx. Symm.	LS-Symm.	Chiral Symm.
Evidence	Bases for PDG	π nonet as NG boson

scheme unifying these two viewpoints. The following is an overview of our attempt, taking an example of the light-quark hadron system:

The left column concerns the NRQM, while the right column does the covariant oscillator quark model (COQM) as a basic kinematical framework of our attempt.

In NRQM the confining force is assumed to be spin-independent and the mass spectra have the $SU(6)_{SF}$ spin-flavor symmetry. In the extended (old) version of COQM the confining force is assumed to be Lorentz-scalar ("boosted-spin" independent) and the mass spectra have the $\tilde{U}(12)_{SF}$ symmetry[1] (boosted $SU(6)_{SF}$ symmetry). The 3-dimensional space-coordinates of constituent quarks and/or anti-quarks, as variables in the meson and baryon wave functions (WF), in NRQM are extended to the 4-dimensional Lorentz-vectors in COQM. Similarly the multi-Pauli-spinors, as spin WF, in NRQM are extended to the covariant multi-Dirac spinors in COQM; where in the old version only the positive-energy spinors $u_+(\mathbf{P})(\bar{v}_+(\mathbf{P}))$ for quarks (anti-quarks) are considered, and in the extended version the negative-energy spinors $u_-(-\mathbf{P})(\bar{v}_-(-\mathbf{P}))$ (the $\mathbf{P}$ is the space-momentum of hadrons as a whole-entity) are also taken into account. Thus, the WF of hadrons in the new level-classification scheme become the tensors in the $O(3,1)_{\text{Lorentz}} \otimes \tilde{U}(4)_{D.S.} \otimes SU(3)_F$-space (, being extended from the ones in the $O(3) \otimes SU(2)_{P.S.} \otimes SU(3)_F$ space of NRQM). The numbers of freedom of spin-flavor WF in NRQM are $\mathbf{6 \times 6^* = \underline{36}}$ for mesons and $(\mathbf{6 \times 6 \times 6})_{\text{Symm.}} = \underline{\mathbf{56}}$ for baryons: These numbers in COQM become $\mathbf{12 \times 12^* = \underline{144}}$ for mesons and $(\mathbf{12 \times 12 \times 12})_{\text{Symm.}} = \underline{\mathbf{364}} = \underline{\mathbf{182}}$ (for baryons) $+ \underline{\mathbf{182}}$ (for anti-baryons).

TABLE 2. Overview of Attempt : example of light quark hadrons

	NRQM	COQM (old)	(extended)
Confining force	spin-indep.	"boosted" spin-indep.	Lorentz scalar
Symmetry	$SU(6)_{SF}$	"boosted" $SU(6)_{SF}$	$\tilde{U}(12)_{SF}$
(Wave Function)		$f(x_1,x_2)\, u_q \bar{v}_{\bar{q}}$ Dirac spinor	
$(q\bar{q})$-meson	$f(\mathbf{x}_1,\mathbf{x}_2)\, \chi_q \bar{\chi}_{\bar{q}}$ Pauli spinor	$u_q = u_+(\mathbf{P})$, $\quad$ $u_-(-\mathbf{P})$ $\bar{v}_{\bar{q}} = \bar{v}_+(\mathbf{P})$, $\quad$ $\bar{v}_-(-\mathbf{P})$ boosted Pauli spin	
(qqq)-baryon	$f(\mathbf{x}_1,\mathbf{x}_2,\mathbf{x}_3)\, \chi_{q_1}\chi_{q_2}\chi_{q_3}$	$f(x_1,x_2,x_3)\, u_{q_1}u_{q_2}u_{q_3}$ $u_q = u_+(\mathbf{P})$, $\quad$ $u_-(-\mathbf{P})$	
$(\bar{q}\bar{q}\bar{q})$-anti-baryon	$f(\mathbf{x}_1,\mathbf{x}_2,\mathbf{x}_3)\, \bar{\chi}_{\bar{q}_1}\bar{\chi}_{\bar{q}_2}\bar{\chi}_{\bar{q}_3}$	$f(x_1,x_2,x_3)\, \bar{v}_{\bar{q}_1}\bar{v}_{\bar{q}_2}\bar{v}_{\bar{q}_3}$ $\bar{v}_{\bar{q}} = \bar{v}_+(\mathbf{P})$, $\quad$ $\bar{v}_-(-\mathbf{P})$	
space(-time)	$O(3) \otimes SU(2)_S \otimes SU(3)_F$	$O(3,1) \otimes \tilde{U}(4)_{D.S.} \otimes SU(3)_F$	
(Spin Wave Function)	**multi-Pauli spinor**	**multi-*boosted*** Pauli spinor	**multi-Dirac spinor**

---

[1] The $\tilde{U}(12)_{SF}$ symmetry was first proposed in 1965 as a generalization of the static $SU(6)_{SF}$ symmetry. However, at that time only the boosted Pauli-spinors are taken as physical components of fundamental representation of $\tilde{U}(4)_{D.S.}$. Now in the extended scheme all general Dirac spinors prove to be physical.

Inclusion of heavy quarks is straightforward: The WF of general q and/or Q hadrons become tensors in $O(3,1) \otimes [\tilde{U}(4)_{D.S.} \otimes SU(3)_F]_q \otimes [SU(2)_{P.S.} \otimes U(1)_F]_Q$.

COVARIANT FRAMEWORK FOR DESCRIBING COMPOSITE HADRONS

As WF of mesons and baryons we set up the following field-theoretical expressions, respectively, as

$$\Phi_A{}^B(x_1,x_2) = \langle 0|\psi_A(x_1)\bar{\psi}^B(x_2)|M\rangle + \langle \bar{M}|\psi_A(x_1)\bar{\psi}^B(x_2)|0\rangle, \quad (1)$$

$$\Phi_{A_1A_2A_3}(x_1,x_2,x_3) = \langle 0|\psi_{A_1}(x_1)\psi_{A_2}(x_2)\psi_{A_3}(x_3)|B\rangle + \langle \bar{B}|\psi_{A_1}(x_1)\psi_{A_2}(x_2)\psi_{A_3}(x_3)|0\rangle, \quad (2)$$

where ψ_A is the quark field ($A = (\alpha, a)$; $\alpha = 1 \sim 4$ (a) denoting Dirac spinor (flavor) indices) and $\bar{\psi}^B$ denotes its Pauli-cpnjugate. We start from the Yukawa-type Klein Gordon equation as a basic wave equation[2].

$$[\partial^2/\partial X_\mu^2 - \mathcal{M}^2(r_\mu, \partial/\partial r_\mu)]\Phi(X,r,\cdots) = 0, \quad (3)$$

where $X_\mu(r_\mu)$ are the center of mass (relative) coordinates of hadron systems. The WF are separated into the positive (negative)-frequency parts concerning the CM plane-wave motion and expanded in terms of eigen-states of the squared-mass operator as

$$\Phi(X,r,\cdots) = \sum_{P_N,N}\left[e^{iP_N\cdot X}\psi_N^{(+)}(P_N,r,\cdots) + e^{-iP_N\cdot X}\psi_N^{(-)}(P_N,r,\cdots)\right], \quad (4)$$

$$\mathcal{M}^2(r_\mu,\partial/i\partial r_\mu,\cdots)\psi_N^{(\pm)} = M_N^2\psi_N^{(\pm)}, \quad (5)$$

$$\mathcal{M}^2 = \mathcal{M}_{\text{conf}}^2 + \delta\mathcal{M}_{\text{pert. QCD}}^2. \quad (6)$$

The $\mathcal{M}^2$ consists of two parts: The confining-force part $\mathcal{M}_{\text{conf}}^2$ is assumed to be Lorentz-scalar and $A, (B)$-independent, leading to the mass spectra with the $\tilde{U}(12)$ symmetry and also with the chiral symmetry. As its concrete model we apply the covariant oscillator in COQM, leading to the straight-rising Regge trajectories. The effects due to perturbative QCD $\delta\mathcal{M}^2$ are neglected in this talk.

The internal WF is, concerning the spinor freedom, expanded in terms of complete set of relevant multi-spinors, Bargmann-Wigner (BW) spinors.

$$\text{meson}: \quad \psi_{N,A}^{(\pm)B}(P_N,r) = \sum_W W_\alpha^{(\pm)\beta}(P_N)M_{N,a}^{(\pm)b}(r,P_N) \quad (7)$$

$$\text{baryon}: \quad \psi_{N,A_1A_2A_3}^{(\pm)}(P_N,r_1,r_2) = \sum_W W_{\alpha_1\alpha_2\alpha_3}^{(\pm)}(P_N)B_{N,a_1a_2a_3}(r_1,r_2,P_N). \quad (8)$$

The BW spinors are defined as multi-Dirac spinor solutions of the relevant local Klein-Gordon equation:

$$(\partial^2/\partial X_\mu^2 - M^2)W_\alpha{}^{\beta\cdots}(X) = 0 \quad (9)$$

TABLE 3. Bargmann-Wigner (BW) Equations and Spinors for mesons. Only the positive frequency parts are given. The negative frequency parts $W^{(-)}$ are obtained from the operation $W^{(-)} = W^{(+)}\{u \leftrightarrow v\}$. For example, $C^{(-)} = C^{(+)}\{u \leftrightarrow v\} = v_\alpha(P)\bar{u}^\beta(-P)$.

[Meson] $W_\alpha^{(+)\beta}(P)$	$M^{(+)}(P)$	BW-Equation	($P_0 \equiv E_P > 0$)
$U_\alpha{}^\beta(P) \equiv u_\alpha(P)\bar{v}^\beta(P);$	$P_s, V_\mu.$	$(iP\cdot\gamma^{(1)}+M)U = 0,$	$U(-iP\cdot\gamma^{(2)}+M) = 0$
$C_\alpha{}^\beta(P) \equiv u_\alpha(P)\bar{v}^\beta(-P);$	$S, A_\mu.$	$(iP\cdot\gamma^{(1)}+M)C = 0,$	$C(iP\cdot\gamma^{(2)}+M) = 0$
$D_\alpha{}^\beta(P) \equiv u_\alpha(-P)\bar{v}^\beta(P);$	$S, A_\mu.$	$(-iP\cdot\gamma^{(1)}+M)D = 0,$	$D(-iP\cdot\gamma^{(2)}+M) = 0$
$V_\alpha{}^\beta(P) \equiv u_\alpha(-P)\bar{v}^\beta(-P);$	$P_s, V_\mu.$	$(-iP\cdot\gamma^{(1)}+M)V = 0,$	$V(iP\cdot\gamma^{(2)}+M) = 0$

$$W_{\alpha\ldots}{}^{\beta\ldots}(X) \equiv \sum_{\mathbf{P},P_0=E}\left(e^{iPX}W_{\alpha\ldots}^{(+)\beta\ldots}(P) + e^{-iPX}W_{\alpha\ldots}^{(-)\beta\ldots}(P)\right). \quad (10)$$

For mesons and baryons BW spinors are bi-Dirac and tri-Dirac spinors, respectively. We further go into more details of BW spinors. First we define the Dirac spinors for constituent quarks and anti-quarks with hadron 4-momentum P_μ as "mono-index" BW spinors:

$$(\partial^2/\partial X_\mu^2 - M^2)\psi_\alpha(X) = 0, \quad (11)$$

$$\psi_{q,\alpha}(X) \equiv \sum_{P_\mu (P_0=\pm E_P)} e^{iPX} u_{q,\alpha}(P_\mu) = \sum_{\mathbf{P},P_0=E_P}(u_+(\mathbf{P})e^{iPX} + u_-(-\mathbf{P})e^{-iPX}), \quad (12)$$

$$\psi_{\bar{q},\alpha}(X) \equiv \sum_{P_\mu (P_0=\pm E_P)} e^{-iPX} v_{\bar{q},\alpha}(P_\mu) = \sum_{\mathbf{P},P_0=E_P}(v_+(\mathbf{P})e^{-iPX} + v_-(-\mathbf{P})e^{iPX}), \quad (13)$$

where the hadron 4 momentum P_μ satisfies the equations

$$P_\mu^2 + M^2 = 0, \; P_0 = \pm E_\mathbf{P}, \; E_\mathbf{P} \equiv \sqrt{\mathbf{P}^2 + M^2}. \quad (14)$$

Here it is to be noted that all 4-independent solution $u_q(P)$ ($v_{\bar{q}}(P)$) with spin $\sigma_3(\sigma_3' = -\sigma_3^T) = \pm 1$ and $P_0 = \pm E_\mathbf{P}$ for quarks(anti-quarks) inside of hadrons.

The BW equations, the BW-spinors as their solutions and their irreducible composite hadrons are summarized in Tables **3** and **4**, respectively, for $q\bar{q}$-mesons and qqq-baryons. It is worthwhile to note that here exist new types of BW spinors for mesons(baryons); $C(P)$, $D(P)$ and $V(P)$ ($V(P)$ and $F(P)$) in addition to the conventional $U(P)$ ($E(P)$), boosted multi-Pauli spinors.

TRANSFORMATION RULE FOR HADRONS AND CHIRAL SYMMETRY

By using the covariant quark representation of composite hadrons given above we can derive automatically their rule for any (relativistic) symmetry transformation from that of constituent quarks. The rules for chiral transformation of mesons and baryons are, respectively,

$$\text{meson}: \quad \psi_A{}^B(P,r) \longrightarrow [e^{i\alpha^a\lambda^a/2}\gamma_5 \psi(P,r) e^{i\alpha^a\lambda^a/2}\gamma_5]_A{}^B, \quad (15)$$

TABLE 4. Bargmann-Wigner (BW) Equations and Spinors for baryons. Only the positive frequency parts are given. The negative frequency parts $W^{(-)}$ are obtained from the operation $W^{(-)} = W^{(+)}\{u \to v\}$. For example, $E^{(-)}_{\alpha_1\alpha_2\alpha_3} = E^{(+)}_{\alpha_1\alpha_2\alpha_3}\{u \to v\} = v_{\alpha_1}(P)v_{\alpha_2}(P)v_{\alpha_3}(P)$.

[Baryon] $W^{(+)}_{\alpha_1\alpha_2\alpha_3}(P)$	$B^{(+)}(P)$	BW-Equation ($P_0 \equiv E_P > 0$)
$E_{\alpha_1\alpha_2\alpha_3}(P) \equiv u_{\alpha_1}(P)u_{\alpha_2}(P)u_{\alpha_3}(P);$	$\psi(\frac{1}{2}), \psi_\mu(\frac{3}{2}).$	$(iP\cdot\gamma^{(1,2,3)} + M)E = 0,$
$G_{\alpha_1\alpha_2\alpha_3}(P) \equiv u_{\alpha_1}(P)u_{\alpha_2}(P)u_{\alpha_3}(-P);$	$\psi(\frac{1}{2}), \psi_\mu(\frac{3}{2}).$	$(iP\cdot\gamma^{(1,2)} + M)G = 0,$ $(-iP\cdot\gamma^{(3)} + M)G = 0,$
$F_{\alpha_1\alpha_2\alpha_3}(P) \equiv u_{\alpha_1}(P)u_{\alpha_2}(-P)u_{\alpha_3}(-P);$	$\psi(\frac{1}{2}), \psi_\mu(\frac{3}{2}).$	$(iP\cdot\gamma^{(1)} + M)G = 0,$ $(-iP\cdot\gamma^{(2,3)} + M)G = 0,$

baryon: $\quad \psi_{A_1A_2A_3}(P, r_1, r_2) \longrightarrow [\Pi^3_{i=1} e^{i\alpha^a\lambda^{a,(i)}/2}\gamma_5^{(i)} \psi(P, r_1, r_2)]_{A_1A_2A_3}.$ (16)

The physical meaning of chiral transformation are clearly seen from the operations:

$$u(P) \xrightarrow{\gamma_5} u'(P) = \gamma_5 u(P) = u(-P); \quad u_\pm(\mathbf{P}) \xleftrightarrow{\gamma_5} u_\mp(-\mathbf{P}), \quad (17)$$
$$v(P) \longrightarrow v'(P) = \gamma_5 v(P) = v(-P); \quad \bar{v}_\pm(\mathbf{P}) \longleftrightarrow \bar{v}_\mp(-\mathbf{P}). \quad (18)$$

That is, the chiral transformation transforms the members of relevant BW-spinors with each other. Accordingly, if $\mathcal{M}^2$ operator is independent of Dirac indices, the hadron mass spectra have effectively the $\tilde{U}(4)$ symmetry and also the chiral symmetry.

For convenience of later discussions we note further on physical meaning of BW equations and introduce the notion of "exciton-quark". That is, the BW spinors with the total hadron momentum P_μ and mass M are equivalent to the product of free Dirac spinors of the exciton quark with momentum $p^{(i)}_\mu \equiv \kappa^{(i)}P_\mu$ and mass $m^{(i)} \equiv \kappa^{(i)}M$ ($\sum_i \kappa^{(i)} = 1$), as is seen from the equations (in an example of the U-type (E-type) BW spinors of meson(baryon) systems).

$$\text{meson} \quad (iP\cdot\gamma^{(1)} + M)U(P) = 0 \xrightarrow{\times\kappa^{(1)}} (ip^{(1)}\cdot\gamma^{(1)} + m^{(1)})U(P) = 0$$
$$U(P)(-iP\cdot\gamma^{(2)} + M) = 0 \xrightarrow{\times\kappa^{(2)}} U(P)(-ip^{(2)}\cdot\gamma^{(2)} + m^{(2)}) = 0 \quad (19)$$
$$p^{(1)}_\mu + p^{(2)}_\mu = P_\mu; \qquad m^{(1)} + m^{(2)} = M, \quad (20)$$
$$\text{baryon} \quad (iP\cdot\gamma^{(i)} + M)E(P) = 0 \xrightarrow{\times\kappa^{(i)}} (ip^{(i)}\cdot\gamma^{(i)} + m^{(i)})E(P) = 0 \quad (21)$$
$$p^{(1)}_\mu + p^{(2)}_\mu + p^{(3)}_\mu = P_\mu; \qquad m^{(1)} + m^{(2)} + m^{(3)} = M. \quad (22)$$

The above consideration is valid through all ground-state and/or excited state hadrons: Accordingly the mass M_N of the N-th excited hadron with the 4-momentum P_N is generally given as a sum of the N-th excited mass m_N of the exciton quark with the 4-momentum $P^{(i)}_N \equiv \kappa^{(i)}P_N$.

$$M_N = m^{(1)}_N + m^{(2)}_N + \cdots, \quad p^{(i)}_N = \kappa^{(i)}P_N \left(\sum_i \kappa^{(i)} = 1\right). \quad (23)$$

TABLE 5. Light exciton-quark mass $m_{n,N}$ for mesons

		$n\bar{n}$	$n\bar{c}$	$n\bar{b}$	
Ω	/GeV	1.1	2.0	4.6	Chiral symm.
$m_{n,N}$	$N=0$	0.38	0.38	0.38	○
(GeV)	$N=1$	0.64	0.70	0.74	△
	$N=2$	0.83	0.95	1.07	×
Chiral	symm.	$N \leq 1$	$N \leq 0$ or 1	$N \leq 0$ or 1	

LEVEL STRUCTURE OF MESONS

(Phenomenological criterion for chiral symmetry) Considering the physical meaning of BW equations (see, Eqs. (19) and (21)) mentioned in the last section, we may set up the phenomenological criterion for chiral symmetry being effective as

$$m_{q,N}^2 \ll \Lambda_{\text{conf.}}^2 (\approx \Lambda_{\chi SB}^2) \approx 1 \text{GeV}^2. \tag{24}$$

We can estimate the values of exciton light-quark mass $m_{q,N}$ by applying the following mass formulas for the light-light $n\bar{n}$-meson ($n = u$ or d) and the light-heavy $n\bar{Q}$- and $Q\bar{n}$-meson systems ($Q = c$ or b).

$$M_N^2 = M_0^2 + N\Omega, \quad M_N = m_{q,N} + m_{q(Q),N} \quad (m_{q,0} = m_q, m_{Q,0} = m_Q) \tag{25}$$

$$M_N^2 = \left\langle \left(\sqrt{m_q^2 + \mathbf{p}^2} + \sqrt{m_{q(Q)}^2 + \mathbf{p}^2}\right)^2 + V \right\rangle_N \equiv \left(\sqrt{m_q^2 + \Lambda_N^2} + \sqrt{m_{q(Q)}^2 + \Lambda_N^2}\right)^2. \tag{26}$$

The equation (25) is the conventional formula in COQM, where the Ω^{-1} is the inverse Regge-slope and the zero-th exciton quark mass $m_{q,0}(m_{Q,0})$ is identified with the corresponding constituent-quark mass $m_q(m_Q)$. The equation (26) comes from the standard bound-state picture of hadrons, where V is the scalar confining potential and Λ_N corresponds to the average value of relative momentum $|\mathbf{p}|$ of constituent quarks in the N-th excited meson rest-frame. The result of values of light-exciton quark masses, thus estimated using the values of Ω and constituent quark masses obtained in the preceding analyses, is collected in Table **5**.

By inspecting the values of $m_{n,N}$ in Table **5** in relation to the criterion Eq. (24) we are able to infer that the chiral symmetry concerning the light quarks is valid (still effective) for the ground (first excited) state of $n\bar{n}$ and $n\bar{Q}$ meson systems, while the symmetry will prove invalid from the N-th ($N \geq 2$) excited hadrons.

(Level structure of ground state mesons) In Table **6** we have summarized the properties of ground state mesons in the light and/or heavy quark systems. It is remarkable that there appear new multiplets of the scalar and axial-vector mesons in the q-$\bar{Q}$ and Q-$\bar{q}$ systems and that in the q-$\bar{q}$ systems the two sets (Normal and Extra) of pseudo-scalar and of vector meson nonets exist. It is also to be noted that the π nonet (ρ nonet) is assigned to the $P_s^{(N)}$ ($V_\mu^{(N)}$) state, whose spin WF is much changed from that in NRQM. We call the new type of particles in the extended COQM (which have never appeared in NRQM) as "chiralons".

TABLE 6. Level structure of ground-state mesons

	mass	Approx. Symm.	Spin WF	$SU(3)$	Meson Type
$Q\bar{Q}$	$m_Q + m_{\bar{Q}}$	LS symm.	$u_Q(P)\bar{v}^Q(P)$	**1**	P_s, V_μ
$q\bar{Q}$	$m_q + m_{\bar{Q}}$	q-Chiral Symm.	$u_q(P)\bar{v}^Q(P)$	**3**	P_s, V_μ
		$\bar{Q}$-HQS	$u_q(-P)\bar{v}^Q(P)$	**3**	S, A_μ
$Q\bar{q}$	$m_Q + m_{\bar{q}}$	$\bar{q}$-Chiral Symm.	$u_Q(P)\bar{v}^q(P)$	**3***	P_s, V_μ
		Q-HQS	$u_q(P)\bar{v}^Q(-P)$	**3***	S, A_μ
$q\bar{q}$	$m_q + m_{\bar{q}}$	Chiral Symm.	$(1/\sqrt{2})(u(P)\bar{v}(P) \pm u(-P)\bar{v}(-P))$		$P_s^{(N,E)}, V_\mu^{(N,E)}$
			$(1/\sqrt{2})(u(P)\bar{v}(-P) \pm u(-P)\bar{v}(P))$		$S^{(N,E)}, A_\mu^{(N,E)}$

TABLE 7. Level structure of Mesons in general

$(q\bar{q})$			
N = all	$P_s^{(N,E)} \otimes \{L,N\}$	$P = (-1)^{L+1}$	$C = (-1)^L$
	$V_\mu^{(N,E)} \otimes \{L,N\}$	$P = (-1)^{L+1}$	$C = (-1)^{L+1}$
$N = 0$(and 1)	$S^{(N,E)} \otimes \{L,N\}$	$P = (-1)^L$	$C = \pm(-1)^L$
	$A_\mu^{(N,E)} \otimes \{L,N\}$	$P = (-1)^L$	$C = \pm(-1)^L$

$(q\bar{Q}$ or $Q\bar{q})$		$(Q\bar{Q})$	
N = all	$P_s \otimes \{L,N\}$	N = all	$P_s \otimes \{L,N\}$
	$V_\mu \otimes \{L,N\}$		$V_\mu \otimes \{L,N\}$
$N = 0$(and 1)	$S \otimes \{L,N\}$		
	$A_\mu \otimes \{L,N\}$		

(Level structure of mesons in general) The mass of the ground and excited state mesons is given by

$$M_N^2 = M_0^2 + N\Omega = m_N^{(1)} + m_N^{(2)}. \tag{27}$$

Their quantum numbers are given in Table 7. Here it is to be noted that some chiralons have the "exotic" quantum numbers from the conventional NRQM viewpoint.

The schematic picture of meson spectroscopy is shown in Fig. **1**.

LEVEL STRUCTURE OF BARYONS

The baryon WF Eq. (2) should be full-symmetric (except for the color freedom) under exchange of constituent quarks: The full-symmetric total WF in the extended scheme is obtained, in the following three ways, as a product of the sub-space WF with respective symmetric properties:

$$|\rho F \sigma\rangle_S = |\rho\rangle_S |F\sigma\rangle_S \ (a); \quad |\rho\rangle_\alpha |F\sigma\rangle_\alpha + |\rho\rangle_\beta |F\sigma\rangle_\beta \ (b);$$
$$|F\rangle_A |\rho\sigma\rangle_A \ (c); \tag{28}$$

where $|\rho\rangle_S$ is the full-symmetric ρ-spin space WF and so on. $\rho \otimes \sigma = \gamma$ is the conventional two, ρ and σ spin, 2 by 2 matrix representation of the 4 by 4 Dirac matrix, and

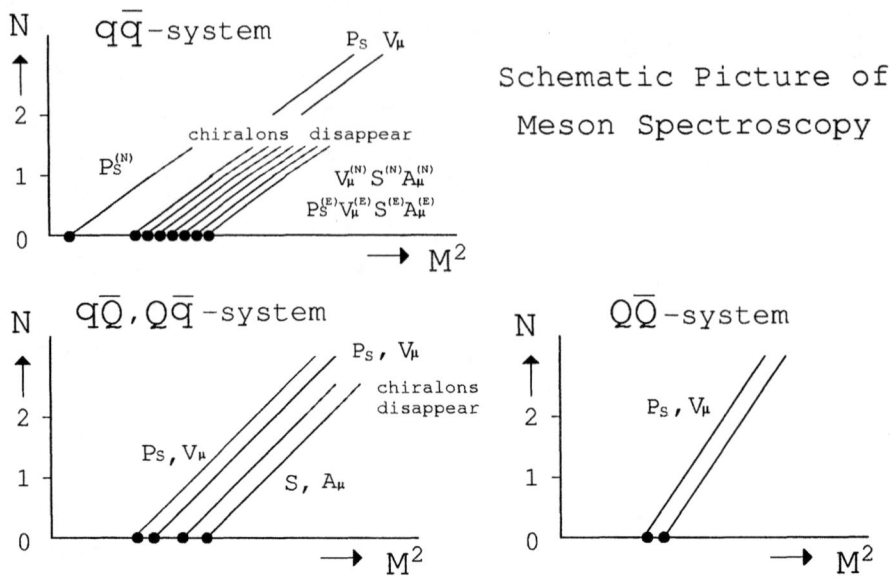

FIGURE 1. Schematic picture of meson spectroscopy

$|\ \rangle_{\alpha(\beta),A}$ mean the $\alpha(\beta)$-type partial symmetric and full anti-symmetric subspace WF, respectively. The intrinsic parity operation is given by $\hat{P} = \Pi_{i=1}^{3}\gamma_{4}^{(i)}$, that is, the parity of $(E^{(+)}, G^{(+)}, F^{(+)})$ BW spinors are $(+,-,+)$ and those of $(E^{(-)}, G^{(-)}, F^{(-)})$ BW spinors are $(-,+,-)$. The symmetry properties of ground state light-quark baryon WF and their level structures thus determined are summarized in Table 8.

Here it is remarkable that there appear chiralons in the ground states. That is, the extra positive parity **56**-multiplet of the static $SU(6)$ and the extra negative parity **70**-multiplet of the $SU(6)$ in the low mass region. It is also to be noted that the chiralons in the first excited states are expected to exist. The above consideration on the light-quark baryons are extended directly to the general light and/or heavy quark baryon systems: The chiralons are expected to exist also in the qqQ and qQQ-baryons, while no chiralons in the QQQ system.

EXPERIMENTAL CANDIDATES FOR CHIRAL PARTICLES

In our level-classification scheme a series of new type of multiplets of the particles, chiralons, are predicted to exist in the ground and the first excited states of $q\bar{q}$ and $q\bar{Q}$ or $Q\bar{q}$ meson systems and of qqq, qqQ and qQQ-baryon systems. Presently we can give only a few experimental candidates or indications for them:

($q\bar{q}$-mesons) One of the most important candidates is the scalar σ nonet to be assigned as $S^{(N)}(^{1}S_{0})$: [$\sigma(600)$, $\kappa(900)$, $a_{0}(980)$, $f_{0}(980)$]. The existence of $\sigma(600)$ seems to be established[1] through the analyses of, especially, $\pi\pi$-production processes. A firm

TABLE 8. Level structure of ground-state qqq-baryon

$W^{(+)}$	spin-flavor wave function	$B^{\mathcal{B}}$	static	$SU(6)$
$E^{(+)}$:	$\|\rho\rangle_S\|F\sigma\rangle_S = \|\rho\rangle_S\|F\rangle_S\|\sigma\rangle_S$	$\Delta_{3/2}^{\oplus}$	$10 \times 4 = 40$	
	$\|\rho\rangle_S(\|F\rangle_\alpha\|\sigma\rangle_\alpha + \|F\rangle_\beta\|\sigma\rangle_\beta)$	$N_{1/2}^{\oplus}$	$8 \times 2 = 16$	**56**
$G^{(+)}$:	$\|\rho\rangle_\alpha\|F\sigma\rangle_\alpha + \|\rho\rangle_\beta\|F\sigma\rangle_\beta$; $\|F\sigma\rangle_{\alpha(\beta)} = \|F\rangle_S\|\sigma\rangle_{\alpha(\beta)}$	$\Delta_{1/2}^{\ominus}$	$10 \times 2 = 20$	
	$\|F\rangle_{\alpha(\beta)}\|\sigma\rangle_S$	$N_{3/2}^{\ominus}$	$8 \times 4 = 32$	
	$\|F\rangle_A\|\rho\sigma\rangle_A = \|F\rangle_A(-\|\rho\rangle_\alpha\|\sigma\rangle_\beta + \|\rho\rangle_\beta\|\sigma\rangle_\alpha)$	$\Lambda_{1/2}^{\ominus}$	$1 \times 2 = 2$	
	$\|\rho\rangle_S\|F\sigma\rangle_S = \|\rho\rangle_S(\|F\rangle_\alpha\|\sigma\rangle_\beta + \|F\rangle_\beta\|\sigma\rangle_\alpha)$	$N_{1/2}^{\ominus}$	$8 \times 2 = 16$	**70**
$F^{(+)}$:	$\|\rho\rangle_S\|F\sigma\rangle_S = \|\rho\rangle_S\|F\rangle_S\|\sigma\rangle_S$	$\Delta_{3/2}^{\oplus}$	$10 \times 4 = 40$	
	$\|\rho\rangle_S(\|F\rangle_\alpha\|\sigma\rangle_\alpha + \|F\rangle_\beta\|\sigma\rangle_\beta)$	$N_{1/2}^{\oplus}$	$8 \times 2 = 16$	**56**

$$_{12}H_3 = \underline{364} \xrightarrow{(\times \frac{1}{2})} 182 = \underline{\mathbf{56}}^{\oplus} \oplus \underline{\mathbf{56}}^{\oplus} \oplus \underline{\mathbf{70}}^{\ominus}$$

$\underline{\mathbf{56}}^{\oplus}$	$N_{1/2}^{\oplus}, \Delta_{3/2}^{\oplus}$	
$\underline{\mathbf{56}}^{\oplus}$	$N_{1/2}^{\prime\oplus}, \Delta_{3/2}^{\prime\oplus}$	chiralons
$\underline{\mathbf{70}}^{\ominus}$	$N_{1/2}^{\ominus}, N_{3/2}^{\ominus}, \Delta_{1/2}^{\ominus}, \Lambda_{1/2}^{\ominus}$	

experimental evidence[3] for $\kappa(800-900)$ through the decay process[4] $D^+ \to K^-\pi^+\pi^+$ was reported at this conference.

In our scheme respective two sets of P_s- and of V_μ-nonets, to be assigned as $P_s^{(N,E)}(^1S_0)$ and $V_\mu^{(N,E)}(^3S_1)$, are to exist: Out of the five vector mesons (stressed[5] as problems with vector mesons, [$\rho'(1450)$, $\rho'(1700)$, $\omega'(1420)$, $\omega'(1600)$, $\phi(1690)$], the lower mass $\rho'(1450)$ and $\omega'(1420)$, and the $\phi(1690)$ are naturally able to be assigned as the members of $V_\mu^{(E)}$-nonets;

Out of the three established η, [$\eta(1295)$, $\eta(1420)$, $\eta(1460)$] at least one extra, plausibly $\eta(1295)$ with the lowest mass, may belong to $P_s^{(E)}(^1S_0)$ nonet.

Recently the existence of two "exotic" particles $\pi_1(1400)$ and $\pi_1(1600)$ with $J^{PC} = 1^{-+}$ and $I = 1$, observed[6] in the $\pi\eta$, $\rho\pi$ and other channels, is attracting strong interests among us. These exotic particles with a mass around 1.5GeV may be naturally assigned as the first excited states $S^{(E)}(^1P_1)$ and $A_\mu^{(E)}(^3P_1)$ of the chiralons.

($q\bar{Q}$ or $Q\bar{q}$-mesons) At this conference some experimental indication for existence of the two chiralons in D- and B-meson systems obtained through the $\Upsilon(4S)$ or Z^0 decay process, were reported, respectively,

$$D_1^\chi = A_\mu(^3S_1), \ J^P = 1^+ \ \text{in} \ D_1^\chi \to D^* + \pi$$
$$B_0^\chi = S(^1S_0), \ J^P = 0^+ \ \text{in} \ B_0^\chi \to B + \pi,$$

by Yamada K[7] and Ishida M[8].

(qqq-baryons) The two facts have been a longstanding problem that the Roper resonance $N(1440)_{1/2^+}$ is too light to be assigned as radial excitation of $N(939)$ and that $\Lambda(1405)_{1/2^-}$ is too light as the $L = 1$ excited state of $\Lambda(1116)$. In our new scheme they are reasonably assigned to the members of ground state chiralons with $[SU(6), SU(3), J^P]$, respectively, as

TABLE 9. Assignment of qqq-baryons: The baryons in the boxes are candidates of chiralons.

SU(6)	SU(3), J^P			SU(3), J^P	
56	**8**, $\frac{1}{2}^+$	$N(939), \Lambda(1116), \Sigma(1192), \Xi(1318)$		**10**, $\frac{3}{2}^+$	$\Delta(1232), \Sigma(1385), \cdots$
56'	**8**, $\frac{1}{2}^+$	$\boxed{N(1440)}$,	$\boxed{\Sigma(1660)}$	**10**, $\frac{3}{2}^+$	$\boxed{\Delta(1600)}$
70	**8**, $\frac{1}{2}^-$	$N(1535)$		**10**, $\frac{1}{2}^-$	$\Delta(1620)$
	1, $\frac{1}{2}^-$	$\boxed{\Lambda(1405)}$			

$$N(1440)_{1/2^+} = F(\mathbf{56}, \mathbf{8}, 1/2^+), \quad \Lambda(1405)_{1/2^-} = G(\mathbf{70}, \mathbf{1}, 1/2^-).$$

The particle $\Delta(1600)_{3/2^+}$ which is lighter than $\Delta(1620)_{1/2^-}$ may also belong to the extra **56** of the ground state chiralons. This situation is shown in Table **9**.

CONCLUDING REMARKS

I have presented an attempt for Level-classification scheme unfying the seemingly contradictory two viewpoints; Non-relativistic one with LS-symmetry and Relativistic one with Chiral symmetry .

As results, I have predicted the existence of New Chiral Particles in the lower mass regions "Chiralons", which had never been appeared in NRQM.

We have several good candidates for chiralons, for example,

σ-nonet { $\sigma(600), \kappa(800), a_0(980), f_0(980)$ } as *"Relativistic" S-wave states of* $(q\bar{q})$.
$\pi_1(1400), \pi_1(1600), (1^{-+})$; as *"Relativistic" P-wave states of* $(q\bar{q})$.
Roper resonance $N(1440)_{1/2^+}$ and SU(3) singlet $\Lambda(1405)_{1/2^-}$; as *"Relativistic" S-wave states of* (qqq).

Further search, both experimental and theoretical, for chiralons is necessary and important.

REFERENCES

1. N. A. Tornqvist, summary talk of "σ-meson 2000", KEK-proceedings 2000-4.
   T. Kunihiro; M. Ishida; T. Komada, this conference.
2. S. Ishida, M. Ishida and T. Maeda, Prog. Theor. Phys. **104** (2000), 785.
   H. Yukawa, Phys. Rev. **91** (1953), 415, 416.
   T. Takabayashi, Nuovo Cim. **33** (1964), 668.
3. S. Ishida et al., Prog. Theor. Phys. **98** (1997), 621.
4. C. Gobel, this conference.
5. Donnachie, this conference
6. S. U. Chung, summary talk of Hadron'99. V. Dorofeev; A. Popov, this conference.
7. K. Yamada, this conference.
8. M. Ishida, this conference.

Systematics of $q\bar{q}$ states, scalar mesons and glueball

V.V. Anisovich

St.Petersburg Nuclear Physics Institute email

Abstract. Basing on the latest results of the PNPI (Gatchina) and QM&W College (London) groups, I discuss systematics of the IJ^{PC} $q\bar{q}$ states in terms of trajectories on the (n, M^2) plane, where n is the radial quantum number and M is its mass. In the scalar sector, which is the most interesting because of the presence of extra states with respect to the $q\bar{q}$ systematics, I discuss: 1) the results of the K-matrix analysis of the spectra $\pi\pi$, $\pi\pi\pi\pi$, $K\bar{K}$, $\eta\eta$, $\eta\eta'$, $\pi\eta$ and characteristics of the resonances in the scalar sector, 2) $q\bar{q}$-nonet classification of scalar bare states, 3) accumulation of widths of the $q\bar{q}$ states by the glueball due to the overlapping of f_0-resonances at 1200–1700 MeV, 4) systematics of scalar $q\bar{q}$ states, both bare states and resonances, on the (n, M^2)-plots, 5) constraints on the quark-gluonium content of the resonances $f_0(980)$, $f_0(1300)$, $f_0(1500)$, $f_0(1750)$, and the broad state $f_0(1420^{+150}_{-70})$ from hadronic decays, 6) radiative decays of the P-wave $q\bar{q}$-resonances: scalars $f_0(980)$, $a_0(980)$, and tensor mesons $a_2(1320)$, $f_2(1270)$, $f_2(1525)$. The analysis proves that in the scalar sector we face two exotic mesons: the light σ-meson, $f_0(450)$, and the broad state $f_0(1420^{+150}_{-70})$, which is the descendant of the glueball.

SYSTEMATICS ON THE (N, M^2)-PLOTS

An important role for the unambiguous interpretation of the data is played by the $q\bar{q}$ systematization of the discovered meson states: this may be a guide for the search for new resonances as well as for establishing signatures of the existing states.

Here, following [1], the systematics of $q\bar{q}$ states is presented in terms of the (n, M^2) trajectories where n is the radial quantum number of the $q\bar{q}$ state and M is its mass. The trajectories on the (n, M^2) planes are drawn for the (IJ^{PC})-states with the positive charge parity $(C = +)$: $\pi(10^{-+})$, $\pi_2(12^{-+})$, $\pi_4(14^{-+})$, $\eta(00^{-+})$, $\eta_2(02^{-+})$, $a_0(10^{++})$, $a_1(11^{++})$, $a_2(12^{++})$, $a_3(13^{++})$, $a_4(14^{++})$, $f_0(00^{++})$, $f_2(02^{++})$, and negative one $(C = -)$: $b_1(11^{+-})$, $b_3(13^{+-})$, $h_1(01^{+-})$, $\rho(11^{--})$, $\rho_3(13^{--})$, $\omega/\phi(11^{--})$, $\omega_3(13^{--})$, see Figs. 1 and 2. Open points stand for the predicted states.

The main bulk of information about the mass region 2000–2400 MeV, which is crucial for drawing the trajectories, came from the analysis of Crystal Barrel data for the $p\bar{p}$ annihilation in flight [2].

The trajectories on the (n, M^2)-plots with a good accuracy are linear:

$$M^2 = M_0^2 + (n-1)\mu^2. \tag{1}$$

M_0 is the mass of basic meson and μ^2 is the trajectory slope parameter: μ^2 is nearly the same for all trajectories: $\mu^2 \simeq 1.2 - 1.3 \text{ GeV}^2$.

Trajectories with the same IJ^{PC} can be created by the states with different orbital momenta, with $J = L \pm 1$; in this way they are doubled: these are the trajectories (11^{--}), (12^{++}), and so on. Isoscalar states are formed by two light flavour components, $n\bar{n} = (u\bar{u}+d\bar{d})/\sqrt{2}$ and $s\bar{s}$. Likewise, this also results in doubling isoscalar trajectories.

The representation of $(C = -)$-trajectories is thus determined. The trajectories are nearly linear with the slope $\mu^2 \simeq 1.3$ GeV^2, with an exception of the b_J sector where the slope $\mu^2 \simeq 1.1 - 1.2$ GeV^2.

For the $C = +$ states, the π_J-sector is decisively fixed with the slope $\mu^2 \simeq 1.2$ GeV^2. The only state which breaks linearity of the trajectory is the pion, that is not surprising because of its specific role in the low-energy physics.

The trajectories in the η_J-sector are not unambigously fixed; in Fig. 1b we show the variant with $\mu^2 = 1.3$ GeV^2. The uncertainties are mainly due to the region 1700–2000 MeV in the wave 00^{-+}: it is the region where one may expect the existence of pseudoscalar glueball. Indeed, a strong production of the 00^{-+} wave is observed in the radiative J/Ψ decay [3] that may be a glueball signature (although one should note that lattice calculations provide us with a higher value, ~ 2300 MeV [4]). For sure, this mass region needs an intensive study.

The sector of a_J states, $J = 0, 1, 2, 3, 4$, demonstrates clearly a set of linear trajectories with $\mu^2 \simeq 1.15 - 1.20$ GeV^2, Figs. 1c, 1e. The same slope is observed for f_2 and f_4 mesons, Fig. 1d.

For f_0 mesons we have $\mu^2 \simeq 1.3$ GeV^2. A superfluous state for $q\bar{q}$-trajectories are the light σ-meson [5, 6, 7] and the broad resonance $f_0(1420^{+150}_{-70})$ observed in the K-matrix analysis [8]: one should consider these states as candidates for the exotics.

In the recently performed study of the reaction $p\bar{p} \to \eta\eta\pi^0\pi^0$ in flight, the resonance with mass 1880 ± 20 in the (12^{-+})-wave has been declared [9]: this state is also beyond the π_2-trajectory and should be considered as a hybrid.

SCALARS

The existence of superfluous, with respect to $q\bar{q}$ systematics, states is a motivation to perform intensive studies on scalar-isoscalar sector.

1) K-matrix analysis and resonances in the scalar-isoscalar sector. In the paper [8], on the basis of experimental data of GAMS group, Crystal Barrel Collaboration and BNL group, the K-matrix solution has been found for the waves $00^{++}, 10^{++}, 02^{++}, 12^{++}$ over the range 450–1900 MeV. Also the masses and total widths of resonances have been determined for these waves. The following states have been seen in the scalar-isoscalar sector: $f_0(980)$, $f_0(1300)$, $f_0(1500)$, $f_0(1420^{+150}_{-70})$, $f_0(1750)$. For the scalar-isovector sector, the analysis [8] points to the presence of the following resonances in the spectra: $a_0(980)$, $a_0(1520)$.

The K-matrix amplitude takes correctly into account the threshold singularities of the 00^{++} amplitude related to the channels $\pi\pi$, $\pi\pi\pi\pi$, $K\bar{K}$, $\eta\eta$, $\eta\eta'$. This circumstance allowed us to reconstruct the analytical amplitude in the complex-mass region shown in Fig. 3 by dashed line. In this area, with correctly restored analytical structure of the amplitude 00^{++}, we find out the resonance characteristics: the amplitude poles and decay coupling constants. Besides, we know the K-matrix characteristics such as the K-matrix poles. The K-matrix poles are not the amplitude poles, these latter being connected with physical resonances, but when the decays are switched off, the resonance poles turn into the K-matrix ones. In the states related to the K-matrix poles there is no cloud of real mesons, that is due to the decay processes. This was the reason to call them as "bare states" [8].

Below the mass scale of the K-matrix analysis [8] there is a pole related to the light σ-meson (or $f_0(450)$); its position is shown in Fig. 3 following the results of the dispersion relation N/D-analysis [6] (the mass region validated by this analysis is also shown in Fig. 3). Above the mass region of the K-matrix analysis there are resonances $f_0(2030)$, $f_0(2100)$, $f_0(2340)$ [2].

2) Classification of scalar bare states. The quark-gluonium systematics of scalar particles, in terms of bare states, has been suggested in [10]. A bare state being a member of the $q\bar{q}$ nonet imposes rigid restrictions upon the K-matrix parameters. The $q\bar{q}$ nonet of scalars consists of two scalar-isoscalar states, $f_0^{bare}(1)$ and $f_0^{bare}(2)$, scalar-isovector meson a_0^{bare} and scalar kaon K_0^{bare}. In the leading order of the $1/N$-expansion the decays of these four states into two pseudoscalars are determined by three parameters only, which are the common constant g, suppression parameter λ for strange quark production (in the limit of a precise $SU(3)_{flavour}$ symmetry $\lambda = 1$) and mixing angle φ for the $n\bar{n} = (u\bar{u}+d\bar{d})/\sqrt{2}$ and $s\bar{s}$ components in f_0^{bare}: $n\bar{n}\cos\varphi + s\bar{s}\sin\varphi$. The mixing angle defines scalar-isoscalar nonet partners $f_0^{bare}(1)$ and $f_0^{bare}(2)$: $\varphi(1) - \varphi(2) = 90°$. Restrictions imposed on coupling constants allow one to fix unambigously basic scalar nonet [8, 10]:

$$1^3P_0 q\bar{q}: f_0^{bare}(720\pm 100),\ a_0^{bare}(960\pm 30),\ K_0^{bare}(1220^{+50}_{-150}),\ f_0^{bare}(1260\pm 30),\quad (2)$$

as well as mixing angle for $f_0^{bare}(720)$ and $f_0^{bare}(1260)$: $\varphi(720) = -70°\,^{+5°}_{-10°}$.

To establish the nonet of first radial excitations, $2^3P_0 q\bar{q}$, appeared to be a more difficult task. The K-matrix analysis [8] gives us two scalar-isoscalar states at 1200–1650 MeV, $f_0^{bare}(1230^{+150}_{-30})$ and $f_0^{bare}(1600\pm 50)$; the decay couplings for both of them satisfy the requirements imposed for the glueball. To resolve this dilemma, we have performed the systematization of the $q\bar{q}$ states on the (n, M^2) plot. Such a systematization definitely proves that $f_0^{bare}(1600\pm 50)$ is an extra state for the $q\bar{q}$ trajectory. In this way, $f_0^{bare}(1230^{+150}_{-30})$ and $f_0^{bare}(1810\pm 30)$ must be the $q\bar{q}$ states.

Then the nonet $2^3P_0 q\bar{q}$ looks as follows:

$$2^3P_0 q\bar{q}: f_0^{bare}(1230^{+150}_{-30}),\ f_0^{bare}(1810\pm 30),\ a_0^{bare}(1650\pm 50),\ K_0^{bare}(1885^{+50}_{-100}).\quad (3)$$

The decay couplings of $f_0^{bare}(1600)$ to channels $\pi\pi, K\bar{K}, \eta\eta, \eta\eta'$ obey the requirements for the glueball decay. This gives us the reason to consider this state as the lightest scalar glueball:

$$0^{++}\ glueball: \qquad f_0^{bare}(1600\pm 50). \qquad (4)$$

The lattice calculations are in reasonable agreement with such a value of the lightest glueball mass.

After the onset of decay channels, the bare states have transformed into real resonances. For scalar–isoscalar sector we observe the following transitions after switching-on the decay channels: $f_0^{bare}(720) \pm 100 \to f_0(980)$, $f_0^{bare}(1260 \pm 30) \to f_0(1300)$, $f_0^{bare}(1230^{+150}_{-30}) \to f_0(1500)$, $f_0^{bare}(1600 \pm 50) \to f_0(1420^{+150}_{-70})$, $f_0^{bare}(1810 \pm 30) \to f_0(1750)$.

The evolution of bare states into real resonances is illustrated by Fig. 4: the shifts of amplitude poles on the complex-M plane correspond to a gradual onset of the decay channels. Technically it is done by replacing the phase space ρ_a for $a = \pi\pi, \pi\pi\pi\pi, K\bar{K}, \eta\eta, \eta\eta'$ in the K-matrix amplitude as follows: $\rho_a \to \xi\rho_a$, the parameter ξ running in the interval $0 \le \xi \le 1$. At $\xi \to 0$ one has pure bare states, while the limit $\xi \to 1$ gives us the position of real resonance.

Note that the broad state is denoted in [8] as $f_0(1530^{+90}_{-250})$ that is the averaged value for three solutions found in [8]; the value of the mass given in Figs. 3 and 4, $M = (1420^{+150}_{-70}) - i(540 \pm 80)$ MeV, corresponds to the solution for which the scalar glueball is located near 1600 MeV.

3) The overlapping of f_0-resonances at 1200–1700 MeV: accumulation of widths of $q\bar{q}$ states by the glueball. The appearance of broad resonance is not at all an occasional phenomenon. It has originated as a result of a mixing of states which are due to the decay processes, namely, transitions $f_0(m_1) \to$ *real mesons* $\to f_0(m_2)$. These transitions result in a specific phenomenon, that is, when several resonances overlap, one of them accumulates the widths of neighbouring resonances and transforms into a broad state.

This phenomenon has been observed in [8] for the scalar-isoscalar states, and the following scheme has been suggested in [11]: the broad state $f_0(1420^{+150}_{-70})$ is the descendant of a pure glueball, which being in the neighbourhood of $q\bar{q}$ states accumulated their widths and transformed into a mixture of the gluonium and $q\bar{q}$ states. In [11] this idea has been applied for four resonances $f_0(1300)$, $f_0(1500)$, $f_0(1420^{+150}_{-70})$ and $f_0(1750)$, by using the language of the $q\bar{q}$ and gg states for consideration of the decays $f_0 \to q\bar{q}, gg$ and mixing processes $f_0(m_1) \to q\bar{q}, gg \to f_0(m_2)$. According to [11], the gluonium component is mainly shared between three resonances, $f_0(1300)$, $f_0(1500)$, $f_0(1420^{+150}_{-70})$, so every state is a mixture of $q\bar{q}$ and gg components, with roughly equal percentage of the gluonium (about 30-40%).

The accumulation of widths of overlapping resonances by one of them is a well-known effect in nuclear physics. In meson physics this phenomenon can play an important role, in particular for exotic states which are beyond the $q\bar{q}$ systematics. Indeed, being among the $q\bar{q}$ resonances, the exotic state creates a group of overlapping resonances. The exotic state, which is not orthogonal to its neighbours, after having accumulated the "excess" of width turns into a broad state. This broad resonance should be accompanied by narrow states which are the descendants of states from which the widths have been taken off. In this way, the existence of a broad resonance accompanied by narrow ones may be a signature of exotics. This possibility, in context of searching for exotic states, has been discussed in [12].

The broad state may be one of the components which form the confinement barrier: the broad states after accumulating the widths of neighbouring resonances play for these latter the role of locking states. Evaluation of the mean radii squared of the broad state $f_0(1420^{+150}_{-70})$ and its neighbours-resonances, performed in [12] on the basis of the GAMS data, argues in favour of this idea, for the radius of $f_0(1420^{+150}_{-70})$ is significantly larger than that of $f_0(980)$ and $f_0(1300)$ thus making it possible for $f_0(1420^{+150}_{-70})$ to be the locking state.

4) Systematics of the $q\bar{q}$ scalar states on the (n,M^2) plot. As is stressed above, the systematics of $q\bar{q}$ states on the (n,M^2) plot argues that the broad state $f_0(1420^{+150}_{-70})$ and its predecessor $f_0^{bare}(1600\pm 50)$ are beyond the $q\bar{q}$ classification. We plot in Fig. 5a the (n,M^2)-trajectories for f_0, a_0 and K_0 states. All trajectories are roughly linear, and they clearly represent the states with dominant $q\bar{q}$ component. It is seen that one of the states, either $f_0(1420^{+150}_{-70})$ or $f_0(1500)$, is superfluous for the $q\bar{q}$ systematics. Looking at the (n,M^2)-trajectories of bare states, Fig. 5b, one can see that just $f_0^{bare}(1600)$ does not fall onto any linear $q\bar{q}$ trajectory. So it would be natural to conclude that the state $f_0^{bare}(1600)$ is an exotic one, i.e. the glueball.

For resonances belonging to linear trajectories (Fig. 5a) the $q\bar{q}$ component is dominanting. The scalar-isoscalar resonances $f_0(1300)$, $f_0(1500)$ contain a considerable gluonium component, and certain gluonium admixture exists in $f_0(1750)$. The location of the $f_0(980)$ pole near $K\bar{K}$ threshold allows one to suspect the existence of an admixture of the $K\bar{K}$-component in this resonance. To investigate this admixture the precise measurements of the $K\bar{K}$ spectra in the interval 1000—1150 MeV are necessary: only these spectra could shed the light on the role of the long-range $K\bar{K}$ component in $f_0(980)$.

5) Quark-gluonium content of resonances $f_0(980)$, $f_0(1300)$, $f_0(1500)$, $f_0(1750)$ and the broad state $f_0(1420^{+150}_{-70})$ from hadronic decays. The K-matrix analysis does not supply us with coupling constants of the resonance decay in a direct way. To find them out, additional calculations are needed to know the residues of amplitude poles related to resonances. Such calculations have been carried out in [13] for the channels $f_0 \to \pi\pi, \pi\pi\pi\pi, K\bar{K}, \eta\eta, \eta\eta'$. The conclusion is as follows [14]: the decays couplings to the channels $\pi\pi, K\bar{K}, \eta\eta, \eta\eta'$ do not provide us with a unique solution for absolute weight of the $n\bar{n}$, $s\bar{s}$ and gluonium components but give us relative weights only. The mixing angle φ which enters the quark wave function $q\bar{q} = n\bar{n}\cos\varphi + s\bar{s}\sin\varphi$ can be evaluated as a function of the decay couplings for gluonium and quarkonium components $G(gg \to hadrons)/g(q\bar{q} \to hadrons)$. The ratio of the couplings squared was conventionally called in [14] as probability for the gluonium component in the f_0-meson: $W \equiv G^2/g^2$. The following relations for φ verus W have been found [14]:

$$\varphi[f_0(980)] \simeq -67° \pm 57°\sqrt{W(980)}, \qquad \varphi[f_0(1300)] \simeq -5° \pm 28°\sqrt{W(1300)},$$
$$\varphi[f_0(1500)] \simeq 8° \pm 16°\sqrt{W(1500)}, \qquad \varphi[f_0(1750)] \simeq -27° \pm 42°\sqrt{W(1750)}. \qquad (5)$$

A large admixture of the gluonium, $W \leq 0.4$, may be expected for $f_0(1300)$, $f_0(1500)$, $f_0(1750)$, but it should be considerably less in $f_0(980)$, $W(980) \leq 0.20$.

The analysis [14] proves that $f_0(1420^{+150}_{-70})$ contains the $q\bar{q}$ in the flavour singlet state only:

$$\varphi[f_0(1420^{+150}_{-70})] \simeq 37°, \qquad (6)$$

that perfectly agrees with its gluonium origin: This value of mixing angle practically does not depend on the percentage of the $(q\bar{q})_{singlet}$ and gluonium components in the broad state.

6) Radiative decays of the P-wave $q\bar{q}$-mesons. The investigation of radiative decays is a powerful tool for establishing the quark structure of hadrons. At the early stage of the quark model, the radiative decays of vector mesons provided strong arguments in favour of the idea of constituent quark, a universal object for mesons and baryons [15]. The radiative decays of the $1^3P_J q\bar{q}$ mesons are equally important for the verification of the P-wave multiplet.

In Ref. [16], partial widths of the decays $f_0(980) \to \gamma\gamma$ and $a_0(980) \to \gamma\gamma$ have been calculated assuming $f_0(980)$ and $a_0(980)$ to be dominantly $q\bar{q}$ states, that is, $1^3P_0 q\bar{q}$ mesons. The results of the calculation agree well with experimental data. On the basis of experimental data for the decays $\phi(1020) \to \gamma f_0(980)$ and $f_0(980) \to \gamma\gamma$ the $n\bar{n}/s\bar{s}$ content of $f_0(980)$ has been found. Assuming the flavour wave function in the form $n\bar{n}\cos\varphi + s\bar{s}\sin\varphi$, the experimental data has been described with two possible values of mixing angle: either $\varphi[f_0(980)] = -48° \pm 6°$ or $\varphi[f_0(980)] = 85° \pm 4°$ (negative value is more preferable), see Fig. 6a where the allowed region of φ versus $R^2_{f_0(980)}$ is shown.

The dominance of the quark-antiquark state does not exclude the existence of other components in $f_0(980)$ on the level $10\% - 20\%$, the glueball or long-range $K\bar{K}$ component. The existence of the long-range $K\bar{K}$ component or that of gluonium in the $f_0(980)$ results in a decrease of the $s\bar{s}$ fraction in the $q\bar{q}$ component: for example, if the long-range $K\bar{K}$ (or gluonium) admixture is of the order of 15%, the data require either $\varphi = -45° \pm 6°$ or $\varphi = 83° \pm 4°$.

There is no problem with the description of the decay $a_0(980) \to \gamma\gamma$ within the hypothesis about $q\bar{q}$ origin of the $a_0(980)$: the data are in a good agreement with the results of the calculation by using $R^2_{a_0(980)} \sim 10 - 17$ GeV^{-2}.

Although direct calculations of widths of radiative decays agree well with the hypothesis that the $q\bar{q}$ component dominates $f_0(980)$ and $a_0(980)$, to determine reliably these mesons as members of the $1^3P_0 q\bar{q}$ multiplet one more step is necessary. We have to prove that radiative decays of tensor mesons $a_2(1320)$, $f_2(1270)$, $f_2(1525)$ can be calculated within the same approach and the same technique as it was carried out for $f_0(980)$ and $a_0(980)$. Tensor mesons $a_2(1320)$, $f_2(1270)$, $f_2(1525)$ are the basic members of the P-wave $q\bar{q}$ multiplet, and the existence of tensor mesons had been used to suggest quark-antiquark classification for four P-wave nonets [17]. Under this motivation, partial widths of the tensor $q\bar{q}$ states $a_2(1320) \to \gamma\gamma$, $f_2(1270) \to \gamma\gamma$ and $f_2(1525) \to \gamma\gamma$ have been calculated [16]: the agreement with data has been reached for all calculated partial widths, with similar radial wave functions, that indicates definitely that both scalar ($f_0(980)$, $a_0(980)$) and tensor ($a_0(1320)$, $f_2(1270)$, $f_2(1525)$) mesons belong to the same P-wave $q\bar{q}$ multiplet. In Fig. 6b one can see the region of magnitudes (φ_T, R_T^2) allowed by data on the decays $f_2(1270) \to \gamma\gamma$ and $f_2(1525) \to \gamma\gamma$; here φ_T is mixing angle for $\psi_{f_2(1270)} = \cos\varphi_T n\bar{n} + \sin\varphi_T s\bar{s}$ and $\psi_{f_2(1525)} = -\sin\varphi_T n\bar{n} + \cos\varphi_T s\bar{s}$ and R_T is the tensor-meson radius.

7) Exotics in scalar-isoscalar sector. The established $q\bar{q}$ systematics of scalar mesons in terms of bare states fixes two nonets: $1^3P_0 q\bar{q}$ and $2^3P_0 q\bar{q}$. The resonances which are the descendants of pure $q\bar{q}$ states are located on linear trajectories in the (n, M^2)-plane. The $q\bar{q}$ systematics reveals two extra states which are the light σ-meson, with mass ~ 450 MeV, or $f_0(450)$, and broad state $f_0(1420^{+150}_{-70})$. The broad state is the descendant of a pure glueball state which accumulated the widths of neighbouring $q\bar{q}$ resonances. The origin of the σ-meson is questionable.

In the paper [18], a hypothesis is discussed that the σ-meson owes its origin to strong singlarity in the confinement amplitude: the large-r behaviour of the confinement scalar potential $V(r) \sim r$ evokes a strong t-channel singularity, $1/t^2$. It was assumed in [18] that the singularity of this kind exists at every colour state of the confinement $q\bar{q}$ ladder; then, in the white state related to the $\pi\pi$ channel, the unitarization of singular block might reduce the singularity strenth, thus providing the pole near the $\pi\pi$ threshold, at $\mathrm{Re}\, s \sim 4\mu_\pi^2$, that corresponds to the σ-meson.

The analysis of data on the decays $D^+ \to \pi^+\pi^+\pi^-$ and $D_s^+ \to \pi^+\pi^+\pi^-$ [7] agrees with such an idea. The $s\bar{s}$ component at $f_0(980)$ has been evaluated in [19] by comparing the branching ratios $D_s^+ \to \pi^+\phi(1020)$ and $D_s^+ \to \pi^+ f_0(980)$ being about 50% (this estimate agrees with hadronic and radiative decays of $f_0(980)$ for the solution with negative mixing angle φ). The ratio of yields of $f_0(450)$ and $f_0(980)$ in the reaction $D^+ \to \pi^+\pi^+\pi^-$ tells us that $f_0(450)$ is dominantly the $n\bar{n}$ system. The confinement ladder should be formed by the light quarks, (u,d), see for example [20]: in this sense, the structure of $f_0(450)$ is just as it was expected, if it originates from the confinement ladder, as it was supposed in [18].

ACKNOWLEDGMENTS

Thanks are due to A.V. Anisovich, D.V. Bugg, L.G. Dakhno, D.I. Melikhov, V.A. Nikonov, A.V. Sarantsev for numerous discussions. The work is suppored by the RFFI grant N 01-02-17861.

REFERENCES

1. A.V. Anisovich, V.V. Anisovich and A.V. Sarantsev, Phys. Rev. **D62**:051502 (2000).
2. A.V. Anisovich, C.A. Baker, C.J. Batty et al., Phys. Lett. **B449**, 114 (1999); **B452**, 173 (1999); **B452**, 180 (1999); **B452**, 187 (1999); **B476**, 15 (2000); **B491**, 47 (2000); **B507**, 23 (2001); **B508**, 6 (2001); Nucl. Phys. **A651**:253 (1999); **A662**:319 (2000).
3. B.V. Bugg and B.S. Zou, privite communication.
4. G.S. Bali et al., Phys. Lett. **B309** 378 (1993); J. Sexton, A. Vaccarino and D. Weingarten, Phys. Rev. Lett. **75** 4563 (1995); C.J. Morningstar, M. Peardon, Phys. Rev. **D56** 4043 (1997).
5. J.L. Basdevant, C.D. Frogatt, J.L. Petersen, Phys. Lett. **B41**, 178 (1972); D. Iagolnitzer, J. Justin, J.B. Zuber, Nucl. Phys. **B60**, 233 (1973); B.S. Zou, D.V. Bugg, Phys. Rev. **D48**, R3942 (1994); **D50**, 591 (1994); G. Janssen, B.C. Pearce, K. Holinde, J. Speth, Phys. Rev. **D52**, 2690 (1995).
6. V.V. Anisovich and V.A. Nikonov, Eur. Phys. J. **A8**, 401 (2000).
7. E.M. Aitala, et al. Phys. Rev. Lett. **86**, 770 (2001); **86**, 779 (2001).
8. V.V. Anisovich, A.A. Kondashov, Yu.D. Prokoshkin, S.A. Sadovsky, A.V. Sarantsev, Yad. Fiz. **60**, 1489 (2000) [Physics of Atomic Nuclei, **60**, 1410 (2000)]; V.V. Anisovich, Yu.D. Prokoshkin and A.V. Sarantsev, Phys. Lett. **B389**, 388 (1996); V.V. Anisovich and A.V. Sarantsev, Phys. Lett. **B382**, 429 (1996).
9. A.V. Anisovich, C.A. Baker, C.J. Batty et al., Phys. Lett. **B500**, 222 (2001);
10. A.V. Anisovich and A.V. Sarantsev, Phys. Lett. **B413**, 137 (1997); V.V. Anisovich, Physics-Uspekhi **41**, 419 (1998).

11. A.V. Anisovich, V.V. Anisovich, Yu.D. Prokoshkin and A.V. Sarantsev, Zeit. Phys. **A357**, 123 (1997); A.V. Anisovich, V.V. Anisovich, A.V. Sarantsev, Phys. Lett. **B395**, 123 (1997); Zeit. Phys. **A359**, 173 (1997).
12. V.V. Anisovich, D.V. Bugg and A.V. Sarantsev, Phys. Rev. **D58**:111503 (1998); V.V. Anisovich, D.V. Bugg and A.V. Sarantsev, Yad. Fiz. **62**, 1322 (1999) [Phys. Atom. Nuclei, **62**, 1247 (1999)].
13. V.V. Anisovich, V.A. Nikonov and A.V. Sarantsev, hep-ph/0102338, Yad. Fiz., in press.
14. V.V. Anisovich, V.A. Nikonov and A.V. Sarantsev, hep-ph/0108188; Yad. Fiz., in press.
15. V.V. Anisovich, A.A. Anselm, Ya.I. Azimov, G.S. Danilov and I.T. Dyatlov, Phys. Lett. **16**, 194 (1965); W.E. Tirring, Phys. Lett. **16**, 335 (1965); L.D. Soloviev, Phys. Lett. **16**, 345 (1965); C. Becchi and G. Morpurgo, Phis. Rev. **140**, 687 (1965).
16. A.V. Anisovich, V.V. Anisovich, D.V. Bugg and V.A. Nikonov, Phys. Lett. **B456**, 80 (1999); A.V. Anisovich, V.V. Anisovich, and V.A. Nikonov, hep-ph/ 0108186, Eur. Phys. J. A, in press.
17. R. Gatto, Phys. Lett. **17**, 124 (1965); V.V. Anisovich, A.A. Anselm, Ya.I. Azimov, G.S. Danilov and I.T. Dyatlov, Pis'ma ZETF **2**, 109 (1965).
18. V.V. Anisovich and V.A. Nikonov, hep-ph/0008163 (2000).
19. V.V. Anisovich, D.V. Bugg, V.A. Nikonov, and A.V. Sarantsev, to be published.
20. V.N. Gribov, Eur. Phys. J. **C10**, 71 (1999).

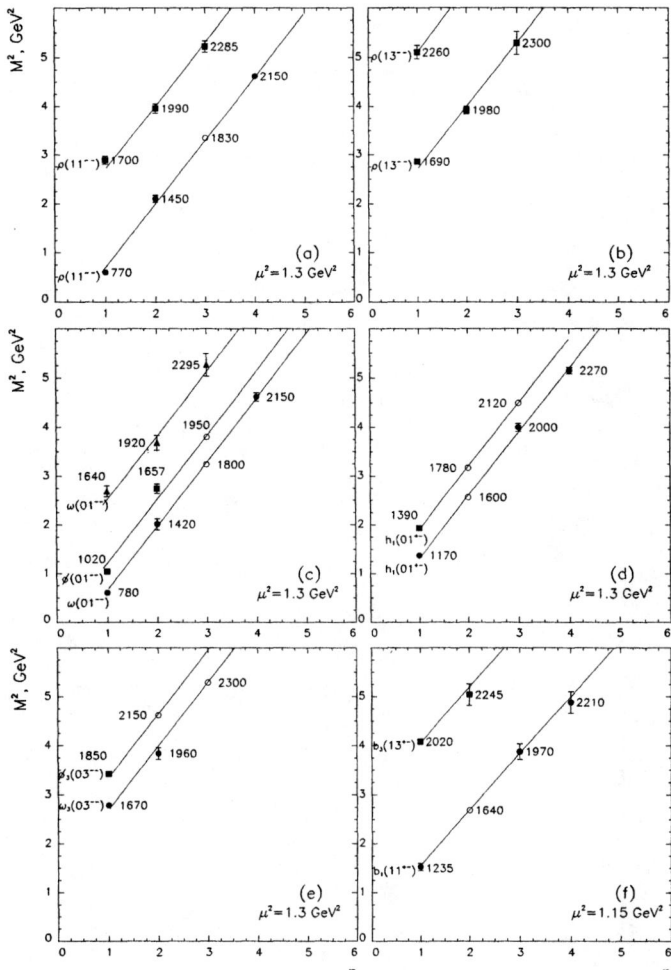

FIGURE 1. Trajectories of the $(C=-)$-states on the (n,M^2) plane.

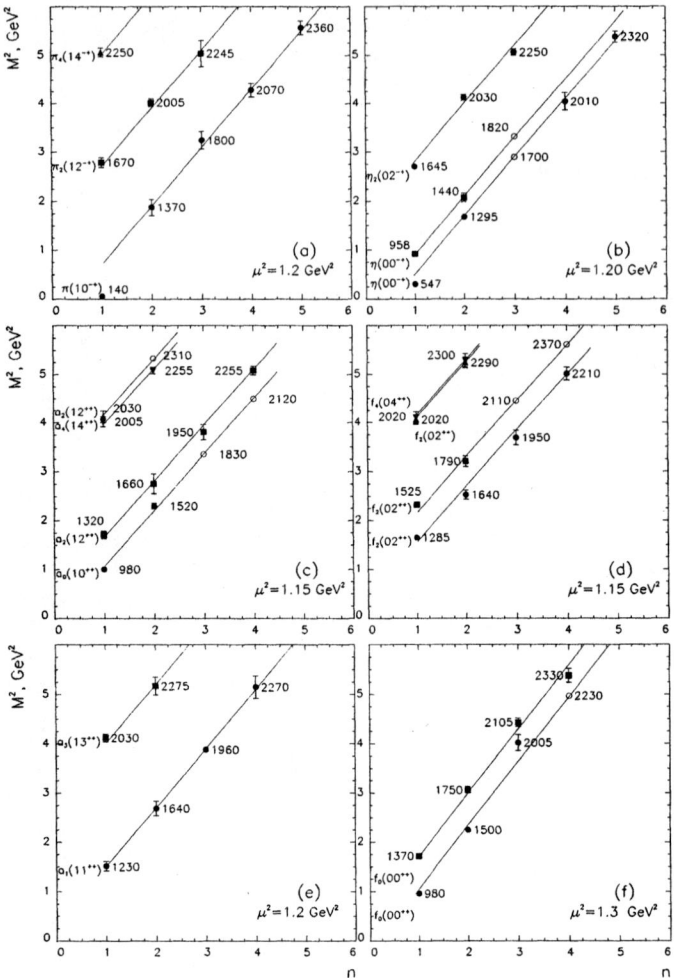

FIGURE 2. Trajectories of the $(C = -)$-states on the (n, M^2) plane.

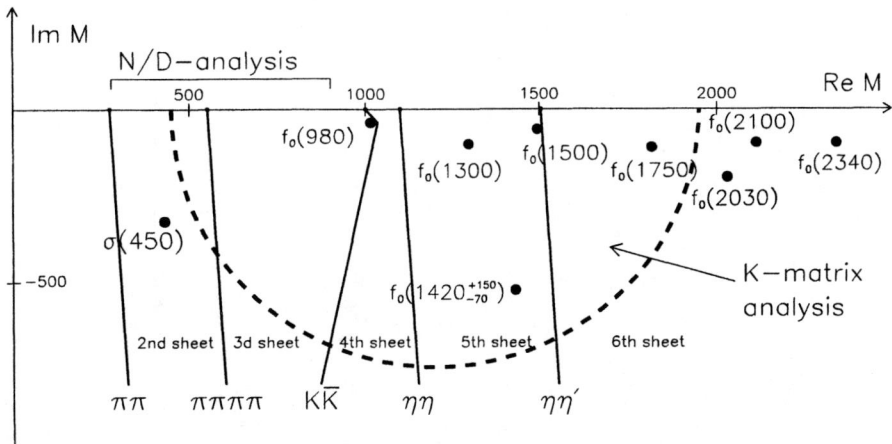

FIGURE 3. Complex M-plane in the ($IJ^{PC} = 00^{++}$) sector. Dashed line encircle the part of the plane where the K-matrix analysis [8] reconstructs the analytic K-matrix amplitude: in this area the poles corresponding to resonances $f_0(980)$, $f_0(1300)$, $f_0(1500)$, $f_0(1750)$ and the broad state $f_0(1420^{+150}_{-70})$ are located. Beyond this area the light σ-meson is located (the position of pole found in the N/D method [6] is shown) as well as resonances $f_0(2030)$, $f_0(2100)$, $f_0(2340)$ [2]. Solid lines stand for the cuts related to the thresholds $\pi\pi, \pi\pi\pi\pi, K\bar{K}, \eta\eta, \eta\eta'$.

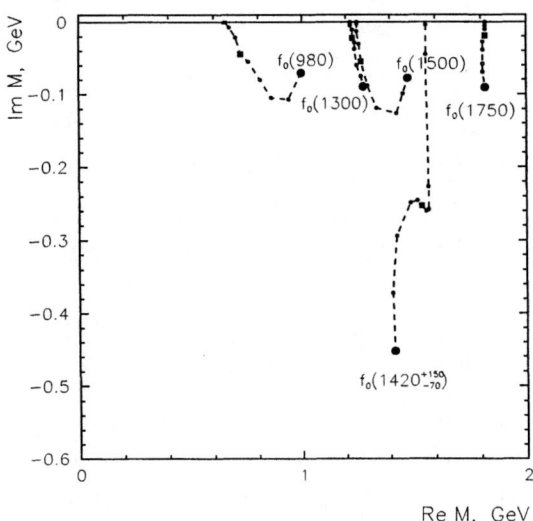

FIGURE 4. Complex M-plane: trajectories of the poles for $f_0(980)$, $f_0(1300)$, $f_0(1500)$, $f_0(1750)$, $f_0(1420^{+150}_{-70})$ during gradual onset of the decay processes.

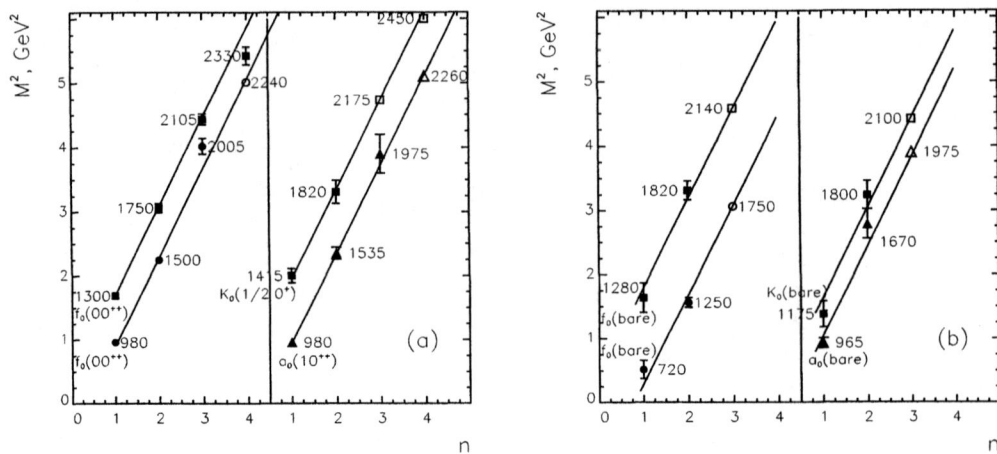

FIGURE 5. Linear trajectories in (n, M^2)-plane for scalar resonances (a) and scalar bare states (b). Open points stand for predicted states.

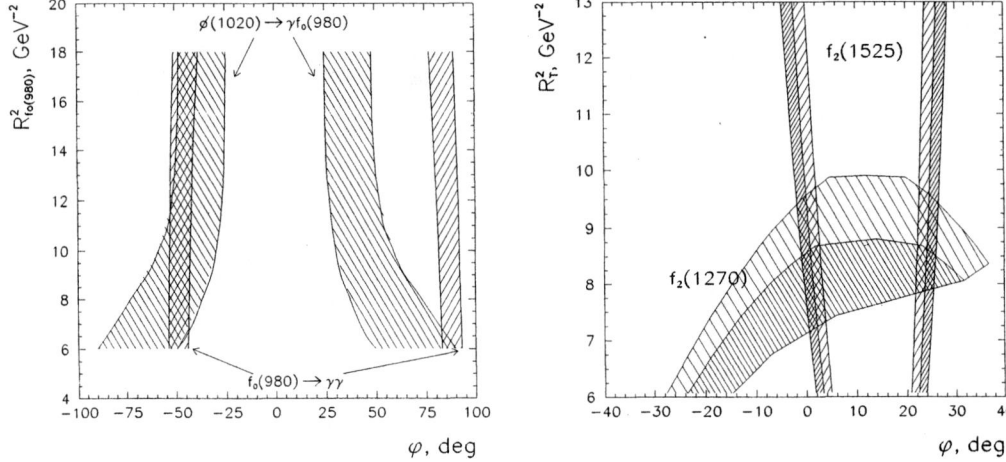

FIGURE 6. a) The $(\varphi, R^2_{f_0(980)})$-plot: the shaded areas are the allowed ones for the reactions $\phi(1020) \to \gamma f_0(980)$ and $f_0(980) \to \gamma\gamma$. b) The (φ_T, R^2_T)-plot for the reactions $f_2(1270) \to \gamma\gamma$ and $f_2(1525) \to \gamma\gamma$; the allowed areas are shaded.

On the dominance of non-exotic meson-meson scattering by s-channel $q\bar{q}$ confinement states and the classification of the scalar mesons

Eef van Beveren[a] and George Rupp[b]

[a] Centro de Física Teórica, Departamento de Física,
Universidade, P3004-516 Coimbra, Portugal, (eef@teor.fis.uc.pt)

[b] Centro de Física das Interacções Fundamentais, Instituto Superior Técnico,
Edifício Ciência, P1049-001 Lisboa Codex, Portugal, (george@ajax.ist.utl.pt)

Abstract. Non-exotic scalar-meson resonances in S-wave meson-meson scattering are studied in the light of a unitarised Schrödinger model. The resulting poles in the scattering matrices, by analytical continuation into the complex-energy plane, are grouped into nonets of isoscalar, isodoublet, and isotriplet resonances. All singularities can be related to *quark-antiquark confinement states*, the light-quark nonet of which has ground states at 1.3 to 1.4 GeV and level spacings of some 300–400 MeV, except for a nonet of light scalar mesons below 1 GeV. All non-exotic S-wave resonances reported by experiment fit into this scheme.

INTRODUCTION

Lattice QCD in principle offers the most direct way to link to experiment what we believe to be the fundamental theory of strong interactions [1]. However, in view of the constant evolution of results from sophisticated lattice QCD solutions [2], it becomes ever more puzzling why mesonic resonances can be described as simple quark-antiquark systems in effective theories [3]. But apparently, such a scenario works!

Seemingly, the perturbative vacuum states of QCD at low energies are not quarks and gluons, but rather confined constituent quarks and residual interactions, a picture that has been successful for several decades by now. What do lattice calculations teach us about constituent quarks? One might think of colour-triplet configurations of quarks, antiquarks, and glue, maybe with admixtures of higher colour multiplets, which mutually feel colour interactions. Can lattice QCD identify such substructures? Moreover, what happens exactly when those substructures suddenly turn colourless and cease to be confined? Does one observe colourless substructures that drift apart on the lattice?

In unitarised meson models one assumes an effective mass for the constituent quark, a confinement force for the remaining colour interactions, and a mechanism for decay [4,5]. However, not knowing how the separation into constituent quarks, confinement, and decay can be derived from QCD, each model is the result of educated guesses, rather than of rigorous derivations starting from QCD. This frustrating state of the affairs has in the past three decades led to a proliferation of effective models and theories. However, no model exists so far which completely describes the resonances of meson-meson scattering to a satisfactory degree of accuracy. Nevertheless, some educated guesses are less successful than others. For example, $q\bar{q}$ models for resonances that do not take meson loops into consideration [3] can never be in agreement with experiment, since the resulting spectrum consists of zero-width bound states, whereas large widths are measured.

NON-EXOTIC MESON-MESON SCATTERING

The assumption that non-exotic meson-meson scattering is dominated by the s-channel states of the confinement mechanism has been worked out in a series of papers [5–8]. Here, we will confine our attention to a toy model that we studied in Ref. [9]. There we obtain for the elastic low-energy partial-wave meson-meson scattering matrix,

$$S_\ell(p) = \exp(2i\delta_\ell(p)) \quad , \qquad (1)$$

the relation

$$\cot g\left(\delta_\ell(p)\right) = \frac{n_\ell(pa)}{j_\ell(pa)} - \left[2\lambda^2 \, \mu \, pa \, j_\ell^2(pa) \sum_{n=0}^{\infty} \frac{|\mathcal{F}_n(a)|^2}{E - E_n}\right]^{-1} , \qquad (2)$$

where p represents the relative momentum in the CM frame of the two mesons, ℓ their relative angular momentum, and μ their reduced mass; E_n ($n = 0, 1, 2, \ldots$) represent the energy eigenvalues of the constituent $q\bar{q}$ system to which the meson pair couples, and $\mathcal{F}_n$ the corresponding $q\bar{q}$ eigenfunctions; n_ℓ and j_ℓ stand for the spherical Bessel and Neuman functions, respectively.

The intensity of the coupling between the meson-meson system and the $q\bar{q}$ system is described by the parameter λ, whereas a stands for the average distance at which the transitions from one system to the other take place [10].

For $\lambda = 0$, we find $S_\ell(p) = 1$, which describes a system of two non-interacting mesons. For small values of λ, the scattering matrix (1) has poles in the lower half of the complex-energy plane, which can approximately be given by

$$E_{\text{pole}} \approx E_n - |\mathcal{F}_n(a)|^2 \left[\sum_{n' \neq n} \frac{|\mathcal{F}_n(a)|^2}{E_n - E_{n'}} - \frac{i}{2\lambda^2 \, \mu \, pa \, j_\ell(pa) \, h_\ell^{(1)}(pa)} \right]^{-1}, \quad (3)$$

indicating that to each value of the radial quantum number n corresponds one singularity, i.e., one meson-meson scattering resonance. For higher values of λ, one can determine the locations of the poles in the scattering matrix by numerical methods.

The real parts of the singularities roughly correspond to the central resonance positions, E_r, whereas the moduli of the imaginary parts approximately equal half the resonance widths, Γ_r. In short,

$$E_{\text{pole}} \approx E_r - i \frac{\Gamma_r}{2}. \quad (4)$$

Singularities may be located on the real energy axis below threshold, representing stable (with respect to strong decay) mesons (e.g. K, J/Ψ, Υ [7]).

For practical purposes, one might truncate the sum in formula (2) and substitute the truncated part by a constant [9].

For $\lambda \to \infty$ one finds

$$\cotg(\delta_\ell(p)) = \frac{n_\ell(pa)}{j_\ell(pa)}, \quad (5)$$

which represents scattering from a hard sphere of radius a. In this case, the interior of the $q\bar{q}$ state becomes unobservable and no resonance spectrum can be deduced from meson-meson scattering.

COMPARISON WITH EXPERIMENT

In Ref. [9] we compare the predictions of formula (1) for $K\pi$ S- and P-wave scattering in $I = 1/2$ with the experimental cross sections. For P waves, in the region of the $K^*(892)$ resonance, we find that the lowest-lying state of the confinement spectrum is at some 945 MeV, whereas the corresponding pole comes out at $(887 - 27i)$ MeV.

In a more refined model [7], which also takes inelasticity into account by considering the coupling to all channels with allowed initial and final states of pseudoscalar and vector mesons, and furthermore employs a more sophisticated mechanism for the coupling of $q\bar{q}$ confinement states to meson-meson scattering channels, it is found that the ground state of the confinement spectrum in this case comes out at 1.19 GeV, some 300 MeV above the position of the $K^*(892)$ pole. It shows that bare states can be several hundreds of MeVs away from the actual central resonance positions, and, moreover, that such conclusions are model dependent. The toy model of Ref. [9] yields a shift of only some 60 MeV. In this perspective, the question as to where a bound-state model should find its *bare* states is hard to be answered.

The latter question becomes even more difficult in the case of S-wave scattering. There we find, in the toy model of Ref. [9], that the ground state of the confinement spectrum is at 1.31 GeV and the corresponding pole at $(1.46 - 0.12i)$ GeV. However, further inspection of the singularity structure of the scattering matrix reveals another pole far below this energy region, namely at $(714 - 228i)$ MeV. The latter singularity has no direct relation to any of the bare states. Hence, the $I = 1/2$, $J^P = 0^+$ ground state of bound-state models should be at some 1.3 GeV and not below 1 GeV. This result is confirmed by the full model [6] (pole at $(727 - 263i)$ MeV), in which also the poles belonging to the two isoscalars $f_0(980)$ and the rather controversial $f_0(470 - 208i)$ (σ meson), as well as to the isovector $a_0(980)$, have no direct relation with the ground states of the corresponding confinement spectrum at about 1.2 GeV.

A MODEL STUDY OF POLES

The relation between the $q\bar{q}$ bare states and the poles in the corresponding scattering matrix can be found in the coupled-channel model (1) by considering the process of stepwise reducing the coupling constant λ. Such a study has been performed in detail in the toy model of Ref. [9], with the following result. All poles move towards a corresponding $q\bar{q}$ bare state on the real axis, as predicted by formula (3), except for the S-wave singularity below 1 GeV. The negative imaginary part of this pole grows inversely proportionally to λ^2, implying that the corresponding "resonance" disappears into the background of $K\pi$ scattering. Unfortunately, such processes cannot be tested in experiment, since Nature corresponds to a fixed value for λ.

CONSTITUENT-QUARK-PAIR CREATION

For P- and higher-wave meson-meson scattering, we do not find singularities other than those which can be related to a $q\bar{q}$ state of the confinement spectrum. The extra poles below 1 GeV, described through *pole doubling* in Ref. [11], exclusively appear in S-wave meson-meson scattering. Hence, we must conclude that the latter "resonances" are a consequence of the mechanism of constituent-quark-pair annihilation and creation, which couples meson-meson initial and final states to the $q\bar{q}$ confinement states.

For P and higher waves, the centrifugal barrier prevents the formation of such resonances in meson-meson scattering. But in the absence of a centrifugal barrier for S waves, resonances are formed that in the cases of the $f_0(980)$ and the $a_0(980)$ are narrow enough to be clearly observable, but which for the $f_0(470 - 208i)$ and $K_0^*(727 - 263i)$ are too broad to be firmly established.

One should note here that, when a pole with a large imaginary part lies close to threshold — close meaning that the distance from threshold to the real part of the singularity is smaller than or of the same order as the imaginary part — then the corresponding cross section has a shape which is very different from a standard Breit-Wigner. In the Argand plot, one finds a resonance motion that rapidly slows down for higher energies. If then, moreover, new thresholds get open and other rapid Breit-Wigner resonances show up, its appearance can hardly be recognised as that of a resonance, within the experimental accuracy.

Nevertheless, whether or not one associates a resonance with the controversial $J^P = 0^+$ singularities is of *no importance*. What is crucial in the above observations is the fact that it settles the classification of the $f_0(980)$ and $a_0(980)$ resonances in a nonet scheme for mesons rather more naturally than other proposals.

RESONANCE SHAPES

The model result that some of the light scalar resonances are broad, while others are narrow, has its origin in the effects of inelasticity. In Ref. [6], Table 1 of the Appendix, a list of inelasticity channels for the three scalar-meson isomultiplets is presented, as well as the intensities of the relative couplings.

We learn from this table that the isotriplet couples twice as strongly to $\eta_n\pi$ as to $K\bar{K}$, other thresholds lying at higher or much higher energies, which makes their effect hardly relevant here. However, only a small part of the $\eta_n\pi$ channel decays into $\eta\pi$, the rest into $\eta'\pi$. This implies that, of the lowest channels, the $K\bar{K}$ and also the $\eta'\pi$ channel are far stronger than the $\eta\pi$ channel (see also Ref. [12]). Elastic S-wave $\eta\pi$ scattering in the absence of inelasticity can be described by the toy model of formula (2). In Ref. [9], formula (2) has been applied to elastic S-wave $K\pi$ scattering. Now, when we substitute there the K mass by the η mass, then we obtain a toy model for elastic S-wave $\eta\pi$ scattering. With this substitution, we find for the model of formula (2) indeed a pole close to the $\eta\pi$ threshold, and with a relatively large imaginary part (763 − 199i MeV). Moreover, the related toy-model prediction for the $\eta\pi$ elastic S-wave scattering cross section does not show a clear resonance, exactly as in the case of the $K_0^*(727-263i)$ pole in $K\pi$ elastic scattering.

Inelasticity, which has been taken into account in Ref. [6], has two consequences here: first, the pole moves close to the $K\bar{K}$ threshold, with a smaller imaginary part (968 − 28i MeV), and, second, since the $\eta\pi$ threshold is far enough below that pole, its resonance shape turns more Breit-Wigner-like (see *e.g.* Fig. 2 of Ref. [6]). Nevertheless, upon reducing the coupling constant, the modulus of the imaginary part of this pole increases in a similar way as does the lower pole, the $K_0^*(727-263i)$, in $K\pi$ elastic S-wave scattering. Consequently, the two poles have a similar origin, not directly related to the bare spectrum. It is only through the strong interference of the $K\bar{K}$ and the $\eta'\pi$ channels that a reasonable Breit-Wigner-like shape appears for the light isotriplet resonance $a_0(980)$.

We also observe from the above-referred table of Ref. [6] that, of the isoscalar complex ($n\bar{n}$ coupled to $s\bar{s}$), the $n\bar{n}$ couples strongly to $\pi\pi$, whereas the $s\bar{s}$ couples strongly to $K\bar{K}$, with, again, other thresholds lying higher or much higher, thus making their influence of little importance here. Furthermore, $n\bar{n}$ and $s\bar{s}$ are coupled to one another through the $K\bar{K}$ channel, which implies that the $s\bar{s}$ component of the isoscalar complex also couples to $\pi\pi$, but quite weakly. Hence, for the $s\bar{s}$ resonance $f_0(980)$ we can now repeat the arguments we gave for the isotriplet resonance shape, with $\eta\pi$ replaced by $\pi\pi$. Moreover, since the $\pi\pi$ threshold in the isoscalar case lies much lower than the $\eta\pi$ threshold in the isotriplet case, we find a more convincing Breit-Wigner-like shape for the $f_0(980)$ in $\pi\pi$ scattering [13] than for the $a_0(980)$ in $\eta\pi$ scattering.

However, the $n\bar{n}$ component of the isoscalar complex, which yields a pole close to the $\pi\pi$ threshold, has no further strong-inelasticity channel to allow for a Breit-Wigner-like shape for the corresponding resonance $f_0(470 - 208i)$. The same happens to the isodoublet, which, according to the table of coupling constants mentioned before, couples strongly to the $K\pi$ channel. No further lower-lying inelasticity channel exists in this case. Consequently, also the $K_0^*(727 - 263i)$ has no Breit-Wigner-like shape.

None of the poles of this nonet of scalar *resonances* has a direct relation to the bare spectrum. By stepwise reducing the model coupling constant, all nine poles stepwise disappear into the complex plane with increasing negative imaginary part, whereas the corresponding structures in the meson-meson scattering cross sections stepwise disappear into the background.

THE K-MATRIX

As one can easily observe from formula (2), we have no poles in the K-matrix at the energy eigenvalues E_n of the confinement spectrum, since the hard-sphere-scattering part in the expression for the cotangent of the phase shift does not vanish at energies E_n. This contradicts the observation of Sarantsev and collaborators (Ref. [14] and references therein) that bare states are the singularities of the K-matrix, so this issue deserves further study.

In the limit of an infinitely strong coupling between the confinement and scattering sectors, formula (2) predicts that no bare spectrum can be observed in meson-meson scattering other than the hard-sphere spectrum. This is reasonable, since in that limit the mesons become impenetrable and thus do not allow the observation of the interior dynamics. However, when the hard-sphere-scattering part of formula (2) is removed, then one just obtains stronger resonances close to the poles of the K-matrix when the coupling constant λ is increased.

For small coupling, both models give similar results, except that the real shifts for formula (2) can be much larger when the hard-sphere-scattering part is present. This is probably the reason why the bare states of Ref. [14] are always close to the central resonance energies.

We may therefore conclude that the behaviour of both models for moderate coupling is very similar, except for the interpretation of the bare states. In Ref. [15], we studied other consequences of the fact that mesons are not point particles, but finite distributions of constituent quarks.

Nevertheless, the fits of Ref. [16] to the data are too good for the corresponding model to be totally wrong. It might be that, with a small modification, the latter model would also yield the extra $J^P = 0^+$ nonet of singularities and no related bare states, without destroying the excellent fits to the data.

$J^P = 0^+$ RESONANCES NONETS

In Table 1, we classify [6] the experimentally observed non-exotic scalar mesons into nonets.

radial excitation	isotriplets	isodoublets	isoscalars
pole doubling	$a_0(980)$	$K_0^*(727 - 263i)$	$f_0(470 - 208i)$ and $f_0(980)$
ground state	$a_0(1470)$	$K_0^*(1430)$	$f_0(1370)$ and $f_0(1500)$
first		$K_0^*(1950)$	$f_0(1710)$ and $f_0(?)$
second			$f_0(2020)$ and $f_0(2200)$

TABLE 1. The nonet classification [6] of the S-matrix poles for $J^P = 0^+$ meson-meson scattering.

The $f_0(470 - 208i)$ comes in the *Tables of Particle Properties*, Ref. [17], under $f_0(400\text{--}1200)$, but is not well established as a resonance, while the $K_0^*(727 - 263i)$ is not even mentioned in Ref. [17], although evidence for the existence of structure in that energy region has been reported [6,18,19]. Moreover, a pole in the S-matrix is not necessarily observable as a clear resonance in meson-meson scattering, as we have argued before. Furthermore, the $s\bar{s}$ assignment of the $f_0(980)$ [20,21,14] hints at the existence of a corresponding, most probably lower-lying, $n\bar{n}$ structure in $\pi\pi$ scattering [22].

The $f_0(1370)$ and $f_0(1500)$ resonances have been studied in many works [21,14,23], with a diversity of explanations as to their nature, out of which the above nonet classification is the most comprehensive.

Of all resonances in Table 1, the $K_0^*(1950)$ does not seem to be well in place: the general level splittings of some 300 – 400 MeV do not agree with the jump of 520 MeV from the ground state to the first radial excitation of the scalar isodoublet. However, in the analysis of Ref. [19] one finds in Table 2 a set of possible singularities (in the third Riemann sheet) related to the $K_0^*(1950)$ resonance, which all have real parts in the energy region 1.7 – 1.77 GeV. Moreover, in Ref. [16] a central resonance position of 1.82±0.04 GeV is reported for this resonance [17].

Furthermore, the $f_0(1710)$ is placed at 1.77 GeV in Ref. [24], which, nevertheless, does not alter the above classification, whereas both the $f_0(2020)$ and the $f_0(2060)$

need confirmation and might very well represent the same resonance. However, if there really exist two resonances in this energy region, then our classification indicates that one of them must be of a nature other than $q\bar{q}$.

In conclusion, one should note that several analyses find *too many* f_0 resonances, whereas in our analysis we *lack* an $f_0(1840)$.

Acknowledgements. One of us (EvB) wishes to thank the organisers of this conference for the kind invitation and warm hospitality at Protvino. We wish to thank Joseph Schechter, H. Noya, David Bugg, Carla Göbel, Brian Meadows, Michael Pennington, Teiji Kunihiro, Shin Ishida, Muneyuki Ishida, Bernard Metsch, and Tadayuki Teshima for valuable exchanges of ideas. This work was partly supported by the *Fundação para a Ciência e a Tecnologia* under contract numbers POCTI/-35304/FIS/2000 and CERN/P/FIS/40119/2000.

REFERENCES

1. F. Butler, H. Chen, J. Sexton, A. Vaccarino, and D. Weingarten, Nucl. Phys. **B430**, 179 (1994) [hep-lat/9405003];
   D. Weingarten, Nucl. Phys. **B215**, 1 (1983);
   H. Hamber, E. Marinari, G. Parisi, and C. Rebbi, Phys. Lett. **B108**, 314 (1982);
   H. Hamber and G. Parisi, Phys. Rev. Lett. **47**, 1792 (1981).
2. A. A. Khan *et al.* [CP-PACS Collaboration], hep-lat/0105015;
   S. Aoki *et al.* [CP-PACS Collaboration], Phys. Rev. Lett. **84**, 238 (2000) [hep-lat/9904012];
   Yoshinobu Kuramashi for the CP-PACS Collaboration, Proceedings of the Meeting of Particles and Fields of the American Physical Society (DPF99), Los Angeles, CA, 5-9 Jan. 1999 [hep-lat/9904003].
3. R. Ricken, M. Koll, D. Merten, B. C. Metsch, and H. R. Petry, Eur. Phys. J. **A9**, 221 (2000) [hep-ph/0008221].
4. G. Fogli and G. Preparata, Nuovo Cim. **A48**, 235 (1978);
   N. A. Törnqvist, Annals Phys. **123**, 1 (1979);
   Ibid., "Bags, Unitarity And Meson Spectroscopy," in *C81-03-15.12*, HU-TFT-81-20, Talk given at the 16th Rencontre de Moriond, Les Arcs, France, March 15-27, 1981.
5. E. van Beveren, C. Dullemond, and G. Rupp, Phys. Rev. **D21**, 772 (1980) (Erratum-ibid. **D22**, 787 (1980)).
6. E. van Beveren, T. A. Rijken, K. Metzger, C. Dullemond, G. Rupp, and J. E. Ribeiro, Z. Phys. **C30**, 615 (1986).
7. E. van Beveren, G. Rupp, T. A. Rijken, and C. Dullemond, Phys. Rev. **D27**, 1527 (1983);
8. C. Dullemond, G. Rupp, T.A. Rijken, and E. van Beveren, Comp. Phys. Comm. **27**, 377 (1982).
9. E. van Beveren and G. Rupp, hep-ex/0106077.
10. E. van Beveren, Z. Phys. **C21**, 291 (1984).

11. E. van Beveren and G. Rupp, Eur. Phys. J. **C10**, 469 (1999) [hep-ph/9806246].
12. E. Oset, J. A. Oller, and U. Meißner, nucl-th/0109050;
    O. Krehl, R. Rapp and J. Speth, Phys. Lett. **B390**,23 (1997) [nucl-th/9609013].
13. J. E. Augustin et al. [DM2 Collaboration], Nucl. Phys. **B320**, 1 (1989).
14. V. V. Anisovich, V. A. Nikonov, and A. V. Sarantsev, hep-ph/0108188.
15. E. van Beveren and G. Rupp, Phys. Lett. **B454**, 165 (1999) [hep-ph/9902301];
    Ibid., Eur. Phys. J. **C11**, 717 (1999) [hep-ph/9806248].
16. A. V. Anisovich and A. V. Sarantsev, Phys. Lett. **B413**, 137 (1997) [hep-ph/9705401].
17. D. E. Groom et al. [Particle Data Group Collaboration], Eur. Phys. J. **C15**, 1 (2000).
18. Carla Göbel, on behalf of the E791 Collaboration, Proceedings of Heavy Quarks at Fixed Target (HQ2K), Rio de Janeiro, October 2000, 373-384, hep-ex/0012009;
    Deirdre Black, Amir H. Fariborz, Sherif Moussa, Salah Nasri, and Joseph Schechter, Phys. Rev. **D64**, 014031 (2001) [hep-ph/0012278];
    Deirdre Black, Amir H. Fariborz, and Joseph Schechter, Proceedings of the YITP Workshop on Possible Existence of the Sigma Meson and its Implications to Hadron Physics, *Sigma Meson 2000*, Kyoto, Japan, 12-14 June 2000 [hep-ph/0008246];
    Ibid., Proceedings of the International Workshop on Hadron Physics: *Effective Theories of Low Energy QCD*, Coimbra, Portugal, 10-15 Sept. 1999, AIP Conference Proceedings **508**, 290 (2000) [hep-ph/9911387];
    Ibid., Phys. Rev. **D61**, 074001 (2000) [hep-ph/9907516];
    M.D. Scadron, Proceedings of the YITP Workshop on Possible Existence of the Sigma Meson and its Implications to Hadron Physics, *Sigma Meson 2000*, Kyoto, Japan, 12-14 June 2000 [hep-ph/0007184];
    Shin Ishida, Muneyuki Ishida, and Tomohito Maeda, Prog. Theor. Phys. **104**, 785 (2000) [hep-ph/0005190];
    L. Babukhadia, Ya. A. Berdnikov, A. N. Ivanov, and M. D. Scadron, Phys. Rev. **D62**, 037901 (2000) [hep-ph/9911284];
    V. E. Markushin and M. P. Locher, Proceedings of the Workshop on Hadron Spectroscopy (WHS 99), Rome, Italy, 8-12 March 1999, *Frascati 1999, Hadron Spectroscopy*, 229 (1999) [hep-ph/9906249];
    J. L. Lucio Martinez and Mendivil Napsuciale, Phys. Lett. **B454**, 365 (1999) [hep-ph/9903234];
    Muneyuki Ishida, Prog. Theor. Phys. **101**, 661 (1999) [hep-ph/9902260];
    Amir H. Fariborz and Joseph Schechter, Phys. Rev. **D60**, 034002 (1999) [hep-ph/9902238];
    Deirdre Black, Amir H. Fariborz, Francesco Sannino, and Joseph Schechter, Phys. Rev. **D59**, 074026 (1999) [hep-ph/9808415];
    Ibid., Phys. Rev. **D58**, 054012 (1998) [hep-ph/9804273];
    J. A. Oller and E. Oset, Phys. Rev. **D60**, 074023 (1999) [hep-ph/9809337];
    J. A. Oller, E. Oset, and J. R. Peláez, Phys. Rev. **D59**, 074001 (1999) (Erratum-ibid. **D60**, 099906 (1999)) [hep-ph/9804209];
    Ibid., Phys. Rev. Lett. **80**, 3452 (1998) [hep-ph/9803242];
    Shin Ishida, Muneyuki Ishida, Taku Ishida, Kunio Takamatsu, and Tsuneaki Tsuru, Prog. Theor. Phys. **98**, 621 (1997) [hep-ph/9705437].

M. D. Scadron, Phys. Rev. **D26**, 239 (1982).
19. Matthias Jamin, José Antonio Oller, and Antonio Pich, Nucl. Phys. **B587**, 331 (2000) [hep-ph/0006045].
20. F. De Fazio and M. R. Pennington, hep-ph/0104289;

    E. van Beveren, G. Rupp and M. D. Scadron, Phys. Lett. **B495**, 300 (2000) (Erratum-ibid. **B509**, 365 (2000)) [hep-ph/0009265];

    E. M. Aitala et al. [E791 Collaboration], Phys. Rev. Lett. **86**, 765 (2001) [hep-ex/0007027].
21. F. Kleefeld, E. van Beveren, G. Rupp, and M. D. Scadron, hep-ph/0109158.
22. T. Kunihiro, Prog. Theor. Phys. Suppl. **120**, 75 (1995) [arXiv:hep-ph/9502305].
23. V. V. Anisovich, V. A. Nikonov, and A. V. Sarantsev, hep-ph/0102338;

    R. Barate et al. [ALEPH Collaboration], Phys. Lett. **B472**, 189 (2000) [hep-ex/9911022];

    D. Li, H. Yu, and Q. Shen, Mod. Phys. Lett. **A15**, 1781 (2000);

    Ibid., Eur. Phys. J. **C19**, 529 (2001) [hep-ph/0011129];

    Y. S. Surovtsev, D. Krupa, and M. Nagy, Acta Phys. Polon. **B31**, 2697 (2000) [hep-ph/0009039];

    F. E. Close and A. Kirk, Phys. Lett. **B483**, 345 (2000) [hep-ph/0004241];

    D. Barberis et al. [WA102 Collaboration], Phys. Lett. **B474**, 423 (2000) [hep-ex/0001017];

    T. Teshima, I. Kitamura and N. Morisita, Nuovo Cim. **A103**, 175 (1990);

    T. Teshima, I. Kitamura and N. Morisita, hep-ph/0105107.
24. A. V. Anisovich et al., Phys. Lett. **B449**, 154 (1999).

NONPERTURBATIVE QCD

IHEP, Protvino

Light hadron spectroscopy from lattice QCD

Takashi Kaneko

High Energy Accelerator Research Organization(KEK), Tsukuba, Ibaraki 305-0801, Japan

Abstract. We review recent developments in lattice QCD calculations of the light hadron spectrum and quark masses.

INTRODUCTION

Lattice QCD is a regularization which allows us to study non-perturbative aspects of the strong interaction from the first principle. One of the main goals of the lattice calculations is to confirm the validity of QCD in the low energy region by comparing its prediction for the hadron spectrum to experiment. This would also give us confidence in our ability to calculate other quantities, such as the hadronic matrix elements, non-perturbatively.

The quenched approximation, which neglects dynamical creation and annihilation of quarks, significantly reduces the amount of CPU time required for numerical simulations and hence it has been widely used in previous calculations. Although the quenched QCD reproduces the experimental spectrum remarkably well, a recent precise calculation[1] revealed a systematic deviation due to this approximation.

In order to see if QCD is correct theory of the strong interaction, it is essential to perform calculations with dynamical quarks (full QCD calculations). As a first step in this direction, a number of large-scale simulations have been executed in QCD with two flavors of dynamical quarks, which are identified with the up and down quarks[2, 3, 4]. Several attempts toward realistic calculations in $N_f = 3$ full QCD were also performed very recently.

Spectrum calculations in lattice QCD allow a simultaneous determination of the quark masses, which are important for phenomenology. Previous works[5] demonstrated that the lattice QCD is the best candidate to provide precise determinations of these quantities. Recent large-scale simulations both in quenched and $N_f = 2$ full QCD led to significant progresses in this subject.

In this review, we summarize the present status of lattice QCD calculations of the light hadron spectrum and quark masses. Other important subjects, such as the glueball spectrum and heavy quark physics, are covered in separate talks[6, 7]. The outline is as follows. In the second (next) section, we discuss recent results of the light hadron spectrum in quenched QCD and examine the validity of the quenched approximation. The third section is devoted to quenched calculations of quark masses. In the fourth section, we discuss search for sea quark effects in the hadron spectrum in full QCD. Recent determinations of quark masses in full QCD are reviewed in the fifth section. A

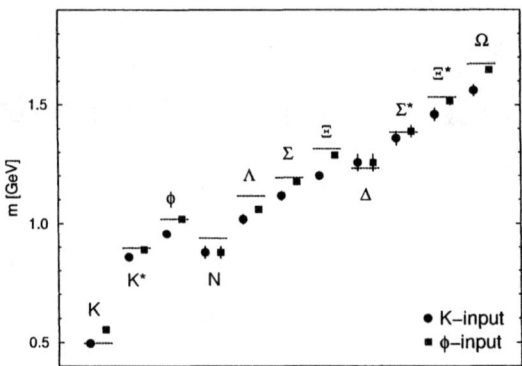

FIGURE 1. Light hadron spectrum in quenched QCD. Circles and squares are results obtained with K- and ϕ-input, respectively. Horizontal dotted lines represent experimental spectrum.

brief conclusion is given in the last section.

LIGHT HADRON SPECTRUM IN QUENCHED QCD

Since the first pioneering studies[8, 9], calculations of the light hadron spectrum have been pursued for a long time in lattice QCD. In the quenched approximation, the large effort has culminated in the large-scale simulations by the CP-PACS collaboration[1]. They use the standard plaquette gauge and the Wilson quark action and simulate four values of lattice spacing in the range of $a \simeq 0.05$–0.10 fm for extrapolation to the limit of zero lattice spacing. Their spatial lattice size is fixed at about 3.0 fm, where finite size effects are expected to be small. The up and down quarks are assumed to be degenerated ($m_{ud} \equiv m_u = m_d$) in their calculation, and π and ρ meson masses are used to fix the light quark mass m_{ud} and the lattice spacing a. They compare two choices of the input meson mass to fix the strange quark mass: one employing the kaon mass (K-input) and other with the ϕ meson mass (ϕ-input).

Their result of the hadron spectrum in the continuum limit is plotted in Fig. 1. While their results show an agreement with experiment with accuracy of 10% level, it is also clear that there exists systematic deviation between the quenched spectrum and experiment. Both of K- and ϕ-inputs give smaller values for meson mass splittings by at most 6 standard deviations. Decuplet baryon splittings are also small. Because all other systematic errors, such as finite size effects and discretization error, are well controlled in their data, this discrepancy can be considered an artifact due to the quenched approximation.

QUARK MASSES IN QUENCHED QCD

Precise determination of quark masses is one of the most important issues in lattice QCD. Several improved techniques have been developed to reduce systematic uncer-

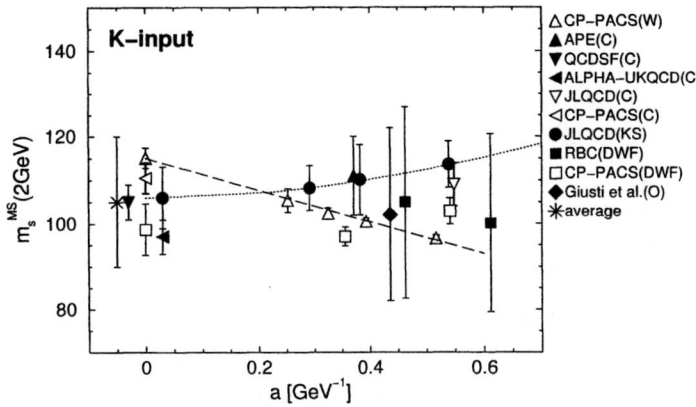

FIGURE 2. Continuum extrapolation of strange quark mass with K-input in $\overline{MS}$ scheme at scale $\mu = 2$ GeV. Filled symbols represent results obtained with non-perturbatively determined matching factors to the $\overline{MS}$ scheme, while the one-loop perturbative matching is used for open symbols. The choice of the quark action is written in brackets, where "W"=Wilson, "C"=clover, [11], KS=Kogut-Susskind, "DWF"=domain-wall,[12, 13, 14] and "O"=overlap fermions[15].

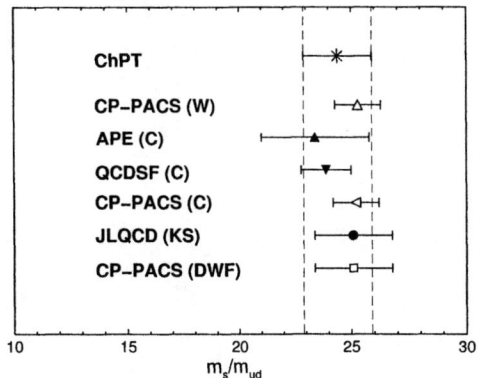

FIGURE 3. Recent results of quark mass ratio m_s/m_{ud} in quenched QCD. The prediction from one-loop ChPT is also plotted.

tainties in the quark mass calculations. An important example is improvement of lattice actions[16], which reduces discretization error of lattice data remarkably. Non-perturbative determination of renormalization constants[10] is also pursued recently in order to remove the large uncertainty associated with the conventional perturbative matching.

Recent results of the strange quark mass m_s with K-input are plotted in Fig. 2[1, 17, 18, 19, 20, 21, 22, 23, 24]. We observe a reasonable agreement among all data within

TABLE 1. Recent simulations in $N_f=2$ full QCD. Values of improvement coefficient c_{SW} for the clover quark action are shown in brackets, where NP and TP stand for non-perturbative and tadpole-improved values, respectively.

group	gauge	quark	a[fm]	L[fm]	ref.
SESAM	plaquette	Wilson	0.08	1.3	[26]
TχL	plaquette	Wilson	0.08	1.9	[27]
SESAM-TχL	plaquette	Wilson	0.09	1.5	[28]
UKQCD	plaquette	clover[11](1.76)	0.12	1.0–1.9	[29]
UKQCD	plaquette	clover (NP)	0.10	1.7	[30]
UKQCD	plaquette	clover (NP)	0.10	1.6	[30]
QCDSF	plaquette	clover (NP)	0.09	2.1,1.5	[31, 32]
JLQCD	plaquette	clover (NP)	0.09	1.1–1.8	[33, 34]
CP-PACS	RG-improved[35]	clover (TP)	0.11–0.22	2.5–2.6	[36]
Columbia	plaquette	KS	0.09	1.5	[37, 38]
MILC	plaquette	KS	0.10–0.32	2.4–3.8	[39, 40]
MILC	Symanzik[41]	improved KS[42, 43]	0.13	2.6	[44]

an accuracy of about 15%. The quenching artifact in the meson spectrum discussed in the previous section, however, indicates that there is an additional uncertainty due to the choice of the input to fix m_s. The CP-PACS collaboration, indeed, observed that there is about 20% discrepancy between $m_s^{\overline{MS}}(2\,\text{GeV})$ with K-input (116(3) MeV) and φ-input (144(6) MeV)[1, 17]. Taking this uncertainty into account, we obtain the average

$$m_s^{\overline{MS}}(2\,\text{GeV}) = 105(26)\,\text{MeV}. \tag{1}$$

For the light quark mass m_{ud}, we consider the ratio m_s/m_{ud}, rather than m_{ud} itself, because systematic uncertainties partially cancel in this ratio. Recent results of m_s/m_{ud} are plotted in Fig. 3. All data are in good agreement with the prediction of chiral perturbation theory (ChPT) at one-loop level: 24.4(1.5)[25]. We therefore apply the ratio from ChPT to Eq. 1 to obtain

$$m_{ud}^{\overline{MS}}(2\,\text{GeV}) = 4.3(1.1)\,\text{MeV}. \tag{2}$$

SEA QUARK EFFECTS IN LIGHT HADRON SPECTRUM

Recent simulations

The observation of the quenching artifact in the hadron spectrum raises a natural question if the difference can be accounted for by the inclusion of dynamical quarks. A number of large-scale simulations, which are listed in Table 1, have been performed recently for the search for the sea quark effect. Most of these simulations were made at a single lattice spacing. An exception is the CP-PACS simulation[36]: they explored the range $a \sim 0.1$–0.2 fm and performed the continuum extrapolation of their data.

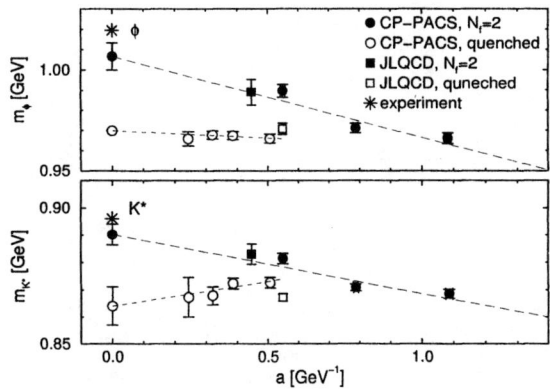

FIGURE 4. Strange vector meson masses with K-input in quenched (open symbols) and $N_f = 2$ full QCD (filled symbols).

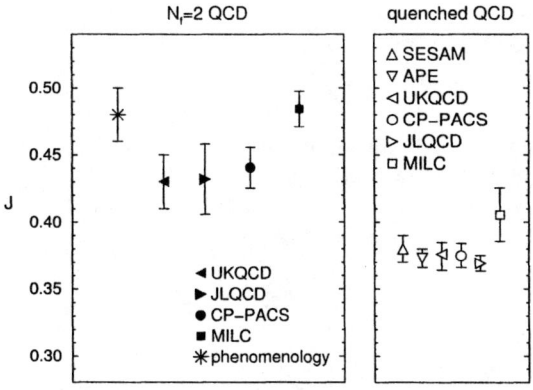

FIGURE 5. Recent results of J parameters in $N_f = 2$ full QCD(left figure) and quenched QCD(right figure). The phenomenological value 0.48(2) is plotted.

Meson spectrum

In Fig.4, we compare strange vector meson masses in quenched and $N_f = 2$ full QCD. While full and quenched results from the CP-PACS simulations take similar values at finite lattice spacing, they appear to have a different slope. The full QCD results extrapolated to a value in the continuum limit which is much closer to the experimental value.

The JLQCD collaboration also observed similar sea quark effects[34]. While continuum extrapolation has not been performed in their data, scaling violation is expected to be small since they use the non-perturbatively $O(a)$ improved lattice action[45, 46].

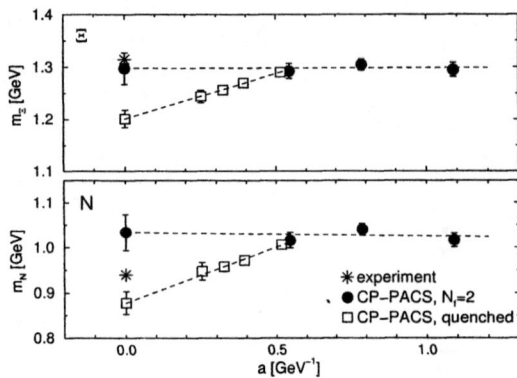

FIGURE 6. Continuum extrapolation of Ξ baryon mass(top figure) and nucleon mass(bottom figure).

The J parameter[47]

$$J \equiv m_V \frac{dm_V}{dm_{PS}^2}\bigg|_{m_{PS}=m_K} \quad (3)$$

can be used as a measure of the sea quark effect in K^*-K hyperfine splitting. Figure 5 shows recent results of J in quenched[26, 30, 34, 36, 44, 48] and $N_f=2$ full QCD[30, 34, 36, 44]. The results in quenched QCD are consistent with each other and significantly smaller than the phenomenological value of J[47]

$$J \equiv m_{K^*} \frac{m_K^* - m_\rho}{m_K^2 - m_\pi^2}. \quad (4)$$

This reflects the small K^*–K hyperfine splitting found in quenched QCD. However, J in full QCD is systematically larger than the quenched results and hence closer to the phenomenological value. This is consistent with the above observation of the sea quark effect that the quenching error in the meson spectrum is remarkably reduced by including effects of dynamical u and d quarks.

Eventually realistic calculations of the light hadron spectrum have to be done in $N_f = 3$ full QCD. The MILC collaboration performed the first study in this direction very recently[44]. Their simulations were performed at a single lattice spacing $a = 0.13$ fm with the spatial lattice size of 2.6 fm using an improved KS quark action. While systematic uncertainties in their data, such as scaling violation, should be studied carefully in future, their result of $J \sim 0.5$ is consistent with the phenomenological value.

Baryon spectrum

The sea quark effect is also expected in the baryon spectrum, as a significant disagreement with experiment was observed in the quenched approximation[1]. In Fig. 6, the

TABLE 2. Recent calculations of quark masses in $N_f = 2$ full QCD. The APE collaboration performed the non-perturbative matching to the $\overline{MS}$ scheme, while the one-loop perturbative matching is used in other studies.

group	gauge	quark	a[fm]	matching to $\overline{MS}$	ref.
SESAM-TχL	plaquette	Wilson	$a \to 0$	perturbative	[28]
APE	plaquette	Wilson	0.08	non-perturbative	[49]
QCDSF	plaquette	clover (NP)	0.09	perturbative	[32]
JLQCD	plaquette	clover (NP)	0.09	perturbative	[33]
CP-PACS	RG-improved	clover (TP)	$a \to 0$	perturbative	[17, 36]

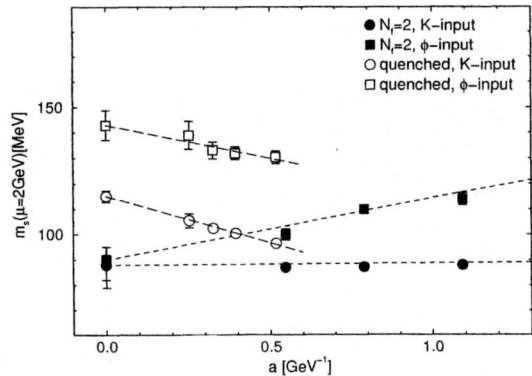

FIGURE 7. CP-PACS result of strange quark mass in $\overline{MS}$ scheme at scale $\mu = 2$ GeV in $N_f = 2$ full QCD (filled symbols) and quenched QCD (open symbols). Circles and squares represent m_s with K- and ϕ-input, respectively.

CP-PACS results for the baryon masses are plotted as a function of the lattice spacing. While the strange baryon masses in full QCD exhibit better agreement with experiment, they found masses of light baryons, nucleon and Δ, to be higher than experiment. It is possible that their spatial lattice size of 2.5 fm is not sufficiently large to avoid finite size effects for the light baryons. Further studies on larger lattice sizes will be necessary to identify sea quark effects on the baryon spectrum.

FULL QCD CALCULATIONS OF QUARK MASSES

Recent calculations of quark masses in $N_f = 2$ full QCD are listed in Table 2. In Fig. 7, we plot the CP-PACS results of m_s in full and quenched QCD as a function of the lattice spacing[17]. They obtained $m_s^{\overline{MS}}(2 \text{ GeV}) = 88^{+4}_{-6}$ MeV or 90^{+5}_{-11} MeV with K- or ϕ-input in the continuum limit. Their important finding is that, in contrast to the quenched results, m_s with K- and ϕ-inputs are consistent with each other in full QCD. This reflects the closer agreement of the strange vector meson masses (m_K and m_ϕ) in full QCD to experiment. Another important results is that the inclusion of dynamical effects of u and

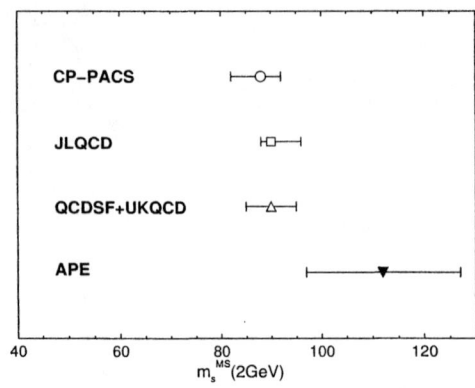

FIGURE 8. Recent result of strange quark mass with K-input in $N_f = 2$ full QCD.

d quarks reduces m_s by about 20%.

We note that the SESAM collaboration also attempted the continuum extrapolation of the quark masses[28]. Their range of the lattice spacing is, however, narrow and far form the continuum limit. This leads to a very large uncertainty in their result $m_s^{\overline{MS}}(2\text{ GeV}) = 92(83)$ MeV.

Other recent calculations of m_s were performed at a single value of lattice spacing[32, 33, 49]. Their results are in reasonable agreement with the CP-PACS result, as shown in Fig. 8. We observe a reasonable agreement among all results. Since only the CP-PACS took the continuum limit, we consider the CP-PACS result the current best estimate:

$$m_s^{\overline{MS}}(2\text{ GeV}) = 88 \left(^{+4}_{-6}\right)(5) \text{ MeV}. \tag{5}$$

The systematic error written in the second bracket includes the difference between results with K- and ϕ-inputs and higher order corrections in the perturbative matching estimated from a naive order counting.

In Fig. 9, we plot recent results of the quark mass ratio m_s/m_{ud} in $N_f = 2$ full QCD. The full QCD results as well as quenched ones are consistent with the prediction from ChPT. By using the ratio from ChPT, we obtain the average

$$m_{ud}^{\overline{MS}}(2\text{ GeV}) = 3.6(0.2)(0.3) \text{ MeV}. \tag{6}$$

CONCLUSION

Recent precise calculations observed that the quenched spectrum clearly and systematically deviates from experiment. This quenching artifact leads to a large uncertainty in the strange quark mass due to the choice of the input meson mass. These findings revealed the limitation of the quenched approximation on a quantitative basis. Therefore,

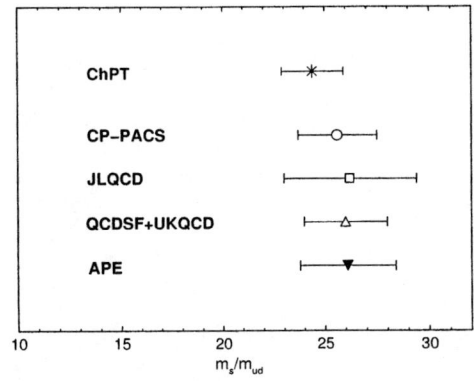

FIGURE 9. Quark mass ratio m_s/m_{ud} in $N_f=2$ full QCD.

high precision calculations with errors down to a few % level have to be done by full QCD simulations.

In $N_f = 2$ full QCD, recent large-scale simulations confirmed that sea quark effects reduce the quenching artifacts on the meson spectrum and the strange quark mass remarkably. It is also found that sea quark effects have significant impact on the quark masses, reducing both of the strange and light quark masses by about 20%.

The direction that followed naturally from these observations has been a pursuit of realistic calculations in $N_f = 3$ full QCD. The first calculation was already performed and further systematic studies will be expected in near future.

ACKNOWLEDGMENTS

I thank R. Burkhalter, S. Hashimoto, and A. Ukawa for useful discussions. This work is supported by the JSPS Research Fellowship.

REFERENCES

1. Aoki, S., et al. (CP-PACS Collaboration), *Phys. Rev. Lett.* **84**, 238-241 (2000).
2. For a recent review, see Aoki, S., *Nucl. Phys. (Proc.Suppl)* **B94**, 3-18 (2001).
3. Kaneko, T., *review in LATTICE 2001*, hep-lat/0111005.
4. Toussaint, D., *review in LATTICE 2001*, hep-lat/0110010.
5. For a recent review, see Lubicz, V., *Nucl. Phys. B (Proc. Suppl.)* **94**, 116-129 (2001).
6. Morningstar, C., *in these proceedings*.
7. Mackenzie, P.B., *in these proceedings*.
8. Hamber, H., and Parisi, G., *Phys. Rev. Lett.* **47**, 1792-1795 (1981).
9. Weingarten, D., *Phys. Lett.* **B109**, 57-62 (1982).
10. For a recent review, see Sint, S., *Nucl.Phys. B (Proc. Suppl.)* **94**, 79-94 (2001).
11. Sheikholeslami, B., and Wohlert, R., *Nucl. Phys.* **B259**, 572-596 (1985).
12. Kaplan, D.B., *Phys. Lett.* **B288**, 342-347 (1992).

13. Shamir, Y., *Nucl. Phys.* **B406**, 90-106 (1993).
14. Furman, V., and Shamir, Y., *Nucl. Phys.* **B439**, 54-78 (1995).
15. Narayanan, R., and Neuberger, H., *Phys. Lett.* **B302**, 62-69 (1993).
16. For reviews, see Wittig, H., *Nucl. Phys. (Proc.Suppl.)* **B63**, 47-52 (1998); Hasenfratz, P., *ibid.* **B63**, 53-58 (1998).
17. Ali Khan, A., *et al.* (CP-PACS Collaboration), *Phys. Rev. Lett.* **85**, 4674-4677 (2000).
18. Becirevic, D., *et al.* (APE Collaboration), *Phys. Rev.* **D61**, 114507(1)-114507(10) (2000).
19. Göckeler, M., *et al.* (QCDSF Collaboration), *Phys. Rev.* **D62**, 054504(1)-054504(11) (2000).
20. Garden, J., *et al.* (ALPHA-UKQCD Collaboration), *Nucl. Phys.* **B571**, 237-256 (2000).
21. Aoki, S., *et al.* (JLQCD Collaboration), *Phys. Rev. Lett.* **82**, 4392-4395 (1999).
22. Wingate, M., *et al.* (RBC Collaboration), *hep-lat/0009022*.
23. Ali Khan, A., *et al.* (CP-PACS Collaboration), *hep-lat/0105020*.
24. Giusti, L., Hoelbling, C., and Rebbi, C., *hep-lat/0108007*.
25. Leutwyler, H., *Phys. Lett.* **B378**, 313-318 (1996).
26. Eicker, N., *et al.* (SESAM Collaboration), *Phys. Rev.* **D59**, 014509(1)-014509(15) (1999).
27. Lippert, T., *et al.* (SESAM-TχL Collaboration), *Nucl. Phys. (Proc.Suppl.)* **B60A**, 311-334 (1998).
28. Eicker, N., *et al.* (SESAM-TχL Collaboration), *proceedings in LATTICE 2001, hep-lat/0110134*.
29. Allton, C.R., *et al.* (UKQCD Collaboration), *Phys. Rev.* **D60**, 034507(1)-034507(15) (1999).
30. Allton, C.R., *et al.* (UKQCD Collaboration), *hep-lat/0107021*.
31. Stüben, H., (QCDSF-UKQCD Collaboration), *Nucl. Phys. (Proc.Suppl.)* **B94**, 273-276 (2001).
32. Pleiter, D., (QCDSF-UKQCD Collaboration), *Nucl. Phys. B (Proc. Suppl.)* **94**, 265-268 (2001).
33. Aoki, S., *et al.* (JLQCD Collaboration), *Nucl. Phys. B (Proc. Suppl.)* **94**, 233-23, (2001).
34. Aoki, S., *et al.* (JLQCD Collaboration), *proceedings in LATTICE 2001, hep-lat/0110179*.
35. Iwasaki, Y., *Univ. of Tsukuba report UTEP-118, unpublished*.
36. Ali Khan, A., *et al.* (CP-PACS Collaboration), *hep-lat/0105015, to be published in Phys.Rev.D*.
37. Sui, C., *et al.*, *Nucl. Phys. (Proc.Suppl.)* **B73**, 228-230 (1999).
38. Mawhinney, R.D., *Nucl. Phys. (Proc.Suppl.)* **B83**, 57-66 (2000).
39. Bernard, C., *et al.* (MILC Collaboration), *Nucl. Phys. (Proc.Suppl.)* **B60A**, 297-305 (1998).
40. Bernard, C., *et al.* (MILC Collaboration), *Nucl. Phys. (Proc.Suppl.)* **B73**, 198-200 (1999).
41. Lüscher, M., and Weisz, P., *Comm. Math. Phys.* **97**, 59-77 (1985).
42. Naik, S., *Nucl. Phys.* **B316**, 238-268 (1989).
43. Orginos, K., Toussaint, D., and Sugar, R.L., *Phys. Rev.* **D60**, 054503(1)-054503(8) (1999).
44. Bernard, C., *et al.* (MILC Collaboration), *Phys. Rev.* **D64**, 054506(1)-054506(15) (2001).
45. Lüescher, M., *et al.*, *Nucl. Phys.* **B478**, 365-400 (1996).
46. Jansen, K., and Sommer, R., (ALPHA Collaboration), *Nucl. Phys.* **B530**, 185-203 (1998).
47. Lacock, P., and Michael, C., *Phys. Rev.* **D52**, 5213-5219 (1995).
48. Becirevic, D., *et al.* (APE Collaboration), *hep-lat/9809129*.
49. Becirevic, D., *et al.* (APE Collaboration), *unpublished*.

Gluonic excitations in lattice QCD: A brief survey

Colin Morningstar

Department of Physics, Carnegie Mellon University, Pittsburgh, PA, USA 15213-3890

Abstract. Our current knowledge about glueballs and hybrid mesons from lattice QCD simulations is briefly reviewed.

INTRODUCTION

Hadronic states bound together by an *excited* gluon field, such as glueballs, hybrid mesons, and hybrid baryons, are a potentially rich source of information concerning the confining properties of QCD. Interest in such states has been recently sparked by observations of resonances with exotic 1^{-+} quantum numbers[1] at Brookhaven. In fact, the proposed Hall D at Jefferson Lab will be dedicated to the search for hybrid mesons and one of the goals of CLEO-c will be to identify glueballs and exotics. Although our understanding of these states remains deplorable, recent lattice simulations have shed some light on their nature. In this talk, I summarize our current knowledge about glueballs and heavy- and light-quark hybrid mesons from lattice QCD simulations.

GLUEBALLS

The glueball spectrum in the absence of virtual quark-antiquark pairs is now well known[2] and is shown in Fig. 1. The use of spatially coarse, temporally fine lattices with improved actions and the application of variational techniques to moderately-large matrices of correlations functions were crucial in obtaining this spectrum. Continuum limit results for the lowest-lying scalar and tensor states are in good agreement with recent previous calculations[3, 4] when expressed in terms of the hadronic scale r_0, the square root of the string tension $\sqrt{\sigma}$, or in terms of each other as a ratio of masses. Disagreements in specifying the masses of these states in GeV arise solely from ambiguities in setting the value of r_0 within the quenched approximation. The glueball spectrum can be qualitatively understood in terms of the interpolating operators of minimal dimension which can create glueball states[5] and can be reasonably well explained[6] in terms of a simple constituent gluon (bag) model which approximates the gluon field using spherical cavity Hartree modes with residual perturbative interactions[7, 8]. The challenge now is to deduce precisely what the spectrum in Fig. 1 is telling us about the long-wavelength properties of QCD. This spectrum provides an important testing ground for

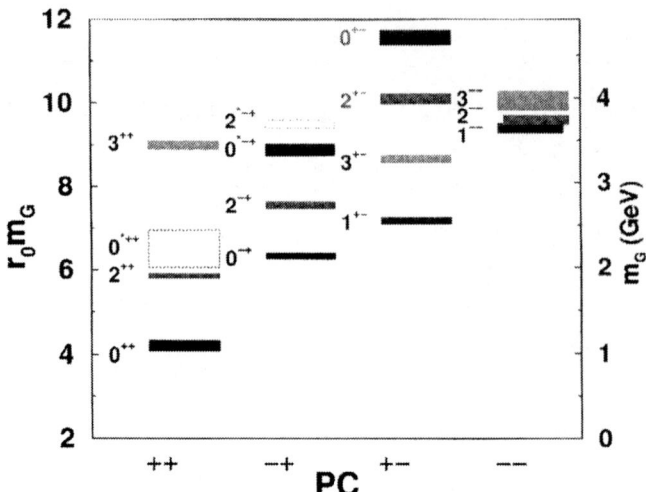

FIGURE 1. The mass spectrum of glueballs in the pure SU(3) gauge theory from Ref. [2]. The masses are given in terms of the hadronic scale r_0 along the left vertical axis and in terms of GeV along the right vertical axis (assuming $r_0^{-1} = 410(20)$ MeV). The mass uncertainties indicated by the vertical extents of the boxes do *not* include the uncertainty in setting r_0. The locations of states whose interpretation requires further study are indicated by the dashed hollow boxes.

models of confined gluons, such as center and Abelian dominance, instantons, soliton knots, instantaneous Coulomb-gauge mechanisms, and so on.

Recently, the two lowest-lying scalar glueballs and the tensor glueball have been studied[9] in $SU(N)$ for $N = 2, 3, 4$, and 5. Their masses have been shown to depend linearly on $1/N^2$, and these masses in the limit $N \to \infty$ do not differ substantially from

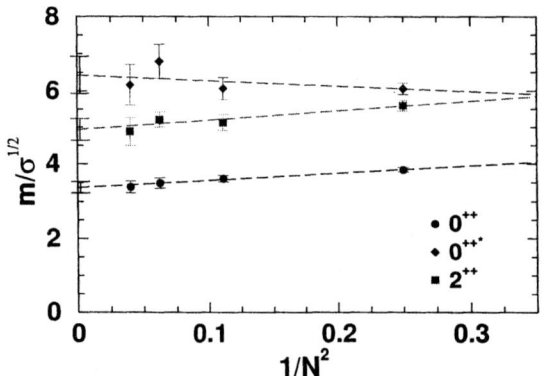

FIGURE 2. Continuum scalar, tensor, and excited scalar $SU(N)$ glueball masses expressed in units of the string tension σ and plotted against $1/N^2$. Linear extrapolations to $N = \infty$ are shown in each case (see Ref. [9]).

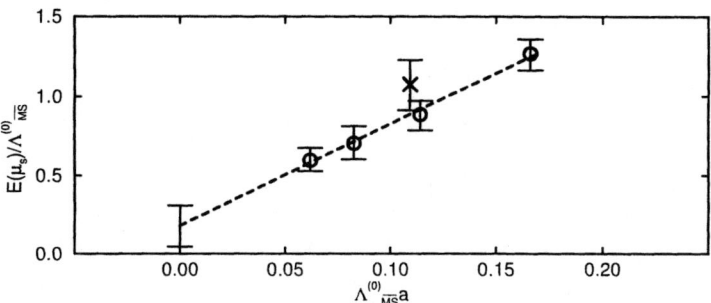

FIGURE 3. Lattice spacing dependence and continuum limit of the glueball-quarkonium mixing energy $E(\mu_s)$ at the strange quark mass in the quenched approximation in terms of $\Lambda_{\overline{MS}}^{(0)} = 234.9(6.2)$ MeV set using the (quenched) mass of the ρ meson. The continuum limit value of $E(\mu_s)$ is 43(31) MeV. The circles indicate results using lattices with spatial extent approximately 1.6 fm; the × indicates a result in a larger volume with spatial extent near 2.3 fm. (see Ref. [13]).

their values for small N (see Fig. 2).

Glueball wavefunctions and sizes have been studied in the past, but much of the early work contains uncontrolled systematic errors, most notably from discretization effects. The scalar glueball is particularly susceptible to such errors for the Wilson gauge action due to the presence of a critical end point of a line of phase transitions (not corresponding to any physical transition found in QCD) in the fundamental-adjoint coupling plane. As this critical point (which defines the continuum limit of a ϕ^4 scalar field theory) is neared, the coherence length in the scalar channel becomes large, which means that the mass gap in this channel becomes small; glueballs in other channels seem to be affected very little. Results in which the scalar glueball was found to be significantly smaller than the tensor were most likely due to contamination of the scalar glueball from the non-QCD critical point. In Ref. [10], an improved gauge action designed to avoid the spurious critical point found that the scalar and tensor glueballs were comparable in size and of typical hadronic dimensions. Operator overlaps obtained from the variational optimizations carried out in Ref. [2], which also used an improved gauge action, concur with such a conclusion.

Vacuum-to-glueball transition amplitudes for operators such as $\langle G|\mathrm{Tr}E^2|0\rangle$ and $\langle G|\mathrm{Tr}B^2|0\rangle$ in the 0^{++} sector, $\langle G|\mathrm{Tr}E\cdot B|0\rangle$ in the 0^{-+} sector, and $\langle G|\mathrm{Tr}E_iE_j|0\rangle$ in the 2^{++} sector, have also been studied in the past[11]. Efforts to revisit these calculations using improved actions and operators with reduced discretization errors is ongoing.

A valiant effort to study the decay of the scalar glueball into two pseudoscalar mesons was made in Ref. [12]. A slight mass dependence of the coupling was found, and a total width of 108(29) MeV for decays to two pseudoscalars was obtained in the quenched approximation. However, these results were obtained with the Wilson gauge action at $\beta = 5.7$, an unfortunate choice since pollution of the scalar glueball from the non-QCD critical point is significant (for example, the mass differs from its continuum limit value by at least 20% and the mixing energy of the glueball with quarkonium differs dramatically from its continuum limit value as discussed below). These large systematic

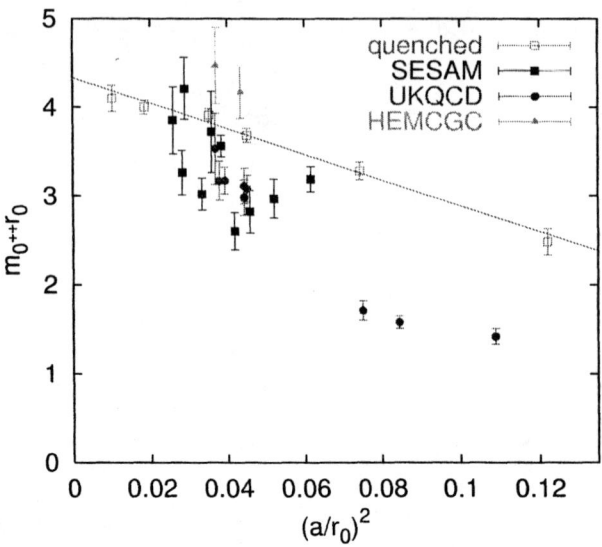

FIGURE 4. The scalar glueball mass with two flavors of virtual quark-antiquark pairs included against the lattice spacing a in terms of the hadronic scale r_0, courtesy of Ref. [18]. Results are shown for three different discretizations of the Dirac action: staggered (HEMCGC, from Ref. [14]), Wilson (SESAM from Ref. [15]), and clover (UKQCD from Refs. [16, 17]). The quarks are all heavier than a third of the strange quark mass. Quenched results are shown for comparison.

uncertainties and the use of the quenched approximation should be kept in mind when considering the value of the width given above.

Mixing of the scalar glueball with quarkonium has been studied in the quenched approximation in Ref. [13]. Several lattice spacings were used to facilitate control of discretization errors. Good control of systematic errors was demonstrated, although extrapolations in the quark mass to μ_n (the u and d quark mass scale) might benefit from the use of chiral perturbation theory. The results for the mixing energy are shown in Fig. 3. Note that the continuum limit is essentially *consistent with zero mixing*. Again, remember that the inclusion of quark loops is incomplete when assessing this conclusion. The large variation of this mixing energy with the lattice spacing is most likely due to the use of the simple Wilson gauge action. The rightmost point corresponds to $\beta = 5.7$ which illustrates the very large lattice artifacts at this spacing. This mixing calculation could benefit enormously from the use of an improved action and anisotropic lattices.

Incorporation of virtual quark-antiquark pairs in calculating the glueball/quarkonium spectrum is a daunting task. The fermion determinant must be included in the Monte Carlo updating, dramatically increasing the computational costs. Mixings with two meson states require all-to-all propagators, further adding to the cost and the stochastic uncertainties. Instabilities of the higher lying states must be properly taken into account (such as with finite volume techniques). Extrapolations to realistically light quark masses must be done carefully, taking decay thresholds into account and possibly requiring sim-

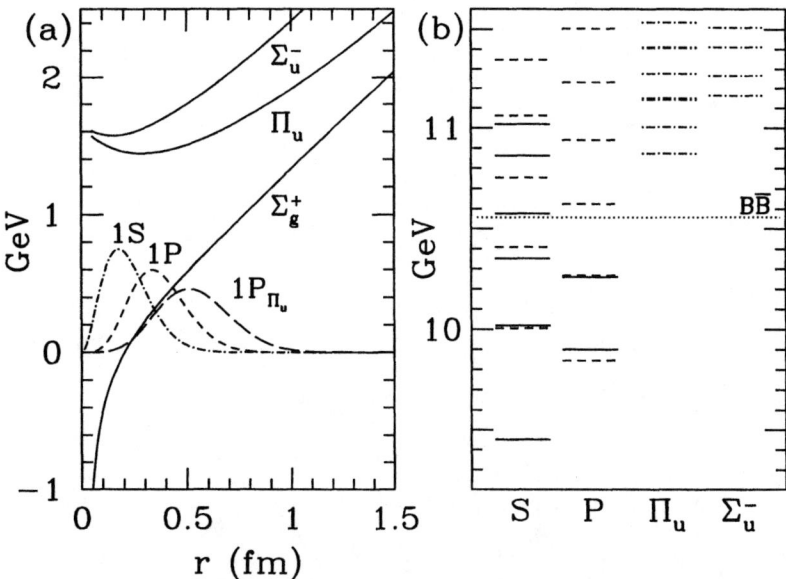

FIGURE 5. (a) Static potentials and radial probability densities against quark-antiquark separation r for $r_0^{-1} = 450$ MeV. (b) Spin-averaged $b\bar{b}$ spectrum in the LBO approximation (light quarks neglected). Solid lines indicate experimental measurements. Short dashed lines indicate the S and P state masses obtained using the Σ_g^+ potential with $M_b = 4.58$ GeV. Dashed-dotted lines indicate the hybrid quarkonium states obtained from the Π_u ($L = 1, 2, 3$) and Σ_u^- ($L = 0, 1, 2$) potentials. These results are from Ref. [22].

ulations to be carried out at quark masses lighter than currently feasible. Nevertheless, a few groups[14, 15, 16, 17] have begun glueball/meson simulations with two flavors of sea quarks, and the results for the scalar glueball mass are shown in Fig. 4. The status of such calculations has recently been reviewed in Ref. [18]. The quarks are still heavier than $m_s/3$, where m_s is the mass of the strange quark. The glueball mass tends to decrease as the light quark mass is reduced, but it increases as the lattice spacing is reduced, making it completely unclear what the end result will be in the continuum limit for realistically light quark masses.

HEAVY-QUARK HYBRID MESONS

One expects that a heavy-quark meson can be treated similar to a diatomic molecule: the slow valence heavy quarks correspond to the nuclei and the fast gluon and light sea quark fields correspond to the electrons[19]. First, the quark Q and antiquark $\overline{Q}$ are treated as static color sources and the energy levels of the fast degrees of freedom are determined as a function of the $Q\overline{Q}$ separation r, each such energy level defining an adiabatic surface or potential. The motion of the slow heavy quarks is then described in the leading Born-Oppenheimer (LBO) approximation by the Schrödinger equation using each of these potentials. Conventional quarkonia are based on the lowest-lying potential;

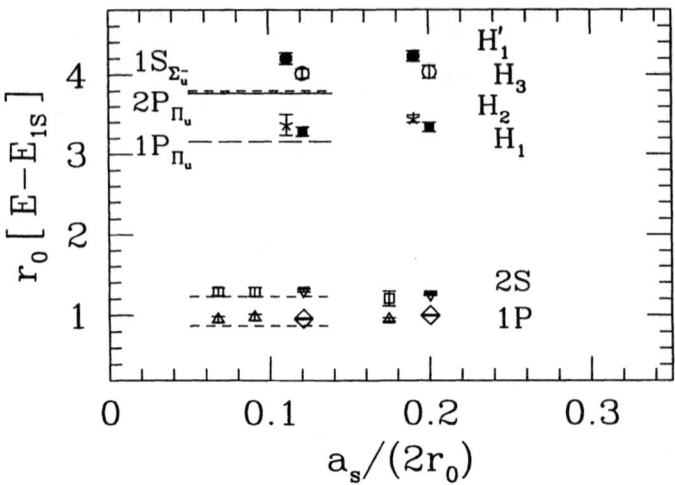

FIGURE 6. Simulation results from Ref. [22] for the heavy quarkonium level splittings (in terms of r_0 and with respect to the 1S state) against the lattice spacing a_s. Results from Ref. [25] using an NRQCD action with higher-order corrections are shown as open boxes and △. The horizontal lines show the LBO predictions. Agreement of these splittings within 10% validates the Born-Oppenheimer approximation.

hybrid quarkonium states emerge from the excited potentials.

The spectrum of the fast gluon field in the presence of a static quark-antiquark pair has been determined in lattice studies[20, 21]. The three lowest-lying levels are shown in Fig. 5. Due to computational limitations, sea quark effects have been neglected in these calculations; their expected impact on the hybrid meson spectrum will be discussed below. The levels in Fig. 5 are labeled by the magnitude Λ of the projection of the total angular momentum $\mathbf{J}_g$ of the gluon field onto the molecular axis, and by $\eta = \pm 1$, the symmetry under the charge conjugation combined with spatial inversion about the midpoint between the Q and $\overline{Q}$. States with $\Lambda = 0, 1, 2, \ldots$ are denoted by $\Sigma, \Pi, \Delta, \ldots$, respectively. States which are even (odd) under the above-mentioned CP operation are denoted by the subscripts g (u). An additional $\pm$ superscript for the Σ states refers to even or odd symmetry under a reflection in a plane containing the molecular axis. The potentials are calculated in terms of the hadronic scale parameter r_0; in Fig. 5, $r_0^{-1} = 450$ MeV has been assumed.

The LBO spectrum[22] of conventional $\bar{b}b$ and hybrid $\bar{b}gb$ states are shown in Fig. 5. Below the $\overline{B}B$ threshold, the LBO results are in very good agreement with the spin-averaged experimental measurements of bottomonium states. Above the threshold, agreement with experiment is lost, suggesting significant corrections either from mixing and other higher-order effects or (more likely) from light sea quark effects. Note from the radial probability densities shown in Fig. 5 that the size of the hybrid state is large in comparison with the conventional 1S and 1P states.

The validity of such a simple physical picture relies on the smallness of higher-order spin, relativistic, and retardation effects and mixings between states based on different adiabatic surfaces. The importance of retardation and leading-order mixings between states based on different adiabatic potentials can be tested by comparing the LBO level splittings with those determined from meson simulations using a leading-order non-relativistic (NRQCD) heavy-quark action. Such a test was carried out in Ref. [22]. The NRQCD action included only a covariant temporal derivative and the leading kinetic energy operator (with two other operators to remove lattice spacing errors). The only difference between the leading Born-Oppenheimer Hamiltonian and the lowest-order NRQCD Hamiltonian was the $\boldsymbol{p}\cdot\boldsymbol{A}$ coupling between the quark color charge in motion and the gluon field. The level splittings (in terms of r_0 and with respect to the $1S$ state) of the conventional $2S$ and $1P$ states and four hybrid states were compared (see Fig. 6) and found to agree within 10%, strongly supporting the validity of the leading Born-Oppenheimer picture.

The question of whether or not quark spin interactions spoil the validity of the Born-Oppenheimer picture for heavy-quark hybrids has been addressed in Ref. [23]. Simulations of several hybrid mesons using an NRQCD action including the spin interaction $c_B \boldsymbol{\sigma}\cdot\boldsymbol{B}/2M_b$ and neglecting light sea quark effects were carried out; the introduction of the heavy-quark spin was shown to lead to significant level shifts (of order 100 MeV or so) but the authors of Ref. [23] argue that these splittings do *not* signal a breakdown of the Born-Oppenheimer picture. First, they claim that no significant mixing of their non-exotic 0^{-+}, 1^{--}, and 2^{-+} hybrid meson operators with conventional states was observed; unfortunately, this claim is not convincing since a correlation matrix analysis was not used. Secondly, the authors argue that calculations using the bag model support their suggestion. These facts are not conclusive evidence that heavy-quark spin effects do not spoil the Born-Oppenheimer picture, but they are highly suggestive. Further evidence to support the Born-Oppenheimer picture has recently emerged in Ref. [24]. The NRQCD simulations carried out in this work examined the mixing of the Υ with a hybrid and found a very small probability admixture of hybrid in the Υ given by $0.0035(1)c_B^2$ where $c_B^2 \sim 1.5 - 3$ is expected.

The dense spectrum of hybrid states shown in Fig. 5 neglects the effects of light sea quark-antiquark pairs. In order to include these effects in the LBO, the adiabatic potentials must be determined fully incorporating the light quark loops. Such computations using lattice simulations are very challenging, but good progress is being made. For separations below 1 fm, the Σ_g^+ and Π_u potentials change very little[15] upon inclusion of the sea quarks (see Fig. 7), suggesting that a few of the lowest-lying hybrid states may exist as well-defined resonances. However, for $Q\overline{Q}$ separations greater than 1 fm, the adiabatic surfaces change dramatically, as shown in Fig. 8 from Ref. [26]. Instead of increasing indefinitely, the static potential abruptly levels off at a separation of 1 fm when the static quark-antiquark pair, joined by flux tube, undergoes fission into two separate $\overline{Q}q$ color singlets, where q is a light quark. Clearly, such potentials cannot support the plethora of conventional and hybrid states shown in Fig. 5; the formation of bound states and resonances substantially extending over 1 fm seems unlikely. Whether or not the light sea quark-antiquark pairs spoil the Born-Oppenheimer picture is currently unknown. Future unquenched simulations should help to answer this question, but it is not unreasonable to speculate that the simple physical picture provided by the Born-Oppenheimer expansion for both the low-lying conventional and hybrid heavy-quark mesons will survive the introduction of the light sea quark effects. Note that the discrep-

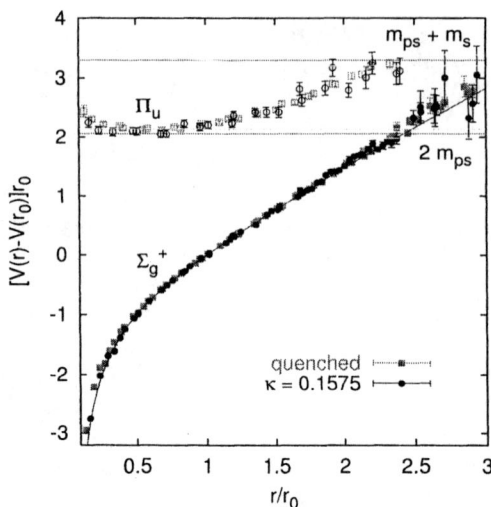

FIGURE 7. Ground Σ_g^+ and first-excited Π_u static quark potentials without sea quarks (squares, quenched) and with two flavors of sea quarks, slightly lighter than the strange quark (circles, $\kappa = 0.1575$). Results are given in terms of the scale $r_0 \approx 0.5$ fm, and the lattice spacing is $a \approx 0.08$ fm. Note that m_S and m_{PS} are the masses of a scalar and pseudoscalar meson, respectively, consisting of a light quark and a static antiquark. These results are from Ref. [15].

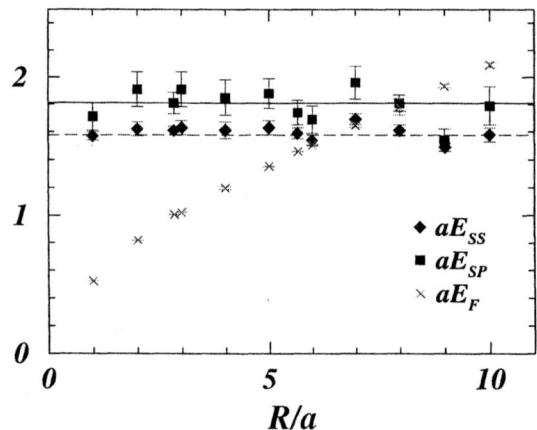

FIGURE 8. Evidence for "string breaking" at quark-antiquark separations $R \approx 1$ fm. E_{SS} is the energy of two S-wave static-light mesons (the light quark bound in an S-wave to the fixed static antiquark), E_{SP} is the energy of an S-wave and a P-wave static-light meson, and E_F is the energy of a static quark-antiquark pair connected by a gluonic flux tube. The distance of separation R refers to the distance between the static quark-antiquark pair. All quantities are measured in terms of the lattice spacing $a \approx 0.16$ fm. Two flavors of light sea quarks are present with masses such that $m_\pi/m_\rho \approx 0.36$. The dashed and solid lines give the asymptotic values $2am_S$ and $a(m_P + m_S)$, where m_S and m_P are the masses of individual S-wave and P-wave static-light mesons, respectively. Mixing between the flux tube and meson-meson channels was found to be very weak. Results are from Ref. [26].

TABLE 1. Recent results for the light quark and charmonium 1^{-+} hybrid meson masses. Method abbreviations: W = Wilson fermion action; SW = improved clover fermion action; NR = nonrelativistic heavy quark action. N_f is the number of dynamical light quark flavors used.

Light quark 1^{-+}			Charmonium $1^{-+} - 1S$		
Ref. & Method	N_f	M (GeV)	Ref. & Method		ΔM (GeV)
UKQCD 97[27]	SW	0	1.87(20)	MILC 97[28]	W	1.34(8)(20)
MILC 97[28]	W	0	1.97(9)(30)	MILC 99[29]	SW	1.22(15)
MILC 99[29]	SW	0	2.11(10)	CP-PACS 99[31]	NR	1.323(13)
LaSch 99[30]	W	2	1.9(2)	JKM 99[22]	LBO	1.19

ancies of the spin-averaged LBO predictions with experiment above the $B\bar{B}$ threshold seen in Fig. 5 most likely arise from the neglect of light sea quark-antiquark pairs.

LIGHT-QUARK HYBRID MESONS

A summary of recent light-quark and charmonium 1^{-+} hybrid mass calculations is presented in Table 1. With the exception of Ref. [30], all results neglect light sea quark loops. The introduction of two flavors of dynamical quarks in Ref. [30] yielded little change to the hybrid mass, but this finding should not be considered definitive due to uncontrolled systematics (unphysically large quark masses, inadequate treatment of resonance properties in finite volume, *etc.*). All estimates of the light quark hybrid mass are near 2.0 GeV, well above the experimental candidates found in the range 1.4-1.6 GeV. Perhaps sea quark effects will resolve this discrepancy, or perhaps the observed states are *not* hybrids. Some authors have suggested that they may be four quark $\bar{q}\bar{q}qq$ states. Clearly, there is still much to be learned about these exotic QCD resonances.

CONCLUSION AND OUTLOOK

Our current understanding of hadronic states containing excited glue is poor, but recent lattice simulations have shed some light on their properties. The glueball spectrum in the pure gauge theory is now well known, and pioneering studies of the mixings of the scalar glueball with scalar quarkonia suggests that these mixings may actually be small. The validity of a Born-Oppenheimer treatment for heavy-quark mesons, both conventional and hybrid, has been verified at leading order in the absence of light sea quark effects, and quark spin interactions do not seem to spoil this. Progress in including the light sea quarks is also being made, and it seems likely that a handful of heavy-quark hybrid states might survive their inclusion. Of course, much more work is needed. The inner structure of glueballs and flux tubes will be probed. Future lattice simulations should provide insight into hybrid meson production and decay mechanisms and the spectrum and nature of hybrid baryons; virtually nothing is known about either of these topics. Glueballs, hybrid mesons, and hybrid baryons, remain a potentially rich source of information (and perhaps surprises) about the confining properties of QCD.

REFERENCES

1. D. Thompson *et al.*, Phys. Rev. Lett. **79**, 1630 (1997); G. Adams *et al.*, Phys. Rev. Lett. **81**, 5760 (1998); A. Abele *et al.*, Phys. Lett. B**423**, 175 (1998).
2. C. Morningstar and M. Peardon, Phys. Rev. D **60**, 034509 (1999).
3. G. Bali *et al.* (UKQCD Collaboration), Phys. Lett. B **309**, 378 (1993).
4. A. Vaccarino and D. Weingarten, Phys. Rev. D **60**, 114501 (1999).
5. R.L. Jaffe, K. Johnson, and Z. Ryzak, Ann. Phys. **168**, 344 (1986).
6. J. Kuti, Nucl. Phys. (Proc. Suppl.) **73**, 72 (1999).
7. T. Barnes, F. Close, and S. Monaghan, Nucl. Phys. **B198**, 380 (1982).
8. C. Carlson, T. Hansson, and C. Peterson, Phys. Rev. D **27**, 1556 (1983).
9. B.Lucini and M. Teper, JHEP 06(2001)050.
10. R. Gupta, A. Patel, C. Baillie, G. Kilcup, and S. Sharpe, Phys. Rev. D **43**, 2301 (1991).
11. Y. Liang *et al.*, Phys. Lett. B**307**, 375 (1993); S.J. Dong *et al.*, Nucl. Phys. (Proc. Suppl.) **63**, 254 (1998).
12. J. Sexton, A. Vaccarino, and D. Weingarten, Phys. Rev. Lett. **75**, 4563 (1995).
13. W.J. Lee and D. Weingarten, Phys. Rev. D **61**, 014015 (2000).
14. K.M. Bitar *et al.*, Phys. Rev. D **44**, 2090 (1991).
15. G.S. Bali *et al.*, Phys. Rev. D **62**, 054503 (2000).
16. C. McNeile and C. Michael, Phys. Rev. D **63**, 114503 (2001).
17. A. Hart and M. Teper, hep-lat/0108022.
18. G. Bali, proceedings of Photon 2001, Ascona, Switzerland, Sep 2001 (hep-ph/0110254).
19. P. Hasenfratz, R. Horgan, J. Kuti, J. Richard, Phys. Lett. B**95**, 299 (1980).
20. K.J. Juge, J. Kuti, and C. Morningstar, Nucl. Phys. B(Proc. Suppl.) **63** 326, (1998); and hep-lat/9809098.
21. S. Perantonis and C. Michael, Nucl. Phys. **B347**, 854 (1990).
22. K.J. Juge, J. Kuti, and C. Morningstar, Phys. Rev. Lett. **82**, 4400 (1999).
23. I. Drummond, N. Goodman, R. Horgan, H. Shanahan, and L. Storoni, Phys. Lett. B**478**, 151 (2000).
24. T. Burch, K. Orginos, and D. Toussaint, Phys. Rev. D **64**, 074505 (2001).
25. C. Davies *et al.*, Phys. Rev. D**58**, 054505 (1998).
26. C. Bernard *et al.*, Phys. Rev. D **64**, 074509 (2001).
27. P. Lacock *et al.*, Phys. Lett. B**401**, 308 (1997).
28. C. Bernard *et al.*, Phys. Rev. D**56**, 7039 (1997).
29. C. Bernard *et al.*, Nucl. Phys. B(Proc. Suppl.) **73**, 264 (1999).
30. P. Lacock and K. Schilling, Nucl. Phys. B(Proc. Suppl.) **73**, 261 (1999).
31. T. Manke *et al.*, Phys. Rev. Lett. **82**, 4396 (1999).

QCD string model for hybrid adiabatic potentials

Yu.S.Kalashnikova and D.S.Kuzmenko

ITEP, 117218, B.Cheremushkinskaya 25, Moscow, Russia

Abstract. Hybrid adiabatic potentials are considered in the framework of the QCD string model. The einbein field formalism is applied to obtain the large-distance behaviour of adiabatic potentials. The calculated excitation curves are shown to be the result of interplay between potential-type longitudinal and string-type transverse vibrations. The results are compared with recent lattice data.

INTRODUCTION

There is no doubts now that hadrons with explicit gluonic degrees of freedom should exist. This idea is supported not only by general arguments from QCD, but also by lattice simulations of pure Yang-Mills theory. In the absence of exact analytical methods of nonperturbative QCD one relies upon models to describe gluonic mesons, so that the challenging question arises of how to introduce effective degrees of freedom for soft (constituent) glue.

There is a lot of indications now that gluonic mesons are already found experimentally, but the conclusive evidences have never been presented; there is no hope that in the nearest future data analyses could shed light on this problem and to offer necessary feedback for model building. The current situation is such that the predictions of different models on hadronic spectra and decays are involved in order to pin-point the signatures for gluonic mesons.

On the other hand, lattice calculations are now accurate enough to provide reliable data on constituent glue and to check model predictions. In this regard recent measurements of gluelump [1] and hybrid adiabatic potentials [2] are of particular interest. These simulations measure the spectrum of the glue in the presence of infinitely heavy adjoint source (gluelump) and in the presence of static quark and antiquark separated by some distance R. These systems are the simplest ones and play the role of hydrogen atom of soft glue studies, as the complicated problem of the centre-of-mass motion separation is not relevant here.

Hybrid adiabatic potentials enter heavy hybrid mass estimations in the Born-Oppenheimer approximation: these potentials are to be inserted into $Q\bar{Q}$ Schroedinger equation in order to obtain spectra of hybrids with heavy quarks. The large R limit is interesting *per se,* as the formation of confining string is expected at large distances, so the direct measurements of the string fluctuations become available and the possibility exists to discriminate between different models of the effective string degrees of freedom.

CONSTITUENT GLUONS AT THE END OF THE STRING

The notion of confinement is usually described in terms of area law asymptotics for the Wilson loop expectation value, defined as an integral along some closed contour C, averaged over gluonic vacuum configurations:

$$\langle W(C) \rangle = \text{Tr} \langle P\exp ig \oint_C A_\mu dz_\mu \rangle, \qquad (1)$$

where trace is taken over colour indices. The area law asymptotics implies that

$$\langle W(C) \rangle \to N_C \exp(-\sigma S), \qquad (2)$$

where N_C is the number of colours, σ is the string tension, and S is the surface bound by the closed contour C. As the initial expression (1) depends only on the contour, the area in (2) should depend only on the contour too, and should be the minimal area.

The area law asymptotics provides the action of the string, and in the case of minimal area this string is also "minimal". The effective string model should be arranged to allow the extra degrees of freedom to populate the string and to be responsible for more complicated string configurations. In what follows these extra degrees of freedom are defined in the framework of the QCD string model. This model deals with quarks and point-like gluons propagating in the confining QCD vacuum, and is based on Vacuum Background Correlators method [3].

The QCD string model for gluons is derived from the perturbation theory in the nonperturbative confining background developed in [4]. The main idea is to split the gauge field as

$$A_\mu = B_\mu + a_\mu, \qquad (3)$$

which allows to distinguish clearly between confining gluonic field configurations B_μ and confined valence gluons a_μ. Confining QCD vacuum is given by the set of gauge invariant field strength correlators made of B_μ, which are responsible for the area law asymptotics (2), while the valence gluons are treated as perturbation at this confining background.

The starting point is the Green function for the gluon propagating in the given external field B_μ [4]:

$$G_{\mu\nu}(x,y) = (D^2(B)\delta_{\mu\nu} + 2igF_{\mu\nu}(B))^{-1}. \qquad (4)$$

The term, proportional to $F_{\mu\nu}(B)$ is responsible for the gluon spin interaction, it can be treated as perturbation [5], and we neglect it for a moment. The next step is to use Feynman-Schwinger representation for the quark-antiquark-gluon Green function, which, for the case of static quark and antiquark, is reduced to the form

$$G(x_g, y_g) = \int ds \int Dz_g \exp(-K_g) \langle \mathcal{W} \rangle_B, \qquad (5)$$

where angular brackets mean averaging over background field. The quantity K_g is the kinetic energy of gluon (to be specified below), and all the dependence on the vacuum background field is contained in the generalized Wilson loop $\mathcal{W}$, depicted in Fig.1.

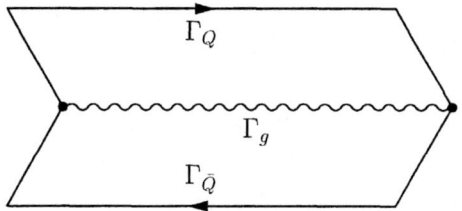

FIGURE 1. Hybrid Wilson loop.

The main assumption of the QCD string model is the minimal area law, which yields for the configuration $\mathcal{W}$ the form [6]

$$\langle \mathcal{W} \rangle_B = \frac{N_C^2 - 1}{2} \exp(-\sigma(S_1 + S_2)), \qquad (6)$$

where S_1 and S_2 are the minimal areas inside the contours formed by quark and gluon and antiquark and gluon trajectories correspondingly.

EINBEIN FIELD FORM OF THE GLUONIC LAGRANGIAN

To define the form of gluon kinetic energy we note that the action of a particle in the external vector field is invariant under reparametrization transformations, and, of course, it remains true after averaging over the background. So, to proceed further we are to fix the gauge, and the most natural way to do this is to identify the proper time τ of the path integral representation with the physical time x_g^0. Then the action of the system can be immediately read out of the representation (5):

$$A = \int d\tau \left\{ -\frac{\mu}{2} + \frac{\mu \dot{r}^2}{2} - \sigma \int_0^1 d\beta_1 \sqrt{(\dot{w}_1 w_1')^2 - \dot{w}_1^2 w_1'^2} \right.$$

$$\left. - \sigma \int_0^1 d\beta_2 \sqrt{(\dot{w}_2 w_2')^2 - \dot{w}_2^2 w_2'^2} \right\}, \qquad (7)$$

where the minimal surfaces S_1 and S_2 are parametrized by the coordinates $w_{i\mu}(\tau, \beta_i)$, $i = 1, 2$, $\dot{w}_{i\mu} = \frac{\partial w_{i\mu}}{\partial \tau}$, $w'_{i\mu} = \frac{\partial w_{i\mu}}{\partial \beta_i}$. In what follows the straight-line ansatz is chosen for the minimal surface:

$$w_{i0} = \tau, \quad \vec{w}_{1,2} = \pm(1-\beta)\frac{\vec{R}}{2} + \beta \vec{r}. \qquad (8)$$

The quantity $\mu = \mu(\tau)$ in the expression (7) is the so-called einbein field [7]; here one is forced to introduce it, as it is the only way to obtain the meaningful dynamics for the massless particle.

Let us introduce another set of einbein fields, $v_i = v_i(\tau, \beta_i)$ to get rid of Nambu-Goto square roots in (7) [8]. The resulting Lagrangian takes the form

$$L = -\frac{\mu}{2} + \frac{\mu \dot{r}^2}{2} - \int_0^1 d\beta_1 \frac{\sigma^2 r_1^2}{2v_1} - \int_0^1 d\beta_1 \frac{v_1}{2}(1 - \beta_1^2 l_1^2) -$$

$$- \int_0^1 d\beta_2 \frac{\sigma^2 r_2^2}{2v_2} - \int_0^1 d\beta_2 \frac{v_2}{2}(1 - \beta_2^2 l_2^2),$$

$$l_{1,2}^2 = \dot{r}^2 - \frac{1}{r_{1,2}^2}(\vec{r}_{1,2}\dot{\vec{r}})^2, \quad \vec{r}_{1,2} = \vec{r} \pm \frac{\vec{R}}{2}, \tag{9}$$

and the corresponding Hamiltonian reads

$$H = H_0 + \frac{\mu}{2} + \int_0^1 d\beta_1 \frac{\sigma^2 r_1^2}{2v_1} + \int_0^1 d\beta_2 \frac{\sigma^2 r_2^2}{v_2} + \int_0^1 d\beta_1 \frac{v}{2} + \int_0^1 d\beta_2 \frac{v_2}{2}, \tag{10}$$

$$H_0 = \frac{p^2}{2(\mu + J_1 + J_2)} +$$

$$\frac{1}{2\Delta(\mu + J_1 + J_2)} \left\{ \frac{(\vec{p}\vec{r}_1)^2}{r_1^2} J_1(\mu + J_1) + \frac{(\vec{p}\vec{r}_2)^2}{r_2^2} J_2(\mu + J_2) + \right.$$

$$\left. \frac{2 J_1 J_2}{r_1^2 r_2^2} (\vec{r}_1 \vec{r}_2)(\vec{p}\vec{r}_1)(\vec{p}\vec{r}_2) \right\}, \tag{11}$$

$$\Delta = (\mu + J_1)(\mu + J_2) - J_1 J_2 \frac{(\vec{r}_1 \vec{r}_2)^2}{r_1^2 r_2^2}, \quad J_i = \int_0^1 d\beta_i \beta_i^2 v_i(\beta_i), \quad i = 1, 2.$$

At first glance, the Hamiltonian (11) looks tractable. Clearly, quantities μ and v_i play the role of gluon constituent mass and energy density distributions along the string respectively. Nevertheless, introducing einbeins does not do miracles for us. These redundant variables are to be found from the conditions

$$\frac{\partial H}{\partial \mu} = 0, \quad \frac{\delta H}{\delta v_i(\beta_i)} = 0, \tag{12}$$

as the equations (12) play the role of second class constraints. One should do it before quantization and substitute the resulting values into the Hamiltonian. Such procedure is hardly possible analytically even at the classical level, and after quantization these extremal values of einbeins would become nonlinear operator functions of coordinates and momenta with inevitable ordering problems arising.

In what follows we use the approximate einbein field method, which treats the einbeins as c-number variational parameters. The eigenvalues of the Hamiltonian (11) are found as functions of μ and v_i and minimized with respect to eibeins to obtain the physical spectrum. Such procedure works surprisingly well in the QCD string model calculations, with the accuracy of about 5-10% [9].

DYNAMICAL REGIMES OF THE GLUONIC HAMILTONIAN

Even with the simplifying assumptions of the approximate einbein field method the problem remains complicated due to the presence of the terms J_i responsible for the string inertia. If one neglect these terms, then the einbeins are eliminated explicitly from the Hamiltonian (11), and one arrives at the potential model Hamiltonian

$$H = \sqrt{p^2} + \sigma r_1 + \sigma r_2. \tag{13}$$

It appears, however, that the neglect of string inertia is justified only for $R \leq 1/\sqrt{\sigma}$ [10]. Indeed, in the einbein field method the potential regime corresponds to the case of v_i independent of β_i. For example, for $R \ll 1/\sqrt{\sigma}$ one has

$$E_n(R) = 2^{3/2}\sigma^{1/2}(n+3/2)^{1/2} + \frac{\sigma^{3/2}R^2}{2^{3/2}(n+3/2)^{1/2}}, \tag{14}$$

$$\mu_n(R) = 2^{1/2}\sigma^{1/2}(n+3/2)^{1/2} - \frac{\sigma^{3/2}R^2}{2^{5/2}(n+3/2)^{1/2}},$$

$$\nu_{1,2n}(R) = \frac{(n+3/2)^{1/2}\sigma^{1/2}}{2^{1/2}} + \frac{3\sigma^{3/2}R^2}{2^{7/2}(n+3/2)^{1/2}},$$

where n is the number of oscillator quanta. The last line in (14) readily gives $J_{1,2}/\mu \approx 1/6$. The situation here is similar to the one in the light quark, glueball and gluelump sectors: the corrections due to the string inertia are sizeable, but not large, and can be taken into account as perturbation [5].

The situation changes drastically for the case of large R, $R \gg 1/\sqrt{\sigma}$. Here one has

$$E_n(R) = \sigma R + \frac{3}{2^{1/3}}\sigma^{1/3}\frac{(n+3/2)^{2/3}}{R^{1/3}}, \tag{15}$$

$$\mu_n(R) = \frac{4\sigma^{1/3}(n+3/2)^{2/3}}{R^{1/3}}, \quad \nu_{1,2n}(R) = \frac{\sigma R}{2}.$$

In this case $J_{1,2} = \frac{1}{6}\sigma R \gg \mu_n$, and the potential regime becomes unadequate.

The case of large R can be treated exactly in the einbein field method. There are two different kinds of excitations, along the $Q\bar{Q}$ axis and in the transverse direction, which are decoupled in the limit of large R. In this case one has

$$E_n(R) = \sigma R + \frac{3}{2^{1/3}}\frac{\sigma^{1/3}(n_z+1/2)^{2/3}}{R^{1/3}} + \frac{2 \cdot 3^{1/2}}{R}(n_\rho + \Lambda + 1), \tag{16}$$

where $\Lambda = \left|\frac{\vec{L}\vec{R}}{R}\right|$ is the projection of orbital momentum onto $Q\bar{Q}$ axis (z axis). Note, that the subleading corrections due to the longitudinal and transverse vibrations are different, the former behaves as $\frac{\sigma}{R}^{1/3}$, like the pure potential regime (15), while the latter displays string-type behaviour $\sim \Lambda/R$.

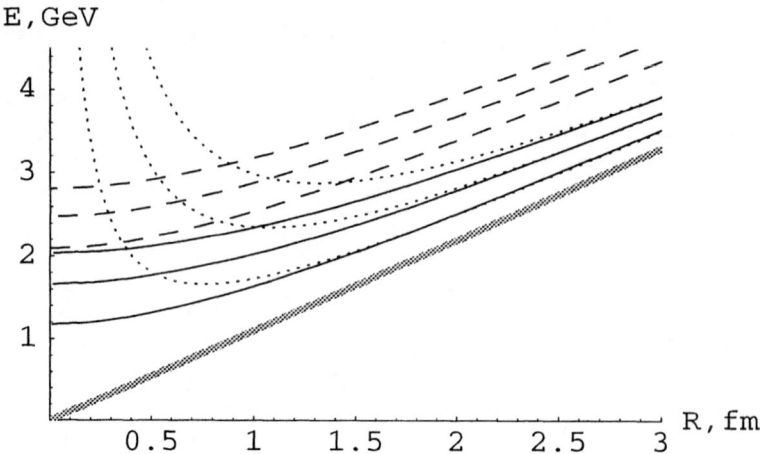

FIGURE 2. Adiabatic hybrid potentials in various regimes. Quasiclassical (solid line), potential (dashed), and flux-tube (dotted) curves for $n_z = n_\rho = 0$ and $\Lambda = 1, 2, 3$. The lowest curve is σR. $\sigma = 0.22 \text{GeV}^2$.

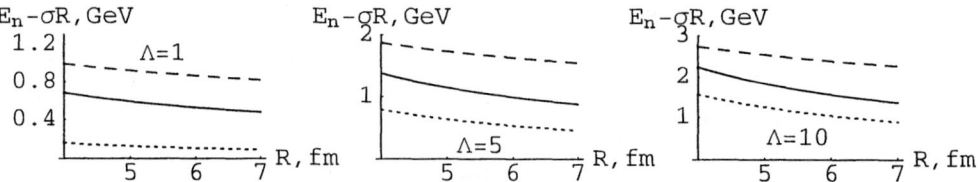

FIGURE 3. Corrections to linear behaviour of potentials. QCD string (solid line), potential (dashed), and flux-tube (dotted) curves; $n_z = n_\rho = 0$; $\sigma = 0.22 \text{GeV}^2$.

The quasiclassical limit of large Λ, where only rotations around z axis are taken into account, was found for the Hamiltonian (11) in [10]. The large R limit reads

$$E(R) = \sigma R + 2\sqrt{3}\frac{\Lambda}{R}, \quad (17)$$

which should be compared with the predictions from naive Nambu-Goto string model

$$E(R) = \sigma R + \frac{\pi \Lambda}{R} \quad (18)$$

in the small oscillation limit. The coefficients π and $2\sqrt{3}$ are close to each other, and differ mainly due to the fact that string configurations differ: there are two straight-line strings in the QCD string model, and one continuous string in the Nambu-Goto case. The quasiclassical excitation curve [10] is shown in Fig.2 together with the Nambu-Goto (18) and potential curves. Note the absence of the unphysical divergent $1/R$ behaviour at small R both for quasiclassical and potential curves.

The dominant subleading regime is defined by longitudinal motion: even if no longitudinal quanta are excited, there is the contribution of zero longitudinal oscillations in

(16). Still, if the distances are not asymptotically large, the potential regime is substantionally contaminated by the string-type transverse vibrations.

Corresponding corrections to linear behaviour for QCD string (16), potential regime (15), and Nambu-Goto string (18) are shown in Fig.3.

There is no QCD string calculations for large distances with proper account of gluon spin yet, so the direct comparison with lattice results is premature. Nevertheless, some preliminary conclusions can be drawn. The behaviour (16) displays the most pronounced difference between QCD string and other models for constituent glue. In the flux tube model [11] the string vibrations are caused by string phonons, so one expects the Nambu-Goto behaviour (18) at large distances. In contrast to this, here the string vibrations are caused by pointlike valence gluons. In the constituent gluon model with linear potential [12] one should have the potential-type behaviour (15), while in the QCD string model the confining force follows from the minimal area law, and, as a consequence, the contributions from the string inertia leave room for the string-type vibrations.

FULL QCD STRING CALCULATIONS IN THE POTENTIAL REGIME

Let us now consider the regime of "small" R relevant to the heavy hybrid mass estimations. Actually these distances are not very small: one expects that in the case of very heavy quarks the hybrid resides in the bottom of the potential well given by the adiabatic curve. The lattice results [2] are not very accurate for small R, but the message is quite clear: the bottom of the potential well is somewhere around 0.25 fm for lowest curves.

If only confining force is taken into account, the QCD string model predicts the oscillator potential (14) with the minimum at $R = 0$. However, the minimum is shifted, if the long range confining force is augmented by the short range Coulomb interaction, which is taken in the form

$$V_c = -\frac{3}{2}\frac{\alpha_s}{r_1} - \frac{3}{2}\frac{\alpha_s}{r_2} + \frac{\alpha_s}{6R}. \tag{19}$$

The coefficients in (19) are in accordance with the colour content of the $Q\bar{Q}g$ system. Note that the $Q\bar{Q}$ Coulomb force is repulsive. It is in contrast to the flux tube model [11], where the string phonons do not carry colour quantum number, so that the $Q\bar{Q}$ pair is in the colour singlet state, and Coulomb interaction is attractive. In the so-called single-bead version of the flux tube model the minimum due the confining force is at $R = 0$, and attractive Coulomb force cannot shift it, so the single-bead version seems to be ruled out by lattice data. As the gluon energies of [2] lie well below the curves (18), the simple Nambu-Goto regime is excluded too.

Below we present the results of the QCD string model with Coulomb force included, for small and intermediate values of R. The calculations were performed in the potential regime, and the string inertia was taken into account perturbatively, which is justified by arguments given above.

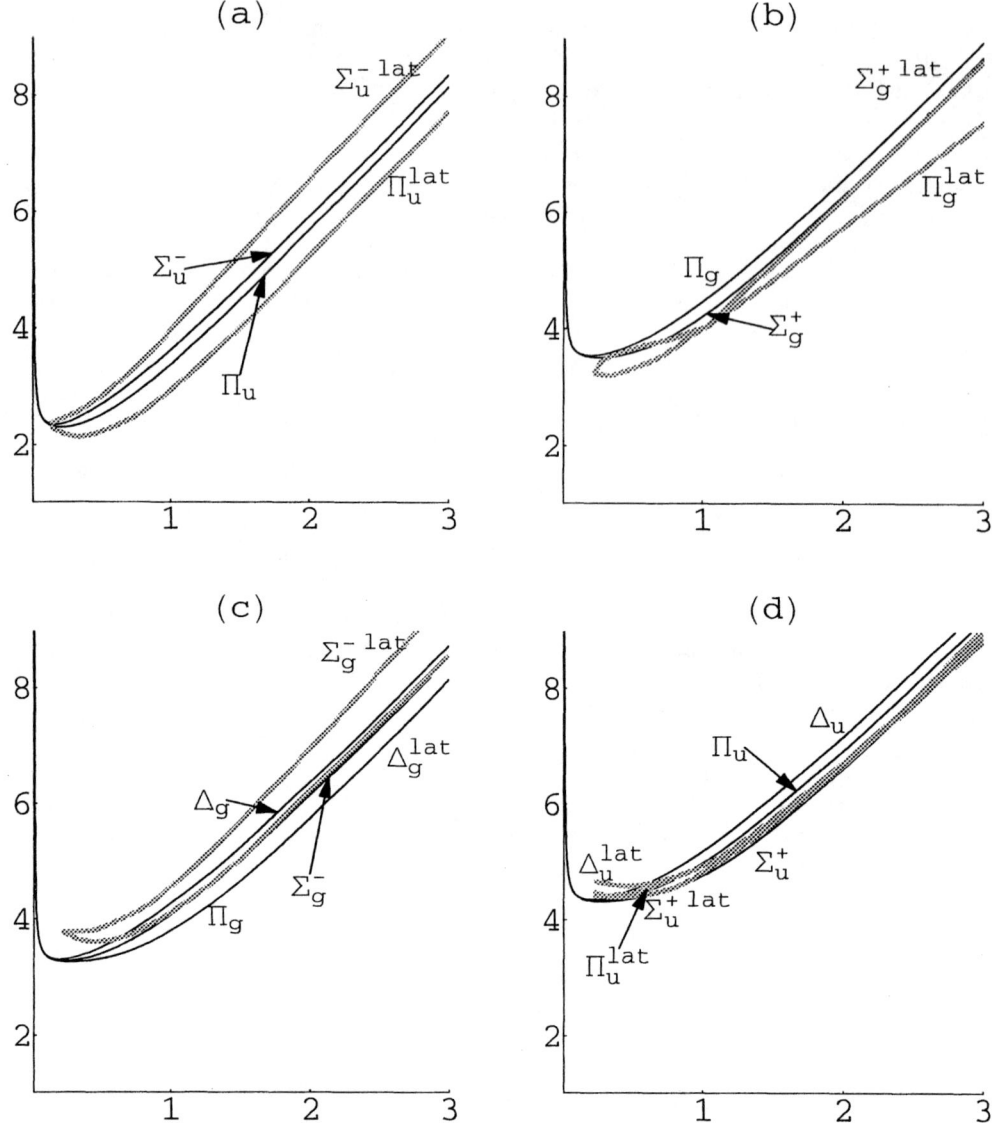

FIGURE 4. Hybrid potentials in full QCD string model (thin solid curves) compared to lattice results (thick light curves). $Q\bar{Q}$ distance R is in units $2r_0 \approx 1$ fm and potentials V are in units $1/r_0$.

Varying over μ_i with result $\mu_{1,2} = \sigma r_{1,2}$, we obtain the Hamiltonian in the form

$$\tilde{H} = \frac{\mu}{2} + \frac{p^2}{2\mu} + \sigma(r_1 + r_2) - \frac{3\alpha_s}{2}\left(\frac{1}{r_1} + \frac{1}{r_2}\right) + \frac{\alpha_s}{6R}. \qquad (20)$$

We calculate unperturbed adiabatic energy levels variationally with Gaussian wave functions used,

$$E_{jl\Lambda}(\mu,R) = \langle \Psi_{jl\Lambda}(\vec{r})|\tilde{H}(\mu,\vec{r},\vec{R})|\Psi_{jl\Lambda}(\vec{r})\rangle \equiv \langle \tilde{H}\rangle_{jl\Lambda}, \quad (21)$$

$$\Psi_{jl\Lambda}(\vec{r}) = \phi_l(\vec{r}) \sum_{\mu_1\mu_2} C^{j\Lambda}_{l\mu_1 l\mu_2} Y_{l\mu_1}\left(\frac{\vec{r}}{r}\right) s_{1\mu_2}, \quad (22)$$

where s is the spin wave function. The unpertubed adiabatic potentials

$$V^0_{jl\Lambda}(R) = E_{jl\Lambda}(\mu^*(R),R), \quad (23)$$

depend on the extremal value μ^* defined from the condition $\frac{\partial E_{jl\Lambda}(\mu,R)}{\partial \mu} = 0$.

The string correction Hamiltonian at $\nu \ll \mu$ after integrating out ν reads

$$H^{\text{string}} = -\frac{\sigma}{6\mu^2}\left(\frac{1}{r_1}L_1^2 + \frac{1}{r_2}L_2^2\right) =$$

$$-\frac{\sigma}{6\mu^2}\left\{\left(\frac{1}{r_1}+\frac{1}{r_2}\right)\left[L^2 + \frac{1}{4}(R_j p_k R_j p_k - R_j p_k R_k p_j)\right] + \right.$$

$$\left. \frac{1}{2}\left(\frac{1}{r_1}-\frac{1}{r_2}\right)[r_j p_k R_j p_k - r_j p_k R_k p_j + R_j p_k r_j p_k - R_j p_k r_k p_j]\right\}, \quad (24)$$

where $\vec{L}_{1,2} = \vec{r}_{1,2} \times \vec{p}$, $\vec{L} = \vec{r} \times \vec{p}$.

Here we estimate the contribution of the string correction (24) taking it at $R=0$, where it reads

$$V^{\text{string}}_l = -\frac{\sigma l(l+1)}{3\mu_l^{*2}}\langle\frac{1}{r}\rangle_l. \quad (25)$$

By the same way we consider spin-orbit correction at $R=0$,

$$V^{SO}_{jl} = -\frac{\sigma \vec{S}\vec{L}}{2\mu_l^{*2}}\langle\frac{1}{r}\rangle_l, \quad (26)$$

where $\vec{S}\vec{L} = \frac{1}{2}(j(j+1) - l(l+1) - 2)$. We also subtract a constant that corresponds to gluon self-energy. From comparison with lattice data [2] its value is found to be $\Sigma_{gl} = 473$ MeV.

In Fig.4 (a)-(d) potential

$$V_{jl\Lambda}(R) = V^0_{jl\Lambda}(R) + V^{\text{string}}_l + V^{SO}_{jl} - \Sigma_{gl} \quad (27)$$

(solid curves) is compared to lattice potentials from [2] (thick grey curves). Parameters $\alpha_s = 0.225$, $\sigma = 0.227$ GeV^2 are taken from fit of the lattice Coulomb+linear ground-state potential. In the Fig. 4 standard notations from the physics of atomic molecules are used.

TABLE 1. Quantum numbers of levels of Fig. 4

(a)	$j=1, l=1, \Lambda=0,1$	Σ_u^-, Π_u
(b)	$j=1, l=2, \Lambda=0,1$	Σ_g^+, Π_g
(c)	$j=2, l=2, \Lambda=0,1,2$	$\Sigma_g^-, \Pi_g, \Delta_g$
(d)	$j=2, l=3, \Lambda=0,1,2$	$\Sigma_u^+, \Pi_u, \Delta_u$

In Table 1 the quantum numbers of levels, shown in Fig.4 (a)-(d), are listed, in terms of j, l, Λ, and atomic notations. Note, that in the QCD string model the gluon is effectively massive and has three polarizations [5], and only two of them are excited with magnetic components of field strength correlators, used in the lattice calculations. It is just these states which are listed in the Table 1. For more details justifying such correspondance see [5].

CONCLUSIONS

The first results of the QCD string model for the hybrid adiabatic potentials look rather encouraging. We have obtained the reasonable agreement with lattice data at small and intermediate interquark distances. The most interesting feature of the QCD string model is the large distance behaviour of the adiabatic potentials, with potential-type longitudinal and string-type transverse vibrations. The full calculations of large distance behaviour with proper account of gluonic spin are in progress now.

ACKNOWLEDGEMENTS

Financial support of RFFI grants 00-02-17836, 00-15-96786, INTAS-RFFI grant IR-97-232 and INTAS CALL 2000-110 is gratefully acknowledged.

REFERENCES

1. Foster, M. and Michael, C., *Phys.Rev.*, **D59**, 094509 (1999)
2. Juge, K.J., Kuti, J., and Morningstar, C., *Proc. LATTICE 98*, Nucl. Phys. Proc. Suppl. **73**, 590 (1999)
3. Simonov, Yu.A., *Lectures at the XVII Int. School of Physics*, Lisbon (1999)
4. Simonov, Yu.A., *Phys.At.Nucl.*, **58**, 107 (1995)
5. Simonov, Yu.A., *Nucl.Phys.*, **B592**, 350 (2000)
   Kaidalov, A.B. and Simonov, Yu.A., *Phys.Lett.* **B477**, 163 (2000)
6. Kalashnikova, Yu.S. and Yufryakov, Yu.B., *Phys.Lett.* **B359**, 175 (1995); *Phys.Atom.Nucl.*, **60**, 307 (1995)
7. Kalashnikova, Yu.S. and Nefediev, A.V., *Phys.At.Nucl.*, **60**, 1389 (1997)
8. Dubin, A.Yu., Kaidalov, A.B. and Simonov, Yu.A., *Phys.Lett.*, **B323**, 41 (1994); *Phys.At.Nucl*, **56**, 1745 (1993)
9. Morgunov V.L., Nefediev, A.V. and Simonov, Yu.A., *Phys.Lett.*, **B459**, 653 (1999)
10. Kalashnikova, Yu.S. and Kuzmenko, D.S., hep-ph/0006073
11. Isgur, N. and Paton, J., *Phys.Rev.*, **D31**, 2910 91985)
12. Horn, D. and Mandula, J., *Phys.Rev.*, **D17**, 537 (1978)

Remembering Nathan

T.Barnes

Physics Division, Oak Ridge National Laboratory, Oak Ridge, TN 37831-6373, USA
Department of Physics and Astronomy, University of Tennessee Knoxville, TN 37996-1501, USA

I was asked to contribute a few remarks as a memorial to our friend and colleague Nathan Isgur, who has made such great contributions to hadron physics, and who sadly is no longer with us. I hope that my recollection of these few personal memories about Nathan and his approach to science will suffice.

In these remarks I would like to recall the topics in hadron physics that Nathan found exciting, as many of us in the hadron community will be interested in furthering these lines of research. I will also do what I can to show why Nathan was so successful as a physicist, what his *modus operandi* was in research, and mention a few topics that he had planned to study, were sufficient time available.

Like many physicists in our field, I first heard of a theorist named Isgur in the context of the baryon spin-orbit problem, around the year 1980. There was an outstanding problem with the nonrelativistic quark potential model, specifically that the one-gluon-exchange (OGE) spin-orbit term predicted much larger spin-orbit splittings in the orbitally-excited baryon spectrum than were observed experimentally. Suddenly the news spread that this Isgur fellow had solved the spin-orbit problem, using what quickly became known as the "Isgur-Karl" quark model [1]. On closer inspection the solution consisted of "tossing the spin-orbit terms out the window", which did not impress me at all when I first heard about it. Some nuclear theorists are still under the misapprehension that this was all that Nathan contributed to the spin-orbit problem. My first meeting with Nathan was at the 1981 Moriond meeting at Les Arcs, which is the sort of venue that all physics conferences should strive for. My impression of Nathan was that this was a very careful, very precise theorist, and that his decision to drop the spin-orbit terms was merely an interim step; indeed, in his subsequent work with Simon Capstick the spin-orbit terms were carefully evaluated with realistic Coulomb plus linear wavefunctions, and the previous problems were found to be an artifact of the use of single Gaussian wavefunctions. These were rather inaccurate in determining short-distance effects, and the discrepancy in spin-orbit splittings was largely resolved when the more accurate wavefunctions were used. This was the actual solution of the spin-orbit problem.

Nathan was an exciting and even "dangerous" participant at conferences, as the following incident will illustrate. The early 1980s was a period in which the first models of glueballs and hybrids were being studied, including the now famous flux-tube model, which Nathan developed in collaboration with Jack Paton (Oxford), and applied to hybrids with Jack and graduate student Rick Kokoski (Toronto). Frank Close and I and several other theorists had earlier published bag model calculations that showed that J^{PC}-exotics should exist, and the I=1 1^{-+} was an especially attractive prospect experimentally, but of course the bag model provided no indication of the decay widths of this or other hybrids; the crucial width question was wide open. Theorists in the hadron community typically met at small conferences or workshops on new facilities, and in 1984 I again met Nathan at one of these, the ELSA workshop in Bad Honnef. Nathan gave a talk about the flux tube model, including some discussion of hybrid mesons. In the question period after his talk I asked a rhetorical question about whether he had done strong decay calculations of hybrid mesons in the flux tube model, knowing full well that he hadn't. (If he *had* done this crucially important calculation, I assumed, he would certainly have told us already.) Nathan replied, in his quiet, measured voice, "Yes, we have calculated strong decay widths." (At this point I was stunned - it hadn't been mentioned in his talk at all - and very worried, since in the worst case the states might be very broad, and hence experimentally inaccessible. His answer could effectively end the entire subject that I had spent so much time and effort working on!) He continued, "Many of the states do come out very broad, but some states are predicted to be relatively narrow and should be observable." (These quotes are to the best of my recollection.) This was the "dangerous" aspect of Nathan; if you asked him a question, he would almost always have the answer, even if he had given no indication that he had already studied that problem, and the audience would later recall this as the definitive result. The answer regarding hybrids, as we all now know, is that the flux tube predicts that hybrids prefer S+P modes, which have less phase space than S+S so that some of the hybrids are expected to be only ca. 100-200 MeV wide. Of course experiment has not yet shown us whether these predictions for hybrids are accurate.

From 1985-1989 I was Nathan's postdoctoral fellow in Toronto, so I was able to see his approach to research at close hand. Nathan's usual technique was to partner with a Toronto graduate student (since UT was Canada's leading university the graduate students were very talented) and assign a sector of Hilbert space to that graduate student to study using the quark potential model. These studies were often encyclopaedic, covering not only the spectrum but also all relevant strong decays and weak and electromagnetic transitions. This completeness has made papers such as Godfrey-Isgur [2] and Capstick-Isgur [3] standard references in meson and baryon spectroscopy, despite the fact that 15 years of experiment and notational changes have since modified these subjects considerably. This graduate student research program had started with Roman Koniuk [4] (assigned qqq, from the baryon spin-orbit problem Nathan and Gabriel Karl at Guelph had treated [1]). This work continued with Steve Godfrey [2] ($q\bar{q}$), Simon Capstick [3] (qqq with improved wavefunctions), John Weinstein [5] ($q^2\bar{q}^2$), Kim Maltman [6] (q^6), Rick Kokoski [7] ($q\bar{q}g$), with Jack Paton, and Daryl Scora [8] ($Q\bar{q}$), with Ben Grinstein and Mark Wise.

One can not only follow Nathan's research career with Toronto graduate students as a systematic progression through Hilbert space, one can also identify "gaps" which Nathan was not able to study, either because there was insufficient time or because the graduate student did not carry the task to completion. These include $q^4\bar{q}$, q^3g (the student encountered difficulties), gg (the flux tube model is probably not realistic here), and configuration mixing, "unquenching the quark model". This was an aspect of Weinstein and Isgur's later work on the 980 MeV states, and the interesting problem of $q\bar{q}$ mixing through meson loops was studied by Nathan and Toronto graduate student Paul Geiger [9]. The "unsolved problems" in Nathan's career which this summary suggests are a more complete study of meson-meson and meson-baryon bound states (which requires configuration mixing of one- and two-hadron states, the "multichannel quark model", which Nathan discussed but did not fully develop); and a QCD-based description of nuclear forces (his 1999 APS talk, which is online at http://www.jlab.org/news/archive/2001/Isgur/index.html, stresses the importance of distinguishing meson exchange from quark-gluon forces, and notes that this "central problem in understanding ordinary matter", the "QCD basis for nuclear physics..." is still "in its infancy"). Nathan's other great interests in hadronic physics, baryon spectroscopy and strong decays, gluonic states and HQET, have already been investigated theoretically at some length using various approaches, but will certainly merit our continued study in future as new data from Jefferson Lab and other facilities becomes available.

The personal humility with which Nathan carried out his work, and the friendly encouragement which he gave to his colleagues and collaborators, were as impressive as his list of research achievements. In reading the comments people have made elsewhere in Nathan's memory, one is struck repeatedly by how often his kindness is mentioned. I would also cite a quiet and slightly amused seriousness as one of his most attractive and characteristic traits. Nathan's low-key confidence in his predictions, which were often amazingly accurate for models of hadronic physics, was something that many of us have encountered. As an example, I once chided Nathan for his surprisingly precise claim of 1.9(1) GeV for the mass of the lightest hybrid meson multiplet. To my surprise Nathan took my comments rather seriously, and replied "At least I can be proven wrong." In hadron physics one more often encounters the habit of "patching up" a model to agree with the latest data points; this has the disadvantage that it can postpone the identification of the best description of the physics, perhaps for a very long time. Nathan's attitude that his model might be incorrect, and that the data really can decide, recommends itself as an approach that is much closer to the scientific ideal.

Nathan's attitude to academic freedom in physics was also exemplary. I can again cite an example from personal experience. When high-Tc superconductivity was discovered in 1987, Eric Swanson (then a Toronto graduate student of Nathan's) and I grew very interested in the subject, and began collaborations with several condensed matter theorists who were also interested in high-Tc and quantum spin systems. The Toronto condensed matter faculty were politically wiser than I, and suggested that I discuss this research topic with Nathan - after all, he was paying us to work on hadrons, and we were instead studying the 2D Heisenberg antiferromagnet. Nathan's reply was that I should "Follow the physics.". This did not surprise me at all at the time, it was the kind of academic freedom that I had always been told encouraged the best research. However I did not then have to write research summaries for grant agencies, or I would have realized how difficult it could be to encourage two young researchers to pursue the "wrong topic" purely because they were interested in it. Nathan was an idealist, even when it was difficult to be one.

Nathan's final contribution to hadrons was as head of the Jefferson Lab theory group through the 1990s. At Jefferson Lab, Nathan emphasized the importance of hadron spectroscopy as a central part of the lab's physics program, and was responsible for the creation of ca. 60 bridge positions between Jefferson Lab and universities in the Southeastern USA. Unfortunately Nathan will not see the realization of the Jefferson Lab physics program, especially the search for "missing baryons" that he had predicted would decay more strongly to vector + nucleon than pion + nucleon; he told me that this was one of his great regrets. He was also a strong supporter of meson spectroscopy at Jefferson Laboratory through the planned energy upgrade and the future meson photoproduction facility, HallD. Nathan was a great influence and source of enthusiasm at Jefferson Lab, and his tenure there showed that he was as effective in an administrative position as he had been as an independent theorist.

Nathan died on 24 July 2001, after five years of treatment for bone marrow cancer. An email I received from Nathan two weeks earlier told me that there had been a sudden and unexpected relapse. He included the admonition "Please carry the torch forward!", which I interpret as his farewell request to us all. Although his medical problem has been of great concern to his colleagues in recent years, to those of us who knew Nathan well it was evident that he personally considered this very serious problem to be an unfortunate but basically irrelevant nuisance. When I first learned of the diagnosis in summer 1996, I immediately telephoned Nathan to discuss his problems and future prospects. We did our best to be serious about the issue for the first few minutes, and discussed the personal problems that we had each encountered recently, but it was soon obvious that this was a very limited set of topics in which neither of us was especially interested. In fairly short order we switched topics to hadrons, and spent over an hour reviewing the current status of meson spectroscopy, hybrids, glueballs, and other really exciting issues. It was clear that the physics of hadrons would remain the paramount interest to Nathan that it was throughout his life. We are very fortunate to have known Nathan, whose life and career as a scientist should be an inspiration to us all.

ACKNOWLEDGMENTS

I would like to thank the organisers of HADRON2001, Prof. Zaitsev in particular, for the kind suggestion that I make a few remarks in Nathan's memory. This work was supported in part by the DOE Division of Nuclear Physics, at ORNL, managed by UT-Battelle, LLC, for the US Department of Energy under Contract No. DE-AC05-00OR22725, and by the US National Science Foundation under Grant No. INT-0004089.

REFERENCES

1. N.Isgur, and G.Karl, *Phys. Rev.*, **D18**, 4187 (1978).
2. S.Godfrey, and N.Isgur, *Phys. Rev.*, **D32**, 189 (1985).
3. S.Capstick, and N.Isgur, *Phys. Rev.*, **D34**, 2809 (1986).
4. R.Koniuk, and N.Isgur, *Phys. Rev.*, **D21**, 1868 (1980).
5. J.Weinstein, and N.Isgur, *Phys. Rev.*, **D27**, 588 (1983).
6. K.Maltman, and N.Isgur, *Phys. Rev.*, **D29**, 952 (1984).
7. N.Isgur, R.Kokoski and J.Paton, *Phys. Rev. Lett.*, **54**, 869 (1985).
8. N.Isgur, D.Scora, B.Grinstein, and M.Wise, *Phys. Rev.*, **D39**, 799 (1989).
9. P.Geiger, and N.Isgur, *Phys. Rev.*, **D41**, 1595 (1990).

HISTORY

IHEP, Protvino

Hadronic Octaves: Symphony in Treble Clef

Yuval Ne'eman*[#]

*Emeritus Professor of Physics
Tel-Aviv University, Tel-Aviv, Israel 69978
and
[#]University of Texas, Austin, TX 78712

Abstract. Pythagoreanism, as derived from the physics of music, an artificially quantized system, involved simple ratios between integers and was conjectured by the Pythagoreans to extend to the whole of physics (the Music of the Spheres). It hit the jackpot in 1895 with Balmer's formula and has dominated XXth Century physics, with its Quantum Foundations. I review the history of Hadron Spectroscopy and my personal role in 1958-1964, ie. (1) my 1960 discovery of SU(3) symmetry with an octet assignment for the $j = \frac{1}{2}$ baryons (independently reached somewhat later by M. Gell-Mann), and (2) in 1961 (with H. Goldberg) my mathematical construction of a structural model which was then developed into the physical quark model by Gell-Mann and Zweig.

Pythagoras to Johann Jakob Balmer

Science, in its precise and full context, starts with the Greeks and more specifically with Pythagoras of Samos and his school in the VIIth century BC, mostly in Southern Italy (Great Greece). Three principles are involved:
(1) It should aim at representing a complete *welt-anschauung*, a self-sufficient representation of *physical reality* in its *entirety*.
(2) It should be formatted as a set of *axioms* from which everything else is derived by *logical inference* and *mathematics* as its extension (to avoid circular arguments, mathematics itself has to be constructed as a basic part of science).
(3) It should allow for *patchwork* construction and a gradual *merger* process.

There was no a priori reason to believe that such a program would succeed and there was special misapprehension relating to some borderlines: *life* as regarded by the vitalists, the *brain-mind* and the hypothetical *body-spirit* tandems, the *time-singularity* at creation. One by one, these lines have been crossed and one's present impression is of no conceptual blank spots left, only unfinished work.

The Pythagoreans went beyond laying the foundations of this edifice. They managed to weave one small patch, namely the physics of *music* [1] and were very impressed (and seemingly carried away) by the elegance of the resulting theory.

Music involves an *artificial quantization*, as holding both ends of a string constrains it to oscillate only at those wavelengths which, when aligned and propagated *n* times along the string's length, will have for *n* an *integer*; in other words, wavelengths corresponding to the string-length divided by an *integer*. The same is true of the oscillating column of air in a wind instrument. The Pythagoreans thus discovered that simple numerical relations such as 1:2:3:4 (a "tetracyte") can describe musical notes and their *harmonics* etc.

At this point, they conjectured that this could not be an exceptional result, specific to the problem they had accidentally selected – and guessed that this is how any chapter in physics should end up – as the *music of the spheres*, namely a set of *pythagoreanisms*, i.e. dimensionless ratios between integers. They were not deterred even after a spectacular defeat of this principle in plain geometry – namely, when, in applying Pythagoras' theorem to the evaluation of the length of the hypotenuse in an isosceles right-angle triangle whose orthogonal sides are of unit length, they proved that *the value of $\sqrt{2}$ cannot be given as the quotient between two integers p and q*, since $\sqrt{2} = (p/q)$ would imply $p^2 = 2q^2$, i.e. just nonsense: the factor 2 would be contained in an even power on the l.h.s. and in an odd power on the r.h.s. The importance of their resulting discovery of the *irrationals* overshadowed the collapse (in mathematics, at least) of their *music of the spheres* conjecture, which continued to dominate the imagination – e.g. of Johannes Kepler and of many others, both before and after him.

These hopes remained unfulfilled and shattered for twenty-five centuries, as long as the physical context consisted of *classical* phenomena. Yet *one day in 1885, the first notes of the music of the spheres were heard*, when Johann Jakob Balmer, a Basel schoolteacher and amateur numerologist, playing with hydrogen spectral data, found a simple formula [2] for these wavelengths. Using the generalized form due to Rydberg [3], these wavelengths are given by (R is Rydberg's constant, k, n are integers and $k > n$, n characterizes a series, k its components), $\lambda^{-1} = R[(n^{-2}) - (k^{-2})]$. Balmer's series had $n = 2$, but the series for all other $n = 1, 2, ..., 5$ soon followed. Finally, the *natural* quantum level had been reached, *atomic spectroscopy* had turned that music on. We take the presently accepted view, according to which, in the Copenhagen interpretation of Quantum Mechanics, the *classical* framework carrying the apparata is only an *effective* result of the *decoherence* process caused by the gigantism in the number of quantum systems involved; in this view, *the fundamental picture of reality is that of a quantized system* – and at least in principle, *the Pythagorean guess was justified*! It was a very wild guess and yet it succeeded.

In a certain context, my contributions in the hadron domain can be considered as having added further parts to that same symphony for which Balmer wrote the Overture. This is a conclusion I had reached in the last five years, while being drawn by the millennium celebrations to look at science throughout the millennia – when I suddenly remembered that back in 1964, after the Omega-minus experiment which validated $SU(3)$ symmetry and the related classification, I had received a letter from Rydberg's daughter, who had noted the similarities and was comparing the excitement...

The "musical" aspect re-surfaced in 1924, now at the *quantum* level, when L. de Broglie introduced the concept of *matter waves*, with wavelength $\lambda = h/p = h/mv$. The above Balmer etc. formulae for wavelengths series in hydrogen atomic spectroscopy had meanwhile been rederived by Niels Bohr in his *ad hoc* atomic "Bohr model", using an ansatz of *quantized angular momenta*, $L = p \wedge r = mv \wedge r \equiv nh/2\pi$. On the other hand, de Broglie treated *musical* orbits, i.e. orbits fitting an integer number of de Broglie wavelengths, $2\pi r = n\lambda_{dB} = nh/mv$ (any other orbit effectively cancels through negative self-interference). It is obvious that division by 2π will yield Bohr's condition, the two quantizations are thus *equivalent*.

Atomic spectroscopy was thus "reduced"; similar treatments have been used for molecular spectra and dynamics. In *nuclear structure*, the puzzle appeared more complicated, though the existence of so-called *magic numbers* finally led to the *shell model*. Further features were described by the *collective model* and by various quadrupolar excitation models (such as Elliott's) fitting the more deformed nuclei. A more comprehensive classification was later provided by the *Interacting Boson Model*.

Now to the hadrons. Initially just a small "community", consisting of the *proton* in 1911, joined by the *neutron* in 1932 and by the 3 *pions* (predicted in 1935, found in 1949) they were now (1960) undergoing an almost exponential growth due to several experimental improvements:
 (a) the development of particle *accelerators*, gradually replacing *cosmic rays* as suppliers of new particles,
 (b) the utilization of *emulsions* to detect particle tracks,
 (c) the invention of *bubble chambers*,
 (d) the method of *inertial mass* plots.

Developments (a) - (c) had mainly brought about new degrees of freedom, e.g. *strange* particles, whereas (d) had opened up a Pandora box involving angular momentum excitations as well. These started in the early fifties with Fermi's "3/3-resonance", produced by scattering pion beams off nucleon targets $\pi + N \to \Delta$. These experimental aspects will be dealt here more comprehensively by my friend and colleague Prof. B. Maglic. I shall focus on the theoretical advances in 1960-1962.

My (meandering) path to London and Particle Physics

The remaining introductory topic to be covered in this review deals with my own path, namely why and how did I sit down in May 1960 at a desk at Imperial College in London, at the ripe age of 35, with a stipend allowing me just one year's leave to be spent on scholarly activities, before returning, after this intellectual escapade, to full service in the Israel Defence Force, with the stated intention (on the part of the Chief of Staff) of putting me in charge of the task of establishing a National Defence College. This is a story I have told several times, especially in my contribution to the *Symmetries in Physics* (1600-1980) conference in 1983 at San Feliu de Guixols, Catalonia (Spain), which is now also included (together with M. Gell Mann's) in our recently republished and thereby extended *Eightfold Way* collection [4].

The main chain of events starts at Tel-Aviv, where I was born in 1925, a grandson of one of the 66 founding couples, with my grandfather and father both mechanical engineers and running a factory producing pumps, both for the water supply of the country's cities, and for irrigation (both deepwell and river pumps) etc., which is why I followed and after matriculating in 1940 at the early age of 15 and having to wait one year because of a minimal admission age of 16, I joined the Technion (Israel Institute of Technology) in Haifa and got my B.Sc. degree and Diploma in Mechanical and Electrical Engineering in 1945. I had already done some original work in Mathematics in High school and was torn between my duty to the family plant and my interest in science, with the further complication relating to my feelings of responsibility with respect to national survival. The scientific interest gradually switched from Mathematics to Physics, mostly under three influences. On the one hand, a popular science lecture series in 1940 on Modern Physics by S. Samboursky, who had taken his doctoral degree under Theodore Kaluza in Berlin and was the first appointed physicist in modern Israel (in 1928 [5], as "Assistant for Physics in the Department of Mathematics" at the newly established – 1925 – Hebrew University, Jerusalem); on the other hand, reading popularization books by Jeans and by Eddington (much later by Gamow too) etc. The third and decisive influence was an undiluted lecture series by F. Ollendorf, nominally professor of electrotechnics, yet on Quantum Physics, in my last year at the Technion (the fall of 1944). After the first two lectures I was the entire residual audience but he went on unperturbed, speaking as if before a full hall. His course included solving the Schroedinger equation for hydrogen and I was overwhelmed by what I viewed as the aesthetical aspects – the features I would now describe as pythagoreanisms. Twenty two years later, when I received an honorary degree from the Technion, Ollendorf commented that his effort for a one-man audience had paid off…

Events on the national scene had meanwhile also come to a head. I had joined the Haganah underground in 1940 and at the Technion, I had been mobilized for training or operational activity about 50% of the school time. I was sent to a non-commissioned officers' school in the summer of 1942 and to the Haganah Officers' School (commanded by Y. Yadin, the archeologist) in the summer of 1945. After finishing my studies in Engineering I did work for about a year (nominally) in the family plant and designed three different models: (1) a horizontal large – flow centrifugal pump representing a model to be used on the upper Jordan River, (2) a vertical deepwell centrifugal pump for orchard irrigation, (3) a large-flow hellical pump for sewage purposes. In mid-1946 I was completely mobilized and participated actively both in the struggle for allowing the Displaced Persons to enter the Palestine Mandate, the protection of the erection of new settlements and sometime an operation blocking an action prepared by one of the other underground organizations, the Zionist leadership being set on avoiding operations that might hurt civilians (or even military personnel, except when unavoidable in the defence of immigration and settlement). In the summers of 1946 and 1947 I commanded NCO courses while getting to know the geographical features of those regions – a fact which helped me later, when these became battlefields. The War of Independence indeed started on November 29th, 1947, and I fought it as an officer in the field, first along the road to Jerusalem, then on the Egyptian front mostly. I established a reputation as a planner, especially after July 1948 when I devised a very original maneuver on our eastern flank, and then in October by developing a new operational technique, adapted to our special conditions. These plans were put into action in October-November 1948, resulting in a successful campaign in the South. After the 1947-49 War, I was appointed Chief of the Operational Section and Vice-Chief of the Operations Department at Defense Headquarters (Gen. Rabin, the future Prime Minister, was the Chief). I attended the French Staff College in 1951-52 and was then appointed Director of Planning, a task which included several civilian components (see my talk at the Seventh Marcel Grossmann meeting in Palo Alto in 1994 [6]) – plus the elaboration of the basic strategy we followed up to and in the 1967 "Six–days War". Meanwhile in 1955, I was transferred to serve as Vice-Chief of Military Intelligence; after the collapse of a network in Egypt and the subsequent changes in command. In this position I developed a few new ideas. The main thrust, however, was in developing a relationship with France, in accordance with French and Israeli interests. This included elements which leaked somehow and underwent imaginative amplification, ending up in making me play roles (under my full name!) in novels about the Middle East or books about Secret Service [7].

Throughout the years I continued to hope that some day I would be able to "quantum tunnel" onto a scientific career. I had made plans for 1948 – and forgot them in fighting. I taught myself General Relativity in 1950 and asked for a leave in 1951-52. I intended to go to France and study Physics under de Broglie; instead, I was prevailed upon by the Vice-Chief of Staff to go there to study Higher Military Studies at the Ecole d'Etat Major, a branch of the famous Ecole Superieure de Guerre where Foch had taught.. Finally in 1957, I felt the situation was less tense and my age such as to make me seriously fear I might miss the last chance I had of moving over to science. Also, I had heard the Technion was starting a Physics Department, under Nathan Rosen (of the Einstein-Rosen cylindrical gravitational wave, the Einstein-Rosen Bridge – a precursor of Black Holes, the Einstein-Podolsky-Rosen entanglement, etc.). I asked Gen. Dayan, the C.o.S., for a two-years leave to return to the Technion. His counter-proposal was "can you do it in London and combine it with the Defense Attache duties?"

Having followed the fates of Steady-State Cosmology, I thought I could fit my studies into this mold, working under Bondi at King's – and accepted. Arriving in London in December 1957, I realized that traffic realities did not allow combining my duties at the Israeli Embassy in Kensington with studies at King's College, East of Trafalgar Square. I found Imperial College in Kensington itself and landed in Salam's Quantum Field Theory program while searching for somebody who might direct me in studying Einstein's latest Unified Field Theory.

Throughout January – June 1958 I made nice progress. Whenever my duties would make me miss a class (and it happened all the time), I would photocopy Ray Streater's notes and study from them. In July 1956, however, there was a revolution in Irak and Gen. Kassem took over after the assassination of the King and of the Prime Minister. The USA sent troops to land in Lebanon, while the UK planned to drop two parachutist battalions in Jordan, to protect King Hussein's regime. I could no more do any physics and spent my time negotiating the purchase of fifty Centurion tanks and two S-class submarines. Next came the task of planning the training program for the weapon systems we were purchasing – these were our first submarines and all ranks had to be trained in parallel and gradually fused into one crew. I applied to the new C.o.S. in Israel, Gen. Laskow, reminding him of the arrangement proposed by his predecessor, under which I had come to the UK, and we reached a new agreement under which I would be replaced as the Defence Attache and get a one-year fellowship – after which I would either resign from the forces and continue on my own in Science – or forget it all and fully return to my military duties. There were some complications – the new Attache who had replaced me suddenly had to be recalled and replaced, which made me have to present a successor to the authorities twice, in "my" five capitals (London, Copenhagen, Oslo, Stockholm and Helsinki). I finally became a free student on May 1st, 1960 – though I still had to go on a further visit to Norway in October, to deliver a lecture I had promised on Israel's strategies, at the Oslo Military Circle, in the presence and with the active participation of H.M. King Olaf VI. By that time I had managed to go deep into my chosen problem in physics and had almost solved it, so that the published text of this lecture [8] appeared at about the same time as my SU(3) paper...[9]

I told Salam about my one-year fellowship, and he suggested a "compact" problem, namely the acquisition of mass by gauge vector-mesons. On the other hand, I had become interested in symmetries and in classification, though I had not yet understood the relationship between the two issues. However, I was coming up with a sequence of new ideas – except that when I would present them to Salam, he would say, "Oh, this is a model which was suggested by Schwinger in 1957 – see etc.." or "Tiomno already tried your new idea etc.". He was becoming impatient while I was gaining confidence, seeing that I was reproducing stuff that some of the best know theorists had come up with and I felt it showed I was still capable of doing serious work.. Before leaving London for the summer, Salam made a last attempt to get me to work on that vector-meson mass problem (it was then taken up and solved by Peter Higgs..) and when I stuck to my symmetry and classification issue, he gave in but said "you are embarking on a highly speculative search; however, if you insist, go ahead and do it, but do it seriously. Don't be satisfied with the little group theory I taught you, which is what I know. Learn it in depth!". He then referred me to "a Russian mathematician named Dynkin", about whom he had heard from Hamermesh – who had mentioned that Dynkin had discovered some subgroups that even Racah had not noticed.."

Philosophy, when applied a priori, may become dogma

We have just seen how the Greek idea of pythagoreanisms had to wait for more than twenty-five centuries before materializing – and in a physical context which could in no way have been guessed in 500 BC – namely modern quantum mechanics, to be eventually understood in terms of the Uncertainty Principle! The same is true of several other guesses in Greek science which were philosophically conceived but could not be realized before certain domains of experience had become accessible. Example: John Dalton's revival and vindication of *atomism*, originally suggested by Democritos of Abdera, had to wait until after the separation of a number of chemical elements had been achieved during the second half of the XVIIIth century (cobalt by Brandt in 1742, nickel by Cronstedt in 1751, manganese by Gahn in 1774, oxygen by Priestley in 1772, molybdenum by Hjelm in 1782, tellurium by Muller, also in 1782, etc.) plus Lavoisier's work on combustion (1772), the separation of the various gases in air (Cavendish, 1781) etc.. Plato's view of a future *geometrization* of physics is yet another example – as it had to wait for Maxwell's equations and for the results of the Michelson-Morley experiment before Einstein's June 1905 paper on the Special Theory of Relativity and Minkowski's subsequent (1908) geometrical interpretation of that paper – plus the General Relativity sequel (1915) which exploited the geometrical interpretion.

The apparent "obviousness" of this comment should not be overstated. In the effort of finding the hadron's *"order"* [10], I have listed four different approaches, two of which were falling precisely into this trap, except that their philosophical ideas were not inspired by Greek science… The *mechanistic* approach of L. de Broglie et al. [11] assumed that all *internal degrees of freedom* are mechanically generated, i.e. that the SU(2) isospin proton-neutron symmetry (with electric charge included) is an *angular momentum* by its nature. There was no experimental data supporting such a view, it was entirely based on *dialectical materialism*. Making this assumption *a priori* was thus not very different from a hypothetical situation in which *Dalton would have assumed Democritos' four elements theory,* rather than applying Democritos' idea as a guideline in analyzing the experimental evidence. Moreover, it is still possible and perhaps even *probable that isospin be an angular momentum after all* – except that this will happen in Kaluza-Klein reality [12], with more dimensions than four (e.g. a total of 11 *bosonic* ones in M-theory [13]), with the isospin angular momentum pointing in a direction within the extra dimensions, which would thus make it impossible to add up spin and isospin.. Materialism would then be vindicated as a philosophy – yet without being forced to concretize its message precisely at the level attained in 1960 in the peeling off of the onion of matter.

The same goes for the *structuralists* [10,14,15]. Their work was also derived from the same gospel of *dialectical materialism* [16]. Sakata and his group were active in constructing a new model, but in their view, *this simply implied the selection of some of the existing particles as being more elementary than the rest*. With philosophy taken as dogma, the treatment seemed to assume features generally encountered only in religious fundamentalism or in messianic movements – namely that this being *revelation*, it applies to the situation at this precise instant in time. Moreover, let us even admit the possibility that they might have been right – nature selecting the layer of the known particles for some deep reason. The dogmatic aspect, however, was so strong among some of the leadership that they wouldn't desist even after the experimental invalidation of their model and the validation of another – thus foregoing the possibility of regarding the experimentally validated model as the expected Messiah (i.e. the hard fundamental little brick of the materialist philosophy). Now to the details.

After the discovery of the pions, Fermi and Yang [14] had suggested that these might be composite particles, with the *proton* and *neutron* set and their antiparticles as fundamental $[(p,n)$ and $(\overline{p,n})]$ with $\pi^+ : p\overline{n}$ etc.,.. After the discovery of strange particles, this was extended by S. Sakata [15] and others to the set p,n,λ which has the correct minimum mix – and the above dialectical notion of having concrete material elementary bodies as building bricks [16].

As to the philosophical point itself, note that, as I explain in the sequel, two years prior to the Ω^- experimental validation [17,18] of the octet model of SU(3) symmetry, I conceived the possibility of making the baryons out of three triplets [19] (the future *quarks* [20]). This meant that, the defeat of its dogmatic application notwithstanding, *materialism had in fact scored again* – by supplying us – in the next layer of matter – with building bricks belonging to the fundamental (or defining) representation.. Again, use the philosophy for post-factum interpretation rather than concretize it and thereby impose a realization fitting a certain state of the phenomenology.

The group pest scores again: the octet model of Unitary Symmetry

The two other approaches used in 1954 – 1961 were not doctrinaire and relied on exploration and mathematical experimentation, with physical experiments as the final test. The difference between the two classes was a matter of one's mathematical toolkit: most particle physicists were familiar with SO(3) and with its double-covering $Spin(3) := \overline{SO}(3) = SU(2)$ which happens to coincide with the unitary-unimodular group in 2 complex dimensions. This should perhaps be rated as an advance, when compared with the state of affairs in the thirties. In the preface to the 2nd edition of his book, *"Group Theory and Quantum Mechanics"*, Hermann Weyl tells us that most physicists were hoping that "the Group Pest" would soon go away – but Weyl warns the physics community that some groups, at least, are here to stay, namely *SO*(3) and the Lorentz group *SL*(2C) and he suggests they learn to live with what they regard as a "pest"... The community must have responded, and particle physicists were now familiar with rotations.. The class of models I have called the *rotational* sequence consists of studies which applied the rotation groups *SO*(*n*) and their double-coverings $\overline{SO}(n) := Spin(n)$ with ever larger dimensionalities, $n = 3$ (Heisenberg's isospin), $n = 4$ (Salam-Polkinghorne), $n = 7$ (Schwinger's and Gell-Mann's "global symmetry", as shown by J. Tiomno), $n = 8, 9$ (Salam and Ward). There could be two reasons for sticking to the orthogonal groups, one ideological (the mechanistic approach we discussed earlier) – and the second practical, i.e. this was the only set of groups with which the physicists were familiar, even those who had obeyed Weyl's injunction..

The more methodical approach, which I followed, after digesting Salam's advice (re the relevant sector in group theory) "learn it in depth", was to start by learning the generalities on group theory and identify that part which was indeed relevant to my search. Clearly, in view of their role in Emmy Noethers' theorems, these were the Lie Groups – and I learned both from the Physics behind Noether's theorems (and even more from their Quantum Field Theory extension by Schwinger) and from the mathematics of Dynkin's treatment – to first focus on the generator algebras. Elie

Cartan had classified all finite dimensional semi-simple Lie algebras in his 1894 thesis [21], and E.B. Dynkin's thesis had added a new diagrammatic methodology [22].

I spent the summer of 1960 tracking down the correct Dynkin paper. The articles mentioned by Hamermesh (which I ordered from the USA, once I had identified the relevant journal) were too advanced and assumed the reader to be familiar with earlier work of Dynkin's, his thesis – which was what I needed. Once I knew what I was after, the question arose whether or not there was a copy in the UK. I was in luck as I finally discovered in the British Museum Library a heliographed copy of the English translation of Dynkin's thesis, published by the USA Office of Naval Research. I enjoyed my reading – the entire treatment of Lie algebras is geometrical and on top of all that, Dynkin's method is entirely geometrical. I have always liked geometry and my intuition is geometrical – which was one reason I had taught myself General Relativity. The particle physics courses in 1958-1960 were very non-geometrical and I now felt like in a reunion with old friends. In a recent volume honoring Dynkin, I have related in some detail my own application of the method, in selecting the global symmetry group of the hadrons and the classification provided by its unitary representations. I have also discussed Dynkin's diagrammatic method and its extensions. Those I described [23] answered new needs after 1974 on the part of particle physicists, namely for Kac-Moody algebras and for Lie Superalgebras.

Returning to my quest in 1960, I realized that what I was after was a Lie Algebra of *rank* $r = 2$. I studied all five candidates A_2, B_2, C_2, D_2, G_2 and selected first A_2, the algebra of *traceless* 3 × 3 matrices, and the mysterious *"exceptional"* G_2 with its Star of David root diagram.. (the existence of the exceptionals is related to the existence of *octonions* as a *non-associative* Hurwitz algebra or number field). As to B_2, C_2, D_2 these three are rotation algebras, respectively in $d = 5^+, 4^-, 4^+$ dimensions, with the upper index denoting *metrics*, respectively symmetrical (e.g. generating the group $SO(5)$), antisymmetrical, i.e. generating the *symplectic* group $Symp(4) = \overline{SO}(5) = Spin(5)$ - and symmetrical again ($SO(4)$). These models had been tried and had generally failed some tests. I finally settled on A_2, dropping G_2 for two main reasons:

(a) its basic representation is 7-dimensional and thus one would have to assign the Λ hyperon to a separate singlet representation, while all the other hyperons make that 7, a separation which seemed to have no experimental justification,

(b) weak currents would include a $(\Delta Q / \Delta S) = -1$ pair, where Q is *electric charge* and S is *strangeness*. Again, there was and still is no support from experiment for such currents being present.

A_2 is the Lie algebra of $SU(3)$. I learned from Salam when I showed him my proposed model, upon his return to London in the late Fall of 1960, that this same *group* had just been postulated by the Sakata group at Nagoya University as the symmetry of the hadrons – but with a basic difference as to the representation selected for the spin $j = ½$ baryons, namely an octet **8** in my model, as against a triplet (p,n,Λ) in the Sakata model – plus another representation to which the Σ triplet was assigned with other missing states; however, there was no room there for a $j = ½\ \Xi$ doublet, which was indeed assigned to a spin $j = 3/2$ representation in that model.

The octet model passed every test I gave it. The only suspicious feature was that *it was not the fundamental* or basic *representation,* it was a higher tensor under $SU(3)$. Did it mean that the proton and neutron were *not the basic constituents* of hadron matter, as we had come to believe? Gradually, I came to adopt this interpretation. The *anomalous magnetic moments* of these particles would not fit a simple Dirac spinor and thus strengthened my doubts with respect to the nucleons' elementarity. At the 1962 International Conference on High Energy Physics at CERN, I had a discussion about this point with David Speiser. He and J. Tarski had launched a similar search through Cartan's classification sometime later in 1961 [24], had also noted the octet in A_2 (and notified Lee and Yang about its existence) but they were unhappy about the implication of non-elementarity of the nucleon and opted for G_2 at the end. The G_2 group was also strongly supported by R. Behrends and A. Sirlin [25], who joined the search, still in 1961. Behrends also participated in a comprehensive repeat effort [26], reviewing the entire algebraic methodology.

I have related elsewhere [4] the sequence of events starting with Salam's return to London in November 1960, after a four months absence ending with the Rochester conference. First, my presenting him with my draft *Derivation of Strong Interactions from a Gauge Invariance* and his suggestion of a joint paper in which he would add a contribution, namely the idea of applying $SU(3)$ as a local gauge symmetry in the Sakata model, with the same resulting vector-meson octet as in my model; then, my naïve rewriting of my draft in terms of the Nagoya group matrices, using the preprint Salam had brought back from Rochester, and giving him that draft; next, waiting for more than a month for him to add that point about gauging $SU(3)$ in the Sakata model and finally asking him about the draft and learning from his answer that he had changed his mind and would publish his point elsewhere, and that I could go ahead and publish my paper as is; submitting it to *Nuclear Physics* ("received 13 February 1961") [9] after having it retyped by my former secretary at the Embassy; Salam greeting me one morning, brandishing a Caltech preprint (CTSL-20) *The Eightfold Way* by M. Gell-Mann [27] and announcing "Gell-Mann has arrived at the same model as you!"; then, my manuscript being returned by the editor (Leon Rosenfeld) with an angry letter to Salam about his student submitting a manuscript without double-spacing it and making life miserable for the journal staff; my resubmission of a properly typed version with two notes added, one about Salam's original point and intention and the other about Gell-Mann's preprint; Salam writing to Gell-Mann and sending him my paper; having decided in May 1960 to wait and see if I was really still capable of doing serious research in theoretical physics, meanwhile withholding until then the submission of my Ph.D. registration forms to London University, I now sent them in; letter from London University's Board of Physics (the result of a meeting where Salam was supposed to present my case, but could not attend..) suggesting I change my Ph.D. program to *Electrical Engineering,* fitting my Technion background; arrival of new edition of Gell-Mann's preprint with added comment acknowledging my contribution; this reprint used by Salam to successfully settle the

London University issue; Gell-Mann submitting his paper for publication in the *Physical Review* ("received March 27, 1961"); false crisis at La Jolla conference (attended by Salam and Gell-Mann) in June 1961, created by wrong preliminary experimental measurement of Σ / Λ relative parity, announcing it as odd and thereby contradicting the octet model; Gell-Mann reacting by withdrawal of manuscript and resubmission (September 20, 1961) to the *Physical Review* of a new and different manuscript [28], now "evenly" presenting two candidate $SU(3)$ models, namely the Sakata **3** and (our) **8**, with the Caltech *Eightfold Way* preprint thus remaining unpublished [27]. Within that same year, Gell-Mann also published with S.L. Glashow (then a Caltech student) [29] a full-fledged algebraic assay of the contents of Cartan's classification of Lie algebras, ending up as a testimonial in favor of the octet model of $SU(3)$ unitary symmetry.

Quarks

I returned to Israel in July 1961, but was back in London in March 1962 to hold my "defence" – the delay corresponding to the required minimum after (my late) registration, as set by London University rules. Meanwhile in Israel, I had been joined by Haim Goldberg, a fresh Racah-tutored Ph.D., whom I acquainted with $SU(3)$ and the two models (I had already similarly briefed Penelope Ionides and Munir Rashid in London, at Salam's request). With Goldberg, we explored the possibility of starting from a basic field[1] in the fundamental representation **3**, making the baryon states in **8** as a higher tensor, a multi-quanta state. We settled on making a baryon as a 3-quanta state, thus being led to adjoin baryon number B and work with $U(3)$ instead of $SU(3)$, assigning $B = 1/3$ to the defining **3**. The fundamental entity would have to carry *fractional electric charges* and it is instructive to read H.J. Lipkin's eyewitness account on p. 180 of his *"Quarks for Pedestrians"* [30]:
"... Goldberg and Ne'eman then pointed that the octet model was consistent with a composite model constructed from a basic triplet with the same isospin and strangeness quantum numbers as the Sakaton, but with baryon number $B = 1/3$. The baryon octet was constructed from three triplets. However, eqns. (1.8) and (1.9) show that particles having third-integral baryon number must also have third-integral electric charge and hypercharge. At that time, the eightfold way was considered to be rather far-fetched and probably wrong. Any suggestion that unitary symmetry was

---

[1] The concept of a relativistic quantized field was undergoing a total devaluation at that time, as a result of the sweeping attacks by Berkeley's G. Chew, following the "news" that the *theory was not unitary off mass shell*. Arriving at Caltech in the fall of 1963, and presenting a model using a Lagrangian, I was asked by Gell-Mann "what is that?" Note that he was soon to find a useful replacement, in the form of *current algebra;* note also that R.P. Feynman did fully restore off-mass-shell unitarity in 1961- 62 by introducing the ghost fields – and yet Chew continued to deny the physical existence of fields at least till the completion in 1971 of 't Hooft's program of renormalization of the Yang-Mills interaction.

based on the existence of particles with third-integral quantum numbers would not have been considered seriously. Thus, the Goldberg-Ne'eman paper presented the triplet as a mathematical device for the construction of the representations in which the particles were classified..."

Our structural paper was sent for publication in *Nuovo Cimento* ("received 22 February 1962") [19] and disseminated as an Israel AEC preprint (in Israel, I had become the Scientific Director of the IAEC's Soreq Research Establishment). More than one and a half years later, Gell-Mann arrived at the same triplet model [20], named the basic entities "quarks", though presenting them now as a *physical* model – though as yet unclear about whether or not these *fractional-charge-quarks* exist as free states (the alternative would then resemble my "*mathematical device*" presentation); another independent introduction of the quarks roughly at that time is due to G. Zweig, who took the view according to which free quarks should be observed [31].

The *quark model* received a major boost in 1964 in a series of algebraic recombinations of *internal su*(3) with *external transformations* (i.e. kinematical, acting on a spacetime background) realized as a tensor-product extension of *SU*(3) charges with the algebra of Dirac gammas, yielding the algebraic kernel of current density components and their space-integrals

$$L^{a\mu} = (1/2)[1(+/-)\gamma_5][1(+/-)\beta] \times \Psi(x)^+ \sigma^\mu \lambda^a \Psi(x)$$

This was initiated by Gürsey and Radicati, [32] who used the *static* spin-unitary spin algebra to derive, for instance the ratio between the magnetic moments of the proton and neutron, a result due to the above and A. Pais [33], namely $\frac{\mu_p}{\mu_n} = -\frac{3}{2}$ clearly a Pythagoreanism. Similarly, using the quark-plus-antiquark canonical $[su(3) \times su(3)]^\beta$ yields the ratio between total cross-sections for the scattering of nucleons versus pions over the same hadron target H (e.g. a nucleon) $r := \sigma(N,H) : \sigma(\pi,H) \to 3/2$ the Levin-Frankfurt [34] relation. Both ratios fit nicely with the measured quantities. Meanwhile, the groups now applying *SU*(3) in Israel and in London had found an experimental test distinguishing between the Sakata and Octet models [35] in proton-antiproton annihilation into two mesons and the Sakata model was in clear contradiction with experiment, while the octet passed the test.. Hundreds of other predictions were being derived and tested [36] including strong-interactions intensity-rules, electromagnetic and weak-interactions hadronic matrix elements (such as magnetic moments or weak transitions), mass formulae (including electromagnetic mass differences). Finally, there was the prediction of the Ω^- hyperon, including the values of its mass and spin by both Gell-Mann and myself at the 1962 CERN Conference (see the two texts by G. Goldhaber in ref. [17]) and its discovery at Brookhaven Lab., early in 1964 [18], which gave the concensus'endorsement seal to the octet. Five years later, it was the SLAC deep-inelastic electron-nucleon scattering result which directly confirmed the presence of (fractionally charged) quarks inside the nucleon, helped by a method suggested and developed by R.P. Feynman and J.D. Bjorken. Three of our experimentalist colleagues were rewarded in 1990 with the Nobel Prize in Physics – namely J. Friedman, H. Kendall and R. Taylor.

Constraints on higher energy regions

Two constraints gradually emerged from the confirmation of (flavor) $SU(3)$ symmetry and of the quark model, namely,
(1) the baryons in $SU(6)$ are observed to be assigned in the **56** *totally-symmetric* representation whose Young tableau is given by a horizontal 3-boxes array. This is incompatible with Fermi-Dirac quantum statistics and can only be so if there is yet *another internal charge* on the quarks. To have it neutralized on hadrons, it should be another ("*color*") $SU(3)$. To fit with the experimenatally quasi-established *confinement* of this charge, this color-$SU(3)$ has to be a Yang-Mills gauge group with *Asymptotic Freedom (Quantum Chromodynamics, =* "*QCD*").
(2) It was argued that a strong interaction cannot have a *broken symmetry*. I answered this argument with an alternative statement of the situation, namely that the *Strong Interaction (= QCD)* is *exactly* flavor-$SU(3)$ invariant, with the *symmetry-breaking* corresponding to a perturbative interaction, somewhere at higher energies. Later formulations assumed that this "*Fifth Interaction*" enters in QCD as different incremental contributions to the *bare quark masses,* $[su(3)_{color}, su(3)_{flavor}] = 0$.

Envoi

I was named *Yuval* by music-loving parents, after the patriarch in *Genesis* 4, 24 "*.Jubal,...father of all such as handle the harp and organ*" (see the illutrations on p. 96 of ref. [1] for "Iubal"'s sponsorship of Pythagoras' work, a medieval superposition of two traditions. The spelling Jubal originates in the Vulgate: (1) Latin "J" was pronounced then as in the present use of "y", and (2) the diacritical mark (*daguesh*) distinguishing in Hebrew between b and v was systematically disregarded (other example: Ya'akov, rendered by the Vulgate as Jacob). My parents may have been disappointed by my passive approach to music (I enjoy it, but avoided the piano lessons they assigned me to). It occurred to me recently that they might have felt somewhat happier, had they been aware of my having contributed to the *Music of the Spheres...* I thus dedicate this explanation to them.

REFERENCES

1. Helm, E.E., *Scientific American*, Dec. 1967 issue, p.93-103. For the *envoi* in the present article, see in particular the illustration on page 96.
2. Balmer, J.J., *Verhandlungen der Naturforschenden Gesellshaft in Basel* **7**, 548-560 (1885).
3. Rydberg, J.R., *Kungliga Vetenskaps Akademiens Handlinger* **23**, #11, 1-155 (1890).
4. Gell-Mann, M. and Ne'eman, Y., eds., *The Eightfold Way*, W.A. Benjamin, Inc., pub. New York (1964),317 pp. Republished (2000) in extended version by Perseus Press, Cambridge, Mass., Advanced Books Classics series, 388 pp.
5. Ne'eman Y., "Seventy Years of Israeli Physics", Proc. Israel Nat. Acad. Sci. and Hum.
6. Ne'eman, Y., "Copernican Humility, Chance and the Creation of Purpose", in Proc. 7th Marcel Grossman Conference (Stanford 1994), R.T. Jaantzen, E. MacKeiser and R. Ruffini, eds., Part B, (Science and Society Public Talks), World Scientific Pub., Singapore (1996) p. 1677-1716.
7. See e.g. F. Forsyth, *The Odessa File,* CORGI books, London (1973) p. 222; G. Pape and T. Aspler, *The Scorpion Sanction,* Bantam books, New York (1981) p. 115.
8. Ne'eman, Y., "The Israeli Army and the Sinai Campaign", *Norsk Militaert Tidsskrift* **12**, 795-814 (1960).
9. Ne'eman, Y., *Nucl. Phys.* **26,** 222-228 (1961) [received 13 Feb 61].
10. Ne'eman, Y., "The Three-Quark Picture", in *The Particle Century*, edited by Gordon Fraser IOP pub., Bristol and Philadelphia, 1998, pp. 34-45.
11. De Broglie, L., Bohm, D., Hillion, P., Halbwachs, F., Takabayasi, T. and Vigier, J.P., *Phys. Rev.* **129**, 438, 451 (1963).
12. Kaluza, T., *Sitz. Preu. Akad. Wiss.* 966 (1921); Klein, O., *Nature* **118**, 516 (1926).
13. Bergshoeff, E. Sezgin E. and Townsend, P.K., *Phys. Letters* **B189**, 75 (1987); Townsend, P.K., *Phys. Letters* **B350**, 184 (1995).
14. Fermi, E., and Yang, C.N., *Phys. Rev.* **76**, 1739 (1949). Goldhaber, M., *Phys. Rev.* **101**, 433 (1956).
15. Sakata, S., *Progr. Theoret. Phys.* (Kyoto) **16**, 686 (1956); see also Okun, L.B., *ICHEP, CERN,* 1958, p. 223.
16. See for example, Y. Fujimoto's address to the 1970 Lenin Symposium *Presentation of S. Sakata's theory of Elementary Particles and Philosophy.*
17. Gell-Mann, M., *Proc. Inter. Conf. High Energy Physics*, CERN, 1962, p. 805. Goldhaber, G., "The Encounter in the Bus", in *From SU(3) to Gravity*, edited by Gotsman, E., and Tauber, G., Cambridge Univ. Press, Cambridge, 1985, pp. 103-106. Cahn, R.N. and Goldhaber, G., *The Experimental Foundations of Particle Physics*, Cambridge Univ. Press, Cambridge, 1988, p. 116.
18. Barnes, V.E., et al. *Phys. Rev. Letters* **12**, 204 (1964).
19. Goldberg, H. and Ne'eman, Y., *Nuovo Cimento* **27**, 1-5 (1963) [received 22 Feb. 62].
20. Gell-Mann, M., *Phys. Letters* **8**, 214-215 (1963).
21. Cartan, E., *Sur la structure des groupes de Transformations Finis et Continus* Vuibert, Paris (1933) [1894 thesis].
22. Dynkin, E.B., *Uspekhi Mat. Nauk* **2** N. 4 (1947); *Am. Math. Soc. Transl.* N. 17 (1950). *Am. Math. Soc. Transl.,* Series 2, 6; 111-224 and 245-368 (1957).
23. Ne'eman, Y., in *Selected Papers of E.B. Dynkin with Commentary*, edited by Yushkevich, A.A., et al., American Math. Soc., Providence, R.I., 2000, pp. 373-382.
24. Speiser, D.R.. and Tarski, J., "Possible Schemes for Global Symmetry", The Institute for Advanced Study, Princeton, New Jersey (1961).
25. Behrends, R.E. and Sirlin, A., *Phys. Rev.* **121**, 324 (1961).
26. Behrends, R.E., Dreitlein, J., Fronsdal, C. and Lee, B.W., *Rev. Mod. Phys.* **34**, 1 (1962).
27. Gell-Mann, M., Caltech preprint CTSL-20, unpub.
28. Gell-Mann, M., *Phys. Rev.* **125**, 1067-1084 (1962) [received Sept. 61].
29. Gell-Mann, M. and Glashow, S.L., *Ann. of Phys. (NY)* **15**, 437 (1961).
30. Lipkin, H.J., *Phys Reports* **8**, 173-268 (1973). See p. 180.
31. Zweig, G., CERN reports (unpublished) TH401, 402 (1964).

32. Gürsey, F. and Radicati, L.A., *Phys. Rev. Lett.* **13**, 173 (1964).
33. Gürsey, F., Pais, A. and Radicati, L.A., *Phys. Rev. Lett.* **13**, 299-301 (1964).
34. Levin, E.M. and Frankfurt, L.L., *Zhur. Eksp. I Teor. Fiz., Ptzma v. Red.* **2**, 105 (1965).
35. Dothan, Y., Goldberg, H., Harari, H. and Ne'eman, Y., *Phys. Lett.* **1**, 308-310 (1962).

BARYONS, QG PLASMA

Open questions in baryon spectroscopy

Volker Credé

Institut für Strahlen- und Kernphysik, University of Bonn, Germany

Abstract. The investigation of the excitation spectrum of nucleons provides important information on many open questions in baryon spectroscopy. The key to any progress is the identification of the effective degrees of freedom leading to a qualitative understanding of strong QCD. The problem of *missing resonances* predicted by quark models and the nature of the nucleon resonance $S_{11}(1535)$ is discussed on the basis of recent experimental results of the CB-ELSA experiment at the e^- accelerator ELSA in Bonn. Among other things, the data show clear structures due to high-mass resonance production. Successive decays via $\Delta(1232)\pi$ are observed. The study of symmetries among negative-parity baryons allows predictions of Δ^* decay properties and, therefore, will shed some light on the rather unknown Δ spectrum.

At present, QCD does not account for quark confinement. However, flavor symmetry seems to play a striking role in the confinement of three quarks. This is well observed for the first (u,d) quark family and predictions are also made for the second family (s). Open questions remain as to what the role of the gluon is, for instance.

INTRODUCTION

Quantum chromodynamics (QCD) is almost without doubt the correct theory of strong interaction. However, the QCD Lagrangian cannot be solved yet in the low-energy regime and for bound states, i.e. to provide the theory of quark confinement. In order to describe and to predict the properties of hadronic states, quark models have been developed instead, which are amazingly successful. It is important to complete the predicted spectra and to classify observed particles as well as to understand their properties within the quark model. Any serious discrepancy between the model and the experimental findings may contribute to elucidate the current uncertainties on QCD.

All constituent quark models can roughly be classified in three different categories:

- **Non-relativistic quark models**: Among others, Gloszman and Riska have developed a model treating quarks as q^3 bound states coupling to mesonic fields. These meson-exchange models are more generally also called Goldstone-boson exchange models. The authors use the exchange of an octet of pseudoscalar mesons [1]. The quarks are assumed to be Pauli spinors, thus obeying the Pauli principle. Consequently, bound states can suitably expanded in a basis of non-relativistic wave functions.

- *Relativised* **quark models**: Another group of models still follows a non-relativistic Ansatz but uses the relativistic terms of the energy for the quarks and, furthermore, applies momentum-dependant relativistic corrections to the confining and short-distance potentials [2].

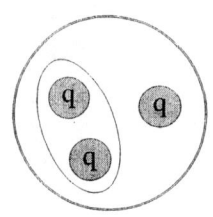

FIGURE 1. Possible quark-diquark structure of the baryon
If one of the internal degree of freedom was frozen, the consequence would be a smaller number of expected baryon resonances.

- **Fully Relativistic quark models**: A relativistically covariant description treats the quarks as full Dirac spinors [3]. The residual interaction is based on instanton-induced forces.

All current models describing the spectrum of baryon resonances predict a series of hitherto unobserved states. This feature would clearly be a big problem for all models as they would have failed to describe physical reality. There are at least two possible explanations which account for this open question. One is that these *missing resonances* have to exist if the 3 valence quarks of the nucleon are not frozen into a quark-diquark substructure. In fact, this would reduce the number of effective degrees of freedom and, therefore, also the number of possible baryon states [4] (Fig. 1).

An alternative explanation is that those *missing resonances* are really missing and have simply not been observed up to now because almost all existing data result from πN elastic-scattering experiments. As a matter of fact, some models focussing on baryon strong decays predict baryon states missing in πN elastic-scattering analyses but showing strong evidence in electromagnetic production [5]. These models create a $q\bar{q}$ pair with vacuum quantum numbers, i.e. quark and antiquark necessarily have relative $L = 1$ and $S = 1$ combined to $J^{PC} = 0^{++}$, and are generally referred to as the $^3P_0$ model [6]. Those *missing resonances* should couple strongly to channels like $\Delta\pi$, for instance [7]. Thus, photoproduction experiments offer a large discovery potential.

Photoproduction of two pseudoscalar mesons off the proton

The observation of baryon resonances in their sequential decay into $p\pi\pi$ is, therefore, particularly suited in photoproduction of $\pi^0\pi^0$ since the latter is not dominated by diffractive scattering of ρ mesons produced by conversion of the incoming photon in the field of the proton. This process dominates the $p\pi^+\pi^-$ channel, especially at high energies. Another dominant contribution in the reaction involving charged pions is the direct production of $\Delta^{++}\pi^-$ (Kroll-Rudermann term). This effect is also excluded in the $p\pi^0\pi^0$ channel.

The study of the reaction $\gamma p \rightarrow p\pi^0\eta$ also allows one to carry out a general search for baryon resonances. These can be Δ excitations decaying into $\Delta(1232)\eta$, for instance. In the following, a deeper interest in the study of this reaction is motivated.

The $N(1535)S_{11}$ has a strong decay mode into $N\eta$ with a branching ratio of (30-55) % while the $N(1650)S_{11}$ hardly decays to $N\eta$. The question arises why this is the case. In Table 1, the negative-parity resonances form isospin groups and are arranged according to their spins. In the quark model, the $N(1535)S_{11}$ and the $N(1520)D_{13}$ are

TABLE 1. Negative-parity baryons and decays into ground state + η
For N^* and Λ^*, a spin flip is required for the decay of interest from states with $s = \frac{3}{2}$. Therefore, the decay is suppressed. See text for details.

$s = \frac{3}{2}$	$N(1650)S_{11}$		$N(1700)D_{13}$		$N(1675)D_{15}$
$s = \frac{1}{2}$	$N(1535)S_{11}$	$\to N\eta$	$N(1520)D_{13}$		

$s = \frac{3}{2}$	$\Lambda(1800)S_{01}$		$\Lambda(????)D_{03}$		$\Lambda(1830)D_{05}$
$s = \frac{1}{2}$	$\Lambda(1670)S_{01}$	$\to \Lambda\eta$	$\Lambda(1690)D_{03}$		

$s = \frac{3}{2}$	$\Sigma(1750)S_{11}$	$\to \Sigma\eta$	$\Sigma(????)D_{13}$		$\Sigma(1775)D_{15}$
$s = \frac{1}{2}$	$\Sigma(1620)S_{11}$		$\Sigma(1670)D_{13}$		

$s = \frac{3}{2}$	$\Delta(1900)S_{31}$		$\Delta(1940)D_{33}$	$\to \Delta\eta$	$\Delta(1930)D_{35}$
$s = \frac{1}{2}$	$\Delta(1620)S_{31}$		$\Delta(1700)D_{33}$		

approximately mass degenerate and form a doublet of states with $s = \frac{1}{2}$. However, the quarks may also couple to $s = \frac{3}{2}$ resulting in three states corresponding to the observed resonances $N(1650)S_{11}$, $N(1700)D_{13}$ and $N(1675)D_{15}$. Different contradictory arguments try to explain the strong coupling of the $N(1535)S_{11}$ to the $N\eta$ decay mode, some even arguing that no genuine three-quark resonance is necessary at all [8, 9]. This conjecture is supported by an amplitude analysis of η production data in which no pole is needed for the $N(1535)S_{11}$ [10]. As a matter of fact, the question as to what the true nature of the resonance is, continues to remain. The systematics of η decays of negative-parity baryons may help to discriminate the proposed models (Tab. 1). The same striking decay pattern as for the nucleons can be found for the Λ resonances where the $\Lambda(1670)$ decays strongly into $\Lambda\eta$ whereas almost no η decay mode can be observed from the state $\Lambda(1800)$. In the case of the Σ baryons, the situation is reversed. The lower-mass state $\Sigma(1620)$ does not decay into $\Sigma\eta$, however, the decay is also forbidden by phase space. Furthermore, the isospin wave functions of Λ and Σ have different symmetries which could be the origin of the observed pattern. In the case of N and Λ resonances, the internal total quark spin of the states at lower mass is dominantly $s = \frac{1}{2}$ whereas of the higher-mass states it is $s = \frac{3}{2}$. This means that for the latter decaying into the ground state, a quark spin flip is required which suppresses the decay. Since no predictions for decays of Δ^* resonances into $\Delta(1232)\eta$ exist, one would expect the $\Delta(1940)D_{33}$ to decay dominantly into $\Delta\eta$ on the basis of this naive picture.

All existing models have failed to describe a group of negative-parity states in the Δ spectrum around 1900 MeV. However, it has to be pointed out that these states cannot be considered well known. In fact, the investigation of these resonances and their decay properties will shed some light on the rather unknown Δ spectrum.

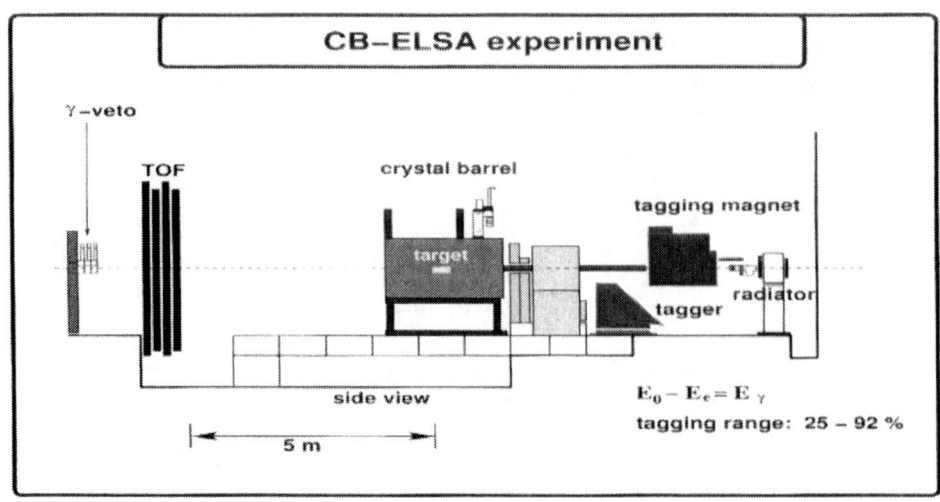

FIGURE 2. Configuration of the CB-ELSA detector for a first series of measurements

THE CRYSTAL-BARREL EXPERIMENT AT ELSA

The ELSA accelerator complex in Bonn provides electron beams up to energies of 3.5 GeV. A LINAC preaccelerates the particles which are then injected into an electron synchrotron. The latter provides electrons up to 1.6 GeV which are finally transferred to the stretcher ring ELSA [11].

The experimental setup of the detector

Electrons extracted from ELSA hit a primary radiation target and produced Bremsstrahlung. The corresponding energy of the photons ($E_\gamma = E_0 - E_{e^-}$) was determined in a tagging system by the deflection of the electrons in a magnetic field. This detector provided a tagged beam in the photon energy range from 25 % up to 95 % of the incoming electron energy. The setup of the CB-ELSA detector used for a first series of experiments is shown in Fig. 2. The calorimeter (Crystal Barrel) consisting of 1380 CsI(Tl) crystals covering about 98 % of 4π solid angle is an ideal detector for photons. The photoproduction target in the center of the Crystal Barrel has a length of 5 cm and was filled with liquid hydrogen. It is surrounded by a scintillating fibre detector which was built to detect and trigger charged particles leaving the target. In addition, it provides an intersection point of a particle's trajectory with the detector and hence helps to identify clusters of charged particles in the barrel. Due to the *in-flight* character of the experiment, the general conception is to combine the calorimeter with suitable forward detectors. In the start configuration, the system was extended by Time-Of-Flight walls of the ELAN experiment previously carried out at ELSA in Bonn.

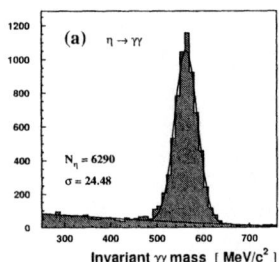 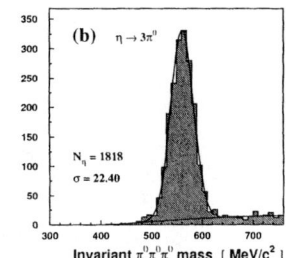 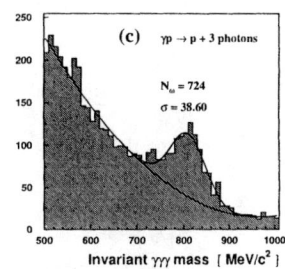

FIGURE 3. Invariant 2γ mass (a), 6γ mass (b) and 3γ mass (c): *beam energy: $E_{e^-} = 2.6\ GeV$*

The latter form together with the tagging system and the inner detector the first-level trigger of the experiment. The second level then consists of a fast cluster encoder which is able to determine the number of clusters in the barrel.

FIRST PRELIMINARY RESULTS OF SPECIFIC REACTIONS

Data has been taken since December 2000 with the whole apparatus fully operating. Measurements at three different ELSA energies have been performed: $E_0 = 1400$, 2600 and 3200 MeV. In the following, first results on the reactions $\gamma p \to p\pi^0\pi^0$ and $\gamma p \to p\pi^0\eta$ are presented. The whole data sample for these reactions comprises about 60 % of the $E_0 = 3200$ MeV data resulting in 150 000 and 20 000 events, respectively. It has to be pointed out that all distributions are neither efficiency corrected nor any flux normalisation has been carried out. However, the reconstruction efficiency is almost flat in $\cos\theta$ (of the proton in the center-of-mass system) and energy. Furthermore, no final tracking has been performed, therefore, improvements can be expected.

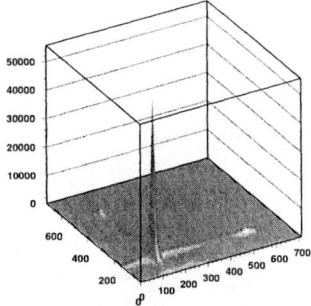 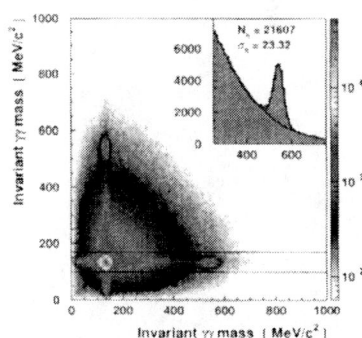

FIGURE 4. proton + 4γ events: invariant $\gamma\gamma$ mass versus $\gamma\gamma$ mass *(all combinations are plotted)*

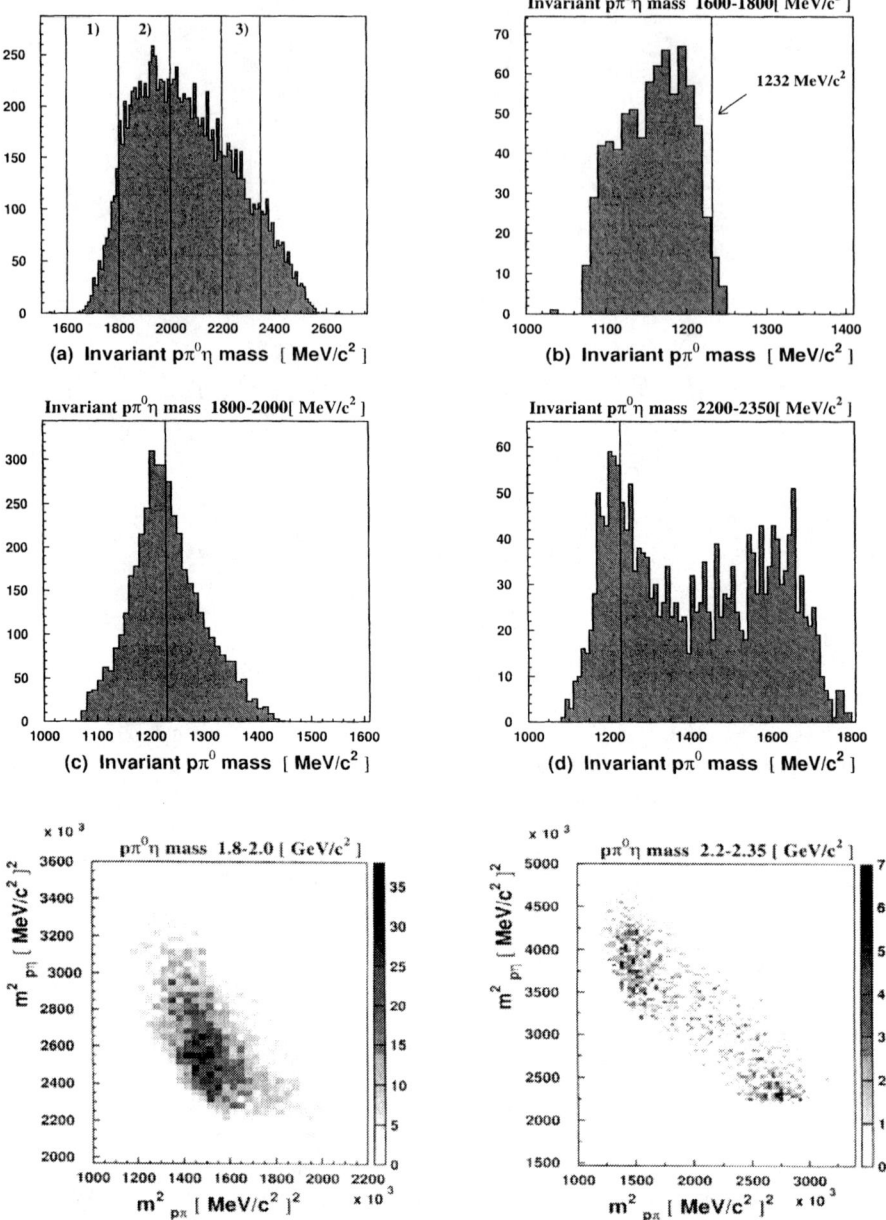

FIGURE 5. Different plots on the reaction $\gamma p \to p\pi^0\eta$. (a) shows the total invariant $p\pi^0\eta$ mass. In (b)-(d), the $p\pi^0$ mass is plotted for the three different $p\pi^0\eta$ mass regions indicated in (a). Clear evidence for the $\Delta(1232)$ can be observed and, thus, hints for resonances decaying via $\Delta\eta$ become obvious. (e) and (f) show Dalitz plots for two different $p\pi^0\eta$ mass regions. Resonance structures become even more transparent.

The good quality of the data can be seen in Fig. 3 (a) and (b) showing the invariant γγ mass as well as the invariant $3\pi^0$ mass, respectively. A clear η signal is visible due to the two neutral decay modes of the reaction γp → pη above a very small background. No additional constraints have been applied to the events but the request for the right multiplicity identifying the proton with the help of the scintillating fibre detector. Fig. 3 (c) shows the invariant 3γ mass. An ω signal can clearly be seen.

Events with four photons in the final state have to be investigated for the reactions of interest. Fig. 4 shows γp → pγγγγ events. In the left picture, clear evidence for $\pi^0\pi^0$ events can be observed. However, a small enhancement in the region of $\pi^0\eta$ events is already visible. The latter become more obvious in the right picture using a logarithmic scale in the third direction. In addition, a horizontal cut is applied and the corresponding slice projected onto the x axis (small inset).

$$\gamma p \rightarrow p\pi^0\eta$$

Fig. 5 (a) shows the total invariant mass for the $p\pi^0\eta$ final state. No structures are visible at first sight. However, if one looks at the invariant $p\pi^0$ mass for different total mass regions, hints for baryon resonances decaying into Δη become visible.

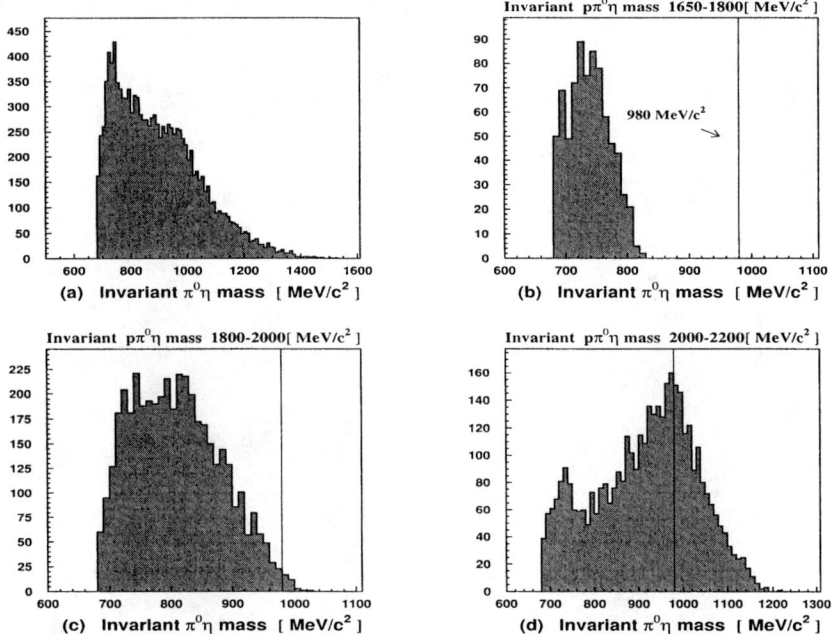

FIGURE 6. Different plots on $\pi^0\eta$ mass distributions of $\gamma p \rightarrow p\pi^0\eta$. *(a) shows the $\pi^0\eta$ mass for all events. A bump at 980 MeV can be seen. In (b)-(d), the $\pi^0\eta$ mass is plotted for the same $p\pi^0\eta$ mass regions as indicated in Fig. 5(a). The $a_0(980)$ threshold is at $E_\gamma = 1449$ MeV.*

Fig. 5 (b), (c) and (d) correspond to the three different mass regions which are indicated in (a). In the total mass region around 1700 MeV, no structure can be seen. However, a clear peak at the Δ mass can already be observed in the mass region around 1900 MeV. This might be a first indication for the $\Delta(1940)D_{33}$ decaying into $\Delta\eta$. For higher $p\pi^0$ masses, further resonance intensity may be hidden in a structure around 1600 MeV. In fact, one has to be careful interpreting structures in the mass projections as those are often reflections of the corresponding Dalitz plots (Fig. 5 (e) and (f)).

Fig. 6 (a) shows the invariant $\pi^0\eta$ mass originating from mass projections of the diagonal bands of the Dalitz plot onto the $\pi^0\eta$ axis. The pictures (b), (c) and (d) represent the same three total mass regions as indicated in the $p\pi^0\eta$ mass spectrum. Possibly, a signal for the $a_0(980)$ can be observed above threshold (d). In any case, further investigation may help to shed some light on the structure of this meson which appears in some theoretical scenarios to be a $K\overline{K}$ molecule whereas it is also discussed as a normal quark-antiquark state.

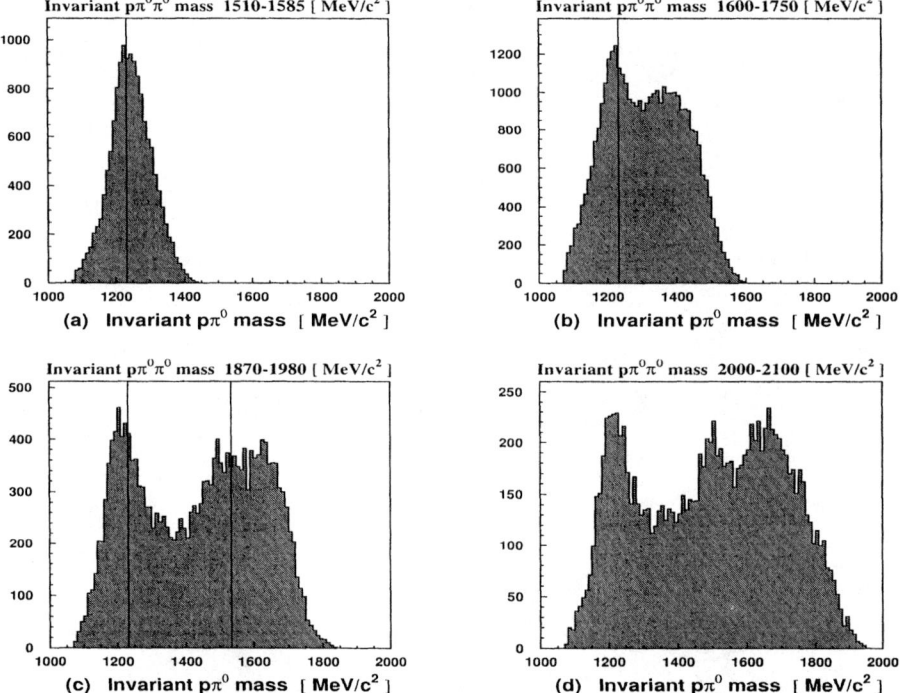

FIGURE 7. Different plots on the reaction $\gamma p \to p\pi^0\pi^0$. *(a)-(d) show the $p\pi^0$ mass for different $p\pi^0\pi^0$ mass regions of the overall distribution (Fig. 8). A clear peak for the $\Delta(1232)$ is observed indicating baryon resonances decaying via $\Delta\pi^0$. However, at higher energies even more structures become visible. These are promising hints for missing resonances.*

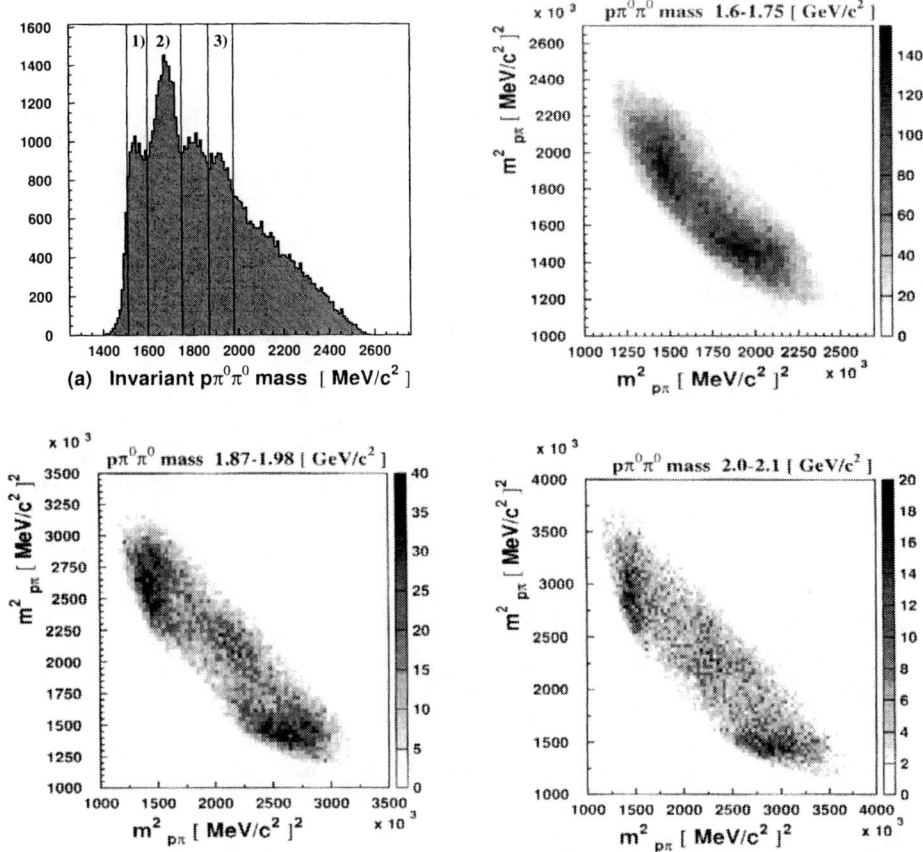

FIGURE 8. Total invariant $p\pi^0\pi^0$ mass (a) and Dalitz plots for the indicated mass regions (b)-(d).

$$\gamma p \to p\pi^0\pi^0$$

In comparison to the reaction discussed above, structures are already visible in the total invariant $p\pi^0\pi^0$ mass spectrum (Fig. 8). A dominant signal around 1500 MeV is suppressed in the 3.2 GeV data set because it lies below the threshold that can be reached with these data. However, it can clearly be seen in data stemming from incoming electrons with $E_0 = 2.6$ GeV. Following the same idea as before, different mass regions are indicated. The corresponding $p\pi^0$ mass spectra are given in Fig. 7 and 8. In the mass region below 1600 MeV only a single signal for the Δ can be observed. Going higher in mass, an additional signal around 1500 MeV and further structures above 1600 MeV become visible. In fact, the structure around 1500 MeV is very likely to be the $D_{13}(1520)$.

SUMMARY AND CONCLUSION

The preliminary results of the first series of measurements with the Crystal-Barrel experiment at ELSA show promising hints that the data will shed some light on the miracle of the *missing resonances*. The good quality of the data in various channels is convincing and resonance structures even at higher masses are visible.

In the $p\pi^0\pi^0$ channel, clear evidence is given for resonance structures decaying via $\Delta(1232)\pi^0$ at different masses between 1500 MeV and 2100 MeV. In addition, the $p\pi^0$ mass distributions (Dalitz plots) give promising hints that decays via $D_{13}(1520)$ and also via higher-mass baryon states (m > 1600 MeV) are observed.

In the $p\pi^0\eta$ channel, decays of baryon resonances via $\Delta(1232)\eta$ with different masses are obvious. Furthermore, decays via $S_{11}(1535)\pi^0$ are seen and even $a_0(980)$ production may have been observed.

Further effort is necessary in order to obtain final results on baryon resonances and their properties, i.e. to improve the reconstruction of data and to perform partial wave analyses. However, the observed baryonic cascades $B^{**} \rightarrow B^*\pi^0 \rightarrow p\pi^0\pi^0$ as well as $B^{**} \rightarrow B^*\pi^0(\eta) \rightarrow p\pi^0\eta$ have never been seen before at this precise level and will reveal a series of new physics results.

REFERENCES

1. L.Y. Gloszman and D.O. Riska: Phys. Rept. **268** (1996) 263
2. S. Capstick and N. Isgur: Phys. Rev. **D34** (1986) 2809
3. U. Löhring, K. Kretzschmar, B.Ch. Metsch and H.R. Petry: EPJ **A10** (2001) 309
4. D.B. Lichtenberg: Phys. Rev. **178** (1969) 2197
5. E.S. Ackleh, T. Barnes and E.S. Swanson: Phys. Rev. **D54** (1996) 6811
6. A.Le Yaouanc, L. Oliver, O. Pene and J.C. Raynal: Phys. Rev. **D18** (1978) 1591
7. S. Capstick and W. Roberts: Phys. Rev. **D49** (1994) 4570; Phys. Rev. **D57** (1998) 4301; Phys. Rev. **D58** (1998) 074011
8. N. Isgur and G. Karl: Phys. Rev. **B72** (1977) 109
9. N. Kaiser, P.B. Siegel and W. Weise: Phys. Lett. **B362** (1995) 23
10. G. Höhler: PiN Newslett. **14** (1998) 168
11. D. Husmann and W.J. Schwille: Phys. Bl. **44** (1988) 40

A Review of Baryon Resonance Analysis and Comparisons with Meson Resonance Analyses

Steven A. Dytman

Department of Physics, University of Pittsburgh, Pittsburgh, PA 15260

Abstract. It is interesting that the fields of meson and baryon spectroscopy have taken divergent paths in the last decade. It is time the fields were more aware of the similarities and this author hopes the fields will benefit from each other's accomplishments. This note will point out the similarities and the differences. Sometimes, the differences are in nomenclature. However, there are also practical reasons for the divergences. The typical meson experiment has a more complicated final state than baryon experiments and the most complete theoretical treatments limit the number of final state particles to three. There are also important differences in sociology.

INTRODUCTION

The field of hadronic structure has become a field that bridges medium energy and high energy physics. The study of the many excited states that nature has given us has become a place where the properties of long-range QCD can be studied, an idea that has become much more exciting with the rapid progress of lattice gauge theory (LGT). There is also the exciting possibility of seeking hybrid mesons and glueballs in the meson sector. Their confirmation or definite denial is equally interesting as an entirely new way to learn about QCD. The study of hadrons was one of the two major subjects presented at a workshop on the future of medium energy physics in 2000 at Sanderling [1]. (The other was structure functions. It is interesting to note that these two subjects have almost no overlap the way they are presently practiced.) To study hadronic structure *many* experiments are required because of the variety of final states possible. There are presently large-scale experiments studying baryons (CLAS at Jlab, GRAAL in Europe, Crystal Barrel at Bonn and Crystal Ball at BNL are represented at this conference) along with a number of smaller experiments. For mesons, the Crystal Barrel experiment has been moved from LEAR to Bonn with a change in emphasis from mesons to baryons. Their data from LEAR is still providing a very quantitative view of meson structure. For the future, a new large experiment (called Hall D at Jlab [2]) is now nearing a formal startup in construction and an important part of the CLEO-c [3] program will be devoted to meson studies. To emphasize the breadth of these collaborations, we note each are healthy mixtures of high energy and medium energy experimentalists. The key to getting physics output from those experiments is a coherent, global model with minimal model dependence. There are a number of models in use, each of which has advantages and shortcomings.

HISTORY

The study of baryons was a significant high energy physics subject in the late 1960's and 1970's. It benefited from the work of numerous luminaries [4, 5] who developed ways to deal with complex data with clever use of algorithms and quantum mechanics. It should be noted that this was before the quark model was recognized as the correct picture of hadronic structure, so the role of quantum mechanics and fundamental principles was extremely important. In developing 'new' methods to extract baryon spectroscopic information from data, this author has benefited greatly from works of that period. Perhaps the ultimate work [5] presented a method to model data relying solely on fundamental properties of unitarity, analyticity, time invariance, and crossing symmetry. Other books [6] take a somewhat more empirical approach. Some of these ideas are of historical interest, but there is much to offer for modern works also. The idea of the pole structure on the Riemann sheets is a way to answer important issues for both mesons and baryons. The major advantage of baryon spectroscopy is the dominance of the s-channel mechanisms where an excited baryon is the sole particle in the intermediate state. Thus, the quantum numbers of the initial and final state match those of the intermediate state. States of strangeness 0 (N^* and Δ^*) are reached through πN and γN and strangeness -1 ($K^- N$) can be easily reached.

The study of mesons was an offshoot of baryons when only fixed-target experiments were available. With at least one baryon in the final state, methods had to be developed to isolate meson excited states. These included central production (GAMS,WA102, and others). More recent πN experiments have generated good statistics at high enough energy that the baryon excitations could be suppressed and t-channel exchanges dominate [7, 8]. The success of these recent experiments is due to a solid understanding of acceptance and careful partial wave analysis (PWA). Additional experiments have been run at e^+e^- colliders (BES, CLEO) and at the LEAR $\bar{p}$ ring where baryons must come in pairs and thus are easily suppressed. Again, understanding of acceptance and a quality PWA is required because the final state has a moderately large number of particles.

While the rate of discovery was large for baryons 20-30 years ago, advances in mesons have dominated the last decade. There is likely another switch underway as the new baryon experiments start to generate high quality data and the study of mesons becomes more specialized.

THE PRESENT SITUATION FOR BARYONS

Since the s channel dominates N^* production, strong resonances will be seen as peaks in the cross section in the appropriate angular momentum state. Baryon resonance production in the s channel is shown in Fig. 1.

There is particular interest in the photocoupling helicity amplitudes ($\gamma N \to N^*$ coupling, $A_{1/2}$, $A_{3/2}$, and $S_{1/2}$) for each resonance because they can be calculated using any quark model. By measuring the Q^2 dependence of these amplitudes, the variation as a function of spatial scale is provided. At large Q^2, the amplitude will agree with perturbative QCD predictions.

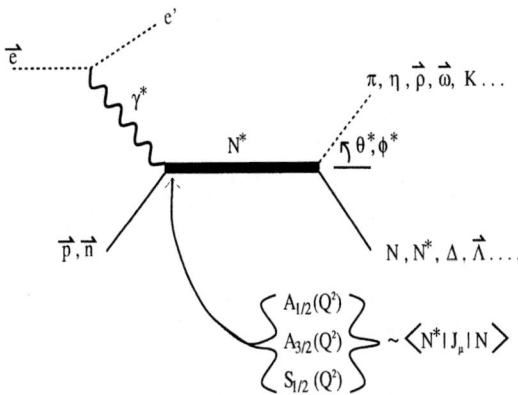

FIGURE 1. The s-channel diagram that dominates reactions such as γN and πN at $\sqrt{s} < 2$ GeV.

Historically, the best information came from πN experiments, a situation that still holds in 2001. However, a major effort is being made to improve the statistics of all relevant electromagnetic reactions by at least an order of magnitude. Jefferson Lab has a beam line that provides electron and photon beams of very low emittance up to 6 GeV coupled with a spectrometer (CLAS) with a single-particle acceptance of about 0.8 for charged particles and about 0.3 for photons. Thus, it can span the entire region of interest ($1.0 < \sqrt{s} < 2.2$ GeV. It runs with instantaneous luminosity of $\sim 10^{34} \text{cm}^{-2}\text{s}^{-1}$ and takes about 0.6 terabytes of data a day for roughly 8 months per year. A large variety of final states (15 so far from $\gamma p \to \pi^+ n$ to $\gamma p \to K^+ K^+ \Xi^-$) have been seen and 6 publications are already in print. It is nicely complemented by the Crystal Barrel experiment at Bonn which specializes in high efficiency detection of photons and can thus reconstruct one or more π^0 and η particles in the final state for photon beams in the same energy range. It has moderate ability to reconstruct charged particles. The Crystal Ball experiment is presently at BNL and has taken data at $\sqrt{s} < 1.6$ GeV for $\pi^- p$ and $\sqrt{s} < 1.8$ GeV for $K^- p$. This is predominantly a high efficiency photon detector with minimal ability to reconstruct charged particles, although upgrades are possible.

The first key point is that experimental capabilities are much improved. The CLAS collaboration has been taking data since early 1998 using mostly a liquid hydrogen target. A few billion events have been taken for $\vec{e}p$, γp, and $\vec{e}\vec{p}$ combinations. A recent run was the first with a linearly polarized photon beam. With a loose trigger, many final states are measured simultaneously for every beam-target combination. The kinematic range is greatly expanded beyond what was possible with old experiments and polarized beams and targets are now common. For example, cross sections for photoproduction of vector mesons were measured at low t where diffractive processes dominate until CLAS began. CLAS has now published papers on ϕ and ρ photoproduction [9] up to t of -5 $(\text{GeV}/c)^2$. The cross section becomes flat at high t, symptomatic of very different processes. An effective 2 gluon model has been successful in describing the data. Another example that directly involves baryons is ω photo- and electroproduction. Preliminary analysis shows relatively flat angular dependence close to threshold. Thus, that data can be analyzed for baryon resonances. The final example is polarized electroproduction with longitudinally polarized beam and target. These data will be used for measurements of the GDH sum rule for virtual photons and asymmetries for specific final states such as π^+ and π^0 have been extracted.

The second key point is that (as for modern meson measurements) with a careful understanding of acceptance and an accompanying PWA, stronger physics conclusions can be made. A correct analysis will also include nonresonant (usually t-channel) mechanisms. Two cases where the nonresonant processes are unimportant are π^0 [10] and η [11] electroproduction which are dominated by the $\Delta(1232)$ and $S_{11}(1535)$ resonances. Thus, simple analyses can give important insights into the underlying physics. The electromagnetic amplitudes for the $\Delta(1232)$ have been sought for many years. The data has often been interpreted in terms of the multipoles (see below). If dominance of the M_{1+} is assumed, the cross section can be fit in terms of 6 complex amplitudes, $|M_{1+}|^2$ and the interference between M_{1+} and the other s- and p-wave amplitudes. They quote values for $\frac{\Re(E_{1+}M_{1+}^*)}{|M_{1+}|^2}$ and $\frac{\Re(S_{1+}M_{1+}^*)}{|M_{1+}|^2}$ at values of $\sqrt{s}$ close to the peak of the resonance as the "E2/M1" and "C2/M1" ratios. There is no separation of resonant and nonresonant features because the nonresonant strength has been found to be close to zero at the resonance peak. Since the values are a few percent up to Q^2 of about 3 $(\text{GeV}/c)^2$, the perturbative QCD limit is far away. To get the full impact of the CLAS data, more complete analyses are required.

GLOBAL ANALYSES FOR BARYONS

The interesting feature of baryon analysis is that the cross section is dominated by production of 1 and 2 mesons in addition to the nucleon in the final state. It is then possible to have a *global* description of *all* baryon resonance data using a coupled channel model. The input data is of the form the T-matrix T_{ab} where a and b are specific initial and final states, e.g. πN or $\eta\Delta$.

$$T(J^\pi) = \begin{pmatrix} T_{\pi N \to \pi N} & T_{\eta N \to \pi N} & T_{\gamma N \to \pi N} & T_{\rho N \to \pi N} & T_{\sigma N \to \pi N} & T_{K\Lambda \to \pi N} \\ T_{\pi N \to \eta N} & T_{\eta N \to \eta N} & T_{\gamma N \to \eta N} & T_{\rho N \to \eta N} & T_{\sigma N \to \eta N} & T_{K\Lambda \to \eta N} \\ T_{\pi N \to \gamma N} & T_{\eta N \to \gamma N} & T_{\gamma N \to \gamma N} & T_{\rho N \to \gamma N} & T_{\sigma N \to \gamma N} & T_{K\Lambda \to \gamma N} \\ T_{\pi N \to \rho N} & T_{\eta N \to \rho N} & T_{\gamma N \to \rho N} & T_{\rho N \to \rho N} & T_{\sigma N \to \rho N} & T_{K\Lambda \to \rho N} \\ T_{\pi N \to \sigma N} & T_{\eta N \to \sigma N} & T_{\gamma N \to \sigma N} & T_{\rho N \to \sigma N} & T_{\sigma N \to \sigma N} & T_{K\Lambda \to \sigma N} \\ T_{\pi N \to K\Lambda} & T_{\eta N \to K\Lambda} & T_{\gamma N \to K\Lambda} & T_{\rho N \to K\Lambda} & T_{\sigma N \to K\Lambda} & T_{K\Lambda \to K\Lambda} \end{pmatrix} \quad (1)$$

A partial wave analysis of the reaction data [12, 13, 14] gives various elements of the matrix shown in Eqn 1 for πN and γN initial states as a function of s. The isobar model provides a simple way to analyze data by fitting the energy dependence to Breit-Wigner shapes. These models are hard to unitarize and cannot be analytic. There are 3 prominent

TABLE 1. Recent results for the detailed properties of $S_{11}(1535)$. * signifies a value that was fixed at the PDG preferred value

Fit ref.	Γ_{full} (MeV)	bf$_{\pi N}$ fraction	$A^p_{1/2}$ $\times 10^{-3}\sqrt{GeV}$	data used (channels)
VPI(96)	105	0.31	60 ± 15	$\pi N, \gamma N$
Drechsel(99)	80	0.40*	67	
Krushe(97)	212	0.45*	120 ± 20	$\gamma p \to \eta p$
Pitt-ANL (00)	126	0.34	92 ± 10	$\pi N, \eta N,$
Feuster (99-00)	151-215	~0.31	93-106	$\pi\pi N, \gamma N$
PDG (00)	100-250	0.35-0.55	90 ± 30	average

codes that analyze a large portion of this matrix using a unitary, time reversal invariant formalism [15, 16, 17]. Vrana, Lee, and this author have resurrected and extended the Carnegie-Mellon-Berkeley analysis [15] to fit all πN and γN data. Like the others, it fits the T_{ab} data with a representation of the resonant and nonresonant processes. Using a Dyson-Schwinger equation, a full description of rescattering in the intermediate state is taken into account. Eight asymptotic channels are presently fit, but the code is easily expanded to include any new data sets. Since this model is analytic, the appropriate Riemann sheet can be searched for the zeroes in the denominator of the T matrix. These are the pole positions of each state; these all correspond to bare poles above πN threshold. The energy dependence of the fit near the pole is then used to determine the Breit-Wigner masses and widths. Since each inelastic threshold is correctly treated, very few resonances have a true Breit-Wigner shape.

An extensive survey of the S_{11} partial wave has been made and will be submitted for publication shortly. This is of interest because the properties of the lowest state (~1535 MeV mass) in this partial wave have been in doubt in recent issues of RPP [18]. In addition, the constituent quark model has a robust prediction of 2 S_{11} states [19] and discovery of a third would be significant. Table 1 shows a sampling of recent results for $S_{11}(1535)$ compared with RPP. Results of a single Breit-Wigner fit using only the η photoproduction data gives a large width. On the other hand, isobar fits using only πN elastic and pion photoproduction data [14, 20] gives widths about half as big. The photocoupling amplitude ($A_{1/2}$) is also large in the former case and small in the latter. The resolution is hopefully obvious that one needs to do a global fit of all the data- the result is an intermediate value.

An extensive study of model dependence was also conducted. A controlled variation of the model, the number of inelastic channels, and the number of resonances was made. The range of values for successful fits matched the results in Table 1 and the estimated error on the coupled channel fits is then seen to be much smaller than that listed by PDG. In a number of cases, we find new values that are outside the PDG ranges. Using only data reported before 2000, we find very little evidence for a 3rd S_{11} state. The situation could change when the new GRAAL data becomes part of the database. It is clear that model dependence must be understood to provide the optimal hadronic spectral information. Modelers must join together to do controlled comparisons to find the best model features and the estimated errors associated with each model.

TRYING TO FIND THE COMMONALITY BETWEEN MESON AND BARYON ANALYSIS

Hopefully, the situation with baryons is now clear. A flood of data of vastly improved quality being analyzed at various labs. Fitting codes already exist that will do *global* analyses of the data. Although all the codes have model dependence, efforts are underway to define these errors. Although not discussed here, there are also efforts in progress to improve the models used. The situation with mesons is less certain to this author. Fitting codes have some of the features of the baryon models, but sometimes the names are different and sometimes further approximations are needed. Global analyses for meson production experiments are much harder to construct because the experiments have a much larger set of final states. For πN experiments, the beam energy used is high enough for the t-channel mechanisms to dominate. Thus, the number of mesons seen in the detectors is large. The E852 code [7] will be used as an example. It unfortunately has a trigger problem (unlike the VES experiment) so that each final state must be separately analyzed in an isobar analysis. The LEAR experiments are analyzed with T-matrix [21] or K-matrix [22] codes that include up to 5 final state mesons. These formalisms are global and unitary, but still have the uncertainty about final states with more than 5 mesons. Unless *all* final states are accounted for, the strength in the states left out will be *incorrectly* attributed to the states included.

The first issue is nomenclature of partial waves. In each case there are complex amplitudes that express the strength for a particular set of quantum numbers. While baryon people label their waves $S_{11}, P_{11}, D_{33}...$, meson people (E852 [7]) use S_0, P_+ and D_+ and talk about reflectivity of the exchanged particle. The baryon notation $(L_{2I,2J})$ comes from the s-channel πN production. The L value is that of πN even in a reaction where pions are never measured. I is the isospin and J is the total angular momentum of the state. The parity is implicit, $P = (-1)^{L+1}$ where the '1' comes from the negative intrinsic parity of the pion. In photoproduction of pseudoscalar mesons, the partial waves (called multipoles) are labelled $E_{0+}^{\frac{1}{2},p}$, $M_{1-}^{\frac{1}{2},n}$, and $M_{2-}^{\frac{3}{2}}$ which really mean S_{11}, P_{11}, and D_{33}. The nomenclature is $\mathcal{M}_{L\pm}^{I,t}$ where I is the isospin, L the πN orbital angular momentum just as for the πN experiments, and $\pm$ denotes the total angular momentum of $L \pm \frac{1}{2}$, i.e. the orbital angular momentum added to the nucleon spin. Because of the lack of definition for photon isospin, the meson-nucleon final state is used to label the isospin and there are separate independent amplitudes for targets (t) of proton (p) and neutron (n). Because the photon is a massless spin 1 particle, its two m projections are split into Electric (E) and Magnetic (M) multipoles $\mathcal{M}$. In an electroproduction experiment, the Scalar (S) multipole is added and the longitudinal multipole is eliminated through the choice of gauge. Thus, a $I = \frac{1}{2}$ partial wave can have up to 4 multipoles and an $I = \frac{3}{2}$ has 2 partial waves for photoproduction and 2 more for electroproduction. For photoproduction of vector mesons, more labels are required [23] and there is no common notation for the general $\pi\pi N$ final state.

The E852 PWA notation assumes a πN initial state and a final state of a nucleon and the meson under study. They use L_m where L labels the meson angular momentum and m labels both the magnetic substate of the meson and the naturality of the t-channel exchanged meson. S_0 means $J^{PC}m^\varepsilon = 0^{++}0^-$, P_- signifies $1^{-+}1^-$ (the exotic wave) and D_+ signifies $2^{++}1^-$ (the a_2) [7]. J is the total angular momentum of the meson, P and C its spatial parity and charge conjugation parity, and ε is the naturality of the exchanged meson (+ signifies exchange of 0^+, 1^-, 2^+....) The naturality is closely linked to the m state of the meson. They plot the magnitude of each partial wave and the phase difference between each pair of partial waves. Analysis of the $\rho\pi$ proposed exotic [8] uses a final state of 3 pions. It requires added labels for the isobar for the paired pions (*isobar*) and the orbital angular momentum (L) between the isobar and the 3rd pion, namely $J^{PC}[isobar]Lm^\varepsilon$.

The meson and baryon labels both have important historical background and are *nontrivial* for the nonexpert to use. The link between the E852 formalism and the baryon formalism is also nontrivial because a t-channel diagram contributes to all s-channel partial waves. Thus, each baryon partial wave would be a sum over many E852 partial waves. Still, the meson and baryon pictures are linked because the most basic picture of a resonance is a pole in the complex plane of the scattering amplitude.

Normally, the resonance is then extracted by doing fits to the partial wave amplitudes. The best treatment of a resonance is as a bare pole that is dressed by coupling to the asymptotic states, e.g. a bare $\Delta(1232)$ state is dressed by πN coupling and a bare f_0 is dressed by $\pi\pi$, $K\overline{K}$, and $\eta\eta$. This is valuable because the position of the bare pole is useful for characterizing the origin of the state. This is the method of our work [15], but not the other 2 methods presently used [16, 13]. For mesons, only the LEAR analysis of Bugg and collaborators [21] uses this picture. It is possible to transform K-matrix fits to a T matrix and search for poles, but the Riemann structure is uncertain. The E852 formalism uses simple Breit-Wigner energy dependence for their 'mass dependent' fits. This is comparable to the isobar fits for baryons [20]. Although isobar fits are very common, this structure cannot satisfy analyticity (i.e. allow searches for poles in the complex plane) and can only satisfy unitarity through ad hoc procedures.

For analysis of mesons, the reaction mechanism is parameterized in simple ways. For example, E852 assumes a t-channel exchange and the experimental trigger is constructed appropriately. For the $\overline{p}p$ reactions, the reaction mechanism is extremely complicated, probably too complicated to model microscopically. E852 assumes the nonresonant processes have *no* angle or energy dependence and don't interfere with the resonant processes, an extreme viewpoint. The VES analysis is more complete, this is one of the primary differences between the two analyses. On the other hand, production and decay are equally important for baryon resonances, as seen in Eqn. 1, and people spend significant time to include the correct resonant and nonresonant diagrams. The spin factors naturally create angular dependence. Because these diagrams interfere, they can create structure in the energy dependence. For photoproduction of pseudoscalar mesons, the number of diagrams to include is small and the strengths are well-understood. Other cases (especially two-pion production) are not as simple; although no nonresonant diagrams are presently included in these cases, they could be added in many cases. Defining values for these parameters will not be simple.

The analyses of $\overline{p}p$ [22, 21] and e^+e^- reactions are similar to baryon analyses because s-channel mechanisms are thought to be dominant. They also have the huge advantage of knowing the quantum numbers of the initial state. $\overline{p}p$ reactions come from atomic orbits that are dominated by $L=0$ for liquid targets and $L=1$ for gas targets. Here the reaction is not known and very likely extremely complicated. Thus, it is parameterized in a simple way similar to the πN case [7]. For e^+e^- reactions, the intermediate state is very often a virtual photon. As a result, $J^{PC} = 1^{--}$ states are predominantly formed, e.g. the Ψ states at $\sqrt{s} \sim 3-4$ GeV and Υ states at $\sqrt{s} \sim 10-11$ GeV. (Lower mass mesons can also be produced by $\gamma\gamma$ interactions. Although the production rates are much smaller, very useful data (e.g the π^0 form factor [24]) have been taken.) Thus, the identification of mesons is more straightforward than for the πN experiments. An important measurement for the new CLEO-c machine will be the radiative decays of the J/Ψ. This partial wave analysis will be very similar to $\bar{p}p$ reactions for the final state.

Decay vertices are common to both fields. They have appropriate spin and isospin operators and phenomenological factors for phase space and sometimes use form factors to account for the composite nature of the particles. All analyses use phase space factors that modify the isobar decay energy dependence. At core, all have factors of channel momenta raised to a power of the orbital angular momentum. Thus, the phase space opens at a rate that depends on the relative angular momentum. The simplest method is to use Blatt-Weisskopf barrier penetration factors. These factors come from decay alpha particles transmitted through a square well potential. Therefore, they are nonrelativistic and have an arbitrary distance scale taken to be about 1 Fm. A more correct method [15] uses relativistic phase space and a dispersion relation to make the entire calculation analytic. Baryon amplitudes commonly use form factors, but meson amplitudes seldom use them. Form factors are important for the Born diagrams used in photoproduction calculations because they assume point particles and the momentum factors in the numerator blow up at high energies. Since meson production amplitudes don't calculate diagrams, this is not an issue there. Almost all of these ad hoc methods introduce undesirable uncertainty in the methods and create doubt in the final result since these are energy dependent factors. Although the energy dependence in an individual diagram is smooth, these diagrams can interfere to produce sharp energy structures. Older formalisms [25] that use renormalization techniques avoid form factors and perhaps provides an avenue for progress.

Finally, there are nontrivial sociological considerations that are important. Baryon data comes in the form of cross sections and polarization observables. They are freely given to the GWU group [12, 14] for analysis and distribution. They have a public web site where the data and various PWA solutions can be seen and used. Historically, only 2-body final states have been covered in their analyses. The GWU group recently received NSF funding for a significant upgrade in their user support, including the handling of $\pi\pi N$ final states. On the other hand, meson production data has more complicated final states. To give one example, the $\omega\pi^0$ final state has 4 pions- 2 charged and 2 neutral. Thus, the data is stored as events and likelihood analyses are used predominantly. A detailed acceptance function is important and must be used in conjunction with the events. This information tends to be with the experimental groups and is never made public. (This trend is changing with high energy experiments. For example, D0 recently opened their database and the new GriPhyN global database for LHC experiments will be open.) As a result, the data is available only to the experimental groups that took it for some time; for example, only the E852 collaboration presents results from their data. This unfortunately inhibits the advance of the field.

CONCLUSIONS

There are many similarities and many differences between baryon and meson analyses. Each is providing important information about the microscopic properties of quark states. These include issues of discovery where new kinds of quark states (hybrid mesons and glueballs) are sought and the detailed spectroscopic information available in the production and decay properties of specific states. With very few exceptions, all of this important information comes from partial wave analyses. A global analysis is required to pull together the disparate reaction data composed of a variety of initial and final states. The analysis is often split into two pieces, the partial wave analysis (PWA) where the result is the strength of a reaction is split into the various quantum states and the resonance analysis where the results are a separation of resonant and nonresonant amplitudes and the spectrum of states reported by PDG. Examples of PWA include the GWU [12, 14]results and the E852 analysis [7, 8]. Examples of the latter include the baryon spectrum results of Ref. [15] which come from fits to the GWU PWA results and the E852 mass dependent fits [7, 8] to their PWA results.

The baryon community has a history of global analyses because the main reaction studied was a single reaction, πN elastic scattering. This focus is rapidly disappearing as modern labs produce new data, but the need for global analyses has increased. Since final states of a baryon and one or two mesons dominate the cross section, the sophisticated models incorporating analyticity, unitarity, and time invariance developed 20-30 years ago can be used with modern treatments and modern computing methods. A wealth of data is required to simultaneously fit all reactions. That data does not exist today, but numerous experiments are rapidly adding to the world's database. The GWU group has maintained the database for many years and has recently received funding to extend their work to $\pi\pi N$ final states. These more complicated final states hold the key to understanding of the spectrum where the so-called missing states are expected to be found. This field is seeing a rebirth of effort. While many new experiments will provide a flood of new data, the physics results will depend on advances to objectively handle the more complicated final states in the new data.

Mesons are produced in the t-channel in πN experiments and in the s-channel in $\bar{p}p$ and e^+e^- reactions. e^+e^- experiments are simpler to interpret than baryon experiments because the reaction mechanism is simpler. Meson production in πN and $\bar{p}p$ experiments have a very complicated and poorly understood reaction mechanism and a very complicated multi-particle final state. This hasn't been an impediment to progress so far. In both $\bar{p}p$ and e^+e^- reactions, the quantum numbers of the initial state are known while baryon and πN meson experiments have a wide range of initial state quantum numbers. While the former are at a lower energy and can cope with this complexity, the latter must use a simple parameterization of the initial state. The solid theoretical basis of baryons allows a much better treatment of background processes which are treated with simple parameterizations or ignored for mesons. The known ways to deal with final states with unitarity and analyticity work best with the smaller number of particles seen in baryon experiments. This advantage is offset by the specific quantum numbers sampled in the s-channel meson production experiments. Thus, every analysis can benefit from more objective ways to cope with complicated final states.

It is hopefully clear that coherent global analyses of global databases with minimal model dependence provide the most objective means to learn about the hadronic spectrum. Sociological issues are very important in that effort. The baryon community has benefitted greatly from mechanisms for global databases and analyses. It is hoped that the meson community can find ways to do this in the future. It is also hoped that the baryon community can do a better job with complicated final states. Perhaps, the two communities would benefit from more mutual discussion.

ACKNOWLEDGMENTS

The author has benefitted greatly from conversations with and the work of David Bugg, Suh-Urk Chung, Paul Eugenio, Curtis Meyer, and Eric Swanson.

REFERENCES

1. Simon Capstick, Steven Dytman, Roy Holt, Xiangdong Ji, Curtis Meyer, John Negele, Eric Swanson et al., "Key Issues in Hadronic Physics", LANL Report hep-ph/0012238.
2. Hall D coll., "Photoproduction of Unusual Mesons", http://dustbunny.physics.indiana.edu/HallD/.
3. CLEO-c coll., "CLEO-c and CESR-c: A New Frontier of Weak and Strong Interactions", report no. CLNS 01/1742, http://www.physics.purdue.edu/Snowmass2001_E2/
4. R.P. Feynman, *Photon-Hadron Interactions*, W.A. Benjamin, 1972.
5. G.F. Chew, *The analytic S matrix; a basis for nuclear democracy*, Benjamin, 1966.
6. A.O. Barut, *The Theory of the Scattering Matrix*, MacMillan, 1967.
7. S.U. Chung, et al., Phys. Rev. **D60**, 092001 (1999).
8. G.S. Adams, et al., Phys. Rev. Lett. **81**, 5760 (1998).
9. E. Anciant, et al.(CLAS Collaboration), Phys. Rev. Lett. **85**, 4682 (2000); M. Battaglieri, et al.(CLAS collaboration), Phys. Rev. Lett. **87**, 172002 (2001).
10. K. Joo, et al.(CLAS collaboration), submitted to Phys. Rev. Lett.
11. R.A. Thompson, et al.(CLAS collaboration), Phys. Rev. Lett. **86**, 1702 (2001).
12. Richard A. Arndt, Igor I. Strakovsky, Ron L. Workman, and Marcello M. Pavan, Phys. Rev. **C52**, 2120 (1995).
13. D.M. Manley, R.A. Arndt, Y. Goradia, and V.L. Teplitz, Phys. Rev. **D30**, 904 (1984).
14. R.A. Arndt, I.I. Strakovsky, and R.L. Workman, Phys. Rev. **C56** 577 (1997).
15. T.P. Vrana, S.A. Dytman and T.-S. H. Lee, Physics Reports **328**, 181 (2000).
16. T. Feuster and U. Mosel, Phys. Rev. **C58**, 457 (1998); **C59**, 461 (1999), A. Waluyo and C. Bennhold, private communication.
17. D.M. Manley and E.M. Saleski, Phys. Rev. **D45**, 4002 (1992), M. Manley, private communication.
18. D.E. Groom, et al.(Particle Physics Group), Review of Particle Physics, Eur. Phys. J. **C15**, 1 (2000).
19. S. Capstick and W. Roberts, Prog. Part. Nucl Phys. **45**, S241 (2000).
20. D. Drechsel, O. Hanstein, S.S. Kamalov, L. Tiator, Nucl. Phys. **A645**, 145 (1999).
21. A.V. Anisovich, V.A. Nikonov, A.V. Sarantsev, V.V. Sarantsev, C.A. Baker, C.J. Batty, D.V. Bugg, A. Hasan, C. Hodd, B.S. Zou, J. Kisiel, Nucl. Phys. **A662**, 319 (2000).
22. A. Abele et al., (The Crystal Barrel Collaboration), Nucl. Phys. **A679**, 563 (2001), Eur. Phys. J. C., (2001).
23. Cetin Savkli, Frank Tabakin, Shin Nan Yang, Phys. Rev. **C53**, 1132 (1996).
24. J. Gronberg et al., CLEO Coll., Phys. Rev. D **57**, 33 (1998).
25. U. Fano, Phys. Rev. **124**, 1866 (1961).

The New Crystal Ball Experimental Program

W.J. Briscoe* and The Crystal Ball Collaboration[†]

*Department of Physics and Center for Nuclear Studies The George Washington University,
Washington, DC 20052, USA
[†]Abilene Christian University, Argonne National Laboratory, Arizona State University,
Brookhaven National Laboratory, University of California at Los Angeles, University of Colorado,
George Washington University, Universität Karlsruhe, Kent State University, University of
Maryland, Petersburg Nuclear Physics Institute, University of Regina, Rudjer Boskovic Institute
and Valparaiso University

Abstract. The Crystal Ball Spectrometer is being used at the Brookhaven National Laboratory Alternating Gradient Synchrotron in a series of experiments that study final states of π^-p and K^-p induced reactions that result in all neutral particles. Data have been obtained on the decays of N^*, Δ, λ, and Σ resonances. Threshold η production has been studied in detail for both π^-p and K^-p. Sequential resonance decays have been studied by studying the $2\pi^o$ production mechanism both in the fundamental interaction and in nuclei. In addition, we have used the ηs produced near threshold to make precision measurements searching in particular for rare and forbidden η decays. The new limits on branching ratios provide stringent constraints on current theoretical models.

INTRODUCTION

The major goal of Nuclear Physics is to understand the strong interaction. The best candidate theory is Quantum Chromodynamics, QCD, which attempts to explain the strong interaction in terms of underlying quark and gluon degrees of freedom. The study of the structure of baryons and their excitations in terms of the elementary quark and gluon constituents is thus pivotal to our understanding of nuclear matter within QCD. Additionally, tests of flavor and other fundamental symmetries are important to understand the nature nuclear matter. Indeed theories such as chiral perturbation theory are becoming very popular and must be tested. Within these goals our motivations for the particular reactions of interest are described briefly in the following paragraphs. We shall expand on areas not covered by other presentation at this conference later in the text.

Physics Interests

The elusive charge-exchange process has been the weakest link in partial-wave and coupled-channel studies. The accurate data that we obtained in this momentum region will help in improving the determinations of the isospin-odd s-wave scattering length, the πNN coupling constant, and the π-N σ term. In addition, better charge-exchange data helps in evaluating the mass splitting of the Δ and the charge splitting of the P_{33}

resonance and may result in new values for the $P_{11}(1440)$ mass and width. The results of these measurements are presented elsewhere in these proceedings. [1]

The radiative decay of a resonance provides the ideal laboratory for testing theories of the strong interaction, gives us insight into the fundamental interactions between mesons and nucleons, and allows us to probe into the structure of the nucleon itself. In particular, these data are important in the study of the radiative decay of the neutral Roper resonance. They can be combined with recent JLab Hall B data for the reactions $\gamma p \to \pi^+ n$ and $\gamma p \to \pi^\circ p$ which study the mesonic decays of the charged Roper in the incident photon energy region from 400 to 700 MeV. In addition, comparison of our data to new JLab data taken on the inverse reaction $\gamma n \to \pi^- p$, using a deuteron target, tests extrapolation techniques for the deuteron correction and study medium effects within the deuteron.

Two pion production provides a means of studying sequential pion resonant decays; with neutral pions we study the $\pi\pi$ interaction in the absence of final-state Coulomb effects and owing to isospin considerations there is no contribution of ρ decay. The study of this process on the proton ($\pi^- p \to \pi^o \pi^o n$) is the subject of a recently completed Ph.D. thesis. [2] We have also published measurements on this process in the nuclear medium. [3]

Near threshold η-production measurements provides data useful in verifying models of η-meson production and are also necessary for extraction of the η-N scattering length. Precise η production data are necessary to resolve ambiguities in the resonance properties of the $S_{11}(1535)$ and in the η photoproduction helicity amplitudes. The results of these measurements are presented elsewhere in these proceedings. [4]

Using the Crystal Ball Detector, we have the ability of selecting pure isospin states. For example in the reactions $K^- p \to \eta \Lambda$ [5] and $K^- p \to \pi^o \Sigma^o$ we select a pure $I = 0$ Λ^* and in the reaction $K^- p \to \pi^o \Lambda$ we select a $I = 1$ Σ^*. Figure 1 shows the production cross section of the former reaction. A significant part of our program is geared toward the study of $K^- p$ reactions are described by Mark Manley in these proceedings [6] and in a recent publication. [7]

The large production cross section of tagged ηs allow us to search for breaking of fundamental symmetries (*e.g.* C and CP invariance), and test Chiral Perturbation Theory as well as other theoretical models. We have published an article on $\eta \to 4\pi^o$ [8] which presents a new upper limit for this branching ratio ($B \leq 6.9 \times 10^{-7}$) of the CP forbidden decay at the 90% confidence level. This value of B puts a 2% limit on CP in quark-family-conserving interactions.

Another article has been published (since my presentation at the conference) on the rare $\eta \to 3\pi^o$ [9] which presents our determination of the quadratic slope parameter α for that decay. The value obtained ($\alpha = -0.031 \pm 0.004$) disagrees significantly with current theory. Since this published material is now readily available, I will discuss our recent and yet unpublished work on the $\pi^o \gamma\gamma$ decay of the η.

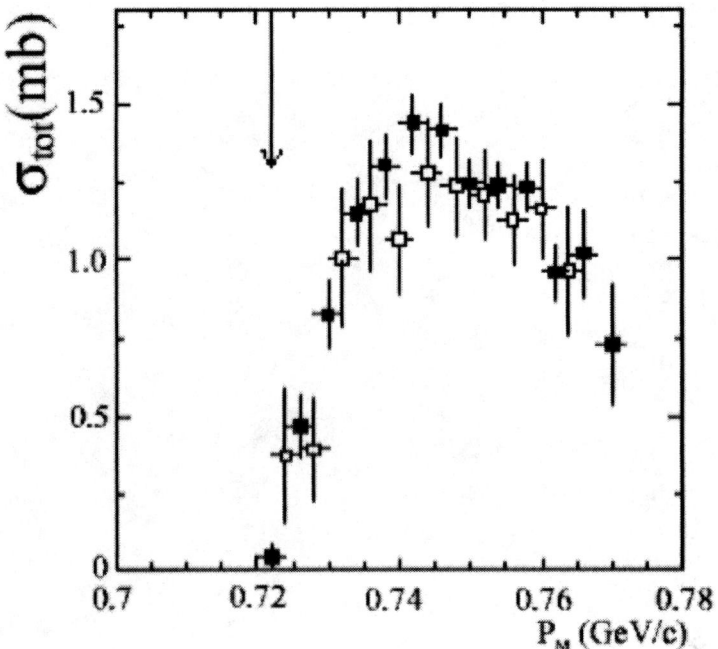

FIGURE 1. Total cross section for the reaction $K^-p \to \eta\Lambda$. Solid squares show cross section derived from $\eta \to \gamma\gamma$ and the open squares are derived from the $3\pi^o$ decay mode. The threshold is indicated by the arrow.

EXPERIMENTAL CONSIDERATIONS

The Crystal Ball

We used the SLAC Crystal Ball to make these measurements at the C6 line at the Brookhaven National Laboratory, BNL, Alternating Gradient Synchrotron, AGS, with incident beam momenta from 147 MeV/c to 760 MeV/c. Data are taken simultaneously on all reactions which helps ensure that background events are accurately subtracted. Data taking using the Crystal Ball began in July 1998 and continued until late November 1998.

The Crystal Ball is a segmented, electromagnetic calorimetric spectrometer, covering 94% of 4π steradians. It was built at SLAC and used for meson spectroscopy measurements there for three years. It was then used at DESY for five years of experiments and put in storage at SLAC from 1987 until 1996 when it was moved to BNL by our collaboration.

The Crystal Ball is constructed of 672 hygroscopic NaI crystals, hermetically sealed inside two mechanically separate stainless steel hemispheres. The crystals are viewed by photomultipliers, PMT. There is an entrance and exit tunnel (see Fig. 2) for the beam, LH$_2$ target plumbing, and veto counters.

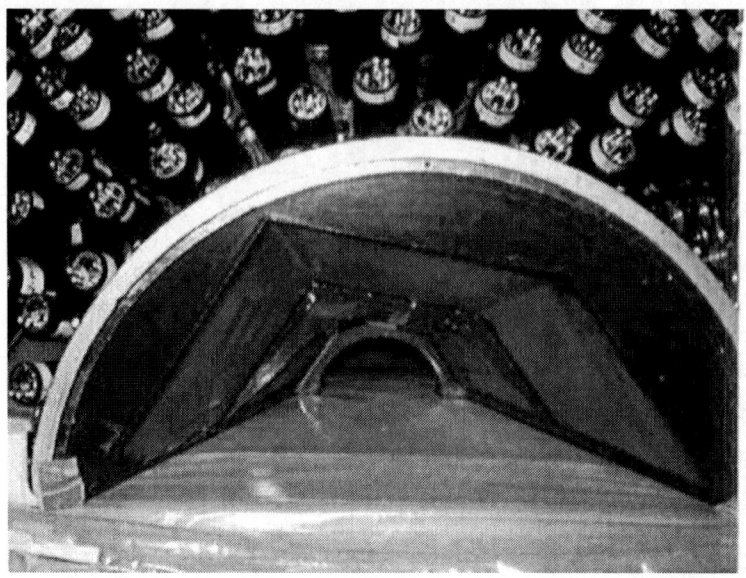

FIGURE 2. A view into the entry tunnel of the Crystal Ball. PMTs can be seen attached to individual crystals.

The crystal arrangement is based on the geometry of an icosahedron (20 triangular faces or "major-triangles" arranged to form a spherical shape). Each "major-triangle" is subdivided into four "minor-triangles", which in turn consist of nine individual crystals. Each crystal is shaped like a truncated triangular pyramid, points towards the interaction point, is optically isolated, and is viewed by a PMT which is separated from the crystal by a glass window. The beam pipe is surrounded by 4 scintillators covering 98% of the target tunnel (these scintillators form the veto-barrel).

This high degree of segmentation provides excellent resolution. Electromagnetic showers in the ball are measured with an energy resolution of $\sigma/E = 2.7\%/E[GeV]^{1/4}$. Shower directions are measured with a resolution in θ of $\sigma = 2°$–$3°$ for energies in the range 50–500 MeV; the resolution in ϕ is $2°/\sin\theta$. Typically, 98% of the deposited energy of each photon is contained in a cluster of thirteen crystals (a crystal with its twelve nearest neighbors). The thickness of the NaI amounts to nearly one hadron interaction length resulting in two-thirds of the charged pions interacting in the detector. The minimum ionization energy deposited is 197 MeV; the length of the counters corresponds to the stopping range of 233 MeV for $\mu^{\pm}$, 240 MeV for $\pi^{\pm}$, 341 MeV for $K^{\pm}$, and 425 MeV for protons. The preliminary energy calibration is performed using the 0.661 MeV γ's from a ^{137}Cs source. The final energy calibration is done using three reactions: i) $\pi^- p \to \gamma n$ at rest, yielding an isotropic, monochromatic γ flux of 129.4 MeV; ii) $\pi^- p \to \pi°n$ at rest, yielding a pair of photons in the energy range 54.3—80 MeV, almost back to back; and iii) $\pi^- p \to \eta n$ at threshold, yielding two photons, about 300 MeV each, in coincidence almost back to back. The PMT analog pulses are sent to ADCs for digitization. Analog sums of the signals from each minor-triangle are available for trigger purposes.

In addition to the expected high efficiency for photons, the Crystal Ball is also fairly responsive to neutrons. We were able to measure the response of the NaI(Tl) to neutrons by using the reaction $\pi^- p \to \pi^0 n$ and kinematics to determine the efficiency (as high as 40%) as a function of energy, see Fig. 3.[10]

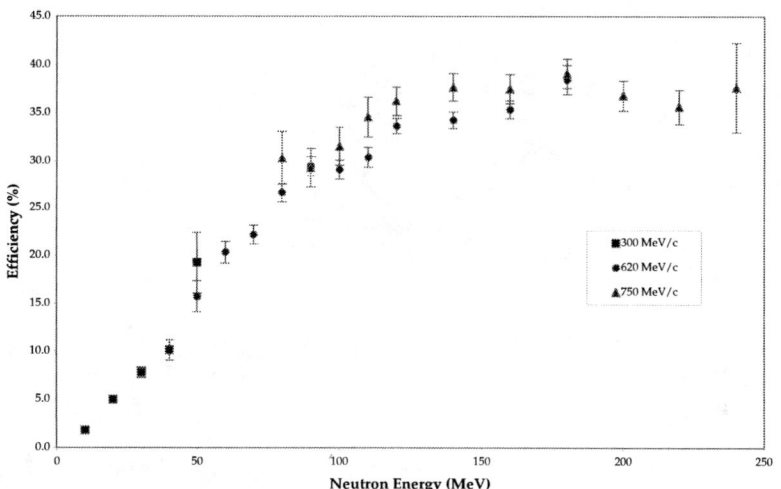

FIGURE 3. Efficiency of NaI(Tl) for neutron detection as a function of neutron energy.

Beam Line

Figure 4 shows the experimental setup at BNL on the C6 beam line. The final stages of the C6 beamline consist of four quadrupoles and a dipole that form a beam momentum spectrometer. Wire chambers are located on both sides of the dipole to track the particles through the spectrometer. The momentum resolution is 0.3%. The scintillators located up and down stream of the dipole provide TOF information and the coincidence trigger for the beam. Scintillators surround the LH2 target to provide a charged particle veto. Two columns of scintillator neutron counters are located downstream of the Crystal Ball. A beam veto scintillator is located further downstream. A concrete shield wall located upstream of the beam stop shields the Crystal Ball from low energy photons from the stop. A Cerenkov counter is located just after this wall to monitor electron contamination in the beam.

The usual trigger consists of: a beam coincidence trigger, no downstream beam veto, and a total energy-over-threshold signal from the Crystal Ball. The Crystal Ball trigger is normally a total energy trigger. A trigger based on the distribution of the energy in

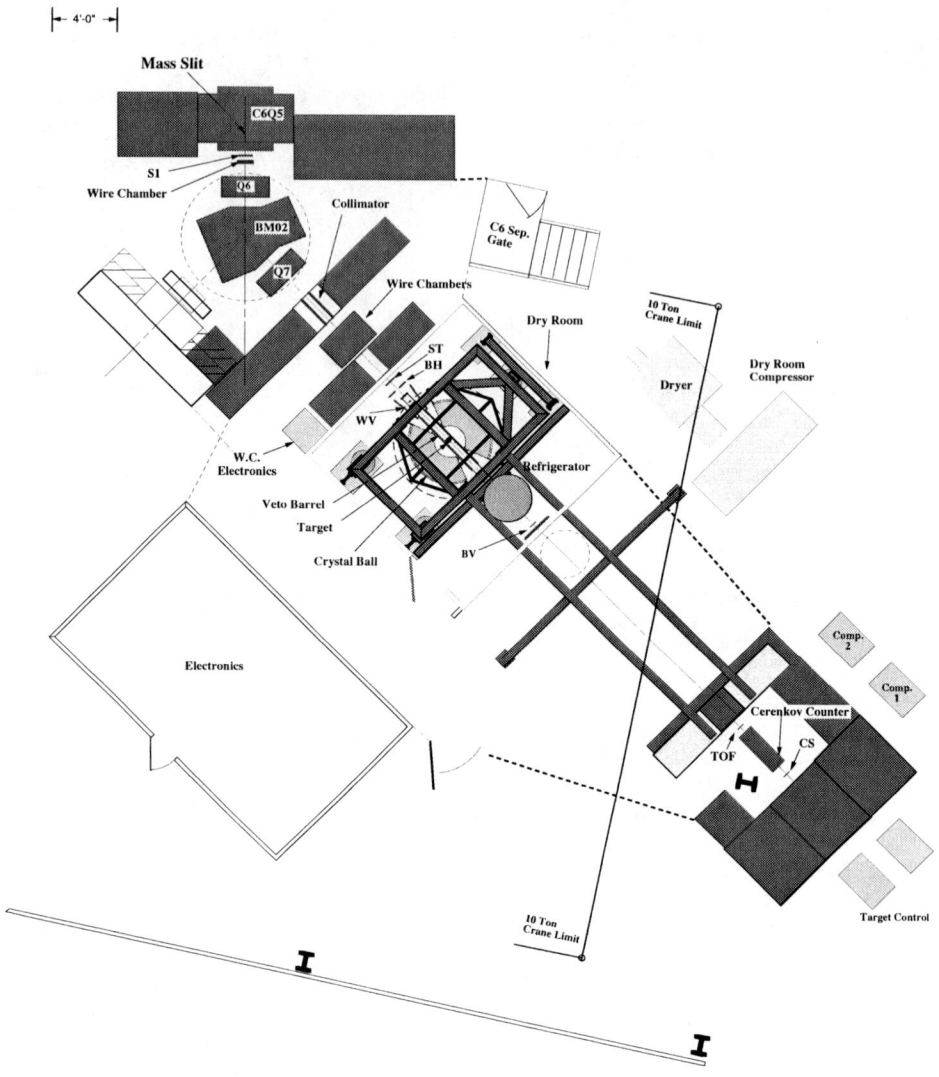

FIGURE 4. Experimental setup at BNL AGS C6 line.

different regions of the Crystal Ball was also used to provide a more restrictive trigger in certain cases.

RARE DECAYS OF THE η MESON

As we have alluded to above, by studying the rare and forbidden decay modes of the η meson, we are able to test the limits of such fundamental symmetries as C and CP invariance and G parity. Additionally, these also provide a laboratory in which we can test chiral perturbation theory and other recently proposed models. As mentioned above some of these measurements are already in the literature.[8] [9]

To measure the η-decay processes, we took a series of dedicated π^-p runs at 720MeV/c which was at the maximum in eta production and yet close enough to threshold that the ηs were essential going forward in the lab. In effect we produced an η beam.

For the decay $\pi^-p \to \eta n \to \pi^o \gamma \gamma n$ we looked at events in which 4 photons were detected. Even with this restriction, we still had backgrounds due to $3\pi^o$ decay and direct $2pi^o$ production. These backgrounds were reduced by a series of kinematic checks which not only required that the 4 photons satisfied the kinematics of the desired final state, but also eliminated any events with even a small probability (>0.1%) of satisfying the kinematics of possible background processes.

In addition to the above kinematic restrictions we made various target and detector cuts that tested our abilities to Monte Carlo the acceptance and detection efficiencies. In all cases we obtained results consistent within our statistical and estimated systematic uncertainties.

Our preliminary result for the $\eta \to \pi^o \gamma \gamma$ decay branching ratio is $3.2 \pm 0.9_{tot} \times 10^{-4}$.[11] This is less than half of the current Particle Data Group value of $7.1 \pm 1.4 \times 10^{-4}$ and disagrees by about 3-4 standard deviations. However, our experimental value does agree with the latest chiral perturbation theory calculations.

Using a similar analysis procedure, one of our colleagues has just reported an upper limit of the $\eta \to \pi^o \pi^o \gamma$ branching ratio of 5×10^{-4} at the 90% confidence level.[12]

SUMMARY

The Crystal Ball Program at BNL has producing a large number of high-quality result in a very short time period after data taking. We has put severe constraints on tests of chiral perturbation theory and the limitations of fundamental symmetries. Full and short reports as well as downloads of publications and conference contributions are available to the public on our Crystal Ball Collaboration web site - URL http://bmkn8.physics.ucla.edu. While hoping to complete our planned experimental program at BNL, members of the collaboration are currently making plans for the future which include bring the Ball back to Europe - in particular to MAMI at Mainz.

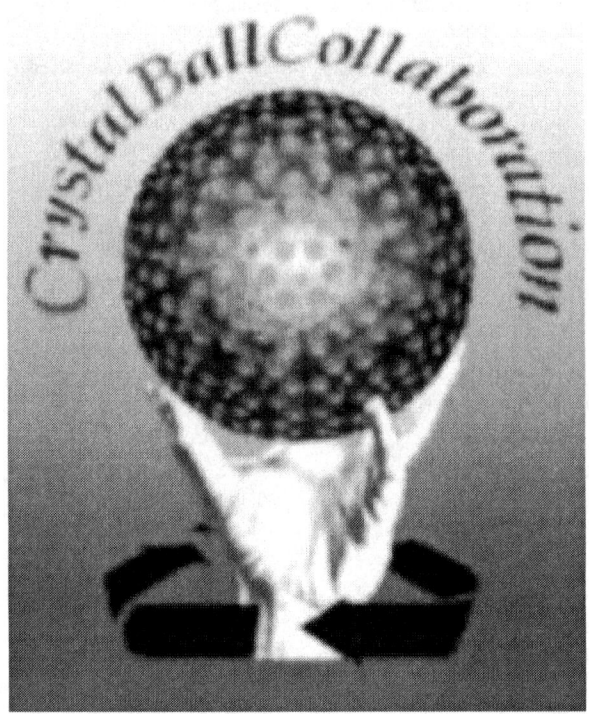

FIGURE 5. The Crystal Ball.

THE COLLABORATION

The new Crystal Ball Collaboration consists of B. Draper, S. Hayden, J. Huddleston, D. Isenhower, C. Robinson and M. Sadler, *Abilene Christian University*, C. Allgower and H. Spinka, *Argonne National Laboratory*, J. Comfort, K. Craig and A. Ramirez, *Arizona State University*, T. Kycia (deceased), *Brookhaven National Laboratory*, M. Clajus, A. Marusic, S. McDonald, B. M. K. Nefkens, N. Phaisangittisakul and W. B. Tippens, *University of California at Los Angeles*, J. Peterson, *University of Colorado*, W. Briscoe, A. Shafi and I. Strakovsky *George Washington University*, H. Staudenmaier, *Universität Karlsruhe*, D. M. Manley and J. Olmsted, *Kent State University*, D. Peaslee, *University of Maryland*, V. Abaev, V. Bekrenev, N. Kozlenko, S. Kruglov, A. Kulbardis, I. Lopatin and A. Starostin, *Petersburg Nuclear Physics Institute*, N. Knecht, G. Lolos and Z. Papandreou, *University of Regina*, I. Supek, *Rudjer Boskovic Institute* and A. Gibson, D. Grosnick, D. D. Koetke, R. Manweiler and S. Stanislaus, *Valparaiso University*.

ACKNOWLEDGEMENTS

The members of the Crystal Ball Collaboration are supported in part by the United States Department of Energy, the United States National Science Foundation, the National Sciences and Engineering Research Council of Canada, the Russian Ministry of Sciences, Volkswagen Stiftung and the George Washington University Research Enhancement Fund and Virginia Campus.

REFERENCES

1. Koulbardis, A., Proceedings of the IXth International Conference on Hadron Spectroscopy, (2001).
2. Craig, K., Ph.D. Thesis, Arizona State University (2001).
3. Starostin, A., *et al.*, Phys. Rev. Lett. **85**, 5539 (2000).
4. Kozlenko, N., Proceedings of the IXth International Conference on Hadron Spectroscopy, (2001).
5. Starostin, A. *et al.*, Phys Rev. C **64**, 055205 (2001).
6. Manley, D.M., Proceedings of the IXth International Conference on Hadron Spectroscopy, (2001).
7. Manley, D.M., **in press**, Phys. Rev. Lett. (2001).
8. Prakhov, S., *et al.*,Phys. Rev. Lett. **84**, 4802 (2000).
9. Tippens, W.B. *et al.*, Phys. Rev. Lett. **87**, 192001 (2001).
10. Stanislaus, T.D.S., *et al.*, Nucl. Instrum. Methods A **462**, 463 (2001).
11. Prakhov, S., Proceedings of the III International Conference on Non-Accelerator New Physics (2001); and Crystal Ball Report CB-01-008 (2001).
12. Prakhov, S., Crystal Ball Report CB-01-009 (2001).

J/ψ ATTENUATION AND THE QUARK-GLUON PLASMA

Kamal K. Seth

Department of Physics, Northwestern University, Evanston, IL 60208, USA

Abstract. J/ψ attenuation is considered to be one of the best tell-tale signatures of Quark-Gluon Plasma which is expected to be formed in the collision of relativistic heavy ions. In order to make sure that the observed J/ψ attenuation does not have other, more benign, explanations, it is necessary to measure J/ψ-Nucleon absorption cross section. It is proposed that the best way to do so is to study J/ψ formation in the annihilation of antiprotons with nuclear protons.

INTRODUCTION

A most exciting physics experiment is currently being done. The New York Times called it "Recreating Creation". It is the attempt to create a plasma of unconfined quarks and gluons (QGP), which are normally bound in hadrons, but which, it is conjectured, were free in the primordial soup just after the Big Bang, when densities and temperatures were very high. It is generally agreed that the best chance for recreating those density and temperature conditions is in the collision of relativistic heavy ions. Will the QGP be created with the heavy ions of the presently available relativistic energies, 100 GeV gold on 100 GeV gold at RHIC at Brookhaven, or will we have to wait for 2.76 TeV lead on 2.76 TeV lead at LHC at CERN? Theoretical conjectures aside[1], we do not really know! And even more importantly, how will we know that we have really succeeded in making QGP, if and when we do succeed? This talk addresses this last question.

CHARMONIUM ATTENUATION BY QGP

In 1986 Matsui and Satz[2] pointed out that in the presence of QGP the quark-antiquark potential gets screened, becoming too weak to bind a $c\bar{c}$ pair into a charmonium state. The conventional quark-antiquark potential

$$V(r) = -(4/3)\alpha_s/r + kr \tag{1}$$

is transformed to

$$V(r) = -(4/3)\alpha_s e^{-\mu r}/r + kr(1 - e^{-\mu r})/\mu r \tag{2}$$

with the Debye screening length $(\mu(T))^{-1} \approx (1.15gT)^{-1}$.

It is claimed that this 'dissociation' or 'melting' condition is reached for J/ψ at a temperature $T \approx 340$ MeV, and for χ_c and ψ' at $T \approx 170$ MeV.

The suppresion of charmonium formation, e.g., that for J/ψ in the collision of two heavy ions, A and B, is measured as:

$$S^{AB} = [\sigma(AB \to J/\psi X)/AB \cdot \sigma(pp \to J/\psi X)] \tag{3}$$

or,

$$= [\sigma(AB \to J/\psi X)/\sigma(AB \to \mu^+\mu^-)_{Drell-Yan}]. \tag{4}$$

S and σ are generally functions of x_f, P_T and $\sqrt{s}$. Gerschel and Hüfner have pointed out that

$$S^{AB}(x_f, AE_p) = S_1^{Ap}(x_f, AE_p) \cdot S_2^{pB}(x_f, E_p) \tag{5}$$

and while S_2^{pB} is known, S_1^{Ap}, which requires an inverse kinematics measurement, is not known. In its absence one could measure S^{pA} at negative x_f, or at least $x_f \approx 0$, because,

$$S_1^{Ap}(x_f, AE_p) = S_1^{pA}(-x_f, E_p) \tag{6}$$

CHARMONIUM ATTENUATION BY CONVENTIONAL MEANS

The conventional, or non-QGP means of J/ψ suppression is measured in terms of an effective absorbtion cross section, σ_{abs}, with J/ψ transforming into other hadrons as a result of its collisions with nucleons or with other hadrons produced in the collisions during the nucleon's passage through a thickness $<t>$ of the nuclear medium (not QGP). For example, for protons incident on a nucleus A

$$S^A = A^{-1}[\sigma(pA \to J/\psi X)/\sigma(pN \to J/\psi X)] \tag{7}$$

$$= A^{-1}exp(-\sigma_{abs} <t>) \approx A^{\alpha-1}, \tag{8}$$

the last being a rather crude parametrization which is often used. With equally crudely estimated

$$<t> = \rho_0 L = [(3/4)r_0 A^{1/3}]/(4/3\pi r_0^3) = (9A^{1/3})/16\pi r_0^2 \tag{9}$$

the cross section can be written as

$$\sigma_{abs}^{pA} = (16\pi/9)r_0^2(1-\alpha)(lnA/A^{1/3}) \tag{10}$$

As an example, for medium to heavy nuclei (A = 30-200), for assumed r_0 = 1.1 fm, a measured value α = 0.91-0.92 would correspond to σ_{abs}^{pA} = 5.6 - 6.7 mb.

CHARMONIUM ATTENUATION - EXPERIMENTAL

Charmonium attenuation in nuclei has been observed in many experiments since 1975. There is a large body of data on J/ψ lepto- and photo-production, summarized, for example, in ref[3], and J/ψ hadro-production with pions and protons, summarized, for example, in refs[1(b),3]. Most of these data were obtained with projectile momenta $\geq$ 100 GeV.

The lepto- and photo-production data were generally analyzed in the Vector Dominance Model, and it was claimed that they implied $\sigma(J/\psi - nucleon) \approx$ 1-2 mb. Recently, it has been shown that these analyses did not use VDM correctly. According to Kopeliovich and Hüfner, a correct application of the VDM leads to $\sigma_{tot}(J/\psi - N) = (4.0 \pm 0.8)$ mb $(\sqrt{s}/10 \text{ GeV})^{0.4}$, so that for incident lepton or photon energies of 100-200 GeV, these data imply $\sigma_{tot}(J/\psi - N) = 5 - 7.5(\pm 20\%)$ mb.

The hadro-production data for 100 - 800 GeV/c protons and pions ($\sqrt{s} \geq 15$ GeV) has been analyzed in terms of the coefficient α of Eq.8. Most recent analyses leads to $<\alpha> = 0.91 \pm 0.01$, which, as mentioned above, corresponds to $\sigma_{abs}^{pA}(J/\psi - N) \approx 6$ mb.[4]

A fair summary of all the above measurements is that for incident particle momenta ≥ 100 GeV/c (or $\sqrt{s} \geq 15$ GeV) all measurments of J/ψ attenuation in nuclear targets are consistent with an absorption cross-section of 5-7 mb.

WHAT DO THE EXPERIMENTAL MEASUREMENTS SAY ABOUT QGP?

The first observations of J/ψ suppression in nucleus-nucleus collisions were made by the NA38 experiment (1986-88) at CERN[5]. These measurements with Oxygen and Sulphur beams of 200 GeV/nucleon incident on Cu and Uranium showed J/ψ suppression which could not be explained in terms of conventional nuclear absorption if $\sigma(J/\psi - N) \approx$ 1-2 mb, as the old VDM analyses of electro- and photo- production experiments was suggesting. This led Satz[6] to cautiously state that is was "theoretically conceivable" that these measurements indicated "quark matter formation". Brodsky and Mueller[7] immediately pointed out that J/ψ "formation time is so long that what passes through the nucleus (in high energy photoproduction or hadroproduction reactions) is not a normal J/ψ, and hence the effective cross section extracted using (the eikonal or Glauber analysis) has little to do with J/ψ scattering on a nucleon." In other words, even the cautious statement about evidence or "quark matter formation" was premature. Further doubts were raised by Gerschel and Hüfner[8] in a paper entitled "Where is QGP", in which they showed that all J/ψ suppression data on γ-A, h-A, and even A-A could be consistently fitted with an empirical $\sigma_{abs}(J/\psi - N) = 6.3 \pm 1.3$ mb. No QGP was needed. [see Figure 1].

The situation changed rather dramatically with NA50's measurement of J/ψ suppression in Pb-Pb collisions at 158 GeV/nucleon[9]. As seen in Fig. 1, the observed J/ψ suppression in the Pb-Pb collisions no longer fitted the universal fit curve of Gerschel and Hüfner; the absorption was much larger. In the simple absorption model it would

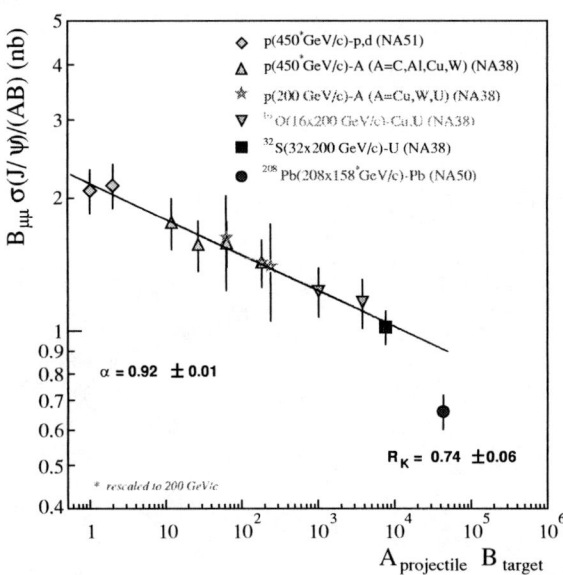

FIGURE 1. J/ψ cross section as a function of A × B from p-p to Pb-Pb interactions

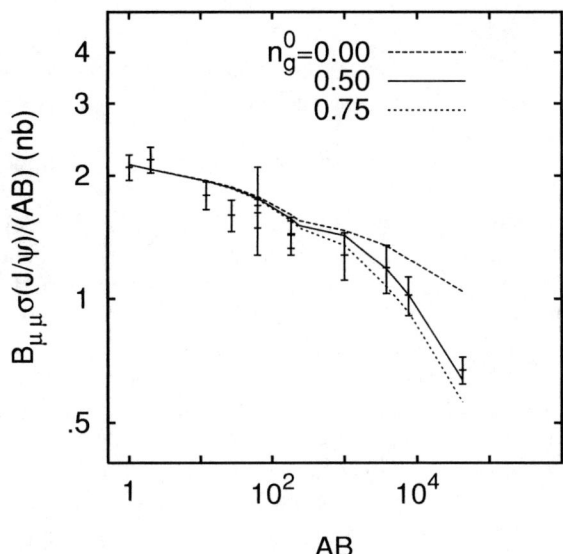

FIGURE 2. Prediction for the suppression of J/ψ in A-A collisions, including Pb-Pb, by Hüfner and Kopeliovich[10]

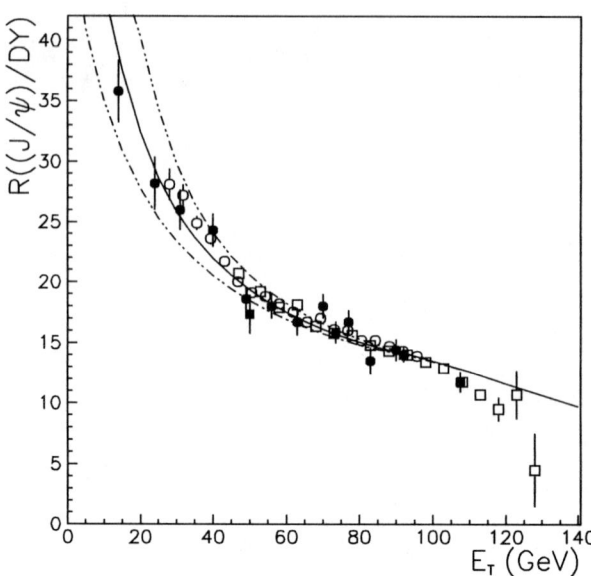

FIGURE 3. Transverse energy distribution for J/ψ production in Pb-Pb collisions. The curves are due to Capella et al.[11].

require $\sigma_{abs}(J/\psi - n) \approx 30$ mb. Clearly something unusual was happening. Could this be the long-sought evidence for QGP formation? The QGP enthusiasts would certainly like to believe that. But then, there are skeptics.

Hüfner and Kopeliovich[10] have proposed a model in which J/ψ suppression results from not only the J/ψ collisions with nucleons, but also the J/ψ collisions with the semi-hard gluons resulting from multiple NN collisions. With this additional contribution they are able to fit all pA and AB collision data including the Pb-Pb data (see Fig. 2). Capella et al.[11] are also able to fit the transverse energy distribution of the latest Pb-Pb suppression data quite satisfactorily in their model of J/ψ interactions with nucleons and hadronic co-movers (see Fig. 3). Similar results are reported by Cassing et al.[12] in their cascade calculations for the co-mover model[12]. None of the above invoke the presence of QGP. However, all of them have some unconstrained parameters, whether they be the J/ψ-nucleon cross section, or prompt gluon or comover density and cross-sections. In order to have confidence in any of these model calculations, it is necessary to provide experimental constraints to any and all parameters. The most important of these is simply the elementary J/ψ-nucleon cross sections. The high energy pA experiments attempting to measure $\sigma(J/\psi - N)$, leave many open questions:

1. Did the $c\bar{c}$ pair produced in the pN collision hadronize to J/ψ while still inside the nucleus?
2. What role does the color-octet state of $c\bar{c}$ play in the process?
3. What role is played by what and how many comovers?
4. What is the contribution of feed-down from higher charmonium/bottomonium resonances?

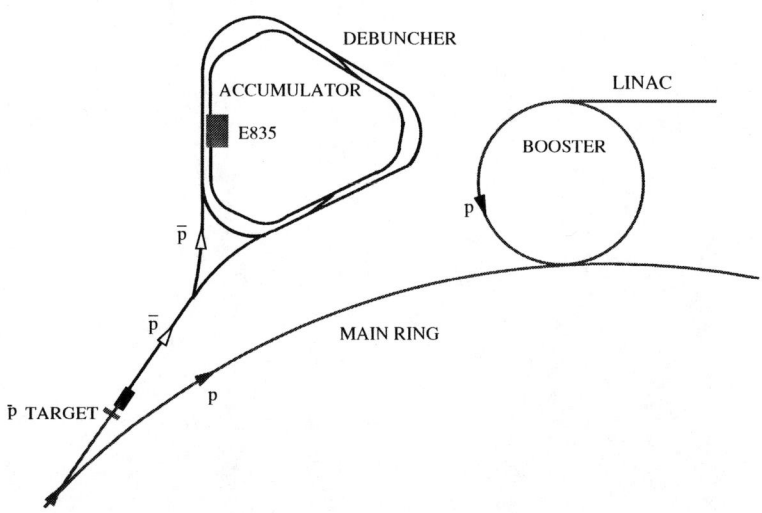

FIGURE 4. Fermilab Experiment E835

Answers to these questions, based on experimental measurements, do not exist. Plenty of conjectures do.

Sometime ago we proposed[13] a unique way to determine $\sigma(J/\psi - N)$ which avoids all the complications posed by the above questions. Here we present this proposal in greater detail.

THE PROPOSED EXPERIMENT

We propose that following the suggestion of Brodsky and Mueller[7] we should measure the attenuation of charmonium resonances in nuclei by forming them in the annihilation of antiprotons with protons bound in nuclear targets. Comparison of these cross sections with formation cross sections on free protons will provide a direct measure of $\sigma(J/\psi - N)$, free from all the problems and ambiguities mentioned earlier. In on-resonance formation of charmonia in $p\bar{p}$ annihilations, there is no feed-down, no comovers, and no color-octet formation. The directly formed $c\bar{c}$ color singlet state hadronizes to the specific charmonium state of choice in a distance of ~ 0.3 fm, travels through the nucleus slowly ($\gamma \approx 1$), and therefore interact with the nucleons in the nucleus as a fully formed charmonium state; there are no color-transparency complications.[14]

How such a measurement may be successfully made is very well illustrated in terms of the Fermilab experiments E760/E835 in which we have been involved for the last 15 years. In these experiments the circulating beam of antiprotons in the Fermilab Accumulator is intersected by a gas-jet of hydrogen, and the reaction products are detected in a surrounding detector system which is optimized for the detection of e^+, e^-,

Charmonium-nucleon cross sections

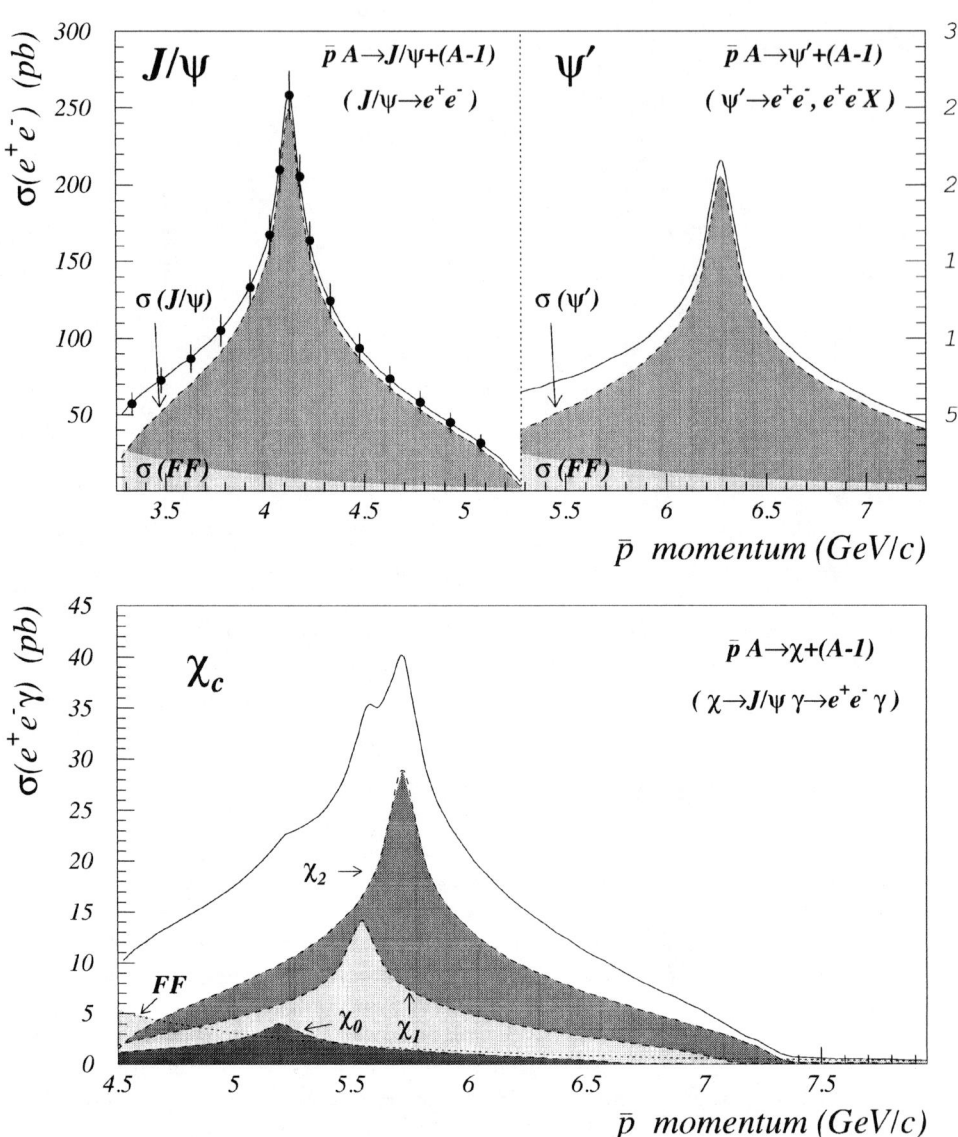

FIGURE 5. Monte-Carlo results for scans of Fermi momentum broadened charmonium resonances measured in antiproton annihilation on nuclear protons

and photons[Fig.4]. The excitation function for a charmonium resonance of choice is obtained by stepping the antiproton energy across the resonance. For example:

$$\bar{p} + p \to J/\psi \to e^+e^- \tag{11}$$

has been measured with high precision[15], and with e^+e^- backgrounds ≤ 1 pb[16]. In the quasi-free reaction

$$\bar{p} + A \to J/\psi + (A-1) \to e^+e^- + (A-1) \tag{12}$$

the J/ψ signal is smeared by the Fermi motion of the target nucleons in the nucleus A. The peak cross section of ≈ 360 nb at the peak of the J/ψ resonance with natural width of 100 keV is smeared to ~ 260 pb with the resonance spread over $\bar{p}$ momentum range of 3.25-5.25 GeV/c, as shown in Fig. 5 (upper left). The only physics background is from the timelike formfactor of the proton at a level of 30 - 5 pb over the same range.

We estimate that the measurement is entirely feasible with a heavy gas jet replacing the hydrogen gas jet of the Fermilab experiments. For example, with 70 mA of antiprotons circulating in the Accumulator at ~ 0.65 MHz, intersected by a Xe gas-jet of 10^{12} atoms/cm^3, a luminosity $\mathcal{L} = 2.5 \times 10^{31}$ cm^{-2}s^{-1} can be acheived. This would lead to a count rate of 200/day reconstucted J/ψ. A 16 point scan in $\Delta p = 125$ MeV steps with $\mathcal{L} = 2$ pb^{-1}/point across the Fermi-broadened J/ψ (3.25 - 5.25 GeV/c) will require ~ 6 weeks of running, and yield $\sigma(J/\psi - N)$ at the level of $\pm 5\%$. In order to rigorously test the assumptions made, several targets, from CH$_4$ to Xe will need to be investigated. In case of the methane jet, because of the hydrogen in CH$_4$, a self-normalized measurement can be made by doing the small step (50-100 keV) scan around 4.07 GeV/c at the same time as the coarse scan for Carbon.

The beauty of the above technique is that $\sigma(\psi' - N)$ can be measured with similar precision by scanning the $\bar{p}$ momentum region 5.2 - 7.2 GeV/c [Fig. 5 (upper right)]. However, the measurement will be a factor two slower even with $\pm 10\%$ statistical uncertainty for $\sigma(\psi' - N)$.

The prospects are a bit better for $\sigma(\Sigma \chi_J \to N)$ measurment in the $\bar{p}$ momentum region 4.7 - 6.7 MeV/c. In this case, the photons from the radiative decay of χ resonances to J/ψ will also need to be detected in order to distinguish from the overlaps with J/ψ and ψ'. The reaction

$$\bar{p}A \to (A-1) + \chi \to (A-1) + \gamma + \psi \to (A-1) + \gamma + (e^+e^-) \tag{13}$$

will have to be measured. A 10% precision can be obtained in the same time as for J/ψ [Fig. 5 (bottom)].

Being not heavy ion experts we have run the proposal presented here by a number of heavy ion experts[17]. They all seem to consider the proposed measurements as extremely important, and we hope that these measurements can be made, if not at Fermilab, at a future antiproton facility like the one which is planned for GSI (see talk by H. Koch in these proceedings).

I would like to thank David Joffe for his help in the preparation of this manuscript.

REFERENCES

[1] Three recent review articles on the subject are: (a) C. Gerschel and J. Hüfner, Annu. Rev. Nucl. Part. Sci. **49**, 255 (1999); (b) R. Vogt, Phys. Reports **310**, 197 (1999); and (c) H. Satz, Rep. Prog. Phys. **63**, 1511 (2000).
[2] T. Matsui and H. Satz, Phys. Lett. **B178**, 416 (1986).
[3] Kamal K. Seth, Proc. RIKEN BNL Workshop on Quarkonium Production in Relativistic Nuclear Collisions, Formal Report, **BNL -52559**, 126 (1998).
[4] There is a very vexing new result from the Fermilab E866 experiment. For 800 MeV protons incident on Be, Fe, and W, this experiment reports [M. Leitch *et al.*, Phys. Rev. Lett. **84**, 3526 (2000)] $\alpha = 0.960(3)$, which corresponds to $\sigma = 2.7 \pm 0.1$ mb. The explanation for this result is not clear, so far.
[5] C. Baglin *et al.* Phys. Lett. **B220**, 471 (1989); *ibid* **B255**, 459 (1991).
[6] H. Satz, Proc. Int. Lepton Photon Symposium, Geneva, 1991, ed. by S. Hagerty (World Scientific, Singapore, 1992) pp. 273-300.
[7] S. J. Brodsky and A. H. Mueller, Phys. Lett. **B206**, 685 (1988).
[8] C. Gerschel and J. Hüfner, Phys. Lett. **B207**, 253 (1988); also Z. Phys C **56**, 171 (1992).
[9] NA50 Collaboration, M. C. Abreu *et al.*, Phys. Lett. **B410**, 337 (1997); *ibid* **B450**, 456 (1999); *ibid* **477**, 28 (2000).
[10] J. Hüfner and B. Z. Kopeliovich, Phys. Lett. **B445**, 223 (1998).
[11] A. Capella *et al.*, Phys. Rev. Lett. **85,** 2080 (2000).
[12] W. Cassing *et al.*, Nucl. Phys. **A674**, 249 (2000).
[13] Kamal K. Seth, Proc SuperLEAR Workshop, Zurich (1991), ed. by C. Amsler and D. Urner, Inst. of Phys. Conf. Series **124**, 261 (1992).
[14] D. Kharzeev, Proc SuperLEAR Workshop, Zurich (1991), ed. by C. Amsler and D. Urner, Inst. of Phys. Conf. Series **124**, 229 (1992).
[15] E760 Collaboration, T. A. Armstrong *et al.*, Phys. Rev. **D47**, 772 (1991).
[16] E835 Collaboration, M. Ambrogiani *et al.*, Phys. Rev. **D60**, 032002 (1999).
[17] The experts contacted include Louis Kluberg (NA50), Jurgen Sthukraft (ALICE), J. Ellis (CERN), H. Satz (Bielefeld), J. Hüfner (Heidelberg), and R. Vogt (LBL).

The Sigma Meson and Chiral Transition in Hot and Dense Matter

Teiji Kunihiro

Yukawa Institute for Theoretical Physics, Kyoto University, Sakyo-ku, Kyoto 606-8502, Japan

Abstract. It is pointed out that the hadron spectroscopy should be a study of the structure of the QCD vacuum, low-energy elementary excitations on top of which are hadrons. Concentrating on the dynamical breaking of the chiral symmetry in the QCD vacuum, we emphasize the importance to clarify what is going on with mesons in the $I = J = 0$-channel, i.e., the sigma meson channel, because it is connected to the quantum fluctuations of the chiral order parameter. After summarizing the significance of the sigma meson in QCD and low-energy hadron phenomenology, we give a review on some theoretical and experimental effort to try to reveal the possible restoration of chiral symmetry in hot and dense nuclear matter including heavy nuclei.

INTRODUCTION

A tricky point in the hadron spectroscopy is that QCD, the fundamental theory of the hadron world, is not written in terms of hadron fields but in terms of quark- and gluon-fields from which hadrons are composed: The quarks and gluons are colored objects which can not exist in the asymptotic state, and low-lying elementary excitations on top of the non-perturbative QCD vacuum are composite and colorless, which we call hadrons. Furthermore, symmetries possessed by the QCD Lagrangian, such as the chiral $SU(3)_L \times SU(3)_R$ symmetry in the massless limit of quarks and the color gauge symmetry are not manifest in our every-day world. This complication of the problem is due to the fact that the true QCD vacuum is completely different from the perturbative one and is actually realized through the phase transitions, i.e., the confinement-deconfinement and the chiral transitions. The notion of such a complicated vacuum structure, i.e., the collective nature of the vacuum and the elementary particles was first introduced by Nambu[1], in analogy with the physics of superconductivity[2]. The nonperturbative nature and the realization of the true QCD vacuum through the phase transitions are being confirmed by the lattice simulations[3]. One may notice that the so called $U_A(1)$ anomaly[4] also characterizes the non-perturbative QCD vacuum. Several rules extracted from the hadron phenomenology such as the vector-meson dominance (VMD)[5, 6] and the Okubo-Zweig-Iizuka (OZI) rule[7] might be also related with some fundamental properties of the QCD vacuum. Thus the hadron spectroscopy can not be failed to be a study of the nature of QCD vacuum including its symmetry properties. In other words, the physics of the hadron spectroscopy is a combination of the condensed matter physics of the QCD vacuum[8, 9] and the atomic physics as played with the constituent quark-gluon model where the vacuum structure is taken for granted[10].

An interesting observation is then that hadrons as elementary excitations on top of the QCD vacuum may change their properties in association with a change or phase transition of the QCD vacuum. What hadrons do change their properties sharply, and how do they in hot and/or dense medium? One should also ask how they are detected in experiment[11, 12, 13, 14, 15]; see also the reviews [8, 16]. For instance, in Table 6.1 in [8], list up are interesting observables and their expected behavior in relation with the chiral transition, possible restoration of the $U_A(1)$-symmetry and precritical deconfinement at finite temperature and/or density. In association with (partial) restoration of chiral symmetry, the mass of the σ meson[13, 19] is expected to decrease. Some people[11, 15, 20] expect that the vector mesons ρ, ω and ϕ also show a decrease of their masses in association with the chiral restoration. The $U_A(1)$ anomaly, which is responsible for lifting the η′ meson mass as high as about 1 GeV and make the η′(η) almost flavor singlet (octet), may be cured at high temperature, which may manifest itself as the decrease of the mass $m_{η'}$[17, 18] for example. The deconfinement may affect the properties of heavy-quark systems such as J/ψ[14, 21] than in light hadrons.

In the present talk, I will focus on the chiral transition in hot and dense hadronic matter and discuss the significance of the scalar and isoscalar meson, the sigma meson and the strength function in its channel in the hadronic medium including heavy nuclei.

THE SIGMA MESON

The order parameter of the chiral transition is the quark condensate $\langle \bar{q}q \rangle \sim \sigma$. There arise two kinds of quantum fluctuations of the order parameter, the modulus and phase fluctuations. The pion corresponds to the latter fluctuations. The particle corresponding to the former fluctuation is a scalar-isoscalar meson, which is traditionally called the σ meson. One may notice that the way of the appearance of the σ meson is analogous to that of the Higgs particle, which comes to exist through the dynamical breaking of the gauge symmetry in the standard model while the corresponding NG boson is absorbed into the longitudinal component of the gauge fields. As one can now see, *the existence of the σ meson is logically related to the fundamental property of the QCD vacuum in which the chiral symmetry is spontaneously broken.* Therefore searching for such a particle in experiment is *not eccentric but as naturally motivated as searching for glue balls in QCD and the Higgs particle in the standard model.* If such a particle or the like could not be identified, one must consider possible dynamical origins to hinder them from appearing.

Here are a short summary of the significance of the sigma meson in hadron physics:

1. The existence of the σ meson as *the quantum fluctuation of the order parameter* of the chiral transition accounts for various phenomena in hadron physics which otherwise remain mysterious[22, 8, 23].
2. There have been accumulation of experimental evidence of a low-mass pole in the σ channel in the pi-pi scattering matrix[25, 26]. It should be emphasized that for obtaining this result, it is essential to respect chiral symmetry, analyticity and crossing symmetry even in an approximate way as in the *N/D* method[27]: The great achievement of the chiral perturbation theory[28] is indispensable for set the precise boundary condition for the scattering matrix in the low-energy region.

3. It is well known that such a scalar meson with the mass range 500 to 700 MeV is responsible for the intermediate range attraction in the nuclear force.[30]
4. The correlation in the scalar channel as summarized by such the sigma meson may account for the enhancement of the $\Delta I = 1/2$ processes in $K^0 \to \pi^+\pi^-$ or $\pi^0\pi^0$ [29]. In fact, the final state interaction for the emitted two pions may include the σ pole, then the matrix element for of the scalar operator $Q_6 \sim \bar{q}_R q_L \bar{q}_L q_R$ is sown to be enhanced dramatically.
5. The collective excitation in the scalar channel as described as the σ meson is essential [32] in reproducing the empirical value of the π-N sigma term [31] $\Sigma_{\pi N} = \hat{m}\langle \bar{u}u + \bar{d}d \rangle$: Empirically, it is known that $\Sigma_{\pi N} \sim$ 40-50 MeV, while the naive quark model only gives as small as $\sim$ 15 MeV. The basic quantities here are the quark contents of baryons $\langle B|\bar{q}_i q_i|B\rangle \equiv \langle \bar{q}_i q_i\rangle_B$ ($i = u,d,s,...$). Actually, it is more adequate to call them the scalar charge of the hadron. The point is such a scalar charge of the nucleon is enhanced with the existence of the sigma meson pole; such an enhancement of charges by collective modes are well known in nuclear physics.

PARTIAL CHIRAL RESTORATION AND THE σ MESON IN HADRONIC MATTER

Although the recent phase shift analyses [26] of the π-πscattering and the identification of the pole in the $I = J = 0$ channel as mentioned in 2 above is a great development in this field, one must say that it is still obscure whether the pole really corresponds to the quantum fluctuation of the chiral order parameter, i.e., our σ. If one were to be able to change the environment freely and trace the possible change of the pole position, the nature of the particle corresponding to the pole can be revealed: A change of the vacuum or the equilibrium state leading to the phase transition will make the mode coupled to the order parameter change. Actually this is the usual strategy in the many-body physics[33] to reveal the nature of elementary excitations. Conversely, an observation of the change of the elementary modes as well as that of other thermodynamic quantities tells us the change of the state of the matter.

Effective theories of QCD[13, 19] show that the sigma meson mass m_σ decreases in association with the chiral restoration in hot and/or dense medium, while the pion mass keeps its value in free space as long as the system is in the Nambu-Goldstone phase. The simulations on the lattice QCD also show a decrease of the *screening mass* m_{sc} and the *generalized mass* m_G in the sigma meson channel; see for instance, [3, 34]. The screening mass which describes the damping of the correlation function in the space direction is not the dynamical mass which is given through the time correlation of the relevant operators. The generalized mass is defined as the inverse of the static susceptibility; $m_G^{-2} = \partial \langle \bar{q}q \rangle / \partial m$ with m being the current quark mass. Nevertheless, it is remarkable that the lattice result is not in contradiction with those in the effective theories.

Then the width of the σ is also expected to decrease due to the depletion of the phase space for the decay $\sigma \to 2\pi$[35]. Thus one can expect a chance to see the σ meson as a sharp resonance at high temperature and/or density.

Some years ago, the present author proposed several nuclear experiments including one using electro-magnetic probes to produce the σ meson in nuclei, thereby have a clearer evidence of the existence of the σ meson and also explore the possible restoration of chiral symmetry in the nuclear medium[23, 36]: As is well known, there arises a scalar-vector mixing in nuclear matter at finite density[37]. To make a veto for the two pions from the rho meson, the produced pions should be neutral ones which may be detected through four γ's.

When a hadron is put in a hadronic medium, the hadron might dissociate into complicated excitations to loose its identity. Then the most informative quantity is the response function or spectral function in the hadron channel of the system. If the coupling of the hadron with the environment is relatively small, then there may remain a peak with a small width in the spectral function corresponding to the hadron. Such a peak is to be identified with an elementary excitation or a quasi particle. Then how will the decrease of m_σ in the nuclear medium affect the spectral function.

It has been shown by using linear chiral models that an enhancement in the spectral function in the σ channel occurs just above the two-pion threshold along with the decrease of m_σ[38]. Subsequently, it has been shown [39] that the spectral enhancement near the $2m_\pi$ threshold takes place in association with partial restoration of chiral symmetry at finite baryon density.

Hatsuda et al[39] started from the following linear sigma model;

$$\mathcal{L} = \frac{1}{4}\text{Tr}[\partial M \partial M^\dagger - \mu^2 M M^\dagger - \frac{2\lambda}{4!}(MM^\dagger)^2 - h(M+M^\dagger)] + \bar{\psi}(i\gamma\cdot\partial - gM_5)\psi + \cdots, \quad (1)$$

where $M = \sigma + i\vec{\tau}\cdot\vec{\pi}$, $M_5 = \sigma + i\gamma_5\vec{\tau}\cdot\vec{\pi}$, ψ is the nucleon field, and Tr is for the flavor index. Consider the propagator of the σ-meson at rest in the medium : $D_\sigma^{-1}(\omega) = \omega^2 - m_\sigma^2 - \Sigma_\sigma(\omega;\rho)$, where m_σ is the mass of σ in the tree-level, and $\Sigma_\sigma(\omega;\rho)$ is the loop corrections in the vacuum as well as in the medium. The corresponding spectral function is given by $\rho_\sigma(\omega) = -\pi^{-1}\text{Im}D_\sigma(\omega)$. One can show that

$$\text{Im}\Sigma_\sigma \propto \theta(\omega - 2m_\pi)\sqrt{1 - \frac{4m_\pi^2}{\omega^2}} \quad (2)$$

near the two-pion threshold in the one-loop order. On the other hand, partial restoration of chiral symmetry implies that m_σ^* defined by $\text{Re}D_\sigma^{-1}(\omega = m_\sigma^*) = 0$ approaches to m_π. Therefore, there exists a density ρ_c at which $\text{Re}D_\sigma^{-1}(\omega = 2m_\pi)$ vanishes even before the complete restoration of chiral symmetry where σ-π degeneracy is realized, namely $\text{Re}D_\sigma^{-1}(\omega = 2m_\pi) = [\omega^2 - m_\sigma^2 - \text{Re}\Sigma_\sigma]_{\omega=2m_\pi} = 0$. At this point, the spectral function is solely given in terms of the imaginary part of the self-energy;

$$\rho_\sigma(\omega \simeq 2m_\pi) = -\frac{1}{\pi\,\text{Im}\Sigma_\sigma} \propto \frac{\theta(\omega - 2m_\pi)}{\sqrt{1 - \frac{4m_\pi^2}{\omega^2}}}, \quad (3)$$

which clearly shows the near-threshold enhancement of the spectral function. This is a general phenomenon correlated with the partial restoration of chiral symmetry.

In [39], the effect of the meson-loop as well as the baryon density was treated as a perturbation to the vacuum quantities. Therefore, our loop-expansion is valid only at

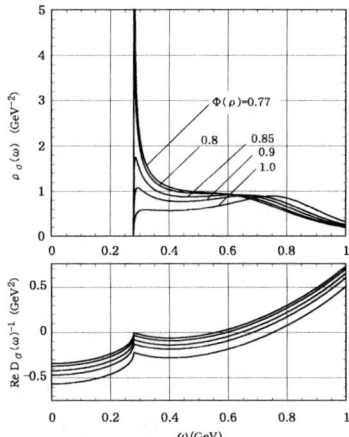

FIGURE 1. The spectral function $\rho_\sigma(\omega)$ (the upper panel) and $\mathrm{Re}D_\sigma^{-1}(\omega)$ (the lower panel) calculated with a linear sigma model. $\Phi(\rho) \equiv \langle\sigma\rangle/\sigma_0$ measures the rate of the partial restoration of the chiral symmetry at the baryonic density ρ.

relatively low densities. When we parameterize the chiral condensate in nuclear matter $\langle\sigma\rangle$ as

$$\langle\sigma\rangle \equiv \sigma_0\, \Phi(\rho), \tag{4}$$

one may take the linear density approximation for small density; $\Phi(\rho) = 1 - C\rho/\rho_0$ with $C = (g_s/\sigma_0 m_\sigma^2)\rho_0$.

The spectral function $\rho_\sigma(\omega)$ together with $\mathrm{Re}D_\sigma^{-1}(\omega)$ calculated with a linear sigma model are shown in Fig.1: The characteristic enhancements of the spectral function is seen just above the $2m_\pi$. It is also to be noted that even before the σ-meson mass m_σ^* and m_π in the medium are degenerate, i.e., the chiral-restoring point, a large enhancement of the spectral function near the $2m_\pi$ is seen.

Is the near-threshold enhancement obtained above specific to the linear representation of the chiral symmetry, where the σ degree of freedom is explicit as in (1). Jido et al[40] showed that the nonlinear realization of the chiral symmetry can also give rise to the near $2m_\pi$ enhancement of the spectral function in nuclear medium as shown in Fig.2.

They begin the discussion with the polar parameterization of the chiral field, $M = \sigma + i\vec{\tau}\cdot\vec{\pi} = (\langle\sigma\rangle + S)U$ with $U = \exp(i\vec{\tau}\cdot\vec{\phi}/f_\pi^*)$. Here f_π^* is a would-be "in-medium pion decay constant".

$$\begin{aligned}\mathcal{L} &= \frac{1}{2}[(\partial S)^2 - m_\sigma^{*2}S^2] - \frac{\lambda\langle\sigma\rangle}{6}S^3 - \frac{\lambda}{4!}S^4 + \frac{(\langle\sigma\rangle+S)^2}{4}\mathrm{Tr}[\partial U\partial U^\dagger] + \frac{\langle\sigma\rangle+S}{4}h\,\mathrm{Tr}[U^\dagger+U] \\ &+ \mathcal{L}_{\pi N}^{(1)} - gS\bar{N}N\,,\end{aligned} \tag{5}$$

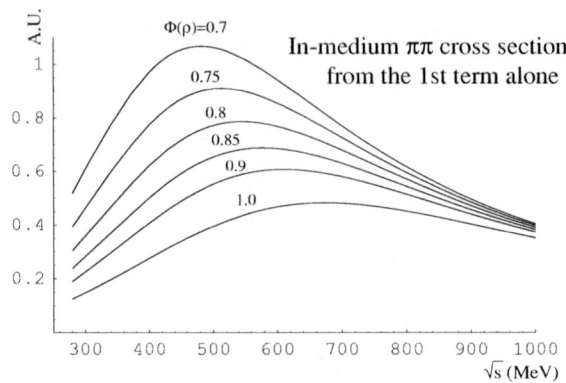

FIGURE 2. In-medium $\pi\pi$ cross section in the $I = J = 0$ channel in the heavy S limit where m_σ^* is taken to be infinity. The cross section is shown in the arbitrary unit (A.U.).

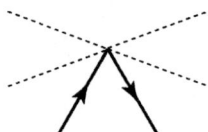

FIGURE 3. The new 4π-N-N vertex generated in the nonlinear realization. The solid line with arrow and the dashed line represent the nucleon and pion, respectively.

with $\mathcal{L}_{\pi N}^{(1)} = \bar{N}(i\gamma\cdot\partial + i\slashed{v} + i\slashed{a}\gamma_5 - m_N^*)N$ and $(v_\mu, a_\mu) = (\xi\partial_\mu\xi^\dagger \pm \xi^\dagger\partial_\mu\xi)/2$, and $m_N^* = g\langle\sigma\rangle$. In this representation, the in-medium $\pi\pi$ amplitude in the tree level reads

$$A(s) = \frac{s - m_\pi^2}{\langle\sigma\rangle^2} - \frac{(s - m_\pi^2)^2}{\langle\sigma\rangle^2} \frac{1}{s - m_\sigma^{*2}}. \tag{6}$$

The first term in (6) comes from the contact 4π coupling generated by the expansion of the second line in (5) with the coefficient proportional to $1/\langle\sigma\rangle^2$. On the other hand, the second term in (6) is from the contribution of the scalar meson S in the s-channel. Fig.2 shows a unitarized in-medium $\pi\pi$ cross section only with the first term in (6), i.e., what is given by the non-linear realization. One sees a clear enhancement of the cross section near the threshold or a softening as chiral symmetry is restored. Although there is no explicit σ-degrees of freedom in this heavy σ approximation, there arises a decrease of the pion decay constant f_π^* in nuclear medium. This is due to a new vertex, i.e., 4πN-N vertex absent in the free space; see Fig. 3. The vertex is responsible for the reduction of f_π^* and hence for the spectral enhancement.

POSSIBLE EXPERIMENTAL EVIDENCE

Interestingly enough, CHAOS collaboration [41] had measured the $\pi^+\pi^\pm$ invariant mass distribution $M^A_{\pi^+\pi^\pm}$ in the reaction $A(\pi^+, \pi^+\pi^\pm)X$ with the mass number A ranging from

2 to 208: They observed that the yield for $M^A_{\pi^+\pi^-}$ near the $2m_\pi$ threshold is close to zero for $A = 2$, but increases dramatically with increasing A. They identified that the $\pi^+\pi^-$ pairs in this range of $M^A_{\pi^+\pi^-}$ is in the $I = J = 0$ state. The A dependence of the the invariant mass distribution presented in [41] near $2m_\pi$ threshold has a close resemblance to our model calculation in Fig.1, which suggests that this experiment may already provide a hint about how the partial restoration of chiral symmetry manifest itself at finite density.

In fact, a state of the art calculation based on the conventional many-body theoretical approach without incorporating the effect of the vacuum change was performed[44]; unfortunately, they all failed in reproducing the sufficient enhancement in the near-threshold region consistently with the other energy region. Once the effect of partial chiral restoration in nuclei is incorporated to the conventional approach[43], as suggested in [39], the agreement of the theory and experiment was remarkable. This is encouraging.

To confirm the threshold enhancement, first of all, more experimental work should be done. Measurement of $2\pi^0$ and 2γ in experiments with hadron/photon beams off the heavy nuclear targets should be done, which is free from the ρ meson meson background inherent in the $\pi^+\pi^-$ measurement. Such an experiment was in fact performed by the Crystal Ball(CB) [45] collaboration at BNL: They claimed that there is no threshold enhancement that was seen in the CHAOS experiment. A reexamination of the data by the CB group has been done by the CHAOS group[46], and emphasized the importance of the combined ratio

$$C^A_{\pi\pi}(M_{\pi\pi}) = \frac{\sigma^A(M_{\pi\pi})/\sigma^A_T}{\sigma^N(M_{\pi\pi})/\sigma^N_T}, \tag{7}$$

where σ^A_T (σ^A_T) is the measured total cross section of the $\pi 2\pi$ process in nuclei (nucleon): This ratio yields the net effect of nuclear matter on the interacting $(\pi\pi)_{I=J=0}$ system. They have shown that the combined ratio grows near the $2m_\pi$ threshold consistently in the two experiments, although the statistics in the CB data is poorer.

Measuring of 2 γ's from the electro-magnetic decay of the σ or $(\pi\pi)_{I=J=0}$ in nuclear matter may be interesting because of the small final state interactions, although the branching ratio is small. One needs also to fight with large background of photons mainly coming from π^0s. Nevertheless, if the enhancement is prominent, there is a chance to find the signal. When σ has a finite three momentum, one can detect dileptons through the scalar-vector mixing in matter: $\sigma \to \gamma^* \to e^+e^-$. The inverse process can be also used to produce the σ or $(\pi\pi)_{I=J=0}$ system by the electro-magnetic probes owing to the scalar-vector mixing in the finite density system where the charge conjugation symmetry is violated. Such an experiment has been planned and being performed in SPRING8[47]. We remark that (d, ^{3}He) or (d, ^{3}He) reactions is also useful to explore the spectral enhancement because of the large incident flux. as in the production of the deeply bound pionic atoms and the possible production of η- or ω- mesic nuclei[48]. The incident kinetic energy E of the deuteron in the laboratory system is estimated to be $1.1\text{GeV} < E < 10 \text{ GeV}$, to cover the spectral function in the range $2m_\pi < \omega < 750$ MeV. A theoretical evaluation of the feasibility of such experiments is now in progress[49].

OTHER POSSIBLE EVIDENCE OF PARTIAL CHIRAL RESTORATION IN NUCLEAR MATTER

It is interesting that there are other possible experimental evidences for partial chiral restoration in nuclear matter than the chiral fluctuations in the sigma meson channel discussed so far. The spectral function deduced from the lepton pairs from the heavy ion collisions shows a softening, which might be an evidence for the partial chiral restoration in nuclear medium[50]: Pisarski[11] was the first who suggested that a decrease of the rho meson mass may be a signature of the chiral restoration in hot hadronic matter. Brown and Rho[15] conjectured also the decrease of the vector meson masses in association with the chiral restoration on the basis of a scaling argument (the so called Brown-Rho scaling). Hatsuda and Lee[20] discussed the vector meson properties using the QCD sum rules. A KEK experiment also shows the softening of the spectral function in the ρ/ω channel in heavy nuclei such as Cu [51]. The deeply bound pionic atom has proved to be a good probe of the properties of the hadronic interaction deep inside of heavy nuclei. Yamazaki[52] suggested that the anomalous energy shift of the pionic atoms (pionic nuclei) owing to the strong interaction could be attributed to the decrease of the effective pion decay constant $f_\pi^*(\rho)$ at finite density ρ which may imply that the chiral symmetry is partially restored deep inside of nuclei.

SUMMARY AND CONCLUDING REMARKS

1. The hadron spectroscopy must be a condensed matter physics of the QCD vacuum, because hadrons are elementary excitations on top of the nontrivial QCD vacuum.
2. The σ meson as the quantum fluctuation of the order parameter of the chiral transition may account for various phenomena in hadron physics which otherwise remain mysterious.
3. There have been accumulation of *experimental evidence of the σ pole* in the pi-pi scattering matrix. Here it has been noticed that the chiral symmetry, analyticity and crossing symmetry are all important.
4. Partial restoration of chiral symmetry in hot and dense medium leads to a peculiar *enhancement in the spectral function in the σ channel near the $2m_\pi$ threshold*.
5. The enhancement is obtained both in the linear and nonlinear realization of chiral symmetry *provided that the possible reduction of the quark condensate or f_π is taken into account*.
6. Such an enhancement has been observed in the reaction $A(\pi^+, (\pi^+\pi^-)_{I=J=0})A'$ by CHAOS group, which might possibly be an experimental evidence of the partial restoration of chiral symmetry in heavy nuclei.
7. It seems that there is no serious contradiction between the CHAOS data $\pi^+\pi^-$ and the Crystal Ball data on $2\pi^0$.
8. Further theoretical and experimental works are needed to confirm that chiral symmetry is partially restored in heavy nuclei.

9. There are other possible experimental evidences which show a partial restoration in dense nuclear matter.

ACKNOWLEDGMENTS

I thank the organizers of this symposium for inviting me to the symposium. Most of this report is based on the works done in collaboration with T. Hatsuda, D. Jido and H. Shimizu, to whom I am grateful. This work is partially supported by the Grants-in-Aid of the Japanese Ministry of Education, Science and Culture (No. 12640263 and 12640296).

REFERENCES

1. Y. Nambu, Phys. Rev. Lett. **4**, 380 (1960); Y. Nambu and G. Jona-Lasinio, Phys. Rev. **122**, 345 (1961), ibid **124**, 246 (1961).
2. Y. Nambu, Phys. Rev. **117**, 648 (1960).
3. F. Karsch, hep-lat/0106019; Z. Fodor and S. D. Katz hep-lat/0106002.
4. G. A. Christos, Phys. Rep. **116**, 219; G. 't Hooft, ibid, **142**, 357 (1986); S. Weinberg, *THE QUANTUM THEORY OF FIELDS. VOL. 2: MODERN APPLICATIONS*, (Cambridge Univ. Press, 1996).
5. J. J. Sakurai, Phys. Rev. Lett. **22**, 981 (1969).
6. M. Bando, T. Kugo, S. Uehara, K. Yamawaki and T. Yanagida, Phys. Rev. Lett. **54**, 1215 (1985); M. Bando, T. Kugo and K. Yamawaki, Phys. Rept. **164**,217 (1988). See also, H. Georgi, Phys.Rev.Lett. **63**, 1917 (1989); Nucl.Phys. **B331**, 311 (1990).
7. See for example, S. Okubo, Prog. Theor. Phys. Suppl. **63**, 1 (1978); H. Lipkin, Int. J. Mod. Phys. **E1**, 603 (1992).
8. T. Hatsuda and T. Kunihiro, Phys. Rep. **247**, 221(1994).
9. K. Rajagopal and F. Wilczek, to appear as Chapter 35 in the Festschrift in honor of B.L. Ioffe, *'At the Frontier of Particle Physics / Handbook of QCD'*, M. Shifman, ed., (World Scientific); hep-ph/0011333.
10. A. De Rujula, H. Georgi and S. L. Glashow, Phys. Rev. **D12** (1975) 149. N. Isgur and G. Karl, Phys. Rev. **D18** (1978) 4187; ibid., **D19** (1979) 2653. F. E. Close, *An Introduction to Quarks and Partons*, (Academic Press, London, 1979). T. Barnes, the summary talk of this conference, these proceedings.
11. R. D. Pisarski, Phys.Lett. **B110**,155 (1982).
12. T. Hatsuda and T. Kunihiro, Phys. Lett. **B145**, 7 (1984).
13. T. Hatsuda and T. Kunihiro, Prog. Theor. Phys. **74**, 765 (1985); Phys. Rev. Lett., 158 **55**.
14. T. Hashimoto, K. Hirose, T. Kanki and O. Miyamura, Phys. Rev. Lett. **57**, 2123 (1986).
15. G. E. Brown and M. Rho, Phys. Rev. Lett. **66**, 2720 (1991).
16. G. E. Brown and M. Rho, Phys.Rep. **269**,333 (1996); hep-ph/0103102.
17. R. D. Pisarski and F. Wilczek, Phys. Rev. **D29**,338 (1984).
18. T. Kunihiro, Phys. Lett. **B219**,363 (1989).
19. V. Bernard, Ulf G. Meissner, I. Zahed, Phys. Rev. Lett. **59**, 966 (1987).
20. T. Hatsuda and S.-H. Lee, Phys. Rev. **C46**,34 (1992); T. Hatsuda, Y. Koike and S.-H. Lee, Nucl. Phys. **B394**, 221 (1993).
21. T. Matsui and H. Satz, Phys. Lett. **B178**, 416 (1986).
22. V. Elias and M. D. Scadron, Phys. Rev. Lett. **53**, 1129 (1984).
23. T. Kunihiro, Prog. Theor. Phys. Supple. **120**, 75 (1995); nucl-th/0006035; nucl-th/9604019; hep-ph/9905262.
24. See for example, T. Kunihiro, hep-ph/0009116, contained in [25].
25. The proceedings of Workshop at Yukawa Institute for Theoretical Physics, "Possible Existence of the σ-meson and Its Implications to Hadron Physics", KEK Proceedings 2000-4, (December, 2000), ed. by S. Ishida et al.

26. N. A. Törnqvist and M. Roos, Phys. Rev. Lett. **76**, 1575 (1996); M. Harada, F. Sannino and J. Schechter, Phys. Rev. **D54**, 1991 (1996); S. Ishida et al., Prog. Theor. Phys.**98**, 1005 (1997); J. A. Oller, E. Oset and J. R. Peláez, Phys. Rev. Lett. **80**, 3452 (1998); K. Igi and K. Hikasa, Phys. Rev. **D59**, 034005 (1999); G. Colangelo, J. Gasser, H. Leutwyler, Nucl. Phys. **B603**,125 (2001); Z. Xiao and H. Zheng, Nucl. Phys. **A695**,273 (2001). See also other contributions to these proceedings.
27. K. Igi and K. Hikasa, Phys. Rev. **D59**, 034005 (1999); J. A. Oller, E. Oset and J. R. Peláez, Phys. Rev. Lett. **80**, 3452 (1998); G. Colangelo, J. Gasser, H. Leutwyler, Nucl. Phys. **B603**,125 (2001); Ż. Xiao and H. Zheng, Nucl. Phys. **A695**,273 (2001).
28. J. Gasser and H. Leutwyler, Nucl. Phys. **B250**, 465 (1985).
29. E. P. Shabalin, Sov. J. Nucl. Phys. **48**,172 (1988); , T. Morozumi, C. S. Lim and A. I. Sanda, Phys. Rev. Lett. **65**, 404 (1990).
30. Prog. Theor. Phys. Suppl. **39** (1967); K. Holinde, Phys. Rep. **68**(1981), 121; G. E. Brown, in "Mesons in Nuclei", ed by M. Rho and D. Wilkinson (North Holland, 1979), p. 329; J. W. Durso, A. D. Jackson and B. J. Verwest, Nucl. Phys. **A345**(1980), 471; K. Machleidt, R. Holinde and Ch. Elster, Phys. Rev. **149** (1987) 1.
31. R. L. Jaffe and C. L. Korpa, Comm. Nucl. Part. Phys. **17** (1987) 163; J. Gasser, H. Leutwyler and M. E. Sainio, Phys. Lett. **253** (1991), 252.
32. T. Kunihiro and T. Hatsuda , Phys. Lett. **B240** (1990) 209; T. Hatsuda and T. Kunihiro , Nucl. Phys. **B387** (1992), 715.
33. See e.g., P. W. Anderson, *Basic Notion of Condensed Matter Physics* (Benjamin, California, 1984).
34. C. DeTar, in *Quark Gluon Plasma 2* ed. by R.C. Hwa (World Scientific 1995); QCD-TARO Collaboration, Nucl. Phys. **B83** Proc.Suppl.,411 (2000).
35. T. Hatsuda and T. Kunihiro , Phys. Lett. **B185**, 304 (1987).
36. T. Kunihiro, invited talk presented at Japan-China joint symposium, " Recent Topics on Nuclear Physics", Tokyo Institute of Technology, 30 Nov - 3 Dec, 1992, (nucl-th/0006035).
37. B. D. Serot and J. D. Walecka, Adv. Nucl. Phys. **16**, 1 (1986); H. A. Weldon, Phys. Lett. **B274**, 133 (1992); T. Kunihiro, Phys. Lett. **B271**, 395 (1991).
38. S. Chiku and T. Hatsuda, Phys. Rev. **D58**, 076001 (1998); M. K. Volkov, E. A. Kuraev, D. Blaschke, G. Roepke and S. M. Schmidt, Phys. Lett. **B424**, 235 (1998).
39. T. Hatsuda, T. Kunihiro and H. Shimizu, Phys. Rev. Lett. **82**, 2840 (1999).
40. D. Jido , T. Hatsuda and T. Kunihiro, Phys.Rev. **D63**,011901 (2000).
41. F. Bonutti et al. (CHAOS Collaboration), Phys. Rev. Lett. **77**, 603 (1996); Nucl. Phys. **A677**, (2000), 213. The experminet was motivated to explore the possible strong pi-pi correlations in the nuclear medium[42].
42. P. Schuck, W. Norönberg and G. Chanfray, Z. Phys. **A330**, 119(1988); G. Chanfray, Z. Aouissat, P. Schuck and W. Nörenberg, Phys. Lett. **B256**, 325 (1991). Z. Aouissat, R. Rapp, G. Chanfray, P. Schuck and J. Wambach, Nucl. Phys. **A581**, 471 (1995); R. Rapp, J. W. Durso and J. Wambach, Nucl. Phys. **A596**, 436 (1996).
43. Z. Aouissat, G. Chanfray, P. Schuck and J. Wambach, Phys. Rev. **C61**, 012202 (2000); D. Davesne, Y. J. Zhang and G. Chanfray, Phys. Rev. **C62**, 024604(2000).
44. R. Rapp et al, Phys. Rev. **C 59**, R1237 (1999); M. J. Vicente-Vacas and E. Oset, Phys. Rev. **C60**, 064621 (1999).
45. A. Starostin et al, Phys. Rev. Lett. **85**, 5539 (2000).
46. P. Camerini et al, nucl-ex/0109007.
47. H. Shimizu, these proceedings.
48. R.S. Hayano, S. Hirenzaki and A. Gillitzer, Eur. Phys. J. **A6**,99 (1999).
49. S. Hirenzaki, T. Hatsuda, K. Kume, T. Kunihiro, H. Nagahiro, Y. Okumura, E. Oset, A. Ramos, H. Toki, Y.Umemoto Nucl.Phys. **A663**, 553 (2000); S. Hirenzaki, H. Nagahiro, T. Hatsuda and T. Kunihiro, in preparation.
50. CERES Collaboration, Phys. Rev. Lett. **75**, 1272 (1995); Phys. Lett. **B422**, 405 (1998); Nucl. Phys. **A638**, 159c, (1998).
51. E325 Collaboration (K. Ozawa et al.), Phys.Rev.Lett. **86**, 5019 (2001).
52. T. Yamazaki et al, Phys.Lett. **B418**, 246 (1998); K. Itahashi et al, Phys. Rev. **C62**, 025202 (2000).

PHENOMENOLOGY

The Constituent Quark Model: a Status Report

Eric S. Swanson

Department of Physics and Astronomy, University of Pittsburgh, Pittsburgh, PA 15260, and Jefferson Lab, 12000 Jefferson Ave, Newport News, VA 23606

Abstract. A brief and biased overview of the status of the constituent quark model is presented. We concentrate on open issues and goals of hadronic phenomenology, rather than specific physics conundrums in the field. Modern attempts at addressing these issues are also presented.

INTRODUCTION

The constituent quark model has a long and distinguished history of service to hadronic physics[1]. However, its utility is restricted to quark-number conserving hadronic processes – it has nothing to say about Fock channel coupling, or about gluonic physics in general. And its connection to QCD is tenuous at best. Indeed, it is clear that present day experiment is outstripping theory and that new reliable and tractable continuum models of QCD are required to interpret and guide the new generation of hadronic experiments.

For example, it is very likely that resonant structure is seen in the exotic $J^{PC} = 1^{-+}$ channel at 1600 MeV; and something is seen at 1400 MeV at BNL, CERN, and VES. Proving that these states are mesonic hybrids will require a substantial improvement in our understanding of the dynamics of soft glue, both in terms of the structure of the putative resonance and in terms of its coupling to 'canonical' mesons. Similarly, it is tempting to interpret the extensive Crystal Barrel data on the $f_0(1500)$ as evidence for a scalar glueball. However, state mixing in the scalar sector is notorious for its strength and its obscurity. This mixing must be thoroughly mastered before we can claim the discovery of a glueball and this task will require a reliable model of soft glue. As a final example, consider the extraction of baryonic resonance parameters from Jefferson Lab, BNL, GRAAL, and Bonn. At modern energies, one must analyse data using coupled channel methods; thus πN, $\pi\pi N$, $\pi\pi\pi N$, ηN, ρN, etc channels become important and one must have a trustworthy method to parameterise the couplings between the different channels. It will also become increasingly important to have reliable estimates of background amplitudes. A moments' reflection reveals that this is a difficult problem in strong QCD: quark exchange diagrams contribute, but may also be present at the effective meson-exchange level. More perplexing is the possibility of quark-antiquark annihilation to intermediate states with excited gluonic content[1].

Although these examples were drawn from the mainstream of hadron spectroscopy, it should be stressed that these issues are rather far reaching. For example, extracting

---

[1] There is compelling evidence of this in meson-meson and meson-baryon scattering data[2].

electroweak phases will require reliable knowledge of strong phases which are generated in the necessarily present hadronic final states. Analyzing forthcoming RHIC data for putative signals of the quark-gluon plasma will require a careful subtraction of hadronic scattering background which may mask the signal. Again, a thorough knowledge of hadronic dynamics is needed.

ISSUES FACING A MODERN CONSTITUENT QUARK MODEL

- **the nature of confinement**

The quark model has long-assumed a linear (or similar) static long range interquark potential. While this is in accord with lattice data, many open issues remain. For example, what is the colour space structure of the long range force? This is required even for heavy quarks. The standard choice is $\lambda \cdot \lambda$ and this has received strong support from the lattice[3]. It is also possible to prove that this is the correct colour structure in the heavy quark limit[4]. The extrapolation to light quark masses remains an open issue. Note that the colour structure is important when one considers hadronic interactions (ie., processes with more than three valence quarks).

The Lorentz structure of confinement is not determined by Wilson loop calculations and is important for those wishing to 'relativise' quark models. Indications from heavy quarkonium spin splittings and from direct lattice computations are that the structure is scalar⊗scalar[5]. However, it should be remembered that this is the form of the *effective* interquark interaction once gluons are integrated out of the theory/model. Indeed, comparison with QCD in Coulomb gauge shows that the Lorentz structure of confinement is vector⊗vector in the heavy quark limit[6], and that the effective scalar interaction arises due to nonperturbative mixing with intermediate hybrids.

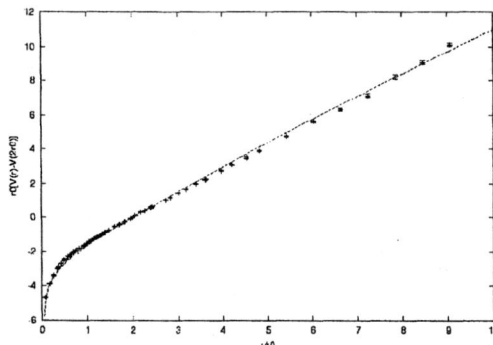

FIGURE 1. The leading Tamm-Dancoff Static Potential (line) compared to the Wilson loop confinement potential (points).

An important, and almost uniformly ignored, aspect of model building is deriving the long range confinement potential. While it is tempting to merely assume that confinement exists and to take its form from the lattice, this risks missing vital aspects of strong QCD because confinement is strongly tied to the vacuum structure of QCD, and therefore is related to the appearance of chiral symmetry breaking and constituent quarks (see 'chiral pions' below). All three of these phenomena are central features of low energy hadronic physics and must be modelled reasonably well before we can have full confidence in the new quark model. That this ambitious goal is possible has recently been demonstrated twice: with the Schwinger Dyson formalism in Landau gauge[7], and with Hamiltonian methods in Coulomb gauge[4]. The result of the latter calculation is shown in Fig. 1.

- **gluodynamics**

Glue, and especially the dynamics of glue, is conspicuously absent from most present day models of strong QCD. It is clear that glue plays a vital role in many aspects of low energy hadronic physics. Indeed, about the only place where it is relatively safe to neglect QCD gluodynamics is when discussing the static properties of mesons and baryons. Thus, for example, a reliable model of gluodynamics is required to address simple questions such as the masses and static properties of glueballs and hybrids. Just as important is the way in which these states couple to 'canonical' matter. This must be known if we are to disentangle exotics from the canonical spectrum. Again the lattice may be of great assistance. Lattice computations of glueball masses[8] serve as a litmus test for any putative models of gluodynamics. Furthermore, high precision computations of the adiabatic excited gluon energies provide our first glimpse into the dynamics of strongly interacting glue[9].

- **unquenching the quark model**

A closely related issue is going beyond the valence approximation in hadronic phenomenology. While it is clear that this is a pressing issue for the investigation of nonvalence physics, such as the strangeness content of the proton[10], or the perplexing robustness of the OZI rule in the face of hadronic loops[11], it is also relevant to spectroscopy. Typical hadronic widths of 150 MeV point to typical hadronic mass shifts of a similar scale. Furthermore, meson loops cause spin splittings which can confound simple attempts at deriving these. It is clear that a quark model which unifies quark-antiquark pair creation with valence physics is required[12].

Although there are a number of technical issues which need to be overcome to achieve this unification (such as efficiently solving coupled channel problems, determining the optimal number of channels to include in a given problem, and accounting for excluded channels), the main issue is the form of the quark creation operator. Certainly this operator is dominated by nonperturbative glue, but a detailed microscopic description is lacking. The most popular model to date is the $^3P_0$ model[13] (first diagram in Fig. 2) which assumes an effective vertex which produces quark pairs with vacuum quantum numbers. This model produces a reasonably reliable phenomenology[14].

Other possible decay mechanisms exist and need to be explored. For example, the naive perturbative diagram (second diagram in Fig. 2) is the leading diagram in perturbation theory. However predictions of the D/S amplitude ratios (which are sensitive to the assumed Lorentz structure of the decay vertex) in $b_1 \to \omega\pi$ and $a_1 \to \rho\pi$ strongly prefer a $^3P_0$ pair creation over $^3S_1$[15].

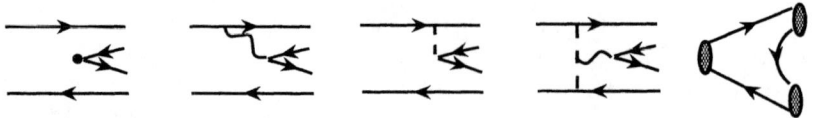

FIGURE 2. Possible decay mechanisms.

Another possible decay mechanism is obtained by isolating the instantaneous portion of diagram 2 (diagram 3 of Fig. 2) – say by working in radiation gauge. However, a model examination of this process reveals that it is strongly suppressed with respect to the $^3P_0$ vertex[16].

A promising approach to hadronic properties is provided by the Schwinger-Dyson formalism. The leading decay mechanism in this approach is the triangle diagram[17] (last diagram of Fig. 2). This method has the benefit of employing the same kernel to describe quark-quark interactions and quark-antiquark pair production. However, this may be too restrictive since vector pair creation appears to be disfavoured by the b_1 and a_1 amplitude ratios. Nevertheless it is possible that the situation may be saved by the relativistic character of the Schwinger-Dyson approach (certainly, this physics is not explored in the quark model calculations which favour the $^3P_0$ mechanism).

The final possibility considered here is the production of a $q\bar{q}$ pair directly from the confinement potential/flux tube (fourth diagram, Fig. 2). This diagram is a leading term in Coulomb gauge QCD[4]. Although the phenomenology of decays in Coulomb gauge QCD has not been explored, it is promising because this diagram yields a vertex with $^3P_0$ quantum numbers, and would be our first microscopic justification of the $^3P_0$ model.

Given the importance of unquenching the quark model and the paucity of our knowledge of the nonperturbative nature of quark pair creation, *a lattice exploration of the form of the decay vertex should be a high priority topic in the near future.*

- **chiral pions**

Pions are an important part of any attempt to understand strong QCD. As the lightest hadrons they dominate nuclear physics; pion cloud effects are important, and pions are ubiquitous in final states of many hadronic experiments. Having a firm grip on their qualities is of central importance to constructing viable models of strong QCD.

It is often said that the constituent quark model view of pions as quark-antiquark bound states is in conflict with their quasiGoldstone boson nature. However, these two world-views need not be at odds. For example, the very existence of the constituent quark model is due to the existence of light pions: the dynamics which causes dynamical symmetry breaking (and Goldstone bosons) also creates quark-like quasiparticle excitations – the constituent quarks[18]. A recent paper[19], show explicitly how the Goldstone, collective, nature of the pion can coexist with the $q\bar{q}$ bound state quark model pion; briefly, both descriptions are correct when the appropriate degrees of freedom are employed (partonic for the Goldstone modes; constituent for the quark model states). Thus it is likely that a good phenomenology may be obtained simply by ignoring the underlying chiral aspects of the pion. However, incorporating the physics of chiral symmetry breaking is important if one wishes to deal with aspects of the QCD vacuum or if the model is strongly constrained (so that pionic fluctuations into many-quark Fock components may not be absorbed into model parameters).

- **relativity**

This is a longstanding and well known problem with the constituent quark model which is a left-over from the early days of hadronic physics. There is really no reason to continue with nonrelativistic approaches (except that they are computationally simple and they work reasonably well!) – and several groups have mounted efforts to construct 'relativised' quark models. These typically fall into two categories, light cone/Bakamjian Thomas models[20] or Schwinger Dyson/Bethe Salpeter models[21]. The latter are closer to field theory (or *are* truncated field theory) and offer great hope.

It is possible to overstate the case for covariance. Any nonperturbative approach must break covariance at some level, for example, the lattice breaks Lorentz invariance by working on a grid and models typically must truncate at some level in Fock space. Both of these problems may be removed in principle – in practice they are *not* removed, but the effects may be checked and are seen to be small (at least in the case of lattice gauge theory). It is perhaps more useful to adopt a practical attitude, for example, it would be useful if the computation of the pion decay constant via the PCAC relation $\langle 0|A_\mu^a(0)|\pi^b(p)\rangle = if^{ab}p_\mu$ did *not* depend on the spacetime index.

- **short distance dynamics**

It is easy to believe that the form of the short distance quark interaction is resolved by QCD; short distance means high Q^2, that means small α_s, and that means perturbative one gluon exchange. The phenomenology of one gluon exchange works extremely well[22] and has the virtue of being universal[23]. However the *evolution* of one gluon exchange to intermediate or large distance is typically ignored. Indeed, in bound state perturbation theory (which is the way all perturbation theory for hadronic physics should be performed), the diagram corresponding to one gluon exchange (first diagram of Fig. 3) corresponds to mixing with intermediate hybrids (second diagram, Fig. 3), and the first diagram does not occur. However, if one is dealing with a field theory, the first diagram reappears as a counterterm which is active at momentum transfer above the renormalization scale. How these two evolve into each other is therefore an issue dealt with by the renormalization group flow of the underlying field theory and should be properly addressed in a new quark model.

This issue is related to a subtlety in most quark models: how are short range and long range dynamics to be merged? If confinement arises from multiple gluon exchange, surely it is not correct to simply add one gluon exchange to an assumed linear potential. Resolving this issue is very difficult in covariant gauges, however, in Coulomb gauge there is a natural separation of instantaneous and transverse potentials which allows the issue to be resolved simply.

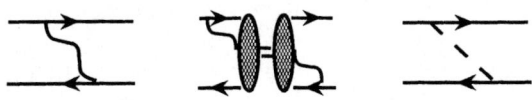

FIGURE 3. Possible short distance interactions.

The last diagram of Fig. 3 represents meson exchange contributions to the quark interaction. If one admits that pion (and meson) loops can affect hadron properties (as we have argued above) then one must allow these sort of diagrams. However, it is an open issue as to how important they are. Robson examined this possibility years ago[24] in the context of the tensor splitting of the $S_{11}(1535)$ and $S_{11}(1650)$ and rejected it. It has since been taken up again[25], although not without criticism[26].

- **topological aspects**

Shortly after the notion of topology (here we focus on instantons) was introduced to QCD, 't Hooft used instantons to resolve the $U_A(1)$ problem[27] – namely that the axial symmetry of (massless) QCD is not realized in the Wigner-Weyl or Nambu-Goldstone modes. It has also been argued that collective effects involving infinitely many instantons may generate the quark condensate, and hence, chiral symmetry breaking. Finally, we mention that instantons induce an effective quark interaction, however it appears that this force does not confine[28].

If we are to accept the instanton resolution to the $U_A(1)$ problem, then instanton field configurations must be accepted as an important subset of the vacuum field configurations, and their effects should be included in a new quark model. Indeed, computations with instanton models indicate that they may successfully describe many properties of light hadrons[28]. There is also lattice evidence that instantons dominate the vacuum. We note, however, that old arguments of Witten against instantons[29] have been resurrected[30]. This paper, in turn, has been criticised[31].

There appears to be little room for instantons in the nonrelativistic constituent quark model, they simply aren't needed to explain the spectrum. However, if instantons do dominate low energy QCD, they must be incorporated into models. The Bonn group (see Refs. [21, 17]) has been developing a model which includes instanton-induced quark interactions in a relativistic Bethe-Salpeter approach, and the resulting phenomenology appears quite successful.

Moving beyond the phenomenological stage requires incorporating the effects of instantons in a way which is consistent with the new quark model's treatment of the vacuum. And this means that a consistent treatment of confinement, chiral symmetry breaking, and instanton effects must be found. I know of no attempts in this direction, and it forms a major challenge for future efforts.

- **hadronic interactions**

Hadronic interactions form an important, if under-appreciated, portion of hadronic physics. They are central to developing a microscopic theory of nuclear physics, to nuclear astrophysics, to the analysis of 'background' in N^* and other resonance (hybrids, glueballs) experiments, and to electroweak experiments (where hadronic final state interactions must be properly accounted for). As such, any new quark model should carry with it a well-defined, tractable, methodology for computing hadronic interactions.

The present state of affairs is less than ideal. Constituent quark model calculations date from the '70's[32], and continue today with resonating group[33] and perturbative[34] methods (see Fig. 4). While it is likely that these quark model calculations provide reasonable guidance at low energies (except for pion-dominated physics where one must hope that the necessary chiral properties are captured in the quark model – see the discussion above), it is less clear how applicable they are at high momentum transfer. It is here that light cone approaches[35] are expected to be applicable (certainly to inclusive reactions, less certainly to exclusive). Of course what is needed is a consistent formalism which allows the computation of the hadronic wavefunctions and hadronic scattering in all energy regimes simultaneously. It is evident that close contact with QCD needs to be maintained if these goals are to be achieved.

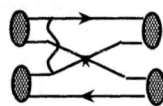 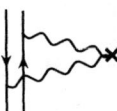

FIGURE 4. Hadronic Interactions. The left figure is a basic diagram in light cone and perturbative quark model computations of hadronic interactions. It also provides the kernel in resonating group methods. The right figure is assumed to dominate the interaction of a small meson with an external colour field.

One attempt in this regard was made many years ago by Peskin[36]. This approach is essentially a multipole expansion of the interaction of a small colour singlet state with an external colour field (see Fig. 4). The resulting dipole interaction is assumed to be applicable to hadron-hadron interactions as well. However, we note that one prediction of this model is that the cross section for ψ' with hadronic matter is 5000 times larger than that of ψs. Indeed, Peskin fears that even the Υ system may be too light for the method to work[37].

The historical litmus test for hadronic models has been a computation of the hadron spectrum. It is becoming increasingly clear that this is inadequate because the extraction of resonance parameters is fraught with ambiguity. Computations which are closer to the data are required – in particular reaction dynamics need to be incorporated into new quark model predictions. This field is in its infancy, however, it has started[38].

CONCLUSIONS

A crucial aspect of the new quark model is a thorough understanding of the QCD vacuum. This is required to meet many of the issues raised above: chiral symmetry breaking, confinement, topology, and gluodynamics. These issues, in turn, are central to developing a viable model of low energy QCD. It will clearly be a stiff challenge to develop a model which adequately addresses all of these issues; however, hadronic physics provides our only window into strongly interacting field theory and is a vital component of nuclear physics, astrophysics, cosmology, and physics beyond the standard model[39]. It will therefore be worth the effort to develop such a model!

An efficient description of hadronic physics will require the identification of appropriate degrees of freedom – constituent quarks, massive gluons, flux tubes, instantons, or something new. However, if the ambitious goals laid out here are to be achieved, a direct connection of these degrees of freedom to QCD must be maintained. We can take heart that progress is being made. Of particular note is the assistance of lattice gauge theory, which promises to be a useful shortcut to the development of new ideas and to testing these ideas.

REFERENCES

1. M. Gell-Mann, Phys. Lett. **8**, 214 (1964); G. Zweig, CERN preprints TH401 and TH412 (1964), Proceedings of Baryon 1980, pg. 439 (Ed. N. Isgur); Morpugo, J. Phys. **2**, 95 (1965); R. Dalitz, *Eighth International Conference on High Energy Physics*, Berkeley, (1966).
2. E. S. Swanson, "Hadron hadron interactions in the constituent quark model: Results and extensions," hep-ph/0102267.
3. G.S. Bali, Phys. Rev. **D62**, 114503 (2000).
4. A. P. Szczepaniak and E. S. Swanson, hep-ph/0107078, to appear Phys. Rev. D.
5. H.J. Schnitzer, Phys. Rev. Lett. **35**, 1540 (1975).
6. A. P. Szczepaniak and E. S. Swanson, Phys. Rev. D **55**, 3987 (1997).
7. D. Atkinson and J.C.R. Bloch, Phys. Rev.d **D58**, 094036 (1998); L. von Smekal, A. Hauck, and R. Alkofer, Ann. Phys. **267**, 1 (1998).
8. C. J. Morningstar and M. J. Peardon, Phys. Rev. D **60**, 034509 (1999).
9. K.J. Juge, J. Kuti, and C.J. Morningstar, Nucl. Phys. Proc. Suppl. **63**, 326 (1998).
10. P. Geiger and N. Isgur, Phys. Rev. D **55**, 299 (1997).
11. P. Geiger and N. Isgur, Phys. Rev. D **47**, 5050 (1993).
12. N. A. Tornqvist and P. Żenczykowski, Phys. Rev. D **29**, 2139 (1984).
13. L. Micu, Nucl. Phys. **B10**, 521 (1969); R. Carlitz and M. Kislinger, Phys. Rev. D **2**, 336 (1970); A. Le Yaouanc, L. Oliver, O. Pene, and J.-C. Raynal, Phys. Rev. D **8**, 2233 (1973); Phys. Lett. **71 B**, 397 (1977); *ibid* **72 B**, 57 (1977).
14. R. Kokoski and N. Isgur, Phys. Rev. D **35**, 907 (1987); S. Capstick and W. Roberts, Phys. Rev. D **47**, 1994; A. LeYaouanc *et al.*, *Hadron Transitions in the Quark Model*, (Gordon Breach, New York, 1988); S. Godfrey and N. Isgur, Phys. Rev. D **32**, 189 (1985); R. Bonnaz and B. Silvestre-Brac, Prog. Part. Nucl. Phys. **44**, 369 (2000).
15. J.W. Alcock, M.J. Burfitt, W.N. Cottingham, Z. Phys. **C25**, 161 (1984); P. Geiger and E. S. Swanson, Phys. Rev. D **50**, 6855 (1994).
16. E. S. Ackleh, T. Barnes and E. S. Swanson, Phys. Rev. D **54**, 6811 (1996).
17. J. C. Bloch, Y. L. Kalinovsky, C. D. Roberts and S. M. Schmidt, Phys. Rev. D **60**, 111502 (1999); M. A. Pichowsky, S. Walawalkar and S. Capstick, Phys. Rev. D **60**, 054030 (1999); R. Ricken, M. Koll, D. Merten, B. C. Metsch and H. R. Petry, Eur. Phys. J. A **9**, 221 (2000).
18. S. L. Adler and A. C. Davis, Nucl. Phys. B **244**, 469 (1984); J. R. Finger and J. E. Mandula, Nucl. Phys. B **199**, 168 (1982); A. Le Yaouanc, L. Oliver, S. Ono, O. Pene and J. C. Raynal, Phys. Rev. D **31**, 137 (1985).
19. A. P. Szczepaniak and E. S. Swanson, Phys. Rev. Lett. **87**, 072001 (2001).
20. S. Capstick and B. D. Keister, Phys. Rev. D **51**, 3598 (1995); L. S. Kisslinger, H. M. Choi and C. R. Ji, Phys. Rev. D **63**, 113005 (2001).
21. P. Maris and C. D. Roberts, Phys. Rev. C **56**, 3369 (1997); U. Loring, B. C. Metsch and H. R. Petry, Eur. Phys. J. A **10**, 395 (2001).
22. A. De Rujula, H. Georgi and S. L. Glashow, Phys. Rev. D **12**, 147 (1975);
23. N. Isgur and G. Karl, Phys. Rev. D **18**, 4187 (1978).
24. D. Robson, *Proceedings of the Topical Conference on Nuclear Chromodynamics*, Argonne National Laboratory (1988), Eds. J. Qiu and D. Sivers (World Scientific), pg. 174.
25. L. Y. Glozman and D. O. Riska, Phys. Rept. **268**, 263 (1996).
26. N. Isgur, Phys. Rev. D **62**, 054026 (2000).
27. G. 't Hooft, Phys. Rev. Lett. **37**, 8 (1976); A.M. Polyakov, Phys. Lett. **59B**, 82 (1975); Nucl. Phys. **B121**, 429 (1977); A.A. Belavin, A.M. Polyakov, A. Schwartz, and Y. Tyupkin, Phys. Lett. **59B**, 85 (1975); R. Jackiw and C. Rebbi, Phys. Rev. Lett. **37**, 172 (1976).
28. See T. Schäfer and E. Shuryak, Rev. Mod. Phys. **70**, 323 (1998).
29. E. Witten, Nucl. Phys. B **149**, 285 (1979); see also J. Kogut and L. Susskind, Phys. Rev. D **11**, 3594 (1975).
30. I. Horvath, N. Isgur, J. McCune and H. B. Thacker, hep-lat/0102003.
31. R. G. Edwards and U. M. Heller, hep-lat/0105004; T. DeGrand and A. Hasenfratz, hep-lat/0103002.
32. D.A. Liberman, Phys. Rev. **D16**, 1542 (1977).
33. M. Oka and K. Yazaki, Phys. Lett. **90B** (1980), 41; Prog. Theor. Phys. **66** (1981), 556; *ibid.*, p.572; A. Faessler, F. Fernandez, G. Lubeck and K. Shimizu, Phys. Lett. **112B**, Y. Suzuki and K.T. Hecht,

Phys. Rev. **C27**. 299 (1983). (1982), 201;
34. T. Barnes and E. S. Swanson, Phys. Rev. D **46**, 131 (1992); E. S. Swanson, Annals Phys. **220**, 73 (1992).
35. S. J. Brodsky, "Hadronic light-front wavefunctions and QCD phenomenology," hep-ph/0102051.
36. G. Bhanot and M. E. Peskin, Nucl. Phys. B **156**, 391 (1979).
37. M. Peksin, private communication.
38. T. S. Lee and T. Sato, Nucl. Phys. A **684**, 327 (2001).
39. S. Capstick *et al.*, "Key issues in hadronic physics," hep-ph/0012238.

Heavy Quark Potential and Mass Spectra of Heavy Mesons

D. Ebert*, R. N. Faustov[†] and V. O. Galkin[†]

*Institut für Physik, Humboldt–Universität zu Berlin, 10115 Berlin, Germany
[†]Russian Academy of Sciences, Scientific Council for Cybernetics, Moscow 117333, Russia

Abstract. The relativistic quark model is presented. The quark-antiquark potential for the Schrödinger-like equation is constructed with the account of retardation effects and one-loop radiative corrections. It consists of the one-gluon exchange part and the confining part which is the mixture of the Lorentz scalar and Lorentz vector contributions. The latter contains both the Dirac and Pauli terms. In the v^2/c^2 approximation the mass spectra of heavy quarkonia (charmonium and bottomonium) are calculated in good agreement with experiment. In the case of heavy-light mesons (B and D) the light quark is treated completely relativistically and only the expansion in the inverse heavy quark mass is used. The mass spectra of the ground and excited states of D, D_s, B, B_s mesons are calculated. They exhibit some features of the so-called "level inversion". The obtained results are generally in accord with experimental data. Still there exist some discrepancies between measurements of different collaborations

INTRODUCTION

The heavy flavour studies lie on the frontiers of elementary particle physics. The investigation of heavy meson mass spectra provides substantial information about the nonperturbative content of quantum chromodynamics (QCD) and fundamental parameters of the relativistic quark model. Since it is impossible to give here any kind of comprehensive review of the subject we present instead the consideration based mainly on two papers [1, 2]. They are extended to include recent experimental data and should be taken as an illustration of theoretical approaches.

RELATIVISTIC QUARK MODEL

In the quasipotential approach the meson is described by the wave function of the bound quark-antiquark state, which satisfies the quasipotential equation of the Schrödinger type in the centre-of-mass frame:

$$\left(\frac{b^2(M)}{2\mu_R} - \frac{\mathbf{p}^2}{2\mu_R}\right)\Psi_M(\mathbf{p}) = \int \frac{d^3q}{(2\pi)^3} V(\mathbf{p},\mathbf{q};M)\Psi_M(\mathbf{q}), \qquad (1)$$

where μ_R is the relativistic reduced mass

$$\mu_R = \frac{M^4 - (m_q^2 - m_Q^2)^2}{4M^3},$$

and $b^2(M)$ denotes the on-mass-shell relative momentum squared

$$b^2(M) = \frac{[M^2 - (m_q + m_Q)^2][M^2 - (m_q - m_Q)^2]}{4M^2}.$$

The kernel $V(\mathbf{p}, \mathbf{q}; M)$ is the quasipotential operator of the quark-antiquark interaction. It is constructed with the help of the off-mass-shell scattering amplitude, projected onto the positive energy states.

We have assumed that the effective interaction is the sum of the usual one-gluon exchange term and the mixture of vector and scalar linear confining potentials. The quasipotential is then defined by

$$V(\mathbf{p}, \mathbf{q}; M) = \bar{u}_q(p)\bar{u}_Q(-p)\left\{\frac{4}{3}\alpha_s D_{\mu\nu}(\mathbf{k})\gamma_q^\mu\gamma_Q^\nu + V_{\rm conf}^V(\mathbf{k})\Gamma_q^\mu\Gamma_{Q;\mu} \right.$$
$$\left. + V_{\rm conf}^S(\mathbf{k})(\mathbf{p},\mathbf{q};M)\right\}u_q(q)u_Q(-q), \tag{2}$$

where α_s is the QCD coupling constant, $D_{\mu\nu}$ is the gluon propagator in the Coulomb gauge and $\mathbf{k} = \mathbf{p} - \mathbf{q}$; the Dirac spinor is given by

$$u^\lambda(p) = \sqrt{\frac{\varepsilon(p) + m}{2\varepsilon(p)}}\begin{pmatrix} 1 \\ \frac{\sigma\mathbf{p}}{\varepsilon(p) + m}\end{pmatrix}\chi^\lambda$$

with $\varepsilon(p) = \sqrt{\mathbf{p}^2 + m^2}$. The effective long-range vector vertex has the form

$$\Gamma_\mu(\mathbf{k}) = \gamma_\mu + \frac{i\kappa}{2m}\sigma_{\mu\nu}k^\nu, \qquad k^0 = 0,$$

where κ is the nonperturbative anomalous chromomagnetic moment of quarks. Vector and scalar confining potentials in the nonrelativistic limit reduce to

$$V_{\rm conf}^V(r) = (1-\varepsilon)(Ar + B), \qquad V_{\rm conf}^S(r) = \varepsilon(Ar + B),$$

reproducing $V_{\rm conf}(r) = V_{\rm conf}^S(r) + V_{\rm conf}^V(r) = Ar + B$, where ε is the mixing coefficient.
The potential parameters are as follows:
$A = 0.18$ GeV^2, $B = -0.30$ GeV, $\alpha_s(m_c) = 0.32$, $\alpha_s(m_b) = 0.22$.
The constituent quark masses are
$m_b = 4.88$ GeV, $m_s = 0.50$ GeV, $m_c = 1.55$ GeV, $m_{u,d} = 0.33$ GeV.
The value of the mixing parameter $\varepsilon = -1$ is fixed by matching the heavy quark expansion and from the consideration of charmonium radiative decays ($J/\psi \to \eta_c\gamma$). The anomalous chromomagnetic quark moment $\kappa = -1$ is determined from the heavy quark expansion and from fitting the fine splitting of heavy quarkonia ^3P_J states. The long range chromomagnetic contribution to the potential is proportional to $(1 + \kappa)$ and vanishes for $\kappa = -1$ in accord with the flux tube model.

HEAVY QUARK-ANTIQUARK POTENTIAL

With the account of retardation effects and one loop radiative corrections the spin-independent potential reads as [1]

$$
\begin{aligned}
V_{SI}(r) = & -\frac{4}{3}\frac{\bar{\alpha}_V(\mu^2)}{r} + Ar + B - \frac{4}{3}\frac{\beta_0\alpha_s^2(\mu^2)\ln(\mu r)}{2\pi}\frac{1}{r} \\
& + \frac{1}{8}\left(\frac{1}{m_a^2} + \frac{1}{m_b^2}\right)\Delta\left[-\frac{4}{3}\frac{\bar{\alpha}_V(\mu^2)}{r} - \frac{4}{3}\frac{\beta_0\alpha_s^2(\mu^2)\ln(\mu r)}{2\pi}\frac{1}{r}\right. \\
& \left. + (1-\varepsilon)(1+2\kappa)Ar\right] + \frac{1}{2m_am_b}\left(\left\{-\frac{4}{3}\frac{\bar{\alpha}_V}{r}\left[\mathbf{p}^2 + \frac{(\mathbf{p}\cdot\mathbf{r})^2}{r^2}\right]\right\}_W\right. \\
& \left. -\frac{4}{3}\frac{\beta_0\alpha_s^2(\mu^2)}{2\pi}\left\{\mathbf{p}^2\frac{\ln(\mu r)}{r} + \frac{(\mathbf{p}\cdot\mathbf{r})^2}{r^2}\left(\frac{\ln(\mu r)}{r}-\frac{1}{r}\right)\right\}_W\right) \\
& + \left[\frac{1-\varepsilon}{2m_am_b} - \frac{\varepsilon}{4}\left(\frac{1}{m_a^2}+\frac{1}{m_b^2}\right)\right]\left\{Ar\left[\mathbf{p}^2 - \frac{(\mathbf{p}\cdot\mathbf{r})^2}{r^2}\right]\right\}_W \\
& - \frac{\varepsilon\lambda_S}{2}\left[\frac{1}{2}\left(\frac{1}{m_a^2}+\frac{1}{m_b^2}\right)+\frac{1}{m_am_b}\right]\left\{Ar\left[\mathbf{p}^2+\frac{(\mathbf{p}\cdot\mathbf{r})^2}{r^2}\right]\right\}_W \\
& + \left[\frac{1}{4}\left(\frac{1}{m_a^2}+\frac{1}{m_b^2}\right)+\frac{1}{m_am_b}\right]B\mathbf{p}^2,
\end{aligned} \quad (3)
$$

where

$$
\bar{\alpha}_V(\mu^2) = \alpha_s(\mu^2)\left[1+\left(\frac{a_1}{4}+\frac{\gamma_E\beta_0}{2}\right)\frac{\alpha_s(\mu^2)}{\pi}\right],
$$

$$
\alpha_s(\mu^2) = \frac{4\pi}{\beta_0\ln(\mu^2/\Lambda^2)}, \quad a_1 = \frac{31}{3} - \frac{10}{9}n_f, \quad \beta_0 = 11 - \frac{2}{3}n_f.
$$

Here n_f is a number of flavours, μ is a renormalization scale and the subscript W means the Weyl ordering of operators.

The spin-dependent part of the quark-antiquark potential for equal quark masses ($m_a = m_b = m$) is given by [1]

$$
V_{SD}(r) = a\,\mathbf{L}\cdot\mathbf{S} + b\left[\frac{3}{r^2}(\mathbf{S}_a\cdot\mathbf{r})(\mathbf{S}_b\cdot\mathbf{r}) - (\mathbf{S}_a\cdot\mathbf{S}_b)\right] + c\,\mathbf{S}_a\cdot\mathbf{S}_b, \quad (4)
$$

$$
\begin{aligned}
a = & \frac{1}{2m^2}\left\{\frac{4\alpha_s(\mu^2)}{r^3}\left(1+\frac{\alpha_s(\mu^2)}{\pi}\left[\frac{1}{18}n_f - \frac{1}{36} + \gamma_E\left(\frac{\beta_0}{2}-2\right) + \frac{\beta_0}{2}\ln\frac{\mu}{m}\right.\right.\right. \\
& \left.\left.\left. + \left(\frac{\beta_0}{2}-2\right)\ln(mr)\right]\right) - \frac{A}{r} + 4(1+\kappa)(1-\varepsilon)\frac{A}{r}\right\}, \\
b = & \frac{1}{3m^2}\left\{\frac{4\alpha_s(\mu^2)}{r^3}\left(1+\frac{\alpha_s(\mu^2)}{\pi}\left[\frac{1}{6}n_f + \frac{25}{12} + \gamma_E\left(\frac{\beta_0}{2}-3\right) + \frac{\beta_0}{2}\ln\frac{\mu}{m}\right.\right.\right. \\
& \left.\left.\left. + \left(\frac{\beta_0}{2}-3\right)\ln(mr)\right]\right) + (1+\kappa)^2(1-\varepsilon)\frac{A}{r}\right\}, \\
c = & \frac{4}{3m^2}\left\{\frac{8\pi\alpha_s(\mu^2)}{3}\left(\left[1+\frac{\alpha_s(\mu^2)}{\pi}\left(\frac{23}{12}-\frac{5}{18}n_f-\frac{3}{4}\ln 2\right)\right]\delta^3(r)\right.\right.
\end{aligned}
$$

$$+\frac{\alpha_s(\mu^2)}{\pi}\left[-\frac{\beta_0}{8\pi}\nabla^2\left(\frac{\ln(\mu/m)}{r}\right)+\frac{1}{\pi}\left(\frac{1}{12}n_f-\frac{1}{16}\right)\nabla^2\left(\frac{\ln(mr)+\gamma_E}{r}\right)\right]\right)$$
$$+(1+\kappa)^2(1-\varepsilon)\frac{A}{r}\bigg\},$$

where **L** is the orbital momentum and $\mathbf{S}_{a,b}$, $\mathbf{S}=\mathbf{S}_a+\mathbf{S}_b$ are the spin momenta. The total angular momentum is $\mathbf{J}=\mathbf{L}+\mathbf{S}$.

HEAVY QUARKONIUM MASS SPECTRA

The heavy quarkonium is similar to the positronium atom and its levels are usually specified by the notation $n^{(2S+1)}L_J$, where n is the radial quantum number. The results of calculations of heavy quarkonium mass spectra on the basis of Eqs. (1), (3), (4) are presented in Tables 1 and 2 [1].

Recently the contribution of the finite c-quark mass to the bottomonium mass spectrum in one loop was considered in our model [4]. The correction to the $b\bar{b}$ static poten-

TABLE 1. Charmonium Mass Spectrum (GeV).

State $(n^{(2S+1)}L_J)$	Particle	Theory [1]	PDG(2000) [3]	BES	CLEO	E835
1^1S_0	η_c	2.979	2.9798(18)	2.9763	2.9804	
1^3S_1	J/Ψ	3.096	3.09687(4)			
1^3P_0	χ_{c0}	3.424	3.4150(8)	3.4141		3.4154
1^3P_1	χ_{c1}	3.510	3.51051(12)			
1^3P_2	χ_{c2}	3.556	3.55618(13)			
2^1S_0	η_c'	3.583	3.594(5)			
2^3S_1	Ψ'	3.686	3.68596(9)			
1^3D_1		3.798	3.7699(25)*			
1^3D_2		3.813				
1^3D_3		3.815				
2^3P_0	χ_{c0}'	3.854				
2^3P_1	χ_{c1}'	3.929				
2^3P_2	χ_{c2}'	3.972				
3^1S_0	η_c''	3.991				
3^3S_1	Ψ''	4.088	4.040(10)*			
2^3D_1		4.194	4.159(20)*			
2^3D_2		4.215				
2^3D_3		4.223				

* Mixture of S and D states

TABLE 2. Bottomonium Mass Spectrum (GeV).

State $(n^{(2S+1)}L_J)$	Particle	Theory [1]	PDG(2000) [3]	CLEO	MD1
1^1S_0	η_b	9.400			
1^3S_1	Υ	9.460	9.46030(26)		9.46051
1^3P_0	χ_{b0}	9.864	9.8599(10)	9.8600	
1^3P_1	χ_{b1}	9.892	9.8927(6)	9.8937	
1^3P_2	χ_{b2}	9.912	9.9126(5)	9.9119	
2^1S_0	η'_b	9.990			
2^3S_1	Υ'	10.020	10.02326(31)		10.0235
1^3D_1		10.151			
1^3D_2		10.157			
1^3D_3		10.160			
2^3P_0	χ'_{b0}	10.232	10.2321(6)		
2^3P_1	χ'_{b1}	10.253	10.2552(5)		
2^3P_2	χ'_{b2}	10.267	10.2685(4)		
3^1S_0	η''_b	10.328			
3^3S_1	Υ''	10.355	10.3552(5)		
2^3D_1		10.441			
2^3D_2		10.446			
2^3D_3		10.450			
3^3P_0	χ''_{b0}	10.498			
3^3P_1	χ''_{b1}	10.516			
3^3P_2	χ''_{b2}	10.529			
4^1S_0	η'''_b	10.578			
4^3S_1	Υ'''	10.604	10.5800(35)		

tial due to $m_c \neq 0$ is approximately given by [5] ($a_0 = 5.2$, $\gamma_E = 0.5772...$)

$$\delta V(r) \cong -\frac{4}{9}\frac{\alpha_s^2}{\pi r}[\ln(\sqrt{a_0}m_c r) + \gamma_E + \mathrm{E}_1(\sqrt{a_0}m_c r)], \qquad \mathrm{E}_1(x) = \int_x^\infty \frac{dt}{t}e^{-t}.$$

Averaging over solutions of Eq. (1) with the Cornell potential yields the following bottomonium mass shifts:

State	1S	1P	2S	1D	2P	3S
$\langle \delta V \rangle$, MeV	-12	-9.3	-8.7	-7.6	-7.5	-7.2

These shifts are within the estimates of theoretical uncertainties (~ 10 MeV) and partially could be adsorbed either in the value of the constituent b quark mass or in the constant term B of the confining potential.

MASS SPECTRA OF B AND D MESONS

The B and D mesons are alike the hydrogen atom with the heavy quark $Q = b, c$ near the centre-of-mass and the light quark $q = u, d, s$ orbiting around it. Thus it is appropriate to make the expansion in the inverse heavy quark mass. In the limit of infinitely heavy quark ($m_Q \to \infty$) its mass and spin decouple and as a result heavy quark symmetry arises. In this limit the meson angular momentum is a sum of the orbital momentum $\mathbf{L}$ and the light quark spin $\mathbf{S}_q$ i.e. $\mathbf{j} = \mathbf{L} + \mathbf{S}_q$. The $1/m_Q$ correction to the $Q\bar{q}$ potential depends on the heavy quark spin $\mathbf{S}_Q$ and leads to the spin-spin interaction. The total angular momentum is $\mathbf{J} = \mathbf{j} + \mathbf{S}_Q$. Thus, for S-wave mesons there is a doublet of $j = 1/2$ states with $J^P = 0^-, 1^-$ (P is the meson parity) which are degenerate in the limit $m_Q \to \infty$. For P-wave mesons there are similarly two doublets of initially degenerate states with $j = 1/2$ ($J^P = 0^+, 1^+$) and $j = 3/2$ ($J^P = 1^+, 2^+$). The $j = 1/2$ levels are expected to be broad because they decay in an S-wave, while the $j = 3/2$ levels should be narrow since they decay in a D-wave.

The heavy-light quark-antiquark potential in configuration space in the limit $m_Q \to \infty$ reads as [2] ($V_{\text{Coul}}(r) = -(4/3)\alpha_s/r$, $\alpha_s = 0.5$ for B, D; $\alpha_s = 0.45$ for B_s, D_s)

$$V_{m_Q \to \infty}(r) = \frac{E_q + m_q}{2E_q}\left[V_{\text{Coul}}(r) + V_{\text{conf}}(r) + \frac{1}{(E_q + m_q)^2}\left\{\mathbf{p}[\tilde{V}_{\text{Coul}}(r)\right.\right.$$
$$\left.+ V^V_{\text{conf}}(r) - V^S_{\text{conf}}(r)]\mathbf{p} - \frac{E_q + m_q}{2m_q}\Delta V^V_{\text{conf}}(r)[1 - (1 + \kappa)] \right.$$
$$\left.\left.+ \frac{2}{r}\left(\tilde{V}'_{\text{Coul}}(r) - V'^S_{\text{conf}}(r) - V'^V_{\text{conf}}(r)\left[\frac{E_q}{m_q} - 2(1+\kappa)\frac{E_q + m_q}{2m_q}\right]\right)\mathbf{LS}_q\right\}\right]. \quad (5)$$

The $1/m_Q$ correction to the potential (5) is given by [2]

$$\delta V_{1/m_Q}(r) = \frac{1}{E_q m_Q}\left\{\mathbf{p}\left[V_{\text{Coul}}(r) + V^V_{\text{conf}}(r)\right]\mathbf{p} + V'_{\text{Coul}}(r)\frac{\mathbf{L}^2}{2r}\right.$$
$$- \frac{1}{4}\Delta V^V_{\text{conf}}(r) + \left[\frac{1}{r}V'_{\text{Coul}}(r) + \frac{(1+\kappa)}{r}V'^V_{\text{conf}}(r)\right]\mathbf{LS}$$
$$+ \frac{1}{3}\left(\frac{1}{r}V'_{\text{Coul}}(r) - V''_{\text{Coul}}(r) + (1+\kappa)^2\left[\frac{1}{r}V'^V_{\text{conf}}(r) - V''^V_{\text{conf}}(r)\right]\right) \quad (6)$$
$$\left.\times \left[-\mathbf{S}_q\mathbf{S}_Q + \frac{3}{r^2}(\mathbf{S}_q\mathbf{r})(\mathbf{S}_Q\mathbf{r})\right] + \frac{2}{3}\left[\Delta V_{\text{Coul}}(r) + (1+\kappa)^2\Delta V^V_{\text{conf}}(r)\right]\mathbf{S}_Q\mathbf{S}_q\right\}.$$

Here the prime denotes differentiation with respect to r, $\mathbf{L}$ is the orbital momentum, $\mathbf{S}_q$ and $\mathbf{S}_Q$ are the spin operators of the light and heavy quarks, $\mathbf{S} = \mathbf{S}_q + \mathbf{S}_Q$ is the total spin.

First Eq. (1) is solved numerically with the complete potential (5) and then the corrections (6) is treated perturbatively. The results of the calculation of heavy-light meson mass spectra are presented in Tables 3-6 and in Figs. 1-4 [2].

From Figs. 1-4 it follows that in the limit $m_Q \to \infty$ the P-wave doublet with $j = 1/2$ lies higher than the one with $j = 3/2$ (abnormal level ordering), i.e. these doublets are inverted. For finite m_Q the hyperfine splitting of the doublet states makes this picture

TABLE 3. Mass Spectrum of D Mesons (GeV).

State	Particle	Theory [2]	PDG(2000)	CLEO	DELPHI
$1S_0$	D	1.875	1.8693(5)		
$1S_1$	D^*	2.009	2.0100(5)		
$1P_2$	D_2^*	2.459	2.459(4)		
$1P_1$	D_1	2.414	2.4222(18)	$2.425 \pm 0.002 \pm 0.002$	
$1P_1$	D_1	2.501		$2.461^{+0.041}_{-0.034} \pm 0.01 \pm 0.032$	
$1P_0$	D_0^*	2.438			
$2S_0$	D'	2.579			
$2S_1$	$D^{*\prime}$	2.629			2.637(9) ?

TABLE 4. Mass Spectrum of D_s Mesons (GeV).

State	Particle	Theory [2]	PDG(2000)	FOCUS
$1S_0$	D_s	1.981	1.9686(6)	
$1S_1$	D_s^*	2.111	2.1124(7)	
$1P_2$	D_{s2}^*	2.560	2.5735(17)	2.5673(13)
$1P_1$	D_{s1}	2.515	2.53535(60)	2.5351(6)
$1P_1$	D_{s1}	2.569		
$1P_0$	D_{s0}^*	2.508		
$2S_0$	D_s'	2.670		
$2S_1$	$D_s^{*\prime}$	2.716		

much more complicated (some of the levels from different doublets overlap). Only for B mesons the pure inversion is preserved.

In the limit $m_Q \to \infty$ heavy quark symmetry predicts simple relations between the spin-averaged masses of B and D states

$$\bar{M}_{B_1} - \bar{M}_{D_1} = \bar{M}_{B_{s1}} - \bar{M}_{D_{s1}} = \bar{M}_{B_s} - \bar{M}_{D_s} = \bar{M}_B - \bar{M}_D = m_b - m_c = 3.33 \text{ GeV}, \quad (7)$$

where $\bar{M}_{B_1} = (3M_{B_1} + 5M_{B_2})/8$, $\bar{M}_B = (M_B + 3M_{B^*})/4$, etc. From Tables 3-6 the following values (in GeV) for the calculated mass differences can be obtained

$$\begin{array}{cccc} \bar{M}_{B_1} - \bar{M}_{D_1} & \bar{M}_{B_{s1}} - \bar{M}_{D_{s1}} & \bar{M}_{B_s} - \bar{M}_{D_s} & \bar{M}_B - \bar{M}_D \\ 3.29 & 3.30 & 3.34 & 3.33 \end{array}$$

Thus relations (7) are satisfied with good accuracy.

The hyperfine mass splittings of the initially degenerate P states

$$\Delta M_B \equiv M_{B_2} - M_{B_1}, \; M_{B_1} - M_{B_0}; \qquad \Delta M_D \equiv M_{D_2} - M_{D_1}, \; M_{D_1} - M_{D_0}$$

should scale with heavy quark masses: $\Delta M_B = (m_c/m_b)\Delta M_D$ and the same for B_s and D_s mesons. Our model predictions for these splittings are displayed in Table 7 [2].

TABLE 5. Mass Spectrum of B Mesons (GeV).

State	Part.	Theor.[2]	PDG(2000)	OPAL	L3	DELPHI	CDF	ALEPH
$1S_0$	B	5.285	5.2790(5)					
$1S_1$	B^*	5.324	5.3250(6)					
$1P_2$	B_2^*	5.733			5.768(8)?	5.732(21)		5.739(13)
$1P_1$	B_1	5.719		5.738(9)			5.71(2)	
$1P_1$	B_1	5.757			5.670(16)?			
$1P_0$	B_0^*	5.738		5.839(14)				
$2S_0$	B'	5.883						
$2S_1$	$B^{*\prime}$	5.898				5.90(2) ?		

TABLE 6. Mass Spectrum of B_s Mesons (GeV).

State	Particle	Theory [2]	PDG(2000)	OPAL
$1S_0$	B_s	5.375	5.3696(24)	
$1S_1$	B_s^*	5.412	5.4166(35)	
$1P_2$	B_{s2}^*	5.844		5.853(15)
$1P_1$	B_{s1}	5.831		
$1P_1$	B_{s1}	5.859		
$1P_0$	B_{s0}^*	5.841		
$2S_0$	B_s'	5.971		
$2S_1$	$B_s^{*\prime}$	5.984		

CONCLUSIONS

So one may conclude that for heavy quarkonium mass spectra most of theoretical calculations are generally in good agreement with high-precision experimental data. This makes possible an accurate determination of fundamental parameters governing the dynamics of the heavy $Q\bar{Q}$ interaction. On the other hand, for heavy-light meson mass spectra the situation is much more indefinite and unclear in both theory and experiment. Not all models predict the P level inversion. The broad $j = 1/2$ levels are very poorly studied due to difficulties in their observation. Much further work is required to remove discrepancies between existing experimental data.

This research was supported in part by the *Deutsche Forschungsgemeinschaft* under Contract Eb 139/2-1, *Russian Foundation for Fundamental Research* under Grant No. 00-02-17768 and *Russian Ministry of Education* under Grant No. E00-3.3-45.

TABLE 7. Hyperfine Splittings of P Levels (MeV).

States	ΔM_D	$\frac{m_c}{m_b}\Delta M_D$	ΔM_B	ΔM_{D_s}	$\frac{m_c}{m_b}\Delta M_{D_s}$	ΔM_{B_s}
$1P_2 - 1P_1$	45	14	14	45	14	13
$1P_1 - 1P_0$	63	20	19	61	19	18

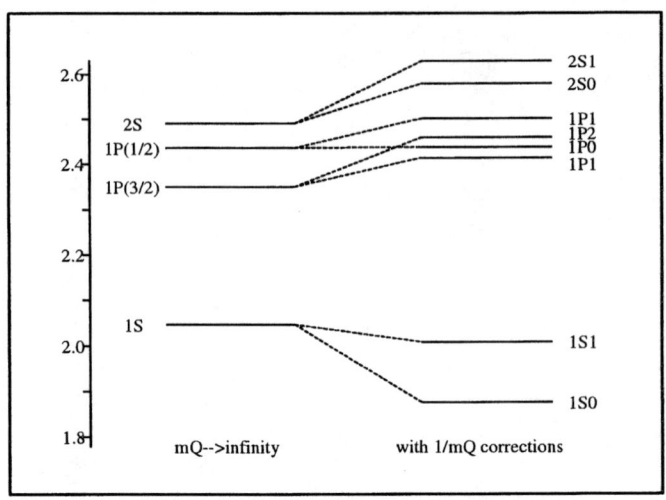

FIGURE 1. The ordering pattern of D meson states. The mass scale is in GeV.

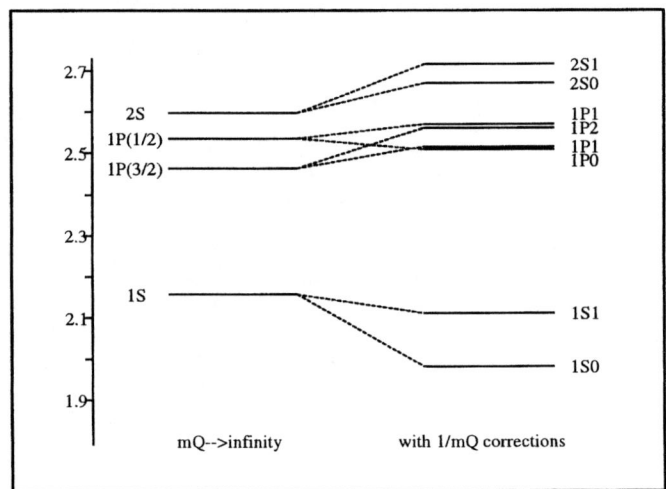

FIGURE 2. The ordering pattern of D_s meson states. The mass scale is in GeV.

REFERENCES

1. Ebert, D., Faustov, R. N., and Galkin, V. O., *Phys. Rev. D* **62**, 034014-1-11 (2000).
2. Ebert, D., Faustov, R. N., and Galkin, V. O., *Phys. Rev. D* **57**, 5663-5669 (1998); **59**, 019902 (1999) (Erratum).

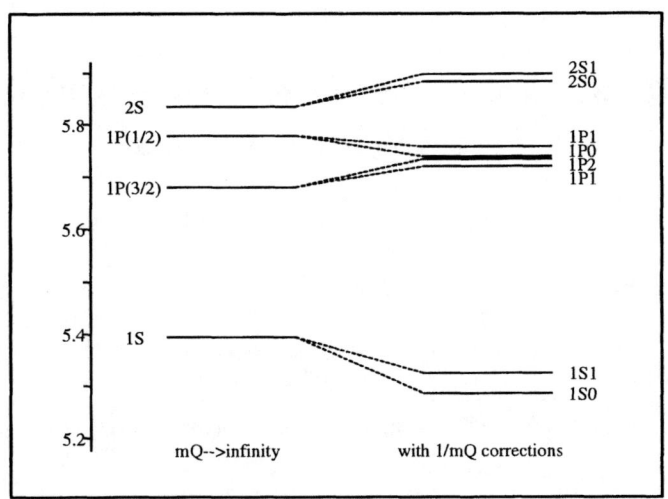

FIGURE 3. The ordering pattern of B meson states. The mass scale is in GeV.

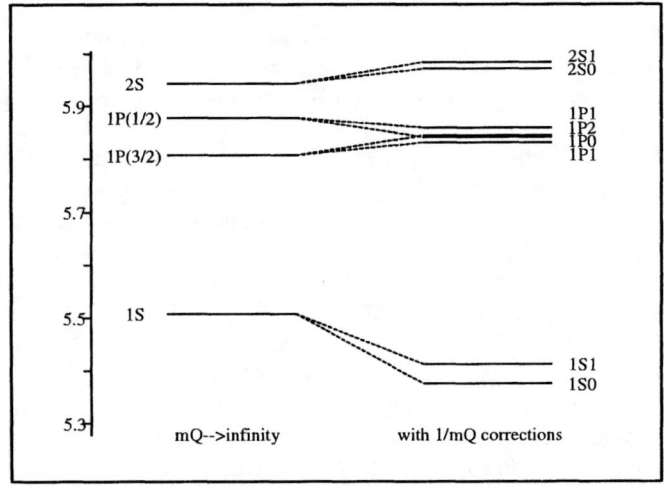

FIGURE 4. The ordering pattern of B_s meson states. The mass scale is in GeV.

3. Particle Data Group (PDG), *Eur. Phys. J. C* **15**, 1-878 (2000).
4. Ebert, D., Faustov, R. N., and Galkin, V. O., in preparation.
5. Melles, M., *Phys. Rev. D* **62**, 074019-1-14 (2000).

Recent Results of $\psi(2S)$ Decay Branching Ratios and Decay Widths from BES

Yongsheng Zhu   (BES Collaboration)

IHEP, Beijing 100039, China

Abstract. The $\psi(2S)$ total decay width, partial widths to hadrons, $\mu^+\mu^-$, $\pi^+\pi^- J/\psi$ final states and corresponding branching ratios are measured with a scan experiment in the vicinity of $\psi(2S)$ resonance at the BES. The preliminary results give much improved accuracies compared with existing data. Also reported are the branching ratios for $\psi(2S)$ to τ-pair and radiative decays.

INTRODUCTION

The BES is a conventional cylindrical detector, which is described in detail in Ref.[1]. A central drift chamber (CDC) surrounding the beam pipe provides trigger information. Charged tracks are reconstructed in a forty-layer main drift chamber (MDC) with a momemtum resolution of $\sim 1.8\%\sqrt{1+p^2}$ (p in GeV/c), and energy loss(dE/dx) resolution of $\sim 8\%$ for bhabha electrons. Scintillation counters provide time-of-flight (TOF) measurements, with the resolutions of ~ 375 ps for bhabha events. A 12-radiation-length, lead-gas barrel shower counter (BSC) measures the energies of electrons and photons over $\sim 80\%$ of the total solid angle. A solenoidal magnet provides a 0.4 T magnetic field in the central tracking region. Three double-layer muon counters (MUC) instrument the magnet flux return, and serve to identify muons with momentum greater than 500 MeV/c.

The BESII [2] is the BES upgrated version, in which, the vertex chamber (VC) replaces the CDC with improved spatial resolution, the new BTOF reaches the time resolution of 180 ps for bhabha events, the deadtime of the new DAQ system is ~ 8 ms per event, in contrast to 20 ms per event for old DAQ system.

In this paper, we report the results on the branching ratios of $\psi(2S)$ radiative decay and $\tau^+\tau^-$ pair final states based on 3.96 million $\psi(2S)$ events collected by BESI [3], and the preliminary results on $\psi(2S)$ decay widths based on scan experiment with BESII detector.

$\psi(2S)$ RADIATIVE DECAY

In perturbative QCD, the dominant process of J/ψ and $\psi(2S)$ hadronic decay is through the $c\bar{c}$ annihilation into three gluons. Since the decay width is proportional to the amplitude of the $c\bar{c}$ wave function at the origing, $|\psi(0)|^2$, it is expected that the branching ratio of J/ψ and $\psi(2S)$ decay into light quark states are related as [6]:

$$Q_h \equiv \frac{B(\psi(2S) \to h)}{B(J/\psi \to h)} = \frac{B(\psi(2S) \to ggg)}{B(J/\psi \to ggg)} \simeq \frac{B(\psi(2S) \to e^+e^-)}{B(J/\psi \to e^+e^-)} = (14.6 \pm 2.2)\% \quad (1)$$

This relation is referred to as the 15% rule. The prediction was originally made for the total decay width into three gluons. Since the partial width of individual channels involving the initial annihilation of $c\bar{c}$ quarks are also functions of the $|\psi(0)|^2$, we expect this rule to be generally valid. The radiative J/ψ and $\psi(2S)$ decays are similar to their hadronic decays into three gluons except one of the gluon lines replaced by a photon line. Thus one power of the coefficient α_S is replaced by α_{QED} in the cross section formula. It is therefore expected the "15%" rule should also work for radiative decay [7].

Previous measurements on the branching ratios or upper limit of $\psi(2S)$ radiative decays into non-charmonium final states have only been carried out by Mark I [8] and BES [9] collaboration.

We studied the events with topologies of $\psi(2S) \to \gamma(\pi\pi, K\bar{K}, \eta\eta) \to 5\gamma, \gamma 2P, \gamma 4P$, here P stands for charged pion or kaon. We use a data set that is a subset of 3.96 million $\psi(2S)$ events recorded by the BESI detector [10]. In this analysis, photons are identified by the BSC. For a π^0 or η decaying into two photons, the invariant mass of two photons are required to be close to the mass of π^0 or η, in addition, the helicity angle of the decay should be flat and hence asymmetric background decays can be removed by requiring $|\cos\theta_{helicity}| < 0.99$. Only charged tracks falling in the region $|\cos\theta| < 0.8$ are used. Charged tracks identified as electrons or positrons by the BSC or identified as μ^+ or μ^- by the MUC are rejected. TOF information, dE/dx information and kinematic fitting are used to identify charged particles. Kinematic fits are also used to improve momentum measurements and mass resolutions, as well as to resolve combinatorial ambiguityes if more than one is possible.

$$\psi(2S) \to \gamma\pi\pi$$

For the candidates of $\psi(2S) \to \gamma\pi^+\pi^-, \gamma\pi^0\pi^0$, the invariant mass of two pions are ploted in Fig.1, the $f_2(1270)$ signals are clearly seen. Both distributions are fitted with a D-wave Breit-Wigner function with the resonant parameters fixed at the PDG values of $f_2(1270)$. The combined result from these two channels is $B(\psi(2S) \to \gamma f_2(1270)) = (2.27 \pm 0.26 \pm 0.39) \times 10^{-4}$. This result gives the $Q_{\gamma f_2(1270)}$ value of $(16.4 \pm 3.8)\%$ using similar branching ratio from J/ψ decay, which is consistent with "15%" rule. A $f_J(1710)$ signal is obeserved in $\pi^+\pi^-$ invariant mass, giving: $B(\psi(2S) \to \gamma f_J(1710)) \times B(f_J(1710) \to \pi^+\pi^-) = (3.38 \pm 0.87 \pm 1.41) \times 10^{-5}$ or $< 5.50 \times 10^{-5} (90\% C.L.)$

The region with a $\pi^0\pi^0$ invariant mass greater than 3 GeV (see Fig. 2a) has signal peak due to χ_{c0} and χ_{c2} states. The mass distribution is fitted with two Breit-Wigners plus a polynomial background and yields : $B(\chi_{c0} \to \pi^0\pi^0) = (2.65 \pm 0.30 \pm 0.58) \times 10^{-3}$, $B(\chi_{c2} \to \pi^0\pi^0) = (8.7 \pm 2.4 \pm 5.0) \times 10^{-4}$.

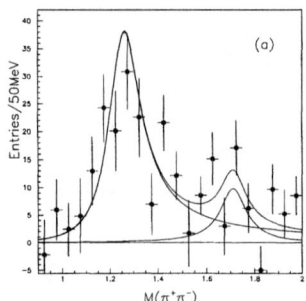
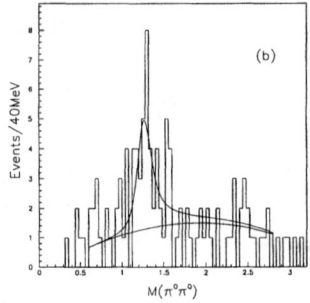

FIGURE 1. Invariant mass of two pions in $\psi(2S) \to \gamma\pi^+\pi^-, \gamma\pi^0\pi^0$

$\psi(2S) \to \gamma\eta\eta$

In the region $M_{\eta\eta} > 3$ GeV in Fig. 2b, a χ_{c0} signal and a weak χ_{c2} signal are observed. Two Breit-Wigners with a polynomial background are used to fit the mass distribution, giving the branching ratios: $B(\chi_{c0} \to \eta\eta) = (1.94 \pm 0.81 \pm 0.59) \times 10^{-3}$, $B(\chi_{c2} \to \eta\eta) < 1.22 \times 10^{-3} (90\% C.L.)$.

Flavor SU(3) symmetry predicts that the branching ratio of χ_{c0} decay into $\pi^0\pi^0$ and $\eta\eta$ should be the same except for a phase space factor and a barrier factor which is $p^{(2s+1)}$, where p is the momentum of the π^0 or η in χ_c's rest frame and s is the spin of χ_c's. Based on the PDG values for χ_{c0}, this predicts $B(\chi_{c0} \to \eta\eta)/B(\chi_{c0} \to \pi^0\pi^0) = 0.95$, consistent with our measurement of $\frac{B(\chi_{c0} \to \eta\eta)}{B(\chi_{c0} \to \pi^0\pi^0)} = 0.73 \pm 0.32 \pm 0.27$ within the error.

$\psi(2S) \to \gamma K\bar{K}$

Fig. 3a shows the invariant mass distribution of $M_{K^+K^-}$, a $f_J(1710)$ signal and a possible $f_2'(1525)$ signal are observed. The distribution is fitted using an S-wave Breit-Wigner and a D-wave Breit-Wigner with mass and width fixed at the PDG value of $f_J(1710)$ and $f_2'(1525)$ respectively. The fit yields a branching ratio of: $B(\psi(2S) \to \gamma f_J(1710)) \times B(f_J(1710) \to K^+K^-) = (5.59 \pm 1.12 \pm 0.96) \times 10^{-5}$ or $< 7.48 \times 10^{-5} (90\% C.L.)$. This result is again consistent with "15%" rule.

For the $\psi(2S) \to \gamma K_S^0 K_S^0 \to \gamma\pi^+\pi^-\pi^+\pi^-$ channel, both K_S^0 vertices are reconstructed and $M_{\pi^+\pi^-}$ must be close to K_S^0 mass. If more than one combination survives, the combination with the smallest value of $\sqrt{(m_{\pi_1^+\pi_2^-} - m_{K_S^0})^2 + (m_{\pi_3^+\pi_4^-} - m_{K_S^0})^2}$ is chosen. An S-wave Breit-Wigner plus a polynomial background are used to fit the invariant mass in Fig. 3b, giving the branching ratio of $B(\psi(2S) \to \gamma f_J(1710)) \times B(f_J(1710) \to K_S^0 K_S^0) = (2.10 \pm 0.96 \pm 1.11) \times 10^{-5}$ or $< 3.98 \times 10^{-5} (90\% C.L.)$.

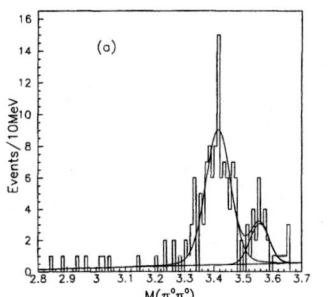
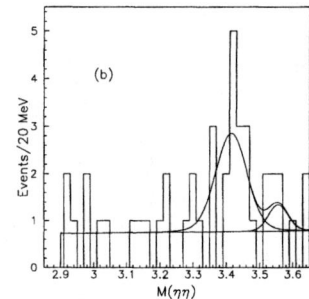

FIGURE 2. Invariant mass of $\pi^0\pi^0$ and $\eta\eta$ in $\psi(2S) \to 5\gamma$ decay

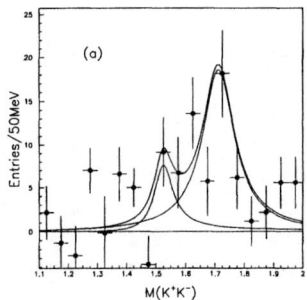
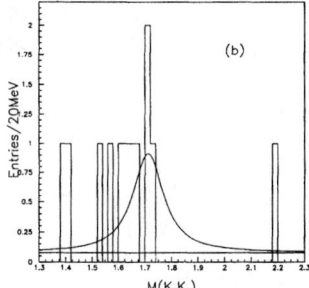

FIGURE 3. Invariant mass of K^+K^- and $K_S^0 K_S^0$ in $\psi(2S) \to \gamma K\overline{K}$

$\psi(2S)$ DECAY TO τ PAIR

The $\psi(2S)$ provides a unique opportunity to compare the three lepton generations by studying the leptonic decays $\psi(2S) \to e^+e^-, \mu^+\mu^-$, and $\tau^+\tau^-$. The sequential lepton hypothesis leads to

$$\frac{B_e}{v_e(\frac{3}{2} - \frac{1}{2}v_e^2)} = \frac{B_\mu}{v_\mu(\frac{3}{2} - \frac{1}{2}v_\mu^2)} = \frac{B_\tau}{v_\tau(\frac{3}{2} - \frac{1}{2}v_\tau^2)} \quad (2)$$

with $v_l = (1 - \frac{4m_l^2}{M_{\psi(2S)}^2})^{\frac{1}{2}}, l = e, \mu, \tau$. Substituting mass values for the leptons and the $\psi(2S)$ gives $B_e \simeq B_\mu \simeq \frac{B_\tau}{0.3885} \equiv B_l$. While the B_e and B_μ have already been reported by some experiments [5], the B_τ is still not measured until now.

The $\tau^+\tau^-$ events are identified by requiring one τ decay via $e\nu\overline{\nu}$ and the other via $\mu\nu\overline{\nu}$. To select candidate events, a cut of acolinearity angle between two well reconstructed tracks with opposite charge is applied to reject bhabhas, μ pairs and cosmic rays. The acoplanar angle between two planes defined by two tracks with beam direction is required to be greater than $20°$ to suppress radiative bhabhas and radiative muon pairs.

The track momentum must be less than the maximum kinematically allowed value for a τ decay. Particle identification is implemented with dE/dx, time-of-flight, energy deposit (from BSC) and hit information in the MUC to ensure $e-\mu$ final state topology.

The same requirements are applied to 5 million events taken at the J/ψ energy to estimate the expected background, which corresponds to $n_{bg} = 0.73$ for the whole $\psi(2S)$ data set. A Monte Carlo study on the two-photon processes shows negligible contamination.

To obtain the resonant τ-pair events number, the QED term including the interference part is subtracted from the total number of $\tau^+\tau^-$ events. B_τ is caculated by

$$B_\tau = \frac{(n_{e\mu} - n_{bg})/(B\varepsilon) - L\sigma_{Q+I}}{N_{\psi(2S)}}. \tag{3}$$

Here B is the fraction of $\tau^+\tau^-$ events yielding the $e\mu$ topology, ε is the candidate events detection efficiency, L is the integrated luminosity determined by large angle bhabha events, and σ_{Q+I} is the QED τ-pair cross section including interference at the center-of-mass energy corresponding to the $\psi(2S)$ resonance [11].

The resultant branching ratio is $B_\tau = (2.71 \pm 0.43 \pm 0.55) \times 10^{-3}$. Our value of B_τ corrected by a factor of 0.3885 is equal to $(7.0 \pm 1.1 \pm 1.4) \times 10^{-3}$, which agrees with $B_e = (8.8 \pm 1.3) \times 10^{-3}$ and $B_\mu = (10.3 \pm 3.5) \times 10^{-3}$ [5] within the error. Assuming lepton universality, the average value B_l is $(8.4 \pm 1.0) \times 10^{-3}$. From the relation $\Gamma_t = \Gamma_e/B_l$ and using PDG's value of $\Gamma_e = (2.12 \pm 0.18)$ keV, we obtain $\Gamma_t = (252 \pm 37)$ keV, which is well consistent with the result of $\Gamma_t = (243.3 \pm 34.2)$ keV given by BES $\psi(2S)$ scan experiment described in next section.

$\psi(2S)$ SCAN EXPERIMENT

Since the discovery of $\psi(2S)$ particle [12] in 1974, a few measurements [13] for its total and partial decay widths $\Gamma_t, \Gamma_h, \Gamma_{\pi^+\pi^-J/\psi}, \Gamma_\mu$, and corresponding branching ratios $B_h, B_{\pi^+\pi^-J/\psi}, B_\mu$ have been carried out. The significant deviations can be seen in previous measurements [14,15]. These parameters are of particular interest, of which, $\psi(2S) \to \mu^+\mu^-$ is used in reconstructing B mesons for CP violation measurements [18], and $\psi(2S) \to \pi^+\pi^-J/\psi$ is usually a tag mode in $\psi(2S)$ any branching ratio measurement because of its large branching ratio and easy tagging.

Twenty-four center-of-mass energy points were scanned in the vicinity of the $\psi(2S)$ peak ranging from 3.67 GeV to 3.71 GeV. In addition, 6 hours of separated-beam-data were taken at the first and the last points for background study. The following reactions have been studied carefully

$$e^+e^- \to \psi(2S) \to e^+e^-, \quad e^+e^- \to e^+e^- \tag{4}$$

$$e^+e^- \to \psi(2S) \to \mu^+\mu^-, \quad e^+e^- \to \mu^+\mu^- \tag{5}$$

$$e^+e^- \to \psi(2S) \to hadrons, \quad e^+e^- \to hadrons \tag{6}$$

$$e^+e^- \to \psi(2S) \to \pi^+\pi^-J/\psi \tag{7}$$

The corresponding cross sections have been determined and fitted to extract the $\psi(2S)$ decay widths and branching ratios.

Event selection

Events from reaction (4,5) are seperated from cosmic ray by requiring that time-of-flight of two particles must fall into the circle $\sqrt{(t_1-5)^2+(t_2-5)^2} < 4.5(ns)$, the charge of two tracks must be opposite for e^+e^- events and the acollinear angle must be less than 10 degrees for $\mu^+\mu^-$ events. In order to reject contaminations from two-body hadronic events such as $\psi(2S) \to \pi^+\pi^-, K^+K^-, \bar{p}p$, the hit information of μ-counter is used, which also seperates the $\mu^+\mu^-$ events from e^+e^- events at the same time. The μ-counter hit cut implies geometric limition of $|\cos\theta_\mu| \leq 0.65$, while for e^+e^- events, the cut is $|\cos\theta_e| \leq 0.72$. To eliminate the background of the lepton-pairs coming from $\psi(2S) \to XJ/\psi, J/\psi \to l^+l^-$, the track momentum must satisfy:

$$(p_1 > 1.75) \cup (p_2 > 1.75) \cup (\sqrt{(p_1-1.85)^2+(p_2-1.85)^2} < 0.25) \ (GeV) \qquad (8)$$

The deposit energies of two tracks in the BSC are used to identify electron pair from muon- and hadron pair with the cut of $\sqrt{(E_{dep1}-1.85)^2+(E_{dep2}-1.85)^2} < 1.2$ GeV.

For hadron-event selection, because there is no particular event topology to follow, we use the "rejection" algolithm, namely reject all possible backgrounds or contaminations, such as cosmic ray, beam-associate background, two-photon processes, mis-identified "hadron" events from $\psi(2S) \to l^+l^-, l = e, \mu, \tau$, and from $\psi(2S) \to \gamma\gamma$ followed by γ conversion, etc., and keep the remained events with charged tracks number greater then 2 as the selected hadron events sample.

After eliminating the selected $e^+e^-, \mu^+\mu^-$ events, the hadron event selection is carried out. We collect only the events with at least 2 well reconstructed charged tracks, which $|\cos\theta|$ are less then 0.8. Only the well isolated neutral clusters with deposit energy greater than 80 MeV are taken as "real photon". The total deposit energy (charged and netrual tracks) must be larger than $0.36E_{beam}$. All tracks in an event should not point to same hemisphere in Z direction to suppress the beam-associate background. For 2 prongs events, the collinear angle must be greater than 10 degrees, the number of real photon must be geater than 1.

The backgrounds from $e^+e^- \to e^+e^-, \mu^+\mu^-, \tau^+\tau^-, \gamma\gamma$ events passing through above selection criteria can be estimated by $N_{ii} = L \cdot \varepsilon_{ii} \cdot \sigma_{ii}$, where σ_{ii} is the theoretical cross section, ε_{ii} is the efficiency of $e^+e^- \to ii$ event ($ii = e^+e^-, \mu^+\mu^-, ...$) selected as "hadron event", which is determined by Monte-Carlo simulation. Then the selected hadron events number is $N_h^{obs} = N_h - N_{ee} - N_{\mu\mu} - N_{\tau\tau} - N_{\gamma\gamma}$, where N_h is the number of events passing the hadron selection cuts.

As for the selection of $\pi^+\pi^- J/\psi$ events, the method [19] is that one selects a pair of low energy pions to determine the mass recoiling against them:

$$m_{\pi^+\pi^-}^{recoil} = [(m_\psi(2S) - E_{\pi^+} - E_{\pi^-})^2 - (\vec{p}_{\pi^+} + \vec{p}_{\pi^-})^2]^{1/2} \qquad (9)$$

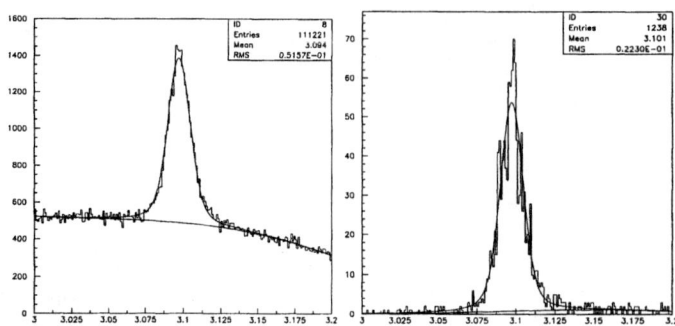

FIGURE 4. Inclusive and exclusive $m^{recoil}_{\pi^+\pi^-}$ distribution of $\psi(2S) \to \pi^+\pi^- J/\psi$ and $\psi(2S) \to \pi^+\pi^- J/\psi, J/\psi \to l^+l^-$ for 24 scan points

The $m^{recoil}_{\pi^+\pi^-}$ distribution shows a strong J/ψ peak, which is fitted with a signal shape plus a polynomial background to obtain the number of $\psi(2S) \to \pi^+\pi^- J/\psi$ events. The very clean $\psi(2S) \to \pi^+\pi^- J/\psi, J/\psi \to l^+l^-$ events are used for the signal shape. The inclusive and exclusive $m^{recoil}_{\pi^+\pi^-}$ distributions for all 24 scan points are shown in Fig. 4.

Acceptance

The acceptances for all the final states are determined by the product of trigger efficiency ε_{trg} and reconstruction-selection efficiency ε_{rs}. The trigger efficiencies are measured by comparing the responses to different trigger requirements in special runs taken at the J/ψ resonance [20]. The trigger efficiencies for $e^+e^-, \mu^+\mu^-$, hadron and $\pi^+\pi^- J/\psi$ final states are 1.0000, 0.9936, 0.9985 and 1.0000 respectively for this scan experiment. The event reconstruction and selection efficiencies are determined by Monte-Carlo simulation. For e^+e^- and $\mu^+\mu^-$ events, the efficiencies of QED process are acquired by RADEE and RADMU generators written by F.A.Berends etal. [21], and the efficiencies of $\psi(2S) \to e^+e^-, \mu^+\mu^-$ are determined by V2LL generators, which are similar to RADEE and RADMU generators, but the initial radiative corrections are erased. As to the hadronic events, the JETSET 7.4 event generator [22] is used to determine detection efficiencies for both QCD process and $\psi(2S)$ hadronic decays. The parameters of the generator are adjusted to reproduce distributions of kinematic variables such as multiplicity, sphericity, transverse momentum, etc. in BEPC energy region data [23]. The efficiency for $\psi(2S) \to \pi^+\pi^- J/\psi$ events is obtained using PPGEN generator, which is a BES code written specially for this process.

The acceptances for different processes with their uncertainties are listed in Table 1, wherein the acceptances for e^+e^- events are for restricted solid angle ($|cos\theta| \leq 0.72$) while the acceptances for hadron, $\mu^+\mu^-, \pi^+\pi^- J/\psi$ final states are for all angles. The errors contain the uncertainties by varying the event selection cuts and Monte Carlo simulation uncertainties including errors due to the imperfection of BES detector simulation and modelling uncertainties.

TABLE 1. Acceptance and systematic error

process	A	$\Delta A/A$
$e^+e^- \to e^+e^-$	0.6548	0.02425
$e^+e^- \to \psi(2S) \to e^+e^-$	0.7033	0.02425
$e^+e^- \to \mu^+\mu^-$	0.3977	0.05647
$e^+e^- \to \psi(2S) \to \mu^+\mu^-$	0.4432	0.05647
$e^+e^- \to hadrons$	0.7381	0.07237
$e^+e^- \to \psi(2S) \to hadrons$	0.7789	0.03137
$e^+e^- \to \psi(2S) \to \pi^+\pi^- J/\psi$	0.4218	0.0159

Fit of observed cross sections and results

The four data sets for $e^+e^- \to$ hadrons, $e^+e^-, \mu^+\mu^-, \pi^+\pi^- J/\psi$ are fitted simultaneously to obtain the partial widths of $\psi(2S)$ to hadrons, muons, and $\pi^+\pi^- J/\psi$, the total width is assumed to be the sum of 4 partial widths $\Gamma_t = \Gamma_h + \Gamma_\mu + \Gamma_e + \Gamma_\tau$ with assumption $\Gamma_e = \Gamma_\mu = \Gamma_\tau/0.3885$.

As the branching ratio of $\psi(2S)$ to e^+e^- is small, the cross section of e^+e^- final state is dominated by QED process, therefore the integrated luminosities for 24 energy points are determined by e^+e^- events numbers at these points by iteration scheme: First we calculate the luminosity by taking total number of e^+e^- events as Bhabbha events at each point to determine the $\psi(2S)$ decay widths, then assuming $e - \mu$ universality ($B_e = B_\mu$), splitting the e^+e^- events into QED part, resonant part and interference part, re-calculate the integral luminosity for each energy point and re-do the fit to obtain new values of $\psi(2S)$ decay widths. Such a recursive iteration is repetedly carried out untill the value of integral luminosity for each point is converged in two succesive iterations.

The theoretical cross section used for the fit of hadron data takes a Breit-Wigner amplitude and a nonresonant amplitude, for the $\pi^+\pi^- J/\psi$ final state only a Breit-Wigner amplitude is considered, while for the $\mu^+\mu^-$ and e^+e^- events, the interference between the $\psi(2S)$ resonant term and QED term is also included. The radiative corrections of these two processes are taken into account by the formulation of ref. [24]. For the e^+e^- final state, where QED t channel photon exchange also contributes, a theoretical cross section with radiative correction is derived following the method of ref. [25]. The vacuum polarization effect on cross sections are also taken into account [26]. These theoretical cross sections are convoluted with the energy distribution of the colliding beams, which is treated as a Gaussian function. We use G.D'Agostini scheme [27] to construct χ^2 to account for correlated data and implement minimization. In the fit following parameters are allowed to vary : the mass M, the partial widths Γ_h, Γ_μ and $\Gamma_{\pi^+\pi^- J/\psi}$, the energy spread of the machine, the nonresonant hadronic cross section, and the scale factors f_h, f_L, f_π for hadron, e^+e^- and $\pi^+\pi^- J/\psi$ final states normalization.

The fitting curves are shown in Fig. 5. The fit gives a χ^2 of 68.1 with the number of degrees of freedom 62. The fitted spread in the center-of-mass energy of the machine is 1.305 ± 0.002 MeV, in agreement with the expectation (~ 1.3 MeV). The resultant R ratio for the nonresonant hadronic cross section near the $\psi(2S)$ resonance is 2.25 ± 0.12, which coincides with the BES R measurements [28] reasonably well. The assumption

TABLE 2. BES preliminary results on $\psi(2S)$ decay widths and branching ratios

Parameter	BES	Mark I	PDG2000
$\Gamma_t (keV)$	243.3 ± 34.2	228 ± 56	277 ± 31
$\Gamma_h (keV)$	237.6 ± 33.7	224 ± 56	
$\Gamma_{\pi\pi J/\psi} (keV)$	78.0 ± 10.6		
$\Gamma_\mu (keV)$	2.36 ± 0.20	2.1 ± 0.3	2.12 ± 0.18
$\Gamma_{\gamma^* h} (keV)$	5.00 ± 0.43	6.7	
$B_h (\%)$	97.69 ± 0.18	98.1 ± 0.3	98.10 ± 0.30
$B_{\pi\pi J/\psi} (\%)$	32.0 ± 1.5	32 ± 4	31.0 ± 2.8
$B_\mu (\%)$	0.97 ± 0.10	0.93 ± 0.16	1.03 ± 0.35
$B_{\gamma^* h} (\%)$	2.06 ± 0.20	2.9 ± 0.4	

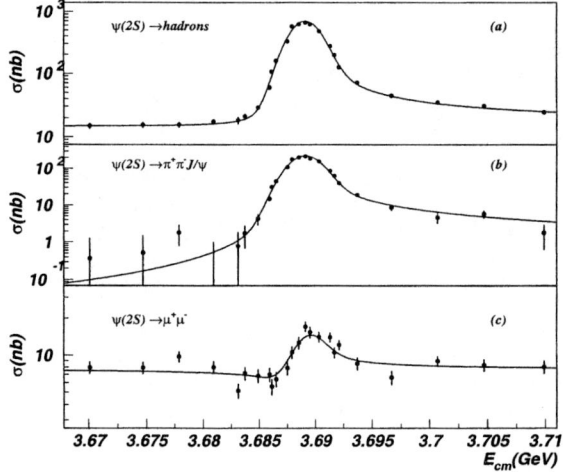

FIGURE 5. The cross sections of hadron, $\pi^+\pi^- J/\psi$ and $\mu^+\mu^-$ final states in the vicinity of $\psi(2S)$ resonance. The solid curves represent the results of the fit to the data

that leptons couple to the $\psi(2S)$ only via an intermediate virtual photon [29] implies the existence of the decay $\psi(2S) \rightarrow \gamma^* \rightarrow hadrons$ with a branching ratio

$$\Gamma_{\gamma^* h}/\Gamma_t = R\Gamma_e(=\Gamma_\mu)/\Gamma_t = 0.0206 \pm 0.0020 \qquad (10)$$

corresponding to a width $\Gamma_{\gamma^* h}$ of 5.00 ± 0.43 keV. Here $R = 2.12 \pm 0.12$ is the fitted value of 2.25 ± 0.12 subtracted by J/ψ resonance tail cross section calculated based on BES determined J/ψ line shape [30]. The Preliminary results of the fit for decay widths and branching ratios are given in Table 2, together with those of Mark I [14] and PDG2000 [5] for comparison. It can be seen that for all these decay widths and branching ratios BES gives the most accurate values from single experiment, in particular the accuracy of $B_{\pi^+\pi^- J/\psi}$ is much improved, and the first direct measurement of $\Gamma_{\pi^+\pi^- J/\psi}$ is given.

ACKNOWLEDGMENTS

This work is supported in part by the National Natural Science Foundation of China under Contract No. 19991480, the Chinese Academy of Sciences under Contract No. KJ95T-03, and by the Department of Energy under Contracts Nos. DE-FG03-93ER40788 (Colorado State University), DE-AC03-76SF00515 (SLAC), DE-FG03-94ER40833 (U Hawaii), DE-FG03-95ER40925 (UT Dallas).

REFERENCES

1. Bai, J.Z., et al., *Nucl. Instr. Meth.* **A344**,319-334(1994).
2. Bai, J.Z., et al., *Nucl. Instr. Meth.* **A458**,627-637(2001).
3. The total number of $\psi(2S)$ recorded by the BES detector is determined by the total number of $\psi(2S) \to \pi^+\pi^- J/\psi$ events divided by the PDG branching ratio value for this mode. This total $\psi(2S)$ number is $(3.96 \pm 0.36) \times 10^6$. [4] [5].
4. Bai, J.Z., et al., *Phys. Rev.* **D58**,092006(1998).
5. Groom, D.E., et al., *Euro. Phys. Jnl.* **C15**(2000).
6. Hou, W.S., and Soni, A., *Phys. Rev. Lett.* **50**,569(1980); Karl, G., and Roberts, W., *Phys. Lett.* **B144**,243(1984); Brodsky, S.J., et al., *Phys. Rev. Lett.* **59**,621(1987); Chaichian, M., et al., *Nucl. Phys.* **B323**,75(1989); Pinsky, S.S., *Phys. Lett.* **B236**,479(1990); Brodsky, S.J., and Karliner, M., *Phys. Rev. Lett.* **78**,4682(1997); Chen, Y.Q., and Braaten, E., *Phys. Rev. Lett.* **80**,5060(1998).
7. Appelquist, T., et al., *Phys. Rev. Lett.* **3.4**,363(1975); Chanowitz, M., *Phys. Rev.* **D12**,918(1975); Okun, L., and Voloshin, M., ITEP-95-1976(unpublished); Brodsky, S.J., et al., *Phys. Lett.* **B73**,203(1978); Koller, K., and Walsh, T., *Nucl. Phys.* **B140**,449(1978).
8. Scharre, D.L., et al., *Phys. Lett.* **B97**,329(1980).
9. Bai, J.Z., et al., *Phys. Rev.* **D58**,097101(1998).
10. In this analysis, only runs of good quality have been used, corresponding to $\sim (3.83 \pm 0.36) \times 10^6 \psi(2S)$ events and $(1.188 \pm 0.003 \pm 0.016) \times 10^6 \pi^+\pi^- J/\psi$ events. [4] [5].
11. The expression for the $\tau^+\tau^-$ cross section, including the center-of-mass energy spread, initial state radiation corrections, vacuum polarization corrections, Coulomb interaction corrections, and final state radiation corrections, is given by Wu, J.M., BIHEP-TH-00/45,2000 (unpublished).
12. Aubert, J.J., et al., *Phys. Rev. Lett.* **33**,1404(1974); Augustin, J.E., et al., *Phys. Rev. Lett.* **33**,1406(1974); Bacci, C., et al., *Phys. Rev. Lett.* **33**,1408(1974).
13. Abrams, G.S., *Stanford Symp* **25**(1975); Boyarski, A.M., et al., *Palermo Conf.* **54**(1975); Abrams, G.S., et al., *Phys. Rev. Lett.* **34**,1181(1975); Boyarski, A.M., et al., *Phys. Rev. Lett.* **34**,1357(1975).
14. Hilger, E., et al., *Phys. Rev. Lett.* **35**,625(1975).
15. Armstrong, T.A., et al., *Phys. Rev. Lett.* **68**,1486(1992); *Phys. Rev.* **D47**,772(1993).
16. Ambrogiani, M., et al., *Phys. Rev.* **D62**,032004(2000).
17. Luth, V., et al., *Phys. Rev. Lett.* **35**,1124(1975).
18. Aubert, B., et al., SLAC-PUB-8909, hep-ex/0107025, submitted to *Phys. Rev.* **D**.
19. Bai, J.Z., et al., *Phys.Rev.* **D58**,092006(1998).
20. Huang, G.S., et al., *HEP & NP* **25**,889-897(2001). (in Chinese).
21. Berends, F.A., et al., *Nucl. Phys.* **B228**,537(1983); **B177**,237(1981); **B57**,381(1973).
22. Sjostrand, T., *Computer Phys. Comm.* **82**,74(1994).
23. Chen, J.C., et al., *Phys. Rev.* **D62**,034003(2000).
24. Kuraev, E.A., and Fadin, V.S., *Sov.J. Nucl. Phys.* **41**,466(1985).
25. Beenakker, W., et al., *Nucl. Phys.* **B349**,323(1991).
26. Tsai, T.S., SLAC-PUB-3129 (1983); Berends, F.A., and Komen, G.J., *Phys. Lett.* **B63**,432(1976).
27. D'Agostini, G., *Nucl. Instr. Meth.* **A346**,306-311(1994).
28. Bai, J.Z., et al., *Phys. Rev. Lett.* **84**,594-597(2000).
29. Kopke, L., and Wermes, N., *Phys. Rep.* **74**,67(1989).
30. Bai, J.Z., et al., *Phys. Lett.* **B355**,374-380(1995).

$q\bar{q}$ and 'Extra' Mesons up to 2400 MeV

D.V. Bugg

Queen Mary, University of London, Mile End Rd, London E1 4NS, UK

Abstract. The spectrum of $q\bar{q}$ states observed in Crystal Barrel data in flight is very simple: a set of parallel straight-line trajectories of mass squared v. radial excitation number. There are 6 extra states, making candidates for 4 glueballs and 2 hybrids. Mass ratios of glueball candidates agree closely with the latest Lattice Gauge calculations. The evidence for 2^{-+} hybrids $\eta_2(1860)$ and $\pi_2(1880)$ is reviewed. Further data from a polarised target are highly desirable in channels $3\pi^0$, $\eta\pi^0$, $\omega\pi^0$ and $\omega\eta$ in order to separate cleanly triplet states with $L = J \pm 1$. These data would provide precise measurements of tensor, spin-orbit and spin-spin splitting of masses.

OUTLINE OF THE ANALYSIS

The essential idea is to study *formation* of s-channel resonances in

$$\bar{p}p \to Resonance \to A + B$$

through 17 channels $A + B$. A full partial wave analysis of both production and decay is made. The analysis uses Crystal Barrel data in flight, and also includes data from the earlier PS172 experiment on $\bar{p}p \to \pi^-\pi^+$ [1]. Crystal Barrel data were taken at 9 momenta from 600 to 1940 MeV/c, i.e. the mass range 1960 to 2410 MeV/c². PS172 data extend down to 360 MeV/c or a mass of 1910 MeV/c². Simultaneous analyses have been carried out of the channels listed in Table 1. In any one channel, significance levels of resonances are broadly similar to those of classic πN analyses of baryon resonances above 1700 MeV; for $I = 0$, $C = +1$, the analysis of 8 channels with consistent parameters leads to extremely secure amplitudes and a complete set of the expected $q\bar{q}$ states, each of which has a statistical significance of $\geq 25\sigma$.

TABLE 1. Channels studied. Decays $\eta \to \gamma\gamma$ and $3\pi^0$ are both used, and $\eta' \to \eta\pi^0\pi^0$.

$I = 0, C = +1$	$I = 1, C = +1$	$I = 0, C = -1$	$I = 1, C = -1$
$\pi^0\pi^0, \pi^-\pi^+, \eta\eta, \eta\eta'$	$\eta\pi^0, \eta'\pi^0$	$\omega\eta$	$\omega\pi^0$
$\eta\pi^0\pi^0, \eta'\pi^0\pi^0, \eta\eta\eta$	$3\pi^0, \eta\eta\pi^0$	$\omega\pi^0\pi^0$	$\omega\eta\pi^0$
$\eta\eta\pi^0\pi^0$	$\eta\pi^0\pi^0\pi^0$		

Historically, the steps in the analysis have been as follows. Two-body channels from the first line of Table 1 were studied first [2-5]. Independently, 3-body channels from the second line were analysed [6-11]. In each case, data from every momentum were analysed separately into partial waves. Many resonances appeared with consistent parameters in the intensities derived from these analyses. Subsequently, data at all momenta were analysed in terms of s-channel resonances [12].

A vital observation for $I = 0, C = +1$ is a large signal from the well-known $f_4(2050)$ in both $\pi\pi$ and $\eta\pi\pi$. It serves as an interferometer, fixing phases of other partial waves. These are found to follow closely the phase of $f_4(2050)$, [see Fig. 13 of Ref. 7], hence requiring nearby resonances in all or most partial waves. The analyses also requires $f_4(2300)$ at higher mass with parameters consistent with earlier work. It acts as an interferometer for the high mass region. In the $\pi^-\pi^+$ data, a very clear and strong $\rho_3(1985)$ is found. This in turn provides an interferometer for $I = 1, C = -1$.

In the final step, all channels are fitted simultaneously at all momenta. Here an important constraint arises from analyticity. All partial waves are fitted using analytic functions of s. Simple Breit-Wigner amplitudes with constant widths are used, plus backgrounds in low partial waves, either constant or linear in s or resonances below the $\bar{p}p$ threshold. No backgrounds are needed for partial waves with $J \geq 3$.

A comparative study of $\omega \to \pi^0\gamma$ and $\omega \to \pi^+\pi^-\pi^0$ demonstrates close agreement between charged and neutral final states with completely different systematics [13]. A similar check comes from comparing $\eta \to \gamma\gamma$ and $\eta \to 3\pi^0$; 4γ and 8γ data for both $\eta\eta$ and $\pi^0\eta$ channels agree very closely [2].

High partial waves are generally easy to determine from their strong angular dependence; low partial waves, particularly for $J^P = 0^-$, 1^- and 1^+ are more difficult. There are two lucky breaks which strenghen the analysis of $I = 0, C = +1$. The channel $\eta'\pi^0\pi^0$ is almost purely $f_2(1270)\eta'$ S-wave, and provides a precise determination of a 2^{-+} resonance at 2248 MeV[14]. The $\eta\eta\eta$ channel is dominated by a 0^- resonance at 2320 MeV[15]. Similar lucky breaks arise for $I = 1, C = +1$, where two 0^- resonances and one 2^- appear strongly in the spinless channel $a_0(980)\eta \to \eta\eta\pi^0$ [9].

THE MASS SPECTRUM

Fig. 1 shows the mass spectrum for $I = 0, C = +1$. For all quantum numbers, states lie close to parallel straight-line trajectories of M^2 (where M is mass) against radial excitation number, n; they are closely related to Regge trajectories. The best determined is the $^3P_2$ trajectory, where $f_2(1270)$ acts as an anchor point at low mass. The mean slope, averaged over all J^P and allowing for errors, is 1.143 GeV^2 per unit of n.

A subjective rating of the reliability of resonances is that 4^+, 3^+ and 2^+ are 4*, as is the upper 2^- state; all are observed accurately in at least two channels. The upper 0^- and 0^+ states and the lower 2^- are 3*; all are observed in at least two channels (the 0^+ independently in $\pi^0\pi^0$ and $\pi^-\pi^+$). Both 1^+ states and the lower 0^- are observed only in one set of data, and should be classified as 2*. From Crystal Barrel data alone, the lower 0^+ state is 2*; it interferes with the strong $f_0(2100)$, making its mass determination poor. It is, however, also observed in WA102 data [16].

Fig. 2 shows the mass spectrum for $I = 1, C = +1$. It is similar to that for $I = 0$, but less precise, mostly because of the absence of polarisation data. Fig. 3 illustrates the intensities of partial waves with these quantum numbers. Squares illustrate the scatter of fits at individual momenta, compared with full curves which show the simultaneous fit to all momenta. A specific remark is that the dominant signal comes from a $^3F_3$ resonance with $M = 2031 \pm 12$ MeV, $\Gamma = 150 \pm 18$ MeV. This is close to the $^3F_4$ and $^3F_2$ states,

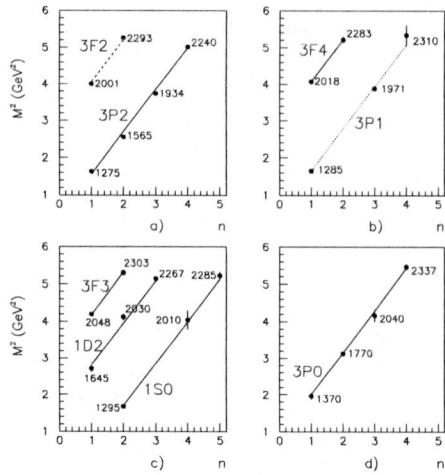

FIGURE 1. Trajectories of $I=0, C=+1$ $q\bar{q}$ states. Numbers indicate masses in MeV.

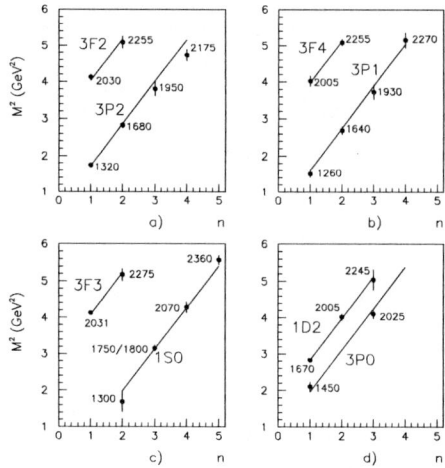

FIGURE 2. Trajectories of $I=1, C=+1$ $q\bar{q}$ states.

which lie respectively at 2005^{+25}_{-45} MeV and 2030 ± 20 MeV. In his summary talk, Barnes quoted VES data giving a much lower 3^+ mass but very large width. It is our opinion that there is unresolved $^3P_2$ mixed into VES data; this large 2^+ signal is visible in Crystal Barrel data at low mass in Fig. 3(i).

Fig. 4 shows mass spectra for $C=-1$ states with $I=1$ and 0. The former is almost

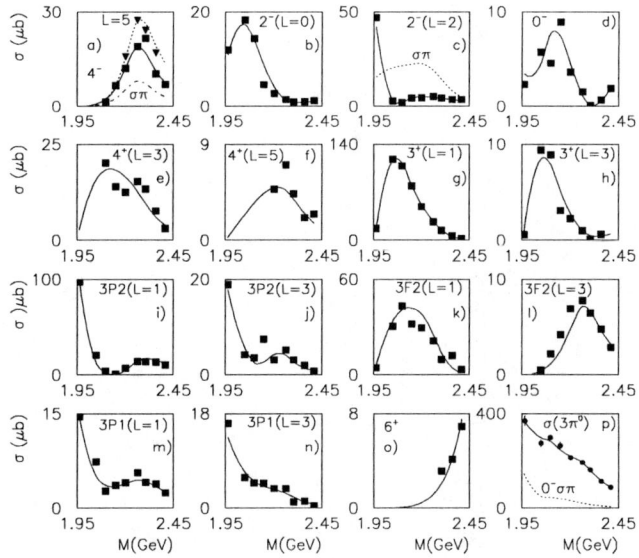

FIGURE 3. Intensities of $f_2(1270)\pi$ partial waves, $I = 1, C = +1$. Full curves show the simultaneous fit to all momenta; squares show results at individual momenta.

complete, but suffers from sizeable errors because of the absence of polarisation data. For that reason, $^3S_1$ and $^3D_1$ states are poorly separated and one is missing. The least accurate spectrum is for $C = -1, I = 0$, where all 1^- states are confused or missing. The problem here is that $\omega\pi^0\pi^0$ is a difficult channel, and the statistics of $\omega\eta$ are a factor 7 lower than for $\omega\pi^0$. Surprisingly, the lower $I = 0$ $J^P = 3^+$ state expected near 2000 MeV in Fig. 4(c) is also missing, although analysis is still in progress of further data for $\omega \to \pi^+\pi^-\pi^0$. Despite the larger errors on Figs. 2 and 4, there is little doubt that they follow a similar pattern to Fig. 1.

EXTRA STATES

For $I = 0, C = +1$, there are definitely two extra 0^+ states. It is arguable whether the lower one is $f_0(1500)$ or $f_0(1370)$, since they lie close together. The upper one is $f_0(2105)$, which has very curious properties. It is observed strongly in many sets of data: $J/\Psi \to \gamma(4\pi)$ [17] (where it decays purely to $\sigma\sigma$), $\bar{p}p \to \eta\eta\pi^0$ [18], $\bar{p}p \to \eta\eta$

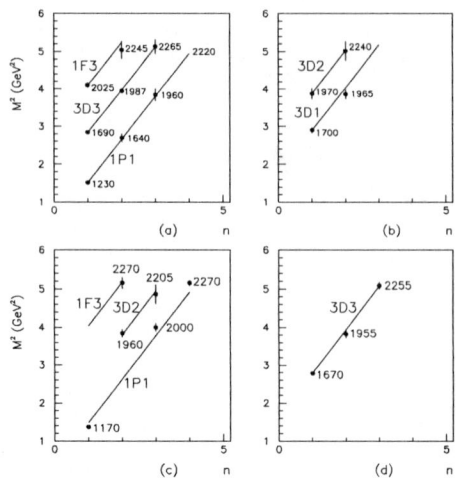

FIGURE 4. Trajectories of (a) and (b) $I = 1, C = -1$ states, (c) and (d) $I = 0, C = -1$.

[19] and $\bar{p}p \to \pi\pi$. Its parameters are well determined, but it is mistakenly listed by the Particle Data Group under $f_2(2150)$ despite clear identification in every publication of quantum numbers 0^+. Its curious property is that it decays more strongly to $\eta\eta$ than $\pi^0\pi^0$. The η is 80% $n\bar{n}$ and 60% $s\bar{s}$, so an $n\bar{n}$ state would be expected to have a branching ratio $\pi^0\pi^0/\eta\eta = 1/(0.8)^4 = 2.45$, compared with the observed ratio 0.71 ± 0.17. If the $f_0(2105)$ is treated as a mixed state $\cos\theta(u\bar{u} + d\bar{d})\sqrt{2} + \sin\theta s\bar{s}$, this requires a mixing angle of $\theta = (65 \pm 6)°$. Its strong production from $\bar{p}p$ but dominant $s\bar{s}$ decay suggests exotic character. All other states in Fig. 1 have $\theta = 0$ to $\pm 15°$.

Another extra state is a broad $J^P = 2^+$ signal with $M = 1980 \pm 50$ MeV, $\Gamma = 500 \pm 100$ MeV. It first appeared in Central Production of 4π [20] and has been confirmed with higher statistics in WA102 data [21]. In between, it has been observed clearly in $\eta\eta$ in $\bar{p}p \to (\eta\eta)\pi$ [18]. It is also required in the partial wave analysis of $\bar{p}p \to \pi\pi$ and $\eta\eta$[3]. It is produced in central production with an azimuthal dependence like that of $f_0(1500)$ but distinctively different to that of well identified $q\bar{q}$ states [22]; this is one reason to consider it an exotic candidate. Other reasons are its unusually large width and its strong appearance in $\eta\eta$.

The latest Lattice Gauge calculations of the glueball mass spectum by Morningstar and Peardon [23] predict that the lowest glueballs will consist of two 0^+ states, one 2^+ and one 0^-. There is a scale error of $\sim 10\%$ on mass predictions, but mass ratios are predicted more accurately. Table 2 [24] shows that mass ratios agree remarkably well with those amongst $f_0(1500)$, $f_0(2105)$, $f_2(1980)$ and $\eta(2190)$[25]. However, observed decay branching ratios demand that glueballs mix strongly with neighbouring $q\bar{q}$ states.

TABLE 2. Glueball mass ratios predicted in Ref. 23, compared with experimental candidates; errors are in parentheses.

Ratio	Prediction	Experiment
$M(2^{++})/M(0^{++})$	1.39(4)	1.32(3)
$M(0^{-+})/M(0^{++})$	1.50(4)	1.46(3)
$M(0^{*++})/M(0^{++})$	1.54(11)	1.40(2)
$M(0^{-+})/M(2^{++})$	1.081(12)	1.043(36)

CANDIDATES FOR 2^{-+} HYBRIDS

There are also extra 2^{-+} states with $I = 0$ and $I = 1$. The trajectory of Fig. 1 accomodates a well identified $\eta_2(2030)$ [26] with the right mass for this trajectory. There is in addition an $\eta_2(1860)$. It was reported in 1996 in Crystal Barrel data for $\eta\pi^0\pi^0\pi^0$ at two beam momenta [27] and subsequently confirmed at all other momenta [26]. It cannot be explained as the high energy tail of $\eta_2(1645)$, since its production cross section is a factor 11-22 too strong. It has been confirmed by WA102 in both $\eta\pi\pi$ and $K\bar{K}\pi$ channels [28]. It decays strongly to $f_2(1270)\eta$ and $a_2(1320)\pi$, favoured decay modes for a hybrid [29,30].

In Crystal Barrel data on $\bar{p}p \to \eta\eta\pi^0\pi^0$ [31], there is evidence at momenta 900–1200 MeV/c for a corresponding $I = 1$ state with $M = 1880 \pm 20$ MeV, $\Gamma = 255 \pm 45$ MeV, decaying to $a_2(1320)\eta$. Fig. 5 shows fits to several mass projections for data at 1200 MeV/c. The evidence for $\pi_2(1880)$ is that there is a strong $a_2(1320)\eta$ threshold enhancement in the $\eta\eta\pi$ mass range above 1800 MeV. It is too strong by a factor 7 to be explained as the high energy tail of $\pi_2(1670)$.

There is also evidence for it in $3\pi^0$ data at the bottom of the available mass range in $[f_2(1270)\pi^0]_{L=2}$ [8], Fig. 3(c). In those data, there is also a $\pi_2(2005)$ in Crystal Barrel data for the $[\bar{p}p \to f_2(1270)\pi]$ S-wave, Fig. 3(b). There is further clearer evidence for it in $\bar{p}p \to a_0(980)\pi$ in $\eta\eta\pi^0$ data [9]. It lies on the expected 2^{-+} trajectory through $\pi_2(1670)$.

There is also earlier evidence from the experiment of Daum et al. for a strong $f_2(1270)\pi$ D-wave enhancement around 1850 MeV [32]. At this meeting, Popov showed confirmation of $\pi_2(1880)$ in E852 data [33]. A search is needed for the expected $s\bar{s}g$ state at ~ 2100 MeV; likely decay modes are to $f_2'(1525)\eta$ and $K_2(1430)K$.

THE NEED FOR POLARISATION DATA

For $I = 0$, $C = +1$, polarisation data from PS172 for $\bar{p}p \to \pi^-\pi^+$ play a vital role in two ways. The polarisation separates cleanly states with $L = J \pm 1$, e.g. $^3F_2$ and $^3P_2$. Secondly, polarisation is phase sensitive. It depends on the imaginary parts of interferences, while differential cross sections contain the real parts of interferences. The two together produce accurate mass determinations for almost all resonances. Masses of $^3F_2$, $^3F_3$ and $^3F_4$ states give a tensor splitting which agrees with one gluon exchange within errors of $\pm 30\%$. Spin-orbit and spin-spin splittings are consistent with zero.

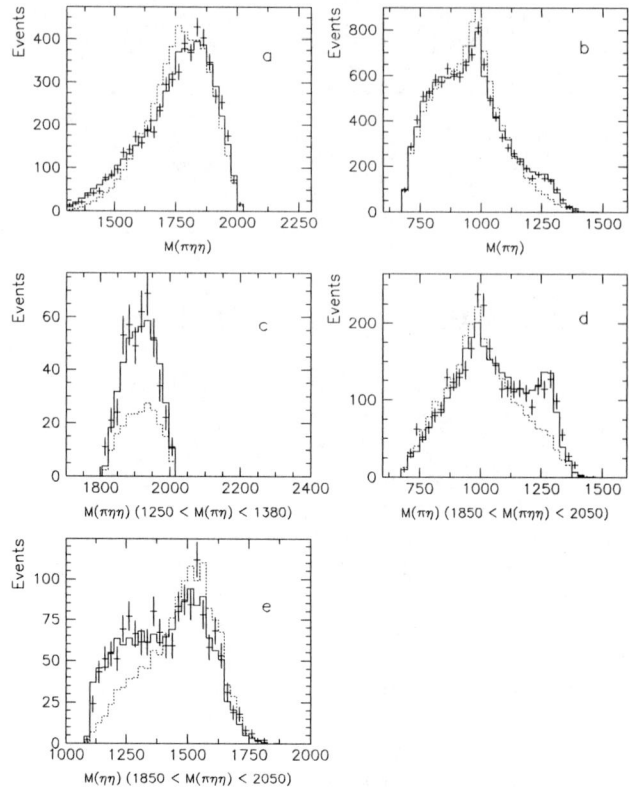

FIGURE 5. Full histograms show fits to $\eta\eta\pi^0\pi^0$ data at 1200 MeV/c compared with data; units of mass are MeV; dotted histograms show fits without $\pi_2(1880)$: (a) $M(\pi\eta\eta)$, (b) $M(\pi\eta)$, (c) $M(\pi\eta\eta)$ for events with a $\pi\eta$ cominbination within ±65 MeV of $a_2(1320)$, (d) and (e) $M(\pi\eta)$ and $M(\eta\eta)$ for events with $M(\pi\eta\eta)$ 1850–2050 MeV.

If polarisation data are dropped from the analysis, errors on many resonance masses increase by a factor 3 to 4, because of poor separation between $^3P_2$ and $^3F_2$ and between $^3F_4$ and $^3H_4$.

It would be very valuable to have data from a polarised target for $I = 1$, $C = +1$ and for $C = -1$, particularly for channels $3\pi^0$, $\eta\pi$, $\omega\pi$, and $\omega\eta$. Such an experiment is technically straightforward and could be carried out with avalable $\bar{p}$ inensities at the Brookhaven AGS - given a polarised target with a dilution refrigerator! Data-taking would be 3 months for a beam intensity $\geq 2 \times 10^4$ $\bar{p}$/s.

For $I = 1$, $C = +1$, there are presently two quite distinct solutions fitting $\eta\pi^0$ data. This ambiguity is almost but not quite resolved by $3\pi^0$ data. There, one solution is better than the other by 1000 in log likelihood; it is also very similar to the solution found for $I = 0$, $C = +1$. It is the solution shown in Figs. 2 and 3. The two solutions differ only for 2^+ and 4^+ states, where mass differences between the two solutions are ~ 30 MeV. However, accurate determinations of spin-splittings will only be possible by resolving this ambiguity. A polarisation experiment down to 400 MeV/c would play a major role in confirming the existence of two neighbouring $I = 1$, $J^{PC} = 2^{-+}$ states at 1880 and 2005 MeV (hence the hybrid candidate), and providing a clear separation of the $^3P_2$ and $^3F_2$ states reported in Crystal Barrel $3\pi^0$ data at ~ 1950 and 2030 MeV respectively.

For channels with $C = -1$, data from a polarised target would separate $^3S_1$ from $^3D_1$ and $^3D_3$ from $^3G_3$. A simulation shows that such data are expected to lead to a complete $q\bar{q}$ spectrum for all quantum numbers, unless resonances fail to couple to the available channels $\omega\pi^0$, $\omega\eta\pi^0$ for $I = 1$ and $\omega\eta$, $\omega\pi^0\pi^0$ for $I = 0$.

There is the prospect that Hall D experiments at Jefferson Lab can contribute to the clean identification of $J^{PC} = 1^{--}$ mesons. What is needed is an experiment on diffraction dissociation of linearly polarised photons. The linear polarisation couples helicity states and separates $^3S_1$ from $^3D_1$. This is where most of the ambiguities presently lie for $C = -1$ states.

REFERENCES

1. A.Hasan et al., Nucl. Phys. B 378 (1992) 3.
2. A.Anisovic et al., Phys. Lett. B468 (1999) 304 and 309, $\pi\pi$, $\eta\eta$, $\eta\pi$.
3. A.Anisovic et al., Phys. Lett. B471 (1999) 271, $I = 0, C = +1$.
4. A.Anisovic et al., Phys. Lett. B507 (1999) 23, $\omega\eta$.
5. A.Anisovic et al., Phys. Lett. B508 (1999) 6, $\omega\pi$.
6. A.Anisovic et al., Phys. Lett. B452 (1999) 273 $\eta\pi\pi$.
7. A.Anisovic et al., Nucl. Phys. A651 (1999) 253, $\eta\pi\pi$.
8. A.Anisovic et al., Phys. Lett. B in press, $3\pi^0$.
9. A.Anisovic et al., Phys. Lett. B, in press, $\eta\eta\pi^0$.
10. A.Anisovic et al., Phys. Lett. B476 (2000) 15, $\omega\pi^0\pi^0$.
11. A.Anisovic et al., Phys. Lett. B513 (2001) 281, $\omega\eta\pi^0$.
12. A.Anisovic et al., Phys. Lett. B491 (2000) 47, $I = 0, C = +1$.
13. V. Nikonov, this meeting.
14. A.Anisovic et al., Phys. Lett. B491 (2000) 40, $\eta'\pi\pi$.
15. A.Anisovic et al., Phys. Lett. B496 (2000) 145, $\eta\eta\eta$.
16. D. Barberis et al., Phys. Lett. B413 (1997) 217.
17. D.V. Bugg et al., Phys. Lett. B353 (1995) 378.
18. A.Anisovic et al., Phys. Lett. B449 (1999) 145, $\eta\eta\pi^0$.
19. A.Anisovic et al., Nucl. Phys. A662 (2000) 319, $\eta\eta$.
20. F. Antinori et al., Phys. Lett. B353 (1995) 589.
21. D. Barberis et al., Phys. Lett. B471 (2000) 440.
22. F.E. Close, A. Kirk and G. Schule, Phys. Lett. B477 (2000) 13.
23. C.J. Morningstar and M. Peardon, Phys. Rev. D60 (1999) 034509.
24. D.V. Bugg, M. Peardon and B.S. Zou, Phys. Lett. B486 (2000) 49.
25. D.V. Bugg, L.Y. Dong and B.S. Zou, Phys. Lett. B458 (1999) 511.
26. A.Anisovic et al., Phys. Lett. B477 (2000) 19.
27. J. Adomeit et al., Zeit. Phys. C 71 (1996) 227.
28. D. Barberis et al., Phys. Lett. B413 (1997) 217.
29. T. Barnes, F.E. Close, P.R. Page and E.S. Swanson, Phys. Rev. D55 (1997) 4157.
30. P.R. Page, E.S. Swanson, and A.P. Szczepaniak, Phys. Rev. D59 (1999) 0334016.
31. A.Anisovic et al., Phys. Lett. B500 (2001) 222, $\eta\eta\pi^0\pi^0$.
32. C. Daum et al., Nucl. Phys. B182 (1981) 269.
33. A. Popov, this conference.

STANDARD MODEL

IHEP, Protvino

A new measurement of Re(ε'/ε) in $K^0 \to 2\pi$ decays by the experiment NA48 at CERN

M. Holder, M. Ziolkowski [1]

Fachbereich Physik, Universität Siegen, D-57068 Siegen, Germany

The goal of the experiment is to improve the error of previous mesurements of Re(ε'/ε) by about a factor 3, because the deviation from zero is only about three standard deviations in one experiment [1] and one standard deviation in the other [2]. Direct CP- violation, that is CP- violation in a $\Delta S = 1$ transition, is expected in the standard model, where CP -violation is the consequence of a nonzero phase in the quark mixing matrix. Both, indirect CP- violation, by $K^0 - \bar{K}^0$ state mixing, characterized by the quantity ε, and direct CP- violation, characterized by ε' can be related to the elements of the quark mixing matrix. The theoretical expectations for Re(ε'/ε) are around 10^{-3}. Because Re(ε'/ε) is related to the $K^0 \to 2\pi$ rates by the double ratio

$$R = \frac{\frac{\Gamma(K_L \to \pi^0\pi^0)}{\Gamma(K_S \to \pi^0\pi^0)}}{\frac{\Gamma(K_L \to \pi^+\pi^-)}{\Gamma(K_S \to \pi^+\pi^-)}} = 1 - 6Re(\varepsilon'/\varepsilon), \tag{1}$$

this ratio has to be measured with an accuracy of about 10^{-3}. This implies statistical errors based on millions of events in all four decay channels, and systematic errors in the relative branching ratios of 10^{-3}; this represents a real challenge.

The experiment NA48 was therefore designed for high intensity beams and for best possible control of systematic errors. All four decay channels are measured concurrently. A brief description of the apparatus has been given in Ref. [3]. The K_L beam is defined by collimators pointing under 2.4 mrad to the K_L target which is hit by 450 GeV protons. A K_S beam is derived from a target near the begin of the decay region, where the vacuum tank opens up to allow for a measurement of the decay products. It is sufficient that the proton intensity on the K_S target is only about 10^{-5} of the intensity on the K_L target, since $K_S \to 2\pi$ are the dominant decay modes, and the decay length is shorter than the fiducial volume. A small fraction of primary protons which have not interacted in the K_L target are sent to the K_S target. First, they are deflected back to the K_L beam axis by means of a

---

[1] on behalf of the NA48 Collaboration: Cagliari, Cambridge, CERN, Dubna, Edinburgh, Ferrara, Firenze, Mainz, Orsay, Perugia, Pisa, Saclay, Siegen, Torino, Warszawa, Wien

bent crystal [4], using the channelling effect. They are then transported to the K_S target by quadrupoles and bending magnets. The K_S target is located 7.2 cm above the center of the K_L beam. The K_S collimator defines a neutral beam which converges to the center of the K_L beam at the position of the photon detector 100m further downstream. This convergence and the small vertical separation of the beams insure that the illumination of the detector is practically the same for K_L and K_S decays. The different decay point distribution of K_L and K_S decays, due to the different lifetimes, are equalized in the analysis, by weighting the K_L events. Late K_L decays obtain a small weight since the corresponding K_S decays are rare. The loss in statistical accuracy is about 20%; the gain in systematic uncertainty due to the reduction of acceptance differences is however substantial.

An observed decay can be attributed to either K_L or K_S beam if the decay products are charged particles. In this case the transverse coordinates of the decay point can be reconstructed. This is not possible for decays into neutrals. Therefore another method is required. The method used in this experiment is based on time measurement. The event time is measured in scintillators, or also by the liquid krypton calorimeter, and is compared to the time of protons in their way to the K_S target. An event is tagged as a K_S decay if it is coincident with a proton to the K_S target, and as a K_L decay if it is not. Because of accidental coincidences there is a certain fraction ($\approx 11\%$) of mistagged events. This fraction can be measured with charged decays. The tagging counter [5] in the proton beam to the K_S target is in fact a series of 12 vertical and 12 horizontal counters, very thin in the direction transverse to the beam, in order to cover only a small fraction of the beam in each counter. The inefficiency of the tagging system is only about $2 \cdot 10^{-4}$. The accuracy of the time measurement is around 0.2 ns, both in the tagger and in the main detector. The time window for a tagging coincidence is defined offline as 2 ns.

The main new development for the NA48 detector is a total absorption liquid krypton calorimeter for the measurement of photon energies and impact points. The calorimeter [6] consists of ≈ 13000 towers of 2 cm*2 cm area and 125 cm (27 radiation lengths) deep, pointing to the begin of the decay volume. A central anode strip in each tower collects the electrons from the ion pairs created in the krypton by photon showers. Both cathode and anode strips are realized as thin Cu-Be strips of 2 cm width, guided between front and back of the calorimeter by a set of four intermediate vetronite frames with suitable windows, such that each individual strip follows a zig-zag path with 48 mrad bends through the windows in these frames. The fluctuation of the energy lost in these frames contibutes a small term to the resolution. From calibrations with electrons and with photons from π^0 decays the resolution is found to be

$$\frac{\sigma(E)}{E} = \frac{(3.2 \pm 0.2)\%}{\sqrt{E}} \oplus \frac{(9 \pm 1)\%}{E} \oplus (0.42 \pm 0.05)\%. \qquad (2)$$

The absolut energy scale is related to geometry using the kinematics of $K_S \to 2\pi^0$ decays. In the K_S beam the beginning of the decay volume is defined by an anticounter (preceeded by a 3 cm thick iridium crystlal [7] for efficient photon conversion). In each event, the distance of the decay point from the calorimeter is calculated from the positions and energies of the photons, using the kaon mass as a constraint. If the energy scale is wrong, the edge of the decay point distribution corresponding to the position of the anticounter is misplaced. The systematic uncertainty in the reconstructed anticounter position is 2 cm, corresponding to an error in the energy scale of $2 \cdot 10^{-4}$. The response of the calorimeter to photon energies between 3 GeV and 100 GeV is linear to very good accuracy; after small corrections with electrons from Ke3 decays the remaining nonlinearity is 10^{-3}. The calorimeter calibration is maintained by test pulse injection (out of burst), by Ke3 decays and by special runs with polyethylen targets at two positions in the decay volume, struck by a pion beam for copious production of π^0 's and η 's.

Charged Kaon decay products are measured in a magnetic spectrometer consisting of a central dipole magnet with a field integral corresponding to a momentum kick of 265 MeV/c, and of four drift chambers, two in front of the magnet and two behind it. Each chamber [8] is in fact composed of eight wire planes; a pair of planes for right/left ambiguity resolution repeated four times with wire orientation rotated by 90^0 or $+- 45^0$ in successive views. The drift space is +- 5 mm; coordinates are measured with 100 μm resolution. The plane efficiency is around 99 %. The track loss by inefficiency in events not affected by overflows is less than 0.1 %; the difference in inefficiencies for K_S and K_L decays is negligeable. Accidentals with too many hits in the chambers, mainly from interactions of photons or electrons in the beam pipe which traverses the whole detector, are suppressed in the readout [9] by an overflow condition. An overflow in a wire plane is generated whenever more than 7 hits are found within 100 ns. The front-end buffers are then cleared and the presence of an overflow is recorded in the data stream. In the offline analysis events are required to be free of overflows in +- 300ns window around the event time, both for charged and neutral decays. Between 10% and 20% of neutral events are lost by this requirement, depending on whether no or at most one plane is allowed to have an overflow. The final result turns out to be the same for both conditions.

The spectrometer resolution is limited by multiple scattering and measurement accuracy. It can be parametrized as

$$\frac{\sigma(p)}{p} = 0.48\% \oplus 0.009 * p\% \qquad (3)$$

(p in GeV/c). The mass resolution for $K \to 2\pi$ decays is 2.5 MeV on average.

A fast reconstruction of two- track events is based on space points in chambers 1,2 and 4. Events with invariant $\pi\pi$ masses less than 475 MeV or decay vertices corresponding to more than 4.5 K_S lifetimes behind the location of the K_S anticounter are rejected online, in the level 2 trigger. The calculation is allowed to take 100 μs at most. The events are distributed to several fast processors in order to minimize dead time. A deadtime condition, if it nevertheless occurs, is then also applied to neutral events in the offline analysis. The efficiency of this trigger is measured with downscaled unbiased events; it is around 98 % for the present data set. The (statistical) error on the efficiency contributes to the systematic error of ε'/ε. Also used in the trigger for charged decays is a hadron calorimeter made of iron- scintillator sandwhiches, located downstream of the liquid krypton calorimeter and used in conjunction with it for total energy measurement. A total energy of at least 35 GeV is required in the trigger. Muon counter arrays behind the hadron calorimeter are used offline for muon rejection.

The trigger for neutral decays is a pipeline system with 3 μs latency. It uses a peak finding algorithm for the number of clusters in x- and y- projections and a lifetime cut, similar to the one used in charged decays, based on first and second moments of the energy deposition.

In the offline reconstruction of charged decays the efficiency of the first chamber near the beampipe of 16 cm diameter is most crucial, since the two beams, which converge on the calorimeter, are still separated vertically by about 1 cm. The chambers are in fact centered midway between the two beams. The standard reconstruction algorithm starts from track candidates in the (x-, y-, u-, v-) projections. An alternative reconstruction based on space points is about 5 % less efficient, because at least 3 hits out of the 4 projections are required for each space point; the reconstructed events are however less affected by accidentals. The two pattern recognitions give the same result for ε'/ε.

Background in charged decays (Fig.1) is evaluated in the $(p_t^2, m_{\pi\pi})$ plane. The transverse momentum p_t of the kaon is calculated from the distance of the reconstructed decay vertex to the line joining the target and the barycenter of the coordinates in the first chamber. In this way the reconstruction errors affect $K_S \to \pi^+\pi^-$ decays and $K_L \to \pi^+\pi^-$ decays in the same way; their p_t distributions should be identical. The background in K_L decays comes mainly from unidentified $K\mu 3$ and Ke3 decays, and to a small extent from kaons produced in the walls of the final collimator by particles emerging from the walls of the beam defining collimator, 80 m upstream. The total background is $(0.17 \pm 0.03)\%$. In K_S decays background from K_S- collimator scattering is essentially removed by a center-of-gravity cut, both in charged and neutral decays.

TABLE 1. Number of selected events after accounting for mistagging.

	Statistic im milions		
	1998	1999	total
$K_L \to \pi^0\pi^0$	1.047	2.243	3.290
$K_S \to \pi^0\pi^0$	1.638	3.571	5.209
$K_L \to \pi^+\pi^-$	4.541	9.912	14.453
$K_S \to \pi^+\pi^-$	6.910	15.311	22.221
Statistical error	$18.0 \cdot 10^{-4}$	$12.2 \cdot 10^{-4}$	$10.1 \cdot 10^{-4}$

Background in neutral events comes mainly from unidentified $K_L \to 3\pi^0$ decays.

In the $M(\gamma_1\gamma_2), M(\gamma_3\gamma_4)$ - plane of invariant $\gamma\gamma$ masses (Fig.2) this background is proportional to the area centered on the π^0 masses. Subdividing the plane into error ellipses of equal area, the background is flat as a function of ellipse number (Fig.3). The background in K_L neutral decays is $(0.06 \pm 0.02)\%$.

Accidentals can affect charged and neutral decays differently. Care is taken that the intensity distribution in the spill is the same for K_S and K_L beams, as much as possilble. The remaining accidental effects due to different sensitivity in charged and neutral decays and due to beam intensity difference are each at a level of 1-2 %. Only their product affects the double ratio. Accidental effects are therefore expected to be small. They are evaluated by overlaying " random " events (taken proportional to beam intensity) with data or with Monte Carlo generated events. In both methods the remaining accidental correction is compatible with zero.

TABLE 2. Corrections and systematic uncertainties on the double ratio. In order to obtain the effect on $\text{Re}(\varepsilon'/\varepsilon)$, the numbers must be divided by a factor of 6.

	in 10^{-4}	
$\pi^+\pi^-$ trigger inefficiency	-3.6	±5.2
AKS inefficiency	+1.1	±0.4
Reconstruction of $\pi^0\pi^0$	-	±5.8
Reconstruction of $\pi^+\pi^-$	+2.0	±2.8
Background to $\pi^0\pi^0$	-5.9	±2.0
Background to $\pi^+\pi^-$	+16.9	±3.0
Beam scattering	-9.6	±2.0
Accidental tagging	+8.3	±3.4
Tagging inefficiency	-	±3.0
Acceptance	+26.7	
statistical		±4.1
systematic		±4.0
Accidental activity	-	±4.4
Long term variation of K_S/K_L	-	±0.6
Total	+35.9	±12.6

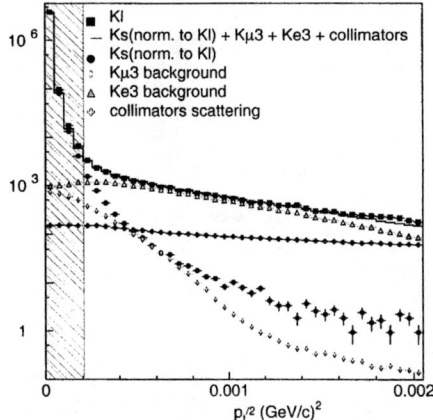

FIGURE 1. Comparison of the $P_T'^2$ tail of the $K_L \to \pi^+\pi^-$ candidates with the sum of all known components.

The statistics of accepted events and various corrections with their uncertainties are listed in Tab. 1 and Tab.2. The analysis is made in bins of kaon energy (Fig.4), since the decay spectra in the K_S and K_L beams are not the same. The variation of the double ratio with cuts in the analysis is shown in Fig.5. The double ratio is measured to be

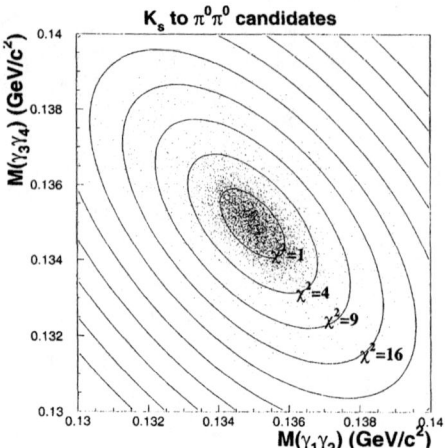

FIGURE 2. Distribution of $K_S \to 2\pi^0$ candidates in the space of two reconstructed $\gamma\gamma$ masses, $M(\gamma_1\gamma_2)$ and $M(\gamma_3\gamma_4)$.

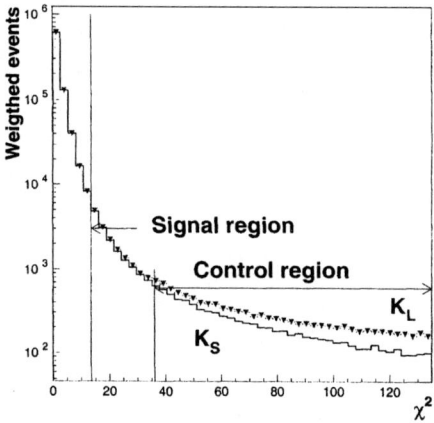

FIGURE 3. The comparison of the χ^2 distribution for K_L and $K_S \to 2\pi^0$ candidates, showing the excess due to the $3\pi^0$ background in the K_L sample.

$$R = (0.99098 \pm 0.00101(stat.) \pm 0.00126(syst.), \qquad (4)$$

or, using (1)

$$Re(\varepsilon'/\varepsilon) = (15.0 \pm 1.7(stat.) \pm 2.1(syst.))10^{-4}. \qquad (5)$$

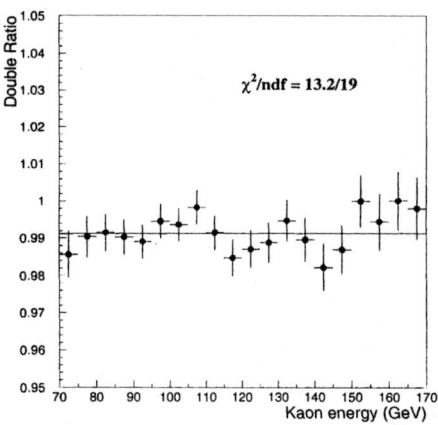

FIGURE 4. Corrected double ratio as a function of the kaon energy.

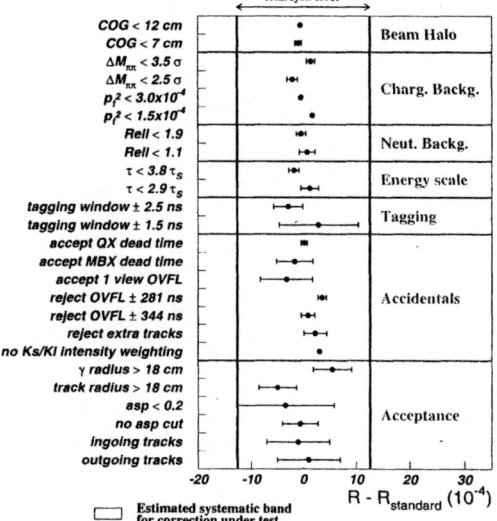

FIGURE 5. Stability of the double ratio with variations of the selection cuts. The grey band shows the uncertainty related to the cut concerned.

Adding the two errors in quadrature gives

$$Re(\varepsilon'/\varepsilon) = (15.0 \pm 2.7(stat.))10^{-4}. \qquad (6)$$

This result establishes a nonzero value of ε'/ε with a significance of nearly 6 standard deviations. It is in rough agreement with model predictions based on the assumption that CP- violation is due to a nonzero phase in the quark mixing matrix; these predictions are however not yet as precise as the experimental result.

REFERENCES

1. G.D.Barr et al., *Phys. Lett.* **B317**(1993),233
2. L.K.Gibbons et al.,*Phys. Rev.Lett.* **70**(1993),1203
3. M. Holder, Proceedings of the Workshop on Kaon Physics, Orsay (ed. L. Iconomidou- Fayard) Editions Frontieres (1996),77 I. Augustin, Proceedings of the 28 th Int. Conf. on High Energy Physics, Warsaw, Poland (1996), 1679
4. N. Doble et al., *Nucl. Instr. and Meth.* **B119** (1996),181
5. P.Grafström et al., *Nucl. Instr. and Meth.* **A344** (1994),487
6. R. Moore et al., *Nucl. Instr. and Meth.* **B119** (1996),149
7. V. Fanti et al., *Nucl. Instr. and Meth.* **A344** (1994),507 G.D.Barr et al., *Nucl. Instr. and Meth.* **A370** (1996),413
8. D.Bederede et al.,*Nucl. Instr. and Meth.* **A367** (1995),88
9. I. Augustin et al.,*Nucl. Instr. and Meth.* **A403** (1998),472

W mass and width determination at LEP II

Franco Ligabue

Scuola Normale Superiore, Pisa

Abstract. The methods used by the four LEP experiments to select W pair events are briefly reviewed, as well as the main methods for extracting the W mass and width. The currently relevant and in some cases still open systematic issues are illustrated. The preliminary results based on approximately 2500 pb^{-1} collected from 1996 to 2000, have been combined to give $M_W = 80.450 \pm 0.039$ GeV/c^2 and $\Gamma_W = 2.150 \pm 0.091$ GeV/c^2

Introduction

The LEP II programme succesfully started in June 1996, when the centre-of-mass energy of e^+e^- collisions was first set to 161 GeV, allowing the first on-shell W^+W^- pairs to be produced and reconstructed by the four LEP experiments. The measurement of the W pair production cross-section led to the first W mass determination at LEP [1]. In the second half of 1996 the centre-of-mass energy was taken to 172 GeV, and increased every year until peak energies of about 208 GeV were reached in 2000, the last year of running. LEP delivered a total integrated luminosity of about 700 pb^{-1} (corresponding approximately to 12,000 W pairs) per experiment[1], allowing a statistically precise determination of the W mass from direct reconstruction. At the present level of accuracy, the measurement is sytematics dominated. The results presented here are preliminary.

Event Selection and Mass Extraction

Several algorithms are used by the four experiments to select W pair events according to the possible W decay channels, which give rise to distinct final state topologies. The complexity of the selection algorithms and the background contamination increase with the number of hadronic jets in the final state.

Final states where both W's decay to a lepton-neutrino pair (*fully leptonic* channel) are characterised by low charged track multiplicity, large missing energy and momentum due to the two energetic neutrinos, and by the presence of two highly acollinear and acoplanar energetic tracks (electrons or muons from the W decay, or tracks from single-prong tau decays) or collimated low multiplicity jets from tau decays. The selection efficiency depends on the lepton species in the final state, ranging from 60-70% (no

---

[1] The results presented here refer to the whole LEPII statistics for all experiments except OPAL which has not yet included the year 2000 data

τ) down to 20-45% for the ττ channel. The typical purities show the same behaviour, ranging from 95% (no τ) to 80% (ττ).

Final states where only one of the W's decays leptonically (*semileptonic* channel) represent 43.7% of W^+W^- decays. The selection is based on the presence of a high energy neutrino, showing as large missing momentum pointing at a large angle with respect to the beam direction, and of a high-energy lepton usually isolated from the two hadronic jets. The main backgrounds come from radiative Z returns $e^+e^- \to q\bar{q}\gamma$ events and from four-fermion $e^+e^- \to q\bar{q}\ell\bar{\ell}$ events.

The overall signal efficiency for the semileptonic selection depends again on the presence of τ leptons in the final state, ranging typically from 70% ($q\bar{q}\tau\nu$) to 85% ($q\bar{q}e\nu$, $q\bar{q}\mu\nu$), with corresponding purities of 85% and 90-95%.

The *fully hadronic* ($W^+W^- \to q\bar{q}q\bar{q}$) channel accounts for 45.9% of W pair decays. These final states are characterised by no missing energy and momentum, and a four-jet topological structure with high sphericity, and high track multiplicity. Due to the comparatively large $e^+e^- \to q\bar{q}(\gamma)$ cross-section, tails in the multijet production distributions from hard gluon emission cause a sizeable amount of background to be effectively irreducible, and make it hard to devise a simple discriminating technique. Therefore, all four experiments make use of multivariate discrimination algorithms (*e.g.* Neural Networks) which combine the information from many event variables. Efficiencies for typical cuts on the discriminating variable range approximately from 78% to 88%, while purities, which are limited by the irreducible QCD background, are around 80%.

Mass measurement from lineshape

The first measurement of M_W at LEPII came in 1996 [1], The working point at 161 GeV was chosen to maximise the sensitivity of the cross-section value to M_W, which is only relevant near the pair-production threshold. Around 10 pb^{-1} per experiment were collected, leading to the following measurement:

$$M_W^{\text{threshold}} = 80.40 \pm 0.20_{\text{stat}} \pm 0.07_{\text{syst}} \pm 0.07_{\text{LEP}} \text{ GeV}/c^2$$

where the first error is statistical, the second systematic; the third comes from the uncertainty on the beam energy as measured by the LEP division, and is obviously correlated among the four experiments. Although the importance of this measurement relies in its complementarity with respect to the one from direct reconstruction, its statistical power is in fact very small compared the latter, and its weight in the final combination is almost negligible [18].

Mass measurement from direct reconstruction

Although the fully leptonic channel contains information on the W mass, the core of the mass measurement from direct reconstruction comes from the fully hadronic and semileptonic channels, both because they comprise most of the statistics, and because their kinematics is completely reconstructable. ALEPH and OPAL however have performed dedicated analyses [2], [10] which exploit kinematic properties of the fully leptonic events, such as the lepton spectrum, to extract M_W. The results are given here after combination with those from the semileptonic channel.

Kinematic Fit and Jet Pairing

Since the energy of hadronic jets is measured with relatively poor precision (compared to the beam energy), a kinematic fitting procedure is used in semileptonic and fully hadronic events to recompute the jet and lepton momenta. Energy and momentum conservation is imposed (four constraints), allowing the energy and direction af each jet to vary within the expected resolution around the measured central value, corrected for expected biases obtained from simulation. In semileptonic events, three of the four constraints are used to determine the energy and direction of the undetected neutrino. Each event provides two invariant masses, one for each decaying W, which can be constrained to be equal (5C fit, 2C for the sempileptonic case).

In the fully hadronic channel, the events are forced into a four-jet topology. In some cases, to account for hard gluon radiation, a five-jet topology is preferred. Only one of the three possible di-jet combinations that can be used in the kinematic fit (ten in the case of five jets in the final state) is the correct one.

Various techniques are employed to define the best combination, using for instance the kinematic fit χ^2, or the probability of the kinematic configuration as given by the Standard Model matrix element. Since these methods are never perfectly efficient, most experiment recover some information from second best combinations. DELPHI [4], for instance, combines the information from all possible jet pairings, after computing a probability for each combination to be the correct one, based on candidate W production angles.

Mass and Width extraction

The event-by-event invariant mass from the kinematic fit provides an estimator for M_W from which the value of the W mass has to be extracted. Figure 1 shows some of the invariant mass distributions obtained by the experiments. The extracton of M_W can be achieved

- by parametrizing the invariant mass distribution (which can be two-dimensional, in case two masses per event are considered) with a functional shape containing M_W as a parameter. Usually a Breit-Wigner shape is used: OPAL [9] for instance uses an "asymmetric" Breit-Wigner shape with different widths on each side of the peak. The value of m_0 obtained with this and similar methods is not in general an unbiased estimator of M_W, essentially because of distorsions due to initial state radiation and detector effects. A simple linear relation between the generated and the fitted mass (calibration curve), obtained from simulation, is used to correct the fit result (and in general to check the calibration of any extraction method);

- by comparing the invariant mass distribution obtained from data to the same distribution from simulation, for different input M_W values, and determining the best value for M_W from a maximum-likelihood fit. Since sufficiently large samples of Monte Carlo events can only be generated for few input mass values, a reweighting techinque is employed to obtain a Monte Carlo distribution for any new value of M_W: each event is assigned a weight which accounts for the changed differential production cross-section: $w = \left|\mathcal{M}(M_{\text{new}})\right|^2 / \left|\mathcal{M}(M_{\text{gen}})\right|^2$. This method is intrinsically unbiased and automatically takes into account initial state radiation and detector effects, provided they are correctly simulated by the Monte Carlo. Fitting the reconstructed mass alone (1-dimensional fit or 2 dimensional in case the two masses from the 4C kinematical fit are used) has the disadvantage of disregarding event-by-event errors. Detailed studies have shown that extending the fit to a multidimensional distribution can sizeably improve the statistical sensitivity of the method. ALEPH, for instance, found that for the semileptonic channel, using a three-dimensional distribution (invariant mass of the hadronic system, mass from the 2C kinematic fit, mass error from the same fit) decreases the expected statistical error by 12%;
- by building an event-by-event based likelihood: DELPHI, for instance, builds a likelihood which convolutes the probability that the observed kinematical configuration comes from the decay of two heavy objects of specified masses, with the probability — determined from theory and from detector simulation – that those two masses are actually produced and reconstructed.

The reliability of the statistical error from the mass fit is checked by generating a large number or Monte Carlo samples of the same size of the data sample, and by looking at the fit error and pull distributions.

All experiments also perform a simultaneous fit for M_W and Γ_W to the invariant mass distribution, leaving both parameters free instead of assuming the Standard Model relation between the W mass and its width. The correlation between the two parameters turns out to be very small.

Systematics

Since at present the measurement is systematics-dominated, the efforts of the collaborations are clearly focussed on the main sources of estimated systematic uncertainty, especially if correlated between years and experiments. It is perhaps worthwhile to recall here that the measured quantity M_W is not strictly a physical observable, but rather a so-called "peudo-observable": basically a semi-theoretical parameter which is used as input in the Monte Carlo generators. What all analyses ultimately do is to determine the one such parameter which best describes the data. All systematic uncertainties can therefore be ascribed to some inadequacy in the Monte Carlo description.

In the following, the currently relevant systematic issues are listed ad briefly discussed.

Detector Effects

This is one of the main sources of uncorrelated systematics. It arises from inadequacies in the simulation of the detector response. In order to investigate and to quantify them, use is made of the so-called "calibration data" collected at the Z peak every year, at the beginning and at the end of the running period. Some of the discrepancies found in these data are generally used to correct the simulated data at higher energy.

The residual discrepancies, which cause the systematic uncertainty, are generally accounted for by altering the MonteCarlo, either at an early stage in the recostruction chain (*e.g.* alteration of the calorimeter response) or at a late stage (*e.g* smearing of jet energies and angles) These two approaches are not independent. Therefore, it is important to avoid double-counting effects if they are both used: studies in this respect are still under way. The current estimates of the effect on M_W range from 8 to 27 MeV.

Fragmentation

The "standard" procedure for the assessment of a systematic uncertainty due to inadequacies in the simulation of hadronisation processes has been the comparison of different Monte Carlo generators (typically JETSET [12] and HERWIG [13]). Since the ultimate goal should rather be a comparison between Monte Carlo and data, extensive studies have been going on to try and identify "fragmentation-sensitive" variables which could be used for such comparison. Other approaches are the variation of fragmentation parameter within a single fragmentation model, and the use of the so called Mixed Lorentz Boosted Z's (MLBZ), which combine pairs of independent events collected at the Z peak (both in data and in Monte Carlo) into fake four-jet topologies mimicking WW events. Preliminary results from these studies give mass uncertainties which are compatible with those from model-model comparison (10 to 30 MeV). This uncertainty is treated as fully correlated between channels, years and experiments, although the role played by detector effects (which might introduce some degree of uncorrelation) is still being investigated.

Radiative corrections

The complete set of $O(\alpha)$ corrections was not present in the Monte Carlo W pair generator (KORALW [11], using GENTLE calculations). This caused the predicted total pair-production cross section to be significantly higher (2.5%) than the one measured at LEPII. The inclusion of such corrections, however, does not only change the overall normalization of the signal, but also its differential cross-section, which might affect the mass determination. Two new Monte Carlo calculations are now available (RACOONWW [14] and YFS [15]) whose predicted total cross-section agrees much better with the data. These calculations are not yet usable at full-generator level, and their effect can only be simulated by event weights for the time being. The full difference in the mass after applying such weight is taken conservatively as an estimate of the systematic error, which is correlated between experiments and years. The current estimates range from 5 from 16 MeV.

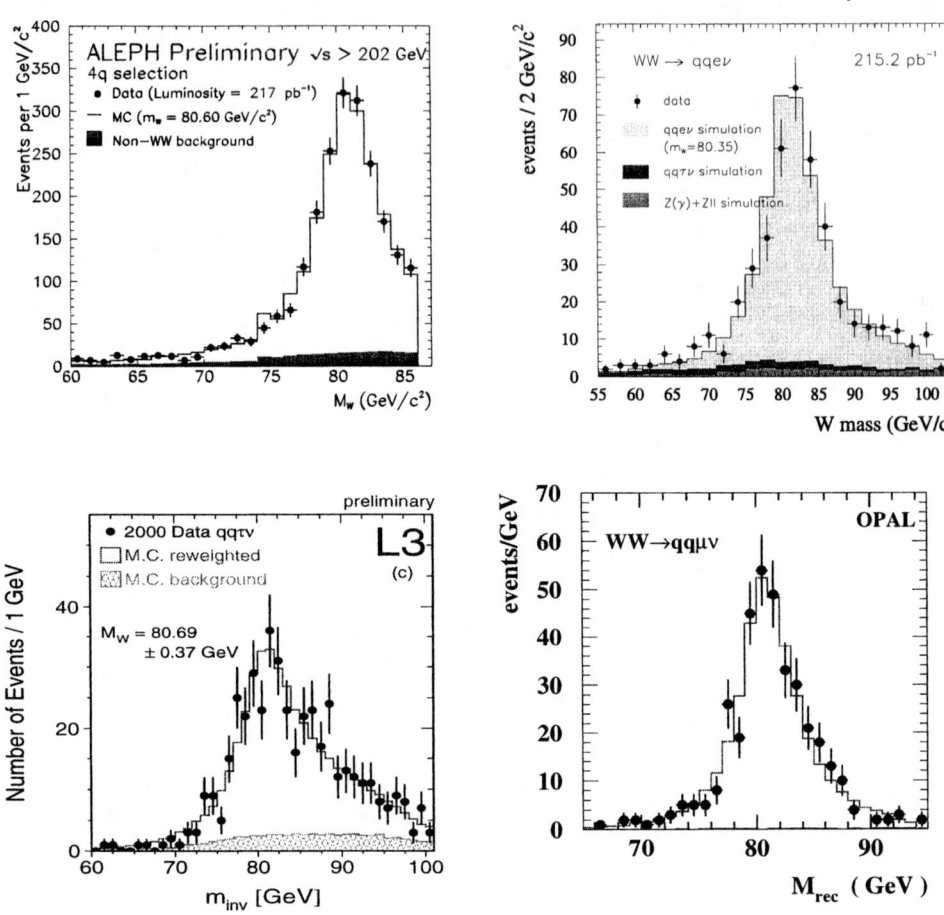

FIGURE 1. Examples of Invariant mass distributions from kinematic fit: top-left: fully hadronic channel (ALEPH); top-right: DELPHI evqq; bottom-left: L3 τνqq; bottom-right: OPAL μνqq (at $\sqrt{s} = 189$ GeV)

Beam Energy

The value of LEP beam energy as provided by the LEP group is explicitly used in the kinematic fitting procedure, which introduces an effective linear dependence on the value of the reconstructed mass: $\Delta M_W / M_W = \Delta E_{\text{LEP}} / E_{\text{LEP}}$.

The beam energy is provided to the collaborations by the Lep Energy Working Group [19], and is obtained from the measurement of the integral of the dipole magnetic field along the beam orbit. The magnetic field is measured by means of NMR probes located inside bending magnets. The NMR measurement is calibrated at LEPI energies ($E_{\text{beam}} < 60$ GeV) by means of the very precise Resonant Depolarisation measurements [19] ($\Delta E/E \approx 2 \times 10^{-5}$). The low-energy calibration has to be extrapolated to LEPII energies. The error on this extrapolation is the main contribution to the estimated uncertainty on E_{beam}. At least three alternative methods are used to cross-check the calibration and extrapolation:

- an alternative measurement of the field integral, achieved by means of flux-loops which sample 96.5% of the bending field along the circumference.
- a spectrometer measuring the beam deflection caused by a very accurately mapped dipole field.
- a mesurement of the beam energy which exploits the well-known energy dependence of the energy loss from synchrotron radiation; this can be achieved by measuring the synchrotron tune Q_s, which depends on both the beam energy and the energy loss rate.

The three cross-check methods give consistent results and confirm the NMR measurement within its uncertainty (from 20 to 25 MeV depending on the year of running, which correspond to $\Delta M_W = 17 - 20$ MeV)). Extensive studies are going on expecially on the spectrometer and Q_s measurement, which look promising and in principle should be able to reduce ΔE_{beam} to 15 MeV.

Final State Interactions

Finally, a very important source of correlated systematics is the possibility of Final State Interactions (FSI) in the fully hadronic channel: Bose-Einstein correlations between same-charge pions, and cross-talk at hadronisation level between the two hadronically decaying W's (Colour Reconnection).

Bose Einstein Correlations. This effect, which has been measured on Z data, can alter the measured mass if it's present for pion pairs coming from different W bosons. It can be is investigated in WW data, essentially by comparing the two-particle density distribution in fully hadronic WW events, with the same distribution in Monte Carlo with and without Bose-Einstein correlations. The most recent results [16] from the four experiments tend to disfavour the presence of correlations between pions from different W's as described in the available Monte Carlo models. For the time being, however, no use is made of these negative results in the assessment of the systematic error, which is based on the mass shift observed when the effect is included in the simulated data, and ranges between 20 and 67 MeV

Colour Reconnection. This non-perturbative QCD effect, which can be roughly described as a non-independence of the hadronisation processes for the decay products of the two W bosons, is described by a variety of phenomenological Monte Carlo models. The expected shift on the measured W mass is very model-dependent: from extensive studies using one particular model called SKI [20], where the reconnection

TABLE 1. Error decomposition for the combined LEP W mass results [18]

Source	Error on M_W (MeV/c^2)		
	$q\bar{q}\ell\bar{\nu}_\ell$	$q\bar{q}q\bar{q}$	Combined
ISR/FSR	8	9	8
Fragmentation	19	17	17
Detector effects	12	8	10
LEP Beam Energy	17	17	17
Colour Reconnection	—	40	11
Bose-Einstein Correlations	—	25	7
Other	4	4	3
Total Systematic	29	54	30
Statistical	33	30	26
Total	44	62	40

probability can be expressed in terms of a single free parameter κ, the four experiments have been shown to have the same sensitivity to Colour Reconnection in terms of ΔM_W. Some of the extreme models which predicted large mass shifts also predicted significant difference in charge multiplicity between semileptonic and hadronic events, which allowed their exclusion based on early LEPII data. Some CR models can be tuned using LEPI data, after which they predict small shifts in the W mass. The current value for the CR systematic error, estimated using a umber of different "realistic" models, ranges from 33 to 66 MeV for the four experiments. Like in the case of Bose-Einstein correlation, recent efforts [17] have been aimed at extracting some information on the amount of Colour Reconnection in data, through the study of some sensitive variables, and particularly the so-called "particle flow" (essentially the particle density) between jets from different W's. This might allow to set some limits on the size of CR effects in the data which would help reduce the FSI systematic uncertainty on the W mass.

Another approach aimed at the reduction of the FSI systematics is the devising of analyses less sensitive to such effects. Since these are basically affecting soft interjet tracks, the explored strategy is the removal of such tracks, by cutting on their momentum or by slightly modifying the jet definition ("cone" algorithm). These procedures have been shown to reduce the systematic mass shift (within a particular CR model) at the expense of part of the statistical sensitivity.

Combined Results for M_W and Γ_W

The preliminary results for the four LEP experiments have been combined [18] taking into account all possible correlations.

The difference between the measurements in the fully hadronic and leptonic channel is 9 ± 44 MeV/c^2, showing no evidence of FSI effects. The weight of the fully hadroinc in the combination is only 0.27, in spite of the lower statistical error, due to the large FSI systematic error.

TABLE 2. Summary of M_W determinations from direct reconstruction at LEP II. Separate result for the hadronic and semileptonic channels are shown, as well as their combination. For all results, the first error is statistical, the second is systematic. See ref. [18] for details on the error breakdown and combination procedures.

	M_W (GeV/c^2)		
	$\ell\nu q\bar{q}$	$q\bar{q}q\bar{q}$	combined
ALEPH	80.456 ± 0.051 ± 0.032	80.507 ± 0.054 ± 0.045	80.477 ± 0.045 ± 0.049
DELPHI	80.414 ± 0.074 ± 0.048	80.348 ± 0.053 ± 0.055	80.399 ± 0.055 ± 0.049
L3	80.314 ± 0.074 ± 0.045	80.478 ± 0.063 ± 0.069	80.389 ± 0.048 ± 0.051
OPAL	80.515 ± 0.067 ± 0.030	80.408 ± 0.066 ± 0.100	80.491 ± 0.053 ± 0.038
Combined	80.448 ± 0.033 ± 0.029	80.457 ± 0.030 ± 0.054	80.450 ± 0.026 ± 0.030

The combined value for M_W, including the measurements from the fully leptonic channel and from the lineshape, is

$$M_W = 80.450 \pm 0.026_{\text{syst}} \pm 0.030_{\text{syst}} \text{ GeV}/c^2$$

The combinination of the width measurements gives

$$\Gamma_W = 2.150 \pm 0.068_{\text{stat}} \pm 0.060_{\text{syst}} \text{ GeV}/c^2.$$

M_W and the Standard Model

A precise measurement of M_W at LEP is important due to the sensitivity of the value of M_W to the Standard Model parameters (m_{top}, m_{Higgs}) through radiative corrections. Figure 2 (left) shows the LEP measurements compared with the Standard Model predictions. Figure 2 (right) compares the direct mass determinations (M_W, m_{top}) with the indirect ones from the Standard Model fit to Electroweak data. As can be seen from both plots, the direct mass measurement prefers light Higgs masses.

Conclusions and Perspectives

The excellent performances of the LEP machine, of the four detectors and of the analysis teams have allowed a precision measurement of M_w which at this preliminary stage is already beyond the design precision goal of a 50 MeV error. Much effort is being spent on systematic studies, especially in order to reduce the FSI uncertainty. Taking into account the statistical improvements from the inclusion of the OPAL 2000 data and from the reprocessing of all the LEPII data, realistic estimates predict a final LEPII $\Delta M_W \approx 30 - 35$ MeV.

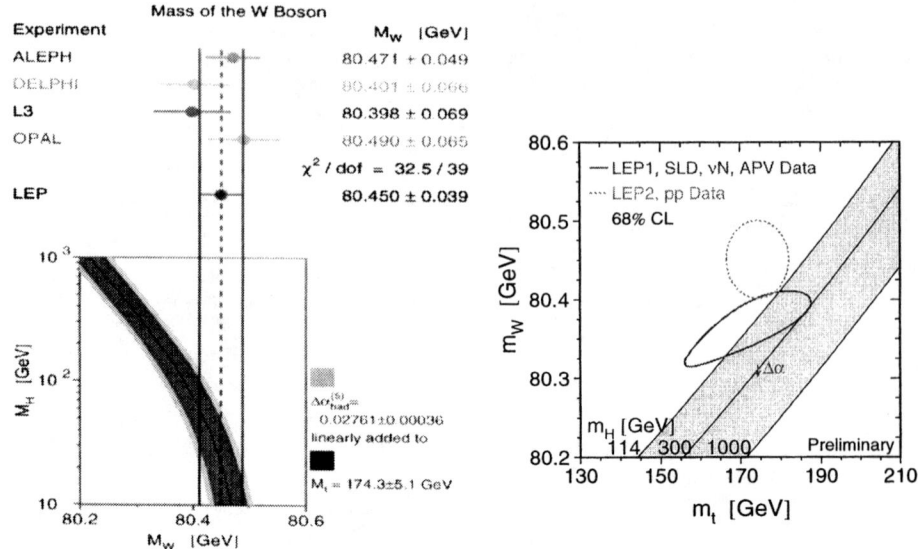

FIGURE 2. *Left*: Summary of M_W measurements at LEPII compared with the Standard Model predictions as a function of m_H. *Right*: comparison of direct and indirect measurements of M_W and m_{top}

REFERENCES

1. ALEPH Collaboration, Phys. Lett. **B401** (1997) 159,
   DELPHI Collaboration, Phys. Lett. **B397** (1997) 159,
   L3 Collaboration, Phys. Lett. **B398** (1996) 159,
   OPAL Collaboration, Phys. Lett. **B398** (1999) 159
2. ALEPH Collaboration, Eur. Phys. J, **C17** (2000) 241
3. ALEPH Collaboration, ALEPH note 2001-020 CONF 2001-017
4. DELPHI Collaboration, Phys. Lett. **B511** (2001) 159
5. DELPHI Collaboration, DELPHI 2001-103 CONF 531
6. L3 Collaboration, Phys. Lett. **B454** (1999) 386
7. L3 Collaboration, L3 note 2377 (1999), L3 note 2575 (2000), L3 note 2637 (2001)
8. OPAL Collaboration, Phys. Lett. **B507** (2001) 29
9. OPAL Collaboration, OPAL Physics Note PN422 (Updated July 2001)
10. OPAL Collaboration, OPAL Physics Note PN480 (2001)
11. M.Skrzypek, S. Jadach, W. Placzek and Z.Wąs, Comp. Phys. Commun. 94 (1996)
12. T. Sjöstrand, Comp. Phys. Commun. 82 (1994) 74
13. G. Corcella et al, CERN-TH 2000-284
14. A.Denner, S. Dittermaier, M. Roth and D. Wackeroth, Phy. Lett **B475** (2000) 127
15. M.Skrzypek, W. Placzek, S. Jadach,M.Skrzypek, B.F.L. Ward and Z.Wąs, Phys. Lett **B417** (1998);
    id., CERN-TH/2001-040 (to be submitted to Comput. Phys. Commun.)
16. O.Pooth, *Bose-Einstein correlations in W-pair events* PRHEP-hep2001/120 to appear in the Proceedings of the EPS HEP 2001 Conference, Budapest, Hungary, 18-20 July 2001
17. D.Duchesneau, *Color reconnection studies in W-pair events* PRHEP-hep2001/119 to appear in the Proceedings of the EPS HEP 2001 Conference, Budapest, Hungary, 18-20 July 2001
18. LEP W Working Group, note LEPEWWG/MASS/2001-02 and references therein
19. LEP Energy Working Group, LEPEWG 01/01, and references therein
20. T. Sjöstrand and V. Khoze, Z. Phys **C62** (1994) 281

Recent QCD Results from CDF

Igor V. Gorelov

Dept. of Physics and Astronomy, Univ. of New Mexico,
800 Yale Blvd. NE, Albuquerque, NM 87131, USA
(on behalf of the CDF Collaboration)

Abstract. Experimental results on QCD measurements obtained in recent analyses and based on data collected with CDF Detector from the Run 1b Tevatron running cycle are presented. The scope of the talk includes major QCD topics: a measurement of the strong coupling constant α_s, extracted from inclusive jet spectra and the underlying event energy contribution to a jet cone. Another experimental object of QCD interest, prompt photon production, is also discussed and the updated measurements by CDF of the inclusive photon cross section at 630 GeV and 1800 GeV, and the comparison with NLO QCD predictions is presented.

INTRODUCTION

Although the topics of the talk outlined in the abstract concern different experimental objects, hadron jets and electromagnetic clusters in photon studies, they probe the common subtleties of perturbative QCD (pQCD) – NLO processes and non-perturbative contributions. Moreover the measurements include similar theoretical uncertainties such as renormalization μ_R and factorization μ_F scales and choice of PDF. From the experimental point of view, a hadron jet is a cone of radius R in η,ϕ over a seed *filled* with energy deposits contrary to a prompt photon object where the similar cone over el.-mag. cluster is required to be *"empty" to isolate* the cluster seed. The topic of underlying events covers specifically the study of the largest experimental (though inevitably connected with the theory) uncertainty in the jet inclusive measurement of α_s.

MEASUREMENT OF THE STRONG COUPLING CONSTANT α_S

The measurement of α_s is extracted from the inclusive jet differential cross section over the jet transverse energy E_T range from 40 GeV to 450 GeV. The measurements are based on a data sample of integrated luminosity $\mathcal{L} = 87 pb^{-1}$ collected by CDF during 1994-95 (Run 1b) at $\sqrt{s} = 1.8 TeV$. The CDF detector is described elsewhere [1]. The details of the inclusive jet cross section measurement can be found in [2]. Briefly, the iterative fixed cone algorithm with $R \equiv \sqrt{\delta_\phi^2 + \delta_\eta^2} = 0.7, \eta \equiv -\ln(\tan(\theta/2))$ is used. Only the central range of $0.1 < |\eta| < 0.7$ is considered. The raw experimental E_T spectrum is corrected bin by bin for the calorimeter response and resolution, for the underlying event energy using an iterative unsmearing procedure. The α_s is determined by comparing the jet cross-section with NLO pQCD, $O(\alpha_s^3)$, calculations realized in

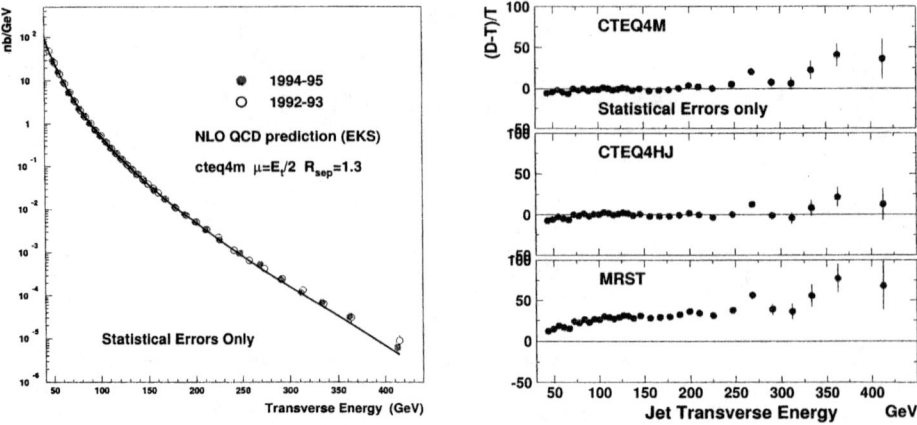

FIGURE 1. The corrected inclusive jet cross section (left) and the comparison to theory for various PDFs (right). Scales $\mu_R = \mu_F$ are set to $E_T^{jet}/2$ and phenomenological $\mathcal{R}_{sep} = 1.3$.

the JETRAD Monte-Carlo program with NLO pQCD contributions incorporated [3]. To match the experimental efficiency of identifying overlapping jets, two partons are required to be separated by more than $\mathcal{R}_{sep} \times R$ with a phenomenological factor $\mathcal{R}_{sep} = 1.3$; otherwise they are merged into a single jet. The corrected inclusive jet cross sections for Run 1b and published Run 1a data are shown in the left picture of Fig. 1 with the scale parameters in JETRAD set as $\mu_R = \mu_F = E_T^{jet}/2$. The right plot shows that the choice of PDF can *accommodate* the discrepancy between theory and data. The statistical precision of data is significantly better than the systematics in measurement and theory. The Run 1b measurement is in impressive agreement with NLO pQCD predictions provided the flexibility allowed by current knowledge of PDFs. This agreement has led to the proposal to use inclusive jet data to determine α_s.

Knowing that in the region $E_T^{jet} \in (40...450)GeV$ and $0.1 < |\eta| < 0.7$, the non-perturbative contributions are estimated to be negligible the NLO pQCD inclusive jet cross section can be parameterized as

$$d\sigma/dE_T = \alpha_s^2(\mu_R)\hat{X}^{(0)}(\mu_F, E_T)[1 + \alpha_s(\mu_R)k_1(\mu_R, \mu_F, E_T)]$$

Here both $\hat{X}(0)(...)$ and $k_1(...)$ are calculated by the JETRAD Monte-Carlo with NLO pQCD included. In this procedure the scale factors $\mu_R = \mu_F$ are set to E_T^{jet} unlike in comparisons of the inclusive jet cross section (see Fig. 1). The choice of PDF is CTEQ4M (see also below a discussion of theoretical systematics). Applying the same cuts and algorithms at the final parton level as are used in the data (JETRAD generator calculates weights) α_s is calculated for each of the 33 bins of the experimentally corrected $d\sigma/dE_T$ spectrum. These bin by bin calculations yield measurements of the *running* α_s presented in Fig. 2. The running $\alpha_s(\mu_R = E_T^{jet})$ for every point is evolved to the mass of Z^0, $\alpha_s(\mu_R = M_Z)$, using the evolution equation (inset plot in Fig. 2). Averaging over

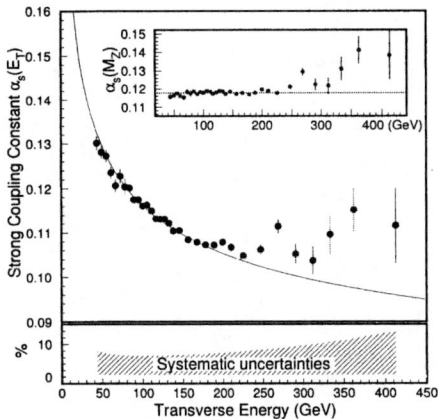

FIGURE 2. The running $\alpha_s(\mu_R = E_T^{jet})$ measurements. The inset plot shows the distribution of already evolved $\alpha_s(M_Z)$. The solid curve corresponds to an NLO pQCD prediction for the evolution $\alpha_s(E_T^{jet})$ using the averaged over (40...250)GeV bins value of $\alpha_s(M_Z) = 0.1178$.

$E_T^{jet} \in (40...250) GeV$ yields:

$$\alpha_s(M_Z) = 0.1178 \pm 0.0001 \, (stat.)$$

The points above 250 GeV increase the average by 0.0001.

As for the inclusive production measurement for bins below 250 GeV, there is good agreement with QCD while at higher E_T^{jet} the discrepancy has the same source as for the excess at the inclusive spectra (see left plot in Fig. 1). It is still not well understood but can be adjusted with the appropriate choice of PDF and its high-x gluon component. To test the behavior of the evolved $\alpha_s(M_Z)$ with an energy E_T below 250 GeV, all 33 measurements have been fit with the linear function $P_0 + P_1 \times (E_T/E_T^0 - 1)$ with E_T^0 set to 92.8GeV. With $\chi^2/N_{d.o.f.} = 1.3$ the fit yielding $P_0 = 0.1176 \pm 0.0003, P_1 = 0.0003 \pm 0.0003$ proves an independence of $\alpha_s(M_Z)$ from E_T; that is pQCD predictions for the evolution of $\alpha_s(\mu_R)$ are *correct*.

Experimental systematic uncertainties (see breakdown of sources on left plots at Fig. 3) are derived from those on the inclusive jet cross section measurement. The dominant source is the calorimeter response to jets. In total the uncertainties propagated to $\alpha_s(M_Z)$ and summed in quadrature yield a total systematic experimental uncertainty of $\pm^{0.0081}_{0.0095}$.

Theoretical systematics includes (see left plots in Fig. 3) the uncertainty of renormalization scales μ_R, μ_F varied independently and reaching largest changes at $\mu_R = \mu_F$. The shift in $\alpha_s(M_Z)$ induced by variation is found to be $\pm^6_4\%$. Examining α_s^{PDF} for other PDF sets including CTEQ4A and MRST yields another uncertainty due to PDF choice to be $\pm 5\%$. The best agreement between data and theory over (40...250)GeV is found for CTEQ4M at $\alpha_s^{PDF} = 0.116$ *used in final fit*. Variation on $\mathcal{R}_{sep} = 1.3...2.0$ results in 5-7% changes in the cross section and induces a 2-3% uncertainty in $\alpha_s(M_Z)$.

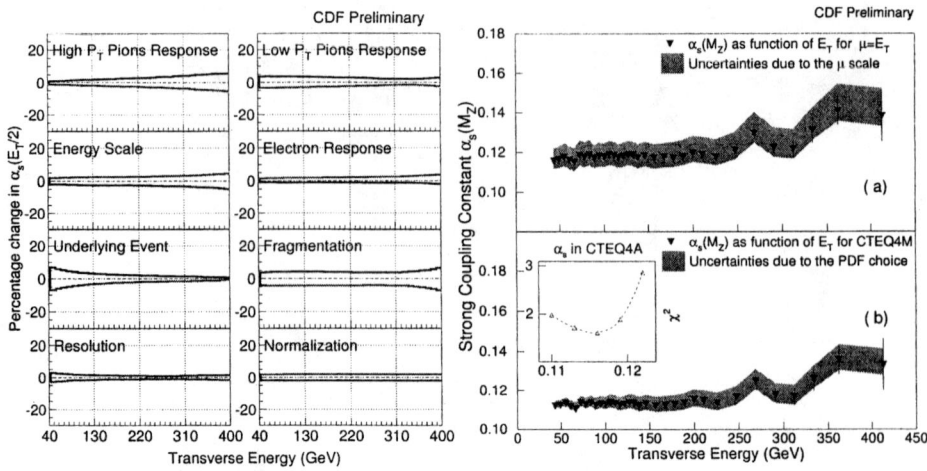

FIGURE 3. Experimental (left) and theoretical (right) systematic uncertainties in α_s.

Finally in conclusion of this topic the analysis results in a number obtained at $\mu_R = \mu_F = E_T$ as

$$\alpha_s(M_Z) = 0.1178 \pm 0.0001(stat.)^{+0.0081}_{-0.0095}(exp.syst.)$$

The theoretical uncertainties from PDF and μ-scale choices of ($\sim 5\% \oplus \pm^{6}_{4}\%$) are *comparable* with the experimental systematics. The value is in *good agreement* with the world average $\alpha_s(M_Z) = 0.1181 \pm 0.0020$[5].

UNDERLYING EVENT ENERGY FLOW

The underlying event is the energy originating from soft spectator parton interactions while the hard partons produce jets. It appears as the energy deposited in a detector by particles from the breakup of interacting hadron beams, from initial state radiations in $2 \rightarrow 2$ and also due to semi-hard processes between spectators like multiple parton scattering. It is an essentially non-perturbative contribution.

As the jet clustering is based on a fixed cone algorithm, there is a contribution due to the underlying event energy flowing into that cone. This energy needs to be subtracted.

The similarity with minimum bias events is used presently as an assumption for energy corrections. Unfortunately the large uncertainty $\sim 30\%$ on the contribution to the energy scale induces the largest uncertainty ($\sim 15...20\%$ at low energy range) in the inclusive jet cross section analysis. Therefore adequate estimates of underlying event energy are of *vital importance*.

In the analysis the jets in a jet event sample are reconstructed with a jet cone radius $R = 0.7$ in the central $|\eta| < 0.7$ region while in a minimum bias data sample a *random* η of a "lead jet" is taken with the same definitions. For each event two cones w.r.t. the leading jet at $\eta = \eta_{LeadJet}$ and having $\phi = \phi_{LeadJet} \pm 90°$ are inspected. These two cones

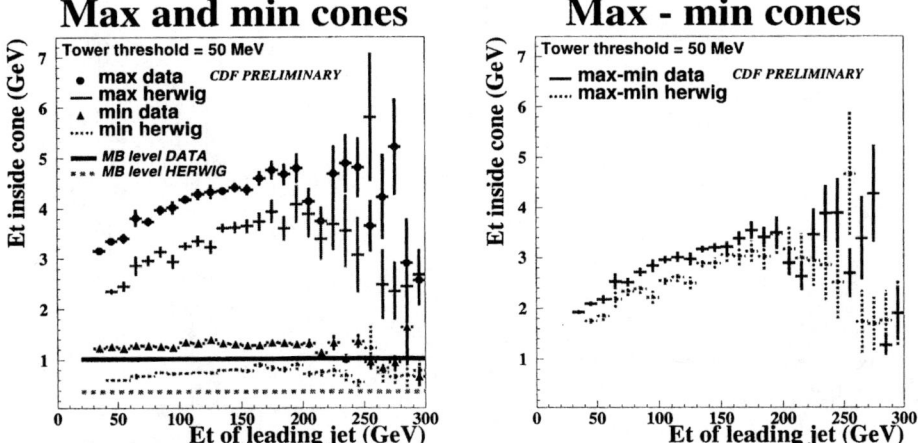

FIGURE 4. The energy flow into Max. and Min. cones (left) and the difference between energy flow into max. and min. cones (right).

at 90° are presumed to be in a semi-quiet region, far away from the two leading jets, though still in central rapidity region. Considering energy E_T flowing into 90° cones, the cones are sorted by energy and identified as "Max. cone" and "Min. cone". The Max. cone most probably will contain NLO corrections to $2 \to 2$ while the most quiet Min. cone gives an indication of the amount of the underlying event. The difference between energies contained in Max. and Min. cones is taken as an indication of NLO contributions. The behavior of E_T energy flow in every cone and their difference versus E_T of the leading jet is shown in Fig. 4 (left and right plots correspondingly) for data taken at $\sqrt{s} = 1800 GeV$. First of all *the flatness* of the E_T energy flowing into the Min. cone versus $E_T^{LeadJet}$ is remarkable while E_T in the Max. cone increases with $E_T^{LeadJet}$. The similarity in shape of the distributions is reproduced in the CDF detector simulated (QFL program) Herwig Monte-Carlo data sample. There is an evident offset between data and Monte-Carlo of $\sim$800MeV for Max. cone and $\sim$500MeV for Min. cone.

For the min. bias data sample (see the same Fig. 4) we consider *a random cone* at $\eta_{\text{``LeadJet''}}$ in the same central $|\eta| < 0.7$ region and examine the $E_T^{\text{``LeadJet''}}$ in that cone to be the level of E_T energy flow. The Min. cone E_T for the jet data sample is higher than the level of E_T in a *random cone* for min. bias events by $\sim$300MeV. The behavior is similar for corresponding Monte-Carlo samples.

The above considerations give a clear indication that underlying events of jet data include an "echo" from the hard interaction and are not adequate to min. bias ambient energy flow and that Herwig model for min. bias does not give a unified description for soft and hard physics.

One has to mention that both in jet and min. bias data there is *an offset* of $\sim$500MeV and $\sim$680MeV between *data and Monte-Carlo* simulation.

It is interesting to have a look at the difference of energies

$$E_T^{Max.cone} - E_T^{Min.cone}$$

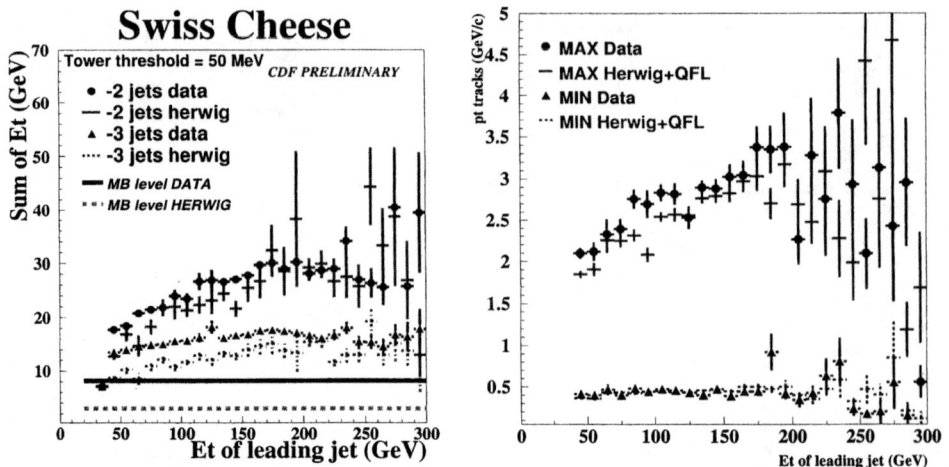

FIGURE 5. Total transverse energy with jets excluded (left plot) and total track P_T flow inside Max. and Min. cones (right plot), track analog of calorimeter E_T flow

– this can be taken as an estimate of the "NLO correction" to $2 \rightarrow 2$ (see a right plot at Fig. 4). Here the underlying event energy expected to be minimized and data agree with Monte-Carlo in shapes having still an offset of $\sim 300 MeV$. The distribution drops at very high E_T due to poor statistics.

The alternative to "Min., Max. cones" and their subtraction approach is a calculation of total transverse energy:

$$E_T^{SwissCheese} = \sum_{towers}^{|\eta|<0.7} E_T^{tower} - \sum_{jet=1}^{2\ or\ 3\ jet} [\sum_{towers} E_T^{tower}]$$

Here we require $E_T^{Jet} > 5GeV$ – see Fig. 5. In the left plot of Fig. 5 the 2 jet subtracted energy should contain mainly NLO (3rd parton) but the 3 jet subtracted energy should have little of NLO; but data and "Herwig + CDF simulation" points are still higher (with a slope) than min. bias data level. This difference indicates possible contributions from hadronization of jets, multiple (double) parton scattering and higher order radiations.

To understand data and Monte-Carlo difference *an independent analysis based on tracks* has been undertaken. Similar quantities have been constructed using tracks instead of calorimeter towers – see the right plot of Fig. 5. The plot shows a total track P_T flow inside Max. and Min. cones again versus calorimeter $E_T^{LeadJet}$ at $\sqrt{s} = 1800 GeV$ with data–Monte-Carlo difference disappearing. The agreement is also good for other plots where tracks substitute towers in Max. and Min. cones at $90°$. Again here the min. bias data level is $\sim 20\%$ smaller than the min. cone jet data level. As we have already observed with calorimeter towers, the min. bias spectra are "softer" than the underlying event spectra from the jet data sample.

Also from track based plots we conclude that the reason for the discrepancy in "calorimeter vs. tracks" lies in the scale at low energies.

In a summary of the "Min., Max. cone" analysis one should mention several points:

- Using "Min., Max. cone" approach we found that energy in Min. cone forms a plateau against energy of a leading jet and exhibits the behavior of underlying events. The height of this plateau is larger than the one observed in min. bias data.
- The total E_T energy with 3 jets contribution subtracted also exhibits the flattening in data and Herwig + "CDF simulation".
- Disagreement in energy flow (Min. cone, "Swiss Cheese") for jet samples with min. bias events indicates the presence of semi-hard multiple hadron interactions and other higher order effects in underlying events which are not present in min. bias samples. Consequently the Herwig model for min. bias events does not give a unified description for soft and hard physics.
- We find the offsets between our data and "Herwig + CDF simulation" at calorimeter level.
- The reason for the offset of data relative to "Herwig + CDF simulation" is still under investigation but may probably be due to the under-estimation of low energy deposits in the calorimeter.
- The Monte-Carlo models should be tuned to better describe the underlying event in $p\bar{p}$ collisions.

ISOLATED PROMPT PHOTON CROSS SECTIONS

Here we discuss another object of QCD interest – prompt photon production in hadronic interactions at Tevatron. The prompt (or direct) photons involve both LO and NLO pQCD processes briefly listed below:

- $gq \to q\gamma$ – *dominating diagram*
- $q\bar{q} \to g\gamma$ – LO annihilation
- $q\bar{q} \to g\gamma g, qg\gamma$ – NLO initial and final state rad. corrections
- photon bremsstrahlung – produced along with hadrons, *suppressed* by photon isolation cuts (see below)

The prompt photon is identified as *an isolated* electromagnetic object, not accompanied by *nearby* energy flow coming from neutral component of hadron jets like π^0, η.

The analysis exploits the electromagnetic clusters from the electromagnetic compartment of CDF central calorimetry (CEM), with lateral profile measured in the proportional chamber CES positioned in the CEM at a depth of shower maximum and the fraction of events with conversions in $1.075 X_0$ of magnet coil counted by proportional chamber CPR installed just in front of the CEM.

The data used in the analysis were taken during 1994-95 (Run 1b). The trigger requires cluster energy $E^{elm}_{cluster} > Thr$ with three thresholds: Thr = 10, 23, 50 GeV corresponding to three event samples. The important cuts and selection criteria are

- *isolation* criteria – require energy deposited in a cone of $R \equiv \sqrt{\delta_\phi^2 + \delta_\eta^2} = 0.4$ around $E^{elm}_{cluster}$ to be $E^{elm}_{isol.cone} < 1 GeV$
- consider only central region $|\eta| < 0.9$
- $|Z_{vertex}| < 60$cm
- require *NO* charged reconstructed track pointing to CPR
- require *NO* other photon above 1GeV in CES to suppress further multiple π^0, η

The signal efficiency of these cuts for *prompt photons* is $\sim 39\%$. The main remaining backgrounds are single π^0, η. This neutral el.-mag. background is subtracted statistically applying two methods – the profile method and the conversion method (see [6] for details). For both methods one determines in every P_T bin the number of signal events as

$$N_{1\gamma} = \left(\frac{\varepsilon_{data} - \varepsilon_{bgr}}{\varepsilon_{1\gamma} - \varepsilon_{bgr}} \right) N_{data}$$

The profile method is based on parameterized shower profile in CES chambers measured in electron test beam. Then the profile (broader for multiphoton background) is fit to test beam data on an event by event basis. The signal and background efficiencies are determined from CDF Monte-Carlo simulation. This method is good for low energy photons with $P_T < 36$ GeV/c.

The conversion method is based on counting with the CPR the photon conversions in the solenoidal coil and used for prompt photons with $P_T > 36$ GeV/c. The background of multiple photons convert more readily than a single prompt photon signal. The signal efficiency of $\varepsilon_{1\gamma} \sim 1 - \exp(-7X_0/9) \approx 60\%$ and the background efficiency of $\varepsilon_{bgr} \sim 1 - (1 - P_{1\gamma})^2 \approx 84\%$ are applied as *weighting factors* for the data on an event by event basis. For both methods ε_{bgr} is measured using the reference data sample of π^0 produced from decay $\rho^\pm \to \pi^+\pi^0, \pi^0 \to \gamma\gamma$ (see Fig. 6). For the conversion method, the weighting functions ε_{bgr} has been *recalibrated* using the experimental $\rho^\pm$ data sample. One has to mention that for the profile method based on $\rho^\pm$ data $\varepsilon_{bgr} = 46.1 \pm 1.0\%$ to be compared with expected $\varepsilon_{bgr} = 46.4 \pm 0.4\%$. For the conversion method based on $\rho^\pm$ data $\varepsilon_{bgr} = 86.8 \pm 0.7\%$ while the expected (based on weighting function) $\varepsilon_{bgr} = 83.2 \pm 0.7\%$. Here the weighting function has been recalibrated to match the data.

Finally we derive the prompt photon inclusive cross section in P_T bins shown in Fig. 7. CDF 1800 GeV and 630 GeV data agree well with the corresponding D0 and UA2 measurements. The comparison of data at $\sqrt{s} = 1800 GeV$ with NLO pQCD calculations shown versus absolute P_T in the right plot of Fig. 8 uses the CTEQ4M PDF, includes NLO fragmentation terms, and uses considerably varied μ-scale choices – *NO* combination has been found that matches the shape of data to within several σ.

The left plot in Fig. 8, showing the difference between data and theory versus scaled momentum $X_T = 2P_T/\sqrt{s}$, exhibits a rise in the measured cross section below 0.1, indicating an enhanced soft gluon contribution which is spoiling the agreement with NLO. Again CDF data points are in agreement with corresponding D0 and UA2 points. But the CDF data points $(D-T)/T$ for $\sqrt{s} = 630 GeV, 1800 GeV$ are different by $\sim 50\%$

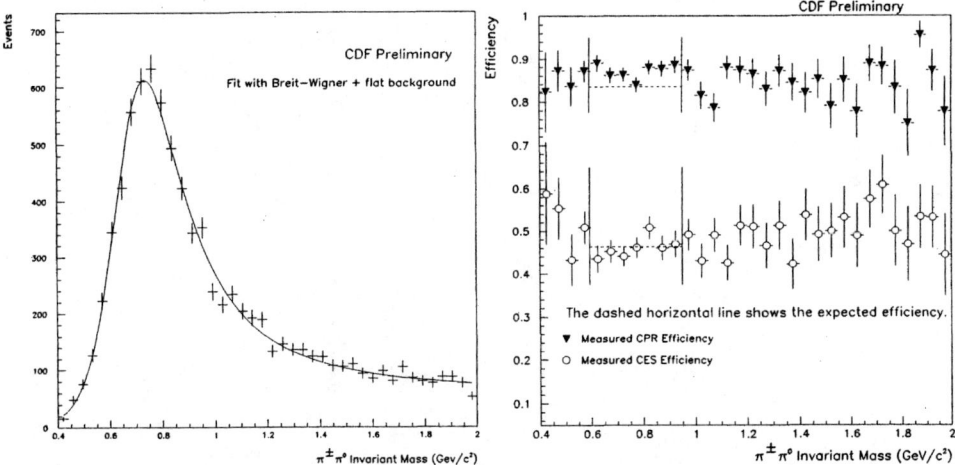

FIGURE 6. $\mathcal{M}(\rho^\pm \to \pi^0\pi^+)$ mass spectrum (left histogram) and background efficiency (right plot). The signal band is $\mathcal{M}(\rho^\pm) \in (0.6, 0.95) GeV/c^2$ and the background band – $\mathcal{M}(\rho^\pm) \in (1.7, 2.0) GeV/c^2$.

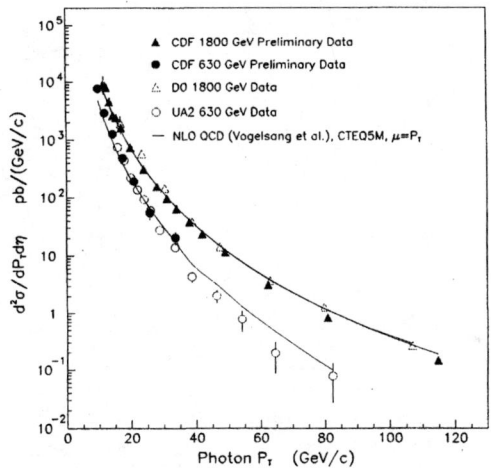

FIGURE 7. The prompt photon inclusive cross section at 1800 GeV and 630 GeV with D0 and UA2 results overlapped.

while the experimental uncertainties are of $\sim 9\%$ only. There is *disagreement* with theory predictions at the *low X_T region which is rich with gluon content*.

One possibility to correct standard NLO pQCD with typical settings to match the data is adding K_t as part of the non-collinear initial state radiation. Both plots in Fig. 9 demonstrate that this works well to describe the shape of the data for both energies. The other curve is a more fundamental attempt by Baer-Reno[7] to add a parton shower to NLO pQCD.

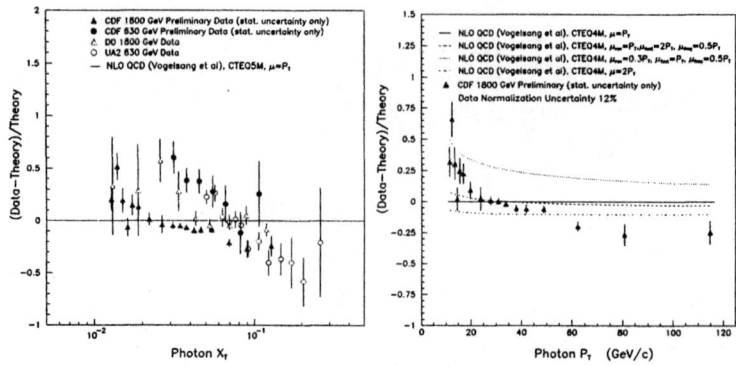

FIGURE 8. Data versus NLO pQCD predictions: X_T spectra (left plot) and P_T spectra (right plot).

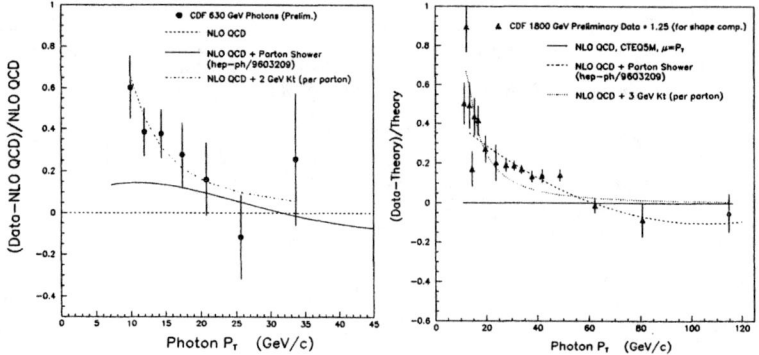

FIGURE 9. Data versus theory for 630 GeV (left) and 1800 GeV (right) with added K_t.

In conclusion of this topic, CDF has updated and improved prompt photon measurements at $\sqrt{s}$ =630GeV and 1800GeV. Data show at low X_T the effect of soft multiple gluon emission. Theory does not describe CDF data at both energies at low X_T. Prompt photon production continues to be a good place to test modern NLO pQCD calculations. *Ad hoc* inclusion of K_t smearing effects in simple Gaussian smearing models works well, though for gluon distribution studies one needs more fundamental approaches.

ACKNOWLEDGMENTS

The author is grateful to his colleagues from the CDF QCD working group for useful suggestions and comments made during preparation of this talk. The author would like to thank Prof. Sally C. Seidel for support of this work, fruitful discussions, and comments.

REFERENCES

1. F. Abe et al., The CDF Collaboration, *Nucl. Instrum. Methods*, **A271**, 387(1988).
2. T. Affolder et al., The CDF Collaboration, *Phys. Rev.*, **D64**, 032001(2001).
3. W. Giele, E.W.N. Glover and J. Yu, *Phys. Rev.*, **D53**, 120(1996).
   W. Giele, E.W.N. Glover and D.A. Kosower, *Phys. Rev. Lett.*, **73**, 2019(1994).
4. T. Affolder et al., The CDF Collaboration, FERMILAB-CONF-01/246-E. Submitted to *Phys. Rev. Lett.*, August 22, 2001.
5. D.E. Groom et al., Particle Data Group, *Eur. Phys. J.*, **C15**, 1(2000).
6. F. Abe et al., The CDF Collaboration, *Phys. Rev. Lett.*, **73**, 2662(1994).
7. H. Baer and M.H. Reno, *Phys. Rev.*, **D54**, 2017(1996).

NEW FACILITIES

Recent J/ψ Physics Results from BES

Zijin Guo*†1

Representing the BES Collaboration

*Institute of High Energy Physics, Academy of Sciences,
Beijing 100039, P. R. China
†China Center for Advanced Science and Technology (CCAST), World Laboratory,
Beijing 100080, P. R. China

Abstract. Here we report partial wave analysis results for $J/\psi \to p\bar{p}\eta$ based on 7.8×10^6 BES I J/ψ events, and through which we determined two N^* resonances $N^*(1535)S_{11}$ and $N^*(1650)S_{11}$. BES just finished collecting 5×10^7 J/ψ events. Some very preliminary results are presented from the 2.4×10^7 J/ψ events, namely the first new data set and showed the high statistics and good data quality of the new J/ψ data. Using the 2.4×10^7 J/ψ events, we performed the partial wave analysis in $J/\psi \to \gamma K^+K^-$, and the results showed that the 0^{++} component is dominant in the mass region around 1.7 GeV. This conclusion is consistent with other experiments.

INTRODUCTION

The Beijing Spectrometer (BES) is a large general purpose solenoidal detector at the Beijing Electron Positron Collider (BEPC). The beam energy of BEPC is in the range from 1.0 to 2.8 GeV with a peak luminosity at J/ψ energy of $5.0 \times 10^{30} cm^{-2} s^{-1}$. The details of BES I was described in ref. [1]. The upgrades of BES I to BES II [2] include the replacement of the central drift chamber with a vertex chamber composed of 12 tracking layers, the installation of a new barrel time-of-flight counter (BTOF) with the time resolution of 180ps and the installation of a new main drift chamber (MDC), which has 10 tracking layers and provides a dE/dx resolution of $\sigma_{dE/dx} = 8.0\%$ for particle identification and $\sigma_p/p = 1.78\%\sqrt{1+p^2}$ (p in GeV) momentum resolution for charged tracks. The barrel shower counter (BSC), which covers 80% of 4π solid angle, has an energy resolution of $\sigma_E/E = 21\%/\sqrt{E}$ (E in GeV) and a spatial resolution of 7.9 mrad in ϕ and 2.3 cm in z, is located outside the TOF. Outermost is a μ identification system, which consists of three double layers of proportional tubes interspersed in the iron flux return of the magnet.

BES I has collected about 7.8 million J/ψ events, 3.96 million ψ' events, and 22.3pb^{-1} of D and D$_s$ events. At the end of 1999, we started a new J/ψ run with the upgraded BES II. Up to the end of April, 2001, we completed the J/ψ run and

---

1 Present E-mail address: guozj@mail.ihep.ac.cn

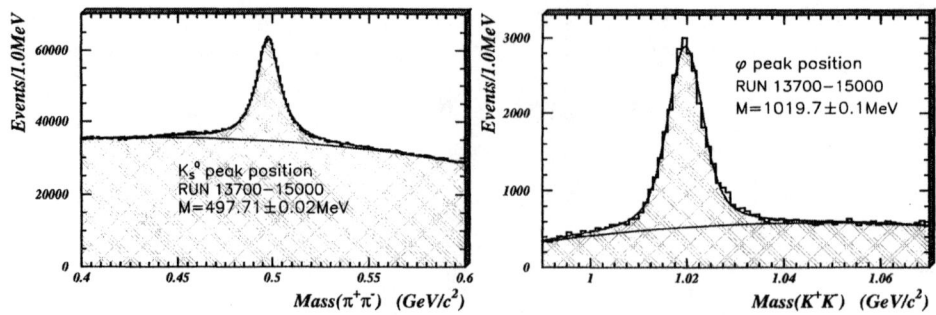

FIGURE 1. Inclusive K_S^0 and ϕ signals

accumulated about 5×10^7 J/ψ events, with the first data set being taken in 1999-2000 run (about 24 million) and the second data set in 2000-2001 (about 27 million). This is the largest J/ψ sample in the world.

Besides the high statistics, the quality of the new J/ψ data is also quite good. The inclusive K_S^0 and ϕ signals and their fitted masses are shown in Fig. 1. The signals are clear and fitted masses are consistent with PDG values. Based on present data, plenty of physics topics are expected, such as glueball and hybrid searches, study of the excited baryon states, precise measurements and rare decay study. Currently, we are working on the calibration of the second new data set.

STUDY OF THE EXCITED BARYON STATES

Nucleons are the most common form of hadronic matter on the earth and probably in the whole universe. To understand the internal quark-gluon structure of nucleon and its excited states N^*'s is one of the most important tasks in nowadays particle and nuclear physics. The main source of information for the nucleon internal structure is their spectrum, various production and decay rates. Our present knowledge on this aspect came almost entirely from the old generation of πN experiments of more than twenty years ago and our understanding on baryon spectroscopy is still poor. Considering its importance for the understanding of the nonpertubertive QCD, a new generation of experiments on N^* physics with electromagnetic probes has recently been started at new facilities such as JLAB, ELSA at Born, GRAAL at Grenoble and Spring8 at JASRI. The J/ψ experiments at BES/BEPC is an excellent place for studying N^* resonances, especially in the mass range 1-2 GeV [3].

Among many interesting channels of J/ψ decays, the $J/\psi \to p\bar{p}\eta$ is a relatively simple one to begin with. According to the information from $\pi N \to \eta N$ and $\gamma N \to \eta N$ experiments, as well as some early quark shell model calculation, only $N^*(1535)S_{11}$ may have a branching ratio to $N\eta$, while $N^*(1650)S_{11}$ may have a branching ratio to $N\eta$ of up to 10% and other N^* resonances below 2.0 GeV have much smaller branching ratios to the $N\eta$. A partial wave analysis using BES I 7.8×10^6 J/ψ events is performed. The $p\eta(\bar{p}\eta)$ invariant mass spectrum is shown in Fig. 2. There is a clear enhancement

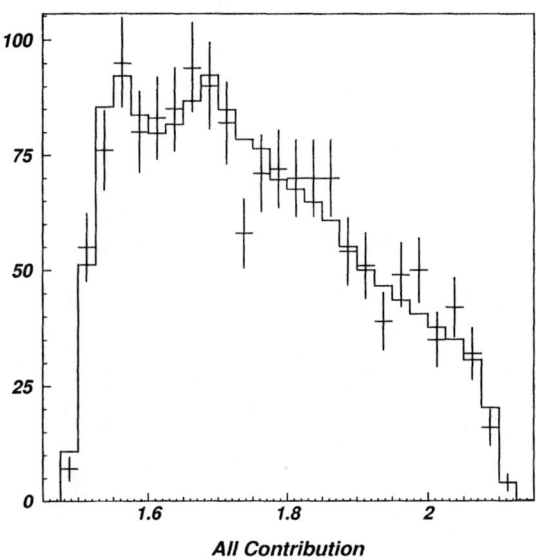

FIGURE 2. $p\eta(\bar{p}\eta)$ invariant mass spectrum for $J/\psi \to p\bar{p}\eta$, crosses are data and histogram the fit

near the $p\eta(\bar{p}\eta)$ threshold. There are also some bumps around 1.65 GeV and 1.8 GeV.

The peak around 1535 MeV near the $p\eta$ threshold optimizes at M= 1530 ± 10 MeV and $\Gamma = 95 \pm 25$ MeV. The data favor $J^P = \frac{1}{2}^-$ over others. It makes the largest contribution $(56 \pm 15)\%$ to the $p\bar{p}\eta$ final states. Our results for $N^*(1535)$ are consistent with the resonance parameters of PDG and a recent detailed analysis of πN S_{11} partial wave by Vrana, Dytman and Lee [4]. The second peak around 1650 MeV is also fitted with $J^P = \frac{1}{2}^-$ resonance $N^*(1650)$. It optimizes at M= 1647 ± 20 MeV, $\Gamma = 145^{+80}_{-45}$ MeV, and contributes $(24^{+5}_{-15})\%$ to the $p\bar{p}\eta$ final states. If we assume no contribution from resonance around 1650 MeV, then the mass and width of $S_{11}(1535)$ optimize around 1570 MeV and 270 MeV, respectively, which are out of the range of PDG values, and the log likelihood is worse by 14. The comparison of BES fitted parameters of $N^*(1535)$ and $N^*(1650)$ with PDG values are summarized in Table 1.

TABLE 1. Comparison of the fit results with PDG values

$N^*(1535)$ parameters	BES	PDG2000
Mass (MeV)	1530 ± 10	1520 – 1555
Γ (MeV)	95 ± 25	100 – 250
$N^*(1650)$ parameters	BES	PDG2000
Mass (MeV)	1647 ± 20	1640 – 1680
Γ (MeV)	145^{+80}_{-45}	145 – 190

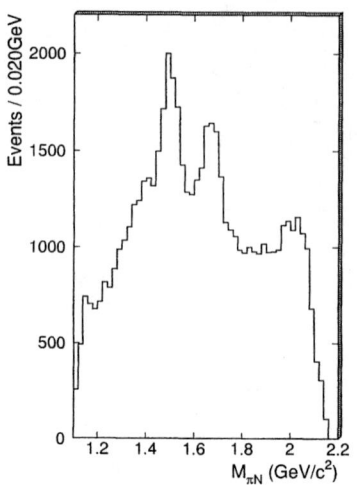

FIGURE 3. $\pi^- p(\pi^- \bar{n})$ invariant mass spectrum for $J/\psi \to p\pi^-\bar{n}$ (preliminary)

With BES II 24M J/ψ events, namely the first data set, we also analyzed the $J/\psi \to p\pi^-\bar{n}$ very preliminarily. The $\pi^- p(\pi^-\bar{n})$ invariant mass spectrum shown in Fig. 3 indicates two narrow peaks around 1535 MeV and 1670 MeV, respectively. Our preliminary partial wave analysis showed that the first peak is consistent with the contributions from $N^*(1535)(\frac{1}{2}^-)$ and $N^*(1520)(\frac{3}{2}^-)$, and the second peak could be from $N^*(1675)(\frac{5}{2}^-)$ and $N^*(1680)(\frac{5}{2}^+)$. PDG gives the width of $N^*(1535)$ as 100-250 MeV, while our value is optimized to about 100 MeV, with a less model dependent method and here the background is less compared with other experiments.

RADIATIVE DECAYS OF J/ψ

Quark model works well in describing most observed states until now, but problems and puzzles still exist, that might signal physics beyond quark model. One of the distinctive features of QCD as a non-Abelian gauge theory is the self-interaction of gluons. The indirect evidence for gluon-gluon interactions has been obtained at high energies. However, glueballs, the bound states of gluons, predicted by QCD, have not been confirmed yet. Therefore, the observation of glueballs is, to some extent, a direct test of QCD.

The J/ψ decays have long been known as the best place for looking for glueballs. The branching ratio for a radiative J/ψ decay is typically between $\sim 10^{-4}$ and 10^{-3}. The previous best experiments produce on the order of several million J/ψ, so only a few thousand accepted events can be expected for each of these states. Consequently the statistical power is meager, and complete partial wave analyses are difficult. Now we hope to use the new J/ψ data to do more work.

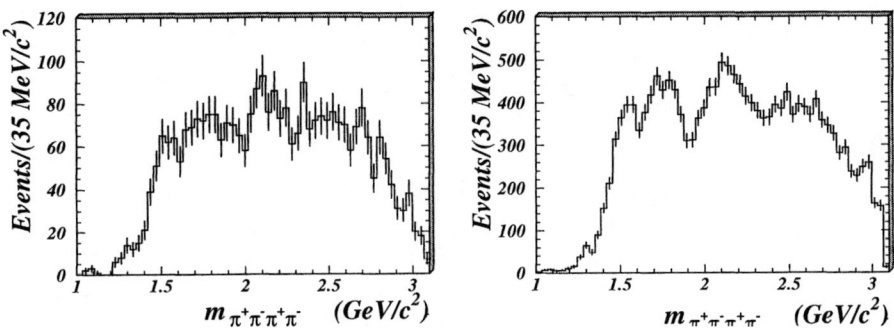

FIGURE 4. The 4π invariant mass spectra of $J/\psi \to \gamma\pi^+\pi^-\pi^+\pi^-$. The left picture is obtained from BES I data; while the right one from BES II 24M J/ψ data (preliminary).

Fig. 4 shows the invariant mass spectra of the four charged pions in the $J/\psi \to \gamma\pi^+\pi^-\pi^+\pi^-$ obtained from BES I 7.8M and BES II 24M J/ψ data separately. The two spectra are similar and the statistics of BES II's is much higher while the signals of the resonances around 1.5 GeV and 1.7 GeV are more significant than BES I's.

The most interesting issue of $J/\psi \to \gamma K^+K^-$ is the study of the $f_0(1710)$. As one of the earliest glueball candidates, since the discovery by the Crystal Ball Collaboration in $J/\psi \to \gamma\eta\eta$ [5], there has been a long history of uncertainty about the properties of $f_0(1710)$, especially its spin-parity, 0^+ or 2^+. The latest analysis of MARK III data by W. Dunwoodie [6] favors $J^P = 0^+$ over an earlier assignment of 2^+. The latest analyses of central production data of WA76, WA102 also favor 0^+ [7, 8]. QCD lattice calculations predict that the ground scalar glueball lies in the mass range 1.5-1.7 GeV [9-12]. It's well known that the J/ψ radiative decay is a glue-rich production mechanism. So the careful study of the properties of $f_0(1710)$ in $J/\psi \to \gamma K^+K^-$ becomes quite important and urgent.

Using BES II 24M J/ψ data we performed the partial wave analysis of $J/\psi \to \gamma K^+K^-$ preliminarily. The K^+K^- invariant mass spectrum after event selection of BES II 24M J/ψ data is shown Fig. 5, where we could see clear $f_2'(1525)$ and $f_0(1710)$ signals. The shadow indicates the background mainly due to feed-through from the $J/\psi \to \pi^0 K^+K^-$ channel. The partial wave analysis is confined to mass less than 2 GeV in which the background is very small relative to the higher mass, and a description in terms of only 0^{++} and 2^{++} amplitudes is appropriate.

Amplitudes are fitted to relativistic covariant tensor expressions [13], and maximum likelihood method is employed in the fit. For spin 0, E1 and E3 transitions from J/ψ are allowed with $L = 0$ and 2 respectively. These two amplitudes differ only by a form factor, a function of the momentum k of the photon. For narrow resonances, this form factor has little effect, so there is effectively only one helicity amplitude; we therefore drop the $L = 2$ amplitude. For spin 2, there are three transitions: E1, M2 and E3.

What we do is called "SLICE FIT", namely the bin-by-bin fit with a 50 MeV bin width to determine the components with spin-parity equals to 0^+ and 2^+ separately in each bin. Fig. 6 shows the fit results after the detector acceptance correction. The remarkable feature of this figure is that: The 2^{++} distribution shows a clear signal corresponding to the $f_2'(1525)$; there is some 2^{++} contribution at about 1.7 GeV, but the region is

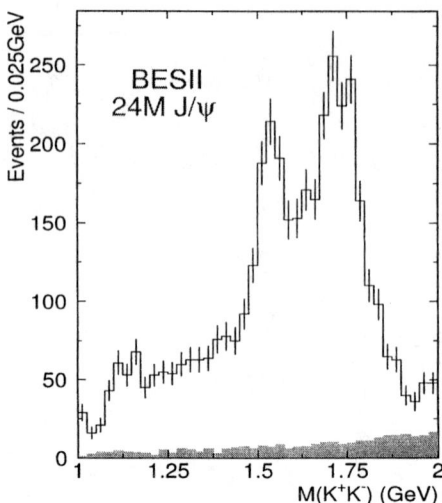

FIGURE 5. The K^+K^- invariant mass spectrum for the $J/\psi \to \gamma K^+K^-$ sample (preliminary)

predominantly 0^{++}. This signal is probably resonant, but it should be emphasized that relative phase motion has not yet been measured in this analysis.

HADRONIC DECAYS OF J/ψ

At the same time, we are also very interested in hadronic decays of J/ψ. It's well known that about 80% of the J/ψ decays are purely hadronic final states and it has been proposed long ago to use this pure initial state to perform refined hadronic spectroscopy including glueball search.

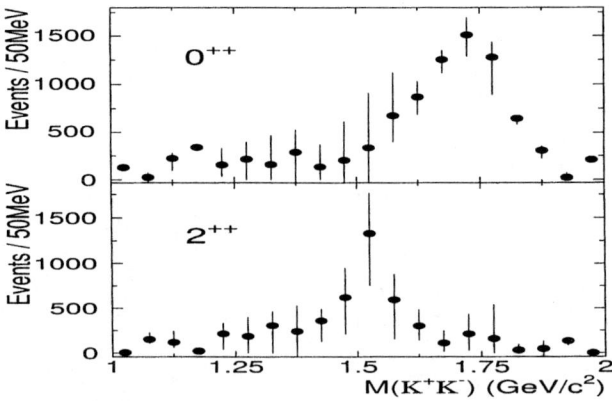

FIGURE 6. The mass dependence of the 0^{++} and 2^{++} components for $J/\psi \to \gamma K^+K^-$ after the detector acceptance correction (very preliminary)

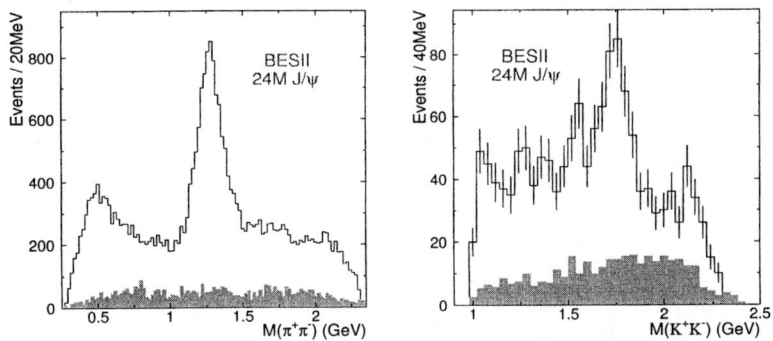

FIGURE 7. The $\pi^+\pi^-$ and K^+K^- invariant mass distributions of the $J/\psi \to \omega\pi^+\pi^-$ and $J/\psi \to \omega K^+K^-$ candidate events, respectively (preliminary).

Both Mark III and DM2 [14, 15] analyzed $J/\psi \to \phi\pi^+\pi^-$, ϕK^+K^-, $\omega\pi^+\pi^-$ and ωK^+K^- for the study of $f_0(980)$, $f_2'(1525)$ and the structure around 1700 MeV. With 2.4×10^7 BES II J/ψ events, BES analyzed $J/\psi \to \omega\pi^+\pi^-$, ωK^+K^-, $\phi\pi^+\pi^-$ and ϕK^+K^- channels preliminarily. Fig. 7 shows the $\pi^+\pi^-$ and K^+K^- invariant mass distributions of the $J/\psi \to \omega\pi^+\pi^-$ and $J/\psi \to \omega K^+K^-$ candidate events, respectively. The shadows indicate the background events estimated from the ω side-bands.

For $J/\psi \to \omega\pi^+\pi^-$, besides its high statistics, clear $f_2(1270)$ and $b_1(1235)$ signals could be observed from the Dalitz plot shown in Fig. 8. DM2, MARK III and BES II all found a broad structure at low $\pi\pi$ mass in $J/\psi \to \omega\pi^+\pi^-$. As we know, the σ is required as $\pi\pi$ S-wave in $\pi\pi$ scattering, and there is a pole at M $\sim 500 - i300$ MeV [16-19]. However, the existence of σ as a resonance has not been generally accepted. Our preliminary partial wave analysis showed that the large peak at low $\pi\pi$ mass fits naturally with spin zero. The two main features of the $J/\psi \to \omega K^+K^-$ data are conspicuous peaks in the K^+K^- mass spectrum at $\sim$ 1750 and 2120 MeV.

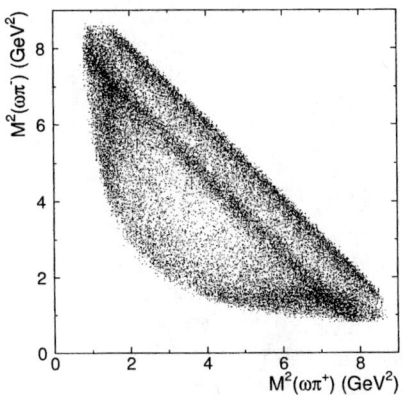

FIGURE 8. Dalitz plot of $J/\psi \to \omega\pi^+\pi^-$ (preliminary)

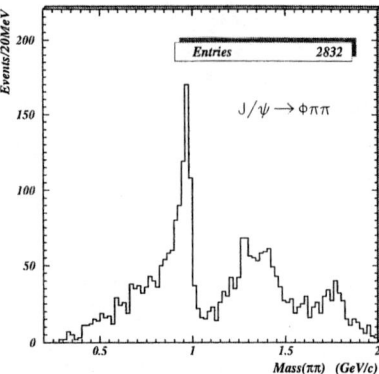 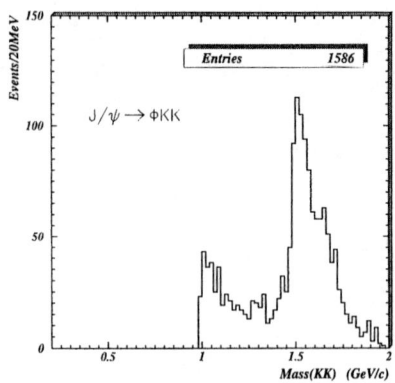

FIGURE 9. The $\pi^+\pi^-$ and K^+K^- invariant mass distributions of the $J/\psi \to \phi\pi^+\pi^-$ and $J/\psi \to \phi K^+K^-$ candidate events, respectively (preliminary).

The $\pi^+\pi^-$ and K^+K^- invariant mass distributions of the $J/\psi \to \phi\pi^+\pi^-$ and $J/\psi \to \phi K^+K^-$ candidate events obtained from BES II 24M J/ψ data are shown in Fig. 9, respectively. There is a beautiful $f_0(980)$ signal in the $\pi^+\pi^-$ spectrum and some possible structures around 1350 and 1770 MeV. Now the complete partial wave analyses of the above four channels are in progress.

SUMMARY

Based on 7.8×10^6 BES I J/ψ data, we have performed the study of excited nucleon states from J/ψ decays and determined two N^* resonances $N^*(1535)S_{11}$ and $N^*(1650)S_{11}$. BES has accumulated 50M J/ψ events until 2001, which is the largest J/ψ data sample in the world now. Some preliminary results of radiative and hadronic J/ψ decays showed the high statistics and good data quality of the BES II new J/ψ data. With 2.4×10^7 BES II J/ψ data, a partial wave analysis has been applied to $J/\psi \to \gamma K^+K^-$ decay and the fit results showed that the 0^{++} component is dominant in the mass region around 1.7 GeV. This conclusion is consistent with the other experiments. We expect many new results from BES in the future.

ACKNOWLEDGMENTS

We acknowledge the staff of the BEPC accelerator and IHEP computing center for their efforts. The work was supported in part by the National Natural Science Foundation of China under Contracts No. 19991480, No. 19825116 and No. 19605007, and by the Department of Energy of US under Contracts No. DE-FG03-93ER40788 (Colorado State University), No. DE-AC03-76SF00515 (SLAC), No. DE-FG03-94ER40833 (University of Hawaii) and No. DE-FG03-95ER40925 (University of Texas at Dallas).

REFERENCES

1. BES Collaboration, Bai, J. Z., et al., *Nucl. Instrum. Meth. A* **344**, 319-334 (1994).
2. BES Collaboration, Bai, J. Z., et al., *Nucl. Instrum. Meth. A* **458**, 627-637 (2001).
3. BES Collaboration, Bai, J. Z., et al., *Phys. Lett. B* **510**, 75-82 (2001).
4. Vrana, T. P., Dytman, S. A., and Lee, T. S. H., *Phys. Rept.* **328**, 181-236 (2000).
5. Edwards, C., et al., *Phys. Rev. Lett.* **48**, 458-461 (1982).
6. Dunwoodie, W., *Hadron Spectroscopy*, AIP Conf. Series 432 (1997) 753.
7. French, B., et al., *Phys. Lett. B* **460**, 213-218 (1999).
8. Barberis, D., et al., *Phys. Lett. B* **453**, 305-315 and 316-324 (1999); *Phys. Lett. B* **462**, 462-470 (1999).
9. Bali, G., Schilling, K., Hulsebos, A., Irving, A., Michael, C., and Stephenson, P., *Phys. Rev. B* **309**, 378-384 (1993).
10. Michael, C., *Hadron Spectroscopy*, AIP Conf. Series 432 (1997) 657.
11. Lee, W., and Weingarten, D., hep-lat/9805029.
12. Morningstar, C., and Peardon, M., *Phys. ReV D* **60** (1999) 034509.
13. Guo, Z. J., Ph.D. thesis, Institute of High Energy Physics, Beijing (2000).
14. Augustin, J. E., et al., *Nucl. Phys. B* **320**, 1-25 (1989).
15. Falvard, A., et al., *Phys. Rev. D* **38**, 2706-2721 (1988).
16. Bugg, D. V., Sarantsev, A. V., and Zou, B. S., *Nucl. Phys. B* **471**, 59-89 (1996).
17. Oller, J. A., Oset, E., and Pelaez, J. R., *Phys. Rev. D* **59** (1999) 074001.
18. Locher, M. P., Markushin, V. E., and Zheng, H. Q., *Phys. Rev. D* **55**, 2894-2901 (1997).
19. Markushin, V. E., *Workshop on Hadron Spectroscopy*, Frascati Physics Series 15, Eds. Bressani, T., Feliciello, A., and Fillippi, A., (1999) 229.

The Hall D Detector at Jefferson Lab

Curtis A. Meyer

Carnegie Mellon University

Abstract. The Hall D experiment at Jefferson Lab is part of the proposed CEBAF upgrade to 12 GeV beam energy. The Experiment will study gluonic excitations of mesons in the 1.5 to 2.5 GeV/c^2 mass region using an 8 to 9 GeV beam of linearly polarized photons.

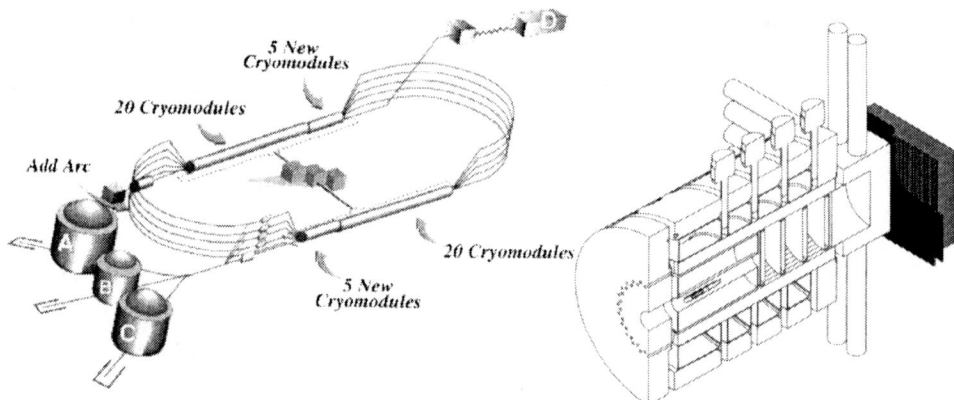

FIGURE 1. The left–hand figure is the CEBAF accelerator after the 12 GeV upgrade. The upgrade is accomplished by adding 10 new cryomodules, and a fifth arc to one end of the accelerator. Hall D extracts electrons at the highest beam energy and delivers them to a tagger complex. Photons are produced via coherent Bremsstrahlung off a thin diamond target, and the deflected electrons are then detected in a tagger. Only photons can be delivered to the Hall D building which houses the Detector (right–hand figure). The detector has nearly 4π coverage for both charged particles and photons and sits in a 2.2 T solenoidal field. It uses a combination of time of flight and threshold cherenkov detectors to do K–π separation.

The Hall D detector [1] is a major new initiative to study light quark mesons and to search for gluonic excitations using a beam of linear polarized photons. The experiment is a key piece of the proposed 12 GeV upgrade to the Jefferson Lab accelerator, (see Figures 1). Current experimental studies have started to show evidence for exotic mesons, both mesons with manifestly exotic quantum numbers, and for the scalar glueball. Understanding the spectrum of these states, and how they are related and mixed with the normal mesons is a critical element in understanding QCD in the non-perturbative regime. This will also provide stringent milestones for theory, in particular the lattice, to be able to explain both the spectrum and decay properties of such states.

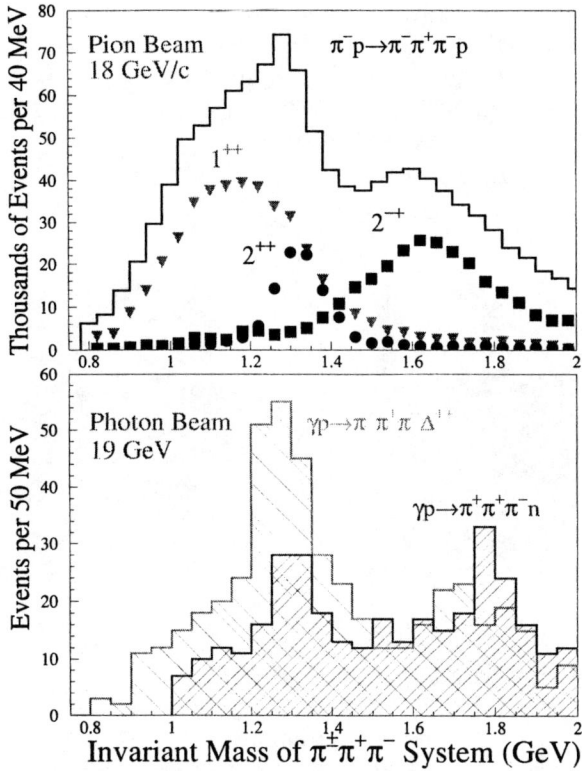

FIGURE 2. The upper figure shows data on the 3π final state from a pion-beam induced reaction, (Brookhaven E852 [6]). The lower figure is bubble chamber data showing the current statistics available in photon beam reactions [7]. It is interesting to note both the three orders of magnitude differences in statistics, as well as the different shapes for the invariant mass spectrum.

Recent progress in the field of meson spectroscopy has been spear-headed by efforts in $p\bar{p}$ annihilation at LEAR [2], central production experiments at CERN [3], and pion scattering experiments at BNL and at VES [4]. These efforts have produced several large, clean data sets that have shown that meson states beyond the naive quark model exist, however, the exact interpretation of these states is not clear. In the exotic meson sector, there appear to be two isovector states with $J^{PC} = 1^{-+}$. However, a hybrid interpretation expects exactly one such state. In addition, both states are lighter than what is expected by current calculations.

While one could in principal try to continue exploiting current reactions, it is useful to step back and look at all production processes. One quickly notices that for historical reasons of either low rates, duty factor, or poor beam definition, there is virtually no data on the photoproduction of mesons. In addition, photoproduction is fairly unique in these processes in that the incident particle carries one unit of intrinsic angular momentum into the reaction. Typically, photoproduction is viewed through the eyes of Vector Meson Dominance, where before a t-exchange process with the nuclear target, the photon is transformed into a vector meson such as a ρ, ω or a ϕ. This is also unique in that these mesons have the $q\bar{q}$ in a Spin 1 configuration, (compared to a spin 0 configuration for π's and K's. This has an added advantage that in many models of hybrid mesons, these gluonic excitations are built from a q-$\bar{q}$ system that is in a spin 1 system. Photoproduction is quite likely to be exactly the place to look to observe many of these reactions [5].

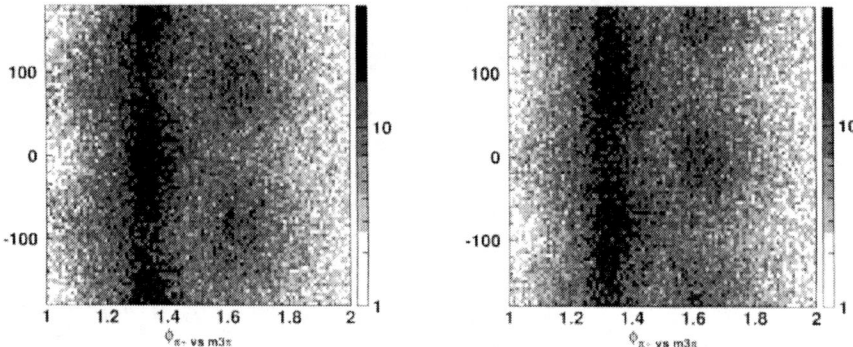

FIGURE 3. Plots of ϕ_{GJ} versus 3π-mass for 100% polarized photons. The figure on the left is for α near 90°, while that on the right is for α near 0°. These two cases correspond to photons polarized along the y axis and the x axis respectively as seen in the Gottfreid-Jackson (GJ) frame. Photons in other directions are coherent mixtures of these two states.

The last piece of this is the need for linearly polarized photons to carry out the analysis of the the final states. Naively, the linear polarization picks out a new direction in addition to the the photon momentum, thus reducing the complications in the analysis by a factor of two. In fact, it can also be used to select the naturality of the exchange particle, and give us additional information on how the mesons are observed. Figure 3 shows how the linear polarization can be used to extract the naturality of the exchanged particle. The angle α is the direction of the polarization vector as seen in the Gottfried-Jackson frame. The angle ϕ_{GJ} is in the same frame but measures the ϕ of the π^-. The case of $\alpha = 0°$ and $\alpha = 90°$ show different ϕ dependences for the same resonances. This dependence allows us to extract the naturality of the exchange. The goal of the Hall D experiment is to use 8 to 9 GeV linearly polarized photons on a liquid hydrogen target to study the meson spectrum from roughly $1\,\text{GeV}/c^2$ up to about $2.5\,\text{GeV}/c^2$. This will be done using a new hermetic detector with excellent charged particle and photon capabilities and very good particle identification.

The new detector would be situated in a new Hall, (Hall D), at the Jefferson Lab CEBAF accelerator. The linearly polarized photons will be produced using a tagged coherent Bremsstrahlung process, and in order to achieve both good rates and sufficient polarization a 12GeV primary electron will be needed. The new Hall will be located at the opposite end of the accelerator as the current Halls, and with the addition of a 5'th arc of magnets, the electrons delivered to the Hall D photoproduction target would have $5\frac{1}{2}$ passes through the CEBAF linacs, rather than the 5 as seen by the other halls. This leads to 12 GeV beam in Hall D, and 11 GeV beam in Halls A, B and C.

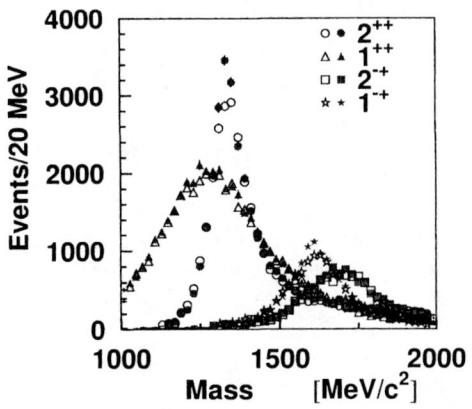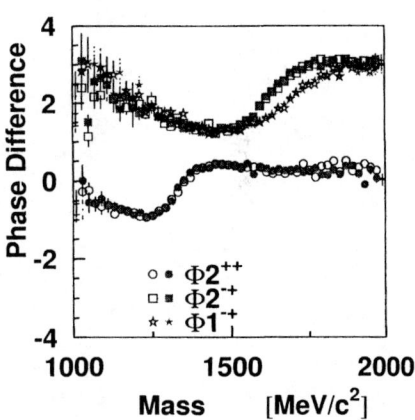

FIGURE 4. Fits to the weighted data. These fits compare generated data (solid shapes) to data that has been run through the HallD Monte Carlo, (open figures). The left figure compares the fits to the intensities of four waves, while the right figure shows the phase differences between the listed waves and the 1^{++} wave.

The Hall D physics program would start with $10^7\,\gamma/s$ tagged linearly polarized photons in the 8 to 9 GeV energy region. This will allow the experiment to study meson states up to masses of $2.5\,GeV/c^2$ with minimal acceptance problems. The ultimate goal will be to push this to $10^8\,\gamma/s$ as the experiment is understood and additional computer power is added to the trigger system. Photoproduction cross sections of interest range from about $1\,\mu b$ down to the nb level. Even at the initial rates, in one year of running the statistics will exceed the best in π scattering experiments by at least a factor of 100, and lead to on the order of 10^7 events from a $1\,\mu b$ cross section. The detector is optimized to do excellent reconstruction for almost all final states, so when coupled with the high statistics, this will allow us not only to find these exotic mesons, but also to map out their decays in detail. It is the decay patterns which are the challenge for QCD to explain, in particular in cases where different states can mix. Figure 2 shows a comparison between the current state of π reactions, where excellent statistics are available, and the state of γ induced reactions. There is currently 3 orders of magnitude less data in the γ system.

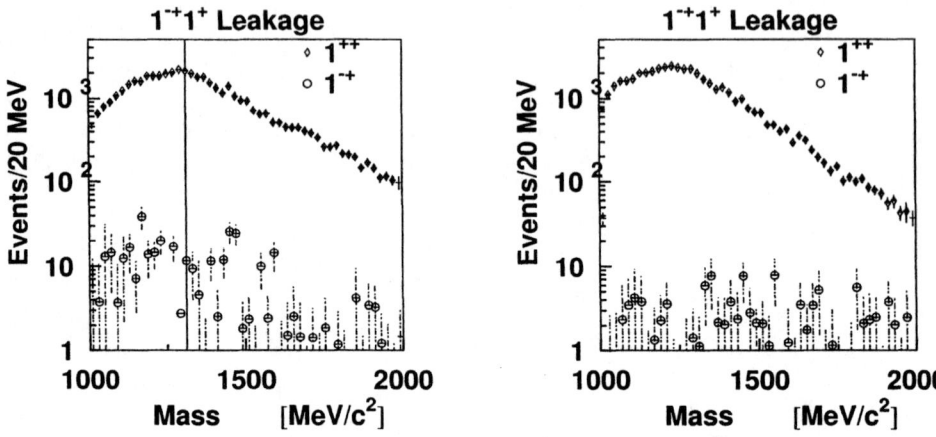

FIGURE 5. Leakage into the exotic wave (1^{-+}) from the a_a (1^{++}) wave. The left hand figure is for a purely charged final state in which the detector position resolution in the chambers has been degraded by a factor of two. The right hand figure is for a neutral final state in which there has been significant degradation to the resolution in the Barrel Calorimeter.

As part of the effort to better design the detector, a Partial Wave Analysis has been started to study both the detector's ability to analyze data, and to try and understand where there will be potential leakage problems in the detector. The reaction $\pi^+\pi^+\pi^-$ final state has been simulated with for intermediate resonances which decay via $\rho\pi$ to the 3π final state. These include the $a_1(1270)$, the $a_2(1320)$, the $\pi^2(1670)$ and an exotic $\pi^1(1600)$ state. Figure 4 shows the results for the intensity of each wave, and the phase difference to the a_1 signal. Both are shown for both a *perfect* detector and for an acceptance corrected detector. The results are very hard to distinguish. In addition to this, we have removed the exotic 1^{-+} wave from the simulated data, and then both worsened, and put in mismatch between the geometry with which we simulate the physics and with which we normalize the PWA. Using this latter technique, we have attempted to generate leakage into the exotic wave. The results of two of these studies are shown in Figure 5. This figure compares the leakage with the strength of the a_1 signal, from which most of it originates. It can be seen that the leakage is under 1% of the a_1 strength. Our current detector design will allow us to measure with high statistics reactions with sub-nanobarn cross sections, and to pull out exotic signals that are as small as a couple of percent of a stronger channel.

Understanding the bound states of QCD in the light-quark sector is an important goal of both theorists and experimentalists trying to understand the strong limit of QCD. Recent lattice calculations [8], [9] have started to address these, and improvements in theory and modeling over the next few years will rely heavily on understanding the role that glue plays in QCD, and information on gluonic excitations will be a crucial test in our understanding. The Hall D detector will fill an important role in this regard, and with high quality, high statistics data available in a previously very poorly studied channel, is very likely to make crucial discoveries in this field.

REFERENCES

1. See http://dustbunny.physics.indiana.edu/HallD/ for additional information. In particular, **The Hall D Design Report, Version 3**, http://dustbunny.physics.indiana.edu/HallD/DR.html.
2. Claude Amsler, Rev. Mod. Phys. **70**, 1293, (1998).
3. D. Barberis, et al., (The WA102 Collaboration), hep-ex/0003033, March 2000.
4. Stephen Godfrey and Jim Napolitano, Rev. Mod. Phys. **71**, 1411, (1999).
5. Nathan Isgur, Phys. Rev. D**60**, 114016, (1999).
6. G. S. Adams et al. (E852 Collaboration), Phys. Rev. Lett. **81**, 5760, (1998).
7. G. Condo, et al., Phys. Rev. D**41**, 3317, (1990).
8. C. Bernard, et al. (MILC Collaboration), Phys. Rev D**56**, 7039, (1997).
9. K, Juge, J. Kuti and C. J. Morningstar, Nucl. Phys. Proc. Suppl., **63**, 326, (1997).

The Antiproton Project at GSI

Helmut Koch

Institut for Experimental Physics I, Ruhr –Universitaet Bochum,
D-44780 Bochum, Germany

Abstract. Experiments with antiprotons at LEAR and FermiLab have started a new era in hadron spectroscopy. Candidates for bound states with gluonic degrees of freedom were found and the spectroscopy in the charmonium region has reached a new level of precision. It is planned to extend measurements of this kind at GSI/Darmstadt. Antiprotons with energies up to 15 GeV will interact with a Hydrogen cluster target in a storage ring with high luminosity. The machine and the detector will be discussed together with the physics program. The main emphasis will be on the search for missing charmonium states and for charmed hybrids, but the large production rate of $D\bar{D}$ -pairs will also allow searches for rare D-decays and for CP-violation in the charm system.

INTRODUCTION

This talk deals with plans to construct a High Energy Storage Ring (HESR) for Antiprotons at the Gesellschaft für Schwerionenforschung (GSI) at Darmstadt, Germany. The project is part of an upgrade program to the existing facilities and is described in detail in the forthcoming proposal [1]. Antiprotons with momenta between 1.5 and 15 GeV/c will collide with protons (Pellet -or Jet-target) or with heavier nuclei (wire-target) allowing cm-energies as high as 5.5 GeV. The minimal luminosity will be $2*10^{32}$ cm^{-2}s^{-1}, and the relative momentum resolution will be as good as 10^{-4}, for lower energies eventually 10^{-5}, if high energy electron cooling can be realized. The measuring program which is outlined in the following deals mainly with all kinds of non-perturbative QCD-effects with emphasis on the Charm sector.

STATUS OF PHYSICS WITH ANTIPROTONS

High energy antiproton beams have considerably contributed to the recent progress in particle physics. The intermediate Vector Bosons ($W^{\pm}, Z^0$) and the TOP–quark have been discovered using antiproton beams. But also at intermediate energies very successful investigations have been taking place. High precision charmonium spectroscopy experiments were performed at FermiLab [2] and a wealth of interesting data has been obtained at LEAR, particularly in the field of light quark spectroscopy, resulting in the discovery of a good candidate for the glueball ground state [3] and the confirmation of two states with spin-exotic quantum numbers [4,5].

Fig.1 shows a comparison between signals for $\chi_{c_{1,2}}$-states obtained in conventional e^+e^--experiments (Crystal Ball) and in $\bar{p}p$-collisions (E760, FermiLab). In contrast to e^+e^--experiments, $\bar{p}p$- reactions allow the direct formation of these states resulting in a much better mass resolution.

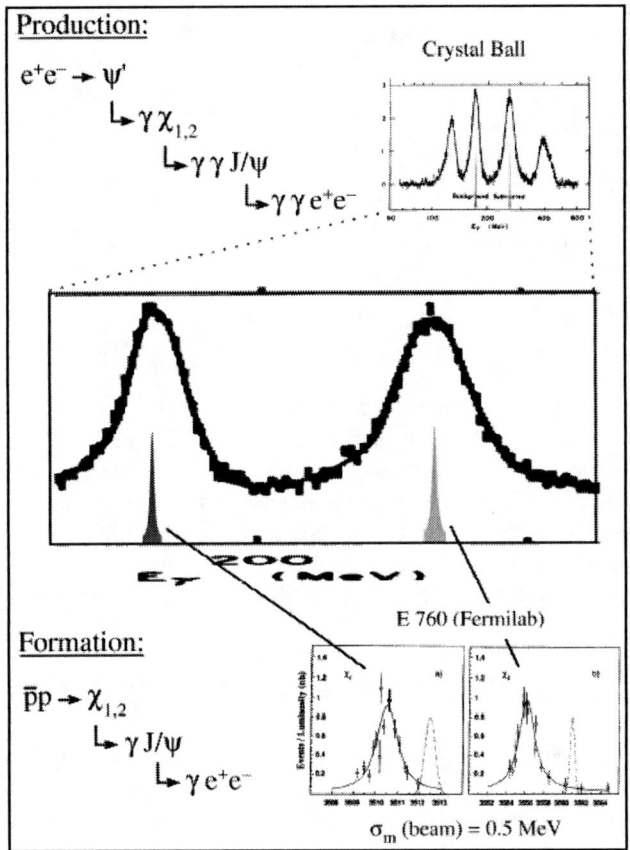

FIGURE 1. Mass spectra for $\chi_{c_{1,2}}$-states as obtained in an e^+e^--production experiment (broad structures) and in a $\bar{p}p$- formation experiment (narrow structures). This figure was kindly provided by U. Wiedner (Uppsala)

The evidence for a spin exotic state $\pi_1(1400)(J^{PC}=1^{-+})$ measured with the Crystal Barrel detector at LEAR is shown in Fig. 2. Of particular importance is, that the exotic π_1–state is produced with a strength similar to the one of a conventional $q\bar{q}$-state ($a_2(1320)$). The same is true for the $f_0(1500)$-state, firstly found at LEAR, being at present the best candidate for the glueball ground state.

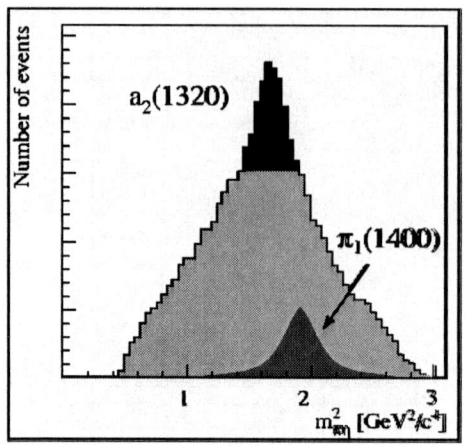 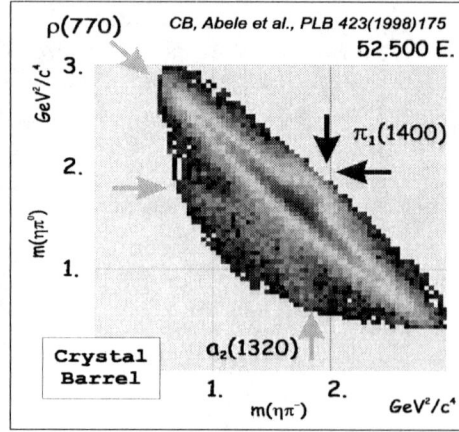

FIGURE 2. Shown on the left is the Dalitz – Plot of the reaction $\bar{p}d \to \eta\pi^-\pi^o + p$ as measured by the Crystal Barrel Collaboration at LEAR. Shown on the right is the $m^2_{\pi\eta}$ - projection.

Both examples show the merits of experiments with antiprotons at medium energies: (1) The cross sections are high facilitating the search for rare particles. (2) Most particles can be directly created in formation processes regardless of their J^{PC} quantum numbers. (3) Antiproton induced reactions have low particle multiplicities, allowing the reconstruction of complete events and thus reliable partial wave analyses. (4) Exotic states are produced with rates similar to those of $q\bar{q}$-and qqq–systems. (5) The experimental conditions are very clean due to cooled, high quality $\bar{p}$ - beams.

ANTIPROTON FACILITY AT GSI

Among other projects dealing with Heavy Ion and Plasma physics the upgrade program of GSI foresees a storage ring (HESR) for antiprotons with a maximal energy of 15 GeV. Fig. 3 shows the outline of the assembly: The antiprotons will be produced in the SIS 200 synchrotron allowing the acceleration of protons up to 60 GeV. Up to 2×10^7 antiprotons per second are produced in a target station very similar to the one at CERN. Accumulation and cooling of the antiprotons is performed in two rings (CR, NESR). Afterwards the antiprotons are injected into the SIS 200 machine in order to accelerate or decelerate them to the desired energy, and they are then ejected into the HESR. The HESR storage ring has two long straight sections for cooling devices and for the set up of a general purpose detector. Using pellet or gas jet targets, luminosities as high as 2×10^{32} cm^{-2}s^{-1} can be reached. In a later stage of the project, a luminosity increase to 5×10^{32} is envisaged allowing high rate D-pair production on a thin wire target.

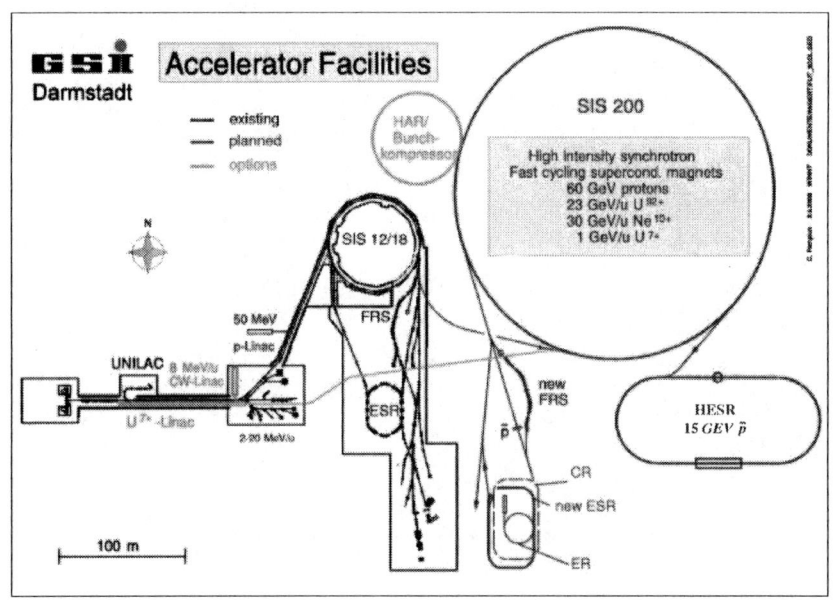

FIGURE 3. Existing and proposed accelerator scenario at GSI.

Fig. 4 shows the general purpose detector, which will be used for most of the experiments discussed in the following. It has a nearly full angle coverage for charged particles and gammas, high rate capability and good particle identification (e, μ, π, K, p) and allows efficient triggering on e, μ, K and D's.

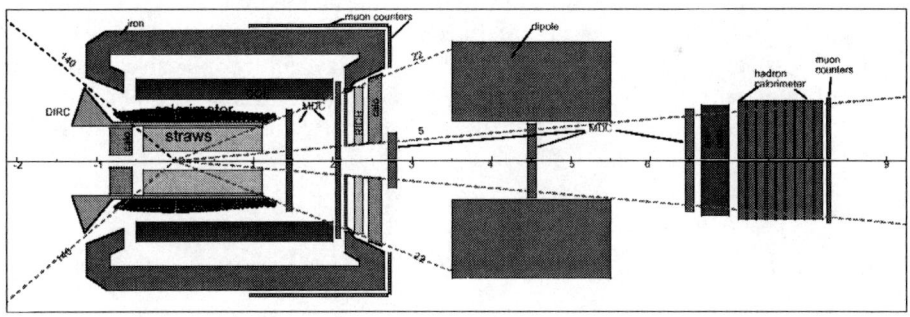

FIGURE 4. General purpose detector for HESR. For special experiments, e.g. the Hypernuclear studies, the area around the interaction point has been modified.

The tracking will be performed using a combination of Pixel-, Straw- and Mini-Drift-Chamber detectors. At present an electromagnetic PbWO$_4$ calorimeter with Avalanche-Photodiode readout is foreseen, but also pure CsI is still in discussion. For particle identification Aerogel-Cerenkovs and a Detector for Internally Reflected Cerenkov light (DIRC) will be used. The Muon Chambers consist of Plastic Scintillator strips. The trigger is based on fast Lepton- and Kaon- identification followed by a sophisticated software part using pipeline techniques.

PHYSICS HIGHLIGHTS

The physics program at such a facility is very rich. Besides the topics outlined in the following it also includes low energy experiments, like the study of the $\bar{p}$p-annihilation process, of antiprotonic atoms and high precision experiments on Antihydrogen.

Charmonium Spectroscopy

The latest results of the FermiLab experiment E835 were summarized at this meeting by A. Tomaradze [2]. They are very impressive as far as masses, widths and decay modes of χ_c-states and of the η_c ground state are concerned. However, a lot of problems remains unsolved after the end of data taking at FermiLab. The $^1P_1$-state is not firmly established and the η_c' was not seen at all. Most of the D-wave states (some of them may be very narrow) and the radially excited P-states above the $D\bar{D}$-threshold have not been found yet. More exclusive decays have to be studied in order to shed light on anomalies found, e.g. the $\pi\rho$ - puzzle.

The high luminosity of HESR and the universal detector, which - in contrast to the E835 detector - can detect leptonic and hadronic decay modes equally well, will allow to extend the program started at FermiLab considerably. E.g., 10^6 and 10^4 reconstructed J/ψ- and χ_2-states, respectively, are expected per day, allowing scans in the energy regions of interest in steps of 10 MeV, in special cases of 1 MeV. Leptonic modes like e^+e^- and $\mu^+\mu^-$ would be registered in parallel to electromagnetic ($\gamma\gamma$) and hadronic modes ($\phi\phi$,etc.)

Search for Charmed Hybrids

In Fig.5 the occurrence of $c\bar{c}$-Hybrids ($c\bar{c}g$) is demonstrated, originating together with the usual $c\bar{c}$ -states from second order LQCD calculations [7]. Similar results were obtained in various model calculations (Bag, Flux-Tube [6], etc.). All predictions indicate that the lowest energy ($c\bar{c}g$) -states have masses between 3.9 and 4.5 GeV/c^2 with the quantum numbers $J^{PC} = 2^{\mp\pm}, 1^{\mp\pm}, 2^{\mp\pm}, 0^{\pm\mp}$, three of them being spin exotic, including the ground state (1^{-+}). At least several of the states may be narrow ($\Gamma\sim$MeV), as they are forbidden to decay to $D\overline{D}$-pairs. E.g., the 0^+ can not decay to $\overline{D}D, \overline{D}^*D^*, \overline{D}_sD_s$, due to CP-conservation, and a dynamical selection rule might forbid decays of the kind $(c\overline{c}g) \to (\overline{Q}q)_{L=0} + (Q\overline{q})_{L=0}$, so that only $(c\overline{c}g)$ states with masses higher than 4.3 GeV/c^2 would decay to $D\overline{D}$. In case that the $D\overline{D}$-mode is forbidden, favorite $(c\overline{c}g)$-decays would contain a J/ψ, e.g. $1^{-+} \to J/\psi + \omega, \phi, \gamma$.

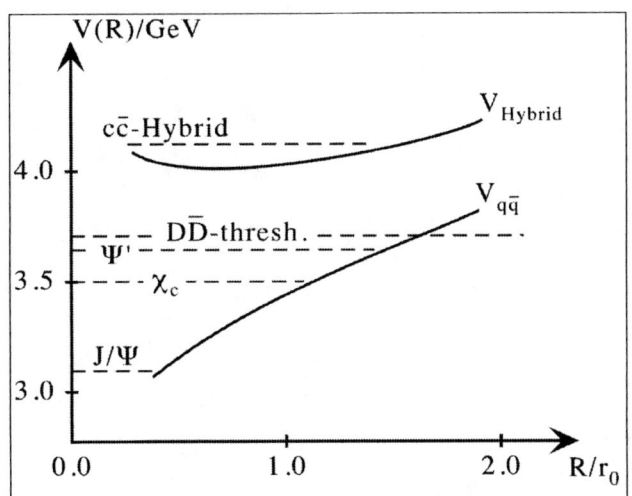

FIGURE 5. $c\overline{c}$- and $c\overline{c}g$-potentials resulting from second order LQCD calculations. Also indicated is the position of the usual $c\overline{c}$-states and of the lowest energy $c\overline{c}$-hybrid.

The chance to find $(c\overline{c}g)$-hybrids seems to be considerably higher than in the light quark sector due to the lower state density and the narrow widths of states in the charm sector. Non spin exotic states would be searched for in scanning experiments, in parallel to the Charmonium measurements. The rates may be as high as 10^4/day. States with spin exotic quantum numbers have to be detected in production experiments of the type $\overline{p}p \to (c\overline{c}g) + \pi^0/\eta$. Here, the rates are lower (10^2/day), but still sufficient for a spin-parity analysis.

In addition, of course, a program for measurements of hybrids in the light quark sector concentrating on the mass region around 1.9 GeV/c^2 can be easily performed.

Search for Heavier Glueballs

LQCD predictions for the glueball mass spectrum are shown in Fig. 6. The spectrum extends up to 5 GeV/c^2, also exhibiting several spin exotic glueballs (odd balls), the lightest one ($J^{PC} = 2^{+-}$) at 4.3 GeV/c^2. Again, at least in special cases, e.g. for odd balls, the widths may be reasonably narrow. The search for such states would proceed in parallel to the measurements discussed before. Favorable decay channels would be $\phi\phi$ or $\phi\eta$, which are easily distinguishable from typical annihilation reactions and exhibit low l-waves facilitating a spin parity-analysis.

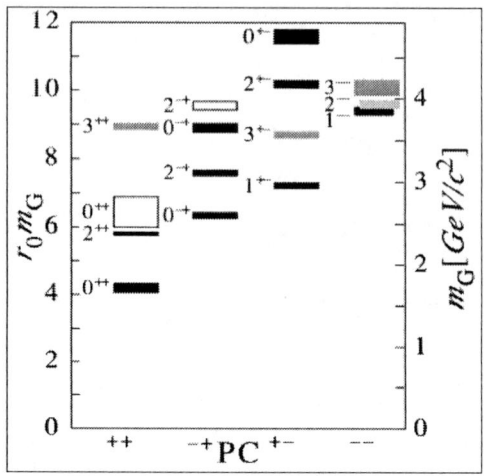

FIGURE 6. Glueball mass spectrum as predicted by LQCD [8].

Experiments with open Charm/ Strangeness

The production rate for reconstructed $D\overline{D}$ pairs (up to 10^7/year) at HESR is comparable to the one of a τ/c-factory. Also charmed and strange Baryon-Antibaryon pairs are produced with high rates, e.g. 10^7 $\Lambda_c \overline{\Lambda}_c$ and 10^9 $\Xi^- \Xi^+$ per year. The experimental conditions for spectroscopy experiments are very favorable, as all pairs will be produced near threshold, resulting in modest particle energies and low multiplicity events. One of the particles can be used for trigger purposes allowing a complete investigation of its partner.

The experimental program incorporates Baryon spectroscopy experiments and the investigation of rare D-decays. An example would be leptonic D-decays ($D^+ \to \mu^+\nu$), the rate of which is predicted by LQCD and is a sensitive test of the D-structure.

In a later stage also CP-violation experiments are feasible [9]. The comparison of the angular decay asymmetries for $\Lambda \to p\pi^-$ and $\overline{\Lambda} \to \overline{p}\pi^+$ is a sensitive test of direct CP-violation in the Baryon sector [10]. Also CP-violation in the $D^0\overline{D}^0$- sector can be investigated. According to the Standard Model CP-violation in mixing is very small [11], but direct CP- violation could be searched for in single Cabibbo suppressed decays.

Antiproton Nucleus Interactions

In recent experiments at GSI it was found that charged Pions [12] and Kaons [13] exhibit effective masses, different from their vacuum values, when these particles are produced in a nuclear medium (see Fig.7). The effects are due to the quark condensate. Similar effects are predicted for D^+ and D^- [14], which, however, depend on the gluon condensate, leading to a decrease of D^-/D^+ - masses in nuclear matter. This would produce very spectacular effects, when ($c\bar{c}$)-states are produced via $\bar{p}p$ - annihilation in nuclear matter. The $D\bar{D}$ - threshold would be lower than its vacuum value leading to a substantial increase of the width of the ψ"- state. e.g.

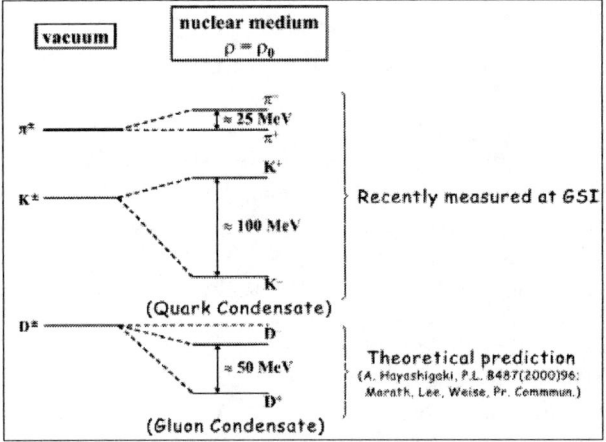

FIGURE 7. Experimental values and predictions for effective masses of Pions, Kaons and D-Mesons inside a nuclear medium.

For the identification of the Quark-Gluon-Plasma the magnitude of the J/ψ - Nucleon absorptive cross section plays a decisive role. It could be measured with high accuracy in a $\bar{p}$-scan between 3.4 and 4.6 GeV/c, observing the reaction $\bar{p} + A \rightarrow J/\psi + (A-1)$. Similar experiments could be performed on other ($c\bar{c}$)-states.

The investigation of Hypernuclei adds a third dimension to the nuclear chart yielding very valuable information on nuclear structure. Of particular interest are double Λ - Hypernuclei, for which only three candidates exist yet. They can be copiously produced at HESR using the two-step reaction

$$\bar{p}\,(2.6\ \text{GeV/c}) + A \rightarrow \Xi^-_{slow} + \bar{\Xi}^-\,Trigger;\ \ \Xi^-_{slow} + A' \rightarrow_{\Lambda\Lambda} A'$$

As a secondary active target (A') high resolution solid state micro-tracking detectors (Diamond, Si) are used together with an efficient, position sensitive Ge-γ-array [15,16] allowing high rate spectrocsopy experiments yielding hundreds of γ-transitions per-day. For these experiments the interior of the general purpose detector has to be modified accordingly

STATUS OF THE PROJECT

The parameters of the SIS 200 synchrotron and of the HESR are worked out in detail. The performance of the general purpose detector was studied using a sophisticated Geant 4 simulation and was found to match the expectations [17]. The proposal for the GSI upgrade will be available soon [1]. Referee Committees have started with the survey of the project and a decision is expected in the middle of the year 2002.

CONCLUSIONS

Antiproton induced reactions which can be studied at the High Energy Storage Ring at GSI exhibit unique features: (1) Gluonic hadrons seem to have high production rates in $\bar{p}p$ - annihilation. (2) High statistics data with low multiplicity events with a symmetric production of particles and antiparticles will be obtained. (3) Many of the interesting states can be directly formed and investigated in a scan-mode with high mass resolution. A rich and unique Physics Program with emphasis on charmed particles can be performed including J/ψ-nucleon interactions, the study of effective hadron masses in nuclear matter, precision charmonium spectroscopy and the search for charmed hybrids and heavier glueballs. In a later stage, even CP-violation experiments are in reach. Additionally, the low energy experiments started at LEAR could be continued, also including further studies on Antihydrogen.

REFERENCES

1. Conceptual Design Report for an International Accelerator Facility for beams of Ions and Antiprotons, GSI, in print
  see also: Letter of Intent: Construction of a Glue / Charm -Factory at GSI (1999)
2. Tomaradze, A., these proceedings
3. Amsler, C., *Rev. of Mod. Phys.* **70** (1998) 1293
4. Abele, A., et al., *Phys. Letters* **B423** (1998) 175
5. Reinnarth, J., Proc. LEAP 2000, *Nucl. Phys.* **A 692** (2001)
6. Isgur, N., Kokoski, R., and Paton, J., *Phys .Rev. Letters* **54** (1985) 869
  Page, P.R., Swanson, E.S., and Szczepaniak, A.P., *Phys. Rev.* **D 59** (1999) 034016
7. Michael, C., Proc. of Heavy Flavours 8, Southampton, UK, 1999
8. Morningstar, C.J., and Peardon, M., *Phys. Rev.* **D60** (1999) 034509
9. Bigi, I.I., and Sanda, A.I., *CP Violation*, Cambridge Monographs on Particle Physics, Nuclear Physics and Cosmology **9** (2000)
10. Hamann, N., Proc. of the Super LEAR Workshop (1991), Inst. of Physics Conference Series, No. 124

11. Burdman, G., Charm Mixing and CP Violation in the Standard Model, hep-ph / 9407378, CHARM 2000 Workshop, FermiLab (1994)
12. Yamazaki, T., et al., *Phys. Letters* **B418** (1998) 246
    Gillitzer, A., Proc. of Int. Workshop "Structure of Hadrons", Hirschegg (2001)
13. Barth, R., et al., *Phys. Rev. Letters* **78** (1997) 4007
14. Morath, Ph., Lee, S.H., and Weise, W., private communication
15. Gerl, J., et. al., VEGA-Versatile and Efficient Gamma Detectors for GSI (1997)
16. Annis, P., et. al., Nucl. Inst. Meth. **A449** (2000) 60
17. Ritmann, J., Ganzhur, S., Hartmann, O., Hejny, V., Schwarz, C., HESR-Simulation-Group

The KLOE Physics Program

The KLOE Collaboration [1]
Presented by P. Valente

Abstract. The KLOE experiment, running at the Frascati DAΦNE φ-factory since 1999, was designed to measure the $\Re(\varepsilon'/\varepsilon)$ ratio to an accuracy of few 10^{-4}. However, its physics program is wider than CP violation, covering both kaon and non-kaon physics items.

During the year 2000 data taking, ~ 20 pb^{-1} were collected and a total integrated luminosity of 200 pb^{-1} is expected by the end of 2001. A wealth of measurements can be addressed with these data and a number of preliminary results have been already obtained using the year 2000 statistics, in particular on rare K_S decays. After a brief description of the detector, the K_S tagging technique is discussed. Preliminary measurements of BR($K_S \to \pi^+\pi^-$)/BR($K_S \to \pi^0\pi^0$) and BR($K_S \to \pi e \nu$) are then presented. Finally, the perspectives for future analyses are discussed.

INTRODUCTION

The KLOE experiment at the Frascati φ-factory, DAΦNE, started data taking in April 1999. The φ(1020) meson decays $\sim 34\%$ of the time into a $K_L K_S$ pair; at peak energy, this corresponds to about 1 million of such decays *per* delivered pb^{-1}. DAΦNE is therefore a source of nearly monochromatic, tagged, neutral K mesons. The detector was primarily designed for measuring $\Re(\varepsilon'/\varepsilon)$, both with quantum interferometry and with the double ratio method, to an accuracy of a few 10^{-4} [1]. However, the KLOE physics program is much wider, covering studies of T and CPT asymmetries, φ radiative decays, measurements of rare kaon decays for precise tests of chiral perturbation theory, and measurement of the e^+e^- hadronic cross section below the φ peak using the radiative return.

---

[1] A. Aloisio, F. Ambrosino, A. Antonelli, M. Antonelli, C. Bacci, G. Barbiellini, F. Bellini, G. Bencivenni, S. Bertolucci, C. Bini, C. Bloise, V. Bocci, F. Bossi, P. Branchini, S. A. Bulychjov, G. Cabibbo, R. Caloi, P. Campana, G. Capon, G. Carboni, M. Casarsa, V. Casavola, G. Cataldi, F. Ceradini, F. Cervelli, F. Cevenini, G. Chiefari, P. Ciambrone, S. Conetti, E. De Lucia, G. De Robertis, P. De Simone, G. De Zorzi, S. Dell'Agnello, A. Denig, A. Di Domenico, C. Di Donato, S. Di Falco, A. Doria, M. Dreucci, O. Erriquez, A. Farilla, G. Felici, A. Ferrari, M. L. Ferrer, G. Finocchiaro, C. Forti, A. Franceschi, P. Franzini, C. Gatti, P. Gauzzi, A. Giannasi, S. Giovannella, E. Gorini, F. Grancagnolo, E. Graziani, S. W. Han, M. Incagli, L. Ingrosso, W. Kluge, C. Kuo, V. Kulikov, F. Lacava, G. Lanfranchi, J. Lee-Franzini, D. Leone, F. Lu, M. Martemianov, M. Matsyuk, W. Mei, A. Menicucci, L. Merola, R. Messi, S. Miscetti, M. Moulson, S. Mueller, F. Murtas, M. Napolitano, A. Nedosekin, M. Palutan, L. Paoluzi, E. Pasqualucci, L. Passalacqua, A. Passeri, V. Patera, E. Petrolo, D. Picca, D. Pirozzi, L. Pontecorvo, M. Primavera, F. Ruggieri, P. Santangelo, E. Santovetti, G. Saracino, R. D. Schamberger, B. Sciascia, A. Sciubba, F. Scuri, I. Sfiligoi, J. Shan, P. Silano, T. Spadaro, E. Spiriti, G. L. Tong, L. Tortora, E. Valente, P. Valente, B. Valeriani, G. Venanzoni, S. Veneziano, A. Ventura, Y. Wu, G. Xu, G. W. Yu, P. F. Zema, Y. Zhou.

About 20 pb^{-1} were collected at the energy of the ϕ peak in the year 2000. Competitive results, in particular on radiative decays of the ϕ (see the two separate reports from KLOE [2, 3]) and rare K_S decays have already been obtained with this data set. In the following, preliminary results from the analysis of two different K_S channels are presented: $K_S \to \pi^+\pi^-/K_S \to \pi^0\pi^0$ and $K_S \to \pi e \nu$.

The measurement of the ratio of branching ratios $K_S \to \pi^+\pi^-/K_S \to \pi^0\pi^0$, is relevant for *CP* violation studies, since it enters the double ratio from which $\Re(\varepsilon'/\varepsilon)$ is derived. In particular it provides a valid bench-test for the study of all the sources of systematic errors. It is also interesting in itself for studies of chiral perturbation theory, especially if the radiation of soft photons in the charged decay is properly taken into account.

The measurement of the branching ratio of the $K_S \to \pi e \nu$ decay is also presented: the branching ratio measurement inclusive of both lepton charges allows a test of the $\Delta S = \Delta Q$ rule. The violation parameter is defined as

$$x = \frac{A\left(K^0 \to \pi^+ e^- \bar{\nu}\right)}{A\left(\overline{K}^0 \to \pi^+ e^- \bar{\nu}\right)}$$

assuming no *CPT* violation[2]: $\Re(x)$ can then be extracted from the K_S semileptonic branching ratio using the known value for that of the K_L. A measurement of the partial width with a relative accuracy of 2% (more than 2000 observed events) would give $\Re(x)$ with an error of 6×10^{-3}, competitive with the present experimental knowledge [4]. Moreover, one may test *CPT* conservation by checking the equality of the charge asymmetry in semileptonic decays of K_L and K_S [5]. Up to now, only one measurement of the branching ratio $K_S \to \pi e \nu$ exists, based on a data sample of 75 ± 13 events [6]. The present KLOE result is based on a sample of about 600 events (~ 17 pb^{-1} from the year 2000 data set) after the subtraction of a background of less than 10%.

THE KLOE DETECTOR

The KLOE detector (see Fig. 1) consists of a large drift chamber, a hermetic electromagnetic calorimeter, and a large magnet surrounding the whole detector. The magnet consists of a superconducting coil and an iron yoke, and provides a solenoidal field of 5.2 kG. At the interaction region (IP), the Al-Be alloy beam pipe has a spherical shape, 10 cm in radius; also the 3 low-β quadrupoles are inside the detector, starting at 40 cm in the longitudinal coordinate (z); they are therefore instrumented with tile electromagnetic calorimeters (QCAL).

The drift chamber [7, 8] is a cylinder of 2 m radius, 3.3 m long, with 52 140 wires and a total number of 12 582 square cells arranged in 58 layers (12 inner layers with 2×2 cm^2 cells, and 46 outer layers with 3×3 cm^2 cells). The chamber is strung in an all-stereo geometry, with constant inward radial displacement of the wires at the chamber center.

---

[2] And neglecting $|\varepsilon|^2$ with respect to 1 and $2\Im(x)\Im(\varepsilon)$ with respect to $\Re(x)$, where ε is the parameter of the *CP* violation in mixing.

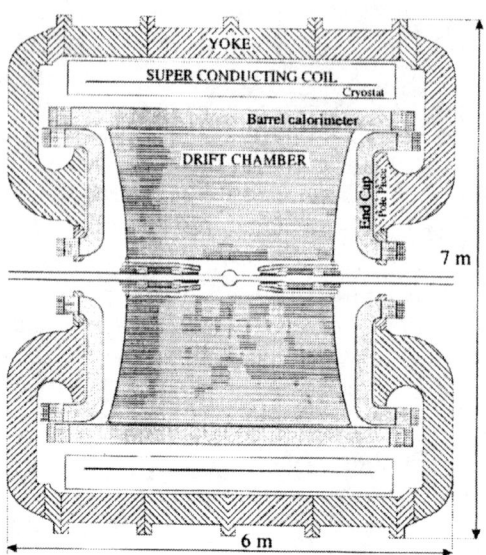

FIGURE 1. Side view of the KLOE detector.

It is made of ultra-light materials (carbon fiber/epoxy walls, tungsten and aluminum wires), and is operated with a low-Z gas mixture (He/iC$_4$H$_{10}$), to minimize multiple scattering of charged particles and regeneration of K_L's. A spatial resolution better than 200 μm is obtained, while the momentum resolution for 510 MeV/c electrons and positrons is 1.3 MeV/c in the angular range $50° < \theta < 130°$.

The electromagnetic calorimeter [9, 10] is a lead/scintillating-fiber sampling calorimeter, divided into a barrel section and two end-caps. The modules are read out at both ends by a total of 4880 photomultipliers. In order to minimize dead zones in the overlap region between barrel and end-caps, the modules of the latter are bent outwards with respect to the decay region.

The calorimeter was designed to detect photons with energy as low as 20 MeV with very high efficiency, and to accurately measure their energy and time-of-flight. Absolute calibrations of energy and time scales are performed using collision data, such as Bhabha and $e^+e^- \to \gamma\gamma$ events. An energy resolution of $5.7\%/\sqrt{E(\text{GeV})}$ is measured throughout the whole calorimeter together with a linearity in energy response better than 1% above 80 MeV and 4% between 20 to 80 MeV. Moreover, γ samples from different processes are selected to measure the time resolution at various energies: the parametrization $\sigma_t = (54 \text{ ps}/\sqrt{(E(GeV)} \oplus 147$ ps was obtained, where the first term agrees with test-beam results, while the second is dominated by the intrinsic time spread due to the bunch length fluctuations.

The trigger uses both calorimeter and drift chamber information, implemented in two-level scheme [11]; for the analyses presented here, only events triggered by the calorimeter were considered, for which the request is basically to have two clusters (barrel-barrel or barrel/end-cap coincidences). The time at which the trigger is issued is synchronized with the DAΦNE radiofrequency (the nominal bunch spacing is T_{bunch} = 2.7 ns). This is relevant for the time-of-flight measurements, for which the correct bunch crossing has to be evaluated in order to estimate the 'time zero' of the event.

At the present luminosity, KLOE acquires data at a rate of $\sim$ 2 kHz resulting in a throughput to the DAQ system of $\sim$ 3 Mbytes/s [12]. Data are then reconstructed quasi-on line by a dedicated farm, divided into different data streams according to a first rough identification of the event type, and stored into a 220 Tbyte tape library accessible by the users for the final physics analyses.

TAGGING OF K_S BEAM

When a ϕ meson decays into two neutral kaons, C-parity invariance forces the two kaons to be in a (antisymmetric) $K_S K_L$ state; moreover, in KLOE, the ϕ meson is produced almost at rest, with a small boost, $p_x \sim$ 13 MeV/c, due to the beam crossing angle 2×12.5 mrad. The two kaons are then produced almost back-to-back in the laboratory with mean decay paths $\lambda_L \sim 3.5$ m and $\lambda_S \sim 0.6$ cm. The observation of a K_L therefore *tags* the presence of the K_S in the opposite hemisphere.

The present analyses use the K_L interaction in the calorimeter (the so called 'K_{crash}') to tag the K_S: one has to identify a cluster due to a slowly moving ($\beta \approx 0.22$) neutral particle. More than 50% of the K_L's produced reach the calorimeter before decaying; so the K_{crash} tag identifies a particularly clean, high statistics K_S sample on the basis of the following simple requests:

1. A calorimeter cluster with energy larger than 50 MeV and transverse distance larger than 60 cm due to the K_S decay; This is needed to determine the t_0 of the event, namely the time at which the ϕ production and decay occurred.
2. A calorimeter cluster (the K_{crash}) in the barrel region with energy larger than 100 MeV and time compatible with the K_L velocity $c\beta^*$ in the ϕ rest frame. The applied cut on the cluster velocity, after having subtracted the t_0 of the event, is therefore:

$$0.195 < \beta^* < 0.2475.$$

The tag efficiency is slightly dependent on the K_S decay type, since the t_0 estimate is determined by the fastest particle in the event: a prompt γ in the case of $K_S \to \pi^0\pi^0$ events, a pion in $K_S \to \pi^+\pi^-$ ones, and a pion or electron for semileptonic decays. The distributions for the reconstructed β^* for charged and neutral two pions decays are shown in Fig. 2: with the velocity cut specified above, the ratio of tagging efficiencies for the two K_S modes is:

$$\varepsilon^{+-}/\varepsilon^{00} = (95.030 \pm 0.005)\%$$

where the error is statistical only. In the following, all events are tagged making use of the K_{crash} selection.

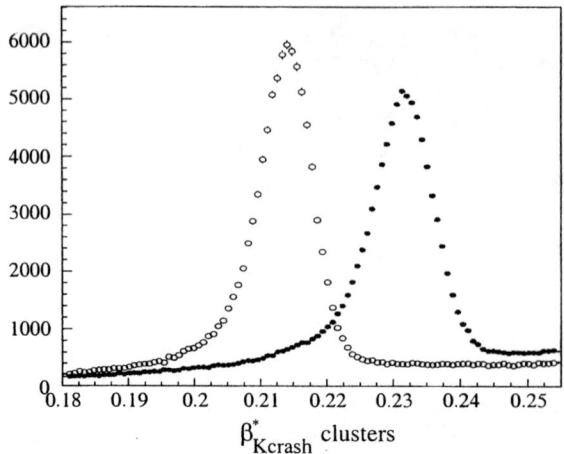

FIGURE 2. Clusters of K_L interacting in the calorimeter: β^* distributions of clusters for $K_S \to \pi^0\pi^0$ (empty dots) or $K_S \to \pi^+\pi^-$ (full dots) events. The distance between the peaks corresponds to 1 bunch time spacing.

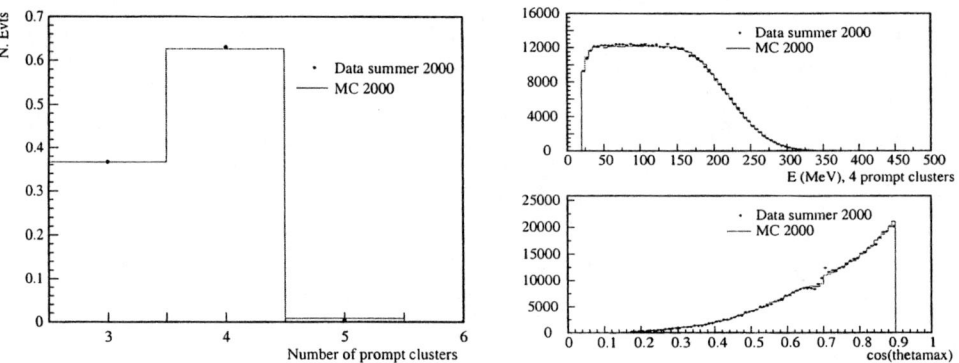

FIGURE 3. Left: Prompt clusters distribution in K_{crash} events for data (dots) and Monte Carlo (line). Right: angular (top) and energy (bottom) distributions of the clusters in 4 γ events. Dots are data, solid line is the Monte Carlo for $K_S \to \pi^0\pi^0$ events.

ANALYSIS OF $K_S \to \pi^0\pi^0$ AND $K_S \to \pi^+\pi^-$ DECAYS

K_S decays into two neutral pions are selected by requiring four calorimeter clusters with a time compatible with the hypothesis of being due to prompt photons (within 5 σ_t) and energy larger than 20 MeV. The distribution of the number of prompt clusters, the energy spectrum and the angular distribution for the events with four prompt clusters (Fig. 3) agree with Monte Carlo (MC) expectations at the percent level.

The photon detection efficiency is estimated from real data using γ's from $\phi \to$

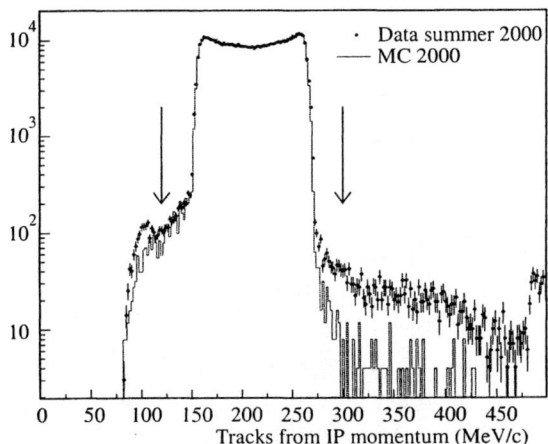

FIGURE 4. Momentum distribution for the tracks originating from the IP in K_{crash} selected events. Black points with error bars are data, the solid histogram is the MC expectation for $K_S \to \pi^+\pi^-$ events. Tracks with P> 300 MeV/c are mostly due to machine background, the small peak at P~ 100 MeV/c is due to $\phi \to K^+ K^-$ events.

$\pi^+\pi^-\pi^0$ decays as a control sample. The final selection efficiency for the $K_S \to \pi^0\pi^0$ decay channel is ε_{00}=(56.7±0.1)% and is dominated by the geometrical acceptance.

The selection of $K_S \to \pi^+\pi^-$ events proceeds via the request of two opposite tracks with polar angle in the interval $30° < \theta < 150°$, originating in a cylinder of 4 cm radius and 10 cm length around the IP. A further request is applied on the measured momenta to remove the residual background due to charged kaon decays (as shown in Fig. 4): 120 MeV/c $<$ p $<$ 300 MeV/c. Both tracks are also required to reach the calorimeter, which enhances the probability for having a good t_0 determination.

The dependence of the acceptance with respect to the energy of the radiated photon is crucial in the analysis of $K_S \to \pi^+\pi^-(\gamma)$ events and is under study: the loose track momentum cut is not expected to produce a significant bias if a soft photon is radiated with energy lower than 60-80 MeV. The track reconstruction efficiency is measured in momentum and polar angle bins from data subsamples. The final selection efficiency is ε_{+-}=(58.5±0.1)%, and is again dominated by the geometrical acceptance.

The trigger efficiency is determined again using real data for both decay types. It is (99.69 ± 0.03)% for the neutral decay and (96.5 ± 0.1)% for the charged one. The above figures also include the requirement of a at least one good cluster to determine the t_0 of the event ('t_0 efficiency'), as explained in the previous paragraph. Background levels are kept well below 1% for both decay types.

Using part of the data acquired in year 2000, about 17 pb^{-1}, 1636457 $K_S \to \pi^+\pi^-$ and 772976 $K_S \to \pi^0\pi^0$ decays have been selected, yielding:

$$\Gamma(K_S \to \pi^+\pi^-)/\Gamma(K_S \to \pi^0\pi^0) = 2.247 \pm 0.005_{stat} \pm 0.034_{syst}$$

to be compared (Fig. 5) with the world average [13]: 2.197 ± 0.026. The systematic error is dominated by residual uncertainties in photon counting and in the estimate of the

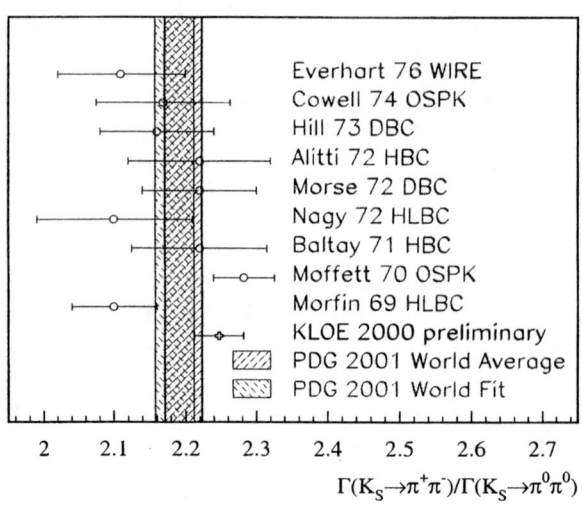

FIGURE 5. Existing measurements of the ratio of partial widths $\Gamma(K_S \to \pi^+\pi^-)/\Gamma(K_S \to \pi^0\pi^0)$; the world average and fit [13] 1-σ intervals are also shown.

difference between the tagging efficiencies for the two channels. More precise studies are presently under way.

ANALYSIS OF K_S SEMILEPTONIC DECAYS

Events with two oppositely-charged tracks from the IP are initially selected; they are rejected if the two tracks' invariant mass (in the pion hypothesis) and the resulting K_S momentum in the ϕ rest frame are compatible with those expected for a $K_S \to \pi^+\pi^-$ decay. The preselection efficiency, given the tag, is $(62.4 \pm 0.3)\%$, as calculated by Monte Carlo.

In order to perform a time-of-flight identification of the charged particles, both tracks are required to be associated with calorimeter clusters. The *geometrical acceptance* for this request is estimated by Monte Carlo to be $(51.1 \pm 0.2)\%$. Using associated cluster times and flight paths, the π-π mass hypothesis is rejected and the correct π-e mass assignment between the two charged tracks is performed. The efficiency, estimated by means of K_L's decaying into $\pi e \nu$ before the internal chamber wall, is $(84.0 \pm 0.6)\%$.

The trigger and t_0 efficiencies, and the efficiency for correctly associating both tracks to calorimeter clusters are measured directly from data as well, making use of $K_L^0 \to \pi^{\pm} e^{\mp} \bar{\nu}(\nu)$, $\phi \to \pi^+\pi^-\pi^0$ and $K_S^0 \to \pi^+\pi^-$ subsamples. The product of these efficiencies turns out to be $(81.7 \pm 0.5)\%$. The overall efficiency is then $(21.8 \pm 0.3)\%$.

Finally the event is kinematically closed: the K_S momentum is estimated using the measured K_L direction and the ϕ meson 4-momentum; the missing energy and momentum of the K_S-π-e system, corresponding to those of the neutrino, are then computed. Their difference is distributed as shown in Fig. 6; it must be peaked around zero for the

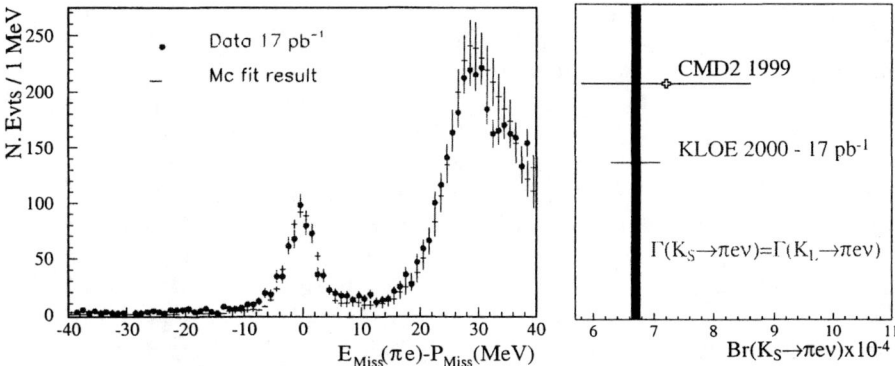

FIGURE 6. (Left) Difference between missing energy and missing momentum for $K_S \to \pi e \nu$ candidates; the peak at zero is due to the signal. The distribution is fit to a linear combination of the MC spectra of signal and background. (Right) Comparison of the presented result with the only existing measurement and with the theoretical value, obtained assuming $\Gamma_S(\pi e \nu) = \Gamma_L(\pi e \nu)$.

signal. Data are fit using MC spectra for both signal and the residual background, due mostly to $K_S \to \pi^+ \pi^-$ events with an early decay of one of the two pions.

Using about 17 pb^{-1} of data, we have observed 627 ± 30 signal events; the normalization of the yield to the number of $K_S \to \pi^+ \pi^-$ events gives

$$BR(K_S \to \pi e \nu) = (6.8 \pm 0.3_{\text{stat}}) \times 10^{-4}.$$

In the ratio, the tagging efficiency, which is the largest source of systematic uncertainty, mostly cancels out. Other systematic effects, presently under study, are estimated to be at the level of a few percent. This result is compared in Fig. 6 with that of the CMD2 Collaboration [6], $BR = (7.2 \pm 1.4) \times 10^{-4}$, and with the prediction obtained assuming $\Gamma_S(\pi e \nu) = \Gamma_L(\pi e \nu)$, $BR = (6.70 \pm 0.07) \times 10^{-4}$.

HADRONIC CROSS SECTION

Even though DAΦNE is operating at a fixed center of mass energy $\sqrt{s} = m_\phi$, the hadronic cross section below the ϕ mass can be measured at KLOE using events with radiation in the initial state (ISR): $e^+ e^- \to \gamma \gamma^* \to \gamma$ hadrons. This "radiative return" allows the measurement of the hadronic cross section in the Q^2 invariant mass range $2m_\pi^2 \leq Q^2 \leq m_\phi^2$ [14].

The precise knowledge of the hadronic cross section – especially at low energies – is of great interest for an improvement of the theoretical error of the muon anomaly a_μ. The hadronic contributions cannot be calculated in the framework of perturbative QCD and are rather obtained through dispersion integrals using experimental cross section data as input [15]. The cross section for the reaction $e^+ e^- \to \rho \gamma \to \pi^+ \pi^- \gamma$ as a function of Q^2 ($\pi\pi$ invariant mass) is used to extract the cross section of the dominating hadronic process at low energies: $\sigma(e^+e^- \to \pi^+\pi^-)$ [3]. This approach is complementary to the conventional energy scan, since with this method many systematic effects (e.g. luminosity, beam energy measurement) enter only once.

An important issue for this measurement is the suppression of Final State Radiation (FSR) with respect to ISR. Monte Carlo simulations [14] showed that cuts on the photon polar angle ($10^o \leq \Theta_\gamma \leq 40^o$) and energy ($E_\gamma \geq 50$ MeV) suppress FSR below 1%.

The signature of $\phi \to \pi^+\pi^-\gamma$ events are two charged tracks from the IP and a prompt photon. The $\pi^+\pi^-\gamma$ sample is hugely contaminated with radiative Bhabhas. An efficient $\pi - e$ separation can be achieved taking advantage of the excellent timing resolution of the calorimeter and by applying a time-of-flight technique, similar to what is done for the K_S semileptonic decays. In this case, the time of the clusters and the shower profile information are combined in a likelihood function. After the likelihood cut, good agreement between data and Monte Carlo Q^2 distributions is found. Preliminary results have been already obtained using the year 2000 data sample. However, to achieve a 1% statistical accuracy on the muon anomaly determination, at least 100 pb^{-1} of integrated luminosity are needed (see [16] for a detailed discussion).

CONCLUSIONS AND PERSPECTIVES

About 20 pb^{-1} were collected at the energy of the ϕ peak in the year 2000. With this data set, competitive results on the measurement of $BR(K_S \to \pi^+\pi^-)/BR(K_S \to \pi^0\pi^0)$ and $BR(K_S \to \pi e \nu)$ have been obtained.

During 2001, the data taking has proceeded with an integrated daily luminosity ranging between 1 and 2 pb^{-1}/day, and a peak luminosity that has reached 4×10^{31} cm^{-2}s^{-1}. Hence, a total integrated luminosity of ~ 200 pb^{-1} will be available by the end of 2001, corresponding to 2×10^8 K_S's (0.8×10^8 of which K_{crash}-tagged).

This will allow improvements of the present analyses and will open new perspectives for studying more rare K_S decays such as $K_S \to \gamma\gamma$, $K_S \to \pi^+\pi^-(\gamma)$, and $K_S \to \pi^0\pi^0\pi^0$ for which there is work currently in progress. The number of expected rare K_S events for 200 pb^{-1} is summarized in Table 1.

TABLE 1. Yield of rare K_S decays (tagged by the K_{crash}) for a total integrated luminosity of 200 pb^{-1}.

Channel	Number of observed events
$K_S \to \pi^+\pi^-$	2×10^7
$K_S \to \pi^0\pi^0$	1×10^7
$K_S \to \pi e \nu$	6 700
$K_S \to \gamma\gamma$	70
$K_S \to \pi^0\pi^0\pi^0$	$BR < 3 \times 10^{-8}$

KLOE can also perform a precision measurement of the hadronic cross section below

---

[3] The $\pi^+\pi^-\gamma$ cross section is enhanced due to the coupling of the virtual photon to the ρ in the case of ISR: $e^+e^- \to \gamma^*\gamma \to \rho\gamma \to \pi^+\pi^-\gamma$

1 GeV using the radiative return method, namely $e^+e^- \to \rho\gamma \to \pi^+\pi^-\gamma$ when the photon is radiated in the initial state, as a function of Q^2 ($\pi\pi$ invariant mass). With this approach, which is complementary to the conventional energy scan, many systematic effects can be reduced. The accuracy on the determination of the hadronic contribution to the muon-anomaly, δa_μ^{hadr}, can be reduced to $\approx 1.5 \cdot 10^{-10}$ in the energy range below the ϕ resonance with an integrated luminosity of the order of 100 pb^{-1}, since the main contribution comes from the $\pi^+\pi^-$ cross section.

REFERENCES

1. C. Bloise et al., The KLOE Collaboration, *KLOE perspectives for ϵ'/ϵ*, Proceedings of KAON 2001 Conference, Pisa (2001).
2. The KLOE Collaboration, presented by C. Di Donato, $\eta,\eta\prime$, these proceedings.
3. The KLOE Collaboration, presented by B. Valeriani, a_0, f_0, these proceedings.
4. A. Apolostakis et al., CPLEAR Collaboration, Phys. Lett. **B 456**, 297-303 (1999); A. Angelopulos et al., CPLEAR Collaboration, Phys. Lett. **B 444**, 43-51 (1998).
5. G. D'ambrosio et al., The second DAΦNE physics handbook, **Volume I**, 63 (1995).
6. R. R. Akhmetshin et al., CMD2 Collaboration, Phys. Lett **B456**, 90-94 (1999).
7. A. Aloisio et al., The KLOE Collaboration, *The KLOE Central Drift Chamber, Addendum to the Technical Proposal*, **LNF-94/028** (1994).
8. M. Adinolfi et al., *The Tracking detector of the KLOE experiment*, **LNF-01/016 (P)** (2001), Submitted to Nucl. Inst. Meth. A.
9. A. Aloisio et al., The KLOE Collaboration, *The KLOE detector, Technical Proposal*, **LNF-93/002** (1993).
10. M. Adinolfi et al., *The KLOE electromagnetic calorimeter*, **LNF-01/017 (P)** (2001), Submitted to Nucl. Inst. Meth. A)
11. A. Aloisio et al., The KLOE Collaboration, *The KLOE Trigger System, Addendum to the Technical Proposal*, **LNF-96/043** (1996).
12. A. Aloisio et al., The KLOE Collaboration, *The KLOE Data Acquisition System, Addendum to the Technical Proposal*, **LNF-95/014** (1995).
13. D.E. Groom et al., The European Physical Journal **C15** (2000).
14. S. Binner et al., Phys. Lett. **B 459**, 279 (1999).
15. S. Eidelman, F.Jegerlehner, Z.f.Physik **C 67**, 585 (1995).
16. A. Aloisio et al., The KLOE Collaboration, *Measuring the hadronic cross-section at KLOE using the radiative return*, Lepton Photon 2001, Roma **hep-ex/0107023** (2001).

Exploring the Charm Sector with CLEO-c

D. Urner

Cornell University, Wilson Lab, Ithaca NY 14853, USA

Abstract. The CLEO collaboration proposes to explore the charm sector starting early 2003. It is foreseen to collect on the order of 6 million $D\bar{D}$ pairs, 300000 $D_s\bar{D}_s$ pairs at threshold and one billion J/ψ decays. High precision charm data will enable us to validate upcoming Lattice QCD calculations that are expected to produce 1-3% errors for some non-perturbative QCD quantities. These can then be used to improve the accuracy of CKM elements. The radiative J/ψ decays will be the first high statistics data set well suited for meson spectroscopy between 1600 and 3000 MeV.

INTRODUCTION

Let's start with a lofty goal: We strive for the mastery of a non-perturbative strongly coupled theory: QCD. Couplings in field theory do not typically have to be weak. Indeed, strong interactions are the expected phenomena if one reaches beyond the Standard Model. It will therefore be of great benefit if we can understand the effects of strong couplings in QCD. High precision predictions of QCD will also remove road blocks for many weak and flavor physics measurements.

Lattice QCD has matured over the last decade. We finally can expect the first non-perturbative QCD results with 1-3% errors. CLEO-c will provide crucial data in a timely fashion to validate them and help guide the theory on its long way from easier predictions to a full understanding of non-perturbative QCD effects. This will result, for example, in improved measurements of V_{cs} and V_{cd} at the 1% level. CLEO-c data will also provide a large number of basic measurements needed in heavy flavor physics and future efforts in understanding physics beyond the standard model. A detailed description can be found in [1].

Data Sets

We plan to acquire the CLEO-c data in a 3 year program. The use of present and future CLEO data sets will be considered in the course of this paper. The expected size of the data sets are shown in Table 1.

CLEO-c intends to accumulate 30×10^6 events at ψ'' and about 1.5×10^6 events at $\psi(4140)$. The number of expected $D_s\bar{D}_s$ events is uncertain within a factor of two because of conflicting earlier measurements. This will be clarified with an early scan, which will determine the point of largest $D_s\bar{D}_s$ production. We expect to collect a total of about one billion J/ψ events. Smaller data sets are considered at the $\tau\tau$ threshold (3557 MeV) at the $\psi'(3686)$ at $\Lambda_c\bar{\Lambda}_c$ threshold (5200 MeV) and a scan over the full 3-7 GeV

TABLE 1. Size of datasets considered in the discussion of this paper. The data at the $\Upsilon(4S)$ represents the CLEOII and CLEOIII data sets. The data sets at the narrow Υ resonances are taken just prior to CLEO-c.

Center of mass energy	Luminosity	Decays/Physics
$\Upsilon(4S)(10580)$	24 fb^{-1}	2-photon physics
$\Upsilon(3S)(10355)$	1 fb^{-1}	η_b
$\Upsilon(1S)(9460)$	1 fb^{-1}	meson spectroscopy
$\Upsilon(2S)(10023)$	1 fb^{-1}	
$\psi''(3770)$	3 fb^{-1}	$1.5 \times 10^6\ D\bar{D}$
$\psi(4140)$	3 fb^{-1}	$3.0 \times 10^5\ D_s\bar{D}_s$
$J/\psi(3097)$	1 fb^{-1}	6.0×10^7 radiative J/ψ decays

region.

In 2002, before lowering the energy to the charm sector, CLEO plans to collect data at the three narrow 1S, 2S and 3S Υ resonances with an integrated luminosity of about 1fb^{-1} each.

The CLEOII and CLEOIII data sets gathered at the $\Upsilon(4S)$ contain a large number of 2-photon events.

Accelerator: Modifications to CESR

For the upgrade to CLEO III new superconducting quadrupoles for the final focusing system were built. They prove to be crucial, since they enable us to lower the beam energy and run in the region of the charm system. Accelerators find that the luminosity typically scales at best with $L \sim E_b^4$. This behavior can be changed if one introduces wigglers, which will cool the beam transversely to ideally a linear correlation of luminosity and beam energy. We plan to build 14 superconducting wiggler modules, each having 1.3 m of length, a peak field of 2T, and a 40 cm period. A 3-pole test module is shown in figure 1. The projected beam spread at the J/ψ will be about: $\Delta E_b \sim 1.2$ MeV. The expected machine performance is shown in Table 2.

TABLE 2. Expected machine performance for CLEO-c.

$\sqrt{s}$	$L(10^{32}\text{cm}^{-2}\text{s}^{-1})$
4.1 GeV	3.6
3.77 GeV	3.0
3.1 GeV	2.0

FIGURE 1. A 3 pole test module for the super conducting wigglers needed in the upgrade to CLEO-c

Detector: CLEO III becomes CLEO-c

The CLEO III detector is a wonderful detector to study the charm system. The tracking system and the calorimeter cover 93% of the solid angle, while the ring-imaging Čerenkov counter (RICH) covers 83% of the solid angle.

The tracking system consists of the main drift chamber [2] for which we find a hit resolution of 88 μm. The 4 layer silicon detector [3] has prematurely degraded, which is observed as a dramatic efficiency loss in the sensors signals. The origin of this problem is yet unknown. Cornell is building a 6-layer high angle stereo drift chamber to replace the silicon detector. For low momentum tracks, such as those typically generated when running at J/ψ energies, the performance of this drift chamber is comparable to a silicon vertex detector, because multiple scattering is the dominatn contribution to the track resolution. We plan to run with a reduced B field of 1T (1.5 Tesla for Υ running). That leads to a resolution of 0.35% at 1 GeV for charged tracks.

The calorimeter [4] consists of 7800 Cesium Iodide crystals and measures the photons with a $\frac{\sigma_E}{E}$ = 2% at 1 GeV and 4% at 100 MeV.

Particle identification can be done with dE/dx with a resolution of 5.7% for minimum ionizing pions. A ring imaging Čerenkov counter [?] has been installed with the CLEOIII upgrade. It has been shown to perform excellently. It covers 83% of the solid angle. For 0.9 GeV particles the kaons are identified with 87% efficiency and a pion fake rate of 0.2%.

Data from the drift chamber and the calorimeter are used in the trigger. It is pipelined with a latency of 2.5 μs. The trigger is fully programmable and can easily be adapted to the new event signatures.

The data acquisition system can accept hardware triggers up to 1 kHz and is designed to write data to tape with a speed of about 300 Hz. The event size is 25 kB and the data throughput is 6Mb. This means that the data acquisition infrastructure will be able to handle the large data rates at the J/ψ peak.

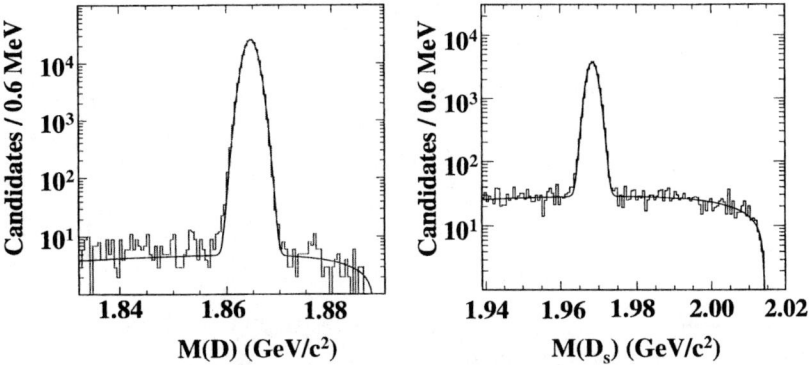

FIGURE 2. MC events equivalent to 1fb^{-1} of data. Left: D→Kπ tags. The width of the D peak is 1.3 MeV/c^2. Right: D_s →KKπ tags. The width of the D_s peak is 1.4 MeV/c^2.

RUNNING AT THRESHOLD

There are important advantages to running at the open flavor thresholds. Large cross sections mean the data can be acquired in one run period rather than over many years. The multiplicity of particles in the final state is smaller, which reduces backgrounds and increases efficiencies. We expect very clean data samples. Single tag events can be used to constrain ν reconstruction, double tags will be used for hadronic measurements. We anticipate 6 million D tags and 300,000 D_s tags. The signal-to-background ratio for the D→Kπ tag is estimated to be S/B: 5000 and for D_s →KKπ tag: S/B ∼ 100, (see Figure 2). Many of the analyses can profit from using kinematic constraints at threshold. This can, for example, lead to a reduced systematic error if one can forego lepton identification. Further advantages are that the initial states are pure (no fragmentation) and that the D's are coherently produced.

Absolute Hadronic Charmed Hadron Branching Fractions

Absolute branching fractions are important since, for a lot of analyses at the highest energies as well as in the *B*-system, an inaccurate knowledge of D, D_s,... decays can result in large systematic errors. Using double-tagged events at threshold leaves only major systematic error contributions from efficiency uncertainties in the tracks and showers. An overview of the expected results is shown in Table 3.

In the case of D decays the statistics are high enough that we can concentrate on the most simple D decays. For one fb^{-1} we expect 1500 D^0 →$K^+\pi^-$ events with no background and 8446 D^+ →$K^-\pi^+\pi^+$ with 25 background events. For D_s, the decays into $K^-K^+\pi^-$, $K^-K^+\pi^-\pi^0$, $\eta\pi^-$, $\eta\rho^-$, and $\eta'\pi^-$ were considered and combinations with less than 20% background were used to generate the numbers shown in Table 3.

The charmed baryon resonances measured with collisions at a large center of mass are typically presented as branching ratios with respect to the Λ_c^+ →$pK^-\pi^+$ decay. With

TABLE 3. Total number of double tag events and precision on absolute charm branching fractions, assuming 3fb^{-1} data at $D\bar{D}$, $D_s\bar{D}_s$ thresholds and 1fb^{-1} at $\Lambda_c\bar{\Lambda}_c$ threshold

Particle	# of double tags	Statistical Error	Systematic Error	Background Error	Total Error
D^0	53,000	0.4%	0.4%	0.06%	0.6%
D^+	60,000	0.4%	0.6%	0.10%	0.7%
D_s^+	6,000	1.3%	1.1%	0.90%	1.9%
Λ_c	17,000	~4%	small	small	~4%

one fb^{-1} at the threshold (4.6 GeV), one expects about 500 $\Lambda_c\bar{\Lambda}_c$ double-tagged events, a rough estimate since neither the cross section nor the $\Lambda_c^+ \to pK^-\pi^+$ branching fraction are well known.

Meson Decay Constant

The hadronic physics for leptonic decays of the D_q and B_q mesons is encapsulated in single non-perturbative QCD parameters f_q. Given the knowledge of their values, one can extract $|V_{cs}|$, $|V_{cd}|$, $|V_{ts}|$ and $|V_{td}|$. Our current knowledge of the uncertainty in f_{D_s} and f_D is 35% and 100%, respectively, while f_B and f_{B_s} will not be measured in the foreseeable future. Lattice QCD should, however, be able to calculate the ratio of $\frac{f_B}{f_D}$ very accurately, so that the determination of f_D will indirectly help in the extraction of $|V_{ts}|$ and $|V_{td}|$.

The leptonic D and D_s decay branching fractions are sizable and enable a direct determination of the charm meson decay constant from the measurement of $D^+ \to \mu^+\nu$, $D_s^+ \to \mu^+\nu$, and $D_s^+ \to \tau^+\nu$. If one assumes unitarity of the CKM matrix to constrain the values of $|V_{cd}|$ and $|V_{cs}|$ one can measure $\frac{\delta f_{D_s}}{f_{D_s}} = 2.1\%$ and $\frac{\delta f_D}{f_D} = 2.6\%$.

Semileptonic Form Factors and Determination of $|V_{cs}|$ and $|V_{cd}|$

The semileptonic from factors $|f_+(q^2)|^2$ encapsulate the hadronic physics of semileptonic decays

$$\frac{d\Gamma}{dq^3} = \frac{G_F^2}{24\pi^3}|V_{cs}|^2 p_K^3 |f_+(q^2)|^2 \quad (1)$$

in the example of a $c \to s$ transition. If we again assume 3 generation unitarity one can extract the semileptonic form factors from processes like $D^0 \to \pi e\nu$ or $D^+ \to K^{*0}e^+\nu$. The estimated precision for the parameters of the pseudoscalar to pseudoscalar and pseudoscalar to vector form factors are shown in Table 4.

TABLE 4. This table contains uncertainties on the branching fractions for several D and D_s decay modes and precision of semileptonic form factor parameters.

Decay Mode	PDG2000 ($\delta B/B$ %)	CLEO-c ($\delta B/B$ %)	Form Factor Type	Form Factor Parameter	expected Precision	CKM Element
$D^0 \to K^- e^+ \nu$	5	0.4				$|V_{cs}|$
$D_s^+ \to \phi^- e^+ \nu$	25	3.1				$|V_{cs}|$
$D^+ \to \pi^0 e^+ \nu$	48	2.0				$|V_{cd}|$
$D^0 \to \pi^- e^+ \nu$	16	1.0	PS $\to$ PS	$\delta f_+(0)/f_+(0)$ slope	$\sim 1\%$ $\sim 4\%$	$|V_{cd}|$
$D^+ \to \overline{K}^{*0} e^+ \nu$	9	0.6	PS $\to$ V	$\delta A_1(0)/A_1(0)$ $\delta A_2(0)/A_2(0)$ $\delta V(0)/V(0)$	$\sim 2\%$ $\sim 5\%$ $\sim 5\%$	$|V_{cs}|$

$|V_{cd}|$ is related to the decay rate Γ, which we can calculate from the absolute branching ratio $B(D^0 \to K^- e^+ \nu)$ and the mean lifetime τ_{D^0} as:

$$\Gamma(D^0 \to K^- e^+ \nu) = \frac{B(D^0 \to K^- e^+ \nu)}{\tau_{D^0}} = T_d |V_{cs}|^2 \quad (2)$$

T_d is taken from theory and requires the knowledge of the PS to PS semileptonic form factor. We can expect lattice QCD calculations with an error of $\delta T_s/T_s = 3\%$ within a few years as discussed further below. This will result in a precision for $|V_{cd}|$ of 1.7%. In a similar way, $|V_{cs}|$ can be determined with a precision of 1.6%. Both CKM matrix elements can also be extracted from leptonic decays. Combining leptonic and semileptonic measurements we expect final precisions of 1.4% and 1.1% on $|V_{cd}|$ and $|V_{cs}|$, respectively.

MAPPING OUT THE Υ AND J/ψ SYSTEMS

Starting late fall 2001 CLEO will collect about 1 fb^{-1} on each of the $\Upsilon(1S)$ - $\Upsilon(3S)$ resonances. Most present theories [5] indicate that the ground state of the Υ system, the η_b, could be discovered via the hindered M1 transition from the $\Upsilon(3S)$ state. One also would expect to see the decay $\Upsilon(3S) \to \pi^+ \pi^- h_b$ [6]. If found, its large predicted decay branching fraction into η_b of 50% opens a further avenue to observe the η_b.

CLEO also should be able to observe the $1^3 D_J$ states. The $b\bar{b}$ system is unique in that it has states with L=2 that lie below the open-flavor threshold. We expect unquenched lattice calculations soon for the center of gravity of the triplets of both D states to about ~ 5 MeV. Current theoretical calculations and our existing data suggest that we can expect 20 - 40 fully reconstructed events in the decay $\Upsilon(3S) \to \gamma_1 \chi_b' \to \gamma_1 \gamma_2 (^3 D_J)$ $\to \gamma_1 \gamma_2 \gamma_3 \chi_b \to \gamma_1 \gamma_2 \gamma_3 \gamma_4 \Upsilon(1S) \to \gamma_1 \gamma_2 \gamma_3 \gamma_4 l^+ l^-$, which should enable the extraction of the center of gravity of the triplet of the $1^3 D_J$ state to about 3 MeV. From this, the mass of the lowest state can be predicted and a scan can directly establish its mass, gaining a measure of the S-D mixing in the $b\bar{b}$ system.

In the J/ψ system only very few states are measured with high precision. Although we know the ground state η_c, its width is measured very poorly. The η_c' and h_c states need confirmation. The $^{3,1}D_J$ and $2^{3,1}P_J$ states still have to be found. The region above the $D\bar{D}$ threshold is generally explored very little, despite the fact that a lot of interesting physics might be extracted if one could identify, for example, charmed hybrid states. CLEO-c will have to make a scan in order to find the energy with the largest $D_s\bar{D}_s$ decay rate. A more detailed scan between 3.6 GeV and 5 GeV is also considered.

THE BRIGHT FUTURE OF LATTICE QCD

Lattice QCD is a full implementation of QCD and can therefore in principle produce accurate results, also for the low energy phenomena that cannot be treated perturbatively. It has, however, failed to make good predictions with well understood errors because of technical difficulties [7]. In the last few years, some real breakthroughs have been achieved [8], so that we can expect some of the theoretically easier calculations to appear in the next few years with sound error estimates of 1-2%.

In general, one was able to incorporate known features of QCD into the lattice calculations. Perturbation theory is used to describe short distance physics and connect the lattice to the continuum. Second order results lead to relative errors of $O(\alpha_s^3)$. One would like to keep lattice spacing as large as possible in order to minimize computer time. Improved discretizations remove errors by adding correction terms. These and other improvements are required in order that one can finally get unquenched results, which include effects from $q\bar{q}$ loops. When this can be done with realistic d,u-quark masses, the errors on the results become quantifiable and errors of order 2% are achievable for many calculations. Examples are masses, decay constants, semileptonic form factors, and mixing amplitudes for D, D_s, D^*, D_s^*, B, B_s, B^*, B_s^*, and corresponding baryons; masses, leptonic widths, electromagnetic form factors, and mixing amplitudes for any meson in ψ and Υ families below the open flavor threshold.

It should be stressed that lattice QCD can make all these predictions using only the quark masses and α_s as input parameters. Today there are not enough measurements in the one percent region for the quantities mentioned above. CLEO-c, however, will be able to provide most of them accurately enough, enabling us to validate the new lattice results and methods. Particularly the measurements in the ψ and Υ region provide an excellent test ground. This validation is needed so that lattice QCD results can be trusted and used to increase the accuracy of theoretical predictions for many different aspects of physics.

Impact of CLEO-c Results and Lattice QCD on Flavor Physics

An immediate application of lattice QCD results by CLEO-c involves the extraction of $|V_{cs}|$ and $|V_{cd}|$ by the use of the meson decay form factor and the measurement of leptonic decay branching fractions of D and D_s, as well as the semileptonic form factors and the measurement of the semileptonic decay branching fractions of D and D_s, described above.

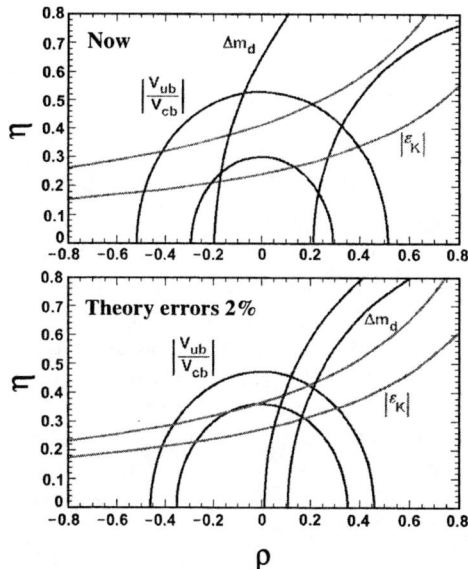

FIGURE 3. Above: Current situation taken from [9], © 2001 Springer-Verlag. Below: Prediction on the limits using 2% errors on form factors and today's data.

Even before a full validation of lattice QCD, measurements of the D-meson form factor will result in better predictions of the B-meson form factor, which will tighten the constraints on the unitarity triangle from $B\bar{B}$ and $B_s\bar{B}_s$ mixing measurements. Using SU(3) and heavy quark symmetry the semileptonic form factor for the process $B \to \pi e^+ \nu$ can be predicted from the measurement of the semileptonic form factor extracted from the process $D^+ \to K^{*0} e^+ \nu$, which will result in an improved $|V_{ub}|$ determination.

Finally, the impact of 2% theoretical errors on form factors produced by a validated lattice QCD theory are shown in Figure 3 using today's measurements. It is obvious that only an increased bound from $|V_{ub}|$ and $B\bar{B}$ mixing results together with the $\sin(2\beta)$ measurements will be able to give a significant answer on the question if the unitarity triangle is indeed closed or if the effect of new physics is observed. The limits from the ε_K measurement will also be improved considerably.

MESON SPECTROSCOPY FROM J/ψ DECAYS

Lattice QCD will have answers for some questions in the near future, however there are many quantities that will remain difficult to calculate for some time. There are only some ideas on how to treat effects like mixing, or the inclusion of gluonic degrees of freedom. The important information that is needed to gain better understanding is the extraction of the relevant degrees of freedom of the strong interaction in the non-perturbative regime.

A much better understanding of the light meson spectrum will be very helpful to gain this information and to guide the lattice.

An important advance will be if one could identify the $s\bar{s}$ states in the light meson sector. There are only a handful of $s\bar{s}$ states that are unambiguously known. However, they are needed in order to study complete nonets, and to discriminate $q\bar{q}$ against non-$q\bar{q}$ resonances. CLEO-c will attempt to collect in the order of one billion J/ψ decays. This data set will contain some 60 million radiative J/ψ decays. This will be the first data set with large statistics well suited to do meson spectroscopy in the region of 1.6 to 2.6 GeV. It should enable us to identify most of the $s\bar{s}$ states, since the initial state for radiative J/ψ decays is well defined and the CLEO detector will measure all final states simultaneously with 93% coverage in solid angle. Another way of identifying $s\bar{s}$ states is via ψ and $\psi' \rightarrow$VF flavor tagging. Lets say the vector state V might be reconstructed as ω, ϕ, or ρ state. The particle F then will be dominantly a state of the same flavor as V due to the suppression of hair-pin diagrams.

It has been said many times that identifying the $q\bar{q}$ states is required first in order to find left over states. Further analysis is needed to determine if they are multiquark states, meson-antimeson molecules, hybrids (qqg) or glueballs. In practice however, these states can mix with the $q\bar{q}$ state. Current data seem to indicate that we have such a case in the scalar sector with the $f_0(1400)$, $f_0(1500)$ and $f_0(1710)$, which are thought to be the mixtures of two $q\bar{q}$ states and the glueball, although there are a fair number of other ideas on how to identify the scalars. Mixing to such a large degree does not have to be the typical case, and the idea of identifying the $q\bar{q}$ nonets should usually be possible.

The scalar sector requires special attention. The CLEO-c data set will be the only data set on the horizon that measures all three resonances simultaneously with high statistics and a well defined initial state. Furthermore, the scalars should be observed in $\Upsilon(1S)$ decays, with lower statistics but well defined initial states. However the data set that will best reveal the nature of the scalar resonances is the 2-photon data collected in the 25fb^{-1} of CLEOII and CLEOIII data, since the coupling to glue is suppressed. Therefore the two photon partial width of the scalar states should give us a handle on their glue content.

One also expects hybrid states containing 2 quarks and a gluon as constituents. These are particularly interesting since there are hybrid states with quantum numbers not realized by $q\bar{q}$ states. The identification of a spectrum of hybrid states is very challenging, but has the reward to make unique measurements to test lattice QCD predictions. CLEO-c has limited access to hybrid production from χ_{c1} decays, enabling us to measure C-odd states as the 1^{-+} exotic hybrid. Since the radiative decay fraction of the $\psi' \rightarrow \chi_{c1}$ is about 9%, a sizeable number of χ_{c1} decays should be recorded.

A MULTITUDE OF OTHER EXCITING MEASUREMENTS

There are a large number of other measurements possible, which are part of the very diverse CLEO-c program. There are just some highlights mentioned here. In τ-physics we expect an improvement in the accuracy of our knowledge of the τ-mass by a factor of three and of the Michel Parameter η by a factor of four. We expect to detect many rare decays of the D^0 and D^+ meson or set limits on the order of a few times 10^{-6}. CLEO has the unique ability to perform an R-scan between 3-7 GeV, which would require about 150 pb^{-1} of data and could be acquired within one week. This is of importance because it is expected that the 3-7 GeV region will otherwise become the region contributing dominantly to uncertainties in the determination of $\alpha(M_Z)$ and the hadronic contribution to $(g-2)_\mu$.

CONCLUSIONS

The CLEO-c program has many exciting aspects. It is made unique, however, by the fact that it provides the data needed by an emerging lattice community to validate their new results to a level of a few percent. This combined effort should lead to a situation that many non-perturbative QCD calculations with well determined errors will be available in flavor physics and wherever they are needed to explore beyond the Standard Model.

REFERENCES

1. CLEO collaboration, *CLNS 01/1742 at www.lns.cornell.edu/public/CLNS/2001/CLEO.html*, CLEO-c and CESR-c: A New Frontier of Weak and Strong Interactions.
2. D. Peterson et al., Proceedings of th 8th Vienna Wire Chamber Conference, 19-23 February, Vienna, Austria, to be published in *Nucl. Instrum. Methods A*.
3. E. von Toerne et al., *hep-ex/0103037*, submitted to *Nucl. Instrum. Methods A*.
4. T. Hill, *Nucl. Instrum. Methods A* **418**, 32 (1998).
5. S. Godfrey and J. L. Rosner, EFI-01-10, April 2001, *hep-ph/0104253*, submitted to *Physical Review D* (Brief Reports).
6. Y.-P. Kuang and T.-M. Yan, *Phys. Rev. D* **24**, 2874 (1981); M. B. Voloshin, *Yad. Fiz.* **43**, 1571 (1986).
7. G. P. Lepage and P. B. Mackenzie, *Phys. Rev. D* **48**, 2250 (1993).
8. See for example: M. Alford et al., *Phys. Lett. B* **361**, 87 (1995); M. Luscher et al, *Nucl. Phys. B* **478**, 365 (1996).
9. A. Hocker et al., "A New Approach to a Global Fit of the CKM Matrix" *Eur. Phys. Jour. C* **21**, 225-259 (2001).

SUMMARY

IHEP, Protvino

Hadron 2001 Conference Summary: Theory

T. Barnes

Physics Division, Oak Ridge National Laboratory, Oak Ridge, TN 37831-6373, USA
Department of Physics and Astronomy, University of Tennessee Knoxville, TN 37996-1501, USA

Abstract. This contribution reviews some of the theoretical issues and predictions that were discussed at HADRON2001. The topics are divided into principle areas, 1) exotics, 2) vectors, 3) scalars, and 4) higher-mass states. The current status of theoretical predictions for each area are summarized, together with a brief description of experiment. New and detailed experimental results are presented in the companion Experimental Summary by Klempt.

INTRODUCTION AND OVERVIEW

Hadron physics is concerned with the questions of what hadrons exist in nature and how these hadrons interact and decay. In each of these areas there are important issues that are poorly understood. Our nominal classification of hadrons as quarkonia, glueballs and hybrids (and perhaps multiquarks) is of course an oversimplification, and it is not yet clear what resemblence the real hadron spectrum has to our expectations for gluonics and other exotica. The most widely used model of open-flavor hadron strong decays, the $^3P_0$ model, is a naive pair-production prescription with no clear connection to QCD. Finally, the nature of the strong force between hadrons in general, which is clearly a very important issue in strong interaction physics, remains controversial.

The year 2001 is a transitional period for hadron physics, as was reflected in the material presented. Two high-statistics experiments using hadron beams, E852 at BNL and the Crystal Barrel at LEAR, ended several years ago. Results from several new final states studied at these experiments were presented here, and some of the results were very interesting indeed; nontheless it is clear that we are near the end of new results from these experiments. Hadron spectroscopy using hadron beams will continue here in Protvino, but will not again be a major world enterprise until new facilities such as GSI and perhaps KEK join this effort.

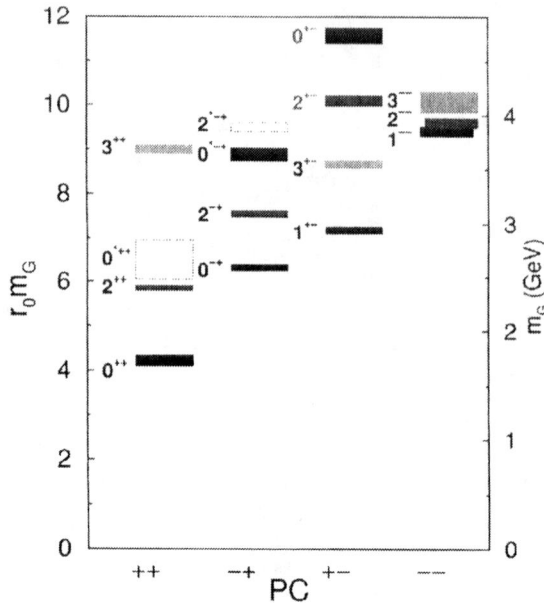

FIGURE 1. The quenched LGT glueball spectrum of Morningstar and Peardon [1, 2].

In the near future we can expect to see exciting new results from electron beam and e^+e^- facilities. For light hadrons this will most noticably involve Novosibirsk (with an energy upgrade to an invariant mass of around 2 GeV) and Frascati (now studying the ϕ but with capabilities for operating at higher mass). These facilities will be complicated by studies of $c\bar{c}$ and charm spectroscopy at BES (very nice results for states above $D\bar{D}$ threshold were shown here), and in the near future, CLEO-c. These facilities can also study the very interesting questions in light meson spectroscopy that can be addressed using two-photon collisions and initial-state radiation.

Hadron spectroscopy of late has also received contributions from machines such as LEP and KEK, which were designed for electroweak physics but can make very interesting contributions to light meson spectroscopy, in this case through two-photon collisions. Experiments that are nominally studies of weak interaction physics, such as charm meson decays, have also rediscovered strong interaction physics in the form of important FSIs. The implications of these FSIs for light scalar mesons led to some interesting interactions between representatives of the "old" and "new" cultures in hadron physics in the course of this meeting.

In theory, we also have seen a mix of "old" and "new" approaches in this meeting. The traditional quark models of hadrons [3, 4] remain the most relevant to experimentalists over the largest part of the $q\bar{q}$ and qqq spectrum, since the results are known to be reasonably accurate numerically, and the radial and orbital excitations of greatest current interest are readily accessible to these methods. In parallel, the "first principles" LGT approach [1, 5, 6] has made great progess in its applications to the spectrum of pure glue and mixed quark-gluon states. In the glueball sector the LGT results [2] (Fig.1) are widely regarded as near definitive (within the quenched approximation), which is why we no longer hear suggestions that the "σ" or $\eta(1440)$ might be glueballs; LGT has eliminated these possibilities in favor of a much higher glueball mass scale. Similarly, the approximate agreement between the predicted LGT scalar glueball mass and the $f_0(1500)$ has been considered to be a very strong argument in favor of a glueball (or mixed glueball-$q\bar{q}$) assignment. Similarly the LGT estimate of the hybrid mass scale reported at this meeting, which is quite similar to the flux-tube model estimate, is considered to be a serious problem for the light exotic candidate $\pi_1(1400)$. Clearly LGT is now the leading theoretical approach for estimating the masses of gluonic states. Although predictions for the masses of the lower-lying excited mesons and baryons can similarly be extracted from LGT, and in some cases should be relatively straightforward since some are the lightest states in their sector, this important application has not yet received sufficient attention from LGT groups. The spectrum of excited $q\bar{q}$ states in LGT is obviously a very important topic, which should be considered by LGT collaborations with improved statistics in future.

The next important step in theoretical technique, both in LGT [1, 5, 6] and in quark models [7], may be the removal of the "quenched approximation" through the incorporation of creation and annihilation of intermediate $q\bar{q}$ pairs. This will lead to several perhaps very important effects, such as large mass shifts due to virtual decays. The reasons for the success of the naive LGT quenched approximation, and the closely related quark-model valence approximation, are important and long-standing questions that can be addressed in this work.

PRINCIPAL TOPICS

Exotica

"Exotica" generically refers to states that are not dominantly $q\bar{q}$ mesons or qqq baryons, to the extent that this can be quantified. In this Hilbert space classification our current expectation is that the possible types of exotica are hybrids, glueballs and multiquark systems, with the latter category including quasinuclear "molecules" and possibly multiquark hadrons. There will of course be configuration mixing between these ideal "conventional" and "exotica" basis states, except in the cases of outright exotic quantum numbers such as I=2 or $J^{PC} = 1^{-+}$. The amount of configuration mixing will be strongly channel-dependent, and in some cases may preclude a separation into exotica and conventional hadronic resonances. One now familiar example is the scalar glueball sector, in which the strong decays of the $f_0(1300)$, $f_0(1500)$ and $f_0(1710)$ are all far from expectations for pure $q\bar{q}$ or glue states, due perhaps to very large $|n\bar{n}\rangle \leftrightarrow |G\rangle \leftrightarrow |s\bar{s}\rangle$ mixing effects. Alternatively, in the cases of exotic flavor or J^{PC} we can be certain that identification of a resonance is an indication of a state beyond the naive quark model of $q\bar{q}$ mesons and qqq baryons. The identification of the spectrum of such states is the most important task for QCD spectroscopy at present.

Theorists derived the expected spectrum of hybrids (including J^{PC}-exotics) in various models beginning in the mid 1970s. It is now widely accepted that hybrid mesons span all J^{PC}, and the lightest hybrid exotic should be a 1^{-+}. In some models such as the flux tube model there are additional exotics present in the lowest multiplet, specifically 0^{+-} and 2^{+-}. These states are also expected in the bag model, but at rather higher mass.) The search for such exotic quantum numbers was given a strong incentive by the flux tube calculations of Isgur, Kokoski and Paton[8], who predicted very characteristic decay modes for hybrids, specifically S+P final states such as $f_1\pi$ and $b_1\pi$. Their mass estimate of ca. 1.9 GeV was somewhat higher than was predicted earlier, for example using the bag model. The restricted S+P decay modes compensated for the increased phase space at the higher flux-tube mass scale, so the flux-tube decay calculations found that some hybrids, notably a $\pi_1(1900)$, should be relatively narrow. Of the other relatively narrow states predicted by this model, the most remarkable are an "extra" ω that would favor K_1K modes and an "extra" π_2 that would decay strongly to $b_1\pi$. (The $b_1\pi$ mode is forbidden to the quark model $\pi_2(1670)$ because the $^1D_2 \to {}^1P_1 + {}^1S_0$ transition is spin singlet to spin singlet, which vanishes in the $^3P_0$ decay model.)

Relatively recent theoretical results on the hybrid mass scale in LGT were presented at this meeting. These results are more accurate at higher quark masses, due to the use of a nonrelativistic expansion of the QCD action; this leads to mass predictions that have much smaller statistical errors for states that incorporate heavy quarks. The masses predicted for the 1^{-+} $b\bar{b}$- and $c\bar{c}$-hybrids in the most recent calculations (reported here by Morningstar [1]) are $M_{b\bar{b}\ hybrid} \approx 10.9\text{-}11.0$ GeV and $M_{c\bar{c}\ hybrid} \approx 4.3$ GeV, which should be very useful as motivation for future studies of the higher-mass $c\bar{c}$ system at CLEO and BES. (Models typically anticipate approximately degenerate 1^{-+} and 1^{--} hybrids, so we expect to see an "extra" $c\bar{c}$ 1^{--} in e^+e^- annihilation at about this mass.) The especially interesting $n\bar{n}$-hybrid with 1^{-+} quantum numbers is predicted to lie at

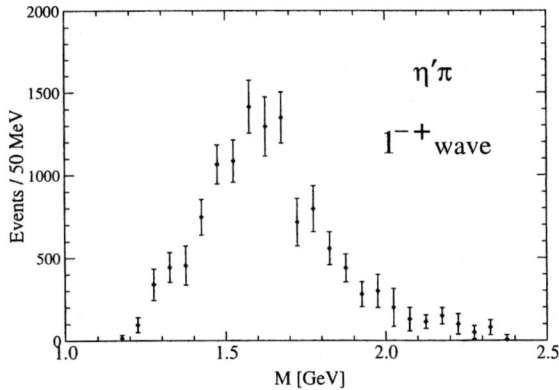

FIGURE 2. The E852 1^{-+} wave in $\eta'\pi^-$, showing a dominant $\pi_1(1600)$ exotic [11].

about 1.9-2.1 GeV [1], quite close to the flux tube model estimate. As a final interesting point, NRQCD is now finding results for the masses of nonexotic hybrids as well; a level ordering of $2^{-+} > 1^{--} > 1^{-+} > 0^{-+}$ found by Drummond *et al* [9] using NRQCD LGT was reported at this meeting [1]; this ordering was predicted by the bag model. In contrast the usual flux tube model results predict these states to be degenerate. This may be another area in which LGT can act as *de facto* theoretical QCD data that can be used to distinguish between different intuitive models, pending experimental results.

Regarding the *de jure* data on exotics, two candidate J^{PC} exotic meson resonances have been proposed, both with I=1, $J^{PC} = 1^{-+}$ quantum numbers; the $\pi_1(1400)$ and $\pi_1(1600)$. Obviously, establishing (or refuting) these candidate exotic resonances is of paramount importance for the future development of spectroscopy, since if confirmed they provide a benchmark for the mass of the lightest exotic resonance and the energy scale of exotic radial excitations. Unfortunately the $\pi_1(1400)$ signal (in $\eta\pi$) is rather weak, so it is difficult to distinguish this resonance interpretation from a nonresonant background phase. (This simple statement summarizes two decades of experiment.)

At this meeting we have heard from the VES collaboration [10] that they now have no clear preference for a $\pi_1(1400)$ resonance interpretation; they find fits of similar quality from a nonresonant signal. Since the favored theoretical methods anticipate a much higher mass of ca. 1.9-2.0 GeV for the lightest hybrid meson multiplet, which includes the lightest expected I=1 1^{-+} exotic, theorists would generally be happier if the $\pi_1(1400)$ were to be reinterpreted as a nonresonant signal, and the very clearly resonant $\pi_1(1600)$ were to replace it as the lightest exotic. Of course we must be cautious here because these predictions are for an unfamiliar system in the quenched approximation; the mass shifts due to couplings to virtual meson loops are currently unclear, and may be rather large. This will be a very important issue for future theoretical studies.

In contrast, the $\pi_1(1600)$, which is already claimed in $\eta'\pi$, $\rho\pi$ and $b_1\pi$ final states, may now be clearer. In their contribution [11], E852 showed results from the $\eta'\pi$ final state, in which the dominant low-energy resonance is the $\pi_1(1600)$ (see Fig.2). The usually dominant $a_2(1320)$ is much weaker in this channel due to small branching fraction of $B(a_2 \to \eta'\pi) \approx 0.5\%$; this leaves a remarkably robust 1^{-+} exotic wave, which if confirmed as resonant (see [10] for a cautionary note) will presumably be a benchmark for future studies of exotics. Note that the fitted width of $\Gamma_{tot}(\pi_1(1600)) = 340 \pm 40 \pm 50$ MeV is rather broader than the earlier estimates from the $\rho\pi$ final state.

Vectors

The conference began with a summary by Donnachie of the status of light vectors [12]. Although this might appear to be a rather specialized topic, in my opinion it merits a special section because much of the future work on light meson spectroscopy will concentrate on the vector sector. This is because the new and upgraded e^+e^- machines at Frascati and Novosibirsk produce vector mesons in e^+e^- annihilation, and future photoproduction facilities such as HallD at Jefferson Lab will also produce 1^{--} states (not uniquely, but vectors should also dominate diffractive photoproduction).

This limitation to 1^{--} states is an advantage in disguise; as usual in the 1-2 GeV mass region we have broad overlapping resonances, but since only 1^{--} is important in e^+e^- annihilation, we expect to produce only a few resonances per flavor channel. Thus it should be possible to establish clearly what states are present in the light meson spectrum, and whether there is indeed an overpopulation of states relative to the naive $q\bar{q}$ quark model.

Application of the quark potential model to the $n\bar{n}$ sectors leads to predictions of 2^3S_1 radial excitations near 1.5 GeV, L=2 $^3D_1$ $n\bar{n}$ states near 1.7 GeV, and a 3S radial excitation near 2.1 GeV. Experiment appears to support the existence of these 2S and D states (Fig.3), with ρ and ω flavor states roughly degenerate, and some evidence for K and ϕ analogues expected about 0.12 and 0.25 GeV higher in mass. Note however that $K^*(1410)$ appears surprisingly light if it is a partner to $\rho(1465)$ and $\omega(1420)$ 2S states. (A parenthetical note: Could this indicate the presence of the 1^{-+} exotic, with 1^{-+}-1^{--} mixing in the kaon sector analogous to the K_1 states?)

Of course only the ρ°, ω and ϕ are accessible to e^+e^-, and again we are fortunate in e^+e^- because the relative flavor cross section ratios for $\rho^\circ : \omega : \phi$ of 9 : 1 : 2 are known. (Some additional suppression of $s\bar{s}$ production is expected, due to the larger m_s.) With only two $q\bar{q}$ states anticipated by theorists per flavor sector between $\approx$ 1.5 GeV and $\approx$ 2.0 GeV, this problem sounds almost too simple!

There are two complications that have left the vector sector in a confused state despite decades of previous study, primarily using e^+e^- and photoproduction facilities. The first and most important problem is that the more accessible ρ° and ω states are quite broad, so we face the famous problem of overlapping resonances. Another difficulty is that we anticipate a 1^{--} hybrid meson multiplet somewhere in this mass region (degenerate with the π_1, in the flux tube model), so we may have not two but three states (2S, D, H) in this mass region. Actually this is again fortunate, since it affords us the opportunity to study conventional $q\bar{q}$ and hybrid states in a very restricted slice through Hilbert space, in a channel in which the important mixing effects can also be investigated. What we learn from excited vectors as an isolated case study may be crucial in helping us to understand the other nonexotic sectors of light meson spectroscopy.

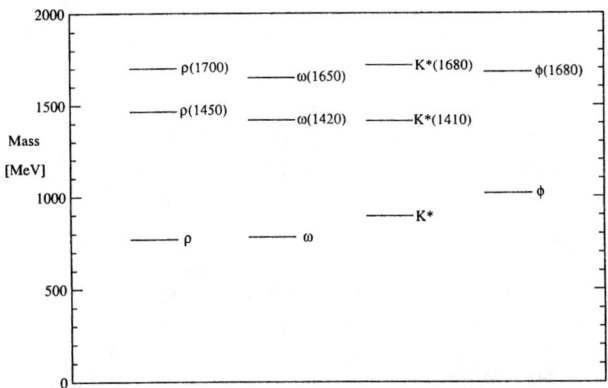

FIGURE 3. Experimental vector mesons below 2 GeV.

TABLE 1. Theoretical partial widths of 2S, 1D and hybrid ρ states.

	$\pi\pi$	$\omega\pi$	$\rho\eta$	$\rho\rho$	KK	K^*K	$h_1\pi$	$a_1\pi$	total
$\rho_{2S}(1465)$	74.	122.	25.	-	35.	19.	1.	3.	279.
$\rho_{1D}(1700)$	48.	35.	16.	14.	36.	26.	124.	134.	435.
$\rho_H(1500)$	0	5	1	0	0	0	0	140	≈ 150

In addition to the location of the individual levels, which may well be quite different from quark model expectations if $q\bar{q} \leftrightarrow$ hybrid mixing is important, the strong decay modes of the vectors will be especially interesting. This is because much of our theoretical "scaffolding" for hadrons and their strong decays relies on the so-called $^3P_0$ model, which assumes that strong decays take place through production of an additional $q\bar{q}$ pair with vacuum quantum numbers ($J^{PC} = 0^{++}$, hence $^3P_0$). Other variant decay models such as the flux tube model of decays are relatively minor variants of the original $^3P_0$ model, introducing for example a smooth spatial modulation of the pair production amplitudes. Although this model has been employed by theorists to reach a broad range of conclusions about hadrons (such as S+P decay modes for hybrids, and a list of "missing baryons" which are purportedly missing because they couple weakly to πN), it has been tested in disturbingly few decays. The most sensitive and well known tests are in decays of axial vectors to vector plus pseudoscalar. This channel allows both S- and D-waves in the VPs final state, so one can determine the relative magnitude *and* sign of S and D through the decay product angular distribution. The D/S ratio is quite sensitive to the quantum numbers of the $q\bar{q}$ pair produced in the decay, and the observed value of $\approx +0.28$ in $b_1 \to \omega\pi$ [11] strongly supports the $^3P_0$ model. The decay $a_1 \to \rho\pi$ is predicted to have a D/S ratio of -1/2 times the $b_1 \to \omega\pi$ ratio, which is also reasonably well satisfied. (Actually the D/S amplitude ratio is complex, since S- and D-wave VPs final states develop different FSI phases. This allows one to determine the phase shift difference $\delta_S - \delta_D$ in the $\omega\pi$ system at the b_1 mass, which has only recently been appreciated and exploited.) These two measurements, and some additional support from other axial vector decays in kaon and charm systems, are the only clear checks of this very widely used decay model.

When applied to these excited vector states, these strong decay models predict markedly different favored modes [13, 14, 15] that may be useful as signatures for the different states. In Table 1 we show results for the three ρ-type excited vectors; evidently these have comparable theoretical widths but very distinct branching fractions. The broad 4π states from $h_1\pi$ and $a_1\pi$ are predicted to arise from the 2D (comparable $h_1\pi$ and $a_1\pi$) and H ($a_1\pi$ only), whereas 2S should couple strongly to neither, instead populating $\pi\pi$ and $\omega\pi$. If this is accurate, it shows the importance of measuring as many final states as possible, especially since mixing of these basis states may be important.

As noted by Donnachie, the existing data has many gaps in energy and final state coverage, but it is clear that the 4π modes do not appear to agree with Table 1. There is evidence for $a_1\pi$ dominance of broad 4π states from the ratio of $\pi^+\pi^-\pi^\circ\pi^\circ/2\pi^+2\pi^-$, which would only be expected from a hybrid! (This assuming the flux-tube model of hybrid decays is accurate.) With accurate measurements of the final states in Table 1, we should be able to distinguish the ρ excitations present in this channel, and should learn about state mixing and strong decay amplitudes in the process.

Table 1 lists only ρ states. Donnachie noted that the flux tube decay model predicts a very narrow ω_H 1^{--} hybrid, coupled strongly only to $K_1 K$ decay modes [15]. If this state is near the $\pi_1(1600)$ mass these modes are closed, and the flux-tube suppressed mode of $\rho\pi$ is expected to lead to a total width of only ~ 20 MeV [12, 15]. This remarkable prediction strongly motivates a simultaneous study of ω-flavor 1^{--} states.

It may be that the $^3P_0$ model is inaccurate outside the 1^+ channel, in which case most of our predictions of hadron strong decays will be inaccurate. Evidence for a failure of the $^3P_0$ model in $\pi_2 \to \rho\omega$ was presented by E852 at this meeting, which I will mention in the section on higher-mass states.

Scalars

Introduction

I will first discuss the famous "980 states", in which there has been clear progress recently, and a close interplay between theory and experiment may have clarified much about the nature of these states. These results were clearly considered by many to be the most interesting presented at this meeting. Next I will briefly discuss the broad "σ" scalar and its purported strange partner, which were discussed at this meeting at some length but (as usual) no clear consensus as to the best description of the physics was evident. Finally I will suggest interesting future possibilities for clarifying the nature of the various scalars in the next round of experiments. Although the scalar sector includes the scalar glueball, and allows one to address the very important question of glueball-quarkonium mixing, little new experimental material was presented at this meeting, so I will not discuss glueballs as a separate topic.

"980" States

The two mesons near 980 MeV, once the S* and δ, now the $f_0(980)$ and $a_0(980)$, have long attracted attention as being anomalous in many of their properties. Although close to degenerate, so that we might expect them to be nonstrange $n\bar{n}$ I=0,1 partners, their very strong coupling to $K\bar{K}$ suggests that these are actually not conventional $n\bar{n}$ quark model states. Other problems are that their strong total widths are much smaller than expectations for $n\bar{n}$ at this mass, their masses are well below those of other P-wave $n\bar{n}$ states and are just below the $K\bar{K}$ threshold, and their electromagnetic couplings (specifically $\gamma\gamma$) are much weaker than we would expect for $n\bar{n}$. This list of problems can be expanded considerably.

Historically three models of these states have been considered by theorists. These suggest that the $f_0(980)$ and $a_0(980)$ might be four-quark clusters (primarily supported by Achasov et al.), weakly-bound kaon-antikaon quasinuclear states (Weinstein and Isgur), or simply $q\bar{q}$ quark model states, whose properties happen to differ from our naive expectations for ordinary mesons. Of course all accessible basis states will mix in physical hadrons, perhaps significantly, so we should more properly regard these models as suggestions regarding which component dominates in the expansion

$$|980\rangle = c_{q\bar{q}}|q\bar{q}\rangle + c_{q^2\bar{q}^2}|q^2\bar{q}^2\rangle + c_{K\bar{K}}|K\bar{K}\rangle + \ldots . \quad (1)$$

Of course the coefficients are actually spatial wavefunctions, so the distinction between $|q^2\bar{q}^2\rangle$ and $|K\bar{K}\rangle$ basis states is rather qualitative.

An important test proposed to distinguish between these descriptions (assuming dominance of one basis state) arises in $\phi(1020)$ radiative decays. In both the four-quark and $K\bar{K}$-molecule models it is assumed that the 980 states are produced in ϕ radiative transitions by photon emission from a virtual $K\bar{K}$ loop, with a direct photon coupling to the K^+K^- loop but not to $K^0\bar{K}^0$. The corresponding decay rate was evaluated by Achasov, Devyanin and Sheshtakov [16] and by Close, Isgur and Kumano [17]. Their result for the branching fractions is

$$B(\phi \to \gamma f_0(980)) = B(\phi \to \gamma a_0(980)) \approx (2.0 \pm 0.5) \cdot 10^{-4} \cdot F(R)^2, \quad (2)$$

where $F(R)$ is a form factor that depends on the spatial wavefunctions of the mesons; $F(R)$ would be unity for a pointlike $K^+K^-m(980)$ coupling. In contrast, a $q\bar{q}$ picture of the $f_0(980)$ and $a_0(980)$ would predict very small branching fractions of perhaps 10^{-6} if $f_0(980) = s\bar{s}$, and even smaller were they $n\bar{n}$.

Klempt will discuss the experimental results for these branching fractions from Novosibirsk and Frascati in his experimental summary. Here I will simple note that they are comparable in scale to the $\approx 1\text{-}3 \cdot 10^{-4}$ quoted above, but the weaker result that the branching fractions are equal,

$$\left. \frac{B(\phi \to \gamma f_0(980))}{B(\phi \to \gamma a_0(980))} \right|_{theory} = 1 \quad (3)$$

is not at all well satisfied! Instead the experimental ratio is

$$\left. \frac{B(\phi \to \gamma f_0(980))}{B(\phi \to \gamma a_0(980))} \right|_{expt.} \approx 4 . \quad (4)$$

If we reconsider the charged-kaon-loop radiative decay models to see what might have gone wrong, we find that the ratio of unity follows from the assumption that both 980 states are isospin eigenstates. It was instead argued long ago in both $q^2\bar{q}^2$ [16] and $K\bar{K}$ [18] models that one should anticipate important isospin violation in these systems. For $q^2\bar{q}^2$ this arises from mixing through nondegenerate K^+K^- and $K^\circ\bar{K}^\circ$ loops, and for $K\bar{K}$ from the fact that these are weakly bound $K\bar{K}$ systems, with zeroth-order K^+K^- and $K^\circ\bar{K}^\circ$ masses that differ by an amount comparable to the binding energy. There was already evidence for isospin mixing in these states, through $\pi\pi \to \pi\eta$ transitions evident in E852 data, and through evidence for central production of both the $f_0(980)$ and $a_0(980)$. The central production data suggests a mixing angle near 15°, which led Close and Kirk [19] to a modified prediction for the radiative transition ratio of

$$\frac{B(\phi \to \gamma f_0(980))}{B(\phi \to \gamma a_0(980))}\bigg|_{theory} = 3.2 \pm 0.8, \tag{5}$$

which is consistent with observation. The absolute scale of the rates suggests a hard form factor $F(R) \approx 1$, which supports the picture of a compact four-quark system. Close and Kirk interpret this as evidence for a combination of a $K\bar{K}$ system with a compact $q^2\bar{q}^2$ core.

In summary, we have clear and consistent evidence of a large isospin mixing angle in these states from three experimental processes, at a level not seen in other hadrons. This is a very interesting result indeed. A future calculation that is immediately suggested by this observation is to determine the mixing angles predicted by the two models of isospin violation, mixing through kaon loops versus mixing due to weak binding of nondegenerate K^+K^- and $K^\circ\bar{K}^\circ$ systems.

This evidence of a large isospin mixing angle between the nominally I=0 $f_0(980)$ and I=1 $a_0(980)$ immediately suggests several interesting measurements, which might check this result and independently determine the mixing angle. These include 1) the $\gamma\gamma$ widths, which were also predicted to be equal for both states because of photon coupling to the charged kaon loop alone, and which we therefore expect to be skewed in favor of the $f_0(980)$ by the same ratio as the radiative transition; 2) the relative annihilation decay rates of $J/\psi \to \phi(\pi\pi)$ and $J/\psi \to \phi(\pi\eta)$ (with isospin eigenstates we would expect to see no 980 signal in $\pi\eta$, since this is driven by an $s\bar{s}$ source; similarly for $D_s \to \pi(\pi\pi)$ and D_s to $\pi(\pi\eta)$). Finally, radiative transitions such as $a_0(980) \to \gamma\omega$ and $f_0(980) \to \gamma\omega$ can be used to quantify the $n\bar{n}$ components in the 980 states, since E1 radiative transition amplitudes of light quarkonia are reliably calculable in the quark model.

Broad Scalars ("Let Sleeping Dragons Lie.")

Discussions of the status of broad scalars have appropriately spanned decades. The contending "camps" in this area have long since settled on favorite explanations of the low-energy "σ" and "κ" effects, and these views are held with the tenacity of religious convictions. This situation makes for bad science, and we may need new, independent experimental information about the light scalar sector before we can make any progress in our understanding of broad scalar states.

At this meeting we have heard discussions of the relatively recent information on the light $\pi\pi$ and $K\pi$ systems that has come from charm decay experiments. In these experiments it was noted that there are clear low-energy enhancements in I=0 $\pi\pi$ and I=1/2 $K\pi$ subsystems, which can be fitted by *very* light scalar resonances. Specifically, masses of ≈ 480 MeV and ≈ 800 MeV were quoted for "σ" and "κ" states [20]. This is probably a premature conclusion, since only the low-energy tails of the purported resonance phase shifts are actually in evidence in the charm data; the crucial observation of a complete Breit-Wigner phase motion through 180° has not been made. It was noted here by Ochs [21] and by Pennington that the elastic $\pi\pi$ and $K\pi$ phase shifts themselves do not show evidence of "complete" low mass scalar resonances, so concluding that these exist based on the charm decay data in isolation, which only covers part of the range of invariant mass that has already been studied in light hadronic processes, is unjustified. The discussions at HADRON2001 following the charm decay presentations suggested that the charm decay analyses should include what is already known about these phase shifts over the full relevant mass range, for example through the parametrization of Au, Morgan and Pennington [22].

Experience suggests that progress may follow from a high-statistics study of a new production mechanism in the relevant mass region, as was provided by ϕ radiative decays for the 980 states. I would suggest that future high-statistics two-photon collisions, especially $\gamma\gamma \to \pi°\pi°$, may be definitive in resolving the resonances present in the light I=0 scalar channel. This reaction is quite simple (only S- and D-waves are produced significantly at low energies), and with high statistics it should be possible to determine the S-wave phase motion through interference with the $f_2(1270)$ D-wave. This reaction was studied earlier by the Crystal Ball collaboration [23], *albeit* with quite limited statistics; their results showed a broad scalar signal under the $f_2(1270)$, but the data was not adequate for a determination of the mass and width. If one could track the phase motion of the S-wave in this process (perhaps augmented by $\gamma\gamma \to \pi^+\pi^-$ and $\gamma\gamma \to \eta\eta$ data) it should be possible to identify the lighter f_0 scalar resonances. We should be aware that slowly-varying background phases are also present, which may significantly modify the fitted resonance parameters in this channel; as an example, the Jülich group note that t-channel ρ exchange in $\pi\pi$ scattering with a realistic $\rho\pi\pi$ coupling strength can explain most of the low-energy $\pi\pi$ phase shifts in both I=0 and I=2 channels [24]. Thus we may not learn where the light scalar resonances lie until we have understood nonresonant "background" phase shifts as well.

LGT predictions for scalar $q\bar{q}$ masses would also be of great interest. Although these would be "quenched" results, these bare numbers actually are used in some models of $\pi\pi$ scattering, and in any case there is so much uncertainty in this field at present that *any* more definitive theoretical result would be important. Just as the large LGT glueball mass scale in Fig.1 [2] has eliminated the "σ" and $\eta(1440)$ from serious contention as glueball candidates, so LGT results for the scalar $q\bar{q}$ spectrum could help to identify the more plausible scenarios in this most obscure and controversial sector of Hilbert space.

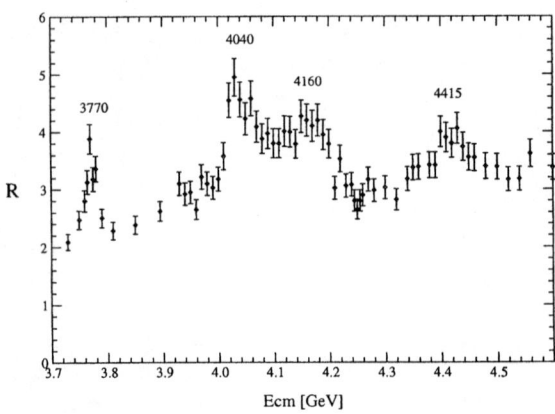

FIGURE 4. The BES measurement of R [25].

Higher-mass States

Heavy Quarkonium

We heard several interesting experimental contributions about heavy quarkonium at HADRON2001, specifically about the charmonium system. Although little new theoretical activity was reported in this field (the exception is heavy-quark hybrid masses from LGT), we will presumably see future theoretical interest in the charmonium system in response to high statistics studies at BES and CLEO-c. For this reason it seems appropriate to at least mention some of the charmonium results reported, and to suggest some possibly interesting questions for future experimental and theoretical investigation.

First, BES has reported results for the inclusive hadron cross section ratio R in the region above open charm threshold [25] (Fig.4). This is an important advance, as the rather noisy previous results from the late 1970s suggested the higher-mass resonances $\psi(3770), \psi(4040), \psi(4160)$ and $\psi(4415)$ but were far from definitive. Only these four $c\bar{c}$ resonances are regarded as established above open charm threshold, and their masses are consistent with potential model expectations for 1^3D_1, 3^3S_1, 2^3D_1 and 4^3S_1 levels (in order of increasing mass).

Despite this agreement of masses, there are serious problems with the properties reported for these states relative to potential model expectations. The $\psi(3770)$ and $\psi(4160)$ should both appear quite weakly in e^+e^- if they are D-wave $c\bar{c}$ states, since the wavefunction at contact vanishes in this case. Instead the e^+e^- width of the $\psi(3770)$ is much larger than expected, and the reported $\psi(4160)$ e^+e^- width is comparable to the nominally 3^3S_1 $\psi(4040)$. Of course this is based on the old, rather noisy, measurements. The $\psi(4160)$ signal in the new BES data appears weaker, and when fitted this new e^+e^- width may be rather smaller than previous estimates.

The exclusive strong branching fractions of these higher-mass $c\bar{c}$ states will also be very important measurements. The existing claims for strong branching fractions include an estimate that the $\psi(4040)$ favors the $D^*\bar{D}^*$ mode over $D\bar{D}$ by about a factor of

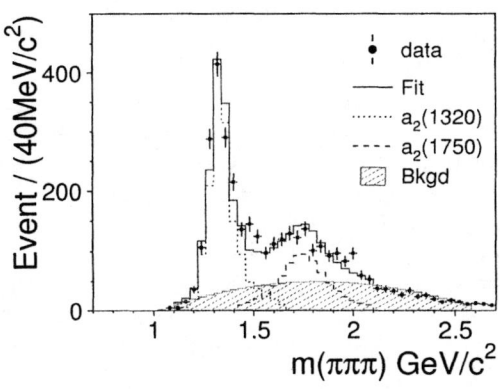

FIGURE 5. An example of $\gamma\gamma$ production of a higher-mass $q\bar{q}$ state, from Belle [28].

~ 500 [26, 27], despite the absence of $D^*\bar{D}^*$ phase space! (Recall $M(D^*) = 2.01$ GeV.) This remarkable result previously led to suggestions that the $\psi(4040)$ might be a $D^*\bar{D}^*$ molecule. The conventional $c\bar{c}$ description nonetheless appears plausible, in view of the agreement with the predicted mass of the 3^3S_1 $c\bar{c}$ level. The $\psi(4040)$ e^+e^- width, which is comparable to the e^+e^- widths of the $\psi(3686)$ and $\psi(4415)$ 2S and 4S radial excitations, also supports a $c\bar{c}$ $\psi(4040)$ assignment. The unusual strong branching fractions may be due to nodes in the strong decay amplitudes; the nodes in the 3S radial wavefunction may well have produced counterintuitive branching fractions for the $\psi(4040)$. Clearly, reasonably accurate measurement of the exclusive branching fractions of the higher $c\bar{c}$ states to all open charm final states will be an extremely interesting set of measurements, which can be used as detailed tests of strong decay models.

One limitation of $e^+e^- \to \gamma \to q\bar{q}$ annihilation is that it produces only 1^{--} states. One can extend these studies to the two-photon collision process $e^+e^- \to e^+e^-\gamma\gamma$, $\gamma\gamma \to q\bar{q}$, to search for states with even C-parity. These two-photon widths are intrinsically interesting to theorists, since they can be calculated in quark models, and may provide sensitive tests of the quark model states. Fig.4 shows a new measurement of a candidate $a_2(1750)$ radial excitation in $\gamma\gamma$, reported here by the BELLE Collaboration [28]. The relative two-photon partial widths of a given J^{PC} flavor multiplet vary with flavor as $f : a : f' = 25 : 9 : 2$, so two-photon couplings can be used to identify flavor partners of a given state, or quantify the level of flavor mixing. (There is some suppression of the heavier $s\bar{s}$-$\gamma\gamma$ coupling.) Two-photon couplings may also be useful in distinguishing different types of scalar states, since we naively expect glueballs and multiquark states to have rather smaller $\gamma\gamma$ couplings than $n\bar{n}$ states. In contrast, in the quark model a light scalar $f_0^{(n\bar{n})}(1300)$ is predicted to have a two-photon width of ≈ 5 KeV, larger than any other light $n\bar{n}$ meson.

Two-photon couplings of charmonia are very interesting in part because they allow us to test calculations of $q\bar{q} \to \gamma\gamma$ widths in a regime in which the nonrelativistic quark model should give reasonably accurate results. Typical theoretical predictions are $\approx$ 5-7 KeV for the $\eta_c(2980)$ and $\approx$ 0.5-2 KeV for the P-wave $c\bar{c}$ states χ_0 and χ_2. The ratio of χ_0/χ_2 partial widths varies over the range $\approx$ 3-10, depending on theoretical assumptions. These measurements of $\gamma\gamma$ charmonium widths have a long history of uncertainty, due to the intrinsically small $O(\alpha^4)$ cross sections. It is now clear that the experimental $\eta_c(2980)$ $\gamma\gamma$ width [29] is not far from theoretical expectations. The P-wave states have somewhat smaller $\gamma\gamma$ widths and less characteristic decays, and so have been more difficult to measure. One competing technique that proved quite successful was to use $p\bar{p}$ annihilation to make the $c\bar{c}$ state, followed by detection of $\gamma\gamma$ against a very large hadronic background. (This was done by E760 and E835 at Fermilab.) A new BELLE measurement of the $\gamma\gamma$ width of the tensor $\chi_2(3556)$ in e^+e^- collisions was reported here [28],

$$\Gamma_{\gamma\gamma}(\chi_2)\Big|_{\text{BELLE}} = 0.84(0.08)(0.07)(0.07) \text{ KeV} \tag{6}$$

which is about a factor of three larger than the Fermilab result

$$\Gamma_{\gamma\gamma}(\chi_2)\Big|_{\text{E835}} = 0.270(0.049)(0.033) \text{ KeV}, \tag{7}$$

presented here by Tomaradze [29]. This is about a 4σ difference, so the discrepancy does appear significant. I am amused to note that a previous $\chi_2(3556) \to \gamma\gamma$ calculation [30] found a value of $\Gamma_{\gamma\gamma}(\chi_2) \approx 0.56$ KeV, comfortably between the two experimental results.

A Striking $^3P_0$ Decay Model Failure

One especially interesting new result reported at this meeting concerned resonances observed in the $\rho\omega$ final state. This is very important theoretically because the VV system can have $S = 0, 1$ and 2, so there is considerable scope for testing strong decay models. (Recall that we have all been using the $^3P_0$ model or variants to predict light meson decays, D meson decays, hybrid decays, missing baryons and so forth for decades, but this model has seen little in the way of sensitive tests of the quantum numbers of the $q\bar{q}$ pair formed in the decay.) The historically convincing angular correlation tests were in 1^+ decays to VPs final states, specifically $b_1 \to \omega\pi$ and $a_1 \to \rho\pi$, in which both S- and D-wave VPs final states are produced. The model does predict these two D/S ratios approximately correctly, but it has seen few sensitive tests in other J^P sectors. When applied to decays into VV final states, the model typically predicts a nontrivial pattern of large, small or identically zero decay amplitudes, which can be compared to these new results on $\rho\omega$.

The $\pi_2(1670)$ is an interesting initial state for these decay model tests; it is a spin singlet ($^1D_2$ in the quark model), so many decay amplitudes are predicted to be zero due vanishing spin matrix elements. For example, the decay $\pi_2 \to b_1\pi$ is strictly forbidden in the $^3P_0$ model, since this would be an S=0 to S=0 transition (the mesons all have S=0); the $^3P_0$ transition operator has S=1 ($\vec{\sigma} \cdot \vec{p}$), so there is no S=0 to S=0 matrix element. The fact that this branching fraction is indeed quite small is one of the few recent decay model tests.

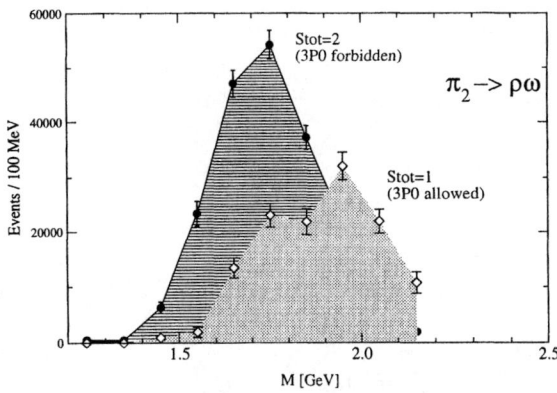

FIGURE 6. Final S_{tot} amplitudes in $\pi_2 \to \rho\omega$, showing violation of $^3P_0$ model expectations [11].

On considering the decay $\pi_2 \to \rho\omega$, one immediately finds a dramatic failure, assuming that the newly reported experimental decay amplitudes are correct. The 2^{-+} $\rho\omega$ system can in general have the quantum numbers $^3P_2$, $^3F_2$, $^5P_2$, and $^5F_2$, but the $^5P_2$ and $^5F_2$ $\rho\omega$ final states are forbidden to $\pi_2 \to \rho\omega$ in the $^3P_0$ model, since we have an S=0 initial state and an S=1 transition operator. We should only find $^3P_2$ and $^3F_2$ $\rho\omega$ final states. Of these we expect the $^3P_2$ $\rho\omega$ wave to dominate $\pi_2 \to \rho\omega$, since there is little phase space.

Experimentally only the S=2 $\rho\omega$ final state is observed to peak in the $\pi_2(1670)$ region, which implies that this decay is dominated by a *spin tensor* transition. This final state might be generated by $q\bar{q}$ pair production from a transverse gluon, but it is certainly not anticipated by the usual $^3P_0$ strong decay model. Subsequent angular analysis of the VV system may provide other interesting results regarding the mechanism of these still poorly understood strong decays.

SUMMARY AND APOLOGIA

In this report I have briefly summarized several interesting topics that were discussed in presentations at HADRON2001. These included evidence of and expectations for exotic mesons, the status of light vector mesons, the very interesting new results on the 980 states, new results for R in the open-charm region, and evidence for a failure of the $^3P_0$ model. Although this is nominally a theory summary, hadron physics is largely driven by experiment, so I have actually cited some new experimental results that seemed of special interest to theorists.

I have been rather selective in this report, due primarily to a lack of time available for completion of this summary. For this reason many of the results presented at HADRON2001, notably relating to heavy quark and quarkonium physics and baryon physics, were not discussed here. The "future facilities" discussions have clearly shown that this concentration on light u, d, s hadrons will change in future meetings, at which time we can expect to see exciting new results on charmonium states, both regarding the states themselves and their decay products. The traditional concentration of the HADRON conference series on meson physics was also discussed at this meeting, and it was suggested that in future there should be a serious effort to include developments in baryons as a major part of the meeting. With new results from facilities such as Jefferson Lab, this will certainly be appropriate, and will make the job of the conference summary speakers even more difficult.

ACKNOWLEDGMENTS

It is a great pleasure to thank Prof.Zaitsev and the organisers of HADRON2001 for their kind invitation to review some of the theoretical aspects of the physics discussed at this meeting. This work was supported in part by the DOE Division of Nuclear Physics, at ORNL, managed by UT-Battelle, LLC, for the US Department of Energy under Contract No. DE-AC05-00OR22725, and by the US National Science Foundation under Grant No. INT-0004089.

REFERENCES

1. C.Morningstar, these proceedings.
2. C.J.Morningstar, and M.J.Peardon, Phys. Rev., D60, 034509 (1999).
3. E.S.Swanson, these proceedings.
4. Yu.S.Kalashnikova, these proceedings.
5. T.Kaneko, these proceedings.
6. P.Mackenzie, these proceedings.
7. E.vanBeveren, these proceedings.
8. N.Isgur, R.Kokoski, and J.Paton, Phys. Rev. Lett. 54, 869 (1985).
9. I.T.Drummond et al., Phys. Lett. B478, 151 (2000).
10. V.Dorofeev, these proceedings.
11. A.Popov, these proceedings.
12. A.Donnachie, these proceedings.
13. A.Donnachie and Yu.S.Kalashnikova, Phys. Rev. D60, 114011 (1999).
14. T.Barnes, F.E.Close, P.R.Page and E.S.Swanson, Phys. Rev. D55, 4157 (1997).
15. F.E.Close and P.R.Page, Nucl. Phys. B443, 233 (1995).
16. N.N.Achasov, S.A.Devyanin, and G.N.Shestakov, Phys. Lett. B88, 367 (1979).
17. F.E.Close, N.Isgur, and S.Kumano, Nucl. Phys. B389, 513 (1993).
18. T.Barnes, Phys. Lett. B165, 434 (1985).
19. F.E.Close, and A.Kirk, Phys. Lett. B515, 13 (2001).
20. C.Göbel, these proceedings.
21. W.Ochs, these proceedings.
22. K.L.Au, D.Morgan, and M.R.Pennington, Phys. Rev. D35, 1633 (1987).
23. H.Marsiske et al., Phys. Rev. D41, 3324 (1990); H.Bienlein, Proc. of the IXth Internatl. Workshop on Photon-photon Collisions, La Jolla, CA, 22-26 March 1992, pp.241-257, eds. D.O.Caldwell and H.P.Paar (World Scientific, 1992).
24. D.Löhse et al, Nucl. Phys. A516, 513 (1990); G.Janssen et al, Phys. Rev. D52, 2690 (1995).
25. W.Li, these proceedings.
26. K.Seth, these proceedings.
27. S.F.Tuan, these proceedings.
28. S.Hou, these proceedings.
29. A.Tomaradze, these proceedings.
30. T.Barnes, Proc. of the IXth Internatl. Workshop on Photon-photon Collisions, La Jolla, CA, 22-26 March 1992, pp.263-282, eds. D.O.Caldwell and H.P.Paar (World Scientific, 1992).

HADRON 2001 Summary: Experiment

Eberhard Klempt

Institut für Strahlen- und Kernphysik, University of Bonn, Germany

Abstract. New data, developments and ideas presented at Hadron2001 are summarized. Emphasis is laid on searches for gluonic excitations of mesons, on hybrids and glueballs. Tests of the Standard Model are shortly reviewed.

INTRODUCTION

A wealth of new and exciting data have been presented to the 9th International Conference on Hadron Spectroscopy. The new results were distributed over more than 140 contributions and it is of course impossible even only to mention them all in this summary. The main focus of this review will be on results on light-quark spectroscopy with emphasis on the search for gluonic excitations. The excuse is that this field is the central issue of this conference series and a large amount of new results were presented. Ted Barnes in his summary [1] has reviewed why it is important to search for mesons beyond the quark model, for hybrids and for glueballs.

Hybrids, mesons in which the gluon flux-tube carrying the forces between quark and antiquark is excited [2], may best be identified in searching for mesons with exotic quantum numbers, with quantum numbers which are not accessible to normal $q\bar{q}$ states. In particular, the $J^{PC} = 1^{-+}$ wave has been studied intensively in a large number of final states. Since 1999, the last hadron conference, greatly improved statistics have been accumulated in Brookhaven and in Protvino. The results of these analyses were presented during the conference and will be reviewed here. Hybrids are not limited to have exotic quantum numbers; they should also show up as additional states not finding their place in one of the meson nonets. This approach requires a careful discussion of light-meson spectroscopy, and of the decay pattern expected for $\bar{q}q$ states and for hybrids.

Glueballs are supposed to carry no constituent quarks; they manifest the new degrees brought into spectroscopy by color. In particular, the lowest-mass glueball, according to lattice gauge calculations a scalar state at about 1700 MeV, has been the object of an intense discussion for about two decades. Morningstar warned us that we should take the results on glueball masses from lattice calculations with proper scientific scepticism [3].

Before entering the field of meson spectroscopy, I would like at least to mention some of the highlights related to the Standard Model even though these results are slightly outside the main road of our field.

THE STANDARD MODEL

To the outstanding results presented at this conference belong of course the different measurements of CP-violation parameters. After a long controversy, the results on ε'/ε obtained at CERN and at FNAL now agree. The results now establish CP-violation in the weak decay amplitude at 10 standard deviations. The new world average is now [4, 5]

$$Re(\varepsilon'/\varepsilon) = (17.2 \pm 1.8) \cdot 10^{-4}.$$

CP-violation is not only accessible for strange mesons but also in the heavy-quark sector. New results were presented from BABAR and BELLE and they both have established that CP-violation can be measured in B-factories. When the results from various sources are combined, CP-violation in the $B^0\bar{B}^0$ system is now established at 3 standard deviations [6]. Of course this is only a start and high precision data are expected in the years to come.

We all followed the difficult decision the CERN management had to take when first candidates for decays of the Higgs particle were reported. Events were found in which four b-quarks were identified indicating that the sequence

$$e^+e^- \to \text{Higgs} + Z^0 \quad \text{with} \quad \text{Higgs} \to \bar{b}b, \, Z^0 \to \bar{q}q$$

was observed [7]. The evidence was reported also here at this conference; it suggests that the Higgs might be just around the corner at a mass of slightly above 115 GeV/c^2. Sadly enough, LEP was closed and we have to wait for Fermilab to find out if the Higgs is really so close.

The standard model continues to resist all attends to find physics beyond its limits. Precise measurement of the weak boson masses [8] and of the Cabibbo-Kobayashi-Maskawa matrix elements were discussed [9]. These measurements over-constrain the unitary triangle but show no evidence for any deviation. Rare Kaon decays also challenge the Standard Model at very large energies [10, 11, 12].

The decays of charmed [13, 14, 15] and beauty [16] hadrons and of J/ψ and ψ' decays [17, 18] provide valuable information on the life times of heavy quarks and on QCD oriented questions like: is the fragmentation of quarks independent of the production mechanism [19]? Is the fraction of b-quarks in the fragmentation process understood [20, 21]? What is the reason for the strange pattern of ψ_{2S} decays to vector and pseudoscalar mesons and can this anomaly be extended to other decay processes [22]?

The new measurement of the anomalous magnetic moment (g-2) of the muon is supposed to challenge the Standard Model. In the new BNL-experiment, 400.000 muons were stored in a storage ring and observed to decay. The decay time distribution shows the known oscillatory behavior; 150 revolutions of the muon spin relative to its moment vector were observed leading to a very high precision in a_μ. The experimental result,

$$a_\mu(99) = (11659202 \pm 14 \pm 6) \cdot 10^{-10}$$

differs from the Standard Model prediction

$$a_\mu(SM) = (11659159.6 \pm 6.7) \cdot 10^{-10}$$

at the 2.5σ level [23]. Does this indicate physics beyond the Standard Model?

A large contribution to the anomaly stems from hadronic loops in radiative corrections. The loops can be calculated by integration of the e^+e^- cross section for annihilation into hadrons (via formation of vector mesons). Very precise new data from Novosibirsk on vector mesons were reported at this conference on e^+e^- annihilation into pionic and kaonic final states, with an impressive reduction of the statistical and systematic errors [24]. The sum of all channels measured contributes to the anomalous magnetic moment $702 \cdot 10^{-10}$. This value is now larger than the value used before, $673.9 \cdot 10^{-10}$, and the discrepancy between $a_\mu(99)$ and $a_\mu(SM)$ below one standard deviation. Obviously, this is one of the many places where the physics of the Standard Model and hadron physics meet, and for me the point where I can turn to a discussion of hadron spectroscopy.

As mentioned in the Introduction, the most important topic in light-meson spectroscopy is to find out which consequences QCD has for the dynamics of quarks in an energy regime in which perturbative approximations fail. There is the exciting possibility that QCD leads to the existence of hybrids and of glueballs. The summary will concentrate on those contributions which contribute to a clarification of this question.

HYBRIDS

Is there a $J^{PC} = 1^{--}$ hybrid?

In his speech opening Hadron 2001, A. Donnachie reviewed the status of vector-meson radial excitations [25]. The status, as presented by the Particle Data Group, is not at all satisfactory. The resonances

| ρ(1450) | ω(1420) | Φ(1680) | K*(1410) |

are assigned to the 1^3S_1 nonet, the states

| ρ(1700) | ω(1650) | K*(1680) |

to the 1^3D_1 nonet. It is clearly surprising that the K*(1410) is lower in mass than the ρ(1450) and ω(1420). Also the decay pattern of the ρ(1450) is by no means consistent with expectations based on the $^3P_0$ model. Table 1 shows theoretical expectations.

The e^+e^- annihilation cross section into two $\pi^+\pi^-$ pairs has the same size as the one for $\pi^+\pi^-2\pi^0$; both cross sections reach a peak value of about 35 nb at 1500 MeV. The isobar h_1 contributes only to the latter and not to the former final state. The 4π final-state is therefore reached only via a_1 and not via h_1. This is a pattern expected for a hybrid!

The Crystal Barrel collaboration reported [27] results of an analysis of various final states in $\bar{p}n$ annihilation at rest (in D_2). Imposing masses and widths for the two known ρ(1450) and ρ(1700), they find a pattern listed in Table 1.

The 4π partial decay-width is again too large (its decomposition into isobars does not agree with the results from e^+e^- annihilation, possibly due to the role of the spectator

TABLE 1. Decay widths in MeV calculated within the framework of the $^3P_0$ model [28] for pure quarkonia states and for a hybrid state calculated within the framework of the flux-tube model [29]. Decays marked with "–" are not part of the calculations (π^* is a shortcut for $\pi(1300)$).

decay mode	$\pi\pi$	$\omega\pi$	$a_2\pi$	$a_1\pi$	$h_1\pi$	$\rho\rho$	$\pi^*\pi$	$\rho\sigma$	$K\bar{K}$
Calculated partial widths in the $^3P_0$ model:									[28]
$2^3S_1\,\rho(1465)$	74	122	0	3	1	0	–	–	35
$1^3D_1\,\rho(1700)$	48	35	2	134	124	0	14	–	36
$3^3S_1\,\rho(1900)$	1	5	46	26	32	70	16	–	1
Calculated partial widths in the flux-tube model:									[29]
Hybrid-$\rho(\sim 1500)$	0	5–10	~ 0	140	0	0	0	–	–
Experimental results from Crystal Barrel:									[27]
	45 ± 13	115 ± 89			$\Gamma_{4\pi}=121\pm 48$				23 ± 7

pion); the pattern of two-body decays supports however the $\bar{q}q$ interpretation of the $\rho(1450)$, and the large widths for decays into $\pi\pi$, $\pi\omega$ and $\bar{K}K$ are incompatible with the prediction - based on the flux tube model - for a hybrid state. On the other hand, the $\rho(1700)$ has a decay pattern suggesting that it is the 1^3D_1 state and not the 2^3S_1 ρ radial excitation. If the $\rho(1450)$ is a hybrid, where is the $\bar{q}q$ state ?

If the masses in the fit to the Crystal Barrel data are not fixed to PDG values, there is a surprise: the $\pi\omega$ phase rises very rapidly at rather low $\pi\omega$ masses: Thus the first resonance above the $\omega(782)\pi$ threshold could be at about 1200 MeV. So the question arises if there is a further multiplet in this mass region.

The very precise Novosibirsk data are limited to energies below 1400 MeV but constrain of course also fits to the full energy range, covered by the DM2 results. The SND collaboration studied ω' radial excitations. M. Achasov [26] reported an analysis of the $\pi^+\pi^-\pi^0$ and $\pi^+\pi^-\omega$ channels where three excitations were required to describe the data. The $\omega(1770)$ is very clear from the $\omega\pi^+\pi^-$ channel, the splitting of the $\omega(1420)$ of the PDG into a state at 1250 and 1400 MeV improves the fit but the evidence for the low-mass state is certainly not overwhelming.

Hence there is evidence that the $\rho(1450)$ and the $\omega(1420)$ might be split; the lower mass particles at around 1250 MeV being the 2^3S_1 radial excitation. The higher mass particle has - at least the $\rho(1450)$ - a decay pattern which follows the flux-tube prediction for a hybrid.

It should be mentioned here that there are new beautiful results on rare decay modes of vector mesons. The reader is referred to contributions [30, 31, 32, 35, 33, 34].

Conclusions on 1^{--} hybrids. At present, I would conclude that there are indications that non-$q\bar{q}$ mesons hide in the spectrum of vector mesons. Likely, we have however to wait for the energy upgrade of the Novosibirsk collider before we come to final conclusions and to a clear view of vector mesons excitations.

$J^{pc} = 1^{-+}$ exotics

Introduction. At Hadron 2001, the compatibility and consistency of the results from different channels and from the experiments at BNL and Protvino were discussed. Hence it seems adequate to point out some differences and similarities between the Brookhaven and the VES Experiment. A more detailed discussion can be found in [36, 37, 38, 39, 41].

The Brookhaven experiment uses a beam of 18 GeV/c while VES uses 28 and 37 GeV/c. Brookhaven uses a hydrogen target, VES uses a nuclear target. Also the analyses are different. Brookhaven fits directly amplitudes to the angular distributions while VES extracts the density matrix elements from the data. The amplitudes are then fitted to the density matrix elements. They have developed a new technique to emphasize the coherent amplitudes by extracting the largest density matrix elements. This new technique works very successfully and reduces the background. It should be mentioned that various sources of incoherent background exist. The nucleon may undergo a spin-flip. At VES, the nuclear target may become excited. Both are processes which are distinguishable and lead to a new incoherent set of amplitudes. Also the use of an incomplete set of amplitudes introduces some apparent incoherencies which need to be taken into account. The similarity of the amplitudes and phases from the two experiments proves that these differences in experimental technology have no significant impact on the physics results.

I would like to explain a basic problem in extracting resonant structures in an exotic partial wave using $\pi\eta$ scattering as an example. The scattering process may receive contributions from two different amplitudes, from formation of resonances in the s-channel or from t-channel exchange contributions. Resonance formation leads to a structure in the cross section and to a rapid phase motion; the decays of a s-channel resonance should not depend on the production mechanism. Exchange processes in the t-channel are background amplitudes. They may be associated with slow phase motions and may also lead to some structure in the cross section. In $\pi\eta$ scattering with orbital angular momentum 2, the scattering amplitude is dominated by the $a_2(1320)$ meson; the contribution of background amplitudes due to t-channel exchange processes is small. In the orbital angular momentum L=1 wave, resonance formation is certainly much weaker and hence the relative background amplitudes may be much larger. With this word of caution we discuss the experimental results on in the $J^{pc} = 1^{-+}$ partial wave.

The $\pi\eta$ system. Both experiments, at Brookhaven and at Protvino, find nearly identical amplitudes; the phases for the $J^{pc} = 1^{-+}$ wave show a rapid phase variation against the 2^{++}-wave. Amplitude and phase can be fitted using a Breit-Wigner ansatz and lead to a description of the data by one resonance at a mass of 1400 MeV or slightly below. This is the simplest explanation of the data and adopted by the BNL-group. The VES group is fully compatible with these findings. A Breit-Wigner fit describes the data very well and gives results fully compatible with the BNL result. Here one has to note that the VES group indeed found phase motion and the amplitude variation many years before this was reported by the Brookhaven experiment.

The VES group also investigated the necessity of a resonance in the partial wave. They tried hard to find background amplitudes in all participating waves which might conspire to mimic a resonance in the $J^{pc} = 1^{-+}$ wave. Indeed they succeed and this gives a warning that also other interpretations might be found which do not necessarily lead to the claim of resonances with exotic quantum numbers.

The question arises if the optimistic approach which assumes that an exotic resonance has been found is justified. Maybe the Crystal Barrel data may help here to resolve this question. In antiproton annihilations at rest on neutrons into $\pi^-\pi^0\eta$, the Crystal Barrel Collaboration observes a clear phase variation in the $\pi\eta$ L=1 partial wave. Hence the optimistic approach seems to be justified, and we may consider the resonant interpretation of the data as very likely.

The $\eta'\pi$ system. Experimentally, the situation is similar in the $\pi\eta$ sector. Evidence for a non vanishing partial wave with a rapid phase motion was first reported by the VES collaboration but not interpreted as exotic resonance. Brookhaven observed this reaction and reported a new analysis with high statistics at this conference. The old data and new data show a very similar pattern and can be interpreted by a resonance with a mass at about 1600 MeV and a width of 340 MeV. The old VES data are experimentally very close to the Brookhaven result, however the new data with much higher sensitivity seem not to support a resonant structure. The Crystal Barrel collaboration reported supportive evidence for this exotic state from proton-antiproton annihilation into $\pi^+\pi^-\eta'$ [42].

$\omega\rho$ and $b_1(1235)\pi$. Both experiments agree that there is a resonant structure at 1600 MeV.

$\rho\pi$. A sizeable background is present in the low mass range at VES and at BNL. It is particularly large at low masses and excludes the possibility to search for $\pi_1(1400) \to \rho\pi$ decays. The method of choosing the largest eigenvalue was used by the VES group. For large momentum transfer ($t' \geq 0.15$ GeV^2), a structure at 1600 MeV appears in the 1^{-+} wave but the group is not fully convinced that the peak requires a resonant interpretation. In a coupled channel analysis of $\eta'\pi$, $\rho\pi$ and $b_1(1235)\pi$, evidence for a $J^{pc} = 1^{-+}$ wave resonating at 1610 MeV was reported. In the BNL data - using the full t' range - a clear peak, also at 1600 MeV, in the exotic wave is observed. Its phase is measured against 6 other partial waves; all show a rapid phase advance by 180^0. Meyer-Wildhagen presented Crystal Barrel data on $\bar{p}n \to \pi^-3\pi^0$ [43]. The exotic 1^{-+} wave is clearly identified; possibly both the $\pi_1(1400)$ and $\pi_1(1600)$ contribute.

Conclusions on 1^{-+} exotics. In my view, there is good evidence that two 1^{-+} exotic mesons have been discovered, the $\pi_1(1400)$ and $\pi_1(1600)$. Ted Barnes had reminded us that hybrids are expected - in the flux tube model - at masses at about 1.9 GeV and above. Low hybrid masses at 1.4 GeV are, however, certainly not fully excluded. So we ask: what is the nature of these two resonances? A striking feature is their decay pattern. The $\pi_1(1400)$ is seen only in the $\pi\eta$ decay mode; the $\pi_1(1600)$ is observed in several channels, including the $\pi\eta'$ decay mode, but not in $\pi\eta$. One may argue that the $\pi\eta'$ decay indicates a gluonic part in the $\pi_1(1600)$ wave function; this could be the reason why it decays into $\pi\eta'$ and not into $\pi\eta$. I do not share this view. In the limit of SU(3) flavor conservation and in the limit that the η is the octet particle, the decay a $J^{PC} = 1^{-+}$ into $\eta\pi$ is forbidden. If the $\pi_1(1600)$ is an octet meson, it cannot decay into $\pi\eta$; it must decay into $\pi\eta'$. The suppression of the $\pi\eta$ decay mode is therefore not surprising. But why does then the $\pi_1(1400)$ decay into $\pi\eta$? It cannot belong to a meson octet; instead it has to be part of a decuplet. In turn, a decuplet cannot decay into $\pi\eta'$. The strange couplings of the $\pi_1(1400)$ and $\pi_1(1600)$ reflect therefore their different flavor structure. The $\pi_1(1400)$ belonging to a decuplet of particles must be a 4-quark state and cannot be of hybrid nature. The $\pi_1(1600)$ however can be both, it could be a hybrid or a multi-quark state. Of course, the closeness in mass and the similarity of the production cross sections may be considered as a hint that both particles have a similar internal structure.

The $\eta(1295)$, radial excitation or non-$q\bar{q}$ state ?

There is another state at comparatively low mass which cannot be a quark anti-quark state and must therefore have a more complicated structure. This is the $\eta(1295)$. The Particle Data Group in the year 2000 edition lists the following nonet of pseudoscalar radial excitations:

| $\pi(1300)$ | $\eta(1295)$ | $\eta(1440)$ | K(1460) |

New data presented at this conference show that this assignment must be wrong. First I show that the $\eta(1440)$ cannot be the pseudoscalar $s\bar{s}$ state. The $\eta(1440)$ is produced in the pion exchange reaction

$$\pi^- p \rightarrow n\,\eta(1440)$$

as a strong signal. No signal is however observed in the reaction

$$K^- p \rightarrow \Lambda\,\eta(1440).$$

States with hidden strangeness are abundantly produced in Kaon-induced reactions while in pion-induced reactions no $s\bar{s}$ states can be produced. The experimental pattern proves therefore that the $\eta(1440)$ cannot be a dominant $s\bar{s}$ state. But then, why is the $\eta(1440)$ decay into $K\bar{K}$ so strong? This can be understood on the basis of the $^3P_0$ model for meson decays. Let us assume that the $\eta(1440)$ is the pseudoscalar radial excitation of the η. Under this assumption the transition amplitude of a radial excitation into $a_0(980)\pi$ vanishes at a mass of the radial excitation of about 1450 MeV. The decay of the $\eta(1440)$ into $a_0(980)\pi$ is then largely suppressed and shifted to low masses. The K^*K decay mode does not suffer from the zero in the transition amplitude and appears un-shifted, and with not reduced strength. The $^3P_0$ model therefore predicts that the radial excitation of the η, if it has a mass in the 1400 to 1500 MeV range, should decay strongly into $K^*\bar{K}$ while the $a_0(980)\pi$ decay mode should be suppressed and shifted to low masses. With its large $n\bar{n}$ component, it should be produced abundantly in pion- induced reactions. All the predictions have been observed experimentally.

Then, what is the $\eta(1295)$? The $\eta(1295)$ is seen in various pion-induced experiments, for instance recently by the Brookhaven group in the reaction proton into neutron plus $\eta\pi\pi$ and $\bar{K}K\pi$ at 18 GeV as reported on this conference [39, 40]. The pseudoscalar intensity is now even stronger than the $f_1(1285)$ intensity, the 1^{++} wave. The 0^{-+} contribution shows peaks at 1.295 and at 1.4 GeV. The 1^{++} wave peaks at 1285 MeV and also at 1.4 GeV but the wave is greatly reduced compared to the pseudoscalar wave. Hence a pseudoscalar state at 1295 MeV is likely to exist even though we note that the properties of the $f_1(1285)$ and the $\eta(1295)$ depend very much on experiment and analysis; obviously there is some feed-through between $f_1(1285)$ and $\eta(1295)$. Also the new BNL data require the properties of the $f_1(1285)$ to be changed.

In any case, the $\eta(1295)$ cannot be a normal $q\bar{q}$ state. The L3 collaboration reported production of pseudoscalar resonances in two photon collisions. Of course, two-photon fusion is an established tool to constrain the flavor structure of produced particles; a few contributions were dedicated to apply this method [44, 45, 46]. At low (transverse) q^2, they observe in the $K\bar{K}\pi$ mass distribution a clear signal at 1450 MeV but no signal at 1295. At large q^2 ($\geq 1 GeV^2$), a second peak shows up at below 1300 MeV. Note that two real photons (or nearly real photons) do couple to pseudoscalar mesons but not to states with spin 1, due to the Yang-Landau theorem. The strong signal at low q^2 must therefore be due to $\gamma\gamma \to \eta(1440)$; the $\eta(1295)$ obviously decouples from 2 photons. The large 2γ coupling of the $\eta(1440)$ excludes any glueball interpretation of the $\eta(1440)$; the small 2γ coupling of the $\eta(1295)$ makes it very unlikely that it is a conventional $q\bar{q}$ state. At (transverse) q^2 larger than 1 GeV^2, a peak at ~ 1.3 GeV shows up. Virtual photons do have coupling to 1^{++} states; the signal has therefore to be assigned to the $f_1(1285)$ and cannot stem from the $\eta(1295)$.

The two-γ coupling of the $\eta(1440)$ and the decoupling of the $\eta(1295)$ supports the conclusion that the $\eta(1440)$ must be a radial excitation while the $\eta(1295)$ requires an exotic interpretation. We note in passing that the early experiments had observed no signal at 1440 MeV for two un-tagged (real) photons while a few events were seen when one photon was tagged. This result was interpreted as evidence for the glueball nature of the $\eta(1440)$ and as evidence that the $f_1(1420)$ really exists. These early results are now proven wrong.

A similar argument has been put forward by the Crystal Barrel Collaboration [47]. They observe in $p\bar{p}$ annihilation at rest a strong signal due to $\eta(1440)$ production while no signature is observed from $\eta(1295)$. The production rate of the $\eta(1440)$ in $p\bar{p} \to \pi^+\pi^-(\pi^+\pi^-\eta)$ is of the same order of magnitude as that for production of the $\pi(1300)$. The rate for $\eta(1295)$ production is however lower by a factor 30 than this naive expectation. Again, the $\eta(1440)$ makes a much better compagnon of the $\pi(1300)$ than the spurious $\eta(1295)$.

Finally, I would like to mention that the OBELIX collaboration reported a scalar resonance at 1420 MeV, with isospin 2 [48]. If confirmed, this would be an explicitely exotic resonance, of the first kind. The identification of exotics in the baryon sector is even more difficult. There are states with possibly anomalously large couplings to final states with strangeness [49]. They could be penta-quarks and contain hidden strangeness.

Conclusions on 0^{-+} hybrids or glueball. I believe that there is only one $\eta(1440)$ in the 1400 to 1500 MeV mass range. Its splitting can be understood within the $^3P_0$ model. The $\eta(1440)$ and not the $\eta(1295)$ is the radial excitation of the η. It is the nature of the $\eta(1295)$ which is unclear.

HYBRID CANDIDATES AT HIGH MASSES

In the flux tube model, hybrids have masses at or above 1.9 GeV and do not need to have exotic partial waves. It is therefore important to identify high-mass meson resonances and to establish the pattern of quarkonia states over the full mass range up to 2.2 GeV or even higher. A large number of resonances, partly new ones, partly known ones, were reported at Hadron 2001, in [50, 51, 52] and references given above.

Of particular interest are the $\pi(1800)$ and the $\pi_2(1900)$. The pseudoscalar isovector resonance at a mass of 1800 MeV had been discovered by VES and further studied at Brookhaven and at VES. The state is now seen in various decay channels; the observations can be grouped into channels where the $\pi(1800)$ has an apparently high mass of about 1870 MeV. These are $\eta\eta\pi$, with two identified isobars $f_0(1500)\pi$ and $a_0(980)\eta$, and $\eta\eta'\pi$. In contrast, the $\omega\rho$ amplitude shows a maximum at 1775 MeV. VES reports that the three-pion channel globally has a resonant π wave at 1775 MeV; BNL separates the 3π mode into one isobar, $(\pi\pi)_s$-wave, at high mass and a low-mass state with $\rho\pi$ and $f_0(980)\pi$ isobars. Hence there is evidence that the $\pi(1800)$ is split into 2 states, a $\pi(1775)$ and a $\pi(1870)$.

Clearly, the quark model cannot accommodate two pionic excitations so close in mass. One of these needs to be of different nature, possibly a hybrid. This idea is supported by decay calculations which predict that a $q\bar{q}$ state and a hybrid should have different decay patterns [1]. The expectations are listed in Table 2.

TABLE 2. Partial width of a ~ 1800 MeV $\bar{q}q$ and resonance and a hybrid with quantum numbers of a pion [1].

Decay	$\rho\pi$	$\rho\omega$	$\rho(1465)\pi$	$f_0(1300)\pi$	$f_2\pi$	$\bar{K}K^*$	tot
3^1S_0	31	73	53	7	28	36	228
hybrid	30	0	30	170	6	5	240

The high-mass component of the $\pi(1800)$ with its strong decay to scalar plus pseudoscalar can thus be identified with a hybrid, the low-mass component with the second π radial excitation.

The π_2 wave also shows an interesting double structure. Fits to the 2^{-+} partial wave require not only the $\pi_2(1670)$ but also a second state at 1900 MeV. Both channels, $\rho\pi$ and $\omega\rho$, cannot be fitted with just one resonance; a high-mass shoulder is seen in addition to the well-established $\pi_2(1670)$. The $\omega\rho$ 2^{-+} partial wave - from which the evidence for the second state is derived - poses a problem: the $\pi_2(1670)$ is seen to decay into $\omega\rho$ via the intrinsic-spin 2 amplitude. This decay mode is incompatible with the $^3P_0$ predictions (using a $\vec{\sigma}\cdot\vec{p}$ operator which creates or destroys only one unit of spin). If the experimental result proves to be correct, the $^3P_0$ model is false and cannot be used to identify non-$q\bar{q}$ objects.

In any case, the occurrence of two resonances in the same partial wave separated in mass only by 200 to 250 MeV is a challenge to the quark model and indicates the presence of dynamics beyond the $q\bar{q}$ system. This claim is supported by the possible observation of two η_2 states, one well known at a mass of 1645 MeV and second one at a mass of 1860 MeV. Likely the $\pi_2(1890)$ and $\eta_2(1860)$, if confirmed, belong to the same particle multiplet. The 2 η_2 states are certainly not a $n\bar{n}$ and a $s\bar{s}$ state since they both are produced with similar yield in $p\bar{p}$ annihilation.

Conclusions on high-mass hybrids. There is good evidence that the $\pi(1800)$ is split into two components, a $\pi(1775)$ $\bar{q}q$ resonance and a $\pi(1870)$ hybrid. Also in the π_2 and η_2 partial waves, two separate resonances were reported. The experimental evidence for two close-by states - where only one is expected in the quark model - is the primary reason for this evidence. In case of the $\pi(1800)$ there is additional support for this interpretation from the observed decay pattern. The decay of the supposedly $\bar{q}q$ $\pi_2(1670)$ into $\omega\rho$ with intrinsic spin 2 is however very intriguing: if confirmed, it invalidates the $^3P_0$ model which is the basis for the identification of resonances as quarkonia or hybrids.

SCALAR MESONS AND THE SEARCH FOR THE SCALAR GLUEBALL

The particle data group assigns the

| $a_0(1450)$ | $f_0(1370)$ | $f_0(1750)$ | $K_0(1460)$ |

to the lowest lying 1^3P_0 meson nonet. The $a_0(980)$ and $f_0(980)$ are interpreted as molecules or four-quark states. The $f_0(1500)$ is the tenth meson, not belonging to the scalar nonet. It is considered as scalar glueball of lowest mass. New data were presented at this conference which shake this interpretation.

The $f_0(980)$ and $a_0(980)$

There is a long standing debate on the nature of the $a_0(980)$ and $f_0(980)$ states. Both are close to the $\overline{K}K$ threshold and their mass is obviously strongly influenced by the threshold. Their unexpected small width and their strong coupling to $\overline{K}K$ is the basis for their interpretation as $\overline{K}K$ molecules. Following Jaffe, there is a strong attraction between qq and $\overline{qq}$ in S-wave and spin singlet; a low-mass nonet can be constructed with $a_0(980)$ and $f_0(980)$ being the two $(n\bar{n}s\bar{s})$ states. Again, the two states are not $\bar{q}q$ states, and can be disregarded when the lowest lying $\bar{q}q$ scalar nonet is constructed. There are, however, also arguments speaking in favor of the two states being normal $\bar{q}q$ mesons.

There is the believe that radiative decays of the Φ meson into $a_0(980)$ and $f_0(980)$ should clarify the internal structure of these two important mesons. Results from Novosibirsk [53] and preliminary data from CLOE [54] were reported at this conference on the branching ratios for these two reactions. The two results agree approximately but not fully within the quoted errors. However, the CLOE result is still preliminary and the small discrepancies do not lead to different conclusions. N. Achasov [53] argued in his contribution that the rates are only consistent with a $\overline{K}K$ molecular interpretation. A. Anisovich [55] presented a calculation based on the hypothesis that the two scalar mesons are $\bar{q}q$ states, and obtained full agreement with data. The wave function at the origin is not calculated but has the same size as other $\bar{q}q$ mesons.

V. Uvarov [56] compared the yields of various mesons in the decay of Z^0 bosons. The fraction of mesons produced in the fragmentation depends on their mass and on the intrinsic number of strange mesons. So the production rates for mesons like ω, ρ, $f_0(980)$ and $a_0(980)$, and $f_2(1270)$ lie on one line which is linear on a logarithmic scale. The K and K* lie on a separate line due to having one strange quark, the Φ and the $f_2(1525)$ lie on a third line as function of their mass. Hence the first evidence favors a $\bar{q}q$ interpretation of the $f_0(980)$ and $a_0(980)$. However, V. Anisovich [57] and Sarantsev [58] assign a bare mass of 720 MeV to the $f_0(980)$. This mass would then fall on the line with two intrinsic unit of strangeness and hence the production rate could be compatible with a $\bar{K}K$ structure. The phase space is of course given by the physical mass; no argument is given why the pole in the K-matrix should be responsible. Also, the argument does not apply to $a_0(980)$ production.

In a very detailed analysis, the Delphi Collaboration has demonstrated that the production from Z^0 decay does not differ in any respect from the production of well known $\bar{q}q$ states and that an interpretation as four-quark state or molecule does not seem plausible. Hence the question if the $f_0(980)$ and $a_0(980)$ are $\bar{q}q$ states, $\bar{K}K$ molecules or four-quark states seems still, from an experimental point of view, still unresolved.

Conclusions on $a_0(980)$ and $f_0(980)$. Generally speaking, we should expect that these mesons have a complex Fock expansion and that a $\bar{q}q$ component, a molecular component and a four-quark component can coexist with open and presently unknown fractional contributions. Personally, I believe that the $\bar{q}q$ component is the largest one. The best possibility to find out the size of the various component might be to search for $a_0(980)$ to $\rho\gamma$ radiative decays. In any case, if these mesons do have a $\bar{q}q$ component, this component then reflects the genuine $\bar{q}q$ state which is attracted by the $\bar{K}K$ threshold and thus acquires a large $\bar{K}K$ component. This view would be inconsistent with leaving these two mesons out of the discussion of ordinary $\bar{q}q$ states.

The σ particle

Low-energy pion-pion scattering is often supposed to be dominated by scalar resonance, the σ-meson, having a mass around 400 to 700 MeV. As chiral partner of the pion it plays an important role in the discussion. From an experimental point of view it is decisive of course to ask for the experimental significance leading to introduce this particle. Since there are no open channels, the phase motion associated with the resonance must be observable. This was not demonstrated so far; quite in contrary, Ochs [59] showed that there is no visible phase motion in data with good phase sensitivity. Clearly, the $\pi\pi$ phase shift rises from threshold to 1000 MeV by 90° (after substracting the $f_0(980)$ phase motion) but there must be a phase advance by 180°. It is claimed that the $\pi\pi$ phase would be repulsive in absence of the σ particle (the isospin 2 interaction is repulsive), but this claim is not supported by quark-model calculations [60].

Scalar mesons with isospin zero

Beautiful results were presented from BABAR [61, 62], BELLE [63] and from Fermilab [64] on the decay of B mesons into different final states. These data may have a large impact on low-mass light mesons which are abundantly reduced in decays. Particularly interesting are decays of D_s mesons to three pions since in this decay there is primary formation of an $s\bar{s}$ state which then decays into non-strange particles. This transition resembles pseudoscalar mesons which also link $n\bar{n}$ components and $s\bar{s}$ components. But also decays of D_s into $K_s^0 K_s^0 \pi$ and D decays into 3 pions, into one Kaon and to two pions, two Kaons and one pion, and into 3 Kaons show very interesting structures. The statistics in these channels is limited at the moment but very high statistics data can be expected in the near future. Also, the analysis methods will partly be needed to become more sophisticated before final conclusions can be drawn.

BES reported a considerable increase in statistics of J/ψ radiative decays [65] in a large variety of final states. Particularly interesting is the decay into $\bar{K}K$, the reaction in which the old θ(1690) was discovered. The new data show that the $f_J(1710)$ as it is called now, clearly has J=0. A small tensor distribution is possible but not really required. This resonance is discussed as possible scalar glueball.

In this context, its two-photon width is very important. The BELLE Collaboration investigated photon-photon fusion to $\bar{K}K$ [6]. They clearly see the $f_2(1525)$ and have further peaks at 1.75 GeV, 2 GeV and 3 GeV. The resonances at 1750 and 2000 MeV favor spin 2. In J/ψ decays, the dominant part had scalar quantum numbers. This part has little coupling to two photons; hence it is not $\bar{q}q$. The small tensor part in J/ψ decays is, in comparison, enhanced in two-photon fusion. That part is $\bar{q}q$.

Unfortunately, the situation is not so clear. The BELLE Collaboration also has data on photon-photon fusion into $K_s^0 K_s^0$ [6]. In this reaction they find the $f_2(1525)$, as before, and a resonance at 1750 MeV. This time the amplitude analysis favours spin zero. Clearly the K^+K^- and $K_s^0 K_s^0$ must have identical partial wave contributions from scalar or tensor mesons and the situation is certainly not well understood.

Unfortunately, the WA102 collaboration is not represented at this conference. But in this context, I have to mention their results on central production of four pions. The scalar part of four pion central production shows a strong peak due to the $f_0(1370)$, a dip at 1500 MeV which is assigned to the $f_0(1500)$, and a wide bump at a mass of about 1800 MeV. The latter is decomposed into the $f_0(1750)$ and a further scalar meson at about 2 GeV. This distribution is seen in the $\pi^+\pi^-2\pi^0$ and in the $2\pi^+2\pi^-$ final states; the $4\pi^0$ final state has contributions only from the $f_0(1500)$.

The picture resembles very much the one in the two-pion sector. The $f_0(980)$ is seen as a dip in a wide distribution [66], called $f_0(1000)$ by Morgan and Pennington, and *red dragon* by Minkowski and Ochs [59]. The wide distribution is of unknown nature; it may be a very wide glueball [59] or generated by t-channel exchange. The strange behavior of the 4π system can be understood assuming that it also generated dominantly by ρ exchange in the t channel. In Pomeron-Pomeron scattering, ρ exchange may lead to ρρ but never to $4\pi^0$. Thus the $f_0(1500)$ is a $\bar{q}q$ state removing intensity from the ρρ scattering process, the broad distribution is due to ρρ interactions via t-channel exchanges. The $f_0(1370)$ and the broad bump at 1800 MeV are a further *red dragon*.

In contrast, the Crystal Barrel Collaboration observes a strong signal from $f_0(1370)$

also in its $4\pi^0$ decay, both in the reaction $\bar{p}p \to 5\pi^0$ and $\bar{p}n \to \pi^- 4\pi^0$, the production process is not limited to Pomeron-Pomeron plus final-state interactions. There is a clear conflict between WA102 and Crystal Barrel data, and this may indicate that the $f_0(1370)$ decay modes are not independent of its production mechanism. This unusual behavior suggests that the $f_0(1370)$ is not a s-channel resonance but rather generated by t-channel exchange processes, in particular by ρ exchange. Theoreticians should be cautious when using the $f_0(1370)$ in mixing scenarios in which $\bar{q}q$ states are mixed with the lowest scalar glueball [67].

Finally I should mention that the t-channel poles which we observe do not necessarily need to be the genuine $\bar{q}q$ resonances as calculated for instance in quark models. The bare states couple to their final states and this may result in grossly shifted resonance positions. Anisovich and collaborators assign the K matrix pole to the bare poles, to the true $\bar{q}q$ states. This pattern of states is very different from the pattern of T-matrix pole positions which are listed by the PDG. Of course, this is a highly theoretical issue but we should have in mind that a straight forward interpretation of meson resonances may lead to wrong conclusions. This warning is particularly true in case of scalar mesons. The shifts are much smaller in cases where the orbital angular momentum barrier is active.

Conclusions on scalar mesons and the scalar glueball. I do not share the optimistic view that the scalar glueball has un-revealed its existence and has been identified by mixing between adjacent $\bar{q}q$ states. Such scenarios have several rather weak points. The $f_0(980)$ belongs, in my view, to the scalar $\bar{q}q$ states. (Admittedly, it may have a large $K\bar{K}$ component). The $f_0(1370)$ is likely generated by t-channel exchanges and is rather a $\rho\rho$ molecule and not a genuine $\bar{q}q$ state. Its decay properties seem to depend on the production mechanism. In the solution offered by Anisovich and collaborators, the $f_0(980)$ is described by two K-matrix poles far apart from the T-matrix pole position. At large momentum transfer to the $\pi\pi$ system, the $f_0(980)$ is seen as clear peak above little background, and it seems unnatural that two K-matrix poles conspire to produce a peak above a small residual background. Thus I believe that the scalar nonet is given by the nine states

$a_0(980)$	$f_0(980)$	$f_0(1500)$	$K_0(1460)$

The $a_0(1450)$, $f_0(1370)$, $f_0(1750)$ and $K_0(1950)$ could form the nonet of scalar radial excitations. The broad scalar background has certainly contributions from t-channel exchange processes; it may comprise contributions from the scalar glueball. But this is speculative, mass and width can certainly not be given.

The tensor glueball

Finally I would like to recall searches for the tenser glueball. There is one famous candidate the so called $\zeta(2220)$. It is observed to decay into $\pi\pi$, $\eta\eta$, and proton antiproton, but in all channels with low statistical significance. In particular, it is unclear if the width is really so small as claimed. The decay of the $\zeta(2220)$ to $\pi\pi$ and to proton antiproton allows to calculate the production cross section with which one should see the state in proton antiproton annihilation in flight. The Crystal Barrel Collaboration has searched for the resonance in a fine scan of antiproton proton annihilation into various final states and no signal was found at the expected height [68]. So there is at least an inconsistency in the decay pattern. Certainly, the statistical significance of this state is not large enough to claim that an anomalous state was discovered which could be identified with the tensor glueball. Another longstanding claim for the tensor glueball was made at Brookhaven from the reaction π^- + proton → neutron + $\Phi\Phi$. New data on $\Phi\Phi$ in central production were shown at this conference [69]. The signal is clearly seen, the mass distribution shows a threshold enhancement which can be fitted using one resonance only. A slightly improved description can be found using two tensor resonances; three are certainly not required. From the quark model, we expect two $s\bar{s}$ tensor states in this mass region. So the claim for a tensor glueball which mixes with the two quarkonia states is no longer justified.

BARYON SPECTROSCOPY

Baryons are hadrons! For a long time, light-baryon spectroscopy played practically no role in the hadron conference series. Now I am very pleased to see that there are several talks related to baryon spectroscopy. The reason for this is of course the chance that the field may get a boost because of the new facilities at Jefferson lab, Spring8, Grenoble, MAMI and ELSA. Surprisingly, the clearest resonant structures came from BES. The preliminary partial wave analysis of the reaction $J/\psi \to p\pi^-\bar{n}$ suggests that the $N^*_{1/2-}(1535)$, $N^*_{3/2-}(1520)$, $N^*_{5/2+}(1675)$, and $N^*_{5/2+}(1680)$ are observed [65]. The phase space ends at slightly above 2 GeV, but the power of the method is established. Study of ψ' decays will open the phase space up to the interesting region up to 2.7 GeV.

Photo- and electroproduction

At Jefferson lab, electroproduction of $K^+\Lambda$ was studied with very high precision, and over a wide energy range [70]. Total and differential cross sections and the polarization transferred to the Λ were measured over a wide range of momentum transfers. The precision of the data is certainly a challenge to any theoretical model aiming at describing the $s\bar{s}$ production mechanism.

From MAMI and ELSA, a test of the Gerasimov-Drell-Hearn sum rule was reported. The summation over all energies over the total photo-absorption cross section $\sigma_{3/2} - \sigma_{1/2}$ for polarised photons and polarised protons is related to the anomalous part of the proton magnetic moment. The high-precision data from MAMI covering the range up to 860 MeV are now augmented by data from ELSA up to 2.4 GeV. The integrated cross section difference starts to level off approaching the GDH sum rule value. If the high-energy part as expected from dispersion relation is added the sum is slightly overshot.

The GRAAL collaboration reported measurements of the η photoproduction. Precise data are available from MAMI but only up to 800 MeV. The new GRAAL data extend the range to 1.1 GeV [72], and (not yet analyzed) data at higher energies are on tape. The

Crystal Ball was used at BNL to study pion and Kaon induced η production at threshold [73, 74, 75]. In both cases the cross section rises steeply; in pion scattering due to the onset of the $NS_{11}(1535)$; in Kaon scattering the $\Lambda S_{01}(1670)$ is observed. These two resonances have large couplings to the η plus ground state; they share this property with the $\Sigma S_{11}(1750)$. These are the only known resonances with large couplings to the η. V. Credé (for the CB-ELSA collaboration) reported first results on photoproduction of $\pi^0 \eta$ where they may see a $\Delta \eta$ threshold enhancement [76]. Data on $2\pi^0$ production also show interesting structures over a wide mass range: baryon spectroscopy has entered a new phase and we may expect a substantial increase in our knowledge.

Analysis problems and a common data base

Dytman demonstrated how refined the analyses have to become to get the best precision out of the data [77]. Combined analyses of several reactions in multi-channel fits are required to identify the exact pole positions. Here I guess we - the meson spectroscopy community - have to learn a lesson: groups working at J-Lab (and elsewhere) have formed BRAG, a baryon resonance analysis group. Data are made publicly available; data are published with fits and reference to an analysis paper describing in details the analysis methods. The Carnegie Mellon University plans to set up a large data-base center for multi-particle production experiments (like we had at Durham for data on cross sections). I firmly believe that this is the way we have to go, and we all should contribute to support such a center.

FUTURE FACILITIES

e^+e^- colliders

We have seen the substantial increase in significance which was obtained by an increase of the statistics in J/ψ decays from a few million events to now $24 \cdot 10^6$ events. Bejing plans an further improvement of the luminosity [81]; in parallel, Cornell has decided to go down in energy and up in luminosity [82]. In a couple of years we will have 10^9 J/ψ and ψ′ decays. These data will have a decisive impact on light-meson spectroscopy. In particular, we can hope that the question if glueballs exist can finally be answered. Does the low-mass scalar glueball manifest itself by mixing with $^3P_0$ $\bar{q}q$ mesons, has it to be identified with the *red dragon* of Minkowski and Ochs [59], or is the life time of glueballs so short that they do not manifest themselves in meson spectroscopy ?

We also will see high-statistics data from KLOE [83] and from Novosibirsk [24]. KLOE will provide not only data on ε'/ε from Φ decays into $\overline{K}K$. In parallel we will get precise information on radiative decays of Φ mesons into light mesons. The energy upgrade in Novosibirsk will provide for precision studies of light vector mesons.

At the high-energy end, we have seen the significant impact B factories will have on the spectroscopy of light mesons. We can anticipate that also D^* resonances will play a major role as bridge from light to heavy mesons.

Photo- and electroproduction

The Jefferson laboratory proposes an energy upgrade to 12 GeV [79]. One of the fascinating options will be to use coherent bremsstrahlung to produce a linearly polarised photon beam of 8 GeV, collimated to accept only the narrow (0.5 GeV) energy window in which the polarization is high. The hope is that the polarised photon beam has a particularly large coupling to mesonic systems with intrinsic quark spin 1, and that the string providing the binding between quark and antiquark can and will be excited.

MAMI in Mainz will receive an upgrade to 1.4 GeV. This will allow precision experiments at different thresholds; the limits of chiral symmetry will be tested at larger energies, e.g. at the strangeness production thresholds. The lower baryon resonances will be mapped precisely, and transition form factors to these states can be determined. Several other facilities extend the energy range all over the baryon resonance region, Spring8, GRAAL, ELSA and, of course, J-lab with its present facility.

The GSI Project

The Gesellschaft für Schwerionenforschung plans a complex facility, with a wide range of experimental possibilities. The core of the facility is a high-intensity 60 GeV proton synchrotron with fast cycling superconducting magnets [80]. The complex allows studies of rare isotopes, plasma physics and hadronic matter at highest baryon densities. Of particular importance for us is the option to produce intense beams of antiprotons. A high-energy (15 GeV) storage ring for antiprotons will support a rich program. There is the chance that hybrids with hidden charm can be formed; some of these hybrids may be below the preferred decay mode, one S-wave and one P-wave D-meson, and could thus be narrow. The potential of such an instrument was demonstrated at Fermilab but certainly not exploited in full.

Intense Kaon beams are not yet part of the GSI proposal but I am sure, the pressure on GSI to install such a beam line will increase once the proton synchrotron is operational. Kaon-induced reactions are mandatory for a proper understanding of low-energy phenomena; the beams can make a significant contribution to meson and baryon physics, to nuclear physics and - through Kaon decays - possible also to physics beyond the Standard Model.

CONCLUDING REMARKS

I would like to conclude in expressing my satisfaction that the study of baryons became again a lively subject in hadron spectroscopy. In light-meson spectroscopy we became used to think in a rather well-defined frame: mesons are described as excitations of constituent quarks, the intrinsic forces are given by a kind of effective gluon exchange. And gluons play an important dynamical role, in creating hybrids and glueballs. The widespread conviction that this picture is correct is however much more driven by theoretical visions than by experimental facts. I believe that baryon spectroscopy can provide very important checks of the understanding of low-energy strong interaction. First, baryons are three-quark systems. There is more freedom in the system and the internal interactions are un-revealed in a more direct way. And, secondly, the community has developed a different language to describe strong interactions. When strong interactions are discussed, the concept of gluon exchange is replaced by quark- and gluon-condensates. And instead of quenched lattice QCD, superconductivity provides a frame of visualizing strong QCD. It is my hope that the study of mesons *and* baryons and joint efforts of both communities will lead to better understanding of strong interactions in the low-energy range. And this is of course our *mission*, certainly not stamp collection but also not just to identify hadronic systems beyond the quark model.

There are several first-class facilities allowing to study strong interaction in the confinement region. Some of them have just started operation, others are being constructed, others are in the planning stage. Even if not all of the new ones will be funded, there is ample room for imaginative new experiments. The future of the field does not depend on others, it depends on us: we have to ask the right questions, we have to find the right answers, and we have to communicate our enthusiasm for the field to others: to our students, to our colleagues and to the general public.

Last not least, it is my privilege as concluding speaker to thank the organizers for the work they did in order to host this exciting conference. We all will memorize the friendly atmosphere, the concerts, the excursion and the forest around the place and, above all, the friendship and hospitality we received at HADRON2001 in Protvino.

REFERENCES

1. Barnes F. E., Summary (theory)
2. Kalashnikova Yu., QCD string model for hybrid adiabatic potentials
3. Morningstar C., Gluonic excitations
4. Wah Y., New KTeV results on ε'/ε
5. Ziolkowsky M., A new measurement of CP violation in two pion decays of the neutral kaon
6. Hou S., Recent results from Belle
7. White R., Search for the Higgs
8. Ligabue F., W mass measurement at LEP
9. Urner D., Measurements of HQET parameters and CKM matrix elements
10. Zintchenko A., New results on rare decays and on future NA48
11. White H. B., Future Kaon physics facility at the Fermilab main injector
12. Wah Y., Precision measurement of Standard Model parameters with rare Kaon decays
13. Vaandering E. Recent results on charmed hadron spectroscopy and charmed lifetimes from FOCUS.
14. Chistov R., $Omega_c^0$ production and decays at Belle
15. Meadows B., Production and Decay of the Λ_c Charmed Baryon from Fermilab E791.
16. Gittelman B., B meson decays from the Upsilon(4S)
17. Tomaradze A., Latest results on charmonium Spectroscopy from Fermilab E835
18. Zhu Y., Recent results of $\psi(2S)$ decay branching ratios and decay widths from BES
19. Chliapnikov P., Do the strange quarks produced in Z decays and valence strange quarks of incident kaon in Kp reactions fragment differently?
20. Hill R., R_b at 192, 196, 200, 202 GeV using ALEPH detector at LEP
21. Karshon U., Heavy Quark production and spectroscopy at HERA

22. Tuan S. F., Aspects of charmonium
23. Grigoriev D., Precise measurement of positive muon anomalous magnetic moment
24. Fedotovich G., Precise measurement of hadronic cross sections with CMD-2 detector at VEPP-2M
25. Donnachie A., Problems with vector mesons
26. Achasov M., Review of experimental results from SND
27. Pick B., Higher vector meson states
28. Barnes T., Close F.E., Page P.R., Swanson E.S., Phys. Rev. **D55** (1997) 4157
29. Close F.E., Page P.R., Nucl. Phys. **B443** (1995) 233
30. Berdyugin A., Study of the $e^+e^- \to \eta\gamma$ process with an SND detector
31. Shwartz B., New results on the rare decays of the light mesons at CMD-2 detector.
32. Dimova T., Conversion $\Phi(1020)$ decays into $\pi^0 e^+e^-, \eta(550)e^+e^-$ and $\eta(550)$ into γe^+e^- from SND experiment at VEPP-2M.
33. Di Donato C., Detection of $\Phi \to \pi^0$ + photon, $\Phi \to \eta$ + photon, $\Phi \to \eta'$ + photon with the KLOE detector at DAPHNE
34. Baratt A., New measurement of the branching ratios of tagged Kaon decays with CMD-2
35. Kozhevnikov A. On the spectroscopy of heavy ρ', ρ'', and ω', ω'' resonances
36. Popov A., Recent results from Brookhaven E852 experiment.
37. Dorofeev V., 1-+ at VES
38. Sarycheva L. Exotics in $\pi^- p$ interactions at 18 GeV/c
39. Eugenio P., Partial wave analysis of $\pi^+\pi^-\eta$ in the reaction $\pi^- p \to \pi^-\pi^+\eta$ n at 18 Gev/c
40. Nikolaenko V., Study of the reaction K^- N to $K^-\pi^+\pi^-$ N at VES
41. Kachaev I., Study of π^- N to $\pi^+\pi^-\pi^-$ N at VES
42. Reinnarth J., Search for the $\pi_1(1400)$ and the $\pi_1(1600)$ in the reaction $p\bar{p} \to \pi^+\pi^-\pi^+\pi^-\eta$ in annihilation at rest
43. Meyer-Wildhagen F., $\rho - \pi$-states in antiproton-neutron annihilation into $\pi^- 3\pi^0$
44. Schegelsky V., Hadronic resonance production in gamma-gamma collisions
45. Sokolov A., Inclusive D-meson and Λ_c production in two-photon collisions at LEP
46. Shapkin M., Inclusive J/ψ production in two-photon collisions at LEPII with the DELPHI detector
47. Suh J.-S., Radial excitations of pseudoscalar mesons
48. Fillippi A., Study of isospin 2 resonant states in $\bar{n}p$ annihilations
49. Kurshetsov V., Search for exotic baryons with hidden strangeness in proton diffractive production processes
50. Singovski A., Meson spectroscopy with CMS detector
51. Bugg D., I=1 C=+1 mesons, 1960-2410 MeV
52. Eugenio P., A Study of the $\eta\eta\pi^-$ system produced in the reaction $\pi^- p \to p \pi^+\pi^-\pi^- 4\gamma$ at 18 GeV/c
53. Achasov N., Analysis of the nature of the $\Phi \to \gamma\pi\eta$ and $\Phi \to \gamma\pi^0\pi^0$ decays
54. Valeriani B., Study of Φ decays to $f_0 + \gamma$ and $a_0 + \gamma$ with KLOE at DAPHNE.
55. Anisovich A., Two-photon partial widths of scalar and tensor mesons and its quark structure
56. Uvarov V., Determination of the strangeness contents of light-flavour isoscalars from its production rates in hadronic Z decays measured at LEP
57. Anisovich V., Quark-antiquark systematics and scalars and glueball
58. Sarantsev A., Systematical study of the mesons: a combined analysis of the data from different experiments
59. Ochs W., Light scalar meson spectrum
60. Barnes F. E., Private communication
61. Palano A., Three-body decays of D_0 and D_s mesons
62. Deppermann T., Dalitz analyses of $D_S \to K_S^0 K_S^0 \pi^\pm$ and $D_S \to \pi^+\pi^-\pi^\pm$
63. Bondar A., Study of three-body charmless B decays at Belle
64. Göbel C., Light meson physics from charm decays at Fermilab E791
65. Guo Z., Recent results from BES J/ψ physics
66. Surovtsev Yu. The f_0 mesons in processes $\pi\pi \to \pi\pi, K\bar{K}$
67. Teshima T. Mixing among scalar mesons and scalar glueball
68. Seth K., A high resolution search for the tensor glueball with the Crystal Barrel detector
69. Reyes M., Preliminary partial wave analysis results of the centrally produced $\Phi\Phi$ system
70. Hicks K., Strangeness production at CLAS
71. Krimmer J., Experimental check of the GDH sum rule at MAMI and ELSA

72. Kouznetsov V., Meson photoproduction and Compton scattering at GRAAL
73. Briscoe W., An overview of the Crystal Ball Program at BNL
74. Manley D. M., New results on baryon spectroscopy with the Crystal Ball spectrometer
75. Kozlenko N., Differential cross section of the reaction $\pi^- p \to \eta n$ using the Crystal Ball detector.
76. Credé V., Open questions in baryon spectroscopy
77. Dytman S., Coupled channel analysis of all N*production data and on eta photoproduction
78. Mokeev V. High lying N* studies in phenomenological analysis of charge double pion production
79. Meyer C., The Hall D Project at Jefferson Lab
80. Koch H., The Antiproton Project at GSI
81. Li W., The status of BEPC/BES and the upgrade program
82. Urner D., Exploring the Charm Sector with CLEO-C
83. Valente P. The KLOE physics program

PARALLEL SESSIONS

SCALARS, THEORY

Mixing among Scalar Mesons and Scalar Glueball

Tadayuki Teshima, Ichijiro Kitamura and Norikazu Morisita

Department of Applied Physics, Chubu University, Kasugai 487-8501, Japan

Abstract. In order to answer to serious issues on scalar mesons; why the high-mass scalar mesons which are considered to be $L = 1$ $q\bar{q}$ nonet are so heavier than the masses predicted from $L = 1$ $q\bar{q}$ 1^{++} and 2^{++} mesons, why the light scalar mesons $a_0(980)$ and $f_0(980)$ have the puzzling mass character as $I = 1$ $a_0(980)$ mass is nearly equal to the $s\bar{s}$ like $I = 0$ $f_0(980)$ mass, we assume that the high mass of the $L = 1$ $q\bar{q}$ scalar mesons is caused by the mixing with the light scalar mesons and the puzzling mass character arises from the $qq\bar{q}\bar{q}$ structure for light scalar mesons. We consider that $a_0(980)$ and $f_0(980)$ constitute the $SU(3)$ nonet together with the re-established $\sigma(600)$ and the $\kappa(900)$. In this context, we analyze the inter-mixing between the light scalar nonet and the high mass $L = 1$ $q\bar{q}$ nonet and the intra-mixing among each nonet including the glueball mixing with the high mass scalar nonet.

INTRODUCTION

For a long time, there have been the following serious issues on scalar mesons: (1) Recent assignment of scalar mesons for $L = 1$ $q\bar{q}$ $SU(3)$ nonet [1] is $a_0(1450)$, $K_0^*(1430)$, $f_0(1370)$ and $f_0(1710)$. If so, what are the low-mass scalars, $a_0(980)$ and $f_0(980)$. (2) $a_0(980)$ and $f_0(980)$ are considered as $qq\bar{q}\bar{q}$ states [2] or $K\bar{K}$ molecules [3]. In mass region lower than 1 GeV, $\sigma(600)$ [4] as a chiral scalar partner of π is recently re-established. Could these and $\kappa(900)$ [5] form a light scalar nonet? (3) Why $L = 1$ $q\bar{q}$ scalar mesons have so heavier than the predicted masses from $L = 1$ $q\bar{q}$ 1^{++} and 2^{++} mesons? (4) $f_0(1500)$ is considered to be the most probable candidate of scalar glueball [6]. This exists near the mass region of $L = 1$ $q\bar{q}$ scalar mesons, then should mixes strongly with $I = 0$ $q\bar{q}$ meson. How rate this glueball $f_0(1500)$ mixes with $q\bar{q}$ scalar mesons? Assuming that the low mass scalar $a_0(980)$, $\kappa(900)$, $\sigma(600)$ and $f_0(980)$ make a scalar nonet and the high mass $L = 1$ $q\bar{q}$ nonet mixes with the low mass nonet [7], we will explain the high mass scalar mesons to have so heavier mass than the masses predicted from $L = 1$ $q\bar{q}$ 1^{++} and 2^{++} meson masses. We assume the $qq\bar{q}\bar{q}$ structure [2] for light scalar nonet. From this assumption we can explain the puzzling mass character for light scalar mesons as $I = 1$ $a_0(980)$ mass is nearly equal to the $s\bar{s}$-like $I = 0$ $f_0(980)$ mass. In this context, we analyze the inter-mixing between the light scalar nonet and the high mass scalar nonet and the intra-mixing among each nonet including the glueball mixing with the high mass $I = 0$ scalar mesons.

STRUCTURE OF LIGHT SCALAR MESONS

We present the scalar meson field as $N_a'^b$ for $L = 1$ $q\bar{q}$ scalar mesons and N_a^b for light scalar mesons. The $N_a'^b$ is the $SU(3)$ nonet represented by the field q_a and unti-triplet quark field $\bar{q}^b$ as $N_a'^b \sim q_a \bar{q}^b$. We assume that the light scalar mesons have $qq\bar{q}\bar{q}$ structure, then a and b in representation N_a^b denote the $SU(3)$ indices of "dual" quark $T_a = \epsilon_{abc}\bar{q}^b\bar{q}^c$ and "dual" anti-quark $\bar{T}^a = \epsilon^{abc}q_b q_c$, respectively [2, 8].

$$N_b^a \sim T_b \bar{T}^a \sim \epsilon_{bde}\bar{q}^d\bar{q}^e \epsilon^{abc} q_b q_c \quad \text{for } qq\bar{q}\bar{q} \text{ light scalar mesons.} \tag{1}$$

The explicit flavor configuration for these scalar nonet are represented as

$$\begin{aligned}
a_0^+ &\sim & \bar{s}dus, & \quad a_0^0 &\sim & \tfrac{1}{\sqrt{2}}(\bar{s}dds - \bar{s}uus), & \quad a_0^- &\sim & \bar{s}uds, \\
\kappa^+ &\sim & \bar{s}dud, & \quad \kappa^0 &\sim & \bar{s}uud, & & & \\
\bar{\kappa}^0 &\sim & \bar{u}dus, & \quad \kappa^- &\sim & \bar{u}dds, & & & \\
f_0 &\sim & \tfrac{1}{\sqrt{2}}(\bar{s}dds + \bar{s}uus), & \quad \sigma &\sim & \bar{u}dud & & &
\end{aligned} \tag{2}$$

in the ideal mixing limit.

We assume that the masses of light scalar mesons are described by the following chiral symmetric effective Lagrangian density

$$L^{eff} = -a\text{Tr}(NN) - b\text{Tr}(NNM) - \frac{1}{2}\lambda \text{Tr}(N)\text{Tr}(N), \tag{3}$$

where M is the "spurion matrix" representing the symmetry breaking effects of strange quark mass. From these expressions, we get the relations

$$m_{f_0}^2 = m_{a_0}^2 + 2\lambda, \quad m_{\sigma_0}^2 = 2m_\kappa^2 - m_{a_0}^2 + \lambda, \quad m_{f_0}^2 = \sqrt{2}l\lambda. \tag{4}$$

If $m_s > m_{u,d}$, we can get the desirable mass order, $m_{f_0}^2 \approx m_{a_0}^2 > m_\kappa^2 > m_{\sigma_0}^2$.

STRENGTH OF INTER-MIXING

The mass values of the 2^{++} and 1^{++} mesons cited in 2000 PDG [1] are

$$\begin{cases}
m_{a_2(1320)} = 1318\text{MeV}, & m_{K_2^*(1430)} = 1429\text{MeV}, \\
\quad m_{f_2(1270)} = 1275\text{MeV}, & m_{f_2'(1525)} = 1525\text{MeV}, \\
m_{a_1(1260)} = 1230\text{MeV}, & m_{K_1(1270/1400)} = 1339\text{MeV}, \\
\quad m_{f_1(1285)} = 1282\text{MeV}, & m_{f_1(1420)} = 1426\text{MeV}.
\end{cases} \tag{5}$$

For the $L = 1$ $q\bar{q}$ bound states, there is the $L \cdot S$ force relation as

$$m^2(2^{++}) - m^2(1^{++}) = 2(m^2(1^{++}) - m^2(0^{++})). \tag{6}$$

From this relation, the masses of $L = 1$ $q\bar{q}$ 0^{++} mesons before inter-mixing denoted by $\overline{a_0(1450)}$, $\overline{f_0(1370)}$, $\overline{K_0^*(1430)}$ and $\overline{f_0(1710)}$ are estimated as follows;

$$m_{\overline{a_0(1450)}} = m_{\overline{f_0(1370)}} = 1236\text{MeV}, \quad m_{\overline{K_0^*(1430)}} = 1307\text{MeV}, \quad m_{\overline{f_0(1710)}} = 1374\text{MeV}, \tag{7}$$

where the ideal (intra-)mixing mass relation $m^2_{\overline{f_0(1710)}} = 2m^2_{\overline{K^*_0(1430)}} - m^2_{\overline{a_0(1450)}}$ is used. The $m_{\overline{a_0(980)}}$ for $a_0(980)$ before inter-mixing is estimated as 1271MeV using the mass relation for 2-body mixing $m^2_{a_0(1450)} - m^2_{\overline{a_0(1450)}} = m^2_{\overline{a_0(980)}} - m^2_{a_0(980)}$. Similarly, the mass $m_{\overline{\kappa(900)}}$ for $I = 1/2$ $\kappa(900)$ before inter-mixing to be 1047MeV is obtained. The mass $m_{\overline{\sigma(600)}}$ for $\sigma(600)$ before inter-mixing is obtained as 760MeV from the ideal (intra-)mixing mass relation $m^2_{\overline{\sigma(600)}} = 2m^2_{\overline{\kappa(900)}} - m^2_{\overline{a_0(980)}}$.

$$m_{\overline{a_0(980)}} = m_{\overline{f_0(980)}} = 1271\text{MeV}, \quad m_{\overline{\kappa_0(900)}} = 1047\text{MeV}, \quad m_{\overline{\sigma(600)}} = 760\text{MeV}. \quad (8)$$

We can easily estimate the strength of inter-mixing for the $I = 1$ $a_0(1450)$ and $a_0(980)$ and $I = 1/2$ $K^*_0(1430)$ and $\kappa(900)$ because these states have no effects from the intra-mixing. If we express the transition strength between $\overline{a_0(980)}$ and $\overline{a_0(1450)}$ as λ_{a_0}, then the mass matrix is written as

$$\begin{pmatrix} m^2_{\overline{a_0(980)}} & \lambda_{a_0} \\ \lambda_{a_0} & m^2_{\overline{a_0(1450)}} \end{pmatrix}, \quad m_{\overline{a_0(980)}} = 1.271\text{GeV}, \quad m_{\overline{a_0(1450)}} = 1.236\text{GeV} \quad (9)$$

and this has the eigenvalues $m_{a_0(980)} = 0.985$GeV and $m_{a_0(1450)} = 1.474$GeV at the $\lambda_{a_0} = 0.600$GeV^2. Mixing angle is evaluated as $\theta_{a_0} = 47.1°$. Similarly, for $I = 1/2$ scalar mesons, mass eigenvalues $m_{\kappa(900)} = 0.900$GeV and $m_{K^*_0(1430)} = 1.412$GeV are obtained at the $\lambda_{K_0} = 0.507$GeV^2 and mixing angle $\theta_{K_0} = 29.5°$.

INTER-, INTRA- AND GLUEBALL MIXING

We assume that the inter-mixing between the light scalar mesons N and $L = 1$ $q\bar{q}$ mesons N' are represented as

$$L^{eff}_{01} = -\lambda_{01} \epsilon^{abc} \epsilon_{dec} N^d_a N'^e_b = \lambda_{01}(\text{Tr}(NN') - \text{Tr}(N)\text{Tr}(N'))$$
$$= \lambda_{01}[a^+_0 a'^-_0 + a^-_0 a'^+_0 + a^0_0 a'^0_0 + \kappa^+ K^{*-}_0 + \kappa^- K^{*+}_0$$
$$+ \kappa^0 \overline{K}^{*0}_0 + \overline{\kappa}^0 K^{*0}_0 - f_N f'_N - \sqrt{2} f_S f'_N - \sqrt{2} f_N f'_S]. \quad (10)$$

For the intra- and glueball mixing, we adopt the mixing analyzed in our previous paper [9], then the overall mixing containing the inter-, intra- and glueball mixing are represented as

$$\begin{pmatrix} f_0(980) \\ \sigma(600) \\ f_0(1370) \\ f_0(1710) \\ f_0(1500) \end{pmatrix} = \begin{pmatrix} m^2_N + 2\lambda_0 & \sqrt{2}\lambda_0 & \lambda_{01} & \sqrt{2}\lambda_{01} & 0 \\ \sqrt{2}\lambda_0 & m^2_S + \lambda_0 & \sqrt{2}\lambda_{01} & 0 & 0 \\ \lambda_{01} & \sqrt{2}\lambda_{01} & m^2_{N'} + 2\lambda_1 & \sqrt{2}\lambda_1 & \sqrt{2}\lambda_G \\ \sqrt{2}\lambda_{01} & 0 & \sqrt{2}\lambda_1 & m^2_{S'} + \lambda_1 & \lambda_G \\ 0 & 0 & \sqrt{2}\lambda_G & \lambda_G & \lambda_{GG} \end{pmatrix} \begin{pmatrix} f_N \\ f_S \\ f'_N \\ f'_S \\ f_G \end{pmatrix}. \quad (11)$$

We estimate the best fit values for λ_{01}, λ_0, λ_1, λ_G and λ_{GG} using the mass values as $m_N = m_{\overline{a_0(980)}} = m_{\overline{f_0(980)}} = 1271$MeV, $m_S = m_{\overline{\sigma(600)}} = 760$MeV,

$m_{N'} = m_{\overline{a_0(1450)}} = m_{\overline{f_0(1370)}} = 1236$MeV and $m_{S'} = m_{\overline{f_0(1710)}} = 1374$MeV. Taking the least χ^2 defined by $\sum_n (m_n - m_{n_0})^2 / \Delta m_n^2$, where m_n and Δm_n represent the experimental mass values and mass errors of the scalar meson n and m_{n_0} represents the value of the mass estimated, we can get the following best-fit values for λ_{01}, λ_0, λ_1, λ_G and λ_{GG}, and $m_{f_0(980)}$, $m_{\sigma(600)}$, $m_{f_0(1370)}$, $m_{f_0(1710)}$ and $m_{f_0(1500)}$:

$$\lambda_{01} = 0.51\text{GeV}^2, \quad \lambda_0 = 0.05\text{GeV}^2, \quad \lambda_1 = 0.05\text{GeV}^2,$$
$$\lambda_G = 0.26\text{GeV}^2, \quad \lambda_{GG} = 1.53\text{GeV}^2,$$
$$m_{f_0(980)} = 0.981(0.980/0.01)\text{GeV}, \quad m_{\sigma(600)} = 0.455(0.600/0.10)\text{GeV},$$
$$m_{f_0(1370)} = 1.376(1.350/0.05)\text{GeV}, \quad m_{f_0(1710)} = 1.715(1.715/0.007)\text{GeV},$$
$$m_{f_0(1500)} = 1.499(1.500/0.01)\text{GeV}, \quad (12)$$

where values in parentheses are (the experimental mass/the experimental error) of scalar mesons. Mixing matrix is obtained for estimated values of λ's as

$$\begin{pmatrix} f_0(980) \\ \sigma(600) \\ f_0(1370) \\ f_0(1710) \\ f_0(1500) \end{pmatrix} = \begin{pmatrix} 0.7129 & -0.3282 & -0.2223 & -0.5548 & 0.1640 \\ 0.1605 & 0.8402 & -0.5056 & -0.0604 & 0.0945 \\ 0.0625 & 0.4027 & 0.7000 & -0.5191 & -0.2729 \\ 0.5024 & 0.1550 & 0.4481 & 0.5221 & 0.5002 \\ -0.4580 & 0.0085 & 0.0639 & -0.3828 & 0.7997 \end{pmatrix} \begin{pmatrix} f_N \\ f_S \\ f_{N'} \\ f_{S'} \\ f_G \end{pmatrix}.$$
(13)

We estimated the $\chi^2 = 2.40$ for this case. On the other hand, when we adopt the inter-mixing expressed by the effective Lagrangian $\text{Tr}(NN')$ instead of $\text{Tr}(NN') - \text{Tr}(N)\text{Tr}(N')$, we get the χ^2 values as $\chi^2 = 249.9$, which cannot be accepted. If we do not include the glueball into the overall mixing, the χ^2 value is $\chi^2 = 3.92$, then the glueball mixing with other scalar mesons is preferred to no glueball mixing in the scalar meson spectrum.

REFERENCES

1. Particle Data Group, *Eur. Phys. Jour.* **C15**, 1–878(2000).
2. Jaffe, R. L., *Phys. Rev. D* **15**, 267–280(1977).
3. Weinstein, J., and Isgur, N., *Phys. Rev. Lett.* **48**, 659–662(1982).
4. Törnqvist, N. A., and Roos, M., *Phys. Rev. Lett.* **76**, 1575–1578(1996); Harada, M., et al., *Phys. Rev.* **D54**, 1991–2004(1996); Ishida, S., et al., *Prog. Theor. Phys.* **98** 1005–1010(1997).
5. Ishida, S., et al., *Phys. Rev.* **98** 621–629(1997); van Beveren, E., et al., *Zeit Phys.* **C30**, 615–620(1986).
6. Amsler, C., and Close, F., *Phys. Lett. B* **353**, 385–390(1995); Anisovich, V. V., Bugg, D. V., and Sarantsev, A. V., *Phys. Rev. D* **58**, 111503–11506(1998); Jaminon, M., and Van den Bosche, B., *Nucl. Phys. A* **619**, 285–294(1997); Lee, W., and Weingarten, D., *Phys. Rev. D* **59**, 094508–094514 (1999).
7. Black, D., Fariborz, A. H., Sannio, F., and Schechter, J., *Phys. Rev. D* **59**, 074026–074035(1999).
8. Achasov, N. N., *Nucl. Phys.* **A675**, 279c–284c(2000); Black, D., Fariborz, A. H., and Schechter, J., *Phys. Rev. D* **61**, 074001–074009(2000).
9. Teshima, T., and Oneda, S., *Phys. Rev. D* **33**, 1974–1979 (1986); Teshima, T., Kitamura, I., and Morisita, N., *Il Nuovo Cim.* **103A**, 175–184(1990).

The f_0 Mesons in Processes $\pi\pi \to \pi\pi, K\overline{K}$

Yu.S. Surovtsev[*], D. Krupa and M. Nagy

[*]*Bogoliubov Laboratory of Theoretical Physics, Joint Institute for Nuclear Research, Dubna 141 980, Moscow Region, Russia*
Institute of Physics, Slov.Acad.Sci., Dúbravská cesta 9, 842 28 Bratislava, Slovakia

Abstract. Combined analysis the experimental data on the processes $\pi\pi \to \pi\pi, K\overline{K}$ in the channel with $I^G J^{PC} = 0^+ 0^{++}$ in a model-independent approach leads to the following results: 1) The $f_0(665)$ state with properties of the σ-meson is proved to exist; 2) It is shown that the $f_0(980)$ and especially $f_0(1370)$ (if exists) have a dominant $s\bar{s}$ component; 3) Indications for the glueball nature of the $f_0(1500)$ and for the considerable $s\bar{s}$ component in the $f_0(1710)$ are obtained; 4) Conclusion on the linear realization of chiral symmetry (χS) is drawn.

INTRODUCTION

Obviously, it is important to have a model-independent information on investigated states and on their QCD nature. It can be obtained only on the basis of the first principles (analyticity, unitarity) immediately applied to analyzing experimental data. Earlier, we have proposed that method for 2- and 3-channel resonances [1]. We apply this below for the 2-channel case of the processes $\pi\pi \to \pi\pi, K\overline{K}$ in the channel with $I^G J^{PC} = 0^+ 0^{++}$. We shall show that the large background, which earlier one has obtained in various analyses of the s-wave $\pi\pi$ scattering [2], hides, in reality, the σ-meson [3] below 1 GeV and the effect of the left-hand branch-point. Furthermore, we shall obtain definite indications about the QCD nature of other f_0 resonances and about the linear χS realization.

Note that recent analyses of old and new experimental data, finding σ-meson below 1 GeV (see, *e.g.*, [2]), use either the Breit – Wigner form, which is insufficiently-flexible even if modified, or specific forms of interactions in the quark models; therefore, there one cannot talk about a model independence of results. Besides, there a large $\pi\pi$-background is obtained.

TWO CHANNEL FORMALISM

The 2-channel S-matrix for the coupled processes $\pi\pi \to \pi\pi, K\overline{K}$ is determined on the 4-sheeted Riemann surface. The elements $S_{\alpha\beta}$, where $\alpha, \beta = 1(\pi\pi), 2(K\overline{K})$, have the right-hand cuts along the real axis of the s-plane, starting at $4m_\pi^2$ and $4m_K^2$, and the left-hand cuts, beginning at $s = 0$ for S_{11} and at $4(m_K^2 - m_\pi^2)$ for S_{22} and S_{12}. The Riemann-surface sheets are numbered according to the signs of analytic continuations of the channel

momenta $k_1 = (s/4 - m_\pi^2)^{1/2}$, $k_2 = (s/4 - m_K^2)^{1/2}$ as follows:
signs $(\text{Im}k_1, \text{Im}k_2) = ++, -+, --, +-$ correspond to the sheets I, II, III, IV.

In the work [4], on the basis of the formulas of analytic continuations of the S-matrix elements to the unphysical sheets, we have shown that 2-channel resonances are represented by compact clusters of poles and zeros on the 4-sheeted Riemann surface of three types, which conveniently are distinguished by a pair of conjugate zeros on sheet I: (a) in S_{11}, (b) in S_{22}, (c) in each of S_{11} and S_{22}. The cluster kind is related to the state nature. The resonance, coupled relatively more strongly to the $\pi\pi$ channel than to the $K\bar{K}$ one, is described by the cluster of type (a); in the opposite case, of type (b) (say, the state with the dominant $s\bar{s}$ component); the flavour singlet (e.g. glueball) must be represented by the cluster of type (c).

For the simultaneous analysis of data on coupled processes we use the Le Couteur-Newton relations [5]. To take into account two right-hand branch-points at $4m_\pi^2$ and $4m_K^2$ and the left-hand one at $s = 0$, the uniformizing variable $v = (m_K \sqrt{s - 4m_\pi^2} + m_\pi \sqrt{s - 4m_K^2})/\sqrt{s(m_K^2 - m_\pi^2)}$ is used. It maps the 4-sheeted Riemann surface onto the plane, divided into two parts by a unit circle centered at the origin. The sheets I (II), III (IV) are mapped onto the exterior (interior) of the unit disk on the upper and lower v-half-plane, respectively. The physical region extends from the point i on the imaginary axis ($\pi\pi$ threshold) along the unit circle clockwise in the 1st quadrant to point 1 on the real axis ($K\bar{K}$ threshold) and then along the real axis to point $b = \sqrt{(m_K + m_\pi)/(m_K - m_\pi)}$ into which $s = \infty$ is mapped on the v-plane. The intervals $(-\infty, -b]$, $[-b^{-1}, b^{-1}]$, $[b, \infty)$ on the real axis are the images of the corresponding edges of the left-hand cut of the $\pi\pi$-scattering amplitude. The type (a) resonance is represented in S_{11} by two pairs of the poles on the images of the sheets II and III, symmetric to each other with respect to the imaginary axis, by zeros, symmetric to these poles with respect to the unit circle.

$S_{11}(v)$ has no cuts; however, $S_{12}(v)$ and $S_{22}(v)$ do have the cuts, arising from the left-hand cut on the s-plane, which further is neglected in the Riemann-surface structure, and the contribution on this cut is taken into account in the $K\bar{K}$ background as a pole on the real s-axis on the physical sheet in the sub-$K\bar{K}$-threshold region.

On v-plane, the Le Couteur-Newton relations are [1, 5]

$$S_{11} = \frac{d(-v^{-1})}{d(v)}, \quad S_{22} = \frac{d(v^{-1})}{d(v)}, \quad S_{11}S_{22} - S_{12}^2 = \frac{d(-v)}{d(v)}. \tag{1}$$

The $d(v)$-function already does not possess branch-points and is taken as $d = d_B d_{res}$, where $d_B = B_\pi B_K$; B_π contains the possible remaining $\pi\pi$-background contribution (the consequent analysis gives $B_\pi = 1$); $B_K = v^{-4}(1 - v_0 v)^4(1 + v_0^* v)^4$ [4] is that part of the $K\bar{K}$ background which does not contribute to the $\pi\pi$-scattering amplitude. $d_{res}(v) = v^{-M} \prod_{n=1}^{M}(1 - v_n^* v)(1 + v_n v)$ represents the contribution of resonances, described by one of three types of the pole-zero clusters, where M is the number of pairs of the conjugate zeros.

ANALYSIS OF THE EXPERIMENTAL DATA

When analyzing simultaneously the experimental data on the processes $\pi\pi \to \pi\pi, K\bar{K}$, we consider four admitted variants, in which the following states are taken into account: Variant 1: The $f_0(665)$ and $f_0(980)$) with the clusters of the type (**a**), and $f_0(1500)$, of the type (**c**); Variant 2: The same three resonances + the $f_0(1370)$ of the type (**b**); Variant 3: The $f_0(665)$, $f_0(980)$) and $f_0(1500)$ + the $f_0(1710)$ of the type (**b**); Variant 4: All the five resonances of the indicated types.

The $\pi\pi$-scattering data are described from the threshold to 1.89 GeV in all the four variants and are taken in this region from analysis [6], and below 1 GeV, from many works [2, 4]. For $\pi\pi \to K\bar{K}$, all accessible data are taken [2, 4], and the description regions extend from the threshold to ~ 1.4 GeV for variant 1, to ~ 1.46 GeV for variant 2, to ~ 1.5 GeV for variants 3 and 4. The quality of fits is illustrated by values of the total χ^2/N_{DF} for both analysed processes: 1.98, 2.45, 1.76 and 2.59 for variants 1,....,4 (the number of fitted parameters is 17 for variant 1, 21 for 2 and 3, 25 for 4; $v_0 = 0.954381 + 0.29859i, 0.97925 + 0.202657i, 0.954572 + 0.29798i, 0.982091 + 0.188405i$ for 1,...,4). We note that two variants (1 and 3) are best – both without the $f_0(1370)$.

Let us indicate the obtained poles of clusters for variant 4 on the complex energy plane ($\sqrt{s_r} = E_r - i\Gamma_r$, [MeV]): for $f_0(665)$: $600 \pm 16 - i(605 \pm 28)$ on sheet II, and $715 \pm 17 - i(59 \pm 6)$ on sheet III; for $f_0(980)$: $985 \pm 5 - i(27 \pm 8)$ on sheet II, and $984 \pm 18 - i(210 \pm 22)$ on sheet III; for $f_0(1370)$: $1310 \pm 22 - i(410 \pm 29)$ on sheet III, and $1320 \pm 20 - i(275 \pm 25)$ on sheet IV; for $f_0(1500)$: $1528 \pm 22 - i(385 \pm 25)$ on sheet II, and $1490 \pm 30 - i(220 \pm 24)$, $1510 \pm 20 - i(370 \pm 30)$ on sheet III, and $1510 \pm 21 - i(308 \pm 30)$ on sheet IV; for $f_0(1710)$: $1700 \pm 25 - i(86 \pm 16)$ on sheet III, and $1700 \pm 20 - i(115 \pm 20)$ on sheet IV.

The coupling constants with the $\pi\pi$ (g_1) and $K\bar{K}$ (g_2) systems are calculated through the residues of amplitudes at the pole on sheet II – for the (**a**) and (**c**) resonances, and on sheet IV – for the (**b**) states. Taking the resonance part of the T-matrix ($S_{ii} = 1 + 2i\rho_i T_{ii}$, $S_{12} = 2i\sqrt{\rho_1\rho_2}T_{12}$, where $\rho_i = \sqrt{(s-4m_i^2)/s}$) as $T_{ij}^{res} = \sum_r g_{ir}g_{rj}D_r^{-1}(s)$ with $D_r(s)$ being an inverse propagator ($D_r(s) \propto s - s_r$), we obtain (in GeV): for $f_0(665)$: $g_1 = 0.652 \pm 0.065$ and $g_2 = 0.724 \pm 0.1$, for $f_0(980)$: $g_1 = 0.167 \pm 0.05$ and $g_2 = 0.445 \pm 0.031$, for $f_0(1370)$: $g_1 = 0.116 \pm 0.03$ and $g_2 = 0.99 \pm 0.05$, for $f_0(1500)$: $g_1 = 0.657 \pm 0.113$ and $g_2 = 0.666 \pm 0.15$.

The $f_0(980)$ and especially the $f_0(1370)$ are coupled essentially more strongly to the $K\bar{K}$ system than to the $\pi\pi$ one, that tells about the dominant $s\bar{s}$ component in these states. The $f_0(1500)$ has the approximately equalled coupling constants with the $\pi\pi$ and $K\bar{K}$ systems, that apparently could point up to its dominant glueball component. The $f_0(1710)$ is represented by the cluster, pointing up to a dominant $s\bar{s}$ component.

We present also the calculated scattering lengths (in $m_{\pi^+}^{-1}$): For the $K\bar{K}$ scattering: $a_0^0 = -1.25 \pm 0.11 + (0.65 \pm 0.09)i$, (1), $a_0^0 = -1.548 \pm 0.13 + (0.634 \pm 0.1)i$, (2), $a_0^0 = -1.19 \pm 0.08 + (0.622 \pm 0.07)i$, (3), $a_0^0 = -1.58 \pm 0.12 + (0.59 \pm 0.1)i$, (4). $\mathrm{Re}a_0^0(K\bar{K})$ is very sensitive to whether this state exists or not.

For $a_0^0(\pi\pi)$, we obtain $0.27 \pm 0.06, 0.267 \pm 0.07, 0.28 \pm 0.05, 0.27 \pm 0.08$ for variants 1,...,4. Compare with the results of some other both theoretical and experimental works: 0.26 ± 0.05 (L. Rosselet et al.[6], analysis of $K \to \pi\pi e\nu$ using Roy's model); 0.24 ± 0.09 (A.A. Bel'kov et al.[6], analysis of $\pi^- p \to \pi^+\pi^- n$ using the effective range formula); 0.23 (S. Ishida et al.[7], modified analysis of $\pi\pi$ scattering using Breit-Wigner forms); 0.16 (S. Weinberg [8], current algebra (non-linear σ-model)); 0.20 (J. Gasser, H. Leutwyler [8], one-loop corrections, non-linear χS realization); 0.217 (J. Bijnens at al.[8], two-loop corrections, non-linear χS realization); 0.26 (M.K. Volkov [8], linear χS realization); 0.28 (A.N. Ivanov, N.I. Troitskaya [8], a variant of theory with linear χS realization.

We have presented model-independent results: the poles, coupling constants and scattering lengths. Masses and widths of these states that should be calculated from the pole positions and coupling constants are highly model-dependent. $E.g.$, supposing, that the $f_0(665)$ is the σ-meson, we obtain $m_\sigma \approx 342$ MeV from the relation $g_{\sigma\pi\pi} = (m_\sigma^2 - m_\pi^2)/\sqrt{2}f_{\pi^0}$ (here $f_{\pi^0} = 93.1$ MeV). If we take the resonance part of amplitude as $T^{res} = \sqrt{s}\Gamma/(m_\sigma^2 - s - i\sqrt{s}\Gamma)$, we obtain $m_\sigma \approx 850$ MeV and $\Gamma \approx 1240$ MeV.

Summary

On the basis of a simultaneous description of the isoscalar s-wave channel of the processes $\pi\pi \to \pi\pi, K\bar{K}$ with a parameterless representation of the $\pi\pi$ background, a model-independent confirmation of the σ-meson below 1 GeV is obtained.

The existence of the low-lying state $f_0(665)$ with the σ-meson properties and the obtained $a_0^0(\pi\pi) \approx 0.27[m_{\pi^+}^{-1}]$ suggest the linear χS realization.

The analysis of the used experimental data evidences that the $f_0(980)$ and especially the $f_0(1370)$ resonance (if exists – variants 2 and 4), have the dominant $s\bar{s}$ component. The $K\bar{K}$ scattering length is very sensitive to whether the $f_0(1370)$ state exists or not.

The $f_0(1500)$ has the approximately equalled coupling constants with the $\pi\pi$ and $K\bar{K}$ systems, that apparently could point up to its dominant glueball component.

The $f_0(1710)$ is represented by the cluster, corresponding the state with dominant $s\bar{s}$ component.

This work has been supported by the Grant Program of Plenipotentiary of Slovak Republic at JINR. Yu.S. and M.N. were supported in part by the Slovak Scientific Grant Agency, Grant VEGA No. 2/7175/20; and D.K., by Grant VEGA No. 2/5085/99.

REFERENCES

1. Krupa, D., Meshcheryakov, V.A., and Surovtsev, Yu.S., *Nuovo Cim.* **A109**, 281 (1996).
2. Review of Particle Physics, *Europ. Phys. J.* **C15**, 1 (2000).
3. Nambu, Y., and Jona-Lasinio, G., *Phys. Rev.* **122**, 345 (1961); Volkov, M.K., *Ann. Phys.* **157**, 282 (1984); Hatsuda, T., and Kunihiro, T., *Phys. Rep.* **247**, 223 (1994); Delbourgo, R., and Scadron, M.D., *Mod. Phys. Lett.* **A10**, 251 (1995).
4. Surovtsev, Yu.S., Krupa, D., and Nagy, M., *Phys. Rev.* **D63**, 054024 (2001).
5. Le Couteur, K.J., *Proc. Roy. Soc.* **A256**, 115 (1960); Newton, R.G., *J. Math. Phys.* **2**, 188 (1961); Kato, M., *Ann. Phys.* **31**, 130 (1965).
6. Hyams, B., et al., *Nucl. Phys.* **B64**, 134 (1973); *ibid.* **B100**, 205 (1975).
7. Ishida, S., et al., *Progr. Theor. Phys.* **95**, 745 (1996).
8. Weinberg, S., *Phys. Rev. Lett.* **17**, 616 (1966); Gasser, J., and Leutwyler, H., *Ann. Phys.* **158**, 142 (1984); Bijnens, J., et al., *Phys. Lett.* **B374**, 210 (1996); Volkov, M.K., *Phys. Elem. Part. At. Nucl.* **17**, part 3, 433 (1986); Ivanov, A.N., Troitskaya, N.I., *Nuovo Cim.* **A108**, 555 (1995).

Chiral Symmetry and Scalars

S.F. Tuan

*Department of Physics, University of Hawaii at Manoa
Honolulu, HI 96822-2219, U.S.A.*

Abstract. The suggestion by Jaffe that if σ is a light $q^2\bar{q}^2$ state 0^{++} then even the fundamental chiral transformation properties of the σ becomes **unclear**, has stimulated much interest. Adler pointed out that in fact the seminal work on chiral symmetry via PCAC consistency, is really quite consistent with the σ being predominantly $q^2\bar{q}^2$. This interpretation was actually backed by subsequent work on effective Lagrangian methods for linear and non linear realizations. More recent work of Achasov suggests that intermediate four-quark states determine amplitudes involving other scalars $a_0(980)$ and $f_0(980)$ below 1 GeV, and the report by Ning Wu that study on σ meson in $J/\psi \to \omega\pi^+\pi^-$ continue to support a non $q\bar{q}$ σ with mass as low as 390 MeV. It is also noted that more recent re-analysis of πK scattering by S. Ishida *et al.* together with the work of the E791 Collaboration, support the existence of the scalar κ particle with comparatively light mass as well.

In an intriguing paper Jaffe [1] pointed out that the QCD "Breit Interaction" summarized by an effective Hamiltonian acting on the quarks' spin and color indices,

$$H_{eff} \propto -\sum_{i\neq j} \overset{\lambda}{\sim}_i \cdot \overset{\lambda}{\sim}_j \, \vec{\sigma}_i \cdot \vec{\sigma}_j$$

affirm earlier work [2] that $f_0(980), a_0(980), \sigma(560)$, and $\kappa(900)$ scalars make a nonet with mass spectrum, decay couplings and widths that look qualitatively like $\bar{q}\bar{q}qq$ system. Alford and Jaffe [3] raised the pertinent question that if light $\bar{q}^2q^2$ states are, in fact, a universal phenomenon below 1 GeV, and if σ is predominantly a $\bar{q}^2q^2$ object, then the chiral transformation properties of the σ have to be re-examined. The π and the σ are usually viewed as members of a (broken) chiral multiplet. In the naive $\bar{q}q$ model both π and σ are in $(1/2, 1/2) \oplus (1/2, 1/2)$ representation of $SU(2)_L \otimes SU(2)_R$ before symmetry breaking. In a $\bar{q}^2q^2$ model, as in the real world, the chiral transformation properties of the σ are **not clear**.

There remains a body of recent literature [4] which retains in essence the $\bar{q}q$ model for the σ meson. For instance Törnqvist *et al.* [4] used chiral-symmetry constraints in their study. Chiral symmetry constraints have been discussed by for instance Oller [5] where it is said that the range of applicability of chiral constraints could be enlarged up to around 0.8 GeV. Because of the model dependence of experimental analysis, current wisdom suggests that σ has mass between 400 to 700 MeV and hence the use of chiral constraints would appear to be valid. However in the context of Törnqvist et al. [4] the 4q $\bar{q}^2 q^2$ scheme is not easy to combine with chiral symmetry constraints, which are crucial to their work. Indeed for weak interactions, their (chiral) results are the same as the strong interaction quark-level linear σ model (LSM) $\bar{q}q$ scheme in one-loop order together with the electromagnetic (LSM) analogue [6]. Törnqvist [6] expressed further concern that 4q or rather 2 meson models and chiral symmetry, the chiral symmetry can of course be imposed in a model like the linear σ model (LSM), but then all states σ, f_0, a_0 and the **pion** would be basically 4q states! We shall return to this concern later on in this paper.

On an optimistic note, Adler [7] pointed out that in the original PCAC Consistency Condition paper [8], when analysed for the pion- pion scattering case, led to the conclusion that there had to be a broad low energy pion-pion scattering resonance. This is then quite consistent with the σ being predominantly $q^2 \bar{q}^2$. Secondly, the numerical estimates of the "sigma term" from current algebra [9], assuming it is $q\bar{q}$ [or $(3,\bar{3}) + (\bar{3},3)$] were always an embarrassment, since they were generally off by a factor of two whereas other things worked much better than that (typically of order 10% or less [8]). This again is quite consistent with the dominant spectral weight not being in the $q\bar{q}$ channel. Third, in Zumino's 1970 Brandeis lectures [10] on effective Lagrangian methods, he discusses nonlinear realizations on pp. 451-454 (see also pp. 481-483, 485); he first describes the linear realization of the σ model, stating that σ..... (is the field) of a scalar isoscalar $\pi - \pi$ resonance. He then shows how by a redefinition the same low energy results arise from a nonlinear transformation involving the redefined pion field only; in this nonlinear transformation, $\vec{\pi}^2$ plays a role analogous to that played by σ in the linear case. So again, it is expected that the σ should be a two pion state, and hence not surprisingly that it is dominantly $q^2 \bar{q}^2$.

Jaffe [11] expanded on his understanding (or lack thereof) of the role of σ in chiral symmetry [3]. Since chiral $SU(2)$ symmetry is spontaneously broken, the physical particles do not have to transform as irreducible representations of $SU(2) \times SU(2)$. There is a prejudice (originating in the quark model?) that the pion transforms like $\bar{q}q$, and an even less well justified prejudice that the σ transforms in the same way as the pion. However, there does not exist any good reason to think that the transformation properties of the σ are linked to those of the pion when $SU(2) \times SU(2)$ is spontaneously broken. Perhaps another way of saying the same thing [12] is that chiral symmetry does not mesh well with either constituent quarks nor with QCD's current quarks, hence chiral symmetry does not require multi-quark states to fuse into a $q\bar{q}$ state as originally thought.

Experimental evidence for the existence of the scalar σ at the low mass value of 390 MeV with total width of order 282 MeV has been recently reported by Ning Wu [13] based on the study of σ particle in $J/\psi \to \omega \pi^+ \pi^-$ from 7.8×10^6 BESI J/ψ data. There is also the newly reported [14] $\sigma(\pi\pi)$ scalar resonance with a σ mass and width of $478 \pm 24 \pm 17 MeV/c^2$ and $342 \pm 42 \pm 21 MeV/c^2$. Indeed recent re-analysis of the $\pi\pi$ scattering data by S. Ishida et al. [15] shows evidence for the existence of σ with comparatively light mass also. This same scattering data [15] for πK also showed evidence for the existence of the κ particle also of relatively light mass. This is corroborated again by the newly reported [14] $\kappa(K\pi)$ scalar resonance with a κ mass and width of 815 ± 30 MeV/c^2 and 560 ± 116 MeV/c^2. However Achasov [16] has cautioned that information on these scalars can be obtained only in strongly model dependent ways up to now. It seems reasonable that together with the status of $f_0(980)$ and $a_0(980)$ rather carefully analysed by Achasov and Gubin [17] **we do nevertheless have a nonet of $q^2\bar{q}^2$ scalars below 1 GeV**, though the mass and width of some of these scalars remain to be pinned down more precisely. Coming back to a more theoretical understanding of the situation, Achasov [16] reassured that Törnqvist's fear [6] that the pion also may end up as a 4q state is strongly overstated. The point is that one can not say that a field contains a fixed number of quarks. It is approximately true only in some energy (virtuality) region. For example, when virtualities of σ states have the order of the pion mass they show themselves as two-quark states, the chiral partners of pions, but when virtualities of σ states are of the order of 1 GeV (remember a σ of mass 700 MeV remains in the acceptable range), they can show themselves as four-quark states. Jaffe [11] elaborated further that the "quark content" of a particular meson is a heuristic concept at best. In some contexts the pion appears to be a $q\bar{q}$ state (for example as a member of an SU(3) meson octet); in others it appears to be a "wave on the chiral vacuum", which would be a coherent state in the Bogoliubov sense, including arbitrarily high numbers of $q\bar{q}$. The point is that the σ, the $f_0(980)$, and $a_0(980)$ has always been that the principal features of their mass spectrum, couplings to pseudoscalars, and to electromagnetic fields, are well described by a dominant $qq\bar{q}\bar{q}$ content. Hence there is agreement with Achasov that the quark content can be regarded as "virtuality" dependent. Jaffe [11] also pointed out that his understanding of Jona Lasinio/Nambu spontaneous symmetry breaking where

$$\sigma = \sqrt{1 - \vec{\pi}^2/f_\pi^2} \tag{1}$$

$$= 1 - \frac{\vec{\pi}^2}{2f_\pi^2} + \cdots \tag{2}$$

is in fact the same as that of Zumino [10] who discussed spontaneous symmetry breaking in the non linear realization case as

$$\delta\vec{\pi} = 2\vec{\alpha}\sqrt{\kappa^2 - \vec{\pi}^2} \tag{3}$$

where $\kappa = (1/2)f_\pi$. Expansion of the r.h.s. of (3) in terms of $[\vec{\pi}^2/\kappa^2]$, one would get something very similar to the r.h.s. of (2) up to a multiplicative factor. Hence Jaffe is in agreement with Adler [7].

We have certainly come a long way from the traditional naive quark model classification of hadron states of some 35 years ago [18]. For some trained in the traditional approach like myself, what is described above comes as a surprise bordering on shock. Hence the opportunity to air out these concerns at Hadron2001 is much appreciated. (During the discussions after this talk, Professor J. Schechter pointed out the work of the Syracuse group [19] which addressed quantitatively some of the issues presented here.)

I wish to thank my scientific colleagues Kolia Achasov, Steve Adler, Bob Jaffe, and Nils Törnqvist for very helpful communications and discussions. This work was supported in part by the U.S. Department of Energy under Grant DE-FG-03-94ER40833 at the University of Hawaii at Manoa.

REFERENCES

1. R.L. Jaffe, hep-ph/0001123, M.I.T. CTP # 2938.
2. R.L. Jaffe, *Phys. Rev.* **D15**, 267 (1977); *Phys. Rev.* **D15**, 281 (1977).
3. M. Alford and R.L. Jaffe, hep-lat/0001023, MIT-CTP-2940.
4. N.A. Törnqvist *et al.*, hep-ph/0005106; M. Volkov *et al.*, hep-ph/0007131.
5. J.A. Oller, hep-ph/0007349.
6. M.D. Scadron and N.A. Törnqvist (private communications).
7. S.L. Adler (private communication).
8. S.L. Adler, *Phys. Rev.* **137**, B1022 (1965).
9. E. Reya, *Rev. Mod. Phys.* **46**, 545 (1974).
10. B. Zumino, Lectures on Elementary Particles and Quantum Field Theory, Volume 2 (1970) (edited by S. Deser, M. Grisaru, and H. Pendleton).
11. R.L. Jaffe (private communication).
12. J.L. Rosner (private communication).
13. Ning Wu, hep-ex/0104050.
14. Carla Gobel, E791 Collaboration, hep-ex/0012009.
15. S. Ishida *et al.*, *Prog. Theor. Phys.* **95**, 745 (1996).
16. N.N. Achasov (private communication).
17. N.N. Achasov and V.V. Gubin, *Phys. Rev.* **D63**, 094007 (2001); see also N.N. Achasov, *these Proceedings*.
18. R.H. Dalitz, in *High Energy Physics* (Gordon and Breach, New York, 1966), p. 253.
19. D. Black *et al*, Phys. Rev. **D64**, 014031 (2001).

The σ-Meson Production in Excited ϒ Decay Processes

T. Komada, M. Ishida[†] and S. Ishida[‡]

Dept. of Engineering Sci., Junior Coll. Funabashi, Nihon Univ., Funabashi 274-8501, JAPAN
[†]*Dept. of Phys., Tokyo Inst. of Technology, Tokyo 152-8551, JAPAN*
[‡]*Atomic Energy Research Inst., Coll. of Sci. and Tech., Nihon Univ., Tokyo 101-0062, JAPAN*

Abstract. We analyze the $\pi\pi$ production amplitudes in the excited ϒ decay processes, $\Upsilon(2S) \to \Upsilon(1S)\pi^+\pi^-$, $\Upsilon(3S) \to \Upsilon(1S)\pi^+\pi^-$ and $\Upsilon(3S) \to \Upsilon(2S)\pi^+\pi^-$, and the $\pi\pi$ and $K\bar{K}$ production amplitudes in the charmonium decay processes, $\psi(2S) \to J/\psi\pi^+\pi^-$ and $J/\psi \to \phi\pi^+\pi^-$, ϕK^+K^-, including the possible effect of light σ production. The amplitudes are parametrized by the sum of Breit-Wigner amplitudes for the σ and the other relevant particles and of the direct 2π-production amplitude, following the VMW method. All the $\pi\pi$ (and $K\bar{K}$) mass spectra are reproduced well with the obtained values of σ-parameters, $m_\sigma = 526^{+48}_{-37}$MeV and $\Gamma_\sigma = 301^{+145}_{-100}$MeV, which is almost consistent with the values in our previous phase shift analyses.

(**Introduction**)   Whether the light σ-meson really exists or does not is an important problem in hadron physics. For many years, its existence had been neglected phenomenologically mainly due to the negative result of the analyses of $I = 0$ S-wave $\pi\pi$ scattering phase shift.

In many $\pi\pi$-production experiments, a large event concentration or a bump structure in the spectra of $\pi\pi$ invariant mass $m_{\pi\pi}$ around 500 MeV had been observed, however, conventionally it was not regarded as σ-resonance, but as a mere $\pi\pi$-background, under influence of the so callled "universality argument."[1] In this argument, it is stated that because of the unitarity of S-matrix and of the analyticity of the amplitudes, the $\pi\pi$ production amplitude $\mathcal{F}$ takes the form $\mathcal{F} = \alpha(s)\mathcal{T}$ ($\mathcal{T}$ being the $\pi\pi$ scattering amplitude), with a slowly varying real function $\alpha(s)$. The pole position of S-matrix is determined solely through the analysis of $\mathcal{T}$, which was believed to have no light σ-pole at that time.

Recently the data of $\pi\pi$-scattering phase shift had been reanalyzed by many groups[2] including ours and the existence of light $\sigma(450 \sim 600)$ was strongly suggested. The result of no σ-existence in the conventional analyses was pointed out[3] to be due to the lack of consideration on the cancellation mechanism guaranteed by chiral symmetry, and shown to be not correct. Furthermore, we have pointed out that the "universality argument" should be revised, taking into account the quark physical picture of hadrons[4]: The essential point is that the strong interaction is a residual interaction of QCD among all color-neutral bound states of quarks, anti-quarks and gluons, and accordingly the requirement of unitarity and analyticity must be made on the S-matrix elements with the right "quark physical bases," including not only the stable π meson but also the "stable" σ meson (which is also considered to be a $q\bar{q}$ bound state). The universality argument (and the conventional application of final state interaction (FSI) theorem) with the S-matrix bases including only stable particles, is not correct: Since this σ-meson has, in principle, an independent production coupling generally with a strong phase from the 2π system.

Thus, the production amplitude $\mathcal{F}$ has "independent" properties from the scattering amplitude $\mathcal{T}$, except that the pole position of resonant particles should be universal through both amplitudes. Accordingly, the data of many ππ-production experiments, which had been analyzed following the universality argument, must be reanalyzed independently from $\mathcal{T}$ by including the effect of σ-production. We parametrize $\mathcal{F}$ as a sum of Breit-Wigner amplitudes for the relevant resonances (including σ-resonance) and of the direct 2π production amplitude, following the VMW method. In the VMW method the above mentioned universal pole singularity is explicitly taken into account, and this method is consistent with the unitarity of the total S-matrix (and also with the FSI theorem, applied correctly) with the above metioned right bases. The VMW method had already been applied to the processes of $J/\psi \to \omega\pi\pi$ decay, $p\bar{p} \to 3\pi^0$ annihilation by us[6], and essentially similar method to the process $D^- \to \pi^-\pi^+\pi^+$ by E791[8]. The large event concentration in low $m_{\pi^+\pi^-}$ region in $J/\psi \to \omega\pi\pi$ and $D^- \to \pi^-\pi^+\pi^+$ was satisfactorily explained by σ-production, while in $p\bar{p} \to 3\pi^0$ clear evidences for σ-meson production were shown. In this paper we apply this method to the analyses of the hadronic decays of excited Υ, Υ(2S, 3S), and ψ(1S, 2S).

(Method of the analyses) We analyze[9] the $m_{\pi\pi}$ spectra of the processes, $\Upsilon(2S) \to \Upsilon(1S)\pi^+\pi^-$, $\Upsilon(3S) \to \Upsilon(1S)\pi^+\pi^-$, $\Upsilon(3S) \to \Upsilon(2S)\pi^+\pi^-$ and $\psi(2S) \to J/\psi\pi^+\pi^-$ following the VMW method in one-channel form. The $\mathcal{F}$ is given by a coherent sum of σ-Breit-Wigner amplitude and of direct 2π-production amplitude as

$$\mathcal{F} = \frac{e^{i\theta_\sigma} r_\sigma}{m_\sigma^2 - s - i\sqrt{s}\Gamma_\sigma(s)} + r_{2\pi}e^{i\theta_{2\pi}}; \quad \Gamma_\sigma(s) = \frac{g_\sigma^2 p_1(s)}{8\pi s}, \quad p_1(s) = \sqrt{\frac{s}{4} - m_\pi^2}, \quad (1)$$

where $r_\sigma(r_{2\pi})$ is the σ(2π) production coupling and $\theta_\sigma(\theta_{2\pi})$ corresponds to the production phase. These parameters are process-dependent. The $J/\psi \to \phi\pi^+\pi^-$, ϕK^+K^- process is analyzed by VMW method in two-channel form, where $f_0(980)$, $f_0(1370)$, $f_0(1500)$, as well as σ, are included. All the relevant spectra are fitted by using common values of the parameters, the mass of m_σ and the σππ coupling $g_{\sigma\pi\pi}$.

(Results and Conclusion) The results of our analysis are shown in Fig. 1. The spectra of six different processes are reproduced well by the interference between the σ-Breit-Wigner amplitude with $m_\sigma = 526^{+48}_{-37}$MeV, $\Gamma_\sigma = 301^{+145}_{-100}$MeV, and the direct 2π production amplitude. The corresponding pole position is $\sqrt{s_{pole}} = (535^{+48}_{-36}) - i(155^{+76}_{-53})$MeV. The total χ^2 is $\chi^2/(N_{data} - N_{param}) = 86.5/(150 - 37) = 0.77$. The contribution of σ and of direct 2π production amplitudes to the spectra are given, respectively, by dot and dot-dashed lines in this figure. It is notable that the destructive interference between σ amplitude and 2π amplitudes explains the suppression of the spectra in the threshold region of $\Upsilon(2S \to 1S)$ and $\psi(2S \to 1S)$ decays, while in $\Upsilon(3S \to 1S)$ decay these two amplitudes interfere constructively, and the steep increase from the threshold is reproduced. These threshold behaviors of the production amplitudes are[3] shown to be consistent

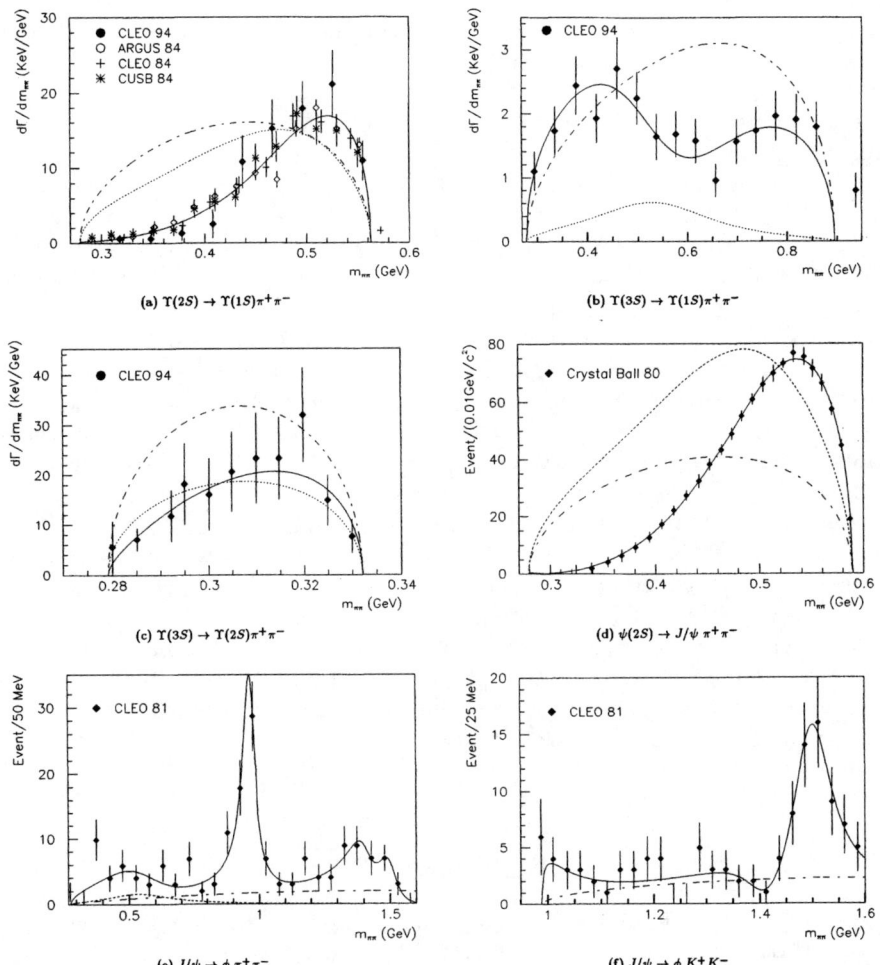

FIGURE 1. The result of the fit to the $\pi\pi$ (or KK) mass spectra of (a) $\Upsilon(2S) \to \Upsilon(1S)\pi\pi$, (b) $\Upsilon(3S) \to \Upsilon(1S)\pi\pi$, (c) $\Upsilon(3S) \to \Upsilon(2S)\pi\pi$, (d) $\psi(2S) \to J/\psi\pi\pi$, (e) $J/\psi \to \phi\pi\pi$, and (f) $J/\psi \to \phi KK$. The bold line represents the fit, while the dotted (dot-dashed) does the contribution of σ(direct 2π)-production. By multiplying appropriate proportional factors, the different data sets are adjusted to the one with the largest number of events. For experimental references, see [9].

with the restriction from chiral symmetry. It is especially interesting that the double peak structure, with the bottom around the σ-peak position, of the spectra in $\Upsilon(3S \to 1S)$ decays is also reproduced well by the above interference between the direct 2π production amplitude with zero phase and the σ-production amplitude with a moving phase. That is, we are observing the very phase motion of the σ-Breit-Wigner formula through the variation of the spectra in this process.

The obtained values of masses and widths of σ-meson given above are almost consistent with the results obtained in our previous $\pi\pi$ phase shift analysis[2], $m_\sigma = 535 \sim 675$ MeV and $\Gamma_\sigma = 385 \pm 70$ MeV. These results give strong evidence for existence of the light $\sigma(450–600)$.

The consistency of our result of analyses with the general constraints from chiral symmetry is discussed in the separate talk[10].

In this conference, there was raised a doubt[7] on the σ-existence along the line of the universality argument;

i) on $J/\psi \to \omega\pi\pi$ process. The experimental $\cos\theta$-distributions show the parabolic shape around $\cos\theta \approx 0$ in the range of $m_{\pi\pi}$, 300 through 700 MeV, coming from the interference between the S and D waves. However, the reverse of the direction of the parabolic shape, which is expected from the relative phase motion, to the background D-wave phase, of the σ-Breit-Wigner amplitude, was not observed. This criticism will become invalid, if we take into account the possible strong phase. For example, $\theta_\sigma = -90°$.

ii) on pp-central collision, $pp \to pp\pi\pi$. The result of the partial wave analysis by WA102 shows that the non-resonant $\pi\pi$-production from one pion exchange is enough to reproduce the S- and D-wave components in low $m_{\pi\pi}$ region. We consider that it is necessary, in addition to this, to consider the effect of the σ-Breit-Wigner amplitude, since the relative phase between S- and D-wave amplitudes shows a structure around $m_{\pi\pi} \sim 500$MeV region, implying the interference between the non-resonant $r_{2\pi}e^{i\theta_{2\pi}}$ and $r_\sigma e^{i\theta_\sigma}$ in our formula Eq. (1). By the way, the experimental relative phase is completely different from that in the $\pi\pi$-scattering. This fact clearly shows the universality-argument (where $\alpha = \mathcal{F}/\mathcal{T}$ is supposed to be real) is not correct.

iii) on $\pi\pi$-scattering. No rapid phase variation due to $\sigma(600)$ has been observed. We had mentioned in many occasions[10] the reason why the σ-Breit-Wigner phase motion is not directly observed in $\pi\pi$-scattering: Because of the constraints from chiral symmetry the σ-amplitude must be strongly cancelled out by the non-resonant repulsive $\pi\pi$-amplitude in $\pi\pi$-scattering.

REFERENCES

1. K. L. Au, D. Morgan and M. R. Pennington, Phys. Rev. **D35** (1987), 1633. M. R. Pennington, in proc. of Hadron95 in Manchester, UK, 1995, (World Scientific).
2. N. A. Tornqvist, summary talk of "σ-meson 2000", KEK-proceedings 2000-4. S. Ishida, M. Ishida, H. Takahashi, T. Ishida, K. Takamatsu and T. Tsuru, Prog. Theor. Phys. **95** (1996), 745; **98** (1997), 1005. E.Beveren, T.A.Rijken,K.Metzger,C.Dullemond,G.Rupp and J.E.Ribeiro, Z.Phys. **C30**(1986), 615. N. N. Achasov and G. N. Schestakov, Phys. Rev. D **49** (1994), 5779. R. Kaminski, L. Lesniak and J. -P. Maillet, Phys. Rev. D **50** (1994), 3145. N. A. Tornqvist, Z. Phys. C **68** (1995), 647. N. A. Tornqvist and M. Roos, Phys. Rev. Lett. **76** (1996), 1575. M. Harada, F. Sannino and J. Schechter, Phys. Rev. D **54** (1996), 1991.
3. M.Ishida, Prog. Theor. Phys. **96** (1996), 853.; M. Ishida in proc. of Hadron97 in BNL, 1997, AIP conf. proc. 432; in proc. of WHS99 in Frascati, 1999, Frascati Physics Series 15, 1999.
4. S. Ishida, in proc of "Possible Existence of σ-Meson and Its Implication to Hadron Physics (σ-meson 2000)," YITP Kyoto, 2000; in proc. of Hadron97 in BNL, 1997, AIP conf. proc. 432.
5. T. Tsuru, in proc. of σ-Meson 2000; K. Takamatsu in proc. of Hadron97.
6. M. Ishida, T. Komada et al., Prog. Theor. Phys. **104** (2000), 203.
7. W. Ochs, this conference.
8. C. Gobel, this conference.
9. T. Komada, M. Ishida and S. Ishida, this conference; Phys. Lett. **B 508** (2001), 31.
10. M. Ishida, S. Ishida, T. Komada and S. I. Matsumoto, this conf.; Phys. Lett. **B 518** (2001), 47.

BARYONS, EXPERIMENT

High lying N* studies in electromagnetic double charged pion production

V. I. Mokeev*, M. Ripani[†], M. Anghinolfi[†], M. Battaglieri[†],
R. De Vita[†], G. V. Fedotov*, E. N. Golovach*, B. S. Ishkhanov*,
M. V. Osipenko*, G. Ricco[†¶], V. Sapunenko[†],
M. Taiuti[†] and CLAS Collaboration

*Institute of Nuclear Physics, Moscow State University,
Vorob'evy gory, Moscow, 119899 Russia*
[†]*Istituto Nazionale di Fisica Nucleare, Sez. di Genova, Genova, Italy*
[¶] *Universita di Genova, via Dodecaneso 33, I-16146 Genova, Italy*

Abstract. A phenomenological model for double charged pion production is presented, aimed to extract N* electromagnetic form factors from measured observables (differential cross-sections, asymmetries). The preliminary results of CLAS data analysis on double charged pion production by virtual photons are discussed, focusing on high lying N* electromagnetic excitation and signals from possible "missing" baryon states.

INTRODUCTION

Investigation of double charged pion production by real and virtual photons on proton represents an important issue in studies of high lying N* states (M > 1.6GeV) and search for missing baryon states now in progress in JLAB,GRAAL,ELSA [1,2].

We developed a phenomenological approach to describe double charged pion production on proton by real and virtual photons [3]-[7] and applied it to the analysis of data the CLAS Collaboration at Jefferson Lab in N* excitation region [1,2]. Our approach relates quantities of physical interest (N* electromagnetic form factors, contributions of different quasi-two-body channels: $\pi^-\Delta^{++}$, $\pi^+\Delta^0$, ρp) with measured observables (differntial cross-sections, asymmetries), allowing to extracting physical parameters from experimental data.

MODEL DESCRIPTION

Complexity of double charge pion production mechanisms was parametrized in terms of meson-baryon degrees of freedom on tree level as coherent sum of quasi-two-body-processes: $\gamma p \to \pi^- \Delta^{++}$, $\gamma p \to \pi^+ \Delta^0$, $\gamma p \to \rho p$, with subsequent decay of unstable intermediate particles to the final state [5]- [7]. The amplitudes of these mechanisms were evaluated in Breit-Wigner ansatz as product of two-body production amplitudes and Breit-Wigner propagator. A 3-body phase space independent from kinematical variables of the final state was added and determined from fit of data. Decay amplitudes were evaluated from effective Lagrangians with form factors dependent from running mass of decay products [6,7].

$\pi - \Delta$ and $\rho - p$ production amplitudes were treated as coherent sum of N^* excitation and non-resonant mechanisms. N^* contributions were treated in Bret-Wigner ansatz [4]. All resonances below 2 GeV with 3 or 4 stars in PDG were included. N^* strong decay amplitudes were taken from analysis [8]. $A_{1/2}$, $A_{3/2}$ electromagnetic form factors were taken from experimental data when available or quark model predictions to select among models for N^* structure description.

In our approach the same quantities could be extracted from experimental data fit for subsequent theoretical analysis and interpretation.

Non-resonant mechanisms for $\pi\Delta$ production were described by a set of gauge invariant Born terms, where the pion propagator in t-channel was substituted by pion Regge trajectory to describe data at W > 1.7GeV [5,9] and gauge invariance was restored after reggeization. Gauge invariant Born terms in conjunction with reggeization allow to describe [11] π^- angular distribution and to overcome shortcomings discussed in [4]. A special procedure for an effective description of interactions in the intial and final state with open inelastic channels was developed [4] for $\pi\Delta$ channel, where such effects appear to be of particular importance.

For the non-resonant rho production a simple diffractive ansatz [10] is used. Such approximation provides reasonable description of data in limeted -t range $< 0.5 \text{GeV}^2$. However in the N^* region resonant contibution becomes dominant at $-t > 0.5 \text{GeV}^2$.

As a test of our approach, we calculated the double charged pion cross-section and evaluated the contributions of $\gamma p \to \pi^- \Delta^{++}$ and $\gamma p \to \rho p$ quasi-two-body channels at the photon point in the whole N^* excitation region (W < 2.0GeV). PDG values for $A_{1/2}$ and $A_{3/2}$ N^* amplitudes were used. The free parameters of the model (pion Regge trajectory coupling, magnitude of ρ diffractive production amplitude and 3-body phase space) were determined from simulataneous fit of $\pi^+\pi^-$ and π^+p invariant mass distributions [6,11]. The model results and the ABBHHM data [11] are shown in Fig. 1. Our approach provides a good reproduction of data both for the overall cross-section and for the contributions of $\pi^- \Delta^{++}$ and ρp channels.

Comprehensive tests [3]- [7] showed that our model can decribe W and Q^2 dependencies in world data on double charged pion production total cross-sections, as well as invariant $(\pi^+\pi^-)$ (π^+p) and angular distribution.

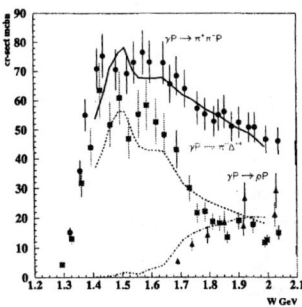

FIGURE 1. Calculated $\gamma p \to \pi^+\pi^-p$ cross-section at the photon point in comparison with ABBHHM data [11]. Complete cross-section is shown by solid line, while $\pi^-\Delta^{++}$ and ρp channel contributions are presented by dashed and dotted-dashed lines respectively.

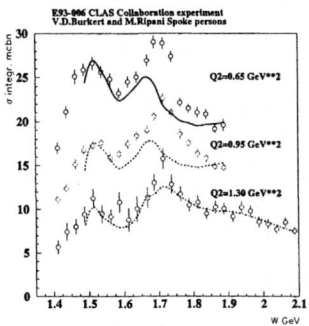

FIGURE 2. The comparison between preliminary CLAS Collaboration data on double charge pion production by virtual photons [2] and our model evaluation with SQTM [12] N* $A_{1/2,3/2}$ form factors.

PRELIMINARY CLAS COLLABORATION DATA ANALYSIS ON DOUBLE CHARGED PION PRODUCTION

We applied our approach to the CLAS Collaboration preliminary data on reaction $\gamma p \to \pi^+\pi^-p$, obtained in E93-006 experiment [2] in order to extract electromagnetic form factors of high lying $N^*(M > 1.6 \text{GeV})$ and search for possible signals from missing baryon states. W dependence of integrated cross-section was studied for three Q^2-bins around 0.65, 0.95 and 1.30 GeV^2. Invariant $(\pi^+\pi^-)$, (π^+p) mass distributions as well as π^- angular distributions were analysed together for all W and Q^2-bins. First, we evaluated differential and total cross-sections, using a Single Quark Transition Model SQTM interpolation of world data [12] for N^* electromagnetic form factors. Free model parameters (Reggeon coupling, etc) were determined from data fit. The comparison of our calculations with preliminary CLAS data [2] is shown in Fig 2. The calculation reproduces the gross features of CLAS data, however at W around 1.7 GeV for all Q^2 bins it is possible to notice an excess of measured cross-section with respect to the calculation. Moreover the calculated $(\pi^+\pi^-)$ invariant mass distribution shows peak mostly due to ρ meson decay, while data show smooth behaviour(Fig 3). It is clear from Fig. 3 that peak is coming from resonant mechainsms. We found that the P13(1720) state is a main contributor in this peak. To reproduce $(\pi^+\pi^-)$ invariant mass distribution data around 1.7 GeV in all Q^2-bins we decrease P13(1720) $A_{1/2,3/2}$ values with respect to SQTM prediction by approximately factor 2. To extract $A_{1/2,3/2}$ N^* electromagnetic form factors from data we fluctuated all of them within SQTM prediction uncertainties (10-20%). For each sample of $A_{1/2,3/2}$ values we calculted $(\pi^+\pi-),(\pi^+p)$ invariant mass and π^- angular distributions in each W, Q^2 bin and evaluated χ^2, picking-up the best solution.

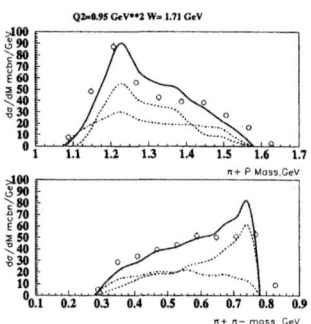

FIGURE 3. The comparison between measured and calculated $(\pi^+\pi^-)$, (π^+p) invariant mass distributions with SQTM [12] N^* $A_{1/2,3/2}$ form factors. The resonant (dashed lines) and non-resonant (dott-dashed lines) process contiburions are also shown.

We found that data could be decribed the above mentioned reduction of $A_{1/2,3/2}$ electromagnetic form factors for the P13(1720) state and $A_{1/2,3/2}$ of all other N^* within SQTM prediction uncertainties [12] for all W, Q^2 bins apart from structure around 1.7 GeV. This structure could be reproduced in two hypothesis:a) increasing by a factor 2 NRQM prediction [12] for conventional D13(1700) $A_{1/2,3/2}$ form factors (SQTM predicts zero electromagnetic couplings for this state); b) implementation of missing P13(1720) state, deriving it's quantum numbers and couplings from data fit. Both hypothesis are currently under investigation and we expect to be able to draw a definite conclusion about this point.

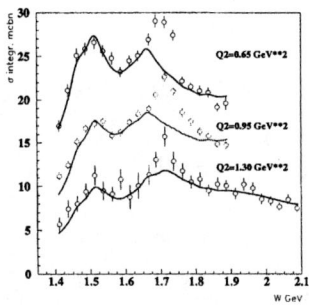

FIGURE 4. The integrated cross-section description with N* $A_{1/2,3/2}$ electromagnetic form factors extracted from data fit (see text).

REFERENCES

1. V.D.Burkert, in Proc. of the Workshop on The Physics of Excited Nucleon NSTAR2001, ed. by D.Drechsel,L.Tiator, p.457.
2. M.Ripani ibid. p.439.
3. V.Mokeev, M.Ripani, e.a, ibid. p.181.
4. M.Ripani, V.Mokeev, e.a. Nucl. Phys. A672 220 (2000).
5. V.Mokeev, M.Ripani, e.a. in Proc. of NSTAR 2000 Conference, ed by V.D.Burkert, L.Elouadrhiri, J.J.Kelly, R.C.Minehart,p. 234.
6. M.Ripani, V.Mokeev.e.a. Physics of Atom. Nucl. 63 1943 (2000).
7. V.Mokeev, M.Ripani, e.a. Physics of Atom. Nucl 64 1 (2001).
8. D.M.Manley, E.M.Salesky, Phys. Rev. D45 4002 (1992).
9. M.Guidal, J.-M.Laget, M.Vanderhaegen, Phys.Lett. B400 6 (1997).
10. D.G.Cassel e.a., Phys. Rev. D24 787 (1981).
11. ABBHHM Collaboration Phys. Rev. 175 1669 (1968).
12. V.D.Burkert, Czech. Journ. Phys. 46 627 (1996).

Strangeness Production at CLAS

K.H. Hicks for the CLAS Collaboration

Ohio University, Athens, OH, 45701, USA

Abstract. The CLAS detector is a large acceptance spectrometer located at the electron accelerator at Thomas Jefferson National Accelerator Facility. Production of K^+ and K^* mesons along with associated hyperons were measured using the CLAS detector at electron beam energies of 2.4-4.4 GeV. In addition, polarization transfer from the electron helicity to the Λ hyperon was measured. These data are compared with theoretical models, where available, based on fits to older, less precise K^+ data. The new data for CLAS indicate that improvements in the models are necessary in order to interpret the results.

INTRODUCTION

Kaon electroproduction is complementary to pion electroproduction, and comparison of both reactions will be helpful to understand the role of various N^* resonances. Since s-quarks are not present as valence quarks within the nucleon, the reaction amplitudes depend on the wavefunction of $s\bar{s}$ quark pairs produced from the vacuum. Most models assume the $s\bar{s}$ are in a $^3P_0$ state. Polarization observables, which are easily measured because the Λ is self-analyzing, provide information about the state of the produced $s\bar{s}$ pair. This in turn can be used in theoretical models to improve our understanding of the reaction mechanism. We may also find that N^* resonances have different decay amplitudes into strange and non-strange channels. Unlike pion production, data for kaon production has been sparse, and the potential for finding "new" resonances is appealing.

In K^+ production, both t-channel and s-channel diagrams contribute. Because of the heavy mass of the s-quark, the t-channel contributes mostly at forward kaon angles. The N^* resonances contribute through the s-channel, and hence will be more visible at backward kaon angles. In addition, only N^* resonances can contribute to the Λ final state, whereas both N^* and Δ^* resonances contribute to the Σ^0 production. Because Λ and Σ^0 hyperons have the same valence quarks (uds) and nearly the same mass, they act as an isospin filter for s-channel production. A comparison of the W-dependence for these two final states helps to untangle the role of N^* and Δ^* resonances in the reaction mechanism.

Other reactions with strange quarks, such as electroproduction of vector mesons like the K^* are also useful to understand the role of N^* resonances in the reaction mechanism. However, the higher threshold for K^* production (above $W = 2.0$ GeV) restricts this channel to resonances above the cluster of $L = 1$ resonances at about 1.7-1.9 GeV. This may be an advantage in a search for higher lying resonances which may be obscured by t-channel processes in other reaction channels [1].

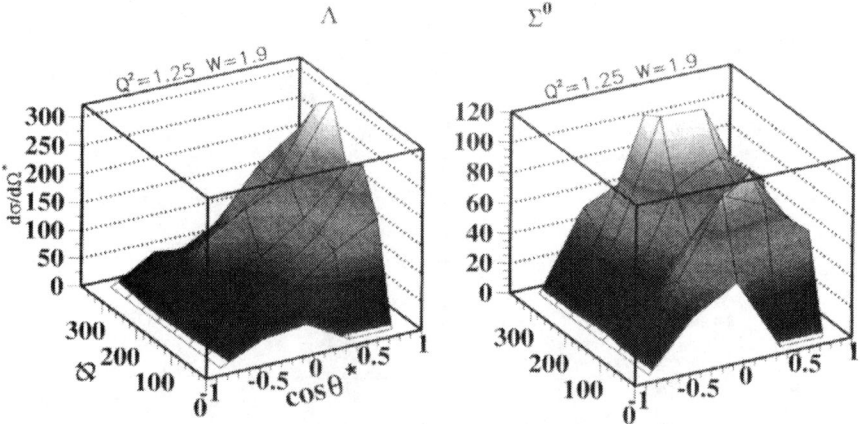

FIGURE 1. Overall angular distribution of cross section to the Λ and Σ^0 final states.

EXPERIMENTAL SETUP

The CLAS detector is a novel design with a torroidal magnetic field. The nearly 4π coverage is segmented into six sectors, each with three sets of drift chambers followed by plastic scintillators for time-of-flight measurement. At forward angles are Cerenkov detectors and electromagnetic calorimeters used in part for detection of the scattered electron. The CLAS detector has been described in detail elsewhere [2].

RESULTS AND DISCUSSION

Fig. 1 shows the angular distributions for kaon electroproduction to Λ and Σ final states. The data for the Λ exhibit a large increase as θ^*, the angle of the kaon relative to the virtual photon direction, becomes small ($\cos\theta^* \simeq 1$) whereas for the Σ data tend to peak at central angles (near $\cos\theta^* = 0$). In addition, the ϕ-dependence of the cross sections is very different for these two final states at a given kaon angle. This difference is perhaps surprising when considering the Λ and Σ^0 are different only in the isospin configuration of the u-d quarks. (The 8effect on kinematics, but kinematics can not explain the differences seen in Fig. 1.) We note that constituent quark models [3] predict a larger coupling constant for t-channel processes for the Λ than for the Σ, which is consistent with the data in Fig. 1.

The unpolarized and interference cross sections for the *Lambda* (circles) and Σ (squares) production are shown in Fig. 2 as a function of $\cos\theta^*$. Also shown in this figure are curves from a hydrodynamic model by Mart and Bennhold [4] where the effective degrees of freedom are baryons and baryon resonances (s-channel) and mesons (t-channel), for the Λ and Σ channels. Although the calculations predict the general magnitude of the unpolarized cross sections ($\sigma_U = \sigma_T + \varepsilon\sigma_L$ where T and L represent the transverse and longitudinal photon polarizations and ε is the standard kinematic vari-

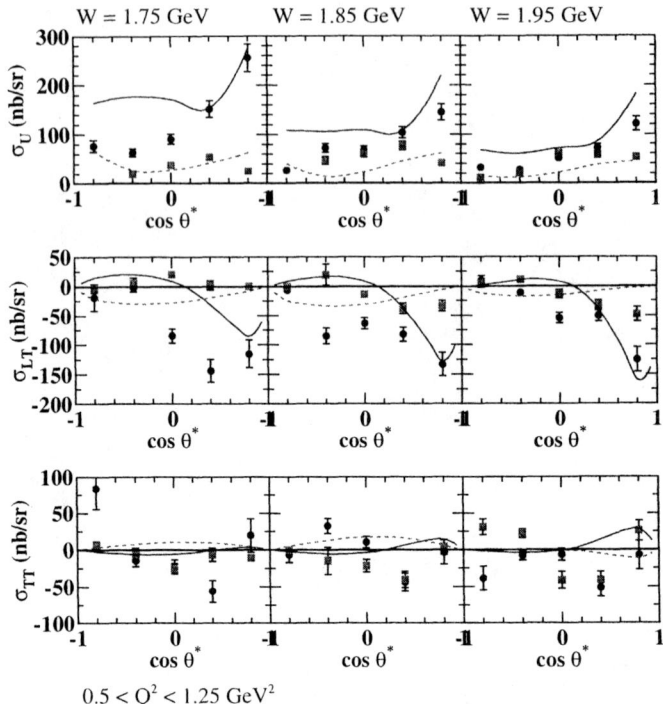

FIGURE 2. Angular distributions of cross sections for different bins in W. The Λ results are given by circles and the Σ results by squares. The curves are calculations for Λ (solid) and Σ (dashed).

able), the interference cross sections for Λ production show structure not present in the solid curves. For Σ production, the interference structure functions are small, but again differ significantly from the preductions shown in the dashed curves. The calculations are based on phenomenological fits to previous data [5?] which had much larger error bars. Clearly, the new precise data from CLAS will force improvements in the theoretical models which in turn could lead to a more fundamental understanding of the production mechanism of strange quarks.

Both K and K^* mesons, with masses of 494 and 892 MeV respectively, contribute to the t-channel in the $H(e,e'K^+)$ reaction. As a result, any model of this reaction will necessarily predict the coupling constants for $K*$ production. By measuring the $H(e,e'K^*)$ reaction in addition to K^+ production, tigher constraints may be placed on theoretical models. Production of the neutral K^* has significant experimental advantages, because it decays almost immediately into K^+ and $\pi-$. The large acceptance of CLAS provides the kinematic phase space necessary to measure K^{0*} production. Also, t-channel contributions are minimized due to the neutral charge, resulting in the prediction that s-channel nucleon resonances are the primary contributions. This reaction has a threshold just above $W = 2.0$ GeV, and so ite may be sensitive to higher-energy N^* resonances. Known N^* resonances above 2 GeV are the $G_{17}(2190)$, $H_{19}(2220)$ and

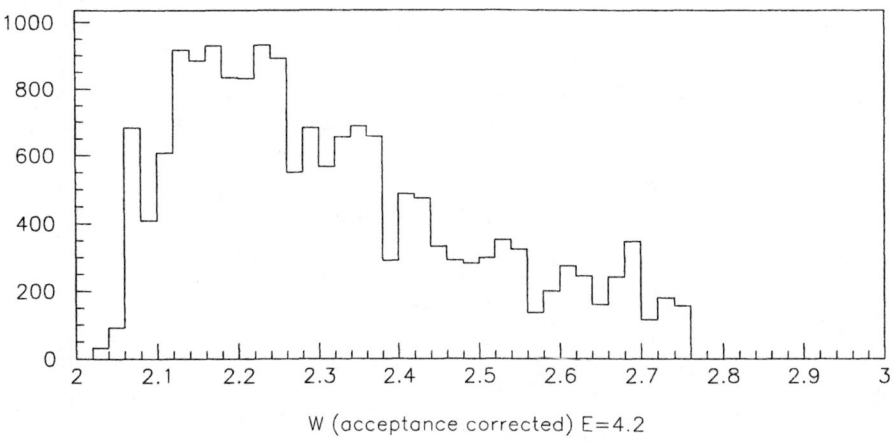

FIGURE 3. Relative yield (vertcal axis) of K^* electroproduction as a function of W, after correction for the detector acceptance. The threshold for the $H(e,e'K^{0*})\Sigma^+$ reaction is $W = 2.08$ GeV.

$G_{19}(2250)$. The W distribution of strength for K^{0*} production are shown in Fig. 3. We note that the acceptance correction for the CLAS detector has only a small affect on the shape of the W-dependence. This suggests that the excess strength at 2.1-2.3 GeV is due to N^* resonance contributions, although it is not clear if the known (high-spin) resonances are contributing strongly here. Calculations are being developed using the same hadrodynamic model as used for K^+ production, and should provide evidence for the structure seen near threshold for K^{0*} production.

In conclusion, new precise data on strangeness production from CLAS is challenging theoretical models and will hopefully lead to a new understanding of the role of N^* resonances in electroproduction reactions.

ACKNOWLEDGMENTS

Gabriel Niculescu provided invaluable assistance in the preparation of this paper. The author is grateful for financial support from the National Science Foundation.

REFERENCES

1. Bennhold, C., "Kaon Photoproduction on the Nucleon", in *Strange Quarks in Hadrons, Nuclei and Nuclear Matter*, edited by K. H. Hicks, World Scientific, Singapore, 2000, pp. 64–73.
2. Brooks, W., *Nucl. Phys. A*, **664**, 1077c (2000).
3. Capstick, S., and Roberts, W., *Phys. Rev. D*, **58**, 1 (1998).
4. Mart, T., and Bennhold, C., *Phys. Rev. C*, **61**, 012201 (1999).
5. Bebek, C., *Phys. Rev. D*, **15**, 3082 (1977).

Experimental check of the GDH sum rule at MAMI and ELSA

J. Krimmer for the GDH-Collaboration

*Physikalisches Institut, Universität Tübingen, Auf der Morgenstelle 14,
D-72072 Tübingen, Germany*

Abstract. The experimental check of the GDH sum rule is being performed at the tagged photon facilities of the electron accelerators MAMI (Mainz) and ELSA (Bonn), using circularly polarized photons impinging on a longitudinally polarized proton target, together with detector systems covering almost the whole solid angle range. Results from the MAMI experiment for the double polarized total photoabsorption cross section in the low energy region ($200\,\text{MeV} < E_\gamma < 800\,\text{MeV}$) will be shown together with their contribution to the GDH sum rule. Furthermore first results from the ELSA experiment in the higher energy region ($680\,\text{MeV} < E_\gamma < 1900\,\text{MeV}$) will be presented.

INTRODUCTION

Improvents in polarized beam and target technologies allow nowadays the investigation of polarization degrees of freedom of the nucleon.

Special interest exists in the experimental check of the Gerasimov-Drell-Hearn (GDH) sum rule [1, 2] which was already derived in the late 1960's.

$$\int_0^\infty d\nu \frac{\sigma_{3/2} - \sigma_{1/2}}{\nu} = \frac{2\pi^2 \alpha}{m^2} \kappa^2 \qquad (1)$$

where $\sigma_{3/2}$ and $\sigma_{1/2}$ are the helicity dependent total photoabsorption cross sections off the nucleon, with photon and nucleon spin parallel or antiparallel, respectively. As weighted by the photon energy ν and integrated over the whole energy range, the l.h.s. represents the complete excitation spectrum of the nucleon, wheras on the r.h.s. just static properties of the nucleon occur, like the mass (m) and the anomalous magnetic moment (κ), α denotes the fine structure constant.

The GDH sum rule can be derived from the Compton forward scattering amplitude, in a model independent way, by using very fundamental physics principles (lorentz and gauge invariance, unitarity, low energy theorems, unsubtracted dispersion relation).

In the past, estimates for the GDH sum rule have been given by multipole analyses of the existing pion photoproduction data (mainly unpolarized experiments), where just [3] gives a very rough estimate of the contribution from the double pion channel. The results are compared to the sum rule value in Table 1.

Apart from [8] the different approaches are fairly in agreement among themselves, but have some deviation to the sum rule value.

TABLE 1. Various theoretical predictions for the GDH sum rule

	$I_p(\mu b)$	$I_n(\mu b)$	$I_p - I_n(\mu b)$
GDH sum rule	204.5	232.8	-28.3
Karliner [3]	261	183	78
Workman '92 [4]	260	157	103
Burkert '93 [5]	203	125	78
Sandorfi '94 [6]	289	160	129
Drechsel '98 [7]	261	180	81
Bianchi '99 [8] (phen. Regge appr.)	207	226	-29

Especially for $I_p - I_n$ the estimates yield to the wrong sign compared to the sum rule.

The lack of double polarized experiments and the theoretical situation shown in Table 1 was the motivation for the formation of the GDH-Collaboration with the goal to experimentally check the GDH sum rule at MAMI (200 MeV $< E_\gamma <$ 800 MeV) and ELSA (680 MeV $< E_\gamma <$ 3.2 GeV) and to investigate the helicity dependence of all open partial reaction channels at MAMI.

THE GDH EXPERIMENT AT MAMI

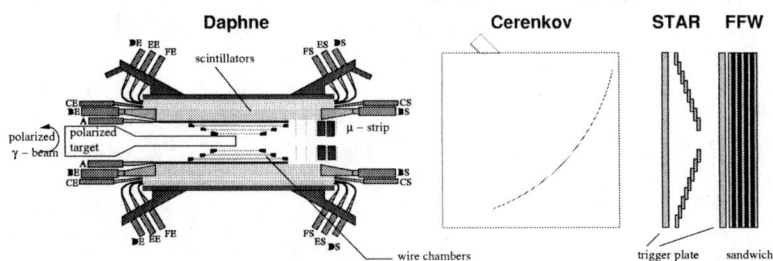

FIGURE 1. The detector setup at MAMI

The experiment was carried out at the tagged photon facility [9] in the A2 hall of the MAMI electron accelerator. Circularly polarized photons were produced by the helicity transfer of longitudinally polarized electrons at the bremsstrahlung process. The MAMI accelerator provides polarized electrons from a source based on a GaAs crystal with a polarization of 75 % or higher.

Polarized nucleons are available by a solid state "frozen spin" target [10], consisting of a horizontal cryostat with an internal superconducting holding coil. It has the advantage, that outgoing particles can be detected in nearly 4π acceptance, wheras the stray field from the holding coil is minimal inside the detector.

Typical operation parameters during data taking are a maximum polarization of $\approx$ 85 % with relaxation times of about 100 hours.

Figure 1 shows the detetor setup used at the MAMI experiment. The central detector DAPHNE [11] allows identification and tracking of charged particles. It is completed with a silicon microstrip detector MIDAS [12]. The threshold Cerenkov detector allows online suppression of electrons coming from atomic processes. The gap in the forward direction is filled by the annular STAR detector [13] and an additional sandwich counter.

THE GDH EXPERIMENT AT ELSA

After the measurements on the proton were finished at MAMI the polarized target was moved to the Electron Stretcher Accelerator ELSA in Bonn.

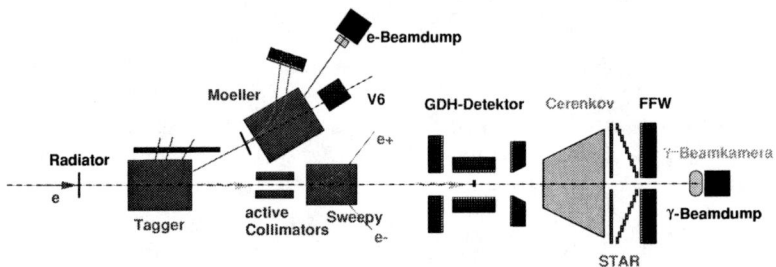

FIGURE 2. The detector setup at ELSA

The detector setup at ELSA is given in Figure 2. The tagging system allows the energy determiniation of the circularly polarized photons produced by bremsstrahlung of longitudinally polarized electrons. The polarization of the electron beam is constantly monitored by a two arm Møller spectrometer.

After passing an active collimator system [14] and a dipole magnet to remove charged particles, the photon beam hits the polarized target. Outgoing hadrons are detected with the GDH Detector [15] which has a detection efficiency for charged particles and decay photons $> 99\%$. The threshold Cerenkov detector suppresses the electromagnetic background. The far-forward (FFW) and STAR [13] detector cover an angular range from $2°$ to $17°$. Therefore more than 99% of the solid angle are covered by the whole detector setup.

A photon-camera [16] right in front of the γ beamdump permanently supervises the position and intensity of the photon beam. The photon flux is measured with a total absorbing leadglass detector in the γ beamdump.

RESULTS AND CONCLUSIONS

The first quarter of Figure 3 shows the result for the difference of the helicity dependent photoabsorption cross sections $\sigma_{3/2}$, $\sigma_{1/2}$ as it was measured at MAMI. The contribution to the GDH integral up to 800 MeV is 226 ± 5 (stat.) ± 12 (syst.) μb [17]. The plot continues with data from ELSA for three different primary electron energies, up to 1.9 GeV, with more positive contributions to the GDH integral in the 2. and 3. resonance region. Data taking at ELSA will continue up to 3 GeV, to find a zero crossing of $\sigma_{3/2} - \sigma_{1/2}$, which is necessary to fulfill the sum rule, and which was predicted by [8].

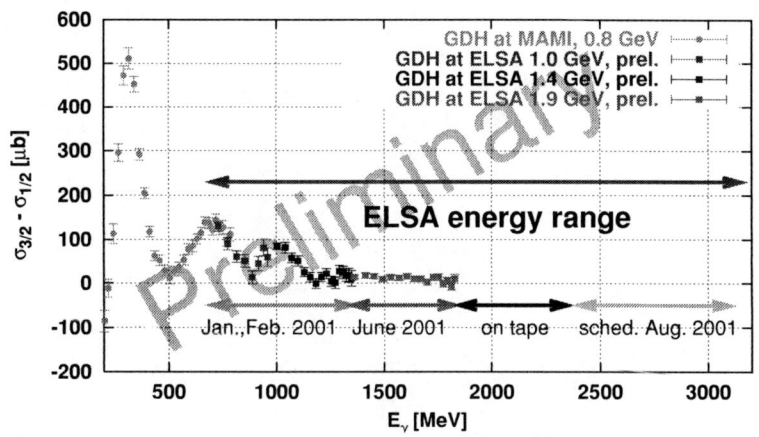

FIGURE 3. Difference of total photoabsorption cross sections $\sigma_{3/2}$, $\sigma_{1/2}$ off the proton

The analysis for all single and double pion partial reaction channels from the MAMI experiment is partly finished (e.g. [18]) and will be completed soon.

REFERENCES

1. Gerasimov, S.B. *Sov. J. Nucl. Phys.* **2**, 430 (1966)
2. Drell, S.D., and Hearn, A.C., *Phys. Rev. Lett.* **16**, 906 (1966)
3. Karliner, I., *Phys. Rev. D* **7**, 2717 (1973)
4. Workman, R., and Arndt, R, *Phys. Rev. D* **45**, 1789 (1992)
5. Burkert, V., and Li, Z., *Phys. Rev. D* **47** 46 (1993)
6. Sandorfi, A., *et al.*, *Phys. Rev. D* **50**, 11 (1994)
7. Drechsel, D., and Krein, G., *Phys. Rev. D* **58**, 116009 (1998)
8. Bianchi, N., and Thomas, T., *Phys. Lett. B* **450**, 439 (1999)
9. Anthony, I., *et al*, *Nucl. Inst. Meth. A* **301**, 103 (1991)
10. Bradke, C., *et al.*, *Nucl. Inst. Meth. A* **436**, 430 (1999)
11. Audit, G., *et al.*, *Nucl. Inst. Meth. A* **301**, 473 (1991)
12. Altieri, S., *et al.*, *Nucl. Inst. Meth. A* **452**, 185 (2000)
13. Sauer, M., *et al.*, *Nucl. Inst. Meth. A* **378**, 143 (1996)
14. Zeitler, G., *et al.*, *Nucl. Inst. Meth. A* **459**, 6, (2001)
15. Helbing, K., *et al.*, *Nucl. Inst. Meth. A* in press
16. Krimmer, J., Grabmayr, P., and Sauer, M., *Nucl. Inst. Meth. A* in press
17. Arends, H.-J., *et al.*, *Phys. Rev. Lett.* **87**, 022003, (2001)
18. Holvoet, H, *http://pit.physik.uni-tuebingen.de/grabmayr/CD-Lund2001/holvoet/holvoet.html*

CENTRAL PRODUCTION, HEAVY IONS

Preliminary Results of a PWA of the Centrally Produced $\phi\phi$ System

M.A.Reyes[a], M.C.Berisso[b], D.C.Christian[c], J.Felix[f], A.Gara[d],
E.E.Gottschalk[c], G.Gutiérrez[c], E.P.Hartouni[e], B.C.Knapp[d],
M.N.Kreisler[b,e], S.Lee[b], K.Markianos[b], G.Moreno[f],
M.H.L.S.Wang[b,e], A.Wehman[c], D.Wesson[b]

[a] *Universidad Michoacana de San Nicolás de Hidalgo, Morelia, Michoacán, México*
[b] *University of Massachusetts, Amherst, Massachusetts, USA*
[c] *Fermilab, Batavia, Illinois, USA*
[d] *Columbia University, Nevis Laboratory, New York, USA*
[e] *Lawrence Livermore National Laboratory, Livermore, California, USA*
[f] *Universidad de Guanajuato, León, Guanajuato, México*

Abstract. We present preliminary results of a Partial Wave Analysis of the centrally produced $\phi\phi$ system at 800 GeV/c in the reaction $pp \to p_{slow}(\phi\phi)p_{fast}$. Our preliminary results with one and two $M=0$ waves, indicate that most of the cross section can be described by two waves, with $J^{PC}LS^\eta = 2^{++}02^{-1}, 0^{++}00^{-1}$.

The first observation of $\phi\phi$ production was made using the BNL–MPS Spectrometer at 22.6 GeV/c [1] in the OZI [2] suppressed reaction,

$$\pi^- p \to \phi\phi n \qquad (1)$$

A Partial Wave Analysis (PWA) of their data showed that only three 2^{++} waves were necessary to fit the data [3]. The larger than expected cross section that was observed indicates that these states may not be conventional $q\bar{q}$ mesons.

We present here our preliminary results of a PWA of the centrally produced $\phi\phi$ system in the 800 GeV/c doubly diffractive reaction,

$$pp \to p_{slow}(\phi\phi)p_{fast}, \quad \phi \to K^+K^- \qquad (2)$$

using events of this reaction selected from the 4×10^9 pp interaction data sample recorded by Fermilab E690 during the 1991 fixed target run.

The E690 apparatus consisted of a high rate, open geometry multiparticle spectrometer used to measure the target system (T) in $pp \to p_{fast}(T)$ reactions, and a beam spectrometer system used to measure the incident 800 GeV/c beam and scattered proton. A liquid hydrogen target was located just upstream of the multiparticle spectrometer. The 96 cell Cherenkov counter located at the downstream

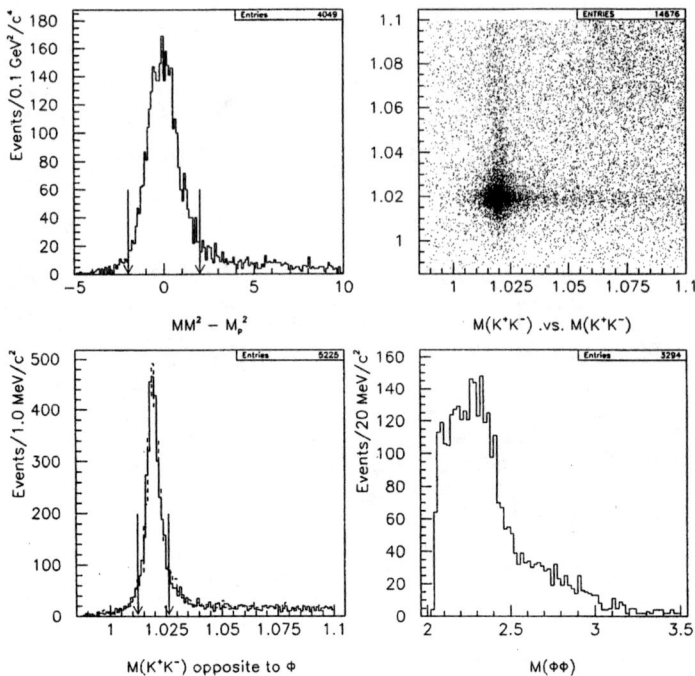

FIGURE 1. Missing mass squared minus proton mass squared for events in reaction (2) (upper left). K^+K^- invariant mass, first vs. second pair (upper right). K^+K^- invariant mass, when the other pair lies in the ϕ–mass band (lower left). $\phi\phi$ invariant mass (lower right).

end of the main spectrometer magnet used Freon 114 as a radiator and had a pion threshold of 2.57 GeV/c. The E690 apparatus has been described elsewhere [4].

After the track and vertex reconstruction stage of the data analysis, final state (2) was selected by requiring a primary vertex in the LH_2 target with two positive and two negative tracks, an incoming beam track and a fast forward proton. Cherenkov particle identification was required for at least one of the four tracks. No direct measurement was made of the slow proton, but a kinematical cut of $p_z < 250$ MeV/c or $\arctan(p_t/p_z) > 30$ for the missing momentum was used to require that it was outside the acceptance of the detector.

The missing mass squared (MM^2) minus proton mass squared shown in Fig.1 has a clear peak around zero for events in reaction (2). The scatter plot shows the first versus second pair mass for the 14678 events selected with $m(K^+K^-) < 1.1$ GeV/c^2 and $-2 < MM^2 - m_p^2 < 2$ GeV^2/c^4. $\phi\phi$ events are predominantly produced over ϕK^+K^- and $K^+K^-K^+K^-$ events, these being the only significant background source. The lower left plot shows the K^+K^- invariant mass, when the other pair lies in the ϕ–mass band of $1.0124 < M(K^+K^-) < 1.0264$ GeV/c^2. The lower right plot shows the $\phi\phi$ invariant mass after all selection cuts, showing a high bump between 2–2.5 GeV/c^2. Only one combination per event enters this plot. 3180

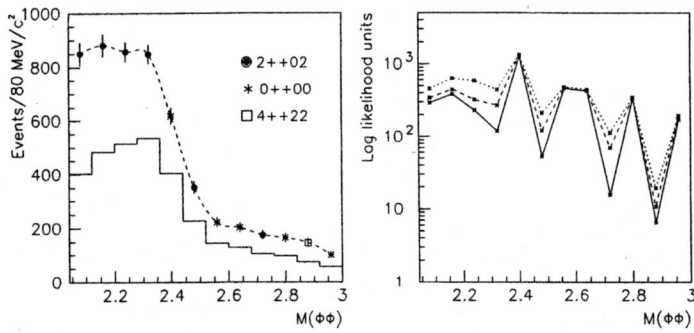

FIGURE 2. Left: $M(\phi\phi)$ data distribution (histogram) and acceptance corrected cross section (markers) from the PWA using only one $M=0$ wave. Right: Log-likelihood difference between the solution with highest likelihood and the next three best solutions, solid, dashed and dotted, respectively, in each bin.

events with $M(\phi\phi) < 3$ GeV/c^2 remain.

Six angles are chosen to specify the spin and angular momentum of the $\phi\phi$ system. Two of them (γ, β) are defined as the Gottfried-Jackson (GJ) angles of one of the ϕ mesons, in the rest frame of the $\phi\phi$ system, with the z-axis in the direction of $\vec{p}_{fast} - \vec{p}_{beam}$, and the y-axis in the direction of the $\vec{p}_{fast} - \vec{p}_{beam} \times \vec{p}_{slow} - \vec{p}_{tgt}$ cross product. The rest are the two pairs of GJ angles $(\alpha_{1,2}, \theta_{1,2})$ for the K^+'s in their parent ϕ rest frames, with the z'-axis in the direction of $\vec{p}_\phi$, and with $y' = \hat{z} \times \hat{z}'$.

The allowable $\phi\phi$ basis vectors in terms of the total angular momentum J, orbital angular momentum L, parity P, and exchange naturality η, are given by [5]

$$G^{J^P L S M \eta}(\gamma, \beta, \alpha_1, \alpha_2, \theta_1, \theta_2) = \text{Real}\left[\frac{(1-i) - \eta(1+i)}{2} \sum_{\mu,\lambda} C(1,1,S|\mu,-\lambda) \times \right.$$
$$\left. C(L,S,J|0,\mu-\lambda) e^{iM\gamma} e^{i\mu\alpha_1} e^{i\lambda\alpha_2} d^J_{M,\mu-\lambda}(\beta) d^1_{\mu,0}(\theta_1) d^1_{\lambda,0}(\theta_2)\right] \quad (3)$$

where $M = |J_z|$. For this system $I = 0$, $C = +$, and $L + S =$ an even number.

We performed a PWA of our data divided in 12 bins of 80 MeV/c^2 beginning at 2.04 GeV/c^2, using only the 14 $M = 0$ waves. The results with one wave only are shown in Fig.2. Only two waves are seen for the data below 2.6 GeV/c^2, $J^{PC}LS\eta = 2^{++}02^{-1}$, $0^{++}00^{-1}$. The acceptance corrected cross section shows high acceptance for this reaction. Monte Carlo events for acceptance corrections were generated flat in all six angles, $\phi\phi$ mass, and x_F of the $\phi\phi$ system. No background subtraction was used for the analysis. The plot on the right shows that the second best solution was at least 3.5 sigmas away ($n_\sigma = \sqrt{2\,\Delta(\ln \mathcal{L})}$).

The results with two waves are shown in Fig.3. We added one wave at a time to the one wave solution, and selected the best solution. Below 2.6 GeV/c^2 the cross section is mainly comprised of only two waves, $J^{PC}LS\eta = 2^{++}02^{-1}$, $0^{++}00^{-1}$. The solid line in the plot on the right in Fig.3 shows that including a second wave greatly improves the significance of the PWA.

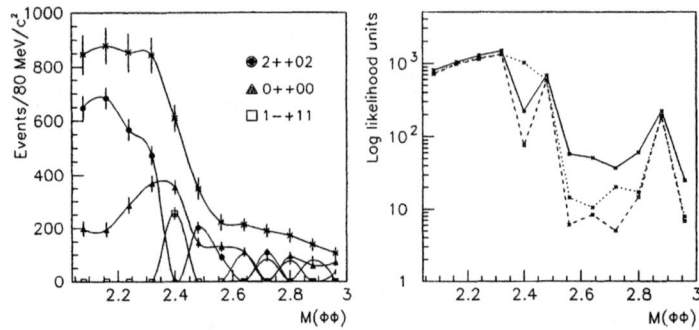

FIGURE 3. Left: PWA results with two $M = 0$ waves. Below $2.6\,\text{GeV}/c^2$ we still found that waves $2^{++}02$ and $0^{++}00$ describe most of the cross section (*). Curves are only used to follow waves from bin to bin. Right: Log-likelihood difference between the one-wave and two-wave solutions (continuous line.) The dotted and dashed lines show the difference to the next two best two-wave solutions in each bin.

CONCLUSIONS

We report preliminary results of a PWA of the centrally produced $\phi\phi$ system. Using only $M = 0$ waves, we find that most of the cross section can be described using only two waves, with $J^{PC}LS^\eta = 2^{++}02^{-1}$, $0^{++}00^{-1}$. Since we are only using $M = 0$ waves at this time, we suspect that results may change when including $M = 1$ waves.

Acknowledgements

This work was funded in part by the Department of Energy under Contracts No. DE-AC02-76CHO3000 and No. DE-AS05-87ER40356, the National Science Foundation under Grants No. PHY89-21320 and No. PHY90-14879, and CONACyT de México under Grants No. 1061-E9201, No. 3793-E9401, and No. CIC2001-4.17.

REFERENCES

1. A.Etkin et al., *Phys.Rev.Lett.* **40**, 422 (1978).
2. S.Okubo, *Phys.Lett.* **5**, 165 (1963); G.Zweig, *CERN Report No. TH-401 and TH-412*, 1964 (unpublished); J.Iizuka, *Prog.Theor.Phys.*, Suppl. **37-38**, 21 (1966).
3. A.Etkin et al., *Phys.Lett.* **B201**, 568 (1988).
4. E.P.Hartouni et al., *Nucl.Instrum.Methods* **A 317**, 161 (1992); J.Uribe et al., *Phys. Rev.* **D49**, 4373 (1994); D.C.Christian et al., *Nucl.Instrum.Methods* **A345**, 62 (1994).
5. R.S.Longacre, *AIP Conf. Proc.* **113**, 0051 (1984).

Glueball candidates production in peripheral Heavy Ion Collisions at ALICE

Mikhail Bashkanov

Interphysics, MEPHI, Kashirskoe-31, Moscow, Russia

Abstarct. To prove the existence of glueball we should check two photon width of all glueball candidates. Peripheral heavy ion collisions provides such possibility. Using some restrictions, which discussed below we can even find glueball in two-photon collision.

INTRODUCTION

From the beginning of the Quantum Chromodinamics there is question: is there a particle, which consist of only gluons, so called glueball. According to the theory a lot of such particles should exist. And the lightest glueball is to have mass nearly 1500 MeV. Next glueball should have mass near 2000 MeV and spin equal to 2. The main problem is that a lot of particles have such properties. So our aim is not to find the glueball, but to prove that our particle is glueball. It seems to be most probable that glueball with quantum numbers 0^{++} is heavily mixed with ordinary mesons. So it should be easier to find tensor glueball, in spite of bigger mass. Most of all such difference in masses is not important both for RHIC (Relativistic Hadron Collider) and for LHC (Large Hadron Collider). There is one more reason to consider tensor glueball. There is a process where we can separate tensor glueball from tensor meson.

To prove the glueball nature of a particle we should detect two-photon width, because other glueball properties are model-dependent. It is well known fact that glueball can not interact with photon directly. So two-photon width of a glueball is small in comparison with two photon width of an ordinary meson. We can directly measure two-photon width only in two processes: glueball production in two-photon collision and glueball decay to two photon. Both these processes have their advantages and disadvantages. It is very convenient to use peripheral collision of heavy ions, because glueballs produced here without hard escort of other particles. If we want to detect glueball decay to two photons we should have good EMC (Electro Magnetic Calorimeter). But there is no EMC at ALICE experiment at LHC. ALICE is detector system at LHC, which will use heavy ions. So we can not detect such decay channel at ALICE. You can say that at RHIC there is a peripheral collision group at STAR detector — Yes. But there are two other problems. First one. EMC at STAR is building now, and it will be build at 2004. There is other deep problem. Trigger that separates peripheral collisions can not work with trigger which select central collisions. As you know the main program at STAR is central collisions so there are no enough time for peripheral collisions. In such situation it is very difficult to get enough statistics to measure two-photon width. So there is only one way to measure two-photon width of a glueball. This way is to use glueball production through two photon collisions. Certainly the situation is not so serene as possible. In peripheral heavy ion collisions glueball can be produced via double-pomeron and two-photon collisions. And two-photon cross section is less than other one. I calculated these cross sections. Two glueball candidates $f_J(1710)$ and $\xi_J(2220)$ will be considered. Most probable spin of these states is two. But for $f_J(1710)$ 0 spin state also possible. And $\xi_J(2220)$ can also have 0 or 4 spins.

TENSOR GLUEBALL PRODUCTION

Glueball can be produced through double-pomeron exchange in peripheral heavy ion collisions. Taking into account assumption that pomeron is two-gluon state, we can say that glueball can be produced both exclusively and inclusively.
In the inclusive process one gluon from each pomeron used to produce glueball while other one fragment into mesons. The average number of mesons in such process is about four for STAR.
In the exclusive process one gluon the same way form glueball whereas other boost it (Fig1).

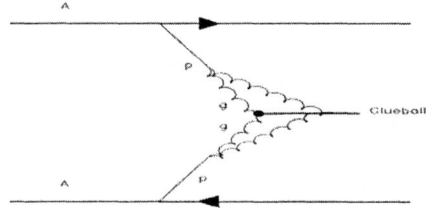

FIGURE 1. Exclusive scalar glueball production through double-pomeron exchange in peripheral heavy ion collision.

In two-photon process two photons form glueball through quark loop.
Fields from nucleons add coherently in a single nucleus, so double-pomeron and two-photon cross sections for heavy ions higher than in proton collisions. Using Shramm's and Natal's works we can calculate such cross sections (Tab. 1). We can see that two photon cross sections are low enough. But we can detect it because of high nucleus luminosity.
As it was mentioned we can separate inclusive double-pomeron process from exclusive, because they have different final states. If we consider scalar gueball production we can not distinguish exclusive pomeron and photon process, because final states are the same. Our situation changes considerably if we discuss tensor glueball. The main difference comes from different spins. Pomeron has spin 0, whereas photon has spin 1. The same way if one gluons from our pomerons produce tensor glueball two other gluons can not boost our glueball, because they carry moment equal to two. Of cause they can fragment into other particles, but we can

distinguish such situation. There is a possibility that our two gluons will be absorbed by nuclei, but in such situation they destroy our nuclei (Fig2.a). There is only one case, which can imitate two-photon collision: if our two gluons fragment into one meson with spin 2 and it meson decay into two photons (Fig2.b.). This situation is next to impossible.

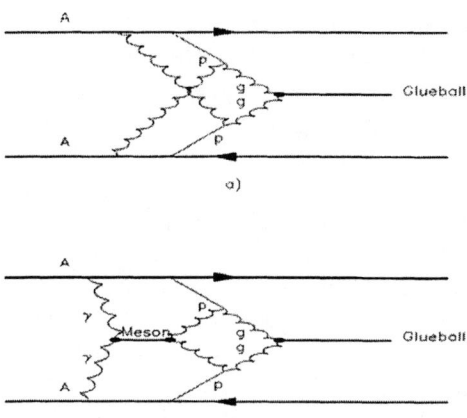

FIGURE 2. a) Tensor glueball production. b) Imitation of tensor glueball production through two photon collision in double-pomeron exchange.

Let's talk now about two-photon collisions. In the begin state there is only two nuclei with spin 0 and orbital moment 0. Then one photons from each nucleus collide and produce glueball with spin 2. So our nuclei also have to carry spin 2. There are 3 possible situation:
1. Nuclei can emits soft photons and relax to the ground state.
2. Nuclei can excite in resonance and relax using neutron emission.
3. Nuclei can carry orbital moment.

Here there is only one bad situation for us. Let's estimate probability of destruction of our nuclei. Here W is the probability that our nuclei excite in the resonance. The probability of excitation with high orbital moments is negligible.

$$W = W_1(J=1)W_2(J=1) + W_1(J=2)W_2(J=0) + W_1(J=0)W_2(J=2) \qquad (1)$$

Spin unit excitation is Giant Dipole Resonance excitation with high probability. For row estimation we can assume that excitation is independent from glueball production. The probability of GDR excitation for lead specimens for ALICE conditions is equal to 37%. So the probability of first term in our formula is about 0.14. The probability of second term in our equation is less than the first so to estimate upper bound of our value we can take the same value for second and third terms. So the probability of destruction of our nuclei is less than 0.5. It is better take into account that the probability of GDR drops rapidly for light ions.

If we will consider pomeron as three gluon state then it is also seems unprobable to pack 6 gluons in one particle to get exclusive glueball production through pomeron-pomeron exchange.

So we should look for final state where there is only one glueball and two nuclei in the beam. In such situation we can consider only our process and not others, and we will not loose a lot in the cross section. It is also very good that tensor glueball produced polarized in two-photon collisions, so it is decay according $\sin^4\theta$ distribution. So the probability of such decay detection is higher than in isotropic situation. It also helps us to determine spin of our glueball candidate. Of cause we can not detect any interesting decays in such situation such as proton-antiproton decay for $\xi_J(2220)$, but we need not it. We need to determine two-photon cross section experimentally to prove that our candidate is glueball. We can investigate interesting decay canals in double-pomeron exchange where we can get high statistics.

SUMMARY

We should do next steps to find our glueball.
1. Check final state, where there is only one particle — glueball candidate.
2. Draw $\cos(\theta)$ dependence for our candidate. If $f(\cos(\theta))=1$ then change particle. If $f(\cos(\theta)) = \sin(\theta)^4$ then spin of our candidate is 2.
3. Determine $\sigma_{\gamma\gamma \to G} \times Br(G \to \pi\pi)$. Glueball dominantly decay to two pions. If this value is rather small then Ok.
4. From double pomeron exchange determine $Br(G \to \pi\pi):Br(G \to KK):Br(G \to \eta\eta):Br(G \to \eta'\eta')$ for our glueball candidate.
5. Using this data estimate $\sigma_{\gamma\gamma \to G}$ and calculate $\Gamma_{\gamma\gamma \to G}$. If $\Gamma_{\gamma\gamma \to G} \approx 1 \div 10 eV$ then our glueball candidate is pure glueball or glueball with small $q\bar{q}$ mixture.

TABLES

TABLE 1. Glueball candidates cross sections in pomeron-pomeron and two photon collisions.

	\multicolumn{6}{c}{LHC}					
	\multicolumn{6}{c}{$f_J(1710)$, J=0}					
	pp $\gamma=7000$		PbPb $\gamma=2750$		CaCa $\gamma=3500$	
	σ, mb	Events for 10^6 sec	σ, mb	Events for 10^6 sec	σ, mb	Events for 10^6 sec
Inclusive	0.55	$1.7 \cdot 10^9$	61.8	$61.8 \cdot 10^5$	33.6	$1.0 \cdot 10^{11}$
Exclusive	$1.1 \cdot 10^{-5}$	$3.3 \cdot 10^4$	0.3	$3.0 \cdot 10^4$	0.1	$3.0 \cdot 10^8$
$\gamma\gamma$	—		$2.6 \cdot 10^{-3}$	269	$1.4 \cdot 10^{-5}$	$4.2 \cdot 10^4$
	\multicolumn{6}{c}{$\xi_J(2220)$, J=2}					
	pp $\gamma=7000$		PbPb $\gamma=2750$		CaCa $\gamma=3500$	
	σ, mb	Events for 10^6 sec	σ, mb	Events for 10^6 sec	σ, mb	Events for 10^6 sec
Inclusive	0.22	$0.7 \cdot 10^9$	23.1	$2.3 \cdot 10^6$	13.3	$4.0 \cdot 10^{10}$
$\gamma\gamma$	—	—	$1.3 \cdot 10^{-3}$	133	$0.7 \cdot 10^{-5}$	$2.1 \cdot 10^4$

J/ψ production in peripheral collisions of heavy ions

S. Timoshenko

Moscow State Engineering Physics Institute(Technical Univercity),
Moscow, 115409, Kashirskoe ave. 31, Russia.

Abstract. J/ψ production in photon-pomeron, pomeron-pomeron and photon-gluon exchange in peripheral nuclear collision at energy of colliding nucleus of 100 GeV per nucleon was calculated. Production and decay J/ψ in these processes for STAR detector was simulated. I try to find such kinematics areas where pomeron processes will dominate above a background (photon-gluon process).

INTRODUCTION

The aim of heavy ions collisions experiments is studying and detection new state of the nuclear matter — quark - gluon plasma (QGP).

Such condition of a nuclear matter from high density of energy and temperature is reached in the central collisions of nucleus. The energy of colliding nucleus at RHIC (Relativistic Heavy Ion Collider) collider is 200 GeV per nucleon for gold nuclei.

Besides the central collisions of heavy ions when QGP can be produced, there are also very interesting physical processes occurring at peripheral interactions of nuclei. In such collisions both strong and electromagnetic interactions are possible. The peripheral processes occurring at impact parameter b > 2R where R is radius of a nucleus is of our special interest.

At such restrictions on impact parameter strong interaction is not dominant more and it compete with coherent photon-photon, photon-pomeron and pomeron-pomeron processes. During mesons production nuclei do not change the initial condition in such processes. For registration of such processes on STAR the special trigger is foresee. In such coherent processes mesons, glueball, higgs boson and other exotic particles can be produced. The purpose of this work is to separate pomeron processes on a background of others.

CALCULATION CROSS SECTIONS.

Pomeron is a carrier of strong interaction, but colorless and has quantum numbers of vacuum. Here use phenomelogical model pomeron suggested Donnachie and Landshoff [1]. In this model pomeron represents as a condition from two or several gluon plus the mixed condition of quarks, antiquarks. We can identify such pomeron process using J/ψ production.

In the pomeron interactions it is necessary to differentiate inclusive and exsclusive processes. Let's consider them on an example of photon-pomeron interaction. At the inclusive process there is an interaction of a photon with gluon which situate in interacting pomeron. The other pomeron component fragment in escorting particles (jet). At exclusive process complete fusion of a photon with pomeron is occurring. It is possible to show, that the probability of inclusive process is more than exclusive process at some orders of magnitude. Because there are no restrictions in the first process. As an example we can use Schramm's results on glueball production [2]. Thus further we will consider only inclusive processes at pomeron interactions. Below we will estimate the cross sections of J/ψ production.

Let's consider photon - gluon process. Heavy nucleus is a powerful source of electromagnetic fields. This field can be described as a virtual flow of photons. We will use equivalent photons method for our calculations [3]. The cross section for photon-gluon fusion into a pair of heavy quarks in a relativistic heavy-ion collision can be cast in the following form [4]

$$\sigma = \int d^2 b \int d\omega_1 \int d\omega_2 g(\omega_2) \int d\vec{x}_\perp N(\vec{r}) n(\omega_1, \vec{b} - \vec{r}) \sigma_{\gamma g \to q\bar{q}}(\omega_1, \omega_2). \quad (1)$$

The integration over the impact parameter is restricted to peripheral collision, where b is large than twice the nuclear radius R. Where r this distance of one gluon to the center of a nucleus. $N(\vec{r})$ is the nucleon distribution of a nucleus projected onto the transverse plane of beams. $g(\omega)$ is stands for gluon distribution of a nucleon. $n(\omega, b)$ is the equivalent photon distribution, which describes the number of photons with energy ω at a distance |b| from the center of the nucleus. $\sigma_{\gamma g}$ is the elementary photon-gluon fusion cross section

The received cross section of pair production is described with the help of the pertubative theory. This a theory can't be applied at big distances. The color evaporation model [5] successfully describes cross sections production of the charm particles at big distances.

This model predicts that cross section of J/ψ production proportionally to the cross section of a cc pair production integrated in the appropriate mass range

$$\sigma = \rho_{J/\psi} \int_{2m_c}^{2m_D} dM_{c\bar{c}} \frac{d\sigma}{dM_{c\bar{c}}}, \quad (2)$$

where $\frac{d\sigma}{dM_{c\bar{c}}}$ is calculable perturbatively, M_c − is the invariant mass of the cc pair, m_c - is the charm quark mass and $2m_d$ is the DD threshold (m_d mass D-meson). The $\rho_{J/\psi}$ factor - free parameter of the theory also is equal 0.025

For photon-pomeron and pomeron-pomeron processes calculations is made similarly. To take into account the impact parameter dependence in our processes it is necessary to consider pomeron spatial distribution [6].

Using color evaporation model we calculated total cross section of J/ψ production in these processes. Also cross sections at impact parameter b more than two R and a more realistic approach accounts for inelastic scattering effects using the Glauber approximation [7]. For heavy nuclei this factor provides for an extremely effective suppression these processes.

We can see that cross section of gluon-photon process with J/psi production approximately three times higher when photon-pomeron cross section. The cross section of pomeron-pomeron process is suppressed.

TABLE 1. Photon-gluon, photon-pomeron, pomeron-pomeron cross section for J/ψ production in Au(197) collisions at RHIC.

	γg	γp	pp
Σ_{tot}, mb	0.385	0.234	52.0
$\Sigma(b>2R)$, mb	0.153	0.063	0.51
$\Sigma_{elast.}$, mb	0.118	0.043	0.005

MONTE CARLO SIMULATION

We simulated J/ψ production at STAR experiment to allocate pomeron processes. And try to find such kinematics areas where pomeron processes will dominate above a photon-gluon.

To calculate J/ψ registration efficiency at STAR it is necessary to know kinematics characteristics of secondary particles from J/ψ decay. Let's consider J/ψ decay to e^-e^+ and $\mu^-\mu^+$.

The detectors designed for registration of secondary particles in experiment STAR (Solenoidal Tracker at RHIC) at RHIC, have technical and geometrical acceptances. So, for the basic detector at STAR i.e. Time Projection Chamber (TPC), rapidity acceptances lays in the interval $-2.0<\eta<2.0$. Peripheral interactions trigger decreases this interval up to $-1.7<\eta<1.7$. Effective measurement of a moment of the charged particles in TPC is Pt>0.15 GeV. At STAR there is an electromagnetic calorimeter (EMC), for measurement of energy of photons and electron. Acceptances for EMC on rapidity lays in the interval $-1.0<\eta<1.0$, that is appreciable, than for TPC. Distribution on transverse momentum J/ψ for TPC show in Fig.1

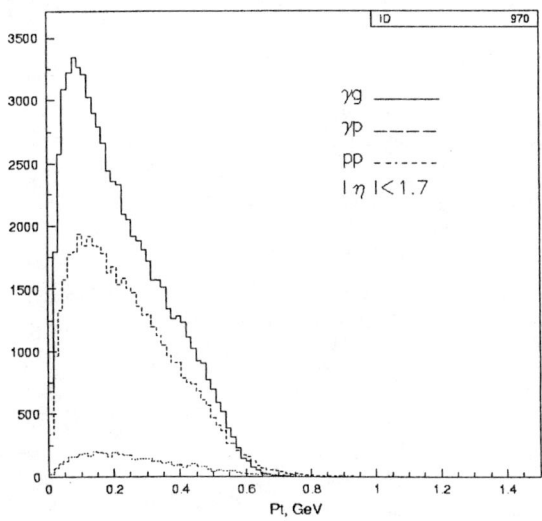

FIGURE 1. The distribution transverse momentum of J/ψ for TPC acceptances. The solid curve for photon-gluon interaction, the dashed curve photon-pomeron interaction and the dotted curve for pomeron-pomeron intetaction.

Taking into account these restrictions we can say, that it is possible to restore about 89% of J/psi events produced in pomeron-pomeron interaction, approximately 90% for a photon-pomeron process and 45% for a photon-gluon interactions in TPC and 56%, 56%, 21% correspondingly at EMC.

Also J/ψ transverse momentum distribution were constructed. Difference in the transverse momentum distribution for different processes can be explained using indeterminancy principle $\Delta r \Delta p \sim h$ It means the less the distance the more momentum. Since gluon in the pomeron has momentum scale about pomeron radius 0.45fm, and gluon in a nucleon is about

1.2fm, hence, the transverse momentum of the particle formed in photon-pomeron and pomeron-pomeron processes, will be more, than in photon-gluon process.

Our task is registration of J/psi production in pomeron interactions. To allocate such events on a photon-gluon background, we can make cut on a transverse momentum. We will select only those events which Pt > 0.6 GeV. As a result of selection for TPC it was received, that in these restrictions get about 3% of events from general number J/psi produced in pomeron processes and less than 0.3% of events from a photon-gluon of process. For EMC 1.5% and 0. % correspondingly. Thus using such selection we reduce a background. So we raises the signal/ background relation of an order.

CONCLUSION

We calculated cross section production J/ψ in photon-gluon, photon-pomeron and pomeron-pomern processes with impact parameter collision nuclei. Simulated production J/ψ in these processes. Let's consider J/ψ decay to e^-e^+ and $\mu^-\mu^+$.

In result of selection at transverse momentum was received, that in these restrictions get about 3% of events from general number J/ψ produced in pomeron processes and less than 0.3% of events from a photon-gluon of process for TPC. For EMC 1.5% and 0.1% correspondingly.

Thus using such selection we reduce a background. So we raises the signal/ background relation of an order.

REFERENCES

1. Donnachie, A., and Landshoft, P. V., *Nucl. Phys. B* **331**, p509 (1989).
2. Schramm, A., *J. Phys. G.* **25**, p1965 (1999).
3. Williams, E. J., *Proc.R.Soc.London Ser. A* **139**, p163 (1933).
4. Greiner, M., Vidovic, M., and Soff, G., *Phys.Rev. C* **47**, p2288 (1993).
5. Gay Ducati, M. B., *Phys. Lett. B* **464**, p286 (1999).
6. Muller, B., and Schramm, A., *Nucl.Phys.A* **523**, p667 (1991).
7. Franco, V., and Glauber, R. J., *Phys. Rev.* **142**, p1195 (1966).

A Comment on the New Low-lying States in the Recent Experimental Results

H.Noya

Institute of Physics, Faculty of Economics, Hosei University at Tama,
Machida, Tokyo 194-0298, Japan
e-mail : hnoya@mt.tama.hosei.ac.jp

and

H.Nakamura

2-25-3, Moridai, Atsugi, Kanagawa 243-0037, Japan

abstract. We present a new interpretation for the low-lying states recently observed by Tatischeff et al. based on the assumption of *the polarization of the baryon number*.

It is well known that many exotic mesons and dibaryons exist which seem to be beyond the limit of conventional theory. The diquark cluster model (DCM) reproduced successfully the mass spectra of these exotic mesons above 1200 Mev [1] and of the dibaryons above 2000 Mev.[2,3] Especially, it was a remarkable success to predict the masses of the scalar meson with I = 2 state which were observed by the OBELIX group.[4]

However, the low-lying groups of the exotic mesons and the dibaryons which were observed by Tatisceff et al. are not suitable for the application of the DCM. In this short article, we present our interpretation for these new states based on *the polarization of the quanta*, a conjecture for the multiquark system which has already been discussed by one of us (H.N.).[7,8]

We assume for a multiquark system that the DCM mass formula holds in the excited state as the system is dominated by a fixed center, while the mass formula based on the two center quark model (TQM) used by Tatischeff et al. holds in the ground state. We think that the polarization of the baryon number (quark number) causes the separation of the quarks and antiquarks in the system as illustrated in the fig.1, and the quark and the antiquark poles in the polarization play the role of the two centers described in the TQM.

Applying the TQM for the q^k-$\bar{q}^h$ system (q = u,d), the mass M of the system is given by

$$M = M_0 + M_1\{i_1(i_1 + 1) + i_2(i_2 + 1) + 1/3[s_1(s_1 + 1) + s_2(s_2 + 1)]\} \quad (1)$$

where i_1 and s_1 represent the isospin and spin of the q^k-cluster while i_2 and s_2 are counter parts in the $\bar{q}^h$-cluster. With $n_T = k + h$ and $n_Q = |k - h|$, we adopt the following very simple approximation for the parameters M_0 and M_1.

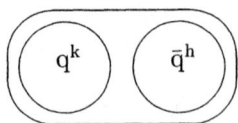

Figure 1. The two center model of the multi-quark system (TQM). (The polarization of the baryon number)

$$M_m = a_m + b_m n_T + c_m n_Q \qquad (m = 0, 1) \qquad (2)$$

Six parameters a_m, b_m and c_m are determined by the use of the data for the $q^3 - \bar{q}^3, q^4 - \bar{q}^4$ and $q^7 - \bar{q}$ system, respectivly. Then one obtains $a_0 = 169, b_0 = 23.5, c_0 = 252.9, a_1 = 39, b_1 = -1.5$ and $c_1 = -3.62$ (in Mev).

Using these parameters, we calculated the mass spectra of the exotic mesons $q^2 - \bar{q}^2, q^3 - \bar{q}^3, q^4 - \bar{q}^4, q^5 - \bar{q}^5$ system and of the dibaryons $q^7 - \bar{q}, q^8 - \bar{q}^2, q^9 - \bar{q}^3$ system, respectivly. The results are illustrated in the fig.2 and the fig.3 together with the experimental mass spectrum of the exotic mesons below 1000 Mev and that of the dibaryons below 2000 Mev by Tatischeff et al.[5,6] The mass spectrum of the narrow $q^2 - \bar{q}^2$ system calculated by the DCM mass formula is also illustrated in the fig.2 with the experimental one by Chiba et al.[9-10] One can see the excellent agreement between the experimental results and calculated ones. We remark that the mass spectrum of the narrow dibaryons above 2000 Mev can be reproduced quite well by the DCM mass formula.[1-3] Also, the masses of the narrow low-lying states of the baryon found by Tatischeff [11] agree well with the prediction of the DCM.[12] We think that these excellent agreements constitute evidence for the validity of our approach to the multiquark system.

Here, we make a comment on our fundamental assumption, the polarization of the baryon number. This assumption was originally introduced by one of us (H.N.) to explain an anomalous cosmic ray event, the centauro event (the multiple production of many hadrons without the π^0).[7] He suggested in a previous article[8] that *the world of antimatter* may exist somewhere in our universe, if the assumption is true. We remark that the incomprehensible property of the γ-ray burst (GRB) may be understood very naturally if one regards the GRB as the vanishing of an anti-star which comes from the world of the antimatter caught by a star in our world. Acutually, one can explain well the peculiar propertiess of the GRB, the production of the extrodinary amount of the γ-ray (10^{54} erg) from a very small region (< 300 km) in a remote galaxy ($>$ several 10^9 light-years), by a simple picture, the vanishing of an anti-neutron star in the galaxy near the border between the antiworld and our world. A detailed discussion is in preparation and will be published elsewhere.

We remark that, in the present picture for the GRB, a considerably large amount of anti-carbon would be scattered around in the burst. Then, if our picture for the GRB is right, there is a possibility anti-carbon will be found in the cosmic rays .

We showed that the ground state of a multiquark system is possibly dominated by a fundamental force which causes the polarization of the baryon number. The existance of the fundamental force may be examined by the use of a high-energy $e^+ - e^-$ collider. If such a force is really present, a centauro event must be observed, although it might be a rather rare event.

We expect our conjecture to be examined with an experiment in the near furture.

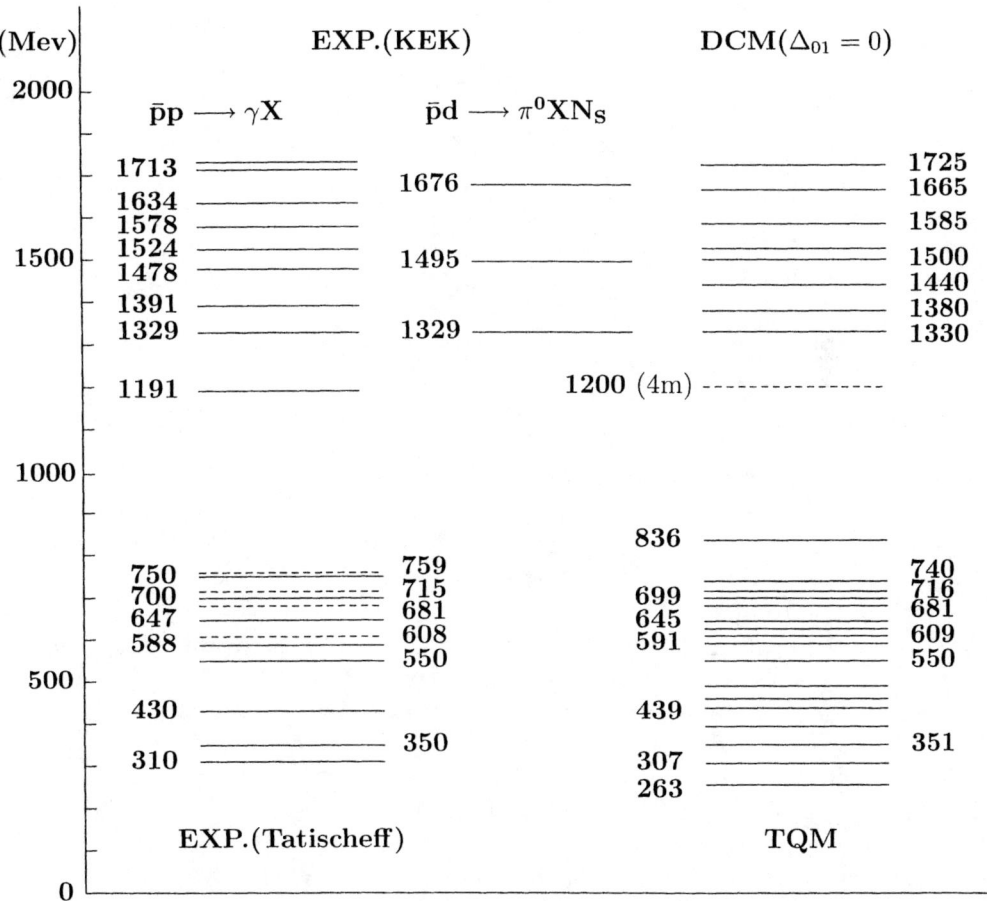

Figure 2. The mass spectrum of the narrow mesonic resonance below 2 Gev calculated by the DCM and the TQM mass formula. The data are taken from Chiba et al.[9-10] and Tatischeff et al.[5]

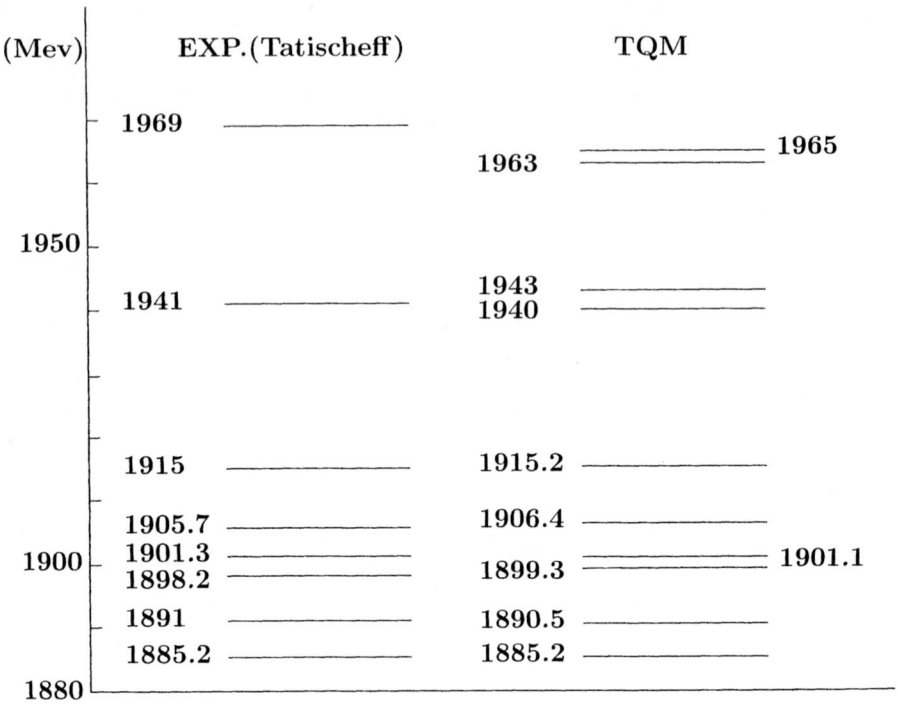

Figure 3. The mass spectrum of the narrow dibaryons calculated by the TQM. The data are taken from Tatischeff et al.[6]

REFERENCES

1. Y.Uehara, N.Konno, H.Nakamura and H.Noya : Nucl.Phys. **A606**, 357 (1996)
2. N.Konno, H.Nakamura and H.Noya : Phys.Rev. **D35**, 239 (1987)
3. H.Noya, H.Nakamura and N.Konno : Nucl.Phys.B (Proc.Suppl.) **21**, 335 (1991
4. A.Filippi : the paper presented at the LEAP 2000 Conference, Venice, Italy
5. J.Yonnet et al. : Phys.Rev. **C63**, 014001 (2000)
6. B.Tatischeff et al. : Phys.Rev. **C62**, 054001 (2000)
7. H.Nakamura, K.Arita and K.Mori : Nuovo Cimento **21**, 33 (1978)
8. H.Nakamura : Nucl.Phys. **A675**, 238c (2000)
9. M.Chiba et al. : Phys.Rev. **D36**, 3321 (1987)
10. M.Chiba et al. : Phys.Rev. **C60**, 035204 (1999)
11. B.Tatischeff et al. : Phys.Rev.Lett. **79**, 601 (1997)
12. N.Konno : Nuovo Cimento **111A**, 1393 (1998)

CHARM AND BEAUTY DECAYS

Measurement of HQET Parameters and CKM Matrix Elements

D. Urner

Cornell University, Wilson Lab, Ithaca NY 14853, USA

Abstract. The determination of CKM matrix elements in the b-sector is discussed, emphasizing the new measurements of $|V_{ub}|$ and $|V_{cb}|$ by the CLEO collaboration.

INTRODUCTION: CKM MATRIX AND UNITARITY TRIANGLE

The mixing of quarks in flavor and their organization into three generations is summarized by the complex Cabibbo-Kobayashi-Maskawa (CKM) [1][2] matrix. The components V_{ij} describe the relative weak couplings between up and down type quarks. Within the Standard Model (SM), the parameters of this matrix are fundamental and have to be determined by experiment.

The SM assumes the unitarity of the CKM matrix, which leaves us with nine parameters, four of which are fundamental: three real mixing angles and one imaginary phase. CP violation can, hence, only occur if this phase does not vanish. The orthogonality of the CKM matrix imposed by unitarity can be expressed geometrically as triangles in which the areas are proportional to the degree of CP violation. What is called The Unitarity Triangle is one of the six possible configurations, for which $V_{ud}V_{ub}^* + V_{cd}V_{cb}^* + V_{td}V_{tb}^* = 0$ and the side $V_{cd}V_{cb}^*$ is used to normalize the other sides. An explicit unitary parameterization that expands around the parameter $\lambda = \sin\Theta_C$ was formulated by Wolfenstein [3] and modified by Buras [4] to attain exact unitarity and higher precision.

Direct angle measurements are hard, since they rely on rare B decay processes. The only accessible angle for the foreseeable future is β for which the asymmetric B factories Babar and Belle have a measurement [5] that combined give a result of $\sin(2\beta)=0.79\pm0.1$, hence establishing a non-zero CP-violating phase. The length of the three sides of the Unitarity Triangle can be determined from non-CP-violating processes which can be used to extract CP violation indirectly. Disagreement between the two methods would suggest new physics beyond the standard model. We expect an accurate measurement of $\sin(2\beta)$ by the B-factories in the next few years. Therefore it is important to improve the accuracy of our knowledge of the length of the sides of the Unitarity Triangle as well, with the goal to maximize the sensitivity to new physics.

Our knowledge of the length of the sides is currently limited by the measurements of V_{ub}, V_{cb} (see below), V_{td}, and V_{ts}. The latter two can be extracted from B_d and B_s mixing [7]. The frequency of the oscillations yields in the B_d system $\Delta m_d = 0.489 \pm 0.008 ps^{-1}$ and in the B_s system $\Delta m_s > 14.6 ps^{-1}$. To extract V_{td} and V_{ts} however one needs the knowledge of f_{B_d} and f_{B_s}, which can be computed by lattice QCD to about

20% with a result of $|V_{td}V_{tb}^*| = (8.3 \pm 1.6) \times 10^{-3}$.

The measurements of V_{ub} and V_{cb} presented below are done with the CLEO detector. The full CLEO II and CLEO II.V data samples containing 9.7×10^6 $B\bar{B}$ events are used for all but the exclusive V_{ub} analysis, which uses 3.3×10^6 $B\bar{B}$ from CLEO II data only.

EXCLUSIVE V_{ub} AND V_{cb} MEASUREMENTS

The determination of of CKM matrix elements from exclusive semileptonic decays requires the knowledge of the semileptonic form factor that encapsulates the hadronic physics between the outgoing quarks.

The extraction of V_{ub} from $B \to \rho l \nu$ and $B \to \pi l \nu$ decays is difficult because of the small decay branching fractions and because it requires neutrino reconstruction. Using only CLEO II data results in a statistical error of only 4%. The largest error contribution comes from the form-factor model uncertainty of 17%. The best measurement currently is: $V_{ub} = (3.25 \pm 0.30_{stat+syst} \pm 0.55_{theory}) \times 10^{-3}$ [8].

Extracted from $B \to D^* l \nu$ decays, V_{cb} can be determined more accurately since the semileptonic form factor for D^* at rest $F(1)$ can be calculated to 4.6% using heavy quark effective theory (HQET). The analysis therefore measures the decay width in 10 bins:

$$\frac{d\Gamma}{dw} = \frac{G_F^2}{48\pi^3}|V_{cb}|^2[F(w)]^2 g(w) \qquad (1)$$

with $w = v_B \dot{v}_{D^*}$ = the D^* boost in the B rest frame separately for charged and neutral B decays. In each case 10 w bins are fitted for $B \to D^* l \nu$, $B \to D^{**} l \nu$, and several background hypotheses. The efficiency corrected $F(w)|V_{cb}|$ distribution shown in figure 1 is then fit to $1-\rho^2$ and extrapolated to $w = 1$ with results of $\rho = 1.51 \pm 0.09 \pm 0.21$ and $F(w)|V_{cb}| = (42.2 \pm 1.3 \pm 1.8) \times 10^{-3}$. The LEP experiments have a combined result [9] using a similar technique of $35.6 \pm 1.7, 1$ which is about 7% consistent with the CLEO result.

DETERMINATION OF HQET PARAMETERS

HQET parameters extracted from $b \to s\gamma$ decays and $B \to X_c l \nu$ hadronic mass moments can be used to reduce the uncertainty in the theoretical prediction of the factor that relates the semileptonic decay width to the square of the CKM matrix element. Inclusive observables can be written as double expansions in powers of α_s and $\frac{1}{M_B}$. We use an expansion in the pole mass [10]. The expansion parameters Λ ($O\frac{1}{M_B}$), λ_1, λ_2 ($O\frac{1}{M_B}$) have to be determined experimentally.

The rare radiative penguin decay $b \to s\gamma$ is interesting because it is sensitive to physics beyond the SM (charged Higgs, ...). The newest CLEO analysis takes the whole E_γ spectrum from 2.0 to 2.7 GeV into account, which leads to small model dependencies. Since the background from continuum photons is two orders of magnitude larger than the signal, a large fraction of off resonance data is the key to this analysis. The $B\bar{B}$ background was estimated from Monte Carlo. From the spectrum shown in figure 1, one

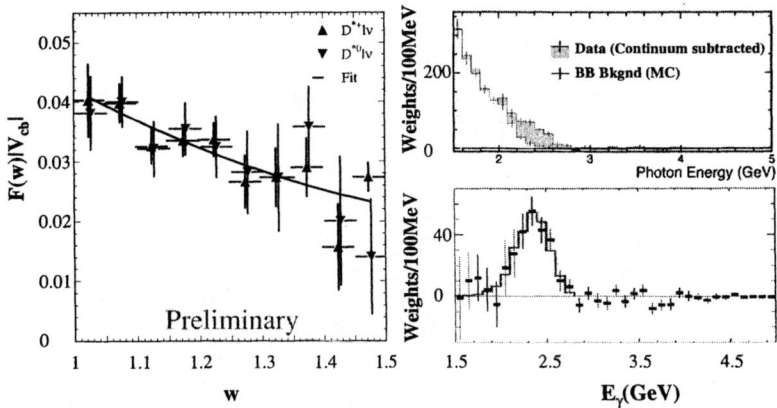

FIGURE 1. Left: Value of $F(w)|V_{cb}|$ versus w with entries from charged and neutral B decays. The line is from a fit to $1-\rho^2$. Upper right: $b \to s\gamma$ after continuum subtraction. Lower right: final $b \to s\gamma$ spectrum. The line shows the expected shape by the Spectator Model.

extracts the following quantities [11]: $B(B \to s\gamma) = (3.19 \pm 0.43 \pm 0.27) \times 10^{-4}$, $\langle E_\gamma \rangle = 2.346 \pm 0.032 \pm 0.011$ GeV, and $\langle (E_\gamma - \langle E_\gamma \rangle)^2 \rangle = 0.0226 \pm 0.0066 \pm 0.0020$ GeV^2.

The hadronic mass moments are extracted from a fit to the $\tilde{M}_X^2 = m_B^2 + m_{l\nu}^2 - 2E_B E_{l\nu}$ distribution. The moments are calculated relative to the spin-averaged D, D* mass $\overline{M}_D$ [12]: $\langle M_X^2 - \overline{M}_D^2 \rangle = 0.251 \pm 0.066$ GeV^2 and $\langle (M_X^2 - \langle M_X^2 \rangle)^2 \rangle = 0.576 \pm 0.17$ GeV^4.

These moments can be related to the HQET parameters such that one can extract $\overline{\Lambda} = 0.35 \pm 0.08 \pm 0.10$ GeV and $\lambda_1 = -0.238 \pm 0.071 \pm 0.078$ GeV^2. $\lambda_2 = 0.128$ GeV^2 can be extracted from the B,B* mass difference. Second moments are not used since the expansion converges slowly, which leads to large theoretical errors. The central values agree, however.

INCLUSIVE V_{ub} AND V_{cb} MEASUREMENTS

Inclusive measurements assume quark-hadron duality, for which there is no consensus on the uncertainty. Using the heavy quark expansion one extracts $|V_{cb}| = (40.4 \pm 0.9_{\Gamma_{exp}} \pm 0.5_{(\overline{\Lambda},\lambda_1)_{exp}} \pm 0.8_{1/M_B^3}) \times 10^{-3}$ [12]. It is important to note, that the same $\overline{MS}$ renormalization scheme and the same order were used for determination of the HQET parameter and the extraction of V_{cb}. A check with a different scheme would be desirable.

V_{ub} is extracted from the inclusive lepton momentum spectrum in the endpoint region between 2.2 and 2.6 GeV for the semileptonic decay channel $B \to X_u l\nu$ (see figure 2). The estimation of the fraction of $b \to u$ transitions in the chosen endpoint region relies on theoretical calculations that have serious model dependencies. Assuming that the effects of the b quark motion are the same for all "massless" partons, one can use the $B \to s\gamma$ shape parameters to reduce the uncertainty in this fraction to 25% in the current analysis. After continuum and background suppression the $B \to X_c l\nu$ yield is subtracted and one extracts $|V_{ub}| = (4.09 \pm 0.14 \pm 0.66) \times 10^{-3}$.

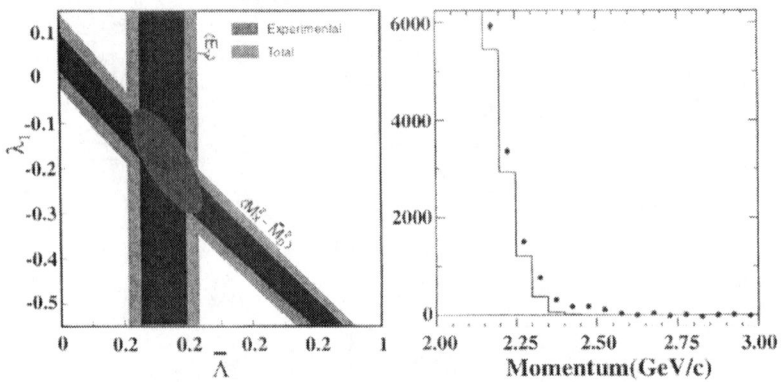

FIGURE 2. Left: The vertical band shows the first moment of the b→sγ photon spectrum, the diagonal band shows the first hadronic moment of B→ $X_c l \nu$ decays. Dark are experimental, light are theoretical errors. The oval shows an area of possible λ_1 and $\bar{\Lambda}$ values. Right: Endpoint region of continuum suppressed data (dots) and b→clν Monte Carlo (histogram)

CONCLUSIONS

New inclusive measurements of V_{cb} and V_{ub} with an accuracy of 3.2% and 17%, respectively, were made possible by the usage of HQET parameters gained from b→sγ and B→ $X_c l \nu$ hadronic mass moment analyses. Exclusive measurements, however, need accurate lattice QCD calculations of the semileptonic form factors. With new techniques and a validation from data planned to be taken with the CLEO-c experiment such results can be expected within a few years; see the talk "Exploring the Charm Sector with CLEO-c" in these proceedings. Together with the expected $\sin(2\beta)$ measurement improvements, the precise measurements of the CKM matrix will test its unitarity and the verity of the SM.

REFERENCES

1. N. Cabibbo, *Phys. Rev. Lett.* **10** (1963) 531.
2. M. Kobayashi and T. Maskawa, *Prog. Theor. Phys.* **49** (1973) 652.
3. L. Wolfenstein, *Phys. Rev. Lett.* **51** 1983) 1945.
4. A. J. Buras, M. E. Lautenbacher and G. Ostermaier, *Phys. Rev. D* **50** (1994) 3433.
5. K. Abe et al. (BELLE Colab.), *Phys. Rev. Lett.* **87**, 091802 (2001).
6. B. Aubert et al. (BABAR Colab.), *Phys. Rev. Lett.* **87**, 091801 (2001).
7. D. Abbaneo et al. (LEP B-Oscillations Working Group), *CERN-EP-2001-50*, June 26, 2001.
8. B. H. Behrens et al. (CLEO Colab.), *Phys. Rev.* **D61** (2000) 052001 [hep-ex/9905056].
9. D. Abbaneo et al. (LEP V_{cb} Work. Group), *lepvcb.web.cern.ch/LEPVCB/Winter01.html*.
10. A. Falk, M. Luke et al. *Phys. Rev.* **D57** (1998) 424.
11. S. Chen et al. (CLEO Colab.), submitted to *Phys. Ref. D*, CLNS 01/1751 at www.lns.cornell.edu/public/CLNS/2001/CLEO.html.
12. D. Cronin-Hennessy et al. (CLEO Colab.), submitted to *Phys. Rev. Lett.*, CLNS 01/1752 at www.lns.cornell.edu/public/CLNS/2001/CLEO.html.

Study of Ω_c^0 Production and New Results on Λ_c^+ Decays at Belle

Ruslan Chistov (on behalf of the *Belle Collaboration*)

ITEP, B.Cheremushkinskaya 25, 117259, Moscow, Russia

Abstract. We present a measurement of $\Omega_c^0 \to \Omega^-\pi^+$ decay. We also present measurements of the Cabibbo-suppressed decays $\Lambda_c^+ \to \Lambda^0 K^+$ and $\Lambda_c^+ \to \Sigma^0 K^+$ (both first observations) and $\Lambda_c^+ \to \Sigma^+ K^+ \pi^-$ (seen with large statistics for the first time). For the Cabibbo-suppressed decays $\Lambda_c^+ \to pK^+K^-$ and $\Lambda_c^+ \to p\phi$, and the W-exchange decays $\Lambda_c^+ \to \Sigma^+ K^+ K^-$ and $\Lambda_c^+ \to \Sigma^+ \phi$, we present measurements with improved accuracy. We found the first evidence for $\Lambda_c^+ \to \Xi(1690)K^+$ and set an upper limit on non-resonant $\Lambda_c^+ \to \Sigma^+ K^+ K^-$ decay. This analysis was performed using 23.6 fb^{-1} of data collected by the Belle detector at the e^+e^- asymmetric collider KEKB.

STUDY OF Ω_C^0 PRODUCTION USING THE $\Omega_c^0 \to \Omega^-\pi^+$ DECAY

The decay $\Omega_c^0 \to \Omega^-\pi^{+1}$ has previously been seen by E687 and CLEO[1]. We reconstruct 23.5 ± 5.4 events for this decay mode at a mass of 2697.3 ± 1.5 MeV: the mass distribution and fit are shown in Figure 1(a). To extract the product of production cross section and branching ratio, we fit the Ω_c^0 signal in five x_p bins. The Peterson et al.[2] fragmentation function was used to fit the resulting x_p spectrum (see Figure 1(b)). We find $\varepsilon_p = 0.18^{+0.27}_{-0.10}$ from which we extract $\sigma(\Omega_c^0) \times \mathcal{B}(\Omega_c^0 \to \Omega^-\pi^+) = 24.2^{+51.4}_{-13.8}$ fb. All quoted errors are statistical only. Our measurements of Ω_c mass and production cross section are consistent with those of CLEO.

OBSERVATION OF THE DECAYS $\Lambda_c^+ \to \Lambda^0 K^+$ AND $\Lambda_c^+ \to \Sigma^0 K^+$

The signal for $\Lambda_c^+ \to \Lambda^0 K^+$ decay is shown in Figure 2(a). The result of the fit is shown by the superimposed curve which takes into account the reflections from $\Lambda_c^+ \to \Lambda^0 \pi^+$ and $\Lambda_c^+ \to \Sigma^0 \pi^+$ decays. We find a yield of 214 ± 30 $\Lambda_c^+ \to \Lambda^0 K^+$ decays, the first observation of this decay mode. For normalization, we use the decay $\Lambda_c^+ \to \Lambda^0 \pi^+$ and extract a branching ratio $\mathcal{B}(\Lambda_c^+ \to \Lambda^0 K^+)/\mathcal{B}(\Lambda_c^+ \to \Lambda^0 \pi^+) = 0.085 \pm 0.012 \pm 0.015$.

Figure 2(b) presents a signal for the $\Lambda_c^+ \to \Sigma^0 K^+$ decay. We find 70 ± 17 $\Lambda_c^+ \to \Sigma^0 K^+$ events, the first observation of this decay mode. For normalization, we use the decay $\Lambda_c^+ \to \Sigma^0 \pi^+$ and extract a branching ratio $\mathcal{B}(\Lambda_c^+ \to \Sigma^0 K^+)/\mathcal{B}(\Lambda_c^+ \to \Sigma^0 \pi^+) = 0.073 \pm 0.018 \pm 0.016$

---

[1] The inclusion of charge-conjugate states is implied.

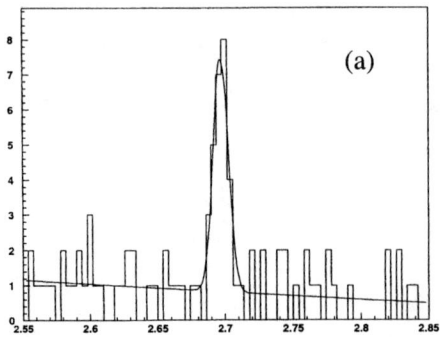 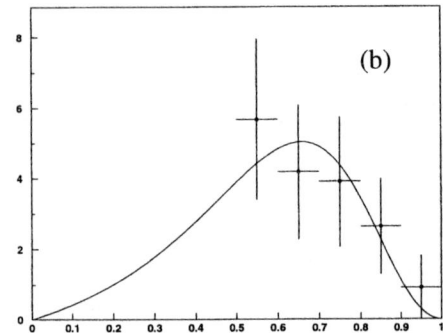

FIGURE 1. (a): The invariant mass spectrum for $\Omega^-\pi^+$ combinations. (b): x_p spectrum for Ω_c^0.

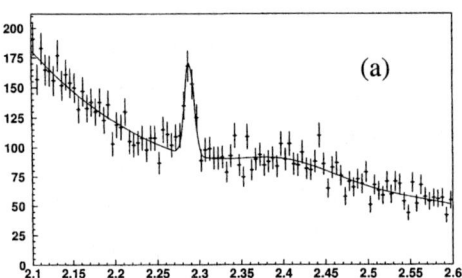 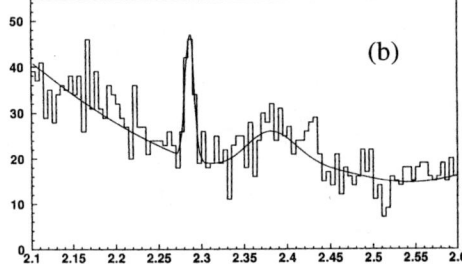

FIGURE 2. The invariant mass spectra of (a) $\Lambda^0 K^+$ and (b) $\Sigma^0 K^+$ combinations

OBSERVATION OF THE $\Lambda_c^+ \to \Sigma^+ K^+ \pi^-$ DECAY

The first evidence for the Cabibbo-suppressed decay $\Lambda_c^+ \to \Sigma^+ K^+ \pi^-$ was published by the NA32 collaboration[3]: they found 2 events. The signal for $\Lambda_c^+ \to \Sigma^+ K^+ \pi^-$ decay is shown in Figure 3(a). We also form $\Sigma^+ K^+ \pi^-$ combinations using "Σ^+" candidates from Σ^+ mass sidebands, shown in Figure 3(a) with the shaded histogram. We find 72 ± 16 $\Lambda_c^+ \to \Sigma^+ K^+ \pi^-$ events. For normalization we reconstruct $\Lambda_c^+ \to \Sigma^+ \pi^+ \pi^-$ decays, and then we extract a branching ratio $\mathcal{B}(\Lambda_c^+ \to \Sigma^+ K^+ \pi^-)/\mathcal{B}(\Lambda_c^+ \to \Sigma^+ \pi^+ \pi^-) = 0.059 \pm 0.014 \pm 0.006$

MEASUREMENT OF THE $\Lambda_c^+ \to \Sigma^+ K^+ K^-$ AND $\Lambda_c^+ \to \Sigma^+ \phi$ DECAYS

The decays $\Lambda_c^+ \to \Sigma^+ K^+ K^-$ and $\Lambda_c^+ \to \Sigma^+ \phi$ proceed dominantly via W-exchange diagrams, and were observed by CLEO[4].

Figure 3(b) shows the invariant mass spectrum for $\Lambda_c^+ \to \Sigma^+ K^+ K^-$ combinations. The fit to this spectrum yields 161 ± 16 $\Lambda_c^+ \to \Sigma^+ K^+ K^-$ decays. For normalization we reconstruct the $\Lambda_c^+ \to \Sigma^+ \pi^+ \pi^-$ decay mode and extract a branching ratio $\mathcal{B}(\Lambda_c^+ \to \Sigma^+ K^+ K^-)/\mathcal{B}(\Lambda_c^+ \to \Sigma^+ \pi^+ \pi^-) = (7.5 \pm 0.8 \pm 1.5) \times 10^{-2}$. For the $\Lambda_c^+ \to \Sigma^+ \phi$ contribution we obtain 93 ± 14 events (Figure 4(a)) and extract a branching ratio $\mathcal{B}(\Lambda_c^+ \to \Sigma^+ \phi)/\mathcal{B}(\Lambda_c^+ \to \Sigma^+ \pi^+ \pi^-) = (9.1 \pm 1.4 \pm 1.8) \times 10^{-2}$.

The $\Sigma^+ K^-$ mass distribution, presented in Figure 4(b), shows evidence for the $\Xi(1690)^0$ resonant state. We obtain 45 ± 15 $\Lambda_c^+ \to \Xi(1690)^0 K^+$ decays and find a combined branching ratio $\frac{\mathcal{B}(\Lambda_c^+ \to \Xi(1690)^0 K^+)}{\mathcal{B}(\Lambda_c^+ \to \Sigma^+ \pi^+ \pi^-)} \times \mathcal{B}(\Xi(1690)^0 \to \Sigma^+ K^-) = (2.1 \pm 0.7 \pm 0.4) \times 10^{-2}$. We also obtain an upper limit on the branching ratio $\mathcal{B}(\Lambda_c^+ \to \Sigma^+ K^+ K^-)_{non-res}/\mathcal{B}(\Lambda_c^+ \to \Sigma^+ \pi^+ \pi^-) < 0.017$ @90% CL

MEASUREMENT OF THE $\Lambda_c^+ \to pK^+K^-$ AND $\Lambda_c^+ \to p\phi$ DECAYS

The first evidence for the $\Lambda_c^+ \to p\phi$ decay was reported by NA32, who claimed a signal of 2.8 ± 1.9 events [5]. The decay $\Lambda_c^+ \to pK^+K^-$ was observed for the first time by E687[6] and the most recent statistically significant resonant analysis was published by CLEO[7].

Figure 5(a) presents the signal for $\Lambda_c^+ \to pK^+K^-$ decay: we find 446 ± 72 events for this mode. For normalization we reconstruct the $\Lambda_c^+ \to pK^-\pi^+$ decay and extract a branching ratio $\mathcal{B}(\Lambda_c^+ \to pK^+K^-)/\mathcal{B}(\Lambda_c^+ \to pK^-\pi^+) = (1.50 \pm 0.25 \pm 0.15) \times 10^{-2}$. For the $\Lambda_c^+ \to p\phi$ contribution, we obtain 205 ± 30 events as shown in Figure 5(b), where the shaded histogram shows the ϕ signal from the Λ_c^+ sidebands. We extract a branching ratio $\mathcal{B}(\Lambda_c^+ \to p\phi)/\mathcal{B}(\Lambda_c^+ \to pK^-\pi^+) = (1.5 \pm 0.23 \pm 0.15) \times 10^{-2}$. For the non-$\phi$ contribution we obtain 222 ± 64 events and calculate a branching ratio $\mathcal{B}(\Lambda_c^+ \to pK^+K^-)_{non-\phi}/\mathcal{B}(\Lambda_c^+ \to pK^-\pi^+) = (0.75 \pm 0.23 \pm 0.08) \times 10^{-2}$.

ACKNOWLEDGMENTS

The authors would like to thank Alexander Bondar, Pavel Pakhlov and Bruce Yabsley for their help in preparing this talk.

REFERENCES

1. P.L.Frabetti *et al.* (E687 Collaboration), *Phys.Lett.* **B300**, 190-194 (1993); D.Cronin-Hennessy *et al.* (CLEO Collaboration), *CLNS 00/1691, CLEO 00-18*.
2. C.Peterson *et al.*, *Phys.Rev.* **D27**, 105-111 (1983)
3. S. Barlag *et al.* (NA32 Collaboration), *Phys. Lett.* **B283**, 465–470, 1992
4. P. Avery *et al.* (CLEO Collaboration), *Phys. Rev. Lett.* **71**, 2391–2395, 1993
5. S. Barlag *et al.* (NA32 Collaboration), *Z. Phys.* **C 48**, 29–45, 1990
6. P.L. Frabetti *et al.* (E687 Collaboration), *Phys. Lett.* **B314**, 477–481, 1993
7. J. Alexander *et al.* (CLEO Collaboration), *Phys. Rev.* **D53**, 1039–1050, 1996

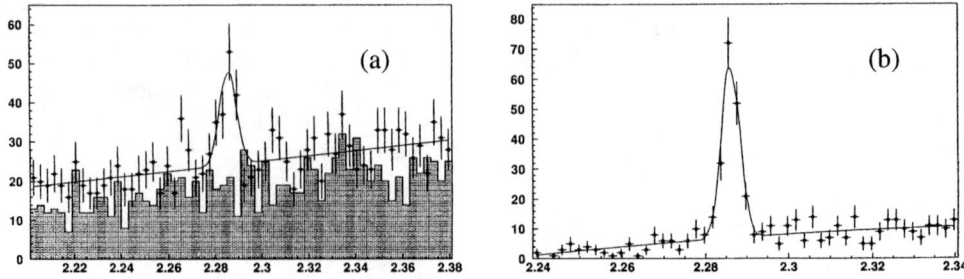

FIGURE 3. The invariant mass spectra of (a) $\Sigma^+ K^+ \pi^-$ and (b) $\Sigma^+ K^+ K^-$ combinations

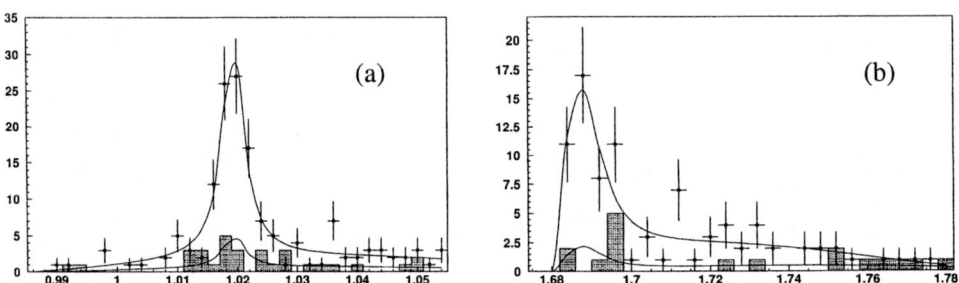

FIGURE 4. The invariant mass spectra of (a) $K^+ K^-$ and (b) $\Sigma^+ K^-$ combinations, both from $\Lambda_c^+ \to \Sigma^+ K^+ K^-$ decay.

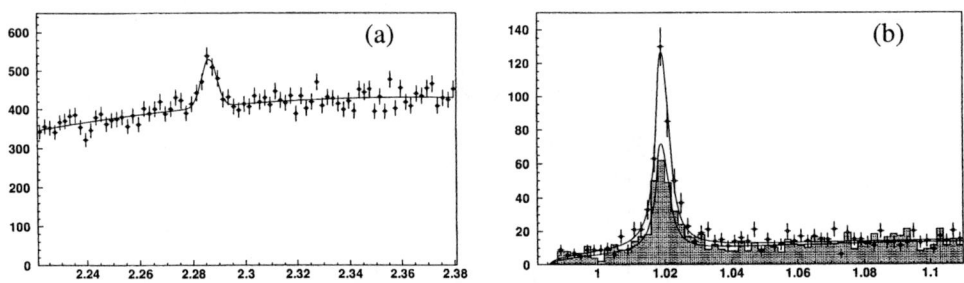

FIGURE 5. The invariant mass spectra of (a) pK^+K^- and (b) K^+K^- combinations from the signal region.

Production and Decay of the Λ_c Charmed Baryon from Fermilab E791.

B. Meadows [1]

University of Cincinnati, Cincinnati, OH, 45221, USA

Abstract. Results are presented for the 500 GeV/c pion production asymmetry and polarization of the Λ_c ($\overline{\Lambda_c}$) charmed baryon from Fermilab experiment E791. An analysis of the decay to the $p\overline{K}\pi$ final state is described. Resonant sub-channel fractions **and phases** are given and possible resonant effects in the low mass $p\overline{K}$ system discussed. Significant decay to $\Lambda_c \to \Delta^{++}K^-$ establishes for the first time the importance of a W exchange mechanism in charmed baryon decay.

Measurements of asymmetry $A_{\mathcal{P}} = (d\sigma_{\mathcal{P}} - d\sigma_{\overline{\mathcal{P}}})/(d\sigma_{\mathcal{P}} + d\sigma_{\overline{\mathcal{P}}})$ in the yield of particle $\mathcal{P}$ and anti particle $\overline{\mathcal{P}}$ can provide information on the production mechanisms involved. Dependences of A on x_F and p_T^2 can distinguish different production models. Several experiments [1, and refs 1-9 therein] have shown that production of charmed mesons is characterized by leading particle effects and that asymmetries can be large. Leading particle behaviour has also been observed in production of strange hyperons in E791 [2] - even in a very central region.

Branching fractions for baryon decays provide information on the relative importance of lowest order decay mechanisms - W exchange or spectator processes. In $\Lambda_c \to p\overline{K}\pi$ decay, [2] W exchange can contribute to $pK^{*\circ}$, $\Lambda^*\pi$, $\Sigma^*\pi$ or $pK\pi$ channels, but for the $\Delta^{++}K^-$ mode it is the *only* low order process possible. Evidence for this decay requires a large sample of $pK\pi$ decays and proper analysis of interference effects in the system.

Reported here are the first published [3] measurements of both x_F and p_T^2 dependence of A for charmed baryon production. We also present [4] the first full analysis of charmed baryon decay, measuring Λ_c branching fractions, relative phases and polarization.

This study is based on a sample of 2×10^{10} events produced from the interaction of 500 GeV/c π^- incident on thin foils, one *Pt* and four *C*. *Pt* target data (unequal numbers of n and p) were not used in the asymmetry study. The detector and data reconstruction are described in [5]. Cuts on geometric and kinematic quantities were made to identify $\Lambda_c \to pK^-\pi^+$ decays. The decay vertex had to be well separated ($> 5\sigma$) from both production vertex and nearest target material. The yield, shown in Fig. 1(a)-(d), was $1,025 \pm 45$ $\Lambda_c \to pK^-\pi^+$ and 794 ± 42 $\overline{\Lambda_c} \to \overline{p}K^+\pi^-$. Events were divided into 5 regions of x_F and 5 of p_T^2 in the overall ranges: $-0.1 < x_F < 0.6$ and $p_T^2 \le 8$ $(GeV/c)^2$ chosen to have clear Λ_c signals in each. Fits similar to those shown in the figure were made to each sample to determine the number $N(\Lambda_c)$ and $\overline{N}(\overline{\Lambda_c})$ of signal events in each range.

---

[1] Representing the E791 Collaboration
[2] Note that charged conjugate states are implied unless stated otherwise.

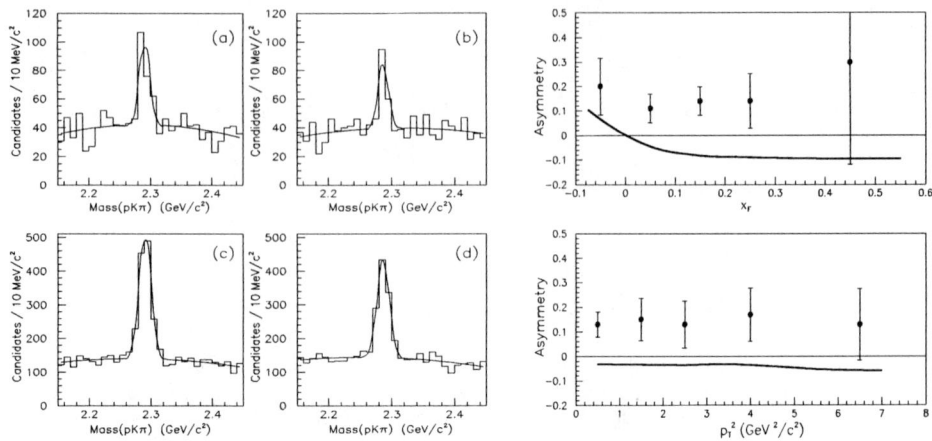

FIGURE 1. $\Lambda_c(\overline{\Lambda_c})$ samples: (a) Λ_c ($x_F < 0$); (b) $\overline{\Lambda_c}$ ($x_F < 0$); (c) Λ_c ($x_F > 0$); (d) $\overline{\Lambda_c}$ ($x_F > 0$). Fits shown in (a)-(d) are Gaussian peaks on 2nd order polynomial backgrounds. Asymmetries: (e) vs x_F (f) vs. p_T^2. The solid curves in (e) and (f) are the prediction of Pythia/Jetset.

$\Lambda_C/\overline{\Lambda_C}$ ASYMMETRIES

Efficiencies ε ($\overline{\varepsilon}$) for $\Lambda_c(\overline{\Lambda_c})$ were not quite equal due to the asymmetric effect of the intense π^- beam on the drift chambers. This effect was greatest at large x_F and low p_T^2. It was necessary therefore to estimate the ratio $r = \varepsilon/\overline{\varepsilon}$ in each of the 5 x_F and 5 p_T^2 ranges using Monte Carlo samples of Λ_c & $\overline{\Lambda_c}$ generated with Pythia/Jetset, projected through a simulated E791 detector and subjected to the same reconstruction code and selection criteria as the data. Corrected asymmetries $A = (N - \overline{N}/r)/(N + \overline{N}/r)$ were then obtained in each range.

The main sources of systematic uncertainty (parametrization of signal and background shapes and precision of r) amounted to less than 50% of the statistical uncertainty in all instances. The results are shown in Figure 1(e) and (f) and compared with earlier $\pi^- N$ studies in Table 1.

TABLE 1. Comparison with asymmetries (%) from earlier $\pi^- N$ experiments.

x_F region	E791	ACCMOR [6]	SELEX [7]
$x_F < 0$	$20 \pm 10 \pm 6$	--	--
$x_F > 0$	$12.3 \pm 3.7 \pm 1.6$	0.5 ± 7.9	25 ± 15

The asymmetry is positive and flat throughout the range. This might result from the additional energy required to produce additional baryons when a $\overline{\Lambda_c}$ is produced, favouring Λ_c production in general. The solid curve in Figure 1 is the prediction of Pythia/Jetset and clearly does not describe the data well. Two component intrinsic charm/coalescence models [8], [9] predict a rising asymmetry beginning at the low end or possibly below the range of this data. Leading particle effects would also result in a rising asymmetry in the entire $x_F < 0$ region. The data do not rule that possibility out.

ANALYSIS OF THE DECAY $\Lambda_C \to PK^-\pi^+$

A cleaner sample was required for this analysis. The length cut was increased to 8σ and a neural net criterion was used to optimize the significance $S/\sqrt{S+B}$ of the signal ($S = 886 \pm 43$) over background ($B \sim 300$) in the fit region.

These decays were defined by five independent variables, e.g. two Dalitz plot coordinates and orientation of the decay plane relative to production plane (z axis). The Λ_c could have polarization $P_{\Lambda_c}\hat{z}$.

Each isobar decay channel $\Lambda_c \to R(\to ab)c$ was assigned an amplitude labeled by the z component of Λ_c spin, m, and proton helicity ($\pm\frac{1}{2}$) in the Λ_c rest frame
$\mathcal{A}^R_{m,\pm\frac{1}{2}} = B^R(M_{ab})(a_\pm e^{i\alpha_\pm}|m,\pm\frac{1}{2},\lambda_{\alpha_\pm}> + b_\pm e^{i\beta_\pm}|m,\pm\frac{1}{2},\lambda_{\beta_\pm}>)$ where $\lambda_{\alpha_\pm}$ & $\lambda_{\beta_\pm}$ are the two possible helicities for R with unknown coefficients $a_\pm e^{i\alpha_\pm}$ & $b_\pm e^{i\beta_\pm}$. The $B^R(M_{ab})$ were Breit Wigner functions. For non-resonant NR decay to $pK\pi$ a similar amplitude with $B=1$ was used. An unbinned, maximum likelihood fit was used to determine $a_\pm,\alpha_\pm,b_\pm,\beta_\pm$ and three values for P_{Λ_c} (one in each of three x_F ranges). The signal probability density function was

$$\mathcal{P}_s = \frac{1}{2} \times \varepsilon \times [(1+P_{\Lambda_c})\left(\left|\sum_R \mathcal{A}^R_{\frac{1}{2},\frac{1}{2}}\right|^2 + \left|\sum_R \mathcal{A}^R_{\frac{1}{2},-\frac{1}{2}}\right|^2\right)$$
$$+ (1-P_{\Lambda_c})\left(\left|\sum_R \mathcal{A}^R_{-\frac{1}{2},\frac{1}{2}}\right|^2 + \left|\sum_R \mathcal{A}^R_{-\frac{1}{2},-\frac{1}{2}}\right|^2\right)]$$

Five dimensional efficiency (ε) and background density were estimated empirically from a MC sample and $M_{pK\pi}$ sidebands. Modes included were $p\bar{K}^{*0}(890)$, $\Delta^{++}(1232)K^-$, $\Lambda(1520)\pi^+$ and NR.

The fit shown in Figure 2 is seen to be good except for the low mass K^-p region where an unmodelled enhancement is seen. Many Y^* exist which could possibly account for this. Each $Y^*\pi$ channel added to our fit requires ≥ 4 more parameters making it difficult, with our limited sample, to include more than one Y^*. Adding $\Lambda(1600)\pi$, $\Sigma(1600)\pi$ or the tail of the $\Sigma(1405)\pi$ alone made no significant improvement.

Isobar fractions f_R were computed by integrating over the five dimensions of the fit $\vec{x}$:

$$f_R = \int \sum_{m,\pm\frac{1}{2}} \left|\mathcal{A}^R_{m,\pm\frac{1}{2}}\right|^2 d\vec{x} / \int \sum_{m,\pm\frac{1}{2}} \left|\sum_R \mathcal{A}^R_{m,\pm\frac{1}{2}}\right|^2 d\vec{x}$$

Branching ratios with respect to $pK^-\pi^+$ are compared with earlier results in Table 2.

Good agreement is seen, but the significance of signals from NA32, where only mass projections were fit is overestimated. NA32 errors are comparable to E791, but E791's sample is much larger. This is because correlations among channels and relative phases were neglected in the NA32 analysis.

The $\Delta^{++}K^-$ mode is comparable to $p\bar{K}^{*0}$ and clearly significant.

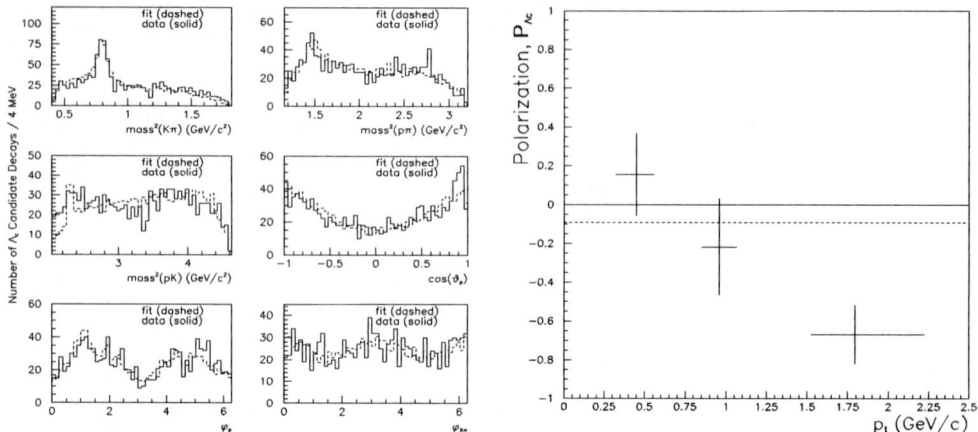

FIGURE 2. Left: Projections of fit (solid lines) onto three mass pair and three angular variables (dotted lines). Data lie in the range $2265 < M(pK^-\pi^+) < 2315$ MeV/c². **Right:** $\Lambda_c/\overline{\Lambda_c}$ Polarization from fit.

TABLE 2. First three columns are branching fractions relative to total $pK\pi$ mode (% corrected for unseen decays). The last four columns are resonant phases, described in the text measured only by E791.

Mode	E791	NA32 [10]	ISR [11]	E791 relative phases (degrees)			
				α_+	β_+	α_-	β_-
$pK^{*0}(890)$	$29\pm4\pm3$	$35^{+6}_{-7}\pm3$	42 ± 24	58 ± 28	135 ± 38	198 ± 24	303 ± 32
$\Delta^{++}(1232)K^-$	$18\pm3\pm3$	$12^{+4}_{-5}\pm5$	40 ± 17	285 ± 23	280 ± 23	$=\alpha_+$	$=\beta_+$
$\Lambda(1520)\pi^+$	$15\pm4\pm2$	$9^{+4}_{-3}\pm2$	–	340 ± 30	-3 ± 32	$=\alpha_+$	$=\beta_+$
NR	$55\pm6\pm4$	$56^{+7}_{-9}\pm5$	–	199 ± 31	0 (fixed)	43 ± 41	65 ± 21

SUMMARY

Λ_c production asymmetry in the range $-0.1 < x_F < 0.6$ and $p_T^2 < 8$ $(GeV/c)^2$ is constant at $\sim +0.15$ favouring Λ_c over $\overline{\Lambda_c}$. Models requiring a rising asymmetry toward negative x_F are not ruled out however. An amplitude analysis of the Λ_c decay shows the $\Lambda_c \to \Delta^{++}K^-$ mode to be large indicating that the W exchange amplitude is important.

REFERENCES

1. E791 Collaboration (E.M. Aitala *et al.*), *Phys. Lett.*, **B411**, 230 (1997).
2. E791 Collaboration (E.M. Aitala *et al.*), *Phys. Lett.*, **B496**, 9 (2000).
3. E791 Collaboration (E.M. Aitala *et al.*), *Phys. Lett.*, **B495**, 42 (2000).
4. E791 Collaboration (E.M. Aitala *et al.*), *Phys. Lett.*, **B471**, 449 (2000).
5. Aitala, E. M., et al., *Eur. Phys. J. direct*, **C4**, 1 (1999).
6. ACCMOR Collaboration (S. Barlag *et al.*), *Phys. Lett.*, **B247**, 113 (1990).
7. M. Iori *et al.* (1999), Proc. of EPS-HEP99 conference, Tampere, Finland, July.
8. Vogt, R., and Brodsky, S. J., *Nucl. Phys.*, **B478**, 311–334 (1996).
9. Herrera, G., and Magnin, J., *Eur. Phys. J.*, **C2**, 477–482 (1998).
10. ACCMOR Collaboration (A. Bozek *et al.*), *Phys. Lett.*, **B312**, 247 (1993).
11. Split Field Magnet Collaboration (M. Basile *et al.*), *Nuovo Cimento*, **62A**, 14 (1981).

Dalitz Analyses of $D_s^\pm \to K_S^0 K_S^0 \pi^\pm$ and $D_s^\pm \to \pi^+ \pi^- \pi^\pm$

T. Deppermann*, K. Peters* and H. Schmücker*

Institut für Experimentalphysik I, Ruhr-Universität Bochum, 44780 Bochum, Germany

Abstract. The high luminosity of PEP-II in combination with the vertexing possibilities of the BABAR-detector [1] offer unique opportunities on light meson spectroscopy. The basic interest in this domain is the search for exotic states. Many of the candidates for conventional mesons are under heavy discussion because of analysis ambiguities, mixing with conventional mesons and some are only seen in selected reactions. Many ambiguities can be resolved using a clean initial state which restricts the final state to specific quantum numbers, such as weak decays of $D_s^\pm$-mesons into three pseudoscalars, allowing only a few resonances to occur.

The selections and the results of the Dalitz plot analyses of the decays $D_s^\pm \to K_S^0 K_S^0 \pi^\pm$ and $D_s^\pm \to \pi^+ \pi^- \pi^\pm$ are presented. These decays are of particular interest for the discussion of the existence and the spin of isoscalar resonances above $1.5\ GeV/c^2$. In this mass region the glueball ground state is conjectured to lie and might mix strongly with other isoscalar $J^{PC} = 0^{++}$-states.

INTRODUCTION

Several candidates for exotics have been identified. One of them is the $f_0(1500)$, which is believed to have a strong glueball component. It was first seen in $p\bar{p}$-annihilation in the Crystal Barrel experiment at LEAR. Unfortunately its interpretation depends on the overall knowledge of the scalar nonet of conventional mesons which may mix with a glueball state of identical quantum numbers. But this nonet is badly known. There are several candidates for the isoscalar sector, like the $f_0(980)$, the $f_0(1370)$ or the $f_J(1710)$ if its spin is zero. However, the spin of the $f_J(1710)$ is not yet settled. In some experiments it turned out to be zero while other experimental results are consistent with a spin of 2 for that resonance. In contrast to many other mesons the $f_J(1710)$ has a small production ratio in $p\bar{p}$ and πp reactions and is poorly known compared to other scalar mesons. Therefore for the interpretation of the light meson spectrum it is important to settle the spin of the $f_J(1710)$ as well as to understand production and decay of the $f_0(980)$.

THE DECAY CHANNEL $D_s^\pm \to K_S^0 K_S^0 \pi^\pm$

Reconstruction

The analysis of this decay channel is based on a data sample of $18.4\ fb^{-1}$. The preselection of the occurring kaons, decaying into two charged pions, is done using a ge-

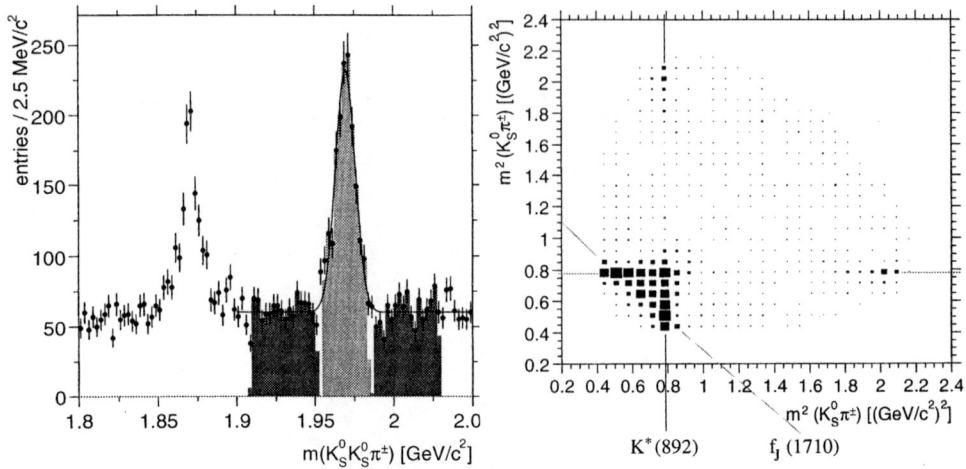

FIGURE 1. $K_s^0 K_s^0 \pi^\pm$ mass histogram and corresponding Dalitz plot

ometric vertex algorithm. K_s^0-pairs on their part are then combined with the remaining charged tracks and kinematically fitted in order to create $D_s^\pm$-candidates. Since background rises exponentially with decreasing $D_s^\pm$-momentum, a minimum momentum cut (2.5 GeV/c) for the $D_s^\pm$-candidates was applied. It is not feasible to extract a sufficient clean $D_s^\pm$-signal at lower momenta. Further background suppression is done be requiring all $D_s^\pm$-mesons having momenta below 3.5 GeV/c to come from the decay $D_s^{*\pm} \to D_s^\pm \gamma$ to be accepted. The selection was optimized by cutting on the vertex probability, lifetime, and mass window of the K_s^0 as well as the vertex probability of the $D_s^\pm$. Finally the compatibility of the $D_s^\pm$-track with the interaction point was checked.

Fitting the $D_s^\pm$-peak in the $K_s^0 K_s^0 \pi^\pm$-mass spectrum (fig.1, left plot) with a Gaussian on a constant background results in a number of events of 1070 ± 47 and a sigma of 6.2 MeV/c^2. The light grey histogram is the one of the signal region while the sideband regions used to estimate the background in the signal region are indicated by dark grey histograms.

Dalitz plot

The corresponding symmetrized, sideband subtracted $K_s^0 \pi^\pm - K_s^0 \pi^\pm$-Dalitz plot is shown in fig. 1, right plot. Clearly visible are the $K^*(892)$ bands in parallel to the axes.

But these structures are not symmetrical. There is an additional diagonal structure in the lower left corner of the plot, which might come from the $f_J(1710)$.

Fitting

Fitting the distribution in the Dalitz plot with an event-based, unbinned loglikelihood-fit and including only well known resonances, results in the grey histograms in the Dalitz

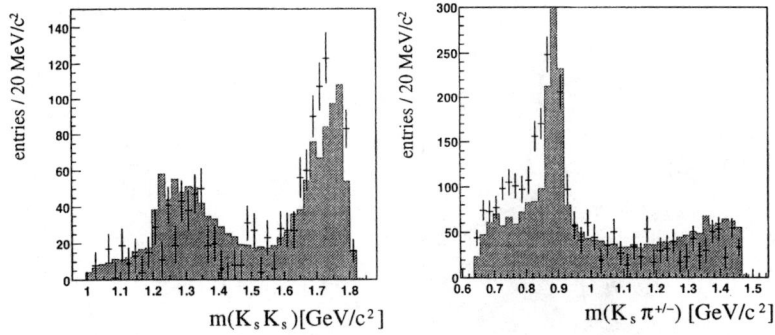

FIGURE 2. Comparison of fit results and measured distributions (error bars)

plot projections of fig.2 to be compared with the measured data indicated by error bars. In this case the resonances $K^*(892)$ and $K_0^*(1430)$ were included in the fit and one can see that the fit does not satisfactorily reproduce the measured distributions. In case of the $K_s^0 K_s^0$-projection the peak at about $1.7\,GeV/c^2$ is not well reproduced while in the $K_s^0\pi^\pm$-projection there is something missing to create the shoulder on the left of the $K^*(892)$-peak. There is need of an additional resonance at a $K_s^0 K_s^0$-mass of about $1700\,MeV/c^2$.

THE DECAY CHANNEL $D_S^\pm \to \pi^+\pi^-\pi^\pm$

Reconstruction

The analysis of the decay channel $D_s^\pm \to \pi^+\pi^-\pi^\pm$ is also based on a data volume of $18.4\,fb^{-1}$. Particle identification for the pions was applied in form of particle selectors working on dE/dx information of BABAR's drift chamber and vertex detector as well as on the information of its Cherenkov detector. All possible 3-pion combinations have to fall into the mass window $[1.8\,GeV/c^2; 2.1\,GeV/c^2]$ in order to be accepted as $D_s^\pm$-candidates and are then kinematically fitted. A minimum momentum cut was applied for the same reason as in the analysis presented previously - there is no chance to get a clean $D_s^\pm$-signal below $2.5\,GeV/c$. But the combinatorial background in this channel is much higher. Therefore the constraint that the $D_s^\pm$-candidates come from the decay $D_s^{*\pm} \to D_s^\pm \gamma$ was applied to all the $D_s^\pm$-candidates. The selection was optimized by cutting on the minimum momentum of the three pions as well as on the vertex probability and the compatibility of the $D_s^\pm$-track with the interaction point.

The $D_s^\pm \to \pi^+\pi^-\pi^\pm$ mass spectrum is shown in fig. 3, left plot. Fitting the peak with a Gaussian on a linear background results in a number of events of 903 ± 64 and a sigma of $7.4\,MeV/c^2$. The light and dark grey histograms again indicate signal and background region, respectively.

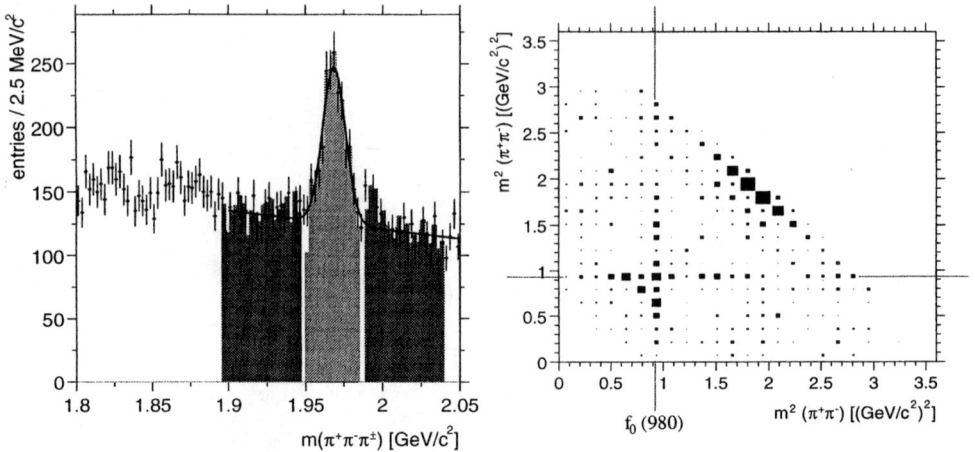

FIGURE 3. $\pi^+\pi^-\pi^\pm$ mass histogram and corresponding Dalitz plot

Dalitz plot

The symmetrized, side-bin subtracted Dalitz plot is shown in fig.3, right plot. Clearly visible are the horizontal and vertical bands of the $f_0(980)$. Work on the determination of the origin of the diagonal structure as well as a spin parity analysis of this channel is going on.

OUTLOOK

First results of a partial wave analysis of the decay channel $D_s^\pm \to K_s^0 K_s^0 \pi^\pm$ indicate a strong contribution of the $f_J(1710)$ being necessary in order to understand this decay. First results of the analysis of the channel $D_s^\pm \to \pi^+\pi^-\pi^\pm$ and the corresponding Dalitz plot were shown.

Work in progress is a completion of the partial wave analysis of the channel $D_s^\pm \to K_s^0 K_s^0 \pi^\pm$ as well as start of it for the channel $D_s^\pm \to \pi^+\pi^-\pi^\pm$. Detailed studies of systematical errors, the sensitivities of the fits and further investigation of the background are going on. In case of the $K_s^0 K_s^0 \pi\pm$-channel we need to study the spin of the $f_J(1710)$ in detail while extending the analysis of the 3-pion channel on more date will allow us to cut on higher $D_s^\pm$-momenta, resulting in a comparable number of events at much lower background.

REFERENCES

[1] BABAR web page: http://www.slac.stanford.edu/BFROOT/

Measurement of Inclusive $f_1(1285)$ and $f_1(1420)$ Production in Z Decays with the DELPHI Detector

Dmitri Ryabchikov for the DELPHI collaboration

Abstract. Inclusive production of two $(K\bar{K}\pi)^0$ states in the mass region 1.22–1.56 GeV in Z decay at LEP I has been observed by the DELPHI Collaboration. The measured masses and widths are 1274 ± 4 and 29 ± 12 MeV for the first peak and 1426 ± 4 and 51 ± 14 MeV for the second. A partial-wave analysis has been performed on the $(K\bar{K}\pi)^0$ spectrum in the mass range; the first peak is consistent with the quantum numbers $I^G(J^{PC}) = 0^+(0^{-+}/1^{++})$ and the second with $I^G(J^{PC}) = 0^+(1^{++})$. These measurements, as well as their total hadronic production rates per hadronic Z decay, are consistent with the mesons of the type $n\bar{n}$, where $n = \{u,d\}$. They are very likely to be the $f_1(1285)$ and the $f_1(1420)$, respectively.

INTRODUCTION

The inclusive production of mesons has been a subject of long-standing study at LEP[1][2], as it provides an insight into the nature of fragmentation of quarks and gluons to hadrons. So far the studies have been done on the S-wave mesons (both $^1S_0$ and $^3S_1$) such as π and ρ, as well as certain P-wave mesons $f_2(1270)$ and $K_2^*(1430)$ (i.e. $^3P_2$) and $f_0(980)$ and $a_0(980)$ ($^3P_0$). Very little is known about the production of mesons belonging to other P-wave (i.e. $^3P_1$ and $^1P_1$). For the first time, we present here a study of the inclusive production of $J^{PC} = 1^{++}$ mesons $f_1(1285)$ and $f_1(1420)$ (i.e. $^3P_1$).

EXPERIMENTAL PROCEDURE

The analysis presented here is based on a data sample of about 3.3 million hadronic Z decays collected from 1992 to 1995 with the DELPHI detector. Detailed description of the DELPHI detector and its performance can be found elsewhere[4][5].

Hadronic events are selected by requiring at least 5 charged particles, with at least 3-GeV energy in each hemisphere of the event—defined with respect to the beam direction—and total energy at least 12% of the center-of-mass energy. The contamination from events due to beam-gas scattering and to γ-γ interactions is estimated to be less than 0.1% and the background from $\tau^+\tau^-$ events less than 0.2% of the accepted events.

$K^\pm$ identification has been provided by the RICH detectors for particles with momenta above 700 MeV/c, while the ionization loss measured in the TPC has been used for momenta above 100 MeV/c. A more detailed description of the identification tags can be found in Ref. [1]. The K_s candidates are detected by their decay in flight into $\pi^+\pi^-$. The details of the method and the various cuts applied are described in Ref. [6].

After all the above cuts, only events with at least one $K_sK^+\pi^-$ or $K_sK^-\pi^+$ combination have been kept in the analysis, resulting in a sample of 705 688 events.

$K_s K^{\pm}\pi^{\mp}$ MASS SPECTRA

Because of big combinatorial background there is no visible enhancement in the mass region between 1.25 to 1.45 GeV both in total $K_s K^{\pm}\pi^{\mp}$ mass spectrum and in that with the K^* cut $0.822 < M(K\pi) < 0.962$ GeV. The key to a successful study of the $f_1(1285)$ and $f_1(1420)$ is to make a mass cut $M(K_s K^{\pm}) \leq 1.04$ GeV, as shown in Fig. 1-a), where two clear peaks are seen. There are two reasons for this: (1) the decay mode $a_0(980)^{\pm}\pi^{\mp}$ is selected by the mass cut, while the general background for the $K\bar{K}\pi$ system is reduced by a factor of $\simeq 7$ at 1.42 GeV or more at higher masses; (2) the interference effect of the two $K^*(892)$ bands on the Dalitz plot at $M(K\bar{K}\pi) \sim 1.4$ GeV is enhanced, if the G-parity is positive[9].

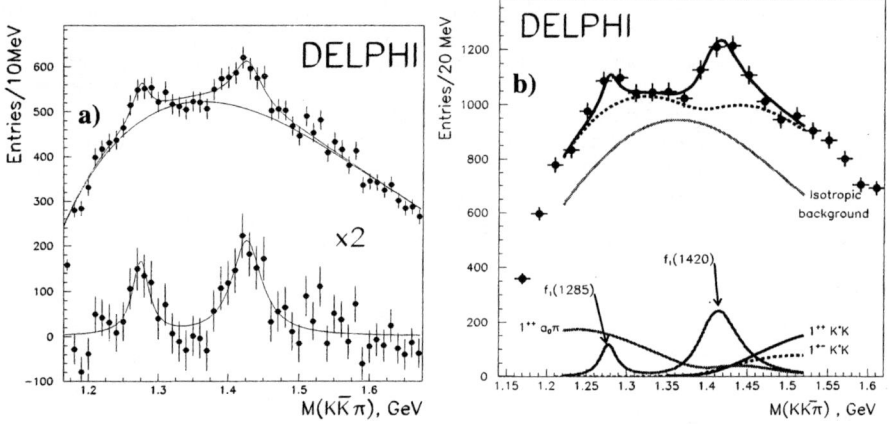

FIGURE 1. a)$M(K_s K^{\pm}\pi^{\mp})$ distribution with a mass cut $M(K_s K^{\pm}) < 1.04$ GeV. The two solid curves in the upper part of the histogram describe Breit-Wigner fits over a smooth background (see text). The lower histogram and the solid curve give the same fits with the background subtracted and amplified by a factor of two. b)distributions per 20 MeV with a breakdown into the partial-waves for the signals and the background. The signals consist of $1^{++} a_0(980)\pi$ for the first peak and $1^{++} K^*(892)\bar{K}$ for the second peak. The background consists of non-interfering superposition of isotropic distribution (1), $1^{++} a_0(980)\pi$ (2), $1^{++} K^*(892)\bar{K}$ (3) and $1^{+-} K^*(892)\bar{K}$ (4).

The results of the fit with smooth background and two S-wave Breit-Wigner forms are shown in Fig. 1-b). The fitted mass and width are (1274 ± 4) MeV and (29 ± 12) MeV for the first peak and (1426 ± 4) MeV and (51 ± 14) MeV for the second one.

The main sources of systematic errors come from the various cuts and selection criteria applied for the V^0 reconstruction and $K^+/-$-mesons identification and also from the conditions of the mass-fit procedure. The first type of error is estimated to be 7% of a given cross section., in the low $K_s K^{\pm}$ mass region. To estimate the second type of error, we have performed a series of fits in different mass intervals and with different background parametrisations. In this way we estimate the fit uncertainty to be 15% for the $f_1(1285)$ and 14% for the $f_1(1420)$. It should be emphasized that the quoted masses and the widths are not intended to be new experimental measurements; rather, they are merely given as an indication that our peaks are consistent with the known parameters.

PARTIAL-WAVE ANALYSIS

We have chosen to employ the so-called Dalitz plot analysis, integrating over the three Euler angles. This entails an essential simplification in the number of parameters required in the analysis, as the decay amplitudes involving the D-functions defined over the three Euler angles and their appropriate decay-coupling constants, are orthogonal for different spins and parities[7]. The actual fitting of the data is done by using the maximum-likelihood method, in which the normalization integrals are evaluated with the accepted Monte Carlo events[8], thus taking into account the finite acceptance of the detector and the event selection.

We assume that the background does not interfere with signals and that it is a non-interfering superposition of a flat distribution (on the Dalitz plot) and the partial waves $I^G(J^{PC}) = 0^+(1^{++})a_0(980)\pi$, $0^+(1^{++})(K^*(892)\bar{K}+c.c.)$ and $0^-(1^{+-})(K^*(892)\bar{K}+c.c.)$.

The signal regions, for $M(K\bar{K}\pi)$ in $1.26 \to 1.30$ and $1.38 \to 1.48$ GeV, have been fitted with a non-interfering superposition of the partial waves $I^G(J^{PC}) = 0^+(1^{++})$, $0^+(1^{+-})$ and $0^-(0^{-+})$, where the decay channels $a_0(980)\pi$ and $K^*(892)\bar{K}+c.c.$ are allowed to interfere within a given J^{PC}. All other possible partial waves have been found to be negligible in the signal regions. Because of a lack of phase space, the two isobars $a_0(980)$ and $K^*(892)$ cannot be distinguished for $M(K\bar{K}\pi)$ below 1.30 GeV, so we have kept the $a_0(980)\pi$ decay mode only. The fit results can be summarized as follows: (1) the maximum likelihood is found to be the same for $I^G(J^{PC}) = 0^+(1^{++})a_0(980)\pi$ and for $0^-(0^{-+})a_0(980)\pi$, i.e. the 1.28- GeV region is equally likely to be the $f_1(1285)$ or the $\eta(1295)$; (2) in the 1.4-GeV region, the maximum likelihood is marginally better (by about 3 for $\Delta \ln L$) for $I^G(J^{PC}) = 0^+(1^{++})f_1(1420)$ than $I^G(J^{PC}) = 0^+(0^{-+})\eta(1440)$; the $I^G(J^{PC}) = 0^+(1^{+-})h_1(1380)$ is excluded in this analysis (by about 13 for $\Delta \ln L$). These results are also shown in Fig. 1-b).

DISCUSSION AND CONCLUSIONS

We have measured the production rate $\langle n \rangle$ per hadronic Z decay for $f_1(1285)/\eta(1295)$ and $f_1(1420)$. We assume for this study that *both have spin 1*. The results are

$$\langle n \rangle = 0.132 \pm 0.034 \quad \text{for} \quad f_1(1285)$$
$$\langle n \rangle = 0.0512 \pm 0.0078 \quad \text{for} \quad f_1(1420) \tag{1}$$

taking a $K\bar{K}\pi$ branching ratio of $(9.0 \pm 0.4)\%$ for the $f_1(1285)$ and 100% for the $f_1(1420)$[3]. The production rate per spin state [i.e. divided by $(2J+1)$] has been studied in Ref. [2]; in Fig. 2 is given all the available data for those mesons with a 'triplet' $q\bar{q}$ structure, i.e. $S = 1$ in the spectroscopic notation $^{2S+1}L_J$. To this figure we have added our two mesons for comparison. It is seen that both $f_1(1285)$ and $f_1(1420)$ come very close to the line corresponding to other mesons whose constituents are thought to be of the type $n\bar{n}$. This is suggestive of two salient facts: (1)the first peak at 1.28 GeV is very likely to be the $f_1(1285)$; (2) both $f_1(1285)$ and $f_1(1420)$ have little $s\bar{s}$ content. Indeed, the two states which are thought to be pure $s\bar{s}$ mesons, the ϕ and the $f'_2(1525)$, are down by a factor $\gamma^k \approx 1/4$ ($\gamma = 0.50 \pm 0.02$ and $k = 2$), as shown in Fig. 2. This is highly unlikely given the production rate (1).

We have studied the inclusive production of $f_1(1285)/\eta(1295)$ and $f_1(1420)$ in Z decays at LEP I. The measured masses and widths are 1274 ± 4 and 29 ± 12 MeV for the first peak and 1426 ± 4 and 51 ± 14 MeV for the second one. For the first time, a partial-wave analysis has been carried out on the $(K\bar{K}\pi)^0$ system. The results show that the first peak is equally likely to be the $f_1(1285)$ or the $\eta(1295)$, while the second peak is consistent with the $f_1(1420)$. However, the hadronic production rate of these two states suggests that their quantum numbers are very probably $I^G(J^{PC}) = 0^+(1^{++})$ and that their quark constituents are mainly of the type $n\bar{n}$, where $n = \{u,d\}$.

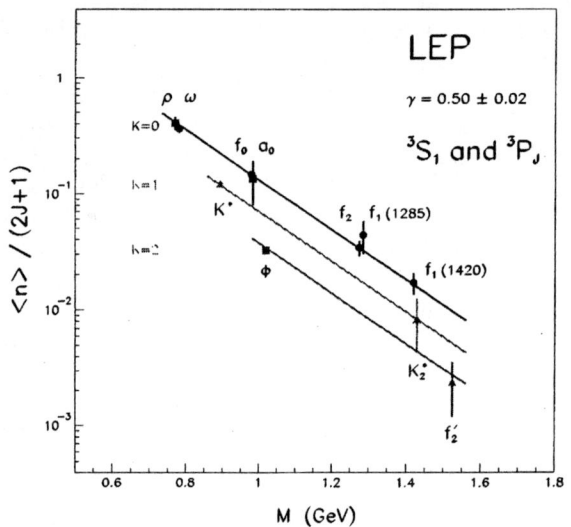

FIGURE 2. Total production rate per spin state and isospin for scalar, vector and tensor mesons as a function of the mass (open symbols). The two solid circles correspond to the $f_1(1285)$ and the $f_1(1420)$.

REFERENCES

1. P. Abreu et al., DELPHI Collab., Phys. Lett. **B449** (1999) 364.
2. V. Uvarov, Phys.Lett. B511 (2001) 136; 'Determination of the strangeness content of light-flavor isoscalars from its production rates in hadronic Z decays measured at LEP,' (arXiv:hep-ph/0105185) May 2001.
3. Review of Particle Physics, Eur. Phys. J. C15, (2000) 1.
4. P. Aarnio et al., DELPHI Collab., Nucl. Inst. Meth. **A303** (1991) 233.
5. P. Abreu et al., DELPHI Collab., Nucl. Inst. Meth. **A378** (1996) 57.
6. P. Abreu et al., DELPHI Collab., Z.Phys. C 65 (1995) 587.
7. See, for example, S. U. Chung, 'Spin Formalisms,' CERN preprint 71-8 (1971).
8. S. U. Chung et al., Phys. Rev. **D60** (1999) 092001.
9. See, for example, S. U. Chung, 'Analysis of $K\bar{K}\pi$ systems (Version I),' BNL-QGS-98-901 (1998), posted on the website: http://cern.ch/suchung/.

EXOTICS

Exotic $J^{PC} = 1^{-+}$ Mesons at the Present Time
(discussion of some problems)

L. I. Sarycheva* and V. L. Korotkikh*

*Scobeltsyn Institute of Nuclear Physics, Moscow State University, Moscow 119899, Russia

Abstract. Some problems and questions on Exotic meson physics are presented. The particular attention is given to the discrepancy between theory and experiment and to the interpretation of experimental results.

INTRODUCTION

Not many physicits doubt that other forms of hadronic matter with gluonic degrees of freedom can exist. More than 170 cites are devoted to so called "Non $q\bar{q}$-candidates"(see PDG [1]). A $q\bar{q}$ meson with orbital momentum l and total spin s has a parity $P = (-1)^{l+1}$ and a charge conjugation $C = (-1)^{l+s}$. This excludes states with $J^{PC} = 0^{--}, 0^{+-}, 1^{-+}, 2^{+-}$ etc. Resonances with these exotic quantum numbers could be Hybrids (bound state of quark, antiquark and gluon) or multiquark states. The lightest state is expected to be a 1^{-+} state. The last years give the experimental evidence of two such states at mass 1.4 GeV/c^2 and 1.6 GeV/c^2 [2, 3, 4, 5, 6, 7, 8, 9].

THEORY AND EXPERIMENT DISCREPANCY

Four experimental group are more active in Exotic state study last time. The results on the 1^{-+} exotics of BNL (6 USA and 2 Russian institutions), VES (IHEP, Russia), GAMS (IHEP, Russia) and Crystall Barrel (CERN) are demonstrated in Table 1.

The results of the experiments of the various reactions, on the different targets and at the different beam energies are remarkably coincident. It removes such kind of questions as the correct calculation of the different apparatus acceptance, some distinctions of partial wave analysis, the different background contributions. So the evidences of 1^{-+} exotic states at mass 1.4 GeV/c^2 ($\pi_1(1400)$) and 1.6 GeV/c^2 ($\pi_1(1600)$) are very weighable.

On the other hand the theoretical predictions (see review [10] and numerical cites in it) are indeterminate because the nonperturbative effects are very strong for light mesons. More exact QCD lattice gauge theory (LGT) predicts lightest hybrid masses between 1.7 GeV/c^2 and 2.1 GeV/c^2. The estimations of famous flux-tube model (FTM) are close to LGT. The lightest gluonic hybrid $J^{PC} = 1^{-+}$ has a mass around 1.9 Gev/c^2. It is larger than the experimental values for $J^{PC} = 1^{-+}$ (Table 1). Other models such as a bag model, a diquark cluster model, QCD sum rules (see cites in [10]) predict the mass of 1^{-+} hybrid in large mass range 1.4 to 2.0 GeV/c^2. The theoretical result is very strong

TABLE 1. Evidence for $J^{PC} = 1^{-+}$ exotics

Experiment	Mass (MeV/c^2)	Width (MeV/c^2)	Decay mode	Reaction
	$\pi_1(1400)$			
BNL-94 [2]	$1370 \pm 16^{+50}_{-30}$	$385 \pm 40^{+65}_{-105}$	$\eta\pi^-$	$\pi^- p \to \eta\pi^- p$
BNL-95 [3]	1359^{+16+10}_{-14-24}	314^{+31+9}_{-29-66}	$\eta\pi^-$	$\pi^- p \to \eta\pi^- p$
CBar [4]	$1400 \pm 20 \pm 20$	$310 \pm 50^{+50}_{-30}$	$\eta\pi^-$	$\bar{p}n \to \pi^-\pi^0\eta$
CBar [5]	1360 ± 25	220 ± 90	$\eta\pi^0$	$\bar{p}p \to \pi^0\pi^0\eta$
GAMS [6]	1370 fixed	300 ± 125	$\eta\pi^0$	$\pi^- p \to \eta\pi^0 n$
	$\pi_1(1600)$			
BNL [7]	$1593 \pm 8^{+29}_{-47}$	$168 \pm 20^{+150}_{-12}$	$\rho\pi^-$	$\pi^- p \to \pi^+\pi^-\pi^- p$
BNL [8]	$1597 \pm 10^{+45}_{-10}$	$340 \pm 40 \pm 50$	$\eta'\pi^-$	$\pi^- p \to \eta'\pi^- p$
VES [9]	1610 ± 20	290 ± 30	$\rho\pi, \eta'\pi, b_1\pi$	$\pi^-\text{Be} \to \rho\pi(\eta'\pi, b_1\pi)X$

model dependent. So, the first problem is that the theory gives larger value of 1^{-+} exotic meson or the accuracy of theoretical predictions is lower than the experimental ones.

The second problem is the discrepancy concerning the branching ratio. For example, the FTM predicts that $J^{PC} = 1^{-+}$ exotic state has a dominant $b_1\pi$ decay with $\rho\pi$ weak and $\eta\pi$ and $\eta'\pi$ very small decay branching [10]. VES result shows the branching of $\pi_1(1600) \to f$, where f is $(b_1\pi), (\eta'\pi), (\rho\pi)$ [11]:

$$\Gamma(\pi_1(1600) \to f) = \begin{cases} 1, & b_1\pi; \\ 1.0 \pm 0.3, & \eta'\pi; \\ 1.6 \pm 0.4, & \rho\pi. \end{cases}$$

Either these three modes are not all due to a hybrid exotic, or our understanding of hybrid decay is inaccurate.

There is another question concerning to the 1^{-+} exotic decay. The state $\pi_1(1400)$ decays to $\eta\pi$ and not to $\eta'\pi$ (Table 1). The state $\pi_1(1600)$ decays quite invert. π and η belong to an octet and η' is a singlet of SU(3). E.Klempt [12] showed that $J^{PC} = 1^{-+}$ state cannot decay into two octet pseudoscalar mesons in the limit of flavor symmetry. But it is possible the 1^{-+} octet to decay to $\eta'\pi$. So, $\pi_1(1600)$ may be the octet state and $\pi_1(1400)$ must be a multiplet of higher order. The easiest choice is a decuplet, which can describe $(q\bar{q} + q\bar{q})$ states. The decuplet cannot possible be a hybrid: gluonic excitations do not contribute to the flavor. The strange phenomenon that the $\pi_1(1600)$ does not decay into $\eta\pi$ thus provides the clue for the interpretation of the $\pi_1(1400)$ as decuplet state [12]. The η-η' mixing can destroy this rule. So, S.U. Chung[13] suggested that $\pi_1(1400)$ could be a complicated mixture $(q\bar{q}+$ gluon) hybrid and a $(q\bar{q} + q\bar{q})$ state.

DATA INTERPRETATION

Background

Are there exotic mesons (see Table 1) the new physical effects or is it a background display? The people ask this question because the data background and exotic signal is comparable in $\eta\pi^-$, $\rho\pi^-$ and $\eta\pi^0$ systems. The situation is more better in $\eta'\pi^-$ system because here the $\pi_1(1600)$ contribution is comparable with $a_2(1300)$ [8]. The ground answer is that the PWA is studying an interference effect of weak (say P-wave) with a strong wave (D-wave). And the interference contribution is larger than the background. But the questions are remained. Let doesn't discuss the incoherent isotropical background $B_{INC}(m)$, which adds to common distribution

$$W(m,\Omega) = |D(m)I_D(\Omega) + P(m)I_P(\Omega)|^2 + B_{INC}(m). \quad (1)$$

Here m is a mass of meson system, Ω are particle decay angles. $I_D(\Omega)$ and $I_P(\Omega)$ is a wave decay amplitudes. Each wave has its own background:

$$\begin{aligned} D(m) &= f_{BW}^{(D)}(m) + B_{COH}^{(D)}(m); \\ P(m) &= f_{BW}^{(P)}(m) + B_{COH}^{(P)}(m). \end{aligned} \quad (2)$$

The authors [2] used the real and mass independent background $B_{COH}^{(D)}$ and no background in P-wave. The description of experimental intensities and relative phase is good with Breit-Wigner amplitudes $f_{BW}^{(D)}(m)$ and $f_{BW}^{(P)}(m)$. Such simplest model is used in other works [3, 6, 7, 8, 9]. The remain questions are the next. Can we describe data without resonant BW in P-wave, but with the complex background $B_{COH}^{(D)}$ in D-wave and with some mass dependence phase of this background? How it could be interpretated? What is the contribution $B_{COH}^{(P)}(m)$ in the P-wave and has it the mass dependent phase? Theory could help, but it has to suggest description of two wave simultaneously at least.

Non-resonant Deck-type background

The authors [14] suggested a new interpretation of $\pi_1(1400)$. They used K-matrix approach, which connects various channels of reaction. Model has many free parameters. It describes approximately the peak at $m = 1.4$ GeV/c^2 by the interference of a Deck-type background with a real hybrid at $m = 1.6$ GeV/c^2. But such kind of approach cannot describe the Exotics of the Crystal Barrel experiment (Table 1) in the annihilation $\bar{p}p$ and $\bar{p}n$.

SOME PROBLEMS OF DATA ANALYSIS

The angular distribution in PWA is a sum

$$W(\Omega) = \sum_{\varepsilon,k} |{}^{\varepsilon}U_k(\Omega)|^2 + B_{INC}. \tag{3}$$

There is no interference between waves with different reflectivities $\varepsilon = \pm 1$, which are coincides with a naturalities η of exchanged particles. Also are for the non-spinflin ($k=1$) and the spinflip ($k=2$) nucleon amplitude. It is direct consequence of parity conservation in the production process. One of the problem is that the ambiguous solutions appear particularly when there are many unnatuarally ($\eta = -1$) parity exchange (UNPE) waves for the quasi-two-body reactioms. If their contributions are small smeared between solutions as it is for $\eta\pi^-$ [2], $\rho\pi^-$ [7], $\eta'\pi-$ [8], then they don't strongly affect the NPR waves. In these cases the people average ambiguous solutions of NPE waves, But the interpretation problem of 1^{-+} UNPE waves in $\rho\pi^-$ system is remained. Its contribution is very unstable and there is no strong partner wave in UNPE sector to set its resonance nature. Exotic 1^{-+} of UNPE is seen in $\rho\pi^-$ BNL data at 18 GeV/c and doesnt seen in $\rho\pi^-$ VES data at 18 GeV/c. But both experiments see 1^{-+} NPE wave.

The interesting suggestion to select the physical solution between ambiguous solutions is made by Sadovsky [6] on the example $\eta\pi^0$ by the help of Gersten's root motion. He reanalysed GAMS data and now he claims that the 1^{-+} exotic state at $m = 1.4 GeV/c^2$ decay to the channel $\eta\pi^0$.

There is also a problem of the spin-density matrix rank. If we set only the spinflip or only non-spinflip amplitudes, then we have rank=1 [2]. It may be not right if the mechanism of π_1 and a_2 production is different. Definite indication is seen in the different t-dependence of $\eta'\pi^-$ events in the region of $\pi_1(1600)$ and a_2 mesoms [8]. The slope of t-dependence of $\pi_1(1600)$ is smaller comparing with the slope of a_2. If the production of π_1 and a_2 is caused by the different Regions or the different Region combinations, then it may be a reason of different the spinflip and non-spinflip amplitudes.

REFERENCES

1. Groom, D.E. et al., (PDG), Eur.Phys.J. **C15**, 1 (2000)
2. Chung, S.U. et al., (E852), Phys. Rev. **D60**, 092001 (1991)
3. Chung, S.U. et al., (E852), Nucl.Phys. **A675**, 453c (2000)
4. Abele, A. et al., (CBar), Phys.Lett. **B423**, 175 (1998)
5. Abele, A. et al., (CBar), Phys.Lett. **B446**, 349 (1999)
6. Sadobsky, S.A., Nucl.Phys. **A655**, 131c (1999)
7. Adams, G.S. et al., (BNL), Phys. Rev.Lett. **81**, 5760 (1998)
8. Ivanov, E.I. et al., (BNL), Phys. Rev.Lett. **86**, 3977 (2001)
9. Khokhlov, Y.V. (VES), Nucl.Phys. **A663**, 596 (2000)
10. Barnes, T. *Exotic Mesons, Theory and Experiment*, hep-ph/00072296, 1999 and this conference
11. Dorofeev, V.,*New results fron VES*, hep-ex/9905002, 1999
12. Klempt, E.,*Meson Spectroscopy:Glueballs, Hebrids, and $q\bar{q} + q\bar{q}$ Mesons*, hep-ex/0101031, 2001
13. Chung, S.U.,*On SU(3) Representation of $q\bar{q} + q\bar{q}$ mesons*, BNL-QGS-00-501, 2000
14. Donnachie, A and Page, P. *Interpretation of Experimental J^{PC} Exotic Signal* , preprint CEBAF, JLAB-THY-98-20, 1998

A Study of the reaction $\pi^- p \to \omega \pi^- \pi^0 p$ at 18 GeV/c

Alexei V. Popov for the E852 Collaboration

Institute for High Energy Physics, Protvino, Russia

Abstract. In this paper the results of the partial-wave analysis of the reaction $\pi^- p \to \omega \pi^- \pi^0 p$ are presented. Both mass-independent and mass-depended analyses of this reaction were performed. The resonance-like structures are observed in the $J^P = 2^+, 0^-, 2^-, 4^+$ waves (the $\omega(782)\rho^-(770)$ intermediate state), and in the $J^P = 3^+$ wave (the $b_1(1235)\pi$ intermediate state). A strong evidence of an exotic resonance in the $J^P = 1^-$ wave (the $b_1(1235)\pi$ intermediate state) is also obtained.

INTRODUCTION

There are two major intermediate states in this reaction: $\omega(782)\rho^-(770)$ and $b_1(1235)\pi$. The first state is interesting for a study of "classical" mesons ($q\bar{q}$ model), $J^P = 0^-, 2^-, 4^+$ are expected, which can be useful for $\pi(1800), \pi_2(2100), a_4(2040)$ mesons study. The last state is more interesting for a hybrid mesons search because, for example, the flux tube model [1, 2, 3] predicts that hybrid mesons will decay preferentially to such final states as $b_1(1235)\pi$ and $f_1(1285)\pi$.

The total number of $2.25 \cdot 10^5$ events of the reaction $\pi^- p \to \omega_{(\pi^+\pi^-\pi^0)}\pi^-\pi^0 p$ was selected by applying some kinematical cuts to the event sample collected during the 1995 data taking run using the Multi-Particle spectrometer (MPS) at the Alternating Gradient Synchrotron (AGS) facility of Brookhaven National Laboratory (BNL).

The production and the decay of the $\omega\pi^-\pi^0$ system were studied by the partial wave analysis (PWA). The $\omega\pi^-\pi^0$ system decaying through intermediate isobars can be described in the Gottfried-Jackson frame by the following set of quantum numbers: $J^P M^\eta LS$, where J is a total spin, P - parity, M - magnetic quantum number, η - naturality, L - angular momentum in the system of two isobars, S - spin of two isobars. The $J^P M^\eta LS$ decay amplitudes for the $\omega\pi^-\pi^0$ system were written using a non-relativistic tensor formalism [4]. The mass-dependent analysis was a global mass-dependent PWA fit with mass and model dependent spin density matrix.

WAVES WITH NON EXOTIC QUANTUM NUMBERS

The waves with quantum numbers $J^P = 0^-, 2^+, 2^-, 4^+$ in $\omega\rho$ sector and $J^P = 3^+$ in $b_1(1235)\pi$ sector are the most significant in the fit. The dominant wave is $2^+1^+ S \omega\rho$. There is a clear signal from $a_2(1320)$-meson and some bump at 1.6 GeV/c^2 which can be described as a threshold effect (Fig. 1a). Fitted parameters (from mass-dependent fit)

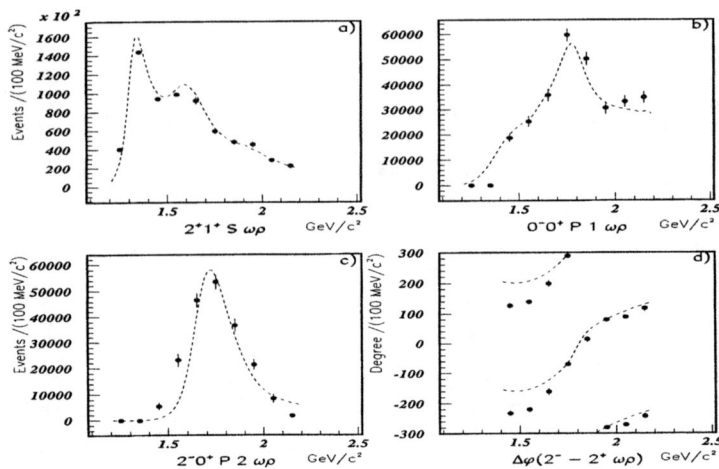

FIGURE 1. a), b), c) Acceptance corrected partial wave intensities for the 2^+1^+ S $\omega\rho$, 0^-0^+ $P1$ $\omega\rho$, 2^-0^+ $P2$ $\omega\rho$ waves; d) relative phase difference between 2^- and 2^+ waves. Phase is plotted repeatedly for each 360° interval.

for $a_2(1320)$ are:
$M = 1313 \pm 1(stat.) \pm 2(syst.)$ MeV/c^2, $\Gamma = 119 \pm 2(stat.) \pm 4(syst.)$ MeV/c^2.

There is also a clear signal from $\pi(1800)$-meson in 0^-0^+ $P1$ $\omega\rho$ wave (Fig. 1b). Fitted parameters for $\pi(1800)$ are:
$M = 1774 \pm 8(stat.) \pm 14(syst.)$ MeV/c^2, $\Gamma = 189 \pm 11(stat.) \pm 21(syst.)$ MeV/c^2.

The biggest wave in the $J^P M^\eta = 2^-0^+$ sector is 2^-0^+ $P2$ $\omega\rho$ (Fig.1c). It looks like a wide bump which peaks at 1.75 GeV/c^2 and has a $\sim 360^o$ phase movement against the 2^+1^+ S $\omega\rho$ wave, which cannot be described by one resonance (Fig. 1d). In the overall 2^-0^+ intensity picture two peaks are clearly seen, one at 1.7 GeV/c^2 which is the well-known $\pi_2(1670)$ and another peak at 1.9 GeV/c^2 (Fig. 2b). The second object dominates in 2^-0^+ $P1$ $\omega\rho$ wave (Fig. 2a). In a mass-dependent analysis model two resonances were used to describe the data. Parameters for the first resonance were fixed on $\pi_2(1670)$ PDG values [5]. Fitted parameters for the second resonance are:
$M = 1890 \pm 10(stat.) \pm 26(syst.)$ MeV/c^2, $\Gamma = 350 \pm 22(stat.) \pm 55(syst.)$ MeV/c^2.

There is a clear resonance-like behavior in the 3^+0^+ F $b_1(1235)\pi$ wave which can be interpreted as a $a_3(1900)$ meson decaying to $b_1(1235)\pi$ (Fig. 2c). Fitted parameters for $a_3(1900)$ are:
$M = 1873 \pm 11(stat.) \pm 18(syst.)$ MeV/c^2, $\Gamma = 259 \pm 18(stat.) \pm 21(syst.)$ MeV/c^2.

The 4^+1^+ $D2$ $\omega\rho$ wave also has a resonant behavior around 2 GeV/c^2 which can be interpreted as a $a_4(2040) \to \omega\rho$ decay (Fig. 2d). Fitted parameters for $a_4(2040)$ are:
$M = 1952 \pm 6(stat.) \pm 14(syst.)$ MeV/c^2, $\Gamma = 231 \pm 11(stat.) \pm 21(syst.)$ MeV/c^2.

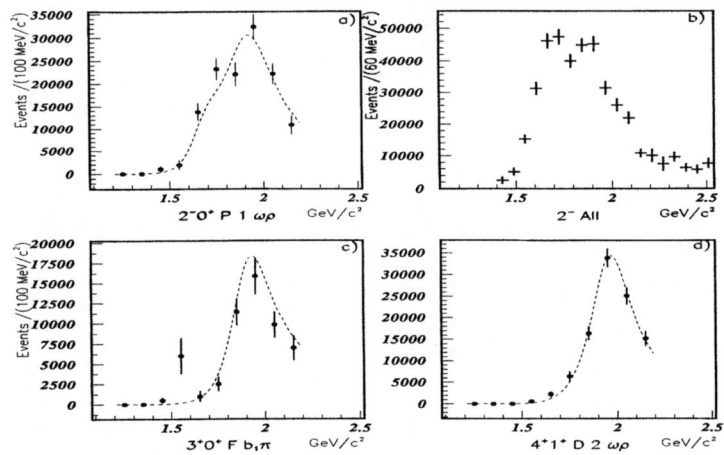

FIGURE 2. a) Acceptance corrected partial wave intensity for the 2^-0^+ $P1$ $\omega\rho$ wave; b) overall intensity for the 2^-0^+ waves; c) and d) acceptance corrected partial wave intensities for the 3^+0^+ F $b_1(1235)\pi$ and 4^+1^+ $D2$ $\omega\rho$ waves.

WAVES WITH EXOTIC QUANTUM NUMBERS

There is a significant 1^- wave in $b_1(1235)\pi$ system (1^-1^+ S $b_1(1235)\pi$) which peaks near 1.6 GeV/c^2 and has a shoulder at 1.9 - 2.0 GeV/c^2. The relative phase between this wave and 2^+1^+ S $\omega\rho$ wave is rising in the interval 1.4 - 1.9 GeV/c^2, which gives a strong evidence of an exotic resonance in this wave ($\pi_1(1600)$ [6, 7, 8]) (Fig. 3a, 3b). The parameters of this object from the mass-dependent analysis:
$M = 1582 \pm 10(stat.) \pm 20(syst.)$ MeV/c^2, $\Gamma = 289 \pm 16(stat.) \pm 27(syst.)$ MeV/c^2.

To describe a shoulder at 1.9 - 2.0 GeV/c^2 the mass-dependent fit with two resonances was performed. This fit gives a better description of the data than a fit with one resonance. The statistical significance of the second object is $2\Delta L/\Delta N \approx 8.9 \gg 1$, where ΔL is a likelyhood difference between two fits and ΔN is a difference between the number of parameters, which gives some evidence of another object existence at 2.0 GeV/c^2, but at this time it is hard to make any conclusions about existence and properties of this object (Fig. 3c, 3d).

REFERENCES

1. N. Isgur, J. Patton, *Phys. Rev.* **D31(11)**, 2910-2929 (1985)
2. R. Kokoski, N. Isgur, *Phys. Rev.* **D35(3)**, 907-933 (1987)
3. F.E. Close, P.R. Page, *Nuclear Physics* **B443**, 233-254 (1995)
4. C. Zemach, *Physical Review* **B1201**, 133 (1964)
5. *Review of Particle Physics, The European Physical Journal* **C15**, 1 (2000)
6. G.S. Adams et al., *Phys. Rew. Lett.* **81**, 5760 (1998)

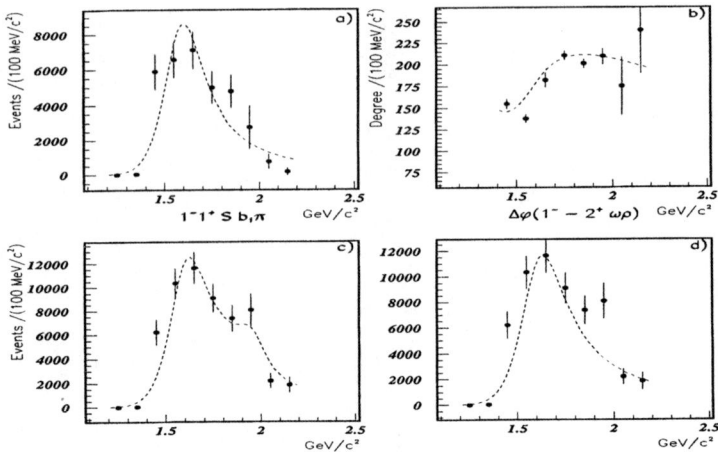

FIGURE 3. a) Acceptance corrected partial wave intensity for the 1^- wave in the $b_1(1235)\pi$ system $(1^-1^+ \, S \, b_1(1235)\pi)$; b) relative phase difference between 1^- and 2^+ waves; c) and d) overall intensity for the 1^-1^+ waves and their mass-dependent fit description in a model with two resonances and in a model with one resonance.

7. E. Ivanov et al., *Phys. Rev. Lett.* **86**, 3977 (2001)
8. D. Amelin et al., *YAF*, **62**, 487-495 (1999)

Partial Wave Analysis of $\pi^-\pi^-\pi^+\eta$ in the Reaction $\pi^- p \to \pi^-\pi^-\pi^+\eta p$ at 18 GeV/c

Joachim Kuhn

for the E852 collaboration

Rensselaer Polytechnic Institute, Dept. of Physics, 110 8th Street, Troy, NY 12180, USA

Abstract. A partial wave analysis of $\pi^-\pi^-\pi^+\eta$ was performed on data taken during the 1995 running period of Experiment 852 (E852) at the Multiple Particle Spectrometer at Brookhaven National Laboratory.

New $J^{PC} = 1^{++}$ and $J^{PC} = 2^{-+}$ states were found in $f_1(1285)\pi^-$, $a_2^-(1320)\eta$, $a_0^-(980)\rho$ and $a_1^-(1265)\eta$ decays. Considerable strength was also observed in the $J^{PC} = 1^{-+}$ waves in the $f_1(1285)\pi^-$ and $a_0(980)\rho$ decay channels, but the process seems not to be dominated by the previously discovered $\pi_1(1600)$.

INTRODUCTION

The search for exotic mesons has been the main focus of meson spectroscopy for many years, but only recently states which are not composed of a $q\bar{q}$ system have been established experimentally[1, 2, 3]. All of these states have been detected looking for resonances with unusual combinations of spin J, parity P and charge conjugation C, J^{PC}, which cannot be accessed by a $q\bar{q}$ meson. Of special interest are $J^{PC} = 1^{-+}$ hybrids, which are predicted to be lowest in mass[4, 5, 6, 7]. According to the flux tube model[4, 5] this state should primarily decay into S- and P-wave mesons, for example $b_1(1235)\pi$ or $f_1(1285)\pi$.

The analysis on the $\pi^-\pi^-\pi^+\eta$ final state presented in this work makes use of the $\pi^-\pi^+\eta$ decay mode of the $f_1(1285)$, in the hope to extract evidence for a $J^{PC} = 1^{-+}$ exotic state.

DATA SELECTION

During the 1995 data taking run of E852 at the MPS facility at Brookhaven National Laboratory a total of 710×10^6 events were recorded, 265×10^6 of which were of the required trigger topology of three downstream charged tracks in coincidence with a large angle recoil. Two clusters were required in a lead glass calorimeter, in agreement with the decay of $\eta \to \gamma\gamma$. After reconstruction, topological and kinematical selection 82,645 events of the type $\pi^- p \to \pi^-\pi^-\pi^+\eta p$ remained. It is noteworthy that, in order to limit the possible number of waves, a cut was performed to remove events of the type $\pi^- p \to \eta'(958)\pi^- p$. The effect of this cut can be seen in figure 1, which also shows the

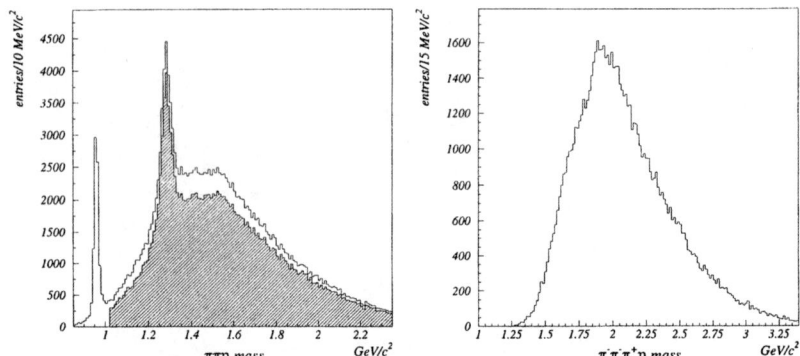

FIGURE 1. **Left:** $\pi^-\pi^+\eta$ mass. The shaded histogram is the distribution for the final data set, the unshaded shows the distribution before removing events from the reaction $\eta' \to \pi\pi\eta$. **Right:** $\pi^-\pi^-\pi^+\eta$ mass for the final data set.

total meson mass spectrum for the final data set.

PARTIAL WAVE ANALYSIS

A full partial wave analysis was performed in order to extract the spin and parity of the meson X^- in the reaction $\pi^- p \to X^- p \to \pi^-\pi^-\pi^+\eta p$. The decay of this state was assumed to populate two body intermediate states. The following possibilities were considered as intermediate states: $f_1(1285)\pi^-$, $\eta(1295)\pi^-$, $a_2^-(1320)\eta$, $a_1^-(1260)\eta$, $a_0^-(980)\rho$ and $a_2^-(1320)\rho$.

The partial wave analysis was performed in the four meson mass range from 1.30 GeV/c^2 to 2.90 GeV/c^2, in bins of 0.08 GeV/c^2. The 4-momentum transfer t was required to be between -0.1 GeV$^2/c^2$ and -1.5 GeV$^2/c^2$. This limited the data set to 68,900 events.

The intensity and phase distributions were fitted pairwise over mass ranges where the coherence between the two waves in consideration was high. The results of these fits can be seen in figures 2 and 3. Figure 4 shows the intensity from the mass-independent analysis for the $J^{PC} = 1^{-+}$ waves in the $f_1(1285)\pi^-$ and $a_0^-(980)\rho$ decay mode. Since there are still studies underway for these two waves the Breit-Wigner fit is not shown. However, the distribution suggests that the mass of the state in $f_1(1285)\pi^-$ is inconsistent with the $\pi_1(1600)$ excitation alone.

CONCLUSIONS

A data set of 68,900 events consistent with the reaction $\pi^- p \to \pi^-\pi^-\pi^+\eta p$ has been selected from the 1995 run period of E852 at BNL. A partial wave analysis of this

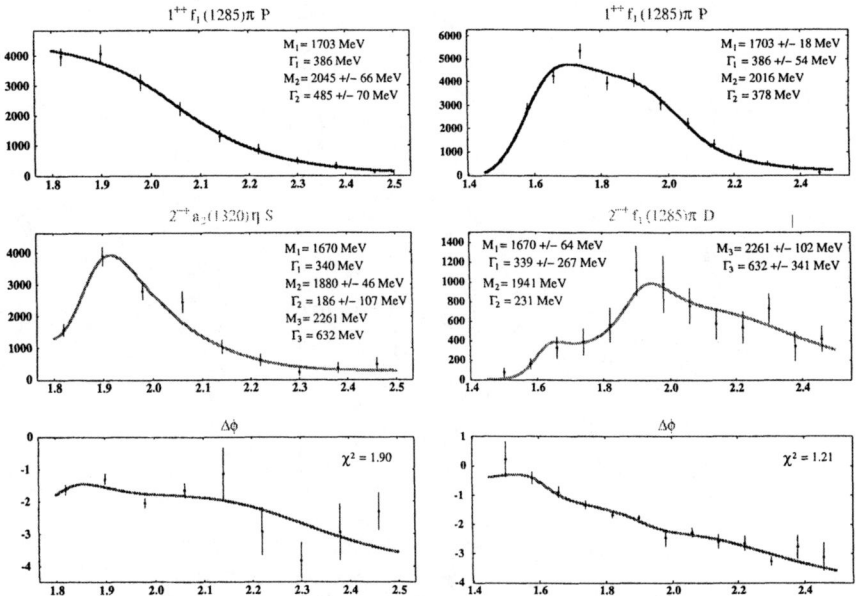

FIGURE 2. Results of the PWA fit and the Breit-Wigner fit. The plots are grouped in three: $\Delta\phi$ is the difference of the phase of the wave in the top and the wave in the middle plot.

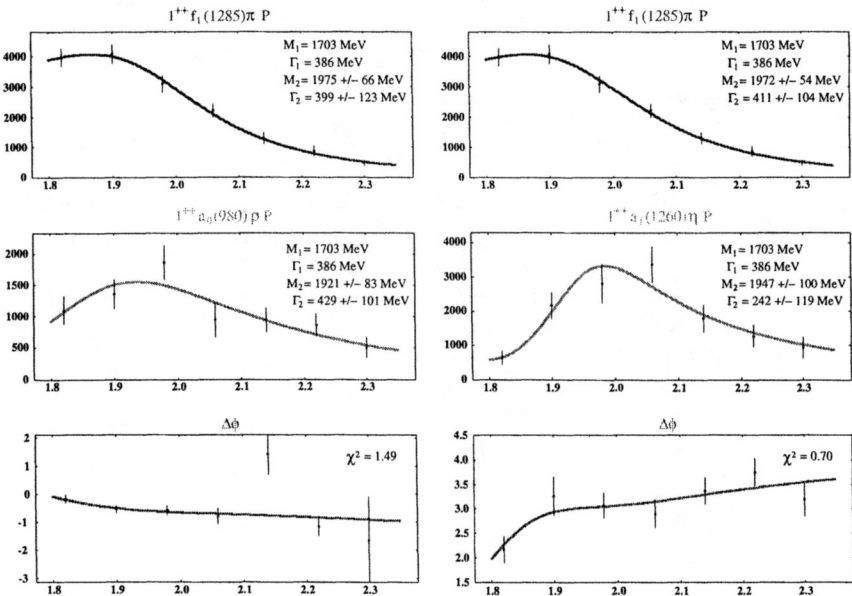

FIGURE 3. Results of the PWA fit and the Breit-Wigner fit. The plots are grouped in three: $\Delta\phi$ is the difference of the phase of the wave in the top and the wave in the middle plot.

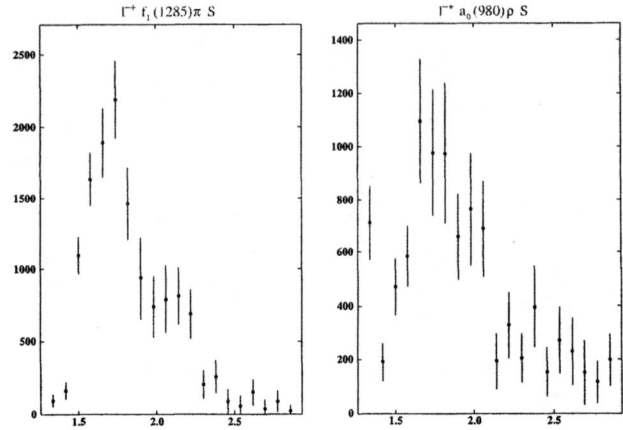

FIGURE 4. Intensities of the $J^{PC} = 1^{-+}$ exotic waves in the PWA fit.

data set shows that most of the intesity is produced via natural parity exchange. The strongest intensities are observed in the waves with $J^{PC} = 1^{++}$ in the channels $f_1(1285)\pi^-$ P, $a_0^-(980)\rho$ P, $a_1^-(1260)\eta$ P and in the waves with $J^{PC} = 2^{-+}$ in the channels $f_1(1285)\pi^-$ D and $a_2^-(1320)\eta S$.

The intensities of theses waves and the phases between them can be well described by Breit-Wigner shapes. There is strong evidence for two $J^{PC} = 1^{++}$ states with masses $^{(1^+)}M_1 = 1.703$ GeV/c^2 and $^{(1^+)}M_2 = 2.016$ GeV/c^2 and widths $^{(1^+)}\Gamma_1 = 0.386$ GeV/c^2 and $^{(1^+)}\Gamma_2 = 0.378$ GeV/c^2. The $J^{PC} = 2^{-+}$ waves can succesfully be fitted to three Breit-Wigner resonances with masses $^{(2^-)}M_1 = 1.670$ GeV/c^2, $^{(2^-)}M_2 = 1.941$ GeV/c^2 and $^{(2^-)}M_3 = 2.261$ GeV/c^2 and widths $^{(2^-)}\Gamma_1 = 0.339$ GeV/c^2, $^{(2^-)}\Gamma_2 = 0.231$ GeV/c^2 and $^{(2^-)}\Gamma_3 = 0.632$ GeV/c^2. $^{(1^+)}M_1$, $^{(2^-)}M_1$ and $^{(2^-)}M_3$ and the corresponding widths are consistent with previously observed states, listed in[8].

There is also considerable strength in the $J^{PC} = 1^{-+}$ waves in the $f_1(1285)\pi^-$ and $a_0^-(980)\rho$ final state. The process however does not seem to be dominated by the previously observed $\pi_1(1600)$ and a Breit-Wigner fit to the intensity alone yields a mass above 1.6 GeV/c^2. Systematic studies are currently underway to determine parameters for the exotic signal.

REFERENCES

1. Adams, G. S. et al., *Phys. Rev. Lett.*, **81**, 5760 (1998).
2. Chung, S. U. et al., *Phys. Rev. D*, **60**, 92001 (1999).
3. Ivanov, E. I. et al., *Phys. Rev. Lett.*, **86**, 3977 (2001).
4. Kokoski, R., and Isgur, N., *Phys. Rev. D*, **35**, 907 (1987).
5. Close, F. E., and Page, P. R., *Nucl. Phys. B*, **443**, 233 (1995).
6. Lacock, P. et al., *Phys. Lett. B*, **401**, 308 (1997).
7. Bernard, C. et al., *Phys. Rev. D*, **56**, 7039 (1997).
8. The Particle Data Group, *Eur. Jou. of Phys. C*, **3**, 1–794 (1998).

A Study of the $\eta\eta\pi^-$ System Produced in the Reaction $\pi^- p \to p\pi^+\pi^-\pi^- 4\gamma$ at $18 GeV/c$

Paul Eugenio
on behalf of the Brookhaven E852 Collaboration

Carnegie Mellon University
Pittsburgh, PA 15206, USA
Florida State University
Tallahassee, FL 32306, USA

Abstract.
Results are reported on the partial wave analysis of the $\eta\eta\pi^-$ system in the reaction $\pi^- p \to p\pi^+\pi^-\pi^- 4\gamma$ at 18 GeV/c where both π^0 and η decay to $\gamma\gamma$. The data were obtained using the MultiParticle Spectrometer at Brookhaven National Laboratory. The $a_0(980)$ and $f_0(1500)$ are observed in the $\eta\pi^-$ and $\eta\eta$ mass spectra, respectively. The accepted $\eta\eta\pi$ mass spectrum exhibits a structure consistent with the $\pi(1800)$. The results of the PWA show the $\pi(1800)$ decaying to both $a_0(980)\eta$ and $f_0(1500)\pi$ decay modes.

INTRODUCTION

An analysis effort of the Brookhaven E852 data is currently under way which focuses on the $\eta\eta\pi^-$ system produced in the reaction $\pi^- p \to p\pi^+\pi^-\pi^- 4\gamma$. The flux-tube model predicts characteristic decays of hybrid mesons to $L = 0$, $L = 1$ meson pairs[1, 2]. Several states are predicted to decay into $\eta\eta\pi$. The $\pi(1800)$ recently reported by the VES Collaboration[4] to have a large decay to $\eta\eta\pi$ has been argued to be a hybrid meson based on its decay to flux-tube favored modes[5].

The data were collected during the 1995 Brookhaven E852 run where 265 million triggers of the type designed to enrich this exclusive final state sample were acquired. A detailed description of the E852 apparatus is given in Reference [3]. Out of these data, 45.6 thousand events were fully reconstructed and kinematically fitted to the hypothesis of $\pi^- p \to p\pi^+\pi^-\pi^-\pi^0\eta$. Events that also fitted a $\pi^- p \to p\pi^+\pi^-\pi^-\pi^0\pi^0$ hypothesis were eliminated.

FEATURES OF THE DATA

$\pi^+\pi^-\pi^-\pi^0\eta$ System

The accepted invariant mass distribution of the $\eta\pi^+\pi^-\pi^-\pi^0$ system is shown in Figure 1a. There are no obvious resonant-like structures seen in the raw distribution. The accepted invariant mass distribution of the $\pi^+\pi^-\pi^-\pi^0$ subsystem is shown in Figure

1b. This distribution exhibits a prominent enhancement at 1300 MeV/c^2 with a width of 200 MeV/c^2. Assuming only isovector contributions, an overall charged 4π system must have an $I^G = 1^+$. The $b_1(1235)$ is the only known candidate for this structure.

Figure 1c displays the accepted $\pi^+\pi^-\pi^\circ$ invariant mass distribution. Two prominent narrow resonances are observed: at 550 MeV/c^2, an η signal; and at 785 MeV/c^2, an ω signal. A study of $\omega\eta\pi^-$ events shows that $b_1(1235)$ observed in the 4π spectrum shows up as an $\omega\pi^-$ intermediate state.

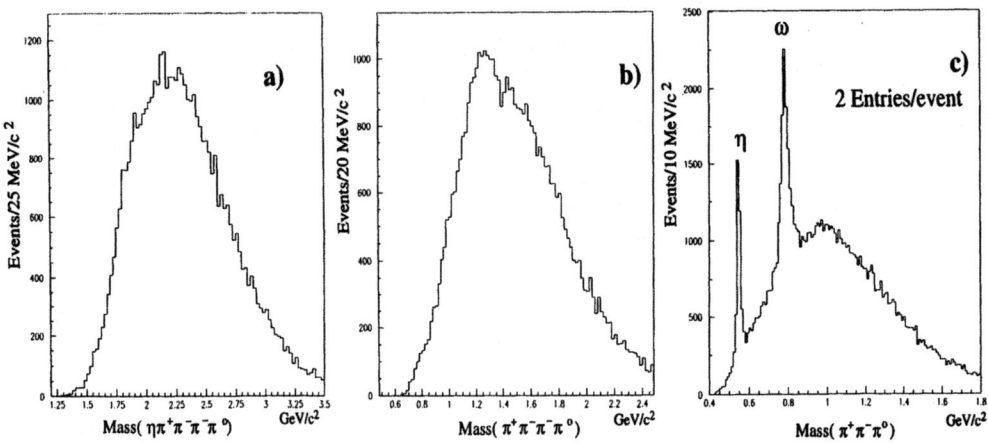

FIGURE 1. Distribution from the reaction $\pi^- p \to p\pi^+\pi^-\pi^-\pi^\circ\eta$: [a] $\eta\pi^+\pi^-\pi^-\pi^\circ$ effective mass, [b] $\pi^+\pi^-\pi^-\pi^\circ$ effective mass, and [c] $\pi^+\pi^-\pi^\circ$ effective mass (2 entries/event).

$\eta\eta\pi^-$ System

For an η selection of $500\,\text{MeV}/c^2 \leq Mass(\pi^+\pi^-\pi^\circ) \leq 580\,\text{MeV}/c^2$, about 5000 events were acquired for the study of the $\eta\eta\pi^-$ system. The non-$\eta \to 3\pi$ background has been estimated to be less than 10%. The accepted invariant mass distribution of the $\eta\pi$ subsystem is shown in Figure 2a. The main feature is a clear $a_\circ(980)$ signal. In addition, the distribution exhibits a shoulder-like structure in the region of 1300 MeV/c^2 suggesting the presence of some $a_2(1320)$ signal. The $\eta\eta$ subsystem can only couple to spin-even positive-parity isoscalar states ($f_0, f_2, f_4, \ldots$). Shown in Figure 2b is the accepted $\eta\eta$ invariant mass distribution. It exhibits a broad structure in the 1400-1500 MeV/c^2 region consistent with a $f_0(1500)$ signal. Figure 2c shows the accepted $\eta\eta\pi^-$ invariant mass distribution. The mass spectrum exhibits a resonant-like structure with a mass of 1800 MeV/c^2 and a width of about 200 MeV/c^2.

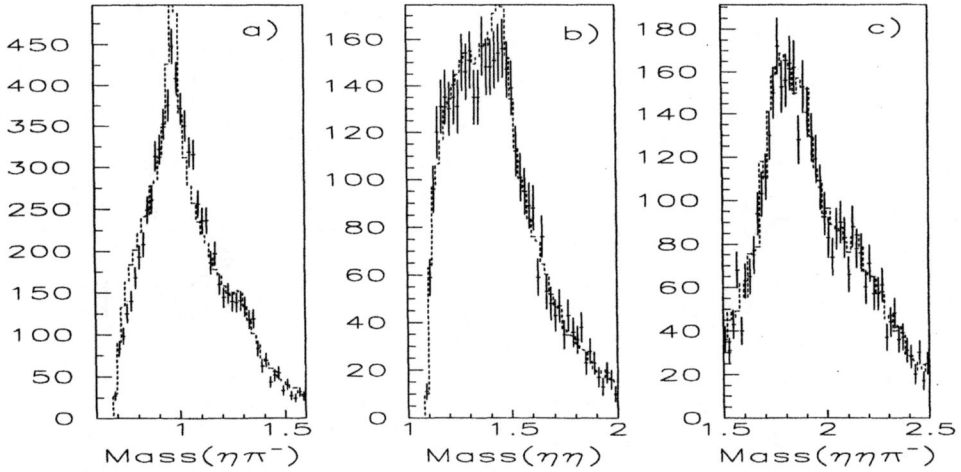

FIGURE 2. The accepted mass spectra for the $\eta\eta\pi^-$ analysis: a) Mass($\eta\pi^-$) [events/ 20 MeV/c^2], Mass($\eta\eta$) [events/ 20 MeV/c^2], and Mass($\eta\eta\pi^-$) [events/ 40 MeV/c^2]. The quality of the fit is shown by a comparison of the experimental data (with error bars) to Monte Carlo data weighted using the PWA fit results(dashed histogram).

PARTIAL WAVE ANALYSIS

A partial wave analysis of the data was performed in $50\,\text{MeV}/c^2$ wide mass bins from $1675\,\text{MeV}/c^2$ to $2175\,\text{MeV}/c^2$ and in the t region: $0 \leq |t| \leq 1.2(\text{GeV}/c^2)^2$ using the Brookhaven PWA program[6]. Partial waves for $L < 4$ and $|M| < 2$ were considered in the analysis. A good description of the data was achieved with the minimum set of 5 partial waves($J^{PC}M^\varepsilon L(isobar\ decay)$): $0^{-+}0^+S\ a_0(980)\eta$, $0^{-+}0^+S\ f_0(1500)\pi$, $2^{-+}0^+S\ a_2(1320)\eta$, $2^{-+}0^+D\ a_0(980)\eta$, and a flat non-interfering background. A comparison of the experimental data and the Monte Carlo data weighted using the fit results is shown in Figure 2. The $\pi(1800)$, $J^{PC} = 0^{-+}$, is observed decaying to $a_0(980)\eta$ and $f_0(1500)\pi$ (See the 0^{-+} intensities in Figures 3a,b). Both 0^{-+} intensities were fitted independently to a relativistic Breit-Wigner mass distribution resulting in: $Mass = 1884 \pm 19(stat)\,\text{MeV}/c^2$, $\Gamma = 222 \pm 39(stat)\,\text{MeV}/c^2$, $\chi^2/(9DoF) = 0.97$ for the $a_0(980)\eta$ decay mode, and $Mass = 1862 \pm 24(stat)\,\text{MeV}/c^2$, $\Gamma = 166 \pm 46(stat)\,\text{MeV}/c^2$, $\chi^2/(9DoF) = 0.76$ for the $f_0(1500)\pi$ decay mode[1]. The $2^{-+}0^+S\ a_2(1320)\eta$ intensity shown in Figure 3c exhibits a structure which peaks near $1900\,\text{MeV}/c^2$. This structure is consistent with the $\pi_2(1670)$ and a $\pi_2(1900)$. This new $\pi_2(1900)$ state has recently been observed via $f_1(1285)\pi$, $\omega\rho$, and $a_2(1320)\eta$ decay modes[7]. The $2^{-+}0^+D\ a_0(980)\eta$ wave is required in the fit and tends to contribute more at higher masses (See Figure 3d).

---

[1] The errors quoted do not take into account systematics.

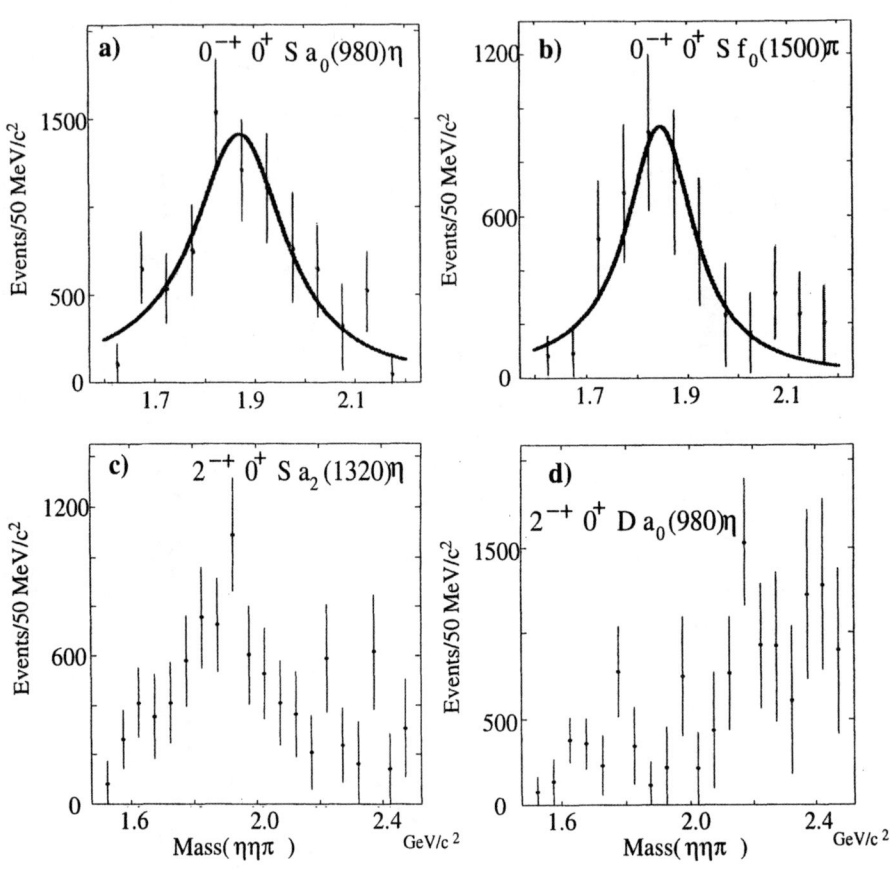

FIGURE 3. The partial wave intensities($J^{PC}M^{\varepsilon}L(isobar\ decay)$)): a) $0^{-+}0^{+}S\ a_0(980)\eta$, b) $0^{-+}0^{+}S\ f_0(1500)\pi$, c) $2^{-+}0^{+}S\ a_2(1320)\eta$, d) $2^{-+}0^{+}D\ a_0(980)\eta$ from the $\eta\eta\pi^{-}$ PWA.

REFERENCES

1. N. Isgur and J. Paton, *Phys. Rev.*, **D31**, 2910, (1985).
2. Close, F., and Page, P., *Nucl. Phys.* **B443** 33, (1995).
3. Brookhaven E852 Collaboration (S.U. Chung *et al.*), *Phys. Rev.*, **D60**,92001, (1999);
4. D. Amelin *et al.*, *Phys. Lett.*, **B356**, 595, (1995).
5. F. Close and P. Page, *Phys. Rev.*, **D56**, 1584, (1997).
6. J.P. Cummings and D.P. Weygand, *Brookhaven Report*, **BNL-64637**, (1997).
7. A. Popov BNL-E852, Hadron 2001 (these proceedings); P. Eugenio & J. Kuhn BNL-E852, Hadron 2001 (these proceedings).

Study of reaction $\pi^- A \to \pi^+\pi^-\pi^- A$ at VES setup.

Igor Kachaev for VES Collaboration. [1]

Department of Hadron Physics, IHEP, Protvino, Russia, 142284

Abstract.
The results on partial wave analysis of 3π system in reaction $\pi^- A \to \pi^+\pi^-\pi^- A$ at the momentum 36.6 GeV/c on the beryllium target are presented. New method of amplitude analysis is suggested — extraction of largest eigenvalue of density matrix. Exotic wave with $J^{PC} = 1^{-+}\rho\pi$ is studied in four t' regions. No narrow object around $M = 1.6$ GeV/c² is found. Unusually steep t' dependence for $\pi(1300)$ object is detected.

INTRODUCTION. THE PARTIAL WAVE ANALYSIS.

The VErtex Spectrometer (VES) setup is a large aperture magnetic spectrometer including the system of proportional and drift chambers, a multichannel threshold Čerenkov counter, beam-line Čerenkov counters, a lead-glass γ-detector (LGD) and trigger hodoscope. This permits full identification of multi-particle final states. The setup runs on the negative particle beam with the momentum of 36.6 GeV/c. The description of the setup can be found in [1].

In this report we present some results of partial wave analysis of the 3π system in the reaction $\pi^- Be \to \pi^+\pi^-\pi^- Be$ for different t' regions $|t'| = 0.01 - 0.07 - 0.15 - 0.30 - 0.80$ GeV²/c². The discussed results are based upon the statistics of about $8.0 \cdot 10^6$ events. Our previous results were published in [2].

The PWA has been performed in the 0.8–2.6 GeV/c² mass region in 50 MeV bins for different t' regions. Modified version of the Illinois PWA program [3] with maximum likehood method has been used for the analysis. Amplitudes were written using isobar model and relativistic covariant helicity formalism according to [4]. Explicit t'-dependence $f(t) = te^{-bt}$ was included for waves with nonzero projection of spin on GJ z-axis. Density matrix of full rank was used to describe final state. The set of 42 partial waves in the form $J^P LM^\eta$ *isobar* [3] was used in the analysis. Full wave set and parameterization of isobars including special treatment of $\pi\pi$ S-wave can be found in [2]. For the channels with $J^{PC} = 0^{-+}, 1^{++}, 2^{-+}$ largest waves are assigned to their own density matrix elements and are enabled to freely interfere with each other.

---

[1] Amelin D.V., Dorofeev V.A., Dzhelyadin R.I., Gouz Yu.P., Kachaev I.A., Karyukhin A.N., Khokhlov Yu.A., Konoplyannikov A.K., Konstantinov V.F., Kopikov S.V., Kostyukhin V.V., Kostyukhina I.V., Matveev V.D., Nikolaenko V.I., Ostankov A.P., Polyakov B.F., Ryabchikov D.I., Solodkov A.A., Solovianov O.V., Zaitsev A.M.

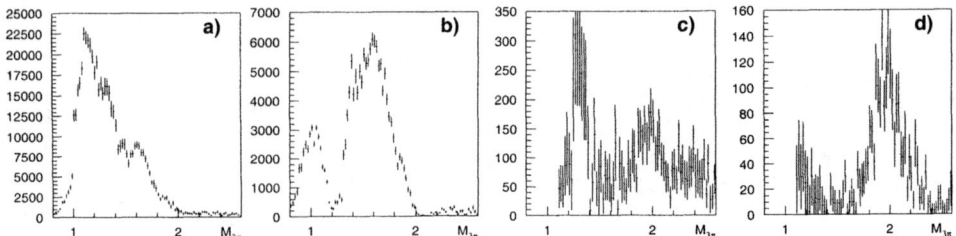

FIGURE 1. Wave $2^-P0^+\rho\pi$ at $|t'| < 0.03$ GeV$^2/c^2$ in a) full density matrix, b) largest eigenvalue; wave $4^+G1^+\rho\pi$ at $|t'| < 0.03$ GeV$^2/c^2$ in c) full density matrix, d) largest eigenvalue.

EXTRACTION OF LARGEST EIGENVALUE.

Results of PWA are represented in general by *density matrix*. For physical analysis *amplitudes* are much more convenient. We present here a new type of amplitude analysis — extraction of largest eigenvalue of density matrix. Density matrix can be represented by its eigenvalues and eigenvectors:

$$\rho = \sum_{k=1}^{d} e_k * V_k * V_k^+ \quad \text{where} \quad \begin{cases} e_k \text{ is k-}th \text{ eigenvalue, } e_1 > e_2 > \ldots > e_d \\ V_k \text{ is k-}th \text{ eigenvector} \end{cases}$$

Single out leading term:

$$\rho = \rho_L + \rho_S, \quad \rho_L = e_1 * V_1 * V_1^+, \quad \rho_S = \sum_{k=2}^{d} e_k * V_k * V_k^+$$

Here ρ_L is coherent part of density matrix and ρ_S is the rest (incoherent part). To be of physical meaning, this decomposition must be *stable* with respect to variations of density matrix elements. This is so if eigenvalues are *well separated* in comparison with errors in ρ matrix: $|e_1 - e_2| \gg \sigma(\rho_{ij})$. This is the case for $\pi^+\pi^-\pi^-$ where $e_1 \sim 1$, $e_2 \sim 0.1$.

Extraction of largest eigenvalue has the following advantages. By construction matrix ρ_L has rank one, so phases are well defined. It quantitatively uses information about coherence factors, which is often ignored. Practical experience shows that resonance structures tend to concentrate in ρ_L and leakage (see below) is suppressed in ρ_L. Nevertheless, ρ_S can contain different non-leading exchanges, albeit it often contains garbage. We can also note that if the wave is small in ρ_L, its phase with respect to largest waves can not be measured.

As a restriction this method requires a lot of data for good fit with small errors. It is not applicable if eigenvalues are not separated. In this case sometimes a group of clustered eigenvalues can be extracted. It is also not applicable if all eigenvalues except one are not statistically significant, as it is the case for unnatural sector for $\pi^+\pi^-\pi^-$ system.

An example of separation of largest eigenvalue is present in figure 1. On sub-figures a) and b) one can see a huge difference between wave $2^-P0^+\rho\pi$ in the full density matrix and in the largest eigenvalue. It was already noted [5, 2] that at low t' region this wave is large and highly incoherent with others at $M_{3\pi} \approx 1.2 - 1.4$ GeV/c^2 and have

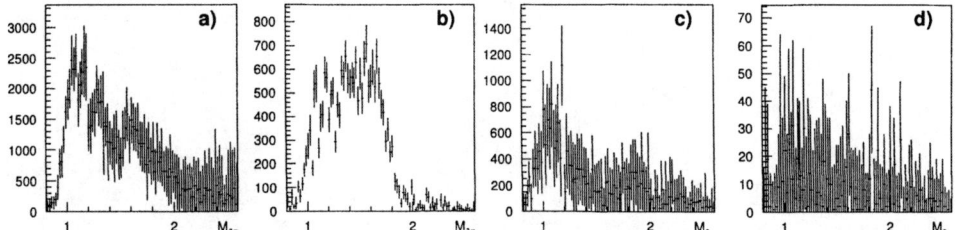

FIGURE 2. Leakage study. Wave $1^{-+}\rho\pi$ at $0.03 < |t'| < 1.0$ GeV$^2/c^2$ in: a) real data, full ρ matrix; b) real data, largest eigenvalue; c) leakage, full ρ matrix; d) leakage, largest eigenvalue.

only a relatively small shoulder at $M_{3\pi} \approx 1.6$ GeV/c^2, which corresponds to well known $\pi_2(1670)$. Our analysis confirms this as shown in figure 1 a). Contrary to this in the largest eigenvalue this wave is dominated by physical $\pi_2(1670)$. The bump at low $M_{3\pi}$ is at least ten times suppressed, which is consistent with its low coherence with other waves. The physical nature of this phenomena is still unknown.

In figures 1 c), d) one can see the difference between full density matrix and largest eigenvalue for the wave $4^+G1^+\rho\pi$ at $|t'| < 0.03$ GeV$^2/c^2$. At both figures we can see at $M_{3\pi} \approx 2.0$ GeV/c^2 a signal for $a_4(2050)$ while in fig.1 c) we can also see a bump at $M_{3\pi} \approx 1.3$ GeV/c^2 which is absent in fig.1 d). This bump is a leakage from $a_2(1320)$ with intensity about 0.2% of total number of events or about 7.5% of 2^+ wave in this low $|t'|$ region. This leakage is at least ten times suppressed in coherent part of density matrix.

LEAKAGE STUDY.

We have used the following method to study a possible leakage effects due to finite setup resolution and limited knowledge of setup acceptance. At first, we fit real data with small but representative wave set — 12 largest waves. The result of this step is a reasonably accurate representation of multidimensional distribution of real events. Next we generate Monte-Carlo events according to density matrix from this fit, smear these events according to modeled setup resolution and fit them as usual using standard wave set with all 42 waves. To study dependence of results of variation of modeled acceptance, we have used Monte-Carlo events without smearing, but used corrupted MC program (with hodoscope trigger logic excluded) for final fits with 42 waves.

The results of leakage study for exotic wave $1^{-+}\rho\pi$ are shown in fig. 2. One can see that exotic wave can contain 30–50% of leakage, but can not be described by leakage. The structure of density matrix in the real data and leakage is quite different — in the coherent part of ρ leakage is 20–50 times suppressed. Most other waves can contain 5–10% of leakage.

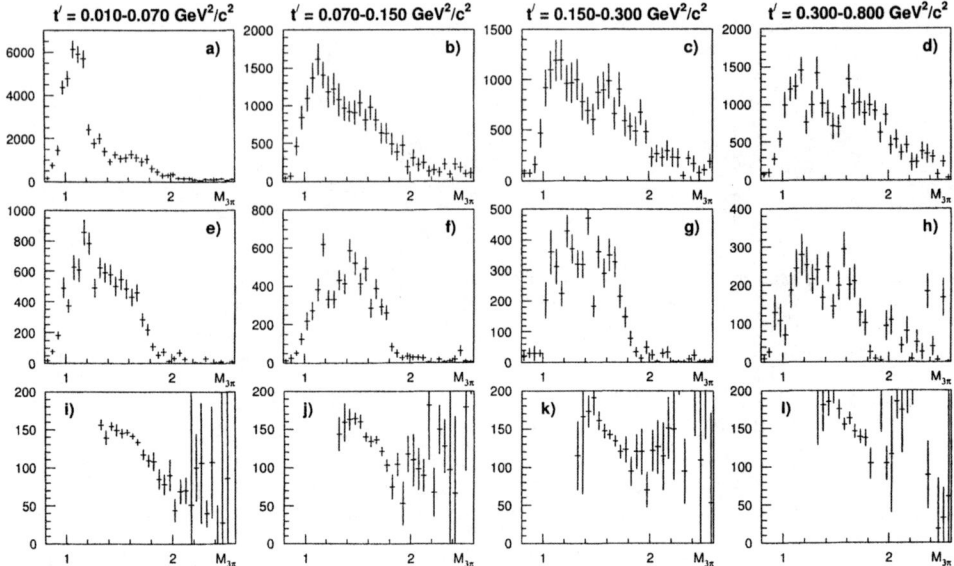

FIGURE 3. Wave $1^-P1^+\rho\pi$ at different $|t'|$ regions: a–d) full density matrix; e–h) largest eigenvalue; i–l) phase difference $\varphi(1^{-+}\rho\pi) - \varphi(2^{-+}f_2\pi)$, full density matrix.

WAVES WITH $J^{PC} = 1^{-+}$.

Waves with exotic quantum numbers $J^{PC} = 1^{-+}\rho\pi$ were included in our analysis with all possible projections $M^\eta = 1^+, 0^-, 1^-$. Waves with $M^\eta = 0^-, 1^-$ are small in comparison with $M^\eta = 1^+$ (except for $M_{3\pi} < 1.2$ GeV/c^2) and are not considered by us as significant. Wave $1^-P1^+\rho\pi$ in four different t' regions is shown in fig. 3. In the full density matrix (fig. 3 a–d) at low t' this wave consists mainly of a bump around $M = 1.0 - 1.2$ GeV/c^2 and a shoulder at $M \approx 1.6$ GeV/c^2. At higher t' regions the bump diminishes while the shoulder remains approximately the same. From figures 3 e–h) one can see that this shoulder corresponds mainly to the coherent part of the density matrix, and this part remains stable over investigated t' region. The bins over t' are selected so that numbers of events in $a_2(1320)$ peak are approximately the same for all bins, namely about 20000 events/50 MeV, so t' distribution for coherent part of exotic wave is roughly the same as for $a_2(1320)$ while its crossection is only 2–3% of it. One can see a sharp drop on the coherent part of exotic wave at $M_{3\pi} \approx 1.8$ GeV/c^2. We can see analogous effect in some other $\rho\pi$ waves and it can be connected with worse description of high $M_{3\pi}$ region. Phase difference $\varphi(1^-P1^+\rho(770)\pi) - \varphi(2^-S0^+f_2(1270)\pi)$ is shown in fig. 3 i–l). This phase difference is not constant, visible drop corresponds to phase raise of $\pi_2(1670)$ resonance. Again the shape of phase variation is more or less stable over inspected t' region. In general we can not see here narrow exotic object with $M \approx 1.6$ GeV/c^2.

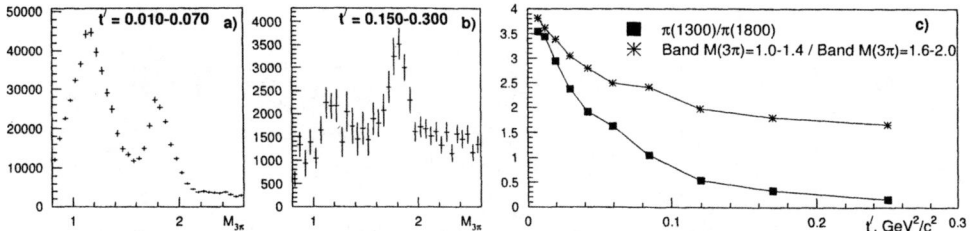

FIGURE 4. Wave $J^{PC} = 0^{-+}\varepsilon\pi$: a) $|t'| = 0.01 - 0.07$ GeV$^2/c^2$ region; b) $|t'| = 0.15 - 0.30$ GeV$^2/c^2$ region; c) relative $|t'|$ dependence of $\pi(1300)/\pi(1800)$ signals and corresponding $M_{3\pi}$ bands.

WAVES WITH $J^{PC} = 0^{-+}$.

Results of PWA for the wave $0^- S0^+ \varepsilon \pi$ in different t' regions are presented in figure 4. At low t' two peaks are visible which corresponds to $\pi(1300)$ and $\pi(1800)$. At higher t' peak for $\pi(1300)$ is clearly suppressed. Relative strength of $\pi(1300)$ and $\pi(1800)$ signals together with relative number of events in $M_{3\pi} = 1.0 - 1.4$ GeV/c^2 and $M_{3\pi} = 1.6 - 2.0$ GeV/c^2 bands is shown in fig 4 c) in ten t' regions. One can see that t' distribution for $\pi(1300)$ is much more steep than for $\pi(1800)$ or even for the total number of events in corresponding $M_{3\pi}$ band. A possible explanation of this phenomena can be given if the $\pi(1300)$ bump is at least partially non-resonant and produced by Deck-type final state scattering.

CONCLUSIONS.

Partial wave analysis of $\pi^+\pi^-\pi^-$ final state at different t' regions was performed on VES data. New type of amplitude analysis was suggested — extraction of the coherent part of density matrix. Its advantages and limitations were briefly discussed.

Wave $J^{PC} = 1^{-+} \rho\pi$ was studied in different t' regions $t' = 0.010$–0.070–0.150–0.300–0.800 GeV$^2/c^2$. Wave shape is broad and more or less the same in all t' regions studied. Clear phase variation with respect to $\pi_2(1670)$ is visible in all t' regions. No narrow object around $M = 1.6$ GeV/c^2 is found.

Waves $J^{PC} = 0^{-+}$ were studied in the same t' regions. Abnormally steep t' distribution for $\pi(1300)$ was established. This phenomena can be understood if $J^{PC} = 0^{-+}$ wave in $\pi(1300)$ region is partially consists of Deck-type background.

This work is supported in part by INTAS-RFBR 97-02-71017, RFBR 00-02-16555, RFBR 00-15-96689 grants.

REFERENCES

1. S.I. Bityukov et al., Phys. Lett., **B268**, 137 (1991).
2. D.I. Amelin et al., Phys. Lett., **B356**, 595 (1995).
3. J.D. Hansen et al., Nucl. Phys., **B81**, 403 (1974).
4. S.U.Chung, Phys. Rev., **D48**, N3, 1225 (1993).
   F.Filippini et al., Phys. Rev., **D51**, N5, 2247 (1995).
5. C. Daum et al., Phys. Lett., **B89**, 285 (1980).

Study of isospin 2 states in n̄p annihilations

A. Filippi[1]

Istituto Nazionale di Fisica Nucleare, sez. di Torino, via P. Giuria, 1, 10125 Torino, Italy

Abstract. The latest results of a study of the possible presence of isospin two states in antineutron-proton annihilation reactions into three and five charged pion exclusive final states are presented.

INTRODUCTION

The only observation of a $I = 2$ state reported by the PDG [1] goes back to the Eighties, when at PETRA several experiments recorded an enhancement in the $\gamma\gamma \to \rho^0\rho^0$ cross section close to threshold [2]. An analogous enhancement was not observed in $\gamma\gamma \to \rho^+\rho^-$, so the observed effect was interpreted as due to the presence of an exotic $q^2\bar{q}^2$ isotensor state, with $J = 2$, interfering with some other isoscalar components of the same tensor multiplet, in the framework of the MIT bag model [3]. According to it, as well as to some other more recent models (like the di-quark cluster one [4]), these states are located in the 1500÷1700 MeV mass region, and should preferentially decay into two vector mesons (VV); the decay into two pseudoscalar mesons (PP) should be suppressed due to an OZI-like mechanism.

The n̄p annihilation data collected by the OBELIX experiment (operational at LEAR, CERN, from 1990 to 1996) into exclusive final states composed by an odd number (3, 5) of charged pions have been studied in order to investigate the possible presence of $I = 2$ states detectable in the two decay modes $\pi^+\pi^+$ (n̄p $\to 2\pi^+\pi^-$ data), and $\rho^0\rho^0 \to 2\pi^+2\pi^-$ (n̄p $\to 3\pi^+2\pi^-$). In the latter case the neutral charge component only of the isotensor multiplet can be observed, as no π^0's are present in the final state.

STUDY OF THE $\pi^+\pi^+$ SYSTEM IN THE n̄p $\to \pi^+\pi^+\pi^-$ REACTION

The addition of a $\pi^+\pi^+$ resonant state in the form of a Breit-Wigner function to the global amplitude describing the n̄p $\to 2\pi^+\pi^-$ Dalitz plot delivers a strong *statistical* indication for the presence of a scalar signal at 1420 MeV, 160 MeV wide[2] [5]: its manifestation is a clean peak in the log(likelihood) ($\mathcal{L}$) trend, as shown in Fig. 1. The intensity of such a signal is very weak, amounting to less than 0.4% over the full Dalitz plot volume. The presence of a resonance with such features is unexpected on the basis

---

[1] OBELIX (PS201) Experiment
[2] in the following, as a shortcut, this scalar isotensor state will be denoted as t_0.

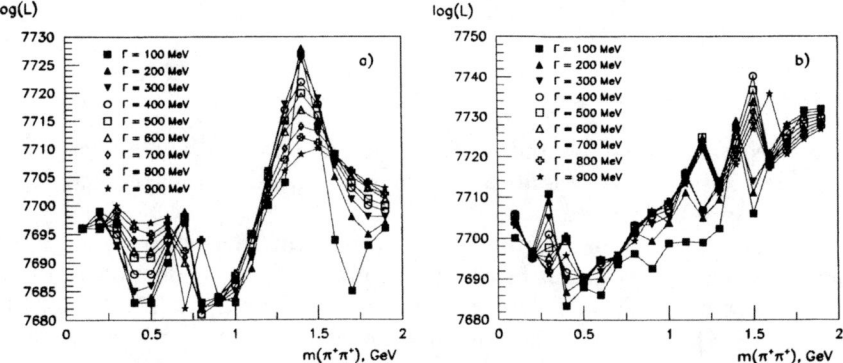

FIGURE 1. $\mathcal{L}$ values for best fit solutions as a function of $\pi^+\pi^+$ invariant mass and for fixed values of the width of the possible isospin two resonant state, in the two hypotheses a) of a scalar resonance, or b) of a tensor one.

of $\pi^+\pi^+$ phase shifts measurements [6]. However, this state could belong to a multiplet of exotic states [7] for which some manifestations, with different spin and parities, have been recently seen by several experiments [8].

STUDY OF THE $\bar{n}p \to \pi^+\pi^+\pi^+\pi^-\pi^-$ REACTION

The whole statistics available for the $\bar{n}p \to 3\pi^+2\pi^-$ channel amounts to 26271 events, with a residual background not exceeding 0.5%. The measured annihilation frequency for this reaction, with $\bar{n}$ momentum from ~ 50 to 405 MeV/c, is $(5.4 \pm 0.6)\%$, in good agreement with the analogous measurement in $\bar{p}d$ annihilations at rest [9]. A complete spin-parity analysis of these data has been performed in order to reproduce the shape of experimental spectra, which deviates quite remarkably from the pure phase-space description. Since the $\bar{n}p$ annihilations occur in flight, partial waves up to P have to be included and summed incoherently in the amplitude, as described in detail in Ref. [10].

Analysis. Several annihilation patterns have been taken into account and introduced in the total amplitude, namely:

- $\bar{n}p \to (4\pi)_A^0 \pi^+$, $(4\pi)_A^0 \to (2\pi)_B^0 (2\pi)_C^0$
  - A: $f_0(1370)$, $f_0(1500)$, $f_2(1270)$, t_0
  - BC: $\rho^0\rho^0$, $\sigma\sigma$ ($t_0 \to \rho^0\rho^0$ only)
- $\bar{n}p \to (4\pi)_A^0 \pi^+$, $(4\pi)_A^0 \to (3\pi)_B^{\pm} \pi^{\mp}$, $(3\pi)_B^{\pm} \to (2\pi)_C^0 \pi^{\pm}$
  - A: $f_0(1370)$, $f_0(1500)$
  - B: $\pi(1300)$
  - C: ρ^0, σ

- $\bar{n}p \to (3\pi)^+_A (2\pi)^0_B$, $(3\pi)^+_A \to (2\pi)^0_C \pi^+$
  - A: $\pi(1300)$, $a_1(1260)$, $a_2(1320)$
  - B: ρ^0, σ
  - C: ρ^0, σ ($a_2 \to \rho^0 \pi$ only)

In the previous list, σ is a shortcut for the $\pi^+\pi^-$ S-wave amplitude. As in this case we are especially interested in the low mass behavior of this interaction, a phenomenological parameterization from [11] is used for it.

No isovector radial excitation ($\rho'(1450)$, $\rho'(1700)$) is inserted in the amplitude as the description of the data by means of direct, uncorrelated $\rho^0\sigma$ production (with spin 1) gives better results. Three body direct production contributions ($\rho^0\rho^0\pi^+$, $\sigma\sigma\pi^+$) are introduced as well. Bose-Einstein correlation factors are moreover implemented to reproduce properly final state interactions between like pions [11].

Differently from previous analyses of $\mathcal{N}\mathcal{N}$ five pions final states [12, 13], a new method to parameterize the resonance widths is used, following Refs. [14, 15]. Since any resonance width is proportional to the phase space available for each of its decay channels, and the phase space rises very rapidly as a function of energy for the $\sigma\sigma$ and $\rho\rho$ channels, some important dynamical effects may appear, unaccounted for if constant widths simply are introduced in the Breit-Wigner functions. The most evident effect is a movement of the peak position, which is shifted to higher masses and broadened. As foreseen in Ref.[14], the $f_0(1350)$ peak actually shows up in the experimental spectra at about 1500 MeV and its contribution is somehow absorbed by the $f_0(1500)$ one, which is found to be dominant.

Results. Clear minima of $\mathcal{L}$ are found for the masses and widths of some of the states, introduced in the amplitude, decaying into four and three pions. In particular, for $f_0(1500)$ the values $m = (1490 \pm 10)$ MeV and $\Gamma = (133 \pm 27)$ MeV are found, and for $\pi(1300)$ the best fit values are $m = (1335 \pm 12)$ MeV and $\Gamma = (339 \pm 26)$ MeV. On the other hand, for the $f_0(1370)$ a minimum for the mass is found at $m = (1359 \pm 8)$ MeV, but no minimum for its width under 700 MeV; for a possible tensor state decaying into four pions, no minimum is found for both mass and width so its presence can be safely discarded.

A good fit of the experimental data ($\langle\chi^2\rangle = 1577/1035$), for which an example is shown in Fig. 2, has been obtained when the main contributions in the amplitude are played by $f_0(1500)\pi^+$ ($\sim 27\%$, switching off interference contributions as in Ref. [15]), by non resonant $\rho^0\sigma\pi^+$ ($\sim 48\%$) and by $\rho^0\rho^0\pi^+$ ($\sim 9\%$). The intensity required by the fit for other two-body contributions is no more than a few percent. Notably, with the energy-dependent width parameterization the weight of $f_0(1370)\pi^+$ is about 2%, a

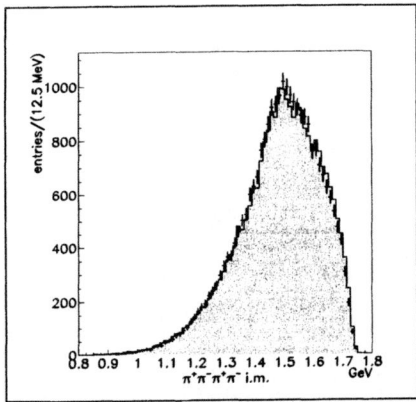

FIGURE 2. Invariant mass of the $\pi^+\pi^+\pi^-\pi^-$ system (points with errors). The grey superimposed histogram is the result from the best fit described in the text: the χ^2 is 1.14 (errors on fit histogram included).

factor of ten smaller as compared to that obtained with constant widths. The $f_0(1370)$ is found to decay with equal strength to $\rho^0\rho^0$ and $\sigma\sigma$; on the contrary, for the $f_0(1500)$ the $\pi(1300)\pi$ decay channel appears to be dominant, while the $\rho^0\rho^0$ one is much weaker, in agreement with Ref. [13]. However, Ref. [13] disagrees with the dominant $\pi(1300) \rightarrow \sigma\pi$ decay mode required by the fit, which, on the other hand, is consistent with the issues of $\bar{p}p \rightarrow 2\pi^+2\pi^-$ channel analysis by OBELIX [16].

On the possible presence of a I=2 state. The insertion in the amplitude of a further scalar state, decaying in $\rho^0\rho^0$ only, has been attempted, weighting it by proper Clebsh-Gordan coefficients to account for its isospin. The tensor hypothesis has been considered as unlikely, on the basis of the results described above. Starting from the mass and width values out of the fits on three pions data, $\mathcal{L}$ reaches its minimum at $m = (1457 \pm 32)$ and $\Gamma = (142 \pm 118)$ MeV. The weight for the $t_0\pi^+$ amplitude, integrated over the full phase space, is about 2%, namely at the same level of $f_0(1370)\pi^+$.

CONCLUSIONS

According to the MIT bag model the the ratio between the couplings of an $I = 2$ state to VV and PP mesons should be about $(0.743/0.041)^2 \simeq 328$ [3]; the phase space correction factor to this ratio for a state at about 1600 MeV is $\simeq 0.4$. Correcting the $t_0\pi^+$ production yields obtained from the fits in the two mentioned cases for the branching ratios for decays into unmeasured channels, and taking into account the measured frequencies for $\bar{n}p$ annihilation in five and three charged pions (3.2%), one gets the frequency ratio $f(\bar{n}p \rightarrow t_0\pi^+ \rightarrow \rho\rho\pi \rightarrow 5\pi)/f(\bar{n}p \rightarrow t_0\pi^+ \rightarrow 3\pi) \simeq 100$. Therefore, the results obtained so far do not rule out the presence of a $I = 2$ state decaying in both $\pi^+\pi^+$ and $\rho^0\rho^0$, even if they are obviously not conclusive.

REFERENCES

1. Particle Data Group, *Review of Particle Physics*, *Eur. Phys. J.* C **15**, 1 (2000).
2. Albrecht, H. et al., *Phys. Lett.* B **217**, 205 (1989); Behrend, H.-J. et al., *Phys. Lett.* B **218**, 493 (1989); Achasov, N.N. and Shestakov, G.N., *Sov. Phys. Usp.* **34**, 471 (1991).
3. Li, B.A. and Liu, K.F., *Phys. Rev.* D **30**, 613 (1984).
4. Noya, H., *these Proceedings*.
5. OBELIX Collaboration, Filippi, A. et al., *Phys. Lett.* B**495**, 284 (2000).
6. Hoogland, W. et al., *Nucl. Phys.* B **69** (1974), 266; *Nucl. Phys.* B **126**, 109 (1977).
7. Chung, S.U., *private communication*.
8. Thompson, D.R. et al., *Phys. Rev. Lett.* **79**, 1630 (1997); Abele, A. et al., *Phys. Lett.* B **423**, 175 (1998); Abele, A. et al., *Phys. Lett.* B **446**, 349 (1999); Klempt, E., *these Proceedings*.
9. Bridges, D. et al., *Phys. Rev. Lett.* **57** 1534 (1986), and references therein.
10. OBELIX Collaboration, Bertin, A. et al., *Phys. Rev.* D **57**, 55 (1998).
11. Gaspero, M., *Nucl. Phys.* A **562**, 407 (1993).
12. OBELIX Collaboration, Adamo, A. et al., *Nucl. Phys.* A **558**, 13c (1993); Ableev, V. et al., "Study of resonances decaying into four pions in $\overline{\mathcal{NN}}$ annihilations", in *HADRON'95*, edited by M.C. Birse, G.D. Lafferty, J.A. McGovern, World Scientific Publishing Co Pte Ltd, Singapore, 1996, pp. 337–339
13. Crystal Barrel Collaboration, Abele, A. et al., *Eur. J.* C **19**, 667 (2001).
14. Achasov, N.N., and Shestakov, G.N., *Phys. Rev.* D **53**, 3559 (1996).
15. Anisovich, A.V. et al., *Nucl. Phys.* B **690**, 567 (2001).
16. OBELIX Collaboration, Salvini, P., *Nucl. Phys.* A **655**, 51c (1999).

MODELS

IHEP, Protvino

Instanton effects in the meson and baryon spectrum

Bernard Metsch

Institut für Theoretische Kernphysik der Universität Bonn,
Nußallee 14–16, D-53115 Bonn, Germany

Abstract. The Bethe-Salpeter equation provides a suitable basis for the formulation of a covariant constituent quark model for both mesons and baryons. With the concept of effective, constituent quarks and instantaneous potentials both for confinement and the spin-dependent residual quark interactions a successful description of the hadron spectra up to the highest angular momenta as well as numerous decays is achieved. This will be exemplified by a discussion of some prominent features in the resulting description of hadron properties.

INTRODUCTION

The starting point for our relativistic quark model is the Bethe-Salpeter equation in momentum space

$$\chi_P(p) = S_1^F(p_1) \int \frac{d^4 p'}{(2\pi)^4} [-iK(P,p,p')\chi_P(p')] S_2^F(p_2), \qquad (1)$$

where $\chi_P(x) = \langle 0|T[\Psi^1(\frac{1}{2}x)\bar{\Psi}^2(-\frac{1}{2}x)]|P\rangle$ is the Bethe-Salpeter amplitude for a bound state of mass $P^2 = M^2$, $p_{1/2} = \frac{1}{2}P +/- p$ denotes the momentum of the (anti)quark, S^F is the Feynman quark propagator and K the irreducible quark interaction kernel. Inspired by the success of the non-relativistic constituent quark model we make the following approximations: The propagators are assumed to be of the free type with an effective constituent quark mass m_i, i.e. $S_i^F(p) = i(\slashed{p} - m_i + i\varepsilon)^{-1}$. The kernel K is assumed to depend only on the components of p and p' perpendicular to P, i.e. $K(P,p,p') = V(p_\perp, p'_\perp)$ with $p_\perp := p - (p \cdot P/P^2)P$. Integrating in the bound state rest frame over the time component p^0 and introducing the Salpeter (or equal-time) amplitude $\Phi(p) = \int dp^0 \chi_P(p^0, \vec{p})|_{P=(M,\vec{0})}$ we obtain the Salpeter equation:

$$\begin{aligned}\Phi(\vec{p}) &= \int \frac{d^3 p}{(2\pi)^3} \frac{\Lambda_1^-(\vec{p})\gamma^0[V(\vec{p},\vec{p}')\Phi(\vec{p}')]\gamma^0\Lambda_2^+(-\vec{p})}{M+m_1+m_2}\\&- \int \frac{d^3 p}{(2\pi)^3} \frac{\Lambda_1^+(\vec{p})\gamma^0[V(\vec{p},\vec{p}')\Phi(\vec{p}')]\gamma^0\Lambda_2^-(-\vec{p})}{M-m_1-m_2},\end{aligned} \qquad (2)$$

with the projectors $\Lambda_i^\pm(\vec{p}) = (\omega_i(\vec{p}) \pm H_i(\vec{p}))/2\omega_i(\vec{p})$, the Dirac Hamiltonian $H_i(\vec{p}) = \gamma^0(\vec{\gamma} \cdot \vec{p} + m_i)$ and where $\omega_i(\vec{p}) = \sqrt{m_i^2 + \vec{p}^2}$. In our quark model the interaction consists

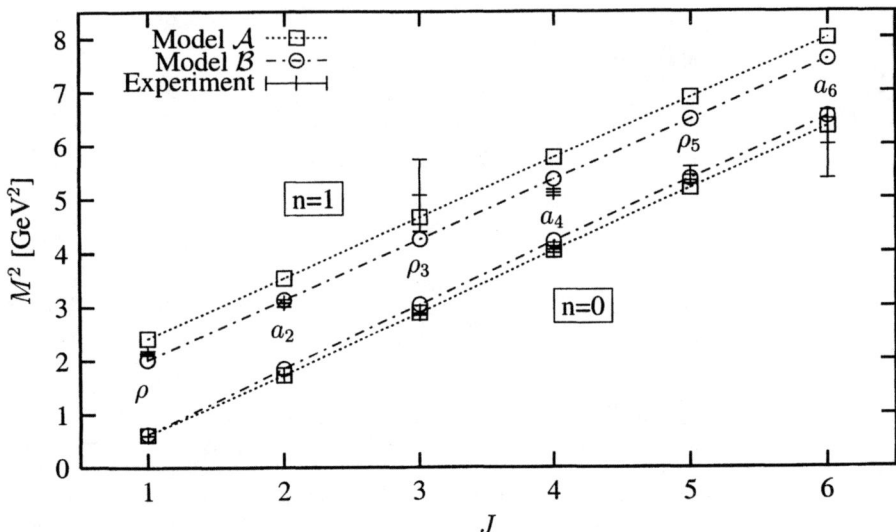

FIGURE 1. Isovector Regge-trajectory for ground states and first radial excitations. Experimental data from [1], apart from a'_4 mass from [2].

of 't Hoofts instanton induced interaction acting exclusively on (pseudo)scalars:

$$\int d^3p' \left[W(\vec{p},\vec{p}')\Phi(\vec{p}') \right] = 4G^{(g,g')} \int d^3p' w_\lambda(\vec{p}-\vec{p}') \left[\gamma^5 \text{tr}\left(\Phi(\vec{p}')\gamma^5\right) + \text{tr}\left(\Phi(\vec{p}')\right) \right], \quad (3)$$

where $G^{(g,g')}$ is a flavor matrix containing effective coupling constants g, g' and w_λ is a regularising Gaussian (see [7] for details) and a confinement potential of the form

$$\left[V(\vec{p},\vec{p}')\Phi(\vec{p}') \right] = v(|\vec{p}-\vec{p}'|)\left[\Gamma_D \Phi(\vec{p}')\Gamma_D \right], \quad (4)$$

where $v(p)$ is the Fourier-transform of a linearly rising potential $v(r) = a + br$ and with a spin structure which minimizes unwanted spin-orbit effects. In particular we studied $\Gamma_D \cdot \Gamma_D = \frac{1}{2}(\mathbf{1} \cdot \mathbf{1} - \gamma^0 \cdot \gamma^0)$ (Model $\mathcal{A}$) and $\Gamma_D \cdot \Gamma_D = \frac{1}{2}(\mathbf{1} \cdot \mathbf{1} - \gamma^5 \cdot \gamma^5 - \gamma^\mu \cdot \gamma_\mu)$ (Model $\mathcal{B}$), which is also $U_A(1)$- as well as $\gamma^5 \gamma^\mu$-invariant and symmetric under Fierz-transformations in Dirac space.

MESONS

Both versions yield an excellent description of ground state Regge-trajectories, see e.g. Fig. 1. For an extensive discussion of meson properties we refer to [5, 6]. Here we merely summarize the main results concerning the spectrum: Both versions can adequately describe the pseudoscalar and vector ground state nonet, including the $\pi - \eta - \eta'$ splitting and the $\eta - \eta'$ mixing due to the instanton induced interaction. In both models there is no explanation for the $\eta(1295)$; there are only four pseudoscalar/isoscalar states

TABLE 1. Scalar ground state and first excited state nonet classification. Experimental data from [1], K-matrix analysis from [3].

	Mod. $\mathcal{A}$	Exp.	Mod. $\mathcal{B}$	K-Mat.		Mod. $\mathcal{A}$	Exp.	Mod. $\mathcal{B}$	K-Mat.
f_0	985	980	665	720	f_0''	1775	1715	1555	1600
a_0	1320	1430	1055	960	a_0'	1930	–	1665	1640
f_0'	1470	1500	1260	1260	f_0'''	2115	2020	1870	1810
K_0^*	1425	1450	1190	1200	$K_0'^*$	2060	1945	1790	1820

predicted to have masses below 1.9 GeV. The main differences are found in the scalar sector, see Table 1. In model $\mathcal{B}$ scalar states are found to be roughly 250 MeV lighter that in model $\mathcal{A}$. Although the results from model $\mathcal{B}$ do not coincide with masses quoted by the Particle Data Group, they do agree rather well with the values given by Anisovich [3] from a K-matrix analysis of the scalar sector. Although these so-called 'bare states' are preferable for a comparison with results from quark models which do not take into account the coupling to decay channels, our classification on the basis of the meson spectra alone should be considered preliminary until such analyses have been performed for all quantum numbers. A remarkable feature of model $\mathcal{B}$ is that it produces an almost linear dependence of the mass squared as a function of the radial quantum number according to $M^2 = M_0^2 + (n-1)\mu^2$, where the value is consistent with $\mu^2 = 1.25 \pm 0.15$ GeV^2 found in [4]. Some examples are displayed in Fig. 2. The amplitudes were tested by a calculation of a multitude of electro-weak observables

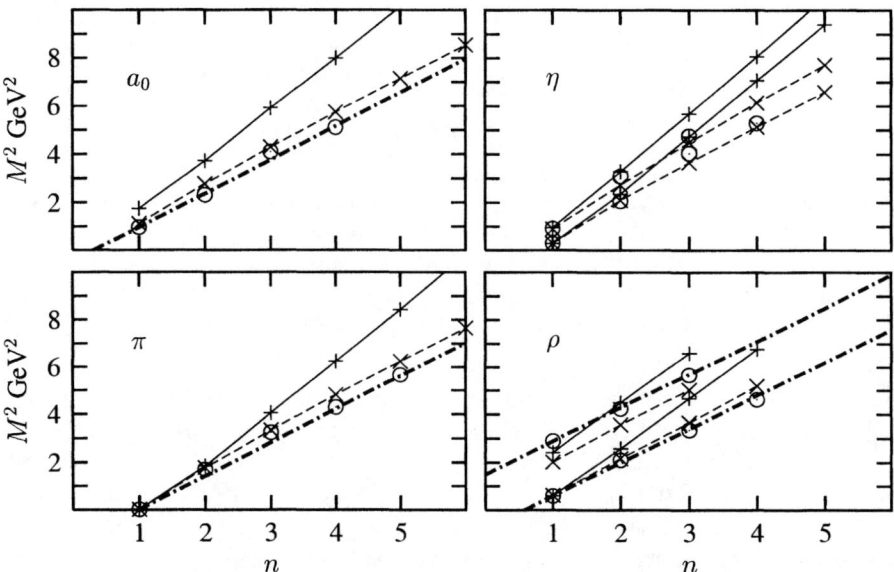

FIGURE 2. (M^2, n)-trajectories for the $\pi(10^{-+}), \rho(11^{--}), a_0(10^{++})$ and $\eta(00^{-+})$ states. The results from model $\mathcal{A}(\mathcal{B})$ are represented by $+(\times)$, experimental data (see [6] for details) by open circles. The thick dashed-dotted line corresponds to the formula $M^2 = M_0^2 + (n-1)\mu^2$ of [4].

within the Mandelstam formalism, again we refer to [5, 6] for examples. Here, we like to point out that the instanton induced interaction, which affected the masses of the (pseudo)scalar mesons only, in fact also plays an important role in strong two particle decays of mesons. In lowest order such decays are given by a simple quark loop diagram, which respects the well-known OZI rule. As described in [8] the three body part of the effective instanton induced Lagrangian contributes an additional, OZI-rule violating part to the decay, provided the spins of all participating mesons vanish. Fixing the strength of this contribution to the $K_0^*(1430) \to K\pi$ decay we find a reasonable description of many other decays, some of which are given in Table 2. More details are given in [11].

TABLE 2. Strong two body decays widths (in MeV) of mesons. The consecutive columns give the contribution calculated with model $\mathcal{B}$ from the lowest order quark loop contribution (Γ_L), from the instanton induced interaction (Γ_I), their sum (Γ_S) and the experimental values from [1], except $\eta''(1440)$, taken from [9] and $f_0'(1500)$, taken from [10].

Decay	Γ_L	Γ_I	Γ_S	$\Gamma_{\text{exp}}^{\text{tot}}$	Decay	Γ_L	Γ_I	Γ_S	Γ_{exp}
$\rho \to \pi\pi$	122	0	122	151 ± 1	$K^* \to K\pi$	42.1	0	42.1	50.8 ± 0.9
$a_2 \to \pi\rho$	28.0	0	28.0	72.8 ± 2.8	$K_0^* \to K\pi$	3.5	167	220	273 ± 53
$\to \pi\eta$	19.3	0	19.3	15.1 ± 1.2	$\to K\eta$	2.6	13.8	28.3	–
$\to K\bar{K}$	5.3	0	5.3	5.10 ± 0.83	$\to$ all			248	294 ± 23
$a_0 \to K\bar{K}$	1.0	34.9	24.3	seen	$K_0^{*'} \to K\pi$	7.0	51.2	95.9	105 ± 103
$\to \eta\pi$	0.01	47.8	46.4	dominant	$\to K\eta$	0.9	4.7	9.6	–
$\to$ all			71	50–100	$\to K\eta'$	0.5	1.0	2.8	–
$f_0/\sigma \to \pi\pi$	165	60	423	600–1000	$\to$ all			108	201 ± 113
$f_0' \to \pi\pi$	3.2	4.4	15.2	44.1 ± 15.4	$\eta'' \to K^*K$	21.8	0	21.8	21.6 ± 5.2
$\to \sigma\sigma$	20.4	0.21	16.5	18.6 ± 12.5	$\to a_0\pi$	29.8	0.01	28.6	26.6 ± 7.0
$\to \pi'\pi$	13.4	0.3	9.9	35.5 ± 29.2	$\to \eta\sigma$	5.2	3.0	16.2	27.0 ± 6.0
$\to$ all			103	130 ± 30	$\to$ all			67	50–80

BARYONS

With essentially the same approximations as discussed above for mesons –i.e. effective constituent quarks and instantaneous potentials which include a linearly rising string-like confinement potential with a suitable spin-structure and the instanton induced interaction to provide the major spin-dependent splittings– we also developed a relativistic quark model of baryons. For the details we refer to [12, 13, 14]. The major results can be summarized as follows:

- Since the instanton induced interaction does not act on baryons which do not contain a flavour antisymmetric quark-pair, the Δ-resonance spectrum is determined by the confinement potential alone. Indeed in this sector an in general satisfactory description is found in this manner, examplified by the description of the Δ-Regge trajectory, see table 3, with the exception of the $\Delta^{\frac{3}{2}+}(1600)$ and some not well established negative parity resonances around 1.9 GeV;

- Once the strengths of the instanton induced interaction is fixed to reproduce successfully the ground state octet-decuplet splittings, we find that we can automatically account for the low position of the Roper resonance $N^{\frac{1}{2}+}(1440)$ as well as

TABLE 3. Masses (in GeV) of states belonging to the positive parity Δ-, N- and Λ-Regge trajectories. Experimental data from [1], calculated results correspond to Model $\mathcal{A}$ of [13, 14].

	cal.	exp.		cal.	exp.		cal.	exp.
$\Delta\frac{3}{2}^+$	1.26	1.23	$N\frac{1}{2}^+$	0.94	0.94	$\Lambda\frac{1}{2}^+$	1.11	1.12
$\Delta\frac{7}{2}^+$	1.96	1.94-1.96	$N\frac{5}{2}^+$	1.72	1.67-1.69	$\Lambda\frac{5}{2}^+$	1.83	1.81-1.83
$\Delta\frac{11}{2}^+$	2.44	2.30-2.50	$N\frac{9}{2}^+$	2.22	2.18-2.31	$\Lambda\frac{9}{2}^+$	2.34	2.34-2.37
$\Delta\frac{15}{2}^+$	2.82	2.75-3.09	$N\frac{13}{2}^+$	2.62	2.57-3.10	$\Lambda\frac{13}{2}^+$	2.75	–

its strange partners $\Lambda\frac{1}{2}^+(1600)$ and $\Sigma\frac{1}{2}^+(1660)$; Although the nucleon states on the nucleon Regge-trajectory are affected by the instanton induced interaction we find in fact a constant shift in the squared mass, see also table 3.

- At the same time we find that in the N- and Λ-sector there are for a given parity a number of states containing flavour-antisymmetric diquark configurations with vanishing spin, which, due to the instanton induced interaction are shifted to a position which is virtually degenerate with states of opposite parity unaffected by this interaction. In this manner we can account for experimentally observed 'parity-doublets' such as e.g. $N\frac{5}{2}^+(1680) - N\frac{5}{2}^-(1675)$, $N\frac{9}{2}^+(2220) - N\frac{9}{2}^-(2250)$ and $\Lambda\frac{5}{2}^+(1820) - N\frac{5}{2}^-(1830)$.

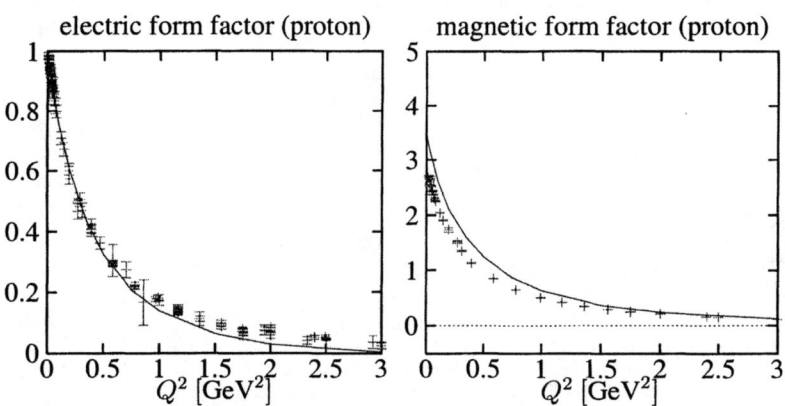

FIGURE 3. Electric (in [e]) and magnetic (in [μ_N]) form factor of the proton.

Also here we tested the amplitudes obtained by calculating various electromagnetic observables [15]. As an example we present the electric and the magnetic form factor of the proton in Fig. 3. In general we can account rather well for the shape of the form factors. Due to an approximation in the treatment of the two-body interaction the magnitude of the magnetic moments turn out to be too large by about 15%. Recently we started with calculating two body hadronic decay widths of baryon resonances, which with the Bethe-Salpeter amplitudes of the baryons and mesons involved determined by

TABLE 4. Partial decay widths (in MeV) of N- and Δ-resonances. Experimental data from [1].

Decay	cal.	exp.	Decay	cal.	exp.
$\Delta\frac{3}{2}^+(1232) \to N\pi$	109	119 ± 5	$N\frac{5}{2}^-(1675) \to N\pi$	7	68 ± 17
$N\frac{1}{2}^+(1440) \to N\pi$	10	228 ± 83	$\to \Delta\pi$	38	83 ± 19
$\to \Delta\pi$	41	88 ± 43	$N\frac{5}{2}^+(1680) \to N\pi$	86	85 ± 13
$N\frac{3}{2}^-(1520) \to N\pi$	86	68 ± 51	$N\frac{3}{2}^-(1700) \to N\pi$	0.6	10 ± 10
$\to \Delta\pi$	25	24 ± 9	$\Delta\frac{3}{2}^-(1700) \to N\pi$	4	45 ± 30
$N\frac{1}{2}^-(1535) \to N\pi$	18	68 ± 51	$N\frac{1}{2}^+(1710) \to N\pi$	3	15 ± 23
$\to \Delta\pi$	1	< 1.5	$\to \Delta\pi$	64	68 ± 36
$\Delta\frac{1}{2}^-(1620) \to N\pi$	5	38 ± 15	$N\frac{3}{2}^+(1720) \to N\pi$	0.1	23 ± 15
$\to \Delta\pi$	64	68 ± 36	$N\frac{5}{2}^+(1905) \to N\pi$	0.8	35 ± 26
$N\frac{1}{2}^-(1650) \to N\pi$	3	109 ± 46	$N\frac{7}{2}^+(1950) \to N\pi$	58	113 ± 19
$\to \Delta\pi$	8	6 ± 6			

the spectrum is a parameter free calculation. Some preliminary results [16] presented in Table 4 are very encouraging for further investigations.

ACKNOWLEDGMENTS

Matthias Koll, Klaus Kretzschmar, Ulrich Löring, Dirk Merten and Ralf Ricken did the actual calculations. Also a longterm, fruitful collaboration with Herbert Petry is gratefully acknowledged.

REFERENCES

1. Particle Data Group, *Eur. Phys. J.* **C 15**, 1 (2000).
2. Anisovich, V. V. et al., *Phys. Lett.* **B 452**, 180 (2000).
3. Anisovich, V. V., *AIP Conf. Proc.* **432**, 421 (1997).
4. Anisovich, A.V. Anisovich, V.V., Sarantsev, A.V., *Phys. Rev.* **D 62**, 051502 (2000).
5. Koll, M., Ricken, R., Merten, D., Metsch, B.C., Petry, H.R., *Eur. Phys. J.* **A 9**, 73 (2000).
6. Ricken, R., Koll, M., Merten, D., Metsch, B.C., Petry, H.R., *Eur. Phys. J.* **A 9**, 221 (2000).
7. Münz, Claus R., Resag, Jörg, Metsch, Bernard C., Petry, Herbert R., *Nucl. Phys.* **A578**, 327 (1994).
8. Ritter, C., Metsch, B.C., Münz, C.R., Patry, H.R., *Phys. Lett.* **B 380**, 431 (1996).
9. Suh, J.-S.,*Untersuchung des E-Mesons in der Proton-Antiproton Vernichtung in Ruhe*, PhD Thesis, Univ. Bonn (1999).
10. Abele, A. et al., *Eur.Phys.J.* **C21**, 261 (2001).
11. Ricken, R., *Properties of Light Mesons in a Relativistic Quark Model*, PhD Thesis, Univ. Bonn (2001).
12. Löring, U., Kretzschmar, K., Metsch, B.Ch., Petry, H.R., *Eur. Phys. J.* **A 10**, 309 (2001).
13. Löring, U., Metsch, B.Ch., Petry, H.R., *Eur. Phys. J.* **A 10**, 395 (2001).
14. Löring, U., Metsch, B.Ch., Petry, H.R., *Eur. Phys. J.* **A 10**, 447 (2001).
15. Kretzschmar, K., *Electroweak Form Factors in a Covariant Quark Model of Baryons*, PhD Thesis, Univ. Bonn (2001).
16. Merten, D., *private communication* (2001).

Light-light and heavy-light mesons in the model of QCD string with quarks at the ends

A.V.Nefediev

ITEP, 117218, B.Cheremushkinskaya 25, Moscow, Russia

Abstract. The variational einbein field method is applied to the model of the QCD string with quarks at the ends for the case of light–light and heavy–light mesons. Special attention is payed to the proper string dynamics. The correct string slope of the Regge trajectories is reproduced for light–light states which comes out from the picture of rotating string. Masses of several low-lying orbitally and radially excited states in the D, D_s, B, and B_s meson spectra are calculated and a good agreement with the experimental data as well as with recent lattice calculations is found. The role of the string correction to the interquark interaction is discussed at the example of the identification of $D^{*\prime}(2637)$ state recently claimed by DELPHI Collaboration. For the heavy–light mesons the standard constants used in Heavy Quark Effective Theory are extracted and compared to the results of other approaches.

INTRODUCTION

One of the most beautiful phenomena observed in QCD — namely, the formation of an extended string between the colour sources, implies that the string degrees of freedom in hadrons should be taken into account in the proper way. In the present contribution the model of the QCD string with quarks at the ends is used to investigate the spectra of the light–light and heavy–light mesons. The variational einbein field method (see [1] and references therein) is used in numerical calculations of the spectra.

LIGHT-LIGHT MESONS

Starting from the gauge invariant Green's function of a $q\bar{q}$ meson, performing integration in the path integral over the fermionic and gluonic fields, and using the minimal area law assumption in the latter case, one can extract the Lagrangian of the spinless quark-antiquark system in the form (see *e.g.* [2])

$$L(t) = -m_1\sqrt{\dot{x}_1^2} - m_2\sqrt{\dot{x}_2^2} - \sigma\int_0^1 d\beta\sqrt{(\dot{w}w')^2 - \dot{w}^2 w'^2}, \quad (1)$$

where we synchronize the quark times, $x_{10} = x_{20} = t$, and choose the minimal-string profile function in the straight-line form, $w_\mu(t,\beta) = \beta x_{1\mu}(t) + (1-\beta)x_{2\mu}(t)$.

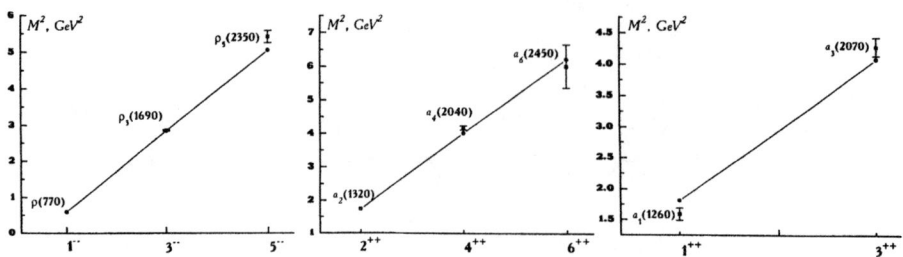

FIGURE 1. The lowest Regge trajectories for the light–light mesons (experimental data are given by boxes with error bars).

If the einbeins $\mu_{1,2}$ and $\nu(\beta)$ are now introduced in (1) to get rid of the square roots, then the centre-of-mass Hamiltonian for the case of massless quarks reads ($\mu_1 = \mu_2 = \mu$)

$$H = \frac{p_r^2}{\mu} + \mu + U(\mu, r), \qquad U(\mu, r) = \frac{\sigma r}{y} \arcsin y + \mu y^2, \qquad (2)$$

where the extremum in ν is already taken at the level of the Hamiltonian yielding $\nu_{ext}(\beta) = \sigma r (1 - 4y^2(\beta - \frac{1}{2})^2)^{-1/2}$, so that y is the solution to the transcendental equation

$$\frac{L}{\sigma r^2} = \frac{1}{4y^2}(\arcsin y - y\sqrt{1-y^2}) + \frac{\mu y}{\sigma r}. \qquad (3)$$

The spectrum of the Hamiltonian (2) is found using the quasiclassical method with the consequent minimization of each eigenvalue with respect to the einbein μ, which is thus treated as a variational parameter, playing the role of the effective constituent mass of the quark. It appears due to the interaction and takes the value of $200 - 300 MeV$ even if one starts with zero current quark mass. The results of numerical calculations with $\sigma = 0.17 GeV^2$ are given in Fig.1. The interested reader can find details in papers [1]. Note, that the Regge trajectories for the light–light mesons remain nearly straight-line up to very low momenta and the only fitting parameter is the overall negative shift $\Delta M^2 \sim -1 GeV$, one and the same for all three trajectories. Another important comment is that the correct string slope of the trajectories, $2\pi\sigma$, appears quite naturally in the given approach as a consequence of the rotating string inertia properly taken into account in the Hamiltonian (2), so that the form of the effective potential $U(\mu, r)$ does not amount to the naive sum of the centrifugal barrier for the quarks and the linearly rising potential σr [1].

HEAVY-LIGHT MESONS

In the calculations of the light-light meson spectra we totally ignored the quark spin, concentrating on the proper account of the string dynamics. Now, to have reliable

TABLE 1. Splittings for the D, D_s, B, and B_s mesons in MeV.

Splitting	D_s-D	$D_s^*-D^*$	D^*-D	$D_s^*-D_s$	B_s-B	$B_s^*-B^*$	B^*-B	$B_s^*-B_s$
Experiment	99	102	141	144	90	91	46	47
Theory	114	115	146	147	100	102	63	65

predictions for the heavy–light meson masses, we supply the Hamiltonian for spinless quarks connected by the string,

$$H_0 = \sum_{i=1}^{2}\left(\frac{\vec{p}^2+m_i^2}{2\mu_i}+\frac{\mu_i}{2}\right)+\sigma r - \frac{\sigma(\mu_1^2+\mu_2^2-\mu_1\mu_2)}{6\mu_1^2\mu_2^2}\frac{\vec{L}^2}{r}, \quad (4)$$

by spin-dependent corrections due to confining and the OGE interaction ($\kappa = -\frac{4}{3}\alpha_s$),

$$V_{sd} = \frac{8\pi\kappa}{3\mu_1\mu_2}(\vec{S}_1\vec{S}_2)|\psi(0)|^2 + \frac{\kappa}{\mu_1\mu_2 r^3}\left(3(\vec{S}_1\vec{n})(\vec{S}_2\vec{n})-(\vec{S}_1\vec{S}_2)\right) - \frac{\sigma}{2r}\left(\frac{\vec{S}_1\vec{L}}{\mu_1^2}+\frac{\vec{S}_2\vec{L}}{\mu_2^2}\right) \quad (5)$$

$$+\frac{\kappa}{r^3}\left(\frac{1}{2\mu_1}+\frac{1}{\mu_2}\right)\frac{\vec{S}_1\vec{L}}{\mu_1}+\frac{\kappa}{r^3}\left(\frac{1}{2\mu_2}+\frac{1}{\mu_1}\right)\frac{\vec{S}_2\vec{L}}{\mu_2}+\frac{\kappa^2}{2\pi\mu^2 r^3}\left(\vec{S}\vec{L}\right)(1.43-\ln(\mu r)),$$

as well as by the Coulomb interaction $-\frac{4}{3}\frac{\alpha_s}{r}$ and the overall negative constant shift $-C_0$. The latter remains the only fitting parameter, whereas we use the standard values for others: $\sigma = 0.17 GeV^2$, $m_u = 5 MeV$, $m_d = 9 MeV$, $\alpha_s = 0.4$ for D mesons and $\alpha_s = 0.39$ for B's. The last term on the r.h.s. of equation (4) is the string correction which is always negative and accounts for the proper string dynamics. The results of numerical calculations for the spectra of orbitally and radially excited D, D_s, B, and B_s mesons, as well as comparison with the lattice data and results of other approaches, can be found in [1, 3]. In Table 1 we give the splittings for the above-mentioned mesons compared to the experimental values. Let us also quote the masses of the radially ($n = 1$) and orbitally ($l = 2$) excited D mesons: $M(0^-) = 2664 MeV$, $M(2^-) = 2663 MeV$, $M(3^-) = 2654 MeV$. Thus the 3^- state is the lightest one and it is the most probably candidate for the resonance $D^{*\prime}(2637)$ recently claimed by the DELPHI Collaboration [4]. Such an identification resolves the problem usually encountered in the framework of the quark models: being too narrow this meson can not be associated with the first radial excitation, whereas model predictions for orbitally excited states lie about $50-60 MeV$ higher than needed. In our approach the proper string dynamics, in the form of the correction to the Hamiltonian, lowers the energy of the orbitally excited state 3^-. It is also instructive to note, that the fitted value of the parameter C_0 is insensitive to the heavy quark and depends only on the light-quark content of the meson, that supports the idea that it is due to the self-energy of the latter.

BRIDGE TO HEAVY QUARK EFFECTIVE THEORY

The suggested approach allows us to find analytic formulae for the constants used in the Heavy Quark Effective Theory, as well as to evaluate them numerically. In the standard

TABLE 2. Standard parameters used in HQET.

	$m_1 \to \infty$ $m_2 \to 0$	Fit (7)	Sum rules [5]	B mesons decays [6]	DS equation [7]
$\bar{\Lambda}$, GeV	0.471	0.485	$0.4 \div 0.5$	0.39 ± 0.11	0.493/0.288
λ_1, GeV^2	-0.506	-0.379	-0.52 ± 0.12	-0.19 ± 0.10	-
λ_2, GeV^2	0.21	0.17	0.12	0.12	-

parameterization the mass of a heavy-light meson is

$$M_{hl} = m_Q + \bar{\Lambda} - (\lambda_1 + d_H \lambda_2)/2m_Q + O\left(1/m_Q^2\right) \quad (6)$$

with d_H being +3 for 0^- states or -1 for 1^- ones. For the idealized case ($m_1 \to \infty$, $m_2 = 0$) our formulae simplify considerably, giving analytical expressions for all three constants [1]. A more reliable way to estimate two of them is to find the best fit of the form

$$M_{fit} = m_Q + \bar{\Lambda} + C_0 - \lambda_1/2m_Q \quad (7)$$

with C_0 fixed by fitting the experimental spectrum, as discussed above, and varying m_Q around the bottom quark mass. The results are listed in Table 2, where they are compared with those of other approaches, demonstrating good agreement with the latter.

CONCLUSIONS

In conclusion let us emphasize that the proper dynamics of the gluonic degrees of freedom in hadrons, taken into account in the form of an effective QCD string between quarks, are of paramount importance in establishing the hadron spectra of mass and their properties. Account for the inertia of the rotating string inside mesons allowed us to reproduce the correct string slope of the Regge trajectories, as well as to fit the spectrum of D and B mesons, and to resolve the problem of identification of the state recently claimed by the DELPHI Collaboration. The variational einbein field method used in the calculations is proved to be efficient and accurate, that allows one to use it in investigations of various relativistic systems.

Financial support of RFFI grants 00-02-17836, 00-15-96786, and 01-02-06273, INTAS-RFFI grant IR-97-232 and INTAS CALL 2000-110 is gratefully acknowledged.

REFERENCES

1. Morgunov, V.L., Nefediev, A.V., and Simonov, Yu.A., *Phys.Lett.* **B459**, 653 (1999).
   Kalashnikova, Yu.S., Nefediev, A.V., Simonov, Yu.A., *Phys.Rev.* **D64**, 014037 (2001).
2. Dubin, A.Yu., Kaidalov, A.B., Simonov, Yu.A., *Phys.Lett.* **B323**, 41 (1994); *ibid.* **B343**, 310 (1995).
3. Kalashnikova, Yu.S., Nefediev, A.V., *Phys.Lett.* **B492**, 91 (2000).
4. DELPHI Collaboration, Abreu, P. *et.al.*, *Phys.Lett.* **B426**, 231 (1998).
5. Ball, P. and Braun, V., *Phys.Rev.* **D49**, 2472 (1994).
6. Gremm, M., Kapustin, A., Ligeti, Z., Wise, M.B., *Phys.Lett.* **B377**, 20 (1996).
7. Simonov, Yu.A., and Tjon J.A., *Phys.Rev.* **D62**, 014501 (2000).

EFFECTIVE LAGRANGIANS INDUCED BY THE ANOMALOUS WESS-ZUMINO ACTION AND $I^G(J^{PC}) = 1^-(1^{-+})$ EXOTIC STATES

N.N. Achasov and G.N. Shestakov

Laboratory of Theoretical Physics, S.L. Sobolev Institute for Mathematics, 630090, Novosibirsk, Russia

Abstract. A simple dynamical model for the exotic waves with $I^G(J^{PC}) = 1^-(1^{-+})$ in the reactions $\rho\pi \to \rho\pi$, $\rho\pi \to \eta\pi$, $\rho\pi \to \eta'\pi$, $\rho\pi \to (K^*\bar{K} + \bar{K}^*K)$, and in the related ones, is constructed beyond the scope of the quark-gluon approach. The model satisfies unitarity and analyticity and uses as a "priming" the anomalous non-diagonal $VPPP$ interaction which couples together the four channels $\rho\pi$, $\eta\pi$, $\eta'\pi$, and $K^*\bar{K} + \bar{K}^*K$. The possibility of the resonance-like behavior of the $I^G(J^{PC}) = 1^-(1^{-+})$ amplitudes belonging to the $\{10\} - \{\overline{10}\}$ and $\{8\}$ representations of $SU(3)$ as well as their mixing is demonstrated explicitly in the 1.3–1.6 GeV mass range which, according to the current experiments, is really rich in exotics.

INTRODUCTION

Phantoms of manifestly exotic π_1 states with $I^G(J^{PC}) = 1^-(1^{-+})$ have more and more agitated the experimental and theoretical communities [1-3]. They were discovered in the 1.3–1.6 GeV mass range in the $\eta\pi$, $\eta'\pi$, $\rho\pi$, $b_1\pi$, and $f_1\pi$ systems produced in π^-p collisions at high energies and in $N\bar{N}$ annihilation at rest in the GAMS, KEK, VES, CB, and BNL experiments [1-3].

The first evidence for the possible existence of an exotic 1^{-+} state coupled to the $\eta\pi$ and $\rho\pi$ channels and belonging to the icosuplet representation of $SU(3)$ was obtained by J. Schechter and S. Okubo about 37 years ago with the bootstrap technique [4].

Recently theoretical considerations concerning the mass spectra and decay properties of exotic hadrons have been based, in the main, on the MIT-bag model, constituent gluon model, flux-tube model, QCD sum rules, lattice calculations, and various selection rules.

Current algebra and effective Lagrangians are also important sources of theoretical information on exotic partial waves. It is sufficient to remember the prediction obtained within the framework of these approaches for the $\pi\pi$ S-wave scattering length with isospin $I = 2$. There also exist a good many of the model constructions which show that the low-energy contributions calculated within the effective chiral Lagrangians framework may in principle transform with increasing energy into resonances with the experimentally established parameters. The important ingredient of all these models is the successfully selected unitarization scheme for the original chiral amplitudes which is used to match the low-energy and resonance regions. Such models are well known, for example, for the $\pi\pi$ scattering channels involving the σ and ρ resonances (see, for example, Ref. [5]). In the present work we continue in this way and construct

A MODEL FOR THE $I^G(J^{PC}) = 1^-(1^{-+})$ WAVES IN THE REACTIONS $VP \to PP$, $PP \to PP$, AND $VP \to VP$

using, as the starting point, the following anomalous effective Lagrangian for point-like $VPPP$ interaction of the vector (V) and pseudoscalar (P) mesons

$$L(VPPP) = ih\varepsilon_{\mu\nu\tau\kappa}\text{Tr}(\hat{V}^\mu \partial^\nu \hat{P} \partial^\tau \hat{P} \partial^\kappa \hat{P}) + i\sqrt{1/3}\, h'\, \varepsilon_{\mu\nu\tau\kappa}\text{Tr}(\hat{V}^\mu \partial^\nu \hat{P} \partial^\tau \hat{P}) \partial^\kappa \eta_0,$$

where h and h' are the coupling constants, $\hat{P} = \sum_{a=1}^{8} \lambda_a P_a/\sqrt{2}$, $\hat{V}^\mu = \sum_{a=0}^{8} \lambda_a V_a^\mu/\sqrt{2}$, and λ_a are the Gell-Mann matrices. This Lagrangian induced by the anomalous Wess-Zumino action and generates the tree exotic amplitudes with $I^G(J^{PC}) = 1^-(1^{-+})$ for the inelastic reactions $\rho\pi \to \eta_8\pi$, $K^*\bar{K} \to \eta_8\pi$, $\bar{K}^*K \to \eta_8\pi$ belonging to the $\{10\}-\{\overline{10}\}$ representation of $SU(3)$ (in this case, there are, at least, the $qq\bar{q}\bar{q}$ states in the s-channel) and the tree amplitudes for the reactions $\rho\pi \to \eta_0\pi$, $K^*\bar{K} \to \eta_0\pi$, $\bar{K}^*K \to \eta_0\pi$ belonging to the $\{8\}$ representation of $SU(3)$ (in this case, there are both $qq\bar{q}\bar{q}$ and $q\bar{q}g$ states in the s-channel).[1] In the next orders, these tree amplitudes induce as well the exotic ones for the elastic processes $\rho\pi \to \rho\pi$, $\eta\pi \to \eta\pi$, and so on. In this connection it is of interest to consider the following 4×4 system of scattering amplitudes for the coupled exotic channels of the reactions $VP \to VP$, $VP \leftrightarrow PP$ and $PP \to PP$:

$$T_{ij} = \begin{bmatrix} T(\rho\pi \to \rho\pi) & T(\rho\pi \to \eta\pi) & T(\rho\pi \to \eta'\pi) & T(\rho\pi \to K^*K) \\ T(\eta\pi \to \rho\pi) & T(\eta\pi \to \eta\pi) & T(\eta\pi \to \eta'\pi) & T(\eta\pi \to K^*K) \\ T(\eta'\pi \to \rho\pi) & T(\eta'\pi \to \eta\pi) & T(\eta'\pi \to \eta'\pi) & T(\eta'\pi \to K^*K) \\ T(K^*K \to \rho\pi) & T(K^*K \to \eta\pi) & T(K^*K \to \eta'\pi) & T(K^*K \to K^*K) \end{bmatrix}.$$

The subscripts $i,j = 1,2,3,4$ are the labels of the $\rho\pi$, $\eta\pi$, $\eta'\pi$, and K^*K channels, respectively (the abbreviation K^*K implies just the $\bar{K}^*K$ and $K^*\bar{K}$ channels).

We consider three natural limiting (in the sense of $SU(3)$ symmetry) cases: (i) $h' = 0$, i.e., when all exotic amplitudes belong to the $\{10\}-\{\overline{10}\}$ representation of $SU(3)$; (ii) $h = 0$, i.e., when all exotic amplitudes belong to the octet representation of $SU(3)$; and (iii) $h' = h$, when the original $VPPP$ interaction possesses nonet symmetry with respect to the 0^- mesons.

To obtain the unitarized amplitudes in coupled channels, we sum up all the possible chains of the s-channel loop diagrams the typical examples of which are given below.

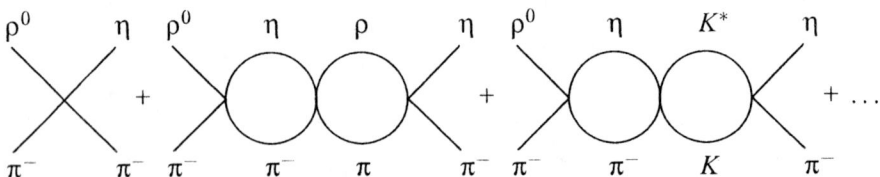

It is the well known field theory way of the unitarization. The relevant summation can be easily carried out by using the matrix equation $\tilde{T}_{ij} = h_{ij} + h_{im}\Pi_{mn}\tilde{T}_{nj}$ for the auxiliary

---

[1] We use the pseudoscalar octet-singlet (η_8-η_0) mixing angle $\theta_P \approx -20°$.

subtracted dispersion integrals

$$F_i = C_{1i} + sC_{2i} + \frac{s^2}{\pi} \int_{m_{i+}^2}^{\infty} \frac{[P_i(s')]^3 \, ds'}{\sqrt{s'} s'^2 (s' - s - i\varepsilon)},$$

where $\sqrt{s}$ is the invariant mass of two-body systems and $P_i(s')$ is the particle momentum in the i intermediate state.

Next, we define $D = \det(\hat{1} - \hat{h}\hat{\Pi})$. It is clear that all the physical amplitudes T_{ij} are proportional to $1/D$. Thus a simplest way to discover "by hand" a possible resonance situation is that to find zero of Re(D) at fixed values of h, h' and $\sqrt{s}$ (for example, at $\sqrt{s} = 1.43$ GeV). Leaving the potentialities of the model almost unchanged, we assume that $C_{11} = C_{14}$ and $C_{21} = C_{24} = 0$ for the VP loops, $C_{12} = C_{13}$ and $C_{22} = C_{23} = 0$ for the PP loops. Thus, in most considered variants, we used as the essential free parameters only two subtraction constants C_{11} and C_{12}. As for the coupling constant h, one may claim [6] that it is not too large in the scale defined by the combination $2g_{\rho\pi\pi}g_{\omega\rho\pi}/m_\rho^2 \approx 284$ GeV^{-3}, namely, that $|\tilde{h} = F_\pi^3 h| \leq 0.4$ (where $F_\pi \approx 130$ MeV), and we are guided by the values of $\tilde{h}$ (and $\tilde{h}' = F_\pi^3 h'$) near 0.1.

Figure 1 shows the typical energy dependences, which occur in our model for cases (i), (ii), and (iii), for the four reaction cross sections $\sigma(\rho^0\pi^- \to \rho^0\pi^-)$, $\sigma(\rho^0\pi^- \to \eta\pi^-)$, $\sigma(\rho^0\pi^- \to \eta'\pi^-)$, and $\sigma(\rho^0\pi^- \to K^{*0}K^-)$ and for the phases of the $\rho\pi \to \rho\pi$ and $\rho\pi \to \eta\pi$ amplitudes. They clearly demonstrate the resonance effects found in the invariant mass region 1.3–1.4 GeV (a similar resonance picture is also obtained for the 1.5–1.6 GeV mass region [7]). Furthermore, the comparison of the obtained cross section values (see Fig. 1) with those of the conventional $a_2(1320)$ resonance production, $\sigma(\rho^0\pi^- \to a_2 \to \rho^0\pi^-) \approx 5.7$ mb and $\sigma(\rho^0\pi^- \to a_2 \to \eta\pi^-) \approx 2.36$ mb at $\sqrt{s} = m_{a_2} = 1.32$ GeV, indicates conclusively that we are certainly dealing with the resonance-like behavior of the $I^G(J^{PC}) = 1^-(1^{-+})$ exotic waves, at least, in the $\rho\pi$, $\eta\pi$, and $\eta'\pi$ channels. Summarizing we conclude that our calculation (see Ref. [7] for details) gives a further new reason in favor of the plausibility of the existence of an explicitly exotic π_1 resonance in the mass range 1.3–1.6 GeV.

REFERENCES

1. Alde, D. et al., *Phys. Lett.* B **205**, 397 (1988). Beladidze, G. M. et al., *Phys. Lett.* B **313**, 276 (1993). Aoyagi, H. et al., *Phys. Lett.* B **314**, 246 (1993). Amsler, C. et al., *Phys. Lett.* B **333**, 277 (1994).
2. Zaitsev, A, HADRON'97, AIP Conference Proceedings **432**, 1998, p. 461. Abele, A. et al., *Phys. Lett.* B **446**, 349 (1999). Alde, D. et al., *Yad. Fiz.* **62**, 462 (1999). Chung, S. U. et al., *Phys. Rev.* D **60**, 092001 (1999). Godfrey, S. and Napolitano, J., *Rev. Mod. Phys.* **71**, 1411 (1999).
3. Popov, A., these proceedings. Dorofeev, V., these proceedings. Reinnarth, J., these prpceedings.
4. Schechter, J. and Okubo, S., *Phys. Rev.* **135**, B1060 (1964).
5. Brown, L. S. and Goble, R. L., *Phys. Rev. Lett.* **20**, 346 (1968). Chan, L.-H. and Haymaker, R. W., *Phys. Rev.* D **10**, 4143 (1974). Dobado, A., Herrero, M. J., and Truong, T. N., *Phys. Lett.* B **235**, 134 (1990). Achasov, N. N. and Shestakov, G. N., *Phys. Rev.* D **49**, 5779 (1994). Harada, M., Sannino, F., and Schechter, J., *Phys. Rev.* D **54**, 1991 (1996). Gómez Nicola, A. and Peláez, J. R., hep-ph/0109056.
6. Jain, P., Johnson, R., Meissner, U.-G., Park, N. W., and Schechter, J., *Phys. Rev.* D **37**, 3252 (1988).
7. Achasov, N. N. and Shestakov, G. N., *Phys. Rev.* D **63**, 014017 (2001).

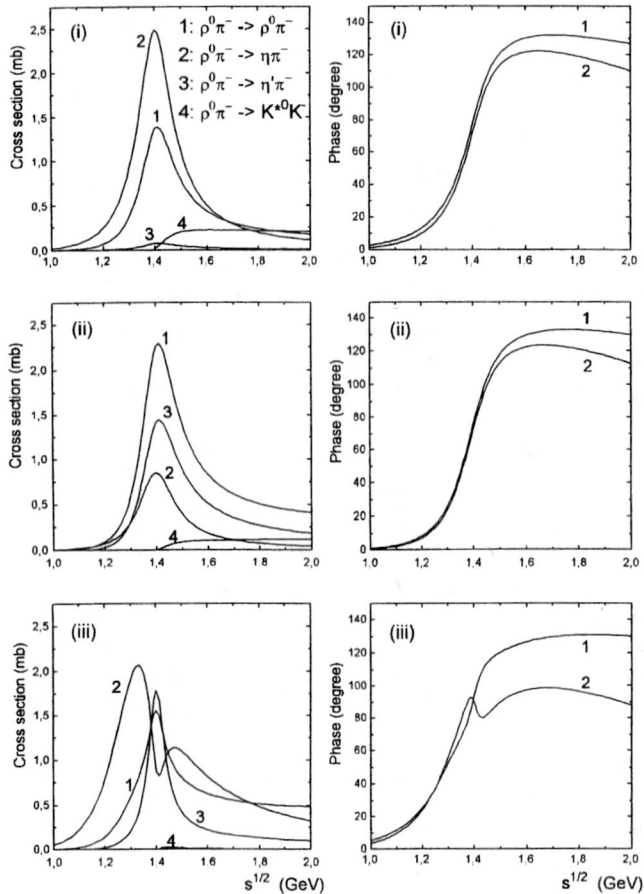

FIGURE 1. The cross sections of the reactions $\rho^0\pi^- \to \rho^0\pi^-$, $\rho^0\pi^- \to \eta\pi^-$, $\rho^0\pi^- \to \eta'\pi^-$, and $\rho^0\pi^- \to K^{*0}K^-$ and the phases of the $\rho\pi \to \rho\pi$ and $\rho\pi \to \eta\pi$ amplitudes. The correspondence between the curve numbers and the reaction channels is shown just in the figure. $\tilde{h} = 0.10746$, $\tilde{h}' = 0$, $C_{11} = 0.17$ GeV^2, $C_{12} = 1.25$ GeV^2 in case (i), $\tilde{h}' = 0.10746$, $\tilde{h} = 0$, $C_{11} = 0.34$ GeV^2, $C_{12} = 0.67$ GeV^2 in case (ii), and $\tilde{h} = \tilde{h}' = 0.10746$, $C_{11} = 0.49$ GeV^2, $C_{12} = 0.5$ GeV^2 in case (iii).

invariant amplitudes $\tilde{T}_{ij}$, the solution of which has the form: $\tilde{T}_{ij} = [(\hat{1} - \hat{h}\hat{\Pi})^{-1}]_{im} h_{mj}$. Here h_{ij} is the matrix of the coupling constants generated by the Lagrangian and $\Pi_{ij} = \delta_{ij}\Pi_j$ is the diagonal matrix of the loops, $\Pi_i = sF_i/(6\pi)$ for the VP loops (i=1,4) and $\Pi_i = F_i/(24\pi)$ for the PP loops (i=2,3), the functions F_i are defined by the doubly

Rare radiative B decays to excited K mesons

D. Ebert*, R. N. Faustov† and V. O. Galkin†

*Institut für Physik, Humboldt–Universität zu Berlin, 10115 Berlin, Germany
†Russian Academy of Sciences, Scientific Council for Cybernetics, Moscow 117333, Russia

Abstract. The exclusive rare radiative B meson decays to orbitally excited axial-vector mesons $K_1^*(1270)$, $K_1(1400)$ and to the tensor meson $K_2^*(1430)$ are investigated in the framework of the relativistic quark model based on the quasipotential approach in quantum field theory. These decays are considered without employing the heavy quark expansion for the s quark. Instead the s quark is treated to be light and the expansion in inverse powers of the large recoil momentum of the final K^{**} meson is used to simplify calculations. It is found that the ratio of the branching fractions of rare radiative B decays to axial vector $K_1^*(1270)$ and $K_1(1400)$ mesons is significantly influenced by relativistic effects. The obtained results for B decays to the tensor meson $K_2^*(1430)$ agree with recent experimental data from CLEO and Belle.

Rare radiative decays of B mesons are induced by flavour changing neutral currents and thus they are sensitive probes of new physics beyond the standard model. Such decays are governed by one-loop (penguin) diagrams with the main contribution from a virtual top quark and a W boson. Therefore, they provide valuable information about the Cabibbo-Kobayashi-Maskawa (CKM) matrix elements V_{ts} and V_{tb}. The statistics of rare radiative B decays considerably increased since the first observation of the $B \to K^*\gamma$ decay in 1993 by CLEO [1]. This allowed a significantly more precise determination of exclusive and inclusive branching fractions [2]. The first observation of the rare B decays to the orbitally excited strange mesons has been reported by CLEO [2]. The branching fraction for the decay to the tensor $K_2^*(1430)$ meson has been measured $\mathcal{B}(B \to K_2^*(1430)\gamma) = (1.66^{+0.59}_{-0.53} \pm 0.13) \times 10^{-5}$, as well as the ratio of exclusive branching fractions $r \equiv \mathcal{B}(B \to K_2^*(1430)\gamma)/\mathcal{B}(B \to K^*(892)\gamma) = 0.39^{+0.15}_{-0.13}$. Recently Belle [3] published the measurement of the corresponding branching fraction $\mathcal{B}(B \to K_2^*(1430)\gamma) = (1.26 \pm 0.66 \pm 0.10) \times 10^{-5}$, which is in good agreement with CLEO result. The data for the other decay channels will be available soon. This significant experimental progress provides a challenge to the theory. Many theoretical approaches have been employed to predict the exclusive $B \to K^*(892)\gamma$ decay rate. Considerably less attention has been payed to rare radiative B decays to excited strange mesons [4, 5, 6, 7]. Most of these theoretical approaches [5, 7] rely on the heavy quark limit both for the initial b and final s quarks and the nonrelativistic quark model. However, the two predictions [5, 7] for the ratio r differ by an order of magnitude, due to a different treatment of the long distance effects and, as a result, a different determination of corresponding Isgur-Wise functions. Only the prediction of Ref. [7] is consistent with the available data. Nevertheless, it is necessary to point out that the s quark in the final K^* meson is not heavy enough, compared to the $\bar{\Lambda} \sim 400$ MeV parameter, which determines the scale of $1/m_Q$ corrections in heavy quark effective theory [8]. Thus the

$1/m_s$ expansion is not appropriate. Notwithstanding, the ideas of heavy quark expansion can be applied to the exclusive $B \to K^*(K^{**})\gamma$ decays. From the kinematical analysis it follows that the final $K^*(K^{**})$ meson bears a large relativistic recoil momentum $|\Delta|$ of order of $m_b/2$ and an energy of the same order. So it is possible to expand the matrix element of the effective Hamiltonian both in inverse powers of the b quark mass for the initial state and in inverse powers of the recoil momentum $|\Delta|$ for the final state [9, 10]. Such an expansion has been realized by us for the $B \to K^*(892)\gamma$ decay in the framework of the relativistic quark model [9]. In Refs. [10] it was shown that in the leading order of this expansion a specific symmetry emerges which imposes several relations between the form factors of semileptonic and rare radiative B decays. It is important to note that rare radiative decays of B mesons require a completely relativistic treatment, because the recoil momentum of the final meson is large compared to the s quark mass. The calculated branching fraction for the decay to the ground K^* state was found [9] in good agreement with experimental data. Here we extend this analysis to the exclusive rare radiative B decays to orbitally excited axial-vector ($K_1^*(1270), K_1(1400)$) and tensor ($K_2^*(1430)$) mesons.

Our relativistic quark model is based on the quasipotential approach in quantum field theory with a specific choice of the quark-antiquark interaction potential. Its main features are described in Ref. [11]. The matrix element of the weak current $J_\mu = \bar{s}\frac{i}{2}k^\nu \sigma_{\mu\nu}(1+\gamma^5)b$ between the states of a B meson and an orbitally excited K^{**} meson has the form [9]

$$\langle K^{**}|J_\mu(0)|B\rangle = \int \frac{d^3p\,d^3q}{(2\pi)^6} \bar{\Psi}_{K^{**}}(\mathbf{p})\Gamma_\mu(\mathbf{p},\mathbf{q})\Psi_B(\mathbf{q}), \tag{1}$$

where $\Gamma_\mu(\mathbf{p},\mathbf{q})$ is the two-particle vertex function and $\Psi_{B,K^{**}}$ are the meson wave functions projected onto the positive energy states of quarks and boosted to the moving reference frame:

$$\Psi_{K^{**}\Delta}(\mathbf{p}) = D_s^{1/2}(R_{L_\Delta}^W) D_q^{1/2}(R_{L_\Delta}^W) \Psi_{K^{**}\,0}(\mathbf{p}). \tag{2}$$

Here $\Psi_{K^{**}\,0}$ is the K^{**} wave function in the rest frame, R^W is the Wigner rotation, L_Δ is the Lorentz boost from the meson rest frame to a moving one, and $D^{1/2}(R)$ is the rotation matrix. We calculated the corresponding decay matrix elements (1) and determined decay form factors which govern rare radiative B decays to orbitally excited K mesons [12]. In order to simplify calculations and get analytical results we used the heavy quark expansion for the b quark from the initial B meson and expansion in inverse powers of large recoil momentum $|\Delta| = (M_B^2 - M_{K^{**}}^2)/(2M_B) \approx M_B/2$ of the final orbitally excited K^{**} meson. The obtained formulas contain all terms up to second order in these expansions. The s quark in the final K^{**} meson was treated fully relativistically. It was shown that resulting formulas in corresponding limits satisfy all constraints imposed by heavy quark symmetry [13] and large energy effective theory [10].

The results of numerical calculations are given in Table 1 [12]. There we also show our previous predictions for the $B \to K^*\gamma$ decay [9]. It is convenient to consider the ratio of exclusive to inclusive branching fractions which depend on form factors and mass

TABLE 1. Theoretical Predictions and Experimental Data for the Branching Fractions (in units of 10^{-5}) and their Ratios $R_{K^*} \equiv \mathcal{B}(B \to K^*\gamma)/\mathcal{B}(B \to X_s\gamma)$, $R_{K_i^{(*)}} \equiv \mathcal{B}(B \to K_i^{(*)}\gamma)/\mathcal{B}(B \to X_s\gamma)$ ($i = 1, 2$), $r \equiv \mathcal{B}(B \to K_2^*\gamma)/\mathcal{B}(B \to K^*\gamma)$ (our values for the $B \to K^*\gamma$ decay are taken from Ref. [9]).

Value	our [12]	[4]	[5]	[7]	CLEO [2]	Belle [3]
$\mathcal{B}(B \to K^*(892)\gamma)$	4.5 ± 1.5	1.35	$1.4 - 4.9$	4.71 ± 1.79	$4.55^{+0.80}_{-0.76}$ *	4.96 ± 0.81 *
					$3.76^{+0.93}_{-0.88}$ †	3.89 ± 1.02 †
R_{K^*} (%)	15 ± 3	4.5	$3.5 - 12.2$	16.8 ± 6.4		
$B \to K_0^*(1430)\gamma$			forbidden			
$\mathcal{B}(B \to K_1^*(1270)\gamma)$	0.45 ± 0.15	1.1	$1.8 - 4.0$	1.20 ± 0.44		
$R_{K_1^*}$ (%)	1.5 ± 0.5	3.8	$4.5 - 10.1$	4.3 ± 1.6		
$\mathcal{B}(B \to K_1(1400)\gamma)$	0.78 ± 0.18	0.7	$2.4 - 5.2$	0.58 ± 0.26		
R_{K_1} (%)	2.6 ± 0.6	2.2	$6.0 - 13.0$	2.1 ± 0.9		
$\mathcal{B}(B \to K_2^*(1430)\gamma)$	1.7 ± 0.6	1.8	$6.9 - 14.8$	1.73 ± 0.80	$1.66^{+0.61}_{-0.55}$	1.26 ± 0.67
$R_{K_2^*}$ (%)	5.7 ± 1.2	6.0	$17.3 - 37.1$	6.2 ± 2.9		
r	0.38 ± 0.08	1.3	$3.0 - 4.9$	0.37 ± 0.10	$0.39^{+0.15}_{-0.13}$	

* $B^0 \to K^{*0}\gamma$
† $B^+ \to K^{*+}\gamma$

ratios only. The recent experimental values for the inclusive decay branching fraction

$$BR(B \to X_s\gamma) = \begin{cases} (3.15 \pm 0.35 \pm 0.32 \pm 0.26) \times 10^{-4} & \text{(CLEO)} \\ (3.39 \pm 0.53 \pm 0.42^{+0.51}_{-0.55}) \times 10^{-4} & \text{(Belle)} \end{cases}$$

are in a good agreement with theoretical calculations [14]

$$BR(B \to X_s\gamma) = (3.29 \pm 0.33) \times 10^{-4}.$$

Our results are confronted with other theoretical calculations [4, 5, 6, 7] and recent experimental data [2]. The QCD sum rules predict (with 20% uncertainty) [15] $\mathcal{B}(B \to K^*\gamma) = 4.4 \times 10^{-5} \times (1 + 8\%)$, where the second term in the brackets is the estimate of the $1/m_c^2$ contribution. We find a good agreement of our predictions for decay rates with the experiment and estimates of Ref. [7] for the decay rates $B \to K^*\gamma$ and $B \to K_2^*\gamma$. Other theoretical calculations substantially disagree with data either for $B \to K^*\gamma$ [4, 6] or for $B \to K_2^*\gamma$ [5] decay rates. Note that one of the main reasons of the too small values for $B \to K^*\gamma$ decay rates in quark models [4, 6] is the use of the nonrelativistic expression for the momentum of the final meson in the argument of the wave function overlap [9]. As a result our predictions and those of Ref. [7] for the ratio r are well consistent with experiment, while the r estimates of [4, 6] and [5] are several times larger than the experimental value (see Table 1). It is not accidental that r values in our and Ref. [7] approaches are close, since the effective expansion due to the relativistic dynamics goes in inverse powers of the s quark energy $\varepsilon_s(p + \Delta) = \sqrt{(\mathbf{p} + \Delta)^2 + m_s^2}$, which is high in one case due to the large s quark mass [7] and in the other one [12] due to the large recoil momentum Δ. The agreement of both predictions for branching fractions could be explained by some specific cancellation of finite s quark mass effects and relativistic corrections which were neglected in Ref. [7]. Though our numerical results for the measured decay rates agree with Ref. [7], we believe that our analysis is more consistent and reliable. We do not use the unphysical limit $m_s \to \infty$, and our quark model consistently takes into account main relativistic effects, for example, the Lorentz transformation (2) of the wave function of the final K^{**} meson to the moving frame. Such a transformation turns out to be very important and leads to the substantial reduction of the $B \to K_1^*(1270)\gamma$ decay rate in our model. We see from Table 1 that our model predicts for the ratio $\mathcal{B}(B \to K_1^*(1270)\gamma)/\mathcal{B}(B \to K_1(1400)\gamma)$ the value 0.7 ± 0.3 while Ref. [7] gives for this ratio a considerably larger value ~ 2, which is the consequence of the nonrelativistic quark model relation between form factors [12]. Thus experimental measurement of $\mathcal{B}(B \to K_1^*(1270)\gamma)$ and $\mathcal{B}(B \to K_1(1400)\gamma)$ can discriminate between these predictions.

This research was supported in part by the *Deutsche Forschungsgemeinschaft* under Contract Eb 139/2-1, *Russian Foundation for Fundamental Research* under Grant No. 00-02-17768 and *Russian Ministry of Education* under Grant No. E00-3.3-45.

REFERENCES

1. CLEO Collaboration, Ammar, R., et al., *Phys. Rev. Lett.* **71**, 674-678 (1993).
2. CLEO Collaboration, Coan, T. E., et al., *Phys. Rev. Lett.* **84**, 5283-5287 (2000).
3. Belle Collaboration, Abe, K., et al., hep-ex/0107065 (2001).
4. Altomari, T., *Phys. Rev. D* **37**, 677-680 (1988).
5. Ali, A., Ohl, T., and Mannel, T., *Phys. Lett. B* **298**, 195-203 (1993).
6. Atwood, D., and Soni, A., *Z. Phys. C* **64**, 241-254 (1994).
7. Veseli, S., and Olsson, M. G., *Phys. Lett. B* **367**, 309-316 (1996).
8. Neubert, M., *Phys. Rep.* **245**, 259-396 (1994).
9. Faustov, R. N., and Galkin, V. O., *Phys. Rev. D* **52**, 5131-5140 (1995).
10. Charles, J., et al., *Phys. Rev. D* **60**, 014001-1-14 (1999); Ebert, D., Faustov, R. N., and Galkin, V. O., *Phys. Rev. D* **65**, 094022-1-15 (2001).
11. Ebert, D., Faustov, R. N., and Galkin, V. O., "Heavy Quark Potential and Mass Spectra of Heavy Mesons", in *these proceedings*.
12. Ebert, D., Faustov, R. N., Galkin, V. O., and Toki, H., *Phys. Lett. B* **495**, 309-316 (2000); *Phys. Rev. D* **64**, 054001-1-11 (2001).
13. Veseli, S., and Olsson, M. G., *Phys. Lett. B* **367**, 302-308 (1996).
14. Ali, A., *Nucl. Instrum. Meth. A* **462**, 11-22 (2001).
15. Ball, P., hep-ph/0010024 (2000).

Wavelet analysis in physics of resonances: application to spectrum of ρ' and ω' excitations and ratio $R_{e^+e^-}$

Victor K. Henner[*], Piter G. Frick[+], Tatyana S. Belozerova[*]

[*] *Perm State University,* [+] *ICMM, Perm Russia*

Abstract. We use the wavelet analysis to separate noise and resonances contributions for some $e^+e^- \to hadrons$ data. With these "cleaned up" data as output, we find ρ' and ω' parameters using the generalized Breit-Wigner method that preserves unitarity in the case of overlapping states.

1. The properties of ρ' and ω' states and even their number are not well defined. The major difficulties in understanding the situation are poor statistics and overlapping of states with the same quantum numbers. In this connection, the goal of our study is to resolve structures in the data (to do that, we give the first, to our knowledge, application of wavelet analysis (WA) for the problems of physics of resonances), and then to find parameters of the excitations related to these structures with a preserving unitary for overlapping states Breit-Wigner (BW) formalism.

We apply the WA, a very efficient multi-scale technique, to resolve structures in e^+e^- annihilation and in p wave $\pi\pi$ scattering. Such a local analysis is very helpful when it is necessary to distinguish between several resonances in data with large errors. Resolving these structures from the experimental noise and background allows us to make more reliable conclusion regarding the resonances. The WA is much less sensitive to the noise than any other analysis and therefore substantially reduces the role of fluctuations. We also apply the WA to smooth out the high frequency noise in the ratio $R_{e^+e^-}$.

Interference of resonances with the same decay channels is the key aspect of any analysis and interpretation. It is often taken into account by relative phases for BW terms, which are treated as free parameters (the most often just 0° or 180°). Whether or not these phases are included, such a sum of BW terms violates unitarity, which is the basic point in the BW description. The BW approach has the advantage of almost complete model independence and contains only masses, widths and branching fractions of the resonances. Because of that, results of any analysis would be desirable to compare with BW. Despite of a common belief that there is no unitary BW multistate method, such a method does exist [1]. The scattering amplitudes are very similar to a standard BW expression:

$$f_{ij}(s) = \sum_r \frac{m_r \Gamma_r g_{ri} g_{rj}}{s - m_r^2 + im_r \Gamma_r} = \sum_r e^{i(\varphi_{ri} + \varphi_{rj})} \frac{m_r \Gamma_r |g_{ri}| |g_{rj}|}{s - m_r^2 + im_r \Gamma_r}, \tag{1}$$

where the phases φ_{ri} are not independent parameters and should be determined in such a way that preserves the unitarity (the number of free parameters is smaller or equal than in a "naive" BW or the K matrix approach). Details of the method and its connection to other preseving unitarity methods can be found in [1].

2. The continuous wavelet transformation of a function of energy $f(E)$ representing the data is defined as:

$$w(a,E) = C_\psi^{-\frac{1}{2}} a^{-\frac{1}{2}} \int_{-\infty}^{+\infty} \psi^*\left(\frac{E'-E}{a}\right) f(E') dE'. \tag{2}$$

Decomposition (2) is performed by convolution of the function $f(E)$ with a bi-parametric family of self-similar functions generated by dilatation and translation of the analyzing function $\psi(E)$, called wavelet: $\psi_{a,b}(t) = \psi((E-b)/a)$, where a scale parameter a characterizes the dilatation, b characterizes the translation, C_ψ is a constant defined through the Fourier transformation of $\psi(E)$. It is a sort of "window function" with a non-constant window's width: high frequency wavelets are narrow (due to the factor $1/a$), while low frequency wavelets are much broader. Contrary to Fourier analysis, the $w(a,E)$ depends both on E and frequency $1/a$. If the objective of the WA is to find the parameters of dominating structures (location and width), then one should use the wavelets with good localization and a small number of oscillations such as "Mexican Hat" (MH), $\psi(E) = (1-E^2) \cdot e^{-E^2/2}$.

The wavelet plane $w(a,E)$ itself illustrates data features indicating the frequency (scale) content of the data as a function of energy. The location of a dark spot on the scale axis, a, corresponds to the width of the maximum. The intensity of spots shows the amplitudes of maxima.

If the reconstructed data are stable under small perturbations, then one can distinguish between the "useful" large scale stains (low frequencies in Fourier space) and contributions of the small-scale features usually generated by noise. The noise is located at the bottom of the wavelet plane (small scale regions, or high frequencies). In order to separate the noise, the wavelet reconstruction is performed for scales greater than a certain boundary scale a_{noise}.

To clear out ρ' and ω' contributions, we perform the WA (with MH wavelet) of $e^+e^- \to hadrons$ data relevant to these excitations, and p-wave $\pi\pi$ scattering data. The main ρ' decay channels are $\pi\pi$, $2\pi^+2\pi^-$, $\pi^+\pi^-2\pi^0$, $\omega\pi^0$, $\eta\pi^+\pi^-$. The WA indicates ρ' states with masses 1.1 - 1.25, 1.4, 1.6 - 1.85, and 2.2 GeV and widths of about 100 - 200 MeV. It makes sense to include these states in further analysis that should be based on some physical models. The first and the last states are sensitive to high frequency noise contribution, which makes them questionable.

The WA shows that some of the considered experimental data are statistically inadequate because they do not allow to separate the noise contribution. The WA gives the criteria for distinguishing between stable and unstable data - the latter do not reproduce the same essential structures when a_{noise} value changes slightly. A high sensitivity to noise might be one of the explanations why the low mass $\rho'(1200)$ is hard to observe and to confirm.

The main ω' decay channels are $\pi^+\pi^-\pi^0$ and $\omega\pi\pi$. The WA gives two ω' states at around 1.4 and 1.6, and possible 1.85 GeV state.

3. For a sake of clarity we give the BW analysis of ω' states only. Note that even the "optimized" data do not directly provide masses and widths - due to overlapping, the observed positions of maxima can differ substantially from the physical resonance masses, and the partial and total widths can differ essentially from WA estimations.

A comparison of the reconstructed with the WA data and BW unitary expression (1) gives good fits for both cases of two and three ω' with the χ^2/n_D about 1.3. Starting from the WA masses and widths we allow big deviations of about 150 MeV from WA numbers. For the case of two ω' the parameters are shown in Table 1 and Table 2. The leptonic widths are $\Gamma_{\omega_1'\to e^+e^-} \approx 0.46$ and $\Gamma_{\omega_2'\to e^+e^-} \approx 0.80$ keV. All these parameters are in a good agreement with the PDG.

For the case of three ω', their masses and total widths are: $m_{\omega_1'}=1.460$, $\Gamma_{\omega_1'}=0.236$, $m_{\omega_2'}=1.615$, $\Gamma_{\omega_2'}=0.139$, $m_{\omega_3'}=1.850$, $\Gamma_{\omega_3'}=0.163$ (in GeV). It is clear that the ω_3' effectively reduces the width of the ω_2'. The ω_3' is questionable because it is sensitive to the cutoff value a_{noise}, but it does not contradict the data.

In this short paper there is a place for two figures only. Fig.1 shows two ρ' maxima around 1.4 and 1.5 - 1.6 GeV with widths about 120 and 150 MeV. These structures are stable: reduction of a_{noise} by a factor of two does not change the locations of the peaks and their shapes. There is also rather sensitive to a_{noise} value maximum at about 1.2 GeV. Fig.2 shows two ω' states stable under variation of a_{noise} value, and the results of BW analysis.

The study of the ratio $R_{e^+e^-}$ demonstrates another application of the WA to high-energy physics. To even out any rapid variations in $R_{e^+e^-}$ we remove the high frequency noise with the WA, that provides the smearing alternative to the procedure proposed in [2]. The restored data keep all main features of $R_{e^+e^-}$ with statistical errors and threshold singularities damped, which makes a direct comparison with the corresponding QCD smeared quantity possible (see the details in work [3]).

TABLE 1. Parameters of the ω' states (in GeV)

Meson	Mass	Width
ω_1'	1.450 ± 0.010	0.199 ± 0.015
ω_2'	1.619 ± 0.005	0.250 ± 0.014

TABLE 2. Branching ratios of the ω' states (in %)

State	ω_1'	ω_2'
e^+e^-	$(23 \pm 1.0)10^{-5}$	$(32 \pm 1.0) 10^{-5}$
$\rho\pi$	69.92 ± 2.85	38.03 ± 1.37
$\omega\pi\pi$	30.08 ± 2.85	61.97 ± 1.36

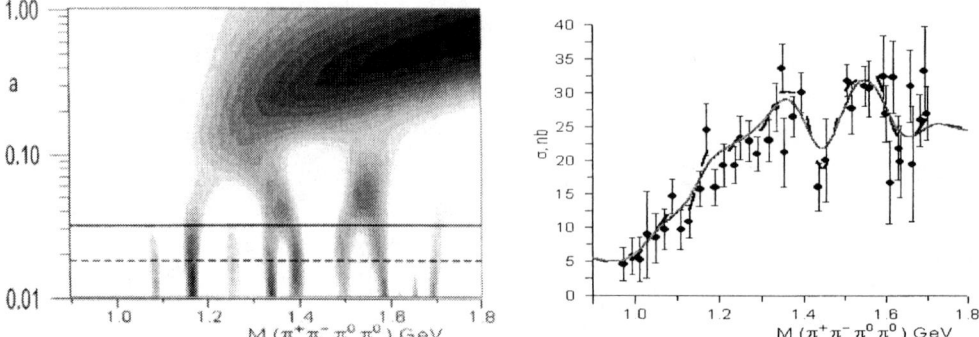

FIGURE 1. Cross-section of $e^+e^- \to \pi^+\pi^-\pi^0\pi^0$. Wavelet plane and two reconstructions for different cutoff values a_{noise}.

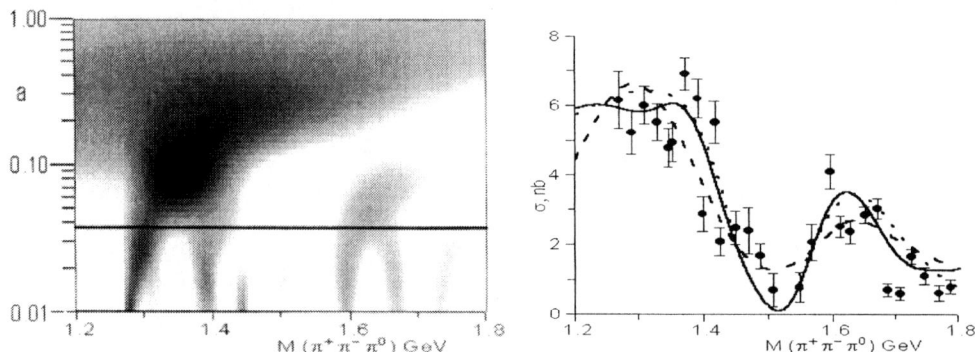

FIGURE 2. Cross-section of $e^+e^- \to \pi^+\pi^-\pi^0$. Wavelet plane and reconstruction (solid line), fits with two ω' (dotted line) and three ω' (dashed line).

4. Numerous applications of wavelets in different fields of mathematics and physics have proved themselves to be a powerful tool for studying fractal signals and data. Due to good scaling properties of wavelets one can consider the data with various resolution which allows us to separate the resonances from noise and from each other. Such a local analysis is very significant when it is necessary to distinguish between several resonances in the data with large errors. The WA can be successfully applied to different problems of nuclear and high-energy physics.

REFERENCES

1. Henner V. and Belozerova T., *Yad.Fizika* **60**,1998 (1997); *Physics of Particles and Nuclei* **29**, 63 (1998).
2. Poggio E., Quinn H. and Weinberg S., *Phys. Rev.* **D13**, 1958 (1976).
3. Henner V., Frick P. and Solovyev V., *Phys.Rev.* (in pess).

Scalar Strong Interaction Hadron Theory- An Alternative to QCD

F. C. Hoh

Dragarbrunnsg. 9B, 75332 Uppsala and associated with Department of Radiation Sciences, Uppsala University, Box 535, SE-751 21 Uppsala, Sweden

Abstract Reasons for not accepting QCD as the correct strong interaction theory given at Hadron '95 and Hadron '97 are expanded and made more precise. The scalar strong interaction hadron theory, recently presented in a book, provides an alternative to QCD. The motivation for and basic steps leading to this theory are outlined.

INTRODUCTION

On the theoretical side of this meeting, the contributions consist largely of the latest versions of phnomenological models, mostly on hadron spectra. These are but a tiny increment to the large body of such models already existing in the literature, or as some put it, on the "market". In his summary talk, Ted Barnes hinted at this situation by putting forth the formula: If there are n theorists, there are $n+1$ theories. Actually, the number $n+1$ is too small and may be replaced by αn, where $\alpha > 1$. This is due to the fact that the same theorist can have different models for different applications. Moreover, there may be several models for the same application area as he updates and improves his earlier models.

There are hundreds of such kinds of models with different assumptions, parameters, form factors, structure functions,... While these models work well within the relatively narrow application angle they are designed for, they provide little understanding of hadronic phenomena in any coherent fashion. This aspect together with the large number of such models makes it very difficult to extract from them eventual guidance to a correct hadron theory. In this respect, the above-mentioned models seem to differ from the models that preceeded quantum mechanics. The atomic models of Bohr and Sommerfeld and the Rydberg formula played impotant roles in the development and verification of the Schrödinger theory.

The great freedom in constructing these models stems from the lack of a working first principle hadron theory. The current strong interaction theory, quantum-chromodynamics(QCD), is nonperturbative at low energies and has not been able to provide useful predictions ever since it was proposed nearly 30 years ago. By contrast, 30 years after the appearance of the Maxwell equations, nearly all basic electromagnetic phenomena were explained. Similarly, 30 years after the publication of the Schrödinger-Dirac equations, most of the atomic phenomena were understood.

Furthermore, Dirac's book appeared already seven years after he published his equation and Schiff's text came out 23 years after the appearance of the Schrödinger equation. Still, there is no book covering QCD.

These circumstances indicate that QCD is not the correct strong interaction theory. Reasons for taking this standpoint have been given in Hadron '95 and Hadron '97[1]. This contribution is a further and more precise development of these references.

QUARKS, FERMIONS, QCD, AND MODELS

QCD as a hadron theory does not include two basic properties of quarks, namely, *property Q1*: *A quark cannot be observed alone* and *property Q2*: *A quark is always accompanied by an antiquark or two other quarks or a diquark.*

Prior to the quark hypothesis of 1963-64, a fermion is defined to be a lepton or a point particle having higher odd half-integer spin. Since the latter has not been observed, it is dropped for the present considerations. A fermion has two basic proper-ties, namely, *property F1*: *A fermion can exist freely and is described by Dirac's wave functions* and *property F2*: *In weakly interacting aggregates, fermions obey Pauli's exclusion principle.*

After the quark hypothesis, quarks were soon embraced into the fermion family. This inclusion is present in QCD and in chiral symmetry models for hadrons because at the times of the proposals of these theories, the nonobservation of free quarks or *property Q1* was not firmly established. Only afterwards was this *property Q1* accepted as an experimental fact.

Accordingly, a quark is not a fermion in the sense of *property F1*. Consequently, there is no reason that *property F2* should hold for quarks. With these observations, the color parts of QCD and chiral symmetry models are no longer needed. Further, quark wave functions similar to those of *property F1* appear in QCD and chiral symmetry models. According to quantum mechanics, these wave functions can be used to form expectation values of dynamical variables such as energy, momentum and position. But such formations would contradict *property Q1*. Therefore, these theories cannot be correct. Also, the concept of spontaneous symmetry breaking in chiral symmetry models is borrowed from solid state physics. But there is no reason why such a mechanism would apply in the totally different case of hadrons.

On par with the chiral symmetry models are the constituent quark models and the Bethe-Salpeter(BS) equation approach[2]. Although no quark wave function appears in the BS approach explicitly, they are present implicitly. This can be seen in the positronium case, which is described by a BS equation with 16 wave function components. The positronium can however disintegrate into a free electron and a free positron and the BS equation decomposes into two Dirac equations describing these free particles. But a meson cannot disintegrate into a free quark and a free antiquark. This difference prevents that the same BS formalism can be applied to a meson. Because of *properties Q1* and *Q2,* quarks in a meson have less degrees of freedom as do the constituents of positronium. Hence the number of meson wave function components should be less than 16(see eq. (3) ff below). Similarly, the number of baryon wave function components should be less than 64(see last §).

The constituent quark models are more phenomenological in nature and have been the most successful ones. The interaction potential among the quarks are usually assumed to be of vector type in order to conform to QCD. In this connection, the spin-orbit terms in the potential have caused quite some controversy[3]. More importantly, confinement potential must be assumed to fit data.

SCALAR STRONG INTERACTION HADRON THEORY

In view of the above considerations, alternatives to QCD are obviously called for. I am as surprised as I was at Hadron '95[1] at that no voice on this issue has been heard also at this meeting. One such alternative was proposed in 1993[4]. Since then, it has been developed in 10 papers which have recently been synthesized and amplified into a book[5].

The approach of this theory is to construct sets of differential equations for hadrons. These sets are then taken as basic postulates and are solved for various applications, just like that we solve the Maxwell and the Schrödinger-Dirac equations for different purposes and get results. They stand or fall with their ability to account for data.

To begin with, it is observed that the conventional form of the Dirac equation involving 4×4 γ matrices is not Lorentz invariant by inspection. van der Warden[6] has transformed this equation into a set of manifestly Lorentz invariant two-spinor equations. Consider the meson case. Quark A at x_I and antiquark B at x_{II} interacting with each other via a massless scalar potential V_S are each described by a set of such two-spinor equations. For quark A, it reads

$$\partial_I^{ab} \chi_{A\dot{b}}(x_I) - iV_{SB}(x_I)\psi_A^a(x_I) = im_A \psi_A^a(x_I)$$
$$\partial_{I\dot{b}c} \psi_A^c(x_I) - iV_{SB}(x_I)\chi_{A\dot{b}}(x_I) = im_A \chi_{A\dot{b}}(x_I)$$
$$\Box_I V_{SB}(x_I) = \tfrac{1}{2} g_s^2 (\psi_B^c(x_I)\chi_{Bc}(x_I) + c.c.) \qquad (1)$$

Here, $a, b, c = 1, 2$ and χ_a(dot over a)$=(\chi_a)^*$, $g_s^2/4\pi$ is the strong quark-quark coupling, I refers to x_I, and m_A is the mass of quark A. The both sets are multiplied together and the basic assumption of the present theory is the generalization of the product wave functions into nonseparable quantities according to

$$\chi_{A\dot{b}}(x_I)\chi_{B\dot{a}}(x_{II}) \to \chi_{\dot{b}\dot{a}}(x_I, x_{II}),, \qquad \psi_A^c(x_I)\psi_B^{\dot{a}}(x_{II}) \to \psi^{c\dot{a}}(x_I, x_{II}) \qquad (2a)$$
$$\chi_{A\dot{b}}(x_I)\psi_{B\dot{e}}(x_{II}) \to \chi_{\dot{b}\dot{e}}(x_I, x_{II}), \qquad \psi_A^c(x_I)\chi_B^f(x_{II}) \to \psi^{cf}(x_I, x_{II}) \qquad (2b)$$
$$V_{SA}(x_{II})V_{SB}(x_I) \to \Phi_m(x_I, x_{II}) \qquad (2c)$$

where Φ_m is the interquark potential. The right sides of (2a) contain mixed spinors of second rank and are assigned to represent mesons. The right sides of (2b) contain unmixed spinors and transform like diquarks. In the so-gneralized equations, these last quantities and unpaired quarks are dropped to conform to *properties Q1* and *Q2*. The resulting meson wave equations read

$$\partial_I^{ab}\partial_{II}^{fe}\chi_{bf}(x_I,x_{II})-(M_m^2-\Phi_m(x_I,x_{II}))\psi^{ae}(x_I,x_{II})=0$$

$$\partial_{Ibc}\partial_{IIed}\psi^{c\dot{e}}(x_I,x_{II})-(M_m^2-\Phi_m(x_I,x_{II}))\chi_{b\dot{d}}(x_I,x_{II})=0$$

$$\Box_I\Box_{II}\Phi_m(x_I,x_{II})=-\tfrac{1}{2}g_s^4\,\mathrm{Re}\,\psi^{b\dot{a}}(x_I,x_{II})\chi_{\dot{a}b}^*(x_I,x_{II}) \tag{3}$$

M_m turns out to be an eigenvalue equal to the average quark mass[5, §2.3.5]. It is obtained after having generalized the quark mass in (1) into a mass operator, similar to that employed in the Gell Mann-Okubo formula.

We note that there is no quark wave function in (3) so that *property Q1* is present. The two indices of the meson wave functions χ and ψ on the right sides of (2a) and in (3) show that a quark and an antiquark appear together in agreement with *property Q2*. On the other hand, *properties F1* and *F2* are absent. There are eight meson wave function components, half the number 16 for the corresponding BS amplitudes. Equations (3) consist of two coupled second order equations containing an interaction potential term that depends nonlinearly upon the meson wave functions.

Equations (3) provide the starting point from which all two quark meson phenomena are to be accounted for, similar to that the Schrödinger-Dirac equations are the basis for understanding all atomic phenomena. So far, (3) has been rather successful when compared to QCD.

A set of baryon wave equations has been analogously constructed. In ground state, a baryon consists of a quark and a diquark and is described by 12 wave function components, far less than 64 in the BS case. However, contact with data has hitherto been hampered by difficulties in solving the differential equations, which are much more complicated than those for the mesons. The present theory has so far only been applied to low energy data. It has not been worked out for application to high energy hadronic phenomena.

REFERENCES

1. Hoh, F. C., "QCD versus Spinor Strong Interaction Theory" in *Hadron '95 The 6th International Conference on Hadron Spectroscopy*, edited by M. C. Birse et al., World Scientific, 1995, pp. 368-370 and "On the Foundation of QCD and an Overview of the Spinor Strong Interaction Theory" in *Hadron Spectroscopy 7th International Conference*, edited by S-U. Chung et al., AIP Conference Proceedings 432, New York, 1997, pp. 181-184.
2. Metsch, B., these proceedings.
3. Barnes, T., these proceedings.
4. Hoh, F. C., *Int. J. Theor. Phys.* **32** 1111-1133 (1993).
5. Hoh, F. C., *Two-Spinor Hadron Theory*, http://www4.tsl.uu.se/~hoh 2001.
6. van der Waerden, B., *Göttinger Nachrichten* 100-109 (1929).

Wavelet Analysis of E852 Experimental Data

V. L. Korotkikh* and L. I. Sarycheva*

*Scobeltsyn Institute of Nuclear Physics, Moscow State University, Moscow 119899, Russia

Abstract. A calculation of background in the $\eta\pi^0$ mass spectrum by the wavelet analysis is presented. The advantages and shortcomings of wavelet method are discussed.

INTRODUCTION

A wavelet analysis [1, 2] became a necessary mathematical tool in many investigations of the nonstationary (in time) or in homogeneous (in space) distributions (see [3] and many references in it). The wavelets can distinct the local characteristics of a signal by changing a scale (dilation) and by translation over the whole region in which it is studied. Due to the completeness of the wavelet functions, they also allow for the inverse transformation to be done. So, there is a possibility to separate a signal (as some local inhomogeneity) and a smooth background.

A large final state statistic of E852 experiment [4, 5, 6] allows to use the wavelet analysis both for one-dimensional and for two-dimensional distributions in the π^-p interactions at 18 GeV/c. For one-dimensional mass distribution it can help to estimate the background under a resonant peak of signal. It is important because the partial wave analysis (PWA) fits the sum of waves and background to the data. If we calculate incorrectly the mass dependence then it can be smeared over some waves and give false effects. It is specially important in a case of the rare exotic effects which are comparable with the background. We consider a mass dependence of $\eta\pi^0$ system [6] produced in the π^-p interaction as an example. One should stress that the wavelet analysis does not pretend to explain the underlying dynamics and physics nature.

BASIC FORMULAE

The discrete wavelets were used. The scaling (φ) and oscillation (ψ) functions are defined as

$$\varphi(x) = \sqrt{2} \sum_{k=0}^{2M-1} h_k \varphi(2x-k), \quad \psi(x) = \sqrt{2} \sum_{k=0}^{2M-1} g_k \varphi(2x-k), \quad (1)$$

where the coefficients h_k and g_k are related by equation $g_k = (-1)^k h_{2M-k-1}$.

The integer M defines the number of coefficients of finite support. The dilated and translated versions of scaling and oscillation function are

$$\varphi_{j,k}(x) = 2^{j/2}\varphi(2^j x - k), \quad \psi_{j,k}(x) = 2^{j/2}\psi(2^j x - k). \quad (2)$$

The integer j sets j-th resolution level and determines the size of cells over x-dependence. The index k is a translation parameter.

Any function $f(x)$ can be decomposed by the wavelet functions

$$f(x) = \sum_{k=0}^{2^{j_n}-1} s_{j,k}\varphi_{j,k}(x) + \sum_{j=j_n}^{j_{max}} \sum_{k=0}^{2^j-1} d_{j,k}\psi_{j,k}(x), \qquad (3)$$

where j_n and j_{max} are a minimum and maximum resolution level to be considered. The coefficients $s_{j,k}$ and $d_{j,k}$ carry information about the content of function $f(x)$ at various scales.

We use a fast wavelet algorithm for direct and reverse transformation [3]. The calculation of $s_{j,k}$ and $d_{j,k}$ is fulfilled by

$$s_{j'+1,k} = \sum_m h_m\, s_{j',2k+m}, \quad d_{j'+1,k} = \sum_m g_m\, s_{j',2k+m}, \qquad (4)$$

where $j' = (j_{max} - j)$ is a reverse resolution index. For start of calculation at $j' = 0$ we take

$$s_{0,k} = f(x_k), \quad k = 1, 2, \ldots, k_{max} \qquad (5)$$

where $k_{max} = 2^{j_{max}}$ at maximum resolution level (at minimum size of cell). All results below are obtained with the orthogonal D^8 Daubechies [1] wavelets ($M = 4$) and $j_{max} = 5$. The coefficient values h_k ($k = 0, 1, \ldots, 7$) are taken from the work [3].

WAVELET ANALYSIS OF BACKGROUND

Let's begin to study the background in a simple example:

$$f(x) = f_{BW}(x) + f_{BG}(x), \qquad (6)$$

where $f_{BW}(x)$ is a Breit-Wigner function (our signal) and $f_{BG}(x)$ is a polynomial of second order. We take the parameters of function which are close to real experimental data (see below). Then we do the wavelet analysis (5) and the reverse transformation (synthesis). The coefficients $d_{j,k}$ give a contribution of oscillations (fluctuations) at various scales. The synthesis with all $d_{j,k}$ gives exact values of function $f(x)$. Let's use the cut of $d_{j,k}$:

$$d'_{j,k} = d_{0,0}, \text{ if } j = 0,\ k = 0. \qquad (7)$$

After calculation of new set of $d'_{j,k}$ with this cut by the fast algorithm of synthesis we get a function $WL(x)$ with minimum fluctuations.

Then we define a background $BG(x)$, calculated by the wavelet analysis, as

$$BG(X) = a\, WL(x) + b, \qquad (8)$$

where a and b are free parameters which are found by fit sum ($f_{BW}(x) + BG(x)$) to the function (7). They are equal to $a = 0.30$, $b = 100$ in our case. The coincidence the calculated $BG(x)$ and the real background $f_{BG}(x)$ is in the limit of 20%.

Let's do the wavelet analysis of only function $f_{BW}(x)$ and calculate the $d_{0,0}^{(BW)}$. The value of parameter a is close to the ratio $d_{0,0}^{(BG)}/d_{0,0}$, which is equal to 0.24. So, the sense of parameter a is the part of wavelet oscillation contribution in the background at minimum resolution level.

Now consider the real experimental data. We have about 20000 events in the reaction $\pi^- p \to \eta \pi^0 n$, $\eta \to \pi^+ \pi^- \pi^0$ at 18 GeV/c (E852 experiment) [6]. Our aim is the estimation of background in the mass spectrum of $\pi^0 \pi^0 \pi^+ \pi^-$-meson system. One of the way is a method of "side bands". Two bands near a peak of η-meson at $m = 0.549$ GeV in mass spectrum of $\pi^0 \pi^+ \pi^-$-system are taken (fig.(1)). The sum of their widths is equal to the central band width of η meson. Then two distributions over mass of four mesons ($\pi^0 \pi^0 \pi^+ \pi^-$) are built (fig.(2)). One is the distribution with events, in which the mass of $\pi^+ \pi^- \pi^0$ mesons is in the η-mass region. Other distribution, called the background, corresponds to the events in which the mass $m(\pi^0 \pi^+ \pi^-)$ is in the side bands regions in fig.(1).

Let's estimate the background in mass spectrum of $(\pi^0 \pi^0 \pi^+ \pi^-)$ system by the wavelet analysis as in the example above. We use the cut (7) of $d_{j,k}$ and calculate $WL(x)$ for the experimental data in fig.(2). Then we fit the sum of background (8) and two Breit-Wigner functions to the data and find the parameters a and b ($a = 0.21$, $b = 135$).

The curve of background, calculated by wavelet analysis, is similar to the background from side band method, but it is more smooth (fig.(2)). The wavelet analysis of background allows to remove a peak near mass $m = 1.32$ GeV/c which can be false events (perhaps, from decay $a_2 \to \pi^0 \pi^+ \pi^-$) in the estimation of background by side bands method. But we have to emphasize that the wavelet analysis doesn't give an absolute value of background.

CONCLUSIONS

A simple one-dimensional experimental distribution was considered. There are no sharp peaks or many peaks with various widths (no strong fluctuations). The region of function argument is limit and short. It is possible to describe the background by the polynomial as in our artificial example (6). The analysis with discrete wavelets restores rather good the smooth background (fig.(2)). There is no necessity to have any additional information besides the distribution in itself.

So, we demonstrate that the wavelet analysis gives a good result. It will be certainly a value analysis in more complex cases with various peak widths and in the large argument region. We suppose that the wavelet analysis will be useful when the background cannot be describe by polynomial.

Let's do some remarks after our experience of wavelet application. It is interesting to continue the present study using other wavelets and more refined cuts of the coefficients $d_{j,k}$. There is a problem how the wavelet analysis can describe the interference effects between Breit-Wigner signal and background. It is useful also to do a combined analysis of the wavelet decomposition and the fit the resonance parameters of signal. There is also a problem of taking into account the experimental errors in the frame of wavelet analysis.

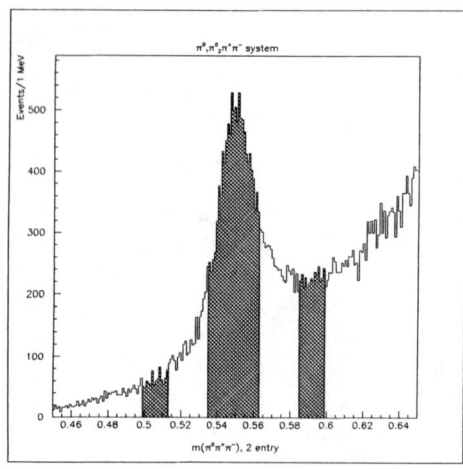
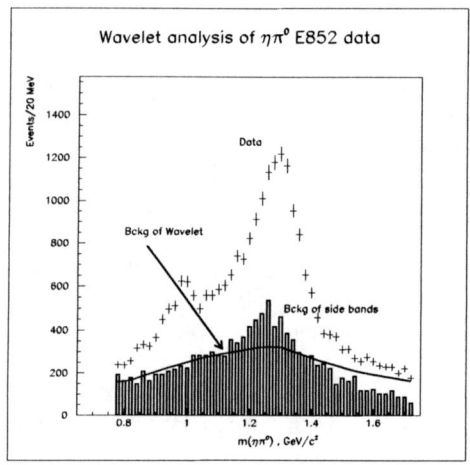

FIGURE 1. Left. Mass distribution of $\pi^0\pi^+\pi^-$-system in the reaction $\pi^- p \to \pi^+\pi^-\pi^0\pi^0 n$ at 18 GeV/c. Central band is a region of η-meson. Left and right bands are used for event selection which corresponds to the background in mass spectrum of $\eta\pi^0$-system, when η decays as $\eta \to \pi^0\pi^+\pi^-$. Mass $m(\pi^0\pi^+\pi^-)$ is given at GeV/c^2.

FIGURE 2. Right. Comparison of two background estimations. Line is the background from wavelet analysis. Histogram is the background from side bands method. Crosses are the $\eta\pi^0$ mass distribution of E852 experiment.

ACKNOWLEDGMENTS

Authors thank I. M. Dremin, O. V. Ivanov and V. A. Nechitailo for very helpful discussion of wavelet method.

REFERENCES

1. Daubechies, I., *Ten Lectures on Wavelets*, SIAM. Philadelphia, 1991.
2. Meyer, Y., *Wavelets: Algorithms and Applications*, SIAM, Philadelphia, 1993.
3. Dremin, I. M., Ivanov O. V., Nechitailo V. A. Usp. Fiz. Nauk, **171**, 465,(2001).
4. Teige, S. et al., Phys. Rev. **D59**, 012001 (1998).
5. Chung, S. U. et al., Phys. Rev. **D60**, 092001 (1999).
6. Korotkikh, V. L. and Sarycheva, L. I. Nucl. Phys. **A675**,413 (2000).

Quasirotational States in Various String Hadron Models and Regge Trajectories

German S. Sharov

Tver State University, Sadovyj per., 35, 170002, Tver, Russia

Abstract. The relativistic string with massive ends describing the meson and four various string baryon models (quark-diquark, linear, three-string and triangle) are considered. For all these systems the rotational motions or planar uniform rotations are used for describing orbitally excited hadron states on the Regge trajectories. We analyze small disturbances of these rotations (quasirotational states) with different methods. This study demonstrates that the classical rotational motions for the meson-like model and the triangle string baryon model are stable, but they are unstable for the linear configuration (q-q-q) and for the three-string (Y). For the latter two models the spectrum of the quasirotational oscillations contains exponentially growing modes. But for the string with massive ends we have two types of these modes describing stationary waves without growth. The obtained quasirotational states are useful for quantization and describing higher radial hadron excitations.

In the string meson model [1] (Fig. 1 a) the relativistic string simulates the strong interaction at large distances between the pair q-$\bar q$ and the QCD confinement mechanism. String models of the baryon are known in the represened in Fig. 1 (b) – (e) four variants with different topology [2, 3]: (b) the meson-like quark-diquark model q-qq, (c) the linear configuration q-q-q, (d) the "three-string" model or Y configuration, and (e) the "triangle" model or Δ configuration [4].

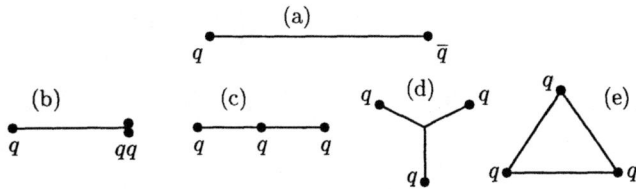

FIGURE 1. String models of the meson (a) and the baryon (b) – (e).

The final choice of the most adequate string baryon model among the four mentioned ones has not been done yet. All these models have a some degree of the QCD motivation [3] and all of them can work in the particle physics and under certain assumptions describe the (quasi)linear Regge trajectories for baryons [5, 6].

For modeling the orbitally excited hadron states on the leading Regge trajectories the rotational motions of all mentioned string models (planar uniform rotations of these systems) are widely used [5, 6]. This motion of the meson-like models q-$\bar q$ or q-qq is the well known rotation of the rectilinear string segment [1] (with the middle quark at the rotational center for the model q-q-q). The world surfaces $X^\mu(\tau,\sigma)$ of these rotational motions may be represented in the form [3, 5]

$$X^\mu(\tau,\sigma) = X^\mu_{rot}(\tau,\sigma) = \Omega^{-1}[\omega\tau e^\mu_0 + \cos(\omega\sigma + \phi_0) \cdot e^\mu(\tau)]. \tag{1}$$

Here Ω is the angular velocity, e^μ_0 is the unit time-like velocity vector of c.m. in Minkowski space with signature $+,-,-,\ldots$, $e^\mu(\tau) = e^\mu_1 \cos\omega\tau + e^\mu_2 \sin\omega\tau$ is the unit ($e^2 = -1$) space-like rotating vector directed along the string, $\sigma \in [0,\pi]$. The parameter ω is connected with the constant speeds v_1, v_2 of the ends: $v_1 = \cos\phi_0$, $v_2 = -\cos(\pi\omega + \phi_0)$, $m_i\gamma^{-1}\Omega = (1-v_i^2)/v_i$, γ is the string tension, m_i are the masses of the endpoints. For the rotational motion the three-string [3] the three rectilinear string segments (1) joined at the angles 120°.

The rotational motion for the baryon model "triangle" describes an uniformly rotating closed string (curvilinear triangle) composed of three segments of a hypocycloid. The corresponding world surface is [4]

$$X^0 = \tau - \frac{T}{D}\sigma, \qquad X^1 + iX^2 = u(\sigma) \cdot e^{i\omega\tau}. \tag{2}$$

Here $u(\sigma) = A_i \cos\omega\sigma + B_i \sin\omega\sigma$, $\sigma \in [\sigma_i, \sigma_{i+1}]$, the complex A_i, B_i and real constants σ_i, D, T are connected by the set of relations [4, 5].

Expressions (1) and (2) are exact solutions of the classic equations of motion and boundary conditions. Both the equations of motion and the boundary conditions result from the action, that has the similar form [3, 5] for all the models in Fig. 1. The equations of motion for all these systems may be linearized

$$\ddot{X}^\mu - X''^\mu = 0, \tag{3}$$

without loss of generality [1, 3] under the orthonormality conditions $(\dot{X} \pm X')^2 = 0$. But the boundary conditions for the massive point at an end

$$m_i \frac{d}{d\tau} U^\mu_i(\tau) \pm \gamma[X'^\mu + \sigma'_i(\tau)\dot{X}^\mu]\Big|_{\sigma=\sigma_i} = 0, \qquad U^\mu_i(\tau) = \frac{\dot{X}^\mu + \sigma'_i X'^\mu}{|\dot{X} + \sigma'_i X'|}\Big|_{\sigma=\sigma_i} \tag{4}$$

or in the middle point (for the models q-q-q or $\triangle$)

$$m_i \frac{d}{d\tau} U^\mu_i(\tau) - \gamma[X'^\mu + \sigma'_i(\tau)\dot{X}^\mu]\Big|_{\sigma=\sigma_i+0} + \gamma[X'^\mu + \sigma'_i(\tau)\dot{X}^\mu]\Big|_{\sigma=\sigma_i-0} = 0, \tag{5}$$

remain essentially nonlinear. Here $\dot{X}^\mu = \partial_\tau X^\mu$, $X'^\mu = \partial_\sigma X^\mu$, $\sigma = \sigma_i(\tau)$ is the i-th quark line on the world surface. The masses m_i make the models much more realistic but they bring additional nonlinearity and (hence) problems with quantization.

The energy E and angular momentum J of the states (1) and (2) are [4, 5]

$$E = E_{st} + \sum_{i=1}^{N} \frac{m_i}{\sqrt{1-v_i^2}} + \Delta E, \qquad J = \frac{1}{2\Omega}\left[E_{st} + \sum_{i=1}^{N} \frac{m_i v_i^2}{\sqrt{1-v_i^2}}\right] + S, \tag{6}$$

where $N = 2$ for the models q-$\bar{q}$, q-qq and $N = 3$ for others, $E_{st} = \gamma\Omega^{-1}\arcsin v_i$ for the motions (1) and $E_{st} = \gamma D(1 - T^2/D^2)$ for the triangle states (2). The quark spins with projections s_i ($S=\sum_{i=1}^{N} s_i$) are taken into account, in particular, as the spin-orbit correction $\Delta E = \Delta E_{SL} = \sum_i \beta(v_i)(\vec{\Omega}\cdot\vec{s}_i)$ to the energy of the classic motion. Here we use $\beta(v_i) = 1 - (1 - v_i^2)^{1/2}$ for this correction [5, 6]. The expression (6) for all string hadron models describes quasilinear Regge trajectories with the similar ultrarelativistic behavior: $J \simeq \alpha' E^2 - \alpha_1 E^{1/2} + \sum_{i=1}^{N} s_i[1 - \beta(v_i)]$, $E \to \infty$. Here the slopes are different: $\alpha' = (2\pi\gamma)^{-1}$ for the meson-like models, $\alpha' = \frac{2}{3}(2\pi\gamma)^{-1}$ for the Y and $\alpha' = \frac{3}{8}(2\pi\gamma)^{-1}$ for the so called simple states (2) of the triangle configuration [4, 5, 6].

All mentioned string baryon models can describe the leading Regge trajectories if we suppose that $\gamma = \gamma_{q-qq} = 0.175$ GeV^2, the effective tension for the Y and "triangle" is to be different $\gamma_Y = \frac{2}{3}\gamma$, $\gamma_\Delta = \frac{3}{8}\gamma$, the effective quark masses $m_u = m_d = 130$ MeV, $m_s = 300$ MeV, $m_c = 1500$ MeV and the spin-orbit correction has the mentioned form. Under these assumptions the orbital excitations were described on the following Regge trajectories: $N(938)$ (with $J^P = 1/2^+, 5/2^+, 9/2^+, \ldots$), $N(1520)$ ($J^P = 3/2^-, 7/2^-, \ldots$), $N(1675)$ ($J^P = 5/2^-, 7/2^+, 9/2^-$), $\Delta(1232)$ ($J^P = 3/2^+, 7/2^+, \ldots$), $\Delta(1930)$ ($J^P = 5/2^+, 9/2^+, \ldots$), $\Delta(1700)$ ($J^P = 3/2^-, 5/2^+, \ldots$); in the strange sector Λ ($J^P = 1/2^+, 3/2^-, 5/2^+, \ldots$), $\Lambda(1405)$ ($J^P = 1/2^-, 3/2^+$), $\Sigma(1193)$ ($J^P = 1/2^+, 3/2^-, 5/2^+, \ldots$), $\Xi(1315)$ ($J^P = 1/2^+, 3/2^-, 5/2^+$), and the charmed trajectory with $\Lambda_c^+(2625)$ [5, 6]. Under the same assumptions in the framework of the string meson model shown in Fig. 1 (a) the Regge trajectories for the light unflavored ($\rho - a$), strange K^*, and charmed mesons D, D^* are also well described. Thus, the comparison of the meson and baryon sectors confirms the adequacy of the considered string models.

Let us study the quasirotational states or small disturbances of the rotational motions (1) and (2) of all these systems. They are interesting due to the following reasons: (a) we are to describe not only orbital, but also higher radial excitations of hadrons; (b) the quasirotational states may help to quantize these nonlinear systems in the linear vicinity of the solutions (1), (2) (if they are stable); (c) they are necessary for testing the stability of rotational motions for all models.

For string with massive ends the quasirotational states were studied in Refs. [3, 7] in the framework of the orthonormality conditions. This let us use the general solution of Eq. (3) $X^\mu(\tau,\sigma) = \frac{1}{2}[\Psi_+^\mu(\tau+\sigma) + \Psi_-^\mu(\tau-\sigma)]$ and reduce the problem (or Eqs. (4)) to the system of nonlinear ordinary differential equations with shifted arguments. Considering small disturbances of the rotation X_{rot}^μ (1) we linearize this system in the vicinity of X_{rot}^μ, solve this linearized system and present arbitrary quasirotational motions for the string with massive ends in the form [7]

$$X^\mu(\tau,\sigma) = X_{rot}^\mu(\tau,\sigma) + \sum_{n=-\infty}^{\infty} \left\{ e_3^\mu \alpha_n \cos(\omega_n\sigma + \phi_n)\exp(-i\omega_n\tau) \right. \\ \left. + \beta_n[e_0^\mu f_0(\sigma) + e_\perp^\mu(\tau)f_\perp(\sigma) + ie^\mu(\tau)f_r(\sigma)]\exp(-i\tilde{\omega}_n\tau) \right\}. \quad (7)$$

Here $e^\mu_\perp(\tau) = \omega^{-1}\frac{d}{d\tau}e^\mu(\tau)$; e_0, e_1, e_2, e_3 is the orthonormal tetrad, $\omega = \omega_1$. Each term in Eq. (7) describes the string oscillation that looks like the stationary wave with n nodes. The two types of these stationary waves (orthogonal and planar) have the frequencies proportional to the roots ω_n and $\tilde\omega_n$ of the equations

$$\frac{\omega^2 - Q_1 Q_2}{(Q_1+Q_2)\omega} = \cot \pi\omega, \qquad \frac{(\tilde\omega^2 - q_1)(\tilde\omega^2 - q_2) - 4Q_1 Q_2 \tilde\omega^2}{2\tilde\omega[Q_1(\tilde\omega^2 - q_2) + Q_2(\tilde\omega^2 - q_1)]} = \cot \pi\tilde\omega, \qquad (8)$$

where $Q_i = \omega_1 v_i/\sqrt{1-v_i^2}$, $q_i = Q_i^2(1+v_i^{-2})$. The roots of Eqs. (8) are real numbers so the rotations (1) of the string with massive ends are stable in the linear approximation.

But the similar quasirotational motions for the string baryon models q-q-q and Y are more complicated, they contain many branches of oscillations and for both models some frequencies in them are complex numbers. For example, for the model three-string these complex frequencies are the roots of the equation

$$2\frac{Q_1\tilde\omega(\omega_1^2 - \tilde\omega^2) - i(\tilde\omega^2 - q_1)(\tilde\omega^2 + \omega_1^2)}{(\tilde\omega^2 - q_1)(\tilde\omega^2 - \omega_1^2) - 4iQ_1\tilde\omega(\tilde\omega^2 + \omega_1^2)} = \cot \pi\tilde\omega, \qquad (9)$$

Imaginary parts of the roots of Eq. (9) are always positive so the disturbances of this class (branch) exponentially grow in time in accordance with the factor $\exp(-i\tilde\omega_n\tau) = \exp(-i\Re\tilde\omega_n\tau)\exp(\Im\tilde\omega_n\tau)$. Arbitrary quasirotational motion may be expanded in the Fourier series similar to Eq. (7) so arbitrary (asymmetric) disturbances will grow exponentially. This means that the rotations of the three-string configuration are unstable even in the linear approximation. The similar picture may be seen for the linear model q-q-q.

But for the string baryon model Δ the classic rotational motions (2) appeared to be stable [3]. This gives some advantages for the string models q-qq and Δ. However the final choice of the most adequate string baryon model among the four existing ones depends not only on the the stability of their classic rotational states and requires more profound investigations of their behavior.

ACKNOWLEDGMENTS

The work is supported by the Russian Foundation of Basic Research (grant 00-02-17359).

REFERENCES

[1] Barbashov B. M., and Nesterenko V. V., *Introduction to the Relativistic String Theory*, Publisher, Singapore, WS, 1990.
[2] Artru X., *Nucl. Phys.* **B85**, 442-460 (1975).
[3] Sharov G. S., *Phys. Rev.* **D62**, 094015 (2000), hep-ph/0004003.
[4] Sharov G. S., *Phys. Rev.* **D58**, 114009 (1998), hep-th/9808099.
[5] Sharov G. S., *Phys. Atom. Nucl.* **62**, 1705 (1999), hep-ph/9809465.
[6] Inopin A., and Sharov G. S., *Phys. Rev.* **D63**, 054023 (2001).
[7] Sharov G. S., hep-ph/0012334.

A Light Meson Translatable Template

C. E. Allgower

IUCF, Indiana University, Bloomington, IN 47405

D. C. Peaslee*

Physics Department, University of Maryland, College Park, MD 20742

Abstract

Recently surveyed (mass)^2 values for I = 0, $J^{PC} = 2^{++}$ light mesons can be assembled into repeating patterns of 4 states, dubbed "templates". Within error, both internal and external template spacings approximate simple multiples of $\Delta m^2 \approx 0.35$ GeV^2. Hopefully, this feature will be useful in predicting the positions of higher isoscalar 2^{++} states.

A recent survey [1] of light mesons suggests $A = dm^2/dn \approx 1.04 \pm .01\ GeV^2$ as a universal slope for the radial trajectories of states with fixed $q\bar{q}$ content and constant quantum numbers IJLPC: confirming the basic notion of the Veneziano model [2], where n=1,2,3... is a sequential label in order of increasing mass for the selected configuration. Examples of such trajectories are shown in Fig. 1, where each dot represents an observed resonance (with error bars where they exceed the dot size). Here the notation in terms of IJLPC is P_0, $a_0 = (0,1)01++$ and P_2, $a_2 = (0,1)21++$. Apparent predictive power in the lower right hand panel indicates a missing a_0 meson at n = 3, m = 1762 ± 19 MeV. For all I=0 states complete separation of $s\bar{s}$ and non-strange mesons is assumed (slight deviations from the ideal mixing angle of 35.3° perturb mass eigenvalues only to second order).

In this more speculative sequel we pursue some implications of assuming that A is totally invariant. Ideally, this would imply that any pattern of m^2 for different mesonic states is directly translatable by units of A to clones at successively higher "radial excitation" numbers n. Any translatable pattern so established can be called a <u>template</u>.

The pursuit of templates would test the universal slope hypothesis more sensitively than in a single survey. What is required are sufficiently many identified states with the same quark content and external quantum numbers.

PACS: 14.40. Keywords: Light mesons, Radial trajectories, Universal slope.

*Speaker

The best documented template consists of 4 states with I=0 and a fixed specification: namely, JPC = 2++. We introduce $F_2 = 023++$ for L=3 and specify the corresponding $s\bar{s}$ mesons by $F_2(s\bar{s})$ and likewise define $P_2(s\bar{s})$. The upper half of Fig. 2 shows this template and its clones as vertical columns corresponding to the radial quantum number n.

Although this template picture is the most direct, it is not unique. The assumption of constant slope A assumes that <u>any</u> selection of four states – comprising one each from the 4 trajectories above – can act as a template.

As an example, define a template as

$$\{n\} = \{nF_2, (n+1)P_2(s\bar{s}), (n+2)P_2, nF_2(s\bar{s})\}$$

Here n, $(n+1)$, $(n+2)$ are radial quantum numbers for states in the nth template. Candidate templates $n = 1,2$ are complete; partial ones only in the case of $n = 3,4$. Comparison of n arrays among different n provides insight into the internal relations among the trajectories. These are shown in the lower half of Fig. 2 (over simplified: equal-spaced pairs are valid only to $\sim 8\%$, which is within current experimental error).

Since universal A reveals no distinction among various trajectories, internal structures like these will supply our first insight into configurational differences among IJLPC states. Because the data are used near the limits of their uncertainty, the results are suggestive rather than conclusive. Nevertheless, within rather generous errors the template structure displays a remarkable simplicity which may be worth noting.

Reference

1. C. E. Allgower and D. C. Peaslee, Phys. Lett. B **513**, 273 (2001).

2. G. Veneziano, Nuovo Cim. **57A**, 190 (1968).

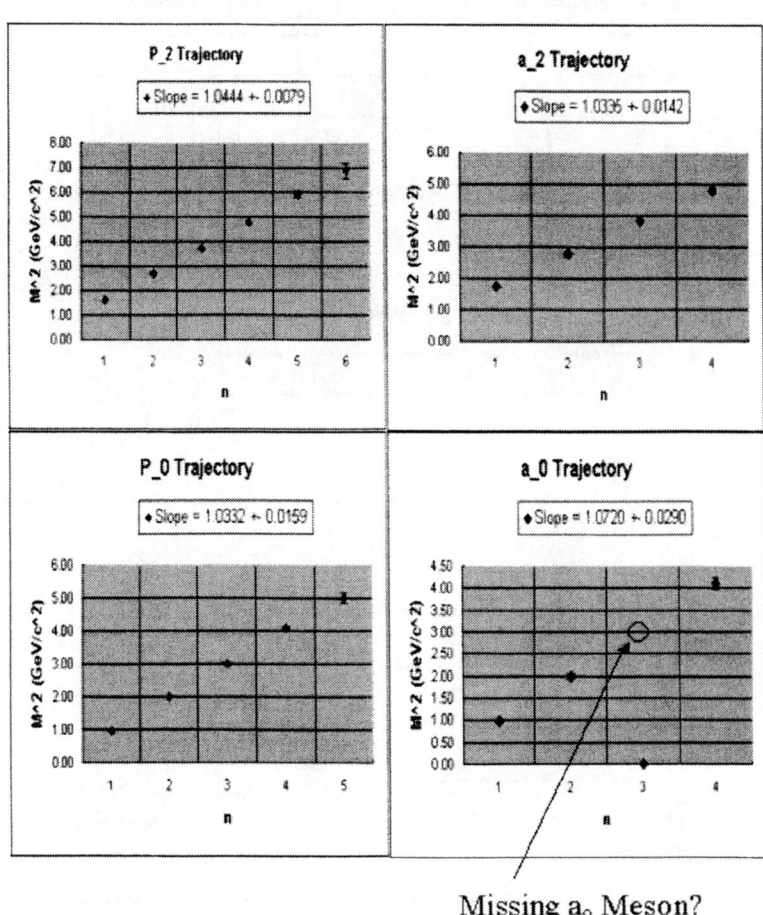

Figure 1: Some Radial Trajectories.

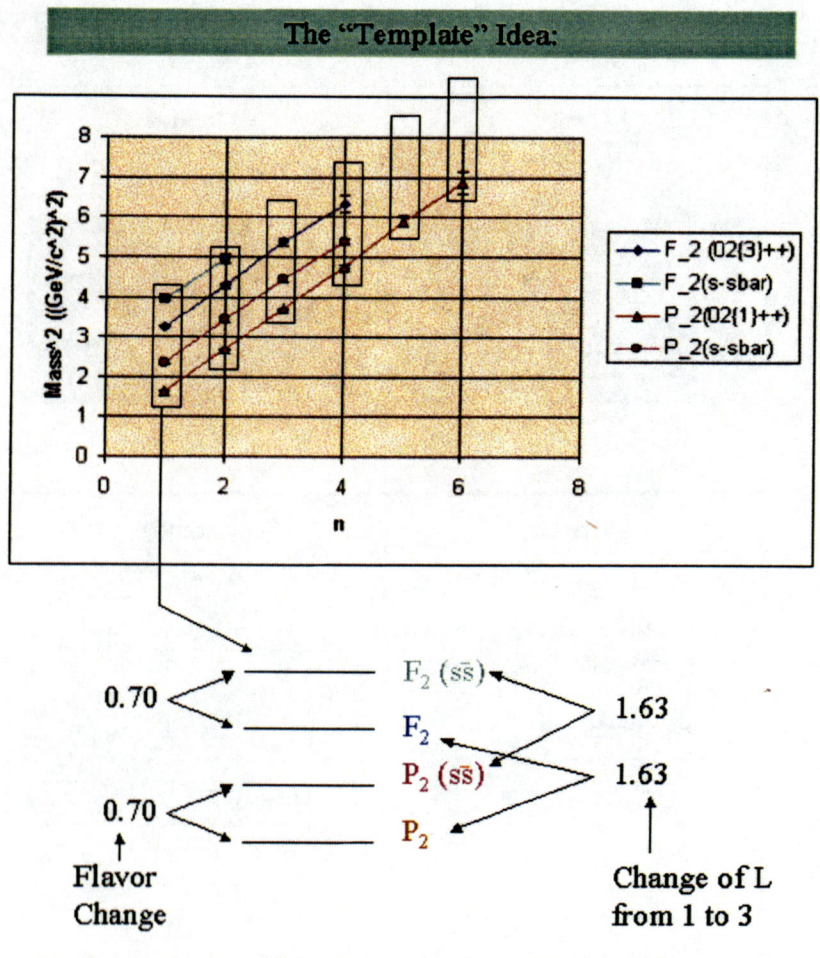

Figure 2: Illustration of "Templates."

STATES WITH OPEN/HIDDEN STRANGENESS

Some Features of $\Lambda\bar{\Lambda}$ system in $\pi^-(p,C) \to \Lambda\bar{\Lambda} + A^*$ at 40 GeV

V. V. Sokolovsky, SERP-E173 experiment[1]

ITEP, Moscow, Russia

Abstract. The results of stydying the $\Lambda\bar{\Lambda}$ system produced in the reaction $\pi^- p \to \Lambda\bar{\Lambda} n$ at a π^--meson energy of 40 GeV are reported. Experimental data ($\sim$ 2300 events) were obtained on the ITEP 6-meter spectrometer with a beam of the IHEP U-70 accelerator. The invariant-mass spectra for the events dominated by the singlet or triplet $\Lambda\bar{\Lambda}$ states were found to differ considerably from each other. The data give evidence for the existance of bumps of the $\Lambda\bar{\Lambda}$ system in the mass regions near 2.3, 2.5 and 2.8 GeV.

The $\Lambda\bar{\Lambda}$ system is also of interest because it may contain states with hidden strangeness, e.g., $s\bar{s}$ states of the ϕ- and f'_2-meson type. A number of broad resonances were found in the $\bar{p}p$ system [1] and in the $\Lambda\bar{p}$ ($\bar{\Lambda}p$) systems with strangeness [2, 3].

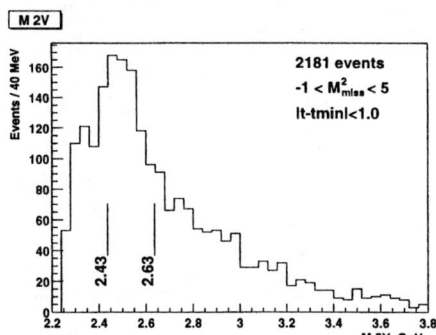

FIGURE 1. $\Lambda\bar{\Lambda}$ effective mass distribution for events of reaction (1).

Relatively high mass of the $\Lambda\bar{\Lambda}$ system and the presence of strange quarks are the main reasons for experimental study of meson decay modes into this system [1, 2, 3]. Some evidence of $\Lambda\bar{\Lambda}$ resonance states was reported in our previous papers [4, 5, 6].

The $\Lambda\bar{\Lambda}$ system consisting of a fermion and an antifermion has spin S = 0 or 1; total angular momentum **J** = **S** + **L**, where **L** is the orbital angular momentum; isospin I = 0; and negative intrinsic parity. The spatial parity is determined as $P = -(-1)^L = (-1)^{L+1}$, the charge parity as $C = (-1)^{L+S}$, and the combined PC parity depends on the spin S of the $\Lambda\bar{\Lambda}$ system. Like the ordinary ($q\bar{q}$) states, the singlet (S = 0) and triplet (S = 1) states have, respectively, opposite and identical signs of the P and C parities.

In this work, the production of pairs of Λ and $\bar{\Lambda}$ hyperons was analyzed on statistics of 2308 events detected in five exposures at 40 GeV measured at 6-meter spectrometer ITEP [7] in experiment SERP-E-173 in reactions

$$\pi^- p \to \Lambda\bar{\Lambda} n, \quad \Lambda \to p\pi^-, \quad \bar{\Lambda} \to \bar{p}\pi^+, \tag{1}$$

---

[1] I.A. Erofeev, O.N. Erofeeva, V.K. Grigoriev, Y.V. Katinov, V.I. Lisin, V.N. Luzin, V.N. Nozdrachev, Y.P. Shkurenko, V.V. Sokolovsky, G.D. Tikhomirov, V.V. Vladimirsky. ITEP, Moscow.

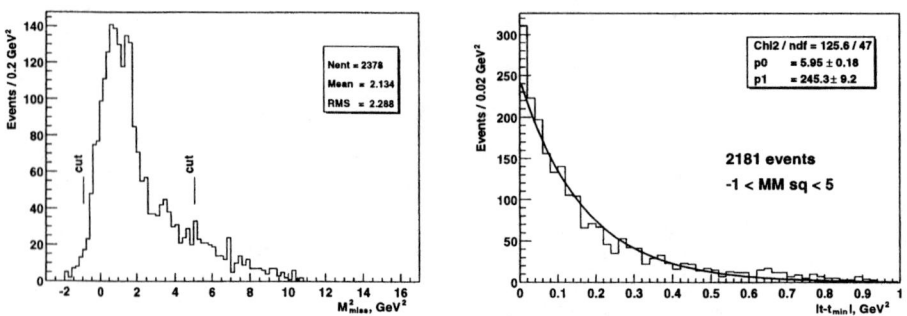

FIGURE 2. Missing mass squared (left) and transverce momentum (right) distribution for reaction (1). Missing mass range used is shown by cut positions.

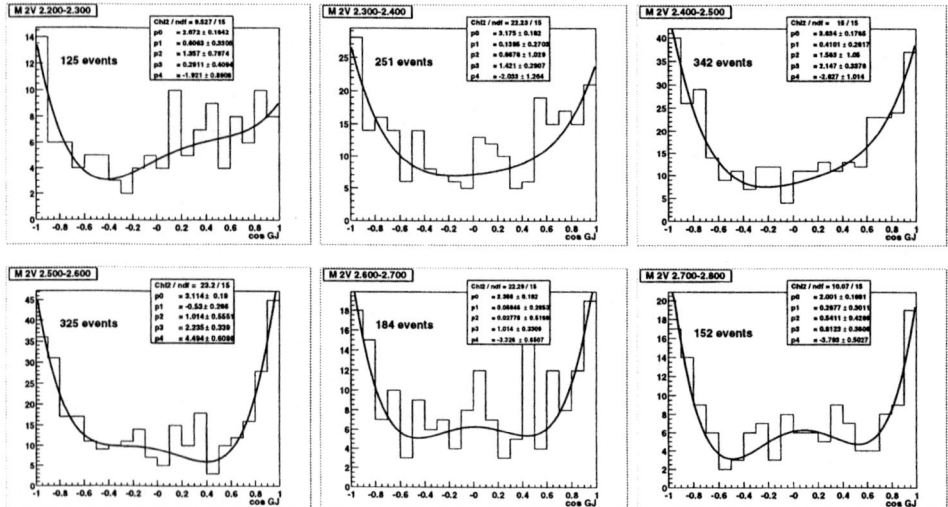

FIGURE 3. Distribution of $\Lambda\bar{\Lambda}$ events over $\cos\theta_{GJ}$ for different ranges of $\Lambda\bar{\Lambda}$ effective mass.

$$\pi^- C \to \Lambda \bar{\Lambda} A^*, \qquad (2)$$

Dozens of resonances are expected in the region 2.23–3.0 GeV/c^2. Our 2300 events of reaction (1) which are considerable part ($\sim$ 70 ev/nb) of fixed target world statistics, is still not enough for PWA.

For some events, K_S and Λ were identified ambiguously, resulting in a background generated by pairs of K_S-mesons. To suppress this background, the invariant masses of both forks were calculated on the assumption that they decay into π-mesons. The events for which both masses fell within the K_S-meson mass range (497±12 MeV) were rejected. As a result, 10±2 % of the events were lost. Nevertheless, the final sample of

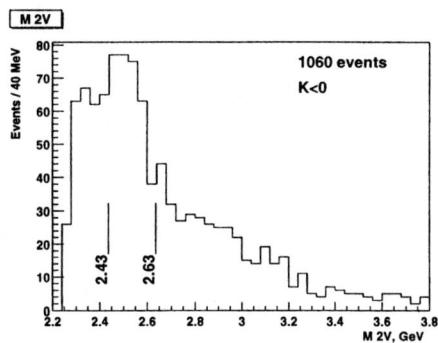 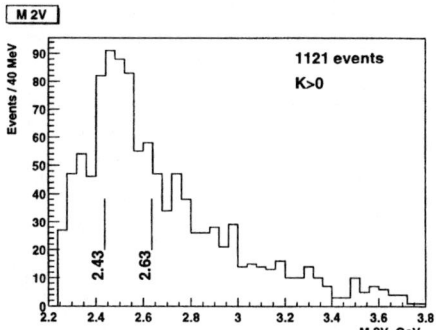

FIGURE 4. $\Lambda\bar{\Lambda}$ effective mass distributions for mainly triplet (left, $C > 0$) and singlet (right, $C < 0$) states.

TABLE 1. Polarization parameters of the $\Lambda\bar{\Lambda}$ system.

	$M_{\Lambda\bar{\Lambda}}$ range	C	S
1	2.28÷2.38	-1.23 ± 0.65	0
2	2.43÷2.55	0.98 ± 0.59	0,1
3	2.75÷2.85	2.02 ± 0.84	1

reaction (1) contained a 12±3% background. This background was estimated from the number of $2V^0$ events fitted to the $\Lambda\Lambda$ and $\bar{\Lambda}\bar{\Lambda}$ pairs forbidden by the conservation laws in the experimental topology.

High precision of coordinate detector of the spectrometer makes it possible [8] to identify $K_S, \Lambda, \bar{\Lambda}$ and γ by kinematical properties of their decays into charged particles without identification of charged particle type. Geometrical and track selection cuts were described in our previous papers [9, 10].

There is a sharp near-threshold rise which can be assigned to the resonances in this region and to the kinematic effect. A maximum at 2.5 GeV and a shoulder near 2.8 GeV may also be manifestations of the resonance structures. In our previous work [6], evidence was obtained for the existence of either resonances or sets of close resonances in these mass regions.

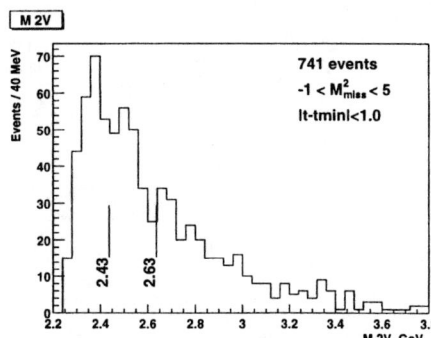

FIGURE 5. $\Lambda\bar{\Lambda}$ effective mass distribution for events of reaction (2).

Weak decays $\Lambda \to p\pi^-$ and $\bar{\Lambda} \to \bar{p}\pi^+$ give information on polarization and spin of the $\Lambda\bar{\Lambda}$ system. Parameter $C = -\frac{9}{\alpha^2}\langle\cos\theta_{p\bar{p}}\rangle$ equals $C = 1$ for triplet state of $\Lambda\bar{\Lambda}$, $C = -3$ for singlet. In this definition α is the polarization parameter $\alpha = \pm 0.647$, the angle $\theta_{p\bar{p}}$ is the angle between the directions of p and $\bar{p}$ momenta.

TABLE 2. Resonance parameters.

Mass (MeV)	Width (MeV)	Cut	A_p(norm.)	σ^* (nb)
2320 ± 20	65 ± 20	1	42 ± 15	1.8 ± 0.4
		2	21 ± 8	
2500 ± 10	205 ± 50	1	71 ± 12	2.9 ± 0.5
		2	41 ± 10	
2805 ± 15	90 ± 25	1	31 ± 10	1.0 ± 0.4
		2	6 ± 4	

The $\Lambda\bar{\Lambda}$ effective mass M distribution has three bumps near 2.3, 2.5 and 2.8 GeV/c^2. Figure 4 shows the same mass distribution, but for $C < 0$ and $C > 0$ samples separately.

CONCLUSION

Like $(q\bar{q})$ states the $\Lambda\bar{\Lambda}$ system has the following quantum numbers: $P=(-1)^{l+1}$, $C=(-1)^{l+s}$, but isospin of the $\Lambda\bar{\Lambda}$ I=0. Bumps reported here can be considered as ordinary $(q\bar{q})$ meson states or exotic one with $(q\bar{q})$ admixture. We report here the preferable spin of $\Lambda\bar{\Lambda}$ only, but other quantum numbers remain unknown.

- Bumps near 2.3, 2.5 and 2.8 GeV/c^2 $\Lambda\bar{\Lambda}$ mass are clearly seen in the sample of 2300 events. Isospin I=0.
- State at 2.3 GeV/c^2 is mainly singlet. $J^{PC} = 0^{-+}, 1^{+-}, ...$
- State at 2.8 GeV/c^2 is mainly triplet.

REFERENCES

1. Rosanska M. et al., *Nucl. Phys.* **B162**, 505 (1980).
2. Baubillier M. et al., *Nucl. Phys.* **B183**, 1 (1981).
3. Armstrong T. et al., *Nucl. Phys.* **B227**, 365 (1983).
4. Barkov B.P. et al., *Yad. Phys.* **22**, 223 (1975); *Sov. J. Nucl. Phys.* **22**, 113 (1975).
5. Bolonkin B.V. et al., *ITEP preprint* ITEP-86, Moscow, 1973.
6. Baloshin O.N. et al., *ITEP preprint* ITEP-2, Moscow, 1982.
7. Baloshin O.N. et al., *ITEP preprint* ITEP-154, Moscow, 1981.
8. Nozdrachev V.N., "The resonance structures of $K_S K_S$ and $\Lambda\bar{\Lambda}$ spectrum at MIS ITEP", this conference.
9. Bolonkin B.V. et al., *Nucl. Phys.* **B309** 426 (1988).
10. Bolonkin B.V. et al., *Yad. Phys.* **58** 1628 (1995).

Observation of narrow states of $K_S^0 K_S^0$ system on 6–m spectrometer and comparison with L3 results

V. I. Lisin, SERP-E173 experiment[1]

ITEP, Moscow, Russia

Abstract. The analysis of the 1200–1300 MeV effective mass region of the $K_S^0 K_S^0$-meson system was carried out for 40000 events obtained with the neutral trigger with 40 cm H_2 target. Irregularity near 1240 MeV and width about 30 MeV is clearly observed. Similar irregularity is observed on another sample of 15000 events on carbon target. The bump in $K_S^0 K_S^0$ mass distribution near 2250 MeV and width 56 MeV is observed under cut $\cos\theta_{GJ} < 0.5$.

INTRODUCTION

We present here recent results on reactions

$$\pi^- p \to K_S^0 K_S^0 n, \tag{1}$$

$$\pi^- C \to K_S^0 K_S^0 A^*, \tag{2}$$

at incident π^- momentum 40 GeV/c on MIS ITEP spectrometer at the 70 GeV IHEP accelerator. The target was surrounded by the system of lead scintillator veto counters. Secondary charged tracks from $K_S^0 \to \pi^+\pi^-$ decays were detected by the system of 62 planes of two-coordinate electrodynamic chambers placed in large volume of magnetic field. K_S^0 mesons were identified by the secondary vertex reconstruction. The co-ordinates of the reconstructed primary vertex were checked to be compatible with the target position. More details on experimental setup and precision are given in [1]. Fig. 1 shows $K_S^0 K_S^0$ effective mass spectrum from the large sample of 40000 $K_S^0 K_S^0$ events (19275 after cuts on missing mass) obtained with neutral trigger on liquid hydrogen target in the reaction (1) starting from the run described in [2] up to our recent runs. Cuts on missing mass (Fig. 2) of the reaction select

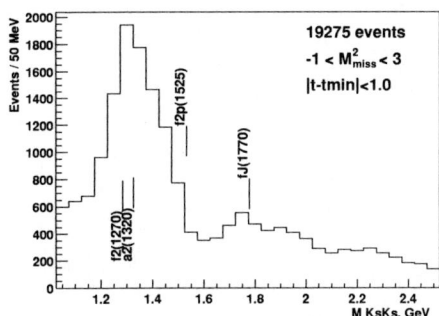

FIGURE 1. $K_S^0 K_S^0$ effective mass distribution for events of reaction (1).

---

[1] I.A. Erofeev, O.N. Erofeeva, V.K. Grigoriev, Y.V. Katinov, V.I. Lisin, V.N. Luzin, V.N. Nozdrachev, Y.P. Shkurenko, V.V. Sokolovsky, G.D. Tikhomirov, V.V. Vladimirsky. ITEP, Moscow.

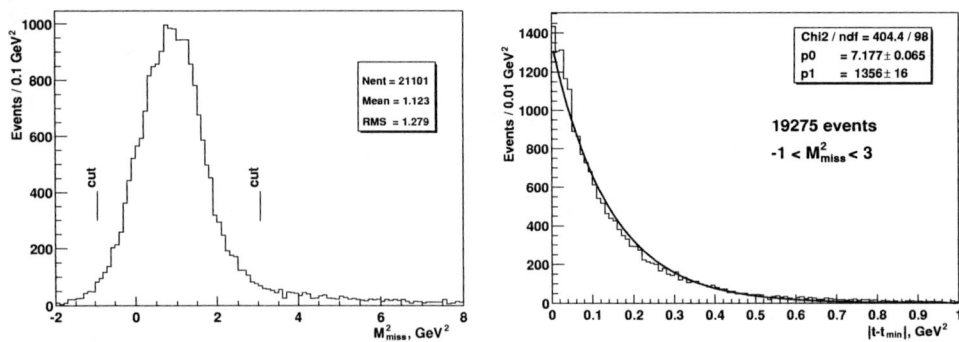

FIGURE 2. Missing mass squared (left) and transverse momentum (right) distributions for reaction (1). Short straight lines indicate cut positions.

events of reaction (1) with small background. Fit of transverse momentum distribution by exponent with two parameters (Fig. 2) gives slope parameter $p_0 = 7.2$ GeV^{-2} favoring OPE exchange domination in this reaction. Threshold behavior of $K_S^0 K_S^0$ distribution may be explained by roughly equal contributions of scalars $f_0(a_0)$. D wave is dominated by the formation of tensor meson $f_2(1270)$ with some contribution of $a_2(1320)$, which has already been investigated in our previous works [3, 4, 5]. Here we analyze the 1200–1300 MeV region of $K_S^0 K_S^0$ effective mass spectrum.

Since the system of two neutral kaons may only be observed in even $J^{PC} = 0^{++}, 2^{++} \ldots$ and higher moments are consistent with zero in this mass region, we used the waves L_m^ε, with $L = S, D$, $m = 0, 1$ and reflectivity $\varepsilon = \pm 1$ (waves with different reflectivity do not interfere):

$$S_0^- = Y_0^0 = 1/\sqrt{4\pi} \tag{3}$$

$$D_0^- = Y_2^0 = \sqrt{5/16\pi}\,(3\cos^2(\theta) - 1) \tag{4}$$

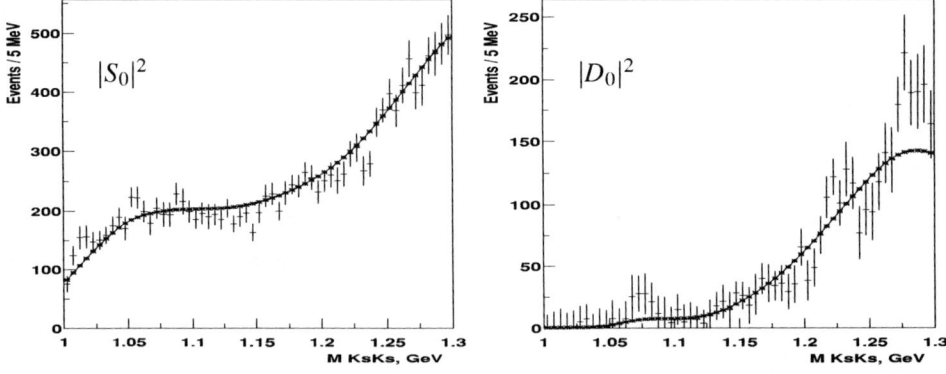

FIGURE 3. $|S_0|^2$ and $|D_0|^2$ as a function of $K_S^0 K_S^0$ effective mass near 1240 MeV with step 5 MeV. Solid lines show results of 30 MeV smoothing.

$$D_1^- = (Y_2^1 - Y_2^{-1})/\sqrt{2} = -\sqrt{15/16\pi} \, \sin(2\theta) \, \cos(\phi) \quad (5)$$

$$D_1^+ = (Y_2^1 + Y_2^{-1})/\sqrt{2} = -i\sqrt{15/16\pi} \, \sin(2\theta) \, \sin(\phi) \quad (6)$$

While D_0 wave is rather small near the threshold, it is rapidly increasing at the left shoulders of $f_2(1270)$ and $a_2^0(1320)$ mesons. Clear interference type signal in D_0 wave near 1240 MeV is shown in Fig. 3. Effective mass resolution for $K_S^0 K_S^0$-system in this region is better than 5 MeV. An interesting interference phenomenon in the region 1200–1300 MeV was also observed in two photon collisions studied with L3 detector at LEP ([6], Fig. 4[2]) which is treated by authors in the text of paper [6] as $f_2(1270)$ and $a_2^0(1320)$ destructive interference, but the mass distribution near 1240 MeV is fitted by the authors [6] by Breit-Wigner function with parameters: mass = 1239±6 MeV, width = 78±19 MeV. On the other hand, our data show rather the interference effect with comparable width about 30 MeV in the D_0 wave where $f_2(1270)$ and $a_2^0(1320)$ dominate. The difference in width may be due to the difference in experimental resolution: 29 MeV in the case of L3 and better than 5 MeV in our case.

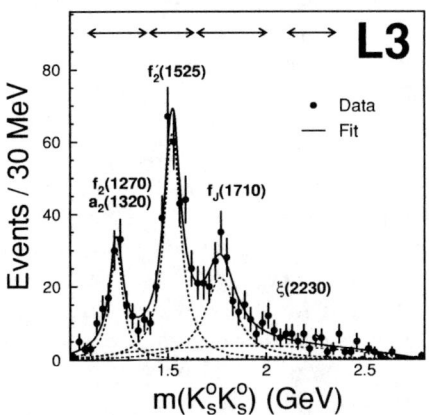

FIGURE 4[2]. The $K_S^0 K_S^0$ mass spectrum: the solid line corresponds to the maximum likelihood fit. The background is fitted by a second order polynomial and the three peaks by Breit-Wigner functions (dashed lines). The arrows correspond to the $f_2(1270)$–$a_2(1320)$, the $f_2'(1525)$, the $f_J(1710)$ and the $\xi(2230)$ mass regions. (The figure and the caption is reprinted from [6]).

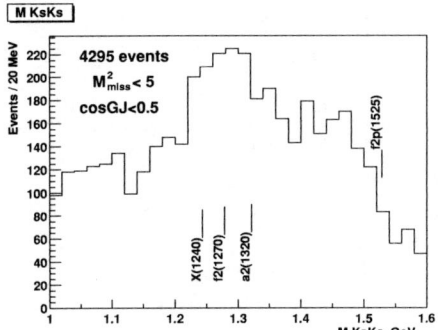 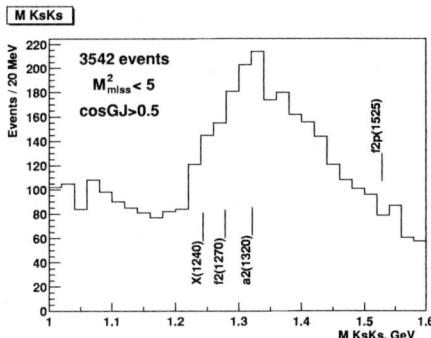

FIGURE 5. $K_S^0 K_S^0$ effective mass distribution for events of reaction (2) with different selections on Gottfried-Jackson angle $\cos\theta_{GJ}$.

---

[2] Fig. 4 is reprinted from *Phys. Lett.* **B501**, L3 Collab., Acciarri M. et al., "$K_S^0 K_S^0$ Final State in Two-Photon Collisions and Implications for Glueballs", Page 178, Fig. 3, Copyright 2001, with permission from Elsevier Science.

Our recent results on reaction (2) show the same irregularity near 1240 MeV on the left shoulder of $f_2(1270)$, $a_2^0(1320)$ which is shown on Fig. 5. This effect is seen even in the mass distribution of $K_S^0 K_S^0$ system. Unfortunately available statistics is not still sufficient for more detailed PWA analysis.

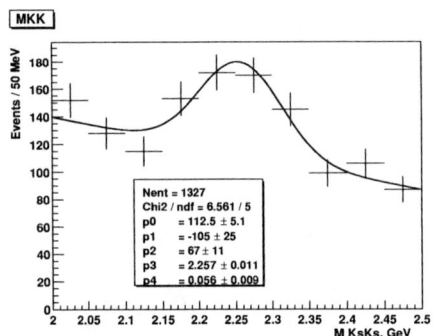

FIGURE 6. $K_S^0 K_S^0$ effective mass distribution for events of reaction (1) near 2250 MeV with selection $\cos\theta_{GJ} < 0.5$.

In hadron collision in our data there is clear enhancement near 2250 Mev (Fig. 6). Gauss fit with three parameters over linear background with two parameters gives mass 2257 ± 11 MeV and relatively small width 56 ± 9 MeV. Our mass resolution in this region is better than 15 MeV. There is no simple way to explain such small width of possible resonance in this region in the terms of quark-antiquark nature of this resonance. Even at mass of 1.3 GeV width of $f_2(1275)$ is 185 MeV, and width should increase with increasing of resonance mass. In $K_S^0 K_S^0$ system produced in two photon collisions ([6], Fig.4) there is no significant signal near 2230 MeV. In centrally produced $K_S^0 K_S^0$ system at 800 GeV/c [7] for $-0.22 < x_F < -0.02$ the $K_S^0 K_S^0$ invariant mass beyond 2 GeV is also smooth. On the other hand, the signal of $\xi(2230)$ is observed [8] in J/ψ radiative decays with mass 2232 ± 15 MeV and width 20 ± 14 MeV.

CONCLUSION

The narrow interference phenomenon in the D_0-wave near $K_S^0 K_S^0$ mass 1240 MeV exists in the reaction $\pi^- p \to K_S^0 K_S^0 n$. The same irregularity is observed in effective mass distribution of $K_S^0 K_S^0$ system from the reaction $\pi^- C \to K_S^0 K_S^0 A^*$ on the left shoulder of $f_2(1270)$, $a_2^0(1320)$.

Statistically significant bump in $K_S^0 K_S^0$ mass distribution near 2250 MeV and relatively small width 56 MeV is observed under cut $\cos\theta_{GJ} < 0.5$.

REFERENCES

1. Nozdrachev V.N., "The resonance structures of $K_S K_S$ and $\Lambda\bar{\Lambda}$ spectrum at MIS ITEP", this conference.
2. Bolonkin B.V. et al., *Nucl. Phys.* **B309**, 426 (1988)
3. Barkov B. P. et al., in *Proc. of HADRON91, College Park*, 47 (1991)
4. Bolonkin B.V. et al., *Phys. Atom. Nucl.* **58**, 1535 (1995)
5. Grigoriev V.K. et al., *Phys. Atom. Nucl.* **59**, 2105 (1996)
6. L3 Collab., Acciarri M. et al., *Phys. Lett.* **B501** 173 (2001).
7. E690 Collab., Reyes M.A. et al., *Phys. Rev. Lett.* **81** 4079 (1998).
8. BES Collab., Bai J.Z. et al., *Phys. Rev. Lett.* **76** 3502 (1996).

Study of the ηη system in the $\pi^- p$ charge exchange reaction at 32 GeV/c with the GAMS-4π spectrometer

A.M. Blick*, F.G. Binon†, A.V. Dolgopolov*, S.V. Donskov*, S. Inaba**, Y. Fujii**, G.V. Khaustov*, V.N. Kolosov ‡, A.A. Kondashov*, A.A. Lednev*, V.A. Lishin*, J.P. Peigneux‡, V.A. Polyakov*, S.A. Sadovsky*, V.D. Samoylenko*, P.M. Shagin*, H. Shimizu§, A.V. Singovsky¶, A.E. Sobol*, J.P. Stroot†, V.P. Sugonyaev*, K. Takamatsu**, T. Tsuru**, Y. Yasu** and A.Yu. Zvyagin*

*Institute for High Energy Physics, 142284, Protvino, Russia
†IISN, Belgium
**High Energy Accelerator Research Organization (KEK), Tsukuba, Ibaraki 305-0801, Japan
‡LAPP-IN2P3, Annecy, France
§Yamagata University, Japan
¶University of Minnesota

Abstract. A study has been made of the ηη system produced in the $\pi^- p$ charge exchange reaction at 32 GeV/c. The experiment was performed at the IHEP 70-GeV proton synchrotron with the GAMS-4π spectrometer. A partial wave analysis has been carried out in the mass range from 1.1 to 3.9 GeV with $-t < 0.2$ (GeV/c)^2 including the waves S, D_0, D_-, D_+, G_0, G_- and G_+. Three states are unambiguously observed in the S-wave: $f_0(1370)$, $f_0(1500)$, and $f_0(1710)$. The large accumulated statistic allows to resolve the $G(1590)$ structure into two separate states the $f_0(1500)$ and the $f_0(1710)$. Two states $f_0(2100)$ and $f_2(1950)$ are observed in one solution. The $f_4(2050)$ is seen in the G_0-wave. A new spin 4 structure observed in the region above 2.7 GeV/c^2. This state needs further investigation due to the possible influence of a spin 6 state.

Introduction. The ηη system has been studied in several experiments searching for exotic states. The $G/f_0(1590)$, a scalar glueball candidate, has been observed during the investigation of ηη systems produced in the $\pi^- p$ charge exchange reaction more than 15 years ago [1, 2]. Listed nowadays in the Particle Data Group tables under the heading $f_0(1500)$, this state has also been observed later in other reactions ($p\bar{p}$-annihilation, J/ψ - radiation decays) where enhanced production of mesons with an enriched gluon component is expected. One more state which could have an exotic nature is the $f_0(1710)$-meson. The clear observation in centrally produced ηη systems of the two scalar states $f_0(1500)$ and $f_0(1710)$[3], as well as the insignificant contribution of the $f_2(1525)$ in the total mass spectrum, allows to say that there are no arguments any more against the splitting of the $G/f_0(1590)$ signal into two states. As a matter of

---

[1] Speaker

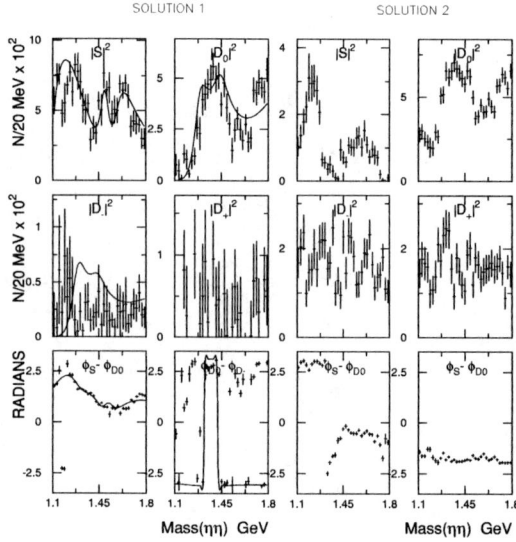

FIGURE 1. PWA in the mass region 1.1 − 1.7 GeV with S, D_0, D_+, D_- waves. Solution 2 is not physical.

fact, the main objections [4] were based on the assumption that the clear peak observed in the $K\bar{K}$ channel by the WA-76 experiment [5] was due to the $f_2(1525)$ meson, in contradiction with recent WA-102 observations[6]. The presence of two narrow scalar states which both fit glueball mass lattice calculations around 1.6 GeV justifies further detailed investigations of the $\eta\eta$ system.

One more important field is the investigation of high spin states. No experimental information on the $\eta\eta$ system is presently available in the high mass region. An investigation of high masses is one of the main goals of the present work.

Mass independent partial wave analysis. A sample of 46500 $\eta\eta$ events with $-t < 0.2(GeV/c)^2$ has been analysed. The partial wave analysis(PWA) has been carried out using mass bins of 20, 40 and 80 MeV in the mass intervals ranging from 1.1 GeV to 1.7 GeV, from 1.7 GeV to 2.5 GeV and from 2.5 GeV to 3.9 GeV, respectively. An event-by-event maximum likelihood method is applied in each mass bin. Within the region below 1.7 GeV, only waves with $J = 0, 2$ contribute. The S, D_0, D_- and D_+ waves are taken into account. Due to the well known ambiguities inherent in PWA, the solution is not unique. The two possible solutions are presented in fig. 1. We consider the second solution to be non-physical due to the high D wave intensity at the reaction threshold. In the physical solution, the waves with projection $m = 0$ dominate over the $m = 1$ waves. Another characteristic feature is phase coherence between the D_0 and D_- waves (phase difference equal to π in our notations) which is also in good agreement with the Ochs-Wagner model predictions.

In the region above 1.7 GeV, only the S, D_0 and G_0 waves (corresponding to $J = 0, 2, 4$ with $m = 0$) have been introduced, as a first step, in the PWA. In this reduced PWA, the number of non-trivial solutions in each mass interval is equal to 2 . As a second step, the

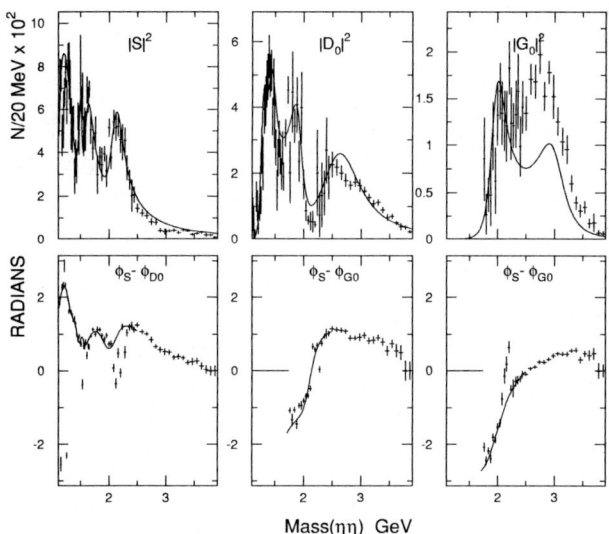

FIGURE 2. First solution of PWA in the mass region $1.1 - 3.9$ GeV with S, D, G−waves. The solid curves are the result of fits with Breit-Wigner resonances.

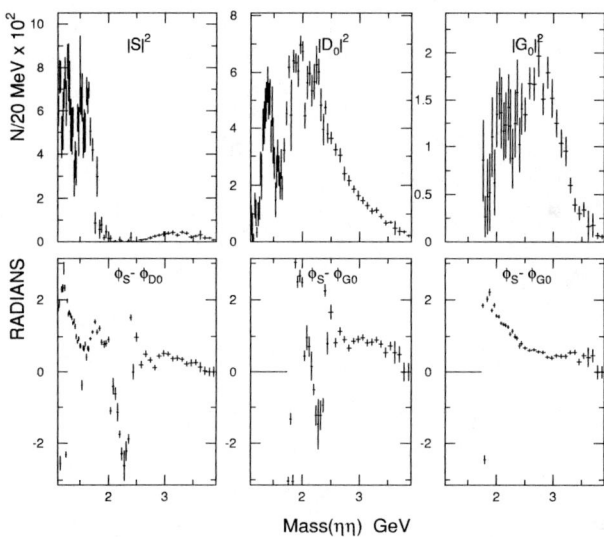

FIGURE 3. Second solution of PWA in the mass region $1.1 - 3.9$ GeV with S, D, G−waves. D_+, D_-, G_+, G_- waves not shown as their contribution is not significant.

waves with $m = 1$, namely D_-, D_+, G_- and G_+, have been added to the found solution and the fitting procedure has been repeated. As a result, solutions have been found with small values of D_-, D_+, G_- and G_+ amplitudes. As the moments with projection $m = 0$ dominate over those with $m = 1$ everywhere, only these solutions are considered in the analysis. Two possible solutions are presented in fig. 2 and fig. 3.

Mass dependent partial wave analysis. To determine the parameters of the resonances produced in the $\eta\eta$ system, an overall fit to the modules of the S, D and G amplitudes and their relative phases has been performed by using the maximum likelihood method. A relativistic Breit-Wigner parametrisation with a Blatt-Weisskopf barrier factor is used to describe the resonances. We impose phase coherence between the D_0 and D_- waves as well as between the G_0 - G_- waves. No absolute normalization of the data has been obtained so far. The sensitivity of the present experiment has been determined by normalisation to the number of observed $f_2(1270)$ mesons in the in the $\pi^0\pi^0$ mass spectrum. The $f_2(1270)$ has a measured cross-section of $(2.61 \pm 0.20\mu b)$ at 38 GeV/c [7] and after proper correction for the energy dependence of the cross section and for the detection efficiency the sensitivity was found to be 2.75 pb. The production cross-section of the different resonances has been evaluated by integrating the corresponding Breit-Wigner functions. To investigate the possible presence of a resonance in the G−wave, a spin 6 resonance is introduced in the fit. This lead to visible discrepancies between the results of mass independent (without spin 6) and mass dependent PWA in the high mass region. In the first solution (fig. 2) the S−wave exhibits a series of bumps separated by dips showing evidence for several scalar resonances. This S−wave behavior is quite similar to that obtained in the $\pi^0\pi^0$ system [8]. In the second solution (fig. 3) one can see broad bump in the D_0−wave in the region of $f_2(2150)$.

The parameters obtained for the resonances in the mass dependent PWA strongly depend on the solution even in the region (below 1.7 GeV) where the solution in the mass independent PWA is unique. The ambiguous spin 0 and 2 structures should be analyzed through other decay channels. A coupled channels analysis of this data is underway.

The obtained branching ratio $BR(f_2(1270) \to \eta\eta)$ is $(2.6 \pm 0.6) \times 10^{-3}$, which is compatible with the PDG value. The branching ratio of $f_4(2050)$ into $\eta\eta$ has been evaluated using the ratio between the $f_4(2050)$ and $f_2(1270)$ production cross-sections measured at 38 GeV/c [8]. It is equal to 0.31 ± 0.08. Using the known values of the $f_4(2050)$ branching ratios, one obtains: $BR(f_4(2050) \to \eta\eta) = (1.8 \pm 0.6) \times 10^{-3}$.

REFERENCES

1. F. Binon et al., *Nuovo Cimento* **78A**, 313 (1983).
2. D. Alde et al., *Yad. Phys.* **44**, 120 (1985).
3. D. Barberis et al., *Phys. Lett.* **B479**, 59 (2000).
4. Yu.D. Prokoshkin, *Soviet Physics Doklady* **36**, 155 (1991).
5. T.A. Armstrong et al., *Phys. Lett.* **B227**, 186 (1989).
6. D. Barberis et al., *Phys. Lett.* **B453** 305 (1999).
7. Yu.D.Prokoshkin and A.A. Kondashov, *Nuovo Cimento* **107A** 1903 (1994).
8. D. Alde et al., *Eur. Phys. J.* **A 3**, 361-371 (1998).

A study of reaction $K^-N \to (K^-\pi^+\pi^-)N$ at 28 GeV/c

VES experiment, presented by V.Nikolaenko

Institute for High Energy Physics, 142281, Protvino, Russia

Abstract. Diffractive-like production of strange meson resonances is studied at the statistics of $\sim 300\,000$ events, which exceeds statistics of previous experiments. Experimental data have been acquired by VES spectrometer [1], exposed in unseparated tagged beam off Serpukhov accelerator on Be target. Preliminary results of the standard 3-meson Partial Wave Analysis (PWA) of $(K^-\pi^+\pi^-)$ system with mass below 2 GeV are presented.

Properties of $(K^-\pi^+\pi^-)$-system have been studied in several experiments, two of them are the most statistically significant: $K^{\pm}p$ at 13 GeV/c [1] (~ 138000 events in total) and K^-p at 63 GeV/c [2] (~ 191000 events). The total mass distribution has two wide bumps, near 1300 and 1800 MeV, called Q- and L-regions, respectively. The wave with spin-parity $J^P = 1^+$ is the dominant one, particularly in the Q-region. There are two well known resonances here, $K_1(1270)$ and $K_1(1400)$, which are interpreted as a mixture of two $(s\bar{u})$-states with parallel and anti-parallel spins of quarks. The second and the 3-rd waves are $J^P = 0^-$ and 2^- waves. A wide resonance with mass $\sim 1400 - 1460$ MeV has been observed in the 0^- wave, which is not yet considered as well established. Events in the 2^- wave are concentrated mainly in the L-region. It would be natural to expect that two 2^- mesons exist in L-region, like two 1^+ resonances in Q-region. There are published fit results for 2^- wave with one or two resonances in the L-region in the $(K^-\pi^+\pi^-)$-system [2], as well as for the $(K\omega)$-system [3]. The χ^2 probabilities are better for two resonance hypothesis, however the difference is not large and this question needs clarification.

The following selection criteria have been applied in this study:

- kaon beam (no signals in two beam Cherenkov detectors tuned below K-threshold; but a signal seen in the 3-rd one, above K-threshold);
- there are 3 charged secondary tracks (one positive, two negative);
- the energy sum of charged tracks is in the range (25.,30.) GeV;
- there is no π^0, η in the event and the total energy deposited in the Electromagnetic Calorimeter is consistent with noise;
- the momentum transfer cut: $|t'| < 0.7\ GeV^2$;

---

[1] supported by grants INTAS-RFBR 97-02-71017, RFBR 00-02-16555 and VNS-RFBR 00-15-96689.

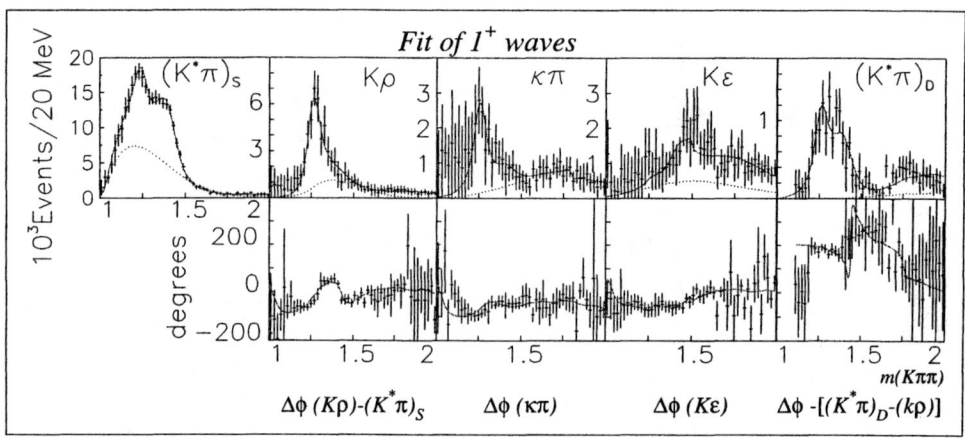

FIGURE 1. PWA results for $J^P M^\eta = 1^+ 0^+$ waves, $1.0 < m(K\pi\pi) < 2.0\,GeV$, $0 < |t| < 0.7\,GeV^2$. Number of events in 20 MeV bins and relative phases are shown. Result of mass-dependent fit is shown by solid lines. Dotted lines show contributions from Deck effect.

- at least one secondary negative track is identified by Cherenkov detector of secondary tracks as K^- or π^-. (The identification works satisfactory in the momentum range from 4.5 to 16.0 GeV/c);
- rejected events with two identified π^- or two K^- or identified K^+.
- rejected events if $\pi^\pm$ or K-meson has $cos\theta_{GJ} < -0.92$ in the Gotfried-Jackson ref. frame, in order to exclude events with $(p\pi)$ or (pK^-) isobars.

Selected sample contains 300895 events with mass $(K^-\pi^+\pi^-)$ in the range $(1.0, 2.1)$ GeV. An analysis of the sample purity has shown that the most serious impurity comes from the $(\pi^-\pi^+\pi^-)$ system produced in K^- beam, this admixture is of order of 5%. Admixtures due to the beam misidentification and from the $(K^-K^+K^-)$ channel are negligible.

At the next step, the 3-meson PWA program was used [4, 5, 6, 7]. Total 30 waves with spin-parity 1^+, 0^-, 2^-, 3^+, 4^-, 2^+ and 1^- were included with intermediate isobars $K^*(890)$, ρ, ε (excluded $f_0(980)$), κ or $K_0(1430)$, $K^{**}(1430)$ or $K_2(1430)$, $f_2(1270)$, $K_3(1780)$, and $f_0(980)$. Like the ACCMOR experiment [2], 5 decay modes were included for 1^+ wave. Constructing the density matrix, all 5 channels for this wave were taken as connected with separate diagonal elements of the ρ-matrix, i.e. we have not assumed that some of 1^+ waves are produced coherently (see [8]). Similarly four large channels in 0^- and three channels in 2^- wave are taken without coherence assumption.

The PWA results for 1^+ channels are shown in Fig. 1. At the next step we fitted the mass spectra for 5 channels and Re/Im parts of non-diagonal elements of density matrix (or the relative phases between channels). The objective was to get a continuous description of all distributions in the mass range from 1.0 to 2.0 GeV. The model contains three resonances: $K_1(1270)$, $K_1(1400)$, $K_1(1750)$ and the Deck amplitudes, all objects are taken as partially coherent. There is a complication with $K_1(1270)$ resonance: the threshold in the dominant decay channel, $K\rho$. This resonance should have a deviation

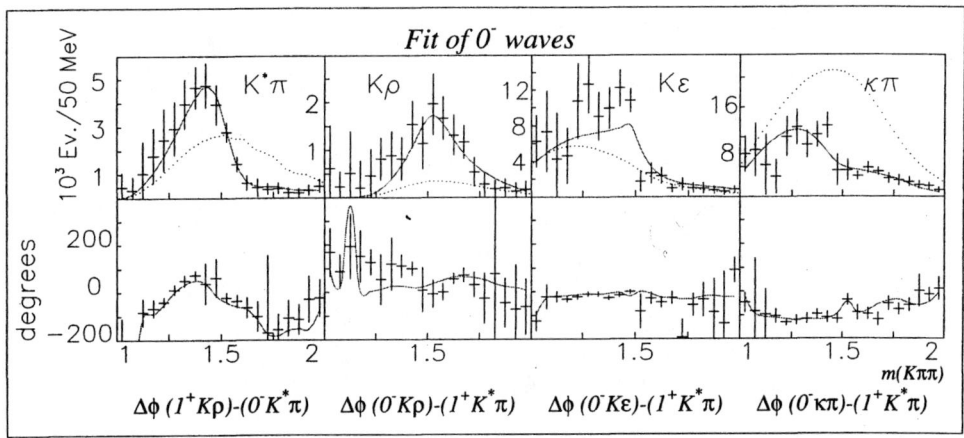

FIGURE 2. PWA results for $J^P M^\eta = 0^- 0^+$ waves, $1.0 < m(K\pi\pi) < 2.0\, GeV$, $0 < |t| < 0.7\, GeV^2$. Number of events in 50 MeV bins and relative phases are shown. Result of mass-dependent fit is shown by solid lines. Dotted lines show contributions from Deck effect.

from the Breight-Wigner shape. This deviation was calculated following the method proposed in [9] for a_1 resonance. A mass shift in the denominator of Breight-Wigner function arises. We estimated the mass shift starting from PDG values of the $K_1(1270)$ mass, width and branching ratios [10] for five measured channels and also $(K\omega)$ channel. Actually, this mass shift amplifies a difference between $K^*\pi$ and $K\rho$ decay channels, which exists already due to the phase space factors. The Deck rescattering amplitudes were simulated by a method proposed in [11]. The simulated shapes of Deck amplitudes were kept, but the overall normalisation and the phase shift were taken as free parameters. Satisfactory fit of all distribution, in the mass range from 1.0 to 2.0 GeV is obtained, with $\chi^2/ND = 217./352$. Fit results are shown by solid lines in Fig. 1. Preliminary parameters of 1^+ resonances are given in Table 1. The branching ratios are corrected for decays of resonances and intermideate isobars into channels with π^0 or K^0. The $(K\omega)$ channel is not included. Errors in Table 1 include systematic uncertainties from inclusion of small waves in PWA and different binning.

A similar procedure was applied for description of four channels in 0^- wave. The phases of 0^- waves were measured with respect to 1^+ waves. The model contains only one resonance, $K_0(1460)$. Fit results are presented in Fig. 2 and in Table 2 ($\chi^2/ND = 130./134$). There is no indication on the 2-nd resonance with a mass of 1800-1900 MeV, which could be expected as a strange partner of the $\pi(1800)$.

2^- wave was measured in channels $(K^*\pi)$, $(K\rho)$, $(K^{**}\pi)$ and (Kf_2) (Fig. 3). Two another decay modes were tried, $(K\varepsilon)$ and $(\kappa\pi)$, but both are negligible. Superimposed curves correspond to the fit of experimental distributions [2] to the model with one resonance and the Deck amplitudes. Fitted parameters are presented in Table 2. Errors

---

[2] Mass spectra in 20 MeV bins and Re/Im parts of non-diagonal elements of the density matrix between 2^- waves and the reference 1^+ waves are fitted.

TABLE 1. Preliminary parameters of resonances in 1^+ wave.

	$K_1(1270)$	$K_1(1410)$	$K_1(1750)$
mass, MeV	1274 ± 9	1407 ± 10	1755 ± 50
width, MeV	45 ± 8	83 ± 10	269 ± 90
$Br(K^*\pi)_S, \%$	16.1 ± 4.1	80.6 ± 9.1	9.7 ± 1.0
$Br(K\rho), \%$	65.7 ± 7.3	6.9 ± 5.7	26.5 ± 11.1
$Br(\kappa\pi), \%$	9.2 ± 0.9	1.4 ± 0.2	4.5 ± 1.8
$Br(\varepsilon K), \%$	1.1 ± 0.4	3.6 ± 1.0	32.2 ± 6.8
$Br(K^*\pi)_D, \%$	7.8 ± 1.2	7.8 ± 1.9	27.0 ± 3.9

TABLE 2. Preliminary parameters of resonances in 0^- and 2^- waves.

0^- wave				$2-$ wave		
	$K_0(1460)$				2^- $K_2(1770)$	
					parabolic err.	MINOS err.
mass, MeV	1450 ± 37		mass, MeV	1718 ± 7	$+5 - 10$	
width, MeV	110 ± 25		width, MeV	97 ± 8	$+8 - 14$	
$Br(K^*\pi)$	38.3 ± 4.1		$Br(K^*\pi), \%$	39.3 ± 6.7	$+8.1 - 4.4$	
$Br(K\rho)$	6.7 ± 0.8		$Br(K\rho)$	13.8 ± 5.1	$+7.1 - 4.6$	
$Br(\varepsilon K)$	7.4 ± 1.2		$Br(K^{**}\pi)$	33.2 ± 9.7	$+8.9 - 7.5$	
$Br(\kappa\pi)$	47.7 ± 5.1		$Br(K f_2)$	13.5 ± 1.9	$+3.0 - 1.4$	

on the resonance mass, width and branching ratios for 2^- wave don't include the shifts which would emerge if one of the channels is excluded from the fit.

One can notice that the shape of the $(K\rho)$ channel demonstrates no resonant-like signal in the mass spectrum. From another side, a clear bump has been observed in LASS experiment in $(K\omega)$ channel [3]. It looks strange, one would expect similar shapes, if the $(K\rho)$ system is produced in pure isospin $I = 1/2$ state.

Like previous experiments [2, 3, 12], we tried to describe the measured 2^- waves by two resonances plus Deck effect amplitudes. Two resonance hypothesis yields $\chi^2/ND = 282/253$ which should be compared with $\chi^2/ND = 326/262$ for the single resonance one. We don't consider the χ^2 difference as significant to justify the introduction of second resonance. Moreover, the fitted mass of 2-nd resonance is lower than the mass of $K_2(1770)$, which disagrees with fit results in previous experiments [2, 3], there the second object is heavier than the $K_2(1770)$.

In conclusion, the main results of this study can be summarised as follows:

- Partial Wave Analysis of $(K^-\pi^+\pi^-)$ system is performed at the statistics of ~ 300000 events. The general pattern of large waves is similar to the pattern observed in K^-p experiment at 63 GeV/c.
- A complex of 5 most significant 1^+ waves in the mass range from 1 to 2 GeV can be described as a superposition of 3 resonances and Deck rescattering amplitudes. The

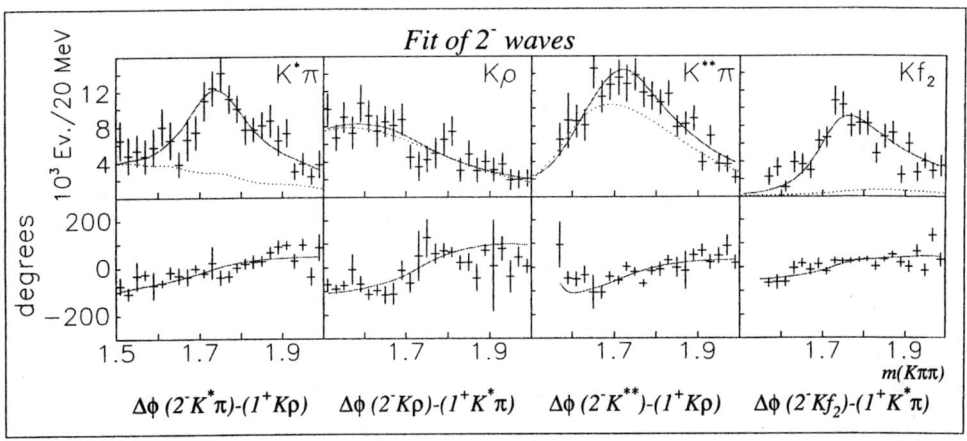

FIGURE 3. PWA results for $J^P M^\eta = 2^- 0^+$ waves, $1.5 < m(K\pi\pi) < 2.0\, GeV$, $0 < |t| < 0.7\, GeV^2$. Number of events in 20 MeV bins and relative phases are shown. Result of mass-dependent fit with one resonance is shown by solid lines. Dotted lines show contributions from Deck effect.

existence of 1^+ resonance with mass close to $1750\, MeV$ is confirmed. Preliminary values of mass, width and Branching ratios of this object are determined.

- A satisfactory description of 4 most significant channels of 0^- wave below $m_{tot}=2\, GeV$ requires one resonance, $K_0(1460)$ and Deck effect. Preliminary measurements of $K_0(1460)$ parameters are presented. There is no evidence for 2-nd resonance in this wave, a strange partner of $\pi(1800)$.

- A satisfactory description of 4 most significant channels of 2^- wave in the mass range from 1.5 to 2.0 GeV is obtained with a single resonance plus Deck effect amplitudes. Two resonance hypothesis gives better χ^2, but the difference is not large enough to justify the introduction of 2-nd object. The shape of the mass spectrum in $(K\rho)$ channel does not agree with the shape obtained for $(K\omega)$ channel in LASS experiment[3].

REFERENCES

1. G.W.Brandenburg et al, Phys. Rev. Lett. 36(1976) 703, 706; Nucl. Phys. B127(1977) 509
2. C.Daum et al., Nucl. Phys. B187(1981)p.1, and ref. therein;
3. D.Aston et al., Phys. Lett. B308(1993)p.186
4. G.Ascoli et al., Phys. Rev. Lett. 25(1970)962;
5. G.Ascoli et al., Phys. Rev. D7(1973)669;
6. G.Ascoli et al., Phys. Rev. D9(1974)1963;
7. D.V.Brockway, University of Illinois report COO-1195-197(1970);
8. J.D.Hansen et al., Nucl. Phys. B81(1974)p.403;
9. N.A.Tornqvist, Z.Phys.C36(1987) 695;
10. Particle Data Group, *Review of Particle Physics* Eur. Phys. J C15(2000), p.530;
11. G.Ascoli et al., Phys. Rev. D8(1973)3894;
12. Particle Data Group, *Review of Particle Physics* Eur. Phys. J C15(2000), p.536-539;

PHENOMENOLOGY

Relative phase between the three-gluon and one-photon amplitudes of the J/ψ decays

N.N. Achasov

Laboratory of Theoretical Physics, Sobolev Institute for Mathematics, Academician Koptiug prospekt, 4, 630090 Novosibirsk, Russia

Abstract. It is shown that the study of the $\omega - \rho^0$ interference pattern in the $J/\psi \to (\rho^0+\omega)\eta \to \pi^+\pi^-\eta$ decay provides evidence for the large (nearly 90°) relative phase between the one-photon and three-gluon decay amplitudes.

In the last few years it has been noted that the single-photon and three-gluon amplitudes in the two-body $J/\psi \to 1^-0^-$ and $J/\psi \to 0^-0^-$ [1, 2, 3] decays appear to have relative phases nearly 90°.

This unexpected result is very important to the observability of CP violating decays as well as to the nature of the $J/\psi \to 1^-0^-$ and $J/\psi \to 0^-0^-$ decays [1, 2, 3, 4, 5, 6, 7]. In particular, it points to a non-adequacy of their description built upon the perturbative QCD, the hypothesis of the factorization of short and long distances, and specified wave functions of final hadrons. Some peculiarities of electromagnetic form factors in the J/ψ mass region were discussed in Ref. [8].

The analysis [1, 2, 3] involved theoretical assumptions relying on the strong interaction $SU_f(3)$-symmetry, the strong interaction $SU_f(3)$-symmetry breaking and the $SU_f(3)$ transformation properties of the one-photon annihilation amplitudes. Besides, effects of the $\rho - \omega$ mixing in the $J/\psi \to 1^-0^-$ decays were not taken into account in Ref. [1] while in Ref. [2] the $\rho - \omega$ mixing was taken into account incorrectly , see the discussion in Ref. [9]. Because of this, the model independent determination of these phases are required.

Fortunately, it is possible to check the conclusion of Refs. [1, 2] at least in one case [9, 10]. We mean the relative phase between the amplitudes of the one-photon $J/\psi \to \rho^0\eta$ and three-gluon $J/\psi \to \omega\eta$ decays.

The point is that the $\rho^0 - \omega$ mixing amplitude is reasonably well studied [11, 12, 13, 14, 15, 16, 17]. Its module and phase are known. The module of the ratio of the amplitudes of the ρ and ω production can be obtained from the data on the branching ratios of the J/ψ-decays. So, the investigation of the $\omega - \rho$ interference in the $J/\psi \to (\rho^0+\omega)\eta \to \rho^0\eta \to \pi^+\pi^-\eta$ decay provides a way of measuring the relative phase of the ρ^0 and ω production amplitudes.

Indeed, the $\omega - \rho$ interference pattern in the $J/\psi \to (\rho^0+\omega)\eta \to \rho^0\eta \to \pi^+\pi^-\eta$ decay is conditioned by the $\rho^0 - \omega$ mixing and the ratio of the amplitudes of the ρ^0 and ω production:

$$\frac{dN}{dm} = N_\rho(m)\frac{2}{\pi}m\Gamma(\rho \to \pi\pi, m)\left|\frac{1}{D_\rho(m)}\left(1 - \varepsilon(m)\left[\frac{N_\omega(m)}{N_\rho(m)}\right]^{\frac{1}{2}}\exp\{i(\delta_\omega - \delta_\rho)\}\right)\right.$$

$$\left. + \frac{1}{D_\omega(m)}\left(\varepsilon(m) + g_{\omega\pi\pi}/g_{\rho\pi\pi}\right)\left[\frac{N_\omega(m)}{N_\rho(m)}\right]^{\frac{1}{2}}\exp\{i(\delta_\omega - \delta_\rho)\}\right|^2 \quad (1)$$

with

$$\varepsilon(m) = -\frac{\Pi_{\omega\rho^0}(m)}{m_\omega^2 - m_\rho^2 + im\left(\Gamma_\rho(m) - \Gamma_\omega(m)\right)}, \quad (2)$$

where m is the invariant mass of the $\pi^+\pi^-$-state, $N_\rho(m)$ and $N_\omega(m)$ are the squares of the modules of the ρ and ω production amplitudes, δ_ρ and δ_ω are their phases, $\Pi_{\omega\rho^0}(m)$ is the amplitude of the $\rho - \omega$ transition, $D_V(m) = m_V^2 - m^2 - im\Gamma_V(m)$, $V = \rho, \omega$. We obtained in Refs. [9, 10]

$$\varepsilon(m_\omega) + g_{\omega\pi\pi}/g_{\rho\pi\pi} = (3.41 \pm 0.24) \cdot 10^{-2}\exp\{i(102 \pm 1)^\circ\}. \quad (3)$$

The branching ratio of the $\omega \to \pi\pi$ decay

$$B(\omega \to \pi\pi) = \frac{\Gamma(\rho \to \pi\pi, m_\omega)}{\Gamma_\omega(m_\omega)} \cdot \left|\varepsilon(m_\omega) + g_{\omega\pi\pi}/g_{\rho\pi\pi}\right|^2. \quad (4)$$

The data [18, 19] were fitted with the function

$$N(m) = L(m) + \left|(N_\rho)^{\frac{1}{2}}F_\rho^{BW}(m) + (N_\omega)^{\frac{1}{2}}F_\omega^{BW}(m)\exp\{i\phi\}\right|^2, \quad (5)$$

where $F_\rho^{BW}(m)$ and $F_\omega^{BW}(m)$ are the appropriate Breit-Wigner terms [18] and $L(m)$ is a polynomial background term.

The results are

$$\phi = (46 \pm 15)^\circ, \quad N_\omega(m_\omega)/N_\rho = 8.86 \pm 1.83 \; [18],$$
$$\phi = -0.08 \pm 0.17 = (-4.58 \pm 9.74)^\circ, \quad N_\omega(m_\omega)/N_\rho = 7.37 \pm 1.72 \; [19]. \quad (6)$$

From Eqs. (1), (4), and (5) it follows

$$N_\rho = N_\rho(m_\rho)\left|1 - \varepsilon(m_\rho)\left[N_\omega(m_\rho)/N_\rho(m_\rho)\right]^{\frac{1}{2}}\exp\{i(\delta_\omega - \delta_\rho)\}\right|^2, \quad (7)$$

$$N_\omega = B(\omega \to \pi\pi)N_\omega(m_\omega), \quad (8)$$

$$\phi = \delta_\omega - \delta_\rho + \arg\left[\varepsilon(m_\omega) + g_{\omega\pi\pi}/g_{\rho\pi\pi}\right] -$$
$$-\arg\left\{1 - \varepsilon(m_\rho)\left[N_\omega(m_\rho)/N_\rho(m_\rho)\right]^{\frac{1}{2}}\exp\{i(\delta_\omega - \delta_\rho)\}\right\} \simeq$$
$$\simeq \delta_\omega - \delta_\rho + \arg\left[\varepsilon(m_\omega) + g_{\omega\pi\pi}/g_{\rho\pi\pi}\right] -$$
$$-\arg\left\{1 - |\varepsilon(m_\omega)|\left[N_\omega(m_\omega)/N_\rho\right]^{\frac{1}{2}}\exp\{i\phi\}\right\}. \quad (9)$$

From Eqs. (3), (6) and (9) we get that

$$\delta_\rho - \delta_\omega = (60 \pm 15)° \quad [18], \tag{10}$$
$$\delta_\rho - \delta_\omega = (106 \pm 10)° \quad [19]. \tag{11}$$

Whereas δ_ρ is the phase of the isovector one-photon amplitude, δ_ω is the phase of the sum of the three-gluon amplitude and the isoscalar one-photon amplitude. But luckily for us the latter is a small correction. Really, it follows from the structure of the electromagnetic current

$$j_\mu(x) = \frac{2}{3}\bar{u}(x)\gamma_\mu u(x) - \frac{1}{3}\bar{d}(x)\gamma_\mu d(x) - \frac{1}{3}\bar{s}(x)\gamma_\mu s(x) + \ldots \tag{12}$$

and the Okubo-Zweig-Iizuka rule the ratio for the amplitudes under consideration (**please image all possible diagrams!**):

$$\frac{A(J/\psi \to \text{the isoscalar photon} \to \omega\eta)}{A(J/\psi \to \text{the isovector photon} \to \rho\eta) \equiv A(J/\psi \to \rho\eta)} = \frac{1}{3}. \tag{13}$$

Taking into account Eqs. (6) and (7) one gets

$$\frac{|A(J/\psi \to \text{the isoscalar photon} \to \omega\eta)|}{|A(J/\psi \to \text{the three-gluon} \to \omega\eta)|} \approx \frac{1}{9}. \tag{14}$$

From Eqs. (10), (11) and (14) one gets easily for the relative phase (δ) between the isovector one-photon and three gluon decay amplitudes

$$\delta = (60 \pm 15)° - 4° \quad [18], \tag{15}$$
$$\delta = (106 \pm 10)° - 6° \quad [19], \tag{16}$$

if the isoscalar and isoscalar one-photon decay amplitudes have the same phase. In case the isoscalar one-photon and three-gluon (isoscalar also!) decay amplitudes have the same phase

$$\delta = (60 \pm 15)° \quad [18], \tag{17}$$
$$\delta = (106 \pm 10)° \quad [19]. \tag{18}$$

So, both the MARK III Collaboration [18] and the DM2 Collaboration [19], see Eqs. (15), (17) and (16), (18), provide support for the large (nearly 90°) relative phase between the isovector one-photon and three-gluon decay amplitudes.

The DM2 Collaboration used statistics only half as high as the MARK III Collaboration, but, in contrast to the MARK III Collaboration, which fitted N_ω as a free parameter, the DM2 Collaboration calculated it from the branching ratio of $J/\psi \to \omega\eta$ using Eq. (8).

In summary I should emphasize that it is urgent to study this fundamental problem once again with KEDR in Novosibirsk and with BES in Beijing.

But I am afraid that only the τ-CHARM factory could solve this problem in the exhaustive way.

I gratefully acknowledge discussions with San Fu Tuan.

The present work was supported in part by the grant INTAS-RFBR IR-97-232.

REFERENCES

1. G. Lopez Castro, J.L. Lucio M. and J. Pestieau, AIP Conf. Proc. **342**, 441 (1995), hep-ph/9902300.
2. M. Suzuki, Phys. Rev. **D 57**, 5717 (1998).
3. M. Suzuki, Phys. Rev. **D 60**, 051501 (1999).
4. J.L. Rosner, Phys. Rev. **D 60**, 074029 (1999).
5. Y.F. Gu and S.F. Tuan, Nucl. Phys. **A 675**, 404c (2000).
6. J.-M. Gerard and J. Weyers, Phys.Lett. **B 462**, 324 (1999).
7. S.F. Tuan, Plenary session, these Proceedings.
8. N.N. Achasov and A.A. Kozhevnikov, Phys. Rev. **D 58**, 097502 (1998); Yad. Fiz. **62**, 364 (1999).
9. N.N. Achasov and V.V. Gubin, Pis'ma v ZhETF **72**, 3 (2000) [JETP Lett. **72**, 1 (2000)].
10. N.N. Achasov and V.V. Gubin, Phys. Rev. **D 61**, 117504 (2000).
11. A.S. Goldhaber, G.S. Fox and C. Quigg, Phys. Lett. **30 B**, 249 (1969).
12. M. Gourdin, L. Stodolsky and F.M. Renard, Phys. Lett. **30 B**, 347 (1969).
13. F.M. Renard, Nucl. Phys. **B 15**, 118 (1970).
14. N.N. Achasov and G.N. Shestakov, Nucl. Phys. **B 45**, 93 (1972); Fiz. Elem. Chastits. At. Yadra **9**, 48 (1978) [Sov. J. Part. Nucl. **9**, 19 (1978)].
15. N.N. Achasov and A.A. Kozhevnikov, Yad. Fiz. **55**, 809 (1992); Int. J. Mod. Phys. **A 7**, 4825 (1992).
16. N.N. Achasov, A.A. Kozhevnikov and G.N. Shestakov, Phys. Lett. **50 B** (1974) 448. N.N. Achasov, N.M. Budnev, A.A. Kozhevnikov and G.N. Shestakov, Yad. Fiz. **23**(1976) 610.
17. Particle Data Group, C. Caso et al., Eur. Phys. J. **C 3**, 1 (1998).
18. D. Coffman *et al.*, Phys. Rev. **D 38**, 2695 (1988).
19. J. Jousset *et al.*, Phys. Rev. **D 41**, 1389 (1990).

Sum rules for total hadronic widths of mesons

Michał Majewski

Dept. of Theoretical Physics, Univ. of Lodz, Pomorska 149/153, 90-236 Łódź, Poland

Abstract. Mass sum rules for meson multiplets derived from exotic commutators may be written for complex masses. Then the real parts give the well known mass formulae (GM-O, Schwinger, Ideal) and the imaginary ones give the corresponding sum rules for total hadronic widths. The masses and widths of the meson nonets submit to a definite orders. It thus follows that tables of the meson nonets should include information about masses, widths and the orders as well as the mixing angle. The width sum rule for the nonet complying with Schwinger mass formula may be depicted as a straight line in the (m, Γ) plane. It is easily verifiable and satisfied better for high mass nonets.

INTRODUCTION

The particle width is one of its main characteristics as much important as mass and discreet quantum numbers. It tells us something different than the mass and sometimes it may tell more: the widths of the particles with similar masses may differ by many orders; then the widths inform us first which interaction—strong, electromagnetic or weak is responsible for their decay. For hadronic decays the differences are not so big, but usually are of the same order as mass differences. Therefore they merit attention.

The difficulty with the widths within the meson multiplet is that they are in a sense accidental. Indeed, selection rules may suppress more or less the decay of a particular particle thus destroying any given regularity. Such an effect should be especially transparent in low mass multiplets where for some particle two-body decays are forbidden and many-body decays are suppressed (e.g. ω-meson). For more massive multiplets, where many decay channels are opened, we may expect better agreement. However the prediction may be interesting in any case.

SUM RULES FOR NONETS

The approach is based on the technique of exotic commutators [1]. The following system of mass sum rules for a nonet has been obtained [2]:

$$l_1^2 + l_2^2 = 1 \tag{1}$$

$$l_1^2 z_1 + l_2^2 z_2 = \frac{1}{3}a + \frac{2}{3}b \; (\equiv z_8) \tag{2}$$

$$l_1^2 z_1^2 + l_2^2 z_2^2 = \frac{1}{3}a^2 + \frac{2}{3}b^2 \tag{3}$$

$$l_1^2 z_1^3 + l_2^2 z_2^3 = \frac{1}{3}a^3 + \frac{2}{3}b^3 \tag{4}$$

Here a, K, z_1, z_2 stand for the mass squared of the isotriplet, isodublet and isoscalar physical mesons respectively ($z_1 < z_2$ by choice), z_8 is the GM–O mass squared, $b = 2K - a$ and the real coefficients l_1, l_2 are introduced by the equation: $|z_8\rangle = l_1|z_1\rangle + l_2|z_2\rangle$. The known mass formulae for a nonet follow from eqs. (1)–(4).

Eqs. (1)–(4) may be considered for complex masses $\hat{M}^2 = \hat{m}^2 - i\hat{m}\hat{\Gamma}$. $\hat{M}^2$ is now non-hermitean, but it can be diagonalized and has orthogonal eigenfunctions. We use for the masses squared the notations: a, K, x_1, x_2, x_8, b and for the appropriate imaginary parts the notations: α, κ, y_1, y_2, y_8, $\beta(=2\kappa - \alpha = \sqrt{(b)}\Gamma_b)$. The coefficients l_1, l_2 are complex numbers, but l_1^2, l_2^2, which are their modula squared, are real.

The real parts of the masses squared satisfy usual mass formula and the imaginary ones give the sum rules for the widths.

For GM-O nonet (follows from eqs. (1), (2)) we find

$$l_1^2 = \frac{x_2 - x_8}{x_2 - x_1}, \quad l_2^2 = \frac{x_8 - x_1}{x_2 - x_1}, \quad \frac{y_2 - y_8}{x_2 - x_8} = \frac{y_8 - y_1}{x_8 - x_1}. \tag{5}$$

For Schwinger nonet (follows from eqs. (1)–(3)) we have

$$\frac{y_2 - y_8}{x_2 - x_8} = \frac{y_8 - y_1}{x_8 - x_1} = \frac{\beta - \alpha}{b - a} \tag{6}$$

and separate equations for the real and imaginary parts of the mass operator:

$$(a - x_1)(a - x_2) + 2(b - x_1)(b - x_2) = 0 \tag{7}$$
$$(\alpha - y_1)(\alpha - y_2) + 2(\beta - y_1)(\beta - y_2) = 0. \tag{8}$$

For these nonets only two mass orders are allowed and two widths orders for each of them:

$$a < x_1 < b < x_2; \quad \alpha > y_1 > \beta > y_2; \quad \text{or} \quad \alpha < y_1 < \beta < y_2 \tag{9}$$
$$x_1 < a < x_2 < b; \quad y_1 > \alpha > y_2 > \beta; \quad \text{or} \quad y_1 < \alpha < y_2 < \beta. \tag{10}$$

Both mass orders are observed [3].

Some of the well established Schwinger nonets are collected in the Table 1.

For the Ideal nonet (folows from eqs. (1)– (4)) we have

$$x_1 = a, \quad x_2 = b, \quad y_1 = \alpha, \quad y_2 = \beta. \tag{11}$$

If we apply eqs. (1)–(3) to the octet states ($l_1^2 = 1$, $l_2^2 = 0$) we get degenerate octet (all mases and widths identical). Degenerate multiplets (octets(?), nonets(?)) do exist. They are shown in the Table 2.

The formula (6) shows that the points $(m^2, m\Gamma)$ of the Schwinger nonet states lie on a straight line in the $(m^2, m\Gamma)$ plane and consequently the poins (m, Γ) lie on the straight line in the plane (m, Γ). The slope of this line is indefinite. Mass-width diagram for the nonet 2^{++} is shown on the Figure 1. For the nonet 1^{--} the agreement is worse, for the nonet 3^{--} it is quite good.

TABLE 1. Some well established nonets of mesons (masses and widths in MeV)

J^{PC} multiplet	m_K Γ_K θ^{GMO}	m_a Γ_a mass order	m_1 Γ_1	m_b $(2\kappa-\alpha)m^{-1}$ width order	m_2 Γ_2
1^{--}					
•$\rho(770)$		769.3 ± 0.8	782.57 ± 0.12	1.0031 ± 0.0011	1019.417 ± 0.014
•$K^*(892)$	893.88 ± 0.26				
•$\omega(782)$	50.7 ± 0.8	150.2 ± 0.8	8.44 ± 0.09	-24.8 ± 2.1	4.458 ± 0.032
•$\Phi(1020)$	$(39.28\pm 0.16)°$	$a<x_1<b<x_2$		$\alpha>y_1>\beta>y_2$	
2^{++}					
•$a_2(1320)$		1318.0 ± 0.6	1275.4 ± 1.2	1532.0 ± 3.1	1525 ± 5
•$K_2^*(1430)$	1429.0 ± 1.4				
•$f_2(1270)$	103.8 ± 4.0	107 ± 5	$185.1^{+3.4}_{-2.6}$	101.5 ± 11.9	76 ± 10
•$f_2(1525)$	$(30.67^{+1.56}_{-1.72})°$	$x_1<a<x_2<b$		$y_1>\alpha>y_2>\beta$	
3^{--}					
•$\rho_3(1690)$		1691 ± 5	1667 ± 4	1857 ± 11	1854 ± 7
•$K_3^*(1780)$	1776 ± 7				
•$\omega_3(1670)$	159 ± 21	161 ± 10	168 ± 10	158 ± 53	87^{+28}_{-23}
•$\Phi_3(1850)$	$(32.0^{+3.5}_{-7.5})°$	$x_1<a<x_2<b$		$y_1>\alpha>y_2>\beta$	
1^{++}					
•$a_1(1260)$		1230 ± 40	1281.9 ± 0.6	1420 ± 0.012	1426.3 ± 1.1
•K_A	1341 ± 6				
•$f_1(1285)$	134 ± 16	$250\div 600$	24.0 ± 1.2	$-447\div 89$	55.5 ± 2.9
•$f_1(1420)$	$35.26°\div 41.00°$	$a<x_1<b<x_2$		$\alpha>y_1>\beta>y_2$	
1^{+-}					
•$b_1(1235)$		1229.5 ± 3.2	1170 ± 20	1414 ± 9	1386 ± 19
•K_B	1322 ± 6				
•$h_1(1170)$	135 ± 17	142 ± 9	360 ± 40	130 ± 40	91 ± 30
•$h_1(1380)$	$0\div 35.26°$	$x_1<a<x_2<b$		$y_1>\alpha>y_2>\beta$	

ACKNOWLEDGMENTS

Valuable discusions with Profs. S.B. Gerasimov, P. Kosiński and V.A. Meshcheryakov are kindly acknowledged.

REFERENCES

1. S.Oneda, K.Terasaki *Progr. Theor. Phys. Suppl.* **82** (1985)
2. M.Majewski and W.Tybor *Acta Physica Polonica* **B15** (1984) 267
3. Particle data Group *Eur. Phys. J.* **C15** (2000) 1

TABLE 2. Degenerate multiplets (masses and widths in MeV)

J^{PC}	m_a Γ_a	m_K Γ_K	m_1 Γ_1	
4^{++} • $a_4(2040)$ • $K_4(2045)$ • $f_4(2050)$	2014 ± 15 361 ± 50	2045 ± 9 198 ± 30	2034 ± 11 222 ± 19	
1^{--} • $\rho(1450)$ • $K^*(1410)$ • $\omega(1420)$	1465 ± 25 310 ± 60	1414 ± 15 232 ± 21	1419 ± 31 174 ± 60	
1^{--} • $\rho(1700)$ • $K^*(1680)$ • $\omega(1650)$ • $\Phi(1680)$	1700 ± 20 240 ± 60	1717 ± 27 322 ± 110	1649 ± 24, 220 ± 35,	1680 ± 20 150 ± 50

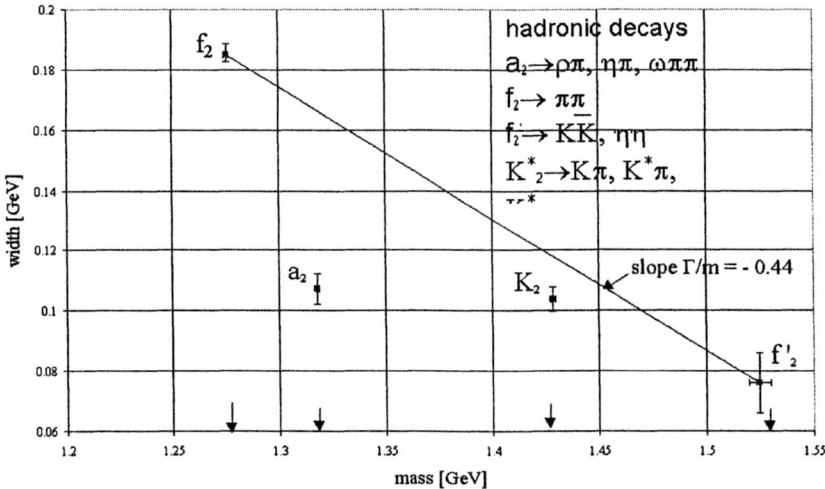

FIGURE 1. Mass-width diagram of 2^{++} mesons—mass order: $x_1 < a < x_2 < b$

Possible Evidence for a Chiral Axial-Vector State in the D Meson System

Kenji Yamada*, Muneyuki Ishida¶, Shin Ishida†, Daiki Ito†, Toshihiko Komada* and Hiroshi Tonooka†

Department of Engineering Science, Junior College Funabashi Campus, Nihon University, Funabashi 274-8501, Japan
¶*Department of Physics, Tokyo Institute of Technology, Tokyo 152-8551, Japan*
†*Atomic Energy Research Institute, College of Science and Technology, Nihon University, Tokyo 101-8308, Japan*

Abstract. We reanalyze the $D^{**}\pi^-$ mass spectrum from CLEO II by the VMW method in order to examine the existence of a chiral axial-vector state, which is predicted in a covariant level-classification scheme recently proposed, other than normal orbitally-excited P-wave states in the D meson system. A result of the present analysis seems to suggest that there exists an extra axial-vector meson, in addition to the two normal ones, in a similar mass region.

INTRODUCTION

In the constituent quark model, together with heavy quark symmetry, the lowest-lying positive parity excitations of heavy-light $Q\bar{q}$ meson systems are expected, in the limit $m_Q \to \infty$, to be two degenerate spin doublets with the total angular momentum j_q=1/2 and 3/2 of the light quark, that is, four orbitally-excited states with L=1 labeled as

$$^{j_q}L_J = {}^{1/2}P_0, \; {}^{1/2}P_1 \quad \text{for the } j_q = \tfrac{1}{2} \text{ doublet,}$$
$$= {}^{3/2}P_1, \; {}^{3/2}P_2 \quad \text{for the } j_q = \tfrac{3}{2} \text{ doublet.}$$

In this limit heavy quark symmetry further requires that the j_q=1/2 states decay to $^{1/2}S_0 + \pi$ or $^{1/2}S_1 + \pi$ only in an S-wave, while the j_q=3/2 states decay only in a D-wave. It is therefore expected that the decay widths of the j_q=1/2 and 3/2 states are broad and narrow, respectively.

On the one hand, a covariant level-classification scheme of quark-antiquark meson systems has been proposed, which gives them a covariant quark representation with definite Lorentz and chiral transformation properties [1]. In this scheme, assuming that chiral symmetry for the light-quark component in heavy-light meson systems is effective, the existence of extra scalar and axial-vector states is predicted, respectively, as chiral partners of the ground-state pseudoscalar and vector mesons. These what we call chiral scalar and axial-vector mesons are distinguished from the above-mentioned P-wave states, since the chiral scalar state is an analogue of the $\sigma(400-600)$ meson, which is difficult to be interpreted as the $^3P_0$ state, as a chiral partner of the π meson in the light-quark system.

In this report we present a possible evidence for the chiral axial-vector state, D_1^χ, in the D meson system.

REANALYSIS OF THE $D^{*+}\pi^-$ MASS SPECTRUM FROM CLEO II

We reanalyze the $D^{*+}\pi^-$ mass spectrum, published by CLEO Collaboration [2], by the VMW method in which the production amplitude is expressed by a sum of Breit-Wigner amplitudes for relevant resonances.

In the present analysis we take into account the four states $D_2^*(^{3/2}P_2)$, $D_1(^{3/2}P_1)$, $D_1^*(^{1/2}P_1)$ and D_1^χ which can decay to $D^*\pi$. Then, following the VMW method, the production amplitude is given by

$$|A(s)|^2 = \left| r_1 e^{i\theta_1} \Delta_{D_1^\chi}(s) + r_2 e^{i\theta_2} \Delta_{D_1^*}(s) \right|^2 + \left| r_3 e^{i\theta_3} \Delta_{D_1}(s) \right|^2 + \left| r_4 e^{i\theta_4} \Delta_{D_2^*}(s) \right|^2,$$

$$\Delta_R(s) = \frac{-m_R \Gamma_R}{s - m_R^2 + i m_R \Gamma_R},$$

where $r_1,...,r_4$ and $\theta_1,...,\theta_4$ are the production couplings and phases of respective resonances, and we assumed that D_1^χ and D_1^* decay only through an S-wave, while D_1 only through a D-wave. The background $D^*\pi$ mass distribution is fit with a five-parameter threshold function given by

$$BG = \alpha(\Delta M)^\beta \exp\left[-\gamma_1 \Delta M - \gamma_2 (\Delta M)^2 - \gamma_3 (\Delta M)^3\right], \quad \Delta M = M(D^*\pi) - m_{D^*} - m_\pi,$$

where the parameters α, β, γ_1, γ_2 and γ_3 are fixed through the fit to the total $D^*\pi$ mass spectrum.

Using the above production amplitude and background, we fit the $D^*\pi$ mass spectrum in the following three cases:

(a) High-mass D_1^* with a mass of $2500 < m_{D_1^*} < 2600$ in MeV,

(b) Low-mass D_1^* with a mass of $2350 < m_{D_1^*} < 2500$ in MeV,

(c) No D_1^χ and D_1^*.

Here the case (c) corresponds to the original analysis by CLEO Collaboration, though the background parametrization is somewhat different. The results of fits are shown in Fig. 1 and obtained values of the resonance parameters are given in Table 1. In both the fits with high- and low-mass D_1^*, it is found that the mass and width of D_1^χ are $\approx$ 2310 MeV and ≈ 20 MeV, respectively, and those of D_1 and D_2^* are similar to the values reported so far. For the mass and width of D_1^* we obtain ≈ 2600 MeV and $\approx$ 200 MeV in the high-mass fit, while ≈ 2420 MeV and ≈ 200 MeV in the low-mass fit.

In all the three cases of fits the χ^2/N_{dof} is best for the high-mass D_1^* fit, though they are not so different from each other. It would be worth while noting that the two cases of fits with D_1^χ seem to describe the data better than the fit without D_1^χ in the mass region 2.15–2.5 GeV, where there appears to be an excess of data events at the mass 2.31–2.33 GeV.

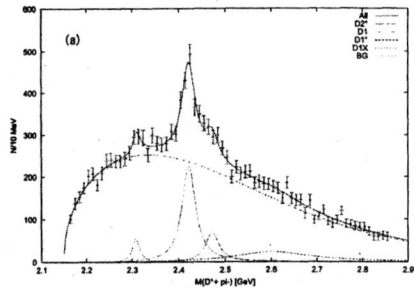
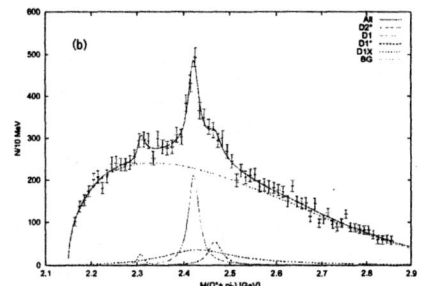
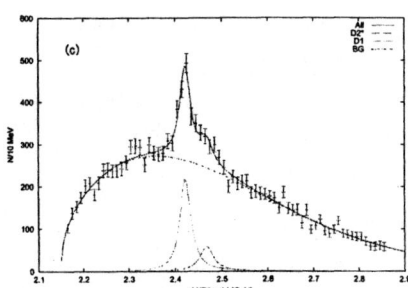

FIGURE 1. The results of the fits to the $D^{*+}\pi^-$ mass spectrum with (a) high-mass D_1^*, (b) low-mass D_1^*, and (c) no D_1^χ and D_1^*.

TABLE 1. Values of the resonance parameters and χ^2/N_{dof} obtained from the respective fits.

State	(a) Fit with high-mass D_1^*		(b) Fit with low-mass D_1^*		(c) Fit without D_1^χ and D_1^*	
	Mass (MeV)	Width (MeV)	Mass (MeV)	Width (MeV)	Mass (MeV)	Width (MeV)
D_1^χ	2308	18.7	2307	17.4	–	–
D_1^*	2596	199	2421	199	–	–
D_1	2421	34.5	2421	27.0	2421	27.5
D_2^*	2472	35.0	2468	35.0	2466	35.0
χ^2/N_{dof}	57.7/52		58.0/52		66.2/59	

THEORETICAL REMARKS ON THE RESULTS

We consider the mass splitting and mixing of P-wave meson multiplets, based on the Breit-Fermi Hamiltonian with vector-gluon and long-range-scalar exchange, where we ignore the P-wave D_1 and D_1^* states mixing with the chiral D_1^χ state. Taking a static potential due to single vector-gluon exchange to be $-4\alpha_s/3r$, the spin-dependent part of the Hamiltonian for P-wave states can be expressed, to first order $1/m_Q$, as

with
$$\delta H = C_q \mathbf{L} \cdot \mathbf{S}_q + C_Q (\mathbf{L} \cdot \mathbf{S}_Q + S_T), \quad S_T = 3(\mathbf{S}_q \cdot \hat{\mathbf{r}})(\mathbf{S}_Q \cdot \hat{\mathbf{r}}) - \mathbf{S}_q \cdot \mathbf{S}_Q$$

$$C_q = \left(\frac{1}{2m_q^2} + \frac{1}{m_q m_Q}\right)\left\langle\frac{4\alpha_s}{3r^3}\right\rangle - \frac{1}{2m_q^2}\left\langle\frac{1}{r}\frac{dV_s}{dr}\right\rangle, \quad C_Q = \frac{1}{m_q m_Q}\left\langle\frac{4\alpha_s}{3r^3}\right\rangle,$$

where $V_s(r)$ is the static potential due to long-range-scalar exchange and the spin-spin interaction is neglected because of its contact nature. The Hamiltonian δH gives rise to the mass splittings among P-wave multiplets and the mixing between the $^{3/2}P_1$ and $^{1/2}P_1$ states. Treating δH as a first-order perturbation and using the mass values of D_2^*, D_1 and D_1^* obtained in the high-mass D_1^* fit, we find ≈ 2470 MeV for the mass of $D_0^*(^{1/2}P_0)$ and $\phi - \phi_{HQ} \approx -3.64°$ for the deviation of the $D_1(^{3/2}P_1) - D_1^*(^{1/2}P_1)$ mixing angle from the heavy-quark-symmetry limit with the parameter values of the unperturbed mass $M_0 = 2490$ MeV common to all four states, $C_q = -73.81$ MeV and $C_Q = 47.15$ MeV, where we have chosen a solution with $C_Q > 0$ in accord with the above definition of C_Q. For the low-mass D_1^* case of fits there is no solution.

CONCLUDING REMARKS

We have reanalyzed the $D^{*+}\pi^-$ mass spectrum published by CLEO Collaboration and found a possible evidence for the chiral axial-vector meson D_1^χ with a mass and width of ≈ 2310 MeV and ≈ 20 MeV, respectively. We have also found the mass and width of D_1^* to be ≈ 2600 MeV and ≈ 200 MeV, together with similar masses and widths of D_1 and D_2^* to those reported so far, among which the spin-dependent Hamiltonian arising from one-vector-gluon and long-range-scalar exchange could account for the mass splittings. To confirm the existence of D_1^χ it goes without saying that further analyses, including other experimental data with high statistics, are necessary in a more precise way.

Furthermore, in establishing the covariant level-classification scheme of meson systems it is important to examine the existence of the chiral scalar meson D_0^χ as well as B_0^χ and B_1^χ in the B meson system. An analysis of the $B\pi$ mass spectrum, to study the existence of B_0^χ, is in progress and its preliminary result has been presented [3].

REFERENCES

1. Ishida, S., Ishida, M., and Maeda, T., *Prog.. Theor. Phys.* **104**, 785-807 (2000).
2. CLEO Collaboration, *Phys. Letters* B **331**, 236-244 (1994); **342**, 453(E) (1995).
3. Ishida, M., and Ishida, S., these proceedings.

Weak decays of heavy mesons in a covariant quark model

D. Merten

Institut für Theoretische Kernphysik, Nußallee 14–16, D-53115 Bonn, Germany

Abstract. Weak decays of heavy mesons will be investigated in the framework of a covariant quark model, which is based on the Bethe-Salpeter equation in instantaneous approximation. Apart from a phenomenological confinement potential, a residual interaction induced by instantons is adopted. An appropiate extension allows a unified description of light and heavy systems. In this model semileptonic and non-leptonic decays of heavy-light mesons are evaluated.

In the last few years, new and improved data on the spectra and the decays of charmed and bottom mesons have become available. Yet many new results will be provided by the B-factories BABAR, BELLE, HERA-B and LHC-B within the next years.

For the theoretical description of meson masses as well as decays over the full kinematic region, constituent quark models, even if the connection to the underlying theory is not quite clear, are still the most successful tool. In previous papers we have developed a relativistic constituent quark model for light mesons with instanton induced forces [2, 3]. In this model, a very good description of the light meson masses has been achieved. Also many decay observables have been calculated in reasonable agreement with the experimental data (see [3] for a recent update). Motivated by this success, the model has been extended for heavy flavours [4]. The resulting spectra together with a brief review of our model are discussed in the following. Knowing the meson amplitudes we have calculated semileptonic and non-leptonic decays of heavy-light mesons [5].

The model is based on the Bethe-Salpeter equation for $q\bar{q}$ bound states. In our ansatz the full quark propagators S_i^F are approximated by free fermion propagators $S_i^F(p) \approx i/(\not{p} - m_i + i\varepsilon)$ with effective constituent quark masses m_i. Furthermore the irreducible interaction kernel is supposed to be instantaneous in the restframe of the meson, leading to the (full) Salpeter equation, which is solved numerically. The interaction kernel is parametrized by a linearly rising confinement potential in configuration space with an adequate Dirac structure. To estimate the influence of the Dirac structure, two possibilities are taken into account: $\mathbb{1} \otimes \mathbb{1} - \gamma^0 \otimes \gamma^0$ ($\mathcal{A}$) which is known to minimize spin-orbit splittings [2], and $\mathbb{1} \otimes \mathbb{1} - \gamma^5 \otimes \gamma^5 - \gamma^\mu \otimes \gamma_\mu$ ($\mathcal{B}$), previously investigated by [6] and [7]. In addition we adopt a residual interaction induced by instantons based on the work of 't Hooft [8] and Shifman *et al.*[9]. This interaction acts on scalar and pseudoscalar mesons only and is flavour dependent. In this way the π-η-η' splitting can be described. The resulting masses of the $J \neq 0$ isovector ground states, in this ansatz determined by the confinement alone, agree very well with the experimental data up to high angular momentum. The spectra of the light isoscalars and kaons are of similar quality. For a detailed discussion we refer to [3].

TABLE 1. Masses of charmed and beauty mesons in [MeV], calculated in model $\mathcal{A}$ and $\mathcal{B}$.

Meson	Exp.[1]	$\mathcal{A}$	$\mathcal{B}$	Meson	Exp.[1]	$\mathcal{A}$	$\mathcal{B}$
$D(0^-)$	1864-1870	1869	1869	$D_s(0^-)$	1968-1969	1969	1969
$D^*(1^-)$	2006-2011	1993	2034	$D_s^*(1^-)$	2112-2113	2049	2116
$D_1(1^+)$	2420-2424	2464	2420	$D_{s1}(1^+)$	2534-2536	2532	2506
$D_2^*(2^+)$	2455-2463	2475	2469	$D_{s2}^*(2^+)$	2572-2575	2541	2552
$B(0^-)$	5279-5280	5279	5279	$B_s(0^-)$	5367-5372	5368	5369
$B^*(1^-)$	5324-5326	5325	5346	$B_s^*(1^-)$	5413-5420	5369	5425

Heavy mesons have been included in this model to achieve a unified description of all mesons. As shown in [4], this is done by naively extending the instanton induced interaction for heavy-light systems. The resulting masses are shown in Table 1. We find good agreement for the heavy-light mesons in both models with small advantages in model $\mathcal{B}$ due to a larger spin-orbit splitting. Thus we think that we have obtained a good estimate for the heavy meson amplitudes. To test these amplitudes further we investigate the semileptonic decays of heavy to light mesons. The relevant current matrix elements are calculated in lowest order according to the prescription of Mandelstam[10].

The effective Lagrangian for the semileptonic decays, e.g. $b \to c$ transitions, after integrating out the W boson, has the usual $V - A$ current-current form. The matrix elements of the vector (V^μ) and axial vector (A^μ) hadronic currents must be parametrized in general by 6 form factors. In the limit of vanishing lepton mass, only 4 of these contribute to the decay rates, which are called in a usual parametrization [11] f_+, V, A_1 and A_2. In the following sections, our results for semileptonic B and charmed meson decays are compared to the experimental data and to the results of the relativised constituent quark model of N. Isgur et al. (ISGW2 [12]).

The decays $B \to D^{(*)} \ell \bar{\nu}_\ell$ have been measured by CLEO [14]. We find good overall agreement with the experimental data for both decays (Fig. 1), using a CKM matrix element of $|V_{cb}| = 0.034 \pm 0.001$ and $|V_{cb}| = 0.035 \pm 0.001$ for model $\mathcal{A}$ and $\mathcal{B}$, respectively, which has been determined by a χ^2 fit. These values are somewhat smaller than the PDG average of $|V_{cb}| = 0.037 - 0.043$. The resulting decay rates are (in $10^{10} s^{-1}$)

model $\mathcal{A}(\mathcal{B})$: $\Gamma(B \to D\ell\bar{\nu}_\ell) = 1.22(1.14)$, $\Gamma(B \to D^*\ell\bar{\nu}_\ell) = 3.21(3.24)$

in satisfying agreement with the current world average and recent new measurements

$B^+ \to \bar{D}^0 \ell^+ \nu_\ell : 1.30 \pm 0.13 [1]$,
$B^0 \to D^- \ell^+ \nu_\ell : 1.36 \pm 0.12 [1]$,
$B^0 \to D^- \ell^+ \nu_\ell : 1.34 \pm 0.07 \pm 0.20 [18]$,

$B^+ \to \bar{D}^{*0} \ell^+ \nu_\ell : 3.21 \pm 0.48 [1]$,
$B^0 \to D^{*-} \ell^+ \nu_\ell : 2.97 \pm 0.17 [1]$,
$B^0 \to D^{*-} \ell^+ \nu_\ell : 3.06 \pm 0.24 \pm 0.26 [18]$,
$B^0 \to D^{*-} \ell^+ \nu_\ell : 3.01 \pm 0.08 ^{+0.23}_{-0.20} [17]$.

For the ($|V_{cb}|$ independent) polarization ratios, we find $\Gamma_L/\Gamma_T = 1.14(1.20)$ for model $\mathcal{A}(\mathcal{B})$, in good agreement with the experimental value of $\Gamma_L/\Gamma_T = 1.24 \pm 0.16$ [14].

The semileptonic decays of charmed mesons have been measured for the $D \to K^{(*)}$ as well as for the $D_s \to \eta/\eta'/\phi$ transitions. The results are shown in Table 2. Our calculation agrees reasonably in both models for the $0^- \to 0^-$ decays $D \to K, D \to \eta/\eta'$. Note that the flavour mixing of η and η' has already been fixed by the mass fit. Although

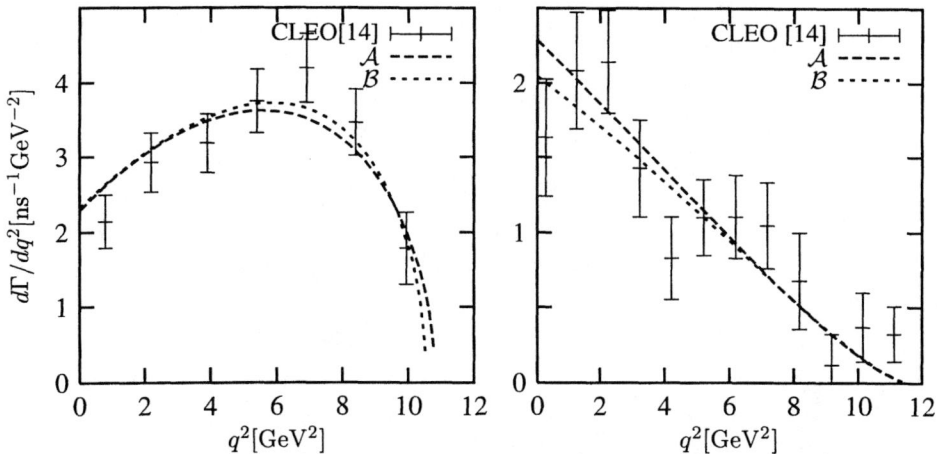

FIGURE 1. The differential decay rate for $B \to D\ell\bar{\nu}$ (top) and $B \to D^*\ell\bar{\nu}$ (bottom).

TABLE 2. $D \to K^{(*)}\ell\bar{\nu}$ and $D_s \to \eta/\eta'/\phi\ell\bar{\nu}$ decay observables and form factors, Γ in $10^{10} s^{-1}$.

Parameter	Exp.[1]	$\mathcal{A}$	$\mathcal{B}$	ISGW2	Parameter	Exp.[1]	$\mathcal{A}$	$\mathcal{B}$	ISGW2
$\Gamma(D \to K)$	7.97 ± 0.36	7.51	7.26	10.0	$\Gamma(D_s \to \eta)$	5.24 ± 1.41	4.05	3.11	3.5
					$\Gamma(D_s \to \eta')$	1.80 ± 0.69	1.27	1.75	3.0
$\Gamma(D \to K^*)$	4.55 ± 0.34	7.64	10.08	5.4	$\Gamma(D_s \to \phi)$	4.03 ± 1.01	7.89	9.67	4.6
Γ_L/Γ_T	1.14 ± 0.08	1.29	1.48	0.94	Γ_L/Γ_T	0.72 ± 0.18	1.20	1.42	0.96
Γ_+/Γ_-	0.21 ± 0.04	0.23	0.34	—	Γ_+/Γ_-	—	0.20	0.33	—
$A_1(0)$	0.56 ± 0.04 [11]	0.69	0.81	—	$A_1(0)$	—	0.66	0.79	—
$V(0)/A_1(0)$	1.82 ± 0.09	1.54	1.18	2.0[15]	$V(0)/A_1(0)$	1.92 ± 0.32	1.77	1.30	2.1[15]
$A_2(0)/A_1(0)$	0.78 ± 0.07	0.81	0.62	1.3[15]	$A_2(0)/A_1(0)$	1.60 ± 0.24	0.85	0.63	1.3[15]

the differences between the results of our two models in these channels are rather large, the experimental data does not allow to prefer one of our parameter sets.

The decays to the pseudovector final states, however, are overestimated by about a factor of 2. The polarization observables on the other hand are comparable with the experimental result, where model $\mathcal{A}$ gives better agreement than model $\mathcal{B}$. This is a well known problem of constituent quark models. With respect to the ISGW2 results it is interesting to note that, whereas the inclusion of relativistic corrections was one of the main ingredients in their model to decrease the $0^- \to 1^-$ decay rate, this problem still exists in our relativistic calculation. The form factor ratios at zero momentum transfer show that this failure can be traced back to too large axial form factors.

To extract further information from our meson amplitudes we finally investigate non-leptonic decays. On tree level, non-leptonic decays are mediated by a single W-boson emission. Contributions by weak annihilation and internal W exchange are neglected.

For these decays the matrix element of a product of currents has to be evaluated. This is usually done using the "factorization approximation", where one assumes that the amplitude is dominated by its factorizable part. Then it is given by the product of two current matrix elements, *e.g.* for the transition $B^0 \to D^-\pi^+$, in lowest order

TABLE 3. Non-leptonic B decay rates Γ in ns^{-1}.

decay mode	Exp.[1] *[16]	$\mathcal{A}$	$\mathcal{B}$	NRSX	decay mode	Exp.[1]	$\mathcal{A}$	$\mathcal{B}$	NRSX
$B^0 \to D^-\pi^+$	1.94 ± 0.26	2.21	1.97	1.94	$B^+ \to \overline{D}^0 D_s^+$	7.9 ± 2.4	8.7	8.3	6.6
$B^0 \to D^-\rho^+$	5.10 ± 0.90	5.68	6.13	4.84	$B^0 \to D^- D_s^+$	5.2 ± 1.9	8.7	8.3	6.6
$B^0 \to D^{*-}\pi^+$	1.78 ± 0.14	2.22	1.88	1.87	$B^+ \to \overline{D}^0 D_s^{*+}$	5.4 ± 2.4	6.1	5.5	6.2
$B^0 \to D^{*-}\rho^+$	4.4 ± 2.2	6.6	6.7	5.5	$B^0 \to D^- D_s^{*+}$	6.5 ± 3.2	6.1	5.5	6.2
$B^0 \to D^- D^+$		0.53	0.49	0.23	$B^+ \to \overline{D}^{*0} D_s^+$	7.3 ± 3.0	6.1	6.5	4.5
$B^0 \to D^- D^{*+}$		0.37	0.34	0.23	$B^0 \to D^{*-} D_s^+$	6.2 ± 2.2	6.1	6.5	4.5
$B^0 \to D^{*-} D^+$		0.39	0.41	0.17	$B^+ \to \overline{D}^{*0} D_s^{*+}$	16.3 ± 6.0	14.7	15.2	15.8
$B^0 \to D^{*-} D^{*+}$	$0.51 \pm 0.10 \pm 0.08^*$	0.86	0.89	0.54	$B^0 \to D^{*-} D_s^{*+}$	12.9 ± 4.5	14.7	15.2	15.8

$$A(B^0 \to D^-\pi^+) = \frac{G_F}{\sqrt{2}} V_{cb} V_{du}^* \langle \pi^+ | h_{\mu du} | 0 \rangle \langle D^- | h_{cb}^\mu | B^0 \rangle.$$

In this way the decay amplitude can be expressed by the decay constant and a form factor of the semileptonic decay at the relevant q^2. Our results for non-leptonic B-decays are shown in table 3, compared with the experimental data from [1] and the calculation NRSX of Neubert et al. [13]. We find good agreement with the data for both our models.

In summary, we find excellent agreement in the description of semileptonic heavy to heavy transitions $B \to D^{(*)}$ over the whole kinematic regime. The results on heavy to light transitions are mostly in agreement with the experimental data, although the common problem of quark models to overestimate the axial form factors is still present. Non-leptonic decay rates of B mesons are well described in factorization approximation.

Acknowledgement: Financial support by funds provided by the Graduiertenkolleg "Die Erforschung subnuklearer Strukturen der Materie" is gratefully acknowledged.

REFERENCES

1. Groom, D. E. et al., *Eur. Phy. J.* **C15**, 1 (2000).
2. Münz, C.R. et al., *Nucl. Phys.* **A578**, 418 (1994); Resag, J. et al., *Nucl. Phys.* **A578**, 379 (1994).
3. Koll, M. et al., *Eur. Phys. J.* **A9**, 73 (2000); Ricken, R. et al., *Eur. Phys. J.* **A9**, 221 (2000).
4. Kaufmann, A., diploma thesis, Universität Bonn TK-99-01 (1999).
5. Merten, D. et al., hep-ph/0104029 (2001).
6. Gross, F. et al., *Phys. Rev.* **D43**, 2401 (1991).
7. Böhm, M. et al., *Nucl. Phys.* **B51**, 397 (1973).
8. 't Hooft, G., *Phys. Rev.* **D14**, 3432 (1976).
9. Shifman, M.A., Vainshtein, A.I., Zakharov, V.I., *Nucl.Phys.* **B163**, 46 (1980).
10. Mandelstam, S., *Proc.Roy.Soc.Lond.* **A233**, 248 (1955).
11. Richman, J.D. et al., *Rev. Mod. Phys.* **67**, 893 (1995).
12. Scora, D. et al., *Phys. Rev.* **D52**, 2783 (1995).
13. Neubert, M., and Stech, B., *Heavy Flavours II*, edited by A.J. Buras and M. Lindner, World Scientific, Singapore, 1998, 294.
14. Avery, P., et al., CLEO CONF 94-7; Barish et al., *Phys. Rev.* **D51**, 1014 (1995); M. Athanas et al., *Phys. Rev. Lett.* **79**, 2208 (1997).
15. Aitala, E.M., et al., *Phys. Lett.* **B440**, 435 (1998); E.M. Aitala et al., *Phys. Lett.* **B450**, 294 (1999).
16. Aubert, B., et al., hep-ex/0107057 (2001).
17. Abreu, P., et al., *Phys.Lett.* **B510**, 55 (2001)
18. Abe, K., et al., Belle-Conf-0121 (2001); Abe, K., et al., Belle-Conf-0122 (2001)

Potential of the Rare Heavy Quark Decay Studies at the ATLAS Experiment

Nikolai Nikitine

Skobeltsyn Institute of Nuclear Physics, Lomonosov Moscow State University, Moscow 119899, Russia

Abstract. In this paper we breafly present the ATLAS detector potential for the studies of the rare b and t quarks decays.

Introduction. The ATLAS detector is one of the four detectors which is planned to work at CERN Large Hadron Collider (LHC). The main ATLAS experiment goals are the Higgs boson and Super Symmetry discoveries. However the b and t physics potential of the ATLAS experiment is high [1].

Rare leptonic $B^0_{d,s} \to \mu^+\mu^-$ and semileptonic $B^0_d \to K^{0*}(\rho^0)\mu^+\mu^-$, $B^0_s \to \phi\mu^+\mu^-$ decays correspond to $b \to s(d)$ quark transitions, which are forbidden at tree level in the SM. These decays provide the precision test of the SM, open the way to study the physics beyond the SM and to estimate the CKM–matrix elements $|V_{ts}|$ and $|V_{td}|$ [2]. For the top quark, the decay $t \to bW$ is absolutly dominante. The other decays can be strongly suppressed. The "radiative" decays $t \to bW(Z,H)$ provide the sensitive probe of the t-quark mass and the decays corresponded to $t \to c(u)$ quark transitions (analogous to the trantitions $b \to s(d)$ for b quark) present the precision test of the SM in top quark sector.

The comparison of b and t quark physics potentials for different experiments are presented in Table 1. The branching ratios of the rare B–meson and top quark decays are presented in the Table 2. It can be seen from the Tables 1 and 2, in the framework of the SM at the LHC one can observe the decay channels $B^0_{d,s} \to \mu^+\mu^-$, $B^0_s \to \phi\mu^+\mu^-$ after 3 year's run at low luminosity and rare radiative decay of the top quark $t \to bWZ$, which could not be detected with other experiments at B factories [3] and Tevatron [4].

Enormous proton–proton collision rate at the LHC produces 90 MHz data flow rate. To reduce this data rate and to select the rare processes the ATLAS use trigger system – special hardware and computing multilevel system selected "interesting" events. It consists of three operational levels: LVL1, LVL2 and Event Filter (EF) [5].

ATLAS trigger simulation. About 1500 signal events for rare leptonic and semileptonic B–meson decays in each channel were simulated and reconstructed in the Inner Detector with minimum bias events. The typical event structure is following:

$\bar{b}b \to$ X including $b \to (c,u)\mu^-\bar{\nu}_\mu$, $p_T(\mu) > 6$ GeV, $\varepsilon_\mu = 0.85$ for **LVL1** Trigger

$\hookrightarrow B^0_{s,d} \to \mu^+\mu^-$ or $B^0_{s,d} \to (V \to h_1 h_2)\mu^+\mu^-$,
$p_T(\mu^\pm) > 6$ GeV, $|\eta(\mu^\pm)| < 2.5$ and $\varepsilon_{\mu^\pm} = 0.95$
$p_T(h_i) > 1$ GeV, $|\eta(h_i)| < 2.5$ and $\varepsilon_{h_i} = 0.9$

where ε_μ is the muon reconstruction efficiency for the LVL1 trigger and $\varepsilon_{\mu^\pm, h_i}$ is reconstruction efficiency for second muon and hadrons, p_T and η is transversal momentum pseudorapidity of particles.

The main background (BG) contribution (13000 events) contains the cascade semileptonic decays of one of b-quarks and non rare semileptonic decays of the both of b-quarks. It was generated using the PYTHIA and ATLFAST (the program for the fast simulation of the ATLAS detector) packages.

To reject BG in the semileptonic channels the following cuts have been used:
1. The mass windows $m(h_1 h_2) = M(K^{*0} or \phi) \pm 2\sigma$, where $\sigma(K^{*0}) = 30$ MeV, $\sigma(\phi) = 3$ MeV or $m(\pi^+\pi^- \to \rho^0) \in [0.60; 0.94]$ GeV;
2. Decay length of $B^0_{s,d} > 0.8$ mm, $\chi^2/\text{ndf} < 10$;
3. Angle between p_T of the reconstructed $B^0_{s,d}$ and the line joinning primary and $B^0_{s,d}$ decay vertices $< 3^o$; $m(\mu^+\mu^-) \notin M(J/\psi \text{ or } \psi')$;
4. Isolation cut: $n_{ch}(p_T > 0.8\,GeV) = 0$ in a cone $\theta < 5^o$;
5. For $B^0_d \to K^{*0}\mu^+\mu^-$: $p_T(K^{*0}) > 5$ GeV;
6. The mass window for $B^0_{s,d}$ – mesons $[-\sigma, +2\sigma]$, where $\sigma \sim 50$ MeV.

In the leptonic channels we use analogous cuts.

In the top-quark sector the sensitivity of the decay $t\bar{t} \to (bWZ)(\bar{b}W)$, where $Z \to \ell^+\ell^-$ and $W \to jj$, has been studied using PYTHIA6.1 and ATLFAST. This decay occurs near the kinematical threshold ($m_t \approx m_b + M_W + M_Z$). Therefore the b-jets p_T spectrum of this decay is soft and efficiency of the t-quark exclusive reconstruction is very low. For the top-quark mass reconstruction we use the "semi-inclusive" technique, where the WZ-pair close to threshold was searched for as evidence of the $t \to bWZ$ decay.

Other rare top quark decays can be observed only at the SM extensions [1].

Expected results. Numbers of expected events for leptonic and semileptonic rare B-meson decays after using all cuts (presented in the preceding section) for 3 years of LHC running at low (30 fb^{-1}) and high (100 fb^{-1}) luminosity with the ATLAS detector have been shown in Table 3. From this table one may see, that the ATLAS can detect rare leptonic and semileptonic decays of the B^0_d and B^0_s mesons at reasonable level.

The Forward–Backward charge asymmetry A_{FB} in the different intervals of the normalised transition momentum $\hat{s} = q^2/M_B^2$ is defined by formula:

$$<A_{FB}>_{[\hat{s}_1,\hat{s}_2]} = \frac{<N_F>_{[\hat{s}_1,\hat{s}_2]} - <N_B>_{[\hat{s}_1,\hat{s}_2]}}{<N_F>_{[\hat{s}_1,\hat{s}_2]} + <N_B>_{[\hat{s}_1,\hat{s}_2]} + <N_{BG}>_{[\hat{s}_1,\hat{s}_2]}},$$

where $<N_F>_{[\hat{s}_1,\hat{s}_2]}$ and $<N_B>_{[\hat{s}_1,\hat{s}_2]}$ the numbers of the positive leptons moving in the forward and backward directions of the $B-$ meson correspondingly in the rest frame of the lepton pair, $<N_{BG}>_{[\hat{s}_1,\hat{s}_2]}$ – the number of the BG events in the corresponding $\hat{s}$ interval.

The predictions of the precision asymmetry mesurments for decay $B^0_d \to K^{*0}\mu^+\mu^-$ at ATLAS in the different $\hat{s}$ intervals after 3 years of LHC running at low luminosity presented in the Table 4. The kinematic limits are given by $\hat{s}_{min} = 4m_\mu^2/M_B^2$ and $\hat{s}_{max} = (M_B - M_{K^*})^2/M_B^2$.

CONCLUSIONS

1. The rare muonic and semimuonic B–meson decays provide the important test of the SM or its extensions and open the way for the estimation of the values of $|V_{ts}|$ and $|V_{td}|$.

2. All of these decays could **be observed at the ATLAS**. Furthermore, the decays $B^0_{s,d} \to \mu^+\mu^-$ and $B^0_s \to \phi\mu^+\mu^-$ could be observed **only at the LHC detectors**.

3. Forward–Backward charge asymmetry in $B^0_{d(s)} \to K^{*0}(\phi)\mu^+\mu^-$ can be measured **only at the LHC detectors**.

4. The rare decays of the t–quark provide the possibility to check the SM one–loop predictions and to measure the top quark mass with high accuracy. These top quark decays could be observed **only at the LHC detectors**.

Acknowledgments. I'm grateful to L.N.Smirnova for useful discussions.

REFERENCES

1. "Proceedings of the Workshop on Standard Model Physics (and more) at the LHC", Edited by G.Altarelli and M.L.Mangano, CERN 2000-004, May 2000; "ATLAS Detector and Physics Performance", CERN/LHCC 99-14/15, ATLAS TDR 14/15, May 1999.
2. A.Ali et al., Z. Phys. C63 (1994), 437; Phys. Rev. D61(2000), 07402; D.Melikhov et al., Phys. Rev. D57 (1998), 681.
3. "The BaBar Physics Book. Physics at an Asymmetric B Factory", Edited by P.F.Harrison and H.R.Quinn, SLAC-R-504, October 1998; "Belle Technical Design Report", KEK Reprint 95-1, 1995; K.Miyabayashi, Acta Phys.Polon. B 32 (2001), 1663.
4. CDF Collaboration, Phys. Rev. D50 (1994), 2966; Phys. Rev. D51 (1995), 4623; Phys. Rev. Lett. 74 (1995), 2626; D0 Collaboration, Phys. Rev. Lett. 74 (1995), 2422.
5. "ATLAS Trigger Performance", CERN/LHCC 98-15, August 1998
6. R.Ammar et al., Phys. Rev. Lett. 71 (1993), 674.

TABLE 1. Today b and top physics experimental potential.

Category	Experiment	$\sqrt{s}$ (GeV)	$\sigma_{q\bar{q}}$ (mb)	L (1/cm^2 s)	$b\bar{b}$ or $t\bar{t}$ (pairs/year)
B–PHYSICS	**LHC–detectors (CERN)** at the low luminosity				
	ATLAS	1.4×10^4	10^{-1}	10^{33}	5×10^{12}
	CMS	1.4×10^4	10^{-1}	10^{33}	5×10^{12}
	LHCb	1.4×10^4	10^{-1}	2×10^{32} *	10^{12}
	B factory				
	B–TeV (FNAL)	2×10^3	10^{-1}	2×10^{32}	2×10^{11}
	Belle (KEK)	10.6	1.1×10^{-6}	10^{34}	10^8
	BaBar (SLAC)	10.6	1.1×10^{-6}	3×10^{33}	3×10^7
TOP–PHYSICS	**LHC (CERN)**				
	a) low luminosity	1.4×10^4	830 pb	10^{33}	8×10^6
	b) high luminosity	1.4×10^4	830 pb	10^{34}	8×10^7
	Tevatron (FNAL)	2×10^3	7pb	2×10^{32}	10^5

TABLE 2. Rare B–meson and top quark decays.

Category	Decay	BR in the SM	LHC experiment
B–PHYSICS	$B_d^0 \to K^{*0}\gamma$	$(4.9 \pm 2.0) \times 10^{-5}$ [6]	LHCb
	$B_d^0 \to K^{*0}\mu^+\mu^-$	1.5×10^{-6}	ATLAS, CMS, LHCb
	$B_s^0 \to \phi\mu^+\mu^-$	$\sim 10^{-6}$	**ATLAS, CMS, LHCb**
	$B_d^0 \to \rho^0\mu^+\mu^-$	$\sim 10^{-7}$	ATLAS, CMS, LHCb
	$B_s^0 \to \mu^+\mu^-$	3.5×10^{-9}	**ATLAS, CMS, LHCb**
	$B_d^0 \to \mu^+\mu^-$	1.5×10^{-10}	ATLAS, CMS, LHCb
TOP–PHYSICS	$t \to bWZ$	2×10^{-6}	ATLAS, CMS
	$t \to gc$	5×10^{-11}	ATLAS, CMS
	$t \to \gamma c$	5×10^{-13}	ATLAS, CMS
	$t \to Zc$	1.3×10^{-13}	ATLAS, CMS

TABLE 3. The expected signal and BG for rare leptonic and semileptonic B–meson decays after 3 years of LHC running at low and high luminosity.

L	Decay	Signal	BG
low	$B_d^0 \to \mu^+\mu^-$	4	93
	$B_s^0 \to \mu^+\mu^-$	27	93
high	$B_d^0 \to \mu^+\mu^-$	14	660
	$B_s^0 \to \mu^+\mu^-$	92	660
low	$B_d^0 \to K^{*0}\mu^+\mu^-$	1995	290
	$B_d^0 \to \rho\mu^+\mu^-$	222	950
	$B_s^0 \to \phi\mu^+\mu^-$	411	140

TABLE 4. The precision for asymmetry measurements in three different $\hat{s}$ intervals was estimated by ATLAS. MSSM is the Minimal Super Symmetry Model.

Interval	$\hat{s}_{min} \div 0.14$	$0.14 \div 0.33$	$0.55 \div \hat{s}_{max}$
ATLAS δA_{FB} (3 years)	5 %	4.5 %	6.5 %
SM A_{FB}	10%	-14%	-29 %
MSSM A_{FB}	$(-17 \div 0.5)\%$	$(-35 \div -13)\%$	$(-33 \div -29)\%$

The heavy baryons in the nonperturbative string approach

I. M. Narodetskii* and M. A. Trusov*

ITEP, Moscow, Russia

Abstract. We present some piloting calculations of the short–range correlation coefficients for the heavy baryons and masses of the doubly heavy baryons $\Xi_{QQ'}$ and $\Omega_{QQ'}$ ($Q, Q' = c, b$) in the framework of the simple approximation within the nonperturbative QCD approach.

The purpose of this talk is to present the results of the calculation [1] of the masses and wave functions of the heavy baryons in a simple approximation within the nonperturbative QCD (see [2] and references therein). The starting point of the approach is the Feynman–Schwinger representation for the three quark Green function in QCD in which the role of the time parameter along the trajectory of each quark is played by the Fock–Schwinger proper time. The proper and real times for each quark related via a new quantity that eventually plays the role of the dynamical quark mass. The final result is the derivation [2] of the Effective Hamiltonian, see Eq. (1) below. In contrast to the standard approach of the constituent quark model the dynamical mass m_i is not a free parameter but it is expressed in terms of the current mass $m_i^{(0)}$ defined at the appropriate scale of $\mu \sim 1$ GeV from the condition of the minimum of the baryon mass M_B as function of m_i: $\frac{\partial M_B(m_i)}{\partial m_i} = 0$. Technically, this has been done using the einbein (auxiliary fields) approach, which is proven to be rather accurate in various calculations for relativistic systems.

This method has been already applied to study baryon Regge trajectories [3] and very recently for computation of magnetic moments of light baryons [4]. The essential point of this talk is that it is very reasonable that the same method should also hold for hadrons containing heavy quarks. In what follows we will concentrate on the masses of double heavy baryons. As in [4] we take as the universal parameter the QCD string tension σ fixed in experiment by the meson and baryon Regge slopes. We also include the perturbative Coulomb interaction with the frozen coupling $\alpha_s(1\text{ GeV}) = 0.4$.

Consider the ground state baryons without radial and orbital excitations in which case tensor and spin-orbit forces do not contribute perturbatively. Then only the spin-spin interaction survives in the perturbative approximation. The EH has the following form

$$H = \sum_{i=1}^{3} \left(\frac{m_i^{(0)2}}{2m_i} + \frac{m_i}{2} \right) + H_0 + V, \tag{1}$$

where H_0 is the non-relativistic kinetic energy operator and V is the sum of the perturbative one gluon exchange potential and the string potential. In Eq. (1) $m_i^{(0)}$ are the current quark masses and m_i are the dynamical quark masses to be found from the minimum condition. Since $m_i \gg m_i^{(0)}$ for light quarks, but $m_i \sim m_i^{(0)}$ for the heavy quarks each light quark contributes to the baryon mass an additional mass $\sim m_q/2$ (not m_q as in the ordinary non–relativistic quark model) whereas each heavy quark contributes $\sim m_i$. The Coulomb–like one–gluon exchange potential is $V_c = -\frac{2\alpha_s}{3} \sum_{i<j} \frac{1}{|\mathbf{r}_{ij}|}$. The string potential has been calculated in [3] as the static energy of the three heavy quarks: $V_{\text{string}}(\mathbf{r}_1, \mathbf{r}_2, \mathbf{r}_3) = \sigma R_{\min}$, where $R_{\min}$ is the sum of the three distances $|\mathbf{r}_i|$ from the string junction point, which for simplicity is chosen as coinciding with the center-of-mass coordinate.

We use the hyper radial approximation (HRA) in the hyper-spherical formalism approach. In the HRA (where only the part of the potential which is invariant under rotation in the six-dimensional space spanned by the Jacobi coordinates is taken into account) the three quark wave function depends only on the hyper-radius $R^2 = \rho^2 + \lambda^2$, where ρ and λ are the three-body Jacobi variables:

$$\rho_{ij} = \sqrt{\frac{\mu_{ij}}{\mu}}(\mathbf{r}_i - \mathbf{r}_j), \qquad \lambda_{ij} = \sqrt{\frac{\mu_{ij,k}}{\mu}} \left(\frac{m_i \mathbf{r}_i + m_j \mathbf{r}_j}{m_i + m_j} - \mathbf{r}_k \right),$$

where

$$\mu_{ij} = \frac{m_i m_j}{m_i + m_j}, \qquad \mu_{ij,k} = \frac{(m_i + m_j) m_k}{m_i + m_j + m_k},$$

and μ is an arbitrary parameter with the dimension of mass which drops off in the final expressions. The confining potential V_{string} has a specific three-body character. However, this potential as well as the Coulomb–like potential) is smooth in the sense that the HRA is already an excellent approximation [5]. Introducing the reduced function $\chi(R) = R^{5/2} \psi(R)$ and averaging $V = V_c + V_{\text{string}}$ over the six-dimensional sphere one obtains the Schrödinger equation

$$\frac{d^2 \chi(R)}{dR^2} + 2\mu \left[E_n + \frac{a}{R} - bR - \frac{15}{8\mu R^2} \right] \chi(R) = 0, \tag{2}$$

where

$$a = \frac{2\alpha_s}{3} \cdot \frac{16}{3\pi} \sum_{i<j} \sqrt{\frac{\mu_{ij}}{\mu}}, \qquad b = \sigma \cdot \frac{32}{15\pi} \sum_{i<j} \sqrt{\frac{\mu(m_i + m_j)}{m_k(m_1 + m_2 + m_3)}}. \tag{3}$$

We use the same parameters as in Ref. [6]: $\sigma = 0.17$ GeV, $\alpha_s = 0.4$, $m_q^{(0)} = 0.009$ GeV, $m_s^{(0)} = 0.17$ GeV, $m_c^{(0)} = 1.4$ GeV, and $m_b^{(0)} = 4.8$ GeV.

The dynamical masses m_i and the ground state eigenvalues E_0 calculated using the described above procedure are given for various baryons in Table 1 of Ref. [1]. For the light baryons the values of light quark masses $m_q \sim 450 - 500$ MeV ($q = u, d, s$) qualitatively agree with the results of Ref. [6] obtained from the analysis of the heavy–light ground meson states, but ~ 60 MeV higher than those of Refs. [3], [4]. This difference is due to the different treatment of the Coulomb and spin–spin interactions. The light quark masses are increased by $\sim 100 - 150$ MeV when going from the light to heavy baryons. For the heavy quarks (c and b) the variation in the values of their dynamical masses is marginal. Note that the masses of the light quarks in baryons are slightly smaller than those in the mesons.

For many applications the quantities $R_{ijk} = \langle \psi_{ijk} | \delta^{(3)}(\mathbf{r}_j - \mathbf{r}_i) | \psi_{ijk} \rangle$ are needed. Note that these quantities depend on the third or 'spectator' quark through the three–quark wave function. To estimate effects related to the baryon wave function we solve Eq. (2) by the variational method. We introduce a simple variational ansätz $\chi(R) \sim R^{5/2} e^{-\mu \beta^2 R^2}$, where β is the variational parameter. For this trial function the quantities R_{ijk} are evaluated explicitly: $R_{ijk} = \left(\frac{2\beta^2 \mu_{ij}}{\pi} \right)^{3/2}$. The results of the variational calculations are given in Table 3 of [1]. Comparing the results with those of Ref. [6] we obtain $R_{ijk} < \frac{1}{2} R_{ij}$, and $R_{ijk} > R_{ijl}$, if $m_k \leq m_l$, where R_{ij} is the corresponding quantity for a meson. Our estimations for the ratios R_{ijk}/R_{ij} (see Table 1) agree with the results obtained using the non–relativistic quark model or the bag model or QCD sum rules which are typically in the range $0.1 - 0.5$. Note also that if i, j are the light quarks, and the quarks k and l are the heavy then $R_{ijk} \approx R_{ijl}$ (i.e. $R_{qqc} \approx R_{qqb}$) in agreement with the limit of the heavy quark effective theory.

TABLE 1. The ratios of the squares of the wave functions determining the probability to find a light quark at the location of the heavy quark inside the heavy baryon and the corresponding meson. The meson wave functions are taken from Ref. [6].

R_{ucd}/R_{uc}	R_{scu}/R_{sc}	$R_{ubd}/R_{\bar{b}d}$	R_{sbu}/R_{sb}
0.436	0.405	0.373	0.340

Note also that the wave function calculated in HRA show the marginal diquark clustering in the doubly heavy baryons. This is principally kinematic effect related to the fact that in the HRA the difference between the various $\bar{r}_{ij}$ in a baryon is due to the factor $\sqrt{1/\mu_{ij}}$ which varies between $\sqrt{2/m_i}$ for $m_i = m_j$ and $\sqrt{1/m_i}$ for $m_i \ll m_j$.

To calculate hadron masses we, as in Ref. [3], first renormalize the string potential: $V_{\text{string}} \to V_{\text{string}} + \sum_i C_i$, where the constants C_i take into account the residual self-energy (RSE) of quarks. In principle, these constants can be expressed in terms of the two scalar functions entering covariant expansion of the bilocal cumulants of gluonic fields in the QCD vacuum. In the present work we treat them phenomenologically. To find C_i we assume, first, that the spin splittings of hadrons with a given quark content arise from the color–magnetic interaction in QCD and, secondly, that the color magnetic interaction can be treated perturbatively. Because the color magnetic interaction between two quarks goes inversely as the product of their masses, the perturbative approximation improves as the quark mass increases. However, this approximation may not be good for the baryons containing light quarks. In what follows we adjust the RSE constants C_i to reproduce the center-of-gravity for baryons with a given flavor. To this end we consider the spin-averaged masses, such as: $\frac{M_N + M_\Delta}{2} = 1.085$ GeV, and $\frac{M_\Lambda + M_\Sigma + 2M_{\Sigma^*}}{4} = 1.267$ GeV and analogous combinations for qqc and qqb states. Then we obtain $C_q = 0.34$, $C_s = 0.19$, $C_c \sim C_b \sim 0$.

We keep these parameters fixed to calculate the masses given in Table 2, namely the spin–averaged masses (computed without the spin–spin term) of the lowest double heavy baryons. In this Table we also compare our predictions with the results obtained using the additive non–relativistic quark model with the power-law potential [7], relativistic quasipotential quark model [8], the Feynman–Hellmann theorem [9] and with the predictions obtained in the approximation of double heavy diquark [10].

TABLE 2. Masses of baryons containing two heavy quarks.

State	present work	Ref. [7]	Ref. [8]	Ref. [9]	Ref. [10]
$\Xi\{qcc\}$	3.69	3.70	3.71	3.66	3.48
$\Omega\{scc\}$	3.86	3.80	3.76	3.74	3.58
$\Xi\{qcb\}$	6.96	6.99	6.95	7.04	6.82
$\Omega\{scb\}$	7.13	7.07	7.05	7.09	6.92
$\Xi\{qbb\}$	10.16	10.24	10.23	10.24	10.09
$\Omega\{sbb\}$	10.34	10.30	10.32	10.37	10.19

In conclusion, we have employed the general formalism for the baryons, which is based on nonperturbative QCD and where the only inputs are the string tension σ, the strong coupling constant α_s and two additive constants, C_q and C_s, the residual self–energies of the light quarks. Using this formalism we have also performed the calculations of the spin–averaged masses of baryons with two heavy quarks. One can see from Table 2 that our predictions are especially close to those obtained in Ref. [7] using a variant of the power–law potential adjusted to fit ground state baryons.

This work was supported in part by RFBR grants Refs. 00-02-16363 and 00-15-96786.

REFERENCES

1. I. M. Narodetskii and M. A. Trusov, Yad. Fiz., in press [hep-ph/0104019].
2. Yu. A. Simonov, Lectures given at the XVII International School of Physics "QCD: Perturbative or Nonperturbative", Lisbon 1999 [hep-ph /9911237].
3. M. Fabre de la Ripelle and Yu. A. Simonov, Ann. Phys. (N.Y.) **212**, 235 (1991).
4. B. O. Kerbikov and Yu. A. Simonov, Phys. Rev. D **62**, 093016 (2000).
5. Yu. S. Kalashnikova, I. M. Narodetskii, and Yu. A. Simonov, Yad. Fiz. **46**, 1181 (1987).
6. Yu. S. Kalashnikova and A. Nefediev, Phys. Lett. B **492**, 91 (2000).
7. E. Bagan, H. G. Dosch, P. Godzinsky, S. Narison, and J.–M. Richard, Z. Phys. C **64**, 57 (1994).
8. D. Ebert et al., Z. Phys. C **76**, 111 (1997).
9. R. Roncaglia, D. B. Lichtenberg, and E. Predazzi, Phys. Rev. D **52**, 1248 (1995).
10. A. K. Likhoded and A. I. Onishchenko, hep-ph/9912425.

Hadron-Hadron Scattering in the Nonrelativistic Quark Model

T. Barnes

Physics Division, Oak Ridge National Laboratory, Oak Ridge, TN 37831-6373, USA
Department of Physics and Astronomy, University of Tennessee Knoxville, TN 37996-1501, USA

Abstract. In this HADRON2001 contribution we summarize the status of our quark-model calculations of hadron-hadron scattering amplitudes in annihilation-free channels. The predictions are in reasonably good agreement with experimentally known S-wave meson-meson and meson-baryon phase shifts, and there are very recent indications that S-wave $\pi\omega$ scattering (extracted from FSIs in b_1 decay) may also be similar to our predictions. Finally, novel applications of this formalism to the dissociation cross sections of charmonia on light hadrons (relevant for QGP studies at RHIC) are discussed.

HADRON-HADRON SCATTERING

Introduction

Early models of strong interhadron forces were constructed by analogy with QED Feynman diagrams, and in the important NN problem it was assumed that these forces were dominated by t-channel meson exchange. At large distances, one-pion-exchange can indeed be confirmed in NN high partial waves. However at short distances one must have serious reservations about this type of model, since exchange of a ca. 1 GeV meson in t-channel implies a range of about 0.2 fm. Since this is much smaller than the extent of a typical hadron, the assumption of t-channel meson exchange appears rather dubious (see Maltman and Isgur [1] for a discussion). The success of meson-exchange models may simply be due to the many parameters available for fitting, and less trivially because it may prove difficult to distinguish short-distance QCD processes such as quark interchange from meson exchange, as these involve the same flavor flow.

Since QCD is a theory of quarks and gluons, and short-ranged scattering probes hadronic wavefunctions, it may be possible to describe hadron-hadron scattering in terms of explicit quark model wavefunctions and interquark forces taken from QCD and hadron spectroscopy. This approach has a long history in the NN problem, and has been applied to many hadronic reactions with much success and the occasional interesting failure. The technique most often used is the resonating-group method, although other variational or nonperturbative methods have also been applied. In this contribution we discuss our results from much simpler Born-order calculations of scattering amplitudes in the quark model, which are more straightfoward to evaluate and are also in reasonable agreement with experimental S-wave scattering amplitudes.

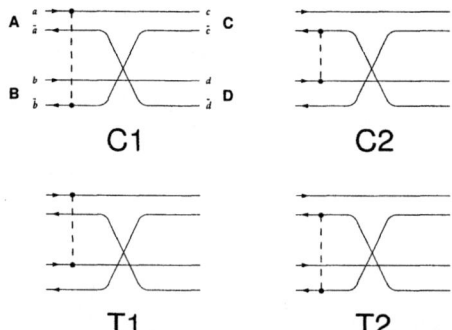

FIGURE 1. The four quark Born diagrams for meson-meson scattering.

Quark Born Diagrams

The usual quark model interaction has $\lambda \cdot \lambda$ color dependence, and in consequence a single interaction between quarks in different hadrons transforms the incident hadronic clusters from color singlets to color octets. Although this makes direct (no quark exchange) scattering zero at Born order, this modified state does have overlap with final color singlet hadrons, provided that we allow quark interchange. We refer to the resulting diagrams as "quark Born diagrams". The four quark Born diagrams one finds for the scattering of two $q\bar{q}$ mesons through this mechanism are shown in Fig.1. (We label these diagrams according to type; if the interacting constituents scatter into the same final hadron this is a "capture" diagram, and if not it is a "transfer" diagram.)

Each diagram has an associated spatial overlap integral, which is weighted by color, spin and flavor multiplicative matrix elements. Detailed evaluation of these diagrams was discussed in Refs.[2, 3], and the "Feynman rules" for the hadron-hadron T-matrices in our current notation are given in Ref.[4]. With Gaussian wavefunctions one may evaluate the T-matrices and phase shifts in closed form. As an example, the I=2 $\pi\pi$ S-wave phase shifts from standard quark model interactions (Ref.[4]) are give in Eq.(1).

$$\delta_0^{I=2\ \pi\pi} = \begin{cases} kE_\pi \frac{\alpha_s}{m^2} \left(-\frac{1}{3^2} \frac{1}{x}\left(1-e^{-2x}\right) - \frac{2^4}{3^{7/2}} e^{-4x/3} \right) & \text{OGE } S \cdot S \\ kE_\pi \frac{\alpha_s}{\beta^2} \left(-\frac{2}{3^2} \frac{1}{x}\left(f_{1,\frac{1}{2}}(-2x) - e^{-2x}\right) - \frac{2^3}{3^{5/2}} f_{1,\frac{3}{2}}(-2x/3)\, e^{-4x/3} \right) & \text{color Cou.} \\ kE_\pi \frac{b}{\beta^4} \left(\frac{1}{3^2} \frac{1}{x}\left(f_{2,\frac{1}{2}}(-2x) - e^{-2x}\right) + \frac{1}{3^{1/2}} f_{2,\frac{3}{2}}(-2x/3)\, e^{-4x/3} \right) & \text{lin. confl.} \end{cases}$$
(1)

where $x = \vec{A}^2/4\beta^2$, $f_{a,c}(x)$ is an abbreviation for the confluent hypergeometric function $_1F_1(a;c;x)$, $|\vec{A}|$ is the pion momentum in the c.m. frame (we assume a relativistic dispersion relation), $\beta = 0.4$ GeV is the standard quark model $q\bar{q}$ wavefunction width parameter, and a conventional quark model parameter set of $\alpha_s = 0.6$, $m_q = 0.33$ GeV and $b = 0.18$ GeV^2 is used to give the curves in Fig.2. The total Born-order S-wave phase shift is the sum of these three contributions. We have confirmed that these results are quite similar to the variational results of Weinstein and Isgur [5], who used essentially

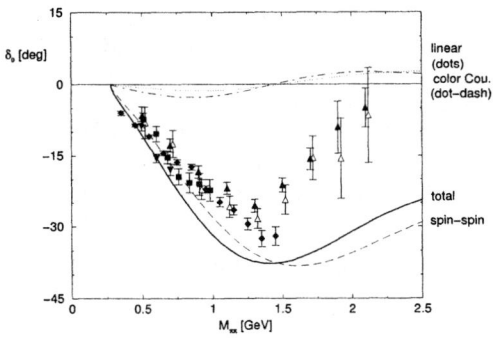

FIGURE 2. I=2 $\pi\pi$ experimental S-wave phase shifts versus Eq.(1), from Ref.[4].

the same interactions but included contributions beyond Born order.

Application of this approach to the scattering of other hadron pairs is straightforward, one need only enumerate the complete set of Born-order scattering diagrams, and evaluate these given a set of external hadron wavefunctions. We have applied this method to S-wave scattering of a wide range of annihilation-free channels, specifically I=3/2 $K\pi$ [6], I=0,1 KN [7], I=0,1 BB [8] (compared to LGT data), and the NN repulsive cores [9], with generally reasonable results. These references consider many additional cases for which we do not have data at present.

More sensitive tests of the hadron scattering mechanism are possible if we consider higher partial waves. Here there is evidence of very interesting physics, for example in the large NN spin-orbit force (referred to by Isgur as the "Holy Grail" of quark-model scattering calculations) and the similarly large KN spin-orbit force. The KN spin-orbit force is apparently not well explained as elastic scattering with quark model forces [10]. (N.Black, unpublished, finds very similar results to this reference.) This discrepancy with experiment may be due to the large inelasticities known experimentally to be present in KN scattering.

We recently considered light vector-pseudoscalar meson scattering as a model spin-orbit problem for the quark Born diagram formalism, and derived the complete set of phase shifts in all partial waves given Gaussian wavefunctions and standard quark model forces [4]. We found that quark-model spin-orbit effects can indeed be quite large, for example in P-wave I=2 $\rho\pi$ elastic scattering we found a phase shift splitting $\delta(^3P_2) - \delta(^3P_0)$ that peaked at about 40°. Although one might suppose vector-pseudoscalar scattering to be experimentally inaccessible, it actually can be measured as a final state interaction in multiamplitude decays. In $b_1 \to \omega\pi$ in particular the S and D amplitudes have FSI phases of $e^{i\delta_S}$ and $e^{i\delta_D}$, so the S-D cross term in the $\omega\pi$ angular distribution is suppressed by $\cos(\delta_S - \delta_D)$ relative to the $|S|^2$ and $|D|^2$ terms. Our prediction is that

$$\delta_S(\omega\pi) - \delta_D(\omega\pi) = -14° \qquad (2)$$

at the b_1 mass. The E852 Collaboration has used this unusual FSI technique to extract this relative phase, and finds a consistent result of $\delta_S(\omega\pi) - \delta_D(\omega\pi) \approx -19°(4°)(8°)$.

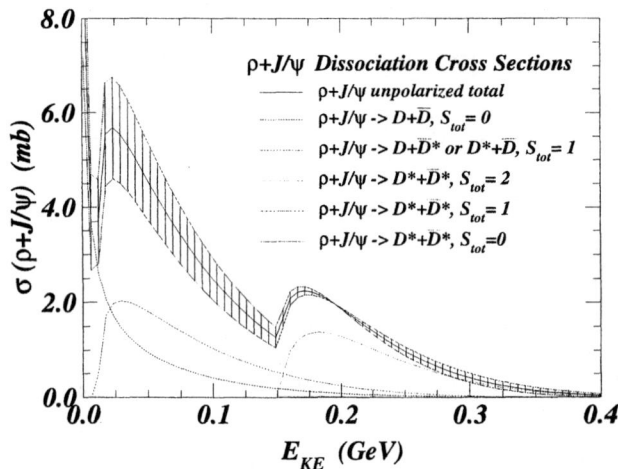

FIGURE 3. The $\rho + J/\psi$ dissociation cross section predicted by quark Born diagrams [12].

($\delta_D - \delta_S = 0.327 \pm 0.061 \pm 0.143$ radians is the tentative, as yet unpublished E852 result; private communication from M.Nozar.)

A New Application: Charmonium Dissociation at RHIC

Recently a novel class of hadronic reactions has attracted the attention of physicists searching for evidence of quark gluon plasma formation in heavy ion collisions. One signature proposed as an indicator of QGP formation is a suppression of the production rate of charmonium bound states such as the J/ψ, since the QGP is expected to screen the linear potential that would normally encourage a $c\bar{c}$ pair produced in the collision to remain bound [11].

If the charmonia that are formed in the collisions can penetrate the cloud of "comoving" light hadrons also produced in the collision, they can be detected through characteristic decays such as $J/\psi \rightarrow \ell^+\ell^-$, and this will be a "clean" experiment. Alternatively, if inelastic charmonium + light hadron cross sections into open-charm final states are sufficiently large, this more conventional $c\bar{c}$ dissociation process may imitate the expected QGP signal and will complicate the interpretation of the experiment. Of course these charmonium + light hadron dissociation cross sections are not at all well known at low energies, and estimates of the scale of these cross sections assuming different theoretical scattering mechanisms cover many orders of magnitude. Here we may have an exciting opportunity to establish the preferred hadron-hadron scattering mechanism in a new regime of QCD.

We have carried out a series of calculations of these charmonium + light meson dissociation cross sections in the constituent interchange model, using the approach described above. We use a standard Coulomb + linear + smeared hyperfine Hamiltonian to determine wavefunctions and scattering amplitudes, which are evaluated using numerical techniques. Our results are that these cross sections at leading order are dissociation to open charm rather than elastic scattering (which is obvious from the flavor flow in Fig.1), and the low-energy cross sections are typically ca. 1 mb in scale [12, 13]. (Earlier work by Martins, Blaschke and Quack [14] using the same scattering formalism found somewhat larger cross sections, due to their assumption of a color-independent confining interaction.) Interestingly, $\rho + J/\psi$ cross sections are much larger than $\pi + J/\psi$, in part because $\rho + J/\psi \to D\bar{D}$ is exothermic and hence diverges as we approach threshold. This process may lead to considerable suppression of the initial J/ψ population, and will require careful consideration in applications of this charmonium-suppression idea to QGP searches.

In future we plan to extend these calculations to a wide range of initial and final states, so that total cross sections (summed over all accessible final states) can be evauated, and more complicated reactions such as $N + J/\psi$ inelastic scattering can be treated. Many other fascinating questions, such as the possibility that charmed mesons might bind to nucleons and nuclei, can also be considered through the application of this model of low-energy interhadron forces.

ACKNOWLEDGMENTS

It is a great pleasure to thank the organisers of HADRON2001 for providing the opportunity to present these results and discuss them with my colleagues in hadron physics. This work was supported in part by the DOE Division of Nuclear Physics, at ORNL, managed by UT-Battelle, LLC, for the US Department of Energy under Contract No. DE-AC05-00OR22725, and by the US National Science Foundation under Grant No. INT-0004089.

REFERENCES

1. K.Maltman, and N.Isgur, *Phys. Rev.*, **D29**, 952 (1984).
2. T.Barnes, and E.S.Swanson, *Phys. Rev.*, **D46**, 131 (1992).
3. E.S.Swanson, *Ann. Phys. (NY)*, **220**, 73 (1992).
4. T.Barnes, N.Black, and E.S.Swanson, *Phys. Rev.*, **C43**, 025204 (2001).
5. J.Weinstein, and N.Isgur, *Phys. Rev.*, **D41**, 2236 (1990).
6. T.Barnes, E.S.Swanson, and J.Weinstein, *Phys. Rev.*, **D46**, 4868 (1992).
7. T.Barnes, and E.S.Swanson, *Phys. Rev.*, **C49**, 1166 (1994).
8. T.Barnes, N.Black, D.J.Dean, and E.S.Swanson, *Phys. Rev.*, **C60**, 045202 (1999).
9. T.Barnes, S.Capstick, M.D.Kovarik, and E.S.Swanson, *Phys. Rev.*, **C48**, 539 (1993).
10. S.Lemaire, J.Labarsouque, and B.Silvestre-Brac, *Nucl. Phys.*, **A** (2002).
11. T.Matsui, and H.Satz, *Phys. Lett.*, **B178**, 416 (1986).
12. C.Y.Wong, E.S.Swanson, and T.Barnes, Heavy Quarkonium Dissociation Cross Sections in Relativistic Heavy-Ion Collisions, Tech. Rep. nucl-th/0106067 (2001).
13. C.Y.Wong, E.S.Swanson, and T.Barnes, *Phys. Rev.*, **C62**, 045201 (2000).
14. K.Martins, D.Blaschke, and E.Quack, *Phys. Rev.*, **C51**, 2723 (1995).

Radial excitations of pseudoscalar mesons

J.S. Suh

Physics Department, Brookhaven National Laboratory, Upton, NY 11973, USA

Abstract. The current status of radial excitations of pseudoscalar mesons is reviewed. From the observed pseudoscalar states, $\pi(1300)$, $K(1460)$, $\eta(1295)$, $\eta(1440)$, and $\eta(1760)$, possible nonets of the first radial excitations are examined, together with open problems.

INTRODUCTION

The pseudoscalar ground state mesons (1^1S_0) π, η, η' and K are well established. In the Particle Data Group (PDG) listing, $\pi(1300)$, $\eta(1295)$, $\eta(1440)$ and $K(1460)$ are suggested as the first radial excitations (2^1S_0) of the pseudoscalar mesons [1]. For the 3^1S_0 states, $\pi(1800)$, $\eta(1760)$ and $K(1860)$ are suggested. In the PDG listing, two pseudoscalar states are reported at the $\eta(1440)$ mass region [1]. However, it is especially hard to find a place for one of the $\eta(1440)$ states within any $q\bar{q}$ model. Therefore, one of the $\eta(1440)$ states is sometimes interpreted as a glueball candidate; as an example see [2]. Expected signatures for glueballs are [3]: (i) enhanced production in gluon rich channels such as radiative J/Ψ decays, (ii) reduced $\gamma\gamma$ couplings, and (iii) decay branching fractions incompatible with SU(3) predictions for $q\bar{q}$ states. In addition, there is no place in $q\bar{q}$ nonets for a glueball.

PSEUDOSCALAR MESONS

$\pi(1300)$

The $\pi(1300)$ has been observed by the VES collaboration in the reaction $\pi^- A \to 3\pi A$ decaying into $\rho\pi$ and $\pi(\pi\pi)_S$ [4]. The state has been confirmed in $\bar{p}N$ annihilation experiments by the Crystal Barrel collaboration [5]. They found a mass of $m = 1375 \pm 40$ MeV/c^2 and a width of $\Gamma = 268 \pm 50$ MeV/c^2 with a dominant decay into $\rho\pi$ from $\bar{p}N$ annihilations at rest [5]. They also found that the $\pi(\pi\pi)_S$ partial decay width of the $\pi(1300)$ is smaller than 15% of its $\rho\pi$ decay width. At this conference the OBELIX collaboration has reported a $\pi(1300)$ with a mass of $m = 1335 \pm 12$ MeV/c^2 and a width of $\Gamma = 339 \pm 26$ MeV/c^2 in the reaction $\bar{n}p \to 3\pi^+ 2\pi^-$ [6]. Their mass and width are compatible with the Crystal Barrel results. However, OBELIX has reported about 100% $\pi(\pi\pi)_S$ decay of the $\pi(1300)$. Recently, the BNL experiment E852 has determined its mass $m = 1343 \pm 15 \pm 24$ MeV/c^2 and width $\Gamma = 449 \pm 39 \pm 47$ MeV/c^2 in the reaction $\pi^- p \to 3\pi p$ [7]. E852 has found the ratio of the branching ratios ($\pi(1300) \to$

$\pi(\pi\pi)_S/(\pi(1300) \to \rho\pi)$ might be dependent on the parametrization of the $(\pi\pi)$S-wave.

K(1460)

The K(1460) has been observed in the $K\pi\pi$ system in the reactions $K^\pm p \to K^\pm \pi^+ \pi^- p$ at 13 GeV/c [8] and by the ACCMOR collaboration in the reaction $K^- p \to K^- \pi^+ \pi^- p$ at 63 GeV/c [9]. At this conference, a K(1460) has been reported by the VES experiment with a mass of $m = 1450 \pm 37$ MeV/c^2 and a width of $\Gamma = 110 \pm 25$ MeV/c^2 [10]. The observed decay modes are $K^*\pi, K\rho, \varepsilon^1 K$ and $\kappa\pi$. They have seen no other 0^- resonance below 2.0 GeV/c in the $K^- \pi^+ \pi^-$ channel [10].

η(1295)

The $\eta(1295)$ has been reported first in the $\eta\pi^+\pi^-$ system in $\pi^- p$ experiments [11][12]. Recently, E852 has observed it in the reaction $\pi^- p \to \eta \pi^+ \pi^- p$ decaying into $a_0\pi$ and $\sigma\eta$ [13] and in the reaction $\pi^- p \to K\bar{K}\pi^0 n$ decaying into $a_0\pi^0$ [14].

η(1440)

The $\eta(1440)$ was discovered in 1963 in a $\bar{p}p$ experiment [15]. Later, it was also observed in J/Ψ radiative decays [16] and in $\pi^- p$ experiments [17][18]. When the $\eta(1440)$ was observed in J/Ψ radiative decays, it was interpreted as a glueball candidate. Upper limits for the two-photon width [24] supported this interpretation, as gluons do not couple directly to photons.

Two pseudoscalar $I = 0$ states have been reported between 1400 MeV/c^2 and 1500 MeV/c^2 in the $K_s^\circ K_s^\circ \pi^\circ$ system produced in $\pi^- p$ interactions [19] and in the $K_s^\circ K^\pm \pi^\mp$ system in J/Ψ radiative decays [20]. Later, these states[2] have also been observed in $\bar{p}p$ experiments at ~ 1405 MeV/c^2 decaying into $a_0(980)\pi$, $\eta\pi\pi$ [21][22] and at ~ 1490 MeV/c^2 decaying into K^*K [23]. The existence of two pseudoscalar states in such a narrow mass range poses severe problems to any quark model. Since the $\eta(1405)$ and the $\eta(1490)$ have been seen, the $\eta(1490)$ has been interpreted as the first radial excitation of the η' because of its decay mode K^*K. The $\eta(1295)$ was already interpreted as the radial excitation of the η. Ever since the $\eta(1405)$ has been interpreted as a glueball.

Later, a decay mode $\pi\pi\gamma$ of the $\eta(1440)$ was reported by the MARK III [26], DM2 [27] and BES collaborations [28] in J/Ψ radiative decays. However, their data did not allow a spin-parity analysis. Hence it was unclear if the signal had to be assigned to $\eta(1440)$ production or to the $f_1(1420)$. Recently, the Crystal Barrel experiment has observed the $\pi\pi\gamma$ decay mode of the $\eta(1405)$ in the $2\pi^+ 2\pi^- \gamma$ final states of $\bar{p}p$

---

[1] where they name the $(\pi\pi)$ S-wave ε. It is also called σ.
[2] called now $\eta(1405)$ and $\eta(1490)$.

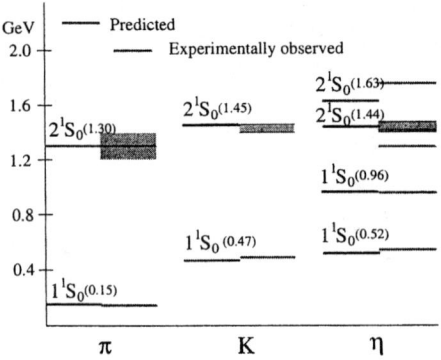

FIGURE 1. Predicted and experimentally observed pseudoscalar mesons. The left lines of each column are the masses of the predicted mesons [35], and the right lines are those of experimentally observed states with errors [1].

annihilation at rest [29]. The signal observed by the Crystal Barrel in $\bar{p}p$ annihilation does not suffer from the $\eta(1405)/f_1(1420)$ uncertainty, and it makes the assumption plausible that the signal from J/Ψ decays originates from the radiative decay $\eta(1405) \to \pi^+\pi^-\gamma$ [29]. They obtained the radiative decay width $\Gamma[\eta(1405) \to \pi^+\pi^-\gamma] \sim 10\%$ of its partial decay width $\Gamma[\eta(1405) \to \pi^+\pi^-\eta]$. Recently, the $\eta(1440)$ has also been observed by the L3 collaboration in untagged $\gamma\gamma$ collisions in the $K^0_s K^\pm \pi^\mp$ decay channel [25]. They obtained a two-photon width $\Gamma_{\gamma\gamma}[\eta(1440)] \cdot BR[\eta(1440) \to K\bar{K}\pi] = (234 \pm 55_{stat} \pm 17_{syst})eV$. The large radiative decay width of the $\eta(1405)$ and its two-photon width make it unlikely that the $\eta(1405)$ or $\eta(1440)$ is a glueball. Furthermore, they do not support the conjecture that $\eta(1440)$ comprises a large glueball fraction. In addition, the latest lattice calculation of the mass of the pseudoscalar glueball is ~ 2.5 GeV/c^2 [30].

$\eta(1760)$

The $\eta(1760)$ has been reported by the MARK III collaboration in $\omega\omega$ [31] and $\rho\rho$ [32] decay modes in J/Ψ radiative decays. It has also been observed by the DM2 collaboration in its $\rho\rho$ decay mode in J/Ψ radiative decays with a mass of $m = 1760 \pm 11$ MeV/c^2 and a width of $\Gamma = 60 \pm 16$ MeV/c^2 [33]. Recently, the BES experiment has reported its $\eta\pi^+\pi^-$ decay mode with a mass of $m = 1760 \pm 35$ MeV/c^2 but without a measurement of the width [34].

NONET ASSIGNMENT OF THE 2^1S_0

Figure 1 shows the predicted pseudoscalar meson spectrum in a relativized quark model with chromodynamics [35] and experimentally observed states [1]. For the radial excitations of the pion and kaon, both are well matched, but there seems to be one (or two if we consider two states at the $\eta(1440)$ region) more observed state(s) than predicted

TABLE 1. Possible nonets of the first radial excitations of the pseudoscalar mesons

$\eta(1295)$ is	$\eta(1440)$ is	Nonet member $\pi(1300), K(1460)$	Mixing angle	Open problems
$q\bar{q}$	two states*	$\eta(1295), \eta(1490)$	$-69° \pm 21°$	$\eta(1295)$,[†] nature of $\eta(1405)$?
$q\bar{q}$	one state	$\eta(1295), \eta(1440)$	$-69° \pm 21°$	$\eta(1295)$[†], splitting of $\eta(1440)$*[†]
not $q\bar{q}$[‡]	two states*	$\eta(1405), \eta(1760)$	$-14° \pm 14°$	Nature of the $\eta(1490)$?
not $q\bar{q}$[‡]	two states*	$\eta(1490), \eta(1760)$	$-19° \pm 19°$	Nature of the $\eta(1405)$?
not $q\bar{q}$[‡]	one state	$\eta(1440)$,[§] $\eta(1760)$	$-19° \pm 19°$	

* The $\eta(1440)$ splitting is considered as two states, $\eta(1405)$ and $\eta(1490)$.
[†] Why is the $\eta(1295)$ seen in the $\pi^- p$ experiment only ?
** How can the splitting of the $\eta(1440)$ be explained ?
[‡] The $\eta(1295)$ is considered as a non $q\bar{q}$ state.
[§] The splitting of the $\eta(1440)$ may provide an evidence that the $\eta(1440)$ should be the radial excitation of the η in the $^3P_0$ model [38].

in the η section. The same discrepancy has been reported in the meson spectrum in a covariant quark model(Fig. 16 in [36]).

From the observed pseudoscalar states, $\pi(1300), K(1460), \eta(1295), \eta(1440)$, and $\eta(1760)$, possible nonets of the first radial excitations are shown in Table 1, together with open problems. Mixing angles are calculated with the help of the Gell-Mann-Okubo formula [1]. Here the same definition of the mixing angle is used for the radial excitations as for the ground states.

The $\eta(1295)$ has been interpreted as the radial excitation of the η since it was first reported [11]. However, the $\eta(1295)$ has been seen neither in J/Ψ radiative decays nor in $\bar{p}p$ experiments[3], while the $\eta(1440)$ has been observed in both reactions which favor the production of pseudoscalar states. A comparison of the branching ratios for production of $\sigma\pi$ and $\sigma\eta$ with those for $\sigma\pi(1300)$ and $\sigma\eta(1295)/\sigma\eta(1440)$ suggests that the $\eta(1440)$ is a better isoscalar companion of the $\pi(1300)$ [38]. Also, the $\eta(1295)$ was suggested not as a $q\bar{q}$ radial excitation but as a hybrid in a bag model [39].

CONCLUSION

The $\pi(1300)$ and $K(1400)$ are most likely the first radial excitations, but the assignment of η's are still open. To establish the "correct" nonet members, the nature of the $\eta(1295)$ and the splitting of the $\eta(1440)$ should be explained.

---

[3] reported by the E771 collaboration at BNL in a $\bar{p}p$ experiment but very weak [37]

ACKNOWLEDGMENTS

I wish to thank the conference organizers for their kind invitation. I would also like to thank Dr. S.U. Chung, Prof. E. Klempt and Dr. H. Willutzki for the opportunity to participate in the conference.

REFERENCES

1. Groom, D.E. et al., *Eur. Phys. Jour.* **C15**, 1 (2000)
2. Minkowski, P. and Ochs, W., *Eur. Phys. Jour.* **C9**, 283-312 (1999)
3. Amsler, C. *Eur. Phys. Jour.* **C15**, 682-683 (2000)
4. Amelin, D.V. et al., *Phys. Lett.* **B356**, 595-600 (1995)
5. Abele, A. et al., *Eur. Phys. Jour.* **C19**, 667-675 (2001)
6. Filippi, A., OBELIX Collaboration, "Study of Isospin 2 Resonant States in $\bar{n}p$ Annihilatios", this conference
7. Chung, S.U. et al., to be submitted to *Phys. Rev.* **D**
8. Brandenburg, G.W. et al., *Phys. Rev. Lett.* **36**, 1239-1242 (1976)
9. Daum, C. et al., *Nucl. Phys.* **B187**, 1-41 (1981)
10. Nikolaenko, V., VES Collaboration, "Study of the reaction $K^-N \to K^-\pi^+\pi^-N$ at VES", this conference
11. Stanton, N.R. et al., *Phys. Rev. Lett.* **42**, 346-349 (1979)
12. Fukui, S. et al., *Phys. Lett.* **B267**, 293-298 (1991)
13. Manak, J.J. et al., *Phys. Rev.* **D62**, 012003-1-8 (2000).
14. Adams, G.S. et al., *Phys. Lett.* **B516**, 264-272 (2001).
15. Baillon, P. et al., *Nuovo Cimento* **50A**, 393-421 (1967)
16. Scharre, D.L. et al., *Phys. Lett.* **97B**, 329-332 (1980)
17. Chung, S.U. et al., *Phys. Rev. Lett.* **55**, 779-782 (1985)
18. Birman, A. et al., *Phys. Rev. Lett.* **61**, 1557-1560 (1988)
19. Rath, M.G. et al., *Phys. Rev.* **D40**, 693-705 (1989)
20. Bai, Z. et al., *Phys. Rev. Lett.* **65**, 2507-2510 (1990)
21. Amsler, C. et al., *Phys. Lett.* **B358**, 389-398 (1995)
22. Suh, J.S. Crystal Barrel Collaboration, *Nucl. Phys.* **A675**, 100-103 (2000)
23. Cicalo, C. et al., *Phys. Lett.* **B462**, 453-461 (1999)
24. Behrend, H.J. et al., *Z. Phys.* **C42**, 367-376 (1989)
25. Vodopianov, I. L3 Collaboration, *Acta Physica Polonica* **B31**, 2453-2458 (2000)
26. Coffman, D. et al., *Phys. Rev.* **D41**, 1410-1413 (1990)
27. Augustin, J.E. et al., *Phys. Rev.* **D42**, 10-19 (1990)
28. Xu, G.F. BES Collaboration, *Nucl. Phys.* **A675**, 337-340 (2000)
29. Klempt, E. et al. Crystal Barrel Collaboration, will be published.
30. Morningstar, C.J. and Peardon, M.J., *Phys. Rev.* **D60**, 034509-1-13 (1999)
31. Baltrusaitis, R.M. et al., *Phys. Rev. Lett.* **55**, 1723-1726 (1985)
32. Baltrusaitis, R.M. et al., *Phys. Rev.* **D33**, 1222-1232 (1986)
33. Bisello, D. et al., *Phys. Rev.* **D39**, 701-712 (1989)
34. Bai, J.Z. et al., *Phys. Lett.* **B446**, 356-362 (1999)
35. Godfrey, S. and Isgur, N., *Phys. Rev.* **D32**, 189-231 (1985)
36. Ricken, R. et al., *Eur. Phys. Jour.* **A9**, 221-244 (2000)
37. Bar-Yam, Z. et al., "Comparison of PWA Results of $K^+K^\circ_S\pi^-$ Final States Produced in 8 GeV/c K^-p, π^-p and $\bar{p}p$ Interactions", in *Hadron 91*, edited by Oneda, S. and Peaslee, D.C., World Scientific, Singapore, 1992, pp. 61-68.
38. Suh, J.S., *Nucl. Phys.* **A675**, 104-107 (2000)
39. Barnes, T. and Close, F.E., *Phys. Lett.* **116B**, 365-368 (1982)

Higher vector meson states

Burkhard Pick representing the Crystal Barrel collaboration

Institut für Strahlen- und Kernphysik, University of Bonn, Germany

Abstract. The partial wave analyses of several data-sets on $\bar{p}d$ at rest taken with the Crystal Barrel detector at LEAR (CERN) show evidence for at least two ρ-like resonances accompanying the $\rho(770)$ ground-state meson. Moreover, the decay channel $\omega\pi$ gives hints for the existence of a ρ' state with a mass around 1180 MeV. This result opens a wide field for interpretations of the spectrum of vector mesons.

INTRODUCTION

The nature of excitations of the $\rho(770)$ ground state has been discussed controversially. In 1986, is was shown that data on e^+e^- annihilation into 2π and 4π require two ρ-like resonances in the 1600 MeV region [1]. These analyses were extended and further data on τ-lepton decays and photoproduction were taken into account [2]. Definite evidence for at least the existence of $\rho(1450)$ and $\rho(1700)$ was confirmed. The existence of a further low-mass ρ' state with a mass around 1250 MeV, proposed by the LASS experiment [3] and former bubble chamber experiments [4], could not be excluded. If such an additional state exists, its electromagnetic coupling will be weak.

The couplings of several decay modes can be interpreted in the light of phenomenological models. Barnes et al. [5] have calculated the partial decay widths within the framework of the $^3P_0$ model assuming the ρ' states to be pure $q\bar{q}$ states (see Tab. 1). For the first radial excitation a decay into $\pi\pi$, $\omega\pi$ and $K\bar{K}$ should be observed whereas the couplings to the 4π decay modes are predicted to be very small. This can be compared to decay widths of a hybrid state calculated within the flux-tube model by Close et al. [6].

TABLE 1. Decay widths in MeV calculated within the framework of the $^3P_0$ model [5] for pure quarkonia states and for a hybrid state calculated within the framework of the flux-tube model [6]. Decays marked with "–" are not part of the calculations (π^* is a shortcut for $\pi(1300)$).

decay mode	$\pi\pi$	$\omega\pi$	$a_2\pi$	$a_1\pi$	$h_1\pi$	$\rho\rho$	$\pi^*\pi$	$\rho\sigma$	$K\bar{K}$
Calculated partial widths in the $^3P_0$ model:									[5]
$2^3S_1\, \rho(1465)$	74	122	0	3	1	0	–	–	35
$1^3D_1\, \rho(1700)$	48	35	2	134	124	0	14	–	36
$3^3S_1\, \rho(1900)$	1	5	46	26	32	70	16	–	1
Calculated partial widths in the flux-tube model:									[6]
Hybrid-$\rho(\sim 1500)$	0	5–10	~ 0	140	0	0	0	–	–

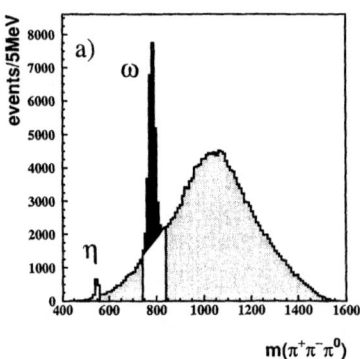
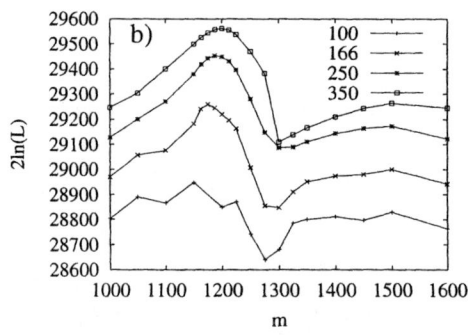

FIGURE 1. a) $\pi^+\pi^-\pi^0$ invariant mass spectrum form $\bar{p}d \to \pi^+\pi^-\pi^-\pi^0\pi^0 p_{spec}$. Each event causes four entries. b) Scan with two ρ' states using Breit-Wigner parameterisation including $\rho^-(770)\omega$, $b_1(1232)\pi$, $\rho(1700)\pi$, $\rho(x)\pi$.

In contrast to the pure quarkonia states, the coupling of a hybrid to $\omega\pi$ is weak and the coupling to $\pi\pi$ vanishes completely. Furthermore, the suppression of $h_1\pi$ compared to $a_1\pi$ is a crucial test for the hybrid interpretation. Data on antiproton annihilation taken with the Crystal Barrel experiment may help to shed more light on the nature of the higher iso-vector vector mesons.

CRYSTAL BARREL DATA

The Crystal Barrel experiment investigated the annihilation of antiprotons on neutrons in liquid deuterium at rest. This is a good environment to study several decay modes of ρ' states into two-body and four-body final states. Explicitly, these are

$$\bar{p}d \to \pi^-\pi^0\pi^0 p_{spectator} \qquad (1)$$

where the partial wave analysis requires the introduction of $\rho(1450)$ and $\rho(1700)$ both decaying into $\pi\pi$ [7]. A 100 MeV cut on the momentum of the spectator proton ensures that the annihilation takes place on a quasi-free neutron. The existence of a further ρ' state did not improve the data description significantly. The ρ' decays into $K\bar{K}$ were investigated in the channel

$$\bar{p}d \to K_S K^-\pi^0 p_{spectator} \qquad (K_S \to \pi^+\pi^-) \qquad (2)$$

Again both states, $\rho(1450)$ and $\rho(1700)$, are required by the data [8]. A combined analysis of the channels

$$\bar{p}d \to \pi^+\pi^-\pi^-\pi^0\pi^0 p_{spectator} \quad \text{and} \quad \bar{p}d \to \pi^-\pi^0\pi^0\pi^0\pi^0 p_{spectator} \qquad (3)$$

allows the observation of 4π decay modes of $\rho(1450)$ and $\rho(1700)$ [9].

The ρ' decay into $\omega\pi$ can be investigated in the channel

$$\bar{p}d \to \omega\pi^-\pi^0 p_{spectator} \qquad (4)$$

Figure 1a) shows the $\pi^+\pi^-\pi^0$ invariant mass spectrum including four entries for each event. If there is an ω decaying into $\pi^+\pi^-\pi^0$, then one entry belongs to the prominent ω signal and three entries form a combinatorial background. In addition, there are events without any ω contribution. A fit yields 21492 ± 204 events including an ω. The distribution of the background originating from $\bar{p}d \to \pi^+\pi^-\pi^-\pi^0\pi^0 p_{spec}$ is well known from the analyses mentioned before. Therefore, these amplitudes are included with fixed parameters in the partial wave analysis.

$$|\mathcal{A}_{total}|^2 = \overbrace{|\mathcal{A}^{4\pi}|^2}^{\text{known background}} + \underbrace{\left[|\mathcal{A}^{\omega}_{1S_0}|^2 + |\mathcal{A}^{\omega}_{3P_0}|^2 + |\mathcal{A}^{\omega}_{3P_1}|^2 + |\mathcal{A}^{\omega}_{3P_2}|^2\right]}_{\omega\pi^-\pi^0 \text{ partial waves}} \quad (5)$$

Since the ω carries spin, its decay angular distribution is also taken into consideration. The contributing partial waves are $\bar{p}n \to \rho(770)\omega$, $\bar{p}n \to b_1(1235)\pi$ and $\bar{p}n \to \rho'\pi$.

Including all established resonances and $\rho(1450)$ ($m = 1435$ MeV, $\Gamma = 325$ MeV) as well as $\rho(1700)$ ($m = 1700$ MeV, $\Gamma = 235$ MeV) in the parameterisation, the partial decay-widths for the $\rho(1450)$ are not in contradiction to the predictions of the $^3P_0$ model for $\pi\pi$, $\omega\pi$ and $K\bar{K}$ decay modes. Only the small 4π decay widths of the $\rho(1450)$ is not consistent with our analyses. Our results disagree with the predictions of the flux-tube model for a hybrid state. The decay into $\pi\pi$ is observed and the suppression of the $h_1\pi$ decay mode in contrast to $a_1\pi$ is not seen. Experimental results for the $\rho(1700)$ and the predictions for the first orbital excitation 1^3D_1 are in good agreement.

A more sophisticated look for the lower mass ρ' states reveals interesting results. Figure 1b) shows the variation of $2\ln L$ for certain masses of one ρ' state where the mass of the second ρ' state is fixed to $m = 1700$ MeV and $\Gamma = 235$ MeV. Around 1180 MeV the data are described well whereas around 1450 MeV no enhancement in $2\ln L$ is visible. This hint at the existence of a low-lying ρ' state decaying into $\omega\pi$ also appears if the parameterisation with two Breit-Wigner resonances is exchanged into a $(\omega\pi)_P$-wave within the $\hat{K}$-matrix formalism. Like before, the variation of $2\ln L$ shows

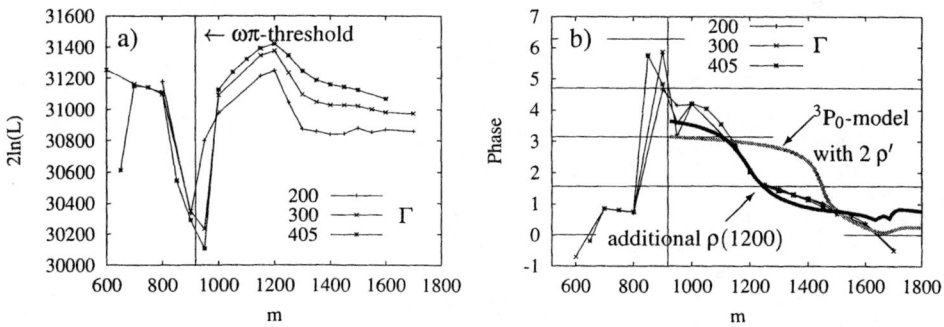

FIGURE 2. 2-pol solution within $\hat{K}$-matrix formalism. a) Variation of the lower mass ρ' pol; b) relative phase between lower and upper ρ' pole assuming the the lower pol to be locate at the indicated mass point. The overlayed curves show theoretically expected phases.

a good description for a pole around a mass of 1180 MeV (see Fig. 2a)). Assuming the lower mass ρ' state to be at a given mass, the relative phase between the poles is shown in Figure 2b). The phase variation is larger than π, there should be a further state, in addition to the well known $\rho(1450)$. The rapid motion cannot be described with the parameters from the $^3P_0$ model. The introduction of a $\rho(1200)$ fits the data much better.

INTERPRETATION: THE SPECTRUM OF VECTOR MESONS

Assuming that a $\rho(1200)$ exists, the field for interesting speculations concerning the spectrum of the vector mesons opens. The physically observed states could mix:

$$(\rho(1200), \rho(1450), \rho(1700)) = \mathcal{M}\left(|2^3S_1\rangle, |H_0\rangle, |1^3D_1\rangle\right) \qquad (6)$$

Furthermore, the $\rho(1200)$ could be assigned to a nonet of $J^{PC} = 1^{--}$ four-quark states:

$$\rho(1200) \qquad \omega(1200) \qquad K^*(1410) \qquad C(1480) \qquad (7)$$

The only accepted state is the $K^*(1410)$. Hints for an $\omega(1200)$ were found by SND [10], the $C(1480)$ was only observed in one experiment at Serpukov [11, 12]. Then $\rho(1450)$ and $\rho(1700)$ could be members of $q\bar{q}$-nonets:

$$2^3S_1: \qquad \rho(1450) \qquad \omega(1420) \qquad K^*(1680) \qquad \phi(1680) \qquad (8)$$

$$1^3D_1: \qquad \rho(1700) \qquad \omega(1650) \qquad K^*(?) \qquad \phi(?) \qquad (9)$$

where some states are still unknown.

SUMMARY AND CONCLUSION

The spectrum of $(I = 1)$ vector mesons in the mass region below 2 GeV may contain more states than just $\rho(770)$, $\rho(1450)$ and $\rho(1700)$. The existence of $\rho(1200)$ seems to be likely.

REFERENCES

1. M. Olsson, C. Erkal: Z. Phys. **C31** (1986) 615; A. Donnachie, H. Mirzaie: Z. Phys. **C33** (1987) 407
2. A.B. Clegg, A. Donnachie: Z. Phys. **C62** (1994) 155
3. D. Aston et al.: Nucl. Phys. (Proc. suppl.) **21** (1991) 105
4. P. Frenkiel et al.: Nucl. Phys. **B76** (1974) 375; J. Ballam et al.: Nucl Phys. **B47** (1972) 61
5. T. Barnes, F.E. Close, P.R. Page, E.S. Swanson Phys. Rev. **D55** (1997) 4157
6. F.E. Close, P.R. Page Nucl. Phys. **B443** (1995) 233
7. A. Abele et al.: Phys. Lett. **B391** (1997) 191
8. K. Wittmack: PhD thesis University of Bonn, 2001
9. A. Abele et al.: Eur. Phys. J. **C21** (2001) 261
10. M.N. Achasov et al.: Phys. Lett. **B462** (1999) 365
11. S.I. Bityukov et al.: Phys. Lett. **B188** (1987) 383
12. N.N. Achasov and A.A. Kozhevnikov: Z. Phys. **C48** (1990) 121

On spectroscopy of ρ', ρ'' and ω', ω'' resonances

N. N. Achasov* and A. A. Kozhevnikov*

Laboratory of Theoretical Physics, Sobolev Institute for Mathematics, 630090, Novosibirsk, Russia

Abstract. Based on coupling constants extracted from fitting various data, the selected branching ratios and full widths of ρ', ρ'' and ω', ω'' resonances are calculated, and some topics of the spectroscopy of these states are discussed

The resonances $\rho' \equiv \rho'_1 \equiv \rho(1450)$, $\rho'' \equiv \rho'_2 \equiv \rho(1700)$ $\left[I^G \left(J^{PC} \right) = 1^+ (1^{--}) \right]$ were observed in the channels $\pi^+\pi^-$, 4π, $\omega\pi$, $\rho\eta$ of the reactions of e^+e^- annihilation, τ lepton and $J/\psi \to \pi^+\pi^-\pi^0$ decays, photoproduction etc. [1]. The resonances $\omega' \equiv \omega'_1 \equiv \omega(1420)$, $\omega'' \equiv \omega'_2 \equiv \omega(1600)$ $\left[I^G \left(J^{PC} \right) = 0^- (1^{--}) \right]$ were observed in the channels $\pi^+\pi^-\pi^0$, $\omega\pi^+\pi^-$, etc. The couplings with the states $K^*\bar{K}+$ c.c. and $K^*\bar{K}\pi+$ c.c. are allowed. Possible indications come from the reactions of e^+e^- annihilation, photoproduction, etc. [1]. The resonances with the same I^G are expected to be strongly mixed via common decay modes. The partial width of each specific mode is strongly energy dependent.

The approach to the treatment of the resonance mixing inspired by field theory is used. It consists in the following steps, assuming, for the sake of brevity, only two mixed states. First, in case of zero width, zero mixing the propagator of the resonance R_a, $a = 1, 2$, is $1/d_a^{(0)}$. Second, in case of finite width, zero mixing the propagator of the resonance R_a becomes

$$\frac{1}{D_a^{(0)}} = \frac{1}{d_a^{(0)}} + \frac{1}{d_a^{(0)}} \Pi_{aa}^{(0)} \frac{1}{d_a^{(0)}} + \cdots = \frac{1}{d_a^{(0)} - \Pi_{aa}^{(0)}},$$

Here $\Pi_{aa}^{(0)}$ are the diagonal polarization operators. At last, in case of finite width, nonzero mixing the propagator and the nondiagonal 'polarization operator' Π_{12} responsible for the mixing are

$$\frac{1}{D_1} = \frac{1}{D_1^{(0)}} + \frac{1}{D_1^{(0)}} \Pi_{12}^{(0)} \frac{1}{D_2^{(0)}} \Pi_{12}^{(0)} \frac{1}{D_1^{(0)}} + \cdots = \frac{D_2^{(0)}}{D_1^{(0)} D_2^{(0)} - \Pi_{12}^{(0)2}} \equiv \left(G^{-1} \right)_{11},$$

analogously for $1/D_2$, and

$$\frac{\Pi_{12}}{D_1^{(0)} D_2^{(0)}} = \frac{\Pi_{12}^{(0)}}{D_1^{(0)} D_2^{(0)}} + \frac{[\Pi_{12}^{(0)}]^3}{[D_1^{(0)} D_2^{(0)}]^2} + \cdots = \frac{\Pi_{12}^{(0)}}{D_1^{(0)} D_2^{(0)} - \Pi_{12}^{(0)2}} \equiv \left(G^{-1} \right)_{12},$$

where

$$G(s) = \begin{pmatrix} D_1^{(0)} & -\Pi_{12}^{(0)} \\ -\Pi_{12}^{(0)} & D_2^{(0)} \end{pmatrix}$$

is the matrix of inverse propagators. The physical states and their couplings are obtained by the diagonalization of G^{-1}. Hereafter $D_a^{(0)} \equiv D_a^{(0)}(s) = m_a^{(0)2} - s - i\sqrt{s}\Gamma_a^{(0)}(s)$, $\Pi_{ab}^{(0)} \equiv \Pi_{ab}^{(0)}(s) = \text{Re}\Pi_{ab}^{(0)} + i\text{Im}\Pi_{ab}^{(0)}(s)$. By unitarity, $\text{Im}\Pi_{aa}^{(0)}(s) = \sqrt{s}\sum_i \Gamma_{a \to i}^{(0)}(s)$, $\text{Im}\Pi_{ab}^{(0)}(s) = \sqrt{s}\sum_i \Gamma_{a \to i}^{(0)}(s)\frac{g_{R_b i}^{(0)}}{g_{R_a i}^{(0)}}$, where $\Gamma_{a \to i}^{(0)}$ is the partial width of the decay $R_a \to i$. Index (0) refers to the unmixed state. The masses $m_a^{(0)}$, coupling constants $g_{R_a i}^{(0)}$, and $\text{Re}\Pi_{ab}^{(0)}$ assumed to be constants should be determined from experiment. The generalization to the case of arbitrary number of states is straightforward. The amplitude for the process $i \to R_1 + R_2 \to f$ is now

$$a_{fi} = (g_{fR_1}^{(0)}, g_{fR_2}^{(0)}) G(s)^{-1} \begin{pmatrix} g_{R_1 i}^{(0)} \\ g_{R_2 i}^{(0)} \end{pmatrix}.$$

There are two essential points related with the resonance mixing and the dependence of partial widths on energy. First, the mass shift of resonances due to their mixing is obtained by the condition of vanishing of the real part of $\det G$. More important is the shift of the resonance position due to energy dependent width. In the case of the single resonance R with the bare mass m_R produced in the reaction $e^+e^- \to R \to f$ whose cross section can be written as

$$\sigma(s) = 12\pi m_R^3 \Gamma_{Rl+l^-}(m_R) g_{Rf}^2 \frac{s^{-3/2} W_{Rf}(s)}{(s - m_R^2)^2 + s\Gamma_R^2(s)},$$

where W_{Rf} is the phase space volume of final state, the peak position is found from the vanishing of derivative of $\sigma(s)$ with respect to s. At sufficiently slow varying phase space the peak shifts to

$$s_R \approx m_R^2 - \frac{1}{2}m_R^2 \Gamma_R^2 \frac{d}{ds_R}\left[\ln\left(s_R^{-3/2} W_{Rf}(s_R)\right)\right] < m_R^2.$$

The data are fitted in the framework of the three-resonance scheme [2] for each channel where the indication on the specific resonance exists, see [3, 4, 5, 6, 7] and references therein [1]. The $V'_{1,2}VP$ coupling constants extracted from the fits turned out to be in the intervals $|g_{\rho'_1\omega\pi}| = 10 - 18$ GeV^{-1}, and $|g_{\rho'_2\omega\pi}| = 2 - 13$ GeV^{-1}. Qualitatively, the relation $|g_{\rho'_{1,2}\omega\pi}| \sim |g_{\omega'_{1,2}\rho\pi}|$ was found to be satisfied. $V'_{1,2}VP$ coupling constants are not suppressed as compared to the VVP ones. Coupling constants with multi-particle states such as $g_{\rho'_{1,2}\rho\pi+\pi^-}$, $g_{\omega'_{1,2}\omega\pi+\pi^-}$, as well as analogous coupling containing strange mesons in final states, should also be included. Taking $g_{\rho'_{1,2}\omega\pi} \sim g_{\omega'_{1,2}\rho\pi} \simeq 10$ GeV^{-1}, and $m_{\rho'_1} \approx m_{\omega'_1} = 1400$ MeV, one can estimate the partial widths to be $\Gamma_{\rho'_1 \to \omega\pi} \sim 280$ MeV, $\Gamma_{\omega'_1 \to \rho\pi} \sim 820$ MeV. Analogously, assuming that $m_{\rho'_2} \approx m_{\omega'_2} = 1750$ MeV, one finds $\Gamma_{\rho'_2 \to \omega\pi} \sim 880$ MeV, $\Gamma_{\omega'_2 \to \rho\pi} \sim 2600$ MeV. VP decay modes are not the only ones to

---

[1] We have used the earlier SND data [3] on the reaction $e^+e^- \to \pi^+\pi^-\pi^0$. Now new SND data [4] on this reaction are available. The question of whether these new data can be described in the three resonance scheme with the only $\omega'_{1,2}$ states in addition to $\omega(782)$ or new states should be invoked is under study.

TABLE 1. Masses, total widths (in the units of MeV), leptonic widths (in the units of keV), and selected branching ratios (percent) of ρ_1' resonance, calculated using the coupling constants extracted from the fits [5] of the specific reaction. The symbol $\sim$ means that only central value is given, while error exceeds it considerably. The ρ_1' resonance does not reveal itself in the $e^+e^- \to \omega\pi^0$ reaction and τ^- decay. Other branching ratios can be found in Ref. [7].

reaction	$e^+e^- \to \pi^+\pi^-$	$e^+e^- \to 2\pi^+2\pi^-$	$e^+e^- \to \pi^+\pi^-2\pi^0$	$K^-p \to \pi^+\pi^-\Lambda$
$m_{\rho_1'}$	1370^{+90}_{-70}	1350 ± 50	1400^{+220}_{-140}	1360^{+180}_{-160}
$B_{\rho_1' \to \pi^+\pi^-}$	1.1 ± 1.1	~ 1.4	~ 8.0	~ 0.7
$B_{\rho_1' \to \omega\pi^0}$	86.5 ± 41.5	93.6 ± 60.0	77.8 ± 62.2	93.3 ± 82.7
$B_{\rho_1' \to 4\pi}$	~ 6.8	~ 0.2	~ 7.2	~ 0.7
$\Gamma_{\rho_1' \to l^+l^-}$	$6.4^{+1.2}_{-1.4}$	$5.4^{+2.6}_{-1.8}$	$6.3^{+3.3}_{-2.5}$	—
$\Gamma_{\rho_1'}$	763 ± 500	~ 518	~ 970	~ 460

TABLE 2. The same as in Table 1, but in the case of the ρ_2' resonance. The latter does not reveal itself in the $K^-p \to \pi^+\pi^-\Lambda$ reaction.

reaction	$e^+e^- \to \pi^+\pi^-$	$e^+e^- \to \omega\pi^0$	$e^+e^- \to 2\pi^+2\pi^-$	$e^+e^- \to \pi^+\pi^-2\pi^0$	$\tau^- \to (4\pi)^-\nu_\tau$
$m_{\rho_2'}$	1900^{+170}_{-130}	1710 ± 90	1851^{+270}_{-240}	1790^{+110}_{-70}	1860^{+260}_{-160}
$B_{\rho_2' \to \pi^+\pi^-}$	~ 0	~ 0	~ 1.2	~ 0.4	1.5 ± 1.4
$B_{\rho_2' \to \omega\pi^0}$	~ 16.7	22.3 ± 8.0	13.4 ± 3.9	31.0 ± 18.6	18.9 ± 2.8
$B_{\rho_2' \to 4\pi}$	~ 68.5	61.2 ± 7.8	74.0 ± 32.1	45.0 ± 18.0	62.6 ± 5.0
$\Gamma_{\rho_2' \to l^+l^-}$	1.8 ± 1.5	5.2 ± 1.5	$4.02^{+0.28}_{-0.27}$	4.5 ± 1.3	9.3 ± 0.6
$\Gamma_{\rho_2'}$	~ 303.9	1886 ± 613	3123 ± 296	3151 ± 1281	3255 ± 388

TABLE 3. Masses, total widths (in the units of MeV), leptonic widths (in the units of eV), and branching ratios (percent) of the $\omega_{1,2}'$ resonances, calculated using the coupling constants extracted from the fits [6] of the specific channel of e^+e^- annihilation. The symbol $\sim$ means that only central value is given, while error exceeds it considerably.

channel	$\pi^+\pi^-\pi^0$	$\omega\pi^+\pi^-$	K^+K^-	$K^0_S K^\pm \pi^\mp$	$K^{*0} K^\mp \pi^\pm$
$m_{\omega_1'}$	1430^{+110}_{-70}	~ 1400	~ 1460	~ 1500	~ 1380
$B_{\omega_1' \to 3\pi}$	~ 21.6	~ 8	~ 67	~ 96	~ 34
$B_{\omega_1' \to K^*\bar{K}+cc}$	~ 0.2	~ 0	~ 1	~ 4	0
$B_{\omega_1' \to K^*\bar{K}\pi}$	~ 0	0	0	0	0
$B_{\omega_1' \to \omega\pi^+\pi^-}$	~ 78.2	~ 92	~ 31.2	~ 0	~ 65.8
$\Gamma_{\omega_1' \to l^+l^-}$	144^{+94}_{-58}	~ 0.2	~ 8	~ 8	~ 48
$\Gamma_{\omega_1'}$	~ 903	~ 129	~ 173	~ 1252	~ 112
$m_{\omega_2'}$	1940^{+170}_{-130}	2000 ± 180	1780^{+170}_{-300}	~ 2120	1880^{+600}_{-1000}
$B_{\omega_2' \to 3\pi}$	~ 22.1	~ 34.2	~ 88.8	~ 91.2	~ 60.1
$B_{\omega_2' \to K^*\bar{K}+cc}$	~ 3.5	~ 5.8	~ 11.2	~ 15.8	~ 8.9
$B_{\omega_2' \to K^*\bar{K}\pi}$	~ 68.2	~ 53.4	0	0	~ 30.9
$B_{\omega_2' \to \omega\pi^+\pi^-}$	~ 6.2	~ 6.6	0	0	~ 0
$\Gamma_{\omega_2' l^+l^-}$	109^{+58}_{-46}	531 ± 225	0	~ 189	1162 ± 922
$\Gamma_{\omega_2'}$	~ 14000	~ 5757	~ 2420	~ 9854	~ 13820

which heavy resonances can decay, and partial widths of the decay into many particle final states are found to be of the same order or even larger as compared to VP ones. The results of the calculation of selected branching ratios and total widths are presented in Tables 1, 2 and 3. One can see that the resonances $\rho_{1,2}'$ and $\omega_{1,2}'$ are broad.

Our conclusions are as follows.

- One should be careful in attributing the specific peak or structure in the cross section to the specific spectroscopy state, because the large width, the rapid growth of the phase space with the energy increase, and the mixing among the resonances result in the shift of the visible peaks in the cross sections. For example, large width of the ω'_2 resonance results in the shift of its peak in the energy dependence of the $e^+e^- \to \omega\pi^+\pi^-$ reaction cross section to $\simeq 1680$ MeV from the bare mass $\simeq 2000$ MeV [6].
- $\rho(1300)$ state observed by LASS team [8] in the reaction $K^-p \to \pi^+\pi^-\Lambda$, revived an old discussion concerning the possible existence of the $\rho(1250)$ meson, in addition to the $\rho(1450)$ claimed to be observed in e^+e^- annihilation. The results presented in Table 1 show that the corresponding peak should be attributed to the same state $\rho(1450)$ as that presented in Reviews of Particle Physics [1].
- The state $\omega(1200)$ observed by SND team [3, 4] in the reaction $e^+e^- \to \pi^+\pi^-\pi^0$ is, in our opinion, the same state as $\omega(1420) \equiv \omega'_1$ presented in PDG [1].
- The states $\rho'_{1,2}$ and $\omega'_{1,2}$ turn out to be the broad resonance structures, as if the conventional quark picture of them as the radial excitations is implied. The present results, having in mind their significant uncertainties, do not contradict to the assignment of ρ'_1 and ω'_1 resonances to the state 2^3S_1. In the meantime, the central values of the ρ'_2 and ω'_2 widths are large, which contradicts to the assigning them to the state 1^3D_1 predicted by Gogfrey and Isgur [9] to be relatively narrow. However, large errors prevent one from drawing final conclusion.
- The very large widths of the resonances may indirectly evidence in favor of some nonresonant contributions to the amplitudes. The accuracy of the existing data is still poor either to isolate such contributions reliably or to specify their form precisely. Considerable increase of experimental statistics hopefully accessible at VEPP-2000 collider aimed at the study of the energy range of e^+e^- annihilation from 1 to 2 GeV will help in resolving the above issues.

REFERENCES

1. Particle Data Group, Groom D. E., *et al.*, *Eur. Phys. J.* **C15**, 1 (2000).
2. Donnachie, A., "Problems with vector mesons", plenary talk, *these Proceedings*.
3. Achasov, M. N., *et al.*, *Phys. Lett.* **B462**, 365 (1999).
4. Achasov, M. N., *et al.*, "Review of experimental results from SND detector", plenary talk, *these Proceedings*, hep-ex/0109035.
5. Achasov, N. N., and Kozhevnikov, A. A., *Phys. Rev.* D **55**, 2663 (1997).
6. Achasov, N. N., and Kozhevnikov, A. A., *Phys. Rev.* D **57**, 4334 (1998).
7. Achasov, N. N., and Kozhevnikov, A. A., *Phys. Rev.* D**62**, 117503 (2000).
8. Aston, D., *et al.*, "$J^P = 1^-$ Radial Excitatios from LASS Data", in *Hadron'91*, edited by S. Oneda and D. C. Peasley, World Scientific, Singapore, 1992, pp. 410-414.
9. Godfrey, S., and Isgur, N., *Phys. Rev.* D**32**, 189 (1985).

LOW ENERGY

IHEP, Protvino

New Results on Baryon Spectroscopy with the Crystal Ball Spectrometer

D. Mark Manley
for the Crystal Ball Collaboration[1]

Department of Physics, Kent State University, Kent, OH 44242 U.S.A.

Abstract. Data on π^-p and K^-p interactions to various neutral final states were measured in 1998 with the Crystal Ball multiphoton spectrometer. The measurements were carried out for beam momenta up to about 760 MeV/c at the C6 beam line of the Alternating Gradient Synchrotron at Brookhaven National Laboratory. This article focuses primarily on the new results from the K^-p measurements. Our data for $K^-p \to \eta\Lambda$ and $K^-p \to \pi^0\Sigma^0$ provide new constraints on the properties of light Λ^* resonances, while our data for $K^-p \to \pi^0\Lambda$ provide new constraints on the properties of light Σ^* resonances. Our preliminary results, for example, on the Λ polarization in $K^-p \to \pi^0\Lambda$ tend to rule out the two-star $\Sigma(1580)3/2^-$. This state, if confirmed, cannot be an ordinary three-quark state because of its very low mass.

INTRODUCTION

Our principal knowledge of the properties of light hyperon (Λ, Σ) resonances is based almost entirely on bubble-chamber experiments that were carried out in the 1970s. Bubble chambers are best suited for studying reactions involving at most a single neutral particle in the final state; consequently, little is known about some K^-p reactions involving all-neutral final states. During the late 1970s, the Crystal Ball multiphoton spectrometer was constructed at the Stanford Linear Accelerator Center (SLAC). This detector contains 672 optically isolated NaI crystals and provides good energy resolution and a solid-angle coverage of 93% of 4π steradians. Between 1978 and 1986, the Crystal Ball was used for meson spectroscopy experiments, initially at SPEAR and later at DESY. Then from 1987 to 1995, the Crystal Ball lay dormant in storage at SLAC. In late 1995, the Crystal Ball was transported to Brookhaven National Laboratory (BNL) where it was installed in the C6 beamline at the Alternating Gradient Synchrotron

---

[1] The Crystal Ball Collaboration consists of: D. Isenhower, M. Sadler (*Abilene Christian University*); C.E. Allgower, H. Spinka (*Argonne National Laboratory*); J.R. Comfort, K. Craig, A.F. Ramirez (*Arizona State University*); M. Clajus, A. Marušić, S. McDonald, B.M.K. Nefkens, N. Phaisangittisakul, S. Prakhov, J.W. Price, A. Starostin, W.B. Tippens (*University of California Los Angeles*); J. Peterson (*University of Colorado*); W.J. Briscoe, A. Shafi, I.I. Strakovsky (*The George Washington University*); H.M. Staudenmaier (*Universität Karlsruhe*); D.M. Manley, J. Olmsted (*Kent State University*); D.C. Peaslee (*University of Maryland*); V.V. Abaev, V. Bekrenev, A.A. Kulbardis, N.G. Kozlenko, S. Kruglov, I.V. Lopatin (*Petersburg Nuclear Physics Institute*); N. Knecht, G. Lolos, Z. Papandreou (*University of Regina*); I. Supek (*Rudjer Bošković Institute*); D. Grosnick, D.D. Koetke, R. Manweiler, T.D.S. Stanislaus (*Valparaiso University*).

(AGS). There it began its "new life" as a detector for baryon spectroscopy experiments. Experiment E913 used the Crystal Ball detector to study N and Δ resonances formed in $\pi^- p$ experiments and experiment E914 used the detector to study Λ and Σ resonances formed in $K^- p$ experiments. This article summarizes some of the results obtained in the $K^- p$ experiments.

SUMMARY OF PRELIMINARY RESULTS

The C6 beamline at the BNL AGS has a maximum laboratory momentum of about 760 MeV/c. Experiment E914 used the Crystal Ball to make simultaneous measurements for $K^- p$ reactions involving all-neutral final states at eight nominal momenta ranging from about 500 to 750 MeV/c. No results were obtained at momenta below 500 MeV/c because beam flux on target was too low for these momenta. For results with adequate statistics, we were able to rebin our data into smaller momentum intervals. We obtained data on several reactions including $K^- p \to \gamma\Lambda$, $K^- p \to \pi^0\Lambda$, $K^- p \to \overline{K^0}n$, $K^- p \to \eta\Lambda$, $K^- p \to \pi^0\pi^0\Lambda$, $K^- p \to \pi^0\overline{K^0}n$, and $K^- p \to \pi^0\pi^0\pi^0\Lambda$. We also obtained data, with one exception,[2] for the corresponding reactions in which the Λ is replaced by Σ^0. Neutral pions were detected by their decays, $\pi^0 \to 2\gamma$, while η mesons were detected by both $\eta \to 2\gamma$ and $\eta \to 3\pi^0 \to 6\gamma$. Short-lived neutral kaons were detected by their decays, $K_S^0 \to 2\pi^0 \to 4\gamma$. We detected the Λ and Σ^0 by their decays, $\Lambda \to \pi^0 n$ and $\Sigma^0 \to \gamma\Lambda$, respectively. In all cases, the final states were reconstructed by detecting the final-state photons. In some cases, final-state neutrons were also detected. Our data span the center-of-mass energy range 1560 to 1680 MeV. Here we present preliminary results for a few selected cases.

We have obtained the world's most precise data [1] in the near-threshold region for the pure isospin ($I = 0$) reaction $K^- p \to \eta\Lambda$. In 1965, Berley et al. [2] discovered that the cross section for this reaction has a strong peak just above threshold. It is now recognized that this peak is associated with S-wave formation of the $\Lambda(1670)1/2^-$ resonance. This feature is analogous to the strong S-wave formation of the $N(1535)1/2^-$ state just above threshold in $\pi^- p \to \eta n$. The $\Lambda(1670)$ and $N(1535)$ are of special interest because they are among the three known baryon resonances with decays having a large η branch.[3] Recently the author carried out a unitary, multichannel fit [3] of S-wave $I = 0$ $\overline{K}N$ amplitudes using the Crystal Ball total cross-section data for $K^- p \to \eta n$ near threshold, which was assumed to be dominantly S-wave. The results of this fit yielded the most reliable resonance parameters to date for $\Lambda(1670)$.

Our data for the pure $I = 1$ reaction $K^- p \to \pi^0\Lambda$ and the pure $I = 0$ reaction $K^- p \to \pi^0\Sigma^0$ are much more precise than older data. These reactions are of interest in part because they selectively excite intermediate Σ^* and Λ^* resonances, respectively. In our momentum range, the most important prior measurements are the 1970 bubble-chamber results of Armenteros et al. [4]. Because the Λ and Σ^0 are self-analyzing, polarizations

---

[2] We were below threshold for the pure $I = 1$ reaction $K^- p \to \eta\Sigma^0$.
[3] The only other known baryon resonance with a large η branch is the $\Sigma(1750)1/2^-$.

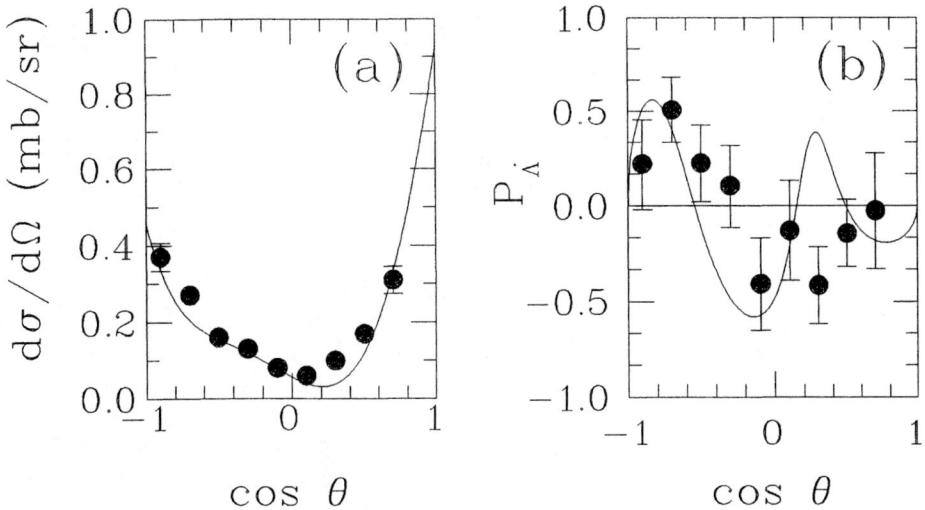

FIGURE 1. (a) Differential cross section and (b) Λ polarization for the reaction $K^-p \to \pi^0\Lambda$ at 750 MeV/c. The data are preliminary results by the Crystal Ball Collaboration. The curves are predictions based on the 1977 PWA of Gopal et al. [5].

as well as differential cross sections can be extracted. Figure 1 shows preliminary Crystal Ball results for the differential cross section and Λ polarization for $K^-p \to \pi^0\Lambda$ at 750 MeV/c. For comparison, we also show curves predicted by the partial-wave analysis (PWA) of Gopal et al. [5]. The curve predicted for the Λ polarization does not agree well with our data. The reason for this is simple: *The older polarization data for these reactions are of such low precision that they made little constraint on PWAs.* The polarization data for $K^-p \to \pi^0\Lambda$ and $K^-p \to \pi^0\Sigma^0$ are sensitive only to P- and D-wave contributions in our momentum range. The P-waves are especially interesting because they will tell us about the $\Lambda(1600)1/2^+$ and $\Sigma(1660)1/2^+$, which are the most likely candidates for SU(3) analogs of the controversial Roper resonance, or $N(1440)1/2^+$. Very little is known about these analog states. For example, the *Review of Particle Physics* [6] lists the totals widths for the $\Lambda(1600)$ and $\Sigma(1660)$ to be 50–250 MeV and 40–200 MeV, respectively.

Another state of great interest is the $\Sigma(1580)3/2^-$. The mass of this state is too low for it to be accommodated as an ordinary 3-quark state. The first evidence for this two-star state was a narrow bump observed in the total $\overline{K}N$ cross section. (See Ref. [7] and references therein.) Later, P. J. Litchfield [8] carried out a PWA of the $K^-p \to \pi^0\Lambda$ data measured by Armenteros et al. [4], which he rebinned into 10-MeV/c intervals. Litchfield concluded including a D_{13} resonance with a mass 1582 ± 4 MeV and width 11 ± 4 MeV made a significant decrease in χ^2. We have made a comparison of our preliminary results for the $K^-p \to \pi^0\Lambda$ Legendre fitting coefficients with Litchfield's predictions. Our results agree much better with his predictions that do *not* include the narrow D_{13} resonance. We conclude that the $\Sigma(1580)$ is not required to describe our data.

We have obtained preliminary Dalitz plots for both $K^-p \to \pi^0\pi^0\Lambda$ and $K^-p \to \pi^0\pi^0\Sigma^0$ at all eight nominal momenta. We expect our data for the former reaction and $K^-p \to \pi^0\Sigma^0$ to provide new information about the Roper analog state, $\Lambda(1600)1/2^+$, through the process $K^-p \to \Lambda(1600)1/2^+ \to \pi^0\Sigma^0(1385) \to \pi^0\pi^0\Lambda$. Indeed, plots of the $\pi^0\Lambda$ invariant mass clearly reveal a strong peak due to the $\Sigma(1385)3/2^+$. Our Dalitz plots for $K^-p \to \pi^0\pi^0\Sigma^0$ reveal the presence of the intermediate state $\pi^0\Lambda(1405)$ although there is a fairly large background from other processes.

CONCLUDING REMARKS

In 1998, we used the Crystal Ball spectrometer to measure new data for several $\overline{K}N$ reactions, including $K^-p \to \pi^0\Lambda$, $K^-p \to \pi^0\Sigma^0$, and $K^-p \to \eta\Lambda$, at the BNL AGS. A multichannel analysis incorporating our $K^-p \to \eta\Lambda$ total cross-section data was carried out and yielded improved resonance parameters for the $\Lambda(1670)1/2^-$. Our preliminary differential cross sections and polarizations for $K^-p \to \pi^0\Lambda$ strongly refute the existence of the $\Sigma(1580)3/2^-$, a state too light to be accommodated as an ordinary 3-quark state if its spin-parity determination is correct. Plans are underway to begin a new PWA of the world data for $\overline{K}N$ reactions, including the new Crystal Ball data. It is anticipated that our new polarization data will effectively add a new constraint to PWAs. A new AGS experiment (E953) to study specific hyperon resonances has been approved. This experiment will provide better statistics for several reactions, including $K^-p \to \pi^0\Sigma^0$ and $K^-p \to \pi^0\Sigma^0(1385) \to \pi^0\pi^0\Lambda$, which should help elucidate the properties of the Roper resonance analog, $\Lambda(1600)1/2^+$.

ACKNOWLEDGMENTS

The author would like to thank the State Research Center of Russia "Institute for High Energy Physics (IHEP)" for partial financial support and hospitality while he was in Protvino, Russia. Partial financial support was also provided by Kent State University's "Center for Nuclear Research". Various members of the Crystal Ball Collaboration are supported in part by DOE, NSF, NSERC of Canada, and the Russian Ministry of Science and Technology. The assistance of BNL is also gratefully acknowledged.

REFERENCES

1. Starostin, A., et al., *Phys. Rev. C* (2001), in press.
2. Berley, D., et al., *Phys. Rev. Letters*, **15**, 641 (1965).
3. Manley, D. M., et al., *Phys. Rev. Letters* (2001), submitted for publication.
4. Armenteros, R., et al., *Nucl. Phys. B*, **21**, 15 (1970).
5. Gopal, G. P., et al., *Nucl. Phys. B*, **119**, 362 (1977).
6. Groom, D. E., et al., *Eur. Phys. J. C*, **15**, 1 (2000).
7. Carroll, A., et al., *Phys. Rev. Letters*, **37**, 806 (1976).
8. Litchfield, P. J., *Phys. Letters B*, **51**, 509 (1974).

Mesurement of differential cross sections of the reaction $\pi^- p \to \eta n$ using the Crystal Ball detector[1]

Nikolai G. Kozlenko
for the Crystal Ball Collaboration [2]

Petersburg Nuclear Physics Institute, Gatchina, Leningrad district, 188350, Russia

Abstract. Preliminary data on the total and differential cross sections of the reaction $\pi^- p \to \eta n$, obtained in 1998 by the Crystal Ball Collaboration, are presented. It is possible to see that the angular distribution at momenta near the threshold (685 MeV/c) is consistent with the S-wave dominance. From the momentum of 720 MeV/c, the P-wave contribution can be found.

Until now experimental information on the cross sections of the reaction $\pi^- p \to \eta n$ has been very scarce and contradictory, especially near the threshold (685 MeV/c). At the same time, obtaining accurate experimental data in the near-threshold region is very important for verifying theoretical models of η-meson production. Such data will also be useful for extracting the ηN scattering length and understanding properties of $S_{11}(1535)$ resonance.

For measuring $\pi^- p \to \eta n$ cross sections it is necessary to detect either the neutron or the photons from the η-meson decay modes $\eta \to 2\gamma$ or $\eta \to 3\pi^0 \to 6\gamma$. An ideal detector for these purposes is a 4π gamma detector, such as the Crystal Ball (CB) [1]. The Crystal Ball detector is shown in Fig. 1. It consists of 672 separate NaI(Tl) crystals covering 93% of 4π steradians, each crystal being 16 radiation lengths thick. The target was surrounded by four veto counters (veto-barrel) to reject charged events.

---

[1] Supported in part by US DOE, NSF, NSERC, Volkswagen Stiftung, the Russian Foundation for Basic Research and the Russian Ministry of Industry, Science and Technology.

[2] The Crystal Ball Collaboration consists of B. Draper, S. Hayden, J. Huddleston, D. Isenhower, C. Robinson and M. Sadler, *Abilene Christian University,* C. Allgower, R. Cadman and H. Spinka, *Argonne National Laboratory,* J. Comfort, K. Craig and A. Ramirez, *Arizona State University,* T. Kycia (deceased), *Brookhaven National Laboratory,* M. Clajus, A. Marusic, S. McDonald, B.M.K. Nefkens, N. Phaisangittisakul, S. Prakhov, J.W. Price and W.B. Tippens, *University of California at Los Angeles,* J. Peterson, *University of Colorado,* W. Briscoe, A. Shafi and I. Strakovsky, *George Washington University,* H. Staudenmaier, *Universität Karlsruhe,* D.M. Manley and J. Olmsted, *Kent State University,* D. Peaslee, *University of Maryland,* V. Abaev, V. Bekrenev, A. Koulbardis, N. Kozlenko, S. Kruglov, I. Lopatin and A. Starostin (now in UCLA,USA), *Petersburg Nuclear Physics Institute,* N. Knecht, G. Lolos and Z. Papandreou, *University of Regina,* I. Supek, *Rudjer Boskovic Institute,* A. Gibson, D. Grosnic, D.D. Koetke, R. Manweiler and S. Stanislaus, *Valparaiso University,* H. Calen, A. Kupse, T. Johanson and U. Wiedner, *Uppsala University.*

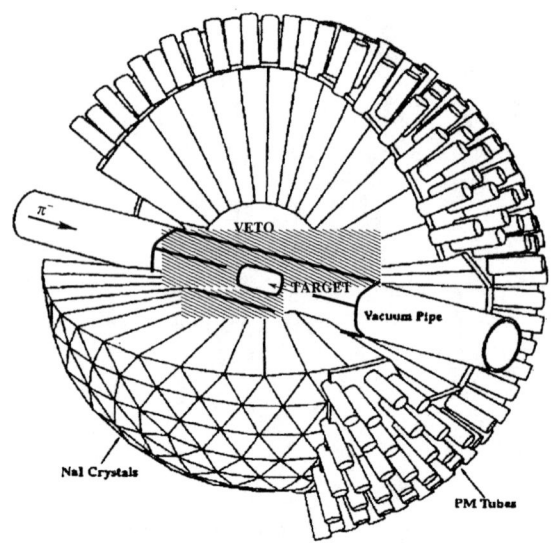

FIGURE 1. Crystal Ball schematic view.

Measurements of the reaction yield were performed in the near-threshold region in the momentum range up to 750 MeV/c.

The central momentum of the incident pions beam was defined with an accuracy of 0.4%, $\Delta P/P$ (FWHM) was 2.3%. The invariant mass of the η meson (for the case when $\eta \to 2\gamma$ and 2γ were detected in the CB) was calculated as $M_{2\gamma}^{inv} = \sqrt{2E_{\gamma_1}E_{\gamma_2}(1-\cos\theta_{\gamma_1\gamma_2})}$. In this formula E_{γ_1} – energy of the first photon; E_{γ_2} – energy of the second photon; $\theta_{\gamma_1\gamma_2}$ – opening angle between directions of the two photons.

A typical invariant mass distribution for 2γ events is shown in Fig. 2. The background

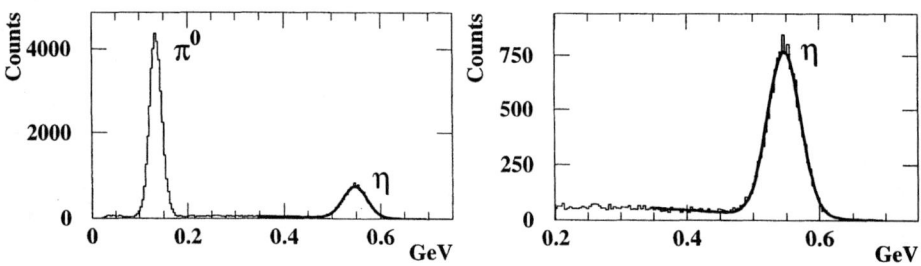

FIGURE 2. Invariant mass spectrum for 2γ. The right histogram is an enlarged version of the left one in the vicinity of the eta.

under the η peak (in Fig. 2) is about 6% and was subtracted when calculating the yield. The background from an empty target was about 2%.

The total cross section σ^{tot} is defined by the formula:

$$\sigma^{tot} \propto \frac{N_\eta}{N_{\pi^-} \, N_p \, A}, \quad \text{where} \quad N_{\pi^-} = N_{tot}\,(1-n_e)\,(1-n_\mu),$$

N_η — number of events under the peak of $M_{2\gamma}^{inv}$ (after background subtraction);
N_{π^-} — number of incident pions;
N_p — number of protons in the target ($1/\text{cm}^2$);
A — acceptance of the Crystal Ball for detection of 2γ (including analysis efficiency), its value is about 0.45 for 720 MeV/c;
N_{tot} — beam monitor (total number of particles including e, μ, π);
n_e, n_μ — fractions of electron and muon contamination, correspondingly.

When obtaining the numbers N_{π^-} and N_p, we used the reconstructed tracks of incident particles from beam proportional chambers. The target has an approximately cylindrical shape, 10 cm length and 10 cm in diameter. Cuts of ± 3 cm in X-coordinate (horizontal) and ± 2 cm in Y-coordinate (vertical) were applied.

The veto barrel efficiency was taken into account by Monte-Carlo calculation and included in the acceptance. The uncertainty in knowledge of the hardware threshold of the veto barrel gave an error of 1% for the value of the total cross section.

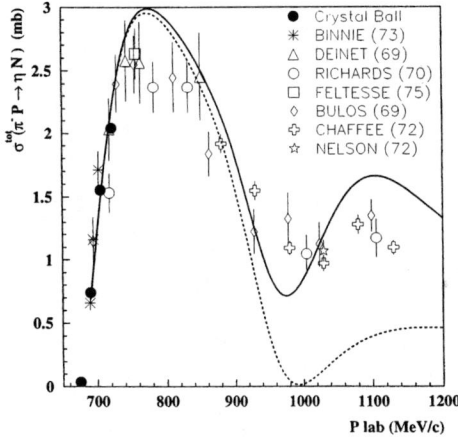

FIGURE 3. Total cross section of $\pi^- p \to \eta n$. The solid line corresponds to calculations [2] including S and P waves. The dashed line corresponds to only S wave contributions alone.

The total cross section obtained in this work is in good agreement with previous experiments but exceeds significantly all existing experimental data in a statistical accuracy – see Fig. 3. The systematic error is about 4%.

Our results do not contradict to the predictions of the K-matrix model of pion-nucleon scattering [2], which takes into account S- and P-wave contributions to the η production.

The differential cross sections for 705, 720, 750 MeV/c are shown in Fig. 4. The angular distributions were fitted with Legendre polynomials. One can see that at the momentum of 705 MeV/c the main contribution comes from S-wave amplitude. But at

720 MeV/c and 750 MeV/c higher waves are needed to describe the differential cross section.

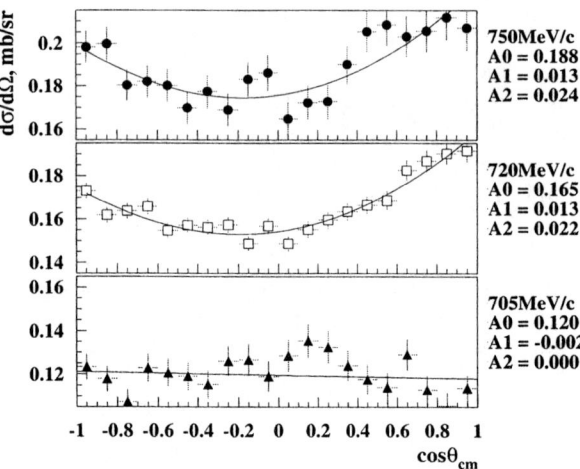

FIGURE 4. The very preliminary differential cross sections of $\pi^- p \to \eta n$ reaction. A_0, A_1, A_2 – the coefficients of Legendre polynomial.

The study of this reaction at momenta higher than 750 MeV/c is necessary. The proposal to do these measurements till 2 GeV/c on D-line at BNL was made by the Crystal Ball Collaboration.

REFERENCES

1. Sadler, M., *πN Newsletter*, **13**, 123 (1997).
2. Gridnev, A., and Kozlenko, N., *Eur. Phys. J.*, **A4**, 187 (1999).

Differential cross sections of the charge exchange reaction $\pi^- p \to \pi^0 n$ in the momentum range from 148 to 323 MeV/c

A. Koulbardis for the Crystal Ball Collaboration[1]

Petersburg Nuclear Physics Institute, Gatchina, Leningrad distr., Russia, 188300

Abstract. Preliminary values of differential cross sections of the reaction $\pi^- p \to \pi^0 n$ at π^- momenta 148, 176, 188, 212, 238, 269, 298, 323 MeV/c are presented. Statistical errors of the differential cross sections are as a rule at a level of 3-5%. The experiment was made using the Crystal Ball detector at the Alternating Gradient Synchrotron (AGS) at Brookhaven National Laboratory (BNL), USA.

INTRODUCTION

Several authors [3] have measured $\pi^- p \to \pi^0 n$ differential cross sections in this momentum range. We are adding 159 new values of differential cross section to 165 existing ones.

THE CRYSTAL BALL DETECTOR

The Crystal Ball (CB) detector (see Fig. 1) was built by SLAC [1] in the 1970s and was used in several experiments. Now the CB detector is installed in beam line C6 of the Alternating Gradient Synchrotron (AGS) at Brookhaven National Laboratory (BNL), USA.

In the center of the detector is a liquid hydrogen target 10 cm in length surrounded by a cylindrical veto barrel to reject charged particles. We mostly register reactions with

---

[1] The Crystal Ball Collaboration consists of B. Draper, S. Hayden, J. Huddleston, D. Isenhower, C. Robinson and M. Sadler, *Abilene Christian University,* C. Allgower, R. Cadman and H. Spinka, *Argonne National Laboratory,* J. Comfort, K. Craig and A. Ramirez, *Arizona State University,* T. Kycia (deceased), *Brookhaven National Laboratory,* M. Clajus, A. Marusic, S. McDonald, B.M.K. Nefkens, N. Phaisangittisakul, S. Prakhov,J.W. Price and W.B. Tippens, *University of California at Los Angeles,* J. Peterson, *University of Colorado,* W. Briscoe, A. Shafi and I. Strakovsky, *George Washington University,* H. Staudenmaier, *Universität Karlsruhe,* D.M. Manley and J. Olmsted, *Kent State University,* D. Peaslee, *University of Maryland,* V. Abaev, V. Bekrenev, A. Koulbardis, N. Kozlenko, S. Kruglov, I. Lopatin and A. Starostin (now in UCLA,USA), *Petersburg Nuclear Physics Institute,* N. Knecht, G. Lolos and Z. Papandreou, *University of Regina,* I. Supek, *Rudjer Boskovic Institute,* A. Gibson, D. Grosnic, D.D. Koetke, R. Manweiler and S. Stanislaus, *Valparaiso University,* H. Calen, A. Kupse, T. Johanson and U. Wiedner, *Uppsala University.*

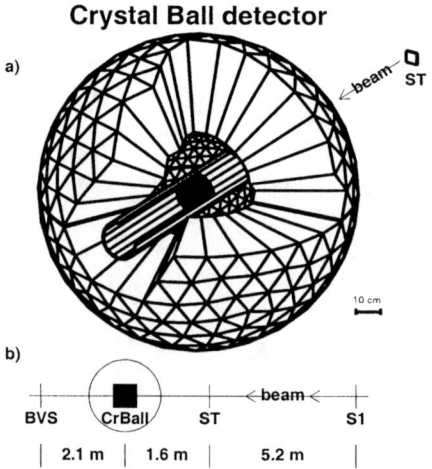

FIGURE 1. The Crystal Ball detector. a) the crystals, the veto barrel and the target are shown with 1/4 cut out, b) schematic picture of the beam line.

only neutral particles in final state. The Crystal Ball detector consists of 672 NaI(Tl) crystals arranged in a 1.3m sphere with holes for beam entrance and exit. The crystals cover 93% of 4π steradians. The sphere is approximated by an icosohedron, a figure consisting of 20 equivalent equilateral triangles called as 'major' triangles. Each major triangle is divided into 4 'minor' triangles and each minor triangle consists of 9 crystals. While majors are similar, there are 2 types of minors - the central minor is slightly different from the corners. The Crystal Ball consists of 11 types of slightly different shaped crystals. The crystals are about 41 cm long truncated triangular pyramids about 14 cm across at the outer and about 5 cm at the inner radius. The inner radius of the sphere of crystals is about 25 cm from the center.

Pion beam momentum and coordinates were measured by a set of 7 drift chambers. Three drift chambers were used to measure the beam horizontal coordinate, three were used to measure the vertical coordinate and one drift chamber situated before the beam bending magnet helped to determine the momentum of the beam particle.

The particles in the beam (π, μ, e) were separated for this work by their time-of-flight on a base of about 9 meters.

THE DATA PROCESSING

The data of $\pi^- p$ charge exchange scattering were taken in October 1998. We are presenting preliminary $\pi^- p \to \pi^0 n$ differential cross sections in π^- momentum range 148-323 MeV/c with momentum spread of $\sigma = 1.5$–1.0%, respectively.

We determined the beam contamination by time-of-flight. We used (see Fig. 1b) scintillator counters S1 and BVS to measure the beam particle's time-of-flight. The distance between the S1 and BVS counters is about 9 meters. The time-of-flight spectra were fitted to get the numbers of π^-, μ^-, and e^- in the beam at the BVS counter position. These numbers were corrected to the center of the target and to the plane just before ST using GEANT [2] to simulate the beam particles starting just before ST separately for π^-, μ^-, and e^- using the momentum and coordinates in the plane just before ST from the actual experimental data.

The π^0 from the reaction $\pi^- p \to \pi^0 n$ decays into two γs immediately and the showers from these two γs give clusters in the Crystal Ball NaI(Tl) crystals. The cluster finding program finds the crystal with maximum deposited energy, adds the energies from adjoining crystals then zeroes the energies in those crystals. It then finds the next crystal with maximum energy to form the next cluster and so on until all the clusters are found. We rejected events containing clusters with the maximum energy in a crystal at the beam entrance or exit holes. The neutron also can give a cluster, but the efficiency of registering the neutron is rather low for these low energies. We used 2 and 3 cluster events to select $\pi^- p \to \pi^0 n$ reaction and to calculate differential cross sections. For 2 cluster events we calculated the invariant mass, assuming that both clusters are from gammas. If the invariant mass coincides with the mass of π^0 and the kinematics of the reaction $\pi^- p \to \pi^0 n$ is fulfilled, then we accept the event. For 3 cluster events, we have the complication that we do not know which cluster results from the neutron and all possible combinations must be examined.

The acceptance of the Crystal Ball for reaction $\pi^- p \to \pi^0 n$ was calculated using GEANT [2] simulation of this reaction. The acceptance is on average about 50% and depends strongly on the scattering angle and beam particle's momentum.

The obtained values of $\pi^- p \to \pi^0 n$ differential cross sections are shown in Fig. 2 together with the results of the WI00 partial-wave analysis of the George Washington group [3].

CONCLUSION

Preliminary values of differential cross sections of the charge exchange reaction $\pi^- p \to \pi^0 n$ are presented. The differential cross sections are in a reasonable agreement with results of partial-wave analysis based on experiments made earlier by other groups.

This work was supported in part by the Russian Foundation for Basic Research, by the Russian State Scientific–Technical Program 'Fundamental Nuclear Physics', U.S. DOE and NSF, NSERC of Canada.

REFERENCES

1. E.D. Bloom and C.W. Peck, Ann. Rev. Nucl. Sci. 33, 143 (1983).
2. GEANT 3.21 CERN Program Library Long Writeup W5013, CERN, Geneva, Switzerland.
3. SAID at http://GWDAC.phys.GWU.EDU

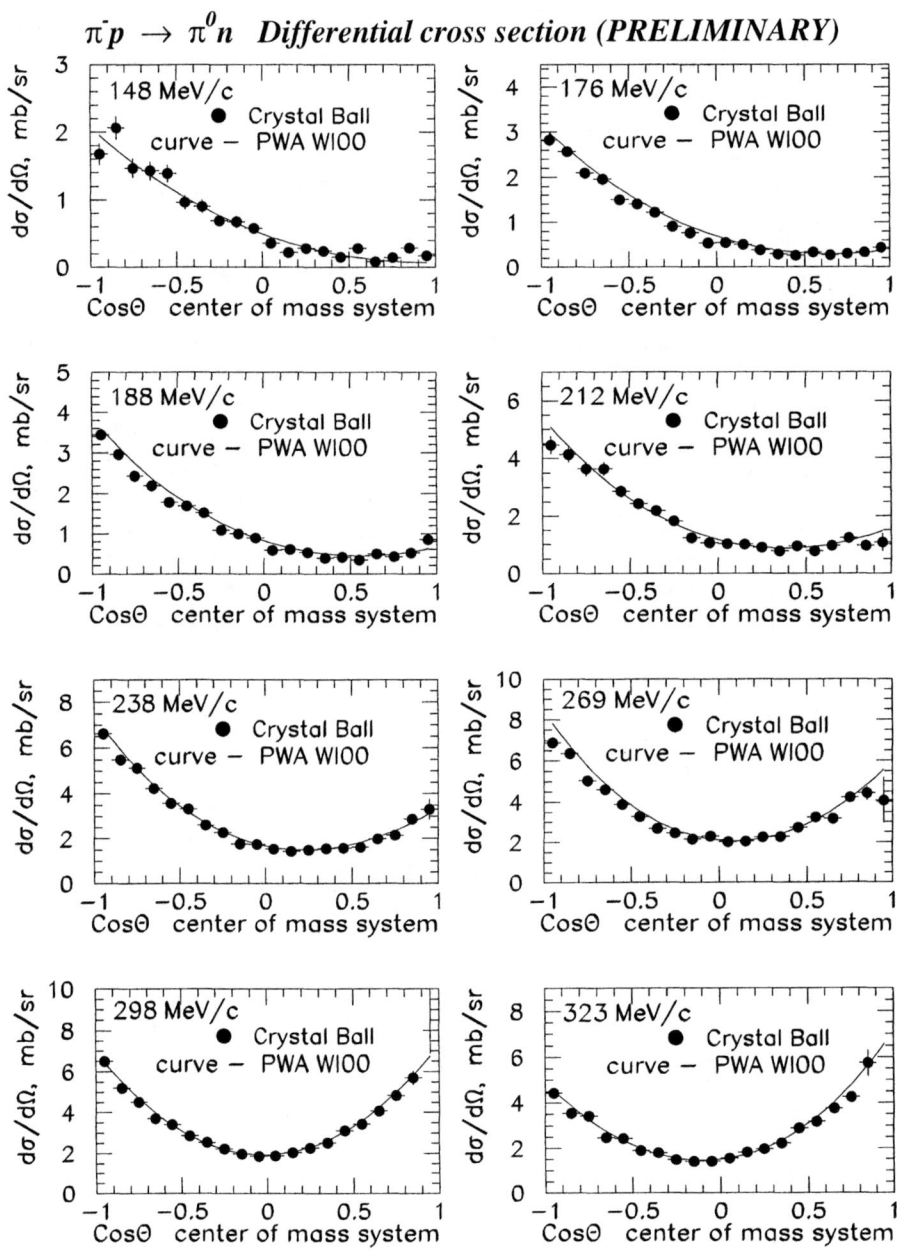

FIGURE 2. Differential cross sections of reaction $\pi^- p \to \pi^0 n$. Black circles are the preliminary values obtained in this experiment. The curve shows the results of the WI00 partial-wave analysis of the George Washington group [3].

LEPS Experiment at SPring-8: detector status and preliminary results

Petr Shagin (for LEPS Collaboration)

IHEP, Protvino, Russia and RCNP, Osaka University, Japan

Abstract.
Photoproduction of ϕ meson on proton is an unique probe to understand the Pomeron exchange process which plays a major role in the high energy particle interactions. There are limited data available for the $\gamma p \to \phi p$ process near the threshold region.

The LEPS experiment of the ϕ photoproduction using linearly polarized photon beam and liquid hydrogen target is described. The tagged photon beam is produced by backward Compton scattering of 351 nm Ar laser light from 8 GeV electron beam in the Spring-8 storage ring. The photon beam linear polarization is close to 100%. The preliminary results of ϕ photoproduction asymmetry is presented.

INTRODUCTION

It is well known that all hadron-hadron total cross sections (including γp) in a wide energy range are reproduced very well in terms of two s-dependent terms with $s^{-0.5}$ and $s^{0.08}$ dependences. The Regge model clearly suggests that the $s^{-0.5}$ term originates from the ρ-meson trajectory. The other term requires the introduction of Pomeron trajectory, whose $\alpha(t=0)$ has to be 1.08. In the Regge phenomenology, one can identify the Pomeron exchange with gluon exchange [4].

At the high energies diffractive photoproduction of a vector mesons from a proton target is well described by Pomeron exchange process. High energy photon converts into the ϕ meson and then scattered from proton by exchange of Pomeron [3].

The polarization data are crucial in distinguishing quark exchange from gluon exchange because of their couplings have different spin dependences.

Precise measurements of differential cross section at small $|t|$ for ϕ photoproduction in the energy range 2~3 GeV may reveal the existence of a 0^{++} glueball trajectory [5, 6]. The contribution of such trajectory falls off rapidly at high energies because of negative value for $\alpha(t=0)$.

The LEPS experiment has measured the ϕ photoproduction from a proton:

$$\gamma + p \to \phi + p, \phi \to K^+ K^-$$

in the energy range from 1.5 to 2.5 GeV, the energy where Pomeron exchange contribution is expected to be dominant. The data were detected from December 2000 to June 2001. This measurements has been unique with use of nearly 100% linearly polarized photons near the production threshold.

Asymmetry of φ photoproduction for the vertically and horizontally polarized photon beams has been measured.

LEPS EXPERIMENT

Laser-Electron-Photon at SPring-8 (LEPS):

High energy photon beam is created by Compton back-scattering (BCS) of laser light from the electron at 8 GeV SPring-8 storage ring. Due to the the relativistic effect with a very large Lorentz factor (16 000), Compton back-scattered photon gains energy of about 5×10^7 times higher than that of incident laser photon [1, 2].

If laser light is polarized and scattered in backward direction, the resulting photon is also polarized due to the angular momentum conservation.

The scattered angle of BCS photons above 1.5 GeV are less than 0.15 mrad which corresponds to a beam size of 1.5 cm at the distance of 100 m from the collision point.

The BCS photon energy is determined by measuring the scattered electron momentum with position sensitive detector.

The tagging counter consist of multi-layers of a 0.1 mm pitch silicon strip detector (SSD) and two plastic scintillator hodoscope planes. The SSD is 10 mm in hight, 64 mm in width, 500μm in depth.

Electrons with the energy of 4.5∼6.5 GeV are detected by the tagger. The corresponding photon energy range is 1.5∼3.5 GeV. The position resolution of the tagger is much better than a required resolution of 1 mm which corresponds to a 30 MeV energy resolution at 3 GeV.

Detector set-up is shown in Fig. 1. The set-up consists of liquid hydrogen target, scintillating veto and start counters, aerogel Cherenkov counters, vertex detector, forward magnetic spectrometer and time-of-flight system.

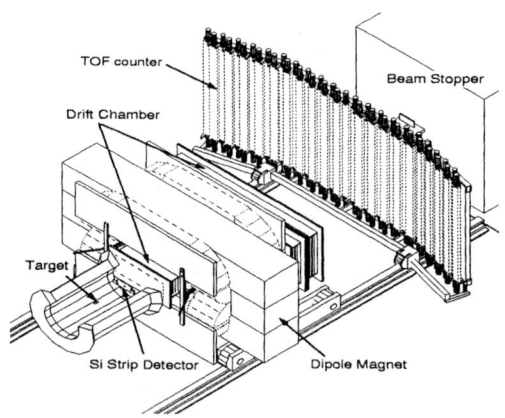

FIGURE 1. LEPS detector setup.

Liquid hydrogen target is a cylindrical volume of 35 mm in diameter and 50 mm in lenght filled with the liquified hydrogen.

Vertex detector measures production angles of charged particles and incident angles of those particles to the dipole magnet. The vertex detector consists of 2 planes (x- and y-) of single-sided SSDs and 6 planes of MWDC (DC_1). The thickness of each SSD is 300μm, the strip pitch is 120μm.

Forward spectrometer consists of dipole magnet and two drift chambers DC_2-DC_3.

Momenta of charged particle pairs are analysed by 0.7 T dipole magnet with a dimension of 135(w)× 55(h) × 60(l) cm^3.

To measure a position of a charged-particle track after the magnet, two drift chambers DC_2 and DC_3 are used. The chambers located about 70 cm downstream of the magnet center. The active area size of both chambers are 200cm(w)×80cm(h). Each chamber has 5 planes: X,X',Y,Y',U(V)in order to solve both left-right and stereo ambiguity. The position resolution is 200 μm for each plane.

Typical overall momentum resolution is 12 MeV at the 2 GeV particle momentum.

The identification of kaon and pion is essentially done by reconstructing its masses from momentum and TOF measurement.

A TOF wall consisiting of 40 2-m long plastic scintillator bars with a cross section of 4cm(t)×12 cm(w). Fast scintillator (BC 406) with attenuation lenth longer than 2 m is used. A typical flight-length of a charged particle is about 4 m. The TOF time resolution of 130 ps is achieved. As shown in Fig.2, pions, kaons, protons and havier hadrons are clearly identified by TOF system.

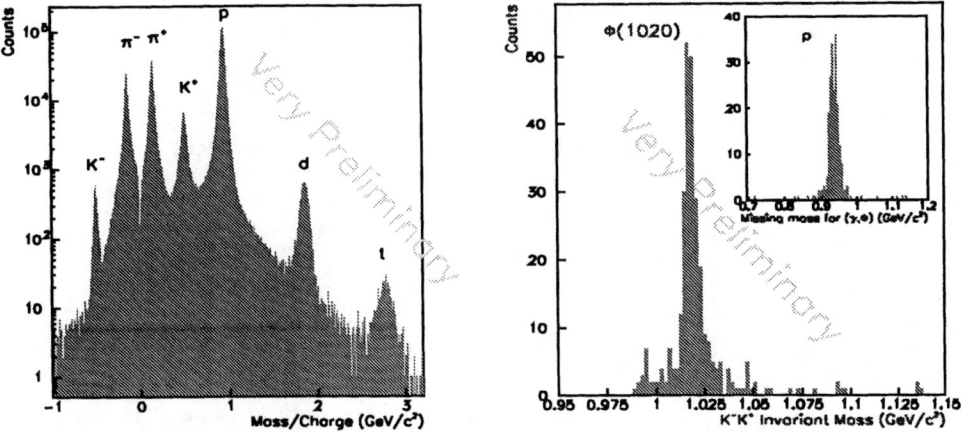

FIGURE 2. Reconstructed mass (left) and invariant mass spectra for K^+K^- reconstructed events (right).

Data Acqusition System included five VME based UNIX processors and event builder was used at LEPS experiment. The UNIX PC was used for the set-up control and data monitoring.

As a trigger signal for the DAQ system the following combination was used:

$$\text{Trigger} = \text{TAG} \times \text{Start} \times \text{TOF} \times \bar{U}\bar{P} \times \bar{A}\bar{C}$$

PRELIMINARY RESULTS

The candidate events for φ into K^+K^- decay were selected using the following criteria: 1) two tracks reconstructed in magnetic spectrometer; 2) masses of both track particles compatible with PDG charged kaon mass; 3) vertex coordinate is compatible with the target location.

The invariant mass distribution of the pair clearly shows φ meson peak (Fig.2).

The preliminary result on φ photoproduction asymmetry is shown in Fig.3. The dependense of K^+K^- events on azimuthal angle indicated the process is dominated by natural parity exchange.

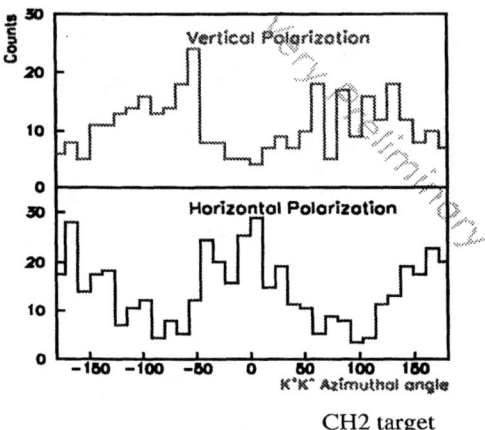

CH2 target

FIGURE 3. The dependence of K^+K^- events on azimuthal angle.

SUMMARY

The LEPS experiment is successfully operated at SPring-8 facility since June 2000. The performance of the detectors are optimized for the detection of φ photoproduction near threshold. The peak of φ meson is clearly seen in K^+K^- invariant mass spectrum. The K^+K^- asymmetry of φ-photoproduction from CH2 target indicates the process is dominates by natural parity exchange. The data analysis is going on. Futher physics results will come soon.

REFERENCES

1. A.M.Sandorfi et al., IEEE Trans. Nucl.Sci. 30, 3083 (1983)
2. C.Shaerf, Nucl. Phys. News 2 (1992) No. 1 7-8
3. T.H.Bauer et al. Rev. Mod. Phys. 50, 261 (1978)
4. A.Donnachie and P.V.Landshoff, Phys. Lett. B296, 227 (1992)
5. T.Nakano and H.Toki, Exciting Physics with New Accelerator Facilities, World Scientific, 1997
6. A.I.Titov, T.S.Lee and H.Toki, Phys. Rev. C 59, 2993 (1999)

RADIATIVE DECAYS

Detection of $\phi \to \pi^0 \gamma$, $\phi \to \eta \gamma$ and $\phi \to \eta' \gamma$ with KLOE detector at DAΦNE

The KLOE Collaboration[1], presented by
Camilla Di Donato

Università degli Studi di Napoli "Federico II", Dipartimento di Scienze Fisiche, Complesso Universitario di Monte S'Angelo Via Cintia ed. G, 80126 Naples, Italy
email=Camilla.DiDonato@na.infn.it

Abstract. KLOE has collected about 30 pb^{-1} in year 2000 at the DAΦNE collider, which corresponds to the largest population of ϕ meson radiative decays studied so far. We present the results for $\phi \to \pi^0 \gamma$, $\phi \to \eta \gamma$ and $\phi \to \eta' \gamma$ obtained analyzing $\int Ldt = 16.6 \, pb^{-1}$. The measurements performed leads to a very accurate determination of the mixing angle in the flavor basis $\varphi_P = (40^{+1.7}_{-1.5})°$.

INTRODUCTION

From the beginning the quark model [1] has found in the radiative decays of light vector mesons to pseudoscalars an useful scenario to probe its validity.
The branching ratio of the decay $\phi \to \eta' \gamma$ is particularly interesting since its value can probe the $|s\bar{s}\rangle$ and gluonium content of the η' [2] and the ratio of its value to that of $\phi \to \eta \gamma$ can be related to $\eta - \eta'$ mixing parameters [3, 4, 5, 6, 7, 8] and determine the mixing angle in the flavor basis φ_P, which has been pointed out as the best suited parameter for a process-independent description of the mixing.
Using about 60% of the luminosity collected in 2000 by the KLOE detector, we are able

---

[1] The KLOE collaboration: A. Aloisio, F. Ambrosino, A. Antonelli, M. Antonelli, C. Bacci, G. Barbiellini, F. Bellini, G. Bencivenni, S. Bertolucci, C. Bini, C. Bloise, V. Bocci, F. Bossi, P. Branchini, S. A. Bulychjov, G. Cabibbo, R. Caloi, P. Campana, G. Capon, M. Carboni, M. Casarsa, V. Casavola, G. Cataldi, F. Ceradini, F. Cervelli, F. Cevenini, G. Chiefari, P. Ciambrone, S. Conetti, E. De Lucia, G. De Robertis, P. De Simone, G. De Zorzi, S. Dell'Agnello, A. Denig, A. Di Domenico, C. Di Donato, S. Di Falco, A. Doria, M. Dreucci, O. Erriquez, A. Farilla, G. Felici, A. Ferrari, M. L. Ferrer, G. Finocchiaro, C. Forti, A. Franceschi, P. Franzini, C. Gatti, P. Gauzzi, A. Giannasi, S. Giovannella, E. Gorini, F. Grancagnolo, E. Graziani, S. W. Han, M. Incagli, L. Ingrosso, W. Kluge, C. Kuo, V. Kulikov, F. Lacava, G. Lanfranchi, J. Lee-Franzini, D. Leone, F. Lu, M. Martemianov, M. Matsyuk, W. Mei, A. Menicucci, L. Merola, R. Messi, S. Miscetti, M. Moulson, S. Müller, F. Murtas, M. Napolitano, A. Nedosekin, F. Nguyen, M. Palutan, L. Paoluzi, E. Pasqualucci, L. Passalacqua, A. Passeri, V. Patera, E. Petrolo, D. Picca, G. Pirozzi, L. Pontecorvo, M. Primavera, F. Ruggieri, P. Santangelo, E. Santovetti, G. Saracino, R. D. Schamberger, B. Sciascia, A. Sciubba, F. Scuri, I. Sfiligoi, J. Shan, P. Silano, T. Spadaro, E. Spiriti, G. L. Tong, L. Tortora, E. Valente, P. Valente, B. Valeriani, G. Venanzoni, S. Veneziano, A. Ventura, Y. Wu, G. Xu, G. W. Yu, P. F. Zema, Y. Zhou.

to give the best measurement of $BR(\phi \to \pi^0 \gamma)$ and $BR(\phi \to \eta' \gamma)$ to date, reducing the statistical error[9, 10]. The accuracy we obtain is significantly better than the current world average and allows us to extract the $\eta - \eta'$ mixing angle in the flavor basis with an error of $\simeq 1.5°$ from the BR's measurement.

ANALYSIS OF $\phi \to \pi^0 \gamma$ AND $\phi \to \eta \gamma$

A consistency test of radiative decay analyses is done using the three photons final state. The events topology is simple: $\phi \to \pi^0 \gamma$ with $\pi^0 \to \gamma\gamma$ and $\phi \to \eta\gamma$ with $\eta \to \gamma\gamma$.

A common strategy is applied to select the events. For both decays we look for three and only three prompt neutral clusters with $21° < \theta_\gamma < 159°$. A *prompt* neutral cluster is defined as a cluster in the Electromagnetic Calorimeter (EmC) with no associated track and satisfying the condition $|t - l/c| < 5 \cdot \sigma_t$, where t is the arrival time on the EmC, l the distance from the interaction point, c is the speed of the light and $\sigma_t = 54ps/\sqrt{E(GeV)} \oplus 147ps$ [11]; the angular cut on θ_γ is due to the presence of magnetic quadrupoles near the interaction point. After the topological selection we perform a kinematic fit constraining global energy-momentum conservation and the speed of light for each photon. The photon assignment is performed minimizing a χ^2 in the $\eta\gamma$ and $\pi^0\gamma$ hypothesis over the possible combinations of the three photons. The final selection is done using $cos\theta_{\gamma\gamma}$ versus $\Delta E_{\gamma\gamma}$ for non radiative photons in order to eliminate QED background. In figure 1.2, there are the two signal bands.

The ratio $R_{\gamma\gamma\gamma} = BR(\phi \to \eta\gamma \to \gamma\gamma\gamma)/BR(\phi \to \pi^0\gamma \to \gamma\gamma\gamma)$ has been evaluated using

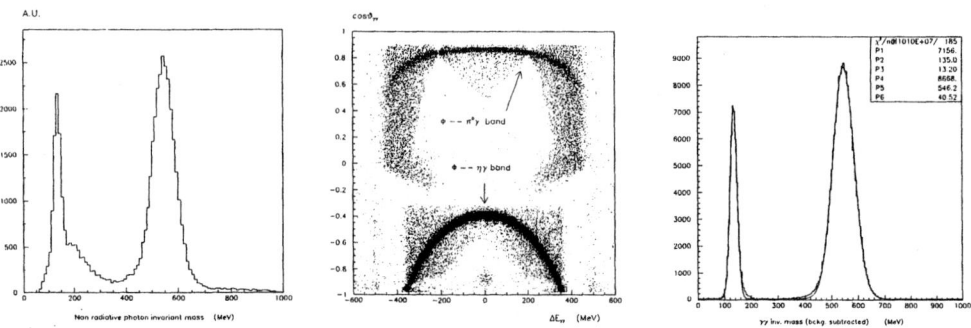

FIGURE 1. 1.Left: non radiative photon invariant mass before background subtraction; 2.Center: $cos\theta_{\gamma\gamma}$ versus $\Delta E_{\gamma\gamma}$; 3.Right: non radiative photon invariant mass after background subtraction.

two samples to check the data of different periods: KLOE'99 ($2.4pb^{-1}$) yields $R_{\gamma\gamma\gamma} = (3.92 \pm 0.05 \pm 0.13)$; KLOE'00 ($16.6pb^{-1}$) yields $R_{\gamma\gamma\gamma} = (3.75 \pm 0.02 \pm 0.09)$. From this analysis [12] we have one of the most accurate determination of the $BR(\phi \to \pi^0\gamma \to \gamma\gamma\gamma)$: $(1.377 \pm 0.007 \pm 0.05) \cdot 10^{-3}$.

ANALYSIS OF $\phi \to \eta'\gamma$ AND $\phi \to \eta\gamma$

The determination of the ratio $R = BR(\phi \to \eta'\gamma)/BR(\phi \to \eta\gamma)$ has been done using the following decays: $\phi \to \eta'\gamma$ with $\eta' \to \pi^+\pi^-\eta$ and $\eta \to \gamma\gamma$; $\phi \to \eta\gamma$ with $\eta \to \pi^+\pi^-\pi^0$ and $\pi^0 \to \gamma\gamma$. The two different decays have the same final state $\pi^+\pi^-\gamma\gamma\gamma$, and thus most of the systematics approximately cancel out when evaluating the ratio R; moreover since the $\phi \to \eta\gamma$ decays can be quite easily selected with small background they constitute a very clean control sample for the analysis. Background events can rise from $\phi \to K_S K_L$ events with one charged vertex where at least one photon is lost and the K_L is decaying near the interaction point (IP) and $\phi \to \pi^+\pi^-\pi^0$ events with an additional photon detected due to accidental photons or splitting of clusters in the electromagnetic calorimeter (EmC).

A first level topological selection for the $\pi^+\pi^-\gamma\gamma\gamma$ channel consists in the request to have 3 and only 3 prompt neutral clusters with $21° < \theta_\gamma < 159°$; opening angle between each pair of photons $> 18°$; one charged vertex inside the cylindrical region $r < 4\ cm$; $|z| < 8\ cm$. This selection is common to both $\phi \to \eta'\gamma$ and $\phi \to \eta\gamma$ events. We perform a kinematic fit constraining global energy-momentum conservation and the speed of light for each photon, without imposing any intermediate particle mass constraint. A loose cut on $\mathcal{P}(\chi^2) > 1\%$ for this fit is imposed to ensure the good reconstruction of the event. Background from $\phi \to \pi^+\pi^-\pi^0$ events is strongly reduced by means of a cut on the charged pions energy end points: $E_{\pi^+} + E_{\pi^-} < 550 MeV$ in the case of $\phi \to \eta\gamma$ events and $E_{\pi^+} + E_{\pi^-} < 430 MeV$ in the case of $\phi \to \eta'\gamma$ events.

Further selection of $\phi \to \eta'\gamma$ events is made via a cut on the total photon energy (to reduce $\phi \to K_S K_L$ background): $\Sigma_\gamma E_\gamma > 540 MeV$.

A useful way to select $\phi \to \eta'\gamma$ events over the $\phi \to \eta\gamma$ background is to look at the kinematic properties of the three photons in both categories of events. The energy spectrum of the photons shows no combinatorial problem, indeed the radiative photon is the most energetic one in $\phi \to \eta\gamma$ events, while it is the less energetic one in $\phi \to \eta'\gamma$ events. The energy of the two most energetic photons after kinematic fit shows a strong correlation for the two photons from η in $\phi \to \eta'\gamma$ events while the $\phi \to \eta\gamma$ events are grouped into two bands around $E_{1(2)} = 363$ MeV, as expected from the presence of the nearly monochromatic radiative photon. The selection of $\phi \to \eta'\gamma$ events is then made cutting on an elliptic shaped region in the $E_1 - E_2$ plane. The $\pi^+\pi^-\gamma\gamma$ invariant mass distribution for the events inside the selected ellipse on data shows a clear peak at the η' mass value with the same σ of the one expected from Monte Carlo, over a small residual background (see fig.2). A "donut shaped" region around the selection ellipse can been used to better understand the background. The signal has been selected in the region $942 MeV/c^2 \leq M_{\pi^+\pi^-\gamma\gamma} \leq 974 MeV/c^2$ and the expected background subtracted. The final number of selected events is then $N_{\eta'\gamma} = 124 \pm 12(stat.) \pm 5(syst.)$, where the statistical and systematic error includes the one from estimating background subtracted. The abundant and pure $\phi \to \eta\gamma$ events can be used as control sample to evaluate systematic effects on the efficiency by comparing data versus Monte Carlo distributions for the variable on which the cuts are set. All comparisons show very good agreement, at percent level or better, between data and Monte Carlo: the overall systematic errors on efficiencies evaluated with Monte Carlo are thus small. The ratio on the number of

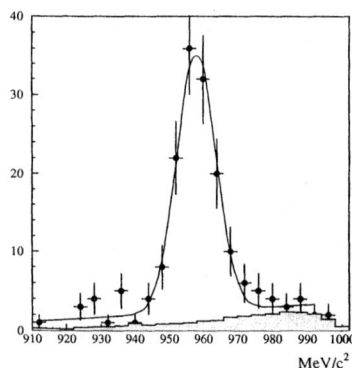

FIGURE 2. The $\pi^+\pi^-\gamma\gamma$ invariant mass for events selected as $\phi \to \eta'\gamma$ candidates. The shaded area is the shape of background obtained selecting events around the elliptical region and normalized to the expected Monte Carlo number of events. The continuous line is the result of a gaussian plus linear fit.

events selected as $\eta'\gamma$ and $\eta\gamma$ respectively, can be related to the ratio of the branching fractions $R = BR(\phi \to \eta'\gamma)/BR(\phi \to \eta\gamma)$ as follows:

$$R = \frac{N_{\eta'\gamma}}{N_{\eta\gamma}} \times \left(\frac{\varepsilon_{\eta\gamma}}{\varepsilon_{\eta'\gamma}}\right) \times \frac{BR(\eta \to \pi^+\pi^-\pi^0)BR(\pi^0 \to \gamma\gamma)}{BR(\eta' \to \pi^+\pi^-\eta)BR(\eta \to \gamma\gamma)}$$

and, thus, we have: $R = (5.3 \pm 0.5(\text{stat.}) \pm 0.3(\text{syst.})) \cdot 10^{-3}$. This value for R can be related directly to the mixing angle in the flavor basis. In the approach by Bramon et al. [5] where SU(3) breaking is taken into account via a constituent quark mass ratio $\frac{m_s}{\bar{m}}$ one has:

$$R = \frac{BR(\phi \to \eta'\gamma)}{BR(\phi \to \eta\gamma)} = \cot^2\varphi_P \left(1 - \frac{m_s}{\bar{m}}\frac{\tan\varphi_V}{\sin 2\varphi_P}\right)^2 \left(\frac{p_{\eta'}}{p_\eta}\right)^3$$

We use the result in the cited paper for all parameters entering the ratio except the mixing angle, in order to estimate the effect of our measurement on the angle φ_P. We use also the approach by Feldmann [7]; we get the same result in extracting the mixing angle in both approaches, i. e.

$$\varphi_P = \left(40\,{}^{+1.7}_{-1.5}\right)^\circ$$

which would result in a mixing angle in the octet-singlet basis $\theta_P = \left(-14.7\,{}^{+1.7}_{-1.5}\right)^\circ$. Moreover, using the value in [9] for the $BR(\phi \to \eta\gamma)$ we can extract the most precise determination of $BR(\phi \to \eta'\gamma)$ to date

$$BR(\phi \to \eta'\gamma) = (6.8 \pm 0.6\,(\text{stat.}) \pm 0.5\,(\text{syst.})) \cdot 10^{-5}$$

This result [13], given also the value of the mixing angle, disfavors large gluonium contents of the η' [2, 3, 14].

Other decay chains, with a different final state, has been used to determine the branching ratio $BR(\phi \to \eta'\gamma)$: one is due to $\eta' \to \pi^+\pi^-\eta$ with $\eta \to \pi^0\pi^0\pi^0$, another is due to $\eta' \to \pi^0\pi^0\eta$ with $\eta \to \pi^+\pi^-\pi^0$, with $\pi^0 \to \gamma\gamma$.

The topology is characterized by seven prompt photons and one charged vertex with two tracks connected. The selection criteria at first level are similar to the one used for the three photons and two tracks final state.

The main background is due to $\phi \to K_S K_L$ when an electromagnetic cluster splitting or accidentals happens. Using event classification we reduce the background. There are no events with the same topology, i.e. alas also no control sample for systematic check. However this is a different approach to measure the $BR(\phi \to \eta'\gamma)$, to be compared with the one obtained using the $\pi^+\pi^-\gamma\gamma\gamma$ final state; the results are consistent in one σ. The number of events selected after background subtraction is $N_{\eta'} = 150 \pm 12$; this is a preliminary result obtained analyzing 16.6 pb^{-1} of the 2000 KLOE data.

CONCLUSIONS

The analysis presented in this paper has been performed on a sample of about 16.6 pb^{-1} of integrated luminosity at the DAΦNE collider. We obtain the best determination of the BR for the process analyzed. The ratio of the BR for $\phi \to \eta'\gamma$ and for $\phi \to \eta\gamma$ can clarify the long-standing $\eta - \eta'$ mixing angle puzzle; our value disfavors a large gluonium content for the η' meson.

REFERENCES

1. Becchi, C., and Morpugno, G., *Phys. Rev*, **B687**, 140 (1965).
2. Close, F. E., *The DAΦNE phyisics handbook*, Maiani, L. Pancheri, G. and Paver, N., Frascati, 1992, vol. 2.
3. Rosner, J. L., *Phys. Rev*, **D27**, 1101 (1983).
4. Bramon, R., A. Escribano, and Scadron, M., *Phys. Lett.*, **B403**, 339 (1997).
5. Bramon, R., A. Escribano, and Scadron, M., *Eur. Phys. J.*, **C7**, 271 (1999).
6. Bramon, R., A. Escribano, and Scadron, M., *Phys. Lett.*, **B503**, 271 (2001).
7. Feldmann, T., *Int. Jou. Mod. Phys.*, **A15**, 159 (2000).
8. Ball, J. M., P. Frere, and Tytgat, M., *Phys. Lett.*, **B365** (1996).
9. The Particle Data Group-: Groom, D. et al., *Eur. Phys. Jou.*, **C15** (2000).
10. The KLOE collaboration : Adinolfi, M. et al., *Contributed paper to ICHEP 2000*, **hep-ex:0006036** (2000).
11. The KLOE Collaboration-: Adinolfi, M. et al., *Nucl. Instrum. Meth.*, **A461**, 344 (2001).
12. *Private conversations with Fabrizio Scuri* (2001).
13. The KLOE collaboration : Aloisio, A. et al., *Contributed paper to Lepton Photon 2001*, **hep-ex:0107022** (2001).
14. Deshpande, N., and Eilam, G., *Phys. Rev.*, **D25**, 270 (1980).

Studies of ϕ decays to $f_0(980)\gamma$ and $a_0(980)\gamma$ with KLOE at DAΦNE

The KLOE Collaboration[1],
presented by B. Valeriani

Institut für Experimentelle Kernphysik der Universität Karlsruhe (TH) im Forschungszentrum Karlsruhe, Postfach 3640, 76021 Karlsruhe, Germany

Abstract. The KLOE experiment [1] has collected $\sim 30\,\mathrm{pb}^{-1}$ by the end of the year 2000. A data sample of $17\,\mathrm{pb}^{-1}$ has been analyzed so far, searching for ϕ radiative decays into $f_0(980)\gamma$ and into $a_0(980)\gamma$. By looking at the five photon final state, corresponding to the two fully neutral decays $f_0(980) \to \pi^0\pi^0$ and $a_0(980) \to \eta\pi^0$, $\eta \to \gamma\gamma$, the following branching ratios have been obtained: $BR(\phi \to f_0(980)\gamma \to \pi^0\pi^0\gamma) = (7.9 \pm 0.2_{stat}) \times 10^{-5}$, $BR(\phi \to a_0(980)\gamma) = (5.8 \pm 0.5_{stat}) \times 10^{-5}$. A preliminary analysis of the decay $\phi \to a_0(980)\gamma \to \eta\pi^0$, $\eta \to \pi^+\pi^-\pi^0$ has given the result: $BR(\phi \to \eta\pi^0\gamma) = (6.7 \pm 0.9_{stat}) \times 10^{-5}$. We find for the ratio of the two rates the value $\frac{BR(\phi \to f_0(980)\gamma)}{BR(\phi \to a_0(980)\gamma)} = 4.1 \pm 0.4_{stat}$, which gives the first significant indication of the structure of these states.

INTRODUCTION

The radiative decays of the ϕ to the scalar mesons f_0 and a_0 are the method to study the nature of these two mesons. Over the years many models have been suggested to describe the internal structure of the two mesons ($qq\bar{q}\bar{q}$ [2], $K\bar{K}$ [3], $q\bar{q}$ [4]). Both measurements of the rates $BR(\phi \to f_0(980)\gamma)$, $BR(\phi \to a_0(980)\gamma)$ and of their ratio precisely allow us to discriminate between them.

---

[1] The KLOE collaboration: A. Aloisio, F. Ambrosino, A. Antonelli, M. Antonelli, C. Bacci, G. Barbiellini, F. Bellini, G. Bencivenni, S. Bertolucci, C. Bini, C. Bloise, V. Bocci, F. Bossi, P. Branchini, S. A. Bulychjov, G. Cabibbo, R. Caloi, P. Campana, G. Capon, G. Carboni, M. Casarsa, V. Casavola, G. Cataldi, F. Ceradini, F. Cervelli, F. Cevenini, G. Chiefari, P. Ciambrone, S. Conetti, E. De Lucia, G. De Robertis, P. De Simone, G. De Zorzi, S. Dell'Agnello, A. Denig, A. Di Domenico, C. Di Donato, S. Di Falco, A. Doria, M. Dreucci, O. Erriquez, A. Farilla, G. Felici, A. Ferrari, M. L. Ferrer, G. Finocchiaro, C. Forti, A. Franceschi, P. Franzini, C. Gatti, P. Gauzzi, A. Giannasi, S. Giovannella, E. Gorini, F. Grancagnolo, E. Graziani, S. W. Han, M. Incagli, L. Ingrosso, W. Kluge, C. Kuo, V. Kulikov, F. Lacava, G. Lanfranchi, J. Lee-Franzini, D. Leone, F. Lu, M. Martemianov, M. Matsyuk, W. Mei, A. Menicucci, L. Merola, R. Messi, S. Miscetti, M. Moulson, S. Müller, F. Murtas, M. Napolitano, A. Nedosekin, F. Nguyen, M. Palutan, L. Paoluzi, E. Pasqualucci, L. Passalacqua, A. Passeri, V. Patera, E. Petrolo, D. Picca, G. Pirozzi, L. Pontecorvo, L. Primavera, F. Ruggieri, P. Santangelo, E. Santovetti, G. Saracino, R. D. Schamberger, B. Sciascia, A. Sciubba, F. Scuri, I. Sfiligoi, J. Shan, P. Silano, T. Spadaro, E. Spiriti, G. L. Tong, L. Tortora, E. Valente, P. Valente, B. Valeriani, G. Venanzoni, S. Veneziano, A. Ventura, Y. Wu, G. Xu, G. W. Yu, P. F. Zema, Y. Zhou.

TABLE 1. Background channels to five photon final state

$\phi \to f_0(980)\gamma \to \pi^0\pi^0\gamma$	S/B	$\phi \to a_0(980)\gamma \to \eta\pi^0\gamma$	S/B
$e^+e^- \to \omega\pi^0 \to \pi^0\pi^0\gamma$	0.6	$e^+e^- \to \omega\pi^0 \to \eta\pi^0\gamma$	71
$\phi \to \rho^0\pi^0 \to \pi^0\pi^0\gamma$	3.7	$\phi \to \rho^0\pi^0 \to \eta\pi^0\gamma$	5.3
$\phi \to a^0\gamma \to \eta\pi^0\gamma$	3.5	$\phi \to f^0\gamma \to \pi^0\pi^0\gamma$	0.27
		$\phi \to \rho^0\pi^0 \to \pi^0\pi^0\gamma$	1
		$e^+e^- \to \omega\pi^0 \to \pi^0\pi^0\gamma$	0.14
$\phi \to \pi^0\gamma$	0.10	$\phi \to \eta\gamma \to \gamma\gamma\gamma$	6.1×10^{-3}
$\phi \to \eta\gamma \to \pi^0\pi^0\pi^0\gamma$	0.02	$\phi \to \eta\gamma \to \pi^0\pi^0\pi^0\gamma$	7.5×10^{-3}

KLOE has obtained some important results on ϕ radiative decays to f_0 and a_0, by the analysis of 17 pb^{-1} of the statistics collected in the year 2000 ($\sim$ 30 pb^{-1}).

$f_0(980) \to \pi^0\pi^0$ and $a_0(980) \to \eta\pi^0$, $\eta \to \gamma\gamma$ decays have been analyzed. The two decays are studied by looking for the five photon final state. The similar kinematics of the two final states minimizes systematic uncertainties in the evaluation of their ratio.

The decay $a_0(980) \to \eta\pi^0$, $\eta \to \pi^+\pi^-\pi^0$ has also been studied.

Selection cuts for the five photon final state

Events with five prompt photons are selected. A prompt photon is defined as an energy release in the calorimeter greater than 7 MeV satisfying the condition $|t - R/c| < 5\sigma_t$, where t and R are respectively time and distance from the interaction point of the cluster, c is the speed of light and σ_t is the time resolution. The acceptance cut on the polar angle $21^0 < \theta < 159^0$ is applied to exclude the blind region around the beam pipe. The expected background processes to $\phi \to f_0(980)\gamma \to \pi^0\pi^0\gamma$ and to $\phi \to a_0(980)\gamma \to \eta\pi^0\gamma$ are listed in Tab.1[2]. The three and seven photon final states can enter the signal selection due to machine background and photon splitting or merging and loss of soft photons.

In order to suppress these background channels, a first kinematic fit is applied to the selected events with the constraints of total energy and momentum conservation and requiring $t - R/c = 0$ for each photon. Using the different hypotheses $\pi^0\pi^0\gamma$, $\eta\pi^0\gamma$, $\omega\pi^0$, $\eta\gamma \to \gamma\gamma\gamma$, $\pi^0\gamma$, a photon pairing is performed to look for the best photon combination. A second kinematic fit is then applied to each of the best combinations: further constraints on π^0 and η masses are required of the assigned $\gamma\gamma$ pairs.

$\phi \to f_0(980)\gamma \to \pi^0\pi^0\gamma$ decay

After the second kinematic fit, $\phi \to a_0\gamma$ and $\phi \to \rho_0\gamma$ backgrounds are reduced by a cut on the angle ψ between the primary photon and the pion flight direction in the $\pi^0\pi^0$

---

[2] The corresponding signal to background ratios have been derived from SND collaboration measurements at VEPP-2M ([5],[6], [7]) and from PDG values

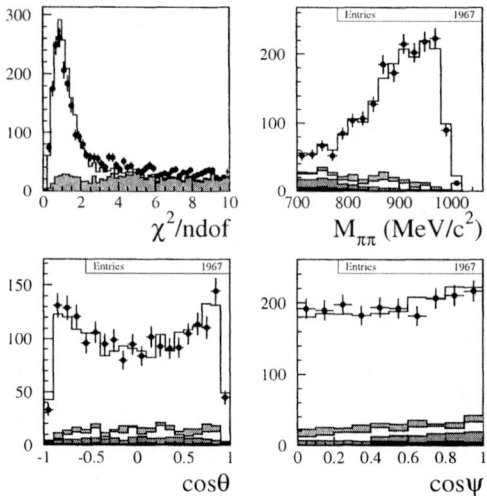

FIGURE 1. $\phi \to \pi^0\pi^0\gamma$ distributions: χ^2/ndf (up-left), $\pi^0\pi^0$ invariant mass (up-right), polar angle of the radiated photon (down-left) and $\cos\psi$ (down-right). Black circles are data, solid histograms are the expected montecarlo signal+background spectra while the background contributions are displayed in colors: green is $e^+e^- \to \omega\pi^0$, yellow is $\phi \to \eta\gamma$, magenta is $\phi \to \rho\pi^0$ and blue is $\phi \to a_0\gamma$.

center of mass. Cuts on the reconstructed ω mass and on the χ^2 from the kinematic fit in the $\omega\pi$ hypothesis are applied to reject $\phi \to \omega\pi$ events.

The data-montecarlo comparison for the most relevant distributions, after all these cuts, is shown in Fig.1. The expected background contributions have also been plotted. The analysis has been performed in a model independent way [8] because of the unknown f_0 spectrum.

After background is subtracted, 1662 ± 48 events are found. Correcting this number by the selection efficiency evaluated from montecarlo and normalizing the result to the $\phi \to \eta\gamma \to \gamma\gamma\gamma$ events [9], we find

$$BR(\phi \to f_0(980)\gamma \to \pi^0\pi^0\gamma) = (7.9 \pm 0.2_{stat}) \times 10^{-5}$$

in the hypothesis that the signal interference with $\phi \to \rho^0\pi^0 \to \pi^0\pi^0\gamma$ is negligible [10].

$\phi \to a_0(980)\gamma \to \eta\pi^0\gamma$, $\eta \to \gamma\gamma$ decay

Most of background contamination due to $\pi^0\pi^0\gamma$ final state (80% of selected events after the second kinematic fit) is rejected by applying a cut on the ψ variable, on the η meson invariant mass distribution and on the difference between the χ^2 values obtained from the kinematic fit in the two different hypotheses of $\pi^0\pi^0\gamma$ and $\eta\pi^0\gamma$. The plot in Fig.2 (left) shows the experimental $M_{\pi\eta}$ distribution with the expected main background

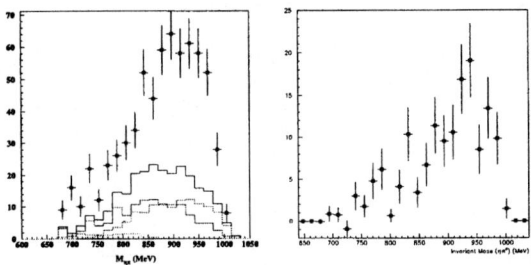

FIGURE 2. $M_{\eta\pi}$ distribution: $\phi \to \eta\pi^0\gamma$, $\eta \to \gamma\gamma$ analysis (left). Black circles are data, solid histograms are the expected montecarlo signal+background spectra while the background contributions are displayed in colors: red is $\pi^0\pi^0\gamma$ final state, green is $e^+e^- \to \omega\pi^0 \to \eta\pi^0\gamma$, blue is $\phi \to \eta\gamma \to \pi^0\pi^0\pi^0\gamma$, black is the total background. $\phi \to \eta\pi^0\gamma$, $\eta \to \pi^+\pi^-\pi^0$ analysis (right).

contributions. At the end of the selection procedure 666 ± 26 events survive, with an expected background population of 253 ± 11 events. The corresponding branching ratio is

$$BR(\phi \to \eta\pi^0\gamma) = (7.4 \pm 0.5_{stat}) \times 10^{-5}.$$

This value includes $\phi \to \rho^0\pi^0 \to \eta\pi^0\gamma$ contribution. The number of expected $\rho^0\pi^0$ events has been evaluated using the average of the recent measurements of $BR(\rho^0 \to \eta\gamma)$ performed by the two VEPP-2M experiments [11], [12]. Subtracting this contribution in the hypothesis of negligible interference with the signal, we obtain:

$$BR(\phi \to a_0\gamma \to \eta\pi^0\gamma) = (5.8 \pm 0.5_{stat}) \times 10^{-5}.$$

From the analysis of the five photon final state a preliminary evaluation of the ratio of the two branching ratios has been obtained. Assuming that $\pi\pi$ and $\eta\pi$ are the dominant decay modes of respectively f_0 and a_0, we obtain:

$$\frac{\phi \to f_0\gamma}{\phi \to a_0\gamma} = \frac{3 \times (\phi \to f_0\gamma \to \pi^0\pi^0\gamma)}{\phi \to a_0\gamma \to \eta\pi^0\gamma} = 4.1 \pm 0.4_{stat}$$

This value is in good agreement with recent theoretical predictions [13].

$\phi \to a_0(980)\gamma \to \eta\pi^0\gamma$, $\eta \to \pi^+\pi^-\pi^0$ decay

$\phi \to a_0(980)\gamma \to \eta\pi^0\gamma$ decay has been checked searching for $\eta \to \pi^+\pi^-\pi^0$ decay. The signal selection requires one vertex in the interaction region with two connected tracks and five energy deposits in the calorimeter as in the previous analyses. A kinematic fit procedure is also in this case applied to reject background events. A preliminary analysis has given the following result:

$$BR(\phi \to \eta\pi^0\gamma) = (6.7 \pm 0.9_{stat}) \times 10^{-5}.$$

The experimental $M_{\eta\pi^0}$ distribution is shown in the plot of Fig.2 (right).

CONCLUSIONS

In KLOE we have performed the analysis of ϕ radiative decays to $f_0(980)$ and $a_0(980)$ with $17\,\text{pb}^{-1}$ collected in 2000. The decay $\phi \to f_0(980)\gamma$ has been studied by searching for $f_0 \to \pi^0\pi^0$ decay and the analysis has given the result $BR(\phi \to f_0(980)\gamma \to \pi^0\pi^0\gamma) = (7.9 \pm 0.2_{stat}) \times 10^{-5}$. The analysis of $\phi \to \eta\pi^0\gamma$, $\eta \to \gamma\gamma$ has found the value $(7.4 \pm 0.5_{stat}) \times 10^{-5}$ for the corresponding branching ratio. By subtracting the expected $\phi \to \rho^0\pi^0 \to \eta\pi^0\gamma$ contribution we obtain $BR(\phi \to a_0(980)\gamma \to \eta\pi^0\gamma) = (5.8 \pm 0.5_{stat}) \times 10^{-5}$. The first analysis of the decay $\phi \to \eta\pi^0\gamma$ with $\eta \to \pi^+\pi^-\pi^0$ has given the result $(6.7 \pm 0.9_{stat}) \times 10^{-5}$.

For the ratio of the two rates $\phi \to f_0\gamma$, $\phi \to a_0\gamma$, we find $4.1 \pm 0.4_{stat}$. This value suggests for the two states a structure seeded by a compact $qq\bar{q}\bar{q}$ core influenced by the S-wave $K\bar{K}$ [13].

The values obtained are already statistically more accurate than present PDG values. A detailed analysis is proceeding concerning systematics. They will not exceed 10%.

REFERENCES

1. KLOE Collaboration, KLOE: a general purpose detector for DAΦNE, *LNF-92/019 (IR)* (1992); KLOE Collaboration, The KLOE detector-Technical Proposal, *LNF-93/002 (IR)* (1993)
2. Jaffe, R. L., *Phys. Rev.*, **D15**, 267 (1977); Alford, M., and Jaffe, R. L., *Nucl. Phys.*, **B578**, 367 (2000)
3. Weinstein, J., and Isgur, N., *Phys. Rev. Lett.*, **48**, 659 (1982)
4. Törnqvist, N. A., *Z. Phys.*, **C68**, 647 (1995)
5. Achasov, M. N., et al., *Phys. Lett.*, **B485**, 349 (2000)
6. Aulchenko, V. M., et al., *JETP Lett.*, **90**, 1067 (2000)
7. Achasov, M. N., et al., *Phys. Lett.*, **B479**, 53 (2000)
8. The KLOE Collaboration, hep-ph/0107024 (2001)
9. KLOE internal memo n° 234, April 2001
10. Achasov, N. N., and Gubin, V. V., *Phys. Rev.*, **D63**, 094007 (2001)
11. Achasov, M. N., et al., *JETP Lett.*, **72**, 282 (2000)
12. Akhmetshin, R. R., et al., *Phys. Lett.*, **B509**, 217 (2001)
13. Close, F. E., and Kirk, A., hep-ph/0106108 (2001)

Study of the process $e^+e^- \to \eta\gamma \to 7\gamma$ in the energy region $\sqrt{s} < 1.4$ GeV[1]

M.N.Achasov*, S.E.Baru*, K.I.Beloborodov*, A.V.Berdyugin*,
A.G.Bogdanchikov*, A.V.Bozhenok*, A.D.Bukin*, D.A.Bukin*,
S.V.Burdin*, T.V.Dimova*, A.A.Drozdetski*, V.P.Druzhinin*,
V.B.Golubev*, V.N.Ivanchenko*, P.M.Ivanov*, I.A.Koop*, A.A.Korol*,
M.S.Korostelev*, S.V.Koshuba*, A.V.Otboev*, E.V.Pakhtusova*,
E.A.Perevedentsev*, A.A.Salnikov*, S.I.Serednyakov*, V.V.Shary*,
Yu.M.Shatunov*, V.A.Sidorov*, Z.K.Silagadze*, A.G.Skripkin[†] and
A.V.Vasiljev*

*Budker Institute of Nuclear Physics, Siberian Branch of the Russian Academy of Sciences,
Lavrentyev 11, Novosibirsk, 630090, Russia
[†]Novosibirsk State University, Novosibirsk, 630090, Russia

Abstract. We present results of studies $e^+e^- \to \eta\gamma$ cross section in the energy region $\sqrt{s} < 1.4$ GeV performed in SND experiment at VEPP-2M e^+e^- collider. The following values for the decay probabilities were obtained $Br(\phi \to \eta\gamma) = (1.341 \pm 0.012 \pm 0.051) \cdot 10^{-2}$, $B(\omega \to \eta\gamma) = (4.22 \pm 0.47 \pm 0.17) \cdot 10^{-4}$, and $B(\rho \to \eta\gamma) = (2.77 \pm 0.26 \pm 0.16) \cdot 10^{-4}$ for $\eta \to 3\pi^0$, $\pi^0 \to \gamma\gamma$ decay mode. It was found that cross section of this process may be described by a sum of ρ, ω and ϕ resonance contributions only.

Introduction

The studies of the radiative decays of light vector mesons (ρ, ω, ϕ) in e^+e^- collisions play an important role in understanding of the electromagnetic structure of $q\bar{q}$-states and low-energy behavior of strong interactions. Although many measurements were carried out for the probabilities of the radiative decays, the achieved accuracy [1] is insufficient for reliable determination of the parameters of phenomenological models [2, 3, 4].

In this work we present the results of studies of the process $e^+e^- \to \eta\gamma$ in the multi-photon final state in the energy region $\sqrt{s} < 1.4$ GeV, which were obtained using the full experimental data set collected in the SND experiment. Since the final state includes seven photons, the background can be substantially suppressed comparing to that in other η-meson decay modes and, therefore, the systematic error can be reduced.

---

[1] Presented by A.V.Berdyugin, e-mail:berdugin@inp.nsk.su

Detector SND and experiment

SND is a general-purpose non-magnetic detector [5] designed and optimized for the studies of processes with neutral particles in the final states. The main part of the detector is an electromagnetic calorimeter consisting of 1632 NaI(Tl) crystals. Full thickness of the calorimeter for the particles originating from the interaction point is $13.5 X_0$. The calorimeter provides a good energy ans angular resolution, which can be approximated as $\Delta E/E = 4.2\%/\sqrt[4]{E(\text{GeV})}$, $\sigma_\varphi = 0.82°/\sqrt{E(\text{GeV})} \oplus 0.63°$.

The experiment with SND detector was carried out at the VEPP-2M collider with the average luminosity $\sim 10^{30} \text{cm}^{-2}\text{sec}^{-1}$. In this work we present the results based on the experimental statistics (Tabl. 1) collected during runs of 1997-2000. Total integrated luminosity accumulated in these runs was 27 pb^{-1}.

TABLE 1. The integrated luminosity statistics.

Experiment	$\sqrt{s}$, MeV	$\int L$, pb^{-1}	energy points
MHAD-97	980 ÷ 1380	5.8	35
PHI-98	984 ÷ 1060	8.7	16
OME-98	360 ÷ 970	3.5	38
MHAD-99	1060 ÷ 1360	3.0	12
OME-00	600 ÷ 940	6.0	27

Analysis

Events of the process $e^+e^- \to \eta\gamma$, $\eta \to 3\pi^0$, $\pi^0 \to \gamma\gamma$ are characterized by the final state with seven photons, some of which may be undetected. Extra photons may also appear due to splitting of showers in the calorimeter, emission of photons by the initial particles at large angles, or stray photons from beam background. The main background process in the ϕ-resonance region is the $\phi \to K_S K_L$ decay with K_S decaying into two neutral pions and K_L, interacting in the calorimeter and producing extra "photons". An additional background comes from by the $e^+e^- \to \omega\pi^0 \to \pi^0\pi^0\gamma$ with addition stray photons and $e^+e^- \to \omega\pi^0\pi^0 \to 3\pi^0\gamma$ in the energy region $\sqrt{s} > 1.2$ GeV. An analysis of the experimental data has demonstrated that the QED process $e^+e^- \to \gamma\gamma\gamma$ can pile up with other events, and result in the required event configuration.

Taking the above-listed background sources into account, we selected events in two steps. The pre-selection was paced on following conditions: (1) six or more reconstructed photons in the event, (2) total energy deposition in the calorimeter (E_{tot}) is in the range from $0.7\sqrt{s}$ to $1.2\sqrt{s}$, where $s = 4E_{beam}^2$, (3) total momentum in the event (p_{tot}) is lower than $0.2E_{tot}/c$, (4) $E_{tot}/\sqrt{s} - p_{tot}c/\sqrt{s} > 0.7$. The last condition was introduced to suppress background from decay $\phi \to K_S K_L$.

Final selection was based on kinematic fit. Employing the 4-momentum conservation it is possible to test for the different intermediate states in the observed events. For each event the following hypotheses were tested:

1. hypothesis $H_{\omega\pi^0}$: five photons in an event are from the process $e^+e^- \to \omega\pi^0 \to \pi^0\pi^0\gamma$

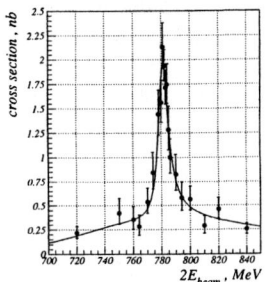 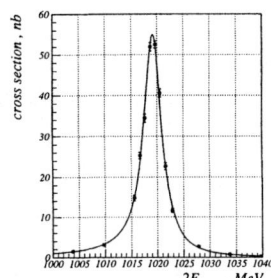 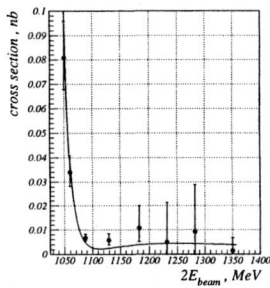

FIGURE 1. The measured total cross section for the $e^+e^- \to \eta\gamma$ process.

2. hypothesis $H_{3\gamma}$: three photons in an event are from the process $e^+e^- \to \gamma\gamma\gamma$ (QED)
3. hypothesis $H_{\omega\pi^0\pi^0}$: seven photons in an event are from the process $e^+e^- \to \omega\pi^0\pi^0 \to 3\pi^0\gamma$
4. hypothesis $H_{\eta\gamma}$: recoil mass ($M_{rec\,\gamma}$) for the photon with maximal energy in process $e^+e^- \to \eta\gamma$ must be around η-meson mass.

The events satisfying hypothesis of background process were rejected. Finally, we selected events satisfying the condition $400 < M_{rec\,\gamma} < 600$ MeV.

In the energy region $\sqrt{s} > 1.06$ GeV the background from $e^+e^- \to \omega\pi^0$ and $e^+e^- \to \omega\pi^0\pi^0$ substantially increase, therefore, we restricted number of particles $N_\gamma = 7$ and used only fully reconstructed $e^+e^- \to \eta\gamma \to 7\gamma$ events.

The number $N(s)$ of the observed events at a given energy point is described by the formula

$$N(s) = L(s)\left[\varepsilon(s)\beta(s)\sigma(s) + \sigma_b(s)\right],$$

where $L(s)$ is an integrated luminosity, ε is an efficiency determined by a simulation, β is a radiative correction factor, σ_b is a cross section of background processes, and σ is a cross section of the process under study. The detection efficiency and factor representing the radiative corrections can be written from a "radiator" function $F(x,s)$ [6] as:

$$\varepsilon(\sqrt{s}) = \frac{\int_0^{\frac{2E_{max}}{\sqrt{s}}} \varepsilon_r(\sqrt{s}, \frac{x\sqrt{s}}{2}) F(x,s)\sigma((1-x)s)dx}{\int_0^{\frac{2E_{max}}{\sqrt{s}}} F(x,s)\sigma((1-x)s)dx} \quad ; \quad \beta = \frac{\int_0^{\frac{2E_{max}}{\sqrt{s}}} F(x,s)\sigma((1-x)s)dx}{\sigma(s)},$$

where ε_r — detection efficiency as a function of the radiative photon energy.

The energy dependence of the resulting cross section (Fig. 1) was parametrized by the vector dominance formulas [7] including the contributions of the ρ, ω and ϕ resonances:

$$\sigma(s) = \frac{F(s)}{s^{3/2}} \left| \sum_{V=\rho,\omega,\phi} \sqrt{\sigma_{V\eta\gamma} \frac{m_V^3}{F(m_V^2)}} \frac{m_V \Gamma_V e^{i\varphi_V}}{D_V(s)} \right|^2,$$

where $F(s) = [(s-m_\rho^2)/\sqrt{s}]^3$, $D_V(s) = m_V^2 - s - i\sqrt{s}\Gamma_V(s)$, $\sigma_{V\eta\gamma} = \frac{12\pi Br_{V\to e^+e^-} \cdot Br_{V\to\eta\gamma}}{m_V^2}$.

The free parameters of the approximations were $\sigma_{V\eta\gamma}$. The relative phases of the resonances were fixed: $\varphi_\rho = \varphi_\omega = 0$, $\varphi_\phi = \pi$. The approximation gives the results listed in Tabl. 2.

TABLE 2. Measured cross section and branching ratios. $B_V = Br_{V\to e^+e^-} \cdot Br_{V\to\eta\gamma}$.

V	ϕ	ω	ρ
B_V	$(3.90 \pm 0.04 \pm 0.12)10^{-6}$	$(2.98 \pm 0.33 \pm 0.09)10^{-8}$	$(1.25 \pm 0.12 \pm 0.04)10^{-8}$
$Br_{V\to\eta\gamma}$	$(1.341 \pm 0.012 \pm 0.051)10^{-2}$	$(4.22 \pm 0.47 \pm 0.17)10^{-4}$	$(2.77 \pm 0.26 \pm 0.16)10^{-4}$
$\sigma_{V\to\eta\gamma}$	$(55.13 \pm 0.50 \pm 1.64)$nb	$(0.715 \pm 0.079 \pm 0.021)$nb	$(0.309 \pm 0.029 \pm 0.009)$nb

Systematic error is determined by contributions of the errors in the determination of the detection efficiency and error in the luminosity measurements. The latest does not exceed 2%. In order to estimate the systematic errors in the detection efficiency we investigated the stability of the results changing the selection conditions. The total systematic error in the efficiency, with the inclusion of all effects, was estimated as 2%.

Conclusion

Our results are in good agreement with previous measurements [8, 9, 10, 11, 12, 13]. The branching ratios for the $\phi, \omega \to \eta\gamma$ decays are measured with an accuracy close to the tabular one [1], and the branching ratio for the $\rho \to \eta\gamma$ decay is determined with double improved accuracy. Note that the quantity $Br_{\phi\to e^+e^-} \cdot Br_{\phi\to\eta\gamma}$ was measured with noticeably higher accuracy than the branching ratio $Br_{\phi\to\eta\gamma}$, because the leptonic width of the ϕ meson is known with accuracy 2.4%, which is worse than the statistical accuracy of our measurement.

REFERENCES

1. D.E.Groom et al., Eur. Phys. J. C 15 (2000) 1.
2. P.O'Donnel, Rev. Mod. Phys. 53 (1981) 673.
3. G.Morpurgo, Phys. Rev. D 42 (1990) 1497.
4. M.Benayoun et al., Phys. Rev. D 59, (1999) 114027.
5. M.N.Achasov et al., Nucl. Instrum. Meth. A 449, (2000) 125.
6. E.A.Kuraev and V.S.Fadin. Sov J. Nucl. Phys. 41 (1985) 466.
7. N.N.Achasov et al., J. Mod. Phys. A 7 (1992) 3187.
8. M.N.Achasov et al., JETP Lett. 68 (1998) 573.
9. R.R.Akhmetshin et al., Phys. Lett. B 460 (1999) 242.
10. M.N.Achasov et al., Eur.Phys.J. C 12 (2000) 25.
11. M.N.Achasov et al., JETP 90 (2000) 17.
12. T.Cose et al., Phys. Rev. D 61 (2000) 032002.
13. R.R.Akhmetshin et al., Phys. Lett. B 509 (2001) 217.

CHIRAL THEORY

Chirality and the Quark Model

Eric S. Swanson* and Adam P. Szczepaniak[†]

*Department of Physics and Astronomy, University of Pittsburgh, Pittsburgh, PA 15260, and
Jefferson Lab, 12000 Jefferson Ave, Newport News, VA 23606
[†]Department of Physics and Nuclear Theory Center, Indiana University, Bloomington, Indiana
47405

Abstract. The relationship of the quark model to the known chiral properties of QCD is a long-standing problem in the interpretation of low energy QCD. In particular, how can the pion be viewed as both a collective Goldstone boson quasiparticle and as a valence quark antiquark bound state? A comparison of the many-body solution of a simplified model of QCD to the constituent quark model demonstrates that the quark model is sufficiently flexible to describe meson hyperfine splitting provided proper renormalization conditions and correct degrees of freedom are employed consistently.

INTRODUCTION

The constituent quark model (CQM) is based on the idea that strong interactions lead to massive quasiparticles (constituent quarks) and that hadronic structure is dominated by the interactions between valence constituent quarks. One argues that constituent quarks are the effective degrees of freedom arising after dynamical chiral symmetry breaking due to bare quark interactions with the quark sea in the chiral noninvariant vacuum. In the CQM, however, properties of the constituent quarks, *e.g.* masses and magnetic moments, are treated as free parameters making the approach insensitive to the underlying chiral structure of QCD. This implies that CQM pions lose their nature as Goldstone bosons and are not much different from, say, ρ mesons. The mass splitting between the pion (a spin-0 constituent $Q\bar{Q}$ state) and the ρ meson (a spin-1 state) is attributed to a residual hyperfine interaction. This interaction is typically associated with the nonrelativistic reduction of single gluon exchange between constituent quarks and yields a vector–pseudoscalar splitting proportional to $\alpha_s/(m_q m_{\bar{q}})$.

Since the notion of gluon exchange is intrinsically perturbative, it is likely a useful concept in heavy quark systems where the large quark mass guarantees its applicability. However its relevance to the light quark sector is unclear. Indeed, one may expect that the π-ρ mass splitting is largely driven by the underlying chiral symmetry rather than by the hyperfine interaction. It is therefore surprising that when the splitting between the flavored quark-antiquark vector and pseudoscalar mesons is plotted against the mass of one of the constituent quarks (the other constituent quark mass is held fixed), it does indeed behave like $1/m_{con}$ typical of a hyperfine interaction all the way down to the light quark sector. This is shown by the squares in Fig. 1. The definition of the CQM masses depends on details of the model, here we use typical values $m_u = m_d = 330$ MeV, $m_s = 550$ MeV, $m_c = 1600$ MeV, and $m_b = 4980$ MeV.

Is the apparent universality of the effective interactions of the constituent quark model

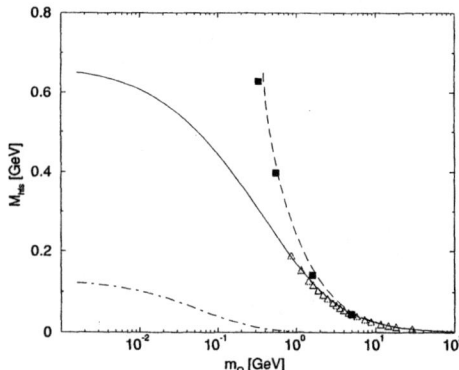

FIGURE 1. Flavored pseudoscalar and vector meson mass splittings as a function of the renormalized heavy quark mass (solid, $m_Q = m_R$) and constituent quark mass (dashed, $m_Q = m_{con}$) together with lattice results[2] (triangles, $m_Q = m_R$) and experiment[3] (squares, $m_Q = m_{con}$).

consistent with the dictates of chiral symmetry in the light meson sector? We address this by explicitly computing the vector-pseudoscalar mass splitting in a simple model of QCD[1] which, nevertheless, captures the main features required to examine the interplay of chiral symmetry and confinement. The model Hamiltonian may be written as $H = H_{can}(\Lambda) + \delta H(\Lambda)$. Here the first term is the canonical Hamiltonian which contains a term which mimics the effects of confinement.

$$H_{can} \to H_c(\Lambda) = \frac{c_c(\Lambda)}{\Lambda^2} \int d\mathbf{x} \left[\psi^\dagger(\mathbf{x}) \mathbf{T} \psi(\mathbf{x}) \psi^\dagger(\mathbf{x}) \mathbf{T} \psi(\mathbf{x}) \right]_\Lambda \qquad (1)$$

Here the subscript Λ indicates that the operators are to be point-split or smeared over a distance $\sim 1/\Lambda$ and $c_c(\Lambda)$ is the coupling which replaces $\alpha_s(\Lambda)$.

The counterterms δH contain relevant, marginal, and irrelevant operators. In the chiral limit, chiral symmetry prevents the occurrence of a relevant operator in the quark sector. For finite quark masses, the relevant operator is absorbed into the definition of the bare quark mass. The effect of transverse gluons eliminated by the cutoff show up through contact operators of dimension six or greater. There are a number of such terms; to illustrate the effect of hyperfine interactions we consider the dominant spin-dependent term,

$$H_h(\Lambda) = \frac{c_h(\Lambda)}{\Lambda^2} \int d\mathbf{x} \left[\psi^\dagger(\mathbf{x}) \mathbf{T} \alpha \psi(\mathbf{x}) \psi^\dagger(\mathbf{x}) \mathbf{T} \alpha \psi(\mathbf{x}) \right]_\Lambda, \qquad (2)$$

where $c_h(\Lambda)$ is proportional $\alpha_s(\Lambda)$. Thus the full model Hamiltonian is given by

$$H = \int \psi^\dagger [-i\alpha \cdot \nabla + \beta m(\Lambda)] \psi + H_c(\Lambda) + H_h(\Lambda). \qquad (3)$$

We note that in the chiral limit Λ is the only scale in the Hamiltonian and that light meson masses will therefore be proportional to this quantity.

Even though one can in principle choose an arbitrary value for the cutoff, there is typically an optimal choice, $\Lambda = \Lambda_R$ which depends on the basis and approximation scheme used to diagonalize the Hamiltonian. Since our goal is to understand the quark model from QCD, we choose Λ_R to match to the quark model. For example, since heavy quark-heavy quark systems are analogous to Coulombic bound states, it is natural to choose Λ_R so that it scales with $m_R \equiv m(\Lambda = \Lambda_R)$ – the heavy quark mass renormalized at the scale Λ_R. In the case of heavy-light systems the dynamics of the "brown muck" of gluons and light quarks is still subject to the full complexity of nonperturbative strong interactions, thus we simply use $\Lambda_R = \langle p \rangle \approx \Lambda_{QCD}$.

Proper determination of the renormalization scale is the first step toward "derivation" of the quark model from QCD. The second, is to implement dynamical chiral symmetry breaking which will generate masses for the light bare quarks thereby creating the constituent quarks of the CQM. The constituent Fock space then provides an efficient basis for diagonalizing the Hamiltonian. This is achieved by solving the gap equation in the BCS Ansatz for the dynamical quark mass, m_{con}.

$$m_{con} = m(\Lambda) + \frac{m_{con}}{\Lambda^2} (\tilde{c}_c(\Lambda) - \tilde{c}_h(\Lambda)) \int^\Lambda q^2 dq \frac{1}{\sqrt{m_{con}^2 + q^2}}, \quad (4)$$

where $\tilde{c}_c = C_F c_c / 2\pi^2$, $\tilde{c}_h = 3 C_F c_h / 2\pi^2$, and the single quasiparticle energies are given by $E(q) = \sqrt{m_{con}^2 + \mathbf{q}^2}$. These resemble energies of the constituent quarks if $m_{con} \sim$ 300 MeV for the light quarks.

The RPA or Bethe-Salpeter equation for the meson bound state, M, is then given by

$$\langle M | [H, Q_M^\dagger] | BCS \rangle = (E_M - E_{BCS}) \langle M | Q^\dagger | BCS \rangle, \quad (5)$$

where $Q_M^\dagger$ is defined in terms of the positive and negative energy wave functions, $Q_M^\dagger = \Sigma_{\alpha\beta} \left[\psi_{\alpha\beta}^+ B_\alpha^\dagger D_\beta^\dagger - \psi_{\alpha\beta}^- D_\beta B_\alpha \right]$ with B and D being the quasiparticle operators. In the simple approximation to H used here, Eq. (5) is an algebraic equation for the bound state masses $E_M = E_M(\Lambda, m(\Lambda), \tilde{c}_c(\Lambda), \tilde{c}_h(\Lambda))$. The cutoff dependence of the couplings is determined by requiring the E_M to be Λ-independent.

The average, renormalized light quark mass, $m_R \equiv m(\Lambda_R)$ is set to 5 MeV, and the two renormalized couplings $\tilde{c}_{R,c,h} \equiv \tilde{c}_{c,h}(\Lambda_R)$ are determined by fitting the π and ρ meson masses, $E_{\pi,\rho} = E_{\pi,\rho}(\Lambda_R, m_R, \tilde{c}_{R,c}, \tilde{c}_{R,h})$.

Solving the gap equation yields a light constituent quark mass of 380 MeV. For a heavy-light meson, the solid line in Fig. 1 shows the dependence of the hyperfine mass splitting $M_{hfs} \equiv E_v - E_{ps}$ as a function of the renormalized heavy quark mass $m_Q = m_R$ with the renormalized light quark mass fixed at 5 MeV. As m_Q increases the splitting falls off as $1/m_Q$; however, as the heavy quark mass approaches the light quark limit the slope changes, reflecting the emergence of chiral symmetry: $E_{ps} \propto \sqrt{m_Q}$. Our predictions reproduce the lattice results (triangles in Fig. 1)[2] available for large quark masses. The lattice results are given for $M_{hfs} \cdot a$ as a function of $m_R \cdot a$ where a is the lattice spacing. Lattice calculations match our results if one takes $a^{-1} = 1.45$ GeV which is very close to that of Ref. [4]. It is nontrivial that our predictions match the lattice over a large range of quark masses since the coupling constants were fixed in the chiral limit.

Is it possible for the quark model to mimic the chiral behavior of the hyperfine splitting shown in Fig. 1? We address this by plotting the hyperfine splitting as a function of the constituent quark mass derived from Eq. 3 (this is shown as a dashed line in Fig. 1). The curve reproduces the observed splitting for $\rho - \pi$, $K^* - K$, $D^* - D$ and $B^* - B$, shown as squares in the figure. We conclude that it is possible for a CQM to mimic the effects of chiral symmetry breaking. This is true because chiral symmetry breaking creates massive quasiparticles which may be used as effective degrees of freedom in model building. Furthermore, the hyperfine $1/m_Q$ behavior which is valid for heavy quarks continues to be valid for lighter constituent quarks.

We have shown how quark model phenomenology may be derived from a simple model of QCD. Elimination of high momentum components from quark-transverse-gluon coupling leads to short range hyperfine interactions. By choosing a renormalization scale which matches the quark model we neglect contributions from long range interactions (due to the exchange of low energy gluons) to the hyperfine interaction – as is consistent with the quark model. Then, studying the quark mass dependence of the pseudoscalar-vector mass splitting we are able to show that interactions between constituent quarks derived from QCD indeed follow that of the naive quark model while respecting chiral symmetry. As a result we have shown how the heavy quark mass limit extrapolates to the chiral limit and have illustrated the interplay between hyperfine interactions and chiral dynamics. Even though the analysis presented here has used a simplified central potential, the key results are independent of this choice. This is because the simplified model captures all features (*i.e.*, the presence of hyperfine interactions, correct momentum scales and dynamical chiral symmetry breaking) of QCD which are relevant.

REFERENCES

1. A. P. Szczepaniak and E. S. Swanson, Phys. Rev. Lett. **87**, 072001 (2001).
2. J. Hein *et al.*, hep-ph/0003130.
3. D.E. Groom *et al.*, Eur. Phys. J. **C15**, 1 (2000).
4. C.T.H. Davies *et al.*, Phys. Rev. **D56**, 2755, (1997).

Property of Chiral Scalar and Axial-Vector Mesons in Heavy-Light Quark Systems

M. Ishida[*] and S. Ishida[†]

[*]*Department of Physics, Tokyo Institute of Technology, Tokyo 152-8551, JAPAN*
[†]*Atomic Energy Research Inst., Coll. of Sci. and Tech., Nihon Univ., Tokyo 101-0062, JAPAN*

Abstract. Recently we have proposed a new level-classification scheme of hadrons with a manifestly covariant framework. In this scheme the requirement of chiral symmetry on the light quark leads to a prediction of existence of new type of scalars X_B, X_D and axial-vectors X_{B^*}, X_{D^*} as the chiral partners of ground state pseudoscalar B, D and vector B^*, D^* mesons, respectively. They belong to "relativistic S-wave states," and are discriminated from the conventional P-wave mesons with $j_q = 1/2$ appearing in the heavy quark effective theory. In this talk we examine the properties of these chiral mesons: The mass-splittings between the respective chiral partners are predicted to be equal, and the decay widths of one pion emission of X_B, X_D, X_{B^*} and X_{D^*} are to take the same value due to both chiral and heavy quark symmetries. Some experimental indications for existence of X_B and X_{D^*} are also given, which are consistent with the above prediction.

Introduction Recently, we have proposed a covariant level classification scheme of hadrons[1], unifying the seemingly contradictory two viewpoints, non-relativistic one with LS-coupling scheme and relativistic one with chiral symmetry. In this scheme, it is expected that the hadron spectra are to show, concerning light quark constituents, the approximate $\tilde{U}(12)_{SF}$ symmetry (including static $SU(6)_{SF}$ as a subgroup) around the lower mass region, and is predicted the existence of many relativistic states, called "chiralons," which are out of the framework of conventional LS coupling scheme in NRQM. Recently, the existence of a light scalar σ meson with the property as partner of π meson in the linear representation of chiral symmetry seems to be confirmed[2, 3]. In our classification scheme, this σ is naturally classified into the relativistic S-wave $q\bar{q}$ state, which is to be discriminated from the $^3P_0$-state appearing in NRQM.

In heavy-light quark $n\bar{b} (= u\bar{b}, d\bar{b})$ systems, the existence of chiralons, new scalar and axial-vector mesons, denoted as $X_B =^t (X_{B^+}, X_{B^0})$ and $X_{B^*} =^t (X_{B^{*+}}, X_{B^{*0}})$, is expected to exist as the chiral partners of the pseudoscalar $B =^t (B^+, B^0)$ and the vector $B^* =^t (B^{*+}, B^{*0})$ mesons, respectively. Similarly, in the $n\bar{c} =^t (u\bar{c}, d\bar{c})$ system, the scalar $X_{\bar{D}} =^t (X_{\bar{D}^0}, X_{D^-})$ and the axial-vector $X_{\bar{D}^*} =^t (X_{\bar{D}^{*0}}, X_{D^{*-}})$ mesons, is expected to exist as the chiral partners of the pseudoscalar $\bar{D} =^t (\bar{D}^0, D^-)$ and the vector $\bar{D}^* =^t (\bar{D}^{*0}, D^{*-})$ mesons, respectively. These chiral scalar and axial-vector mesons are classified to the relativistic S-wave states, which are discriminated from the P-wave mesons appearing in NRQM, or the scalars B_0^*, D_0^* and axial-vectors B_1^*, D_1^* with $j_q = 1/2$ appearing in HQET.

In this talk we investigate the properties of these chiral scalar and axial-vector mesons by taking into account chiral symmetry for the light quark component, as well as HQS for the heavy quark component. A similar approach has already been done in Refs. [4, 5]. However, they assigned as the respective chiral partners of S-wave state mesons to the

$j_q = 1/2$ P-wave mesons. This assignment is crucially different from ours.

Constraints on one pion emission process from HQS We consider general constraints from HQS on the processes of one-pion emission $X_B \to B\pi$, $X_{B^*} \to B^*\pi$, $X_D \to D\pi$ and $X_{D^*} \to D^*\pi$, which are expected to be the main decay modes of the relevant mesons.

The HQ spin symmetry relates B, X_B, D and X_D, respectively, to B^*, X_{B^*}, D^* and X_{D^*}, and the HQ flavor symmetry relates B, B^*, X_B and X_{B^*}, respectively, to $\bar{D}$, $\bar{D}^*$, $X_{\bar{D}}$ and $X_{\bar{D}^*}$ with the same velocity. Thus, the S-matrix elements of the relevant four decay modes are related with one another, and are represented by one universal amplitude ξ as.[1]

$$-i\langle\pi B^*(\mathbf{0},\varepsilon^{(0)})|U_I|X_{B^*}(\mathbf{0},\varepsilon^{(0)})\rangle = -i\langle\pi D^*(\mathbf{0},\varepsilon^{(0)})|U_I|X_{D^*}(\mathbf{0},\varepsilon^{(0)})\rangle = \langle\pi B(\mathbf{0})|U_I|X_B(\mathbf{0})\rangle$$

$$= \langle\pi D(\mathbf{0})|U_I|X_D(\mathbf{0})\rangle = -\xi\frac{1}{(2\pi)^3}\sqrt{\frac{1}{(2\pi)^3 2E_\pi}}i(2\pi)^4\delta^{(4)}(P_{X_M} - P_M - p_\pi), \quad (1)$$

where U_I is the translational operator of time from $-\infty$ to $+\infty$, $P_{X_M}(P_M)$ being the initial (final) heavy meson momentum, p_π being the emitted pion momentum, and $\mathbf{0}$ represents the three velocity $\mathbf{v} = \mathbf{0}$, the longitudinally polarized states of B^*, D^* and X_{B^*}, X_{D^*} appear, and a common factor of $\frac{1}{(2\pi)^3}\sqrt{\frac{1}{(2\pi)^3 2E_\pi}}$ has been introduced.

The decay widths of $X_M = X_B, X_{B^*}, X_D, X_{D^*}$ are given by

$$\Gamma_{X_M} = 3\frac{1}{2m_{X_M}}\frac{|\mathbf{p}|}{4\pi m_{X_M}}(2\xi\sqrt{m_{X_M}m_M})^2 \approx 3\frac{\sqrt{(\Delta m_M)^2 - m_\pi^2}}{8\pi}(2\xi)^2, \quad (2)$$

where we use the approximation $m_{X_M} \approx m_M$, $|\mathbf{p}| = \sqrt{(\Delta m_M)^2 - m_\pi^2}$ is the pion CM momentum, and the factor 3 comes from the isospin degree of freedom of the final $|\pi M\rangle$ state. As expressed by Eq. (2), the decay widths of the relevant processes are dependent only upon the corresponding mass difference Δm_M.

Chiral and Heavy Quark Symmetric Lagrangian Here we construct the chiral and heavy quark symmetric Lagrangian of B and $\bar{D}$ systems, by using the quark bi-spinor representation in the new classification scheme: $U_\alpha^{(\pm)\beta}(v) = \sum_\phi (1/2\sqrt{2})\Gamma_\phi \hat{\phi}(1 + iv\cdot\gamma)_\alpha^\beta$, where $\alpha(\beta)$ are light quark (heavy antiquark) spinor index, and the summation is taken for $\phi = B, X_B, B_\mu^*, X_{B_\mu^*}; \bar{D}, X_{\bar{D}}, \bar{D}_\mu^*, X_{\bar{D}_\mu^*}$; $\Gamma_B = \Gamma_{\bar{D}} = i\gamma_5$, $\Gamma_{X_B} = \Gamma_{X_{\bar{D}}} = \pm 1$, $\Gamma_{B_\mu^*} = \Gamma_{\bar{D}_\mu^*} = i\gamma_\mu$, $\Gamma_{X_{B_\mu^*}} = \Gamma_{X_{\bar{D}_\mu^*}} = \pm\gamma_5\gamma_\mu$. The field $\hat{\phi}$ is related with the ordinarily normalized field ϕ as $\phi(X) = (1/\sqrt{2m_Q})e^{im_Q v\cdot X}\hat{\phi}(X)$, where $m_Q = m_b(m_c)$ for $B(\bar{D})$ system. The chiral $U_A(1)$ transformation for the light quark is given by $U^{(\pm)} \to e^{\pm i\alpha\gamma_5}U^{(\pm)}$. By using the conjugate bispinor, defined by $\bar{U}^{(\pm)} \equiv \gamma_4 U^{(\pm)\dagger}\gamma_4$, the free Lagrangian is given by

$$\mathcal{L}^{\text{free}} = \langle \bar{U}^{(-)}\left(iv\cdot\overleftrightarrow{\partial}/2 - m_q\right)U^{(+)}\rangle = \sum_\phi \hat{\phi}^\dagger\left(iv\cdot\overleftrightarrow{\partial}/2 - m_q\right)\hat{\phi}, \quad (3)$$

where $\langle\ \rangle$ means the trace on spinor indices, and the light quark mass m_q is common for all B and $\bar{D}$ systems. The total meson mass m_M ($M=B,D$) is given by $m_M = m_Q + m_q$, thus $m_B = m_{X_B} = m_{B^*} = m_{X_{B^*}} = m_b + m_q$ and $m_D = m_{X_D} = m_{D^*} = m_{X_{D^*}} = m_c + m_q$ in symmetric limit.

The σ and π fields are transformed as $(\sigma + i\gamma_5\tau\cdot\pi) \to e^{i\alpha\gamma_5}(\sigma + i\gamma_5\tau\cdot\pi)e^{i\alpha\gamma_5}$, thus the chiral symmetric Yukawa coupling is given by

$$\mathcal{L}^{\text{Yukawa}} = -\eta\langle\bar{U}^{(-)}(\sigma + i\gamma_5\tau\cdot\pi)U^{(-)}\rangle. \quad (4)$$

---

[1] Here the normalization of states $|B(\mathbf{0})\rangle \equiv a_{B(\mathbf{0})}^\dagger|0\rangle$, etc. are used, where $[a_{B(\mathbf{p})}, a_{B(\mathbf{p}')}^\dagger] = \delta^{(3)}(\mathbf{p} - \mathbf{p}')$.

Through the spontaneous breaking of chiral symmetry the σ acquires vacuum expectation value $\langle\sigma\rangle \equiv \sigma_0(= f_\pi$ in SU(2) linear σ model), which induces the mass splittings Δm between chiral partners. The Δm are universal in B and $\bar{D}$ systems:

$$\Delta m \equiv m_{X_B} - m_B = m_{X_{B^*}} - m_{B^*} = m_{X_D} - m_D = m_{X_{D^*}} - m_{D^*}, \tag{5}$$

which is given by $\Delta m = 2\eta\sigma_0 = 2\eta f_\pi$ in Lagrangian (4). Even in the case, considering the contribution from all the other possible forms of effective chiral symmetric Lagrangian, the universality of mass-splittings are shown to be preserved. Thus, following the argument given in the last sub-section, the decay widths of one pion emission also become universal,

$$\Gamma(\Delta m) \equiv \Gamma_{X_B \to B\pi} = \Gamma_{X_D \to D\pi} = \Gamma_{X_{B^*} \to B^*\pi} = \Gamma_{X_{D^*} \to D^*\pi}, \tag{6}$$

although the magnitude of $\Gamma(\Delta m)$ cannot be predicted in the present framework.[2]

Experimental Evidence for X_B and X_{D^*}   In order to examine phenomenologically whether these chiral mesons really exist or not, we analyze the $B\pi$ mass spectra[6] in 5.4GeV $< m_{B\pi} <$ 5.9GeV, obtined through Z^0-boson decay by L3[7] and ALEPH[8] collaborations. In the relevant mass region of $B\pi$ channel, the X_B, and the P-wave mesons, $B_2^*(j_q = 3/2)$ and $B_0^*(j_q = 1/2)$, are expected to be observed directly. We use the following forms of squared amplitude $|\mathcal{M}|^2$ and background $|\mathcal{M}|^2_{\text{BG}}$,

$$|\mathcal{M}|^2 = |r_1 e^{i\theta_1} \Delta_{X_B}(s) + r_2 e^{i\theta_2} \Delta_{B_0^*}(s)|^2 + |r_3 e^{i\theta_3} \Delta_{\text{other } B}(s)|^2,$$

$$|\mathcal{M}|^2_{\text{BG}} = P_1(m_{B\pi} - P_2)^{P_3} e^{P_4(m_{B\pi}-P_2)+P_5(m_{B\pi}-P_2)^2+P_6(m_{B\pi}-P_2)^3}, \tag{7}$$

where P_1–P_6 are parameters and $\Delta_R(s) = -m_R\Gamma_R/(s - m_R^2 + im_R\Gamma_R)$; $\Delta_{X_B}(\Delta_{B_0^*})$ are the Breit-Wigner amplitude for $X_B(B_0^*)$ mesons, and $\Delta_{\text{other }B}$ represents contributions from all the other possible resonances including B_2^*.[3] Preliminary results of the fit are given in FIGURE 1. Both data show a dip at the same energy $m_{B\pi} \approx$ 5.55GeV, which is reproduced by the interference between the X_B with a narrow width and the B_0^* with a wide width. The mass and width of X_B is given by $m_{X_B} =$ 5540MeV and $\Gamma_{X_B} =$ 21MeV, and the obtained values of χ^2 is $\tilde{\chi}^2 = 22.92/20 = 1.15$. We also tried the fit without X_B, and obtain almost the same value of $\tilde{\chi}^2 = 26.280/24 = 1.10$, thus, only by this analysis, we cannot obtain the definite conclusion on existence of X_B.

Similar analysis is also done[9] on the $D^*\pi$ mass spectra by CLEO and DELPHI collaborations, and we obtained the preliminary result on existence of X_{D^*} with $m_{X_{D^*}} =$ 2306MeV and $\Gamma_{X_{D^*}} =$ 21MeV.

---

[2] In Lagrangian (4), ξ is given by $\xi = \eta = \Delta m/(2f_\pi)$. By taking $\Delta m \approx$ 300MeV in Eq. (2) as an example, $\Gamma(\Delta m = 300\text{MeV})$=331MeV. We can also consider, as one of the possible forms of the effective Lagrangian, the $\mathcal{L}^{(d)} = -k\langle \bar{U}^{(-)}[iv_\mu\partial_\mu(\sigma + i\tau \cdot \pi)]U^{(-)}\rangle$, where k is a coupling constant of $O(1/m_q)$. In this case $\xi = \eta + 2kv \cdot p_\pi \approx \eta - k\Delta m = (1 - 2kf_\pi)\Delta m/(2f_\pi)$. By taking a natural value of $k \approx 2/m_\rho$ as an example, the $\Gamma(\Delta m = 300\text{MeV})$=88 MeV. As is seen in these examples, the magnitude of $\Gamma(\Delta m)$ itself is largely dependent upon the value of k.

[3] In these experiments, the final photon was not detected, and so the following states decaying into $B^*\pi$ are also to be seen in $B\pi$ spectra indirectly through the successive $B^* \to B\gamma$ decay: $B_2^*(\to B^*\pi \to \gamma B\pi)$, $B_1(j_q = 3/2)(\to B^*\pi \to \gamma B\pi)$, $B_1^*(j_q = 1/2)(\to B^*\pi \to \gamma B\pi)$ and $X_{B^*}(\to B^*\pi \to \gamma B\pi)$. The observed $m_{B\pi}$-values of these resonances become smaller than their real values by the missing photon energy $E_\gamma = m_{B^*} - m_B$. In Eq. (7) $\Delta_{\text{other }B}$ are meant as including, in addition to direct $B_2^*(\to B\pi)$, the above mentioned indirect $B_2^*(\to B^*\pi)$ and $B_1^*(\to B^*\pi)$, which are D-wave decays. The indirect $B_1^*(\to B^*\pi)$ and $X_{B^*}(\to B^*\pi)$, which are S-wave decays interfering with each other, are considered to be included in $\Delta_{B_0^*}$ and Δ_{X_B}, respectively.

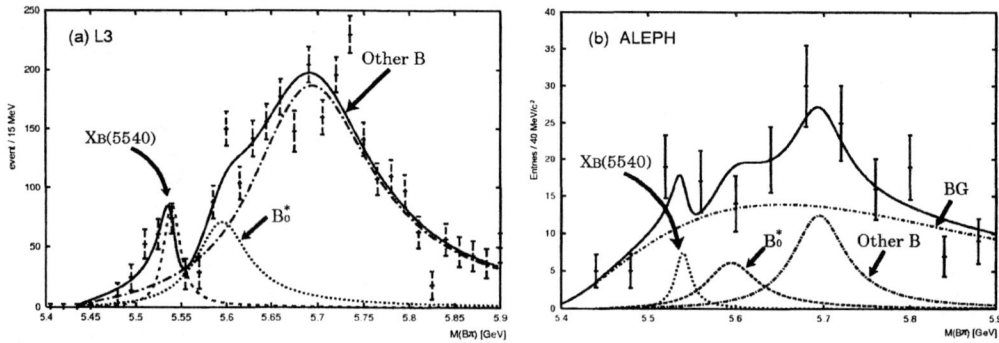

FIGURE 1. $B\pi$ mass spectra obtained in (a)L3[7] and ALEPH[8] collaborations. Contributions from the individual Breit-Wigner amplitudes are shown by dotted lines. In (a) the background contribution is subtracted. The m_{X_B} and Γ_{X_B} are determined only through CLEO data, since ALEPH data have much less statistics.

Here it may be worthwhile to note that the preliminary values above obtained on preperties of X_B and X_{D^*} are consistent with our predictions by heavy quark symmetry and chiral symmetry, Eqs. (5) and (6).
$m_{X_B} - m_B = 261\text{MeV} \approx m_{X_{D^*}} - m_{D^*} = 296\text{MeV}, \quad \Gamma_{X_B} = 21\text{MeV} = \Gamma_{X_{D^*}} = 21\text{MeV}.$
<u>Concluding Remarks</u>   In the new classification-scheme, the chiral scalars X_B, X_D and axial-vectors X_{B^*}, X_{D^*}, which are the chiral partners of B, D and B^*, D^*, respectively, are predicted to exist, which are to be discriminated from the conventional P-wave state mesons. Preliminary analyses of experimental data give some indications of possible existence of both X_B and X_{D^*}, besides B_0^* and D_1^*, with the properties consistent with the theoretical prediction.

This fact seems to suggest that the new level-classification scheme is actually realized in heavy-light quark meson systems.

REFERENCES

1. S. Ishida and M. Ishida, this conf. S. Ishida, M. Ishida, T. Maeda, Prog.Theor.Phys.**104**(2000),785.
2. N. A. Tornqvist, summary talk of "σ-meson 2000", KEK-proceedings 2000-4. T. Kunihiro; M. Ishida; T. Komada, this conference.
3. S. Ishida, M. Ishida, H. Takahashi, T. Ishida, K. Takamatsu and T. Tsuru, Prog. Theor. Phys. **95** (1996), 745; **98** (1997), 1005. E.Beveren, T.A.Rijken,K.Metzger,C.Dullemond,G.Rupp and J.E.Ribeiro, Z.Phys. **C30**(1986), 615. N. N. Achasov and G. N. Schestakov, Phys. Rev. D **49** (1994), 5779. R. Kaminski, L. Lesniak and J. -P. Maillet, Phys. Rev. D **50** (1994), 3145. N. A. Tornqvist, Z. Phys. C **68** (1995), 647. N. A. Tornqvist and M. Roos, Phys. Rev. Lett. **76** (1996), 1575. M. Harada, F. Sannino and J. Schechter, Phys. Rev. D **54** (1996), 1991.
4. W. A. Bardeen and C. T. Hill, Phys. Rev. D **49** (1994), 409.
5. D. Ebert, T. Feldmann, R. Friedrich, H. Reinhardt, Nucl.Phys.B**434**(1995), 619;hep-ph/9409298.
6. The collaboration by D. Ito, T. Komada, T. Maeda, S. Ishida, M. Ishida, I. Yamauchi and K. Yamada.
7. S. Goldfarb, L3 collab, on proc. of WHS99, Frascati Phys. series XV (1999), p.317.
8. S.Monteil, ALEPH collaboration, in proc. of WHS99, p.311.
9. K. Yamada et al., this conference.

Confirmation of σ(450–600)-Meson in $\Upsilon' \to \Upsilon \pi\pi$ & Other $\pi\pi$-Production Processes

M. Ishida, S. Ishida[†], T. Komada[‡] and S. I. Matsumoto[†]

Department of Physics, Tokyo Institute of Technology, Tokyo 152-8551, JAPAN
[†]*Atomic Energy Research Inst., Coll. of Sci. and Tech., Nihon Univ., Tokyo 101-0062, JAPAN*
[‡]*Dept. of Engineering Sci., Junior Coll. Funabashi Campus, Nihon Univ. Funabashi 274-8501, JAPAN*

Abstract. Applying the effective amplitude, which is evidently consistent with general constraints from chiral symmetry, the $\pi\pi$ spectra in the relevant processes are analyzed, leading to a strong evidence for existence of the light σ meson. It is also pointed out that the $\pi\pi$ scattering process, which had been one of the main sources for PDG table for these many years, is, in principle, exceptionally difficult to investigate the property of σ-meson.

(**Introduction**) In the previous work referred to as I[1], we have analyzed systematically the $\pi\pi$ production amplitudes in the various excited Υ and J/ψ decay processes, and reproduced successfully the experimental behaviors: The production amplitude $\mathcal{F}$ is given by a coherent sum of σ Breit-Wigner amplitude $\mathcal{F}_\sigma$ and direct 2π amplitude $\mathcal{F}_{2\pi}$ in the VMW method.

$$\mathcal{F} = \mathcal{F}_\sigma^I + \mathcal{F}_{2\pi}^I; \quad \mathcal{F}_\sigma^I = \frac{r_\sigma e^{i\theta_\sigma}}{m_\sigma^2 - s - i\sqrt{s}\Gamma_\sigma(s)}, \quad \mathcal{F}_{2\pi}^I = r_{2\pi} e^{i\theta_{2\pi}}, \tag{1}$$

where r_σ ($r_{2\pi}$) is production coupling constant of σ-state (2π-state) and $e^{i\theta_\sigma}$ ($e^{i\theta_{2\pi}}$) is a strong phase factor. Here (r_σ, θ_σ) and ($r_{2\pi}, \theta_{2\pi}$) are taken to be process-dependent, free parameters, since the $|\sigma\rangle$ state and $|2\pi\rangle$ state are, from quark physical picture[2], independent bases of S-matrix with an independent vertex, in principle.

However, we must give special attention on the threshold behaviors. Because of the property of π meson as Nambu-Goldstone boson in the case of chiral symmetry breaking, the $|\mathcal{F}|^2$ was widely believed to be suppressed in the $\pi\pi$ threshold. In the $\pi\pi$ scattering, the observed spectrum is actually suppressed near the threshold. However, in the relevant excited-Υ decay processes this threshold suppression is also observed experimentally in $\Upsilon(2S \to 1S)$ and $\psi(2S \to 1S)$, while, in $\Upsilon(3S) \to \Upsilon(1S)\pi\pi$, the steep increase from the $\pi\pi$ threshold is observed.

In the following we examine the consistency of our results in I, especially the threshold behavior of $\Upsilon(3S \to 1S)$, with the constraint from chiral symmetry.

(**Threshold Behavior of Production Amplitude and Chiral Symmetry**)
(*Effective amplitude by linear σ model*) First we consider an effective chiral symmteric Lagrangian[3, 4] including σ and π mesons, $\mathcal{L}^{(n)} = \xi^{(n)} \Upsilon'_\mu \Upsilon_\mu (\sigma^2 + \pi^2)$, where $\Upsilon'(\Upsilon)$ is the initial (final) $b\bar{b}$ quarkonium. Through the spontaneous breaking of chiral symmetry, σ acquires vacuum expectation value, $\langle \sigma \rangle_0 \equiv \sigma_0 = f_\pi$, and $\mathcal{L}^{(n)}$ is rewritten as $\mathcal{L}^{(n)} = \xi^{(n)} \Upsilon'_\mu \Upsilon_\mu (f_\pi^2 + 2f_\pi \sigma' + \sigma'^2 + \pi^2)$, producing the coupling terms; $\mathcal{L}_\sigma = \xi_\sigma \Upsilon'_\mu \Upsilon_\mu \sigma'$ ($\xi_\sigma = 2f_\pi \xi^{(n)}$), $\mathcal{L}_{\pi\pi} = \xi_{2\pi} \Upsilon'_\mu \Upsilon_\mu \pi^2$ ($\xi_{2\pi} = \xi^{(n)}$). The $\pi\pi$ production amplitude $\mathcal{F}^{L\sigma M}$ is given

by sum of $\mathcal{F}_\sigma^{L\sigma M}$ and $\mathcal{F}_{2\pi}^{L\sigma M}$, canceling with each other in $O(p^0)$ level.

$$\mathcal{F}^{L\sigma M} = \mathcal{F}_\sigma^{L\sigma M} + \mathcal{F}_{2\pi}^{L\sigma M} = \frac{\xi_\sigma(-2g_{\sigma\pi\pi})}{m_\sigma^2 - s} + 2\xi^{(n)} = 2\xi^{(n)}\frac{m_\pi^2 - s}{m_\sigma^2 - s}, \quad (2)$$

where the relation of SU(2)LσM, $g_{\sigma\pi\pi} = (m_\sigma^2 - m_\pi^2)/(2f_\pi)$, is used, $s = -(p_1 + p_2)^2$ ($p_{1,2}$ being the pion momenta), and the factor of polarization vectors of initial and final $b\bar{b}$ quarkonia, $\varepsilon(P') \cdot \tilde{\varepsilon}(P)$, is omitted. The final amplitude takes $O(p^2)$ form, being consistent with the derivative coupling property of π meson. In the limit $p_{1\mu} \to 0_\mu$, $s \to m_\pi^2$ and $\mathcal{F}^{L\sigma M} \to 0$. Thus the amplitude has Adler 0, which leads to the threshold suppression in conformity with the observed threshold behavior in $\Upsilon(2S) \to \Upsilon(1S)\pi\pi$.

(Effective amplitude through intermediate glueball states) Next we consider the effective amplitude through the intermediate production of the ground-state scalar and tensor glueballs. By using the framework of new classification scheme[5], the effective interaction is given by $\mathcal{L}^G \approx (\xi^G/M'M)\partial_\mu \Upsilon'_\lambda \partial_\nu \Upsilon_\lambda (\partial_\mu \sigma \partial_\nu \sigma + \partial_\mu \pi \cdot \partial_\nu \pi)$, where $M'(M)$ is the mass of initial(final) Υ. ξ^G is coupling constant. It is notable that, after spontaneous symmetry breaking, it produces no σ-amplitude cancelling the direct 2π-amplitude. Then the production amplitude is given by

$$\mathcal{F}_{2\pi}^G = -\varepsilon(P') \cdot \tilde{\varepsilon}(P)(\xi^G/M'M)(P' \cdot p_1 P \cdot p_2 + P' \cdot p_2 P \cdot p_1), \quad (3)$$

where $P'(P)$ is the momentum of initial(final) Υ.

The $\mathcal{F}_{2\pi}^G$ vanishes when $p_{1\mu} \to 0_\mu$. Thus, it has Adler 0, satisfying the general constraint from chiral symmetry. However, it does not vanish at $s = m_\pi^2$. In the relevant process, the Adler limit $p_{1\mu} \to 0_\mu$ corresponds to neglect $\Delta M \equiv M' - M$ in comparison with m_π. At $s = 4m_\pi^2$ the pion four-momenta should be $p_{1\mu} = p_{2\mu}$, leading to $p_{10} = p_{20} \approx \Delta M/2 = 450\text{MeV} \gg m_\pi$ in $\Upsilon(3S) \to \Upsilon(1S)$ transition. Actually $\mathcal{F}_{2\pi}^G$ can be approximated as $\mathcal{F}_{2\pi}^G \approx -2\xi^G p_{10} p_{20}$, which is almost s-independent in all the physical region and has no zero close to the threshold. Correspondingly, there is no suppression near the threshold.

Quantitative Analysis In the actual analysis we modify $\mathcal{F}^{L\sigma M}$ in Eq. (2), with inclusion of width of the σ-meson, $\Gamma_\sigma(s)(= g_\sigma^2 p_1(s)/(8\pi s))$ and of strong phase, as $\mathcal{F}_{\sigma+2\pi}^{\text{phen}} = (\varepsilon(P') \cdot \tilde{\varepsilon}(P))2\xi^{(n)} e^{i\theta_\sigma} \frac{m_\pi^2 - s}{m_\sigma^2 - s - i\sqrt{s}\Gamma_\sigma(s)}$. The results of the fit by using the semi-phenomenological amplitude $\mathcal{F}^{\text{phen}} \equiv \mathcal{F}_{\sigma+2\pi}^{\text{phen}} + \mathcal{F}_{2\pi}^G$ are depicted in Fig. 1. The $\pi\pi$ mass spectra in $\Upsilon(2S) \to \Upsilon(1S)$, $\Upsilon(3S) \to \Upsilon(1S)$ and $\Upsilon(3S) \to \Upsilon(2S)$ transitions are fitted simultaneously with common values of m_σ and g_σ, leading to the σ-pole position, $\sqrt{s_{\text{pole}}}(\approx m_\sigma - i\Gamma_\sigma/2) = (580^{+79}_{-30}) - i(190^{+107}_{-49})$, which is consistent with the ones obtained in our preceding works: [1] and the analysis[6] of $\pi\pi$ scattering phase shift. The total χ^2 is $\chi^2/(N_{\text{data}} - N_{\text{param}}) = 54.5/(74 - 11) = 0.87$. The $\mathcal{F}_{2\pi}^G$ contribution becomes dominant in $\Upsilon(3S) \to \Upsilon(1S)$ transition, and it explains the steep increase from the threshold.

Now we can see our treatment in the previous analysis[1] is to be consistent with general chiral constraints: Threshold suppression in $\Upsilon(2S) \to \Upsilon(1S)$, required from chiral symmetry, is reproduced phenomenologically by the cancellation between $\mathcal{F}_\sigma^I$ and $\mathcal{F}_{2\pi}^I$ of I (quoted in Eq. (1)). Dominant $\mathcal{F}_{2\pi}^G$-contribution, which is almost s-independent in all physical region, in $\Upsilon(3S) \to \Upsilon(1S)$ is reproduced mainly by constant $\mathcal{F}_{2\pi}^I$ of I.

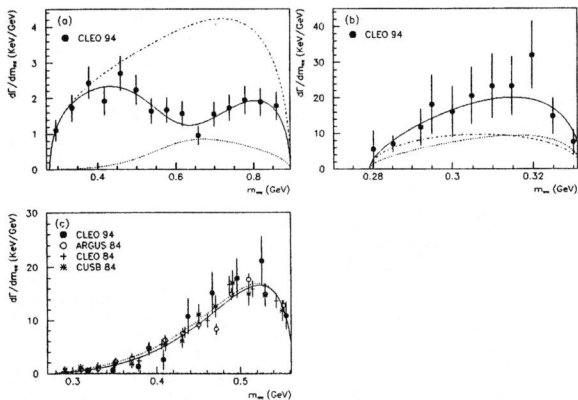

FIGURE 1. Fit to the $\pi\pi$ invariant mass spectrum[4] by the amplitude explicitly consistent with chiral constraint: (a) $\Upsilon(3S) \to \Upsilon(1S)\pi\pi$, (b) $\Upsilon(3S) \to \Upsilon(2S)\pi\pi$ and (c) $\Upsilon(2S) \to \Upsilon(1S)\pi\pi$. The respective contributions from $\mathcal{F}^{phen}_{\sigma+2\pi}$ and from $\mathcal{F}^{G}_{2\pi}$ are shown by dotted and dot-dashed lines.

Features of $\pi\pi$ production amplitudes The $\pi\pi$ production processes have generally much the larger energy release ΔE than m_π. Thus, the momentum of emitted pion becomes large, and the derivative type amplitude, as Eq.(3), may play an important role. The Adler 0 generally does not imply the suppression at the small s region, and the $\pi\pi$ spectrum shows possibly steep increase from threshold. There is no mechanism cancelling the effect of σ production, and the direct σ-peak structure is expected to be observed. Actually,[1] in $J/\psi \to \omega\pi\pi$ decay[8, 9], $p\bar{p} \to 3\pi^0$,[10] and $D^- \to \pi^-\pi^+\pi^+$[11], where ΔE is very large, the σ-peak structure is observed directly. In the relevant Υ decay processes ΔE in $\Upsilon(3S \to 1S)$ is the largest, and only this process shows the steep increase from threshold.

On the contrary in the $\pi\pi$ scattering process, the pion momentum itself becomes small near threshold region, and the spectrum close to threshold is suppressed. Because of this chiral constraints, the σ-amplitude must be cancelled out by the non-resonant repulsive amplitude, and the direct σ-peak cannot be seen in $\pi\pi$-scattering, in principle.

(**Concluding Remarks**) Through the above discussion it proves not adequate to determine the pole-position of σ only through the $\pi\pi$-scattering. However, the present estimation of the σ pole position in PDG table has been done mainly through the analyses of $\pi\pi$ scattering. This is one of the reasons why the present label of σ, "$f_0(400-1200)$ or σ", is largely uncertain and the σ is still regarded as controversial.

The σ-pole position should be determined regarding the $\pi\pi$-production processes with large energy release, which are free from the chiral constraints. The pole positions of σ obtained through various $\pi\pi$-production processes are shown in Fig. 2. They are almost process-independent, and in the range m_σ=450–600MeV. Basing on these results we

---

[1] Concerning the pp-central collision experiment $pp \to pp(\pi^0\pi^0)$ by GAMS[7], which also seems to suggest the σ-existence, see the comment in the talk by T. Komada, this conference[1].

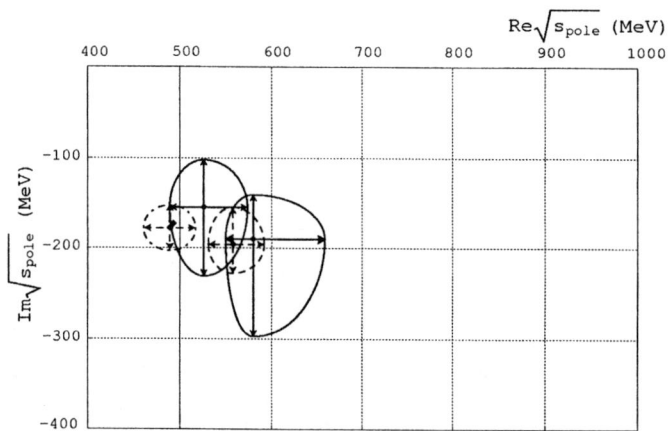

FIGURE 2. Pole positions $\sqrt{s_{pole}} \approx m_\sigma - i\Gamma_\sigma/2$ of σ and their error regions, obtained from the various $\pi\pi$-production processes: Υ decays [1] and [4], $p\bar{p} \to 3\pi^0$[10], $D^- \to \pi^-\pi^+\pi^+$[11] and $J/\psi \to \omega\pi\pi$[9] are shown by the points with crosses. Respective regions (except for J/ψ decay) are surrounded by solid lines, dashed line and dot-dashed line.

conclude that *the existence of $\sigma(450–600)$ is confirmed, and the present label of PDG table should be corrected accordingly.*

In this connection it is notable that, in this conference, a firm experimental evidence for existence of the $I = 1/2$ scalar κ-meson, to be the flavor partner of σ-meson, is reported. This κ is observed[11] as a peak in the $K\pi$-mass spectra in the decay process $D^- \to K^-\pi^+\pi^+$. This process is actually a $K\pi$-production process with the above mentioned large energy release, being free from the chiral constraints. The criticism[12] that no such pole is observed in $K\pi$-scattering process is disregarding the property of K meson as Nambu-Goldstone boson in the case of chiral symmetry breaking. This is the same situation as the direct σ-peak being not observed in $\pi\pi$-scattering. Actually by taking into account this constraint from chiral symmetry, the $K\pi$-scattering data were shown to be consistent with the exsitence of κ[13].

REFERENCES

1. T. Komada, M. Ishida and S. Ishida, this conference; Phys. Lett. **B 508** (2001), 31.
2. S. Ishida, in proc. of Hadron97 in BNL, 1997, AIP conf. proc. 432; WHS99 in Frascati, 1999, Frascati Physics Series 15, 1999. M. Ishida, S. Ishida, T. Ishida, Prog. Theor. Phys. **99** (1998), 1031.
3. L. S. Brown and R. N. Cahn, Phys. Rev. Lett. **35** (1975), 1.
4. M. Ishida, S. Ishida, T. Komada and S. I. Matsumoto, Phys. Lett. **B 518** (2001), 47.
5. S. Ishida, this conference. S. Ishida, M. Ishida, T. Maeda, Prog. Theor. Phys. **104** (2000), 785.
6. S. Ishida et al., Prog. Theor. Phys. **95** (1996), 745; **98** (1997), 1005.
7. D. Alde et al., Phys. Lett. **B397** (1997), 350. T. Ishida, Doctor thesis, Univ. of Tokyo 1996.
8. J. E. Augustin et al., Nucl. Phys. **B320** (1989), 1.
9. K. Takamatsu, in proc. of Hadron97 at BNL, AIP conf. proc. 432, Upton NY 1997.
10. M. Ishida,T. Komada,S.Ishida,T. Ishida,K. Takamatsu,T. Tsuru, Prog. Theor. Phys. **104** (2000), 203.
11. C. Gobel, this conference.
12. S. N. Cherry and M. R. Pennington, Nucl. Phys. **A 688** (2001), 823.
13. S. Ishida, M. Ishida, T. Ishida, K. Takamatsu and T. Tsuru, Prog. Theor. Phys. **98** (1997), 621.

Complete meson-meson scattering within one loop in Chiral Perturbation Theory: Unitarization and resonances

José R. Peláez and A. Gómez Nicola

Departamento de Física Teórica II, Universidad Complutense, 28040 Madrid, SPAIN

Abstract. We review our recent one-loop calculation of all the two meson scattering amplitudes within SU(3) Chiral Perturbation Theory, i.e. with pions, kaons and etas. By unitarizing these amplitudes we are able to generate dynamically the lightest resonances in meson-meson scattering. We thus obtain a remarkable description of the meson-meson scattering data right from threshold up to 1.2 GeV, in terms of chiral parameters in good agreement with previous determinations.

Chiral Perturbation Theory (ChPT) [1] provides a remarkable description of the dynamics of pions, kaons and the eta, which are the pseudo-Goldstone bosons associated to the spontaneous $SU(3)_L \times SU(3)_R$ chiral symmetry breaking down to $SU(3)_{R+L}$. The ChPT Lagrangian contains the most general terms compatible with the symmetry breaking pattern, organized in a derivative and mass expansion (generically p), which, for the amplitudes becomes an expansion in powers of the external momenta and the masses over a scale of $O(1\,\text{GeV})$. The loop divergences appearing at a given order can be absorbed by a finite number of constants in the Lagrangian to the same order. Thus, order by order, the theory is finite and predictive. This approach has been very successful at low energies (less than 500 MeV). However, it has been shown that applying a coupled channel generalization of the Inverse Amplitude Method (IAM) [2, 3] one gets a very good description of meson-meson scattering up to 1.2 GeV, generating dynamically seven light resonances [4]: the ρ, K^*, f_0, a_0, the octet ϕ, the σ and the κ. The properties of the last two, and even their existence, are subject to intense debate within the hadron spectroscopy community (see these proceedings). Remarkably, the chiral amplitudes unitarized with the IAM generate poles associated to these two wide structures [4]. In principle, since this method is built from the perturbative ChPT results, it should respect the good low energy constraints. However, not all the one loop ChPT meson-meson scattering amplitudes were known. Indeed, only $\pi\pi \to \pi\pi$ [5], $K\pi \to K\pi$ [5], $\eta\pi \to \eta\pi$ [5] and $K^+K^- \to K^+K^-$, $K^+K^- \to K^0\bar{K}^0$ [6] were available in the literature. Therefore additional approximations had to be done [4], which spoiled partially the low energy regime and did not allow for a direct comparison with the standard ChPT parameters.

Very recently [7], we have completed the one-loop meson-meson scattering calculation. There are three new amplitudes: $K\eta \to K\eta$, $\eta\eta \to \eta\eta$ and $K\pi \to K\eta$, but we have recalculated the other five amplitudes unifying the notation, ensuring exact perturbative unitarity and also correcting some misprints in the literature. Next, we have applied the coupled channel IAM to these amplitudes. Our results allow for a direct comparison with the standard low-energy chiral parameters, which we find in very good agreement

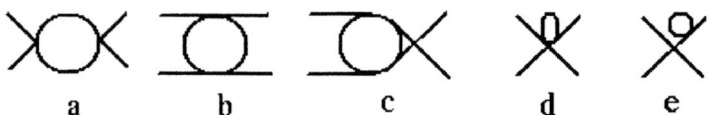

FIGURE 1. Generic one-loop Feynman diagrams that have to be evaluated in meson-meson scattering.

with previous determinations from low-energy data. The main differences with [4] are: i) we consider the full calculation of all the one-loop amplitudes in dimensional regularization, ii) we are able to describe simultaneously the low energy and the resonance regions, and iii) we pay special attention to the estimation of uncertainties.

The $O(p^2)$ scattering amplitudes (low energy theorems) are obtained from the lowest order Lagrangian at tree level, whereas the $O(p^4)$ calculation has the following contributions: First, the one-loop diagrams in Fig.1, which are divergent. Second, the tree level graphs with the second order Lagrangian, which depend on the chiral parameters L_i, that absorb the previous divergences through renormalization. In Table I, we list some determinations of the $L_i^r(\mu)$ renormalized in the usual $\overline{MS}-1$ scheme of ChPT, so that they depend on a scale (except L_3 and L_7), customarily chosen at $\mu = M_\rho$.

TABLE 1. Several sets of chiral parameters ($\times 10^3$) in the literature as well as those from an IAM fit [7], with the uncertainty due to different systematic error used on different fits.

Parameter	$O(p^6)$ K_{l4} decays [8]	$O(p^4)$ K_{l4} decays [8]	ChPT [1, 9]	IAM fits [7]
$L_1^r(M_\rho)$	0.53 ± 0.25	0.46	0.4 ± 0.3	0.56 ± 0.10
$L_2^r(M_\rho)$	0.71 ± 0.27	1.49	1.35 ± 0.3	1.21 ± 0.10
L_3	-2.72 ± 1.12	-3.18	-3.5 ± 1.1	-2.79 ± 0.14
$L_4^r(M_\rho)$	0 (input)	0 (input)	-0.3 ± 0.5	-0.36 ± 0.17
$L_5^r(M_\rho)$	0.91 ± 0.15	1.46	1.4 ± 0.5	1.4 ± 0.5
$L_6^r(M_\rho)$	0 (input)	0(input)	-0.2 ± 0.3	0.07 ± 0.08
L_7	-0.32 ± 0.15	-0.49	-0.4 ± 0.2	-0.44 ± 0.15
$L_8^r(M_\rho)$	0.62 ± 0.2	1.00	0.9 ± 0.3	0.78 ± 0.18

In order to compare with experiment, we use partial waves t_{IJ} of definite isospin I and angular momentum J. Thus, omitting the I,J subindices, the chiral expansion becomes $t \simeq t_2 + t_4 + ...$, with t_2 and t_4 of $O(p^2)$ and $O(p^4)$, respectively. The unitarity relation is rather simple for the partial waves t_{ij}, where i, j denote the different available states. For example, when two states, "1" and "2", are accessible, the partial waves satisfy

$$\text{Im} T = T \Sigma T^* \quad \Rightarrow \quad \text{Im} T^{-1} = -\Sigma \quad \Rightarrow \quad T = (\text{Re}\, T^{-1} - i\Sigma)^{-1} \quad (1)$$

with
$$T = \begin{pmatrix} t_{11} & t_{12} \\ t_{12} & t_{22} \end{pmatrix}, \quad \Sigma = \begin{pmatrix} \sigma_1 & 0 \\ 0 & \sigma_2 \end{pmatrix}, \quad (2)$$

where $\sigma_i = 2q_i/\sqrt{s}$ and q_i is the C.M. momentum of the state i. It can be readily noted that *we only need to know the real part of the Inverse Amplitude*. The imaginary part is fixed by unitarity. Similar expressions can be obtained with n accessible states. Note that, since the unitarity relations are non-linear, they will never be satisfied exactly with a perturbative expansion like that of ChPT. Still, unitarity holds perturbatively, i.e,

$$\text{Im}\, T_2 = 0 + O(p^4), \qquad \text{Im}\, T_4 = T_2 \Sigma T_2^* + O(p^6). \quad (3)$$

A simple way to unitarize ChPT amplitudes is to use in eq.(1) the chiral expansion of $\operatorname{Re} T^{-1} \simeq T_2^{-1}(1 - (\operatorname{Re} T_4) T_2^{-1} + ...)$. Taking into account eq.(3), we find

$$T \simeq T_2(T_2 - T_4)^{-1} T_2, \qquad (4)$$

which is the coupled channel IAM, which we have used to unitarize simultaneously all the one-loop ChPT meson-meson scattering amplitudes [7].

Let us remark that since we have the complete amplitudes renormalized in the $\overline{MS} - 1$ scheme, we can use the IAM with previous L_i^r determinations. Still we find the correct resonant behavior. Nevertheless we have carried out a fit (using MINUIT [10]) of the available data on meson-meson scattering. Since there are incompatibilities between different experiments, a 1%, 3% and 5% systematic error has been added, which introduces an additional source of error. We give in Table 1 the resulting chiral parameters from the fit, whose errors correspond to those of MINUIT combined with those from the systematic uncertainty. Note that they are compatible with previous determinations.

In Fig.2 we show the IAM fit (see [7] for details). The gray error bands cover the uncertainties in the L_i due to MINUIT, and are obtained from a Monte-Carlo gaussian sampling of the parameters. Similarly, the area between the dotted lines covers the errors due to the different choice of systematic error. Note that all the resonant features are reproduced. However, since we have used the full one-loop amplitudes we are able to obtain simultaneously values for the threshold parameters (they have not been fitted) given in table 2. Note the good agreement with the experimental values.

TABLE 2. Meson-meson scattering lengths a_{IJ} and slope parameters b_{IJ}. For experimental references see [7]. Note that our one-loop IAM results are very similar to those of two-loop ChPT.

Threshold parameter	Experiment	IAM fit [7]	ChPT $O(p^4)$ [5, 3]	ChPT $O(p^6)$ [8]
a_{00}	0.26 ±0.05	$0.231^{+0.003}_{-0.006}$	0.20	0.219±0.005
b_{00}	0.25 ±0.03	0.30± 0.01	0.26	0.279±0.011
a_{20}	-0.028±0.012	$-0.0411^{+0.0009}_{-0.001}$	-0.042	-0.042±0.01
b_{20}	-0.082±0.008	-0.074±0.001	-0.070	-0.0756±0.0021
a_{11}	0.038±0.002	0.0377±0.0007	0.037	0.0378±0.0021
$a_{1/20}$	0.13...0.24	$0.11^{+0.06}_{-0.09}$	0.17	
$a_{3/20}$	-0.13...-0.05	$-0.049^{+0.002}_{-0.003}$	-0.5	
$a_{1/21}$	0.017...0.018	0.016±0.002	0.014	
a_{10}		$0.15^{+0.07}_{-0.11}$	0.0072	

Acknowledgments. Work partially supported from the Spanish CICYT projects FPA2000-0956, PB98-0782 and BFM2000-1326.

REFERENCES

1. S. Weinberg, Physica A96, (1979) 327. J. Gasser and H. Leutwyler, Ann. Phys. 158, (1984) 142 and Nucl. Phys. B250, (1985) 465,517,539.
2. T. N. Truong, Phys. Rev. Lett. 661, (1988) 2526 ;Phys. Rev. Lett. 67, (1991) 2260; A. Dobado, M.J.Herrero and T.N. Truong, Phys. Lett. B235, (1990) 134;
3. A. Dobado and J.R. Peláez, Phys. Rev. D47, (1993) 4883; Phys. Rev. D56, (1997) 3057. J. Nieves, M. Pavón Valderrama and E. Ruiz Arriola, hep-ph/0109077.
4. J. A. Oller, E. Oset and J. R. Peláez, Phys. Rev. Lett. 80, (1998) 3452; Phys. Rev. D59, (1999) 074001; Erratum-ibid. D60, (1999) 099906.

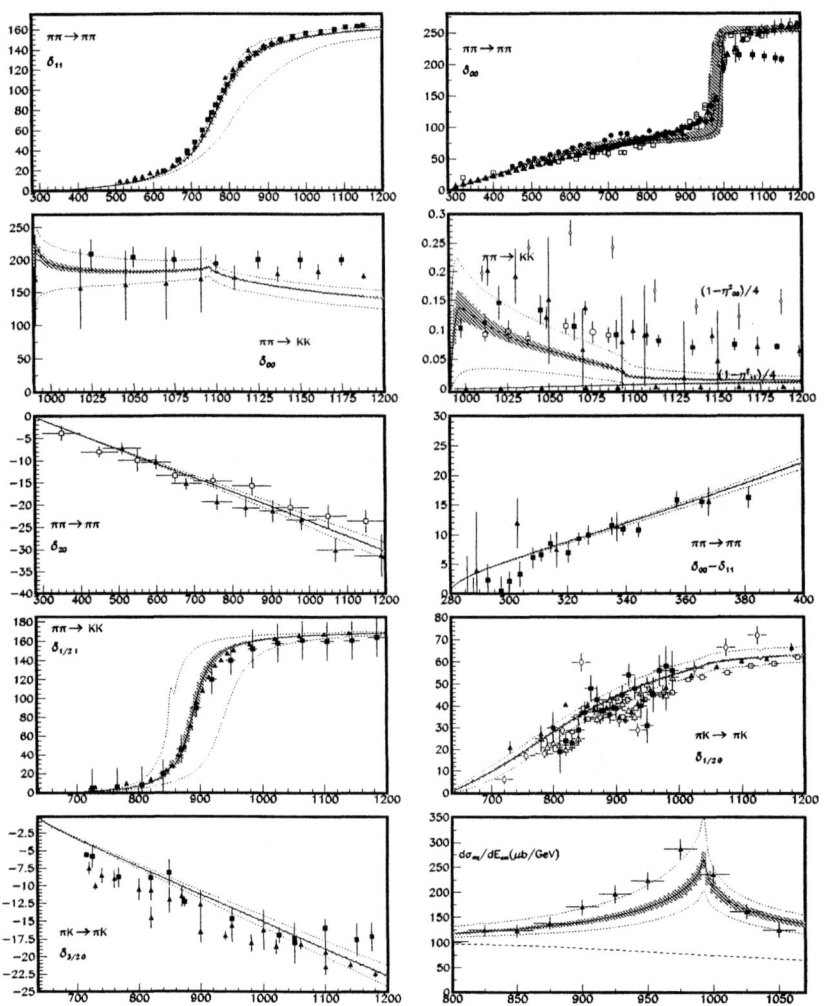

FIGURE 2. Result of the coupled channel IAM fit to meson-meson scattering data (see [7] for references). The shaded area covers the uncertainty due to MINUIT errors. The area between the dotted lines corresponds to the uncertainty in the L_i^r due to the use of different systematic errors on the fits. The dashed line in the last plot is the continuous background underneath the resonant contribution.

5. V. Bernard, N. Kaiser, U.G. Meissner, Phys. Rev. D43 (1991), 2757; Nucl. Phys. B357 (1991), 129; Phys. Rev. D44 (1991), 3698; Nucl. Phys. B364 (1991), 283.
6. F. Guerrero and J. A. Oller, Nucl. Phys. B537, (1999) 459.
7. A. Gómez Nicola and J. R. Peláez, hep-ph/0109056.
8. G. Amorós, J. Bijnens and P. Talavera, Nucl. Phys. B602(2001),87. Nucl.Phys.B585:293-352,2000, Erratum-ibid.B598:665-666,2001.
9. J. Bijnens, G. Colangelo and J. Gasser, Nucl. Phys. B427, (1994) 427.
10. F. James, Minuit Reference Manual D506 (1994).

VERY LOW ENERGY, DEUTERON

IHEP, Protvino

Detection of atoms consisting of π^+ and π^- mesons at PS CERN

L. Afanasyev

JINR Dubna, Russia

on behalf of DIRAC Collaboration

B. Adeva[o], L. Afanasyev[l], M. Benayoun[d], Z. Berka[b], V. Brekhovskikh[n],
G. Caragheorgheopol[m], T. Cechak[b], M. Chiba[j], S. Constantini[p], S. Constantinescu[m], A. Doudarev[l],
D. Dreossi[f], D. Drijard[a], M. Ferro-Luzzi[a], T. Gallas Torreira[a,o], J. Gerndt[b], R. Giacomich[f], P. Gianotti[e],
F. Gomez[o], A. Gorin[n], O. Gortchakov[l], C. Guaraldo[e], M. Hansroul[a], R. Hosek[b], M. Iliescu[e,m],
M. Jabitski[l], N. Kalinina[r], V. Karpukhin[l], J. Kluson[b], M. Kobayashi[g], P. Kokkas[k], V. Komarov[l],
A. Koulikov[l], A. Kouptsov[l], V. Krouglov[l], L. Krouglova[l], K.-I. Kuroda[i], A. Lamberto[f], A. Lanaro[a],
V. Lapshin[n], R. Lednicky[c], P. Leruste[d], P. Levi Sandri[e], A. Lopez Aguera[o], V. Lucherini[e], T. Maki[i],
I. Manuilov[n], L. Montanet[a], J.-L. Narjoux[d], L. Nemenov[a,l], M. Nikitin[l], T. Nunez Pardo[o], K. Okada[h],
V. Olchevskii[l], A. Pazos[o], M. Pentia[m], A. Penzo[f], J.-M. Perreau[a], C. Petrascu[e,m], M. Plo[o], T. Ponta[m],
D. Pop[m], G.F. Rappazzo[f], A. Riazantsev[n], J.M. Rodriguez[o], A. Rodriguez Fernandez[o], V. Rykalin[n],
C. Santamarina[o], J. Saborido[o], J. Schacher[q], C. Schuetz[p], A. Sidorov[n], J. Smolik[c], F. Takeutchi[h],
A. Tarasov[l], L. Tauscher[p], M.J. Tobar[o], S. Trousov[r], P. Vazquez[o], S. Vlachos[p], V. Yazkov[r],
Y. Yoshimura[g], P. Zrelov[l]

[a] *CERN, Geneva, Switzerland*
[b] *Czech Technical University, Prague, Czech Republic*
[c] *Institute of Physics ASCR, Prague, Czech Republic*
[d] *LPNHE des Universites Paris VI/VII, IN2P3-CNRS, France*
[k] *Ioannina University, Greece*
[e] *INFN - Laboratori Nazionali di Frascati, Frascati, Italy*
[f] *Trieste University and INFN-Trieste, Italy*
[g] *KEK, Tsukuba, Japan*
[h] *Kyoto Sangyou University, Japan*
[i] *UOEH-Kyushu, Japan*
[j] *Tokyo Metropolitan University, Japan*
[l] *JINR Dubna, Russia*
[m] *National Institute for Physics and Nuclear Engineering IFIN-HH, Bucharest, Romania*
[n] *IHEP Protvino, Russia*
[o] *Santiago de Compostela University, Spain*
[p] *Basel University, Switzerland*
[q] *Bern University, Switzerland*
[r] *Skobeltsyn Institute for Nuclear Physics of Moscow State Univeristy*

Abstract. The DIRAC experiment aims to measure the lifetime of $\pi^+\pi^-$ atoms in the ground state with 10% precision, using the 24 GeV/c proton beam of the CERN Proton Synchrotron. As the value of the above lifetime of order 10^{-15}s is dictated by a strong interaction at low energy, the precise measurement of this quantity enables to determine a combination of S-wave pion scattering lengths to 5%. Pion scattering lengths have been calculated in the framework of chiral perturbation theory with high precision. Thus the accurate measurement of these values would submit the understanding of chiral symmetry breaking of QCD to a crucial test. Some preliminary results from the analysis of a 2000 data sample are presented.

INTRODUCTION

Pionium ($A_{2\pi}$) is the hydrogen-like atom consisting of π^+ and π^- mesons. Its lifetime is dominated by the charge-exchange process $\pi^+\pi^- \to \pi^0\pi^0$ and decay width is proportional to the square of the atom wave function at origin and to the square of $\Delta = a_0 - a_2$, the difference between the isoscalar and isotensor S-wave $\pi\pi$-scattering lengths. The predicted value for the $A_{\pi\pi}$ lifetime in the ground state, according to the Chiral Perturbation Theory at leading and next-to-leading order in isospin breaking, is of the order of $3 \cdot 10^{-15}$ s [1] [2] [3].

The goal of the DIRAC Experiment [4] is to measure the $A_{\pi\pi}$ lifetime with 10% accuracy in order to determine $|a_0 - a_2|$ with 5% precision in a model independent way [5]. The determination of the S-wave $\pi\pi$-scattering lengths will yield the size of the two-flavour quark condensate [6].

EXPERIMENTAL METHOD

The production cross section of $A_{2\pi}$ in the high-energy inclusive processes have been obtained by L.Nemenov [5]. Relativistic $A_{2\pi}$ produced in hadron-nucleus interaction will move in the target material before their decay. The electro-magnetic interactions with the atoms of the medium might, however, lead to the pionium excitation or breakup (ionization). From the knowledge of the atomic interaction cross sections, for a given target material and thickness, one can calculate the breakup probability P_{br} for arbitrary values of the atom momentum and lifetime [8].

"Atomic pairs" produced from the breakup of $A_{2\pi}$ have very small relative momentum in their c.m. system $Q < 3$ MeV/c that allows to observe they over the background of pions pairs produced in the free state. Moreover the number of the produced $A_{2\pi}$ can be obtained through the number of free pions pairs. Thus comparing the ratio of detected "atomic pairs" to the number of produced $A_{2\pi}$ with the calculated dependence of P_{br} on τ, one can determine the atom lifetime.

THE DIRAC APPARATUS

The DIRAC setup is located at CERN, at the ZT8 beam area of the PS East Hall. It became operational at the end of 1998 and has been collecting data since the summer of 1999, using the 24 GeV/c proton beam extracted from the PS accelerator.

DIRAC is a fixed-target experiment and its apparatus is designed to detect charged pion pairs with high resolution over the pair relative momentum (see Fig.1). It consists of a straight detector section between the target station and the analyzing magnet (2.3 Tm bending power), and of a double arm spectrometer downstream the magnet.

A multi-level trigger system has been designed to select atomic pairs from the overwhelming background of uncorrelated $\pi^+\pi^-$ pairs. The Cherenkov, preshower and muon detectors ensure rejection of e^+e^-, $\mu^+\mu^-$ pairs at trigger level. Pairs of $p\pi^-$ are identified by offline analysis of time-of-flight measurements.

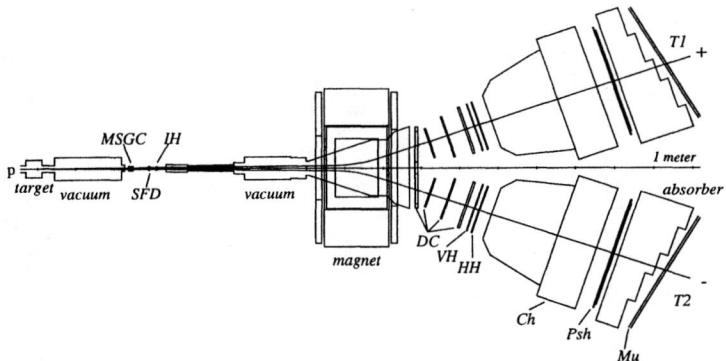

FIGURE 1. Schematic top view of the DIRAC spectrometer. Moving from the target station toward the magnet there are: four MicroStrip Gas Chambers (*MSGC*), two Scintillating Fiber Detectors (*SFD*) and two Ionization Hodoscopes (*IH*). Downstream the dipole magnet, on each arm of the spectrometer, are located: four modules of Drift Chambers (*DC*), the Vertical and Horizontal Hodoscopes (*VH, HH*), the Cherenkov counter (*Ch*), the Preshower detector (*PSh*) and, behind the iron absorber, the Muon detector (*Mu*).

EXPERIMENTAL RESULTS

The number of ionized $A_{2\pi}$ is measured directly from the excess of $\pi^+\pi^-$ pairs with very low relative momentum ($Q < 3$ MeV/c) compared to the expected number of "free" pairs simulated using pairs of accidental coincidence. The Q-distribution of $\pi^+\pi^-$ pairs obtained with the nickel and titanium targets in 2000 are presented on the Fig.2. The first preliminary result on the lifetime of $\pi^+\pi^-$ atom obtained with the Ni2000 data are given on the Fig.3.

The DIRAC Experiment is at present collecting statistics to achieve the final goal of measuring the lifetime of pionium with 10% accuracy. The quality of the data is excellent and a preliminary analysis of data collected two last years with three targets shows the excess of $\pi^+\pi^-$ pairs coming from the breakup of $A_{2\pi}$. By now the total number of the observed "atomic pairs" amounts 5500. At the present rate of $\sim$ 3-4 detected atomic pairs per million triggers we expect to fulfil our experimental programme within the year 2002.

REFERENCES

1. A. Gall et al., *Phys. Lett.* B **462**, 335 (1999).
2. J. Gasser et al., *Phys.Lett.* B **471**, 244-250 (1999).
3. G. Colangelo et al., *Phys.Lett.* B **488**, 261-268 (2000).
   G. Colangelo et al., *Nucl.Phys.B* **603**, 125-179 (2001).
4. B. Adeva et al., Proposal to the SPSLC, *CERN/SPSLC 95-1, SPSLC/P 284*, (1994).
5. L.L. Nemenov, *Sov. J. Nucl. Phys.* **41**, 629 (1985).
6. N.H. Fuchs et al., *Phys. Lett.* B **269**, 183 (1991); *Phys. Rev.* D **47**, 3814 (1993).
7. E. Hartouni et al., *Phys. Rev. Lett.* **72**, 1322 (1994).
8. L.G.Afanasyev and A.V.Tarasov: *Yad.Fiz.* **59**, 2212 (1996); *Phys.Atom.Nucl.* **59**, 2130 (1996).

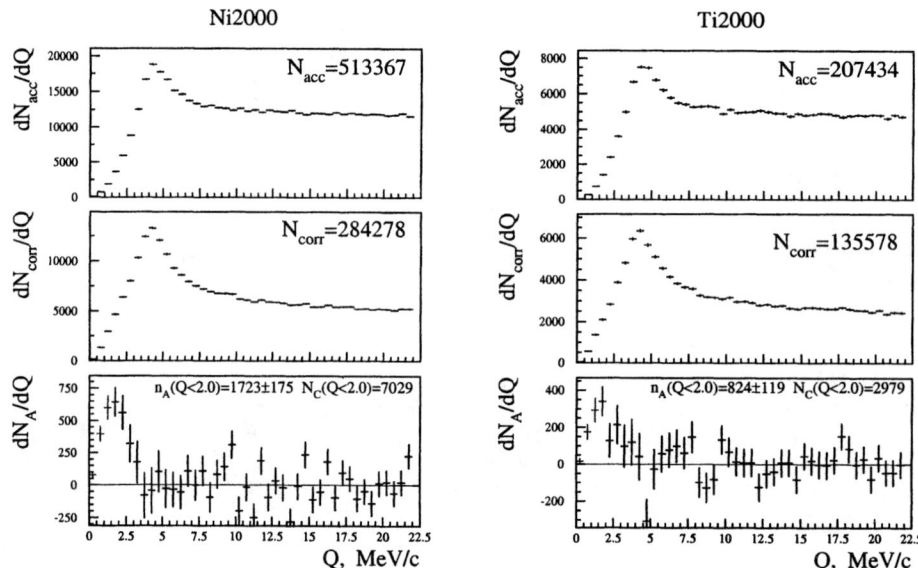

FIGURE 2. Distributions over $\pi^+\pi^-$ pair relative momentum for Ni and Ti targets. The accidentals are shown on the upper plots, the correlated in time pairs on the middle and their normalized difference on the lower one. The "atomic pairs" are observed as the peaks at low Q on the latter plots.

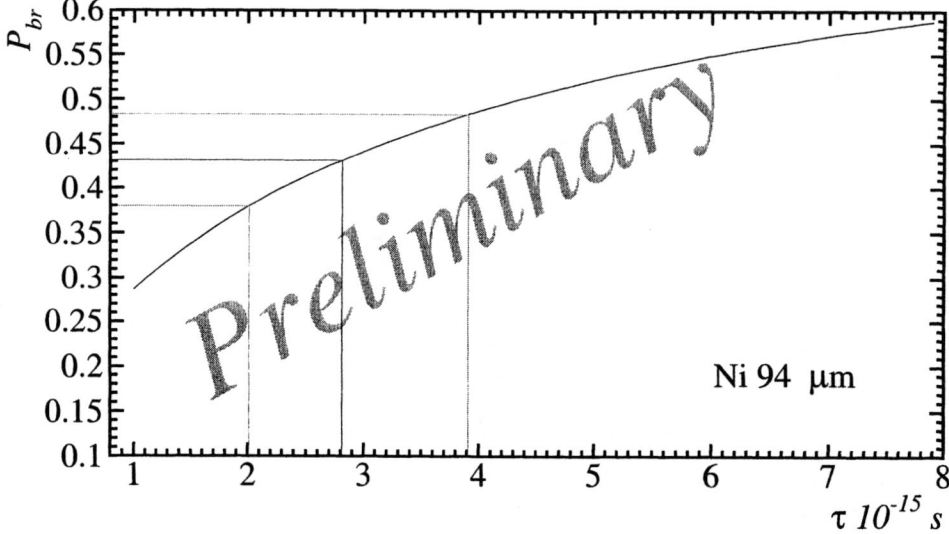

FIGURE 3. The calculated probability of pionium breakup as a function of its lifetime for the Ni target of 94 μm thick (curve). The measured probability of the pionium breakup obtained form Ni2000 data is shown with its statistical errors as the horizontal lines. The crossing gives the lifetime value of $\tau = (2.8^{+1.1}_{-0.8}) \cdot 10^{-15}$ s

Deeply bound $1s$ and $2p$ pionic states in ^{205}Pb and effective pion mass in nuclear matter

Hans Geissel[a], Hansjörg Gilg[b], Albrecht Gillitzer[c], Ryugo S. Hayano[d],
Satoru Hirenzaki[e], Kenta Itahashi[f], Masahiko Iwasaki[f], Paul Kienle[b],
Matthias Münch[b], Gottfried Münzenberg[a], Wolfgang Schott[b],
Ken Suzuki[d], Dai Tomono[f], Helmut Weick[a], Toshimitsu Yamazaki[g], and
Tetsu Yoneyama[f]

[a]*Gesellschaft für Schwerionenforschung, D-64291 Darmstadt, Germany*
[b]*Physik-Department, Technische Universität München, D-85747 Garching, Germany*
[c]*Institut für Kernphysik, Forschungszentrum Jülich, D-52425 Jülich, Germany*
[d]*Department of Physics, University of Tokyo, Hongo, Bunkyo-ku, Tokyo 113-0033, Japan*
[e]*Department of Physics, Nara Women's University, Kita-Uoya Nishimachi, Nara 630-8506, Japan*
[f]*Department of Physics, Tokyo Institute of Technology, Meguro-ku, Tokyo 152-8551, Japan*
[g]*RI Beam Science Laboratory, RIKEN, 2-1 Hirosawa, Wako-shi, Saitama, 351-0198 Japan*

Abstract. We have succeeded in observing the $1s$ and $2p$ bound pionic states in ^{205}Pb nucleus as well separated peaks in the excitation spectrum of the ^{206}Pb$(d,^3$He$)$ reaction. The observed spectrum agreed quite well with the theoretical prediction where the formation cross section was calculated for the deeply bound pionic states. These states are accommodated in proximity to the nuclear surface as halo states in a potential provided by the attractive Coulomb and the repulsive core. The binding energies and the widths were precisely determined, and the values were used to deduce the unknown s-wave part of the π–nucleus interaction and to translate it to the effective pion mass in nuclear matter.

EXPERIMENTAL PROCEDURE AND ANALYSIS

The recent experiment to measure the excitation spectrum of ^{208}Pb$(d,^3$He$)$ reaction near the pion emission threshold revealed that the deeply bound π^- states indeed exist and that their formation and observation is feasible by the nuclear reaction [1]. The spectrum indicated the formation of the bound $2p$ state as a distinct peak and the $1s$ as a clear excess on the tail of the $2p$. This first successful observation led us to concentrate on the investigation of necessary improvements to observe the $1s$ state as a separate peak. We carefully examined the experimental conditions and came to conclude that the replacement of the ^{208}Pb target with ^{206}Pb enhances the isolation of the $1s$ contribution from the $2p$ tail [2].

The experiment using the ^{206}Pb target was performed using the Fragment Separator (FRS) [3] and the SIS-18 synchrotron of GSI, Darmstadt. Detailed descriptions of the experimental procedures and analyses are found in Refs. [4, 5, 6]. We employed a $T_d = 604.3$ MeV deuteron beam with the intensity 1.5×10^{11}/spill and a 1.5 mm wide 25 mg/cm^2 thick ^{206}Pb target. The ^{3}He momentum was analyzed by the FRS and the excitation spectrum of the ^{206}Pb$(d,^3$He$)$ reaction covering the bound π^- region

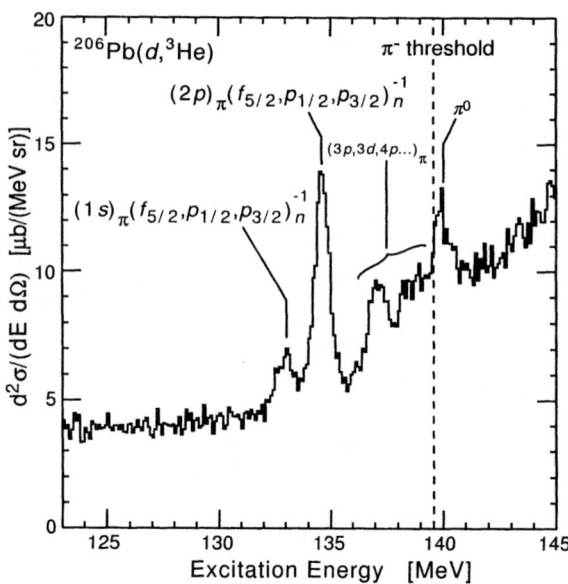

FIGURE 1. Double differential cross section versus excitation energy of the ^{206}Pb$(d,^3$He$)$ reaction measured at the incident deuteron energy (T_d) of 604.3 MeV. The π^- emission threshold is shown by the vertical broken line. The two peaks near the center of the figure correspond to the formation of 1s and 2p bound pionic states.

was measured with the experimental resolution of 0.317 ± 0.048 MeV (FWHM). The systematic uncertainty in the absolute energy scale is estimated to be 0.040 MeV.

The observed excitation spectrum, namely, the double differential cross section versus the excitation energy referring to the ground state of ^{205}Pb (E_x), is shown in Fig. 1. The spectrum can be simply divided into three regions. The right-hand side of the π^- emission threshold ($E_x = 139.570$ MeV, shown by the vertical broken line) from $E_x = 141$ MeV to 145 MeV is a continuum due to the quasi-free pion production. The small peak near the threshold ($E_x \sim 140$ MeV) arises from the $p(d,^3$He$)\pi^0$ reaction by hydrogen contamination in the target. The left-hand side from $E_x = 120$ MeV to 130 MeV is a nearly constant structureless background of about 4 μb/(sr MeV) due to the nuclear excitation without pion production. The central part between 130 MeV and 140 MeV has structures associated with the pionic state formation. The largest peak at $E_x = 135$ MeV corresponds to the bound $(2p)_\pi$ state coupled with mostly the three neutron hole states of $(2f_{5/2}, 3p_{1/2}, 3p_{3/2})_n^{-1}$. The well separated peak at $E_x \sim 133$ MeV near the $(2p)_\pi$ peak is $(1s)_\pi$, which is also a composite with the three hole states. The region between the $(2p)_\pi$ peak and the threshold ($E_x = 136 \sim 139$ MeV) is due to the formation of higher states $(3p, 3d, 4p...)_\pi$. The overall structure agrees remarkably well with the theoretical calculation [2].

In order to determine the 1s and 2p binding energies and widths, we decomposed the region of interest into $(1s)_\pi$ and $(2p)_\pi$ components and a linear background by a least squares fit in a similar manner as in Ref. [6]. The result of this fitting is excellent as

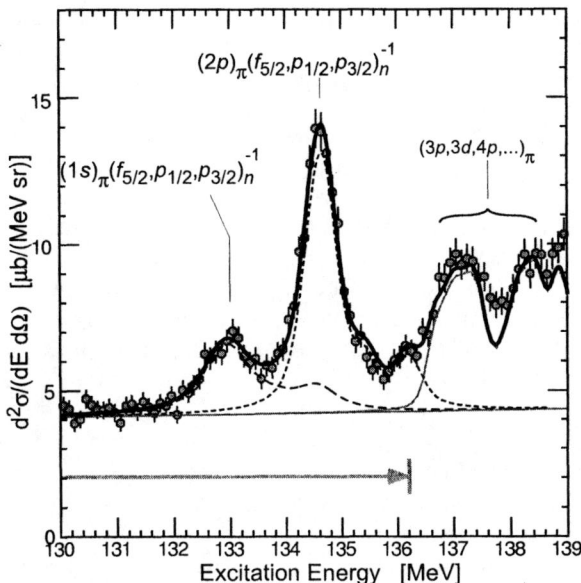

FIGURE 2. The data points of the excitation spectrum, fitted by the solid curve, which was obtained in the fit range indicated by the gray arrow. The broken and dotted curves are the $(1s)_\pi$ and $(2p)_\pi$ components, respectively.

shown in Fig. 2 (the χ^2 is 97.4 for n.d.f. = 123). The solid curve shows the obtained fit function, the broken and dotted curves are the $(1s)_\pi$ and $(2p)_\pi$ contributions, respectively. The peak intensity of $(1s)_\pi$ relative to $(2p)_\pi$ is 69 % larger than the theoretical prediction. Sum of the contribution from shallow states of $(3p,3d,4p...)_\pi$ is not considered in the fitting procedure but was scaled with the same factor used for the $(2p)_\pi$, and is shown as the gray curve. This contribution is added to the fit result to show the solid curve and to clarify the behaviour of the fit function outside the fitting region.

Our final values for the binding energies and widths of the 1s and 2p states are:

$$B_{1s} = 6.768 \pm 0.044 \, (stat.) \pm 0.041 \, (syst.) \text{ MeV}$$
$$\Gamma_{1s} = 0.778^{+0.150}_{-0.130} \, (stat.) \pm 0.055 \, (syst.) \text{ MeV}$$
$$B_{2p} = 5.110 \pm 0.015 \, (stat.) \pm 0.042 \, (syst.) \text{ MeV}$$
$$\Gamma_{2p} = 0.371 \pm 0.037 \, (stat.) \pm 0.048 \, (syst.) \text{ MeV}.$$

DISCUSSION AND CONCLUSION

Since the $(1s)_\pi$ binding energy and width depend nearly entirely on the s-wave potential of the π-nucleus interaction $(U_s(r) = V_s(r) + iW_s(r))$, we can uniquely determine its real part $V_s(0)$ and imaginary part $W_s(0)$, irrespective of the choice of the p-wave parameters.

Fixing the p-wave parameters to the well known values [7] we vary the representative s-wave parameters, $V_s(0)$ and $W_s(0)$, to obtain their theoretical relations with B_{1s} and Γ_{1s}. The obtained relations are compared to the experimental values of B_{1s} and Γ_{1s} yielding

$$V_s(0) = 26.9^{+1.7}_{-1.6} \text{ MeV} \tag{1}$$

$$W_s(0) = -13.8^{+3.4}_{-3.7} \text{ MeV}. \tag{2}$$

Because the binding energies and widths are determined primarily by the strength of $V_s(r)$ at the half density radius, we have to allow an ambiguity as much as 2–8 MeV to extrapolate and to determine the potential depth $V_s(0)$ at *the center of the nucleus*. If there is a neutron skin, the above value is lowered accordingly [8].

The value of $V_s(0)$ indicates a magnitude of the repulsive mass shift of π^- in the center of the present nucleus [9, 6]. This strength is significantly larger than $V_s(0) \sim 16$ MeV which we expect from the free πN scattering lengths after the correction for double scattering ($b_0 = -0.017$ and $b_1 = -0.090$) [10]. The inconsistency between the present value and the above value can be explained by the increase of the isovector strength $|b_1|$ expected from the chiral symmetry restoration of the quark condensate [11, 12].

In conclusion, we have clearly observed the ground $1s$ and $2p$ states of π^- in ^{205}Pb. The deduced binding energy and width of $(1s)_\pi$ were used to determine the s-wave potential strength, which indicates a ~ 27 MeV repulsive mass shift of π^- in this nucleus, significantly larger than that expected from the free πN scattering lengths after double scattering correction.

ACKNOWLEDGMENTS

The authors thank Prof. H. Toki and Dr. Y. Umemoto for theoretical support, and the staff of GSI for providing superb experimental conditions. Some of the authors (K.S., H.G., and K.I.) express their gratitude for the receipt of the fellowship to the Japan Society for the Promotion of Science. This work is supported by the Grant-in-Aid for Scientific Research of Monbusho (Japan) and by the Bundesministerium für Bildung, Wissenschaft, Forschung und Technologie (Germany).

REFERENCES

1. T. Yamazaki *et al.*, Z. Phys. **A355**, 219 (1996).
2. S. Hirenzaki and H. Toki, Phys. Rev. C **55**, 2719 (1997).
3. H. Geissel *et al.*, Nucl. Inst. Meth. **B70**, 286 (1992).
4. K. Suzuki, *Master thesis of the University of Tokyo, unpublished* (2000).
5. H. Gilg *et al.*, Phys. Rev. C **62**, 025201 (2000).
6. K. Itahashi *et al.*, Phys. Rev. C **62**, 025202 (2000).
7. R. Seki and K. Masutani, Phys. Rev. C **27**, 2799 (1983).
8. Y. Umemoto, S. Hirenzaki, K. Kume, and H. Toki, Phys. Rev. C **62**, 024606 (2000).
9. T. Yamazaki *et al.*, Phys. Lett. **418B**, 246 (1998).
10. T.E.O. Ericson, B. Loiseau and A.W. Thomas, Phys. Rev. C (2000).
11. W. Weise, Acta Phys. Pol. **B 31**, 2715 (2000).
12. P. Kienle and T. Yamazaki, Phys. Lett. B., in press.

Search for Supernarrow Dibaryons Production in $pd \to p + pX_1$ and $pd \to p + dX_2$ Reactions

L.V. Fil'kov*, V.L. Kashevarov*, E.S. Konobeevski[†],
M.V. Mordovskoy[†], S.I. Potashev[†], V.A. Simonov[†] and
V.M. Skorkin[†]

*Lebedev Physical Institute, Moscow, Russia; e-mail:filkov@sci.lebedev.ru
[†] Institute for Nuclear Research, Moscow, Russia; e-mail:konobeev@sci.lebedev.ru

Abstract. We study a production of supernarrow dibaryons, the decay of which into two nucleons is forbidden by the Pauli exclusion principle, in the reactions $pd \to p+pX_1$ and $pd \to p + dX_2$ at Linear Accelerator of INR (Moscow). Dibaryons with masses 1904 ± 2, 1926 ± 2 and 1942 ± 2 MeV have been observed in missing mass M_{pX_1} spectra. In missing mass M_{X_1} spectra, the peaks at $M_{X_1} = 966\pm2$, 986 ± 2, and 1003 ± 2 MeV have been found. The analysis of the data obtained leads to the conclusion that the observed dibaryons are supernarrow dibaryons. The possible interpretation of "exotic baryon states" with small masses is discussed.

I INTRODUCTION

Usually one looks for dibaryons in the NN channel (see for review ref. [1]). Such dibaryons have decay widths from a few up to a hundred MeV. Their relative contributions are small enough but the background contribution is big and uncertain as a rule. All this leads often to contradictory results.

In the present work we will consider supernarrow dibaryons (SNDs), a decay of which into two nucleons is forbidden by the Pauli exclusion principle [2-5]. These dibaryons satisfy the following condition:

$$(-1)^{I+S}P = +1 \qquad (1)$$

where I is the isospin, S is the internal spin, and P is the dibaryon parity. In the NN channel, such dibaryons correspond to the following forbidden NN states: even singlets and odd triplets with the isotopic spin $I = 0$ as well as odd singlets and even triplets with $I = 1$. These dibaryons with the masses $M < 2m_N + m_\pi$ ($m_N(m_\pi)$ is the nucleon (pion) mass) can mainly decay into two nucleons by emitting a photon. This is a new class of dibaryons with the decay widths $\leq$ 1keV. For isovector dibaryons, the decay widths are expected to be equal to a few eV or even smaller [4]. The experimental discovery of such states would have important consequences for particle and nuclear physics and astrophysics.

We study the reactions $pd \to p+pX_1$ and $pd \to p+dX_2$ with the aim of searching for SNDs. In the process $pd \to pD$, SNDs can be produced only if the nucleons in the deuteron overlap sufficiently, such that a 6-quark state with deuteron quantum numbers can be formed. In this case, an interaction of a meson or another particle with this state can change its quantum numbers so that a metastable state is formed. Therefore, the probability of the production of such dibaryons is proportional to the probability η of the 6-quark state existing in the deuteron. As indicated above, the SNDs decay mainly emitting a photon. Therefore, if we limit ourselves to the investigation of pd interactions with the photon in the final state, then the contribution of the background processes would be suppressed essentially. On the other hand, the special choice of the kinematics allows to allocate the area where the contribution of the SNDs dominates, even without the detection of the final photon.

It was shown in [3] that the decay of the SND into γNN had to be characterized in the rest frame by a narrow peak near the maximum photon energy in the probability distributions of the dibaryon decay over an emitted photon energy. It leads to an essential limitation of the outgoing nucleon angles. If such a dibaryon decays into γd, the emitted deuteron angles are limited by the following condition: $\sin\theta_d \leq Mp_d^*/(m_d p_D)$, where p_D is the momentum (in the lab. system) of the dibaryon, p_d^* is the momentum (in the c.m. system) of the deuteron, and M is the dibaryon mass. So the nucleons and the deuteron from the decay of SND into γNN and γd have to be emitted in a narrow angle cone with respect to the direction of motion of the dibaryon. On the other hand, if a dibaryon decays mainly into two nucleons, then the expected angular cone of emitted nucleons must be more than 50°. Therefore, a detection of the scattered proton in coincidence with the proton (or the deuteron) from the decay of the dibaryon at correlated angles allowed to suppress essentially the contribution of the background processes and to increase the

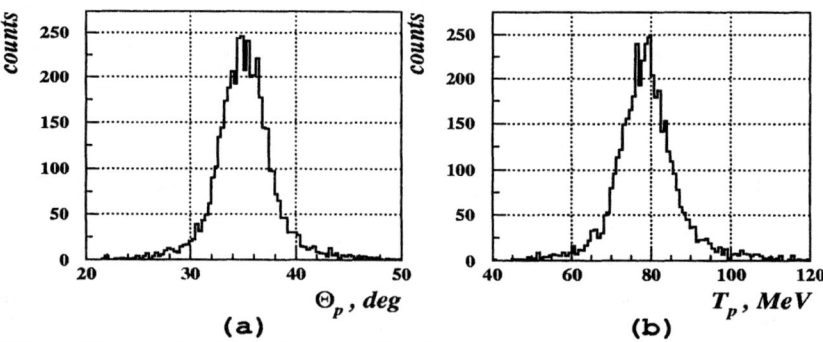

FIGURE 1. The angular (a) and energy (b) distributions of the protons from the isovector SNDs decay.

relative contribution of a possible SND production. The results of the Monte-Carlo simulation of the kinematics of the protons from the decay of the isovector SND with mass 1904 MeV, at the condition that the proton beam energy is equal to 305 MeV and the scattered proton is detected at 70°, are presented in Fig. 1. Fig. 1a demonstrates the expected angular distributions of the protons under consideration. As seen from this figure, these protons are emitted mainly at $\theta_p = 35°$ and the width of the peak is equal to a few degrees. The distributions of these protons over their kinetic energy are presented in Fig. 1b. This figure shows that we should take into account the protons with the energy $60 < T_p < 100$ MeV.

Such cuts are very important as they allow to suppress essentially the contribution from the reaction $p+d \to p+p+n$. Main part of this contribution is determined by the one-nucleon pole diagram with the neutron (and the proton) as a spectator and is proportional to $1/(m_d - 2E_N)^2$ where E_N is the full nucleon energy. The calculation of the kinematics for the other SND masses and quantum numbers have been done too. In order to suppress the background from the reaction $p+{}^{12}C$ in M_{dX_2} spectra we omitted events with the deuteron energy higher than the energy of the elastic scattered deuteron in the $pd \to pd$ reaction at appropriate angles.

II EXPERIMENTAL RESULTS

In the present paper we give the results of an analysis of the experimental data respecting study of the $pd \to p + pX_1$ and $p \to p + dX_2$ processes at the Linear Accelerator of INR with 305 MeV proton beam using the spectrometer TAMS. The properties of this spectrometer are described elsewhere [4]. In this experiment CD_2 and ${}^{12}C$ were used as targets. The scattered proton was detected in the left arm of the spectrometer TAMS at the angle $\theta_L = 70°$. The second charged particle (either p or d) was detected in the right arm by three telescopes located at $\theta_R = 34°, 36°$, and $38°$. A trigger was generated by four-fold coincidence of the two ΔE detector signals of the left arm combined with those of any telescope of the right arm. The selected events contained an information about time of flights and full energies of two particles detected in coincidence in the left and right arms of the spectrometer. An energy resolution was 3 MeV (7 MeV) for the left (right) arm. A time resolution better than 0.5 ns was achieved. Each useful event including two time of flight and two energies were stored event by event and then analyzed off line.

An off line identification of different particles was performed by means of their energies and time of flights. In this way, protons and other charged particles were identified via characteristic loci observed in two-dimensional diagrams of the time of flight vs energy. For example, such an experimental $E - t$ distribution for the right arm detector is displayed in Fig. 2. This figure shows a proton locus, presented a dependence of the proton time of flight on its energy, and a deuteron spot corresponding to a registration of the elastic pd scattering.

Invariant mass spectra of M_{pX_1} and M_{dX_2} was determined using the expression

$$M^2_{pX_1(dX_2)} = m_d^2 + m_p^2 + 2m_d(E_1 - E_2) - 2E_1E_2 + 2p_1p_2\cos\theta_L \qquad (2)$$

at an additional condition of a detection of the proton (or the deuteron) from the decay of the pX_1 (or dX_2) system in the right arm of the detector. In Eq. (1), $m_{p(d)}$ is the proton (deuteron) mass, $E_1(p_1)$ and $E_2(p_2)$ are the energies (momenta) of the incident and scattered protons, respectively. The invariant mass M_{dX_2} spectrum of the reaction $pd \to pdX_2$ for the angle $\theta_R = 38°$ has a peak at the deuteron mass. This peak corresponds to the elastic pd scattering. A measurement of this effect at different angles of the right and left arms was used to calibrate the spectrometer. The mass resolution of the spectrometer was 4 MeV; the angular resolution was 1°.

Let us consider the reaction $pd \to p + pX_1$. Figs. 3a-3c demonstrate the experimental invariant mass M_{pX_1} spectra obtained with the CD_2 (the points with the statistical errors) and ^{12}C (the bars) targets, where (3a), (3b), and (3c) correspond to a detection of the second proton in the right arm detector at $\theta_R = 34°$, $36°$, and $38°$, respectively. The background in these spectra are interpolated by polynomials (dashed curves). Three peaks at $M_{pX_1} = 1904 \pm 2$, 1926 ± 2, and 1942 ± 2 MeV are observed in these spectra. The first two of them confirmed the values of the dibaryon mass obtained by us earlier [4-7] and the resonance at 1942 MeV is a new one.

The experimental invariant mass spectra, obtained with the carbon target, are rather smooth. This smoothness is caused by both an essential increase of the con-

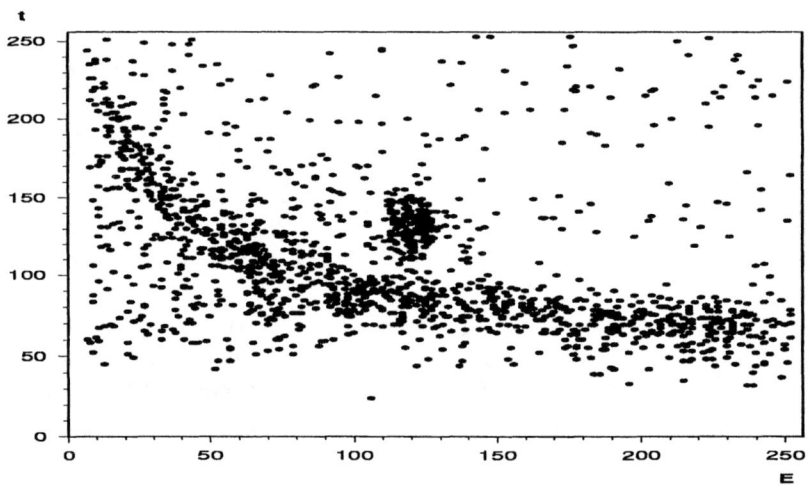

FIGURE 2. E-t distribution for the right arm detector at $\theta_R = 38°$

tribution of background reactions in the interaction of the proton with the carbon and Fermi motion of nucleons in the nucleus. The latter increases essentially the angular cone size of emitted nucleons. In consequence, it is not possible to see peaks of SNDs in the present experiment on the carbon target. As the experiment with the carbon target resulted in the rather smooth spectra, all structures, appearing in the experiment with the CD_2 target, may be explained by an interaction of the proton with the deuteron.

The experimental spectra in Figs. 3a-3c are compared with the prediction of the theoretical model of SNDs $D(T = 1, J^P = 1^{\pm})$ production (the solid curves) constructed in the one meson exchange approach [4] and normalized to the values of peaks in Fig. 3a. These predictions for the SNDs are in agreement with our experimental data within the errors. If the observed states are NN-coupled dibaryons

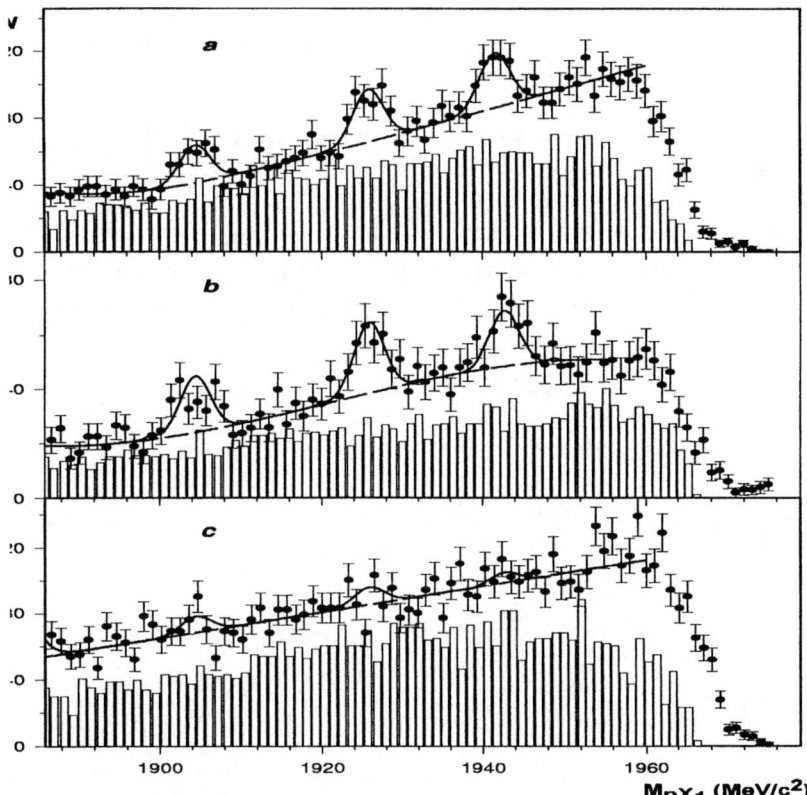

FIGURE 3. The invariant mass M_{pX_1} spectra obtained with CD_2 (the points with the statistical errors) and ^{12}C (the bars) targets; (a) – $\theta_R = 34°$, (b) – $\theta_R = 36°$, (c) – $\theta_R = 38°$. The solid curves are normalized theoretical predictions. The dashed curves correspond to the background interpolated by polynomials.

decaying mainly into two nucleons then the expected angular cone size of emitted nucleons must be more than 50°. Therefore, their contributions to the invariant mass spectra in Fig. 3a-3c would be nearly the same and would not exceed a few events, even assuming that the dibaryon production cross section is equal to that of elastic pd scattering ($\sim 40\mu$b/sr). Hence, the peaks found most likely correspond to SNDs. However, the analysis of the reaction $pd \to p + pX_1$ only in the considered angle range does not allow to determine an isotopic spin of the SNDs. The invariant mass M_{dX_2} spectrum of the reaction $pd \to p + dX_2$, for the sum of angles $\theta_R = 34°$ and $36°$ is shown in Fig. 4.

As seen from this figure, the reaction $pd \to p + dX_2$ gives very small contribution into the production of the dibaryons under study. On the other hand, it is expected [5,4] that isoscalar SNDs contribute mainly into γd channel and isovector SNDs do into γNN one. As the main contribution of the found dibaryons is observed in pX_1 channel, it is possible to expect that $X_1 = \gamma + n$, and all found states are isovector SNDs. The more precise conclusion about the value of the isotopic spin of the observed SNDs could be obtained by the study of the reaction $pd \to n + pX$.

The summary spectrum of the reaction $pd \to p + pX_1$ over angles $\theta_R = 34°$ and 36°, at which where the contribution of the SNDs is maximum, is presented in Fig. 5a. This spectrum was interpolated by a polynomial (for the background) plus Gaussians (for the peaks). The number of standard deviations (SD) is 6.0, 7.0, and 6.3 SD for the resonances at 1904, 1926, and 1942 MeV, respectively. The widths of these resonances are equal to the experimental resolution of ~ 4 MeV.

An additional information about the nature of the observed states was obtained by studying the missing mass M_{X_1} spectra of the reaction $pd \to p + pX_1$. If the

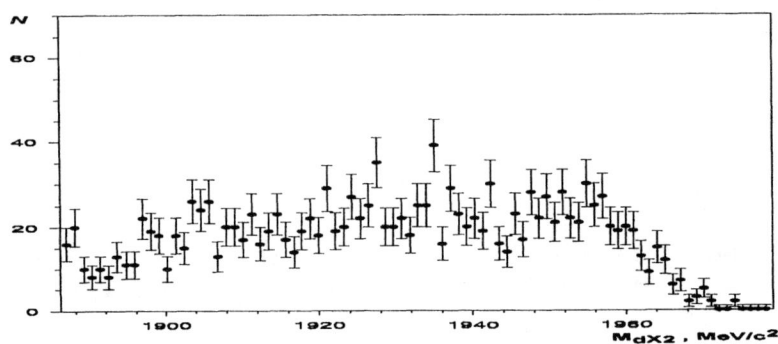

FIGURE 4. The invariant mass M_{dX_2} spectra for the reaction $pd \to p + dX_2$ for the sum of the angles $\theta_R = 34°$ and $36°$.

state found is a dibaryon decaying mainly into two nucleons then X_1 is a neutron and the mass M_{X_1} is equal to the neutron mass m_n. If the value of M_{X_1}, obtained from the experiment, differs essentially from m_n then $X_1 = \gamma + n$ and we have the additional indication that the observed dibaryon is SND. The simulation of missing mass spectra for the reaction $pd \to p + pX_1$, where pX_1 are decay products of the SNDs with the masses 1904, 1926, and 1942 MeV, gave peaks at $M_{X_1} = 965$, 987, and 1003 MeV, respectively. Fig. 5b demonstrates the missing mass M_{X_1} spectrum obtained from the experiment for the sum of the angles $\theta_R = 34°$ and $36°$. As is seen from this figure, besides the peak at neutron mass, which caused by the process $pd \to p + pn$, a resonancelike behavior of the spectrum is observed at 966 ± 2, 986 ± 2, and 1003 ± 2 MeV. These values of M_{X_1} coincide with the ones obtained from the simulation and differ essentially from the value of the neutron mass (939.6 MeV). Hence, for all states under study, we have $X_1 = \gamma + n$ in support

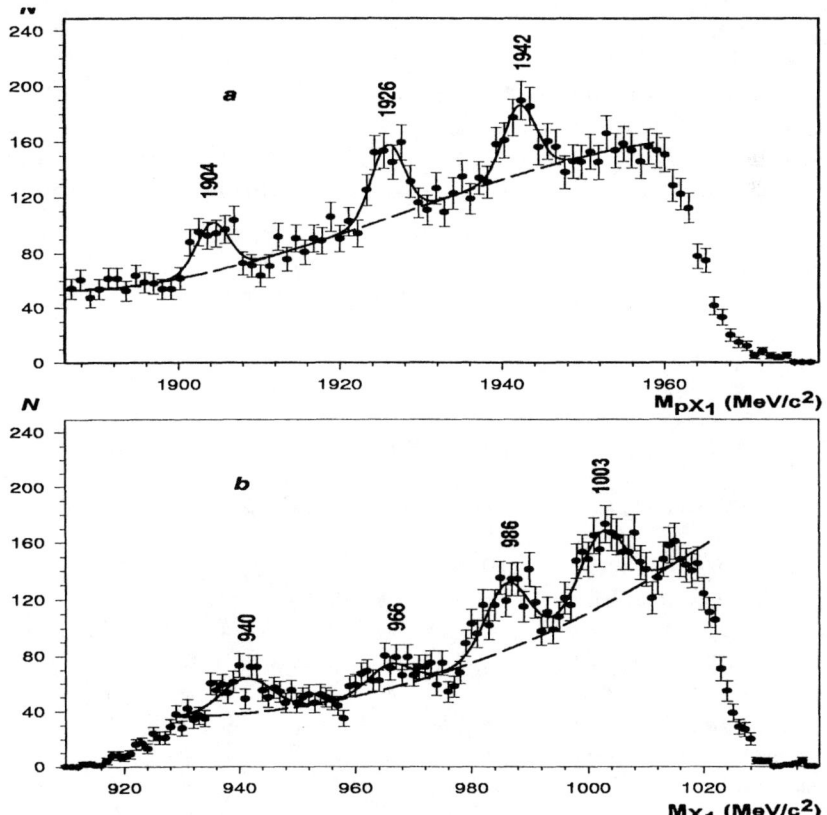

FIGURE 5. The invariant mass M_{pX_1} (a) and the missing mass M_{X_1} (b) spectra for the sum of angles of $\theta_R = 34°$ and $\theta_R = 36°$. The dashed and solid curves are results of interpolation by polynomials (for the background) and Gaussians (for the peaks), respectively

of a statement that the dibaryons found are SNDs.

It should be noted that the peak at $M_{X_1} = 1003 \pm 2$ MeV corresponds to the resonance found in [8] and attributed to an "exotic baryon state" N^*. In this work, the authors brought out three such states with masses 1004, 1044, and 1094 MeV. In principle, SND could decay into NN^*. A possibility of the production of NN^*-coupled dibaryons was considered in [1]. Taking into account the found connection between the SNDs and the resonancelike states X_1, it is possible to assume that the peaks, observed in [8] at 1004 and 1044 MeV, are not "exotic baryon states", but they are the resonancelike states $X_1 = \gamma + n$ caused by possible existence and decay of the SNDs with the masses 1942 and 1982 MeV, respectively.

III CONCLUSIONS

The following conclusion can be made. As a result of the study of the reaction $pd \to p + pX_1$ three narrow peaks at 1904, 1926, and 1942 MeV have been observed in the invariant mass M_{pX_1} spectra. The analysis of the angular distributions of the protons from the decay of pX_1 states and the data on the reaction $pd \to p + dX_2$ showed that the peaks found can be explained as a manifestation of the isovector SNDs, the decay of which into two nucleons is forbidden by the Pauli exclusion principle. The observation of the peaks in the missing mass M_{X_1} spectra at 966, 985, and 1003 MeV is an additional confirmation that the dibaryons found are the SNDs.

ACKNOWLEDGMENTS

Work supported by RFBR grant number 01-02-17398.

REFERENCES

1. Tatischeff, B., Yonnet, J., Boivin, et al., *Phys. Rev.* C **59**, 1878–1889 (1999).
2. Fil'kov, L.V., *Sov.Physics–Lebedev Inst. Reports* **No.11**, 49–55 (1986); *Sov.J.Nucl.Phys.* **47**, 437–439 (1988).
3. Akhmedov, D.M., and Fil'kov, L.V., *Nucl.Phys.* **A544**, 692–712 (1992).
4. Fil'kov, L.V., Kashevarov, V.L., Konobeevski, E.S., Mordovskoy, M.V., Potashev, S.I., and Skorkin, V.M., *Phys. Rev.* C **61**, 044004–044010 (2000).
5. Fil'kov, L.V., Kashevarov, V.L., Konobeevski, E.S., Mordovskoy, M.V., Potashev, S.I., and Skorkin, V.M., *Phys. Atom. Nucl.* **62**, 2021–2023 (1999).
6. Konobeevski, E.S., Mordovskoy, M.V., Potashev, S.I., Skorkin, V.M., Zuev, S.K., Simonov, V.A. and Fil'kov, L.V. *Izv. Ross. Akad. Nauk*, Ser. Fiz. **62**, 2171 (1998).
7. Fil'kov, L.V., Kashevarov, V.L., Konobeevski, E.S., Mordovskoy, M.V., Potashev, S.I., and Skorkin, V.M., *Bulletin of Lebedev Phys. Inst.* No **11**, 36 (1998).
8. Tatischeff, B., Yonnet, J., Willis, N., Boivin, M., Comets, M.P., Courtat, P., Gacougnolle, R., Le Bornec, Y., Loireleux, E., and Reide F. *Phys. Rev. Lett.* **79**, 601 (1997).

Study of two photon production process in proton-proton collisions at 216 MeV

A.S.Khrykin

Joint Institute for Nuclear Research, Dubna, 141980 Russia
Email: Khrykin@nusun.jinr.dubna.su

Abstract. The energy spectrum for high energy γ-rays ($E_\gamma \geq 10$ MeV) from the process $pp \to \gamma\gamma X$ emitted at 90^0 in the laboratory frame has been measured at 216 MeV. The resulting photon energy spectrum extracted from $\gamma-\gamma$ coincidence events consists of a narrow peak (5.3σ) at a photon energy of about 24 MeV and a relatively broad peak (3.5σ) in the energy range of (50 - 70) MeV. This behavior of the photon energy spectrum is interpreted as a signature of the exotic dibaryon resonance d_1^* with a mass of about 1956 MeV which is assumed to be formed in the radiative process $pp \to \gamma d_1^*$ followed by its electromagnetic decay via the $d_1^* \to pp\gamma$ mode. The experimental spectrum is compared with those obtained by means of Monte Carlo simulations.

INTRODUCTION

The process $pp \to pp\gamma\gamma$ at energy below the pion production threshold (πNN) is still poorly explored both theoretically and experimentally. It was first suggested not long ago as a sensitive probe of the possible existence of NN-decoupled nonstrange dibaryon resonances[1, 2]. These are two-baryon states ^2B with zero strangeness and exotic quantum numbers $I(J^P)$ [I is the isospin, J is the total spin, and P is the parity of a dibaryon state], for which the strong decay $^2B \to pp$ is either strictly forbidden by the Pauli principle [for the states with $I(J^P) = 1(1^+, 3^+, etc.)$] or is strongly suppressed by the isospin selection rules (for the states with $I = 2$). Such dibaryon states cannot be simply bound systems of two nucleons, and a proof of their existence would have consequences of fundamental significance for the theory of strong interactions[3, 4, 5].

If the NN-decoupled dibaryons exist in nature, then the $pp\gamma\gamma$ process may proceed, at least partly, through the mechanism that directly involves the radiative excitation $pp \to \gamma\, ^2B$ and decay $^2B \to \gamma pp$ modes of these states. In pp collisions at energies below the πNN threshold, these production and decay modes of the NN-decoupled dibaryon resonances with masses $M_R \leq 2m_p + m_\pi$ would be unique or dominant. Since such dibaryons may decay mainly into the $pp\gamma$ state, their widths should be very narrow ($\leq 1 keV$). The simplest and clear way of revealing them is to measure the photon energy spectrum of the reaction $pp\gamma\gamma$. The presence of an NN-decoupled dibaryon resonance would reveal itself in this energy spectrum as a narrow peak associated with the formation of the resonance and a relatively broad peak originating from its three-particle decay. In the center-of-mass system, the position of the narrow peak (E_R) is determined by the energy of colliding nucleons ($W = \sqrt{s}$) and the mass of this dibaryon resonance as $E_R = (W^2 - M_R^2)/2W$. An essential feature of the $pp\gamma\gamma$ process at an energy below the πNN threshold is that, apart from the resonant mechanism in question, there should only be one more source of photon pairs. This is the double pp bremsstrahlung reaction. But this reaction is expected to play a minor role. Indeed, it involves two electromagnetic vertices, so that one may expect that the $pp\gamma\gamma$-to-$pp\gamma$ cross section ratio should be of the order of the fine structure constant α. However, the cross section for $pp\gamma$ is already small (the total cross section for the $pp\gamma$ reaction at energies of interest is a few μb).

The preliminary experimental studies of the reaction $pp \to \gamma\gamma X$ at an energy of about 200 MeV [6, 7] showed that the photon energy spectrum of this reaction had a peculiar structure ranging from about 20 MeV to about 60 MeV. This structure was interpreted as an indication of the possible existence of an NN-decoupled dibaryon resonance (later called d_1^*) that is produced in the process $pp \to \gamma d_1^*$ and subsequently decays via the $d_1^* \to pp\gamma$ channel. Unfortunately, a relatively coarse energy resolution and low statistics did not allow us to distinguish the narrow γ peak associated with the d_1^* production from the broad γ peak due to its decay and, hence, to determine the resonance mass exactly. To clarify the situation with the dibaryon resonance d_1^*, we have decided to measure the energy spectrum of the $pp \to pp\gamma\gamma$ reaction more carefully.

THE EXPERIMENT AND RESULTS

The experiment was performed using the variable energy proton beam from the phasotron at the Joint Institute for Nuclear Research (JINR). The pulsed proton beam with an energy of about 216 MeV, an energy spread of about 1.5%, and an intensity of about $3.6 \cdot 10^8$ protons/s bombarded a liquid hydrogen target. Both γ quanta of the reaction $pp\gamma\gamma$ were detected by two γ-ray detectors placed in a horizontal plane, symmetrically on either side of the beam at a laboratory angle of 90^0 with respect to the beam direction. The solid angles covered by the detectors were 43 msr and 76 msr, respectively. To reject events induced by charged particles, plastic scintillators were put in front of each γ detector. The electronics associated with the γ detectors and the plastic scintillators together with the data acquisition system provided $\gamma - \gamma$ coincidence candidate events to be recorded on the hard disk of the computer. A further selection of events associated with the process $pp \to \gamma\gamma X$ was done during off-line data processing. The energy threshold for both the γ detectors was set at about 7 MeV.

Measurements were done both for the target filled with liquid hydrogen and for the empty one. Data for the full and empty target were taken in two successive runs for ~31 h and ~21 h, respectively. The integrated luminosity of about 8.5 pb^{-1} was accumulated for the measurement with the full target. The photon energy spectrum of the process $pp \to \gamma\gamma X$ obtained after subtraction of the empty-target contribution from the spectrum measured with the full target is shown in Fig. 1. As can be seen, this spectrum consists of a narrow peak at a photon energy of about 24 MeV and a relatively broad peak in the energy range from about 50 to about 70 MeV. The statistical significances for the narrow and the broad peaks are 5.3σ and 3.5σ, respectively. The width (FWHM) of the narrow peak was found to be about 8 MeV. This width is comparable with that of the energy resolution of the experimental setup. The observed behavior of the photon energy spectrum agrees with a characteristic signature of the sought dibaryon resonance d_1^* that is formed and decays in the radiative process $pp \to \gamma d_1^* \to pp\gamma\gamma$. In that case the narrow peak should be attributed to the formation of this dibaryon, while the broad peak should be assigned to its three-particle decay. Using the value for the energy of the narrow peak $E_R \sim 24$ MeV, we obtained the d_1^* mass $M_R \sim 1956$ MeV. The differential cross section for the resonance production of two photons emitted symmetrically at $\theta_{lab} = \pm 90^0$ from the process $pp \to \gamma d_1^* \to pp\gamma\gamma$ at an energy of 216 MeV was estimated to be ~ 9 nb/sr^2.

Having assumed that the $pp \to \gamma d_1^* \to pp\gamma\gamma$ process with the d_1^* mass of 1956 MeV is the only mechanism of the reaction $pp \to pp\gamma\gamma$, we calculated the photon energy spectra of this reaction for a proton energy of 216 MeV. It was also assumed that the radiative decay of the d_1^* is a dipole $E1(M1)$ transition from the two-baryon resonance state to a pp state in the continuum. The calculations were carried out with the help of Monte Carlo simulations which included the geometry and the energy resolution of the actual experimental setup. The photon energy spectra were calculated for two different scenarios of the d_1^* decay. The difference between them was that one of these scenarios took into account the final state interaction (FSI) of two outgoing protons whereas in the other that interaction was switched off. Each of the scenarios imposed some restrictions on possible quantum numbers of the dibaryon state in question. The scenario including the FSI implies that the final pp-system is in the singlet $^1S_0$ state and consequently it should take place, in particular, for the isovector 1^+ dibaryon state (the simplest exotic quantum numbers), namely, $1^+ \xrightarrow{M1} 0^+$. Moreover such a scenario can take place for any isotensor dibaryon state with the exception of the 0^+ or 0^- state. At the same time, the scenario in which the FSI is switched off, is most likely to occur for the isotensor $0^\pm$ dibaryon state. The spectra calculated for these two decay scenarios and normalized to the total number of $\gamma - \gamma$ events observed in the present experiment are shown in Fig. 1. Comparison of these spectra with the experimental spectrum indicates that both the calculated spectra are in reasonable agreement with the experimental one within experimental uncertainties. In other words, the statistics of the experiment is insufficient to draw any firm conclusions in favor of one of these scenarios and thereby to limit possible quantum numbers of the observed dibaryon state.

Here it is important to note that the d_1^* with any possible set of quantum numbers $I(J^P)$ with the exception of $2(0^\pm)$ should mainly decay to the singlet $^1S_0$ pp state. At the same time the outgoing protons from the process $pp \to \gamma d_1^* \to \gamma\gamma^1S_0 pp$ would mainly be concentrated in a narrow angular cone near the direction of motion of incident protons [8]. We believe that this is why that this process was not found in the Uppsala pp bremsstrahlung data [9].

CONCLUSION

The γ-ray energy spectrum for the $pp \to \gamma\gamma X$ reaction at a proton energy below the pion production threshold has been measured for the first time. The spectrum measured at an energy of about 216 MeV for coincident photons emitted at an angle of 90^0 in the laboratory frame clearly evidences the existence of the NN-decoupled dibaryon

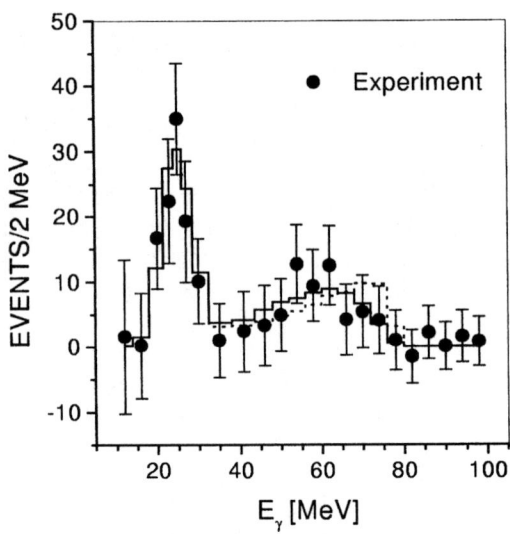

FIGURE 1. Experimentally observed energy spectrum for photons from the $pp\gamma\gamma$ process and energy spectra for photons from the process $pp \to \gamma d_1^* \to \gamma\gamma pp$ calculated with the help of Monte Carlo simulations for two d_1^* decay scenarios: without the FSI (solid line) and with the FSI (dashed line).

resonance d_1^* with a mass of ~ 1956 MeV that is formed and decays in the process $pp \to \gamma\, d_1^* \to pp\gamma\gamma$. The data we have obtained, however, are still incomplete, and additional careful studies of the reaction $pp \to pp\gamma\gamma$ are needed to get proper parameters (mass, width, spin, etc.) of the observed dibaryon state.

REFERENCES

1. S.B. Gerasimov and A.S. Khrykin, Mod. Phys. Lett. **A8**, 2457(1993).
2. S.B. Gerasimov, S.N. Ershov, A.S. Khrykin, Phys. At. Nucl. **58**, 844(1995).
3. P.J. Mülders, A.T. Aerts, and J.J. de Swart, Phys. Rev. **D 21**, 2653(1980).
4. L.A. Kondratyuk, B.V. Martem'yanov, and M.G. Shchepkin, Sov. J. Nucl. Phys. **45**, 776(1987).
5. V.B. Kopeliovich, Phys. At. Nucl. **58**, 1237(1995).
6. A.S.Khrykin, in*Proceeding of the XIV International conference on particles and nuclei (PANIC96),* Willianisburg, Virginia,USA,1996, ed. by Carl E.Carlson and John J.Domingo, (World Scientific, Singapore), p.533.
7. A.S.Khrykin, in *Proceeding of the Seventh International Symposium on Meson-Nucleon Physics and the Structure of the Nucleon,* Vancouver, British Colombia, Canada, 1997, ed. by D.Drechsel, G.Höhler, W.Kluge, and B.M.K.Nefkens, (TRIUMF, Vancouver, 1997), p.250.
8. A. S. Khrykin et al., Phys. Rev. **C64**,034002(2001).
9. H.Calén et al., Phys. Lett. **B427**, 248(1998).

Are the exotic mesons and baryons, recently observed, a signature of quark-hadron duality ?

Boris Tatischeff[*]

Institut de Physique Nucléaire, CNRS/IN2P3, F-91406 Orsay Cedex, France

Abstract. Narrow low mass exotic hadronic structures were recently observed in mesons, baryons and dibaryons. Narrow mesons, in the mass range 300≤M≤ 750 MeV, were observed using the pp → ppX reaction. Narrow baryons in the mass range 1000≤M≤ 1400 MeV were observed using the pp → pπ^+X and dp → ppX reactions. The statistical significances of these structures vary up to 4.6 standard deviations (S.D.) for mesons and up to 16.9 S.D. for baryons. These exotic states are associated with precursor quark deconfinement.

INTRODUCTION

It is often believed than the observation of quark degrees of freedom cannot start at energies lower than several GeV (≈ 10 GeV). However such limit is somewhat unprecise. A relation was proposed by Baldin [1] some years ago. In this theory, the deconfinement is determined as being a process allowing particle creation for $b_{ij} \geq 5$, where

$$b_{ij} = -(P_i/m_i - P_j/m_j)^2 \tag{1}$$

P_i, P_j are the four momentum of both particles involved in the reaction, and m_i, m_j their masses. If we consider a N-N reaction, the previous relation reduces in the laboratory system to $T_i \geq 2.35$ GeV. Following the previous relation, the search for possible signatures of the beginning of the quark deconfinement, could be significant at energies close to 2 GeV. These correspond to the energies of the experiments which results are presented in this paper.

The measurements were performed at Saturne, with proton and deuteron beams. Their energies varied from 1.52 up to 2.1 GeV. The large momentum range SPES3 spectrometer (600 ≤ pc ≤ 1400 MeV) and its detection were settled at different angles from 0^0 up to 17^0 (lab.). Two positively charged particles (pp or pπ^+) were detected and identified in the same detection. A MIT type drift chamber was located in the focal plane of the spectrometer (σ_x=90 μm), (σ_θ=18 mrd). Its efficiency was smooth, without displaying any structure. Two multidrift CERN

---
[*] e-mail: tati@ipno.in2p3.fr

type drift chambers were located perpendicular to the mean particle trajectories. Each of these two chambers consisted of three wire planes (x, u, v). The trigger consisted of four planes of plastic scintillator hodoscopes (20 scintillator/plane). A first time of flight between both extreme trigger planes were used to identify the detected particles. The time resolution for each scintillator was $\sigma \approx 180$ ps. A second time of flight between both particles overdetermined the reaction under study. For T_p=1.52 GeV protons at 0^0, the events from the pp$\rightarrow$pπ^+n reaction came across all the momentum range in the p_p versus p_π scatter plot, and the missing mass histogram was smooth without structure.

The σ of the invariant masses was close to 1 MeV, when the σ of the missing masses was close to 2 MeV at forward angles.

Regular empty target measurements allowed us to observe that the target windows were not a source of noticeable contamination. They also allowed us to conclude that the data were not contaminated by any hot area of incident beam which could have been scattered by some mechanical piece at the entrance of the spectrometer. A simulation code was written in order to perform the corrections for the lost events and for the acceptance. It showed that we did not have any narrow structure in the simulated histograms, and that we have a good control of both the spectrometer and the detection. The simulation code was also used to check that the structures were not produced by eventual particles emitted vertically outside the solid angle and partially absorbed by the lead diaphragm and other mechanical pieces between the target and the spectrometer. The randoms histograms - a few % of the total spectra - did not exhibit any structure. We conclude on the lack of possible contamination, and on the genuine existence of the small and narrow observed structures.

NARROW MESONS

The missing mass of the pp $\rightarrow$ ppX reaction was studied at T_p=1.52, 1.805 and 2.1 MeV. A broad enhancement was observed between 300 and 500 MeV, which is usually called the ABC effect. The corresponding histograms displayed an oscillatory pattern, compatible with the presence of several unresolved structures [2]. The position of the extracted peaks did not depend on the multipion phase space, which exhibits a smooth behaviour. Four peaks were extracted at 313, 353, 426 and 495 MeV. When different incident proton energies and angles were considered, the stability of the two first masses was better than the stability of the two last masses.

Between the η and the ω mesons, narrow structures were observed at 588, 608, 647, 681, 700, (715)[1], and (750) MeV [3].

There is no room for these new mesons within $q\bar{q}$ models. Indeed there is an excellent agreement between the complete classical meson spectrum and the calculations using a relativistic quark model for mesons [4]. Therefore more complex quark configurations are to be considered. We call these states: exotic mesons.

---

[1]between parenthesis since its existence is less well defined

Figure 1 shows an attempt to describe the experimental masses using the following phenomenological mass formula [5] for two coloured constituent quark cluster configurations:

$$M = M_0 + M_1[i_1(i_1+1) + i_2(i_2+1) + (1/3)s_1(s_1+1) + (1/3)s_2(s_2+1)], \quad (2)$$

where M_0 and M_1 are parameters and $i_1(i_2)$, $s_1(s_2)$ are isospin and spin of the first and second quark clusters respectively.

The first four meson masses, extracted from the ABC broad bump, are well described using the $q^2 - \bar{q}^2$ configurations. These states are "broad" (σ(width)$\approx$ 20 MeV) and highly excited. The mesons between 480 and 620 MeV which correspond to narrow and weakly excited peaks, are described using the $q^3 - \bar{q}^3$ configurations. In both cases ($q^2 - \bar{q}^2$ and $q^3 - \bar{q}^3$ configurations) the same parameters: M_0=310 MeV and M_1=30 MeV are used. Finally the mass range between 620 and 750 MeV is described by $q^4 - \bar{q}^4$ configurations with slightly different parameters (by 10%): M_0=357 MeV and M_1=27 MeV.

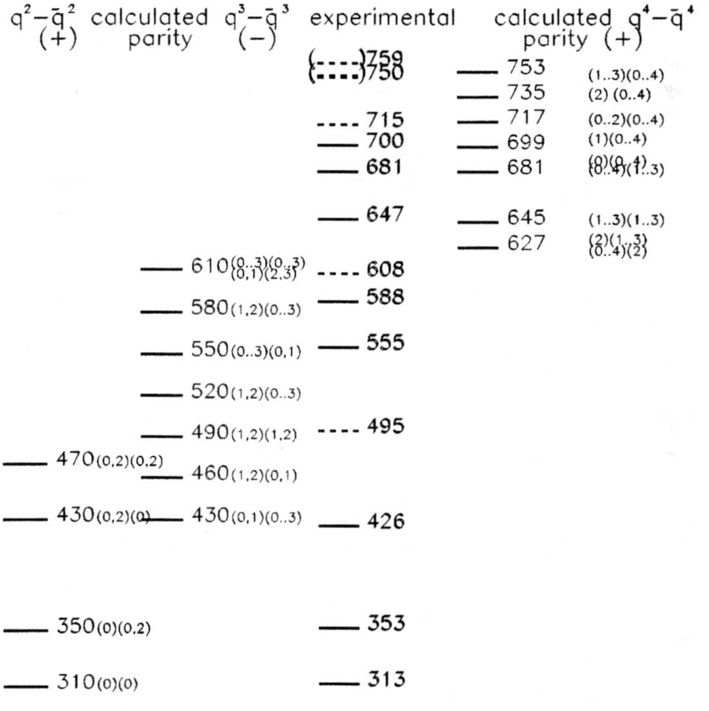

FIGURE 1. Compared experimental and calculated exotic narrow mesonic mass spectra. Into parenthesis, calculated possible (Spin) and (Isospin) respectively.

NARROW BARYONS

The first results of narrow exotic low mass baryons were obtained using the missing mass M_X of the pp $\to$ pπ^+X reaction [6]. Three baryons were observed, with S.D. as large as 10, at 1004, 1044 and 1094 MeV. The corresponding structures were observed at nearly all incident proton energies and scattering angles.

Other baryons - which are less highly excited - were also recently extracted from the same reaction at the following masses: 1136, 1173, 1249, 1277, and (1384) MeV. Since they were each observed in only some experimental spectra corresponding to different angle and-or incident proton energy, their existence is less firmly established than was the existence of the first three lowest narrow baryons. In the mass range $1.1 \leq M \leq 1.45$ GeV they were observed in the missing mass M_X or-and in the invariant mass $M_{p\pi^+}$ of the previous reaction. They were also observed in the missing mass M_X of the dp $\to$ ppX reaction, although with smaller S.D. These last results cannot be used to establish firmly the existence of these states. But since the masses of the structures observed in the dp $\to$ ppX experiment lie close to those found in the first reactions, these results seem to confirm the previous ones. Figure 2 illustrates some of these peak extractions and table 1 describes the conditions of each histogram.

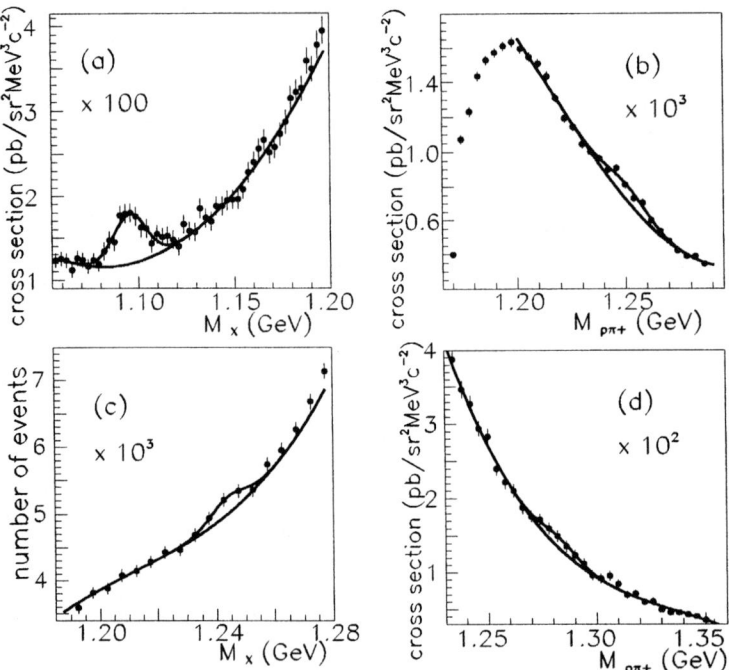

FIGURE 2. A selection of four experimental spectra, showing small and narrow baryonic structures.

	Mass (MeV)	S.D.	T_p (MeV)	θ^0	reaction	observable
(a)	1094	10.9	1805	3.7	pp→pπ^+X	M_X
(b)	1249	8.3	1520	2	pp→pπ^+n	$M_{p\pi+}$
(c)	1249	3.2	2100	17	dp→ppX	M_X
(d)	1277	3.5	1805	6.7	pp→pπ^+X	$M_{p\pi+}$

TABLE 1. Description of the data of Figure 2 showing new structures as observed in different reactions, different observables, incident energies and spectrometer angles.

Only those structures extracted several times at masses less than M±5 MeV were taken into account. The widths of the structures - about a few MeV - are not precise enough to discriminate the experimental from the physical widths.

As it was the case for mesons, there is no room for new baryons in this mass range within the many quark models [7]. Missing baryons are baryons which were calculated by different constituent quarks models, but which were not seen experimentally through elastic πN scattering. They are predicted for masses larger than 1700 MeV, and not in the mass range studied here. Moreover there is - a priori - no reason for them to be narrow since they are calculated within the three quarks assumption. We have therefore several reasons not to associate our narrow baryons to missing baryons, but to associate them to an exotic (not q^3) origin. We considered the same phenomenological mass formula as the one already used for mesons. The two parameters were adjusted in order to get the mass, and different possible (because of large degenerascy) spin and isospin values of nucleon and Roper resonance at 1440 MeV. We got the calculated spectra shown in Figure 3 and observed a very good agreement between measured and calculated masses - mainly for the four lower baryons - although no adjustable parameter was used here.

DISCUSSION

Thanks to good resolution and statistics, narrow structures were observed in hadrons using the pp → pπ^+X, pp → ppX and dp → ppX reactions. Many checks were performed to make sure that these structures were not produced by artefacts. Their small width and their mass stability, whatever the experiment, were used to conclude that they are genuine hadronic structures which were not produced by dynamical rescatterings. Since they cannot be associated to classical quark configurations: $q\bar{q}$ for mesons and q^3 for baryons, we tentatively associated them to coloured cluster quark configurations. The agreement between measured and calculated masses is outstanding since both spectra were obtained quite close although only two (zero) free parameters are used for mesons (baryons). It remains that our attempt to identify the experimental narrow structures to partial quark deconfinement, in these experiments of low temperature and small baryochemical potential (low baryonic number density), although being quite impressive, is not understood today and needs a theoretical explanation.

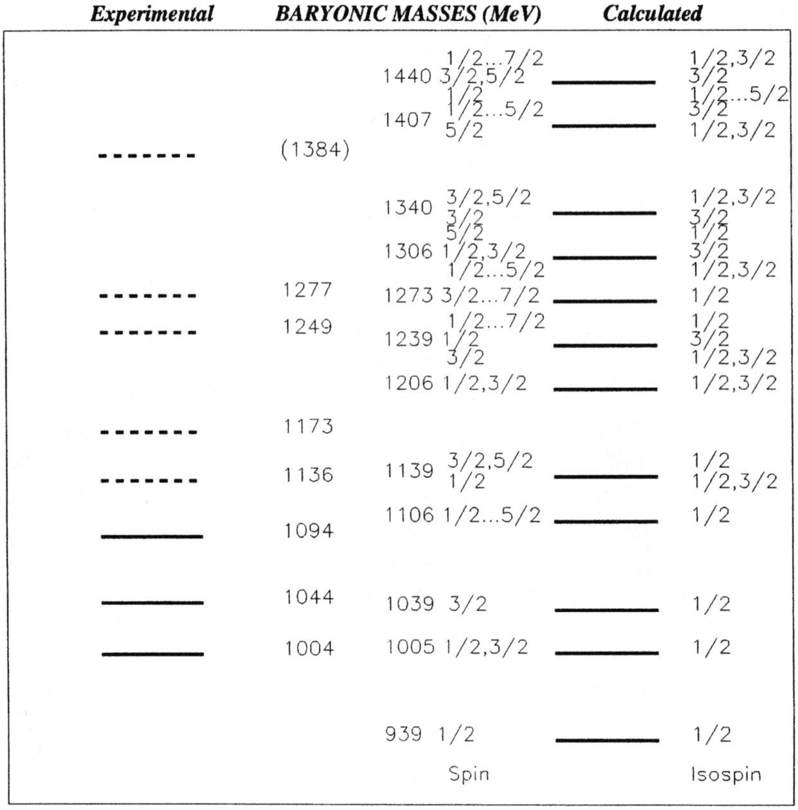

FIGURE 3. Compared experimental and calculated masses of exotic narrow baryonic structures.

REFERENCES

1. Baldin, A.M., *Dubna 1992 JINR*, E1-92-487.
2. Yonnet, J., Tatischeff, B., et al., *Phys. Rev.* **C63**, 014001 (2000).
3. Tatischeff, B., et al., *Phys. Rev.* **C62**, 054001 (2000).
4. Koll, M., et al., *Eur. Phys. J.* **A9**, 73 (2000).
5. Mulders, P.J., Aerts, A.T., de Swart,J.J., *Phys. Rev.* **D21** (1980) 2653; *Phys. Rev.* **D19** (1979) 2635; *Phys. Rev. Lett.* **40** (1978) 1543.
6. Tatischeff, B. et al., *Phys. Rev. Lett.* **79**, 601 (1997).
7. Capstick, S. and Roberts, W., *Progress in Particle and Nuclear Physics* **45**, S241 (2000).

ON A MANIFESTATION OF DIBARYON RESONANCES IN THE STRUCTURE OF PROTON-PROTON TOTAL CROSS-SECTION AT LOW ENERGIES

A.A. Arkhipov

Institute for High Energy Physics, 142280 Protvino, Moscow Region, Russia

Abstract. A manifestation of narrow diproton resonances in the early discovered global structure of proton-proton total cross section (see [8, 9]) at low energies is discussed. It is also discussed the existence of new particle with the mass $1.833\,MeV$ predicted early.

INTRODUCTION

We are all know an incessant interest to the physics of dibaryons and this session at the Conference is an additional confirmation of that [1, 2, 3]. It is well known fact that diprotons have experimentally been observed as a narrow structures in the distributions over invariant mass of proton-proton system in the processes of proton-nucleus interaction [4, 5, 6]. The physical origin of these narrow structures is high interest because it has fundamental importance which is related to the nature of fundamental nucleon-nucleon forces and not only to this one. However the experimental and theoretical understanding of the dibaryon physics is far from desired.

From experimental point of view it would be well done to obtain a strong statement concerning the observation of narrow dibaryons regardless of their origin. However at present time there are many experiments where we can find quite an opposite results: some authors state that they have observed such narrow dibaryons [4, 5, 6] but the others make the contrary conclusion. It is usually supposed that the main reason for these disagreements is the weakness of dibaryons signatures compared to the physical background of a given process under experimental study. In that case there are needed the experiments with a high precision. Of course we can always explain a contradiction between the different experiments by a poor energy resolution and statistics or by kinematically unfavourable conditions and all somethings like that. Therefore it would be desirable to have the measurements with one and the same positive signals coming from different kinds of experiments.

Certainly it's bad that at present time there is no theory which can explain the existence of the dibaryons and describe them.

I will adress here a nontrivial physical phenomenon related to a manifestation of diproton resonances in the proton-proton total cross sections at low energies. We faced with the phenomenon in our study of global structure for the nucleon-nucleon total cross

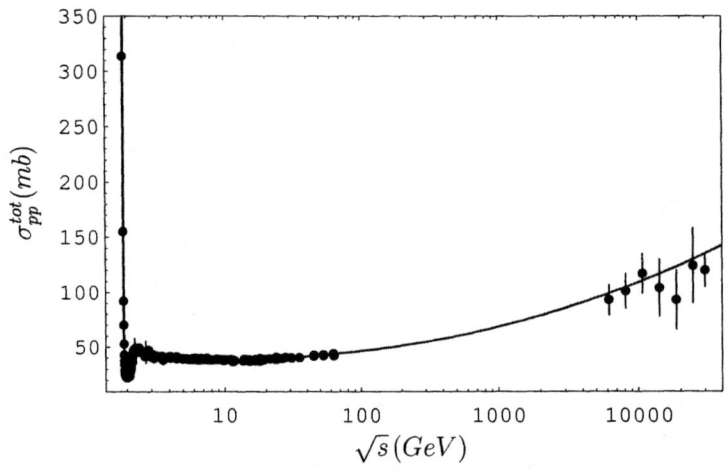

FIGURE 1. The proton-proton total cross-section versus $\sqrt{s}$ with the cosmic rays data points from Akeno Observatory and Fly's Eye Collaboration. Solid line corresponds to our theory predictions.

section.

Mabe it should be emphasized a common experimental point of view that the total cross section is not a suitable characteristic to study the resonance physics. Nevertheless we will show that the existing experimental data set on proton-proton total cross sections allowed us to find a clear signatures for diproton resonances. Let me remind you what was the beginning on.

GLOBAL STRUCTURE OF PROTON-PROTON(ANTIPROTON) TOTAL CROSS SECTIONS

Recently a simple theoretical formula describing the global structure of pp and $p\bar{p}$ total cross-sections in the whole range of energies available up today has been derived. The fit to the experimental data with the formula was made, and it was shown that there is a very good correspondence of the theoretical formula to the existing experimental data obtained at the accelerators [7, 8].

Moreover it turned out there is a very good correspondence of the theory to all existing cosmic ray experimental data as well. The predicted values for σ_{tot}^{pp} obtained from theoretical description of all existing accelerators data are completely compatible with the values obtained from cosmic ray experiments [9]. The global structure of proton-proton total cross section is shown in Fig. 1 extracted from paper [9].

The theoretical formula describing the global structure of proton-proton total cross section is written below

$$\sigma_{pp}^{tot}(s) = \sigma_{asmpt}^{tot}(s) \left[1 + \left(\frac{c_1}{\sqrt{s - 4m_N^2} R_0^3(s)} - \frac{c_2}{\sqrt{s - s_{thr}} R_0^3(s)} \right) (1 + d(s)) + Resn(s) \right],$$

$$R_0^2(s) = \left[0.40874044\sigma_{asmpt}^{tot}(s)(mb) - B(s)\right](GeV^{-2}),$$

$$\sigma_{asmpt}^{tot}(s) = 42.0479 + 1.7548\ln^2(\sqrt{s}/20.74),$$

$$B(s) = 11.92 + 0.3036\ln^2(\sqrt{s}/20.74),$$

$$c_1 = (192.85 \pm 1.68)GeV^{-2}, \quad c_2 = (186.02 \pm 1.67)GeV^{-2},$$

$$s_{thr} = (3.5283 \pm 0.0052)GeV^2,$$

$$d(s) = \sum_{k=1}^{8} \frac{d_k}{s^{k/2}}, \quad Resn(s) = \sum_{i=1}^{N} \frac{C_R^i s_R^i {\Gamma_R^i}^2}{\sqrt{s(s-4m_N^2)}[(s-s_R^i)^2 + s_R^i {\Gamma_R^i}^2]}.$$

For the numerical values of the parameters $d_i(i = 1,...8)$ see original paper [8]. It should be pointed out that the mathematical structure of the formula is very simple and physically transparent: the total cross section is represented in a factorized form. One factor describes high energy asymptotics of total cross section and it has the universal energy dependence predicted by the general theorems in local quantum field theory (Froissart theorem). The other factor is responsible for the behaviour of total cross section at low energies and it has a complicated resonance structure. However this factor has also the universal asymptotics at elastic threshold. It is a remarkable fact that the low energy asymptotics of total cross section at elastic threshold is dictated by high energy asymptotics of three-body (three-nucleon in that case) forces. The appearance of new threshold $s_{thr} = 3.5283\,GeV^2$ in the proton-proton channel, which is near the elastic threshold, is nontrivial fact too.

Some experimental information concerning the diproton resonances is collected in Table 1.

The positions of resonances and their widths, listed in Table 1, were fixed in our fit, and only relative contributions of the resonances C_R^i have been considered as free fit parameters. Fitted parameters C_R^i obtained by the fit are listed in Table 1 too. It should be

TABLE 1. Diproton resonances.

$m_R(MeV)$	$\Gamma_R(MeV)$	Reference	$C_R(GeV^2)$
1937 ± 2	7 ± 2	[6]	0.058 ± 0.018
$1947(5) \pm 2.5$	8 ± 3.9	[4]	0.093 ± 0.028
1955 ± 2	9 ± 4	[6]	0.158 ± 0.024
1965 ± 2	6 ± 2	[6]	0.138 ± 0.009
1980 ± 2	9 ± 2	[6]	0.310 ± 0.051
1999 ± 2	9 ± 4	[6]	0.188 ± 0.070
2008 ± 3	4 ± 2	[6]	0.176 ± 0.050
$2027 \pm ?$	$10-12$		0.121 ± 0.018
2087 ± 3	12 ± 7	[6]	-0.069 ± 0.010
2106 ± 2	11 ± 5	[6]	-0.232 ± 0.025
$2127(9) \pm 5$	4 ± 2	[6]	-0.222 ± 0.056
$2180(72) \pm 5$	7 ± 3	[6]	0.131 ± 0.015
$2217 \pm ?$	$8-10$		0.112 ± 0.031
2238 ± 3	22 ± 8	[6]	0.221 ± 0.078
2282 ± 4	24 ± 9	[6]	0.098 ± 0.024

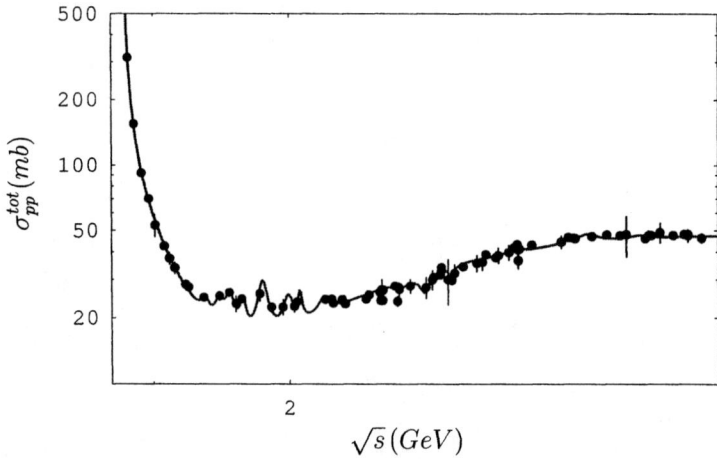

FIGURE 2. The proton-proton total cross-section versus $\sqrt{s}$ at low energies. Solid line corresponds to our theory predictions.

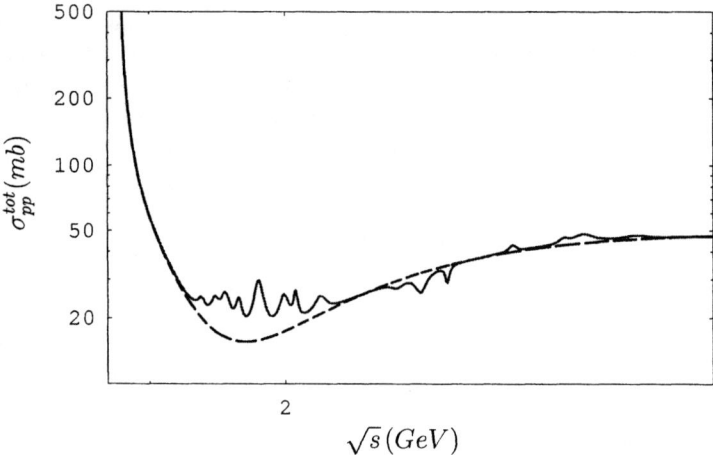

FIGURE 3. The resonance structure for the proton-proton total cross-section versus $\sqrt{s}$ at low energies. Solid line is our theory predictions. Dashed line corresponds to the "background" where all resonances are switched off.

remarked that the experimental data set on proton-proton total cross sections revealed the existence of two unknown resonances with the masses $\sim 2027\,MeV$ and $\sim 2217\,MeV$. These resonances were also included in our fit. Some known diproton resonances are not included in the list by the reason of our computer allowance. We plan to make a more extended analysis in the future.

Our fitting curve is shown in Fig. 2. We also plotted in Fig. 3 the resonance structure for proton-proton total cross section at low energies without the experimental points but with dashed line corresponding the "background" where all resonances are switched off. As it is seen from this Figure there is a clear signature for the diproton resonances.

CONCLUSION

- It appears the diproton resonances are confirmed by the data set for proton-proton total cross section at low energies from statistical point of view (good fit!).
- There is a big bag ("bol'shoi korob") with many dibaryon resonances. This korob-bag is not completely filled yet till now! How many dibaryon resonances are there?
- There are many questions???...There are no answers!!!...What is the physical nature and dynamical origin of dibaryon resonances? What are the quantum numbers: spin, isospin, and so on. A nontrivial fact in our fitting games is the observation that three resonances with the mass 2087, 2106, 2127 MeV have an odd parity.
- Without any doubt the physics of dibaryon resonances is very interesting, very exciting, very promising, very..., very... part of elementary particle and nuclear physics.
- From the global structure it follows that new threshold, which is near the elastic one, looks like a manifestation of a new unknown particle:

$$\sqrt{s_{thr}} = 2m_p + m_\mathcal{L}, \qquad m_\mathcal{L} = 1.833\, MeV.$$

We predicted the position of new threshold with a high accuaracy.
- It seems $\mathcal{L}$-particle may have many faces. We could take a refreshing thought that $\mathcal{L}$-particle may be a bound state of photons–"photoball", or a bound state of electron-positron pairs embedded in continuum, or very deeply bounded system of pions. A very intriguing idea that $\mathcal{L}$-particle is a Higgs particle, which is well known theoretically but it is not observed experimentally, is admissible one as well. Is $\mathcal{L}$-particle a photonium, positronium, pionium, and so on x-onium?
- Could one make an experiment to search $\mathcal{L}$-particle? It is very probably that $\mathcal{L}$-particle has been observed in Darmstadt. We find in the abstract of paper [10]: "The most pronounced line appears at a sum energy of $\sim 810\, keV$, corresponding to an invariant mass of $\sim 1.83\, MeV/c^2$." This result was confirmed by the other group a year later [11]. Now we can understand an independence of Darmstadt effect on the content of beam and target nuclei because this is a manifestation of fundamental nucleon-nucleon dynamics. It's a pity, the present status of Darmstadt efect is not so stable. That is why, it would be very desirable to make new experiments to search $\mathcal{L}$-particle.
- Could one measure a missing mass spectra in one-particle $pp \to pX$ and in two-particle $pp \to ppX$ inclusive reactions with a high precision and with a high resolution in missing mass?
Such measurements will shed more light on the questions surrounding the nature of diproton resonances.
- Surely, it is very important to perform systematic studies and precise calculations using quantum field theoretical methods. In this respect we hope that the discovery of quasicrystal structure of the vacuum in quantum field theory [12] will help us to understand the new sites of the fundamental dynamics.

ACKNOWLEDGMENTS

I am grateful to A.M. Zaitsev provided me the possibility to attend the Conference HADRON 2001 and to present the Report there.

REFERENCES

1. B. Tatischeff, this Conference.
2. E.S. Konobeevski, this Conference.
3. L.V. Filkov, this Conference.
4. B. Tatischeff et al., Phys Rev. C**45** 2005 (1992).
5. Yu.A. Troyan, V.N. Pechenov, Sov. J. Yad. Phys. **56**, 191 (1993).
6. Yu.A. Troyan, Sov. J. Physics of Element. Part. and Atomic Nuclei **24**, 683 (1993).
7. A.A. Arkhipov, *What Can we Learn from the Study of Single Diffractive Dissociation at High Energies?* – in Proceedings of VIIIth Blois Workshop on Elastic and Diffractive Scattering, Protvino, Russia, June 28–July 2, 1999, World Scientific, Singapore, 2000, pp. 109-118; REPORT IHEP 99-43, Protvino, 1999; e-print hep-ph/9909531.
8. A.A. Arkhipov, *On Global Structure of Hadronic Total Cross Sections*, preprint IHEP 99-45, Protvino, 1999; e-print hep-ph/9911533.
9. A.A. Arkhipov, *Proton-Proton Total Cross Sections from the Window of Cosmic Ray Experiments*, preprint IHEP 2001-23, Protvino, 2001; e-print hep-ph/0108118; to be published in Proceedings of IXth Blois Workshop on Elastic and Diffractive Scattering, Pruhonice near Prague, June 9-15, 2001.
10. W. Koenig et al., *On the momentum correlation of (e^+e^-) pairs observed in $U+U$ and $U+Pb$ collisions*, Phys. Lett. **B 218** 12 (1989); T. Covan et al., *Observation of Correlated Narrow-Peak Structures in Positron and Electron Spectra from Superheavy Collision Systems*, Phys. Rev. Lett. **56** 444 (1986)..
11. P. Salabura et al., *Correlated (e^+e^-) peaks observed in heavy-ion collisions*, Phys. Lett. **B 245** 153 (1990).
12. A.A. Arkhipov, *Quark-Quark Forces in Quantum Chromodynamics*, Contribution to the International Conference on High Energy Physics "ICHEP94", Glasgow, 20-27 July, 1994, Ref. gls0064 and references therein; *Single-Time Formalism and Quasicrystal Structure of the Vacuum in Quantum Chromodynamics*, in Proceedings of the XI workshop on soft physics "Hadrons-95", eds. G.Bugrij, L.Jenkovsky, Kiev 1995, p.136-154; Ukrainian Journ. of Phys. V. 41, P. 373-384 (1996).

$N\overline{N}$

A High Resolution Search for the Tensor Glueball Candidate $\xi(2230)$

Kamal K. Seth (for the Crystal Barrel Collaboration)

Department of Physics, Northwestern University, Evanston, IL 60208, USA

Abstract. We report results of a high resolution search for the tensor glueball candidate $\xi(2230)$ in a $\bar{p}p$ formation experiment. $\pi^0\pi^0$ and $\eta\eta$ decay channels were measured in a scan of the mass region 2220 MeV to 2240 MeV. No evidence for the existence of $\xi(2230)$ was found. 95% confidence upper limits for the possible existence of ξ are presented.

INTRODUCTION

As soon as the non-Abelian nature of the quark-gluon field theory, which we now call Quantum Chromodynamics (QCD), was recognized in 1972, Fritzsch and Gell-Mann[1] predicted that "there must exist glue states in hadron spectrum", and in 1975 Fritzsch and Minkowski[2] presented a detailed discussion of the phenomenology of the spectrum of 'glue states', which we now call 'glueballs'. For example, one of the most sophisticated lattice-gauge calculations (which are still made in the quenched approximation) predicts glueball masses, $M(0^{++}) \approx 1730 \pm 50 \pm 80$ MeV, $M(2^{++}) \approx 2400 \pm 25 \pm 120$ MeV[3]. Most recent experimental efforts have been directed to the search and identification of these two, the scalar and the tensor glueballs. In this letter we present the results of a high resolution search for the 2^{++} tensor glueball in a $\bar{p}p$ annihilation experiment, PS197, made with the Crystal Barrel detector[4] at the LEAR facility at CERN.

In order to provide the appropriate perspective for our choice of the method of search for the tensor glueball, it is necessary to briefly review the past searches and their conclusions.

In 1986, Mark III at SLAC, investigating radiative decays of $5.8 \cdot 10^6$ J/Ψ, reported[5] the observation of an abnormally narrow enhancement, dubbed ξ, in the K^+K^- and $K_S K_S$ decay channels, with $M(\xi) \approx 2230$ MeV, and $\Gamma \approx 20$ MeV. However, a similar measurement by DM2 failed to find any narrow enhancement[6].

Several searches for the narrow $\xi(2230)$ in the formation reactions $\bar{p}p \to K^+K^-$ and $K_S K_S$ were also unsuccessful in finding any evidence for it.

However, in 1996, the BES detector at BEPC rekindled the interest in the narrow $\xi(2230)$ by reporting its observation in their sample of the radiative decay of 8 million J/Ψ[7]. They reported the observation of $\xi(2230)$ not only in K^+K^- and $K_S K_S$ decay channels, but also in $\pi^+\pi^-$ and $\bar{p}p$ channels, with consistent masses and widths, $M(\xi) \approx 2232$ MeV and $\Gamma(\xi) = 15 - 20$ MeV.

The BES report of ξ decay into $\bar{p}p$ revived hopes of making a definitive observation of ξ(2230) in high resolution $\bar{p}p$ formation experiments. In this report we present the results of a search for ξ(2230) by the LEAR (CERN) experiment PS197 in a $\bar{p}p$ formation experiment. The all neutral decay channels $\bar{p}p \to \pi^0\pi^0$, $\bar{p}p \to \eta\eta$ and $\bar{p}p \to \eta\pi^0$ were measured.

The measurements reported here were made at LEAR with the Crystal Barrel detector, which has been described in detail elsewhere[4].

The antiproton beam extracted from LEAR had a momentum uncertainty of ±0.3 MeV/c, and an overall, nearly uniform, $\sqrt{s}$ spread of ±0.3 MeV in the momentum range 1412-1461 MeV/c of the present measurements. Data were taken at 9 nearly equally spaced beam momenta in the range 1412-1461 MeV/c, which corresponds to $\sqrt{s} = 2222 - 2240$ MeV. Between 0.5 and 1.2 million all-neutral triggers were taken at each beam momentum.

After initial cuts on E_{tot} and P_{tot} these events were then subjected to the hypotheses $\bar{p}p \to \pi^0\pi^0$, $\eta\eta$ or $\eta\pi^0$, with the confidence level set at > 5%. For example, for $P_{beam} = 1412.3$ MeV/c, the following assignments resulted: 10,567 $\pi^0\pi^0$, 407 $\eta\eta$ and 3,571 $\eta\pi^0$. The overall acceptance and efficiency for event selection was estimated by Monte Carlos simulation to be 33% for $\pi^0\pi^0$, 36% for $\eta\eta$ and 35% for $\eta\pi^0$.

Both the differential cross sections and the integrated cross sections in the range $\cos\theta = 0$ to 0.85 were determined for $\pi^0\pi^0$, $\eta\eta$ and $\eta\pi^0$ final states. The statistical errors in cross sections were ≤ 1.7%, 6.5% and 2.8% for $\pi^0\pi^0$, $\eta\eta$ and $\eta\pi^0$, respectively.

The cross sections integrated in the region $\cos\Theta_{cm} = 0 - 0.85$ for $\bar{p}p \to \pi^0\pi^0$, $\eta\eta$ and $\eta\pi^0$ are shown in Fig. 1.

In the small region of $\sqrt{s}$ investigated in our measurements, in absence of a resonance, the cross sections can be reasonably fitted with straight lines. These straight line fits to the cross sections, $\sigma = A + B\sqrt{s}$, are shown in Fig. 1.

Fig. 1 shows that in none of the integrated cross sections there is any indication of structure. Since the $\eta\pi^0$ decay channel, with isospin 1, cannot have an I=0 glueball, we can remove all luminosity uncertainties by considering $\sigma(\pi^0\pi^0)/\sigma(\pi^0\eta)$ and $\sigma(\eta\eta)/\sigma(\pi^0\eta)$. These are also shown in the bottom two panels of Fig. 1. In both cases essentially perfect fits to straight line are obtained with $\chi^2 = 0.31$, and 0.50, respectively. To summarize, there is no evidence in our integrated cross section data for any structure.

In order to put quantitative limits for the possible presence of a resonance in these data, we have made the following analysis of the $\pi^0\pi^0$ and $\eta\eta$ integrated cross sections. The measured cross sections are considered as the sum of a linear background as fitted above plus a J=2 Breit-Wigner resonance for which $B_{in}B_{out}$ is determined for trial values of width ($\Gamma = 10, 20$ MeV) and resonance energy (variable in small steps from 2222 to 2240 MeV). In this manner 95% confidence upper limits, shown in Fig. 2 are obtained. We note that these imply that for Γ_R between 10 and 20 MeV,

$$B(\bar{p}p \to \xi)B(\xi \to \pi^0\pi^0) < 6 \cdot 10^{-5}, \quad 95\% \text{ CL}$$
$$B(\bar{p}p \to \xi)B(\xi \to \eta\eta) < 4 \cdot 10^{-5}, \quad 95\% \text{ CL}$$

As mentioned earlier, differential cross sections for all three reactions were also

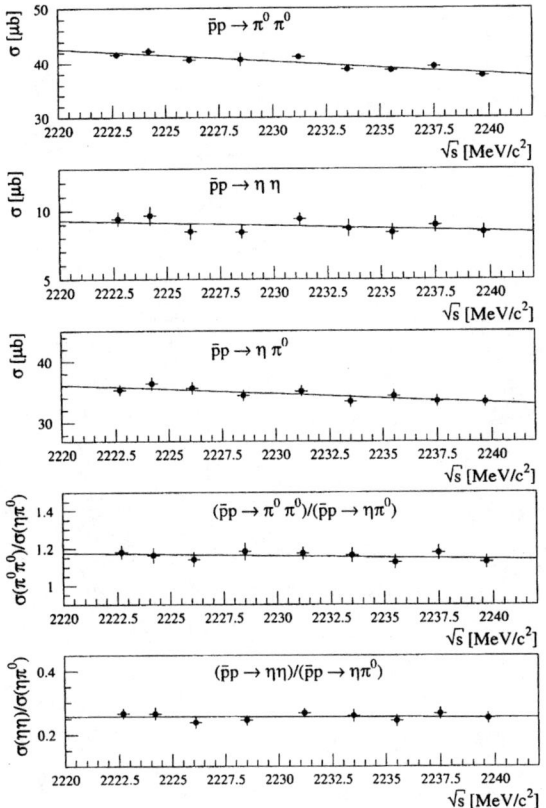

FIGURE 1. The upper three panels show measured cross sections integrated in the region $\cos\theta \leq 0.85$ for $\bar{p}p \to \pi^0\pi^0$, $\eta\eta$, and $\eta\pi^0$, and the straight line fits through them. The lower two panels show $\pi^0\pi^0$ and $\eta\eta$ cross sections divided point by point by the $\eta\pi^0$ cross sections.

measured. A simple analysis of the differential cross sections in terms of Legendre polynomials was made for both $\pi^0\pi^0$ and $\eta\eta$. It was found that the coefficients up to the order of 8 were required, and they all showed smooth evolution from $\sqrt{s} = 2222$ to 2240 MeV. No indication of any behavior indicative of a resonance was found.

Our measured product branching ratio can be combined with the branching ratios reported by BES to yield some other limits. For example, by combining BES product branching ratios for ξ decay to $\pi^0\pi^0$ and $p\bar{p}$ with our product branching ratio for ξ decay to $\pi^0\pi^0$, we obtain

$$B(J/\Psi \to \gamma\xi) > 0.29\%, \quad 95\%\,CL,$$

which would be comparable to the largest measured radiative widths of J/ψ. Upper limits can also be derived by using the above result in the other product branching ratios

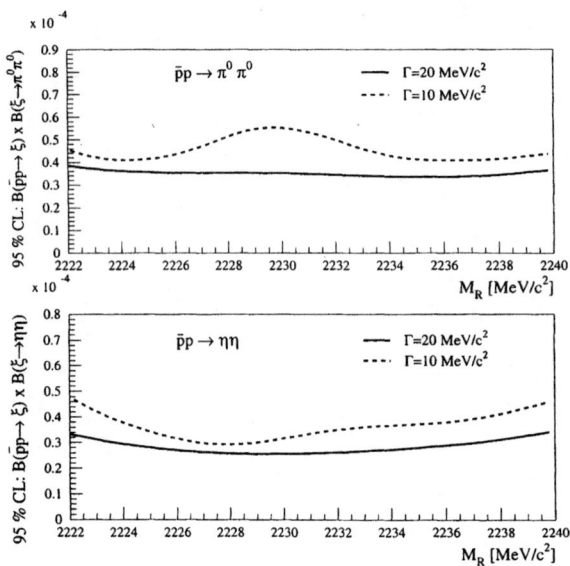

FIGURE 2. 95% confidence limits for the product branching ratios, $B(\bar{p}p \to \xi) B(\xi \to \pi^0\pi^0)$, and $B(\bar{p}p \to \xi) B(\xi \to \eta\eta)$, as function of assumed mass of ξ. Two different values of width of ξ, $\Gamma(\xi) = 10$, and 20 MeV were assumed.

reported by BES. These are 0.51%, 1.2%, 2.3%, 1.1% and 0.9% for the branching ratios for $\xi \to \bar{p}p$, $\pi^0\pi^0$, $\pi^+\pi^-$, K^+K^- and $K_S K_S$, respectively.

To summarize, the results of our measurements of $\bar{p}p \to \pi^0\pi^0$ and $\eta\eta$, we find no evidence for the existence of the narrow glueball candidate $\xi(2230)$ anywhere in the mass range 2222 to 2240 MeV. We have established 95% confidence upper limits for the product branching ratios of $6 \cdot 10^{-5}$ and $4 \cdot 10^{-5}$ for the formation of $\xi(2230)$ in $\bar{p}p$ annihilation and its decay into $\pi^0\pi^0$ or $\eta\eta$, respectively. We are lead to conclude that either the $\xi(2230)$ does not exist in the mass range of our scan, or its coupling to the $\bar{p}p$ channel is much weaker than implied by the BES measurements.

REFERENCES

[1] H. Fritzsch and M. Gell-Mann, Proc. XVI Int. Conf. on High Energy Physics, Fermilab, 1972, vol. 2, p. 135.
[2] H. Fritzsch and P. Minkowski, Nuovo Cimento, **30A**, (1975) 393.
[3] C. Morningstar and M. Peardon, Phys. Rev. **D60**, (1999) 034509.
[4] E. Aker et al., Crystal Barrel Collaboration, Nucl. Instr. Meth. **A321**, (1992) 69.
[5] R. M. Baltrusaitis et al., Mark III Collaboration, Phys. Rev. Lett. **56**, (1986) 107.
[6] J. E. Augustin et al., DM2 Collaboration, Phys. Rev. Lett. **60**, (1988) 2238.
[7] J. Z. Bai et al., BES Collaboration, Phys. Rev. Lett. **76**, (1996) 3502.

Antiproton annihilation on nuclei

K. Protasov* and R. Duperray*

*Institut des Sciences Nucléaires, 53, Avenue des Martyrs, F-38026 Grenoble Cedex, France

Abstract. Recent experimental data on the antiproton in-flight annihilation on light nuclei are analysed. The reasons of the observed equality of the p̄-hydrogen, p̄-deuterium, and p̄-helium annihilation cross sections at low energies are discussed.

The calculations within a simple optical model confirm that the S-wave contribution to the antiproton-nucleus annihilation cross section is approximately the same for all nuclei.

INTRODUCTION

The OBELIX experiment on measurement of the p̄p, p̄d and p̄^{4}He annihilation cross section at low antiproton momentum (from 35 to 100 MeV/c in the laboratory frame) gave quite unexpected result. For lowest antiproton momentum, these annihilation cross sections are approximately equal to each other [1]. This result seems to be in direct contradiction with naive geometrical picture of annihilation, which would suggest that the annihilation cross section increases with the size of nucleus. This observation was confirmed by another LEAR experiment (PS207) [2] where the width of the 1S state of the p̄d atom was found to be approximately equal to that for the p̄p.

The phenomenological analysis of the data [3] shown that the low energy antiproton annihilation on these light nuclei is dominated by the S-wave. The imaginary parts of the S-wave scattering lengths for these nuclei were also extracted:

$$\text{Im } a_{sc}(\bar{p}p) = -[0.69 \pm 0.01(\text{stat}) \pm 0.03(\text{sys})] \text{ fm},$$
$$\text{Im } a_{sc}(\bar{p}D) = -[0.62 \pm 0.02(\text{stat}) \pm 0.05(\text{sys})] \text{ fm},$$
$$\text{Im } a_{sc}(\bar{p}^4\text{He}) = -[0.36 \pm 0.03(\text{stat})^{+0.19}_{-0.11}(\text{sys})] \text{ fm}.$$

In [4], a mechanism explaining the relation between the imaginary parts of the p̄p and p̄d scattering lengths was proposed. This explanation illustrated by three-body calculations with quite realistic interactions was based on the fact that, in case of strong annihilation, the imaginary part of the scattering length is mostly sensitive to the diffuseness of the annihilation potential. This fact leads to approximate equality of the imaginary parts of the scattering amplitudes for all light nuclei. To do the absolute value of Im $a_{sc}(\bar{p}D)$ system bigger than that for the p̄p, one needs to enhance the last one. The two-body attractive antinucleon-nucleon interaction produces S-wave poles and thus can increase the values of the scattering length. This enhancement is more important in the p̄p system than in the p̄d one. These S-wave resonances are necessarily present in all realistic models but their experimental observation is a delicate task because of their large width.

The aim of this contribution is to calculate the antiproton-nucleus annihilation cross sections for heavier nuclei where contributions of higher partial waves can be important but for which the experimental data at low energies are unfortunately absent.

Remind that, in the systems with Coulomb interaction, the behavior of the inelastic cross sections near the threshold is very different from that for systems with neutral particles. Within the scattering length approximation valid at low energies, the annihilation cross section σ_{ann}^l for a given orbital momentum l takes the form [5, 6]:

$$q^2 \sigma_{ann}^l = (2l+1) 4\pi \frac{g_l(\eta) q^{2l+1} C_0^2(\eta) \operatorname{Im}(-a_l^{sc})}{|1 - ig_l(\eta) q^{2l+1} w(\eta) a_l^{sc}|^2}. \tag{1}$$

where q is the center-of-mass momentum, a_l^{sc} the scattering lengh, B the Bohr radius, $\eta = 1/qB$

$$g_0(\eta) = 1, \qquad g_l(\eta) = \prod_{m=1}^{l} \left(1 + \frac{\eta^2}{m^2}\right),$$

$$C_0^2(\eta) = \frac{2\pi\eta}{1 - \exp(-2\pi\eta)}, \qquad h(\eta) = \frac{1}{2}[\Psi(i\eta) + \Psi(-i\eta)] - \frac{1}{2}\ln(\eta^2)$$

with the digamma function Ψ.

One can see from (1) that, at low momentum, σ_{ann}^l for any partial wave has the same $1/q^2$ behavior:

$$\lim_{q \to 0} q^2 \sigma_{ann}^l = \text{const.}$$

The value of this constant depends on the size of nuclei and determines the number of partial waves giving non negligible contribution to the total annihilation cross section.

OPTICAL MODEL CALCULATIONS

To investigate the behaviour of the σ_{ann} for different nuclei, the optical model approach was used. The potentials which describe very well the existing experimental data at intermediate energies were taken from [7].

In figure 1, the antiproton-nucleus annihilation cross section is presented as a function of atomic number A for different nuclei. Three lowest points (for hydrogen, deuterium, and helium) are taken from [3]. For these three nuclei, the S-wave annihilation is dominant and the the cross section is approximately the same. The points for heavier nuclei are the optical model predictions [8]. One can see that, for these nuclei, where many partial waves give important contributions to σ_{ann}, the usual geometrical picture of annihilation is restored (σ_{ann} is proportional to the $A^{2/3}$). Let us emphasize once more that these higher partial waves give finite contributions at practically zero momentum due to Coulomb forces presented in these systems.

Note that the corrections due to the finite size of nuclei are not very important for σ_{ann}: their contribution is important for the real part of the scattering length whereas the σ_{ann} is mostly determined by the imaginary part of the scattering length.

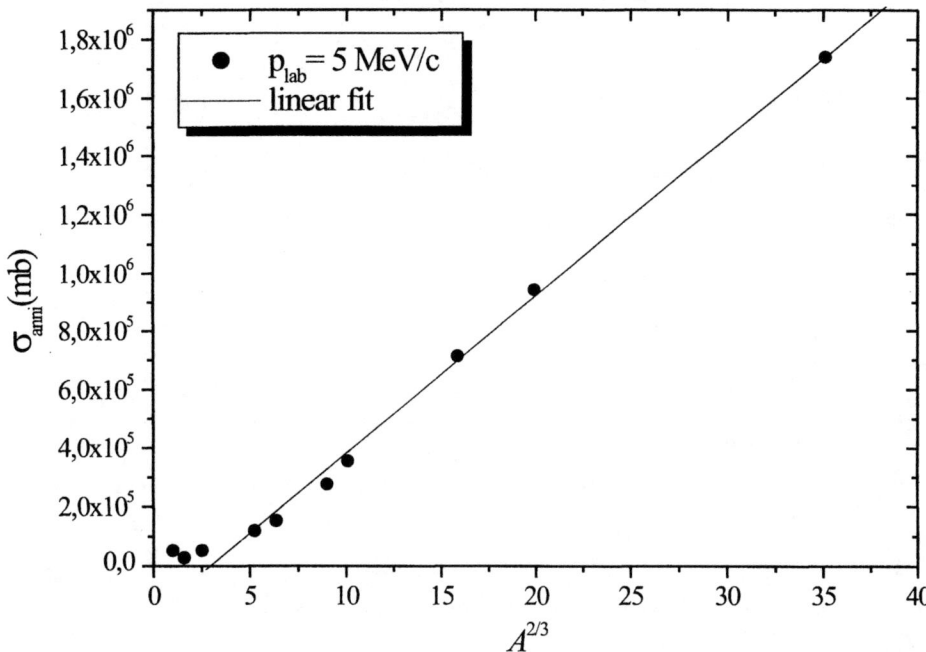

FIGURE 1. Values of the total $\bar{p}$-nucleus annihilation cross section calculated within an optical model approach.

It is very interesting to emphasize the phenomenon discussed in [4], that the S-wave annihilation cross section is sensitive mostly to a diffuseness of the interaction (and therefore the S-wave annihilation should be the same for all nuclei). This pediction is confirmed by our calculations within the optical model. In figure 2, $\sigma_{ann}^{l=0}$ as a function of the laboratory momentum is presented for different nuclei.

This fact allows to do an interesting prediction for the low energy antineutron-nucleus annihilation. In this case, Coulomb interaction is absent and the contributions of higher partial waves are not important. Therefore, the antineutron-nucleus annihilation cross section should be approximately the same for all nuclei.

Note that the wrong conclusion given in [9] about a contradiction between the results of phenomenological analysis [3] and values of the imaginary part of the scattering length for the antiproton-helium system comes from the fact that the authors of [9] completely forgot the experimental errors of the data. One can easily see from the value given in the introduction that the imaginary part of the scattering length is extracted from the experiment with quite important systematic error.

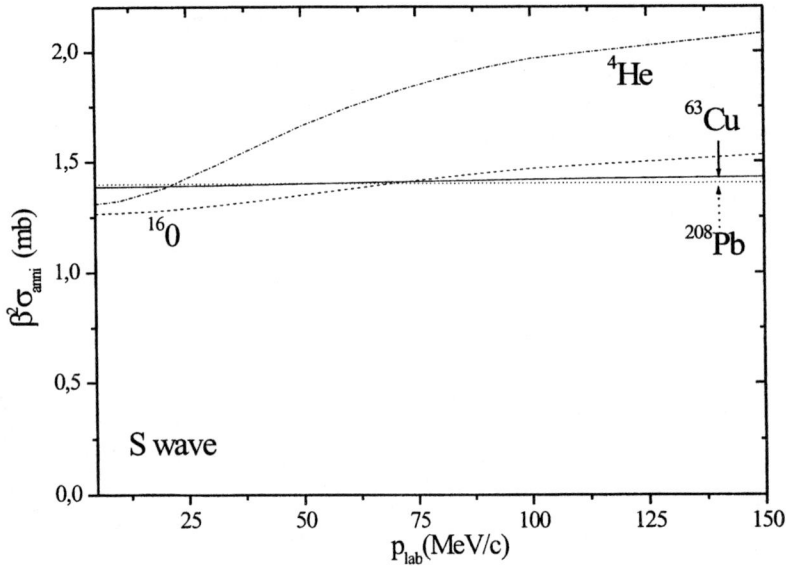

FIGURE 2. The S-wave contribution to the p̄-nucleus annihilation cross section for different nuclei.

CONCLUSIONS

The analysis of the recent experimental data on the antiproton annihilation on nuclei is presented. The calculations performed within the optical model confirm the prediction that and therefore the S-wave annihilation cross sections are approximately the same for all nuclei. This fact gives a quite strong experimental prediction: the low energy antineutron-nucleus annihilation cross-sections should be approximately the same for all nuclei. Due to Coulomb interaction, the antiproton-nucleus cross sections for intermediate and heavy nuclei have usual $A^{2/3}$ dependence.

REFERENCES

1. A. Zenoni et al., *Phys. Lett.* **B461**, 405, (1999); A. Zenoni et al., *Phys. Lett.* **B461**, 413, (1999).
2. M. Augsburger et al., *Phys. Lett.* **B461**, 417, (1999).
3. K.V. Protasov, G. Bonomi, E. Lodi Rizzini, A. Zenoni, Eur. Phys. J. **A7**, 429, (2000).
4. V.A. Karmanov, K.V. Protasov, A.Yu. Voronin, Eur. Phys. J. **A8**, 429, (2000).
5. J. Carbonell, K.V. Protasov, *Hyp. Int.* **76**, 327, (1993).
6. J. Carbonell, K.V. Protasov, *Nucl. Phys. B* (Proc. Suppl.) **56**, Addendum, (1997).
7. G.R. Satchler et al., Phys. Rev. **C29**, 574, (1984); V. Ashford et al., Phys. Rev. **C31**, 663, (1985).
8. R. Duperray, "Annihilation antiproton-noyau à basse énergie", ISN, Grenoble (2001).
9. C.J. Batty, E. Friedman, A. Gal, Nucl. Phys. **A 689**, 721, (2001).

$\rho\pi$–states in the Antiproton–Neutron–Annihilation into $\pi^-3\pi^\circ$

F. Meyer-Wildhagen on behalf of the Crystal Barrel Collaboration

Sektion Physik, Universität München, D-85748 Garching, Germany

Abstract. The preliminary results of the analysis of $\bar{p}n \to \pi^-3\pi^\circ$ annihilation at rest are presented. About 205.520 events were taken with the Crystal Barrel Detector with a liquid deuterium target. A full four–body–partial wave analysis with the maximum likelihood method was applied for a part of this sample. The dominant intermediate states in the data consist of a ρ–meson combined with the $(\pi\pi)$ S–wave or with the D–wave $(f_2(1270))$. Beneath the well established resonances the partial wave analysis yields a broad resonant structure in the exotic $(\rho\pi)$ P–wave at about $1500 MeV/c^2$. In addition, the $(\rho\pi)$–system is found in the $a_1(1260)$, $a_1(1640)$, $a_2(1320)$, $a_2(1650)$, $\pi(1300)$ and $\pi(1800)$ states.

INTRODUCTION

The study of the reaction $\bar{p}n \to \pi^-3\pi^\circ$ is very promising for the search of resonances decaying into $\rho^-\pi^\circ$. The $(\rho\pi)$–system in this channel may have the following quantum numbers: $J^{PC} = 0^{-+}, 1^{++}, 1^{-+}, 2^{++}, \ldots$, Isospin $I = 1$, G–parity $G = -, (C = +)$, Parity $P = (-1)^l$. Notice that the exotic quantum numbers $J^{PC} = 1^{-+}$ are also possible.

About $8.2 \cdot 10^6$ 1–prong events were taken with the Crystal Barrel Detector located at the Low Energy Antiproton Ring (LEAR) at CERN [1]. The antiproton annihilates with the neutron of the deuterium at rest. After the event reconstruction 824909 events are available for the kinematic fit. After the kinematic fit with a confidence level cut of $CL \geq 0.05$ and a cut on the momentum of the spectator proton of $p \leq 100 MeV/c$, 205520 events remain for the partial wave analysis.

In Fig. 1 are plotted the four possible invariant mass spectra for the data (error bars) and phase space distributed Monte Carlo events (solid line). The $(\pi^-\pi^\circ)$–spectrum shows a very prominent $\rho(770)$ peak, wherefore the $\rho(770)$ is involved in the most decay modes. In the $(\pi^\circ\pi^\circ)$ mass distribution the $(\pi\pi)$ S–wave ("σ") shows up and there is a deep incision due to the destructive interference with $f_\circ(980)$. Fig. 1d indicates some structures and some enhancement in the $1500 MeV/c^2$ region.

Since the $\rho(770)$ is involved in most decay modes, it is useful to create a pseudo "Dalitzplot". It gives some information about intermediate $(\rho\pi)$–states, which is not easily seen in the projections for this 4–body channel. In Fig. 2 the $\rho(770)$ is defined by a window cut of $675 MeV/c^2$ to $800 MeV/c^2$. This cut is slightly asymmetric because the mass of the $\rho(770)$ is shifted to a lower value. The reason is the strong decay mode of the $\bar{p}n$ system into $\rho(770)f_2(1270)$ which is near to the edge of phase space. The Dalitzplot for phase space distributed Monte Carlo events is completely flat (Fig. 2b). In contrast the Dalitzplot for the data shows some remarkable structure. The strong contribution of the $f_2(1270)$ with the $\rho(770)$ is seen as a prominent band in the left corner. The dip at its edges is due to $\rho(770) \times f_\circ(980)$. $(\rho\pi)$–resonances are indicated in the region around $2.25 GeV^2/c^4$.

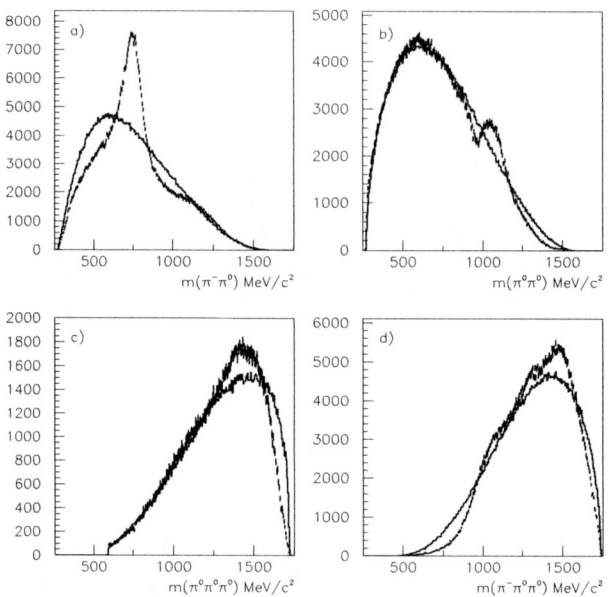

FIGURE 1. All four possible invariant mass spectra of the 205520 data events (with error bars) and the 405638 phase space distributed Monte Carlo events (solid line). Figure a), b) and d) have 3 entries per event due to combinatorics.

PARTIAL WAVE ANALYSIS

The fit of the data is made with the unbinned maximum likelihood method for the four–body channel. Most of the fits are applied in this stage of the analysis only to parts of the statistics in order to save cpu time (mostly 50000 data events, 150000 Monte Carlo events). The angular distribution is calculated with the helicity formalism [2][3]. For parametrization of intermediate states the normal relativistic Breit Wigner formula is used. A 2×2–K–matrix is used for the $(\pi\pi)$ S–wave [4] and a 1×1–K–matrix for two overlapping resonances, eg. $\pi_1(1400)$, $\pi_1(1650)$. The atomic states of the $\bar{p}n$–system are: $^3S_1$ and $^1P_1$. Other states are forbidden because the G–parity of 4 pions is always positive. Table 1 gives an overview of the possible resonances.

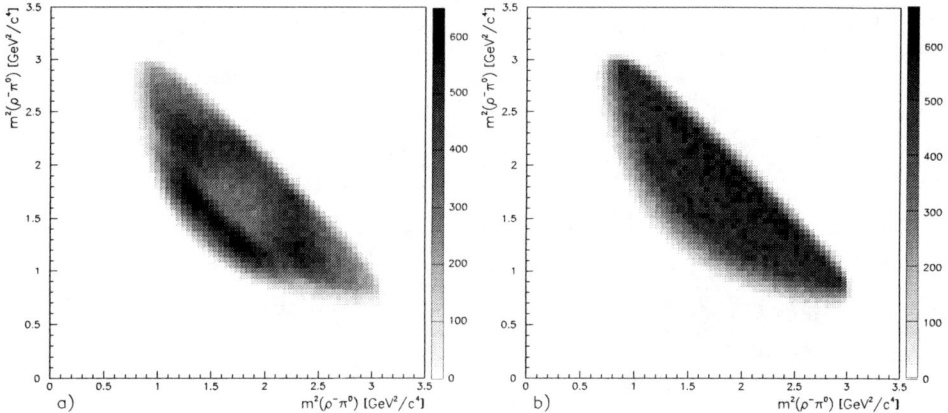

FIGURE 2. Pseudo Dalitzplots with 3 entries per event for the present 4–body channel for the data (a) and phase space (b).

TABLE 1. Combinations of resonances allowed in the present case. The order of the relative contributions according to the fit is indicated by the numbers in the third column.

combinations	resonances	ranking of strength
$(\pi^-\pi^\circ) + (\pi^\circ\pi^\circ)$:	$\rho(770)^- + \sigma, f_2(1270)$	(4,1)
	$\rho(1450)^- + \sigma$	(3)
$[(\pi^-\pi^\circ)\pi^\circ] + \pi^\circ$:	$a_1(1260, 1640)$	(10, 2)
	$a_2(1320, 1650)$	(9, 11)
	$\pi(1300, 1800)$	(7, 5)
	$\pi_1(1405, 1650^*)$	(6)
	$\pi_2(1670)$	(8)

* not included in the shown fit

Figure 3 gives an impression of the status of the analysis. Little problems are seen in the $(\pi^\circ\pi^\circ)$–spectrum. However, at present only fixed poles [4] were used for the 2×2–K–matrix, and only the total strength is fitted. For the exotic $(\rho\pi)$ P–wave, a single resonance of width $310 MeV/c^2$ was assumed, and a mass of $1450 MeV/c^2$ was fitted (see below). The $(\pi^-\pi^\circ\pi^\circ)$–spectrum shows more structure in this region. A solution could be that there are two exotic resonances. A K–matrix–fit with two resonances gave the following very preliminary poles: $m_1 = 1460 MeV/c^2, \Gamma_1 = 314 MeV/c^2, m_2 = 1650 MeV/c^2, \Gamma_2 = 150 MeV/c^2$. The contribution of the $\pi_1(1400)$ comes dominantly from the $^1P_1$–state which is puzzling, as compared with the $\eta\pi^-\pi^\circ$ result [5]. A mass scan (Fig. 4b) of log likelihood shows the need for an exotic resonance. The possibility of 2 resonances is not excluded.

It should be stressed that the invariant mass spectra do not give the full information.

Clear differences between data and fit are apparent in a comparison of the experimental and fitted pseudo Dalitzplots (Figs. 2a and 4a). It is clear that the region around $1500 MeV/c^2$ is not described satisfactorily and improvements of the fit must be sought for.

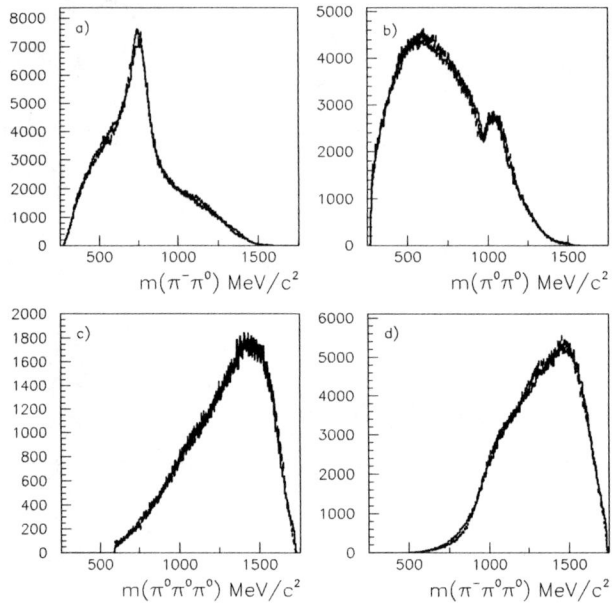

FIGURE 3. Comparison of data (error bars) and fit (solid line) for all projections.

PRELIMINARY RESULTS FOR a_1, a_2, π AND π_2 RESONANCES

This analysis allows the study of different decay modes of one resonance, eg.: $\pi(1300) \to \rho\pi$, $\sigma\pi$. Table 2 shows three branching ratios for the $\pi(1300)$. Our result is in clear contradiction with the PDG but in agreement with another Crystal Barrel work [7]. The difference could result from different $(\pi\pi)$ S–wave parametrizations. It is interesting to compare the results with theory. In [8] it is predicted for the radial excitation of the π–meson, that its decay branching should be stronger into $(\rho\pi)$ than into $(\sigma\pi)$, in agreement with our observation.

Another interesting outcome is the $a_1(1640)$ as a $(\rho\pi)$–resonance. Until now only the decay modes $a_1(1640) \to f_2(1270)\pi$, $\sigma\pi$ were known [6]. The partial wave analysis yields evidence for three new decay modes: $a_1(1640) \xrightarrow{I=0} \rho(770, 1450)^-\pi^\circ$,

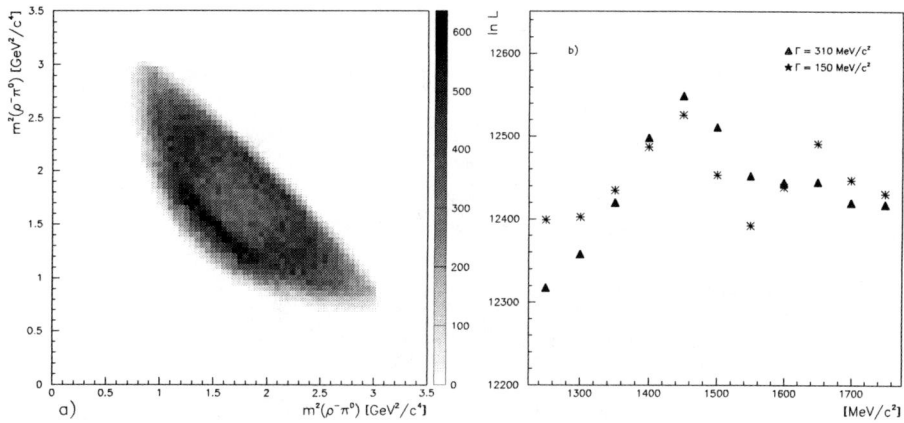

FIGURE 4. a) Theoretical pseudo Dalitzplot, as obtained from the present preliminary fit. b) Mass scan of the π_1 for two different widths.

TABLE 2. Comparison of Braching Ratios for the $\pi(1300)$.

BR $\pi(1300)$	PDG [6]	CB [7]	this work
$\frac{\Gamma(\sigma\pi)}{\Gamma(\rho\pi)}$	= 2.12	< 0.15	< 0.1

$a_1(1640) \xrightarrow{l=2} \rho(770)^-\pi^\circ$. We find that the first two decay modes are stronger than the second, and also stronger than $(f_2\pi)$ and $(\sigma\pi)$. The $a_2(1650)$ appears as a $\rho\pi$–resonance with similar intensitiy as the $a_2(1320)$.

At present, we may draw the conclusion that the exotic π_1–wave around $1500 MeV/c^2$ is needed. Whether more complex resonance structure than a single resonance is present remains to be seen. The decay modes of the $\pi_2(1670) \to \rho\pi, f_2\pi$ are in agreement with the PDG [6]. The results for $\pi(1300)$ support the interpretation as a radial excitation. The partial wave analysis requires the $(\rho\pi)$–resonances listed in Table 1, including in particular the radial excitation of the a_1 which is more intense than the a_1 ground state.

REFERENCES

1. E. Aker et al., Nucl. Instr. and Meth. A 321 (1992) 69
2. S. U. Chung, Spin Formalism, CERN 71-8 (1971)
3. M. Jacob and G. C. Wick, Ann. of Phys. 7 (1959) 404
4. Crystal Barrel Collaboration, Phys. Lett. B 342 (1995) 433
5. A. Abele et al., Crystal Barrel Collaboration, Phys. Lett. B 423 (1998) 175
6. Particle Data Group, Review of Particle Physics, Euro. Phys. J. C15 (2000) 1
7. Crystal Barrel Collaboration, Euro Phys. J. C19 (2001) 667
8. T. Barnes, F. Close, P. Page, E. Swanson, Phys. Rev. D55 (1997) 4157

Search for an exotic partial wave in $\pi\eta'$

Jörg Reinnarth representing the Crystal Barrel collaboration

Institut für Strahlen- und Kernphysik, University of Bonn, Germany

Abstract. To investigate the possible existence of an exotic $J^{PC} = 1^{-+}$-wave decaying into $\pi\eta'$ in $p\bar{p}$-annihilation at rest, the reaction $p\bar{p} \to \pi^+\pi^-\pi^+\pi^-\eta$ was analysed. The data was taken with the Crystal Barrel experiment. The preliminary partial wave analysis requires a resonant $(\eta'\pi)$-P-wave, with a mass of 1555 ± 50 MeV/c^2 and a width of 200 ± 100 MeV/c^2.

INTRODUCTION

Mesons, which are composed of a quark and an antiquark, are not allowed to carry following quantum numbers: $J^{PC}_{exotic} = 1^{-+}, 0^{+-}, 2^{+-}, 0^{--}$, etc. Various models including lattice QCD predict states carrying so called 'exotic' quantum numbers. [1].

The $\pi\eta$ or $\pi\eta'$ system is attractive for the search for exotics since its P-wave carries the non-$q\bar{q}$ quantum numbers $I^G J^{PC} = 1^- 1^{-+}$. A glueball interpretation of an exotic state with these quantum numbers is excluded because of isospin 1.

In different experiments at least two states with the exotic quantum number $I^G J^{PC} = 1^- 1^{-+}$ have been observed in the mass region which is accessible for $p\bar{p}$–annihilation at rest (see Table 1). One of these states was found at a mass of 1400 MeV/c^2 and the other at a mass of 1600 MeV/c^2. In pion-induced reactions on nucleons, evidence for a resonant behaviour in the $(\pi\eta)$-P-wave was claimed by the BNL collaboration [2, 3, 6, 7, 8] and also by the VES collaboration in various decay modes [9]. The $\pi_1(1400)$ has been observed in different Crystal Barrel analyses [4, 5].

TABLE 1. Evidence for $J^{PC} = 1^{-+}$ exotics.

Experiment	mass (MeV/c^2)	width (MeV/c^2)	decay mode	reaction
BNL [2]	$1370 \pm 16 ^{+\ 50}_{-\ 30}$	$385 \pm 40 ^{+\ 65}_{-105}$	$\eta\pi$	$\pi^- p \to \eta\pi^- p$
BNL [3]	$1359 ^{+\ 16}_{-\ 14} ^{+\ 10}_{-\ 24}$	$314 ^{+31}_{-29} ^{+\ 9}_{-66}$	$\eta\pi$	$\pi^- p \to \eta\pi^- p$
CBar [4]	$1400 \pm 20 \pm 20$	$310 \pm 50 ^{+\ 50}_{-\ 30}$	$\eta\pi$	$\bar{p}n \to \pi^-\pi^0\eta$
CBar [5]	1360 ± 25	220 ± 90	$\eta\pi$	$\bar{p}p \to \pi^0\pi^0\eta$
BNL [6]	$1593 \pm 8 ^{+\ 29}_{-\ 47}$	$168 \pm 20 ^{+150}_{-\ 12}$	$\rho\pi$	$\pi^- p \to \pi^+\pi^-\pi^- p$
BNL [7, 8]	$1595 \pm 10 \pm 50$	$395 \pm 30 \pm 50$	$\eta'\pi$	$\pi^- p \to \pi^-\eta' p$
VES [9]	1610 ± 20	290 ± 30	$\rho\pi, \eta'\pi, b_1\pi$	$\pi^- N \to \pi^-\eta' N$

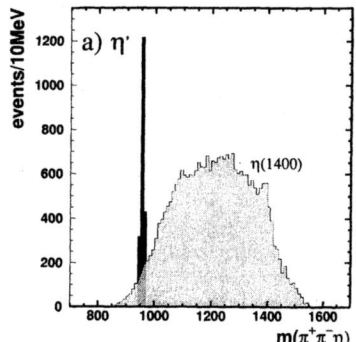
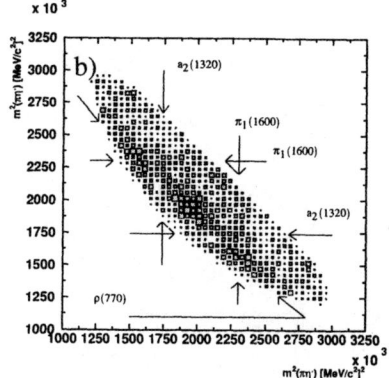

FIGURE 1. a) $\pi^+\pi^-\eta$ invariant mass spectrum from $\bar{p}p \to \pi^+\pi^-\pi^+\pi^-\eta$. Each event causes four entries. b) Dalitzplot of the black coloured events. The contribution of $\rho(770)\eta'$, $a_2(1320)\pi$ and $\pi_1(1600)\pi$ are clearly visible.

CRYSTAL BARREL DATA

The data were recorded by stopping antiprotons from LEAR in a liquid hydrogen target. Out of 7.3 million events taken with a trigger on 4 charged particles, 6026 $p\bar{p} \to \pi^+\pi^-\pi^+\pi^-\eta(\eta \to \gamma\gamma)$ events were selected [10]. Figure 1a) shows the $\pi^+\pi^-\eta$ invariant mass spectrum including four entries for each event. If there is an η' decaying into $\pi^+\pi^-\eta$, then one entry belongs to the prominent η' signal and three entries form a combinatorial background. In addition, there are events without any η' contribution. A fit yields 1327 events including an η'. The distribution of the background originating from $\bar{p}p \to \pi^+\pi^-\pi^+\pi^-\eta$ is well known from the analysis mentioned before [10]. Therefore, these amplitudes are included with fixed parameters in the partial wave analysis and added up incoherently to the η'-amplitudes:

$$|\mathcal{A}_{\text{total}}|^2 = \overbrace{|\mathcal{A}^{4\pi\eta}|^2}^{\text{known background}} + \underbrace{\left[|\mathcal{A}^{\eta'}_{^1S_0}|^2 + |\mathcal{A}^{\eta'}_{^3S_1}|^2\right]}_{\pi^+\pi^-\eta' \text{ partial waves}} \quad (1)$$

The partial wave analysis is based on the isobar model decomposing the reaction into a series of two-body decays. For the reaction $\bar{p}p \to \pi^+\pi^-\eta'(\eta' \to \pi^+\pi^-\eta)$, the model space of possible intermediate states is strongly restricted due to conservation laws. The final state $\pi^+\pi^-\eta'$ restricts the initial state to a positive G-parity. Resonances decaying into $\pi\eta'$ are restricted to isospin 1. The only possible quantum numbers are:
$I^G J^{PC} = 0^+0^{++}[(\pi\pi)_S\text{-wave}]$, $1^-0^{++}[a_0(1450)]$, $1^-2^{++}[a_2]$, $1^-1^{-+}[\pi_1]$ and $1^+1^{--}[\rho(770)]$.

The $a_2(1320)$ is tested with mass and width m = 1318 MeV/c^2, Γ = 107 MeV/c^2. From previous analyses and from SU(3) we expect the following contributions:
$\rho(770)\eta' \sim (22 \pm 5)\%$, $a_0(1450)\pi \sim (3.1 \pm 0.7)\%$, $a_2(1320)(^1S_0)\pi \sim (1.6 \pm 0.5)\%$ [4, 5]. In the partial wave analysis, we accept only solutions which do not disagree strongly with these expectations. Only annihilation from S-states has been taken into

TABLE 2. Partial wave analysis.

Step	$(\pi\pi)_S$	$\rho(770)$	$a_2(1320)$		$a_0(1450)$	$\pi_1(1600)$		$-2\cdot\ln\mathcal{L}$
	$^1S_0$	$^3S_1$	$^1S_0$	$^3S_1$	$^1S_0$	$^1S_0$	$^3S_1$	
SU(3)	-	22±7%	1.6%	-	3.1%	(1560,120)		
1	70.4%	29.6%	-	-	-	-	-	0
2	67.7%	29.4%	1.4%	1.5%	-	-	-	31
3	63.8%	30.4%	1.2%	2.6%	1.9%	-	-	37
4	58.4%	26.3%	0.9%	5.9%	0.4%	2.8%	5.2%	45
2*	67.7%	29.4%	1.4%	1.5%	-	-	-	31
3*	60.3%	25.7%	1.0%	4.9%	-	2.5%	5.5%	44
4*	58.4%	26.3%	0.9%	5.9%	0.4%	2.8%	5.2%	45
$\pi_1(1400)$ included in step 5: $(m,\Gamma) = (1400,200)$ $^1S_0 = 0.5\%$ and $^3S_1 = 1.0\%$								
5	61.7%	20.4%	0.1%	5.8%	0.9%	2.5%	7.7%	47

account. Table 1 shows the different combinations of resonances for the description of the data. The numbers in the table represent the relative contributions of the resonances to the $\pi^+\pi^-\eta'$ data set. In the second line their contributions as expected from SU(3) considerations are listed.

In a first step of the analysis only the $\rho(770)$ and the $(\pi\pi)_S$-wave were introduced into the unbinned loglikelihood fit. A relative branching ratio of 30% is determined for the $\rho(770)$ and 70% for the $(\pi\pi)_S$-wave. The $-2\ln\mathcal{L}$ is normalized to 0.

We now include the isovector tensor state $a_2(1320)$. The fit optimizes for a relative branching ratio of the $\rho(770)$ of 30% and a relative branching ratio of the $(\pi\pi)_S$-wave of 68%. The $a_2(1320)$ contributes with a relative rate of 1.4% to the $^1S_0$-state and 1.5% to the $^3S_1$-state. The likelihood change of $\Delta 2\cdot\ln\mathcal{L}=31$ is significant.

In the next step (3) the $a_0(1450)$ is included. The change of the $-2\cdot\ln\mathcal{L}$ shows a small improvement in the description of the data. The $a_0(1450)$ contributes with 2% in this description. This is the best description of the data which was achieved without introducing a resonance with exotic quantum numbers.

Finally we include a Breit-Wigner resonance in the exotic partial wave. Mass and width are fitted freely and are determined to m=1555±50 MeV/c^2 and Γ=200±100 MeV/c^2. These values are independent of the start values of the minimization procedure. In this step already a problem becomes visible. The fit is not sensitiv to the width of the π_1. The $\rho(770)$ contributes 26% and the $(\pi\pi)_S$-wave 59% to the final state. The contribution of the $a_2(1320)$ with 1% to the $^1S_0$-state and with 6% to the $^3S_1$-state are in the order of magnitude as expected from SU(3). The relativ branching ratio of the $a_0(1450)$ of 0.4% shows no importance of an $a_0(1450)$ for the description of the data. The π_1 is produced with 3% from the $^1S_0$-state and from the $^3S_1$-state with 6%.

To test the importance of the $a_0(1450)$, the order in step 3* and 4* are changed. We now first include the π_1 and observe an improvement of the $\Delta 2\ln\mathcal{L}$ by 13. Including now the $a_0(1450)$ delivers the same result as in step 4 and the increasing likelihood about 1 shows no improvement of the fit.

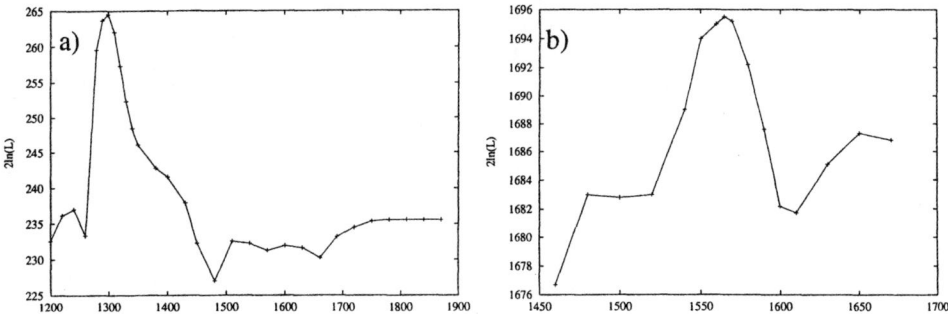

FIGURE 2. a.) Scan of the $a_2(1320)$ mass, $\Gamma=108$ MeV/c^2. b.) Scan of the $\pi_1(1600)$ mass, $\Gamma=200$ MeV/c^2.

To confirm this evidence, scans of the $I^G J^{PC} = 1^- 2^{++}$- and $1^- 1^{-+}$-partial waves are carried out. The scan of the 2^{++}-wave is shown in Figure 2.a). A clear maximum at the $a_2(1320)$ mass is visible. No further maximum can be seen at a mass of 1660 MeV/c^2. The scan of the $\pi_1(1600)$ shows a maximum at a mass of m=1565 MeV/c^2 (2.b). This indicates, that the best description of the data is achieved by including a π_1 at a mass of 1555 MeV/c^2 and a width of 200 ± 100 MeV/c^2.

SUMMARY AND CONCLUSION

Preliminary results of the analysis of the reaction $p\bar{p} \to \pi^+\pi^-\pi^+\pi^-\eta$ show that the description of the data is improved if a $\pi_1(1600)$ is included. The relative branching ratios of different resonances are in the order of magnitude as expected from SU(3) relations using the results from analysis of $\bar{p}p \to \pi^0\pi^0\eta$ and $\bar{p}p \to \pi^0\pi^0\eta'$. A scan of the π_1-mass shows a maximum at a mass of m=1565 MeV/c^2 and a width of $\Gamma=200$ MeV/c^2. A free fit determines the mass and width to m=1555 ± 50 MeV/c^2 and $\Gamma=200\pm100$ MeV/c^2. Unhappily the fit is not sensitiv on the width of the π_1.

REFERENCES

1. C. Morningstar, Nucl. Phys. Proc. Suppl. **B90** (2000) 214
2. D. R. Thompson *et al.* [E852], Phys. Rev. Lett. **Vol.79 Nr.9** (1997) 1630
3. S. U. Chung *et al.* [E852 Collab.], Phys. Rev. **D60** (1999) 092001
4. A. Abele *et al.* [Crystal Barrel Collab.], Phys. Lett. **B423** (1998) 175.
5. A. Abele *et al.* [Crystal Barrel Collab.], Phys. Lett. **B446** (1999) 349.
6. G. S. Adams *et al.* [E852 Collab.], Phys. Rev. Lett. **81** (1998) 5760.
7. K.K. Seth, Proc.PANIC99, Nucl. Phys. **A663 & 664** (2000) 113c
8. S. U. Chung, In *Frascati 1999, Hadron spectroscopy*, 603-607.
9. Y. Khokhlov [VES Collab.], Nucl. Phys. **A663**, (2000) 596.
10. J.S. Suh, Investigation of the E-Meson in Proton-Antiproton-Annihilation at Rest, Dissertation, University Bonn (1999)
11. E. Klempt, Meson Spectroscopy: Glueballs, Hybrids and q$\bar{\text{q}}$ Mesons, Summary of the lessons for the summerschool in Zuoz.

γγ AND INCLUSIVE REACTIONS

IHEP, Protvino

Inclusive D*-meson production in two-photon collisions at LEP

A. A. Sokolov

Institute for High Energy Physics, 142280 Protvino, Moscow region, Russia

Abstract. The inclusive production of D^{*+} is measured by DELPHI in photon-photon collisions at LEP-II energies. The measured cross sections are compatible with the QCD calculations having the contributions from the resolved processes sensitive to the gluon density in photon. The total cross section of the charm quark production in two-photon collisions at LEP-II energies is estimated.

INTRODUCTION

Hadron production in two-photon collisions is described by the Vector meson Dominance Model (VDM) [1], the Quark Parton Model (direct process) (QPM) [2], and the hard scattering of hadronic constituents of quasi-real photons (resolved photon process)[3]. Studies of charm meson production might provide usefull information about the gluon density in the photon.

In this paper we report on measurement of charm production in $\gamma\gamma$ collisions at LEP-II energies via the inclusive production of D^{*+} mesons[1]. D^{*+} mesons were detected by their decay to $D^0\pi^+$, with the D^0 observed in the decay modes $K^-\pi^+$ and $K^-\pi^+\pi^-\pi^+$.

EXPERIMENTAL PROCEDURE

The analysis presented here is based the data taken with the DELPHI detector [4, 5] during the years 1996–2000. The integrated luminosity used in the analysis is 617 pb^{-1}.

After the requirement for the TPC and RICH detectors to have a good-quality operation for the selected events the hadronic two-photon events have been extracted by applying the following cut on the full sample: at least 1 charged track in the barrel region ($45° < \theta < 145°$ with $p_t > 1.2$ GeV/c^2).

The trigger efficiency for the events which passed the above requirement is bigger than 98%.

The following cuts were applied also: visible invariant mass is $W_{vis} < 35$ GeV/c^2; number of charged tracks $4 \leq N_{ch} \leq 16$; the sum of the transverse energy components with respect to the beam direction of all charged particles is $\sum E_T^{vis} > 3$ GeV/c^2.

---

[1] Throughout this paper charge conjugate decays are implicitly included.

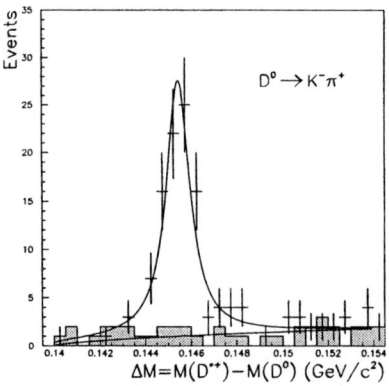 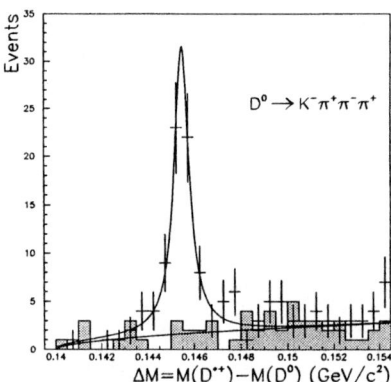

FIGURE 1. The distributions of the mass difference between the D^{*+} and D^0 for the $D^0 \to K^-\pi^+$ (left), $K^-\pi^+\pi^-\pi^+$ (right). The histograms are fitted by the sum of a Breit-Wigner function and the background function $f(\Delta M) = \alpha \cdot (\Delta M - m_\pi)^\beta$. The distributions for "wrong sign" combinations of tracks are shown by hatched histograms.

A total of 215 352 events remain in the data sample after applying all these cuts. The main background comes from the process $e^+e^- \to Z^0\gamma$ and amounts to $\sim 1.2\%$ of the selected $\gamma\gamma$ events.

In order to compare the data with the theoretical predictions, samples of $\gamma\gamma \to q\bar{q}$ events simulated by PYTHIA 6.143 [6] program was used.

D^{*+} PRODUCTION

D^{*+} mesons are detected by their decay into $D^0\pi^+$. The signal is displayed by plotting of a mass difference between D^{*+} and D^0 meson candidates ΔM. D^0 is identified via its decay to $K^-\pi^+$ and $K^-\pi^+\pi^-\pi^+$.

A neural networks based MACRIB package [7], which combines a RICH and dE/dx information, used for charged kaon selection. To classify D^0 candidates the mass windows used in each decay mode were set according to the mass resolution.

In Fig.1 the mass difference ΔM distribution for the $D^0 \to K^-\pi^+$, $K^-\pi^+\pi^-\pi^+$ decay channels are shown. A clear peak around the D^{*+} and D^0 mass difference is observed. The histograms are fitted by the sum of a Breit-Wigner function and the background function $f(\Delta M) = \alpha \cdot (\Delta M - m_\pi)^\beta$. The distributions for "wrong sign" combinations of tracks are also shown. The numbers of the signal events $N_{D^{*+}}$ are 96 ± 11.6, 67 ± 9.6 for $D^0 \to K^-\pi^+$, $K^-\pi^+\pi^-\pi^+$ decay modes respectively.

To get more reliable results we studied a D^{*+} production in the restricted kinematical region (2 GeV/c $< p_T^{D^{*+}} <$ 12 GeV/c, $|\eta^{D^{*+}}| < 1$), where $p_T^{D^{*+}}$, $\eta^{D^{*+}}$ are transverse momentum and pseudorapidity of D^{*+} respectively. To determine the relative fraction of direct and single resolved processes we used the distribution of $x_T^{D^{*+}} = 2p_T^{D^{*+}}/W_{\text{vis}}$ in

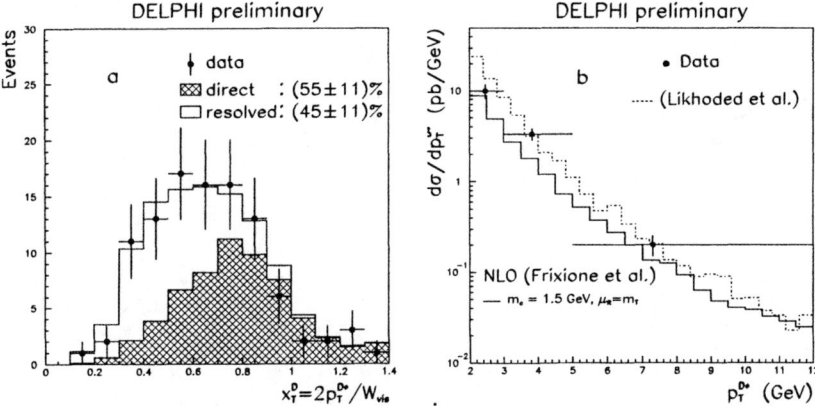

FIGURE 2. $x_T^{D^{*+}} = 2p_T^{D^{*+}}/W_{vis}$ distribution (a) in the restricted kinematical region $2 \text{ GeV/c} < p_T^{D^{*+}} < 12 \text{ GeV/c}$, $|\eta^{D^{*+}}| < 1$ and $p_T^{D^{*+}}$ distribution (b) in the kinematical region $|\eta^{D^{*+}}| < 1$.

this kinematical region (fig.2a). The fit to the $x_T^{D^{*+}}$ distribution yields $f = (55 \pm 11)\%$ direct and $(1-f) = (45 \pm 11)\%$ single resolved contributions (errors are statistical).

A comparison of the measured differential cross-section $d\sigma/dp_T^{D^{*+}}$ with a next-to-leading order (NLO) calculation by Frixione et al. [8] and a model [9] is shown in Fig.2b. The experimental distribution is consistent with the model [9], where a fragmentation of charm quarck near the threshold is described in detail. The NLO calculation with the massive charm quark matrix element underestimates the experimental distribution.

The cross section of inclusive D^{*+} production in the considered kinematical region $2 \text{ GeV/c} < p_T^{D^{*+}} < 12 \text{ GeV/c}$, $|\eta^{D^{*+}}| < 1$ is then

$$\sigma_{mes}^{D^{*+}}(e^+e^- \to e^+e^- D^{*+}X) = N_{exp}^{D^*}/(\varepsilon_{D^*} \cdot \mathcal{L} \cdot Br) = 18.0 \pm 1.8(\text{stat.}) \pm 1.6(\text{syst.}) \text{ pb},$$

where $N_{exp}^{D^*}$ is the observed number of D^{*+} mesons, ε_{D^*} is the reconstruction efficiency of D^{*+} in the corresponding mode, $\mathcal{L}$ is the total integrated luminosity and Br is the branching ratio of D^{*+} decay to the corresponding decay chain [10]. The D^{*+} reconstruction efficiencies $\varepsilon_{D^{*+}}$ for each D^0 decay mode are calculated using PYTHIA.

The main sources of systematic uncertainties are due to the limited number of MC events (~6-8%), K/π selection procedure (~3-5%) and the branching fractions uncertainties (~2.4-4.2%). The systematic uncertainties are added in quadrature.

To obtain the total cross-section of charm production we extrapolated the measured cross section $\sigma_{mes}^{D^{*+}}$ to the full kinematical range. For calculation of the extrapolation factor the PYTHIA based MC was used. The extrapolation factors for the direct and single-resolved processes are calculated separately and then summarized with the factors f and $(1-f)$ respectively. The total cross section of inclusive D^{*+} production is corrected then by the probability of charm quarks to fragment into a D^{*+} [10]. The obtained value of the total cross section is

$$\sigma(e^+e^- \to e^+e^-c\bar{c}) = 783 \pm 78(\text{stat.}) \pm 70(\text{syst.}) \pm 190(\text{extr.}) \text{ pb}$$

agrees with the NLO calculation [8]. The extrapolation procedure introduces large uncertainties to the total cross section value.

CONCLUSIONS

The inclusive production of D^{*+} in $\gamma\gamma$ collisions at LEP-II energies has been measured.

The relative fractions of contributions from direct and single resolved processes in the kinematical region $2 \text{ GeV/c} < p_T^{D^{*+}} < 12 \text{ GeV/c}$, $|\eta^{D^{*+}}| < 1$ was measured. The measured differential cross section $d\sigma/dp_T^{D^{*+}}$ is consistent with the model [9]. The NLO calculation with the massive charm quark matrix element underestimates the experimental distribution.

The extrapolation of the cross section of inclusive D^{*+} production measured in the restricted kinematical region to the total charm cross section gives the value which agrees with the NLO calculation.

REFERENCES

1. Sakurai, J. J., and Schildknecht, D., *Phys. Lett.* B **40**, 121 (1979);
   Ginzburg, I. F., and Serbo, V. G., *Phys. Lett.* B **109**, 231 (1982).
2. Brodsky, S. J., Kinoshita, T., and Terazawa, H., *Phys. Rev.* D **4**, 1532 (1971).
3. Brodsky, S. J., DeGrand, T. A., Gunion, J. F., and Weis, J. H., *Phys. Rev. Lett.* **41**, 672 (1978); *Phys. Rev.* D **19**, 1418 (1979).
4. Aarnio, P., *et al.*, DELPHI Collab., *Nucl. Instrum. Methods* A **303**, 233 (1991).
5. Abreu, P., *et al.*, DELPHI Collab., *Nucl. Instrum. Methods* A **378**, 57 (1996).
6. Sjöstrand, T., *Comput. Phys. Comm.* **82**, 74 (1994).
7. Albrecht, Z., Feindt, M., and Moch, M., *MACRIB. High efficiency - high purity hadron identification for DELPHI*, DELPHI/99-150 (October 1999).
8. Frixione, S., Krämer, M., and Laenen, E., *Nucl. Phys.* B **571**, 169 (2000).
9. Berezhnoy, A. V., Kiselev, V. V., and Likhoded, A. K., *Photoproduction and electroproduction of charm at high energies*, hep-ph/9905555 (Jun 2000);
   Likhoded, A. K., and Berezhnoy, A. V., *private communication*.
10. Review of Particle Physics, *Eur. Phys. J.* C **15**, 1 (2000).

Inclusive J/ψ production in two-photon collisions at LEP II with the DELPHI detector

Mikhail Chapkine

Institute for High Energy Physics, 142280 Protvino, Moscow region, Russia

Abstract. Inclusive J/ψ production in photon-photon collisions has been observed by the DELPHI collaboration at LEP II beam energies. A clean signal from the reaction $\gamma\gamma \to J/\psi + X$ is seen. Number of observed events, $N(J/\psi \to \mu^+\mu^-) = 36 \pm 7$ for the integrated luminosity 617 pb^{-1}, yielding a cross section of $\sigma(J/\psi \to \mu^+\mu^-) = 25.2 \pm 10.2$ pb. Based on a study of the event shapes of different types of $\gamma\gamma$ processes in the PYTHIA program, we conclude that $(74\pm22)\%$ of the observed J/ψ events are due to the 'resolved' photons, the dominant contribution of which is evidently a single color-octet gluon within the photon.

INTRODUCTION

An important component of the e^+e^- collisions at LEP II energies is the two-photon fusion process. It has been pointed out that two-photon production of inclusive J/ψ's $e^+ + e^- \to e^+ + e^- + \gamma_1 + \gamma_2$ followed by $\gamma_1 + \gamma_2 \to J/\psi + X$ is a sensitive tool for the gluon distribution in the photon [1].

There are two important processes leading to inclusive J/ψ production. The first process is undoubtedly attributable to the vector-meson dominance (VMD) model [2]. The second process is due to the color-octet model [3]. It proceeds through the so-called 'resolved' contribution of the photons, in which the intermdediate photons are 'resolved' into its constituent partons.

The purpose of this paper is to study the inclusive J/ψ production, in order to assess the relative importance of the production processes discussed above.

EXPERIMENTAL PROCEDURE

The analysis presented here is based the data taken with the DELPHI detector [4, 5] during the years 1996–2000. The integrated luminosity used in the analysis is 617 pb^{-1}.

After the requirement for the TPC detector to have a good-quality operation for the selected events the hadronic two-photon events have been extracted by applying the following cut on the full sample: there is either (i) at least 1 charged track in the barrel region ($40° < \theta < 140°$ with $p_t > 1.2$ GeV/c^2) or (ii) at least 1 neutral track in Forward Electromagnetic Calorimeter (FEMC) ($10° < \theta < 36°$ and $144° < \theta < 170°$) with energy greater than 10 GeV/c^2 or (iii) sum of number of charged tracks in barrel with $p_t > 1$ GeV/c^2 and charged tracks in forward region ($10° < \theta < 40°$ or $140° < \theta < 170°$) with $p_t > 2$ GeV/c^2 and neutrals in FEMC with $E > 7$ GeV/c^2 greater than one or (iv)

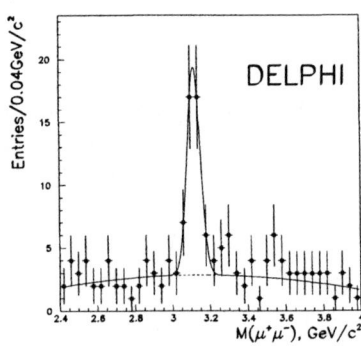

FIGURE 1. $M(\mu^+\mu^-)$ distribution from the DELPHI data.

sum of number of charged tracks in barrel with $p_t > 0.5$ GeV/c and charged tracks in forward with $p_t > 1$ GeV/c and neutral tracks in FEMC with $E > 5$ GeV/c² greater than four. The trigger efficiency for the events which passed the above requirement is bigger than 98%.

Finally the following cuts were applied: visible invariant mass is $W_{vis} < 35$ GeV/c²; number of charged tracks $4 \leq N_{ch} \leq 30$; the sum of the transverse energy components with respect to the beam direction of all charged particles is $\sum E_T^{vis} > 3$ GeV/c².

A total of 274 510 events remain in the data sample after applying all these cuts. The main background comes from the process $e^+e^- \to Z^0\gamma$ and amounts to $\sim 1.2\%$ of the selected $\gamma\gamma$ events.

J/ψ candidates have been selected using the $\mu^+\mu^-$ decay channel. For the muon selection the following criteria have been imposed: track should satisfy the DELPHI standard muon-tagging algorithm [5] or identified as muons by hadronic calorimeter [6]; track should not come from any reconstructed secondary vertex or be identified as kaon, proton or electron by standard DELPHI identification packages. At least two charged particles with zero summary charge should be identified as a muon candidates.

INCLUSIVE J/ψ PRODUCTION

We give in Fig. 1 the invariant mass of $\mu^+\mu^-$ from the DELPHI data selected as outlined in the precious section. It is seen that the J/ψ produced with a little background. A mass-dependent fit to the $M(\mu^+\mu^-)$ distribution with gaussian for the signal and polinomial for the backgrond gives the following results: J/ψ mass $M = 3119 \pm 8$ MeV/c², width $\Gamma(obs) = 35 \pm 7$ MeV/c².

The number of observed events is $N(J/\psi) = 36 \pm 7$ from the fit.

For efficiency estimation we used PYTHIA 6.156 generator [7]. The generated events

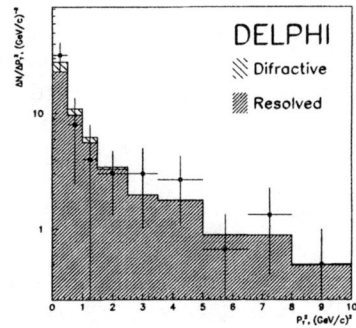 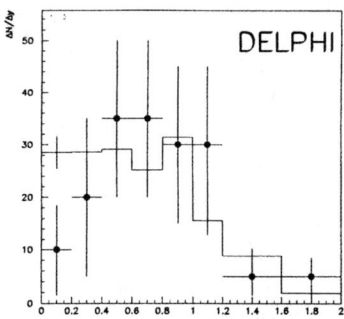

FIGURE 2. $p_T^2(J/\psi)$ distribution from the DELPHI data (left) and $|y|$ distribution for the J/ψ from the DELPHI data (right)

were passed through the simulation of the DELPHI detector [5] and than processed with the same reconstruction and analysis programs as the real data. There is a substantial fraction of PYTHIA events where J/ψ are produced just as a simple fusion of two photons because there is not enough phase space to produce additional particles. We checked that all such events are produced when both the colliding photons are direct or one photon is anomalous and the other one is DIS (we use here the PYTHIA notation [7]). That is why for the efficiency estimation we did not use the events with direct-direct or DIS-anomalous photon interactions. Among the rest of the PYTHIA events about 93% J/ψ are produced when at least one photon is a VDM photon.

Fig. 2 shows the $p_T^2(J/\psi)$ distribution and $|y|$ distribution from the DELPHI data.

As expected, the PYTHIA prediction for the $p_T^2(J/\psi)$ distribution is sharply peaked near zero for the diffractive MC events, while the 'resolved' MC events are very much spread out. We fitted the experimental $p_T^2(J/\psi)$ distribution as a function of the two categories of MC events

$$\frac{dN}{dp_T^2} = f \cdot \left.\frac{dN}{dp_T^2}\right|_{\text{Diffractive}} + (1-f) \cdot \left.\frac{dN}{dp_T^2}\right|_{\text{Resolved}},$$

which gives $f = (26.0 \pm 22.0)\%$. The PYTHIA study tells us that the experimental efficiencies are very different for the two categories: $\varepsilon(\text{Diffractive})=(1.79\pm0.07)\%$, $\varepsilon(\text{Resolved})=(6.79\pm0.16)\%$. From this we deduce that the overall experimental efficiency must be $1/\varepsilon = f/\varepsilon(\text{Diffractive}) + (1-f)/\varepsilon(\text{Resolved})$.

The cross section of inclusive J/ψ production is then

$$\sigma = N \cdot (Br \cdot \mathcal{L} \cdot \varepsilon)^{-1} = 25.2 \pm 10.2 \text{ pb}$$

where $Br = (5.88\pm0.10)\%$ is the branching ratio for $J/\psi \to \mu^+\mu^-$ [8] and $\mathcal{L} = 617\,\text{pb}^{-1}$ is the total integrated luminosity.

Finally, the rapidity distribution for the J/ψ is shown in right side of Fig. 2. The PYTHIA MC events have been combined using the measured fraction f and then normalized to the observed number of events in $0 < |y| < 2$. It is seen that the MC events are in fair agreement with the experimental rapidity dsitribution, although the data tend to have some deficiency below $|y| = 0.4$

CONCLUSIONS

We have studied the inclusive J/ψ production from $\gamma\gamma$ collisions. The data have been taken by the DELPHI collaboration during the LEP II phase, i.e. $\sqrt{s}$ of the LEP machine ranged from 161 to 207 GeV/c^2. A clean signal from the reaction $\gamma\gamma \to J/\psi + X$ is seen.

The preliminary value for inclusive cross section is $\sigma(J/\psi \to \mu^+\mu^-) = 25.2 \pm 10.2$ pb. Based on a study of the event shapes of different types of $\gamma\gamma$ processes in the PYTHIA program, we conclude that some $(74\pm22)\%$ of the observed J/ψ events are due to the 'resolved' photons, the dominant contribution of which is evidently derived from a single color-octet gluon within the photon.

The $p_T^2(J/\psi)$ and y distributions are presented.

REFERENCES

1. *Physics ar LEP2*, edited by G. Altarelli, T. Sjöstrand and F. Zwirner, CERN96-01 (Vol. 1), p. 330 (Feb 1996).
2. J.J. Sakurai and D. Schildknecht, *Phys. Lett.* B **40**, 121 (1979); I. F. Ginzburg and V.G. Serbo, *Phys. Lett.* B **109**, 231 (1982).
3. M. Klasen et al., DESY 01-039 (April 2001); R. M. Godbole et al., LC-TH-2001-019 (February 2001); E. L. Berger and D. Jones, *Phys. Rev.* D **23**, 1521 (1981); H. Jung, G. A. Schuler and J. Terrón, *Int. J. Mod. Phys.* A **7**, 7955 (1992); B. Naroska, *Nucl. Phys. B (Proc.Suppl.)* **82**, 187 (2000).
4. P. Aarnio et al., *Nucl. Instrum. Methods* A **303**, 233 (1991).
5. P. Abreu et al., *Nucl. Instrum. Methods* A **378**, 57 (1996).
6. J. Ridky, V. Vrba, J. Chudoba, *ECTANA. User's Guide*, DELPHI/99-181 TRACK 96 (17 November 1999).
7. T. Sjöstrand, *Comput. Phys. Comm.* **82**, 74 (1994).
8. Review of Particle Physics, *Eur. Phys. J.* C **15**, 1 (2000).

Relating production and masses of the vector and P-wave mesons for light and heavy flavours

P.V. Chliapnikov

Institute for High Energy Physics, Protvino, 142284, Russia

Abstract. The production rates of P-wave mesons and promptly produced vector mesons for light and heavy flavours are analyzed using results of the LEP experiments. It is shown that the mass dependence of production rates for the bottom B^*, B_1^* and B_2^*, charm D^{*+}, D_1^0 and D_2^{*0}, strange charm D_s^{*+} and D_{s1}^0, and light-flavour ρ^0 and $f_2(1275)$, $K^{*0}(892)$ and $K_2^{*0}(1430)$, ϕ and $f_2'(1525)$ mesons is very very similar. This allows to relate the relative production rates for mesons with different flavours and, surprisingly, their masses.

In this report, an attempt is made to relate the production of the vector and P-wave mesons with the light and heavy flavours in Z^0 hadronic decays, using the data from the LEP experiments.

The production rates of the promptly produced light-flavour vector and P-wave mesons were determined by multiplying their total rates, obtained by averaging the results of the LEP experiments [1], by the fractions of primary mesons taken from the JETSET model [2]. For the tensor mesons, only the DELPHI results were used. The mass dependence of the vector and tensor meson rates presented in Fig. 1 for the ρ^0, ω and $f_2(1275)$, the $K^{*0}(892)$ and $K_2^{*0}(1430)$, the ϕ and $f_2'(1525)$ for one spin projection, $\langle n \rangle/(2J+1)$, is very similar. The fit of the corresponding data to three exponentials $a \exp(-bM)$ with different normalization parameters a for the three meson families but the *same* slope parameter yields $b = 4.11 \pm 0.27$ $(\text{GeV}/c^2)^{-1}$.

The charm fragmentation fractions into the vector D^{*+} and D_s^{*+}, and P-wave $D_2^*(2460)$, $D_1(2420)$ and $D_{s1}^0(2536)$ mesons measured in the LEP experiments can be found in [3]. For the determination of a charm fragmentation fraction into the primary D^{*+} meson, the effects of the $D_2^*(2460)$, $D_1(2420)$ and $D_{s1}^0(2536)$ decays into $D^{*+}\pi$ have to be taken into account. With the averaged value of $f(c \to D^{*+}) = 0.238 \pm 0.010$, using the results given in [3] and assuming isospin invarians we obtain $f(c \to D^{*+}_{prompt}) = 0.183 \pm 0.018$. The values of the charm fragmentation fractions into the primary vector and P-wave mesons, divided by the corresponding spin counting factors $2J+1$, are presented in Fig. 1. As one can see their mass dependence is very similar to the one observed for the light-flavour mesons.

The relative B^* production rate in b-quark jet, $\sigma_{B^*}/\sigma_{b-jet}$, was measured by the LEP experiments [4,5] for a mixture of the states B_d^*, B_u^* and B_s^*, with the averaged result of 0.667 ± 0.037. Assuming that $0.3 B_s^*$ are produced for each B_d^*, we obtain $\sigma_{B^*(u,d)}/\sigma_{b-jet} = 0.580 \pm 0.032$. Evidence for the $B^{(*)}\pi$ resonant structure has been clearly observed by the LEP experiments [4,6-9]. The averaged relative rate of all spin

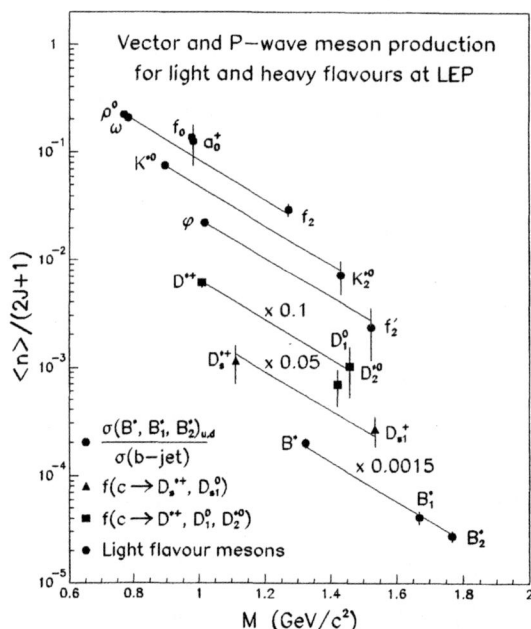

FIGURE 1. The mass dependence of direct production rates, $\langle n \rangle$, for light-flavour vector and tensor mesons, fragmentation fractions $f(c \to D^{*+}, D_1^0, D_2^{*0})$ and $f(c \to D_s^{*+}, D_{s1}^+)$ for charm mesons, and ratios $\sigma(B^*, B_1^*, B_2^*)_{u,d}/\sigma(b-jet)$ for bottom mesons, all divided by the spin counting factor $2J+1$. The data points for charm, strange charm and bottom mesons have been scaled by factors of 0.1, 0.05 and 0.0015 respectively and shifted by 1 GeV/c^2 for charm and by 4 GeV/c^2 for bottom mesons for clarity. The solid lines represent the result of the fit of the data to six exponentials with the *same* slope.

states corresponding to the observed signal amounts to $\sigma_{B_{u,d}^{**}}/\sigma_{b-jet} = 0.258 \pm 0.027$. In the framework of HQS and under constraints common to several theoretical models, attempts have been made [7,9,10] to determine the masses and widths of at least one of these states. The obtained masses are shown (as bold numbers) in Table 1, together with the masses of the states used as constraints in the fits. If the mass dependence of production rates on one spin projection for bottom mesons is similar to the one observed for light-flavour and charm mesons, the production rates on one spin projection for the states with very close masses must be almost the same. For the states with the different masses, such as the B_1^* and B_2^*, violation of the spin counting can be accounted for by the coefficient $\varepsilon = 5B_1^*/(3B_2^*) = \exp(b(M_{B_2^*} - M_{B_1^*}))$ where b is the slope parameter in the exponential mass dependence assumed to be the same as for the light flavours. Thus the relative production rates of the four different states have been set according to the proportion $B_0 : B_1^* : B_1 : B_2^* = \varepsilon : 3\varepsilon : 3 : 5$. With the B_1^* and B_2^* masses from L3 [7], this yields $\sigma_{B_1^*(u,d)}/\sigma_{b-jet} = 0.083 \pm 0.012$ and $\sigma_{B_2^*(u,d)}/\sigma_{b-jet} = 0.092 \pm 0.010$. For determining the rate of the promptly produced $B^*(ud)$, the decays of P-wave mesons

TABLE 1. Masses of the P-wave bottom mesons from the LEP experiments.

Meson	$J^P_{j_q}$	ALEPH [9]	L3 [7]	OPAL [10]	Br($B^*\pi$)
B_0	$0^+_{1/2}$	$5627 \pm^8_{11} \pm^6_4$	$5658 \pm 10 \pm 13$	$\mathbf{5839 \pm^{13}_{14} \pm^{34}_{42}}$	0.0
B^*_1	$1^+_{1/2}$	$5639 \pm^8_{11} \pm^6_4$	$\mathbf{5670 \pm 10 \pm 13}$	$5859 \pm^{13}_{14} \pm^{34}_{42}$	1.0
B_1	$1^+_{3/2}$	$5727 \pm^8_{11} \pm^6_4$	$5756 \pm 5 \pm 6$	$5738 \pm^5_6 \pm 7$	1.0
B^*_2	$2^+_{3/2}$	$\mathbf{5739 \pm^8_{11} \pm^6_4}$	$\mathbf{5768 \pm 5 \pm 6}$	$5750 \pm^5_6 \pm 7$	0.5

into $B^*\pi$ were taken into account using the branching fractions in Table 1. This gives $Br(B_J \to B^*\pi) = 0.714 \pm 0.032 \pm 0.068$, where additional systematic error accounts to the half of difference between this value and $Br(B_J \to B^*\pi(X))$ found by OPAL [10]. Thus we obtained: $\sigma_{B^*(u,d)prompt}/\sigma_{b-jet} = 0.396 \pm 0.042$. The resulting B^*, B^*_1 and B^*_2 relative production rates weighted with the $1/(2J+1)$ factor are shown in Fig. 1 at the B^*_1 and B^*_2 masses measured by L3. The fit of the data to six exponentials with different normalization parameters, but the *same* slope $b = 4.17 \pm 0.21$ $((GeV/c^2)^{-1}$ describes the data well (solid lines in Fig. 1). The slope parameter is practically identical to the value obtained for the light-flavour mesons. Thus the mass dependence of the production rates on one spin projection is indeed very similar for the light-flavour, charm and bottom mesons. For the B^*_1 and B^*_2, the same mass dependence as for the light-flavour mesons is, of course, simply imposed by the coefficient ε. The important result is that the rate of the promptly produced B^* follows the same mass dependence.

The observed universality of the mass dependence allows to relate not only the production rates of the mesons with different flavours, but as well their masses. Fig. 1 suggests the following mass rescaling formulae

$$B_i = B^*_2 - (B^*_2 - B^*)\frac{T - P_i}{T - V}, \quad D_i = D^*_2 - (D^*_2 - D^*)\frac{T - P_i}{T - V} \quad (1)$$

where V, T and P_i are the masses of the vector, tensor and P-wave (with $J^P = 1^+$ or 0^+) light-flavour mesons corresponding to the masses of their respective charm, D^*, D^*_2 and D_i, and bottom, B^*, B^*_2 and B_i, partners. From equations (1), with the K^{*0}, K^{*0}_2, $K_1(1402)$, D^*, D^*_2 and B^* masses from PDG [11] and the B^*_2 mass, 5752 ± 15 MeV/c^2, taken as the average of the masses determined by ALEPH, L3 and OPAL (Table 1), with the error equal to a half of spread between the ALEPH and L3 values, one obtains: $M_{B_1} = 5728 \pm 16$ MeV/c^2 and $M_{D^0_1} = 2433 \pm 6$ MeV/c^2. With the $K_1(1273)$ instead of the $K_1(1402)$, equations (1) give: $M_{B^*_1} = 5625 \pm 16$ MeV/c^2 and $M_{D^{*0}_1} = 2325 \pm 6$ MeV/c^2. The obtained B_1 mass agrees within errors with the averaged value of this mass, 5740 ± 15 MeV/c^2, from ALEPH, L3 and OPAL. The obtained B^*_1 mass is consistent within 2 standard deviations with $M_{B^*_1} = 5670 \pm 16$ MeV/c^2 from L3, but significantly smaller than the OPAL value. This supports the models suggesting that the masses of the broad B^*_1 and B_0 states lie below the B^*_2 mass. The obtained D^0_1 mass is in good agreement with the PDG value of 2422.2 ± 1.8 MeV/c^2 [11]. The D^{*0}_1 mass represents the prediction for the mass of the broad, not yet established state. It is smaller than the measured D^0_1 mass by 97 ± 6 MeV/c^2. Notice that these results identify the $K_1(1402)$ and $K_1(1273)$ as the $1^+(3/2)$ and $1^+(1/2)$ levels, respectively. This is consistent with their masses provided that the states belonging to the $j_q = 3/2$ doublets are heavier for all flavours but not with their widths.

Due to the $K_2^*(1430)$ and $K_0^*(1430)$ mass degeneracy, an attempt to apply equations (1) for determination of the B_0 and D_0 masses gives $M(B_0) \approx M(B_2^*)$ and $M(D_0) \approx M(D_2^*)$. The last relation agrees with the CLEO preliminary result [12] on the mass of the broad $D_J(1/2)$ state. However, this is not consistent with the calculated values of the B_1^* and D_1^{*0} masses given above, as well as with the B_1^* mass determined by L3 and constrained by ALEPH, if the mass difference between the B_1^* and B_0, and the D_1^{*0} and D_0 is small as expected. In order to reproduce the B_0 masses from L3, the K_0^* mass must be 1299 ± 22 MeV/c^2. Such value of the K_0^* mass is consistent with identification of the $K_1(1273)$ as the $1^+(1/2)$ level, since the $0^+(1/2)$ and $1^+(1/2)$ levels should be nearly mass degenerate in all heavy-light systems. A smaller K_0^* mass is also required in the description of the light-flavour P-wave mesons in the nonrelativistic quark model [13]. As suggested in [13], this can be explained if the observable $K_0^*(1430)$ mass is replaced by its "bare" $q\bar{q}$ mass corresponding to the K-matrix pole. In the K-matrix analysis of the 0^{++}-wave [14], the "bare" K_0^* mass, in one of the two possible solutions, is 1220 ± 70 MeV/c^2, consistent within errors with the value of 1299 ± 22 MeV/c^2.

In conclusion, we have shown that the mass dependence of the production rates for six families of mesons: the directly produced vector and tensor light-flavour mesons, the vector and P-wave charm, strange charm and bottom mesons obtained from results of the LEP experiments is very similar. This allowed to relate not only the production rates of mesons with different flavours, but as well their masses, thus showing interesting connection between hadron production properties and their masses.

REFERENCES

1. ALEPH Collab., Barate R. et al., Phys. Rep. **294** 1-165 (1998). DELPHI Collab., Abreu P. et al., Z. Phys. **C73** 61-72 (1996), Phys. Lett. **B449** 364-382 (1999). L3 Collab., Acciarri M. et al., Phys. Lett. **B393** 465-476 (1997). OPAL Collab., Akers R. et al., Z. Phys. **C68** 1-12 (1995); Ackerstaff K. et al., Eur. Phys. J. **C4** 19-28 (1998), **C5** 411-432 (1998).
2. Sjöstrand T., Comp. Phys. Comm. **82** 74-89 (1994).
3. ALEPH Collab., Barate R. et al., Eur. Phys. J. **C16** 597-611 (2000). DELPHI Collab., Abreu P. et al., Eur. Phys. J. **C12** 209-224 (2000). OPAL Collab., Ackerstaff K. et al., Z. Phys. **C76** 425-440 (1997), Eur. Phys. J. **C1** 439-459 (1998).
4. ALEPH Collab., Buskulic D. et al., Z. Phys. **C69** 393-404 (1996).
5. DELPHI Collab., Abreu P. et al., Z. Phys. **C68** 353-362 (1995). L3 Collab., Acciarri M. et al., Phys. Lett. **B345** 589-597 (1995). OPAL Collab., Ackerstaff K. et al., Z. Phys. **C74** 413-422 (1997).
6. DELPHI Collab., Abreu P. et al., Phys. Lett. **B345** 598-608 (1995).
7. L3 Collab., Acciarri M. et al., Phys. Lett. **B465** 323-334 (1999).
8. OPAL Collab., Akers K. et al., Z. Phys. **C66** 19-30 (1995).
9. ALEPH Collab., Barate R. et al., Phys. Lett. **B425** 215-226 (1998).
10. OPAL Collab., G. Abbiendi et al., CERN-EP/2000-125 (submitted to Eur. Phys. J. C).
11. Particle Data Group, Eur. Phys. J. **C15** 1-878 (2000).
12. CLEO Collab., Anderson S. et al., Nucl. Phys. **A663** 647-650 (2000).
13. Chliapnikov P.V., Phys. Lett. **B496** 129-136 (2000).
14. Anisovich A.V. and Sarantsev A.V., Phys. Lett. **B413** 137-146 (1997).

Do strange quarks in Z^0 decays and kaon valence $\bar{s}$ quark in K^+p reactions fragment differently?

P.V. Chliapnikov

Institute for High Energy Physics, Protvino, 142284, Russia

Abstract. The ratios of the production rates $K^{*0}(892)/K$, ϕ/K, ρ^0/π, ω/π, $\Delta^{++}(1232)/p$, $\Sigma^{*+}(1385)/\Lambda$, Ξ^-/Λ and their x_p-dependence obtained from results of the LEP and SLD experiments in Z^0 hadronic decays are analyzed. The corresponding ratios for the promptly produced mesons are estimated at $x_p \to 1$. A comparison of the LEP results with those from the Mirabelle and BEBC bubble chamber K^+p experiments at 32 and 70 GeV/c shows striking similarity in fragmentation of the $\bar{s}$ valence quark of the incident K^+ and strange quarks produced in Z^0 decays.

The K^0, $K^{*0}(892)$, ϕ, $\bar{\Lambda}$, $\overline{\Sigma^*}(1385)$ and $\overline{\Xi^-}$ in the Mirabelle [1] and BEBC [2] K^+p experiments at 32 and 70 GeV/c have been shown to be dominantly produced on the $\bar{s}$ valence quark of the incident K^+. This offered good possibilities to trace the flow of the incident strange valence flavour among the reaction debris and allowed to obtain the reliable estimates of the vector-to-pseudoscalar (V/P) and decuplet-to-octet (D/O) ratios for the promptly produced hadrons. Therefore an interesting test of the LEP results, as well as model predictions, on V/P and D/O ratios for the promptly produced hadrons is feasible using the results of K^+p experiments.

In this report, the ratios of the production rates[1] $K^*(892)/K$, ϕ/K, ρ^0/π, ω/π, $\Delta^{++}(1232)/p$, $\Sigma^*(1385)/\Lambda$ and Ξ^-/Λ obtained from results of the LEP [3-5] and SLD [6] experiments are analyzed. The values of these ratios for the promptly produced hadrons are estimated by studying their x_p-dependence and comparing the fragmentation of strange quarks in Z^0 decays and K^+p reactions. The differential cross-sections, $1/\sigma_h \cdot d\sigma/dx_p$, for the resonances and Ξ^- were taken directly from the LEP experiments. For the K, π, p or Λ they were taken either from the same experiments or from combined x_p-spectra of corresponding particles measured at LEP [3-5] and SLD [6]. These spectra were fitted by a sum of two exponentials, and cross-sections in the corresponding x_p-intervals were calculated using the results of these fits.

The $K^{*0}(892)/(3K^+)$ ratios measured in Z^0 hadronic decays [3-6] are presented as a function of x_p in Fig. 1a. They increase with x_p, presumably approaching the value for the promptly produced mesons at $x_p \to 1$. The corresponding ratios for the promptly produced mesons from the Mirabelle and BEBC K^+p experiments are also shown in Fig. 1a. They are in excellent agreement with a trend of the LEP data for $x_p \to 1$ and predictions of the JETSET [7] and Pei [8] models (also shown in Fig. 1a) for the promptly produced mesons at LEP. The simplest spin models of fragmentation, such

---

[1] The charge conjugates and antiparticles are not included into definition of the rates.

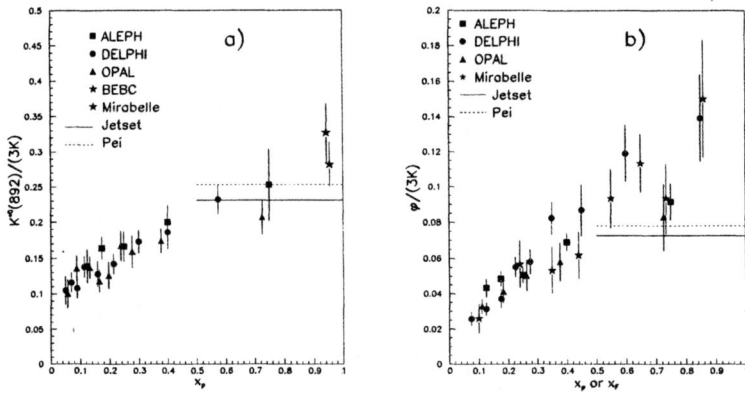

FIGURE 1. The $K^{*0}(892)/(3K^+)$ (a) and $\phi/(3K^+)$ (b) ratios obtained from the LEP experiments as a function of x_p; the $K^{*0}(892)/(3K^0)$ ratios for the promptly produced particles from the Mirabelle and BEBC K^+p experiments (a); the $\phi/(3K_s^0)$ ratio as a function of Feynman-x_F from Mirabelle (b). The JETSET and Pei model predictions for the promptly produced particles at LEP are also shown. Here and in Fig. 2 some data points were slightly shifted to avoid overlap.

as [9], based on quark combinatorics (QC) predict $V/P = 3$ and $D/O = 2$. Recently Anisovich et al. [10] compared these predictions with the ALEPH data [3] and arrived to surprising conclusion that the experimental results are consistent with $V/P = 3$. It is surprising because the results of the LEP experiments on hadron production agree well with the JETSET model, where the probability to produce a vector meson is controlled by the parameter $V/(V+P)$ which is much smaller than 0.75, the value expected from QC. Contrary to [10], we see that QC is ruled out by the LEP, Mirabelle and BEBC data. The experimental values at the largest x_p are a factor of 3 lower than the prediction, $K^*(892)/(3K) = 1$, and deviate from it by at least 11 standard deviations.

The $\phi/(3K^+)$ ratios [3-6] are presented as a function of x_p in Fig. 1b. They exhibit a clear rise with increasing x_p, similar to the one in Fig. 1a. The ϕ/K_s^0 ratio as a function of Feynman variable x_F for $x_F > 0$ from Mirabelle is also shown in Fig. 1b. It is strikingly similar to the behaviour of the LEP data. The JETSET and Pei model predictions are somewhat lower than the experimental data at large x_p, but consistent with them within one-two errors. QC [9, 10] predicts $\phi/(3K) = \lambda$, where λ is the strangeness suppression parameter. For usually accepted value $\lambda \approx 0.3$, it is a factor of more than two higher than the experimental values at large x_p and deviates from the DELPHI value at $x_p = 0.85$ by 7 standard deviations or even more for $\lambda \approx 0.5$ suggested in [10].

The $\rho^0/(3\pi^+)$ and $\omega/(3\pi)$ ratios obtained from [3-6] are presented as a function of x_p in Fig. 2a. For $x_p < 0.5$, all results are consistent within errors. For $x_p > 0.5$, the $\rho^0/(3\pi^+)$ ratio measured by ALEPH is higher than other data points. The $\rho^0/(3\pi^-)$ ratio as a function of Feynman-x_F from Mirabelle (convoluted around $x_F = 0$) is well consistent with the LEP results for $x_p < 0.7$ and lies between the ALEPH and DELPHI data points at $x_p = 0.7$-1.0. Thus, in spite of some inconsistency of the LEP results at the largest

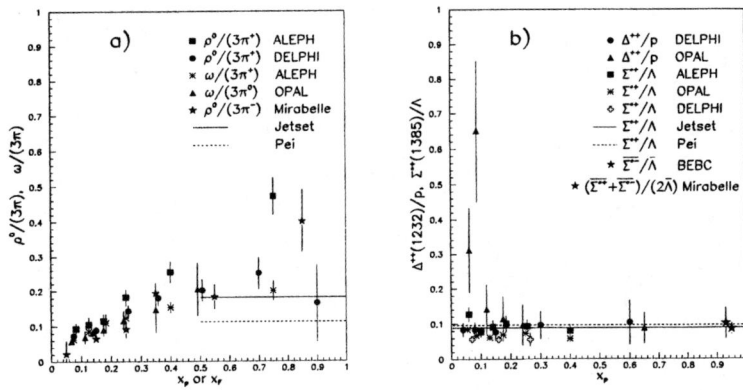

FIGURE 2. The $\rho^0/(3\pi^+)$, $\omega/(3\pi^+)$ and $\omega/(3\pi^0)$ ratios as a function of x_p from the LEP experiments, together with the $\rho^0/(3\pi^-)$ ratio as a function of Feynman-x_F from the Mirabelle experiment and $\rho^0/(3\pi^+)$ ratios for the promptly produced mesons predicted by the JETSET and Pei models (a). The $\Delta^{++}(1232)/p$ and $\Sigma^{*+}(1385)/\Lambda$ ratios as a function of x_p from the LEP experiments, $\Sigma^{*+}(1385)/\Lambda$ ratios of the total rates predicted by the JETSET and Pei models and ratios of the total rates $(\overline{\Sigma^{*+}} + \overline{\Sigma^{*-}})/(2\bar{\Lambda})$ from Mirabelle and $\overline{\Sigma^{*-}}/\bar{\Lambda}$ from BEBC (b).

x_p, the general tendency of the LEP and Mirabelle data in Fig. 2a is similar to the one observed in Fig. 1. However, in this case the Pei model prediction for the ratios of promptly produced mesons at LEP appears to be underestimated. The QC prediction, $\rho^0/(3\pi) = 1$, is not consistent with the data, contrary to conclusion of ref. [10]. It is again a factor of 2 to 5 higher than the experimental values at large x_p. The DELPHI and ALEPH data points at $x_p = 0.6$-0.8 and $x_p = 0.5$-1.0 deviate from this prediction by 16 and respectively 11 standard deviations.

The V/P values at $x_p \to 1$ could be reconciled with the QC predictions in case of strong spin alignment of vector mesons. However, the values of ρ_{00} for the ρ^0, $\rho^\pm$ and ω measured at LEP [4,5] were found compatible, over the entire x_p range, with 1/3 corresponding to a statistical mix of helicity -1, 0 and +1 states. Some preference for occupation of the helicity zero state was possibly observed only for the K^{*0} and ϕ at large x_p [4,5]. Thus this can not explain the failure of the QC model.

The $\Delta^{++}(1232)/p$ and $\Sigma^{*+}(1385)/\Lambda$ ratios based on the results of the LEP experiments (Fig. 2b) are essentially flat as a function of x_p as expected, apart from two suspicious $\Delta^{++}(1232)/p$ OPAL data points at the smallest x_p. The ratios of the total rates $(\overline{\Sigma^{*+}} + \overline{\Sigma^{*-}})/(2\bar{\Lambda})$ and $\overline{\Sigma^{*-}}/\bar{\Lambda}$ from Mirabelle and BEBC as well as the JETSET and Pei model predictions for the ratio Σ^{*+}/Λ of the total rates are well consistent with the LEP data. With the Ξ^- and Λ total rates averaged over results of all LEP experiments, one gets $(\Xi^-/\Lambda)_{tot} = 0.072 \pm 0.004$. From the $\overline{\Xi^-}$ and $\bar{\Lambda}$ total rates in Mirabelle and BEBC, one obtains respectively $(\overline{\Xi^-}/\bar{\Lambda})_{tot} = 0.086 \pm 0.021$ and 0.14 ± 0.04, again in agreement with the LEP value. The JETSET and Pei model predictions, 0.090 and 0.072, respectively, agree also with the LEP, Mirabelle and BEBC results. The $\overline{\Sigma^{*+}}$, $\overline{\Sigma^{*-}}$ and $\overline{\Xi^-}$ in the Mirabelle K^+p experiment can presumably be safely considered as promptly produced. The cross-section of the promptly produced $\bar{\Lambda}$ in Mirabelle can be estimated from the $\bar{\Lambda}$ total rate after subtraction of $\bar{\Lambda}$ from the $\overline{\Sigma^{*+}}$, $\overline{\Sigma^{*-}}$ and $\overline{\Sigma^{*0}}$ decays and small fraction of the centrally produced $\bar{\Lambda}$. This gives $[\overline{\Xi^-}/(\bar{\Lambda}+\overline{\Sigma^0})]_{prompt} = 0.14 \pm 0.04$ and $[(\overline{\Sigma^{*-}}+\overline{\Sigma^{*+}})/2(\bar{\Lambda}+\overline{\Sigma^0})]_{prompt} = 0.14 \pm 0.03$ where we took into account that one can not separate the prompt $\bar{\Lambda}$ and $\overline{\Sigma^0}$. The JETSET and Pei models predict $[\Xi^-/(\Lambda+\Sigma^0)]_{prompt} = 0.11$ and 0.14, respectively, in good agreement with the Mirabelle result. For the $[\Sigma^{*+}/(\Lambda+\Sigma^0)]_{prompt}$, the same models predict 0.13 and 0.41, respectively. The JETSET estimate is in good agreement with the Mirabelle result. The Pei model prediction is significantly higher thus indicating that the fractions of promptly produced octet baryons in this model might be underestimated. The QC [9] predicts $[\Sigma^{*+}/(\Lambda+\Sigma^0)]_{prompt} = 1$, a factor of 7 higher than the Mirabelle value deviating from it by 29 standard deviations.

In conclusion, we have shown that a study of the vector-to-pseudoscalar ratios at LEP as a function of x_p allows to obtain estimates of these ratios for promptly produced mesons at $x_p \to 1$. These estimates, as well as the $\overline{\Xi^-}/(\bar{\Lambda}+\overline{\Sigma^0})$ and $(\overline{\Sigma^{*-}}+\overline{\Sigma^{*+}})/2(\bar{\Lambda}+\overline{\Sigma^0})$ ratios for the total rates, have been found in very good agreement with the results of the Mirabelle and BEBC K^+p experiments. This shows that the fragmentation properties of the $\bar{s}$ valence quark of the incident kaon and strange quarks produced in e^+e^- annihilation are very similar. This interesting experimental observation is not unexpected. In the Lund String model [11], a soft hadronic collision is considered as a colour separation mechanism whereby valence quarks of incident meson act as borders of colour strings, analogous to the $q-\bar{q}$ field in e^+e^- annihilation. Therefore it is not surprising that many features in fragmentation of the $\bar{s}$ valence quark of the incident K^+ and strange quarks in e^+e^- annihilation must be similar.

REFERENCES

1. Ajinenko I.V. et al., *Nucl. Phys.* **B176** 51-60 (1980), *Z. Phys.* **C23** 307-331 (1984); Chliapnikov P.V. et al., *Nucl. Phys.* **B176** 303-320 (1980), *Phys. Lett.* **B130** 432-438 (1983); Kniazev V.V. et al., *Yad. Fiz.* **40** 1460-1476 (1984); Garuchava Z. Sh., Uvarov V.A. and Chliapnikov P.V., *Pis'ma Zh. Eksp. Teor. Fiz.* **40** 121-124 (1988).
2. Barth M. et al., *Nucl. Phys.* **B191** 39-62 (1981), **B246** 431-461 (1984); De Wolf E.A. et al., *Z. Phys.* **C31** 13-19 (1986).
3. ALEPH Collab., Barate R. et al., *Phys. Rep.* **294** 1-165 (1998).
4. DELPHI Collab., Abreu P. et al., *Z. Phys.* **C67** (1995) 543-553, **C73** (1996) 61-72; *Phys. Lett.* **B361** 207-220 (1995), **B406** 271-286 (1997), **B449** 364-382 (1999); *Eur. Phys. J.* **C5** 585-620 (1998).
5. OPAL Collab., Akers R. et al., *Z. Phys.* **C63** 181-196 (1994), **C68** 1-12 (1995); Ackerstaff K. et al., *Z. Phys.* **C74** 437-449 (1997), *Phys. Lett.* **B412** 210-224 (1997), *Eur. Phys. J.* **C4** 19-28 (1998), **C5** 411-437 (1998); Alexander G. et al., *Phys. Lett.* **B358** 162-172 (1995), *Z. Phys.* **C73** 569-586 (1997); Abbiendi G. et al., *Eur. Phys. J.* **C16** 61-70 (2000).
6. SLD Collab., Abe K. et al., *Phys. Rev.* **D59** 052001 (1999).
7. Sjöstrand T., *Comp. Phys. Comm.* **82** 74-89 (1994).
8. Pei Yi-Jin, *Z. Phys.* **C72** 39-46 (1996).
9. Anisovich V.V. and Shekhter V.M., *Nucl. Phys.* **B55** 455-473 (1973).
10. Anisovich V.V., Nikonov V.A. and Nuiri J., *Yad. Fiz.* **64** 877-898 (2001), *Phys. Atom. Nucl.* **64** 812-833 (2001).
11. Andersson B., Gustafson G., Ingelman G. and Sjöstrand T., *Phys. Rep.* **C97** 31 (1983).

Determination of the strangeness content of light-flavour isoscalars from their production rates in hadronic Z decays at LEP

Vladimir A. Uvarov

Institute for High Energy Physics, RU-142284, Protvino, Russia

Abstract. A new phenomenological approach is suggested for determining the strangeness content of light-flavour isoscalars. This approach is based on phenomenological laws of hadron production related to the spin, isospin, strangeness content and mass of the particles. From the total production rates per hadronic Z decay of all light-flavour hadrons, measured so far at LEP, the values of the nonstrange-strange mixing angles are found to be $|\varphi_P| = 42.3° \pm 3.5°$, $|\varphi_V| = 10° \pm 8°$, $|\varphi_T| = 16° \pm 11°$ and $|\varphi_S| = 13° \pm 9°$. Our results on the η–η', ω–ϕ and f_2–f_2' isoscalar mixing are consistent with the present experimental evidence. The strangeness content obtained for the $f_0(980)$ scalar/isoscalar is not consistent with the values supported by recent model studies and is discussed further in the framework of our approach and the K-matrix analysis.

The quark contents of pseudoscalar (P), vector (V), tensor (T) and scalar (S) mesons have been discussed many times since the discovery of unitary SU(3)-flavour symmetry. This is still quite an interesting question because the quark contents of the lightest isoscalars differ from the predictions of the SU(3) quark model and are very important SU(3)-breaking hadronic parameters.

In terms of the $n\bar{n} = (u\bar{u} + d\bar{d})/\sqrt{2}$ and $s\bar{s}$ quark basis, the strangeness contents of the physical isoscalars are given by the *nonstrange-strange mixing angles* φ_P, φ_V, φ_T and φ_S. Assuming orthogonality of the isoscalar partners and no mixing with other states and glueballs, the flavour wave functions of the η–η' pseudoscalars are defined to be $\eta = n\bar{n} \cdot \cos\varphi_P - s\bar{s} \cdot \sin\varphi_P$ and $\eta' = n\bar{n} \cdot \sin\varphi_P + s\bar{s} \cdot \cos\varphi_P$. The flavour wave functions of the ω–ϕ vectors (f_2–f_2' tensors) are defined in a way analogous to the η–η' case, replacing $\eta \to \omega(f_2)$, $\eta' \to \phi(f_2')$ and $\varphi_P \to \varphi_V(\varphi_T)$. The flavour wave function of the $f_0(980)$ scalar is written as $f_0(980) = n\bar{n} \cdot \cos\varphi_S + s\bar{s} \cdot \sin\varphi_S$.

The values of these mixing angles have been estimated from different phenomenological and theoretical analyses (see Refs. [1-13] and references therein). Estimates of the mixing angle φ_P were obtained from the available world data on different decay processes. The corresponding average value was found to be $\varphi_P = 39.2° \pm 1.3°$ [2, 3]. For the mixing angles φ_V and φ_T, most theoretical and phenomenological analyses (see, e.g., Refs. [4-6, 13]) predict values which are close to the "ideal" mixing: $\varphi_V \simeq +3.4°$ and $\varphi_T \simeq -7.3°$. The interpretation of the $f_0(980)$ scalar is one of the most controversial ones in meson spectroscopy [7]. The question is whether the $f_0(980)$ consists mostly of $n\bar{n}$ or of $s\bar{s}$ states. The phenomenological analyses of the experimental data on different decay processes [9-12] favour the $s\bar{s}$ dominance of the $f_0(980)$.

Recently the strangeness contents of light-flavour isoscalars have been obtained for the first time [14] from their total production rates per hadronic Z decay at LEP. Here only some key points and the final results of that analysis [14] are discussed.

It has been shown [15-19] that the total production rates per hadronic Z decay ($\langle n \rangle$) of all light-flavour mesons (M) and baryons (B), measured so far at LEP, follow phenomenological laws related to the spin (J), isospin (I), strangeness content (k) and mass (m) of the particles. These regularities can be combined into one empirical formula:

$$\langle n \rangle = A \cdot \beta_H \cdot (2J+1) \cdot \gamma^k \cdot \exp[-b_H(m/m_0)^{N_H}], \qquad (1)$$

where $H = M$ or B, $m_0 = 1$ GeV, γ is the strangeness suppression factor with a value of $\gamma \simeq 0.5$ for all hadrons, k is the number of s and $\bar{s}$ quarks in the hadron, and b_H is the slope of the m dependence. The values of the power N_H and of the coefficient β_H are different for mesons and baryons: $N_M = 1$, $N_B = 2$ and $\beta_M = 1$, $\beta_B = 4/(C_{\pi/p}\lambda_{QS})$, where $\lambda_{QS} = (2J+1)(2I+1)$ can be interpreted as a fermion suppression factor originating from the quantum statistics properties of bosons and fermions, and $C_{\pi/p}$ is the π/p ratio at the zero mass limit with a value of $C_{\pi/p} \simeq 3$ which could be expected from quark combinatorics. The slope b_M has two values: one for vector, tensor and scalar mesons and the other for pseudoscalar mesons. In spite of this "splitting", Eq. (1) assumes the validity of the relation $V/P = 3$ at the zero mass limit. This difference in slopes can probably be explained by the influence of the spin-spin interaction between the quarks of the meson. However, there is no influence of the spin-orbital interaction of the quarks.

The *purpose of this analysis* [14] is to determine the mixing angles of the η–η', ω–ϕ, f_2–f_2' and $f_0(980)$ isoscalars from the *simultaneous* fit of Eq. (1) to the total production rates per hadronic Z decay of all light-flavour hadrons measured so far at LEP, assuming that the mixing angle φ and the strangeness contents k_1 and k_2 of the isoscalar partners are related by: $k_1 = 2\sin^2\varphi$ and $k_2 = 2\cos^2\varphi$. The strangeness contents of baryons and $I \neq 0$ mesons are taken from the predictions of the SU(3) quark model.

The total production rates *per isospin state* $\langle n \rangle$, used in the fit, were obtained by averaging the total production rates $\bar{n}$ of hadrons belonging to the same isomultiplet. The rates $\bar{n}$ themselves were obtained for at least one state of a given isomultiplet as a weighted-average of the measurements of the four LEP experiments (see details in Ref. [14] and references therein).

Firstly we tested the sensitivity of Eq. (1) to the values of the power N_H. In the test fit the following parameters were fixed: $\varphi_P = 44.7°$, $\varphi_V = 3.7°$, $\varphi_T = -7.3°$ predicted by the quadratic Gell-Mann–Okubo mass formula [13], $\varphi_S = 0$ suggested as an ad hoc value in [15], and $C_{\pi/p} = 3$ predicted by quark combinatorics for production of direct or massless particles. The values obtained, $N_M = 1.04\pm0.06$ and $N_B = 2.07\pm0.05$, strongly suggest the use of the fixed values of $N_M = 1$ and $N_B = 2$ in our approach.

In the final fit only the values of $N_M = 1$ and $N_B = 2$ are fixed, all other parameters are free. This fit (with χ^2/dof = 8.1/12) is illustrated in Fig. 1, where the three curves are the result of the fit for baryons, for mesons with net spin $S=0$ (pseudoscalars), and for mesons with net spin $S=1$ (vectors, tensors and scalars). The values of the strangeness suppression factor and of the π/p ratio at the zero mass limit are found to be $\gamma = 0.50\pm0.02$ and $C_{\pi/p} = 2.8\pm0.2$, in good agreement with our previous results [17-19]. The values of the isoscalar mixing angles are found to be $|\varphi_P| = 42.3°\pm3.5°$, $|\varphi_V| = 10°\pm8°$, $|\varphi_T| = 16°\pm11°$ and $|\varphi_S| = 13°\pm9°$. Quite remarkably, our values

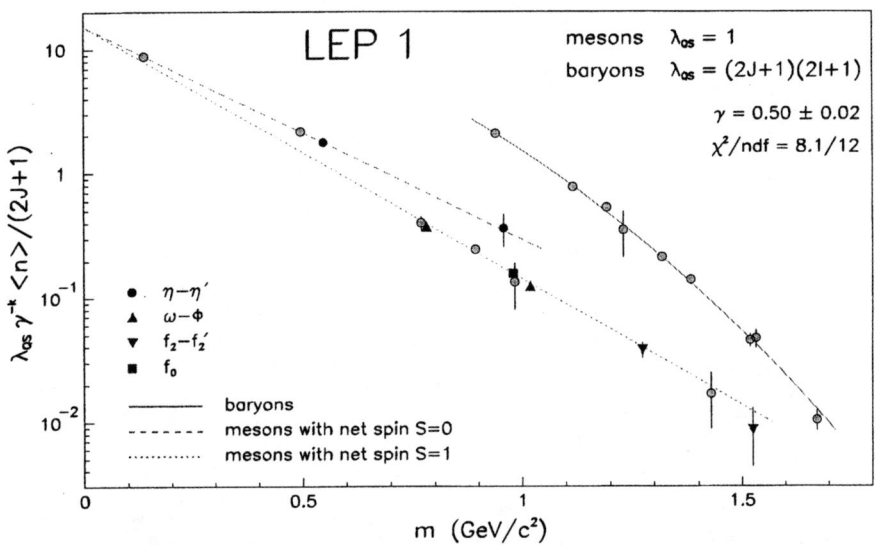

FIGURE 1. Mass dependences of the total production rates per spin and isospin state for hadrons in hadronic Z decays weighted by a factor $\lambda_{QS}\gamma^{-k}$, where $\lambda_{QS} = 1$ for mesons and $\lambda_{QS} = (2J+1)(2I+1)$ for baryons. The references for all data points can be found in Ref. [14].

of the η–η', ω–ϕ and f_2–f_2' mixing angles are compatible within the errors with the predictions of theoretical and phenomenological analyses (see, e.g., Refs. [2-6,13]). However, our value of the mixing angle φ_S is not consistent with the results of recent phenomenological studies which favour the $s\bar{s}$ dominance of the $f_0(980)$ scalar [9-12].

The disagreement obtained for the $f_0(980)$ can probably be related to a question which is still open. This is whether the $f_0(980)$ belongs to the scalar $q\bar{q}$ nonet 1^3P_0 or whether it should be considered as an exotic state. The scalar nonet classification can be performed in terms of so-called "bare states" (the K-matrix poles) [8, 20]. In this way the scalar/isoscalars have been found to be $f_0^{bare}(720\pm100)$ and $f_0^{bare}(1260\pm30)$ with the mixing angle $\varphi_S[f_0^{bare}(720)] = -70°^{+5°}_{-16°}$.

As was observed by Montanet [21], the $f_0(980)$ data point (Fig. 2) shifted from the real mass to the "bare mass" (720 MeV) is close to the line corresponding to the mesons with $k = 2$. This interesting relation between our phenomenology and the K-matrix analysis can probably be considered as a *speculative* argument for re-determining the scalar mixing angle, replacing in the fit the real mass of the $f_0(980)$ by the "bare" one. Such a fit gives a scalar mixing angle of $|\varphi_S^{bare}| = 73°\pm7°\pm24°$, where the second error is due to the uncertainty (±100 MeV) of the "bare mass". This value is well consistent with the predictions of recent phenomenological studies [8, 10-12].

In conclusion, a new phenomenological approach has been suggested for determining the mixing angles of light-flavour isoscalars. For the first time, the total production rates per hadronic Z decay of all light-flavour hadrons, measured so far at LEP, have been used for this purpose. The following values of the pseudoscalar, vector and tensor mixing angles have been obtained: $|\varphi_P| = 42.3°\pm3.5°$, $|\varphi_V| = 10°\pm8°$ and $|\varphi_T| = 16°\pm11°$, in

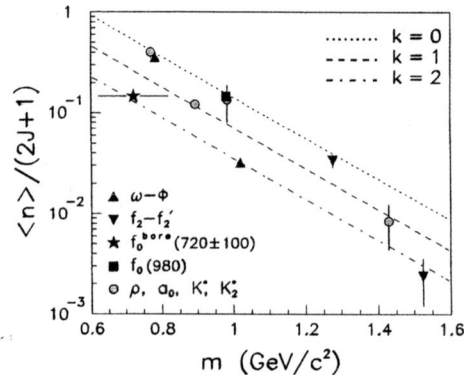

FIGURE 2. Total production rate per spin and isospin state as a function of m for vector, tensor and scalar mesons in hadronic Z decays. Curves are the result of the final fit with $k = 0$, $k = 1$ and $k = 2$.

good agreement with the present experimental evidence. Also two values of the scalar mixing angle have been obtained: $|\varphi_S| = 13°\pm9°$ if the $f_0(980)$ is taken with the real mass, but $|\varphi_S^{bare}| = 73°\pm7°\pm24°$ if the $f_0(980)$ is taken with the "bare mass" of the $f_0^{bare}(720\pm100)$ state. Only the second value is consistent with recent phenomenological analyses. *If their conclusions are correct*, it means that in the framework of our approach the total production rates of scalar mesons are probably given by the "bare masses".

REFERENCES

1. Gilman, F.J., Kauffman, R., *Phys. Rev.* D **36**, 2761 (1987).
2. Bramon, A., Escribano, R., Scadron, M.D., *Eur. Phys. J.* C **7**, 271 (1999).
3. Feldmann, Th., *Int. J. Mod. Phys.* A **15**, 159 (2000).
4. Jones, H.F., Scadron, M.D., *Nucl. Phys.* B **155**, 409 (1979).
5. Dillon, G., Morpurgo, G., *Z. Phys.* C **64**, 467 (1994).
6. Li, D.M., Yu, H., Shen, Q.-X., *J. Phys.* G **27**, 807 (2001).
7. Montanet, L., *Nucl. Phys.* B *(Proc. Suppl.)* **86**, 381 (2000).
8. Anisovich, V.V., et al., *Phys. Atom. Nucl.* **63**, 1410 (2000).
9. Törnqvist, N.A., Roos, M., *Phys. Rev. Lett.* **76**, 1575 (1996).
10. Delbourgo, R., Liu, D., Scadron, M.D., *Phys. Lett.* B **446**, 332 (1999).
11. Anisovich, V.V., Montanet, L., Nikonov, V.N., *Phys. Lett.* B **480**, 19 (2000).
12. van Beveren, E., Rupp, G., Scadron, M.D., *Phys. Lett.* B **495**, 300 (2000).
13. Particle Data Group, Groom, D.E., et al., *Eur. Phys. J.* C **15**, 1 (2000).
14. Uvarov, V., *Phys. Lett.* B **511**, 136 (2001).
15. Uvarov, V., in *Proc. of the 15th Int. Conf. on Particles and Nuclei, Uppsala, 1999*, edited by G. Fäldt et al., *Nucl. Phys.* A **663**, 633 (2000).
16. DELPHI Collaboration, Abreu, P., et al., *Phys. Lett.* B **475**, 429 (2000).
17. Uvarov, V., *Phys. Lett.* B **482**, 10 (2000).
18. Uvarov, V., in *Proc. of the 35th Rencontre de Moriond on QCD and High Energy Hadronic Interactions, Les Arcs, 2000*, hep-ex/0011058.
19. Uvarov, V., in *Proc. of the 30th Int. Symp. on Multiparticle Dynamics, Tihany, 2000*, edited by T. Csörgő et al., World Scientific, Singapore, 2001, p. 190.
20. Anisovich, V.V., Prokoshkin, Yu.D., Sarantsev, A.V., *Phys. Lett.* B **389**, 388 (1996).
21. Montanet, L., talk given at the *Gribov-70 Workshop*, Orsay, 2000.

Measurements of the Vector and Tensor Analyzing Powers of the Inelastic Scattering of Deuterons on Nuclei in the Vicinity of Baryonic Resonances Excitation

L.S. Azghirey*, V.P. Ladygin*, S.V. Afanasiev*, V.V. Arkhipov*,
V.K. Bondarev*, G. Filipov[†], A.Yu. Isupov*, V.I. Ivanov*,
A.A. Kartamyshev**, V.A. Kashirin*, A.N. Khrenov*,
V.I. Kolesnikov*, V.A. Kuznetsov*, N.B. Ladygina*,
A.G. Litvinenko*, S.G. Reznikov*, P.A. Rukoyatkin*,
A.Yu. Semenov*, I.A. Semenova*, G.D. Stoletov*, V.N. Zhmyrov*
and L.S. Zolin*

*JINR, 141980 Dubna, Moscow Region, Russia
[†]Institute of Nuclear Research and Nuclear Energy, 1784 Sofia, Bulgaria
**Russian Scientific Center "Kurchatov Institute", 123182 Moscow, Russia

Abstract. Vector A_y and tensor A_{yy} analyzing powers of the inelastic scattering of deuterons with the momenta of 4.5 GeV/c on ^9Be and 9 GeV/c on ^1H and ^{12}C at an angle of ~80 mr in the vicinity of the excitation of baryonic resonances with masses between 1200 and 2200 MeV/c^2 have been measured. The values of parameter A_y suggest that the role of the spin-dependent part of the $NN \to NN^*$ elementary processes is significant. The new A_{yy} data are in good agreement with the previous data at 4.5, 5.5 and 9 GeV/c, obtained at a zero angle, in an overlapping region of t. The results of the experiments are compared with the calculations within the PWIA and ω-meson exchange model. The data obtained show that polarization observables of (d,d') reaction in the vicinity of the excitation of baryonic resonances may be used to study their properties.

Inelastic scattering of relativistic deuterons on nuclei, $A(d,d')X$, have extensively been studied in the last years [1]-[4]. When the coherent interaction between deuteron and nucleus involves high momentum transfers one may expect it to be sensitive to the deuteron structure and, possibly, to the manifestation of the non-nucleonic degrees of freedom. On the other hand, as the deuteron is an isoscalar probe, inelastic scattering of deuterons is selective to the isospin of the unobserved system X, which is bound to be equal to the isospin of target A. Therefore inelastic scattering of deuterons on hydrogen can be used to obtain information on the excitation of barionic resonances.

The experiments on the inelastic deuteron scattering have been made using a polarized deuteron beam of Dubna Synchrophasotron and the SPHERE setup described elsewhere [4]. The polarized deuterons were produced by the ion source POLARIS [5]. Along with the inelastically scattered deuterons, the apparatus detected the protons originated from deuteron breakup. The time-of-flight (TOF) information with a base line ~ 34 m was used for particle identification in the off-line analysis.

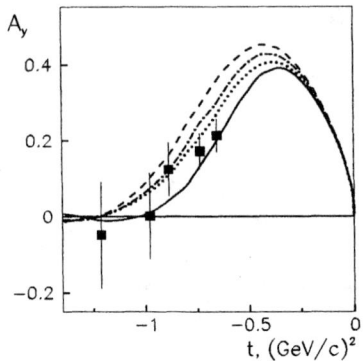

FIGURE 1. Vector analyzing power A_y of the inelastic scattering of 9 GeV/c deuterons on ^{12}C at an angle of 85 mr (full squares). The solid, dashed, dash-dotted and dotted curves are calculated in PWIA using DWFs for Paris, Bonn A, B and C potentials of NN-scattering, respectively.

Here we report data on A_y and A_{yy} of (d,d') reaction on ^1H and ^{12}C at the incident deuteron momentum of 9 GeV/c and at an detection angle of 85 mr (corresponding to the undetected system mass of $M_X \sim 2.2$ GeV/c^2) and similar data for beryllium target at 4.5 GeV/c and ~ 80 mr (corresponding to the $M_X \sim 1.2 \div 1.8$ GeV/c^2). These data, along with the previously obtained data on T_{20} of $^1H(d,d')X$ reactions at 4.5, 5.5 and 0° (corresponding to the Roper resonance ($P_{11}(1440)$) excitation) [2], and data on T_{20} in the inelastic 9 GeV/c-deuteron scattering at 0° on hydrogen [3] are compared with calculations within existing theoretical models.

In Fig. 1 the data on the vector analyzing power A_y of the inelastic scattering of 9 GeV/c deuterons on ^{12}C at an angle of 85 mr (full squares) shown as a function of the 4-momentum t, are compared with calculations in the plane wave impulse approximation (PWIA). Within this approach A_y is expressed through the parameter $R(|t|)$, the ratio of spin-dependent and spin-independent parts of the elementary amplitude $NN \to NN^*$, and a combination of the charge and quadrupole form factors of the deuteron. The data shown correspond mainly to the excitation of $N^*(2190)$; the parameter R was taken in the form $R = 0.4\sqrt{|t|}$. The solid curve in Fig.1 is calculated with the deuteron wave function (DWF) for Paris potential [6], while the dashed, dotted and dash-dotted curves correspond to the DWFs for Bonn A, B and C potentials [7], respectively. As a whole the calculation results are in a reasonable agreement with the data.

Tensor analyzing power A_{yy} of the inelastic deuteron scattering within PWIA, in contrast to A_y, depends almost not at all on the elementary amplitude $NN \to NN^*$

and is defined by the form factors of the deuteron. Calculations have shown marked deviations of the data on A_{yy} from predictions of the PWIA indicating the sensitivity of A_{yy} to the baryonic resonances excitation. Such a sensitivity has been pointed out within the t-channel ω-meson exchange model [8, 9]. According to this approach cross section and the polarization observables can be calculated from known electromagnetic properties of the deuteron and baryonic resonances N^* through the vector dominance model.

In this approach [8, 9] A_{yy} is expressed through the ratio of the cross sections of absorption of virtual isoscalar photons with longitudinal and transversal polarizations by nucleons $r = \sigma_L/\sigma_T$. In the case of N^* resonance excitation the ratio r can be expressed through the longitudinal form factor of N^* excitation on proton or neutron, and two, isoscalar and isovector, transversal form factors. The results for the transversal and longitudinal amplitudes may be obtained from a collective string model of baryons [10]. As to the $P_{11}(1440)$, $S_{11}(1535)$, $D_{13}(1520)$ and $S_{11}(1650)$ resonances, only the Roper resonance $P_{11}(1440)$ has a nonzero isoscalar longitudinal form factor, while other three resonances cannot be excited by isoscalar longitudinal virtual photons.

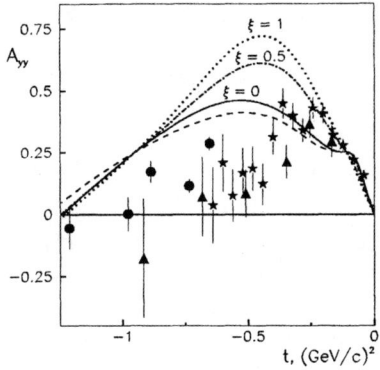

FIGURE 2. Tensor analyzing power A_{yy} of the inelastic deuteron scattering on Be at 4.5 GeV/c and 80 mr (full triangles) and on C at 9 GeV/c and 85 mr (full circles) from this experiment, and on H at 9 GeV/c and 0° (stars) [3] as a function of the 4-momentum t. The solid and dashed curves are calculated in the ω-meson exchange model [9] without and with taking into account the resonances widths, respectively, with $\xi = 0$. The dash-dotted and dotted curves correspond to the values of $\xi = 0.5$ and 1, respectively.

In Fig. 2 the data on the parameter A_{yy} of the inelastic scattering of 9 GeV/c deuterons on ^{12}C at an emission angle of 85 mr (full circles) and 4.5 GeV/c deuterons on ^9Be at an angle of 80 mr (full triangles) are shown as a function of the 4-momentum t. The data on A_{yy} obtained on ^1H at a zero emission angle at 9 GeV/c [3] (stars) are also given. One can see good agreement of the present and previous data in the region where they overlap. The data show an approximate independence on the incident deuteron momentum, and there is no significant dependence on the A-value of the target.

In Fig. 2 the data on A_{yy} are compared with the calculations within the ω-meson exchange model [9]. As inputs of the calculations, the deuteron form factors [11] and standard parameters of the collective string model [10], namely, the constituent quark mass $m = 0.366$ GeV, magnetic moment $\mu = 0.127$ GeV^{-1} and a scale parameter of the distribution $a = 0.232$ fm were used. The solid and dashed curves represent results of calculations neglecting and taking into account the finite widths of the resonances, respectively. There is a reasonable agreement of the calculations with the data up to $|t| \sim 0.3$ (GeV/c)^2, whereas at larger $|t|$ the discrepancy is observed. Such a deviation can be related with neglecting the contribution from the resonances $F_{15}(1680)$ and $P_{13}(1720)$. Moreover, the exchanges by other mesons (σ, η etc.) may play a role at large $|t|$.

In a string model of hadrons one expects that a string elongates with increasing excitation energy. The latter can be studied by introducing the stretchability of the string ξ, which may change between 0 and 1 [10]. The solid, dash-dotted and dotted curves in Fig. 3 are calculated with the values of ξ equal to 0, 0.5, and 1, respectively. The present data and ones obtained previously at a zero angle can be described for $|t| \leq 0.3$ (GeV/c)^2 using $\xi \sim 0.2$.

In conclusion, the behaviour of the A_{yy} data in the vicinity of the $P_{11}(1440)$, $S_{11}(1535)$ and $D_{13}(1520)$ resonances does not contradict to the predictions of the ω-meson exchange model [8, 9], while at higher excited masses this model may require taking into account the additional baryonic resonances and additional mechanisms like exchanges by other mesons [12].

Acknowledgments. Authors express their gratitude to the Director of LHE A.I.Malakhov and vice-Director of LHE V.N.Penev for their permanent help. Authors are grateful to the LHE accelerator staff and POLARIS team for providing good conditions for the experiment. They thank to L.V.Budkin, V.P.Ershov, V.V.Fimushkin, A.S.Nikiforov, Yu.K.Pilipenko, V.G.Perevozchikov, E.V.Ryzhov, A.I.Shirokov, and O.A.Titov for their assistance during the experiment. Authors are indebted to E. Tomasi-Gustafsson for permanent help and fruitful discussions. The research described in this report was supported in part by grant $N°$ 01-02-17299 of the Russian Foundation for Fundamental Research.

REFERENCES

1. L.S. Azhgirey et al., Yad.Fiz. **48**, 1758 (1988) [Sov.J.Nucl.Phys. **48**, 1058 (1988)]
2. L.S. Azhgirey et al., Phys.Lett. B**361**, 21 (1995)
3. L.S. Azhgirey et al., JINR Rapid Comm. **2**[88]-98, 17 (1998)
4. L.S. Azhgirey et al., Yad.Fiz. **62**, 1796 (1999) [Phys.Atom.Nucl. **62**, 1673 (1999)]
5. N.G. Anishchenko et al., in Proc. of the 5-th Int. Symp. on High En. Spin Phys., Brookhaven, 1982 (AIP Conf. Proc. N95, N.Y., 1983) p.445
6. M. Lacombe et al., Phys.Lett. B**101**, 139 (1981)
7. R. Machleidt et al., Phys.Reports **149**, 1 (1987)
8. M.P. Rekalo, E. Tomasi-Gustafsson, Phys.Rev. C**54**, 3125 (1996)
9. E. Tomasi-Gustafsson et al., Phys.Rev. C**59**, 1526 (1999)
10. R. Bijker, F. Iachello, A. Levitan, Ann.Phys. (N.Y.), **236**, 69 (1994); Phys.Rev. C**54**, 1935 (1996); Phys.Rev. D**55**, 2862 (1997)
11. P.L. Chung et al., Phys.Rev. C**37**, 2000 (1988)
12. S. Hirenzaki, E. Oset, C. Djalali, M. Morlet, Phys.Rev. C**61**, 044605 (2000)

RARE DECAYS,
SEARCHES FOR NEW PHENOMENA

New results on rare decays and on future NA48

A. Zinchenko
(for the NA48 Collaboration)

Particle Physics Laboratory, Joint Institute for Nuclear Research, Joliot Curie 6, 141980, Dubna, Moscow region, Russia

Abstract. The decay rate of $K_L \to \pi^0 \gamma\gamma$ has been measured with the NA48 detector at the CERN SPS. The branching ratio is determined to be $(1.36 \pm 0.03_{(stat)} \pm 0.03_{(syst)} \pm 0.03_{(norm)}) \times 10^{-6}$ and the vector coupling constant $a_V = -0.46 \pm 0.03_{(stat)} \pm 0.03_{(syst)} \pm 0.02_{(theo)}$. Using data collected during a special high intensity run at the end of 1999, the following preliminary branching fractions of the weak radiative decays Ξ^0: $Br(\Xi^0 \to \Sigma^0 \gamma) = (3.7 \pm 0.5) \times 10^{-3}$ and $Br(\Xi^0 \to \Lambda\gamma) = (1.9 \pm 0.1 \pm 0.2) \times 10^{-3}$ have been obtained. Future prospects aiming to study rare K_S, Ξ^0 decays and direct CP violation with charged kaon beams are also presented.

EXPERIMENTAL SETUP

The NA-48 experiment has been designed to measure the parameter $R(\varepsilon'/\varepsilon)$ of direct CP violation in the decay of neutral kaons to two pions. In the usual data taking mode, a 450 GeV/c proton beam from CERN-SPS is directed to two different targets. The first ("K_L") target, around 120 m upstream of the decay volume, is exposed to the full proton beam intensity and produces a pure K_L beam. A small fraction of the protons which pass the K_L target is redirected to a second ("K_S") target - about 6 m upstream of the fiducial volume - to generate a neutral beam consisting mainly of K_S mesons and neutral hyperons. The detector components most relevant to measure $K_L \to \pi^0 \gamma\gamma$ and Ξ^0 decays are the magnetic spectrometer [1] with the momentum resolution $\delta p/p = 0.5\% \oplus 0.009\% \times p[GeV/c]$ and the liquid krypton calorimeter [2] with the energy resolution $\delta E/E = 3.2\%/\sqrt{E[GeV]} \oplus 9\%/E \oplus 0.42\%$.

STUDY OF $K_L \to \pi^0 \gamma\gamma$ DECAY

The measurement of the $K_L \to \pi^0 \gamma\gamma$ is useful to test the prediction of the Chiral Perturbation Theory (χPT) together with models based on Vector Meson (VMD) exchange and to get constraints to the CP conserving amplitude of the decay $K_L \to \pi^0 e^+ e^-$ via two photon intermediate state.

The $K_L \to \pi^0 \gamma\gamma$ branching ratio is obtained from the ratio of the number of observed $\pi^0 \gamma\gamma$ events to the number $\pi^0 \pi^0$ events. In the ratio most of the systematic uncertainties cancel. The $K_L \to \pi^0 \gamma\gamma$ events are collected by the same trigger that is used to select the $K_L \to \pi^0 \pi^0$ decays. The obtained invariant mass distributions are shown in fig.1. The number of $K_L \to \pi^0 \gamma\gamma$ candidates is 2588 in the signal region defined as $132 MeV/c^2 <$

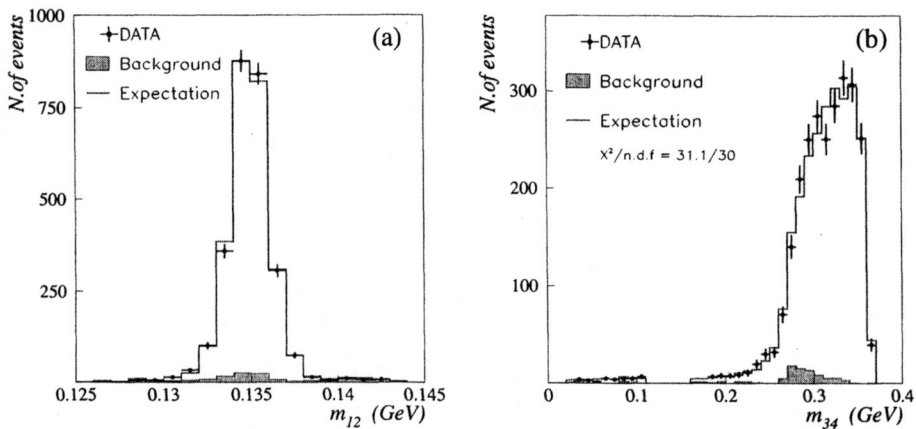

FIGURE 1. $m_{12} - m_{\gamma\gamma}$ for the π^0 candidates (a) and $m_{34} - m_{\gamma\gamma}$ for the non-π^0 candidates (b)

$m_{12} < 138 MeV/c^2$. The overall background is of the order of 3.3%.

To extract a value for the vector coupling constant we perform a fit of the bidimensional distribution of the two relevant kinematic variables m_{34} and $y = |E_3 - E_4|/m_K$ by corresponding Monte Carlo distribution. For the best fit, we obtain

$$a_v = -0.46 \pm 0.03_{(stat)} \pm 0.03_{(syst)} \pm 0.02_{(theo)}$$

with the systematic error dominated by our uncertainty in the background evaluation. We quote as a separate error the uncertainty from the $O(p^4)$ χPT parametrization of the $K_L \to 3\pi$ vertex.

From the whole $\pi^0\gamma\gamma$ candidates sample and with the obtained value of a_v we find the branching ratio to be:

$$BR(K_L \to \pi^0\gamma\gamma) = (1.36 \pm 0.03_{(stat)} \pm 0.03_{(syst)} \pm 0.03_{(norm)}) \times 10^{-6}.$$

The uncertainty related to the knowledge of the experimental $K_L \to \pi^0\pi^0$ branching ratio [3] is quoted separately. The systematic error comes from the residual background estimate (1.5%), the a_v fit (2.2%) and the acceptance calculation (3.2%).

HYPERON DECAYS

Study of radiative decays like $\Xi^0 \to \Lambda\gamma$ and $\Xi^0 \to \Sigma^0\gamma$ gives information about the hyperon structure related to SU(3) violation. Theoretical predictions for the branching ratios of Ξ^0 radiative decays range over almost two orders of magnitude. Experimentally, the situation is unclear, too: Branching ratio measurements agree for $\Xi^0 \to \Sigma^0\gamma$ [4, 5], but do not for $\Xi^0 \to \Lambda\gamma$ [4, 6].

The data used for the analysis described here were taken with the next conditions: The K_L target was removed and a proton beam with about 200 times the usual intensity was directed to the K_S target.

The abundant decay $\Xi^0 \rightarrow \Lambda\pi^0$ is used for normalization. We find about 115,000 accepted candidates. As a dominant Ξ^0 decay, it is virtually free of any backgrounds. An acceptance, derived from Monte Carlo simulation, is 4.7%.

The $\Xi^0 \rightarrow \Sigma^0\gamma$ decay has a signature similar to $\Xi^0 \rightarrow \Lambda\pi^0$. Z-coordinate of Ξ^0 decay vertex has been calculated using the nominal Ξ^0 mass, therefore the number of $\Sigma^0 \rightarrow \Lambda\gamma$ decays was measured. The invariant mass spectrum of the $\Lambda\gamma$ system is presented in fig.2a. Σ^0 signal contains roughly 380 peak events on a background of about 50 events. The background is consistent with the hypothesis of being due exclusively to $\Xi^0 \rightarrow \Lambda\pi^0$ decays. The detection efficiency has been determined to be 4.1%. We obtain

$$Br(\Xi^0 \rightarrow \Sigma^0\gamma) = (3.7 \pm 0.5) \times 10^{-3}$$

(with statistical and systematic error combined).

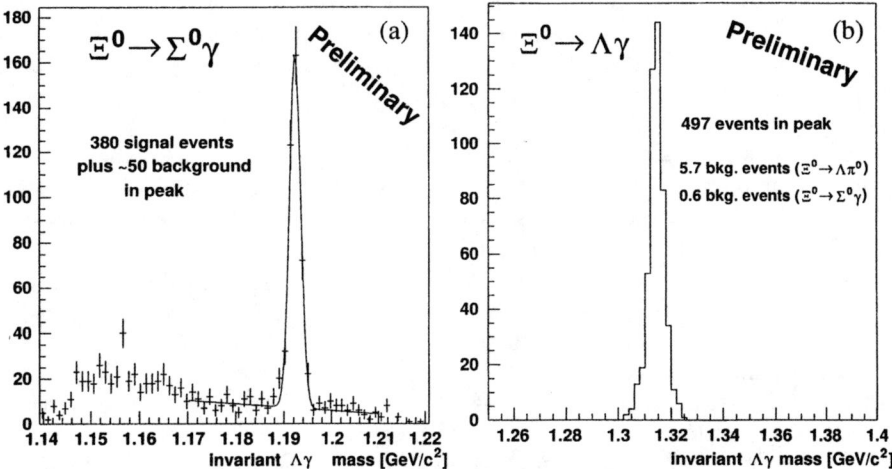

FIGURE 2. Invariant $\Lambda\gamma$ mass for $\Xi^0 \rightarrow \Sigma^0\gamma$ candidates (a): the solid line is the fit to signal (Gaussian function) and background (linear function). Invariant $\Lambda\gamma$ mass for $\Xi^0 \rightarrow \Lambda\gamma$ candidates (b).

We observe 497 candidates for the decay $\Xi^0 \rightarrow \Lambda\gamma$ (fig.2b). The background estimate is 5.7 events from $\Xi^0 \rightarrow \Lambda\pi^0$ and 0.6 events from $\Xi^0 \rightarrow \Sigma^0\gamma$ decays with missing photons in the signal region. The detection efficiency is 10.5%. From this, we obtain the preliminary branching ratio of

$$Br(\Xi^0 \rightarrow \Lambda\gamma) = (1.9 \pm 0.1 \pm 0.2) \times 10^{-3}.$$

The systematic error is dominated by uncertainties in the Ξ^0 polarization and the $\Xi^0 \rightarrow \Lambda\gamma$ decay asymmetry.

FUTURE PROSPECTS

The NA48 Collaboration has recently proposed in 2002 to use a modified K_S beam line on order to investigate K_S and neutral hyperon decays with high sensitivities [7].

One of the main goals of this physics programme is the search for the $K_S \to \pi^0 e^+ e^-$ decay. The expected signal in this case is 7 events/year for a branching fraction of 5×10^{-9}. The $K_S \to \pi^0 \mu^+ \mu^-$ decay will also be investigated but its branching ratio is about five times smaller than the one of $K_S \to \pi^0 e^+ e^-$.

As far as the $K_S \to \pi^0 \pi^0 \pi^0$ channel is concerned, the experiment sensitivity should allow to put a bound on η_{000} of about 1% and provide a further test of CPT.

$K_S \to \gamma\gamma$, $K_S \to \pi^+\pi^-\gamma$ and $K_S \to \pi^+\pi^- e^+ e^-$ decays will be measured with an expect statistical gain of about 50 with respect to previous NA48 measurement.

The β decay of the Ξ^0 hyperon, which has a branching ratio of 2.5×10^{-4}, can be detected in NA48 through the decay $\Xi^0 \to \Sigma^+ e^- \bar{\nu}$ followed by $\Sigma^+ \to p\pi^0$. About 25k events are expected to be collected in the proposed run. These events will be used for a consistency check of SU(3) symmetry and the Cabbibo model in hyperon decays.

In 2003 the NA48 Collaboration has proposed to look for a manifestation of CP violation through the measurement of the Dalitz plot decay parameters in $K^\pm \to \pi^+\pi^-\pi^\pm$ using an extended NA48 setup [8]. The matrix element for the decays $K^\pm \to \pi^+\pi^-\pi^\pm$ can be parametrized by $|M(u,v)|^2 \sim 1 + gu + hu^2 + kv^2$, where $u = (s_3 - s_0)/m_\pi^2$, $v = (s_1 - s_2)/m_\pi^2$, $s_0 = (s_1 + s_2 + s_3)/3$ and $s_i = (P_K - P_i)^2$, P_K and P_i are the four-momenta of the kaon and of the pion (i=3 for the odd pion). A measurement of direct CP violation can be obtained through the observation of non-zero value for the asymmetry $A_g = (g^+ - g^-)/(g^+ + g^-)$. Theoretical predictions for A_g in the framework of the Standard Model are in the $O(10^{-6})$-$O(10^{-4})$ range [9].

The proposed experiment would use simultaneous K^+ and K^- beams, would be selected with the same geometrical acceptance. With an average proton beam intensity of about 1×10^{12}/spill on the production target, NA48 could collect $\approx 10^{10}$ $K^\pm \to \pi^+\pi^-\pi^\pm$ decays per year and measure A_g with a precision better than 10^{-4}.

The use of intense charged kaon beams would also allow to study in a clean way the $\pi^+\pi^-$ interaction at low energy via the measurement of the $K^\pm \to \pi^\pm e^\pm \nu(\bar{\nu})$ decays (BR=3.9×10^{-5}). As no other hadron is present in the final state, the extraction of the $\pi^+\pi^-$ elastic scattering lengths a_0^0 and a_0^2 can be performed. The NA48 experiment aims at a precise measurement of a_0^0 with a statistical precision of about 0.007.

REFERENCES

1. Augustin I. et al. *Nucl.Instr.Meth.*, **A403**, 472(1998).
2. Barr G.D. et al. *Nucl.Instr.Meth.*, **A370**, 413(1996).
3. Groom D.E. et al. *Eur.Phys.Jour.*, **C15**, 1(2000).
4. Fanti V. at al. *Eur.Phys.Jour.*, **C12**, 69(2000).
5. Teige, S. at al. *Phys.Rev.Lett.*, **63**, 2717(1989).
6. James C. et al. *Phys.Rev.Lett.*, **64**, 843(1990).
7. Batley R. et al. *Addendum 2 to P253*, **CERN/SPSC/2000-002**(1999).
8. Batley R. et al. *Addendum 3 to P253*, **CERN/SPSC/2000-003**(1999).
9. Belkov A. et al. *Phys.Lett.B*, **300**, 283(1993).

Search for the Standard Model Higgs Boson at the ALEPH Detector

R. R. White

The Blackett Laboratory, Imperial College, London, SW7 2BW

Abstract. The ALEPH experiment, in the year 2000 recorded an excess of Higgs signal-like events of $\sim 3\sigma$ over the background-only hypothesis, for a Higgs mass of $\sim 115\,\mathrm{GeV}$. This note presents the first results after the final data taking, which were published in November 2000.

INTRODUCTION

ALEPH collected $217\mathrm{pb}^{-1}$ of data with $202 < \sqrt{s} < 209\,\mathrm{GeV}$ during the year 2000. During this final year of LEP running, the machine was operated at its limit in order to maximise the Higgs boson discovery potential.

Two parallel analysis streams are used by ALEPH: selections based on artificial neural networks (NN); and selections based on sequential cuts (cuts). All analyses were frozen before the start of the data taking period.

The major production mechanism is *Higgsstrahlung*: $e^+e^- \rightarrow Z^* \rightarrow ZH$. This process has a kinematic limit for Higgs production in association with an on mass-shell Z^0 of $M_H \leq \sqrt{s} - M_Z$. Other relevant production mechanisms are *vector boson fusion*: $e^+e^- \rightarrow e^+e^- ZZ \rightarrow e^+e^- H$ and $e^+e^- \rightarrow \nu_e \bar{\nu}_e W^+ W^- \rightarrow \nu_e \bar{\nu}_e H$, but their contribution is very small at these energies. Taking into account the efficiency of the analyses, we would expect 7 Higgs events to be selected for a Higgs mass of 114 GeV.

STATISTICAL METHOD

The likelihood ratio, Q, is defined as:

$$Q = \frac{L_{s+b}}{L_b} = \frac{e^{-(s+b)}(s+b)^n \prod_{i=1}^n f_{s+b}(\mathbf{x}_i)}{e^{-b} b^n \prod_{i=1}^n f_b(\mathbf{x}_i)} = e^{-s} \prod_{i=1}^n \left[\left(1 + \frac{s}{b}\right) \frac{f_{s+b}(\mathbf{x}_i)}{f_b(\mathbf{x}_i)} \right] \quad (1)$$

where $f_b(\mathbf{x})$ and $f_{s+b}(\mathbf{x})$ are the probability density functions (PDF) of the discriminating variables $\mathbf{x}^1$, for the background and signal plus background hypotheses, and s and b are the total expected signal and background respectively. The product is

---

[1] The vector **x** always has the reconstructed mass as one of its dimensions and in some channels has another variable as the second dimension.

over the n observed events. The signal plus background PDF can be expressed as $f_{s+b} = (sf_s + bf_b)/(s+b)$, where f_s is the signal only PDF. Then an *event weight*, w_i, can be defined from

$$Q' \equiv -2\ln(Q) = 2s - 2\sum_{i=1}^{n} w_i \quad \text{where} \quad w_i = \ln\left(1 + \frac{sf_s(\mathbf{x}_i)}{bf_b(\mathbf{x}_i)}\right) \quad (2)$$

This makes combining channels simple: one just sums over the event weights for all the candidates from all the channels. The distributions of the test statistic, $-2\ln(Q)$, are found for the background and signal plus background hypothesis (ρ_b and ρ_{s+b} respectively) using a method of fast Fourier Transforms [1]. The confidence level in the hypothesis x is then:

$$c_x(M_H) = \int_{Q'_{obs}(M_H)}^{+\infty} \rho_x(Q', M_H) dQ' \quad (3)$$

The number c_b is measure of discovery significance or incompatibility with the background only hypothesis. When interpreted as the tail of a Gaussian, $1 - c_b$ must be less than 1.3×10^{-3} and 3×10^{-7} for 3 and 5 sigma discoveries respectively. c_s is a measure of consistency with the signal only hypothesis: $c_s < 0.05$ for a 95% signal exclusion (the value of M_H where $c_s(M_H) = 0.05$ gives the limit). However c_s is not directly available since signal and background can not be completely separated. The method adopted by ALEPH is *The Signal Estimator Limit Setting Method*[2]. In this case:

$$c_s \approx CL_{SE} = c_{s+b} + (1 - c_b)e^{-s} \quad (4)$$

SEARCH TOPOLOGIES

There are four SM Higgs decay topologies that are specifically searched for by ALEPH. For $M_H = 114$ GeV the dominant Higgs decays are $H \to b\bar{b}(\sim 74\%)$ and $H \to \tau^+\tau^-(\sim 7\%)$. Therefore the decay channels are mainly determined by the Z^0 decay.

4-jets. This is the dominant channel, with $BR(H \to b\bar{b}, Z^0 \to q\bar{q}) \sim 51\%$. The major backgrounds are $q\bar{q}$ (with either two hard gluons or gluon splitting) and ZZ events.

The missing energy channel. This channel has the second largest branching ratio, $BR(H \to q\bar{q}, Z^0 \to \nu\bar{\nu}) \sim 15\%$. The WW fusion process makes a non-negligible contribution near threshold. The major background, other than ZZ, is $q\bar{q}$ with two ISR photons escaping (in opposite directions) undetected at low angles.

Leptonic channels. This is split into final states with electron or muon pairs and final states with tau pairs. The former has $BR(H \to b\bar{b}, Z^0 \to l^+l^-) \sim 5\%$, where $l = e, \mu$. The signal is distinctive and offers good reconstructed mass resolution.

The tau channel has two topologies depending on whether the Higgs decays to b or τ pairs: $H \to b\bar{b}, Z^0 \to \tau^+\tau^-$ and $H \to \tau^+\tau^-, Z^0 \to q\bar{q}$. Together these account for $\sim 7\%$ of the Higgsstrahlung final states.

RESULTS

Three out of the four channels select events that have a large weight for $M_H \sim 115$ GeV and are subsequently signal like in this region[3]. Seven events, in either the NN or cuts stream, have an event weight greater than 0.4 at $M_H = 115$ GeV: five 4-jet events; one tau event; and one leptonic event. The five 4-jet candidates are detailed in table 1 below.

TABLE 1. Candidate selected by the cuts or NN stream with event weight greater than 0.4 at $M_H = 115$ GeV.

Candidate	$\sqrt{s}$	M_{REC}	b-tagging				Stream
(Run/Event)	(GeV)	(GeV/c^2)	jet 1	jet 2	jet 3	jet 4	
a (56698/7455)	206.6	110.0	0.999	0.836	0.999	0.214	both
b (56065/3253)	206.7	112.9	0.994	0.776	0.993	0.999	both
c (54698/4881)	206.7	114.3	0.136	0.012	0.999	0.999	both
d (56366/0955)	206.4	114.5	0.238	0.052	0.998	0.948	both
e (55982/6125)	206.6	114.6	0.088	0.293	0.895	0.998	cuts

An excess of high mass candidates reduces the expected exclusion limit. Figure 1 shows the signal estimator confidence level for the cuts stream with 98, 99 and 2000 data combined. The NN (cuts) stream can set a 95% confidence level exclusion on the Standard Model Higgs of $M_H < 111.1(110.4)$ GeV, while the median expected is $M_H < 114.1(113.6)$ GeV.

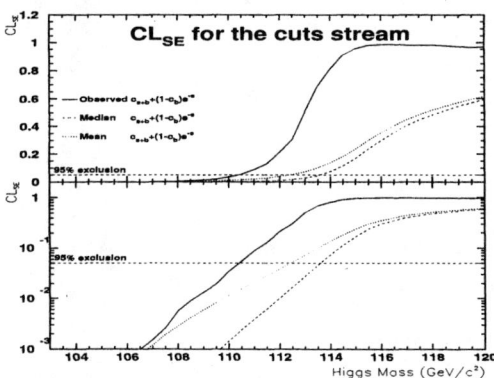

FIGURE 1. CL_{se} for the cuts combined stream, shown on linear and logarithmic scales.

Figure 2a shows the observed test statistic, $-2\ln Q$ from the NN stream. The dashed line is that expected for the null hypothesis, shown with 1 and 2 sigma bands. The minimum of $-2\ln Q$ occurs at $M_H \sim 114.5$ GeV, while the maximum statistical departure from the null hypothesis occurs at $M_H \sim 116$ GeV. Figure 2b shows $1 - c_b$ for the NN stream. The minimum is $3.0(3.1)\sigma$ for the NN(cuts) stream.

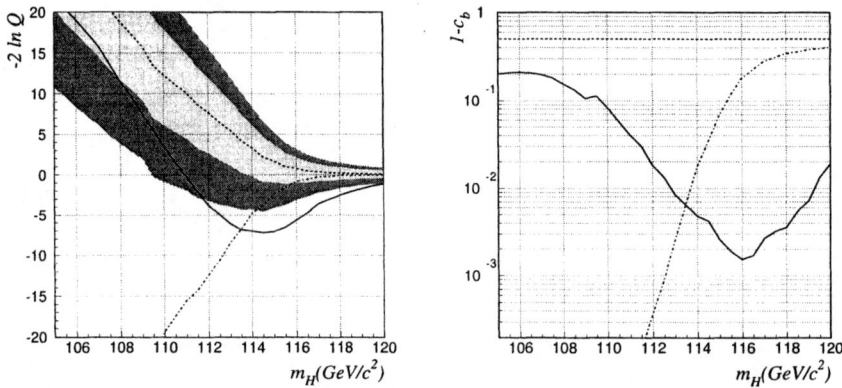

FIGURE 2. Results from the NN stream. a) The test statistic, $-2\ln Q$. The black line is the observed value, while the dashed line is that expected for the null hypothesis (shown with 1 and 2σ bands). The dot-dashed line is the expected for a signal at M_H. b) $1 - c_b$. The null hypothesis is at 0.5, while the minimum in the observed value is 3.0σ.

CONCLUSION

The ALEPH detector has recorded an excess of signal-like events consistent with $M_H \sim 115\,\text{GeV}$. The excess is $\sim 3\sigma$ in both the cuts and NN streams. When the search results of all four LEP experiments are combined[4], a LEP wide excess of $\sim 2.1\sigma$ is observed. This is in agreement with the Standard Model Higgs cross section for $M_H = 115^{+1.3}_{-0.9}\,\text{GeV}$.

ACKNOWLEDGMENTS

We thank the accelerator divisions for the very successful running of LEP at high energies.

REFERENCES

1. H. Hu and J. Nielsen, "Analytic confidence level calculation using the likelihood ratio and fourier transform." ALEPH 99-031 PHYSIC 99-010, April, 1999.
2. S. Jin and P. McNamara, "The signal estimator limit setting method," *Nucl. Phys.* **A462** (2001) 561–567.
3. **ALEPH** Collaboration, R. Barate *et al.*, "Observation of an excess in the search for the standard model Higgs boson at ALEPH," *Phys. Lett.* **B495** (2000) 1–17, hep-ex/0011045.
4. **LEP Higgs Working Group for Higgs boson searches** Collaboration, "Search for the standard model Higgs boson at LEP," arXiv:hep-ex/0107029.

RADIATIVE DECAYS

Radiative Decay Width of the $a_2(1320)$ Meson

Vladimir V. Molchanov

IHEP, Protvino, 142284, Russia

On behalf of the SELEX Collaboration[1]

Abstract. Coherent $\pi^+\pi^-\pi^-$ production in the interactions of a beam of 600 GeV π^- mesons with C, Cu and Pb nuclei has been studied with the SELEX facility (Experiment E781 at Fermilab). The $a_2(1320)$ meson signal has been detected in the Coulomb (low q^2) region. The Primakoff formalism used to extract radiative decay width of this meson yields $\Gamma = 284 \pm 25 \pm 25$ keV, which is the most precise measurement to date.

INTRODUCTION

Radiative decays of mesons and baryons, as well as other electromagnetic processes, are important tools for studying internal structure of these particles and for testing unitary symmetry schemes and quark models of hadrons.

Direct observation and study of rare radiative decays of hadrons is often very difficult to carry out because of high background from hadronic processes with $\pi^0(\eta)$ production and subsequent loss of one decay photon. An alternative way of measuring radiative decay width $\Gamma(h \to a\gamma)$ consists in investigation of coherent production reaction $a + (A,Z) \to h + (A,Z)$. Coulomb contribution to the cross section of this reaction is directly proportional to $\Gamma(h \to a\gamma)$. This fact is commonly known as Primakoff effect [1].

Here we present results on $\Gamma(a_2^- \to \pi^-\gamma)$, measured in Primakoff reaction

$$\pi^- + (A,Z) \to a_2(1320)^- + (A,Z) \tag{1}$$

on C, Cu and Pb nuclei at a beam energy of approximately 600 GeV in an experiment using the SELEX spectrometer (E781) at Fermilab. The a_2 meson has been detected in the decay to three charged pions, thus the basic reaction to be analysed is

$$\pi^- + A \to \pi^+\pi^-\pi^- + A \tag{2}$$

More detailed description of the analysis will soon be published [2].

---

[1] Carnegie-Mellon University, Fermilab, University of Iowa, University of Rochester, University of Michigan-Flint, Ball State University, Petersburg Nuclear Physics Institute, ITEP (Moscow), IHEP (Protvino), Moscow State University, University of São Paulo, Centro Brasileiro de Pesquisas Fisicas (Rio de Janeiro), Universidade Federal de Paraiba, IHEP (Beijing), University of Bristol, Tel Aviv University, Max-Planck-Institut für Kernphysik (Heidelberg), University of Trieste, University of Rome "La Sapienta", INFN, Universidad de San Luis Potosí, Bogazici University.

EXPERIMENTAL APPARATUS

The SELEX facility is a 3-stage magnetic spectrometer, designed mainly to study production and decays of charm baryons in a hyperon beam. It emphasized the forward ($x_F > 0.1$) region and, consequently, had high acceptance for exclusive low multiplicity processes. Only part of the SELEX facility was needed for the study of Reaction (2).

The beam contained approximately equal fractions of Σ^- and π^-, which were reliably identified with the beam transition-radiation detector. A segmented target with 2 Cu and 3 C foils, each separated by 1.5 cm, was used for most of the data taking. During brief periods of running a thin Pb target was used as well. Silicon strip detectors (most of which had 4 μm transverse position resolution) measured parameters of the beam and secondary tracks in the target region. After deflection by two analyzing magnets, tracks were measured in 14 planes of 2 mm proportional wire chambers (PWC). Three lead glass detectors measured photons.

Reaction (2) was singled out with the help of a special exclusive trigger. This trigger used scintillation counters to define beam time and to suppress interactions upstream of the target. Located downstream of the target interaction counters (IC) and located near the PWC hodoscope selected events with three charged tracks. To reduce the background trigger rate, the aperture was limited by veto counters, which had little effect on efficiency for Reaction (2).

DATA ANALYSIS

Events for Reaction (2) were selected by requiring a reconstructed beam track and three charged tracks in the final state. These tracks were required to form a good vertex in the vicinity of one of the targets. The beam particle had to be identified as a pion by the beam TRD. However, there was no identification for the produced particles. To suppress inclusive ($\pi^+\pi^-\pi^- + X$) background, the energy sum of the observed particles was required to be within ± 17.5 GeV of the beam energy. For further supression of background, the most upstream photon detector was used as a guard system, requiring that any registered energy be less than 2 GeV.

Most of the ensuing analysis will be described using the data from the Cu target. The distribution in the square of the transverse momentum (p_T^2) of the 3π-system in Reaction (2) is shown in Fig. 1a. This distribution can be fitted by the sum of two falling exponentials, one with slope parameter $b_1 \approx 180$ GeV^{-2}, which is characteristic of coherent diffractive production on a copper nucleus, and the other with a slope parameter $b_2 \sim 1500$ GeV^{-2}, which is consistent with the estimation for Coulomb production folded in with the experimental resolution in transverse momentum.

Two p_T^2 regions are defined for extracting the mass distribution for the Coulomb production process, as shown in Fig. 1a. The first one ($p_T^2 < 0.001$ GeV^2) contains most of the Coulomb contribution, the second one ($0.0015 < p_T^2 < 0.0035$ GeV^2) has very little of it. But even the first region is dominated by diffractive production. The mass spectra $M(3\pi)$ for these two regions are presented in Fig. 1b. Using results of the fit to Fig. 1a, the mass distribution for events in the second p_T^2 region was normalized to

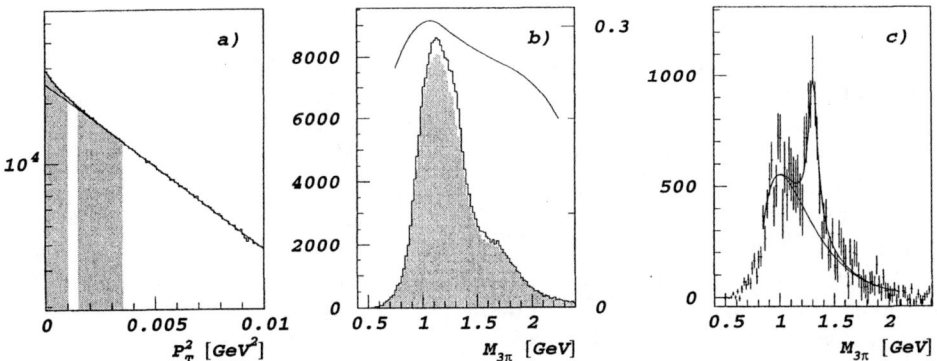

FIGURE 1. Primakoff production of the a_2 meson on Cu target. a) — Transverse momentum distribution for Reaction (2). b) — Mass distribution for events with $p_T^2 < 0.001\,\mathrm{GeV}^2$ (histogram) and $0.0015 < p_T^2 < 0.0035\,\mathrm{GeV}^2$, after normalization for background subtraction (shaded) according to Fig. 1a. The curve shows the efficiency for observing a $\rho\pi$ in a 1^+S0^+ wave, which is dominant in the shown mass spectrum. c) — Result of subtraction. The curve shows fit with a sum of pure Coulomb contribution and smooth background.

the expected number of diffractive events in the first region. Then, the mass distribution from the second region was subtracted from the distribution for the first p_T^2 region. The resulting mass spectrum is shown in Fig. 1c. The $a_2(1320)$ signal stands out clearly.

The differential cross section for Coulomb production of a broad resonance in a pion beam is given by the expression (see Ref. [3] and references therein):

$$\frac{d\sigma}{dM\,dq^2} = 16\alpha Z^2(2J+1)\left(\frac{M}{M^2-m_\pi^2}\right)^3 \frac{m_0^2\Gamma(\pi\gamma)\Gamma(\text{final})}{(M^2-m_0^2)^2+m_0^2\Gamma(\text{all})^2} \frac{q^2-q_{\min}^2}{q^4}|F(q^2)|^2 \quad (3)$$

where α is the fine structure constant, Z is the charge of the nucleus, J and m_0 are spin and mass of the produced resonance, M is the effective mass of the produced system, the Γ are the decay widths for the corresponding modes, q^2 is the square of the momentum transfer, and $q_{\min}^2$ is its minimal value. The Coulomb form factor $F(q^2)$ in Eq. (3) accounts for the nuclear charge distribution, initial and final state absorption, as well as the Coulomb phase. It was calculated in the framework of the optical model described in Ref. [4].

To describe mass spectrum in Fig. 1c with Eq. (3) requires taking into account luminosity of the exposure and efficiency, which includes trigger, acceptance, reconstruction, as well as effects of applied cuts. Cuts on the measured p_T^2 value are especially important for the analysis and thus require precise knowledge of the transverse momentum resolution.

Evaluation of the trigger performance is very difficult because of accidental veto rates and uncertainties in the discrimination of IC analog amplitudes. That is why we chose to normalize the measurement to the diffractive three-pion production process, which dominates Reaction (2) in the region of low q^2. This cancels uncertainties in luminosity and trigger efficiency. The diffractive production was measured using special runs with a so-called "beam" trigger, which used no information from detectors downstream of the targets. Thus, it selected a completely unbiased set of interactions. Cross section of this process on C nucleus (we refer to the number of events in the first diffractive exponential of the p_T^2 distribution for $0.8 < M(3\pi) < 1.5$ GeV region) was found to be 2.57 ± 0.13 mb.

GEANT-based MC was used to calculate acceptance and reconstruction efficiencies for Coulomb and diffractive production on all the targets, and to find the transverse momentum resolution. To verify the transverse momentum resolution simulation we studied decays of Ξ^-, present in the beam. We had about 6800 $\Xi^- \to \Lambda\pi^-$, $\Lambda \to p\pi^-$ decays, with both vertices lying within the target region. These events are topologically similar to those of Reaction (2), and correspond to no momentum transfer ($p_T = 0$). Consequently, the measured momentum transfer gives the resolution. Comparison of measured values with MC showed that the transverse momentum resolution in the MC is better than in the data. The difference in quadrature in the resolution between data and MC was used to correct MC resolutions for the a_2 production. Obtained values vary from 16.2 to 19.3 MeV, depending on the data set, target and transverse direction, and are known with accuracy of $\approx 2\%$.

To obtain the expected shape of the $a_2(1320)$ signal, Eq. (3) for Coulomb production was multiplied by efficiency, convoluted with the p_T-resolution, and integrated over the relevant region of p_T^2. The a_2 mass and full width were fixed in the fit to their known PDG values [5]. To check the stability of the result, we varied the regions of p_T^2 and employed two different fitting procedures. Results for different p_T^2 regions and both fitting procedures were similar.

Because our fitting procedure accounts only for the Coulomb production, the results of the fit must be corrected for strong production contribution. To describe this effect, we extrapolated data on the a_2 production on nucleons and nuclei at lower energies to our energy of 600 GeV. Glauber-like model was used to describe production on nucleus [4].

Corrected $\Gamma(a_2^- \to \pi^- \gamma)$ values and their statistical uncertainties are 350 ± 121, 270 ± 38 and 291 ± 36 keV for C, Cu and Pb targets correspondingly. Net correction for strong production was estimated as $\approx 3 \pm 1.5\%$. It is impossible to correct for interference of the two production mechanisms because the phase difference is not known. This contributes to a systematic uncertainty of $\approx 4.5\%$ in the analysis. Other systematic uncertainties include absolute normalization (5%), transverse momentum resolution (1.8%), accuracy in $F(q^2)$ calculation (1%), and uncertainties in the PDG parameters of the $a_2(1320)$ resonance mass (0.35%), width (3.4%), and branching to $\rho\pi$ (3.8%). The final combined result

$$\Gamma\left[a_2(1320)^- \to \pi^- \gamma\right] = 284 \pm 25 \pm 25 \text{ keV} \qquad (4)$$

has total relative uncertainty of 12.5%. This is the best measurement to date. It is compatible with the world average [5].

REFERENCES

1. Primakoff, H., *Phys. Rev.*, **81**, 899–899 (1951).
2. Molchanov, V., et al., *Phys. Lett. B* (in print), hep-ex/0109016.
3. Huston, J., et al., *Phys. Rev.*, **D33**, 3199–3202 (1986).
4. Bemporad, C., et al., *Nucl. Phys.*, **B51**, 1–15 (1973).
5. Groom, D., et al., *Eur. Phys. J.*, **C15**, 1 (2000).

Radiative transitions in mesons within a non relativistic quark model

R. Bonnaz, B. Silvestre-Brac, C. Gignoux

Institut des Sciences Nucléaires, 53 Avenue des martyrs, F38026 Grenoble-Cedex, France

Abstract. An exhaustive study of radiative transitions in mesons is performed in a non relativistic quark model. Three different types of mesons wave functions are tested. The effect of some usual approximations is commented. Overall agreement with experimental data is obtained

INTRODUCTION

It is believed to day that QCD is a good theory for strong interactions and that mesons and baryons are good laboratories for exploring its properties. In particular the mesonic sector is specially interesting because a meson is a very simple object made of only 2 particles : a quark and an antiquark. But we are in a non perturbative regime of QCD and a basic description is very complicated. This is why so many approximations have been developed to explain the mesonic properties.

Among them, the non relativistic quark model (NRQM) is specially attractive because it is very simple, it allows a proper treatment of the center of mass motion and also because it is very successful in many domains, even when it should in principle fail. The model is based on a Schroedinger equation involving some interquark potential.

The first job is to propose a potential V in order to obtain a good spectrum; then one should test the resulting wave functions on reliable observables. Very interesting are the electromagnetic decays such as radiative transitions. In that case the transition operator is exactly known, so that no free parameter is needed. In consequence they are specially interesting to put severe limitation to the model by testing the wave functions.

The theory of radiative transitions is known for many years [1], both in atomic and nuclear physics, but, in those domains, it is simplified by various approximations. Let us mention: the dipole approximation for E1 transitions, the use of a non relativistic phase space or the long wave length approximation (LWLA). If those approximations are fully justified in atomic or nuclear physics, they are questionable in the mesonic sector. The transition energies E are quite large, and the size of the emitting source R is about 1 fm, so that kR is in general not negligible as compared to unity.

In this work, we present an exhaustive study of all non exotic radiative transitions, with no free parameter and no approximation, except NRQM. We have in mind, of course to test very severely the wave functions at our disposal.

DESCRIPTION OF MESONS

The first job to do is to obtain a good description of mesons and, in particular to rely on good potentials. In this study we present results obtained with 3 different types of potentials. In any case, the potentials contain a central and a hyperfine term.

In our first generation [2] (AL1 and AP1) the central part is a sum of a coulomb term plus a confining term whose power is either 1 (in AL1) or 2/3 (in AP1) while the hyperfine term is a gaussian with a mass dependent size. The spectra are already good but there exists a drawback : the eta sector is degenerate with the pion one.

In order to cure this problem, our second generation potential DNR takes into account instanton induced effects [3]. Moreover it takes care correctly of the asymptotic freedom by adjusting $\alpha_s(q^2)$ to the experimental values. Finally quarks acquire finite size; the final potential is a convolution of the bare potential with the quark gluonic density. The form of the potential is now more complicated and can be found in ref [3]. The spectra are now nicely reproduced both in the light and the heavy quark sectors.

The wave function contains an isospin dependent term and the coupling of spin and space degrees of freedom. For radiative transitions, the formalism is much easier in momentum representation. Moreover every thing can be made analytical if the radial wave function is expanded on gaussian terms.

$$R_{nl}(p) = p^L \sum_{i=1}^{N} c_i \exp(-\frac{A_i}{2} p^2) \qquad (1)$$

The convergence with the number of terms in the expansion is very fast and, with $N = 5$ terms, one can consider that we obtain the exact wave functions in any case.

RADIATIVE TRANSITIONS

The electromagnetic transition is very classic. After non relativistic reduction and quantification in a box of volume V, it takes the form

$$H_I = -\sqrt{\frac{2\pi\alpha}{VE}} \vec{\varepsilon}(\vec{k}, \lambda) \cdot \vec{M} \qquad (2)$$

where α is the fine structure constant, $E = |\vec{k}|$ is the photon energy; $\vec{\varepsilon}$ is the polarisation vector for a real photon with momentum $\vec{k}$ and helicity λ and $\vec{M}$ is the quark operator :

$$\vec{M} = \sum_{j=1,2} \frac{e_j}{2m_j} (2\vec{p}_j - i\vec{\sigma}_j \times \vec{k}) \qquad (3)$$

The first term in this expression is the electric term and the second one the magnetic term. The transtion amplitude from resonance A to meson B is calculated in the rest frame of A. The quark operator is composed of 2 contributions, one $\vec{M}^1$ coming from the emission of the photon by the quark, the other $\vec{M}^2$ by the antiquark. Their expression in terms of the initial and final waves functions are given by :

$$\vec{M}^1 = \frac{e_1}{2m_1} \int d^3 p \phi_B^*(\vec{p} - \frac{m_2}{m_1+m_2}\vec{k})[2\vec{p} - i\vec{\sigma}_1 \times \vec{k}]\phi_A(\vec{p}) \quad (4)$$

$$\vec{M}^2 = \frac{e_2}{2m_2} \int d^3 p \phi_B^*(\vec{p} + \frac{m_1}{m_1+m_2}\vec{k})[-2\vec{p} - i\vec{\sigma}_2 \times \vec{k}]\phi_A(\vec{p})$$

Let us stress that the argument entering the final wave function takes care of the recoil due to the photon emission. The treatment of the phase space is classic. Since we are interested only by the partial widths, one must sum over final states (photon helicity λ and meson polarisation M_b) and average on initial states (meson polarisation M_a). Thus one can define a quantity $X(E)$ like this

$$X(E)\delta_{\vec{K}_b, -\vec{k}} = \frac{1}{2J_a+1} \sum_{\lambda, M_a, M_b} |\langle B\gamma | H_I | A \rangle|^2 \quad (5)$$

Introducing it in the golden rule expression we get the partial width for the transition (the quantification volume V disappears, as it should):

$$\Gamma_{A \to B} = \Phi(E_0) X(E_0) \quad (6)$$

Depending on how we treat the Dirac term imposing energy conservation, one obtains two possibilities for the phase space factors (m_a and m_b are the masses of resonances A and B):

$$\text{Non} - \text{Relativistic}: \quad E_0 = m_a - m_b; \quad \Phi(E_0) = \frac{E_0^2}{\pi} \quad (7)$$

$$\text{Relativistic}: \quad E_0 = (m_a^2 - m_b^2)/(2m_a); \quad \Phi(E_0) = \frac{E_0^2}{\pi} \frac{E_b(E_0)}{m_a} \quad (8)$$

There exist essentially two differences. First the photon energy E_0 differs in both cases. This means that the quantity $X(E_0)$ is different, and we call this a dynamical effect. Second, the relativistic phase space factor (RPS) is obtained from the non relativistic one (NRPS) by a multiplication by a factor $\frac{E_b(E_0)}{m_a}$ that is different from unity; we call this a kinematical effect.

In the equation giving the quark operator (4), we saw that the wave function of the final meson should be calculated with a momentum taking into account the recoil due to photon emission. The long wave length approximation consists precisely in neglecting this recoil. In consequence, the formulae are much simpler and it leads also, in some cases, to forbidden transistions.

To go beyond LWLA, one sees the advantage of chosing wave functions expressed as sum of gaussians. Limiting ourselves to only one gaussian in the initial and final mesons for simplicity, the term appearing in ϕ_A is $\exp(-Ap^2/2)$, the one in ϕ_B is $\exp(-B(\vec{p}-u\vec{k})^2/2)$. By a very simple change of variables, it is easy to diagonalize the quadratic form appearing in the argument of the resulting exponential. The transition operator for the quark takes now the following form :

$$\vec{M}^1_{A \to B} = \frac{e_1}{2m_1} \int d^3q e^{-Dq^2 - Fk^2} [\mathcal{Y}^*_{L_b}(\vec{q} - z\vec{k})\chi_b]_{J_b}$$
$$[2\vec{q} - i\vec{\sigma}_1 \times \vec{k}][\mathcal{Y}_{L_a}(\vec{q} + x\vec{k})\chi_a]_{J_a} \tag{9}$$

and a similar expression for the antiquark transition operator. The exponential is now simple, but the complication is reported on the solid spherical harmonics. But this is not really a problem since $\mathcal{Y}_L(a\vec{q} + b\vec{k})$ is easy to express in term of the more basic quantities $[\mathcal{Y}_{l_1}(\vec{q})\mathcal{Y}_{l_2}(\vec{k})]_L$ with help of well known geometrical coefficients. The rest is only a question of Racah algebra and, since all the integrals appearing in the formalism are analytical, the final expression for the width is itself analytical.

One ends up with a transition operator

$$\vec{M}_{A \to B} = \vec{E} l_{A \to B} + \vec{M} a_{A \to B} \tag{10}$$

composed of two terms, one coming from the electric part and the other from the magnetic part. The final partial width :

$$\Gamma_{A \to B} = \Phi(E_0)[EE + EM + MM] \tag{11}$$

finally contains the phase space factor (7-8), and the square of the M operator which gives rise to terms electric-electric EE, electric-magnetic EM and magnetic-magnetic MM. So, we see that there exist interference effects between electric and magnetic terms; moreover the total theory is able to describe transitions that are forbidden in the LWLA.

RESULTS

Let us come to the results obtained with the previous formalism. All masses used to calculate the phase space factor must be the experimental ones. This is very important to get good values for the photon momentum.

As a first study, one can have a glance on how the things change with the quality of the wave function. Calculating the partial widths with a wave function developed on $N = 1$ gaussian is an approximation valid to 10%; the approximation $N = 2$ is much better, valid to 1%. The approximations with $N = 3$ or more reproduce the exact result. This is why we consider that $N = 5$ used here gives the exact result in practice.

In a second step, we examine the influence of the phase space factor. The conclusion is mainly independent on the treatment and on the interquark potential. In any case, the RPS is much better; the differences between RPS and NRPS can reach a factor 10. In fact, the discrepancy is in part due to the kinematical effect (which is limited between 1/2 and 1); but essentially it is due to a bad determination of the photon momentum which leads to a bad determination of the dynamical factor $X(E_0)$.

The last approximation that we want to comment is LWLA. Some transitions observed experimentally are forbidden in this scheme, but not in the exact treatment; obviously the latter is superior. For other transitions, there is no clear emerging conclusion. It can be shown that the LWLA results are always larger than the exact ones; in some cases this improves the result and in others this deteriorates it. So one cannot draw any systematic conclusion concerning this type of approximation.

Now we come to the results, presented in Table 1, obtained without any approximation, except the NRQM, and with the 3 different interquark potentials presented previously. We do not want to comment each one in particular but try to draw general conclusions. First the 3 different types of wave functions give results of similar quality. Generally speaking, DNR gives the lower values, AP1 the higher ones and AL1 values in between. This is a general trend. Second, the theoretical values have always the right order of magnitude as compared to experimental data. The difference can reach 50%, sometimes a factor 2, but never a factor 10. Moreover, when a potential gives a too large (or small) result, so do the others. In the case of electric transitions there exist interference effects between electric and magnetic terms, but the electric-electric part always gives the dominant contribution, while interference effects can contribute, in some cases, to 25%. For the transitions forbidden by LWLA, the complete theory also gives results with the right order of magnitude.

SUMMARY

We applied the NRQM and made an exhaustive study of radiative transitions in mesons with no approximation and no free parameter, using the wave functions as they come from the Schroedinger equation with quark potentials that give satisfactory spectra.

We stressed the importance of using the exact wave function and a relativistic phase space. The conclusions are essentially independent of the potential under consideration and the theoretical results have always the right order of magnitude. This indicates that NRQM is basically safe and serious for this description. However, the fact that there still exist some differences between the theoretical and the experimental data and the fact that, in some cases, an approximation (LWLA), in principle not justified in this sector, gives better results than an exact treatment indicate that probably some physical ingredients are still missing in the theory.

We think in particular that relativistic corrections both at the level of the wave functions and at the level of the transition operator should have some effects. We also think that using a form factor for the coupling of a constituent quark to the photon should be the key for a correct overall description of radiative transitions.

REFERENCES

1. S. Godfrey, N. Isgur, Phys. Rev. **D32**, 189 (1985).
2. B. Silvestre-Brac and C Semay, ISN **93-69** (unpublished); B. Silvestre-Brac, Few-Body Syst. **20**, 1 (1996); C. Semay and B. Silvestre-Brac, Z. Phys. **C61**, 271 (1994).
3. C. Semay and B. Silvestre-Brac, Nucl. Phys. **A618**, 455 (1997); **A647**, 72 (1999).

TABLE 1. Decay widths for the potentials AL1, AP1 and DNR compared to the experimental results.

decay	AL1	AP1	DNR	Exp.[keV]	decay	AL1	AP1	DNR	Exp.[keV]
$\psi(2S) \to \chi_{c0}(1P)\gamma$ *	14.12	14.06	18.52	25.76± 3.81	$\rho^+ \to \pi^+\gamma$ †	48.48	60.41	44.64	67.82± 7.55
$\psi(2S) \to \chi_{c1}(1P)\gamma$	34.25	34.23	43.34	24.10± 3.49	$\rho^0 \to \pi^0\gamma$	48.66	60.64	44.81	102.48±25.69
$\psi(2S) \to \chi_{c2}(1P)\gamma$	46.39	46.43	57.88	21.61± 3.28	$\rho^0 \to \eta\gamma$	47.73	60.63	51.53	36.18±13.57
$\Upsilon(2S) \to \chi_{b0}(1P)\gamma$	0.41	0.54	0.82	1.89± 0.53	$\omega \to \pi^0\gamma$	459.30	571.79	423.19	714.85±42.74
$\Upsilon(2S) \to \chi_{b1}(1P)\gamma$	1.02	1.35	2.04	2.95± 0.61	$\omega \to \eta\gamma$	6.08	7.72	6.56	5.47± 0.84
$\Upsilon(2S) \to \chi_{b2}(1P)\gamma$	1.49	1.97	2.94	2.90± 0.61	$\phi \to \eta\gamma$	41.27	44.12	31.95	55.82± 2.73
$\Upsilon(3S) \to \chi_{b0}(2P)\gamma$	0.66	0.73	1.09	1.42± 0.25	$\phi \to \eta'\gamma$	0.30	0.32	0.27	0.53± 0.31
$\Upsilon(3S) \to \chi_{b1}(2P)\gamma$	1.65	1.84	2.71	2.97± 0.43	$K^{*0} \to K^0\gamma$	98.28	116.41	85.93	116.15±10.19
$\Upsilon(3S) \to \chi_{b2}(2P)\gamma$	2.44	2.71	3.98	3.00± 0.45	$K^{*+} \to K^+\gamma$	79.07	104.46	66.99	50.29± 4.66
$f_1(1285) \to \rho^0\gamma$	1232.83	1376.96	1160.31	1296.00±295.20	$D^{*0} \to D^0\gamma$	33.60	41.74	28.22	< 800.10 ± 60.90
$\chi_{c0}(1P) \to J/\psi(1S)\gamma$	255.40	260.24	267.75	92.40± 41.52	$D^{*+} \to D^+\gamma$	2.48	3.58	1.84	< 1.44 $^{+2.75}_{-0.92}$
$\chi_{c1}(1P) \to J/\psi(1S)\gamma$	306.63	312.43	322.27	240.24± 40.73	$D_s^{*+} \to D_s^+\gamma$	0.26	0.31	0.18	< 1789.80 ± 47.50
$\chi_{c2}(1P) \to J/\psi(1S)\gamma$	262.05	266.99	279.01	270.00± 32.78	$B^{*+} \to B^+\gamma$	0.97	1.26	0.78	seen
$\chi_{b0}(1P) \to \Upsilon(1S)\gamma$	30.10	30.85	30.06	seen	$B^{*0} \to B^0\gamma$	0.28	0.36	0.23	seen
$\chi_{b1}(1P) \to \Upsilon(1S)\gamma$	31.51	32.26	31.35	seen	$J/\Psi \to \eta_c\gamma$	1.85	1.87	1.75	1.13 ± 0.35
$\chi_{b2}(1P) \to \Upsilon(1S)\gamma$	30.39	31.03	29.91	seen	$\psi(2S) \to \eta_c(1S)\gamma$	4.97	6.34	7.13	0.78 ± 0.19
$\chi_{b0}(2P) \to \Upsilon(1S)\gamma$	14.01	11.80	8.07	seen	$\eta'(958) \to \rho^0\gamma$	112.90	143.62	107.50	61.31± 5.51
$\chi_{b0}(2P) \to \Upsilon(2S)\gamma$	13.31	13.52	14.50	seen	$\eta'(958) \to \omega\gamma$	10.50	13.36	10.01	6.11± 0.78
$\chi_{b1}(2P) \to \Upsilon(1S)\gamma$	13.53	11.38	7.80	seen	$D_{s1}(2536)^{*+} \to D_s^{*+}\gamma$	10.97	11.99	8.63	probably seen
$\chi_{b1}(2P) \to \Upsilon(2S)\gamma$	14.51	14.72	15.79	seen	$a_1(1260)^+ \to \pi^+\gamma$	179.53	229.90	171.45	seen
$\chi_{b2}(2P) \to \Upsilon(1S)\gamma$	11.80	9.90	6.78	seen	$a_1(1260)^0 \to \pi^0\gamma$	-	-	-	seen
$\chi_{b2}(2P) \to \Upsilon(2S)\gamma$	14.63	14.82	15.89	seen	$a_2(1320)^+ \to \pi^+\gamma$	142.01	179.27	136.64	299.60 ± 65.71
$b_1(1235)^+ \to \pi^+\gamma$ **	148.68	152.76	118.27	227.20 ± 58.60					

* the electric, magnetic and interference terms all contribute to the total amplitude
† purely magnetic transition
** purely electric transition

HADRON 2001 participants list

Achasov	Mikhail	Budker Institute of Nuclear Physics
Achasov	Nikolai	Sobolev Institute for Mathematics
Afanasev	Leonid	PNPI
Allaby	James	CERN
Anisovich	Vladimir	PNPI
Anisovich	Alexei	PNPI
Arhipov	Andrei	IHEP
Azhgirey	Leonid	JNRI
Barnes	Ted	Oak Ridge National Labs / Univ. Tennessee
Bashkanov	Mikhail	Moscow State Engineering Physics Institute
Berdyugin	Alexey	Budker Institute of Nuclear Physics
Bondar	Alexandre	Budker Institute of Nuclear Physics
Briscoe	William	The George Washington University
Bugg	David	Queen Mary, University of London
Chapkine	Mikhail	IHEP, Protvino
Chistov	Ruslan	ITEP
Chliapnikov	Pavel	IHEP, Protvino
Chung	Suh-Urk	BNL
Crede	Volker	Institut fur Strahlen- und Kernphysik
Denisov	Serguei	IHEP, Protvino
Deppermann	Thomas	Institute for Experimental Physics I, Bochum University
Di Donato	Camilla	Naples University, Federico II
Donnachie	Alexander	University of Manchester
Dorofeev	Valery	IHEP, Protvino
Dytman	Steven	University of Pittsburgh
Eugenio	Paul	Carnegie Mellon University
Faustov	Rudolf	Sci. Council for Cybernetics (SSC), RAS
Fedotovich	Guennady	Institute of Nuclear Physics
Filippi	Alessandra	INFN sez. di Torino
Fil'kov	Lev	Lebedev Physical Institute
Galkin	Vladimir	Scientific Council for Cybernetics, RAS
Gershtein	Semen	IHEP, Protvino
Gittelman	Bernard	Cornell University
Gobel	Carla	Instituto de Fisica, Univ.de la Republica Uruguay
Gorelov	Igor	Univ. of New Mexico
Grigoriev	Dmitry	Budker Institute of Nuclear Physics
Grigoriev	Vadim	ITEP
Guo	Zijin	IHEP, Beijing
Henner	Victor	Perm State University
Hicks	Kenneth	Ohio University
Hill	Richard	Imperial College, London
Hoh	Frank	Uppsala University
Hou	Suen	National Taiwan University
Ishida	Muneyuki	Tokyo Institute of Technology
Ishida	Shin	College of Science and Technology, Nihon University
Itahashi	Kenta	Tokyo Institute of Technology

Kachaev	Igor	IHEP, Protvino
Kalashnikova	Ylia	ITEP
Kaneko	Takashi	KEK
Karshon	Uri	Weizmann Institute of Science
Khrykin	Anatoly	JINR
Klempt	Eberhard	ISKP, Bonn
Koch	Helmut	Ruhr-Universitaet Bochum
Kolosov	Vladimir	IHEP, Protvino
Komada	Toshihiko	Nihon University
Konobeevski	Eugene	Institute for Nuclear Research, Moscow
Korotkhih	Vladimir	SINP MSU
Koulbardis	Arnis	PNPI
Kouznetsov	Viatchelav	INP
Kozhevnikov	Arkadii	Institute for Mathematics, Novosibirsk
Kozlenko	Nikolai	PNPI
Krimmer	Jochen	Physikalisches Institut, Universitaet Tuebingen
Kunihiro	Teiji	Yukawa Institute for Theoretical Physics
Kurshetsov	Victor	IHEP, Protvino
Landsberg	Leonid	IHEP, Protvino
Li	Weiguo	IHEP, Beijing
Ligabue	Franco	Scuola Normale Superiore, Pisa
Likhoded	Anatoly	IHEP, Protvino
Lisin	Vladimir	ITEP
Mackenzie	Paul	Fermilab
Maglich	Bogdan	Serbian Academy of Sciences and Arts
Majewski	Michal	University of Lodz, Institute of Physics
Manley	D. Mark	Kent State University
Meadows	Brian	University of Cincinnati
Merten	Dirk	Institute for Theoretical Nuclear Physics
Metsch	Bernard	Institut fuer Theoretische Kernphysik der Universitaet Bonn
Meyer	Curtis	Carnegie Mellon University
Meyer-Wildhagen	Frank	Muenchen University
Mokeev	Victor	Moscow State University
Molchanov	Vladimir	IHEP, Protvino
Morningstar	Colin	Carnegie Mellon University
Narodeckij	Ilja	ITEP
Ne'eman	Yuval	Tel-Aviv University
Nefediev	Alexei	ITEP
Nikitine	Nikolai	Skobeltsyn Institute of Nuclear Physics
Nikolaenko	Vladimir	IHEP, Protvino
Nikonov	Victor	PNPI
Noya	Hiroshi	Institute of Physics, Hosei University at Tama
Nozdrachev	Victor	ITEP
Ochs	Wolfgang	Max Planck Institut fuer Physik
Palano	Antimo	INFN and University of Bari
Peaslee	David C.	University of Maryland
Pelaez	Jose	Universidad Complutense, Madrid
Pennington	Michael	University of Durham
Petrov	Vladimir	IHEP, Protvino

Pick	Burkhard	Institut fuer Strahlen und Kernphysik
Pokrovsky	Yury	RRC "Kurchatov Institute"
Popov	Alexei	IHEP, Protvino
Protasov	Konstantin	Institut des Sciences Nucleaires
Reinnarth	Joerg	Institut fuer Strahlen und Kernphysik - Bonn
Reyes	Marco	Instituto de Fisica, Universidad Michoacana
Roinishvili	Vladimir	IHEP, Protvino
Ryabchikov	Dmitry	IHEP, Protvino
Sadilov	Serguei	IHEP, Protvino
Sarantsev	Andrei	PNPI
Sarycheva	Ludmila	SINP, Moscow State University
Schechter	Joseph	Syracuse University
Seth	Kamal	Northwestern University
Shagin	Petr	IHEP, Protvino
Sharov	German	Tver State University
Shestakov	Georgii	Institute for Mathematics, Novosibirsk
Shimizu	Hajime	Yamagata University
Silvestre-Brac	Bernard	Institut des Sciences Nucleaires
Singovski	Alexander	University of Minnesota
Sokolov	Anatoli	IHEP, Protvino
Sokolovsky	Vladimir	ITEP
Solovev	Lev	IHEP, Protvino
Suh	Jun Suhk	Brookhaven National Laboratory
Surovtsev	Yurii	Bogoliubov Laboratory of Theoretical Physics, JINR
Swanson	Eric	University of Pittsburgh
Takamatsu	Kunio	KEK
Tatischeff	Boris	IPN (Orsay)
Teshima	Tadayuki	Chubu University
Timoshenko	Sergey	Moscow State Engineering Physics Institute
Tomaradze	Amiran	Northwestern University
Tsuru	Tsuneaki	KEK
Tuan	San Fu	University of Hawaii at Manoa
Urner	David	Cornell University
Uvarov	Vladimir	IHEP, Protvino
Vaandering	Eric	Vanderbilt University
Valente	Paolo	INFN-LNF
Valeriani	Barbara	IEKP-Universitaet Karlsruhe
van Beveren	Eef	Physics Department, Coimbra University
Vladimirsky	Vasili	ITEP
Wah	Yau	University of Chicago
White	Richard	Imperial College, London
Yamada	Kenji	Nihon University
Zaitsev	Alexander	IHEP, Protvino
Zhu	Yongsheng	IHEP, Beijing
Zintchenko	Andrei	JINR
Ziolkowski	Michal	Univ. of Siegen

AUTHOR INDEX

A

Abaev, V. V., 693, 697, 701
Achasov, M. N., 30, 721
Achasov, N. N., 112, 599, 649, 687
Adeva, B., 745
Afanasiev, S. V., 819
Afanasyev, L., 745
Akhmetshin, R. R., 15
Allgower, C. E., 623, 693, 697, 701
Aloisio, A., 424, 711, 716
Ambrosino, F., 424, 711, 716
Amelin, D. V., 143, 577
Anashkin, E. V., 15
Anghinolfi, M., 505
Anisovich, V. V., 197
Antipov, Y. M., 125
Antonelli, A., 424, 711, 716
Antonelli, M., 424, 711, 716
Arkhipov, A. A., 771
Arkhipov, V. V., 819
Artamonov, A. V., 125
Aulchenko, V. M., 15, 30
Azhgirey, L. S., 819

B

Bacci, C., 424, 711, 716
Bagnasco, S., 95
Baldini, W., 95
Banzarov, V. S., 15
Barbiellini, G., 424, 711, 716
Barkov, L. M., 15
Barnes, T., 251, 447, 673
Baru, S. E., 15, 721
Bashkanov, M., 525
Bashtovoy, N. S., 15
Batarin, V. A., 125
Battaglieri, M., 505
Bekrenev, V., 693, 697, 701
Belkoborodov, K. I., 721
Bellini, F., 424, 711, 716
Beloborodov, K. I., 30
Belozerova, T. S., 607
Benayoun, M., 745
Bencivenni, G., 424, 711, 716

Berdyugin, A. V., 30, 721
Berisso, M. C., 521
Berka, Z., 745
Bertolucci, S., 424, 711, 716
Bettoni, D., 95
Bini, C., 424, 711, 716
Binon, F. G., 637
Black, D., 178
Blick, A. M., 637
Bloise, C., 424, 711, 716
Bocci, V., 424, 711, 716
Bogdanchikov, A. G., 30, 721
Bondar, A. E., 15
Bondarev, D. V., 15
Bondarev, V. K., 819
Bonnaz, R., 839
Borreani, G., 95
Bossi, F., 424, 711, 716
Bozhenok, A. V., 30, 721
Bragin, A. V., 15
Branchini, P., 424, 711, 716
Brekhovskikh, V., 745
Briscoe, W. J., 297, 693, 697, 701
Bugg, D. V., 356
Bukin, A. D., 30, 721
Bukin, D. A., 30, 721
Bulychjov, S. A., 424, 711, 716
Burdin, S. V., 30, 721
Buzzo, A., 95

C

Cabibbo, G., 424, 711, 716
Cadman, R., 697, 701
Calabrese, R., 95
Calen, H., 697, 701
Caloi, R., 424, 711, 716
Campana, P., 424, 711, 716
Capon, G., 424, 711, 716
Caragheorgheopol, G., 745
Carboni, G., 424, 711, 716
Casarsa, M., 424, 711, 716
Casavola, V., 424, 711, 716
Cataldi, G., 424, 711, 716
Cechak, T., 745
Ceradini, F., 424, 711, 716

Cervelli, F., 424, 711, 716
Cester, R., 95
Cevenini, F., 424, 711, 716
Chapkine, M., 803
Chernyak, D. V., 15
Chiba, M., 745
Chiefari, G., 424, 711, 716
Chistov, R., 543
Chliapnikov, P. V., 807, 811
Christian, D. C., 521
Ciambrone, P., 424, 711, 716
Cibinetto, G., 95
Clajus, M., 693, 697, 701
Comfort, J. R., 693, 697, 701
Conetti, S., 424, 711, 716
Constantinescu, S., 745
Constantini, S., 745
Craig, K., 693, 697, 701
Credé, V., 277

D

Dalpiaz, P., 95
Dell'Agnello, S., 424, 711, 716
De Lucia, E., 424, 711, 716
Denig, A., 424, 711, 716
Deppermann, T., 551
De Robertis, G., 711, 716
De Simone, P., 424, 711, 716
De Vita, R., 505
De Zorzi, G., 424, 711, 716
Di Domenico, A., 424, 711, 716
Di Donato, C., 424, 711, 716
Di Falco, S., 424, 711, 716
Dimova, T. V., 30, 721
Dolgopolov, A. V., 637
Donnachie, A., 5
Donskov, S. V., 637
Doria, A., 424, 711, 716
Dorofeev, V. A., 143, 577
Doudarev, A., 745
Draper, B., 697, 701
Dreossi, D., 745
Dreucci, M., 424, 711, 716
Drijard, D., 745
Drozdetski, A. A., 30, 721
Druzhinin, V. P., 30, 721
Duperray, R., 783

Dytman, S. A., 287
Dzhelyadin, R. I., 143, 577

E

Ebert, D., 336, 603
Eidelman, S. I., 15
Erofeev, I. A., 155, 629, 633
Erofeeva, O. N., 155, 629, 633
Eroshin, O. V., 125
Erriquez, O., 424, 711, 716
Eugenio, P., 573

F

Fariborz, A. H., 178
Farilla, A., 424, 711, 716
Faustov, R. N., 336, 603
Fedotov, G. V., 505
Fedotovitch, G. V., 15
Felici, G., 424, 711, 716
Felix, J., 521
Ferrari, A., 424, 711, 716
Ferrer, M. L., 424, 711, 716
Ferro-Luzzi, M., 745
Filipov, G., 819
Filippi, A., 582
Fil'kov, L. V., 753
Finocchiaro, G., 424, 711, 716
Forti, C., 424, 711, 716
Franceschi, A., 424, 711, 716
Franzini, P., 424, 711, 716
Frick, P. G., 607
Fujii, Y., 637

G

Gabyshev, N. I., 15
Galkin, V. O., 336, 603
Gallas Torreira, T., 745
Ganushin, D. I., 30
Gara, A., 521
Garzoglio, G., 95
Gatti, C., 424, 711, 716
Gauzzi, P., 424, 711, 716
Geissel, H., 749
Gerndt, J., 745

Giacomich, R., 745
Giannasi, A., 424, 711, 716
Gianotti, P., 745
Gibson, A., 697, 701
Gignoux, C., 839
Gilg, H., 749
Gillitzer, A., 749
Giovannella, S., 424, 711, 716
Gittelman, B., 85
Göbel, C., 63
Gollwitzer, K., 95
Golovach, E. N., 505
Golovkin, S. V., 125
Golubev, V. B., 30, 721
Gomez, F., 745
Gómez Nicola, A., 739
Gorelov, I. V., 385
Gorin, A., 745
Gorin, Y. P., 125
Gorini, E., 424, 711, 716
Gortchakov, O., 745
Gottschalk, E. E., 521
Gou, Z., 399
Gouz, Y. P., 143, 577
Graham, M., 95
Grancagnolo, F., 424, 711, 716
Graziani, E., 424, 711, 716
Grebeniuk, A. A., 15
Grigoriev, D. N., 15
Grigoriev, V. K., 155, 629, 633
Grosnick, D., 693, 697, 701
Guaraldo, C., 745
Gutiérrez, G., 521

H

Han, S. W., 424, 711, 716
Hansroul, M., 745
Hartouni, E. P., 521
Hayano, R. S., 749
Hayden, S., 697, 701
Henner, V. K., 607
Hicks, K. H., 510
Hirenzaki, S., 749
Hoh, F. C., 611
Holder, M., 367
Hosek, R., 745
Hou, S. R., 43
Hu, M., 95

Huddleston, J., 697, 701
Hughes, V. W., 15

I

Iliescu, M., 745
Inaba, S., 637
Incagli, M., 424, 711, 716
Ingrosso, L., 424, 711, 716
Isenhower, D., 693, 697, 701
Ishida, M., 187, 499, 657, 731, 735
Ishida, S., 187, 499, 657, 731, 735
Ishkhanov, B. S., 505
Isupov, A. Y., 819
Itahashi, K., 749
Ito, D., 657
Ivanchenko, V. N., 30, 721
Ivanov, P. M., 15, 30, 721
Ivanov, V. I., 819
Iwasaki, M., 749

J

Jabitski, M., 745
Joffe, D., 95
Johanson, T., 697, 701

K

Kachaev, I. A., 143, 577
Kalashnikova, Y. S., 5, 241
Kalinina, N., 745
Kaneko, T., 221
Karpov, S. V., 15
Karpukhin, V., 745
Karshon, U., 73
Kartamyshev, A. A., 819
Karyukhin, A. N., 143, 577
Kashevarov, V. L., 753
Kashirin, V. A., 819
Kasper, J., 95
Katinov, Y. V., 155, 629, 633
Kazanin, V. F., 15
Khaustov, G. V., 637
Khazin, B. I., 15
Khokhlov, Y. A., 143, 577
Khrenov, A. N., 819

Khrykin, A. S., 761
Kienle, P., 749
Kitamura, I., 487
Klempt, E., 463
Kluge, W., 424, 711, 716
Kluson, J., 745
Knapp, B. C., 521
Knecht, N., 693, 697, 701
Kobayashi, M., 745
Koch, H., 414
Koetke, D. D., 693, 697, 701
Kokkas, P., 745
Kolesnikov, V. I., 819
Kolganov, V. Z., 125
Kolosov, V. N., 637
Komada, T., 499, 657, 735
Komarov, V., 745
Kondashov, A. A., 637
Konobeevski, E. S., 753
Konoplyannikov, A. K., 143, 577
Konstantinov, V. F., 143, 577
Koop, I. A., 15, 721
Kopikov, S. V., 143, 577
Korol, A. A., 30, 721
Korostelev, M. S., 15, 721
Korotkikh, V. L., 561, 615
Koshuba, S. V., 30, 721
Kostyukhin, V. V., 143, 577
Koulbardis, A., 697, 701
Koulikov, A., 745
Kouptsov, A., 745
Kozhevnikov, A. A., 687
Kozhevnikov, A. P., 125
Kozlenko, N. G., 693, 697, 701
Kreisler, M. N., 521
Krimmer, J., 514
Krokovny, P. P., 15
Krouglov, V., 745
Krouglova, L., 745
Kruglov, S., 693, 697, 701
Krupa, D., 491
Kubarovsky, V. P., 125
Kuhn, J., 569
Kulbardis, A. A., 693
Kulikov, V., 424, 711, 716
Kunihiro, T., 315
Kuo, C., 424, 711, 716
Kupse, A., 697, 701
Kurdadze, L. M., 15
Kuroda, K.-I., 745

Kurshetsov, V., 125
Kuzmenko, D. S., 241
Kuzmin, A. S., 15
Kuznetsov, V. A., 819
Kycia, T., 697, 701

L

Lacava, F., 424, 711, 716
Ladygin, V. P., 819
Ladygina, N. B., 819
Lamberto, A., 745
Lanaro, A., 745
Landsberg, L. G., 125
Lanfranchi, G., 424, 711, 716
Lapshin, V., 745
Lasio, G., 95
Lednev, A. A., 637
Lednicky, R., 745
Lee, S., 521
Lee-Franzini, J., 424, 711, 716
Leone, D., 424, 711, 716
Leruste, P., 745
Levi Sandri, P., 745
Ligabue, F., 375
Lishin, V. A., 637
Lisin, V. I., 155, 629, 633
Litvinenko, A. G., 819
Logashenko, I. B., 15
Lolos, G., 693, 697, 701
Lomkatsi, G. S., 125
Lopatin, I. V., 693, 697, 701
Lopez Aguera, A., 745
LoVetere, M., 95
Lu, F., 424, 711, 716
Lucherini, V., 745
Lukin, P. A., 15
Luppi, E., 95
Luzin, V. N., 155, 629, 633
Lysenko, A. P., 15

M

Macrì, M., 95
Majewski, M., 653
Maki, T., 745
Mandelkern, M., 95
Manley, D. M., 693, 697, 701

Manuilov, I., 745
Manweiler, R., 693, 697, 701
Marchetto, F., 95
Marinelli, M., 95
Markianos, K., 521
Martemianov, M., 424, 711, 716
Marušić, A., 693, 697, 701
Matsumoto, S. I., 735
Matsyuk, M., 424, 711, 716
Matveev, V. D., 143, 577
McDonald, S., 693, 697, 701
Meadows, B., 547
Medovikov, V. A., 125
Mei, W., 424, 711, 716
Menichetti, E., 95
Menicucci, A., 424, 711, 716
Merola, L., 424, 711, 716
Merten, D., 661
Messi, R., 424, 711, 716
Metreveli, Z., 95
Metsch, B., 589
Meyer, C. A., 408
Meyer-Wildhagen, F., 787
Mikhailov, K. Y., 15
Miscetti, S., 424, 711, 716
Mokeev, V. I., 505
Molchanov, V. V., 125, 835
Montanet, L., 745
Mordovskoy, M. V., 753
Moreno, G., 521
Morisita, N., 487
Morningstar, C., 231
Moulson, M., 424, 711, 716
Moussa, S., 178
Mueller, S., 424
Mukhin, V. A., 125
Müller, S., 711, 716
Münch, M., 749
Münzenberg, G., 749
Murtas, F., 424, 711, 716
Mussa, R., 95

N

Nagy, M., 491
Nakamura, H., 533
Napolitano, M., 424, 711, 716
Narjoux, J.-L., 745
Narodetski, I. M., 669

Nasri, S., 178
Nedosekin, A., 424, 711, 716
Ne'eman, Y., 259
Nefediev, A. V., 595
Nefkens, B. M. K., 693, 697, 701
Negrini, M., 95
Nemenov, L., 745
Nesterenko, I. N., 15, 30
Nguyen, F., 424, 711, 716
Nikitin, M., 745
Nikitine, N., 665
Nikolaenko, V. I., 143, 577, 641
Nilov, A. F., 125
Noya, H., 533
Nozdrachev, V. N., 155, 629, 633
Nunez Pardo, T., 745

O

Obertino, M., 95
Ochs, W., 167
Okada, K., 745
Okhapkin, V. S., 15
Olchevskii, V., 745
Olmsted, J., 693, 697, 701
Osipenko, M. V., 505
Ostankov, A. P., 143, 577
Otboev, A. V., 15, 721

P

Pakhtusova, E. V., 30, 721
Palano, A., 53
Pallavicini, M., 95
Palutan, M., 424, 711, 716
Paoluzi, L., 424, 711, 716
Papandreou, Z., 693, 697, 701
Pasqualucci, E., 424, 711, 716
Passalacqua, L., 424, 711, 716
Passeri, A., 424, 711, 716
Pastrone, N., 95
Patalakha, D. I., 125
Patera, V., 424, 711, 716
Patrignani, C., 95
Pazos, A., 745
Peaslee, D. C., 623, 693, 697, 701
Peigeneux, J. P., 637

Peláez, J. R., 739
Pentia, M., 745
Penzo, A., 745
Perevedentsev, E. A., 15, 721
Perreau, J.-M., 745
Peters, K., 551
Peterson, J., 693, 697, 701
Petrascu, C., 745
Petrenko, S. V., 125
Petrolo, E., 424, 711, 716
Petrukhin, A. I., 125
Phaisangittisakul, N., 693, 697, 701
Picca, D., 424, 711, 716
Pick, B., 683
Pirozzi, G., 424, 711, 716
Plo, M., 745
Polunin, A. A., 15, 30
Polyakov, B. F., 143, 577
Polyakov, V. A., 637
Ponta, T., 745
Pontecorvo, L., 424, 711, 716
Pop, D., 745
Popov, A. S., 15
Popov, A. V., 135, 565
Pordes, S., 95
Potashev, S. I., 753
Prakhov, S., 693, 697, 701
Price, J. W., 693, 697, 701
Primavera, M., 424, 711, 716
Protasov, K., 783
Purlatz, T. A., 15

R

Ramirez, A. F., 693, 697, 701
Rappazzo, G. F., 745
Reinnarth, J., 792
Reyes, M. A., 521
Reznikov, S. G., 819
Riazantsev, A., 745
Ricco, G., 505
Ripani, M., 505
Robertis, G. De, 424
Robinson, C., 697, 701
Robutti, E., 95
Rodriguez, J. M., 745
Rodriguez Fernandez, A., 745
Roethel, W., 95
Root, N. I., 15

Rosen, J., 95
Ruban, A. A., 15
Ruggieri, F., 424, 711, 716
Rukoyatkin, P. A., 819
Rumerio, P., 95
Rupp, G., 209
Rusack, R., 95
Ryabchikov, D. I., 143, 555, 577
Rykalin, V., 745
Ryskulov, N. M., 15

S

Saborido, J., 745
Sadler, M., 693, 697, 701
Sadovsky, S. A., 637
Salnikov, A. A., 30, 721
Samoylenko, V. D., 637
Santamarina, C., 745
Santangelo, P., 424, 711, 716
Santovetti, E., 424, 711, 716
Santroni, A., 95
Sapunenko, V., 505
Saracino, G., 424, 711, 716
Sarycheva, L. I., 561, 615
Schacher, J., 745
Schamberger, R. D., 424, 711, 716
Schechter, J., 178
Schmüker, H., 551
Schott, W., 749
Schuetz, C., 745
Schultz, J., 95
Sciascia, B., 424, 711, 716
Sciubba, A., 424, 711, 716
Scuri, F., 424, 711, 716
Semenov, A. Y., 819
Semenova, I. A., 819
Seo, S. H., 95
Serednyakov, S. I., 30, 721
Seth, K. K., 95, 306, 779
Sfiligoi, I., 424, 711, 716
Shafi, A., 693, 697, 701
Shagin, P. M., 637, 705
Shamov, A. G., 15
Shan, J., 424, 711, 716
Sharov, G. S., 619
Shary, V. V., 30, 721
Shatunov, Y. M., 15, 30, 721
Shekhtman, A. I., 15

Shestakov, G. N., 599
Shimizu, H., 637
Shkurenko, Y. P., 155, 629, 633
Shwartz, B. A., 15
Sibidanov, A. L., 15
Sidorov, A., 745
Sidorov, V. A., 15, 30, 721
Silagadze, Z. K., 30, 721
Silano, P., 424, 711, 716
Silvestre-Brac, B., 839
Simonov, V. A., 753
Singovsky, A. V., 637
Skorkin, V. M., 753
Skrinsky, A. N., 15, 30
Skripkin, A. G., 30, 721
Smakhtin, V. P., 15
Smolik, J., 745
Smolyankin, V. T., 125
Snopkov, I. G., 15
Sobol, A. E., 637
Sokolov, A. A., 799
Sokolovsky, V. V., 155, 629, 633
Solodkov, A. A., 143, 577
Solodov, E. P., 15
Solovianov, O. V., 143, 577
Spadaro, T., 424, 711, 716
Spinka, H., 693, 697, 701
Spiriti, E., 424, 711, 716
Stancari, G., 95
Stancari, M., 95
Stanislaus, S., 697, 701
Stanislaus, T. D. S., 693
Starostin, A., 693, 697, 701
Staudenmaier, H., 697, 701
Staudenmaier, H. M., 693
Stepanov, P. Y., 15
Stoletov, G. D., 819
Strakovsky, I. I., 693, 697, 701
Stroot, J. P., 637
Sugonyaev, V. P., 637
Suh, J. S., 678
Sukhanov, A. I., 15
Supek, I., 693, 697, 701
Surovtsev, Y. S., 491
Suzuki, K., 749
Swanson, E. S., 327, 727
Szczepaniak, A. P., 727

T

Taiuti, M., 505
Takamatsu, K., 637
Takeutchi, F., 745
Tarasov, A., 745
Tatischeff, B., 765
Tauscher, L., 745
Teshima, T., 487
Thompson, J. A., 15
Tikhomirov, G. D., 155, 629, 633
Timoshenko, S., 529
Tippens, W. B., 693, 697, 701
Titov, V. M., 15
Tobar, M. J., 745
Tomaradze, A., 95
Tomono, D., 749
Tong, G. L., 424, 711, 716
Tonooka, H., 657
Tortora, L., 424, 711, 716
Trousov, S., 745
Trusov, M. A., 669
Tsuru, T., 637
Tuan, S. F., 105, 495

U

Urner, D., 434, 539
Usov, Y. V., 30
Uvarov, V. A., 815

V

Valente, E., 424, 711, 716
Valente, P., 424, 711, 716
Valeriani, B., 424, 711, 716
Valishev, A. A., 15
van Beveren, E., 209
Vaniev, V. S., 125
Vasiljev, A. V., 30, 721
Vavilov, D. V., 125
Vazquez, P., 745
Venanzoni, G., 424, 711, 716
Veneziano, S., 424, 711, 716
Ventura, A., 424, 711, 716
Victorov, V. A., 125
Vidnovic III, T., 95

Vlachos, S., 745
Vladimirsky, V. V., 155, 629, 633

W

Wang, M. H. L. S., 521
Wehman, A., 521
Weick, H., 749
Werkema, S., 95
Wesson, D., 521
White, R. R., 829
Wiedner, U., 697, 701
Wu, Y., 424, 711, 716

X

Xu, G., 424, 711, 716

Y

Yamada, K., 657
Yamazaki, T., 749

Yasu, Y., 637
Yazkov, V., 745
Yoneyama, T., 749
Yoshimura, Y., 745
Yu, G. W., 424, 711, 716
Yudin, Y. V., 15

Z

Zaitsev, A. M., 143, 577
Zema, P. F., 424, 711, 716
Zhmyrov, V. N., 819
Zhou, Y., 424, 711, 716
Zhu, Y., 346
Zimin, S. A., 125
Zinchenko, A., 825
Ziolkowsky, M., 367
Zolin, L. S., 819
Zrelov, P., 745
Zverev, S. G., 15
Zvyagin, A. Y., 637
Zweber, P., 95